area A
perimeter P
length l

width w
surface area S
altitude (height) h

base b
circumference C
radius r

Rectangle

$$A = lw \qquad P = 2l + 2w$$

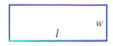

Triangle

$$A = \frac{1}{2}bh$$

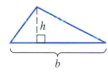

Square

$$A = s^2 \qquad P = 4s$$

Parallelogram

$$A = bh$$

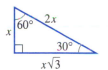

Trapezoid

$$A = \frac{1}{2}h(b_1 + b_2)$$

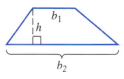

Circle

$$A = \pi r^2 \qquad C = 2\pi r$$

30°–60° Right Triangle

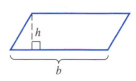

Right Triangle

$$a^2 + b^2 = c^2$$

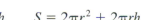

Isosceles Right Triangle

Right Circular Cylinder

$$V = \pi r^2 h \qquad S = 2\pi r^2 + 2\pi rh$$

Sphere

$$S = 4\pi r^2 \qquad V = \frac{4}{3}\pi r^3$$

Right Circular Cone

$$V = \frac{1}{3}\pi r^2 h \qquad S = \pi r^2 + \pi rs$$

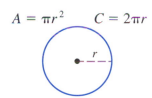

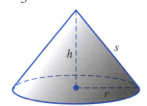

Pyramid

$$V = \frac{1}{3}Bh$$

Prism

$$V = Bh$$

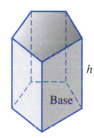

ELEMENTARY AND INTERMEDIATE ALGEBRA

A Combined Approach

SIXTH EDITION

ELEMENTARY AND INTERMEDIATE ALGEBRA

A Combined Approach

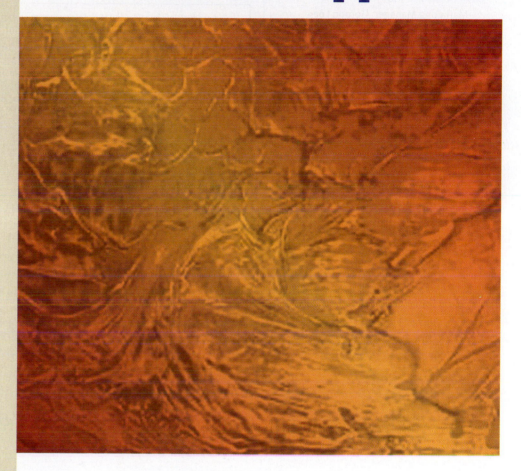

Jerome E. Kaufmann

Karen L. Schwitters
Seminole State College of Florida
Nicolet College

BROOKS/COLE
CENGAGE Learning™

Australia • Brazil • Japan • Korea • Mexico • Singapore • Spain • United Kingdom • United States

BROOKS/COLE
CENGAGE Learning™

Elementary and Intermediate Algebra:
A Combined Approach, **Sixth Edition**
Jerome E. Kaufmann and
Karen L. Schwitters

Editor: Marc Bove

Assistant Editor: Stefanie Beeck

Editorial Assistant: Zack Crockett

Media Editor: Guanglei Zhang and
   Heleny Wong

Marketing Manager: Gordon Lee

Marketing Assistant: Angela Kim

Marketing Communications Manager:
   Darlene Macanan

Sr. Content Project Manager:
   Tanya Nigh

Design Director: Rob Hugel

Art Director: Vernon Boes

Print Buyer: Rebecca Cross

Rights Acquisitions Specialist:
   Don Schlotman

Production Service: Susan Graham

Text Designer: Diane Beasley

Photo Researcher: PreMedia Global

Copy Editor: Susan Graham

Illustrator: Network Graphics

Cover Designer: Irene Morris

Cover Image: Fotolia

Compositor: MPS Limited, a Macmillan
   Company

For product information and technology assistance, contact us at
Cengage Learning Customer & Sales Support, 1-800-354-9706.

For permission to use material from this text or product, submit all requests online at **www.cengage.com/permissions.**
Further permissions questions can be e-mailed to
permissionrequest@cengage.com.

Library of Congress Control Number: 2010936597

Student Edition:
ISBN-13: 978-0-8400-5314-5
ISBN-10: 0-8400-5314-2

Brooks/Cole
20 Davis Drive
Belmont, CA 94002-3098
USA

Cengage Learning is a leading provider of customized learning solutions with office locations around the globe, including Singapore, the United Kingdom, Australia, Mexico, Brazil, and Japan. Locate your local office at **www.cengage.com/global.**

Cengage Learning products are represented in Canada by Nelson Education, Ltd.

To learn more about Brooks/Cole, visit
www.cengage.com/brooks/cole

Purchase any of our products at your local college store or at our preferred online store **www.cengagebrain.com.**

Printed in the United States of America
1 2 3 4 5 6 7 14 13 12 11 10

CONTENTS

8 A Transition from Elementary Algebra to Intermediate Algebra   339

9 Rational Expressions   401

10 Exponents and Radicals   457

11 Quadratic Equations and Inequalities 507

12 Coordinate Geometry: Lines, Parabolas, Circles, Ellipses, and Hyperbolas 561

13 Functions 615

14 Exponential and Logarithmic Functions 667

15 Systems of Equations: Matrices and Determinants 731

Elementary and Intermediate Algebra: *A Combined Approach,* Sixth Edition, presents the basic topics of both elementary and intermediate algebra. By combining these topics in one text, we present an organizational format that allows for the frequent reinforcement of concepts but eliminates the need to completely reintroduce topics, as when two separate texts are used. At any time, if necessary, one can return to the original introduction of a particular topic.

The basic concepts of elementary and intermediate algebra are developed in a logical sequence but in an easy-to-read manner without excessive technical vocabulary and formalism. Whenever possible, the algebraic concepts are allowed to develop from their arithmetic counterparts. The following are two specific examples of this development.

Manipulation with simple algebraic fractions begins early (Sections 2.1 and 2.2) when reviewing operations with rational numbers.

Manipulation with monomials, without any of the formal vocabulary, is introduced in Section 2.4 when working with exponents.

There is a common thread throughout the text, which is *learn a skill, use the skill to solve equations and inequalities,* and *then use equations and inequalities as problem solving tools.* This thread influenced some other decisions.

1. Approximately 800 word problems are scattered throughout the text. These problems deal with a large variety of applications and reinforce the connections between mathematics and the world around us.
2. Many problem-solving suggestions are offered throughout, with special discussions in several sections. Problem-solving suggestions are demonstrated in more than 100 worked-out examples.
3. Newly acquired skills are used as soon as possible to solve equations and inequalities. Equations and inequalities are introduced early in the text and then used throughout in a large variety of problem-solving situations.

In preparing this edition, a special effort was made to incorporate the improvements suggested by reviewers and users of the previous editions, while at the same time, preserving the book's many successful features.

New in This Edition

- For every Example in the text, a corresponding Classroom Example problem was added in the margin. A conscious decision was made to include these classroom examples in the student edition so that the instructor does not have to write the problem on the board in lecture classes. The instructor's edition includes the answer and the student edition does not.

- All chapter summaries have been rewritten in a grid format. The grid contains an objective, a summary of the key concepts for that objective, and an example problem of that objective with a complete solution shown.

- Some users of previous editions made suggestions regarding the problem sets. Wherever possible, we have incorporated those suggestions. For example, in Problem Set 3.3, we moved equations with solutions of the null set and all real numbers from the Further Investigations section to the body of the problem set. In Problem Set 8.4, we added eight additional polynomial problems. Altogether we added or updated approximately 80 problems.

New Features

Design

The new design creates a spacious format that allows for continuous and easy reading, as color and form guide students through the concepts presented in the text. Page size has been slightly enlarged, enhancing the design to be visually intuitive without increasing the length of the book.

OBJECTIVES
1. Solve first-degree equations by simplifying both sides and then applying properties of equality
2. Solve first-degree equations that are contradictions
3. Solve first-degree equations that are identities
4. Solve word problems that represent several quantities in terms of the same variable
5. Solve word problems involving geometric relationships

◀ **Learning Objectives**

Found at the beginning of each section, Learning Objectives are mapped to Problem Sets and to the Chapter Summary.

▶ **Classroom Examples**

To provide the instructor with more resources, a Classroom Example is written for every example. Instructors can use these to present in class or for student practice exercises. These classroom examples appear in the margin, to the left of the corresponding example, in both the Annotated Instructor's Edition and in the Student Edition. Answers to the Classroom Examples appear only in the Annotated Instructor's Edition, however.

Classroom Example
Find the measures of the three angles of a triangle if the second angle is twice the first angle, and the third angle is half the second angle.

45°, 90°, 45°

EXAMPLE 9

Find the measures of the three angles of a triangle if the second is three times the first and the third is twice the second.

Solution

If we let a represent the measure of the smallest angle, then $3a$ and $2(3a)$ represent the measures of the other two angles. Therefore, we can set up and solve the following equation:

$$a + 3a + 2(3a) = 180$$
$$a + 3a + 6a = 180$$
$$10a = 180$$
$$\frac{10a}{10} = \frac{180}{10}$$
$$a = 18$$

If $a = 18$, then $3a = 54$ and $2(3a) = 108$. So the angles have measures of 18°, 54°, and 108°.

Concept Quiz 3.1

For Problems 1–10, answer true or false.

1. Equivalent equations have the same solution set.
2. $x^2 = 9$ is a first-degree equation.
3. The set of all solutions is called a solution set.
4. If the solution set is the null set, then the equation has at least one solution.
5. Solving an equation refers to obtaining any other equivalent equation.
6. If 5 is a solution, then a true numerical statement is formed when 5 is substituted for the variable in the equation.
7. Any number can be subtracted from both sides of an equation, and the result is an equivalent equation.
8. Any number can divide both sides of an equation to obtain an equivalent equation.
9. By the reflexive property, if $y = 2$ then $2 = y$.
10. By the transitive property, if $x = y$ and $y = 4$, then $x = 4$.

◀ **Concept Quiz**

Every section has a Concept Quiz that immediately precedes the problem set. The questions are predominantly true/false questions that allow students to check their understanding of the mathematical concepts and definitions introduced in the section before moving on to their homework. Answers to the Concept Quiz are located at the end of the Problem Set.

▶ Chapter Summary

The new grid format of the Chapter Summary allows students to review material quickly and easily. Each row of the Chapter Summary includes a learning objective, a summary of that objective, and a worked-out example for that objective.

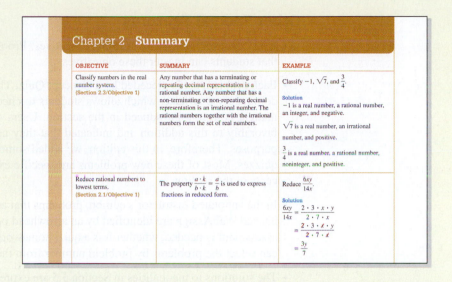

Continuing Features

Explanations

Annotations in the examples and text provide further explanations of the material.

Examples

More than 500 worked-out Examples show students how to use and apply mathematical concepts. Every example has a corresponding Classroom Example for the teacher to use.

Thoughts Into Words

Every problem set includes Thoughts Into Words problems, which give students an opportunity to express in written form their thoughts about various mathematical ideas.

Further Investigations

Many problem sets include Further Investigations, which allow students to pursue more complicated ideas. Many of these investigations lend themselves to small group work.

Graphing Calculator Activities

Certain problem sets contain a group of problems called Graphing Calculator Activities. In this text, the use of a graphing calculator is optional.

Problem Sets

Problems Sets contain a wide variety of skill-development exercises.

Chapter Review Problem Sets and Chapter Tests

Chapter Review Problem Sets and Chapter Tests appear at the end of every chapter.

Cumulative Review Problem Sets

Cumulative Review Problem Sets help students retain skills introduced earlier in the text.

Answers

The answer section at the back of the text provides answers to the odd-numbered exercises in the problem sets and to all exercises in the Chapter Review, Chapter Test, Cumulative Review, and Cumulative Practice Test Problem Sets.

Other Special Features

- Every section opens with learning objectives. Problem sets contain ample problems so that students can master these objectives.

- Each problem set is preceded by a Concept Quiz. The problems in these quizzes are predominately true/false, which allows students to check their understanding of the mathematical concepts introduced in the section. Users of the previous edition reacted very favorably to this addition and indicated that they used the problems for many different purposes. Therefore, in this edition, we added some additional problems to many of the quizzes. Most of these new problems are specific examples that reinforce the basic concepts of the section.

- In the annotated instructor's edition, problems that are available in electronic form in Enhanced WebAssign are identified by an arrowhead pointing to the problem number. If an assessment is needed, whether it is a quiz, homework assignment, or a test, the instructor can select the problems by problem number from the identified problems in the text.

- The solutions to inequalities in Section 3.5 are expressed in both set notation and interval notation. Both notations are used until Chapter 8 when interval notation becomes the typical format. This progression allows the instructors to choose their preferred format for beginning algebra.

- As recommended by the American Mathematical Association of Two-Year Colleges, many basic geometric concepts are integrated in problem-solving settings. *Approximately 20 worked-out examples and 200 problems in this text connect algebra, geometry, and our world.* (For examples, see Problems 17 and 21 on page 164.) The following geometric concepts are presented in problem-solving situations: complementary and supplementary angles; the sum of measures of angles of a triangle equals 180°; area and volume formulas; perimeter and circumference formulas; ratio and proportion; the Pythagorean theorem; isosceles right triangle; and 30°–60° right triangle relationships.

- The graphing calculator is introduced in Chapter 9 and is used primarily as a teaching tool. The students do not need a graphing calculator to study from this text. Starting in Chapter 9, graphing calculators are used at times to enhance the learning of specific algebraic concepts. These examples are written so that students without a graphing calculator can read and benefit from them.

- Beginning with Problem Set 9.1, a group of problems called "Graphing Calculator Activities" is included in many of the problem sets. These activities, which are especially good for small group work, are designed to reinforce concepts (see, for example, Problem Set 12.2), as well as lay groundwork for concepts about to be discussed (see, for example, Problem Set 12.1). Some of these activities ask students to predict shapes and locations of graphs based on previous graphing experiences and then use the graphing calculator to check their predictions (see, for example, Problem Set 12.5). The graphing calculator is also used as a problem solving tool (see, for example, Problem Set 9.6).

- Problems called "Thoughts into Words" are included in every problem set except the review exercises. These problems are designed to encourage students to express in written form their thoughts about various mathematical ideas. For examples, see Problem Sets 2.1, 3.2, 4.4, and 8.1.

- Problems called "Further Investigations" appear in many of the problem sets. These are "extras" that lend themselves to individual or small group work. These problems encompass a variety of ideas: some exhibit different approaches to topics covered in the text, some are proofs, some bring in supplementary topics and relationships, and some are more challenging problems. These problems add variety and flexibility to the problem sets, but they could be omitted entirely without disrupting the continuity of the text. For examples, see Problem Sets 1.2, 4.2, 6.3, 7.3, and 8.3.

- Problem solving is a key issue throughout the text. Not only are equations, systems of equations, and inequalities used as problem solving tools, but many other concepts are used as well. For example, functions, the distance formula, and slope concepts are all developed and then used to solve problems.

- Specific graphing ideas (intercepts, symmetry, restrictions, asymptotes, and transformations) are introduced and used in Chapters 5, 12, 13, and 14. In Section 13.3 the work with parabolas introduced in Chapter 12 is used to motivate definitions for translations, reflections, stretchings, and shrinkings. These transformations are then applied to the graphs of

$$f(x) = x^3, \qquad f(x) = \frac{1}{x}, \qquad f(x) = x, \qquad \text{and} \qquad f(x) = |x|$$

- All answers for Chapter Review Problem Sets, Chapter Tests, Cumulative Review Problem Sets, and Cumulative Practice Tests appear in the back of the text, along with answers to all odd-numbered problems.

- Please note the exceptionally pleasing design features of this text, including the functional use of color. The open format makes for a continuous and easy reading of the material, unlike working through a maze of flags, caution symbols, reminder symbols, and so on.

Ancillaries

For the Instructor

Annotated Instructor's Edition

In the AIE, answers are printed next to all respective exercises. Graphs, tables, and other answers appear in an answer section at the back of the text.

Power Lecture with ExamView®

This CD-ROM provides the instructor with dynamic media tools for teaching. Create, deliver, and customize tests (both print and online) in minutes with ExamView® Computerized Testing Featuring Algorithmic Equations. Easily build solution sets for homework or exams using Solution Builder's online solutions manual. Microsoft® PowerPoint® lecture slides and figures from the book are also included on this CD-ROM.

Complete Solutions Manual

The Complete Solutions Manual provides worked-out solutions to all of the problems in the text.

Enhanced WebAssign   ENHANCED WebAssign

Enhanced as in WebAssign, used by over one million students at more than 1100 institutions, allows you to assign, collect, grade, and record homework assignments via the web. This proven and reliable homework system includes thousands of algorithmically generated homework problems, an eBook, links to relevant textbook sections, video examples, problem-specific tutorials, and more. Note that the WebAssign problems for this text are highlighted by a ▶.

Instructor's Resource Binder

NEW! Each section of the main text is discussed in uniquely designed Teaching Guides, which contain instruction tips, examples, activities, worksheets, overheads, assessments, and solutions to all worksheets and activities.

Website (www.cengagebrain.com)

To access additional course materials and companion resources, please visit www.cengagebrain.com. At the CengageBrain.com home page, search for the ISBN of your title (from the back cover of your book) using the search box at the top of the page. This will take you to the product page where free companion resources can be found.

Text-Specific DVDs

These text-specific DVDs, available at no charge to qualified adopters of the text, feature 10- to 20-minute problem-solving lessons that cover each section of every chapter.

For the Student

Student Solutions Manual

The Student Solutions Manual provides worked-out solutions to the odd-numbered problems in the text.

Interactive Video Skillbuilder CD-ROM

The Interactive Video Skillbuilder CD-ROM contains video instruction covering each chapter of the text. The problems worked during each video lesson are shown first so that students can try working them before watching the solution. To help students evaluate their progress, each section contains a 10-question Web quiz (the results of which can be emailed to the instructor), and each chapter contains a chapter test with answers to each problem on each test. This dual-platform CD-ROM also includes MathCue tutorial and quizzing software, featuring a Skill Builder that presents problems to solve and evaluates answers with step-by-step explanations; a Quiz function that enables students to generate quiz problems keyed to problem types from each section of the book; a Chapter Test that provides many problems keyed to problem types from each chapter; and a Solution Finder that allows students to enter their own basic problems and receive step-by-step help as if they were working with a tutor. Also, English/Spanish closed caption translations can be selected to display along with the video instruction.

Enhanced WebAssign   ENHANCED WebAssign

Enhanced WebAssign, used by over one million students at more than 1100 institutions, allows you to do homework assignments and get extra help and practice via the Web. This proven and reliable homework system includes thousands of algorithmically generated homework problems, an eBook, links to relevant textbook sections, video examples, problem-specific tutorials, and more.

Student Workbook

NEW! Get a head-start! The Student Workbook contains all of the Assessments, Activities, and Worksheets from the Instructor's Resource Binder for classroom discussion, in-class activities, and group work.

Acknowledgments

We would like to take this opportunity to thank the following people who served as reviewers for the fifth edition of *Elementary and Intermediate Algebra: A Combined Approach*:

Ramendra Bose
University of Texas – Pan American

Nam Ngyuen
University of Texas – Pan American

Zhixiong Chen
New Jersey City University

Leticia Oropesa
University of Miami

Betty Larson
South Dakota State University

Ahmed Zayed
DePaul University

We are very grateful to the staff of Brooks/Cole, especially to Marc Bove, for his continuous assistance throughout this project. We would also like to express our sincere gratitude to Susan Graham and to Tanya Nigh. They continue to carry out the details of production in a dedicated and caring way. Additional thanks are due to Laurel Fischer who wrote the solutions manuals.

Jerome E. Kaufmann
Karen L. Schwitters

1 Some Basic Concepts of Arithmetic and Algebra

© photogolfer

Golfers are familiar with positive and negative integers.

Karla started 2011 with $500 in her savings account, and she planned to save an additional $15 per month for all of 2011. Without considering any accumulated interest, the numerical expression 500 + 12(15) represents the amount in her savings account at the end of 2011.

The numbers +2, −1, −3, +1, and −4 represent Woody's scores relative to par for five rounds of golf. The numerical expression 2 + (−1) + (−3) + 1 + (−4) can be used to determine how Woody stands relative to par at the end of the five rounds.

The temperature at 4 A.M. was −14°F. By noon the temperature had increased by 23°F. The numerical expression −14 + 23 can be used to determine the temperature at noon.

In the first two chapters of this text the concept of a *numerical expression* is used as a basis for reviewing addition, subtraction, multiplication, and division of various kinds of numbers. Then the concept of a *variable* allows us to move from numerical expressions to *algebraic expressions*; that is, to start the transition from arithmetic to algebra. Keep in mind that algebra is simply a generalized

Video tutorials based on section learning objectives are available in a variety of delivery modes.

1

approach to arithmetic. Many algebraic concepts are extensions of arithmetic ideas; your knowledge of arithmetic will help you with your study of algebra.

| **1.1** | ## Numerical and Algebraic Expressions |

OBJECTIVES

1 Recognize basic vocabulary and symbols associated with sets

2 Simplify numerical expressions according to the order of operations

3 Evaluate algebraic expressions

In arithmetic, we use symbols such as 4, 8, 17, and π to represent numbers. We indicate the basic operations of addition, subtraction, multiplication, and division by the symbols $+$, $-$, $\cdot$, and $\div$, respectively. With these symbols we can formulate specific **numerical expressions**. For example, we can write the indicated sum of eight and four as $8 + 4$.

In algebra, **variables** allow us to generalize. By letting x and y represent *any* number, we can use the expression $x + y$ to represent the indicated sum of *any two* numbers. The x and y in such an expression are called variables and the phrase $x + y$ is called an **algebraic expression**. We commonly use letters of the alphabet such as x, y, z, and w as variables; the key idea is that they represent numbers. Our review of various operations and properties pertaining to numbers establishes the foundation for our study of algebra.

Many of the notational agreements made in arithmetic are extended to algebra with a few slight modifications. The following chart summarizes the notational agreements that pertain to the four basic operations. Notice the variety of ways to write a product by including parentheses to indicate multiplication. Actually, the ab form is the simplest and probably the most used form; expressions such as abc, $6x$, and $7xyz$ all indicate multiplication. Also note the various forms for indicating division; the fractional form, $\dfrac{c}{d}$, is usually used in algebra, although the other forms do serve a purpose at times.

Operation	Arithmetic	Algebra	Vocabulary
Addition	$4 + 6$	$x + y$	The **sum** of x and y
Subtraction	$7 - 2$	$w - z$	The **difference** of w and z
Multiplication	$9 \cdot 8$	$a \cdot b$, $a(b)$, $(a)b$, $(a)(b)$, or ab	The **product** of a and b
Division	$8 \div 2$, $\dfrac{8}{2}$, $2\overline{)8}$	$c \div d$, $\dfrac{c}{d}$, or $d\overline{)c}$	The **quotient** of c and d

As we review arithmetic ideas and introduce algebraic concepts, it is important to include some of the basic vocabulary and symbolism associated with sets. A **set** is a collection of objects, and the objects are called **elements** or **members** of the set. In arithmetic and algebra the elements of a set are often numbers. To communicate about sets, we use set braces, { }, to enclose the elements (or a description of the elements), and we use capital letters to name sets. For example, we can represent a set A, which consists of the vowels of the alphabet, as

$A = \{$Vowels of the alphabet$\}$ Word description

$A = \{a, e, i, o, u\}$ List or roster description

We can modify the listing approach if the number of elements is large. For example, all of the letters of the alphabet can be listed as

$$\{a, b, c, \ldots, z\}$$

We begin by simply writing enough elements to establish a pattern, then the three dots indicate that the set continues in that pattern. The final entry indicates the last element of the pattern. If we write

$$\{1, 2, 3, \ldots\}$$

the set begins with the counting numbers 1, 2, and 3. The three dots indicate that it continues in a like manner forever; there is no last element.

A set that consists of no elements is called the **null set** (written $\varnothing$). Two sets are said to be *equal* if they contain exactly the same elements. For example,

$$\{1, 2, 3\} = \{2, 1, 3\}$$

because both sets contain the same elements; the order in which the elements are written does not matter. The slash mark through the equality symbol denotes *not equal to*. Thus if $A = \{1, 2, 3\}$ and $B = \{1, 2, 3, 4\}$, we can write $A \neq B$, which we read as "set A is not equal to set B."

Simplifying Numerical Expressions

Now let's simplify some numerical expressions that involve the set of **whole numbers**, that is, the set $\{0, 1, 2, 3, \ldots\}$.

Classroom Example
Simplify $2 + 6 - 3 + 7 + 11 - 9$.

EXAMPLE 1   Simplify $8 + 7 - 4 + 12 - 7 + 14$.

Solution

The additions and subtractions should be performed from left to right in the order that they appear. Thus $8 + 7 - 4 + 12 - 7 + 14$ simplifies to 30. ∎

Classroom Example
Simplify $5(8 + 6)$.

EXAMPLE 2   Simplify $7(9 + 5)$.

Solution

The parentheses indicate the product of 7 and the quantity $9 + 5$. Perform the addition inside the parentheses first and then multiply; $7(9 + 5)$ thus simplifies to $7(14)$, which becomes 98. ∎

Classroom Example
Simplify $(5 + 11) \div (8 - 4)$.

EXAMPLE 3   Simplify $(7 + 8) \div (4 - 1)$.

Solution

First, we perform the operations inside the parentheses; $(7 + 8) \div (4 - 1)$ thus becomes $15 \div 3$, which is 5. ∎

We frequently express a problem like Example 3 in the form $\dfrac{7 + 8}{4 - 1}$. We don't need parentheses in this case because the fraction bar indicates that the sum of 7 and 8 is to be divided by the difference, $4 - 1$. A problem may, however, contain parentheses and fraction bars, as the next example illustrates.

Classroom Example
Simplify $\dfrac{(6-3)(4+1)}{5} + \dfrac{8}{13-5}$.

EXAMPLE 4   Simplify $\dfrac{(4+2)(7-1)}{9} + \dfrac{7}{10-3}$.

Solution

First, simplify above and below the fraction bars, and then proceed to evaluate as follows.

$$\frac{(4+2)(7-1)}{9} + \frac{7}{10-3} = \frac{(6)(6)}{9} + \frac{7}{7}$$

$$= \frac{36}{9} + 1 = 4 + 1 = 5$$

Classroom Example
Simplify $4 \cdot 7 + 3$.

EXAMPLE 5   Simplify $7 \cdot 9 + 5$.

Solution

If there are no parentheses to indicate otherwise, multiplication takes precedence over addition. First perform the multiplication, and then do the addition; $7 \cdot 9 + 5$ therefore simplifies to $63 + 5$, which is 68.

Remark: Compare Example 2 to Example 5, and note the difference in meaning.

Classroom Example
Simplify $6 - 10 \div 5 + 4 \cdot 5$.

EXAMPLE 6   Simplify $8 + 4 \cdot 3 - 14 \div 2$.

Solution

The multiplication and division should be done first in the order that they appear, from left to right. Thus $8 + 4 \cdot 3 - 14 \div 2$ simplifies to $8 + 12 - 7$. We perform the addition and subtraction in the order that they appear, which simplifies $8 + 12 - 7$ to 13.

Classroom Example
Simplify $3 \cdot 8 \div 4 + 20 \div 2 \cdot 5$.

EXAMPLE 7   Simplify $18 \div 3 \cdot 2 + 8 \cdot 10 \div 2$.

Solution

When we perform the multiplications and divisions first in the order that they appear and then do the additions and subtractions, our work takes on the following format.

$$18 \div 3 \cdot 2 + 8 \cdot 10 \div 2 = 6 \cdot 2 + 80 \div 2 = 12 + 40 = 52$$

Classroom Example
Simplify $9 + 3[5(2 + 4)]$.

EXAMPLE 8   Simplify $5 + 6[2(3 + 9)]$.

Solution

We use brackets for the same purpose as parentheses. In such a problem we need to simplify *from the inside out*; perform the operations inside the innermost parentheses first.

$$5 + 6[2(3 + 9)] = 5 + 6[2(12)]$$
$$= 5 + 6[24]$$
$$= 5 + 144$$
$$= 149$$

Let's now summarize the ideas presented in the previous examples regarding **simplifying numerical expressions**. When simplifying a numerical expression, use the following order of operations.

Order of Operations

1. Perform the operations inside the symbols of inclusion (parentheses and brackets) and above and below each fraction bar. Start with the innermost inclusion symbol.

2. Perform all multiplications and divisions in the order that they appear, from left to right.

3. Perform all additions and subtractions in the order that they appear, from left to right.

Evaluating Algebraic Expressions

We can use the concept of a variable to generalize from numerical expressions to algebraic expressions. Each of the following is an example of an algebraic expression.

$$3x + 2y \qquad 5a - 2b + c \qquad 7(w + z)$$

$$\frac{5d + 3e}{2c - d} \qquad 2xy + 5yz \qquad (x + y)(x - y)$$

An algebraic expression takes on a numerical value whenever each variable in the expression is replaced by a specific number. For example, if x is replaced by 9 and z by 4, the algebraic expression $x - z$ becomes the numerical expression $9 - 4$, which simplifies to 5. We say that $x - z$ "has a value of 5" when x equals 9 and z equals 4. The value of $x - z$, when x equals 25 and z equals 12, is 13. The general algebraic expression $x - z$ has a specific value each time x and z are replaced by numbers.

Consider the following examples, which illustrate the process of finding a value of an algebraic expression. We call this process **evaluating algebraic expressions**.

Classroom Example
Find the value of $5x + 4y$, when x is replaced by 3 and y by 13.

EXAMPLE 9   Find the value of $3x + 2y$, when x is replaced by 5 and y by 17.

Solution

The following format is convenient for such problems.

$$\begin{aligned} 3x + 2y &= 3(5) + 2(17) \quad \text{when } x = 5 \text{ and } y = 17 \\ &= 15 + 34 \\ &= 49 \end{aligned}$$

Note that in Example 9, for the algebraic expression $3x + 2y$, the multiplications "3 times x" and "2 times y" are implied without the use of parentheses. Substituting the numbers switches the algebraic expression to a numerical expression, and then parentheses are used to indicate the multiplication.

Classroom Example
Find the value of $11m - 5n$, when $m = 4$ and $n = 7$.

EXAMPLE 10   Find the value of $12a - 3b$, when $a = 5$ and $b = 9$.

Solution

$$\begin{aligned} 12a - 3b &= 12(5) - 3(9) \quad \text{when } a = 5 \text{ and } b = 9 \\ &= 60 - 27 \\ &= 33 \end{aligned}$$

Classroom Example
Evaluate $6xy - 3xz + 5yz$,
when $x = 2$, $y = 5$, and $z = 3$.

EXAMPLE 11   Evaluate $4xy + 2xz - 3yz$, when $x = 8$, $y = 6$, and $z = 2$.

Solution

$$4xy + 2xz - 3yz = 4(8)(6) + 2(8)(2) - 3(6)(2) \quad \text{when } x = 8, y = 6, \text{ and } z = 2$$
$$= 192 + 32 - 36$$
$$= 188$$

Classroom Example

Evaluate
$$\frac{6a + b}{4a - b}$$
for $a = 4$ and $b = 6$.

EXAMPLE 12   Evaluate $\dfrac{5c + d}{3c - d}$ for $c = 12$ and $d = 4$.

Solution

$$\frac{5c + d}{3c - d} = \frac{5(12) + 4}{3(12) - 4} \quad \text{for } c = 12 \text{ and } d = 4$$
$$= \frac{60 + 4}{36 - 4} = \frac{64}{32} = 2$$

Classroom Example
Evaluate $(4x + y)(7x - 2y)$,
when $x = 2$ and $y = 5$.

EXAMPLE 13   Evaluate $(2x + 5y)(3x - 2y)$, when $x = 6$ and $y = 3$.

Solution

$$(2x + 5y)(3x - 2y) = (2 \cdot 6 + 5 \cdot 3)(3 \cdot 6 - 2 \cdot 3) \quad \text{when } x = 6 \text{ and } y = 3$$
$$= (12 + 15)(18 - 6)$$
$$= (27)(12)$$
$$= 324$$

Concept Quiz 1.1

For Problems 1–10, answer true or false.

1. The expression "ab" indicates the sum of a and b.

2. Any of the following notations, $(a)b$, $a \cdot b$, $a(b)$, can be used to indicate the product of a and b.

3. The phrase $2x + y - 4z$ is called "an algebraic expression."

4. A set is a collection of objects, and the objects are called "terms."

5. The sets $\{2, 4, 6, 8\}$ and $\{6, 4, 8, 2\}$ are equal.

6. The set $\{1, 3, 5, 7, \ldots\}$ has a last element of 99.

7. The null set has one element.

8. To evaluate $24 \div 6 \cdot 2$, the first operation that should be performed is to multiply 6 times 2.

9. To evaluate $6 + 8 \cdot 3$, the first operation that should be performed is to multiply 8 times 3.

10. The algebraic expression $2(x + y)$ simplifies to 24 if x is replaced by 10, and y is replaced by 0.

Problem Set 1.1

For Problems 1–34, simplify each numerical expression. **(Objective 2)**

1. $9 + 14 - 7$

2. $32 - 14 + 6$

3. $7(14 - 9)$

4. $8(6 + 12)$

5. $16 + 5 \cdot 7$

6. $18 - 3(5)$

7. $4(12 + 9) - 3(8 - 4)$

8. $7(13 - 4) - 2(19 - 11)$

9. $4(7) + 6(9)$

10. $8(7) - 4(8)$

11. $6 \cdot 7 + 5 \cdot 8 - 3 \cdot 9$

12. $8(13) - 4(9) + 2(7)$

13. $(6 + 9)(8 - 4)$

14. $(15 - 6)(13 - 4)$

15. $6 + 4[3(9 - 4)]$

16. $92 - 3[2(5 - 2)]$

17. $16 \div 8 \cdot 4 + 36 \div 4 \cdot 2$

18. $7 \cdot 8 \div 4 - 72 \div 12$

19. $\dfrac{8 + 12}{4} - \dfrac{9 + 15}{8}$

20. $\dfrac{19 - 7}{6} + \dfrac{38 - 14}{3}$

21. $56 - [3(9 - 6)]$

22. $17 + 2[3(4 - 2)]$

23. $7 \cdot 4 \cdot 2 \div 8 + 14$

24. $14 \div 7 \cdot 8 - 35 \div 7 \cdot 2$

25. $32 \div 8 \cdot 2 + 24 \div 6 - 1$

26. $48 \div 12 + 7 \cdot 2 \div 2 - 1$

27. $4 \cdot 9 \div 12 + 18 \div 2 + 3$

28. $5 \cdot 8 \div 4 - 8 \div 4 \cdot 3 + 6$

29. $\dfrac{6(8 - 3)}{3} + \dfrac{12(7 - 4)}{9}$

30. $\dfrac{3(17 - 9)}{4} + \dfrac{9(16 - 7)}{3}$

31. $83 - \dfrac{4(12 - 7)}{5}$

32. $78 - \dfrac{6(21 - 9)}{4}$

33. $\dfrac{4 \cdot 6 + 5 \cdot 3}{7 + 2 \cdot 3} + \dfrac{7 \cdot 9 + 6 \cdot 5}{3 \cdot 5 + 8 \cdot 2}$

34. $\dfrac{7 \cdot 8 + 4}{5 \cdot 8 - 10} + \dfrac{9 \cdot 6 - 4}{6 \cdot 5 - 20}$

For Problems 35–54, evaluate each algebraic expression for the given values of the variables. **(Objective 3)**

35. $7x + 4y$ for $x = 6$ and $y = 8$

36. $8x + 6y$ for $x = 9$ and $y = 5$

37. $16a - 9b$ for $a = 3$ and $b = 4$

38. $14a - 5b$ for $a = 7$ and $b = 9$

39. $4x + 7y + 3xy$ for $x = 4$ and $y = 9$

40. $x + 8y + 5xy$ for $x = 12$ and $y = 3$

41. $14xz + 6xy - 4yz$ for $x = 8$, $y = 5$, and $z = 7$

42. $9xy - 4xz + 3yz$ for $x = 7$, $y = 3$, and $z = 2$

43. $\dfrac{54}{n} + \dfrac{n}{3}$ for $n = 9$

44. $\dfrac{n}{4} + \dfrac{60}{n} - \dfrac{n}{6}$ for $n = 12$

45. $\dfrac{y + 16}{6} + \dfrac{50 - y}{3}$ for $y = 8$

46. $\dfrac{w + 57}{9} + \dfrac{90 - w}{7}$ for $w = 6$

47. $(x + y)(x - y)$ for $x = 8$ and $y = 3$

48. $(x + 2y)(2x - y)$ for $x = 7$ and $y = 4$

49. $(5x - 2y)(3x + 4y)$ for $x = 3$ and $y = 6$

50. $(3a + b)(7a - 2b)$ for $a = 5$ and $b = 7$

51. $6 + 3[2(x + 4)]$ for $x = 7$

52. $9 + 4[3(x + 3)]$ for $x = 6$

53. $81 - 2[5(n + 4)]$ for $n = 3$

54. $78 - 3[4(n - 2)]$ for $n = 4$

For Problems 55–60, find the value of $\dfrac{bh}{2}$ for each set of values for the variables b and h. **(Objective 3)**

55. $b = 8$ and $h = 12$

56. $b = 6$ and $h = 14$

57. $b = 7$ and $h = 6$

58. $b = 9$ and $h = 4$

59. $b = 16$ and $h = 5$

60. $b = 18$ and $h = 13$

For Problems 61–66, find the value of $\dfrac{h(b_1 + b_2)}{2}$ for each set of values for the variables h, b_1, and b_2. (Subscripts are used to indicate that b_1 and b_2 are different variables.)

61. $h = 17$, $b_1 = 14$, and $b_2 = 6$

62. $h = 9$, $b_1 = 12$, and $b_2 = 16$

63. $h = 8$, $b_1 = 17$, and $b_2 = 24$

64. $h = 12$, $b_1 = 14$, and $b_2 = 5$

65. $h = 18$, $b_1 = 6$, and $b_2 = 11$

66. $h = 14$, $b_1 = 9$, and $b_2 = 7$

67. You should be able to do calculations like those in Problems 1–34 *with* and *without* a calculator. Be sure that you can do Problems 1–34 *with* your calculator, and make use of the parentheses key when appropriate.

Thoughts Into Words

68. Explain the difference between a numerical expression and an algebraic expression.

69. Your friend keeps getting an answer of 45 when simplifying $3 + 2(9)$. What mistake is he making and how would you help him?

Further Investigations

Grouping symbols can affect the order in which the arithmetic operations are performed. For the following problems, insert parentheses so that the expression is equal to the given value.

70. Insert parentheses so that $36 + 12 \div 3 + 3 + 6 \cdot 2$ is equal to 20.

71. Insert parentheses so that $36 + 12 \div 3 + 3 + 6 \cdot 2$ is equal to 50.

72. Insert parentheses so that $36 + 12 \div 3 + 3 + 6 \cdot 2$ is equal to 38.

73. Insert parentheses so that $36 + 12 \div 3 + 3 + 6 \cdot 2$ is equal to 55.

Answers to the Concept Quiz

1. False **2.** True **3.** True **4.** False **5.** True **6.** False **7.** False **8.** False **9.** True **10.** False

1.2 Prime and Composite Numbers

OBJECTIVES

1 Identify whole numbers greater than one as prime or composite

2 Factor a whole number into a product of prime numbers

3 Find the greatest common factor of two or more whole numbers

4 Find the least common multiple of two or more whole numbers

Occasionally, terms in mathematics are given a special meaning in the discussion of a particular topic. Such is the case with the term "divides" as it is used in this section. We say that 6 *divides* 18, because 6 times the whole number 3 produces 18; but 6 *does not divide* 19, because there is no whole number such that 6 times the number produces 19. Likewise, 5 *divides* 35, because 5 times the whole number 7 produces 35; 5 *does not divide* 42, because there is no whole number such that 5 times the number produces 42. We present the following general definition.

Definition 1.1

Given that a and b are whole numbers, with a not equal to zero, a *divides* b if and only if there exists a whole number k such that $a \cdot k = b$.

Remark: Notice the use of variables, a, b, and k, in the statement of a *general* definition. Also note that the definition merely generalizes the concept of *divides*, which was introduced in the specific examples prior to the definition.

The following statements further clarify Definition 1.1. Pay special attention to the italicized words, because they indicate some of the terminology used for this topic.

1. 8 *divides* 56, because $8 \cdot 7 = 56$.

2. 7 *does not divide* 38, because there is no whole number, k, such that $7 \cdot k = 38$.

3. 3 is a *factor* of 27, because $3 \cdot 9 = 27$.

4. 4 is *not a factor* of 38, because there is no whole number, k, such that $4 \cdot k = 38$.

5. 35 is a *multiple* of 5, because $5 \cdot 7 = 35$.

6. 29 is *not a multiple* of 7, because there is no whole number, k, such that $7 \cdot k = 29$.

We use the *factor* terminology extensively. We say that 7 and 8 are factors of 56 because $7 \cdot 8 = 56$; 4 and 14 are also factors of 56 because $4 \cdot 14 = 56$. The **factors** of a number are also divisors of the number.

Now consider two special kinds of whole numbers called *prime numbers* and *composite numbers* according to the following definition.

> **Definition 1.2**
>
> A **prime number** is a whole number, greater than 1, that has no factors (divisors) other than itself and 1. Whole numbers, greater than 1, which are not prime numbers, are called **composite numbers**.

The prime numbers less than 50 are 2, 3, 5, 7, 11, 13, 17, 19, 23, 29, 31, 37, 41, 43, and 47. Notice that each of these has no factors other than itself and 1. The set of prime numbers is an infinite set; that is, the prime numbers go on forever, and there is no *largest* prime number.

We can express every composite number as the indicated product of prime numbers—also called the **prime factored form** of the number. Consider the following examples.

$$4 = 2 \cdot 2 \qquad 6 = 2 \cdot 3 \qquad 8 = 2 \cdot 2 \cdot 2 \qquad 10 = 2 \cdot 5 \qquad 12 = 2 \cdot 2 \cdot 3$$

In each case we expressed a composite number as the indicated product of prime numbers.

There are various procedures to find the prime factors of a given composite number. For our purposes, the simplest technique is to factor the given composite number into any two easily recognized factors and then to continue to factor each of these until we obtain only prime factors. Consider these examples.

$$18 = 2 \cdot 9 = 2 \cdot 3 \cdot 3 \qquad\qquad 27 = 3 \cdot 9 = 3 \cdot 3 \cdot 3$$
$$24 = 4 \cdot 6 = 2 \cdot 2 \cdot 2 \cdot 3 \qquad 150 = 10 \cdot 15 = 2 \cdot 5 \cdot 3 \cdot 5$$

It does not matter which two factors we choose first. For example, we might start by expressing 18 as $3 \cdot 6$ and then factor 6 into $2 \cdot 3$, which produces a final result of $18 = 3 \cdot 2 \cdot 3$. Either way, 18 contains two prime factors of 3 and one prime factor of 2. The order in which we write the prime factors is not important.

Greatest Common Factor

We can use the prime factorization form of two composite numbers to conveniently find their *greatest common factor*. Consider the following example.

$$42 = 2 \cdot 3 \cdot 7$$
$$70 = 2 \cdot 5 \cdot 7$$

Notice that 2 is a factor of both, as is 7. Therefore, 14 (the product of 2 and 7) is the **greatest common factor** of 42 and 70. In other words, 14 is the largest whole number that divides both 42 and 70. The following examples should further clarify the process of finding the greatest common factor of two or more numbers.

Classroom Example
Find the greatest common factor
of 45 and 150.

EXAMPLE 1   Find the greatest common factor of 48 and 60.

Solution

$$48 = 2 \cdot 2 \cdot 2 \cdot 2 \cdot 3$$
$$60 = 2 \cdot 2 \cdot 3 \cdot 5$$

Since two 2s and one 3 are common to both, the greatest common factor of 48 and 60 is $2 \cdot 2 \cdot 3 = $ **12**.

Classroom Example
Find the greatest common factor
of 50 and 105.

EXAMPLE 2   Find the greatest common factor of 21 and 75.

Solution

$$21 = 3 \cdot 7$$
$$75 = 3 \cdot 5 \cdot 5$$

Since only one 3 is common to both, the greatest common factor is **3**.

Classroom Example
Find the greatest common factor
of 18 and 35.

EXAMPLE 3   Find the greatest common factor of 24 and 35.

Solution

$$24 = 2 \cdot 2 \cdot 2 \cdot 3$$
$$35 = 5 \cdot 7$$

Since there are no common prime factors, the greatest common factor is **1**.

The concept of *greatest common factor* can be extended to more than two numbers, as the next example demonstrates.

Classroom Example
Find the greatest common factor
of 70, 175, and 245.

EXAMPLE 4   Find the greatest common factor of 24, 56, and 120.

Solution

$$24 = 2 \cdot 2 \cdot 2 \cdot 3$$
$$56 = 2 \cdot 2 \cdot 2 \cdot 7$$
$$120 = 2 \cdot 2 \cdot 2 \cdot 3 \cdot 5$$

Since three 2s are common to the numbers, the greatest common factor of 24, 56, and 120 is $2 \cdot 2 \cdot 2 = $ **8**.

Least Common Multiple

We stated earlier in this section that 35 is a *multiple of* 5 because $5 \cdot 7 = 35$. The set of all whole numbers that are multiples of 5 consists of 0, 5, 10, 15, 20, 25, and so on. In other words, 5 times each successive whole number ($5 \cdot 0 = $ **0**, $5 \cdot 1 = $ **5**, $5 \cdot 2 = $ **10**, $5 \cdot 3 = $ **15**, and so on) produces the multiples of 5. In a like manner, the set of multiples of 4 consists of 0, 4, 8, 12, 16, and so on.

It is sometimes necessary to determine the smallest common *nonzero* multiple of two or more whole numbers. We use the phrase **least common multiple** to designate this nonzero number. For example, the least common multiple of 3 and 4 is 12, which means that 12 is the smallest nonzero multiple of both 3 and 4. Stated another way, 12 is the smallest nonzero whole number that is divisible by both 3 and 4. Likewise, we say that the least common multiple of 6 and 8 is 24.

If we cannot determine the least common multiple by inspection, then the prime factorization form of composite numbers is helpful. Study the solutions to the following examples very carefully so that we can develop a systematic technique for finding the least common multiple of two or more numbers.

EXAMPLE 5   Find the least common multiple of 24 and 36.

Solution

Let's first express each number as a product of prime factors.

$$24 = 2 \cdot 2 \cdot 2 \cdot 3$$
$$36 = 2 \cdot 2 \cdot 3 \cdot 3$$

Since 24 contains three 2s, the least common multiple must have three 2s. Also, since 36 contains two 3s, we need to put two 3s in the least common multiple. The least common multiple of 24 and 36 is therefore $2 \cdot 2 \cdot 2 \cdot 3 \cdot 3 = $ **72**.

If the least common multiple is not obvious by inspection, then we can proceed as follows.

Step 1   Express each number as a product of prime factors.

Step 2   The least common multiple contains each different prime factor as many times as the *most* times it appears in any one of the factorizations from step 1.

EXAMPLE 6   Find the least common multiple of 48 and 84.

Solution

$$48 = 2 \cdot 2 \cdot 2 \cdot 2 \cdot 3$$
$$84 = 2 \cdot 2 \cdot 3 \cdot 7$$

We need four 2s in the least common multiple because of the four 2s in 48. We need one 3 because of the 3 in each of the numbers, and one 7 is needed because of the 7 in 84. The least common multiple of 48 and 84 is $2 \cdot 2 \cdot 2 \cdot 2 \cdot 3 \cdot 7 = $ **336**.

EXAMPLE 7   Find the least common multiple of 12, 18, and 28.

Solution

$$12 = 2 \cdot 2 \cdot 3$$
$$18 = 2 \cdot 3 \cdot 3$$
$$28 = 2 \cdot 2 \cdot 7$$

The least common multiple is $2 \cdot 2 \cdot 3 \cdot 3 \cdot 7 = $ **252**.

EXAMPLE 8   Find the least common multiple of 8 and 9.

Solution

$$8 = 2 \cdot 2 \cdot 2$$
$$9 = 3 \cdot 3$$

The least common multiple is $2 \cdot 2 \cdot 2 \cdot 3 \cdot 3 = $ **72**.

Concept Quiz 1.2

For Problems 1–10, answer true or false.

1. Every even whole number greater than 2 is a composite number.
2. Two is the only even prime number.
3. One is a prime number.
4. The prime factored form of 24 is 2 · 2 · 6.
5. Some whole numbers are both prime and composite numbers.
6. The greatest common factor of 36 and 64 is 4.
7. The greatest common factor of 24, 54, and 72 is 8.
8. The least common multiple of 9 and 12 is 72.
9. The least common multiple of 8, 9, and 18 is 72.
10. 161 is a prime number.

Problem Set 1.2

For Problems 1–20, classify each statement as true or false.

1. 8 divides 56
2. 9 divides 54
3. 6 does not divide 54
4. 7 does not divide 42
5. 96 is a multiple of 8
6. 78 is a multiple of 6
7. 54 is not a multiple of 4
8. 64 is not a multiple of 6
9. 144 is divisible by 4
10. 261 is divisible by 9
11. 173 is divisible by 3
12. 149 is divisible by 7
13. 11 is a factor of 143
14. 11 is a factor of 187
15. 9 is a factor of 119
16. 8 is a factor of 98
17. 3 is a prime factor of 57
18. 7 is a prime factor of 91
19. 4 is a prime factor of 48
20. 6 is a prime factor of 72

For Problems 21–30, fill in the blanks with a pair of numbers that has the indicated product and the indicated sum. For example, $\underline{8} \cdot \underline{5} = 40$ and $\underline{8} + \underline{5} = 13$.

21. ___ · ___ = 24   and   ___ + ___ = 11
22. ___ · ___ = 12   and   ___ + ___ = 7
23. ___ · ___ = 24   and   ___ + ___ = 14
24. ___ · ___ = 25   and   ___ + ___ = 26
25. ___ · ___ = 36   and   ___ + ___ = 13
26. ___ · ___ = 18   and   ___ + ___ = 11
27. ___ · ___ = 50   and   ___ + ___ = 15
28. ___ · ___ = 50   and   ___ + ___ = 27
29. ___ · ___ = 9   and   ___ + ___ = 10
30. ___ · ___ = 48   and   ___ + ___ = 16

For Problems 31–40, classify each number as prime or composite. **(Objective 1)**

31. 53   32. 57
33. 59   34. 61
35. 91   36. 81
37. 89   38. 97
39. 111   40. 101

For Problems 41–50, familiarity with a few basic divisibility rules will be helpful for determining the prime factors. The divisibility rules for 2, 3, 5, and 9 are as follows.

Rule for 2

A whole number is divisible by 2 if and only if the units digit of its base-ten numeral is divisible by 2. (In other words, the units digit must be 0, 2, 4, 6, or 8.)

EXAMPLES   68 is divisible by 2 because 8 is divisible by 2.

57 is not divisible by 2 because 7 is not divisible by 2.

Rule for 3

A whole number is divisible by 3 if and only if the sum of the digits of its base-ten numeral is divisible by 3.

EXAMPLES   51 is divisible by 3 because $5 + 1 = 6$, and 6 is divisible by 3.

144 is divisible by 3 because $1 + 4 + 4 = 9$, and 9 is divisible by 3.

133 is not divisible by 3 because $1 + 3 + 3 = 7$, and 7 is not divisible by 3.

Rule for 5

A whole number is divisible by 5 if and only if the units digit of its base-ten numeral is divisible by 5. (In other words, the units digit must be 0 or 5.)

EXAMPLES   115 is divisible by 5 because 5 is divisible by 5.

172 is not divisible by 5 because 2 is not divisible by 5.

Rule for 9

A whole number is divisible by 9 if and only if the sum of the digits of its base-ten numeral is divisible by 9.

EXAMPLES   765 is divisible by 9 because $7 + 6 + 5 = 18$, and 18 is divisible by 9.

147 is not divisible by 9 because $1 + 4 + 7 = 12$, and 12 is not divisible by 9.

Use these divisibility rules to help determine the prime factored form of the following numbers. **(Objective 2)**

41. 118   **42.** 76

43. 201   **44.** 123

45. 85   **46.** 115

47. 117   **48.** 441

49. 129   **50.** 153

For Problems 51–62, factor each composite number into a product of prime numbers. For example, $18 = 2 \cdot 3 \cdot 3$. **(Objective 2)**

51. 26   **52.** 16

53. 36   **54.** 80

55. 49   **56.** 92

57. 56   **58.** 144

59. 120   **60.** 84

61. 135   **62.** 98

For Problems 63–74, find the greatest common factor of the given numbers. **(Objective 3)**

63. 12 and 16   **64.** 30 and 36

65. 56 and 64   **66.** 72 and 96

67. 63 and 81   **68.** 60 and 72

69. 84 and 96   **70.** 48 and 52

71. 36, 72, and 90   **72.** 27, 54, and 63

73. 48, 60, and 84   **74.** 32, 80, and 96

For Problems 75–86, find the least common multiple of the given numbers. **(Objective 4)**

75. 6 and 8   **76.** 8 and 12

77. 12 and 16   **78.** 9 and 12

79. 28 and 35   **80.** 42 and 66

81. 49 and 56   **82.** 18 and 24

83. 8, 12, and 28   **84.** 6, 10, and 12

85. 9, 15, and 18   **86.** 8, 14, and 24

Thoughts Into Words

87. How would you explain the concepts of *greatest common factor* and *least common multiple* to a friend who missed class during that discussion?

88. Is it always true that the greatest common factor of two numbers is less than the least common multiple of those same two numbers? Explain your answer.

Further Investigations

89. The numbers 0, 2, 4, 6, 8, and so on are multiples of 2. They are also called *even* numbers. Why is 2 the only even prime number?

90. Find the smallest nonzero whole number that is divisible by 2, 3, 4, 5, 6, 7, and 8.

91. Find the smallest whole number, greater than 1, that produces a remainder of 1 when divided by 2, 3, 4, 5, or 6.

92. What is the greatest common factor of x and y if x and y are both prime numbers, and x does not equal y? Explain your answer.

93. What is the greatest common factor of x and y if x and y are nonzero whole numbers, and y is a multiple of x? Explain your answer.

94. What is the least common multiple of x and y if they are both prime numbers, and x does not equal y? Explain your answer.

95. What is the least common multiple of x and y if the greatest common factor of x and y is 1? Explain your answer.

Answers to the Concept Quiz
1. True   **2.** True   **3.** False   **4.** False   **5.** False   **6.** True   **7.** False   **8.** False   **9.** True   **10.** False

---

1.3   Integers: Addition and Subtraction

OBJECTIVES

1. Know the terminology associated with sets of integers

2. Add and subtract integers

3. Evaluate algebraic expressions for integer values

4. Apply the concepts of adding and subtracting integers to model problems

"A record temperature of 35° *below* zero was recorded on this date in 1904." "The PO stock closed *down* 3 points yesterday." "On a first-down sweep around the left end, Moser *lost* 7 yards." "The Widget Manufacturing Company reported *assets* of 50 million dollars and *liabilities* of 53 million dollars for 2010." These examples illustrate our need for negative numbers.

The number line is a helpful visual device for our work at this time. We can associate the set of whole numbers with evenly spaced points on a line as indicated in Figure 1.1. For each nonzero whole number we can associate its *negative* to the left of zero; with 1 we associate −1, with 2 we associate −2, and so on, as indicated in Figure 1.2. The set of whole numbers along with −1, −2, −3, and so on, is called the set of **integers**.

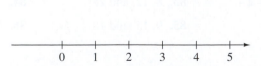

Figure 1.1

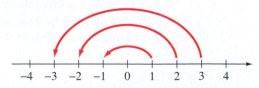

Figure 1.2

The following terminology is used in reference to the integers.

$$\{\ldots, -3, -2, -1, 0, 1, 2, 3, \ldots\} \qquad \text{Integers}$$

$$\{1, 2, 3, 4, \ldots\} \qquad \text{Positive integers}$$

$\{0, 1, 2, 3, 4, \ldots\}$ Nonnegative integers

$\{\ldots, -3, -2, -1\}$ Negative integers

$\{\ldots, -3, -2, -1, 0\}$ Nonpositive integers

The symbol -1 can be read as "negative one," "opposite of one," or "additive inverse of one." The opposite-of and additive-inverse-of terminology is very helpful when working with variables. The symbol $-x$, read as "opposite of x" or "additive inverse of x," emphasizes an important issue: Since x can be any integer, $-x$ (the opposite of x) can be zero, positive, or negative. If x is a positive integer, then $-x$ is negative. If x is a negative integer, then $-x$ is positive. If x is zero, then $-x$ is zero. These statements are written as follows and illustrated on the number lines in Figure 1.3.

If $x = 3$,
then $-x = -(3) = -3$.

If $x = -3$,
then $-x = -(-3) = 3$.

If $x = 0$,
then $-x = -(0) = 0$.

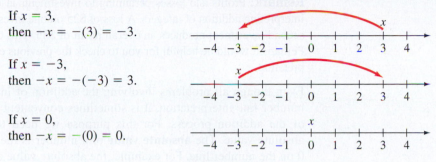

Figure 1.3

From this discussion we also need to recognize the following general property.

Property 1.1

If a is any integer, then

$$-(-a) = a$$

(The opposite of the opposite of any integer is the integer itself.)

Addition of Integers

The number line is also a convenient visual aid for interpreting the addition of integers. In Figure 1.4 we see number line interpretations for the following examples.

Problem	Number line interpretation	Sum
$3 + 2$	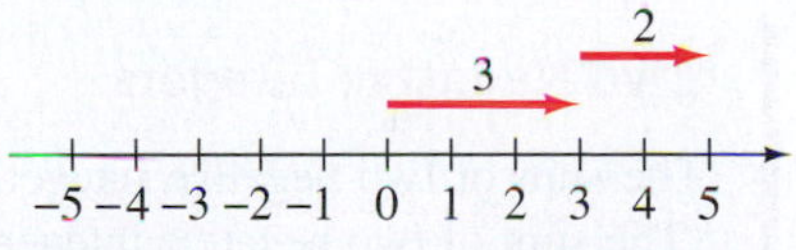	$3 + 2 = 5$
$3 + (-2)$	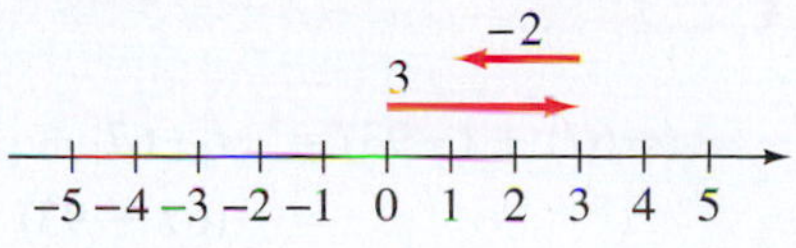	$3 + (-2) = 1$
$-3 + 2$	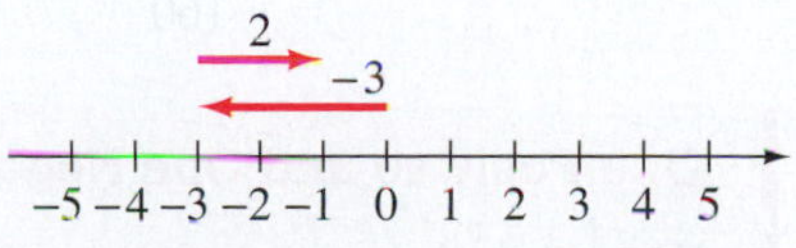	$-3 + 2 = -1$
$-3 + (-2)$	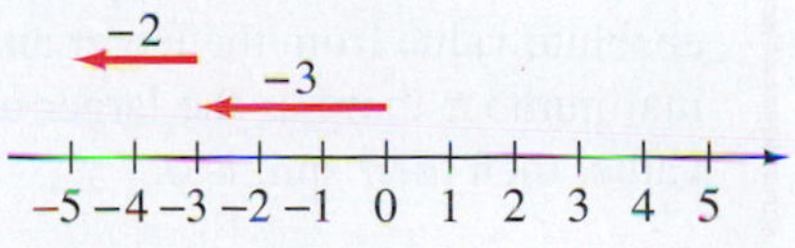	$-3 + (-2) = -5$

Figure 1.4

Once you acquire a feeling of movement on the number line, a mental image of this movement is sufficient. Consider the following addition problems, and mentally picture the number line interpretation. Be sure that you agree with all of our answers.

$$5 + (-2) = 3 \qquad -6 + 4 = -2 \qquad -8 + 11 = 3$$

$$-7 + (-4) = -11 \qquad -5 + 9 = 4 \qquad 9 + (-2) = 7$$

$$14 + (-17) = -3 \qquad 0 + (-4) = -4 \qquad 6 + (-6) = 0$$

The last problem illustrates a general property that should be noted: **Any integer plus its opposite equals zero**.

Remark: Profits and losses pertaining to investments also provide a good physical model for interpreting addition of integers. A loss of $25 on one investment along with a profit of $60 on a second investment produces an overall profit of $35. This can be expressed as $-25 + 60 = 35$. Perhaps it would be helpful for you to check the previous examples using a profit and loss interpretation.

Even though all problems involving the addition of integers could be done by using the number line interpretation, it is sometimes convenient to give a more precise description of the addition process. For this purpose we need to briefly consider the concept of absolute value. The **absolute value** of a number is the distance between the number and 0 on the number line. For example, the absolute value of 6 is 6. The absolute value of -6 is also 6. The absolute value of 0 is 0. Symbolically, absolute value is denoted with vertical bars. Thus we write

$$|6| = 6 \qquad |-6| = 6 \qquad |0| = 0$$

Notice that the absolute value of a positive number is the number itself, but the absolute value of a negative number is its opposite. Thus the absolute value of any number except 0 is positive, and the absolute value of 0 is 0.

We can describe precisely the **addition of integers** by using the concept of absolute value as follows.

Two Positive Integers

The sum of two positive integers is the sum of their absolute values. (The sum of two positive integers is a positive integer.)

$$43 + 54 = |43| + |54| = 43 + 54 = 97$$

Two Negative Integers

The sum of two negative integers is the opposite of the sum of their absolute values. (The sum of two negative integers is a negative integer.)

$$(-67) + (-93) = -(|-67| + |-93|)$$
$$= -(67 + 93)$$
$$= -160$$

One Positive and One Negative Integer

We can find the sum of a positive and a negative integer by subtracting the smaller absolute value from the larger absolute value and giving the result the sign of the original number that has the larger absolute value. If the integers have the same absolute value, then their sum is 0.

$$82 + (-40) = |82| - |-40| = 82 - 40 = 42$$
$$74 + (-90) = -(|-90| - |74|)$$
$$= -(90 - 74) = -16$$
$$(-17) + 17 = |-17| - |17|$$
$$= 17 - 17 = 0$$

Zero and Another Integer

The sum of 0 and any integer is the integer itself.

$$0 + (-46) = -46$$
$$72 + 0 = 72$$

The following examples further demonstrate how to add integers. Be sure that you agree with each of the results.

$$-18 + (-56) = -(|-18| + |-56|) = -(18 + 56) = -74$$
$$-71 + (-32) = -(|-71| + |-32|) = -(71 + 32) = -103$$
$$64 + (-49) = |64| - |-49| = 64 - 49 = 15$$
$$-56 + 93 = |93| - |-56| = 93 - 56 = 37$$
$$-114 + 48 = -(|-114| - |48|) = -(114 - 48) = -66$$
$$45 + (-73) = -(|-73| - |45|) = -(73 - 45) = -28$$
$$46 + (-46) = 0 \qquad -48 + 0 = -48$$
$$(-73) + 73 = 0 \qquad 0 + (-81) = -81$$

It is true that this absolute value approach does precisely describe the process of adding integers, but don't forget about the number line interpretation. Included in the next problem set are other physical models for interpreting the addition of integers. You may find these models helpful.

Subtraction of Integers

The following examples illustrate a relationship between addition and subtraction of whole numbers.

$$7 - 2 = 5 \quad \text{because } 2 + 5 = 7$$
$$9 - 6 = 3 \quad \text{because } 6 + 3 = 9$$
$$5 - 1 = 4 \quad \text{because } 1 + 4 = 5$$

This same relationship between addition and subtraction holds for *all integers*.

$$5 - 6 = -1 \quad \text{because } 6 + (-1) = 5$$
$$-4 - 9 = -13 \quad \text{because } 9 + (-13) = -4$$
$$-3 - (-7) = 4 \quad \text{because } -7 + 4 = -3$$
$$8 - (-3) = 11 \quad \text{because } -3 + 11 = 8$$

Now consider a further observation:

$$5 - 6 = -1 \qquad \text{and} \qquad 5 + (-6) = -1$$
$$-4 - 9 = -13 \qquad \text{and} \qquad -4 + (-9) = -13$$
$$-3 - (-7) = 4 \qquad \text{and} \qquad -3 + 7 = 4$$
$$8 - (-3) = 11 \qquad \text{and} \qquad 8 + 3 = 11$$

The previous examples help us realize that we can state the subtraction of integers in terms of the addition of integers. A general description for the **subtraction of integers** follows.

> ### Subtraction of Integers
>
> If a and b are integers, then $a - b = a + (-b)$.

It may be helpful for you to read $a - b = a + (-b)$ as "a minus b is equal to a plus the opposite of b." Every subtraction problem can be changed to an equivalent addition problem as illustrated by the following examples.

$$6 - 13 = 6 + (-13) = -7$$
$$9 - (-12) = 9 + 12 = 21$$
$$-8 - 13 = -8 + (-13) = -21$$
$$-7 - (-8) = -7 + 8 = 1$$

It should be apparent that the addition of integers is a key operation. The ability to effectively add integers is a necessary skill for further algebraic work.

Evaluating Algebraic Expressions

Let's conclude this section by evaluating some algebraic expressions using negative and positive integers.

Classroom Example
Evaluate each algebraic expression for the given values of the variables.
(a) $m - n$ for $m = -10$ and $n = 23$
(b) $-x + y$ for $x = -11$ and $y = -2$
(c) $-c - d$ for $c = -57$ and $d = -4$

EXAMPLE 1

Evaluate each algebraic expression for the given values of the variables.

(a) $x - y$ for $x = -12$ and $y = 20$

(b) $-a + b$ for $a = -8$ and $b = -6$

(c) $-x - y$ for $x = 14$ and $y = -7$

Solution

(a) $x - y = -12 - 20$ when $x = -12$ and $y = 20$
$$= -12 + (-20) \quad \text{Change to addition}$$
$$= -32$$

(b) $-a + b = -(-8) + (-6)$ when $a = -8$ and $b = -6$
$$= 8 + (-6) \quad \text{Note the use of parentheses when substituting the values}$$
$$= 2$$

(c) $-x - y = -(14) - (-7)$ when $x = 14$ and $y = -7$
$$= -14 + 7 \quad \text{Change to addition}$$
$$= -7$$

Concept Quiz 1.3

For Problems 1–4, match the letters of the description with the set of numbers.

1. $\{\ldots, -3, -2, -1\}$ A. Positive integers
2. $\{1, 2, 3, \ldots\}$ B. Negative integers
3. $\{0, 1, 2, 3, \ldots\}$ C. Nonnegative integers
4. $\{\ldots, -3, -2, -1, 0\}$ D. Nonpositive integers

For Problems 5–10, answer true or false.

5. The number zero is considered to be a positive integer.

6. The number zero is considered to be a negative integer.

7. The absolute value of a number is the distance between the number and one on the number line.

8. The $|-4|$ is -4.

9. The opposite of -5 is 5.

10. a minus b is equivalent to a plus the opposite of b.

Problem Set 1.3

For Problems 1–10, use the number line interpretation to find each sum. **(Objective 2)**

1. $5 + (-3)$ **2.** $7 + (-4)$

3. $-6 + 2$ **4.** $-9 + 4$

5. $-3 + (-4)$ **6.** $-5 + (-6)$

7. $8 + (-2)$ **8.** $12 + (-7)$

9. $5 + (-11)$ **10.** $4 + (-13)$

For Problems 11–30, find each sum. **(Objective 2)**

11. $17 + (-9)$ **12.** $16 + (-5)$

13. $8 + (-19)$ **14.** $9 + (-14)$

15. $-7 + (-8)$ **16.** $-6 + (-9)$

17. $-15 + 8$ **18.** $-22 + 14$

19. $-13 + (-18)$ **20.** $-15 + (-19)$

21. $-27 + 8$ **22.** $-29 + 12$

23. $32 + (-23)$ **24.** $27 + (-14)$

25. $-25 + (-36)$ **26.** $-34 + (-49)$

27. $54 + (-72)$ **28.** $48 + (-76)$

29. $-34 + (-58)$ **30.** $-27 + (-36)$

For Problems 31–50, subtract as indicated. **(Objective 2)**

31. $3 - 8$ **32.** $5 - 11$

33. $-4 - 9$ **34.** $-7 - 8$

35. $5 - (-7)$ **36.** $9 - (-4)$

37. $-6 - (-12)$ **38.** $-7 - (-15)$

39. $-11 - (-10)$ **40.** $-14 - (-19)$

41. $-18 - 27$ **42.** $-16 - 25$

43. $34 - 63$ **44.** $25 - 58$

45. $45 - 18$ **46.** $52 - 38$

47. $-21 - 44$ **48.** $-26 - 54$

49. $-53 - (-24)$ **50.** $-76 - (-39)$

For Problems 51–66, add or subtract as indicated. **(Objective 2)**

51. $6 - 8 - 9$ **52.** $5 - 9 - 4$

53. $-4 - (-6) + 5 - 8$ **54.** $-3 - 8 + 9 - (-6)$

55. $5 + 7 - 8 - 12$ **56.** $-7 + 9 - 4 - 12$

57. $-6 - 4 - (-2) + (-5)$

58. $-8 - 11 - (-6) + (-4)$

59. $-6 - 5 - 9 - 8 - 7$

60. $-4 - 3 - 7 - 8 - 6$

61. $7 - 12 + 14 - 15 - 9$

62. $8 - 13 + 17 - 15 - 19$

63. $-11 - (-14) + (-17) - 18$

64. $-15 + 20 - 14 - 18 + 9$

65. $16 - 21 + (-15) - (-22)$

66. $17 - 23 - 14 - (-18)$

The horizontal format is used extensively in algebra, but occasionally the vertical format shows up. Some exposure to the vertical format is therefore needed. Find the following sums for Problems 67–78. **(Objective 2)**

67. $\begin{array}{r} 5 \\ -9 \\ \hline \end{array}$ **68.** $\begin{array}{r} 8 \\ -13 \\ \hline \end{array}$

69. $\begin{array}{r} -13 \\ -18 \\ \hline \end{array}$ **70.** $\begin{array}{r} -14 \\ -28 \\ \hline \end{array}$

71. $\begin{array}{r} -18 \\ 9 \\ \hline \end{array}$ **72.** $\begin{array}{r} -17 \\ 9 \\ \hline \end{array}$

73. $\begin{array}{r} -21 \\ 39 \\ \hline \end{array}$ **74.** $\begin{array}{r} -15 \\ 32 \\ \hline \end{array}$

75. $\begin{array}{r} 27 \\ -19 \\ \hline \end{array}$ **76.** $\begin{array}{r} 31 \\ -18 \\ \hline \end{array}$

77. $\begin{array}{r} -53 \\ 24 \\ \hline \end{array}$ **78.** $\begin{array}{r} 47 \\ -28 \\ \hline \end{array}$

For Problems 79–90, do the subtraction problems in vertical format. **(Objective 2)**

79. 5
    12

80. 8
    19

81. 6
    −9

82. 13
    −7

83. −7
    −8

84. −6
    −5

85. 17
    −19

86. 18
    −14

87. −23
    16

88. −27
    15

89. −12
    12

90. −13
    −13

For Problems 91–100, evaluate each algebraic expression for the given values of the variables. **(Objective 3)**

91. $x - y$ for $x = -6$ and $y = -13$

92. $-x - y$ for $x = -7$ and $y = -9$

93. $-x + y - z$ for $x = 3$, $y = -4$, and $z = -6$

94. $x - y + z$ for $x = 5$, $y = 6$, and $z = -9$

95. $-x - y - z$ for $x = -2$, $y = 3$, and $z = -11$

96. $-x - y + z$ for $x = -8$, $y = -7$, and $z = -14$

97. $-x + y + z$ for $x = -11$, $y = 7$, and $z = -9$

98. $-x - y - z$ for $x = 12$, $y = -6$, and $z = -14$

99. $x - y - z$ for $x = -15$, $y = 12$, and $z = -10$

100. $x + y - z$ for $x = -18$, $y = 13$, and $z = 8$

A game such as football can be used to interpret addition of integers. A gain of 3 yards on one play followed by a loss of 5 yards on the next play places the ball 2 yards behind the initial line of scrimmage; this could be expressed as $3 + (-5) = -2$. Use this football interpretation to find the following sums for Problems 101–110. **(Objective 4)**

101. $4 + (-7)$

102. $3 + (-5)$

103. $-4 + (-6)$

104. $-2 + (-5)$

105. $-5 + 2$

106. $-10 + 6$

107. $-4 + 15$

108. $-3 + 22$

109. $-12 + 17$

110. $-9 + 21$

For Problems 111–120, refer to the Remark on page 16 and use the profit and loss interpretation for the addition of integers. **(Objective 4)**

111. $60 + (-125)$

112. $50 + (-85)$

113. $-55 + (-45)$

114. $-120 + (-220)$

115. $-70 + 45$

116. $-125 + 45$

117. $-120 + 250$

118. $-75 + 165$

119. $145 + (-65)$

120. $275 + (-195)$

121. The temperature at 5 A.M. was $-17°$F. By noon the temperature had increased by $14°$F. Use the addition of integers to describe this situation and to determine the temperature at noon (see Figure 1.5).

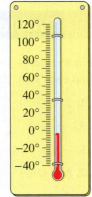

122. The temperature at 6 P.M. was $-6°$F, and by 11 P.M. the temperature had dropped $5°$F. Use the subtraction of integers to describe this situation and to determine the temperature at 11 P.M. (see Figure 1.5).

Figure 1.5

123. Megan shot rounds of 3 over par, 2 under par, 3 under par, and 5 under par for a four-day golf tournament. Use the addition of integers to describe this situation and to determine how much over or under par she was for the tournament.

124. The annual report of a company contained the following figures: a loss of $615,000 for 2007, a loss of $275,000 for 2008, a loss of $70,000 for 2009, and a profit of $115,000 for 2010. Use the addition of integers to describe this situation and to determine the company's total loss or profit for the four-year period.

125. Suppose that during a five-day period, a share of Dell's stock recorded the following gains and losses:

Monday	**Tuesday**	**Wednesday**
lost $2	gained $1	gained $3

Thursday	**Friday**	
gained $1	lost $2	

Use the addition of integers to describe this situation and to determine the amount of gain or loss for the five-day period.

126. The Dead Sea is approximately thirteen hundred ten feet below sea level. Suppose that you are standing eight hundred five feet above the Dead Sea. Use the addition of integers to describe this situation and to determine your elevation.

127. Use your calculator to check your answers for Problems 51–66.

Thoughts Into Words

128. The statement $-6 - (-2) = -6 + 2 = -4$ can be read as "negative six minus negative two equals negative six plus two, which equals negative four." Express in words each of the following.

(a) $8 + (-10) = -2$

(b) $-7 - 4 = -7 + (-4) = -11$

(c) $9 - (-12) = 9 + 12 = 21$

(d) $-5 + (-6) = -11$

129. The algebraic expression $-x - y$ can be read as "the opposite of x minus y." Express in words each of the following.

(a) $-x + y$

(b) $x - y$

(c) $-x - y + z$

Answers to the Concept Quiz

1. B   **2.** A   **3.** C   **4.** D   **5.** False   **6.** False   **7.** False   **8.** False   **9.** True   **10.** True

---

1.4   Integers: Multiplication and Division

OBJECTIVES

1. Multiply and divide integers

2. Evaluate algebraic expressions involving the multiplication and division of integers

3. Apply the concepts of multiplying and dividing integers to model problems

Multiplication of whole numbers may be interpreted as repeated addition. For example, $3 \cdot 4$ means the sum of three 4s; thus, $3 \cdot 4 = 4 + 4 + 4 = 12$. Consider the following examples that use the repeated addition idea to find the product of a positive integer and a negative integer:

$$3(-2) = -2 + (-2) + (-2) = -6$$
$$2(-4) = -4 + (-4) = -8$$
$$4(-1) = -1 + (-1) + (-1) + (-1) = -4$$

Note the use of parentheses to indicate multiplication. Sometimes both numbers are enclosed in parentheses so that we have $(3)(-2)$.

When multiplying whole numbers, the order in which we multiply two factors does not change the product: $2(3) = 6$ and $3(2) = 6$. Using this idea we can now handle a negative number times a positive integer as follows:

$$(-2)(3) = (3)(-2) = (-2) + (-2) + (-2) = -6$$
$$(-3)(2) = (2)(-3) = (-3) + (-3) = -6$$
$$(-4)(3) = (3)(-4) = (-4) + (-4) + (-4) = -12$$

Finally, let's consider the product of two negative integers. The following pattern helps us with the reasoning for this situation:

$$4(-3) = -12$$
$$3(-3) = -9$$
$$2(-3) = -6$$
$$1(-3) = -3$$
$$0(-3) = 0 \qquad \text{The product of 0 and any integer is 0}$$
$$(-1)(-3) = ?$$

Certainly, to continue this pattern, the product of -1 and -3 has to be 3. In general, this type of reasoning helps us to realize that the product of any two negative integers is a positive integer.

Using the concept of absolute value, these three facts precisely describe the **multiplication of integers**:

> **1.** The product of two positive integers or two negative integers is the product of their absolute values.
>
> **2.** The product of a positive and a negative integer (either order) is the opposite of the product of their absolute values.
>
> **3.** The product of zero and any integer is zero.

The following are examples of the multiplication of integers:

$$(-5)(-2) = |-5| \cdot |-2| = 5 \cdot 2 = 10$$
$$(7)(-6) = -(|7| \cdot |-6|) = -(7 \cdot 6) = -42$$
$$(-8)(9) = -(|-8| \cdot |9|) = -(8 \cdot 9) = -72$$
$$(-14)(0) = 0$$
$$(0)(-28) = 0$$

These examples show a step-by-step process for multiplying integers. In reality, however, the key issue is to remember whether the product is positive or negative. In other words, we need to remember that **the product of two positive integers or two negative integers is a positive integer; and the product of a positive integer and a negative integer (in either order) is a negative integer**. Then we can avoid the step-by-step analysis and simply write the results as follows:

$$(7)(-9) = -63$$
$$(8)(7) = 56$$
$$(-5)(-6) = 30$$
$$(-4)(12) = -48$$

Dividing Integers

By looking back at our knowledge of whole numbers, we can get some guidance for our work with integers. We know, for example, that $\dfrac{8}{2} = 4$, because $2 \cdot 4 = 8$. In other words, we can find the quotient of two whole numbers by looking at a related multiplication problem. In the following examples we use this same link between multiplication and division to determine the quotients.

$$\frac{8}{-2} = -4 \quad \text{because } (-2)(-4) = 8$$

$$\frac{-10}{5} = -2 \quad \text{because } (5)(-2) = -10$$

$$\frac{-12}{-4} = 3 \quad \text{because } (-4)(3) = -12$$

$$\frac{0}{-6} = 0 \quad \text{because } (-6)(0) = 0$$

$$\frac{-9}{0} \quad \text{is undefined because no number times 0 produces } -9$$

$$\frac{0}{0} \quad \text{is indeterminate because any number times 0 equals 0}$$

The following three statements precisely describe the **division of integers**:

> 1. The quotient of two positive or two negative integers is the quotient of their absolute values.
>
> 2. The quotient of a positive integer and a negative integer (or a negative and a positive) is the opposite of the quotient of their absolute values.
>
> 3. The quotient of zero and any nonzero number (zero divided by any nonzero number) is zero.

The following are examples of the division of integers:

$$\frac{-8}{-4} = \frac{|-8|}{|-4|} = \frac{8}{4} = 2 \qquad \frac{-14}{2} = -\left(\frac{|-14|}{|2|}\right) = -\left(\frac{14}{2}\right) = -7$$

$$\frac{0}{-4} = 0 \qquad \frac{15}{-3} = -\left(\frac{|15|}{|-3|}\right) = -\left(\frac{15}{3}\right) = -5$$

For practical purposes, when dividing integers the key is to remember whether the quotient is positive or negative. Remember that **the quotient of two positive integers or two negative integers is positive; and the quotient of a positive integer and a negative integer or a negative integer and a positive integer is negative**. We can then simply write the quotients as follows without showing all of the steps:

$$\frac{-18}{-6} = 3 \qquad \frac{-24}{12} = -2 \qquad \frac{36}{-9} = -4$$

Remark: Occasionally, people use the phrase "two negatives make a positive." We hope they realize that the reference is to multiplication and division only; in addition the sum of two negative integers is still a negative integer. It is probably best to avoid such imprecise statements.

Simplifying Numerical Expressions

Now we can simplify numerical expressions involving any or all of the four basic operations with integers. Keep in mind the order of operations given in Section 1.1.

Classroom Example
Simplify $4(-5) - 3(-6) - 7(4)$.

EXAMPLE 1   Simplify $-4(-3) - 7(-8) + 3(-9)$.

Solution

$$
\begin{aligned}
-4(-3) - 7(-8) + 3(-9) &= 12 - (-56) + (-27) \\
&= 12 + 56 + (-27) \\
&= 41
\end{aligned}
$$

Classroom Example
Simplify $\dfrac{-3 + 6(-7)}{-5}$.

EXAMPLE 2   Simplify $\dfrac{-8 - 4(5)}{-4}$.

Solution

$$\frac{-8 - 4(5)}{-4} = \frac{-8 - 20}{-4}$$

$$= \frac{-28}{-4}$$

$$= 7$$

Evaluating Algebraic Expressions

Evaluating algebraic expressions will often involve the use of two or more operations with integers. We use the final examples of this section to represent such situations.

Classroom Example
Find the value of $5m + 4n$, when $m = 3$ and $n = -7$.

EXAMPLE 3   Find the value of $3x + 2y$ when $x = 5$ and $y = -9$.

Solution

$$3x + 2y = 3(5) + 2(-9) \quad \text{when } x = 5 \text{ and } y = -9$$
$$= 15 + (-18)$$
$$= -3$$

Classroom Example
Evaluate $-3x + 11y$ for $x = 5$ and $y = -2$.

EXAMPLE 4   Evaluate $-2a + 9b$ for $a = 4$ and $b = -3$.

Solution

$$-2a + 9b = -2(4) + 9(-3) \quad \text{when } a = 4 \text{ and } b = -3$$
$$= -8 + (-27)$$
$$= -35$$

Classroom Example
Find the value of $\dfrac{3a - 7b}{5}$, when $a = -2$ and $b = -3$.

EXAMPLE 5   Find the value of $\dfrac{x - 2y}{4}$ when $x = -6$ and $y = 5$.

Solution

$$\frac{x - 2y}{4} = \frac{-6 - 2(5)}{4} \quad \text{when } x = -6 \text{ and } y = 5$$
$$= \frac{-6 - 10}{4}$$
$$= \frac{-16}{4}$$
$$= -4$$

Concept Quiz 1.4

For Problems 1–10, answer true or false.

1. The product of two negative integers is a positive integer.
2. The product of a positive integer and a negative integer is a positive integer.
3. When multiplying three negative integers the product is negative.
4. The rules for adding integers and the rules for multiplying integers are the same.
5. The quotient of two negative integers is negative.
6. The quotient of a positive integer and zero is a positive integer.
7. The quotient of a negative integer and zero is zero.
8. The product of zero and any integer is zero.
9. The value of $-3x - y$ when $x = -4$ and $y = 6$ is 6.
10. The value of $2x - 5y - xy$ when $x = -2$ and $y = -7$ is 17.

Problem Set 1.4

For Problems 1–40, find the product or quotient (multiply or divide) as indicated. (Objective 1)

1. $5(-6)$

2. $7(-9)$

3. $\dfrac{-27}{3}$

4. $\dfrac{-35}{5}$

5. $\dfrac{-42}{-6}$

6. $\dfrac{-72}{-8}$

7. $(-7)(8)$

8. $(-6)(9)$

9. $(-5)(-12)$

10. $(-7)(-14)$

11. $\dfrac{96}{-8}$

12. $\dfrac{-91}{7}$

13. $14(-9)$

14. $17(-7)$

15. $(-11)(-14)$

16. $(-13)(-17)$

17. $\dfrac{135}{-15}$

18. $\dfrac{-144}{12}$

19. $\dfrac{-121}{-11}$

20. $\dfrac{-169}{-13}$

21. $(-15)(-15)$

22. $(-18)(-18)$

23. $\dfrac{112}{-8}$

24. $\dfrac{112}{-7}$

25. $\dfrac{0}{-8}$

26. $\dfrac{-8}{0}$

27. $\dfrac{-138}{-6}$

28. $\dfrac{-105}{-5}$

29. $\dfrac{76}{-4}$

30. $\dfrac{-114}{6}$

31. $(-6)(-15)$

32. $\dfrac{0}{-14}$

33. $(-56) \div (-4)$

34. $(-78) \div (-6)$

35. $(-19) \div 0$

36. $(-90) \div 15$

37. $(-72) \div 18$

38. $(-70) \div 5$

39. $(-36)(27)$

40. $(42)(-29)$

For Problems 41–60, simplify each numerical expression. (Objective 2)

41. $3(-4) + 5(-7)$

42. $6(-3) + 5(-9)$

43. $7(-2) - 4(-8)$

44. $9(-3) - 8(-6)$

45. $(-3)(-8) + (-9)(-5)$

46. $(-7)(-6) + (-4)(-3)$

47. $5(-6) - 4(-7) + 3(2)$

48. $7(-4) - 8(-7) + 5(-8)$

49. $\dfrac{13 + (-25)}{-3}$

50. $\dfrac{15 + (-36)}{-7}$

51. $\dfrac{12 - 48}{6}$

52. $\dfrac{16 - 40}{8}$

53. $\dfrac{-7(10) + 6(-9)}{-4}$

54. $\dfrac{-6(8) + 4(-14)}{-8}$

55. $\dfrac{4(-7) - 8(-9)}{11}$

56. $\dfrac{5(-9) - 6(-7)}{3}$

57. $-2(3) - 3(-4) + 4(-5) - 6(-7)$

58. $2(-4) + 4(-5) - 7(-6) - 3(9)$

59. $-1(-6) - 4 + 6(-2) - 7(-3) - 18$

60. $-9(-2) + 16 - 4(-7) - 12 + 3(-8)$

For Problems 61–76, evaluate each algebraic expression for the given values of the variables. (Objective 2)

61. $7x + 5y$ for $x = -5$ and $y = 9$

62. $4a + 6b$ for $a = -6$ and $b = -8$

63. $9a - 2b$ for $a = -5$ and $b = 7$

64. $8a - 3b$ for $a = -7$ and $b = 9$

65. $-6x - 7y$ for $x = -4$ and $y = -6$

66. $-5x - 12y$ for $x = -5$ and $y = -7$

67. $\dfrac{5x - 3y}{-6}$ for $x = -6$ and $y = 4$

68. $\dfrac{-7x + 4y}{-8}$ for $x = 8$ and $y = 6$

69. $3(2a - 5b)$ for $a = -1$ and $b = -5$

70. $4(3a - 7b)$ for $a = -2$ and $b = -4$

71. $-2x + 6y - xy$ for $x = 7$ and $y = -7$

72. $-3x + 7y - 2xy$ for $x = -6$ and $y = 4$

73. $-4ab - b$ for $a = 2$ and $b = -14$

74. $-5ab + b$ for $a = -1$ and $b = -13$

75. $(ab + c)(b - c)$ for $a = -2$, $b = -3$, and $c = 4$

76. $(ab - c)(a + c)$ for $a = -3$, $b = 2$, and $c = 5$

For Problems 77–82, find the value of $\dfrac{5(F - 32)}{9}$ for each of the given values for F. (Objective 2)

77. $F = 59$

78. $F = 68$

79. $F = 14$

80. $F = -4$

81. $F = -13$

82. $F = -22$

For Problems 83–88, find the value of $\dfrac{9C}{5} + 32$ for each of the given values for C. (Objective 2)

83. C = 25

84. C = 35

85. C = 40

86. C = 0

87. C = −10

88. C = −30

For Problems 89–92, solve the problem by applying the concepts of adding and multiplying integers. (Objective 3)

89. On Monday morning, Thad bought 800 shares of a stock at $19 per share. During that week, the stock went up $2 per share on one day and dropped $1 per share on each of the other four days. Use multiplication and addition of integers to describe this situation and to determine the value of the 800 shares by closing time on Friday.

90. In one week a small company showed a profit of $475 for one day and a loss of $65 for each of the other four days. Use multiplication and addition of integers to describe this situation and to determine the company's profit or loss for the week.

91. At 6 P.M. the temperature was 5°F. For the next four hours the temperature dropped 3° per hour. Use multiplication and addition of integers to describe this situation and to find the temperature at 10 P.M.

92. For each of the first three days of a golf tournament, Jason shot 2 strokes under par. Then for each of the last two days of the tournament he shot 4 strokes over par. Use multiplication and addition of integers to describe this situation and to determine how Jason shot relative to par for the five-day tournament.

93. Use a calculator to check your answers for Problems 41–60.

Thoughts Into Words

94. Your friend keeps getting an answer of 27 when simplifying the expression −6 + (−8) ÷ 2. What mistake is she making and how would you help her?

95. Make up a problem that could be solved using 6(24) = 224.

96. Make up a problem that could be solved using (−4)(−3) = 12.

97. Explain why $\dfrac{0}{4} = 0$ but $\dfrac{4}{0}$ is undefined.

Answers to the Concept Quiz
1. True   **2.** False   **3.** True   **4.** False   **5.** False   **6.** False   **7.** False   **8.** True   **9.** True   **10.** True

1.5   Use of Properties

OBJECTIVES

1 Recognize the properties of integers

2 Apply the properties of integers to simplify numerical expressions

3 Simplify algebraic expressions

We will begin this section by listing and briefly commenting on some of the basic properties of integers. We will then show how these properties facilitate manipulation with integers and also serve as a basis for some algebraic computation.

Commutative Property of Addition

If a and b are integers, then

$$a + b = b + a$$

Commutative Property of Multiplication

If a and b are integers, then

$$ab = ba$$

Addition and multiplication are said to be commutative operations. This means that the order in which you add or multiply two integers does not affect the result. For example, $3 + 5 = 5 + 3$ and $7(8) = 8(7)$. It is also important to realize that subtraction and division *are not* commutative operations; order does make a difference. For example, $8 - 7 \neq 7 - 8$ and $16 \div 4 \neq 4 \div 16$.

Associative Property of Addition

If a, b, and c are integers, then

$$(a + b) + c = a + (b + c)$$

Associative Property of Multiplication

If a, b, and c are integers, then

$$(ab)c = a(bc)$$

Our arithmetic operations are binary operations. We only operate (add, subtract, multiply, or divide) on two numbers at a time. Therefore, when we need to operate on three or more numbers, the numbers must be grouped. The associative properties can be thought of as grouping properties. For a sum of three numbers, changing the grouping of the numbers does not affect the final result. For example, $(-8 + 3) + 9 = -8 + (3 + 9)$. This is also true for multiplication as $[(-6)(5)](-4) = (-6)[(5)(-4)]$ illustrates. Addition and multiplication are associative operations. Subtraction and division *are not* associative operations. For example, $(8 - 4) - 7 = -3$, whereas $8 - (4 - 7) = 11$ shows that subtraction is not an associative operation. Also, $(8 \div 4) \div 2 = 1$, whereas $8 \div (4 \div 2) = 4$ shows that division is not associative.

Identity Property of Addition

If a is an integer, then

$$a + 0 = 0 + a = a$$

We refer to zero as the identity element for addition. This simply means that the sum of any integer and zero is exactly the same integer. For example, $-197 + 0 = 0 + (-197) = -197$.

Identity Property of Multiplication

If a is an integer, then

$$a(1) = 1(a) = a$$

We call one the identity element for multiplication. The product of any integer and one is exactly the same integer. For example, $(-573)(1) = (1)(-573) = -573$.

Additive Inverse Property

For every integer a, there exists an integer $-a$ such that

$$a + (-a) = (-a) + a = 0$$

The integer $-a$ is called the additive inverse of a or the opposite of a. Thus 6 and -6 are additive inverses, and their sum is 0. The additive inverse of 0 is 0.

Multiplication Property of Zero

If a is an integer, then

$$(a)(0) = (0)(a) = 0$$

The product of zero and any integer is zero. For example, $(-873)(0) = (0)(-873) = 0$.

Multiplication Property of Negative One

If a is an integer, then

$$(a)(-1) = (-1)(a) = -a$$

The product of any integer and -1 is the opposite of the integer. For example, $(-1)(48) = (48)(-1) = -48$.

Distributive Property

If a, b, and c are integers, then

$$a(b + c) = ab + ac$$

The distributive property involves both addition and multiplication. We say that **multiplication distributes over addition**. For example, $3(4 + 7) = 3(4) + 3(7)$. Since $b - c = b + (-c)$, it follows that **multiplication also distributes over subtraction**. This could be stated as $a(b - c) = ab - ac$. For example, $7(8 - 2) = 7(8) - 7(2)$.

Let's now consider some examples that use these properties to help with various types of manipulations.

Classroom Example
Find the sum [37 + (−18)] + 18.

EXAMPLE 1   Find the sum [43 + (−24)] + 24.

Solution

In this problem it is much more advantageous to group −24 and 24. Thus

$$[43 + (-24)] + 24 = 43 + [(-24) + 24] \quad \text{Associative property for addition}$$
$$= 43 + 0$$
$$= 43$$

Classroom Example
Find the product [(−26)(5)](20).

EXAMPLE 2   Find the product [(−17)(25)](4).

Solution

In this problem it is easier to group 25 and 4. Thus

$$[(-17)(25)](4) = (-17)[(25)(4)] \quad \text{Associative property for multiplication}$$
$$= (-17)(100)$$
$$= -1700$$

Classroom Example
Find the sum
(−32) + 11 + (−15) + 16 + 27 +
(−23) + 52.

EXAMPLE 3   Find the sum 17 + (−24) + (−31) + 19 + (−14) + 29 + 43.

Solution

Certainly we could add in the order that the numbers appear. However, since addition is *commutative* and *associative,* we could change the order and group any convenient way. For example, we could add all of the positive integers and add all of the negative integers, and then add these two results. In that case it is convenient to use the vertical format as follows.

$$
\begin{array}{rrr}
17 & & \\
19 & -24 & \\
29 & -31 & 108 \\
43 & -14 & -69 \\
\hline
108 & -69 & 39
\end{array}
$$

For a problem such as Example 3 it might be advisable to first add in the order that the numbers appear, and then use the rearranging and regrouping idea as a check. Don't forget the link between addition and subtraction. A problem such as 18 − 43 + 52 − 17 − 23 can be changed to 18 + (−43) + 52 + (−17) + (−23).

Classroom Example
Simplify (−25)(−3 + 20).

EXAMPLE 4   Simplify (−75)(−4 + 100).

Solution

For such a problem, it is convenient to apply the *distributive property* and then to simplify.

$$(-75)(-4 + 100) = (-75)(-4) + (-75)(100)$$
$$= 300 + (-7500)$$
$$= -7200$$

Classroom Example
Simplify 24(−16 + 18).

EXAMPLE 5   Simplify $19(-26 + 25)$.

Solution

For this problem we are better off *not* applying the distributive property, but simply adding the numbers inside the parentheses first and then finding the indicated product. Thus

$$19(-26 + 25) = 19(-1) = -19$$

Classroom Example
Simplify 33(6) + 33(−106).

EXAMPLE 6   Simplify $27(104) + 27(-4)$.

Solution

Keep in mind that the distributive property allows us to change from the form $a(b + c)$ to $ab + ac$ or from $ab + ac$ to $a(b + c)$. In this problem we want to use the latter conversion. Thus

$$27(104) + 27(-4) = 27(104 + (-4))$$
$$= 27(100) = 2700$$

Examples 4, 5, and 6 demonstrate an important issue. Sometimes the form $a(b + c)$ is the most convenient, but at other times the form $ab + ac$ is better. A suggestion in regard to this issue—as well as to the use of the other properties—is to think first, and then decide whether or not the properties can be used to make the manipulations easier.

Combining Similar Terms

Algebraic expressions such as

$$3x \qquad 5y \qquad 7xy \qquad -4abc \qquad \text{and} \qquad z$$

are called terms. A **term** is an indicated product, and it may have any number of factors. We call the variables in a term **literal factors**, and we call the numerical factor the **numerical coefficient**. Thus in $7xy$, the x and y are literal factors, and 7 is the numerical coefficient. The numerical coefficient of the term $-4abc$ is -4. Since $z = 1(z)$, the numerical coefficient of the term z is 1. Terms that have the same literal factors are called **like terms** or **similar terms**. Some examples of similar terms are

$$3x \quad \text{and} \quad 9x \qquad\qquad 14abc \quad \text{and} \quad 29abc$$

$$7xy \quad \text{and} \quad -15xy \qquad 4z \quad 9z \quad \text{and} \quad -14z$$

We can simplify algebraic expressions that contain similar terms by using a form of the distributive property. Consider the following examples.

$$3x + 5x = (3 + 5)x$$
$$= 8x$$

$$-9xy + 7xy = (-9 + 7)xy$$
$$= -2xy$$

$$18abc - 27abc = (18 - 27)abc$$
$$= (18 + (-27))abc$$
$$= -9abc$$

$$4x + x = (4 + 1)x \qquad \text{Don't forget that } x = 1(x)$$
$$= 5x$$

More complicated expressions might first require some rearranging of terms by using the commutative property:

$$7x + 3y + 9x + 5y = 7x + 9x + 3y + 5y$$
$$= (7 + 9)x + (3 + 5)y$$
$$= 16x + 8y$$

$$9a - 4 - 13a + 6 = 9a + (-4) + (-13a) + 6$$
$$= 9a + (-13a) + (-4) + 6$$
$$= (9 + (-13))a + 2$$
$$= -4a + 2$$

As you become more adept at handling the various simplifying steps, you may want to do the steps mentally and go directly from the given expression to the simplified form as follows.

$$19x - 14y + 12x + 16y = 31x + 2y$$

$$17ab + 13c - 19ab - 30c = -2ab - 17c$$

$$9x + 5 - 11x + 4 + x - 6 = -x + 3$$

Simplifying some algebraic expressions requires repeated applications of the distributive property as the next examples demonstrate.

$$5(x - 2) + 3(x + 4) = 5(x) - 5(2) + 3(x) + 3(4)$$
$$= 5x - 10 + 3x + 12$$
$$= 5x + 3x - 10 + 12$$
$$= 8x + 2$$

$$-7(y + 1) - 4(y - 3) = -7(y) - 7(1) - 4(y) - 4(-3)$$
$$= -7y - 7 - 4y + 12 \qquad \text{Be careful with this sign}$$
$$= -7y - 4y - 7 + 12$$
$$= -11y + 5$$

$$5(x + 2) - (x + 3) = 5(x + 2) - 1(x + 3) \qquad \text{Remember } -a = -1a$$
$$= 5(x) + 5(2) - 1(x) - 1(3)$$
$$= 5x + 10 - x - 3$$
$$= 5x - x + 10 - 3$$
$$= 4x + 7$$

After you are sure of each step, you can use a more simplified format.

$$5(a + 4) - 7(a - 2) = 5a + 20 - 7a + 14$$
$$= -2a + 34$$

$$9(z - 7) + 11(z + 6) = 9z - 63 + 11z + 66$$
$$= 20z + 3$$

$$-(x - 2) + (x + 6) = -x + 2 + x + 6$$
$$= 8$$

Back to Evaluating Algebraic Expressions

To simplify by combining similar terms aids in the process of evaluating some algebraic expressions. The last examples of this section illustrate this idea.

Classroom Example
Evaluate $9s - 4t + 3s + 7t$ for $s = 2$ and $t = -5$.

EXAMPLE 7   Evaluate $8x - 2y + 3x + 5y$ for $x = 3$ and $y = -4$.

Solution

Let's first simplify the given expression.

$$8x - 2y + 3x + 5y = 11x + 3y$$

Now we can evaluate for $x = 3$ and $y = -4$.

$$11x + 3y = 11(3) + 3(-4)$$
$$= 33 + (-12) = 21$$

Classroom Example
Evaluate $6x + 3yz - 4x - 7yz$ for $x = 4$, $y = -6$, and $z = 3$.

EXAMPLE 8   Evaluate $2ab + 5c - 6ab + 12c$ for $a = 2$, $b = -3$, and $c = 7$.

Solution

$$2ab + 5c - 6ab + 12c = -4ab + 17c$$
$$= -4(2)(-3) + 17(7) \quad \text{when } a = 2, b = -3, \text{ and } c = 7$$
$$= 24 + 119 = 143$$

Classroom Example
Evaluate $4(m + 5) - 9(m - 3)$ for $m = 7$.

EXAMPLE 9   Evaluate $8(x - 4) + 7(x + 3)$ for $x = 6$.

Solution

$$8(x - 4) + 7(x + 3) = 8x - 32 + 7x + 21$$
$$= 15x - 11$$
$$= 15(6) - 11 \quad \text{when } x = 6$$
$$= 79$$

Concept Quiz 1.5

For Problems 1–10, answer true or false.

1. Addition is a commutative operation.

2. Subtraction is a commutative operation.

3. $[(2)(-3)](7) = (2)[(-3)(7)]$ is an example of the associative property for multiplication.

4. $[(8)(5)](-2) = (-2)[(8)(5)]$ is an example of the associative property for multiplication.

5. Zero is the identity element for addition.

6. The integer $-a$ is the additive inverse of a.

7. The additive inverse of 0 is 0.

8. The numerical coefficient of the term $-8xy$ is 8.

9. The numerical coefficient of the term ab is 1.

10. $6xy$ and $-2xyz$ are similar terms.

Problem Set 1.5

For Problems 1–12, state the property that justifies each statement. For example, $3 + (-4) = (-4) + 3$ because of the commutative property for addition. (Objective 1)

1. $3(7 + 8) = 3(7) + 3(8)$

2. $(-9)(17) = 17(-9)$

3. $-2 + (5 + 7) = (-2 + 5) + 7$

4. $-19 + 0 = -19$

5. $143(-7) = -7(143)$

6. $5(9 + (-4)) = 5(9) + 5(-4)$

7. $-119 + 119 = 0$

8. $-4 + (6 + 9) = (-4 + 6) + 9$

9. $-56 + 0 = -56$

10. $5 + (-12) = -12 + 5$

11. $[5(-8)]4 = 5[-8(4)]$

12. $[6(-4)]8 = 6[-4(8)]$

For Problems 13–30, simplify each numerical expression. Don't forget to take advantage of the properties if they can be used to simplify the computation. (Objective 2)

13. $(-18 + 56) + 18$

14. $-72 + [72 + (-14)]$

15. $36 - 48 - 22 + 41$

16. $-24 + 18 + 19 - 30$

17. $(25)(-18)(-4)$

18. $(2)(-71)(50)$

19. $(4)(-16)(-9)(-25)$

20. $(-2)(18)(-12)(-5)$

21. $37(-42 - 58)$

22. $-46(-73 - 27)$

23. $59(36) + 59(64)$

24. $-49(72) - 49(28)$

25. $15(-14) + 16(-8)$

26. $-9(14) - 7(-16)$

27. $17 + (-18) - 19 - 14 + 13 - 17$

28. $-16 - 14 + 18 + 21 + 14 - 17$

29. $-21 + 22 - 23 + 27 + 21 - 19$

30. $24 - 26 - 29 + 26 + 18 + 29 - 17 - 10$

For Problems 31–62, simplify each algebraic expression by combining similar terms. (Objective 3)

31. $9x - 14x$

32. $12x - 14x + x$

33. $4m + m - 8m$

34. $-6m - m + 17m$

35. $-9y + 5y - 7y$

36. $14y - 17y - 19y$

37. $4x - 3y - 7x + y$

38. $9x + 5y - 4x - 8y$

39. $-7a - 7b - 9a + 3b$

40. $-12a + 14b - 3a - 9b$

41. $6xy - x - 13xy + 4x$

42. $-7xy - 2x - xy + x$

43. $5x - 4 + 7x - 2x + 9$

44. $8x + 9 + 14x - 3x - 14$

45. $-2xy + 12 + 8xy - 16$

46. $14xy - 7 - 19xy - 6$

47. $-2a + 3b - 7b - b + 5a - 9a$

48. $-9a - a + 6b - 3a - 4b - b + a$

49. $13ab + 2a - 7a - 9ab + ab - 6a$

50. $-ab - a + 4ab + 7ab - 3a - 11ab$

51. $3(x + 2) + 5(x + 6)$

52. $7(x + 8) + 9(x + 1)$

53. $5(x - 4) + 6(x + 8)$

54. $-3(x + 2) - 4(x - 10)$

55. $9(x + 4) - (x - 8)$

56. $-(x - 6) + 5(x - 9)$

57. $3(a - 1) - 2(a - 6) + 4(a + 5)$

58. $-4(a + 2) + 6(a + 8) - 3(a - 6)$

59. $-2(m + 3) - 3(m - 1) + 8(m + 4)$

60. $5(m - 10) + 6(m - 11) - 9(m - 12)$

61. $(y + 3) - (y - 2) - (y + 6) - 7(y - 1)$

62. $-(y - 2) - (y + 4) - (y + 7) - 2(y + 3)$

For Problems 63–80, simplify each algebraic expression and then evaluate the resulting expression for the given values of the variables. (Objective 3)

63. $3x + 5y + 4x - 2y$ for $x = -2$ and $y = 3$

64. $5x - 7y - 9x - 3y$ for $x = -1$ and $y = -4$

65. $5(x - 2) + 8(x + 6)$ for $x = -6$

66. $4(x - 6) + 9(x + 2)$ for $x = 7$

67. $8(x + 4) - 10(x - 3)$ for $x = -5$

68. $-(n + 2) - 3(n - 6)$ for $n = 10$

69. $(x - 6) - (x + 12)$ for $x = -3$

70. $(x + 12) - (x - 14)$ for $x = -11$

71. $2(x + y) - 3(x - y)$ for $x = -2$ and $y = 7$

72. $5(x - y) - 9(x + y)$ for $x = 4$ and $y = -4$

73. $2xy + 6 + 7xy - 8$ for $x = 2$ and $y = -4$

74. $4xy - 5 - 8xy + 9$ for $x = -3$ and $y = -3$

75. $5x - 9xy + 3x + 2xy$ for $x = 12$ and $y = -1$

76. $-9x + xy - 4xy - x$ for $x = 10$ and $y = -11$

77. $(a - b) - (a + b)$ for $a = 19$ and $b = -17$

78. $(a + b) - (a - b)$ for $a = -16$ and $b = 14$

79. $-3x + 7x + 4x - 2x - x$ for $x = -13$

80. $5x - 6x + x - 7x - x - 2x$ for $x = -15$

81. Use a calculator to check your answers for Problems 13–30.

Thoughts Into Words

82. State in your own words the associative property for addition of integers.

83. State in your own words the distributive property for multiplication over addition.

84. Is $2 \cdot 3 \cdot 5 \cdot 7 \cdot 11 + 7$ a prime or composite number? Defend your answer.

Further Investigations

For Problems 85–90, state whether the expressions in each problem are equivalent and explain why or why not.

85. $15a(x + y)$ and $5a(3x + 3y)$

86. $(-6a + 7b) + 11c$ and $7b + (11c - 6a)$

87. $2x - 3y + 4z$ and $2x - 4z + 3y$

88. $a + 5(x + y)$ and $(a + 5)x + y$

89. $7x + 6(y - 2z)$ and $6(2z - y) + 7x$

90. $9m + 8(3p - q)$ and $8(3p - q) + 9m$

Answers to the Concept Quiz

1. True   **2.** False   **3.** True   **4.** False   **5.** True   **6.** True   **7.** True   **8.** False   **9.** True   **10.** False

Chapter 1   Summary

OBJECTIVE	SUMMARY	EXAMPLE
List the elements of a set. (Section 1.1/Objective 1)	Elements of a set can be shown by a word description or a list.	List the set of even numbers greater than 4. **Solution** $A = \{6, 8, 10, 12, \ldots\}$
Determine if sets are equal. (Section 1.1/Objective 1)	Equal sets have the same members.	Is $A = \{1, 3, 5, 7\}$ equal to $B = \{1, 5\}$? **Solution** $A \neq B$
Use the order of operations to simplify numerical expressions. (Section 1.1/Objective 2)	1. Perform the operations inside grouping symbols and above and below each fraction bar. 2. Perform all multiplications and divisions in order from left to right. 3. Perform all additions and subtractions in order from left to right.	Simplify $\dfrac{7 + 8}{5 - 2} + 3[2(6 + 4)]$. **Solution** $\dfrac{7 + 8}{5 - 2} + 3[2(6 + 4)]$ $= \dfrac{15}{3} + 3[2(10)]$ $= 5 + 3(20)$ $= 5 + 60$ $= 65$
Identify numbers as prime or composite. (Section 1.2/Objective 1)	A prime number is a whole number greater than 1 that only has itself and 1 as factors. Any whole number that is not prime is called a composite number.	List the prime numbers less than 50. **Solution** 2, 3, 5, 7, 11, 13, 17, 19, 23, 29, 31, 37, 41, 43, 47
Factor a number into a product of primes. (Section 1.2/Objective 2)	Start by picking any two factors of the number. If those numbers are not prime, then continue by picking factors of those numbers until every factor is a prime number.	Prime factor 36. **Solution** $36 = 6 \cdot 6$ $\quad = 2 \cdot 3 \cdot 2 \cdot 3$ $\quad = 2 \cdot 2 \cdot 3 \cdot 3$
Find the greatest common factor. (Section 1.2/Objective 3)	The greatest common factor (GCF) of two numbers is the largest divisor of both numbers. Prime factor each number and select each common factor the least number of times it appears in the factorizations.	Find the greatest common factor (GCF) of 60 and 210. **Solution** $60 = 2 \cdot 2 \cdot 3 \cdot 5$ $210 = 2 \cdot 3 \cdot 5 \cdot 7$ $GCF = 2 \cdot 3 \cdot 5 = 30$
Find the least common multiple. (Section 1.2/Objective 4)	The least common multiple (LCM) is the smallest nonzero common multiple of each number. Prime factor each number and select each factor the most number of times it appears in the factorizations.	Find the least common multiple (LCM) of 12 and 54. **Solution** $12 = 2 \cdot 2 \cdot 3$ $54 = 2 \cdot 3 \cdot 3 \cdot 3$ $LCM = 2 \cdot 2 \cdot 3 \cdot 3 \cdot 3 = 108$

(continued)

OBJECTIVE	SUMMARY	EXAMPLE
Add integers. (Section 1.3/Objective 2)	The number line is a convenient visual aid for interpreting the addition of integers. See page 16 for the formal rules of addition.	Find the sums: **(a)** $(-36) + (-14)$ **(b)** $36 + (-14)$ **(c)** $-36 + 14$ **Solution** **(a)** $(-36) + (-14) = -50$ **(b)** $36 + (-14) = 22$ **(c)** $-36 + 14 = -22$
Subtract integers. (Section 1.3/Objective 2)	Subtraction is defined in terms of addition. To subtract a number you can add its opposite.	Find the differences: **(a)** $-4 - 6$ **(b)** $5 - (-2)$ **Solution** **(a)** $-4 - 6$ $= -4 + (-6)$ $= -10$ **(b)** $5 - (-2)$ $= 5 + 2$ $= 7$
Multiply and divide integers. (Section 1.4/Objective 1)	The product (or quotient) of two positive or two negative integers is positive. The product (or quotient) of a positive integer and a negative integer is negative.	Perform the indicated operations: **(a)** $(-3)(-4)$ **(b)** $(3)(-4)$ **(c)** $\dfrac{-12}{-3}$ **(d)** $\dfrac{-12}{3}$ **Solution** **(a)** $(-3)(-4) = 12$ **(b)** $(3)(-4) = -12$ **(c)** $\dfrac{-12}{-3} = 4$ **(d)** $\dfrac{-12}{3} = -4$
Solve application problems involving integers. (Section 1.3/Objective 4) (Section 1.4/Objective 3)	Positive and negative integers can be used to represent many real world problems.	A stock selling for $38.70 went up $3.52 one day and the next day fell $1.82. What is the price of the stock after these two days? **Solution** $38.70 + 3.52 + (-1.82) = 40.40$ The price is $40.40.
Know the properties of integers. (Section 1.5/Objective 1)	The commutative property states that the order in which you add or multiply two numbers does not affect the result. The associative property states that for a sum or product of three numbers, changing the grouping of the numbers does not affect the final result. See pages 27–28 for a full listing of the properties.	State the property demonstrated by $15 + (-10) = -10 + 15$ **Solution** Commutative property for addition

OBJECTIVE	SUMMARY	EXAMPLE
Combine similar terms (Section 1.5/Objective 3)	Use the distributive property to combine similar terms.	Simplify $3x + 12y - 5x + 8y$. **Solution** $3x + 12y - 5x + 8y$ $= 3x - 5x + 12y + 8y$ $= (3 - 5)x + (12 + 8)y$ $= -2x + 20y$
Evaluate algebraic expressions. (Section 1.1/Objective 3) (Section 1.3/Objective 3) (Section 1.4/Objective 2)	First simplify the algebraic expression. Next, replace the variables with the given values. It is a good practice to use parentheses when substituting the values. Be sure to follow the order of operations as you simplify the resulting numerical expression.	Evaluate $4x - 2y$, when $x = 5$ and $y = 6$ **Solution** $4x - 2y$ $= 4(5) - 2(6)$ when $x = 5$ and $y = 6$ $= 20 - 12$ $= 8$

Chapter 1 Review Problem Set

In Problems 1–10, perform the indicated operations.

1. $7 + (-10)$

2. $(-12) + (-13)$

3. $8 - 13$

4. $-6 - 9$

5. $-12 - (-11)$

6. $-17 - (-19)$

7. $(13)(-12)$

8. $(-14)(-18)$

9. $(-72) \div (-12)$

10. $117 \div (-9)$

In Problems 11–15, classify each of the numbers as *prime* or *composite*.

11. 73

12. 87

13. 63

14. 81

15. 91

In Problems 16–20, express each of the numbers as the product of prime factors.

16. 24

17. 63

18. 57

19. 64

20. 84

21. Find the greatest common factor of 36 and 54.

22. Find the greatest common factor of 48, 60, and 84.

23. Find the least common multiple of 18 and 20.

24. Find the least common multiple of 15, 27, and 35.

For Problems 25–38, simplify each of the numerical expressions.

25. $(19 + 56) + (-9)$

26. $43 - 62 + 12$

27. $8 + (-9) + (-16) + (-14) + 17 + 12$

28. $19 - 23 - 14 + 21 + 14 - 13$

29. $3(-4) - 6$

30. $(-5)(-4) - 8$

31. $(5)(-2) + (6)(-4)$

32. $(-6)(8) + (-7)(-3)$

33. $(-6)(3) - (-4)(-5)$

34. $(-7)(9) - (6)(5)$

35. $\dfrac{4(-7) - (3)(-2)}{-11}$

36. $\dfrac{(-4)(9) + (5)(-3)}{1 - 18}$

37. $3 - 2[4(-3 - 1)]$

38. $-6 - [3(-4 - 7)]$

39. A record high temperature of 125°F occurred in Laughlin, Nevada on June 29, 1994. A record low temperature of −50°F occurred in San Jacinto, Nevada on January 8, 1937. Find the difference between the record high and low temperatures.

40. In North America the highest elevation, which is on Mt. McKinley, Alaska, is 20,320 feet above sea level. The lowest elevation in North America, which is at Death Valley, California, is 282 feet below sea level. Find the absolute value of the difference in elevation between Mt. McKinley and Death Valley.

41. As a running back in a football game, Marquette carried the ball 7 times. On two plays he gained 6 yards on each play; on another play he lost 4 yards; on the next three plays he gained 8 yards per play; and on the last play he lost 1 yard. Write a numerical expression that gives Marquette's overall yardage for the game, and simplify that expression.

42. Shelley started the month with $3278 in her checking account. During the month she deposited $175 each week for 4 weeks but had debit charges of $50, $189, $160, $20, and $115. What is the balance in her checking account after these deposits and debits?

In Problems 43–54, simplify each algebraic expression by combining similar terms.

43. $12x + 3x - 7x$

44. $9y + 3 - 14y - 12$

45. $8x + 5y - 13x - y$

46. $9a + 11b + 4a - 17b$

47. $3ab - 4ab - 2a$

48. $5xy - 9xy + xy - y$

49. $3(x + 6) + 7(x + 8)$

50. $5(x - 4) - 3(x - 9)$

51. $-3(x - 2) - 4(x + 6)$

52. $-2x - 3(x - 4) + 2x$

53. $2(a - 1) - a - 3(a - 2)$

54. $-(a - 1) + 3(a - 2) - 4a + 1$

In Problems 55–68, evaluate each of the algebraic expressions for the given values of the variables.

55. $5x + 8y$ for $x = -7$ and $y = -3$

56. $7x - 9y$ for $x = -3$ and $y = 4$

57. $\dfrac{-5x - 2y}{-2x - 7}$ for $x = 6$ and $y = 4$

58. $\dfrac{-3x + 4y}{3x}$ for $x = -4$ and $y = -6$

59. $-2a + \dfrac{a - b}{a - 2}$ for $a = -5$ and $b = 9$

60. $\dfrac{2a + b}{b + 6} - 3b$ for $a = 3$ and $b = -4$

61. $5a + 6b - 7a - 2b$ for $a = -1$ and $b = 5$

62. $3x + 7y - 5x + y$ for $x = -4$ and $y = 3$

63. $2xy + 6 + 5xy - 8$ for $x = -1$ and $y = 1$

64. $7(x + 6) - 9(x + 1)$ for $x = -2$

65. $-3(x - 4) - 2(x + 8)$ for $x = 7$

66. $2(x - 1) - (x + 2) + 3(x - 4)$ for $x = -4$

67. $(a - b) - (a + b) - b$ for $a = -1$ and $b = -3$

68. $2ab - 3(a - b) + b + a$ for $a = 2$ and $b = -5$

For Problems 1–10, simplify each of the numerical expressions.

1. $6 + (-7) - 4 + 12$

2. $7 + 4(9) + 2$

3. $-4(2 - 8) + 14$

4. $5(-7) - (-3)(8)$

5. $8 \div (-4) + (-6)(9) - 2$

6. $(-8)(-7) + (-6) - (9)(12)$

7. $\dfrac{6(-4) - (-8)(-5)}{-16}$

8. $-14 + 23 - 17 - 19 + 26$

9. $(-14)(4) \div 4 + (-6)$

10. $6(-9) - (-8) - (-7)(4) + 11$

11. It was reported on the 5 o'clock news that the current temperature was 7°F. The forecast was for the temperature to drop 13 degrees by 6:00 A.M. If the forecast is correct, what will the temperature be at 6:00 A.M.?

For Problems 12–17, evaluate each of the algebraic expressions for the given values of the variables.

12. $7x - 9y$ for $x = -4$ and $y = -6$

13. $-4a - 6b$ for $a = -9$ and $b = 12$

14. $3xy - 8y + 5x$ for $x = 7$ and $y = -2$

15. $5(x - 4) - 6(x + 7)$ for $x = -5$

16. $3x - 2y - 4x - x + 7y$ for $x = 6$ and $y = -7$

17. $3(x - 2) - 5(x - 4) + 6(x - 1)$ for $x = -3$

18. Classify 79 as a prime or composite number.

19. Express 360 as a product of prime factors.

20. Find the greatest common factor of 36, 60, and 84.

21. Find the least common multiple of 9 and 24.

22. State the property of integers demonstrated by
$[-3 + (-4)] + (-6) = -3 + [(-4) + (-6)]$.

23. State the property of integers demonstrated by
$8(25 + 37) = 8(25) + 8(37)$.

24. Simplify $-7x + 9y - y + x - 2y - 7x$ by combining similar terms.

25. Simplify $-2(x - 4) - 5(x + 7) - 6(x - 1)$ by applying the distributive property and combining similar terms.

2 Real Numbers

People that watch the stock market are familiar with rational numbers expressed in decimal form.

© Stephen Coburn

Caleb left an estate valued at $750,000. His will states that three-fourths of the estate is to be divided equally among his three children. The numerical expression $\left(\dfrac{1}{3}\right)\left(\dfrac{3}{4}\right)(750{,}000)$ can be used to determine how much each of his three children should receive.

When the market opened on Monday morning, Garth bought some shares of a stock at $13.25 per share. The rational numbers 0.75, -1.50, 2.25, -0.25, and -0.50 represent the daily changes in the market for that stock for the week. We use the numerical expression $13.25 + 0.75 + (-1.50) + 2.25 + (-0.25) + (-0.50)$ to determine the value of one share of Garth's stock when the market closed on Friday.

The width of a rectangle is w feet, and its length is four feet more than three times its width. The algebraic expression $2w + 2(3w + 4)$ represents the perimeter of the rectangle.

Again in this chapter we use the concepts of numerical and algebraic expressions to review some computational skills from arithmetic and to continue

Video tutorials based on section learning objectives are available in a variety of delivery modes.

the transition from arithmetic to algebra. However, the set of rational numbers now becomes the primary focal point. We urge you to use this chapter to review and improve your arithmetic skills so that the algebraic concepts in subsequent chapters can build upon a solid foundation.

2.1   Rational Numbers: Multiplication and Division

OBJECTIVES

1 Reduce rational numbers in fractional form to lowest terms

2 Multiply and divide rational numbers

3 Solve application problems involving multiplication and division of rational numbers

Any number that can be written in the form $\dfrac{a}{b}$, where a and b are integers and b is not zero, we call a **rational number**. (We call the form $\dfrac{a}{b}$ a fraction or sometimes a common fraction.) The following are examples of rational numbers:

$$\frac{1}{2} \qquad \frac{7}{9} \qquad \frac{15}{7} \qquad \frac{-3}{4} \qquad \frac{5}{-7} \quad \text{and} \quad \frac{-11}{-13}$$

All integers are rational numbers, because every integer can be expressed as the indicated quotient of two integers. Some examples follow.

$$6 = \frac{6}{1} = \frac{12}{2} = \frac{18}{3} \quad \text{and so on}$$

$$27 = \frac{27}{1} = \frac{54}{2} = \frac{81}{3} \quad \text{and so on}$$

$$0 = \frac{0}{1} = \frac{0}{2} = \frac{0}{3} \quad \text{and so on}$$

Our work in Chapter 1 with division involving negative integers helps with the next three examples.

$$-4 = \frac{-4}{1} = \frac{-8}{2} = \frac{-12}{3} \quad \text{and so on}$$

$$-6 = \frac{6}{-1} = \frac{12}{-2} = \frac{18}{-3} \quad \text{and so on}$$

$$10 = \frac{10}{1} = \frac{-10}{-1} = \frac{-20}{-2} \quad \text{and so on}$$

Observe the following general properties.

Property 2.1

$$\frac{-a}{b} = \frac{a}{-b} = -\frac{a}{b} \qquad \text{and} \qquad \frac{-a}{-b} = \frac{a}{b}$$

Therefore, a rational number such as $\dfrac{-2}{3}$ can also be written as $\dfrac{2}{-3}$ or $-\dfrac{2}{3}$. (However, we seldom express rational numbers with negative denominators.)

Multiplying Rational Numbers

We define multiplication of rational numbers in common fractional form as follows:

> ### Definition 2.1
>
> If a, b, c, and d are integers, and b and d are not equal to zero, then
>
> $$\frac{a}{b} \cdot \frac{c}{d} = \frac{a \cdot c}{b \cdot d}$$

To **multiply rational numbers** in common fractional form we simply multiply numerators and multiply denominators. Furthermore, we see from the definition that the commutative and associative properties, with respect to multiplication, apply to the set of rational numbers. We are free to rearrange and regroup factors as we do with integers. The following examples illustrate Definition 2.1:

$$\frac{1}{3} \cdot \frac{2}{5} = \frac{1 \cdot 2}{3 \cdot 5} = \frac{2}{15}$$

$$\frac{3}{4} \cdot \frac{5}{7} = \frac{3 \cdot 5}{4 \cdot 7} = \frac{15}{28}$$

$$\frac{-2}{3} \cdot \frac{7}{9} = \frac{-2 \cdot 7}{3 \cdot 9} = \frac{-14}{27} \quad \text{or} \quad -\frac{14}{27}$$

$$\frac{1}{5} \cdot \frac{9}{-11} = \frac{1 \cdot 9}{5(-11)} = \frac{9}{-55} \quad \text{or} \quad -\frac{9}{55}$$

$$-\frac{3}{4} \cdot \frac{7}{13} = \frac{-3}{4} \cdot \frac{7}{13} = \frac{-3 \cdot 7}{4 \cdot 13} = \frac{-21}{52} \quad \text{or} \quad -\frac{21}{52}$$

$$\frac{3}{5} \cdot \frac{5}{3} = \frac{3 \cdot 5}{5 \cdot 3} = \frac{15}{15} = 1$$

The last example is a very special case. **If the product of two numbers is 1, the numbers are said to be reciprocals of each other**.

Using Definition 2.1 and applying the multiplication property of one, the fraction $\frac{a \cdot k}{b \cdot k}$, where b and k are nonzero integers, simplifies as shown.

$$\frac{a \cdot k}{b \cdot k} = \frac{a}{b} \cdot \frac{k}{k} = \frac{a}{b} \cdot 1 = \frac{a}{b}$$

This result is stated as Property 2.2.

> ### Property 2.2 The Fundamental Principle of Fractions
>
> If b and k are nonzero integers, and a is any integer, then
>
> $$\frac{a \cdot k}{b \cdot k} = \frac{a}{b}$$

We often use Property 2.2 when we work with rational numbers. It is called the fundamental principle of fractions and provides the basis for equivalent fractions. In the following examples, the property will be used for what is often called "reducing fractions to lowest terms" or "expressing fractions in simplest or reduced form."

Classroom Example
Reduce $\frac{21}{35}$ to lowest terms.

EXAMPLE 1 Reduce $\frac{12}{18}$ to lowest terms.

Solution

$$\frac{12}{18} = \frac{2 \cdot 6}{3 \cdot 6} = \frac{2}{3}$$

EXAMPLE 2   Change $\dfrac{14}{35}$ to simplest form.

Solution

$$\frac{14}{35} = \frac{2 \cdot 7}{5 \cdot 7} = \frac{2}{5} \qquad \text{A common factor of 7 has been divided out of both numerator and denominator}$$

EXAMPLE 3   Express $\dfrac{-24}{32}$ in reduced form.

Solution

$$\frac{-24}{32} = -\frac{3 \cdot 8}{4 \cdot 8} = -\frac{3}{4} \cdot \frac{8}{8} = -\frac{3}{4} \cdot 1 = -\frac{3}{4} \qquad \text{The multiplication property of 1 is being used}$$

EXAMPLE 4   Reduce $-\dfrac{72}{90}$.

Solution

$$-\frac{72}{90} = -\frac{2 \cdot 2 \cdot 2 \cdot 3 \cdot 3}{2 \cdot 3 \cdot 3 \cdot 5} = -\frac{4}{5} \qquad \text{The prime factored forms of the numerator and denominator may be used to help recognize common factors}$$

The fractions may contain variables in the numerator or the denominator (or both), but this creates no great difficulty. Our thought processes remain the same, as these next examples illustrate. Variables appearing in the denominators represent *nonzero* integers.

EXAMPLE 5   Reduce $\dfrac{9x}{17x}$.

Solution

$$\frac{9x}{17x} = \frac{9 \cdot x}{17 \cdot x} = \frac{9}{17}$$

EXAMPLE 6   Simplify $\dfrac{8x}{36y}$.

Solution

$$\frac{8x}{36y} = \frac{2 \cdot 2 \cdot 2 \cdot x}{2 \cdot 2 \cdot 3 \cdot 3 \cdot y} = \frac{2x}{9y}$$

EXAMPLE 7   Express $\dfrac{-9xy}{30y}$ in reduced form.

Solution

$$\frac{-9xy}{30y} = -\frac{9xy}{30y} = -\frac{3 \cdot 3 \cdot x \cdot y}{2 \cdot 3 \cdot 5 \cdot y} = -\frac{3x}{10}$$

Classroom Example
Reduce $\dfrac{-3xyz}{-8yz}$.

EXAMPLE 8   Reduce $\dfrac{-7abc}{-9ac}$.

Solution

$$\frac{-7abc}{-9ac} = \frac{7abc}{9ac} = \frac{7ab\cancel{c}}{9a\cancel{c}} = \frac{7b}{9}$$

We are now ready to consider multiplication problems with the agreement that the final answer should be expressed in reduced form. Study the following examples carefully; we use different methods to handle the various problems.

Classroom Example
Multiply $\dfrac{3}{8} \cdot \dfrac{5}{9}$.

EXAMPLE 9   Multiply $\dfrac{7}{9} \cdot \dfrac{5}{14}$.

Solution

$$\frac{7}{9} \cdot \frac{5}{14} = \frac{7 \cdot 5}{9 \cdot 14} = \frac{\cancel{7} \cdot 5}{3 \cdot 3 \cdot 2 \cdot \cancel{7}} = \frac{5}{18}$$

Classroom Example
Find the product of $\dfrac{6}{7}$ and $\dfrac{21}{30}$.

EXAMPLE 10   Find the product of $\dfrac{8}{9}$ and $\dfrac{18}{24}$.

Solution

$$\frac{\overset{1}{\cancel{8}}}{\underset{1}{\cancel{9}}} \cdot \frac{\overset{2}{\cancel{18}}}{\underset{3}{\cancel{24}}} = \frac{2}{3}$$

A common factor of 8 has been divided out of 8 and 24, and a common factor of 9 has been divided out of 9 and 18

Classroom Example
Multiply $\left(-\dfrac{4}{9}\right)\left(\dfrac{33}{40}\right)$.

EXAMPLE 11   Multiply $\left(-\dfrac{6}{8}\right)\left(\dfrac{14}{32}\right)$.

Solution

$$\left(-\frac{6}{8}\right)\left(\frac{14}{32}\right) = -\frac{\overset{3}{\cancel{6}} \cdot \overset{7}{\cancel{14}}}{\underset{4}{\cancel{8}} \cdot \underset{16}{\cancel{32}}} = -\frac{21}{64}$$

Divide a common factor of 2 out of 6 and 8, and a common factor of 2 out of 14 and 32

Classroom Example
Multiply $\left(-\dfrac{10}{3}\right)\left(-\dfrac{12}{18}\right)$.

EXAMPLE 12   Multiply $\left(-\dfrac{9}{4}\right)\left(-\dfrac{14}{15}\right)$.

Solution

$$\left(-\frac{9}{4}\right)\left(-\frac{14}{15}\right) = \frac{3 \cdot 3 \cdot 2 \cdot 7}{2 \cdot 2 \cdot 3 \cdot 5} = \frac{21}{10}$$

Immediately we recognize that a negative times a negative is positive

Classroom Example
Multiply $\left(\dfrac{6m}{5n}\right)\left(\dfrac{15n}{26}\right)$.

EXAMPLE 13   Multiply $\dfrac{9x}{7y} \cdot \dfrac{14y}{45}$.

Solution

$$\frac{9x}{7y} \cdot \frac{14y}{45} = \frac{9 \cdot x \cdot \overset{2}{\cancel{14}} \cdot \cancel{y}}{\cancel{7} \cdot \cancel{y} \cdot \underset{5}{\cancel{45}}} = \frac{2x}{5}$$

Classroom Example
Multiply $\dfrac{-3z}{8xy} \cdot \dfrac{16x}{9z}$.

EXAMPLE 14   Multiply $\dfrac{-6c}{7ab} \cdot \dfrac{14b}{5c}$.

Solution

$$\frac{-6c}{7ab} \cdot \frac{14b}{5c} = -\frac{2 \cdot 3 \cdot c \cdot 2 \cdot 7 \cdot b}{7 \cdot a \cdot b \cdot 5 \cdot c} = -\frac{12}{5a}$$

Dividing Rational Numbers

The following example motivates a definition for division of rational numbers in fractional form.

$$\frac{\dfrac{3}{4}}{\dfrac{2}{3}} = \left(\frac{\dfrac{3}{4}}{\dfrac{2}{3}}\right)\left(\frac{\dfrac{3}{2}}{\dfrac{3}{2}}\right) = \frac{\left(\dfrac{3}{4}\right)\left(\dfrac{3}{2}\right)}{1} = \left(\frac{3}{4}\right)\left(\frac{3}{2}\right) = \frac{9}{8}$$

↑
Notice that this is a form of 1, and $\dfrac{3}{2}$ is the reciprocal of $\dfrac{2}{3}$

In other words, $\dfrac{3}{4}$ divided by $\dfrac{2}{3}$ is equivalent to $\dfrac{3}{4}$ times $\dfrac{3}{2}$. The following definition for division should seem reasonable:

Definition 2.2

If b, c, and d are nonzero integers and a is any integer, then

$$\frac{a}{b} \div \frac{c}{d} = \frac{a}{b} \cdot \frac{d}{c}$$

Notice that to divide $\dfrac{a}{b}$ by $\dfrac{c}{d}$, we multiply $\dfrac{a}{b}$ times the reciprocal of $\dfrac{c}{d}$, which is $\dfrac{d}{c}$. The following examples demonstrate the important steps of a division problem.

$$\frac{2}{3} \div \frac{1}{2} = \frac{2}{3} \cdot \frac{2}{1} = \frac{4}{3}$$

$$\frac{5}{6} \div \frac{3}{4} = \frac{5}{6} \cdot \frac{4}{3} = \frac{5 \cdot 4}{6 \cdot 3} = \frac{5 \cdot 2 \cdot 2}{2 \cdot 3 \cdot 3} = \frac{10}{9}$$

$$-\frac{9}{12} \div \frac{3}{6} = -\frac{\overset{3}{9}}{\underset{2}{12}} \cdot \frac{\overset{1}{6}}{\underset{1}{3}} = -\frac{3}{2}$$

$$\left(-\frac{27}{56}\right) \div \left(-\frac{33}{72}\right) = \left(-\frac{27}{56}\right)\left(-\frac{72}{33}\right) = \frac{\overset{9}{27} \cdot \overset{9}{72}}{\underset{7}{56} \cdot \underset{11}{33}} = \frac{81}{77}$$

$$\frac{6}{7} \div 2 = \frac{6}{7} \cdot \frac{1}{2} = \frac{\overset{3}{6}}{7} \cdot \frac{1}{\underset{1}{2}} = \frac{3}{7}$$

$$\frac{5x}{7y} \div \frac{10}{28y} = \frac{5x}{7y} \cdot \frac{28y}{10} = \frac{5 \cdot x \cdot \overset{\overset{2}{4}}{28} \cdot y}{7 \cdot y \cdot \underset{2}{10}} = 2x$$

Classroom Example
Lynn purchased 24 yards of fabric for her sewing class. If $\frac{3}{4}$ of a yard is needed for each pillow, how many pillows can be made?

EXAMPLE 15

Frank has purchased 50 candy bars to make s'mores for the Boy Scout troop. If he uses $\frac{2}{3}$ of a candy bar for each s'more, how many s'mores will he be able to make?

Solution

To find how many s'mores can be made, we need to divide 50 by $\frac{2}{3}$.

$$50 \div \frac{2}{3} = 50 \cdot \frac{3}{2} = \frac{50}{1} \cdot \frac{3}{2} = \frac{\overset{25}{\cancel{50}}}{1} \cdot \frac{3}{\underset{1}{\cancel{2}}} = \frac{75}{1} = 75$$

Frank can make 75 s'mores.

Concept Quiz 2.1

For Problems 1–10, answer true or false.

1. 6 is a rational number.

2. $\frac{1}{8}$ is a rational number.

3. $\frac{-2}{-3} = \frac{2}{3}$

4. $\frac{-5}{3} = \frac{5}{-3}$

5. The product of a negative rational number and a positive rational number is a positive rational number.

6. If the product of two rational numbers is 1, the numbers are said to be reciprocals.

7. The reciprocal of $\frac{-3}{7}$ is $\frac{7}{3}$.

8. $\frac{10}{25}$ is reduced to lowest terms.

9. $\frac{4ab}{7c}$ is reduced to lowest terms.

10. To divide $\frac{m}{n}$ by $\frac{p}{q}$, we multiply $\frac{m}{n}$ by $\frac{q}{p}$.

Problem Set 2.1

For Problems 1–24, reduce each fraction to lowest terms.
(Objective 1)

1. $\frac{8}{12}$

2. $\frac{12}{16}$

3. $\frac{16}{24}$

4. $\frac{18}{32}$

5. $\frac{15}{9}$

6. $\frac{48}{36}$

7. $\frac{-8}{48}$

8. $\frac{-3}{15}$

9. $\frac{27}{-36}$

10. $\frac{9}{-51}$

11. $\dfrac{-54}{-56}$

12. $\dfrac{-24}{-80}$

13. $\dfrac{24x}{44x}$

14. $\dfrac{15y}{25y}$

15. $\dfrac{9x}{21y}$

16. $\dfrac{4y}{30x}$

17. $\dfrac{14xy}{35y}$

18. $\dfrac{55xy}{77x}$

19. $\dfrac{-20ab}{52bc}$

20. $\dfrac{-23ac}{41c}$

21. $\dfrac{-56yz}{-49xy}$

22. $\dfrac{-21xy}{-14ab}$

23. $\dfrac{65abc}{91ac}$

24. $\dfrac{68xyz}{85yz}$

For Problems 25–58, multiply or divide as indicated, and express answers in reduced form. **(Objective 2)**

25. $\dfrac{3}{4} \cdot \dfrac{5}{7}$

26. $\dfrac{4}{5} \cdot \dfrac{3}{11}$

27. $\dfrac{2}{7} \div \dfrac{3}{5}$

28. $\dfrac{5}{6} \div \dfrac{11}{13}$

29. $\dfrac{3}{8} \cdot \dfrac{12}{15}$

30. $\dfrac{4}{9} \cdot \dfrac{3}{2}$

31. $\dfrac{-6}{13} \cdot \dfrac{26}{9}$

32. $\dfrac{3}{4} \cdot \dfrac{-14}{12}$

33. $\dfrac{7}{9} \div \dfrac{5}{9}$

34. $\dfrac{3}{11} \div \dfrac{7}{11}$

35. $\dfrac{1}{4} \div \dfrac{-5}{6}$

36. $\dfrac{7}{8} \div \dfrac{14}{-16}$

37. $\left(-\dfrac{8}{10}\right)\left(-\dfrac{10}{32}\right)$

38. $\left(-\dfrac{6}{7}\right)\left(-\dfrac{21}{24}\right)$

39. $-9 \div \dfrac{1}{3}$

40. $-10 \div \dfrac{1}{4}$

41. $\dfrac{5x}{9y} \cdot \dfrac{7y}{3x}$

42. $\dfrac{4a}{11b} \cdot \dfrac{6b}{7a}$

43. $\dfrac{6a}{14b} \cdot \dfrac{16b}{18a}$

44. $\dfrac{5y}{8x} \cdot \dfrac{14z}{15y}$

45. $\dfrac{10x}{-9y} \cdot \dfrac{15}{20x}$

46. $\dfrac{3x}{4y} \cdot \dfrac{-8w}{9z}$

47. $ab \cdot \dfrac{2}{b}$

48. $3xy \cdot \dfrac{4}{x}$

49. $\left(-\dfrac{7x}{12y}\right)\left(-\dfrac{24y}{35x}\right)$

50. $\left(-\dfrac{10a}{15b}\right)\left(-\dfrac{45b}{65a}\right)$

51. $\dfrac{3}{x} \div \dfrac{6}{y}$

52. $\dfrac{6}{x} \div \dfrac{14}{y}$

53. $\dfrac{5x}{9y} \div \dfrac{13x}{36y}$

54. $\dfrac{3x}{5y} \div \dfrac{7x}{10y}$

55. $\dfrac{-7}{x} \div \dfrac{9}{x}$

56. $\dfrac{8}{y} \div \dfrac{28}{-y}$

57. $\dfrac{-4}{n} \div \dfrac{-18}{n}$

58. $\dfrac{-34}{n} \div \dfrac{-51}{n}$

For Problems 59–74, perform the operations as indicated, and express answers in lowest terms. **(Objective 2)**

59. $\dfrac{3}{4} \cdot \dfrac{8}{9} \cdot \dfrac{12}{20}$

60. $\dfrac{5}{6} \cdot \dfrac{9}{10} \cdot \dfrac{8}{7}$

61. $\left(-\dfrac{3}{8}\right)\left(\dfrac{13}{14}\right)\left(-\dfrac{12}{9}\right)$

62. $\left(-\dfrac{7}{9}\right)\left(\dfrac{5}{11}\right)\left(-\dfrac{18}{14}\right)$

63. $\left(\dfrac{3x}{4y}\right)\left(\dfrac{8}{9x}\right)\left(\dfrac{12y}{5}\right)$

64. $\left(\dfrac{2x}{3y}\right)\left(\dfrac{5y}{x}\right)\left(\dfrac{9}{4x}\right)$

65. $\left(-\dfrac{2}{3}\right)\left(\dfrac{3}{4}\right) \div \dfrac{1}{8}$

66. $\dfrac{3}{4} \cdot \dfrac{4}{5} \div \dfrac{1}{6}$

67. $\dfrac{5}{7} \div \left(-\dfrac{5}{6}\right)\left(-\dfrac{6}{7}\right)$

68. $\left(-\dfrac{3}{8}\right) \div \left(-\dfrac{4}{5}\right)\left(\dfrac{1}{2}\right)$

69. $\left(-\dfrac{6}{7}\right) \div \left(\dfrac{5}{7}\right)\left(-\dfrac{5}{6}\right)$

70. $\left(-\dfrac{4}{3}\right) \div \left(\dfrac{4}{5}\right)\left(\dfrac{3}{5}\right)$

71. $\left(\dfrac{4}{9}\right)\left(-\dfrac{9}{8}\right) \div \left(-\dfrac{3}{4}\right)$

72. $\left(-\dfrac{7}{8}\right)\left(\dfrac{4}{7}\right) \div \left(-\dfrac{3}{2}\right)$

73. $\left(\dfrac{5}{2}\right)\left(\dfrac{2}{3}\right) \div \left(-\dfrac{1}{4}\right) \div (-3)$

74. $\dfrac{1}{3} \div \left(\dfrac{3}{4}\right)\left(\dfrac{1}{2}\right) \div 2$

For Problems 75–81, solve the word problems. **(Objective 3)**

75. Maria's department has $\dfrac{3}{4}$ of all of the accounts within the ABC Advertising Agency. Maria is personally responsible for $\dfrac{1}{3}$ of all accounts in her department. For what portion of all of the accounts at ABC is Maria personally responsible?

76. Pablo has a board that is $4\frac{1}{2}$ feet long, and he wants to cut it into three pieces of the same length (see Figure 2.1). Find the length of each of the three pieces.

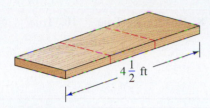

$4\frac{1}{2}$ ft

Figure 2.1

77. A recipe for a birthday cake calls for $\frac{3}{4}$ cup of sugar. How much sugar is needed to make 3 cakes?

78. Jonas left an estate valued at $750,000. His will states that three-fourths of the estate is to be divided equally among his three children. How much should each receive?

79. One of Arlene's recipes calls for $3\frac{1}{2}$ cups of milk. If she wants to make one-half of the recipe, how much milk should she use?

80. The total length of the four sides of a square is $8\frac{2}{3}$ yards. How long is each side of the square?

81. If it takes $3\frac{1}{4}$ yards of material to make curtains for one window, how much material is needed for 5 windows?

82. If your calculator is equipped to handle rational numbers in $\frac{a}{b}$ form, check your answers for Problems 1–12 and 59–74.

Thoughts Into Words

83. State in your own words the property

$$-\frac{a}{b} = \frac{-a}{b} = \frac{a}{-b}$$

84. What mistake was made in the following simplification process?

$$\frac{1}{2} \div \left(\frac{2}{3}\right)\left(\frac{3}{4}\right) \div 3 = \frac{1}{2} \div \frac{1}{2} \div 3 = \frac{1}{2} \cdot 2 \cdot \frac{1}{3} = \frac{1}{3}$$

How would you correct the error?

Further Investigations

85. The division problem $35 \div 7$ can be interpreted as "how many 7s are there in 35?" Likewise, a division problem such as $3 \div \frac{1}{2}$ can be interpreted as, "how many one-halves in 3?" Use this *how-many* interpretation to do the following division problems.

(a) $4 \div \frac{1}{2}$ **(b)** $3 \div \frac{1}{4}$

(c) $5 \div \frac{1}{8}$ **(d)** $6 \div \frac{1}{7}$

(e) $\frac{5}{6} \div \frac{1}{6}$ **(f)** $\frac{7}{8} \div \frac{1}{8}$

86. Estimation is important in mathematics. In each of the following, estimate whether the answer is larger than 1 or smaller than 1 by using the *how-many* idea from Problem 86.

(a) $\frac{3}{4} \div \frac{1}{2}$ **(b)** $1 \div \frac{7}{8}$

(c) $\frac{1}{2} \div \frac{3}{4}$ **(d)** $\frac{8}{7} \div \frac{7}{8}$

(e) $\frac{2}{3} \div \frac{1}{4}$ **(f)** $\frac{3}{5} \div \frac{3}{4}$

87. Reduce each of the following to lowest terms. Don't forget that we reviewed some divisibility rules in Problem Set 1.2.

(a) $\frac{99}{117}$ **(b)** $\frac{175}{225}$

(c) $\dfrac{-111}{123}$ (d) $\dfrac{-234}{270}$ (g) $\dfrac{91}{143}$ (h) $\dfrac{187}{221}$

(e) $\dfrac{270}{495}$ (f) $\dfrac{324}{459}$

Answers to the Concept Quiz

1. True **2.** True **3.** True **4.** True **5.** False **6.** True **7.** False **8.** False **9.** True **10.** True

---

2.2	Rational Numbers: Addition and Subtraction

OBJECTIVES

1 Add and subtract rational numbers in fractional form

2 Combine similar terms whose coefficients are rational numbers in fractional form

3 Solve application problems that involve the addition and subtraction of rational numbers in fractional form

---

Suppose that it is one-fifth of a mile between your dorm and the student center, and two-fifths of a mile between the student center and the library along a straight line as indicated in Figure 2.2. The total distance between your dorm and the library is three-fifths of a mile, and we write $\dfrac{1}{5} + \dfrac{2}{5} = \dfrac{3}{5}$.

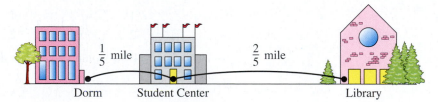

Figure 2.2

A pizza is cut into seven equal pieces and you eat two of the pieces. How much of the pizza (Figure 2.3) remains? We represent the whole pizza by $\dfrac{7}{7}$ and then conclude that $\dfrac{7}{7} - \dfrac{2}{7} = \dfrac{5}{7}$ of the pizza remains.

Figure 2.3

These examples motivate the following definition for addition and subtraction of rational numbers in $\dfrac{a}{b}$ form:

> **Definition 2.3**
>
> If a, b, and c are integers, and b is not zero, then
>
> $$\frac{a}{b} + \frac{c}{b} = \frac{a+c}{b} \qquad \text{Addition}$$
>
> $$\frac{a}{b} - \frac{c}{b} = \frac{a-c}{b} \qquad \text{Subtraction}$$

We say that rational numbers *with common denominators* can be added or subtracted by adding or subtracting the numerators and placing the results over the common denominator. Consider the following examples:

$$\frac{3}{7} + \frac{2}{7} = \frac{3+2}{7} = \frac{5}{7}$$

$$\frac{7}{8} - \frac{2}{8} = \frac{7-2}{8} = \frac{5}{8}$$

$$\frac{2}{6} + \frac{1}{6} = \frac{2+1}{6} = \frac{3}{6} = \frac{1}{2} \qquad \text{Reduce the result}$$

$$\frac{3}{11} - \frac{5}{11} = \frac{3-5}{11} = \frac{-2}{11} \qquad \text{or} \qquad -\frac{2}{11}$$

$$\frac{5}{x} + \frac{7}{x} = \frac{5+7}{x} = \frac{12}{x}$$

$$\frac{9}{y} - \frac{3}{y} = \frac{9-3}{y} = \frac{6}{y}$$

In the last two examples, the variables x and y cannot be equal to zero in order to exclude division by zero. It is always necessary to restrict denominators to nonzero values, although we will not take the time or space to list such restrictions for every problem.

How do we add or subtract if the fractions do not have a common denominator? We use the fundamental principle of fractions, $\dfrac{a}{b} = \dfrac{a \cdot k}{b \cdot k}$, and obtain equivalent fractions that have a common denominator. **Equivalent fractions** are fractions that name the same number. Consider the following example, which shows the details.

Classroom Example
Add $\dfrac{1}{4} + \dfrac{1}{5}$.

EXAMPLE 1   Add $\dfrac{1}{2} + \dfrac{1}{3}$.

Solution

$$\frac{1}{2} = \frac{1 \cdot 3}{2 \cdot 3} = \frac{3}{6} \qquad \frac{1}{2} \text{ and } \frac{3}{6} \text{ are equivalent fractions naming the same number}$$

$$\frac{1}{3} = \frac{1 \cdot 2}{3 \cdot 2} = \frac{2}{6} \qquad \frac{1}{2} \text{ and } \frac{2}{6} \text{ are equivalent fractions naming the same number}$$

$$\frac{1}{2} + \frac{1}{3} = \frac{3}{6} + \frac{2}{6} = \frac{3+2}{6} = \frac{5}{6}$$

Notice that we chose 6 as the common denominator, and 6 is the least common multiple of the original denominators 2 and 3. (Recall that the least common multiple is the smallest nonzero whole number divisible by the given numbers.) In general, we use the least common multiple of the denominators as the **least common denominator** (LCD).

Recall from Section 1.2 that the least common multiple may be found either by inspection or by using prime factorization forms of the numbers. Let's consider some examples involving these procedures.

Classroom Example
Add $\frac{1}{3} + \frac{4}{7}$.

EXAMPLE 2   Add $\frac{1}{4} + \frac{2}{5}$.

Solution

By inspection we see that the LCD is 20. Thus both fractions can be changed to equivalent fractions that have a denominator of 20.

$$\frac{1}{4}+\frac{2}{5}=\frac{1\cdot 5}{4\cdot 5}+\frac{2\cdot 4}{5\cdot 4}=\frac{5}{20}+\frac{8}{20}=\frac{13}{20}$$

Use of the fundamental
principle of fractions

Classroom Example
Subtract $\frac{7}{9} - \frac{4}{15}$.

EXAMPLE 3   Subtract $\frac{5}{8} - \frac{7}{12}$.

Solution

By inspection it is clear that the LCD is 24.

$$\frac{5}{8}-\frac{7}{12}=\frac{5\cdot 3}{8\cdot 3}-\frac{7\cdot 2}{12\cdot 2}=\frac{15}{24}-\frac{14}{24}=\frac{1}{24}$$

If the LCD is not obvious by inspection, then we can use the technique from Chapter 1 to find the least common multiple. We proceed as follows.

Step 1 Express each denominator as a product of prime factors.

Step 2 The LCD contains each different prime factor as many times as the *most* times it appears in any one of the factorizations from step 1.

Classroom Example
Add $\frac{7}{12} + \frac{6}{15}$.

EXAMPLE 4   Add $\frac{5}{18} + \frac{7}{24}$.

Solution

If we cannot find the LCD by inspection, then we can use the prime factorization forms.

$$\left.\begin{array}{l} 18 = 2\cdot 3\cdot 3 \\ 24 = 2\cdot 2\cdot 2\cdot 3 \end{array}\right\} \longrightarrow \text{LCD} = 2\cdot 2\cdot 2\cdot 3\cdot 3 = 72$$

$$\frac{5}{18}+\frac{7}{24}=\frac{5\cdot 4}{18\cdot 4}+\frac{7\cdot 3}{24\cdot 3}=\frac{20}{72}+\frac{21}{72}=\frac{41}{72}$$

Classroom Example
Subtract $\dfrac{3}{10} - \dfrac{11}{15}$.

EXAMPLE 5   Subtract $\dfrac{3}{14} - \dfrac{8}{35}$.

Solution

$$\left.\begin{array}{l} 14 = 2 \cdot 7 \\ 35 = 5 \cdot 7 \end{array}\right\} \longrightarrow \text{LCD} = 2 \cdot 5 \cdot 7 = 70$$

$$\frac{3}{14} - \frac{8}{35} = \frac{3 \cdot 5}{14 \cdot 5} - \frac{8 \cdot 2}{35 \cdot 2} = \frac{15}{70} - \frac{16}{70} = \frac{-1}{70} \quad \text{or} \quad -\frac{1}{70}$$

Classroom Example
Add $\dfrac{-7}{9} + \dfrac{7}{15}$.

EXAMPLE 6   Add $\dfrac{-5}{8} + \dfrac{3}{14}$.

Solution

$$\left.\begin{array}{l} 8 = 2 \cdot 2 \cdot 2 \\ 14 = 2 \cdot 7 \end{array}\right\} \longrightarrow \text{LCD} = 2 \cdot 2 \cdot 2 \cdot 7 = 56$$

$$\frac{-5}{8} + \frac{3}{14} = \frac{-5 \cdot 7}{8 \cdot 7} + \frac{3 \cdot 4}{14 \cdot 4} = \frac{-35}{56} + \frac{12}{56} = \frac{-23}{56} \quad \text{or} \quad -\frac{23}{56}$$

Classroom Example
Add $-2 + \dfrac{4}{9}$.

EXAMPLE 7   Add $-3 + \dfrac{2}{5}$.

Solution

$$-3 + \frac{2}{5} = \frac{-3 \cdot 5}{1 \cdot 5} + \frac{2}{5} = \frac{-15}{5} + \frac{2}{5} = \frac{-15 + 2}{5} = \frac{-13}{5} \quad \text{or} \quad -\frac{13}{5}$$

Denominators that contain variables do not complicate the situation very much, as the next examples illustrate.

Classroom Example
Add $\dfrac{4}{m} + \dfrac{5}{n}$.

EXAMPLE 8   Add $\dfrac{2}{x} + \dfrac{3}{y}$.

Solution

By inspection, the LCD is xy.

$$\frac{2}{x} + \frac{3}{y} = \frac{2 \cdot y}{x \cdot y} + \frac{3 \cdot x}{y \cdot x} = \frac{2y}{xy} + \frac{3x}{xy} = \frac{2y + 3x}{xy}$$

Commutative property

Classroom Example
Subtract $\dfrac{3}{4x} - \dfrac{7}{18y}$.

EXAMPLE 9   Subtract $\dfrac{3}{8x} - \dfrac{5}{12y}$.

Solution

$$\left.\begin{array}{l} 8x = 2 \cdot 2 \cdot 2 \cdot x \\ 12y = 2 \cdot 2 \cdot 3 \cdot y \end{array}\right\} \longrightarrow \text{LCD} = 2 \cdot 2 \cdot 2 \cdot 3 \cdot x \cdot y = 24xy$$

$$\frac{3}{8x} - \frac{5}{12y} = \frac{3 \cdot 3y}{8x \cdot 3y} - \frac{5 \cdot 2x}{12y \cdot 2x} = \frac{9y}{24xy} - \frac{10x}{24xy} = \frac{9y - 10x}{24xy}$$

Classroom Example
Add $\dfrac{7}{6x} + \dfrac{-4}{9yz}$.

EXAMPLE 10   Add $\dfrac{7}{4a} + \dfrac{-5}{6bc}$.

Solution

$$\left.\begin{array}{l} 4a = 2 \cdot 2 \cdot a \\ 6bc = 2 \cdot 3 \cdot b \cdot c \end{array}\right\} \quad \longrightarrow \quad \text{LCD} = 2 \cdot 2 \cdot 3 \cdot a \cdot b \cdot c = 12abc$$

$$\frac{7}{4a} + \frac{-5}{6ac} = \frac{7 \cdot 3bc}{4a \cdot 3bc} + \frac{-5 \cdot 2a}{6bc \cdot 2a}$$

$$= \frac{21bc}{12abc} + \frac{-10a}{12abc}$$

$$= \frac{21bc - 10a}{12abc}$$

Let's now consider simplifying numerical expressions that contain rational numbers. As with integers, multiplications and divisions are done first, and then the additions and subtractions are performed. In these next examples only the major steps are shown, so be sure that you can fill in all of the other details.

Classroom Example
Simplify $\dfrac{3}{5} - \dfrac{1}{3} \cdot \dfrac{3}{4} + \dfrac{1}{4} \cdot \dfrac{1}{2}$.

EXAMPLE 11   Simplify $\dfrac{3}{4} + \dfrac{2}{3} \cdot \dfrac{3}{5} - \dfrac{1}{2} \cdot \dfrac{1}{5}$.

Solution

$$\frac{3}{4} + \frac{2}{3} \cdot \frac{3}{5} - \frac{1}{2} \cdot \frac{1}{5} = \frac{3}{4} + \frac{2}{5} - \frac{1}{10} \qquad \text{Perform the multiplications}$$

$$= \frac{15}{20} + \frac{8}{20} - \frac{2}{20}$$

$$= \frac{15 + 8 - 2}{20} = \frac{21}{20} \qquad \text{Change to equivalent fractions and combine numerators}$$

Classroom Example
Simplify
$\dfrac{1}{3} \div \dfrac{2}{3} + \left(\dfrac{-3}{4}\right)\left(\dfrac{1}{2}\right) + \dfrac{5}{6}$.

EXAMPLE 12   Simplify $\dfrac{3}{5} \div \dfrac{8}{5} + \left(-\dfrac{1}{2}\right)\left(\dfrac{1}{3}\right) + \dfrac{5}{12}$.

Solution

$$\frac{3}{5} \div \frac{8}{5} + \left(-\frac{1}{2}\right)\left(\frac{1}{3}\right) + \frac{5}{12} = \frac{3}{5} \cdot \frac{5}{8} + \left(-\frac{1}{2}\right)\left(\frac{1}{3}\right) + \frac{5}{12} \qquad \text{Change division to multiply by the reciprocal}$$

$$= \frac{3}{8} + \frac{-1}{6} + \frac{5}{12}$$

$$= \frac{9}{24} + \frac{-4}{24} + \frac{10}{24}$$

$$= \frac{9 + (-4) + 10}{24}$$

$$= \frac{15}{24} = \frac{5}{8} \qquad \text{Reduce!}$$

The distributive property, $a(b + c) = ab + ac$, holds true for rational numbers and, as with integers, can be used to facilitate manipulation.

Classroom Example
Simplify $18\left(\dfrac{1}{2} + \dfrac{1}{6}\right)$.

EXAMPLE 13   Simplify $12\left(\dfrac{1}{3} + \dfrac{1}{4}\right)$.

Solution

For help in this situation, let's change the form by applying the distributive property.

$$12\left(\frac{1}{3} + \frac{1}{4}\right) = 12\left(\frac{1}{3}\right) + 12\left(\frac{1}{4}\right)$$
$$= 4 + 3$$
$$= 7$$

Classroom Example
Simplify $\dfrac{5}{7}\left(\dfrac{1}{6} + \dfrac{1}{4}\right)$.

EXAMPLE 14   Simplify $\dfrac{5}{8}\left(\dfrac{1}{2} + \dfrac{1}{3}\right)$.

Solution

In this case it may be easier not to apply the distributive property but to work with the expression in its given form.

$$\frac{5}{8}\left(\frac{1}{2} + \frac{1}{3}\right) = \frac{5}{8}\left(\frac{3}{6} + \frac{2}{6}\right)$$
$$= \frac{5}{8}\left(\frac{5}{6}\right)$$
$$= \frac{25}{48}$$

Examples 13 and 14 emphasize a point we made in Chapter 1. Think first, and decide whether or not the properties can be used to make the manipulations easier. Example 15 illustrates how to combine similar terms that have fractional coefficients.

Classroom Example
Simplify $\dfrac{1}{3}m - \dfrac{2}{5}m + \dfrac{1}{2}m$ by combining similar terms.

EXAMPLE 15   Simplify $\dfrac{1}{2}x + \dfrac{2}{3}x - \dfrac{3}{4}x$ by combining similar terms.

Solution

We can use the distributive property and our knowledge of adding and subtracting rational numbers to solve this type of problem.

$$\frac{1}{2}x + \frac{2}{3}x - \frac{3}{4}x = \left(\frac{1}{2} + \frac{2}{3} - \frac{3}{4}\right)x$$
$$= \left(\frac{6}{12} + \frac{8}{12} - \frac{9}{12}\right)x$$
$$= \frac{5}{12}x$$

EXAMPLE 16

Brian brought 5 cups of flour along on a camping trip. He wants to make biscuits and cake for tonight's supper. It takes $\frac{3}{4}$ of a cup of flour for the biscuits and $2\frac{3}{4}$ cups of flour for the cake. How much flour will be left over for the rest of his camping trip?

Solution

Let's do this problem in two steps. First add the amounts of flour needed for the biscuits and cake.

$$\frac{3}{4} + 2\frac{3}{4} = \frac{3}{4} + \frac{11}{4} = \frac{14}{4} = \frac{7}{2}$$

Then to find the amount of flour left over, we will subtract $\frac{7}{2}$ from 5.

$$5 - \frac{7}{2} = \frac{10}{2} - \frac{7}{2} = \frac{3}{2} = 1\frac{1}{2}$$

So $1\frac{1}{2}$ cups of flour are left over.

Concept Quiz 2.2

For Problems 1–10, answer true or false.

1. To add rational numbers with common denominators, add the numerators and place the result over the common denominator.

2. When adding $\frac{2}{c} + \frac{6}{c}$, c can be equal to zero.

3. Fractions that name the same number are called equivalent fractions.

4. The least common multiple of the denominators can always be used as a common denominator when adding or subtracting fractions.

5. To simplify $\frac{3}{8} - \frac{1}{5}$, we need to find equivalent fractions with a common denominator.

6. To multiply $\frac{5}{7}$ and $\frac{2}{3}$, we need to find equivalent fractions with a common denominator.

7. Either 20, 40 or 60 can be used as a common denominator when adding $\frac{1}{4}$ and $\frac{3}{5}$, but 20 is the least common denominator.

8. When adding $\frac{2x}{ab}$ and $\frac{3y}{bc}$, the least common denominator is ac.

9. $36\left(\frac{1}{2} - \frac{4}{9}\right)$ simplifies to 2.

10. $\frac{2}{3}x - \frac{1}{4}x + \frac{5}{6}x$ simplifies to $\frac{13}{12}x$.

Problem Set 2.2

For Problems 1–64, add or subtract as indicated, and express your answers in lowest terms. **(Objective 1)**

1. $\frac{2}{7} + \frac{3}{7}$

2. $\frac{3}{11} + \frac{5}{11}$

3. $\frac{7}{9} - \frac{2}{9}$

4. $\frac{11}{13} - \frac{6}{13}$

5. $\frac{3}{4} + \frac{9}{4}$

6. $\frac{5}{6} + \frac{7}{6}$

7. $\dfrac{11}{12} - \dfrac{3}{12}$

8. $\dfrac{13}{16} - \dfrac{7}{16}$

9. $\dfrac{1}{8} - \dfrac{5}{8}$

10. $\dfrac{2}{9} - \dfrac{5}{9}$

11. $\dfrac{5}{24} + \dfrac{11}{24}$

12. $\dfrac{7}{36} + \dfrac{13}{36}$

13. $\dfrac{8}{x} + \dfrac{7}{x}$

14. $\dfrac{17}{y} + \dfrac{12}{y}$

15. $\dfrac{5}{3y} + \dfrac{1}{3y}$

16. $\dfrac{3}{8x} + \dfrac{1}{8x}$

17. $\dfrac{1}{3} + \dfrac{1}{5}$

18. $\dfrac{1}{6} + \dfrac{1}{8}$

19. $\dfrac{15}{16} - \dfrac{3}{8}$

20. $\dfrac{13}{12} - \dfrac{1}{6}$

21. $\dfrac{7}{10} + \dfrac{8}{15}$

22. $\dfrac{7}{12} + \dfrac{5}{8}$

23. $\dfrac{11}{24} + \dfrac{5}{32}$

24. $\dfrac{5}{18} + \dfrac{8}{27}$

25. $\dfrac{5}{18} - \dfrac{13}{24}$

26. $\dfrac{1}{24} - \dfrac{7}{36}$

27. $\dfrac{5}{8} - \dfrac{2}{3}$

28. $\dfrac{3}{4} - \dfrac{5}{6}$

29. $-\dfrac{2}{13} - \dfrac{7}{39}$

30. $-\dfrac{3}{11} - \dfrac{13}{33}$

31. $-\dfrac{3}{14} + \dfrac{1}{21}$

32. $-\dfrac{3}{20} + \dfrac{14}{25}$

33. $-4 - \dfrac{3}{7}$

34. $-2 - \dfrac{5}{6}$

35. $\dfrac{3}{4} - 6$

36. $\dfrac{5}{8} - 7$

37. $\dfrac{3}{x} + \dfrac{4}{y}$

38. $\dfrac{5}{x} + \dfrac{8}{y}$

39. $\dfrac{7}{a} - \dfrac{2}{b}$

40. $\dfrac{13}{a} - \dfrac{4}{b}$

41. $\dfrac{2}{x} + \dfrac{7}{2x}$

42. $\dfrac{5}{2x} + \dfrac{7}{x}$

43. $\dfrac{10}{3x} - \dfrac{2}{x}$

44. $\dfrac{13}{4x} - \dfrac{3}{x}$

45. $\dfrac{1}{x} - \dfrac{7}{5x}$

46. $\dfrac{2}{x} - \dfrac{17}{6x}$

47. $\dfrac{3}{2y} + \dfrac{5}{3y}$

48. $\dfrac{7}{3y} + \dfrac{9}{4y}$

49. $\dfrac{5}{12y} - \dfrac{3}{8y}$

50. $\dfrac{9}{4y} - \dfrac{5}{9y}$

51. $\dfrac{1}{6n} - \dfrac{7}{8n}$

52. $\dfrac{3}{10n} - \dfrac{11}{15n}$

53. $\dfrac{5}{3x} + \dfrac{7}{3y}$

54. $\dfrac{3}{2x} + \dfrac{7}{2y}$

55. $\dfrac{8}{5x} + \dfrac{3}{4y}$

56. $\dfrac{1}{5x} + \dfrac{5}{6y}$

57. $\dfrac{7}{4x} - \dfrac{5}{9y}$

58. $\dfrac{2}{7x} - \dfrac{11}{14y}$

59. $-\dfrac{3}{2x} - \dfrac{5}{4y}$

60. $-\dfrac{13}{8a} - \dfrac{11}{10b}$

61. $3 + \dfrac{2}{x}$

62. $\dfrac{5}{x} + 4$

63. $2 - \dfrac{3}{2x}$

64. $-1 - \dfrac{1}{3x}$

For Problems 65–80, simplify each numerical expression and express your answers in reduced form. **(Objective 1)**

65. $\dfrac{1}{4} - \dfrac{3}{8} + \dfrac{5}{12} - \dfrac{1}{24}$

66. $\dfrac{3}{4} + \dfrac{2}{3} - \dfrac{1}{6} + \dfrac{5}{12}$

67. $\dfrac{5}{6} + \dfrac{2}{3} \cdot \dfrac{3}{4} - \dfrac{1}{4} \cdot \dfrac{2}{5}$

68. $\dfrac{2}{3} + \dfrac{1}{2} \cdot \dfrac{2}{5} - \dfrac{1}{3} \cdot \dfrac{1}{5}$

69. $\dfrac{3}{4} \cdot \dfrac{6}{9} - \dfrac{5}{6} \cdot \dfrac{8}{10} + \dfrac{2}{3} \cdot \dfrac{6}{8}$

70. $\dfrac{3}{5} \cdot \dfrac{5}{7} + \dfrac{2}{3} \cdot \dfrac{3}{5} - \dfrac{1}{7} \cdot \dfrac{2}{5}$

71. $4 - \dfrac{2}{3} \cdot \dfrac{3}{5} - 6$

72. $3 + \dfrac{1}{2} \cdot \dfrac{1}{3} - 2$

73. $\dfrac{4}{5} - \dfrac{10}{12} - \dfrac{5}{6} \div \dfrac{14}{8} + \dfrac{10}{21}$

74. $\dfrac{3}{4} \div \dfrac{6}{5} + \dfrac{8}{12} \cdot \dfrac{6}{9} - \dfrac{5}{12}$

75. $24\left(\dfrac{3}{4} - \dfrac{1}{6}\right)$ Don't forget the distributive property!

76. $18\left(\dfrac{2}{3} + \dfrac{1}{9}\right)$

77. $64\left(\dfrac{3}{16} + \dfrac{5}{8} - \dfrac{1}{4} + \dfrac{1}{2}\right)$

78. $48\left(\dfrac{5}{12} - \dfrac{1}{6} + \dfrac{3}{8}\right)$

79. $\dfrac{7}{13}\left(\dfrac{2}{3} - \dfrac{1}{6}\right)$

80. $\dfrac{5}{9}\left(\dfrac{1}{2} + \dfrac{1}{4}\right)$

For Problems 81–96, simplify each algebraic expression by combining similar terms. **(Objective 2)**

81. $\dfrac{1}{3}x + \dfrac{2}{5}x$

82. $\dfrac{1}{4}x + \dfrac{2}{3}x$

83. $\dfrac{1}{3}a - \dfrac{1}{8}a$

84. $\dfrac{2}{5}a - \dfrac{2}{7}a$

85. $\dfrac{1}{2}x + \dfrac{2}{3}x + \dfrac{1}{6}x$

86. $\dfrac{1}{3}x + \dfrac{2}{5}x + \dfrac{5}{6}x$

87. $\dfrac{3}{5}n - \dfrac{1}{4}n + \dfrac{3}{10}n$

88. $\dfrac{2}{5}n - \dfrac{7}{10}n + \dfrac{8}{15}n$

89. $n + \dfrac{4}{3}n - \dfrac{1}{9}n$

90. $2n - \dfrac{6}{7}n + \dfrac{5}{14}n$

91. $-n - \dfrac{7}{9}n - \dfrac{5}{12}n$

92. $-\dfrac{3}{8}n - n - \dfrac{3}{14}n$

93. $\dfrac{3}{7}x + \dfrac{1}{4}y + \dfrac{1}{2}x + \dfrac{7}{8}y$

94. $\dfrac{5}{6}x + \dfrac{3}{4}y + \dfrac{4}{9}x + \dfrac{7}{10}y$

95. $\dfrac{2}{9}x + \dfrac{5}{12}y - \dfrac{7}{15}x - \dfrac{13}{15}y$

96. $-\dfrac{9}{10}x - \dfrac{3}{14}y + \dfrac{2}{25}x + \dfrac{5}{21}y$

For Problems 97–104, solve using addition and subtraction of rational numbers. **(Objective 3)**

97. Beth wants to make three sofa pillows for her new sofa. After consulting the chart provided by the fabric shop, she decides to make a 12″ round pillow, an 18″ square pillow, and a 12″ × 16″ rectangular pillow. According to the chart, how much fabric will Beth need to purchase?

Fabric shop chart

10″ round	$\dfrac{3}{8}$ yard
12″ round	$\dfrac{1}{2}$ yard
12″ square	$\dfrac{5}{8}$ yard
18″ square	$\dfrac{3}{4}$ yard
12″ × 16″ rectangular	$\dfrac{7}{8}$ yard

98. Marcus is decorating his room and plans on hanging three prints that are each $13\dfrac{3}{8}$ inches wide. He is going to hang the prints side by side with $2\dfrac{1}{4}$ inches between the prints. What width of wall space is needed to display the three prints?

99. From a board that is $12\dfrac{1}{2}$ feet long, a piece $1\dfrac{3}{4}$ feet long is cut off from one end. Find the length of the remaining piece of board.

100. Vinay has a board that is $6\dfrac{1}{2}$ feet long. If he cuts off a piece $2\dfrac{3}{4}$ feet long, how long is the remaining piece of board?

101. Mindy takes a daily walk of $2\dfrac{1}{2}$ miles. One day a thunderstorm forced her to stop her walk after $\dfrac{3}{4}$ of a mile. By how much was her walk shortened that day?

102. Blake Scott leaves $\dfrac{1}{4}$ of his estate to the Boy Scouts, $\dfrac{2}{5}$ to the local cancer fund, and the rest to his church. What fractional part of the estate does the church receive?

103. A triangular plot of ground measures $14\dfrac{1}{2}$ yards by $12\dfrac{1}{3}$ yards by $9\dfrac{5}{6}$ yards. How many yards of fencing are needed to enclose the plot?

104. For her exercise program, Lian jogs for $2\dfrac{1}{2}$ miles, then walks for $\dfrac{3}{4}$ of a mile, and finally jogs for another $1\dfrac{1}{4}$ miles. Find the total distance that Lian covers.

105. If your calculator handles rational numbers in $\dfrac{a}{b}$ form, check your answers for Problems 65–80.

Thoughts Into Words

106. Give a step-by-step description of the best way to add the rational numbers $\dfrac{3}{8}$ and $\dfrac{5}{18}$.

107. Give a step-by-step description of how to add the fractions $\dfrac{5}{4x}$ and $\dfrac{7}{6x}$.

108. The will of a deceased collector of antique automobiles specified that his cars be left to his three children. Half were to go to his elder son, $\dfrac{1}{3}$ to his daughter, and $\dfrac{1}{9}$ to his younger son. At the time of his death, 17 cars were in the collection. The administrator of his estate borrowed a car to make 18. Then he distributed the cars as follows:

Elder son: $\dfrac{1}{2}(18) = 9$

Daughter: $\dfrac{1}{3}(18) = 6$

Younger son: $\dfrac{1}{9}(18) = 2$

This takes care of the 17 cars, so the administrator then returned the borrowed car. Where is the error in this solution?

2.3	**Real Numbers and Algebraic Expressions**

OBJECTIVES

1 Classify real numbers

2 Add, subtract, multiply, and divide rational numbers in decimal form

3 Combine similar terms whose coefficients are rational numbers in decimal form

4 Evaluate algebraic expressions when the variables are rational numbers

5 Solve application problems that involve the operations of rational numbers in decimal form

We classify decimals—also called decimal fractions—as terminating, repeating, or nonrepeating. Here are examples of these classifications:

Terminating decimals	Repeating decimals	Nonrepeating decimals
0.3	0.333333 . . .	0.5918654279 . . .
0.26	0.5466666 . . .	0.26224222722229 . . .
0.347	0.14141414 . . .	0.145117211193111148 . . .
0.9865	0.237237237 . . .	0.645751311 . . .

Technically, a **terminating decimal** can be thought of as repeating zeros after the last digit. For example, $0.3 = 0.30 = 0.300 = 0.3000$, and so on.

A **repeating decimal** has a block of digits that repeats indefinitely. This repeating block of digits may contain any number of digits and may or may not begin repeating immediately after the decimal point.

In Section 2.1 we defined a rational number to be any number that can be written in the form $\dfrac{a}{b}$, where a and b are integers and b is not zero. A rational number can also be defined as **any number that has a terminating or repeating decimal representation**. Thus we can express rational numbers in either common-fraction form or decimal-fraction form, as the next examples illustrate. A repeating decimal can also be written by using a bar over the digits that repeat; for example, $0.\overline{14}$.

Terminating decimals	Repeating decimals
$\dfrac{3}{4} = 0.75$	$\dfrac{1}{3} = 0.3333 . . .$

$$\frac{1}{8} = 0.125 \qquad\qquad \frac{2}{3} = 0.66666\ldots$$

$$\frac{5}{16} = 0.3125 \qquad\qquad \frac{1}{6} = 0.166666\ldots$$

$$\frac{7}{25} = 0.28 \qquad\qquad \frac{1}{12} = 0.08333\ldots$$

$$\frac{2}{5} = 0.4 \qquad\qquad \frac{14}{99} = 0.14141414\ldots$$

The **nonrepeating decimals** are called "irrational numbers" and do appear in forms other than decimal form. For example, $\sqrt{2}$, $\sqrt{3}$, and π are irrational numbers; a partial representation for each of these follows.

$$\left.\begin{array}{l} \sqrt{2} = 1.414213562373\ldots \\[6pt] \sqrt{3} = 1.73205080756887\ldots \\[6pt] \pi = 3.14159265358979\ldots \end{array}\right\} \quad \text{Nonrepeating decimals}$$

(We will do more work with the irrational numbers in Chapter 9.)

The rational numbers together with the irrational numbers form the set of **real numbers**. The following tree diagram of the real number system is helpful for summarizing some basic ideas.

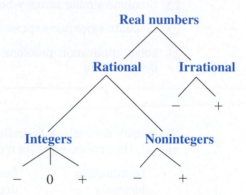

Any real number can be traced down through the diagram as follows.

5 is real, rational, an integer, and positive

−4 is real, rational, an integer, and negative

$\dfrac{3}{4}$ is real, rational, a noninteger, and positive

0.23 is real, rational, a noninteger, and positive

−0.161616 . . . is real, rational, a noninteger, and negative

$\sqrt{7}$ is real, irrational, and positive

$-\sqrt{2}$ is real, irrational, and negative

In Section 1.3, we associated the set of integers with evenly spaced points on a line as indicated in Figure 2.4. This idea of associating numbers with points on a

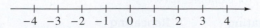

Figure 2.4

line can be extended so that there is a one-to-one correspondence between points on a line and the entire set of real numbers (as shown in Figure 2.5). That is to say, to each real number there corresponds one and only one point on the line, and to each point on the line there corresponds one and only one real number. The line is often referred to as the **real number line**, and the number associated with each point on the line is called the **coordinate** of the point.

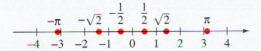

Figure 2.5

The properties we discussed in Section 1.5 pertaining to integers are true for all real numbers; we restate them here for your convenience. The multiplicative inverse property was added to the list; a discussion of that property follows.

Commutative Property of Addition

If a and b are real numbers, then

$$a + b = b + a$$

Commutative Property of Multiplication

If a and b are real numbers, then

$$ab = ba$$

Associative Property of Addition

If a, b, and c are real numbers, then

$$(a + b) + c = a + (b + c)$$

Associative Property of Multiplication

If a, b, and c are real numbers, then

$$(ab)c = a(bc)$$

Identity Property of Addition

If a is any real number, then

$$a + 0 = 0 + a = a$$

Identity Property of Multiplication

If a is any real number, then

$$a(1) = 1(a) = a$$

Additive Inverse Property

For every real number a, there exists a real number $-a$, such that

$$a + (-a) = (-a) + a = 0$$

Multiplication Property of Zero

If a is any real number, then

$$a(0) = 0(a) = 0$$

Multiplicative Property of Negative One

If a is any real number, then

$$a(-1) = -1(a) = -a$$

Multiplicative Inverse Property

For every nonzero real number a, there exists a real number $\dfrac{1}{a}$, such that

$$a\left(\frac{1}{a}\right) = \frac{1}{a}(a) = 1$$

Distributive Property

If a, b, and c are real numbers, then

$$a(b + c) = ab + ac$$

The number $\dfrac{1}{a}$ is called the **multiplicative inverse** or the **reciprocal** of a. For example, the reciprocal of 2 is $\dfrac{1}{2}$ and $2\left(\dfrac{1}{2}\right) = \dfrac{1}{2}(2) = 1$. Likewise, the reciprocal of $\dfrac{1}{2}$ is $\dfrac{1}{\frac{1}{2}} = 2$.

Therefore, 2 and $\dfrac{1}{2}$ are said to be reciprocals (or multiplicative inverses) of each other.

Also, $\dfrac{2}{5}$ and $\dfrac{5}{2}$ are multiplicative inverses, and $\left(\dfrac{2}{5}\right)\left(\dfrac{5}{2}\right) = 1$. Since division by zero is undefined, zero does not have a reciprocal.

Basic Operations with Decimals

The basic operations with decimals may be related to the corresponding operation with common fractions. For example, $0.3 + 0.4 = 0.7$ because $\dfrac{3}{10} + \dfrac{4}{10} = \dfrac{7}{10}$, and

$0.37 - 0.24 = 0.13$ because $\dfrac{37}{100} - \dfrac{24}{100} = \dfrac{13}{100}$. In general, to add or subtract decimals, we add or subtract the hundredths, the tenths, the ones, the tens, and so on. To keep place values aligned, we line up the decimal points.

Addition		Subtraction	
	₁ ₁₁	⁶ ¹⁶	⁸ ¹¹¹³
2.14	5.214	7̶.6̶	9.2̶3̶5
3.12	3.162	4.9	6.781
5.16	7.218	2.7	2.454
10.42	8.914		
	24.508		

The following examples can be used to formulate a general rule for multiplying decimals.

Because $\dfrac{7}{10} \cdot \dfrac{3}{10} = \dfrac{21}{100}$ then $(0.7)(0.3) = 0.21$

Because $\dfrac{9}{10} \cdot \dfrac{23}{100} = \dfrac{207}{1000}$ then $(0.9)(0.23) = 0.207$

Because $\dfrac{11}{100} \cdot \dfrac{13}{100} = \dfrac{143}{10,000}$ then $(0.11)(0.13) = 0.0143$

In general, to multiply decimals we (1) multiply the numbers and ignore the decimal points, and then (2) insert the decimal point in the product so that the number of digits to the right of the decimal point in the product is equal to the sum of the number of digits to the right of the decimal point in each factor.

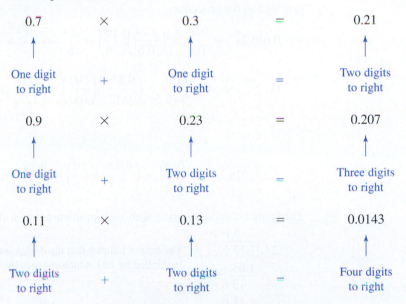

We frequently use the vertical format when multiplying decimals.

41.2	One digit to right	0.021	Three digits to right
0.13	Two digits to right	0.03	Two digits to right
1236		0.00063	Five digits to right
412			
5.356	Three digits to right		

Notice that in the last example we actually multiplied $3 \cdot 21$ and then inserted three 0s to the left so that there would be five digits to the right of the decimal point.

Once again let's look at some links between common fractions and decimals.

Because $\dfrac{6}{10} \div 2 = \dfrac{\overset{3}{\cancel{6}}}{10} \cdot \dfrac{1}{2} = \dfrac{3}{10}$ then $2\overline{)0.6}^{\,0.3}$

Because $\dfrac{39}{100} \div 13 = \dfrac{\overset{3}{\cancel{39}}}{100} \cdot \dfrac{1}{\cancel{13}} = \dfrac{3}{100}$ then $13\overline{)0.39}^{\,0.03}$

Because $\dfrac{85}{100} \div 5 = \dfrac{\overset{17}{\cancel{85}}}{100} \cdot \dfrac{1}{\cancel{5}} = \dfrac{17}{100}$ then $5\overline{)0.85}^{\,0.17}$

In general, to divide a decimal by a nonzero whole number we (1) place the decimal point in the quotient directly above the decimal point in the dividend

$$\left(\text{Divisor}\overline{)\text{Dividend}}^{\,\text{Quotient}} \right)$$

and then (2) divide as with whole numbers, except that in the division process, zeros are placed in the quotient immediately to the right of the decimal point in order to show the correct place value.

$$
\begin{array}{r}
0.121 \\
4\overline{)0.484}
\end{array}
\qquad
\begin{array}{r}
0.24 \\
32\overline{)7.68} \\
\underline{6\ 4} \\
1\ 28 \\
\underline{1\ 28}
\end{array}
\qquad
\begin{array}{r}
0.019 \\
12\overline{)0.228} \\
\underline{12} \\
108 \\
\underline{108}
\end{array}
\qquad
\text{Zero needed to show the correct place value}
$$

Don't forget that division can be checked by multiplication. For example, since $(12)(0.019) = 0.228$ we know that our last division example is correct.

Problems involving division by a decimal are easier to handle if we change the problem to an equivalent problem that has a whole number divisor. Consider the following examples in which the original division problem was changed to fractional form to show the reasoning involved in the procedure.

$$0.6\overline{)0.24} \ \rightarrow \ \dfrac{0.24}{0.6} = \left(\dfrac{0.24}{0.6}\right)\left(\dfrac{10}{10}\right) = \dfrac{2.4}{6} \ \rightarrow \ 6\overline{)2.4}^{\,0.4}$$

$$0.12\overline{)0.156} \ \rightarrow \ \dfrac{0.156}{0.12} = \left(\dfrac{0.156}{0.12}\right)\left(\dfrac{100}{100}\right) = \dfrac{15.6}{12} \ \rightarrow \
\begin{array}{r}
1.3 \\
12\overline{)15.6} \\
\underline{12} \\
3\ 6 \\
\underline{3\ 6}
\end{array}$$

$$1.3\overline{)0.026} \ \rightarrow \ \dfrac{0.026}{1.3} = \left(\dfrac{0.026}{1.3}\right)\left(\dfrac{10}{10}\right) = \dfrac{0.26}{13} \ \rightarrow \
\begin{array}{r}
0.02 \\
13\overline{)0.26} \\
\underline{26}
\end{array}$$

The format commonly used with such problems is as follows.

$$
\begin{array}{r}
5.6 \\
{}_{\llcorner}21.\overline{)1{}_{\llcorner}17.6} \\
\underline{1\ 05} \\
12\ 6 \\
\underline{12\ 6}
\end{array}
\qquad
\begin{array}{l}
\text{The arrows indicate that the divisor and dividend were} \\
\text{multiplied by 100, which changes the divisor to a whole number}
\end{array}
$$

$$
\begin{array}{r}
0.04 \\
3{}_{\llcorner}7.\overline{)1{}_{\llcorner}.48} \\
\underline{1\ 48}
\end{array}
\qquad
\text{The divisor and dividend were multiplied by 10}
$$

Our agreements for operating with positive and negative integers extend to all real numbers. For example, the product of two negative real numbers is a positive real number. Make sure that you agree with the following results. (You may need to do some work on scratch paper since the steps are not shown.)

$$0.24 + (-0.18) = 0.06 \qquad\qquad (-0.4)(0.8) = -0.32$$

$$-7.2 + 5.1 = -2.1 \qquad (-0.5)(-0.13) = 0.065$$
$$-0.6 + (-0.8) = -1.4 \qquad (1.4) \div (-0.2) = -7$$
$$2.4 - 6.1 = -3.7 \qquad (-0.18) \div (0.3) = -0.6$$
$$0.31 - (-0.52) = 0.83 \qquad (-0.24) \div (-4) = 0.06$$
$$(0.2)(-0.3) = -0.06$$

Numerical and algebraic expressions may contain the decimal form as well as the fractional form of rational numbers. We continue to follow the agreement that multiplications and divisions are done *first* and then the additions and subtractions, unless parentheses indicate otherwise. The following examples illustrate a variety of situations that involve both the decimal form and fractional form of rational numbers.

Classroom Example
Simplify
$5.6 \div (-8) + 3(4.2)$
$- (0.28) \div (-0.7)$.

| **EXAMPLE 1** | Simplify $6.3 \div 7 + (4)(2.1) - (0.24) \div (-0.4)$. |

Solution

$$6.3 \div 7 + (4)(2.1) - (0.24) \div (-0.4) = 0.9 + 8.4 - (-0.6)$$
$$= 0.9 + 8.4 + 0.6$$
$$= 9.9$$

Classroom Example
Evaluate $\frac{2}{3}x - \frac{1}{5}y$ for $x = \frac{3}{7}$
and $y = -2$.

| **EXAMPLE 2** | Evaluate $\frac{3}{5}a - \frac{1}{7}b$ for $a = \frac{5}{2}$ and $b = -1$. |

Solution

$$\frac{3}{5}a - \frac{1}{7}b = \frac{3}{5}\left(\frac{5}{2}\right) - \frac{1}{7}(-1) \quad \text{for } a = \frac{5}{2} \text{ and } b = -1$$

$$= \frac{3}{2} + \frac{1}{7}$$

$$= \frac{21}{14} + \frac{2}{14}$$

$$= \frac{23}{14}$$

Classroom Example
Evaluate $\frac{3}{4}a + \frac{1}{3}a - \frac{1}{2}a$
for $a = -\frac{5}{14}$.

| **EXAMPLE 3** | Evaluate $\frac{1}{2}x + \frac{2}{3}x - \frac{1}{5}x$ for $x = -\frac{3}{4}$. |

Solution

First, let's combine similar terms by using the distributive property.

$$\frac{1}{2}x + \frac{2}{3}x - \frac{1}{5}x = \left(\frac{1}{2} + \frac{2}{3} - \frac{1}{5}\right)x$$

$$= \left(\frac{15}{30} + \frac{20}{30} - \frac{6}{30}\right)x$$

$$= \frac{29}{30}x$$

Now we can evaluate.

$$\frac{29}{30}x = \frac{29}{30}\left(-\frac{3}{4}\right) \quad \text{when } x = -\frac{3}{4}$$

$$= \frac{29}{\overset{10}{30}}\left(-\frac{\overset{1}{3}}{4}\right) = -\frac{29}{40}$$

Classroom Example
Evaluate $4a + 5b$ for $a = 2.3$ and $b = 1.4$.

EXAMPLE 4   Evaluate $2x + 3y$ for $x = 1.6$ and $y = 2.7$.

Solution

$$2x + 3y = 2(1.6) + 3(2.7) \quad \text{when } x = 1.6 \text{ and } y = 2.7$$
$$= 3.2 + 8.1 = 11.3$$

Classroom Example
Evaluate $1.5d - 0.8d + 0.5d + 0.2d$ for $d = 0.4$.

EXAMPLE 5   Evaluate $0.9x + 0.7x - 0.4x + 1.3x$ for $x = 0.2$.

Solution

First, let's combine similar terms by using the distributive property.

$$0.9x + 0.7x - 0.4x + 1.3x = (0.9 + 0.7 - 0.4 + 1.3)x = 2.5x$$

Now we can evaluate.

$$2.5x = (2.5)(0.2) \quad \text{for } x = 0.2$$
$$= 0.5$$

Classroom Example
A stain glass artist is putting together a design. She has five pieces of glass whose lengths are 2.4 cm, 3.26 cm, 1.35 cm, 4.12 cm, and 0.7 cm. If the pieces are set side by side, what will be their combined length?

EXAMPLE 6

A layout artist is putting together a group of images. She has four images whose widths are 1.35 centimeters, 2.6 centimeters, 5.45 centimeters, and 3.2 centimeters. If the images are set side by side, what will be their combined width?

Solution

To find the combined width, we need to add the widths.

$$\begin{array}{r} 1.35 \\ 2.6 \\ 5.45 \\ +3.2 \\ \hline 12.60 \end{array}$$

The combined width would be 12.6 centimeters.

Concept Quiz 2.3

For Problems 1–10, answer true or false.

1. A rational number can be defined as any number that has a terminating or repeating decimal representation.

2. A repeating decimal has a block of digits that repeat only once.

3. Every irrational number is also classified as a real number.

4. The rational numbers along with the irrational numbers form the set of natural numbers.

5. 0.141414. . . is a rational number.

6. $-\sqrt{5}$ is real, irrational, and negative.

7. 0.35 is real, rational, integer, and positive.

8. The reciprocal of c, where $c \neq 0$, is also the multiplicative inverse of c.

9. Any number multiplied by its multiplicative inverse gives a result of 0.

10. Zero does not have a multiplicative inverse.

Problem Set 2.3

For Problems 1–8, classify the real numbers by tracing down the diagram on p. 60. **(Objective 1)**

1. -2

2. $1/3$

3. $\sqrt{5}$

4. $-0.09090909\ldots$

5. 0.16

6. $-\sqrt{3}$

7. $-8/7$

8. 0.125

For Problems 9–40, perform the indicated operations. **(Objective 2)**

9. $0.37 + 0.25$

10. $7.2 + 4.9$

11. $2.93 - 1.48$

12. $14.36 - 5.89$

13. $(7.6) + (-3.8)$

14. $(6.2) + (-2.4)$

15. $(-4.7) + 1.4$

16. $(-14.1) + 9.5$

17. $-3.8 + 11.3$

18. $-2.5 + 14.8$

19. $6.6 - (-1.2)$

20. $18.3 - (-7.4)$

21. $-11.5 - (-10.6)$

22. $-14.6 - (-8.3)$

23. $-17.2 - (-9.4)$

24. $-21.4 - (-14.2)$

25. $(0.4)(2.9)$

26. $(0.3)(3.6)$

27. $(-0.8)(0.34)$

28. $(-0.7)(0.67)$

29. $(9)(-2.7)$

30. $(8)(-7.6)$

31. $(-0.7)(-64)$

32. $(-0.9)(-56)$

33. $(-0.12)(-0.13)$

34. $(-0.11)(-0.15)$

35. $1.56 \div 1.3$

36. $7.14 \div 2.1$

37. $5.92 \div (-0.8)$

38. $-2.94 \div 0.6$

39. $-0.266 \div (-0.7)$

40. $-0.126 \div (-0.9)$

For Problems 41–54, simplify each of the numerical expressions. **(Objective 2)**

41. $16.5 - 18.7 + 9.4$

42. $17.7 + 21.2 - 14.6$

43. $0.34 - 0.21 - 0.74 + 0.19$

44. $-5.2 + 6.8 - 4.7 - 3.9 + 1.3$

45. $0.76(0.2 + 0.8)$

46. $9.8(1.8 - 0.8)$

47. $0.6(4.1) + 0.7(3.2)$

48. $0.5(74) - 0.9(87)$

49. $7(0.6) + 0.9 - 3(0.4) + 0.4$

50. $-5(0.9) - 0.6 + 4.1(6) - 0.9$

51. $(0.96) \div (-0.8) + 6(-1.4) - 5.2$

52. $(-2.98) \div 0.4 - 5(-2.3) + 1.6$

53. $5(2.3) - 1.2 - 7.36 \div 0.8 + 0.2$

54. $0.9(12) \div 0.4 - 1.36 \div 17 + 9.2$

For Problems 55–68, simplify each algebraic expression by combining similar terms. **(Objective 3)**

55. $x - 0.4x - 1.8x$

56. $-2x + 1.7x - 4.6x$

57. $5.4n - 0.8n - 1.6n$

58. $6.2n - 7.8n - 1.3n$

59. $-3t + 4.2t - 0.9t + 0.2t$

60. $7.4t - 3.9t - 0.6t + 4.7t$

61. $3.6x - 7.4y - 9.4x + 10.2y$

62. $5.7x + 9.4y - 6.2x - 4.4y$

63. $0.3(x - 4) + 0.4(x + 6) - 0.6x$

64. $0.7(x + 7) - 0.9(x - 2) + 0.5x$

65. $6(x - 1.1) - 5(x - 2.3) - 4(x + 1.8)$

66. $4(x + 0.7) - 9(x + 0.2) - 3(x - 0.6)$

67. $5(x - 0.5) + 0.3(x - 2) - 0.7(x + 7)$

68. $-8(x - 1.2) + 6(x - 4.6) + 4(x + 1.7)$

For Problems 69–82, evaluate each algebraic expression for the given values of the variables. Don't forget that for some problems it might be helpful to combine similar terms first and then to evaluate. **(Objective 4)**

69. $x + 2y + 3z$ for $x = \dfrac{3}{4}$, $y = \dfrac{1}{3}$, and $z = -\dfrac{1}{6}$

70. $2x - y - 3z$ for $x = -\dfrac{2}{5}$, $y = -\dfrac{3}{4}$, and $z = \dfrac{1}{2}$

71. $\dfrac{3}{5}y - \dfrac{2}{3}y - \dfrac{7}{15}y$ for $y = -\dfrac{5}{2}$

72. $\dfrac{1}{2}x + \dfrac{2}{3}x - \dfrac{3}{4}x$ for $x = \dfrac{7}{8}$

73. $-x - 2y + 4z$ for $x = 1.7$, $y = -2.3$, and $z = 3.6$

74. $-2x + y - 5z$ for $x = -2.9$, $y = 7.4$, and $z = -6.7$

75. $5x - 7y$ for $x = -7.8$ and $y = 8.4$

76. $8x - 9y$ for $x = -4.3$ and $y = 5.2$

77. $0.7x + 0.6y$ for $x = -2$ and $y = 6$

78. $0.8x + 2.1y$ for $x = 5$ and $y = -9$

79. $1.2x + 2.3x - 1.4x - 7.6x$ for $x = -2.5$

80. $3.4x - 1.9x + 5.2x$ for $x = 0.3$

81. $-3a - 1 + 7a - 2$ for $a = 0.9$

82. $5x - 2 + 6x + 4$ for $x = -1.1$

83. Tanya bought 400 shares of one stock at $14.35 per share, and 250 shares of another stock at $16.68 per share. How much did she pay for the 650 shares?

84. On a trip Brent bought the following amounts of gasoline: 9.7 gallons, 12.3 gallons, 14.6 gallons, 12.2 gallons, 13.8 gallons, and 15.5 gallons. How many gallons of gasoline did he purchase on the trip?

85. Kathrin has a piece of copper tubing that is 76.4 centimeters long. She needs to cut it into four pieces of equal length. Find the length of each piece.

86. On a trip Biance filled the gasoline tank and noted that the odometer read 24,876.2 miles. After the next filling the odometer read 25,170.5 miles. It took 13.5 gallons of gasoline to fill the tank. How many miles per gallon did she get on that tank of gasoline?

87. The total length of the four sides of a square is 18.8 centimeters. How long is each side of the square?

88. When the market opened on Monday morning, Garth bought some shares of a stock at $13.25 per share. The daily changes in the market for that stock for the week were 0.75, -1.50, 2.25, -0.25, and -0.50. What was the value of one share of that stock when the market closed on Friday afternoon?

89. Victoria bought two pounds of Gala apples at $1.79 per pound and three pounds of Fuji apples at $0.99 per pound. How much did she spend for the apples?

90. In 2005 the average speed of the winner of the Daytona 500 was 135.173 miles per hour. In 1978 the average speed of the winner was 159.73 miles per hour. How much faster was the average speed of the winner in 1978 compared to the winner in 2005?

91. Andrea's automobile averages 25.4 miles per gallon. With this average rate of fuel consumption, what distance should she be able to travel on a 12.7-gallon tank of gasoline?

92. Use a calculator to check your answers for Problems 41–54.

Thoughts Into Words

93. At this time how would you describe the difference between arithmetic and algebra?

94. How have the properties of the real numbers been used thus far in your study of arithmetic and algebra?

95. Do you think that $2\sqrt{2}$ is a rational or an irrational number? Defend your answer.

Further Investigations

96. Without doing the actual dividing, defend the statement, "$\dfrac{1}{7}$ produces a repeating decimal." [*Hint*: Think about the possible remainders when dividing by 7.]

97. Express each of the following in repeating decimal form.

(a) $\dfrac{1}{7}$ **(b)** $\dfrac{2}{7}$

(c) $\dfrac{4}{9}$ **(d)** $\dfrac{5}{6}$

(e) $\dfrac{3}{11}$ **(f)** $\dfrac{1}{12}$

98. (a) How can we tell that $\dfrac{5}{16}$ will produce a terminating decimal?

(b) How can we tell that $\dfrac{7}{15}$ will not produce a terminating decimal?

(c) Determine which of the following will produce a terminating decimal: $\dfrac{7}{8}$, $\dfrac{11}{16}$, $\dfrac{5}{12}$, $\dfrac{7}{24}$, $\dfrac{11}{75}$, $\dfrac{13}{32}$, $\dfrac{17}{40}$, $\dfrac{11}{30}$, $\dfrac{9}{20}$, $\dfrac{3}{64}$.

2.4   Exponents

OBJECTIVES

1. Know the definition and terminology for exponential notation

2. Simplify numerical expressions that involve exponents

3. Simplify algebraic expressions by combining similar terms

4. Reduce algebraic fractions involving exponents

5. Add, subtract, multiply, and divide algebraic fractions

6. Evaluate algebraic expressions that involve exponents

We use exponents to indicate repeated multiplication. For example, we can write $5 \cdot 5 \cdot 5$ as 5^3, where the 3 indicates that 5 is to be used as a factor 3 times. The following general definition is helpful:

> **Definition 2.4**
>
> If n is a positive integer, and b is any real number, then
>
> $$b^n = \underbrace{bbb \cdots b}_{n \text{ factors of } b}$$

We refer to the b as the **base** and n as the **exponent**. The expression b^n can be read as "b to the nth **power**." We frequently associate the terms **squared** and **cubed** with exponents of 2 and 3, respectively. For example, b^2 is read as "b squared" and b^3 as "b cubed." An exponent of 1 is usually not written, so b^1 is written as b. The following examples further clarify the concept of an exponent.

$$2^3 = 2 \cdot 2 \cdot 2 = 8 \qquad\qquad (0.6)^2 = (0.6)(0.6) = 0.36$$

$$3^5 = 3 \cdot 3 \cdot 3 \cdot 3 \cdot 3 = 243 \qquad \left(\frac{1}{2}\right)^4 = \frac{1}{2} \cdot \frac{1}{2} \cdot \frac{1}{2} \cdot \frac{1}{2} = \frac{1}{16}$$

$$(-5)^2 = (-5)(-5) = 25 \qquad -5^2 = -(5 \cdot 5) = -25$$

We especially want to call your attention to the last two examples. Notice that $(-5)^2$ means that -5 is the base, which is to be used as a factor twice. However, -5^2 means that 5 is the base, and after 5 is squared, we take the opposite of that result.

Exponents provide a way of writing algebraic expressions in compact form. Sometimes we need to change from the compact form to an expanded form as these next examples demonstrate.

$$x^4 = x \cdot x \cdot x \cdot x \qquad\qquad (2x)^3 = (2x)(2x)(2x)$$

$$2y^3 = 2 \cdot y \cdot y \cdot y \qquad\qquad (-2x)^3 = (-2x)(-2x)(-2x)$$

$$-3x^5 = -3 \cdot x \cdot x \cdot x \cdot x \cdot x \qquad -x^2 = -(x \cdot x)$$

$$a^2 + b^2 = a \cdot a + b \cdot b$$

At other times we need to change from an expanded form to a more compact form using the exponent notation.

$$3 \cdot x \cdot x = 3x^2$$

$$2 \cdot 5 \cdot x \cdot x \cdot x = 10x^3$$

$$3 \cdot 4 \cdot x \cdot x \cdot y = 12x^2y$$

$$7 \cdot a \cdot a \cdot a \cdot b \cdot b = 7a^3b^2$$

$$(2x)(3y) = 2 \cdot x \cdot 3 \cdot y = 2 \cdot 3 \cdot x \cdot y = 6xy$$

$$(3a^2)(4a) = 3 \cdot a \cdot a \cdot 4 \cdot a = 3 \cdot 4 \cdot a \cdot a \cdot a = 12a^3$$

$$(-2x)(3x) = -2 \cdot x \cdot 3 \cdot x = -2 \cdot 3 \cdot x \cdot x = -6x^2$$

The commutative and associative properties for multiplication allowed us to rearrange and regroup factors in the last three examples above.

The concept of *exponent* can be used to extend our work with combining similar terms, operating with fractions, and evaluating algebraic expressions. Study the following examples very carefully; they will help you pull together many ideas.

Classroom Example
Simplify $5y^3 + 4y^3 - 3y^3$ by combining similar terms.

EXAMPLE 1   Simplify $4x^2 + 7x^2 - 2x^2$ by combining similar terms.

Solution

By applying the distributive property, we obtain

$$4x^2 + 7x^2 - 2x^2 = (4 + 7 - 2)x^2$$

$$= 9x^2$$

Classroom Example
Simplify $6m^2 - 5n^3 + 2m^2 - 12n^3$ by combining similar terms.

EXAMPLE 2

Simplify $-8x^3 + 9y^2 + 4x^3 - 11y^2$ by combining similar terms.

Solution

By rearranging terms and then applying the distributive property we obtain

$$-8x^3 + 9y^2 + 4x^3 - 11y^2 = -8x^3 + 4x^3 + 9y^2 - 11y^2$$

$$= (-8 + 4)x^3 + (9 - 11)y^2$$

$$= -4x^3 - 2y^2$$

Classroom Example
Simplify $6a^4 - 7a - 7a^4 + 5a$.

EXAMPLE 3   Simplify $-7x^2 + 4x + 3x^2 - 9x$.

Solution

$$-7x^2 + 4x + 3x^2 - 9x = -7x^2 + 3x^2 + 4x - 9x$$

$$= (-7 + 3)x^2 + (4 - 9)x$$

$$= -4x^2 - 5x$$

As soon as you feel comfortable with this process of combining similar terms, you may want to do some of the steps mentally. Then your work may appear as follows.

$$9a^2 + 6a^2 - 12a^2 = 3a^2$$

$$6x^2 + 7y^2 - 3x^2 - 11y^2 = 3x^2 - 4y^2$$

$$7x^2y + 5xy^2 - 9x^2y + 10xy^2 = -2x^2y + 15xy^2$$

$$2x^3 - 5x^2 - 10x - 7x^3 + 9x^2 - 4x = -5x^3 + 4x^2 - 14x$$

The next two examples illustrate the use of exponents when reducing fractions.

Classroom Example
Reduce $\dfrac{15m^3n}{18m^2n}$.

EXAMPLE 4   Reduce $\dfrac{8x^2y}{12xy}$.

Solution

$$\frac{8x^2y}{12xy} = \frac{2 \cdot 2 \cdot 2 \cdot x \cdot x \cdot y}{2 \cdot 2 \cdot 3 \cdot x \cdot y} = \frac{2x}{3}$$

Classroom Example
Reduce $\dfrac{12x^4y}{20x^2y^3}$.

EXAMPLE 5   Reduce $\dfrac{15a^2b^3}{25a^3b}$.

Solution

$$\frac{15a^2b^3}{25a^3b} = \frac{3 \cdot 5 \cdot a \cdot a \cdot b \cdot b \cdot b}{5 \cdot 5 \cdot a \cdot a \cdot a \cdot b} = \frac{3b^2}{5a}$$

The next three examples show how exponents can be used when multiplying and dividing fractions.

Classroom Example
Multiply $\left(\dfrac{6m^2}{5n^2}\right)\left(\dfrac{8n^2}{9m^3}\right)$ and express the answer in reduced form.

EXAMPLE 6   Multiply $\left(\dfrac{4x}{6y}\right)\left(\dfrac{12y^2}{7x^2}\right)$ and express the answer in reduced form.

Solution

$$\left(\frac{4x}{6y}\right)\left(\frac{12y^2}{7x^2}\right) = \frac{4 \cdot \overset{2}{12} \cdot x \cdot y \cdot y}{6 \cdot 7 \cdot y \cdot x \cdot x} = \frac{8y}{7x}$$

Classroom Example
Multiply and simplify
$\left(\dfrac{6x^5}{14y^3}\right)\left(\dfrac{10y^4}{12x^3}\right)$.

EXAMPLE 7   Multiply and simplify $\left(\dfrac{8a^3}{9b}\right)\left(\dfrac{12b^2}{16a}\right)$.

Solution

$$\left(\frac{8a^3}{9b}\right)\left(\frac{12b^2}{16a}\right) = \frac{8 \cdot \overset{4}{12} \cdot a \cdot a \cdot a \cdot b \cdot b}{\underset{3}{9} \cdot \underset{2}{16} \cdot b \cdot a} = \frac{2a^2b}{3}$$

Classroom Example
Divide and express in reduced form,
$\dfrac{-3a^2}{4b^3} \div \dfrac{7}{12ab}$.

EXAMPLE 8   Divide and express in reduced form, $\dfrac{-2x^3}{3y^2} \div \dfrac{4}{9xy}$.

Solution

$$\frac{-2x^3}{3y^2} \div \frac{4}{9xy} = -\frac{2x^3}{3y^2} \cdot \frac{9xy}{4} = -\frac{\overset{1}{2} \cdot \overset{3}{9} \cdot x \cdot x \cdot x \cdot x \cdot y}{3 \cdot \underset{2}{4} \cdot y \cdot y} = -\frac{3x^4}{2y}$$

The next two examples demonstrate the use of exponents when adding and subtracting fractions.

Classroom Example
Add $\dfrac{3}{y} + \dfrac{8}{y^2}$.

EXAMPLE 9   Add $\dfrac{4}{x^2} + \dfrac{7}{x}$.

Solution

The LCD is x^2. Thus,

$$\frac{4}{x^2} + \frac{7}{x} = \frac{4}{x^2} + \frac{7 \cdot x}{x \cdot x} = \frac{4}{x^2} + \frac{7x}{x^2} = \frac{4 + 7x}{x^2}$$

Classroom Example
Subtract $\dfrac{2}{mn} - \dfrac{5}{n^2}$.

EXAMPLE 10   Subtract $\dfrac{3}{xy} - \dfrac{4}{y^2}$.

Solution

$$\left. \begin{array}{l} xy = x \cdot y \\ y^2 = y \cdot y \end{array} \right\} \longrightarrow \text{The LCD is } xy^2$$

$$\frac{3}{xy} - \frac{4}{y^2} = \frac{3 \cdot y}{xy \cdot y} - \frac{4 \cdot x}{y^2 \cdot x} = \frac{3y}{xy^2} - \frac{4x}{xy^2}$$

$$= \frac{3y - 4x}{xy^2}$$

Remember that exponents are used to indicate repeated multiplication. Therefore, to simplify numerical expressions containing exponents, we proceed as follows.

1. Perform the operations inside the symbols of inclusion (parentheses and brackets) and above and below each fraction bar. Start with the innermost inclusion symbol.
2. Compute all indicated powers.
3. Perform all multiplications and divisions in the order that they appear from left to right.
4. Perform all additions and subtractions in the order that they appear from left to right.

Keep these steps in mind as we evaluate some algebraic expressions containing exponents.

Classroom Example
Evaluate $5a^2 - 3b^2$ for $a = 4$ and $b = -6$.

EXAMPLE 11   Evaluate $3x^2 - 4y^2$ for $x = -2$ and $y = 5$.

Solution

$$3x^2 - 4y^2 = 3(-2)^2 - 4(5)^2 \quad \text{when } x = -2 \text{ and } y = 5$$

$$= 3(-2)(-2) - 4(5)(5)$$

$$= 12 - 100$$

$$= -88$$

Classroom Example
Find the value of $x^2 - y^2$ for $x = \dfrac{1}{4}$
and $y = -\dfrac{1}{2}$.

EXAMPLE 12   Find the value of $a^2 - b^2$ when $a = \dfrac{1}{2}$ and $b = -\dfrac{1}{3}$.

Solution

$$a^2 - b^2 = \left(\frac{1}{2}\right)^2 - \left(-\frac{1}{3}\right)^2 \quad \text{when } a = \frac{1}{2} \text{ and } b = -\frac{1}{3}$$

$$= \frac{1}{4} - \frac{1}{9}$$

$$= \frac{9}{36} - \frac{4}{36}$$

$$= \frac{5}{36}$$

Classroom Example
Evaluate $7c^2 + 2cd$ for $c = 0.5$ and $d = -0.2$.

EXAMPLE 13   Evaluate $5x^2 + 4xy$ for $x = 0.4$ and $y = -0.3$.

Solution

$$5x^2 + 4xy = 5(0.4)^2 + 4(0.4)(-0.3) \quad \text{when } x = 0.4 \text{ and } y = -0.3$$

$$= 5(0.16) + 4(-0.12)$$

$$= 0.80 + (-0.48)$$

$$= 0.32$$

Concept Quiz 2.4

For Problems 1–10, answer true or false.

1. Exponents are used to indicate repeated additions.
2. In the expression b^n, b is called "the base," and n is called "the number."
3. The term "cubed" is associated with an exponent of three.
4. For the term $3x$, the exponent on the x is one.
5. In the expression $(-4)^3$, the base is 4.
6. In the expression -4^3, the base is 4.
7. Changing from an expanded notation to an exponential notation, $5 \cdot 5 \cdot 5 \cdot a \cdot b \cdot b = 5^3ab$.
8. When simplifying $2x^3 + 5x^3$, the result would be $7x^6$.
9. The least common multiple for xy^2 and x^2y^3 is x^2y^3.
10. The term "squared" is associated with an exponent of two.

Problem Set 2.4

For Problems 1–20, find the value of each numerical expression. For example, $2^4 = 2 \cdot 2 \cdot 2 \cdot 2 = 16$. **(Objective 2)**

1. 2^6
2. 2^7
3. 3^4
4. 4^3
5. $(-2)^3$
6. $(-2)^5$
7. -3^2
8. -3^4
9. $(-4)^2$
10. $(-5)^4$
11. $\left(\dfrac{2}{3}\right)^4$
12. $\left(\dfrac{3}{4}\right)^3$
13. $-\left(\dfrac{1}{2}\right)^3$
14. $-\left(\dfrac{3}{2}\right)^3$
15. $\left(-\dfrac{3}{2}\right)^2$
16. $\left(-\dfrac{4}{3}\right)^2$
17. $(0.3)^3$
18. $(0.2)^4$

19. $-(1.2)^2$ **20.** $-(1.1)^2$

For Problems 21–40, simplify each numerical expression. **(Objective 2)**

21. $3^2 + 2^3 - 4^3$ **22.** $2^4 - 3^3 + 5^2$

23. $(-2)^3 - 2^4 - 3^2$

24. $(-3)^3 - 3^2 - 6^2$

25. $5(2)^2 - 4(2) - 1$

26. $7(-2)^2 - 6(-2) - 8$

27. $-2(3)^3 - 3(3)^2 + 4(3) - 6$

28. $5(-3)^3 - 4(-3)^2 + 6(-3) + 1$

29. $-7^2 - 6^2 + 5^2$

30. $-8^2 + 3^4 - 4^3$

31. $-3(-4)^2 - 2(-3)^3 + (-5)^2$

32. $-4(-3)^3 + 5(-2)^3 - (4)^2$

33. $\dfrac{-3(2)^4}{12} + \dfrac{5(-3)^3}{15}$

34. $\dfrac{4(2)^3}{16} - \dfrac{2(3)^2}{6}$

35. $\dfrac{3(4 - 2)^2}{4} + \dfrac{2(-1 + 3)^3}{4}$

36. $\dfrac{4(-2 - 3)^2}{5} - \dfrac{5(-1 - 5)^2}{6}$

37. $\dfrac{-5(2 - 3)^3}{2} - \dfrac{4(2 - 4)^3}{3}$

38. $\dfrac{-4(2 - 5)^2}{5} + \dfrac{3(-1 - 6)}{2}$

39. $\dfrac{2 - 3[(4 - 5)^2 + 1]}{(3 - 1)^2}$

40. $\dfrac{-3 - 2[(3 - 5)^2 - 2]}{(-2 - 1)^2}$

For Problems 41–52, use exponents to express each algebraic expression in a more compact form. For example, $3 \cdot 5 \cdot x \cdot x \cdot y = 15x^2y$ and $(3x)(2x^2) = 6x^3$. **(Objective 2)**

41. $9 \cdot x \cdot x$ **42.** $8 \cdot x \cdot x \cdot x \cdot y$

43. $3 \cdot 4 \cdot x \cdot y \cdot y$

44. $7 \cdot 2 \cdot a \cdot a \cdot b \cdot b \cdot b$

45. $-2 \cdot 9 \cdot x \cdot x \cdot x \cdot x \cdot y$

46. $-3 \cdot 4 \cdot x \cdot y \cdot z \cdot z$

47. $(5x)(3y)$ **48.** $(3x^2)(2y)$

49. $(6x^2)(2x^2)$ **50.** $(-3xy)(6xy)$

51. $(-4a^2)(-2a^3)$ **52.** $(-7a^3)(-3a)$

For Problems 53–64, simplify each expression by combining similar terms. **(Objective 3)**

53. $3x^2 - 7x^2 - 4x^2$ **54.** $-2x^3 + 7x^3 - 4x^3$

55. $-12y^3 + 17y^3 - y^3$ **56.** $-y^3 + 8y^3 - 13y^3$

57. $7x^2 - 2y^2 - 9x^2 + 8y^2$

58. $5x^3 + 9y^3 - 8x^3 - 14y^3$

59. $\dfrac{2}{3}n^2 - \dfrac{1}{4}n^2 - \dfrac{3}{5}n^2$

60. $-\dfrac{1}{2}n^2 + \dfrac{5}{6}n^2 - \dfrac{4}{9}n^2$

61. $5x^2 - 8x - 7x^2 + 2x$

62. $-10x^2 + 4x + 4x^2 - 8x$

63. $x^2 - 2x - 4 + 6x^2 - x + 12$

64. $-3x^3 - x^2 + 7x - 2x^3 + 7x^2 - 4x$

For Problems 65–74, reduce each fraction to simplest form. **(Objective 4)**

65. $\dfrac{9xy}{15x}$ **66.** $\dfrac{8x^2y}{14x}$

67. $\dfrac{22xy^2}{6xy^3}$ **68.** $\dfrac{18x^3y}{12xy^4}$

69. $\dfrac{7a^2b^3}{17a^3b}$ **70.** $\dfrac{9a^3b^3}{22a^4b^2}$

71. $\dfrac{-24abc^2}{32bc}$ **72.** $\dfrac{4a^2c^3}{-22b^2c^4}$

73. $\dfrac{-5x^4y^3}{-20x^2y}$ **74.** $\dfrac{-32xy^2z^4}{-48x^3y^3z}$

For Problems 75–92, perform the indicated operations and express your answers in reduced form. **(Objective 5)**

75. $\left(\dfrac{7x^2}{9y}\right)\left(\dfrac{12y}{21x}\right)$ **76.** $\left(\dfrac{3x}{8y^2}\right)\left(\dfrac{14xy}{9y}\right)$

77. $\left(\dfrac{5c}{a^2b^2}\right) \div \left(\dfrac{12c}{ab}\right)$

78. $\left(\dfrac{13ab^2}{12c}\right) \div \left(\dfrac{26b}{14c}\right)$

79. $\dfrac{6}{x} + \dfrac{5}{y^2}$ **80.** $\dfrac{8}{y} - \dfrac{6}{x^2}$

81. $\dfrac{5}{x^4} - \dfrac{7}{x^2}$ **82.** $\dfrac{9}{x} - \dfrac{11}{x^3}$

83. $\dfrac{3}{2x^3} + \dfrac{6}{x}$ **84.** $\dfrac{5}{3x^2} + \dfrac{6}{x}$

85. $\dfrac{-5}{4x^2} + \dfrac{7}{3x^2}$ **86.** $\dfrac{-8}{5x^3} + \dfrac{10}{3x^3}$

87. $\dfrac{11}{a^2} - \dfrac{14}{b^2}$ **88.** $\dfrac{9}{x^2} + \dfrac{8}{y^2}$

89. $\dfrac{1}{2x^3} - \dfrac{4}{3x^2}$

90. $\dfrac{2}{3x^3} - \dfrac{5}{4x}$

91. $\dfrac{3}{x} - \dfrac{4}{y} - \dfrac{5}{xy}$

92. $\dfrac{5}{x} + \dfrac{7}{y} - \dfrac{1}{xy}$

For Problems 93–106, evaluate each algebraic expression for the given values of the variables. **(Objective 6)**

93. $4x^2 + 7y^2$ for $x = -2$ and $y = -3$

94. $5x^2 + 2y^3$ for $x = -4$ and $y = -1$

95. $3x^2 - y^2$ for $x = \dfrac{1}{2}$ and $y = -\dfrac{1}{3}$

96. $x^2 - 2y^2$ for $x = -\dfrac{2}{3}$ and $y = \dfrac{3}{2}$

97. $x^2 - 2xy + y^2$ for $x = -\dfrac{1}{2}$ and $y = 2$

98. $x^2 + 2xy + y^2$ for $x = -\dfrac{3}{2}$ and $y = -2$

99. $-x^2$ for $x = -8$

100. $-x^3$ for $x = 5$

101. $-x^2 - y^2$ for $x = -3$ and $y = -4$

102. $-x^2 + y^2$ for $x = -2$ and $y = 6$

103. $-a^2 - 3b^3$ for $a = -6$ and $b = -1$

104. $-a^3 + 3b^2$ for $a = -3$ and $b = -5$

105. $y^2 - 3xy$ for $x = 0.4$ and $y = -0.3$

106. $x^2 + 5xy$ for $x = -0.2$ and $y = -0.6$

107. Use a calculator to check your answers for Problems 1–40.

Thoughts Into Words

108. Your friend keeps getting an answer of 16 when simplifying -2^4. What mistake is he making and how would you help him?

109. Explain how you would simplify $\dfrac{12x^2y}{18xy}$.

Answers to the Concept Quiz

1. False   **2.** False   **3.** True   **4.** True   **5.** False   **6.** True   **7.** False   **8.** False   **9.** True   **10.** True

2.5   Translating from English to Algebra

OBJECTIVES

1 Translate algebraic expressions into English phrases

2 Translate English phrases into algebraic expressions

3 Write algebraic expressions for converting units of measure within a measurement system

In order to use the tools of algebra for solving problems, we must be able to translate back and forth between the English language and the language of algebra. In this section we want to translate algebraic expressions to English phrases (word phrases) and English phrases to algebraic expressions. Let's begin by translating some algebraic expressions to word phrases.

Algebraic expression	Word phrase
$x + y$	The sum of x and y
$x - y$	The difference of x and y
$y - x$	The difference of y and x
xy	The product of x and y
$\dfrac{x}{y}$	The quotient of x and y

(table continues)

Algebraic expression	Word phrase
$3x$	The product of 3 and x
$x^2 + y^2$	The sum of x squared and y squared
$2xy$	The product of 2, x, and y
$2(x + y)$	Two times the quantity x plus y
$x - 3$	Three less than x

Now let's consider the reverse process, translating some word phrases to algebraic expressions. Part of the difficulty in translating from English to algebra is that different word phrases translate into the same algebraic expression. So we need to become familiar with different ways of saying the same thing, especially when referring to the four fundamental operations. The following examples should help to acquaint you with some of the phrases used for the basic operations.

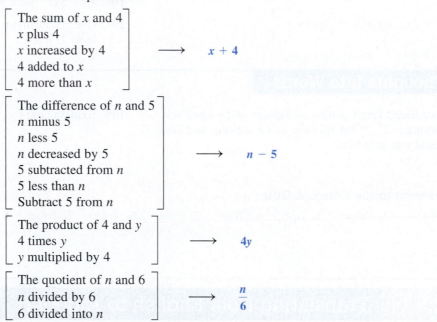

$$\left.\begin{array}{l}\text{The sum of } x \text{ and 4} \\ x \text{ plus 4} \\ x \text{ increased by 4} \\ 4 \text{ added to } x \\ 4 \text{ more than } x\end{array}\right\} \longrightarrow \quad x + 4$$

$$\left.\begin{array}{l}\text{The difference of } n \text{ and 5} \\ n \text{ minus 5} \\ n \text{ less 5} \\ n \text{ decreased by 5} \\ 5 \text{ subtracted from } n \\ 5 \text{ less than } n \\ \text{Subtract 5 from } n\end{array}\right\} \longrightarrow \quad n - 5$$

$$\left.\begin{array}{l}\text{The product of 4 and } y \\ 4 \text{ times } y \\ y \text{ multiplied by 4}\end{array}\right\} \longrightarrow \quad 4y$$

$$\left.\begin{array}{l}\text{The quotient of } n \text{ and 6} \\ n \text{ divided by 6} \\ 6 \text{ divided into } n\end{array}\right\} \longrightarrow \quad \dfrac{n}{6}$$

Often a word phrase indicates more than one operation. Furthermore, the standard vocabulary of sum, difference, product, and quotient may be replaced by other terminology. Study the following translations very carefully. Also remember that the commutative property holds for addition and multiplication but not for subtraction and division. Therefore, the phrase "x plus y" can be written as $x + y$ or $y + x$. However, the phrase "x minus y" means that y must be subtracted from x, and the phrase is written as $x - y$. So be very careful of phrases that involve subtraction or division.

Word phrase	Algebraic expression
The sum of two times x and three times y	$2x + 3y$
The sum of the squares of a and b	$a^2 + b^2$
Five times x divided by y	$\dfrac{5x}{y}$
Two more than the square of x	$x^2 + 2$
Three less than the cube of b	$b^3 - 3$
Five less than the product of x and y	$xy - 5$
Nine minus the product of x and y	$9 - xy$
Four times the sum of x and 2	$4(x + 2)$
Six times the quantity w minus 4	$6(w - 4)$

Suppose you are told that the sum of two numbers is 12, and one of the numbers is 8. What is the other number? The other number is $12 - 8$, which equals 4. Now suppose you are told that the product of two numbers is 56, and one of the numbers is 7. What is the other number? The other number is $56 \div 7$, which equals 8. The following examples illustrate the use of these addition-subtraction and multiplication-division relationships in a more general setting.

Classroom Example
The sum of two numbers is 57, and one of the numbers is y. What is the other number?

EXAMPLE 1

The sum of two numbers is 83, and one of the numbers is x. What is the other number?

Solution

Using the addition and subtraction relationship, we can represent the other number by $83 - x$.

Classroom Example
The difference of two numbers is 9. The smaller number is f. What is the larger number?

EXAMPLE 2

The difference of two numbers is 14. The smaller number is n. What is the larger number?

Solution

Since the smaller number plus the difference must equal the larger number, we can represent the larger number by $n + 14$.

Classroom Example
The product of two numbers is 42, and one of the numbers is r. Represent the other number.

EXAMPLE 3

The product of two numbers is 39, and one of the numbers is y. Represent the other number.

Solution

Using the multiplication and division relationship, we can represent the other number by $\dfrac{39}{y}$.

In a word problem, the statement may not contain key words such as sum, difference, product, or quotient; instead, the statement may describe a physical situation, and from this description you need to deduce the operations involved. We make some suggestions for handling such situations in the following examples.

Classroom Example
Sandy can read 50 words per minute. How many words can she read in w minutes?

EXAMPLE 4

Arlene can type 70 words per minute. How many words can she type in m minutes?

Solution

In 10 minutes she would type $70(10) = 700$ words. In 50 minutes she would type $70(50) = 3500$ words. Thus in m minutes she would type $70m$ words.

Notice the use of some specific examples: $70(10) = 700$ and $70(50) = 3500$, to help formulate the general expression. This technique of first formulating some specific examples and then generalizing can be very effective.

Classroom Example
Jane has d dimes and q quarters. Express, in cents, this amount of money.

EXAMPLE 5

Lynn has n nickels and d dimes. Express, in cents, this amount of money.

Solution

Three nickels and 8 dimes are $5(3) + 10(8) = 95$ cents. Thus n nickels and d dimes are $(5n + 10d)$ cents.

EXAMPLE 6

A train travels at the rate of *r* miles per hour. How far will it travel in 8 hours?

Solution

Suppose that a train travels at 50 miles per hour. Using the formula *distance equals rate times time*, it would travel $50 \cdot 8 = 400$ miles. Therefore, at *r* miles per hour, it would travel $r \cdot 8$ miles. We usually write the expression $r \cdot 8$ as $8r$.

EXAMPLE 7

The cost of a 5-pound box of candy is *d* dollars. How much is the cost per pound for the candy?

Solution

The price per pound is figured by dividing the total cost by the number of pounds. Therefore, the price per pound is represented by $\dfrac{d}{5}$.

An English statement being translated into algebra may contain some geometric ideas. For example, suppose that we want to express in inches the length of a line segment that is *f* feet long. Since 1 foot = 12 inches, we can represent *f* feet by 12 times *f*, written as $12f$ inches.

Tables 2.1 and 2.2 list some of the basic relationships pertaining to linear measurements in the English and metric systems, respectively. (Additional listings of both systems are located on the inside back cover.)

Table 2.1

English system
12 inches = 1 foot
3 feet = 36 inches = 1 yard
5280 feet = 1760 yards = 1 mile

Table 2.2

Metric system
1 kilometer = 1000 meters
1 hectometer = 100 meters
1 dekameter = 10 meters
1 decimeter = 0.1 meter
1 centimeter = 0.01 meter
1 millimeter = 0.001 meter

EXAMPLE 8

The distance between two cities is *k* kilometers. Express this distance in meters.

Solution

Since 1 kilometer equals 1000 meters, we need to multiply *k* by 1000. Therefore, the distance in meters is represented by $1000k$.

EXAMPLE 9

The length of a line segment is *i* inches. Express that length in yards.

Solution

To change from inches to yards, we must divide by 36. Therefore $\dfrac{i}{36}$ represents, in yards, the length of the line segment.

Classroom Example
The width of a rectangle is
x centimeters, and the length is
4 centimeters more than three times
the width. What is the length of the
rectangle? What is the perimeter of
the rectangle? What is the area of
the rectangle?

EXAMPLE 10

The width of a rectangle is w centimeters, and the length is 5 centimeters less than twice the width. What is the length of the rectangle? What is the perimeter of the rectangle? What is the area of the rectangle?

Solution

We can represent the length of the rectangle by $2w - 5$. Now we can sketch a rectangle as in Figure 2.6 and record the given information. The perimeter of a rectangle is the sum of the lengths of the four sides. Therefore, the perimeter is given by $2w + 2(2w - 5)$, which can be written as $2w + 4w - 10$ and then simplified to $6w - 10$. The area of a rectangle is the product of the length and width. Therefore, the area in square centimeters is given by $w(2w - 5) = w \cdot 2w + w(-5) = 2w^2 - 5w$.

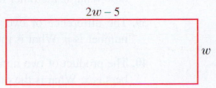

Figure 2.6

Classroom Example
The length of a side of a square is
y yards. Express the length of a side
in feet. What is the area of the
square in square feet?

EXAMPLE 11

The length of a side of a square is x feet. Express the length of a side in inches. What is the area of the square in square inches?

Solution

Because 1 foot equals 12 inches, we need to multiply x by 12. Therefore, $12x$ represents the length of a side in inches. The area of a square is the length of a side squared. So the area in square inches is given by $(12x)^2 = (12x)(12x) = 12 \cdot 12 \cdot x \cdot x = 144x^2$.

Concept Quiz 2.5

For Problems 1–10, match the English phrase with its algebraic expression.

1. The product of x and y **A.** $x - y$
2. Two less than x **B.** $x + y$
3. x subtracted from 2   **C.** $\dfrac{x}{y}$
4. The difference of x and y **D.** $x - 2$
5. The quotient of x and y **E.** xy
6. The sum of x and y **F.** $x^2 - y$
7. Two times the sum of x and y **G.** $2(x + y)$
8. Two times x plus y **H.** $2 - x$
9. x squared minus y **I.** $x + 2$
10. Two more than x **J.** $2x + y$

Problem Set 2.5

For Problems 1–12, write a word phrase for each of the algebraic expressions. For example, lw can be expressed as "the product of l and w." **(Objective 1)**

1. $a - b$

2. $x + y$

3. $\frac{1}{3}Bh$

4. $\frac{1}{2}bh$

5. $2(l + w)$

6. πr^2

7. $\frac{A}{w}$

8. $\frac{C}{\pi}$

9. $\frac{a + b}{2}$

10. $\frac{a - b}{4}$

11. $3y + 2$

12. $3(x - y)$

For Problems 13–36, translate each word phrase into an algebraic expression. For example, "the sum of x and 14" translates into $x + 14$. **(Objective 2)**

13. The sum of l and w

14. The difference of x and y

15. The product of a and b

16. The product of $\frac{1}{3}$, B, and h

17. The quotient of d and t

18. r divided into d

19. The product of l, w, and h

20. The product of π and the square of r

21. x subtracted from y

22. The difference "x subtract y"

23. Two larger than the product of x and y

24. Six plus the cube of x

25. Seven minus the square of y

26. The quantity, x minus 2, cubed

27. The quantity, x minus y, divided by four

28. Eight less than x

29. Ten less x

30. Nine times the quantity, n minus 4

31. Ten times the quantity, n plus 2

32. The sum of four times x and five times y

33. Seven subtracted from the product of x and y

34. Three times the sum of n and 2

35. Twelve less than the product of x and y

36. Twelve less the product of x and y

For Problems 37–72, answer the question with an algebraic expression. **(Objectives 2 and 3)**

37. The sum of two numbers is 35, and one of the numbers is n. What is the other number?

38. The sum of two numbers is 100, and one of the numbers is x. What is the other number?

39. The difference of two numbers is 45, and the smaller number is n. What is the other number?

40. The product of two numbers is 25, and one of the numbers is x. What is the other number?

41. Janet is y years old. How old will she be in 10 years?

42. Hector is y years old. How old was he 5 years ago?

43. Debra is x years old, and her mother is 3 years less than twice as old as Debra. How old is Debra's mother?

44. Jack is x years old, and Dudley is 1 year more than three times as old as Jack. How old is Dudley?

45. Donna has d dimes and q quarters in her bank. How much money in cents does she have?

46. Andy has c cents, which is all in dimes. How many dimes does he have?

47. A car travels d miles in t hours. How fast is the car traveling per hour (i.e., what is the rate)?

48. If g gallons of gas cost d dollars, what is the price per gallon?

49. If p pounds of candy cost d dollars, what is the price per pound?

50. Sue can type x words per minute. How many words can she type in 1 hour?

51. Larry's annual salary is d dollars. What is his monthly salary?

52. Nancy's monthly salary is d dollars. What is her annual salary?

53. If n represents a whole number, what is the next larger whole number?

54. If n represents an even number, what is the next larger even number?

55. If n represents an odd number, what is the next larger odd number?

56. Maria is y years old, and her sister is twice as old. What is the sum of their ages?

57. Willie is y years old, and his father is 2 years less than twice Willie's age. What is the sum of their ages?

58. Harriet has p pennies, n nickels, and d dimes. How much money in cents does she have?

59. The perimeter of a rectangle is y yards and f feet. What is the perimeter in inches?

60. The perimeter of a triangle is m meters and c centimeters. What is the perimeter in centimeters?

61. A rectangular plot of ground is f feet long. What is its length in yards?

62. The height of a telephone pole is f feet. What is the height in yards?

63. The width of a rectangle is w feet, and its length is three times the width. What is the perimeter of the rectangle in feet?

64. The width of a rectangle is w feet, and its length is 1 foot more than twice its width. What is the perimeter of the rectangle in feet?

65. The length of a rectangle is l inches, and its width is 2 inches less than one-half of its length. What is the perimeter of the rectangle in inches?

66. The length of a rectangle is l inches, and its width is 3 inches more than one-third of its length. What is the perimeter of the rectangle in inches?

67. The first side of a triangle is f feet long. The second side is 2 feet longer than the first side. The third side is one foot more than twice the first side. What is the perimeter of the triangle in inches?

68. The first side of a triangle is y yards long. The second side is one yard longer than the first side. The third side is twice as long as the first side. Express the perimeter in feet?

69. The width of a rectangle is w yards, and the length is twice the width. What is the area of the rectangle in square yards?

70. The width of a rectangle is w yards and the length is 4 yards more than the width. What is the area of the rectangle in square yards?

71. The length of a side of a square is s yards. What is the area of the square in square feet?

72. The length of a side of a square is y centimeters. What is the area of the square in square millimeters?

Thoughts Into Words

73. What does the phrase "translating from English to algebra" mean to you?

74. Your friend is having trouble with Problems 61 and 62. For example, for Problem 61 she doesn't know if the answer should be $3f$ or $\dfrac{f}{3}$. What can you do to help her?

Answers to the Concept Quiz

1. E   **2.** D   **3.** H   **4.** A   **5.** C   **6.** B   **7.** G   **8.** J   **9.** F   **10.** I

OBJECTIVE	SUMMARY	EXAMPLE
Classify numbers in the real number system. (Section 2.3/Objective 1)	Any number that has a terminating or repeating decimal representation is a rational number. Any number that has a non-terminating or non-repeating decimal representation is an irrational number. The rational numbers together with the irrational numbers form the set of real numbers.	Classify -1, $\sqrt{7}$, and $\dfrac{3}{4}$. **Solution** -1 is a real number, a rational number, an integer, and negative. $\sqrt{7}$ is a real number, an irrational number, and positive. $\dfrac{3}{4}$ is a real number, a rational number, noninteger, and positive.
Reduce rational numbers to lowest terms. (Section 2.1/Objective 1)	The property $\dfrac{a \cdot k}{b \cdot k} = \dfrac{a}{b}$ is used to express fractions in reduced form.	Reduce $\dfrac{6xy}{14x}$. **Solution** $\dfrac{6xy}{14x} = \dfrac{2 \cdot 3 \cdot x \cdot y}{2 \cdot 7 \cdot x}$ $= \dfrac{2 \cdot 3 \cdot \cancel{x} \cdot y}{2 \cdot 7 \cdot \cancel{x}}$ $= \dfrac{3y}{7}$
Multiply fractions. (Section 2.1/Objective 2)	To multiply rational numbers in fractional form, multiply the numerators and multiply the denominators. Always express the result in reduced form.	Multiply $\left(\dfrac{6}{7}\right)\left(\dfrac{21}{4}\right)$. **Solution** $\left(\dfrac{6}{7}\right)\left(\dfrac{21}{4}\right) = \dfrac{6 \cdot 21}{7 \cdot 4}$ $= \dfrac{2 \cdot 3 \cdot 3 \cdot 7}{7 \cdot 2 \cdot 2}$ $= \dfrac{2 \cdot 3 \cdot 3 \cdot \cancel{7}}{\cancel{7} \cdot 2 \cdot 2}$ $= \dfrac{9}{2}$
Divide fractions. (Section 2.1/Objective 2)	To divide rational numbers in fractional form, change the problem to multiplying by the reciprocal of the divisor. Always express the result in reduced form.	Divide $\dfrac{5}{7} \div \dfrac{6}{11}$. **Solution** $\dfrac{5}{7} \div \dfrac{6}{11} = \dfrac{5}{7} \cdot \dfrac{11}{6}$ $= \dfrac{55}{42}$

OBJECTIVE	SUMMARY	EXAMPLE
Add and subtract rational numbers in fractional form. (Section 2.2/Objective 1)	When the fractions have a common denominator, add (or subtract) the numerators and place over the common denominator. If the fractions do not have a common denominator, then use the fundamental principle of fractions, $\dfrac{a}{b} = \dfrac{a \cdot k}{b \cdot k}$, to obtain equivalent fractions that have a common denominator.	Add $\dfrac{7}{12} + \dfrac{1}{15}$. **Solution** $\dfrac{7}{12} + \dfrac{1}{15} = \dfrac{7 \cdot 5}{12 \cdot 5} + \dfrac{1 \cdot 4}{15 \cdot 4}$ $= \dfrac{35}{60} + \dfrac{4}{60}$ $= \dfrac{39}{60}$ $= \dfrac{13}{20}$
Combine similar terms whose coefficients are rational numbers in fractional form. (Section 2.2/Objective 2)	To combine similar terms, apply the distributive property and follow the rules for adding rational numbers in fractional form.	Simplify $\dfrac{3}{5}x + \dfrac{1}{2}x$. **Solution** $\dfrac{3}{5}x + \dfrac{1}{2}x = \left(\dfrac{3}{5} + \dfrac{1}{2}\right)x$ $= \left(\dfrac{3 \cdot 2}{5 \cdot 2} + \dfrac{1 \cdot 5}{2 \cdot 5}\right)x$ $= \left(\dfrac{6}{10} + \dfrac{5}{10}\right)x$ $= \dfrac{11}{10}x$
Add or subtract rational numbers in decimal form. (Section 2.3/Objective 2)	To add or subtract decimals, write the numbers in a column so that the decimal points are lined up. Then add or subtract the numbers. It may be necessary to insert zeros as placeholders.	Perform the indicated operations: **(a)** $3.21 + 1.42 + 5.61$ **(b)** $4.76 - 2.14$ **Solution** **(a)** $\begin{array}{r} 3.21 \\ 1.42 \\ +5.61 \\ \hline 10.24 \end{array}$ **(b)** $\begin{array}{r} 4.76 \\ -2.14 \\ \hline 2.62 \end{array}$
Multiply rational numbers in decimal form. (Section 2.3/Objective 2)	1. Multiply the numbers and ignore the decimal points. 2. Find the sum of the number of digits to the right of the decimal points in each factor. 3. Insert the decimal point in the product so that the number of decimal places to the right of the decimal point is the same as the above sum. It may be necessary to insert zeros as placeholders.	Multiply $(3.12)(0.3)$. **Solution** $\begin{array}{r} 3.12 \\ 0.3 \\ \hline 0.936 \end{array}$ two digits to the right one digit to the right three digits to the right

(continued)

OBJECTIVE	SUMMARY	EXAMPLE
Divide a rational number in decimal form by a whole number. (Section 2.3/Objective 2)	To divide a decimal number by a nonzero whole number, divide the numbers and place the decimal point in the quotient directly above the decimal point in the dividend. It may be necessary to insert zeros in the quotient to the right of the decimal point.	Divide $13\overline{)0.182}$. **Solution** $$\begin{array}{r} 0.014 \\ 13\overline{)0.182} \\ -13 \\ \hline 52 \\ -52 \\ \hline 0 \end{array}$$
Divide a rational number in decimal form by another rational number in decimal form. (Section 2.3/Objective 2)	To divide by a decimal number, change to an equivalent problem that has a whole number divisor.	Divide $1.7\overline{)0.34}$. **Solution** $$1.7\overline{)0.34} = \left(\frac{0.34}{1.7}\right)\left(\frac{10}{10}\right)$$ $$= \frac{3.4}{17}$$ $$= \begin{array}{r} 0.2 \\ 17\overline{)3.4} \\ -3.4 \\ \hline 0 \end{array}$$
Combine similar terms whose coefficients are rational numbers in decimal form. (Section 2.3/Objective 3)	To combine similar terms, apply the distributive property and follow the rules for adding rational numbers in decimal form.	Simplify $3.87y + 0.4y + y$. **Solution** $3.87y + 0.4y + y$ $= (3.87 + 0.4 + 1)y$ $= 5.27y$
Evaluate algebraic expressions when the variables are rational numbers. (Section 2.3/Objective 4)	Algebraic expressions can be evaluated for values of the variable that are rational numbers.	Evaluate $-\frac{3}{4}y + \frac{1}{3}y$, when $y = -\frac{2}{5}$. **Solution** $$-\frac{3}{4}\left(-\frac{2}{5}\right) + \frac{1}{3}\left(-\frac{2}{5}\right) = \frac{3}{10} - \frac{2}{15}$$ $$= \frac{3 \cdot 3}{10 \cdot 3} - \frac{2 \cdot 2}{15 \cdot 2}$$ $$= \frac{9}{30} - \frac{4}{30} = \frac{5}{30} = \frac{1}{6}$$
Simplify numerical expressions involving exponents. (Section 2.3/Objective 2)	Expressions of the form b^n are read as "b to the nth power"; b is the base, and n is the exponent. Expressions of the form b^n tell us that the base, b, is used as a factor n times.	Evaluate: **(a)** 2^5 **(b)** $(-3)^4$ **(c)** $\left(\frac{2}{3}\right)^2$ **Solution** **(a)** $2^5 = 2 \cdot 2 \cdot 2 \cdot 2 \cdot 2 = 32$ **(b)** $(-3)^4 = (-3)(-3)(-3)(-3) = 81$ **(c)** $\left(\frac{2}{3}\right)^2 = \frac{2}{3} \cdot \frac{2}{3} = \frac{4}{9}$

OBJECTIVE	SUMMARY	EXAMPLE
Evaluate algebraic expressions that involve exponents. (Section 2.3/Objective 2)	Algebraic expressions involving exponents can be evaluated for specific values of the variable.	Evaluate $3x^2y - 5xy^2$ when $x = -2$ and $y = 4$. **Solution** $3(-2)^2(4) - 5(-2)(4)^2$ $= 3(4)(4) - 5(-2)(16)$ $= 48 + 160$ $= 208$
Simplify algebraic expressions involving exponents by combining similar terms. (Section 2.4/Objective 3)	Similar terms involving exponents can be combined by using the distributive propery.	Simplify $2x^2 - 3x^2 + 5x^2$. **Solution** $2x^2 - 3x^2 + 5x^2$ $= (2 - 3 + 5)x^2$ $= 4x^2$
Solve application problems involving rational numbers in fractional or decimal form. (Section 2.1/Objective 3) (Section 2.2/Objective 3) (Section 2.3/Objective 5)	Rational numbers are used to solve many real world problems.	To obtain a custom hair color for a client, Marti is mixing $\frac{3}{8}$ cup of brown color with $\frac{1}{4}$ cup of blonde color. How many cups of color are being used for the custom color? **Solution** To solve, add $\frac{3}{8} + \frac{1}{4}$. $\frac{3}{8} + \frac{1}{4} = \frac{3}{8} + \frac{2}{8} = \frac{5}{8}$ So $\frac{5}{8}$ cup of color is being used.
Translate English phrases into algebraic expressions. (Section 2.5/Objective 2)	To translate English phrases into algebraic expressions, you should know the algebraic vocabulary for "addition," "subtraction," "multiplication," and "division."	Translate the phrase "four less than a number" into an algebraic expression. **Solution** Let n represent the number. The algebraic expression is $n - 4$.

Chapter 2 Review Problem Set

For Problems 1–14, find the value of each of the following.

1. 2^6

2. $(-3)^3$

3. -4^2

4. 5^3

5. $-\left(\frac{1}{2}\right)^2$

6. $\left(\frac{3}{4}\right)^2$

7. $\left(\frac{1}{2} + \frac{2}{3}\right)^2$

8. $(0.6)^3$

9. $(0.12)^2$

10. $(0.06)^2$

11. $\left(-\frac{2}{3}\right)^3$

12. $\left(-\frac{1}{2}\right)^4$

13. $\left(\frac{1}{4} - \frac{1}{2}\right)^3$

14. $\left(\frac{1}{2} + \frac{1}{3} - \frac{1}{6}\right)^2$

For Problems 15–24, perform the indicated operations, and express your answers in reduced form.

15. $\frac{3}{8} + \frac{5}{12}$

16. $\frac{9}{14} - \frac{3}{35}$

17. $\dfrac{2}{3} + \dfrac{-3}{5}$

18. $\dfrac{7}{x} + \dfrac{9}{2y}$

19. $\dfrac{5}{xy} - \dfrac{8}{x^2}$

20. $\left(\dfrac{7y}{8x}\right)\left(\dfrac{14x}{35}\right)$

21. $\left(\dfrac{6xy}{9y^2}\right) \div \left(\dfrac{15y}{18x^2}\right)$

22. $\left(\dfrac{-3x}{12y}\right)\left(\dfrac{8y}{-7x}\right)$

23. $\left(\dfrac{-4y}{3x}\right)\left(-\dfrac{3x}{4y}\right)$

24. $\left(\dfrac{6n}{7}\right)\left(\dfrac{9n}{8}\right)$

For Problems 25–36, simplify each of the following numerical expressions.

25. $\dfrac{1}{6} + \dfrac{2}{3} \cdot \dfrac{3}{4} - \dfrac{5}{6} \div \dfrac{8}{6}$

26. $\dfrac{3}{4} \cdot \dfrac{1}{2} - \dfrac{4}{3} \cdot \dfrac{3}{2}$

27. $\dfrac{7}{9} \cdot \dfrac{3}{5} + \dfrac{7}{9} \cdot \dfrac{2}{5}$

28. $\dfrac{4}{5} \div \dfrac{1}{5} \cdot \dfrac{2}{3} - \dfrac{1}{4}$

29. $\dfrac{2}{3} \cdot \dfrac{1}{4} \div \dfrac{1}{2} + \dfrac{2}{3} \cdot \dfrac{1}{4}$

30. $0.48 + 0.72 - 0.35 - 0.18$

31. $0.81 + (0.6)(0.4) - (0.7)(0.8)$

32. $1.28 \div 0.8 - 0.81 \div 0.9 + 1.7$

33. $(0.3)^2 + (0.4)^2 - (0.6)^2$

34. $(1.76)(0.8) + (1.76)(0.2)$

35. $(2^2 - 2 - 2^3)^2$ **36.** $1.92(0.9 + 0.1)$

For Problems 37–42, simplify each of the following algebraic expressions by combining similar terms. Express your answers in reduced form when working with common fractions.

37. $\dfrac{3}{8}x^2 - \dfrac{2}{5}y^2 - \dfrac{2}{7}x^2 + \dfrac{3}{4}y^2$

38. $0.24ab + 0.73bc - 0.82ab - 0.37bc$

39. $\dfrac{1}{2}x + \dfrac{3}{4}x - \dfrac{5}{6}x + \dfrac{1}{24}x$

40. $1.4a - 1.9b + 0.8a + 3.6b$

41. $\dfrac{2}{5}n + \dfrac{1}{3}n - \dfrac{5}{6}n$

42. $n - \dfrac{3}{4}n + 2n - \dfrac{1}{5}n$

For Problems 43–48, evaluate the following algebraic expressions for the given values of the variables.

43. $\dfrac{1}{4}x - \dfrac{2}{5}y$ for $x = \dfrac{2}{3}$ and $y = -\dfrac{5}{7}$

44. $a^3 + b^2$ for $a = -\dfrac{1}{2}$ and $b = \dfrac{1}{3}$

45. $2x^2 - 3y^2$ for $x = 0.6$ and $y = 0.7$

46. $0.7w + 0.9z$ for $w = 0.4$ and $z = -0.7$

47. $\dfrac{3}{5}x - \dfrac{1}{3}x + \dfrac{7}{15}x - \dfrac{2}{3}x$ for $x = \dfrac{15}{17}$

48. $\dfrac{1}{3}n + \dfrac{2}{7}n - n$ for $n = 21$

For Problems 49–56, answer each of the following questions with an algebraic expression.

49. The sum of two numbers is 72, and one of the numbers is n. What is the other number?

50. Joan has p pennies and d dimes. How much money in cents does she have?

51. Ellen types x words in an hour. What is her typing rate per minute?

52. Harry is y years old. His brother is 3 years less than twice as old as Harry. How old is Harry's brother?

53. Larry chose a number n. Cindy chose a number 3 more than 5 times the number chosen by Larry. What number did Cindy choose?

54. The height of a file cabinet is y yards and f feet. How tall is the file cabinet in inches?

55. The length of a rectangular room is m meters. How long in centimeters is the room?

56. Corinne has n nickels, d dimes, and q quarters. How much money in cents does she have?

For Problems 57–66, translate each word phrase into an algebraic expression.

57. Five less than n

58. Five less n

59. Ten times the quantity, x minus 2

60. Ten times x minus 2

61. x minus three

62. d divided by r

63. x squared plus nine

64. x plus nine, the quantity squared

65. The sum of the cubes of x and y

66. Four less than the product of x and y

1. Find the value of each expression.
   (a) $(-3)^4$ (b) -2^6 (c) $(0.2)^3$

2. Express $\dfrac{42}{54}$ in reduced form.

3. Simplify $\dfrac{18xy^2}{32y}$.

For Problems 4–7, simplify each numerical expression.

4. $5.7 - 3.8 + 4.6 - 9.1$

5. $0.2(0.4) - 0.6(0.9) + 0.5(7)$

6. $-0.4^2 + 0.3^2 - 0.7^2$

7. $\left(\dfrac{1}{3} - \dfrac{1}{4} + \dfrac{1}{6}\right)^4$

For Problems 8–11, perform the indicated operations, and express your answers in reduced form.

8. $\dfrac{5}{12} \div \dfrac{15}{8}$

9. $-\dfrac{2}{3} - \dfrac{1}{2}\left(\dfrac{3}{4}\right) + \dfrac{5}{6}$

10. $3\left(\dfrac{2}{5}\right) - 4\left(\dfrac{5}{6}\right) + 6\left(\dfrac{7}{8}\right)$

11. $4\left(\dfrac{1}{2}\right)^3 - 3\left(\dfrac{2}{3}\right)^2 + 9\left(\dfrac{1}{4}\right)^2$

For Problems 12–17, perform the indicated operations, and express your answers in reduced form.

12. $\dfrac{8x}{15y} \cdot \dfrac{9y^2}{6x}$ 13. $\dfrac{6xy}{9} \div \dfrac{y}{3x}$

14. $\dfrac{4}{x} - \dfrac{5}{y^2}$ 15. $\dfrac{3}{2x} + \dfrac{7}{6x}$

16. $\dfrac{5}{3y} + \dfrac{9}{7y^2}$ 17. $\left(\dfrac{15a^2b}{12a}\right)\left(\dfrac{8ab}{9b}\right)$

For Problems 18 and 19, simplify each algebraic expression by combining similar terms.

18. $3x - 2xy - 4x + 7xy$ 19. $-2a^2 + 3b^2 - 5b^2 - a^2$

For Problems 20–23, evaluate each of the algebraic expressions for the given values of the variables.

20. $x^2 - xy + y^2$ for $x = \dfrac{1}{2}$ and $y = -\dfrac{2}{3}$

21. $0.2x - 0.3y - xy$ for $x = 0.4$ and $y = 0.8$

22. $\dfrac{3}{4}x - \dfrac{2}{3}y$ for $x = -\dfrac{1}{2}$ and $y = \dfrac{3}{5}$

23. $3x - 2y + xy$ for $x = 0.5$ and $y = -0.9$

24. David has n nickels, d dimes, and q quarters. How much money, in cents, does he have?

25. Hal chose a number n. Sheila chose a number 3 less than 4 times the number that Hal chose. Express the number that Sheila chose in terms of n.

For Problems 1–12, simplify each of the numerical expressions.

1. $16 - 18 - 14 + 21 - 14 + 19$

2. $7(-6) - 8(-6) + 4(-9)$

3. $6 - [3 - (10 - 12)]$

4. $-9 - 2[4 - (-10 + 6)] - 1$

5. $\dfrac{-7(-4) - 5(-6)}{-2}$

6. $\dfrac{5(-3) + (-4)(6) - 3(4)}{-3}$

7. $\dfrac{3}{4} + \dfrac{1}{3} \div \dfrac{4}{3} - \dfrac{1}{2}$

8. $\left(\dfrac{2}{3}\right)\left(-\dfrac{3}{4}\right) - \left(\dfrac{5}{6}\right)\left(\dfrac{4}{5}\right)$

9. $\left(\dfrac{1}{2} - \dfrac{2}{3}\right)^2$

10. -4^3

11. $\dfrac{0.0046}{0.000023}$

12. $(0.2)^2 - (0.3)^3 + (0.4)^2$

For Problems 13–20, evaluate each algebraic expression for the given values of the variables.

13. $3xy - 2x - 4y$ for $x = -6$ and $y = 7$

14. $-4x^2y - 2xy^2 + xy$ for $x = -2$ and $y = -4$

15. $\dfrac{5x - 2y}{3x}$ for $x = \dfrac{1}{2}$ and $y = -\dfrac{1}{3}$

16. $0.2x - 0.3y + 2xy$ for $x = 0.1$ and $y = 0.3$

17. $-7x + 4y + 6x - 9y + x - y$ for $x = -0.2$ and $y = 0.4$

18. $\dfrac{2}{3}x - \dfrac{3}{5}y + \dfrac{3}{4}x - \dfrac{1}{2}y$ for $x = \dfrac{6}{5}$ and $y = -\dfrac{1}{4}$

19. $\dfrac{1}{5}n - \dfrac{1}{3}n + n - \dfrac{1}{6}n$ for $n = \dfrac{1}{5}$

20. $-ab + \dfrac{1}{5}a - \dfrac{2}{3}b$ for $a = -2$ and $b = \dfrac{3}{4}$

For Problems 21–24, express each of the numbers as a product of prime factors.

21. 54

22. 78

23. 91

24. 153

For Problems 25–28, find the greatest common factor of the given numbers.

25. 42 and 70

26. 63 and 81

27. 28, 36, and 52

28. 48, 66, and 78

For Problems 29–32, find the least common multiple of the given numbers.

29. 20 and 28

30. 40 and 100

31. 12, 18, and 27

32. 16, 20, and 80

For Problems 33–38, simplify each algebraic expression by combining similar terms.

33. $\dfrac{2}{3}x - \dfrac{1}{4}y - \dfrac{3}{4}x - \dfrac{2}{3}y$

34. $-n - \dfrac{1}{2}n + \dfrac{3}{5}n + \dfrac{5}{6}n$

35. $3.2a - 1.4b - 6.2a + 3.3b$

36. $-(n - 1) + 2(n - 2) - 3(n - 3)$

37. $-x + 4(x - 1) - 3(x + 2) - (x + 5)$

38. $2a - 5(a + 3) - 2(a - 1) - 4a$

For Problems 39–46, perform the indicated operations and express your answers in reduced form.

39. $\dfrac{5}{12} - \dfrac{3}{16}$

40. $\dfrac{3}{4} - \dfrac{5}{6} - \dfrac{7}{9}$

41. $\dfrac{5}{xy} - \dfrac{2}{x} + \dfrac{3}{y}$

42. $-\dfrac{7}{x^2} + \dfrac{9}{xy}$

43. $\left(\dfrac{7x}{9y}\right)\left(\dfrac{12y}{14}\right)$

44. $\left(-\dfrac{5a}{7b^2}\right)\left(-\dfrac{8ab}{15}\right)$

45. $\left(\dfrac{6x^2y}{11}\right) \div \left(\dfrac{9y^2}{22}\right)$ **46.** $\left(-\dfrac{9a}{8b}\right) \div \left(\dfrac{12a}{18b}\right)$

For Problems 47–50, answer the question with an algebraic expression.

47. Hector has p pennies, n nickels, and d dimes. How much money in cents does he have?

48. Ginny chose a number n. Penny chose a number 5 less than 4 times the number chosen by Ginny. What number did Penny choose?

49. The height of a flagpole is y yards, f feet, and i inches. How tall is the flagpole in inches?

50. A rectangular room is x meters by y meters. What is its perimeter in centimeters?

3 Equations, Inequalities, and Problem Solving

The inequality
$$\frac{95 + 82 + 93 + 84 + s}{5} \geq 90$$
can be used to determine that Ashley needs a 96 or higher on her fifth exam to have an average of 90 or higher for the five exams if she got 95, 82, 93, and 84 on her first four exams.

© Laurence Gough

Tracy received a cell phone bill for $136.74. Included in the $136.74 were a monthly-plan charge of $39.99 and a charge for 215 extra minutes. How much is Tracy being charged for each extra minute? If we let c represent the charge per minute, then the equation $39.99 + 215c = 136.74$ can be used to determine that the charge for each extra minute is $0.45.

John was given a graduation present of $25,000 to purchase a car. What price range should John be looking at if the $25,000 has to include the sales tax of 6% and registration fees of $850? If we let p represent the price of the car, then the inequality $p + 0.06p + 850 \leq 25,000$ can be used to determine that the car can cost $22,783 or less.

Throughout this book, you will develop new skills to help solve equations and inequalities, and you will use equations and inequalities to solve applied problems. In this chapter you will use the skills you developed in the first chapter to solve equations and inequalities and then move on to work with applied problems.

Video tutorials based on section learning objectives are available in a variety of delivery modes.

3.1	Solving First-Degree Equations

OBJECTIVES

1 Solve first-degree equations using the addition-subtraction property of equality

2 Solve first-degree equations using the multiplication-division property of equality

These are examples of **numerical statements**:

$$3 + 4 = 7 \qquad 5 - 2 = 3 \qquad 7 + 1 = 12$$

The first two are true statements, and the third one is a false statement.

When we use x as a variable, the statements

$$x + 3 = 4 \qquad 2x - 1 = 7 \qquad \text{and} \qquad x^2 = 4$$

are called **algebraic equations** in x. We call the number a a **solution** or **root** of an equation if a true numerical statement is formed when a is substituted for x. (We also say that a satisfies the equation.) For example, 1 is a solution of $x + 3 = 4$ because substituting 1 for x produces the true numerical statement $1 + 3 = 4$. We call the set of all solutions of an equation its **solution set**. Thus the solution set of $x + 3 = 4$ is $\{1\}$. Likewise the solution set of $2x - 1 = 7$ is $\{4\}$ and the solution set of $x^2 = 4$ is $\{-2, 2\}$. **Solving an equation** refers to the process of determining the solution set. Remember that a set that consists of no elements is called the empty or null set and is denoted by $\varnothing$. Thus we say that the solution set of $x = x + 1$ is $\varnothing$; that is, there are no real numbers that satisfy $x = x + 1$.

In this chapter, we will consider techniques for solving **first-degree equations of one variable**. This means that the equations contain only one variable, and this variable has an exponent of one. Here are some examples of first-degree equations of one variable:

$$3x + 4 = 7 \qquad 8w + 7 = 5w - 4$$

$$\frac{1}{2}y + 2 = 9 \qquad 7x + 2x - 1 = 4x - 1$$

Equivalent equations are equations that have the same solution set. The following equations are all equivalent:

$$5x - 4 = 3x + 8$$

$$2x = 12$$

$$x = 6$$

You can verify the equivalence by showing that 6 is the solution for all three equations.

As we work with equations, we can use the properties of equality.

Property 3.1 Properties of Equality

For all real numbers, a, b, and c,

1. $a = a$ Reflexive property

2. If $a = b$, then $b = a$.   Symmetric property

3. If $a = b$ and $b = c$, then $a = c$.   Transitive property

4. If $a = b$, then a may be replaced by b, or b may be replaced by a, in any statement, without changing the meaning of the statement.   Substitution property

---

The general procedure for solving an equation is to continue replacing the given equation with equivalent but simpler equations until we obtain an equation of the form **variable = constant** or **constant = variable**. Thus in the earlier example, $5x - 4 = 3x + 8$ was simplified to

$2x = 12$, which was further simplified to $x = 6$, from which the solution of 6 is obvious. The exact procedure for simplifying equations becomes our next concern.

Two properties of equality play an important role in the process of solving equations. The first of these is the **addition-subtraction property of equality**.

Property 3.2   Addition-Subtraction Property of Equality

For all real numbers, a, b, and c,

1. $a = b$ if and only if $a + c = b + c$.
2. $a = b$ if and only if $a - c = b - c$.

Property 3.2 states that any number can be added to or subtracted from both sides of an equation and an equivalent equation is produced. Consider the use of this property in the next four examples.

Classroom Example
Solve $d - 7 = 11$.

EXAMPLE 1   Solve $x - 8 = 3$.

Solution

$$x - 8 = 3$$
$$x - 8 + 8 = 3 + 8 \qquad \text{Add 8 to both sides}$$
$$x = 11$$

The solution set is $\{11\}$.

Remark: It is true that a simple equation such as Example 1 can be solved *by inspection*; for instance, "some number minus 8 produces 3" yields an obvious answer of 11. However, as the equations become more complex, the technique of solving by inspection becomes ineffective. So it is necessary to develop more formal techniques for solving equations. Therefore, we will begin developing such techniques with very simple types of equations.

Classroom Example
Solve $m + 12 = -4$.

EXAMPLE 2   Solve $x + 14 = -8$.

Solution

$$x + 14 = -8$$
$$x + 14 - 14 = -8 - 14 \qquad \text{Subtract 14 from both sides}$$
$$x = -22$$

The solution set is $\{-22\}$.

Classroom Example
Solve $a - \dfrac{1}{5} = \dfrac{1}{2}$.

EXAMPLE 3   Solve $n - \dfrac{1}{3} = \dfrac{1}{4}$.

Solution

$$n - \frac{1}{3} = \frac{1}{4}$$
$$n - \frac{1}{3} + \frac{1}{3} = \frac{1}{4} + \frac{1}{3} \qquad \text{Add } \frac{1}{3} \text{ to both sides}$$
$$n = \frac{3}{12} + \frac{4}{12} \qquad \text{Change to equivalent fractions with a denominator of 12}$$

$$n = \frac{7}{12}$$

The solution set is $\left\{\frac{7}{12}\right\}$.

Classroom Example
Solve $0.63 = n + 0.49$.

EXAMPLE 4   Solve $0.72 = y + 0.35$.

Solution

$$0.72 = y + 0.35$$
$$0.72 - 0.35 = y + 0.35 - 0.35 \qquad \text{Subtract 0.35 from both sides}$$
$$0.37 = y$$

The solution set is $\{0.37\}$.

Note in Example 4 that the final equation is $0.37 = y$ instead of $y = 0.37$. Technically, the **symmetric property of equality** (if $a = b$, then $b = a$) permits us to change from $0.37 = y$ to $y = 0.37$, but such a change is not necessary to determine that the solution is 0.37. You should also realize that the symmetric property could be applied to the original equation. Thus $0.72 = y + 0.35$ becomes $y + 0.35 = 0.72$, and subtracting 0.35 from both sides produces $y = 0.37$.

One other comment pertaining to Property 3.2 should be made at this time. Because subtracting a number is equivalent to adding its opposite, we can state Property 3.2 only in terms of addition. Thus to solve an equation such as Example 4, we add -0.35 to both sides rather than subtract 0.35 from both sides.

The other important property for solving equations is the **multiplication-division property of equality**.

> **Property 3.3   Multiplication-Division Property of Equality**
>
> For all real numbers, a, b, and c, where $c \neq 0$,
>
> **1.** $a = b$ if and only if $ac = bc$.
>
> **2.** $a = b$ if and only if $\dfrac{a}{c} = \dfrac{b}{c}$.

Property 3.3 states that an equivalent equation is obtained whenever both sides of a given equation are multiplied or divided by the same nonzero real number. The next examples illustrate how we use this property.

Classroom Example
Solve $\frac{4}{7}y = 8$.

EXAMPLE 5   Solve $\frac{3}{4}x = 6$.

Solution

$$\frac{3}{4}x = 6$$

$$\frac{4}{3}\left(\frac{3}{4}x\right) = \frac{4}{3}(6) \qquad \text{Multiply both sides by } \frac{4}{3} \text{ because } \left(\frac{4}{3}\right)\left(\frac{3}{4}\right) = 1$$

$$x = 8$$

The solution set is $\{8\}$.

Classroom Example
Solve $4c = 33$.

EXAMPLE 6   Solve $5x = 27$.

Solution

$$5x = 27$$

$$\frac{5x}{5} = \frac{27}{5} \qquad \text{Divide both sides by 5}$$

$$x = \frac{27}{5} \qquad \frac{27}{5} \text{ can be expressed as } 5\frac{2}{5} \text{ or } 5.4$$

The solution set is $\left\{\dfrac{27}{5}\right\}$.

Classroom Example
Solve $-\dfrac{4}{5}m = \dfrac{2}{3}$.

EXAMPLE 7   Solve $-\dfrac{2}{3}p = \dfrac{1}{2}$.

Solution

$$-\frac{2}{3}p = \frac{1}{2}$$

$$\left(-\frac{3}{2}\right)\left(-\frac{2}{3}p\right) = \left(-\frac{3}{2}\right)\left(\frac{1}{2}\right) \qquad \text{Multiply both sides by } -\frac{3}{2} \text{ because } \left(-\frac{3}{2}\right)\left(-\frac{2}{3}\right) = 1$$

$$p = -\frac{3}{4}$$

The solution set is $\left\{-\dfrac{3}{4}\right\}$.

Classroom Example
Solve $18 = -8b$.

EXAMPLE 8   Solve $26 = -6x$.

Solution

$$26 = -6x$$

$$\frac{26}{-6} = \frac{-6x}{-6} \qquad \text{Divide both sides by } -6$$

$$-\frac{26}{6} = x \qquad \frac{26}{-6} = -\frac{26}{6}$$

$$-\frac{13}{3} = x \qquad \text{Don't forget to reduce!}$$

The solution set is $\left\{-\dfrac{13}{3}\right\}$.

Look back at Examples 5–8, and you will notice that we divided both sides of the equation by the coefficient of the variable whenever the coefficient was an integer; otherwise, we used the multiplication part of Property 3.3. Technically, because dividing by a number is equivalent to multiplying by its reciprocal, Property 3.3 could be stated only in terms of multiplication. Thus to solve an equation such as $5x = 27$, we could multiply both sides by $\dfrac{1}{5}$ instead of dividing both sides by 5.

Classroom Example
Solve $0.4t = 28$.

EXAMPLE 9   Solve $0.2n = 15$.

Solution

$$0.2n = 15$$

$$\frac{0.2n}{0.2} = \frac{15}{0.2} \qquad \text{Divide both sides by 0.2}$$

$$n = 75$$

The solution set is $\{75\}$.

Concept Quiz 3.1

For Problems 1–10, answer true or false.

1. Equivalent equations have the same solution set.
2. $x^2 = 9$ is a first-degree equation.
3. The set of all solutions is called a solution set.
4. If the solution set is the null set, then the equation has at least one solution.
5. Solving an equation refers to obtaining any other equivalent equation.
6. If 5 is a solution, then a true numerical statement is formed when 5 is substituted for the variable in the equation.
7. Any number can be subtracted from both sides of an equation, and the result is an equivalent equation.
8. Any number can divide both sides of an equation to obtain an equivalent equation.
9. By the reflexive property, if $y = 2$ then $2 = y$.
10. By the transitive property, if $x = y$ and $y = 4$, then $x = 4$.

Problem Set 3.1

For Problems 1–72, use the properties of equality to help solve each equation. **(Objectives 1 and 2)**

1. $x + 9 = 17$
2. $x + 7 = 21$
3. $x + 11 = 5$
4. $x + 13 = 2$
5. $-7 = x + 2$
6. $-12 = x + 4$
7. $8 = n + 14$
8. $6 = n + 19$
9. $21 + y = 34$
10. $17 + y = 26$
11. $x - 17 = 31$
12. $x - 22 = 14$
13. $14 = x - 9$
14. $17 = x - 28$
15. $-26 = n - 19$
16. $-34 = n - 15$
17. $y - \frac{2}{3} = \frac{3}{4}$
18. $y - \frac{2}{5} = \frac{1}{6}$
19. $x + \frac{3}{5} = \frac{1}{3}$
20. $x + \frac{5}{8} = \frac{2}{5}$
21. $b + 0.19 = 0.46$
22. $b + 0.27 = 0.74$
23. $n - 1.7 = -5.2$
24. $n - 3.6 = -7.3$
25. $15 - x = 32$
26. $13 - x = 47$
27. $-14 - n = 21$
28. $-9 - n = 61$
29. $7x = -56$
30. $9x = -108$
31. $-6x = 102$
32. $-5x = 90$
33. $5x = 37$
34. $7x = 62$
35. $-18 = 6n$
36. $-52 = 13n$
37. $-26 = -4n$
38. $-56 = -6n$
39. $\frac{t}{9} = 16$
40. $\frac{t}{12} = 8$
41. $\frac{n}{-8} = -3$
42. $\frac{n}{-9} = -5$
43. $-x = 15$
44. $-x = -17$
45. $\frac{3}{4}x = 18$
46. $\frac{2}{3}x = 32$

47. $-\dfrac{2}{5}n = 14$

48. $-\dfrac{3}{8}n = 33$

59. $-\dfrac{5}{12} = \dfrac{7}{6}x$

60. $-\dfrac{7}{24} = \dfrac{3}{8}x$

49. $\dfrac{2}{3}n = \dfrac{1}{5}$

50. $\dfrac{3}{4}n = \dfrac{1}{8}$

61. $-\dfrac{5}{7}x = 1$

62. $-\dfrac{11}{12}x = -1$

51. $\dfrac{5}{6}n = -\dfrac{3}{4}$

52. $\dfrac{6}{7}n = -\dfrac{3}{8}$

63. $-4n = \dfrac{1}{3}$

64. $-6n = \dfrac{3}{4}$

53. $\dfrac{3x}{10} = \dfrac{3}{20}$

54. $\dfrac{5x}{12} = \dfrac{5}{36}$

65. $-8n = \dfrac{6}{5}$

66. $-12n = \dfrac{8}{3}$

55. $\dfrac{-y}{2} = \dfrac{1}{6}$

56. $\dfrac{-y}{4} = \dfrac{1}{9}$

67. $1.2x = 0.36$

68. $2.5x = 17.5$

69. $30.6 = 3.4n$

70. $2.1 = 4.2n$

57. $-\dfrac{4}{3}x = -\dfrac{9}{8}$

58. $-\dfrac{6}{5}x = -\dfrac{10}{14}$

71. $-3.4x = 17$

72. $-4.2x = 50.4$

Thoughts Into Words

73. Describe the difference between a numerical statement and an algebraic equation.

74. Are the equations $6 = 3x + 1$ and $1 + 3x = 6$ equivalent equations? Defend your answer.

Answers to the Concept Quiz

1. True **2.** False **3.** True **4.** False **5.** False **6.** True **7.** True **8.** False **9.** False **10.** True

---

3.2   Equations and Problem Solving

OBJECTIVES

1 Solve first-degree equations using both the addition-subtraction property of equality and the multiplication-division property of equality

2 Declare variables and write equations to solve word problems

We often need more than one property of equality to help find the solution of an equation. Consider the following examples.

Classroom Example
Solve $2n + 5 = 13$.

EXAMPLE 1   Solve $3x + 1 = 7$.

Solution

$$3x + 1 = 7$$
$$3x + 1 - 1 = 7 - 1 \qquad \text{Subtract 1 from both sides}$$
$$3x = 6$$
$$\dfrac{3x}{3} = \dfrac{6}{3} \qquad \text{Divide both sides by 3}$$
$$x = 2$$

The potential solution can be *checked* by substituting it into the original equation to see whether a true numerical statement results.

✓ **Check**

$$3x + 1 = 7$$
$$3(2) + 1 \overset{?}{=} 7$$
$$6 + 1 \overset{?}{=} 7$$
$$7 = 7$$

Now we know that the solution set is $\{2\}$.

Classroom Example
Solve $7m - 3 = 11$.

EXAMPLE 2   Solve $5x - 6 = 14$.

Solution

$$5x - 6 = 14$$
$$5x - 6 + 6 = 14 + 6 \qquad \text{Add 6 to both sides}$$
$$5x = 20$$
$$\frac{5x}{5} = \frac{20}{5} \qquad \text{Divide both sides by 5}$$
$$x = 4$$

✓ **Check**

$$5x - 6 = 14$$
$$5(4) - 6 \overset{?}{=} 14$$
$$20 - 6 \overset{?}{=} 14$$
$$14 = 14$$

The solution set is $\{4\}$.

Classroom Example
Solve $5 - 2c = 13$.

EXAMPLE 3   Solve $4 - 3a = 22$.

Solution

$$4 - 3a = 22$$
$$4 - 3a - 4 = 22 - 4 \qquad \text{Subtract 4 from both sides}$$
$$-3a = 18$$
$$\frac{-3a}{-3} = \frac{18}{-3} \qquad \text{Divide both sides by } -3$$
$$a = -6$$

✓ **Check**

$$4 - 3a = 22$$
$$4 - 3(-6) \overset{?}{=} 22$$
$$4 + 18 \overset{?}{=} 22$$
$$22 = 22$$

The solution set is $\{-6\}$.

Notice that in Examples 1, 2, and 3, we used the addition-subtraction property first, and then we used the multiplication-division property. In general, this sequence of steps provides the easiest method for solving such equations. Perhaps you should convince yourself of that fact by doing Example 1 again, but this time use the multiplication-division property first and then the addition-subtraction property.

EXAMPLE 4   Solve $19 = 2n + 4$.

Solution

$$19 = 2n + 4$$

$$19 - 4 = 2n + 4 - 4 \qquad \text{Subtract 4 from both sides}$$

$$15 = 2n$$

$$\frac{15}{2} = \frac{2n}{2} \qquad \text{Divide both sides by 2}$$

$$\frac{15}{2} = n$$

✓ Check

$$19 = 2n + 4$$

$$19 \stackrel{?}{=} 2\left(\frac{15}{2}\right) + 4$$

$$19 \stackrel{?}{=} 15 + 4$$

$$19 = 19$$

The solution set is $\left\{\dfrac{15}{2}\right\}$.

Word Problems

In the last section of Chapter 2, we translated English phrases into algebraic expressions. We are now ready to expand that concept and translate English sentences into algebraic equations. Such translations allow us to use the concepts of algebra to solve word problems. Let's consider some examples.

EXAMPLE 5

A certain number added to 17 yields a sum of 29. What is the number?

Solution

Let n represent the number to be found. The sentence "A certain number added to 17 yields a sum of 29" translates into the algebraic equation $17 + n = 29$. We can solve this equation.

$$17 + n = 29$$

$$17 + n - 17 = 29 - 17$$

$$n = 12$$

The solution is 12, which is the number asked for in the problem.

We often refer to the statement "let n represent the number to be found" as **declaring the variable**. We need to choose a letter to use as a variable and indicate what it represents for a specific problem. This may seem like an unnecessary exercise, but as the problems become more complex, the process of declaring the variable becomes more important. We could solve a problem like Example 5 without setting up an algebraic equation; however, as problems increase in difficulty, the translation from English into an algebraic equation becomes a key issue. Therefore, even with these relatively simple problems we need to concentrate on the translation process.

EXAMPLE 6   Six years ago Bill was 13 years old. How old is he now?

Solution

Let y represent Bill's age now; therefore, $y - 6$ represents his age 6 years ago. Thus

$$y - 6 = 13$$
$$y - 6 + 6 = 13 + 6$$
$$y = 19$$

Bill is presently 19 years old.

EXAMPLE 7

Betty worked 8 hours Saturday and earned $60. How much did she earn per hour?

Solution A

Let x represent the amount Betty earned per hour. The number of hours worked times the wage per hour yields the total earnings. Thus

$$8x = 60$$
$$\frac{8x}{8} = \frac{60}{8}$$
$$x = 7.50$$

Betty earned $7.50 per hour.

Solution B

Let y represent the amount Betty earned per hour. The wage per hour equals the total wage divided by the number of hours. Thus

$$y = \frac{60}{8}$$
$$y = 7.50$$

Betty earned $7.50 per hour.

Sometimes we can use more than one equation to solve a problem. In Solution A, we set up the equation in terms of multiplication; whereas in Solution B, we were thinking in terms of division.

EXAMPLE 8

Kendall paid $244 for a CD player and six CDs. The CD player cost ten times as much as one CD. Find the cost of the CD player and the cost of one CD.

Solution

Let d represent the cost of one CD. Then the cost of the CD player is represented by $10d$, and the cost of six CDs is represented by $6d$. The total cost is $244, so we can proceed as follows:

Cost of CD player + Cost of six CDs = $244

$$10d \qquad + \qquad 6d \quad = 244$$

Solving this equation, we have

$$16d = 244$$
$$d = 15.25$$

The cost of one CD is \$15.25, and the cost of the CD player is $10(15.25)$ or \$152.50.

EXAMPLE 9

The cost of a five-day vacation cruise package was \$534. This cost included \$339 for the cruise and an amount for 2 nights of lodging on shore. Find the cost per night of the lodging.

Solution

Let n represent the cost for one night of lodging; then $2n$ represents the total cost of lodging. Thus the cost for the cruise and lodging is the total cost of \$534. We can proceed as follows

Cost of cruise + Cost of lodging = \$534

$$339 \quad + \quad 2n \quad = 534$$

We can solve this equation:

$$339 + 2n = 534$$
$$2n = 195$$
$$\frac{2n}{2} = \frac{195}{2}$$
$$n = 97.50$$

The cost of lodging per night is \$97.50.

Concept Quiz 3.2

For Problems 1–5, answer true or false.

1. Only one property of equality is necessary to solve any equation.

2. Substituting the solution into the original equation to obtain a true numerical statement can be used to check potential solutions.

3. The statement "let x represent the number" is referred to as checking the variable.

4. Sometimes there can be two approaches to solving a word problem.

5. To solve the equation, $\frac{1}{3}x - 2 = 7$, you could begin by either adding 2 to both sides of the equation or by multiplying both sides of the equation by 3.

For Problems 6–10, match the English sentence with its algebraic equation.

6. Three added to a number is 24.

7. The product of 3 and a number is 24.

8. Three less than a number is 24.

9. The quotient of a number and three is 24.

10. A number subtracted from 3 is 24.

A. $3x = 24$

B. $3 - x = 24$

C. $x + 3 = 24$

D. $x - 3 = 24$

E. $\dfrac{x}{3} = 24$

Problem Set 3.2

For Problems 1–40, solve each equation. (Objective 1)

1. $2x + 5 = 13$ **2.** $3x + 4 = 19$

3. $5x + 2 = 32$ **4.** $7x + 3 = 24$

5. $3x - 1 = 23$ **6.** $2x - 5 = 21$

7. $4n - 3 = 41$ **8.** $5n - 6 = 19$

9. $6y - 1 = 16$ **10.** $4y - 3 = 14$

11. $2x + 3 = 22$ **12.** $3x + 1 = 21$

13. $10 = 3t - 8$ **14.** $17 = 2t + 5$

15. $5x + 14 = 9$ **16.** $4x + 17 = 9$

17. $18 - n = 23$ **18.** $17 - n = 29$

19. $-3x + 2 = 20$ **20.** $-6x + 1 = 43$

21. $7 + 4x = 29$ **22.** $9 + 6x = 23$

23. $16 = -2 - 9a$ **24.** $18 = -10 - 7a$

25. $-7x + 3 = -7$ **26.** $-9x + 5 = -18$

27. $17 - 2x = -19$ **28.** $18 - 3x = -24$

29. $-16 - 4x = 9$ **30.** $-14 - 6x = 7$

31. $-12t + 4 = 88$ **32.** $-16t + 3 = 67$

33. $14y + 15 = -33$ **34.** $12y + 13 = -15$

35. $32 - 16n = -8$ **36.** $-41 = 12n - 19$

37. $17x - 41 = -37$ **38.** $19y - 53 = -47$

39. $29 = -7 - 15x$ **40.** $49 = -5 - 14x$

For each of the following problems, (a) choose a variable and indicate what it represents in the problem, (b) set up an equation that represents the situation described, and (c) solve the equation. (Objective 2)

41. Twelve added to a certain number is 21. What is the number?

42. A certain number added to 14 is 25. Find the number.

43. Nine subtracted from a certain number is 13. Find the number.

44. A certain number subtracted from 32 is 15. What is the number?

45. Suppose that two items cost $43. If one of the items costs $25, what is the cost of the other item?

46. Eight years ago Rosa was 22 years old. Find Rosa's present age.

47. Six years from now, Nora will be 41 years old. What is her present age?

48. Chris bought eight pizzas for a total of $50. What was the price per pizza?

49. Chad worked 6 hours Saturday for a total of $39. How much per hour did he earn?

50. Jill worked 8 hours Saturday at $8.50 per hour. How much did she earn?

51. If 6 is added to three times a certain number, the result is 24. Find the number.

52. If 2 is subtracted from five times a certain number, the result is 38. Find the number.

53. Nineteen is 4 larger than three times a certain number. Find the number.

54. If nine times a certain number is subtracted from 7, the result is 52. Find the number.

55. Dress socks cost $2.50 a pair more than athletic socks. Randall purchased one pair of dress socks and six pairs of athletic socks for $21.75. Find the price of a pair of dress socks.

56. Together, a calculator and a mathematics textbook cost $85 in the college bookstore. The textbook price is $45 more than the price of the calculator. Find the price of the textbook.

57. The rainfall in June was 11.2 inches. This was 1 inch less than twice the rainfall in July. Find the amount of rainfall in inches for July.

58. Lunch at Joe's Hamburger Stand costs $1.75 less than lunch at Jodi's Taco Palace. A student spent his weekly lunch money, $24.50, eating four times at Jodi's and one time at Joe's. Find the cost of lunch at Jodi's Taco Palace.

59. If eight times a certain number is subtracted from 27, the result is 3. Find the number.

60. Twenty is 22 less than six times a certain number. Find the number.

61. A jeweler has priced a diamond ring at $550. This price represents $50 less than twice the cost of the ring to the jeweler. Find the cost of the ring to the jeweler.

62. Todd is following a 1750-calorie-per-day diet plan. This plan permits 650 calories less than twice the number of calories permitted by Lerae's diet plan. How many calories are permitted by Lerae's plan?

63. The length of a rectangular floor is 18 meters. This length is 2 meters less than five times the width of the floor. Find the width of the floor.

64. An executive is earning $145,000 per year. This is $15,000 less than twice her salary 4 years ago. Find her salary 4 years ago.

65. In the year 2000 it was estimated that there were 874 million speakers of the Chinese Mandarin language. This was 149 million less than three times the speakers of the English language. By this estimate how many million speakers of the English language were there in the year 2000?

66. A bill from the limousine company was $510. This included $150 for the service and $80 for each hour of use. Find the number of hours that the limousine was used.

67. Robin paid a $454 bill for a car DVD system. This included $379 for the DVD player and $60 an hour for installation. Find the number of hours it took to install the DVD system.

68. Tracy received a bill with cell phone use charges of $136.74. Included in the $136.74 was a charge of $39.99 for the monthly plan and a charge for 215 extra minutes. How much is Tracy being charged for each extra minute?

Thoughts Into Words

69. Give a step-by-step description of how you would solve the equation $17 = -3x + 2$.

70. What does the phrase "declare a variable" mean when solving a word problem?

71. Suppose that you are helping a friend with his homework, and he solves the equation $19 = 14 - x$ like this:

$$19 = 14 - x$$
$$19 + x = 14 - x + x$$
$$19 + x = 14$$
$$19 + x - 19 = 14 - 19$$
$$x = -5$$

The solution set is $\{-5\}$.

Does he have the correct solution set? What would you say to him about his method of solving the equation?

Answers to the Concept Quiz

1. False   **2.** True   **3.** False   **4.** True   **5.** True   **6.** C   **7.** A   **8.** D   **9.** E   **10.** B

| **3.3** | More on Solving Equations and Problem Solving |

OBJECTIVES

1 Solve first-degree equations by simplifying both sides and then applying properties of equality

2 Solve first-degree equations that are contradictions

3 Solve first-degree equations that are identities

4 Solve word problems that represent several quantities in terms of the same variable

5 Solve word problems involving geometric relationships

As equations become more complex, we need additional tools to solve them. So we need to organize our work carefully to minimize the chances for error. We will begin this section with some suggestions for solving equations, and then we will illustrate a *solution format* that is effective.

We can summarize the process of solving first-degree equations of one variable with the following three steps.

Step 1 Simplify both sides of the equation as much as possible.

Step 2 Use the addition or subtraction property of equality to isolate a term that contains the variable on one side and a constant on the other side of the equal sign.

Step 3 Use the multiplication or division property of equality to make the coefficient of the variable one.

The next examples illustrate this step-by-step process for solving equations. Study these examples carefully and be sure that you understand each step taken in the solution process.

Classroom Example
Solve $7m + 1 - 3m = 21$.

EXAMPLE 1   Solve $5y - 4 + 3y = 12$.

Solution

$$5y - 4 + 3y = 12$$
$$8y - 4 = 12 \qquad \text{Combine similar terms on the left side}$$
$$8y - 4 + 4 = 12 + 4 \qquad \text{Add 4 to both sides}$$
$$8y = 16$$
$$\frac{8y}{8} = \frac{16}{8} \qquad \text{Divide both sides by 8}$$
$$y = 2$$

The solution set is $\{2\}$. You can do the check alone now!

Classroom Example
Solve $8w + 5 = 2w - 3$.

EXAMPLE 2   Solve $7x - 2 = 3x + 9$.

Solution

Notice that both sides of the equation are in simplified form; thus we can begin by applying the subtraction property of equality.

$$7x - 2 = 3x + 9$$
$$7x - 2 - 3x = 3x + 9 - 3x \qquad \text{Subtract } 3x \text{ from both sides}$$
$$4x - 2 = 9$$
$$4x - 2 + 2 = 9 + 2 \qquad \text{Add 2 to both sides}$$
$$4x = 11$$
$$\frac{4x}{4} = \frac{11}{4} \qquad \text{Divide both sides by 4}$$
$$x = \frac{11}{4}$$

The solution set is $\left\{\dfrac{11}{4}\right\}$.

Classroom Example
Solve $3d + 1 = 4d - 3$.

EXAMPLE 3   Solve $5n + 12 = 9n - 16$.

Solution

$$5n + 12 = 9n - 16$$
$$5n + 12 - 9n = 9n - 16 - 9n \qquad \text{Subtract } 9n \text{ from both sides}$$
$$-4n + 12 = -16$$
$$-4n + 12 - 12 = -16 - 12 \qquad \text{Subtract 12 from both sides}$$

$$-4n = -28$$

$$\frac{-4n}{-4} = \frac{-28}{-4} \qquad \text{Divide both sides by } -4$$

$$n = 7$$

The solution set is $\{7\}$.

Classroom Example
Solve $2m + 5 = 2m - 3$.

EXAMPLE 4   Solve $3x + 8 = 3x - 2$.

Solution

$$3x + 8 = 3x - 2$$

$$3x + 8 - 3x = 3x - 2 - 3x \qquad \text{Subtract } 3x \text{ from both sides}$$

$$8 = -2 \qquad \text{False statement}$$

Since we obtained an equivalent equation that is a false statement, there is no value of x that will make the equation a true statement. When the equation is not true under any condition, then the equation is called a **contradiction**. The solution set for an equation that is a contradiction is the empty or null set, and it is symbolized by $\varnothing$.

Classroom Example
Solve $3n + 4 = 8n + 4 - 5n$.

EXAMPLE 5   Solve $4x + 6 - x = 3x + 6$.

Solution

$$4x + 6 - x = 3x + 6$$

$$3x + 6 = 3x + 6 \qquad \text{Combine similar terms on the left side}$$

$$3x + 6 - 3x = 3x + 6 - 3x \qquad \text{Subtract } 3x \text{ from both sides}$$

$$6 = 6 \qquad \text{True statement}$$

Since we obtained an equivalent equation that is a true statement, any value of x will make the equation a true statement. When an equation is true for any value of the variable, the equation is called an **identity**. The solution set for an equation that is an identity is the set of all real numbers. We will denote the set of all real numbers as $\{\text{all reals}\}$.

Word Problems

As we expand our skills for solving equations, we also expand our capabilities for solving word problems. No one definite procedure will ensure success at solving word problems, but the following suggestions can be helpful.

Suggestions for Solving Word Problems

1. Read the problem carefully, and make sure that you understand the meanings of all the words. Be especially alert for any technical terms in the statement of the problem.

2. Read the problem a second time (perhaps even a third time) to get an overview of the situation being described and to determine the known facts as well as what is to be found.

3. Sketch any figure, diagram, or chart that might be helpful in analyzing the problem.

(continued)

4. Choose a meaningful variable to represent an unknown quantity in the problem (perhaps *t* if time is an unknown quantity); represent any other unknowns in terms of that variable.

5. Look for a **guideline** that you can use to set up an equation. A guideline might be a formula such as *distance equals rate times time* or a statement of a relationship such as *the sum of the two numbers* is 28. A guideline may also be indicated by a figure or diagram that you sketch for a particular problem.

6. Form an equation that contains the variable and that translates the conditions of the guideline from English into algebra.

7. Solve the equation and use the solution to determine all facts requested in the problem.

8. **Check all answers by going back to the original statement of the problem and verifying that the answers make sense.**

If you decide not to check an answer, at least use the reasonableness-of-answer idea as a partial check. That is to say, ask yourself the question: Is this answer reasonable? For example, if the problem involves two investments that total $10,000, then an answer of $12,000 for one investment is certainly not reasonable.

Now let's consider some examples and use these suggestions as you work them out.

Consecutive Number Problems

Some problems involve consecutive numbers or consecutive even or odd numbers. For instance, 7, 8, 9, and 10 are consecutive numbers. To solve these applications, you must know how to represent consecutive numbers with variables. Let *n* represent the first number. For consecutive numbers, the next number is 1 more and is represented by $n + 1$. To continue, we add 1 to each preceding expression, obtaining the representations shown here:

$$
\begin{array}{cccc}
7 & 8 & 9 & 10 \\
\downarrow & \downarrow & \downarrow & \downarrow \\
n & n+1 & n+2 & n+3
\end{array}
$$

The pattern is somewhat different for consecutive even or odd numbers. For example, 2, 4, 6, and 8 are consecutive even numbers. Let *n* represent the first even number; then $n + 2$ represents the next even number. To continue, we add 2 to each preceding expression, obtaining these representations:

$$
\begin{array}{cccc}
2 & 4 & 6 & 8 \\
\downarrow & \downarrow & \downarrow & \downarrow \\
n & n+2 & n+4 & n+6
\end{array}
$$

Consecutive odd numbers have the same pattern of adding 2 to each preceding expression because consecutive odd numbers are two odd numbers with one and only one whole number between them.

Classroom Example
Find two consecutive even numbers whose sum is 42.

EXAMPLE 6   Find two consecutive even numbers whose sum is 74.

Solution

Let *n* represent the first number; then $n + 2$ represents the next even number. Since their sum is 74, we can set up and solve the following equation:

$$n + (n + 2) = 74$$
$$2n + 2 = 74$$

$$2n + 2 - 2 = 74 - 2$$
$$2n = 72$$
$$\frac{2n}{2} = \frac{72}{2}$$
$$n = 36$$

If $n = 36$, then $n + 2 = 38$; thus, the numbers are 36 and 38.

✓ Check

To check your answers for Example 6, determine whether the numbers satisfy the conditions stated in the original problem. Because 36 and 38 are two consecutive even numbers, and $36 + 38 = 74$ (their sum is 74), we know that the answers are correct.

The fifth entry in our list of problem-solving suggestions is to look for a *guideline* that can be used to set up an equation. The guideline may not be stated explicitly in the problem but may be implied by the nature of the problem. Consider the following example.

Classroom Example
Tyron sells appliances on a salary-plus-commission basis. He receives a monthly salary of $550 and a commission of $95 for each appliance that he sells. How many appliances must he sell in a month to earn a total monthly salary of $1,120.

EXAMPLE 7

Barry sells bicycles on a salary-plus-commission basis. He receives a weekly salary of $300 and a commission of $15 for each bicycle that he sells. How many bicycles must he sell in a week to earn a total weekly salary of $750?

Solution

Let b represent the number of bicycles to be sold in a week. Then $15b$ represents his commission for those bicycles. The guideline "fixed salary plus commission equals total weekly salary" generates the following equation:

Fixed salary + Commission = Total weekly salary

$$\$300 \quad + \quad 15b \quad = \quad \$750$$

Solving this equation yields

$$300 + 15b - 300 = 750 - 300$$
$$15b = 450$$
$$\frac{15b}{15} = \frac{450}{15}$$
$$b = 30$$

Barry must sell 30 bicycles per week. (Does this number check?)

Geometric Problems

Sometimes the guideline for setting up an equation to solve a problem is based on a geometric relationship. Several basic geometric relationships pertain to angle measure. Let's state some of these relationships and then consider some examples.

1. Two angles for which the sum of their measure is 90° (the symbol ° indicates degrees) are called **complementary angles**.

2. Two angles for which the sum of their measure is 180° are called **supplementary angles**.

3. The sum of the measures of the three angles of a triangle is 180°.

Classroom Example
One of two supplementary angles is 26° smaller than the other. Find the measure of each of the angles.

EXAMPLE 8

One of two complementary angles is 14° larger than the other. Find the measure of each of the angles.

Solution

If we let a represent the measure of the smaller angle, then $a + 14$ represents the measure of the larger angle. Since they are complementary angles, their sum is 90°, and we can proceed as follows:

$$a + (a + 14) = 90$$
$$2a + 14 = 90$$
$$2a + 14 - 14 = 90 - 14$$
$$2a = 76$$
$$\frac{2a}{2} = \frac{76}{2}$$
$$a = 38$$

If $a = 38$, then $a + 14 = 52$, and the angles measure 38° and 52°.

Classroom Example
Find the measures of the three angles of a triangle if the second angle is twice the first angle, and the third angle is half the second angle.

EXAMPLE 9

Find the measures of the three angles of a triangle if the second angle is three times the first angle, and the third angle is twice the second angle.

Solution

If we let a represent the measure of the smallest angle, then $3a$ and $2(3a)$ represent the measures of the other two angles. Therefore, we can set up and solve the following equation:

$$a + 3a + 2(3a) = 180$$
$$a + 3a + 6a = 180$$
$$10a = 180$$
$$\frac{10a}{10} = \frac{180}{10}$$
$$a = 18$$

If $a = 18$, then $3a = 54$ and $2(3a) = 108$. So the angles have measures of 18°, 54°, and 108°.

Concept Quiz 3.3

For Problems 1–10, answer true or false.

1. If n represents a whole number, then $n + 1$ would represent the next consecutive whole number.

2. If n represents an odd whole number, then $n + 1$ would represent the next consecutive odd whole number.

3. If n represents an even whole number, then $n + 2$ would represent the next consecutive even whole number.

4. The sum of the measures of two complementary angles is 90°.

5. The sum of the measures of two supplementary angles is 360°.

6. The sum of the measures of the three angles of a triangle is 120°.

7. When checking word problems, it is sufficient to check the solution in the equation.

8. For a word problem, the reasonableness of an answer is appropriate as a partial check.
9. For a conditional equation, the solution set is the set of all real numbers.
10. The solution set for an equation that is a contradiction is the null set.

Problem Set 3.3

For Problems 1–32, solve each equation. (Objectives 1–3)

1. $2x + 7 + 3x = 32$

2. $3x + 9 + 4x = 30$

3. $7x - 4 - 3x = -36$

4. $8x - 3 - 2x = -45$

5. $3y - 1 + 2y - 3 = 4$

6. $y + 3 + 2y - 4 = 6$

7. $5n - 2 - 8n = 31$

8. $6n - 1 - 10n = 51$

9. $-2n + 1 - 3n + n - 4 = 7$

10. $-n + 7 - 2n + 5n - 3 = -6$

11. $3x + 4 = 2x - 5$

12. $5x - 2 = 4x + 6$

13. $5x - 7 = 6x - 9$

14. $7x - 3 = 8x - 13$

15. $6x + 1 = 3x - 8$

16. $4x - 10 = x + 17$

17. $7y - 3 = 5y + 10$

18. $8y + 4 = 5y - 4$

19. $8n - 2 = 11n - 7$

20. $7n - 10 = 9n - 13$

21. $-2x - 7 = -3x + 10$

22. $-4x + 6 = -5x - 9$

23. $-3x + 5 = -5x - 8$

24. $-4x + 7 = -6x + 4$

25. $-7 - 6x = 9 - 9x$

26. $-10 - 7x = 14 - 12x$

27. $2x - 1 - x = 3x - 5$

28. $3x - 4 - 4x = 5 - 5x + 3x$

29. $5n - 4 - n = -3n - 6 + n$

30. $4x - 3 + 2x = 8x - 3 - x$

31. $-7 - 2n - 6n = 7n - 5n + 12$

32. $-3n + 6 + 5n = 7n - 8n - 9$

33. $7x - 3 = 4x - 3$

34. $-x - 4 + 3x = 2x - 7$

35. $-3x + 9 - 2x = -5x + 9$

36. $5x - 3 = 6x - 7 - x$

37. $7x + 4 = -x + 4 + 8x$

38. $3x - 2 - 5x = 7x - 2 - 5x$

39. $-8x + 9 = -8x + 5$

40. $-6x - 8 = 6x + 4$

Solve each word problem by setting up and solving an algebraic equation. (Objectives 4 and 5)

41. The sum of a number plus four times the number is 85. What is the number?

42. A number subtracted from three times the number yields 68. Find the number.

43. Find two consecutive odd numbers whose sum is 72.

44. Find two consecutive even numbers whose sum is 94.

45. Find three consecutive even numbers whose sum is 114.

46. Find three consecutive odd numbers whose sum is 159.

47. Two more than three times a certain number is the same as 4 less than seven times the number. Find the number.

48. One more than five times a certain number is equal to eleven less than nine times the number. What is the number?

49. The sum of a number and five times the number equals eighteen less than three times the number. Find the number.

50. One of two supplementary angles is five times as large as the other. Find the measure of each angle.

51. One of two complementary angles is 6° smaller than twice the other angle. Find the measure of each angle.

52. If two angles are complementary and the difference between their measures is 62°, find the measure of each angle.

53. If two angles are supplementary and the larger angle is 20° less than three times the smaller angle, find the measure of each angle.

54. Find the measures of the three angles of a triangle if the largest is 14° less than three times the smallest, and the other angle is 4° larger than the smallest.

55. One of the angles of a triangle has a measure of 40°. Find the measures of the other two angles if the difference between their measures is 10°.

56. Jesstan worked as a telemarketer on a salary-plus-commission basis. He was paid a salary of $300 a week and a $12 commission for each sale. If his earnings for the week were $960, how many sales did he make?

57. Marci sold an antique vase in an online auction for $69.00. This was $15 less than twice the amount she paid for it. What price did she pay for the vase?

58. A set of wheels sold in an online auction for $560. This was $35 more than three times the opening bid. How much was the opening bid?

59. Suppose that Bob is paid two times his normal hourly rate for each hour he works over 40 hours in a week. Last week he earned $504 for 48 hours of work. What is his hourly wage?

60. Last week on an algebra test, the highest grade was 9 points less than three times the lowest grade. The sum of the two grades was 135. Find the lowest and highest grades on the test.

61. At a university-sponsored concert, there were three times as many women as men in attendance. A total of 600 people attended the concert. How many men and how many women attended?

62. Suppose that a triangular lot is enclosed with 135 yards of fencing (see Figure 3.1). The longest side of the lot is 5 yards longer than twice the length of the shortest side. The other side is 10 yards longer than the shortest side. Find the lengths of the three sides of the lot.

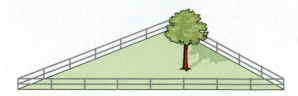

Figure 3.1

63. The textbook for a biology class cost $15 more than twice the cost of a used textbook for college algebra. If the cost of the two books together is $129, find the cost of the biology book.

64. A nutrition plan counts grams of fat, carbohydrates, and fiber. The grams of carbohydrates are to be 15 more than twice the grams of fat. The grams of fiber are to be three less than the grams of fat. If the grams of carbohydrate, fat, and fiber must total 48 grams for a dinner meal, how many grams of each would be in the meal?

65. At a local restaurant, $275 in tips is to be shared between the server, bartender, and busboy. The server gets $25 more than three times the amount the busboy receives. The bartender gets $50 more than the amount the busboy receives. How much will the server receive?

Thoughts Into Words

66. Suppose your friend solved the problem *find two consecutive odd integers whose sum is 28* as follows:

$$x + (x + 1) = 28$$
$$2x = 27$$
$$x = \frac{27}{2} = 13\frac{1}{2}$$

She claims that $13\frac{1}{2}$ will check in the equation. Where has she gone wrong, and how would you help her?

67. Give a step-by-step description of how you would solve the equation $3x + 4 = 5x - 2$.

Further Investigations

68. Make up an equation whose solution set is the null set and explain why.

69. Make up an equation whose solution set is the set of all real numbers and explain why.

Equations Involving Parentheses and Fractional Forms

OBJECTIVES

1 Solve first-degree equations that involve the use of the distributive property

2 Solve first-degree equations that involve fractional forms

3 Solve a variety of word problems involving first-degree equations

We will use the distributive property frequently in this section as we add to our techniques for solving equations. Recall that in symbolic form the distributive property states that $a(b + c) = ab + ac$. Consider the following examples, which illustrate the use of this property to remove parentheses. Pay special attention to the last two examples, which involve a negative number in front of the parentheses.

$$3(x + 2) = \qquad 3 \cdot x + 3 \cdot 2 \qquad = 3x + 6$$

$$5(y - 3) = \qquad 5 \cdot y - 5 \cdot 3 \qquad = 5y - 15 \qquad a(b - c) = ab - ac$$

$$2(4x + 7) = \qquad 2(4x) + 2(7) \qquad = 8x + 14$$

$$-1(n + 4) = \qquad (-1)(n) + (-1)(4) \qquad = -n - 4$$

$$-6(x - 2) = \qquad (-6)(x) - (-6)(2) \qquad = -6x + 12$$

↓

Do this step mentally!

It is often necessary to solve equations in which the variable is part of an expression enclosed in parentheses. We use the distributive property to remove the parentheses, and then we proceed in the usual way. Consider the next examples. (Notice that we are beginning to show only the major steps when solving an equation.)

Classroom Example
Solve $2(d - 5) = 5$.

EXAMPLE 1   Solve $3(x + 2) = 23$.

Solution

$$3(x + 2) = 23$$

$$3x + 6 = 23 \qquad \text{Applied distributive property to left side}$$

$$3x = 17 \qquad \text{Subtracted 6 from both sides}$$

$$x = \frac{17}{3} \qquad \text{Divided both sides by 3}$$

The solution set is $\left\{\dfrac{17}{3}\right\}$.

Classroom Example
Solve $3(m - 1) = 6(m + 2)$.

EXAMPLE 2   Solve $4(x + 3) = 2(x - 6)$.

Solution

$$4(x + 3) = 2(x - 6)$$

$$4x + 12 = 2x - 12 \qquad \text{Applied distributive property on each side}$$

$$2x + 12 = -12 \qquad \text{Subtracted } 2x \text{ from both sides}$$
$$2x = -24 \qquad \text{Subtracted 12 from both sides}$$
$$x = -12 \qquad \text{Divided both sides by 2}$$

The solution set is $\{-12\}$.

It may be necessary to remove more than one set of parentheses and then to use the distributive property again to combine similar terms. Consider the following two examples.

Classroom Example
Solve $2(t + 4) + 6(t + 5) = 32$.

EXAMPLE 3   Solve $5(w + 3) + 3(w + 1) = 14$.

Solution

$$5(w + 3) + 3(w + 1) = 14$$
$$5w + 15 + 3w + 3 = 14 \qquad \text{Applied distributive property}$$
$$8w + 18 = 14 \qquad \text{Combined similar terms}$$
$$8w = -4 \qquad \text{Subtracted 18 from both sides}$$
$$w = -\frac{4}{8} \qquad \text{Divided both sides by 8}$$
$$w = -\frac{1}{2} \qquad \text{Reduced}$$

The solution set is $\left\{-\dfrac{1}{2}\right\}$.

Classroom Example
Solve $4(w - 3) - 8(w - 2) = 9$.

EXAMPLE 4   Solve $6(x - 7) - 2(x - 4) = 13$.

Solution

$$6(x - 7) - 2(x - 4) = 13 \qquad \text{Be careful with this sign!}$$
$$6x - 42 - 2x + 8 = 13 \qquad \text{Distributive property}$$
$$4x - 34 = 13 \qquad \text{Combined similar terms}$$
$$4x = 47 \qquad \text{Added 34 to both sides}$$
$$x = \frac{47}{4} \qquad \text{Divided both sides by 4}$$

The solution set is $\left\{\dfrac{47}{4}\right\}$.

In a previous section, we solved equations like $x - \dfrac{2}{3} = \dfrac{3}{4}$ by adding $\dfrac{2}{3}$ to both sides. If an equation contains several fractions, then it is usually easier to clear the equation of all fractions by multiplying both sides by the least common denominator of all the denominators. Perhaps several examples will clarify this idea.

Classroom Example
Solve $\dfrac{1}{3}y + \dfrac{3}{4} = \dfrac{7}{12}$.

EXAMPLE 5   Solve $\dfrac{1}{2}x + \dfrac{2}{3} = \dfrac{5}{6}$.

Solution

$$\frac{1}{2}x + \frac{2}{3} = \frac{5}{6}$$

$$6\left(\frac{1}{2}x + \frac{2}{3}\right) = 6\left(\frac{5}{6}\right)$$ 6 is the LCD of 2, 3, and 6

$$6\left(\frac{1}{2}x\right) + 6\left(\frac{2}{3}\right) = 6\left(\frac{5}{6}\right)$$ Distributive property

$$3x + 4 = 5$$ Note how the equation has been *cleared of all fractions*

$$3x = 1$$

$$x = \frac{1}{3}$$

The solution set is $\left\{\frac{1}{3}\right\}$.

Classroom Example
Solve $\dfrac{5x}{12} - \dfrac{2}{3} = \dfrac{4}{9}$.

EXAMPLE 6   Solve $\dfrac{5n}{6} - \dfrac{1}{4} = \dfrac{3}{8}$.

Solution

$$\frac{5n}{6} - \frac{1}{4} = \frac{3}{8}$$

$$24\left(\frac{5n}{6} - \frac{1}{4}\right) = 24\left(\frac{3}{8}\right)$$ 24 is the LCD of 6, 4, and 8

$$24\left(\frac{5n}{6}\right) - 24\left(\frac{1}{4}\right) = 24\left(\frac{3}{8}\right)$$ Distributive property

$$20n - 6 = 9$$

$$20n = 15$$

$$n = \frac{15}{20} = \frac{3}{4}$$

The solution set is $\left\{\frac{3}{4}\right\}$.

We use many of the ideas presented in this section to help solve the equations in the next examples. Study the solutions carefully and be sure that you can supply reasons for each step. It might be helpful to cover up the solutions and try to solve the equations on your own.

Classroom Example
Solve $\dfrac{s + 5}{4} + \dfrac{s + 2}{5} = \dfrac{11}{20}$.

EXAMPLE 7   Solve $\dfrac{x + 3}{2} + \dfrac{x + 4}{5} = \dfrac{3}{10}$.

Solution

$$\frac{x + 3}{2} + \frac{x + 4}{5} = \frac{3}{10}$$

$$10\left(\frac{x + 3}{2} + \frac{x + 4}{5}\right) = 10\left(\frac{3}{10}\right)$$ 10 is the LCD of 2, 5, and 10

$$10\left(\frac{x + 3}{2}\right) + 10\left(\frac{x + 4}{5}\right) = 10\left(\frac{3}{10}\right)$$ Distributive property

$$5(x + 3) + 2(x + 4) = 3$$

$$5x + 15 + 2x + 8 = 3$$

$$7x + 23 = 3$$

$$7x = -20$$

$$x = -\frac{20}{7}$$

The solution set is $\left\{-\frac{20}{7}\right\}$.

Classroom Example
Solve $\dfrac{v-5}{4} - \dfrac{v-3}{5} = \dfrac{1}{2}$.

EXAMPLE 8   Solve $\dfrac{x-1}{4} - \dfrac{x-2}{6} = \dfrac{2}{3}$.

Solution

$$\frac{x-1}{4} - \frac{x-2}{6} = \frac{2}{3}$$

$$12\left(\frac{x-1}{4} - \frac{x-2}{6}\right) = 12\left(\frac{2}{3}\right) \qquad \text{12 is the LCD of 4, 6, and 3}$$

$$12\left(\frac{x-1}{4}\right) - 12\left(\frac{x-2}{6}\right) = 12\left(\frac{2}{3}\right) \qquad \text{Distributive property}$$

$$3(x-1) - 2(x-2) = 8$$

$$3x - 3 - 2x + 4 = 8 \qquad \text{Be careful with this sign!}$$

$$x + 1 = 8$$

$$x = 7$$

The solution set is $\{7\}$.

Word Problems

We are now ready to solve some word problems using equations of the different types presented in this section. Again, it might be helpful for you to attempt to solve the problems on your own before looking at the book's approach.

Classroom Example
Ian has 23 coins (dimes and nickels) that amount to $1.45. How many coins of each kind does he have?

EXAMPLE 9

Loretta has 19 coins (quarters and nickels) that amount to $2.35. How many coins of each kind does she have?

Solution

Let q represent the number of quarters. Then $19 - q$ represents the number of nickels. We can use the following guideline to help set up an equation:

Value of quarters in cents + Value of nickels in cents = Total value in cents

$$25q \qquad + \qquad 5(19 - q) \qquad = \qquad 235$$

Solving the equation, we obtain

$$25q + 95 - 5q = 235$$

$$20q + 95 = 235$$

$$20q = 140$$

$$q = 7$$

If $q = 7$, then $19 - q = 12$, so she has 7 quarters and 12 nickels.

Classroom Example
Find a number such that 6 more than three-fourths the number is equal to two-thirds the number.

EXAMPLE 10

Find a number such that 4 less than two-thirds of the number is equal to one-sixth of the number.

Solution

Let n represent the number. Then $\frac{2}{3}n - 4$ represents 4 less than two-thirds of the number,

and $\frac{1}{6}n$ represents one-sixth of the number.

$$\frac{2}{3}n - 4 = \frac{1}{6}n$$

$$6\left(\frac{2}{3}n - 4\right) = 6\left(\frac{1}{6}n\right)$$

$$4n - 24 = n$$

$$3n - 24 = 0$$

$$3n = 24$$

$$n = 8$$

The number is 8.

Classroom Example
John is paid $1\frac{1}{2}$ times his normal hourly rate for each hour he works over 40 hours in a week. Last week he worked 48 hours and earned $962. What is his normal hourly rate?

EXAMPLE 11

Lance is paid $1\frac{1}{2}$ times his normal hourly rate for each hour he works over 40 hours in a week. Last week he worked 50 hours and earned $462. What is his normal hourly rate?

Solution

Let x represent Lance's normal hourly rate. Then $\frac{3}{2}x$ represents $1\frac{1}{2}$ times his normal hourly rate. We can use the following guideline to set up the equation:

Regular wages for first 40 hours + Wages for 10 hours of overtime = Total wages

$$40x \quad + \quad 10\left(\frac{3}{2}x\right) \quad = \quad 462$$

Solving this equation, we obtain

$$40x + 15x = 462$$

$$55x = 462$$

$$x = 8.40$$

Lance's normal hourly rate is $8.40.

Classroom Example
Find two consecutive whole numbers such that the sum of the first plus four times the second is 179.

EXAMPLE 12

Find three consecutive whole numbers such that the sum of the first plus twice the second plus three times the third is 134.

Solution

Let n represent the first whole number. Then $n + 1$ represents the second whole number and $n + 2$ represents the third whole number.

$$n + 2(n + 1) + 3(n + 2) = 134$$
$$n + 2n + 2 + 3n + 6 = 134$$
$$6n + 8 = 134$$
$$6n = 126$$
$$n = 21$$

The numbers are 21, 22, and 23.

Keep in mind that the problem-solving suggestions we offered in Section 3.3 simply outline a general algebraic approach to solving problems. You will add to this list throughout this course and in any subsequent mathematics courses that you take. Furthermore, you will be able to pick up additional problem-solving ideas from your instructor and from fellow classmates as problems are discussed in class. Always be on the alert for any ideas that might help you become a better problem solver.

Concept Quiz 3.4

For Problems 1–10, answer true or false.

1. To solve an equation of the form $a(x + b) = 14$, the associative property would be applied to remove the parentheses.

2. Multiplying both sides of an equation by the common denominator of all fractions in the equation clears the equation of all fractions.

3. If Jack has 15 coins (dimes and quarters), and x represents the number of dimes, then $x - 15$ represents the number of quarters.

4. The equation $3(x + 1) = 3x + 3$ has an infinite number of solutions.

5. The equation $2x = 0$ has no solution.

6. The equation $4x + 5 = 4x + 3$ has no solution.

7. The solution set for the equation $3(2x - 1) = 2x - 3$ is $\{0\}$.

8. The solution set for the equation $5(3x + 2) = 4(2x - 1)$ is $\{2\}$.

9. The answer for a word problem must be checked back into the statement of the problem.

10. If q represents the number of quarters, then $25q$ represents the value of quarters in cents.

Problem Set 3.4

For Problems 1–60, solve each equation. **(Objectives 1 and 2)**

1. $7(x + 2) = 21$
2. $4(x + 4) = 24$
3. $5(x - 3) = 35$
4. $6(x - 2) = 18$
5. $-3(x + 5) = 12$
6. $-5(x - 6) = -15$
7. $4(n - 6) = 5$
8. $3(n + 4) = 7$
9. $6(n + 7) = 8$
10. $8(n - 3) = 12$
11. $-10 = -5(t - 8)$
12. $-16 = -4(t + 7)$
13. $5(x - 4) = 4(x + 6)$
14. $6(x - 4) = 3(2x + 5)$
15. $8(x + 1) = 9(x - 2)$
16. $4(x - 7) = 5(x + 2)$
17. $8(t + 5) = 4(2t + 10)$
18. $7(t - 5) = 5(t + 3)$
19. $2(6t + 1) = 4(3t - 1)$
20. $6(t + 5) = 2(3t + 15)$
21. $-2(x - 6) = -(x - 9)$
22. $-(x + 7) = -2(x + 10)$
23. $-3(t - 4) - 2(t + 4) = 9$
24. $5(t - 4) - 3(t - 2) = 12$
25. $3(n - 10) - 5(n + 12) = -86$

26. $4(n + 9) - 7(n - 8) = 83$

27. $3(x + 1) + 4(2x - 1) = 5(2x + 3)$

28. $4(x - 1) + 5(x + 2) = 3(x - 8)$

29. $-(x + 2) + 2(x - 3) = -2(x - 7)$

30. $-2(x + 6) + 3(3x - 2) = -3(x - 4)$

31. $5(2x - 1) - (3x + 4) = 4(x + 3) - 27$

32. $3(4x + 1) - 2(2x + 1) = -2(x - 1) - 1$

33. $-(a - 1) - (3a - 2) = 6 + 2(a - 1)$

34. $3(2a - 1) - 2(5a + 1) = 4(3a + 4)$

35. $3(x - 1) + 2(x - 3) = -4(x - 2) + 10(x + 4)$

36. $-2(x - 4) - (3x - 2) = -2 + (-6x + 2)$

37. $3 - 7(x - 1) = 9 - 6(2x + 1)$

38. $8 - 5(2x + 1) = 2 - 6(x - 3)$

39. $\dfrac{3}{4}x - \dfrac{2}{3} = \dfrac{5}{6}$

40. $\dfrac{1}{2}x - \dfrac{4}{3} = -\dfrac{5}{6}$

41. $\dfrac{5}{6}x + \dfrac{1}{4} = -\dfrac{9}{4}$

42. $\dfrac{3}{8}x + \dfrac{1}{6} = -\dfrac{7}{12}$

43. $\dfrac{1}{2}x - \dfrac{3}{5} = \dfrac{3}{4}$

44. $\dfrac{1}{4}x - \dfrac{2}{5} = \dfrac{5}{6}$

45. $\dfrac{n}{3} + \dfrac{5n}{6} = \dfrac{1}{8}$

46. $\dfrac{n}{6} + \dfrac{3n}{8} = \dfrac{5}{12}$

47. $\dfrac{5y}{6} - \dfrac{3}{5} = \dfrac{2y}{3}$

48. $\dfrac{3y}{7} + \dfrac{1}{2} = \dfrac{y}{4}$

49. $\dfrac{h}{6} + \dfrac{h}{8} = 1$

50. $\dfrac{h}{4} + \dfrac{h}{3} = 1$

51. $\dfrac{x + 2}{3} + \dfrac{x + 3}{4} = \dfrac{13}{3}$

52. $\dfrac{x - 1}{4} + \dfrac{x + 2}{5} = \dfrac{39}{20}$

53. $\dfrac{x - 1}{5} - \dfrac{x + 4}{6} = -\dfrac{13}{15}$

54. $\dfrac{x + 1}{7} - \dfrac{x - 3}{5} = \dfrac{4}{5}$

55. $\dfrac{x + 8}{2} - \dfrac{x + 10}{7} = \dfrac{3}{4}$

56. $\dfrac{x + 7}{3} - \dfrac{x + 9}{6} = \dfrac{5}{9}$

57. $\dfrac{x - 2}{8} - 1 = \dfrac{x + 1}{4}$

58. $\dfrac{x - 4}{2} + 3 = \dfrac{x - 2}{4}$

59. $\dfrac{x + 1}{4} = \dfrac{x - 3}{6} + 2$

60. $\dfrac{x + 3}{5} = \dfrac{x - 6}{2} + 1$

Solve each word problem by setting up and solving an appropriate algebraic equation. **(Objective 3)**

61. Find two consecutive whole numbers such that the smaller number plus four times the larger number equals 39.

62. Find two consecutive whole numbers such that the smaller number subtracted from five times the larger number equals 57.

63. Find three consecutive whole numbers such that twice the sum of the two smallest numbers is 10 more than three times the largest number.

64. Find four consecutive whole numbers such that the sum of the first three numbers equals the fourth number.

65. The sum of two numbers is 17. If twice the smaller number is 1 more than the larger number, find the numbers.

66. The sum of two numbers is 53. If three times the smaller number is 1 less than the larger number, find the numbers.

67. Find a number such that 20 more than one-third of the number equals three-fourths of the number.

68. The sum of three-eighths of a number and five-sixths of the same number is 29. Find the number.

69. Mrs. Nelson had to wait 4 minutes in line at her bank's automated teller machine. This was 3 minutes less than one-half of the time she waited in line at the grocery store. How long in minutes did she wait in line at the grocery store?

70. Raoul received a $30 tip for waiting on a large party. This was $5 more than one-fourth of the tip the head-waiter received. How much did the headwaiter receive for a tip?

71. Suppose that a board 20 feet long is cut into two pieces. Four times the length of the shorter piece is 4 feet less than three times the length of the longer piece. Find the length of each piece.

72. Ellen is paid time and a half for each hour over 40 hours she works in a week. Last week she worked 44 hours and earned $391. What is her normal hourly rate?

73. Lucy has 35 coins consisting of nickels and quarters amounting to $5.75. How many coins of each kind does she have?

74. Suppose that Julian has 44 coins consisting of pennies and nickels. If the number of nickels is two more than twice the number of pennies, find the number of coins of each kind.

75. Max has a collection of 210 coins consisting of nickels, dimes, and quarters. He has twice as many dimes as nickels, and 10 more quarters than dimes. How many coins of each kind does he have?

76. Ginny has a collection of 425 coins consisting of pennies, nickels, and dimes. She has 50 more nickels than pennies and 25 more dimes than nickels. How many coins of each kind does she have?

77. Maida has 18 coins consisting of dimes and quarters amounting to $3.30. How many coins of each kind does she have?

78. Ike has some nickels and dimes amounting to $2.90. The number of dimes is one less than twice the number of nickels. How many coins of each kind does he have?

79. Mario has a collection of 22 specimens in his aquarium consisting of crabs, fish, and plants. There are three times as many fish as crabs. There are two more plants than crabs. How many specimens of each kind are in the collection?

80. Tickets for a concert were priced at $8 for students and $10 for nonstudents. There were 1500 tickets sold for a total of $12,500. How many student tickets were sold?

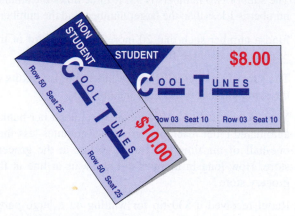

Figure 3.2

81. The supplement of an angle is 30° larger than twice its complement. Find the measure of the angle.

82. The sum of the measure of an angle and three times its complement is 202°. Find the measure of the angle.

83. In triangle *ABC*, the measure of angle *A* is 2° less than one-fifth of the measure of angle *C*. The measure of angle *B* is 5° less than one-half of the measure of angle *C*. Find the measures of the three angles of the triangle.

84. If one-fourth of the complement of an angle plus one-fifth of the supplement of the angle equals 36°, find the measure of the angle.

85. The supplement of an angle is 10° smaller than three times its complement. Find the size of the angle.

86. In triangle *ABC*, the measure of angle *C* is eight times the measure of angle *A*, and the measure of angle *B* is 10° more than the measure of angle *C*. Find the measure of each angle of the triangle.

Thoughts Into Words

87. Discuss how you would solve the equation

$3(x - 2) - 5(x + 3) = -4(x + 9)$.

88. Why must potential answers to word problems be checked back into the original statement of the problem?

89. Consider these two solutions:

$$3(x + 2) = 9$$
$$\frac{3(x + 2)}{3} = \frac{9}{3}$$
$$x + 2 = 3$$
$$x = 1$$

$$3(x - 4) = 7$$
$$3x - 12 = 7$$
$$3x = 19$$
$$x = \frac{19}{3}$$

Are both of these solutions correct? Comment on the effectiveness of the two different approaches.

Further Investigations

90. Solve each equation.

(a) $-2(x - 1) = -2x + 2$

(b) $3(x + 4) = 3x - 4$

(c) $5(x - 1) = -5x - 5$

(d) $\frac{x - 3}{3} + 4 = 3$

(e) $\frac{x + 2}{3} + 1 = \frac{x - 2}{3}$

(f) $\frac{x - 1}{5} - 2 = \frac{x - 11}{5}$

(g) $4(x - 2) - 2(x + 3) = 2(x + 6)$

(h) $5(x + 3) - 3(x - 5) = 2(x + 15)$

(i) $7(x - 1) + 4(x - 2) = 15(x - 1)$

91. Find three consecutive integers such that the sum of the smallest integer and the largest integer is equal to twice the middle integer.

Answers to the Concept Quiz

1. False   **2.** True   **3.** False   **4.** True   **5.** False   **6.** True   **7.** True   **8.** False   **9.** True   **10.** True

3.5 Inequalities

OBJECTIVES

1 Solve first-degree inequalities

2 Write the solution set of an inequality in set-builder notation or interval notation

3 Graph the solution set of an inequality

Just as we use the symbol = to represent is *equal to*, we use the symbols < and > to represent *is less than* and *is greater than*, respectively. Here are some examples of **statements of inequality**. Notice that the first four are true statements and the last two are false.

$$6 + 4 > 7 \qquad \text{True}$$
$$8 - 2 < 14 \qquad \text{True}$$
$$4 \cdot 8 > 4 \cdot 6 \qquad \text{True}$$
$$5 \cdot 2 < 5 \cdot 7 \qquad \text{True}$$
$$5 + 8 > 19 \qquad \text{False}$$
$$9 - 2 < 3 \qquad \text{False}$$

Algebraic inequalities contain one or more variables. These are examples of algebraic inequalities:

$$x + 3 > 4$$
$$2x - 1 < 6$$
$$x^2 + 2x - 1 > 0$$
$$2x + 3y < 7$$
$$7ab < 9$$

An algebraic inequality such as $x + 1 > 2$ is neither true nor false as it stands; it is called an **open sentence**. Each time a number is substituted for x, the algebraic inequality $x + 1 > 2$ becomes a numerical statement that is either true or false. For example, if $x = 0$, then $x + 1 > 2$ becomes $0 + 1 > 2$, which is false. If $x = 2$, then $x + 1 > 2$ becomes $2 + 1 > 2$, which is true. **Solving an inequality** refers to the process of finding the numbers that make an algebraic inequality a true numerical statement. We say that such numbers, called the **solutions of the inequality**, satisfy the inequality. The set of all solutions of an inequality is called its **solution set**. We often state solution sets for inequalities with set builder notation. For example, the solution set for $x + 1 > 2$ is the set of real numbers greater than 1, expressed as $\{x \mid x > 1\}$. The **set builder notation** $\{x \mid x > 1\}$ is read as "the set of all x such that x is greater than 1." We sometimes graph solution sets for inequalities on a number line; the solution set for $\{x \mid x > 1\}$ is pictured in Figure 3.3.

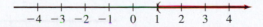

Figure 3.3

The left-hand parenthesis at 1 indicates that 1 is *not* a solution, and the red part of the line to the right of 1 indicates that all real numbers greater than 1 are solutions. We refer to the red portion of the number line as the *graph* of the solution set $\{x \mid x > 1\}$.

The solution set for $x + 1 \leq 3$ ($\leq$ is read "less than or equal to") is the set of real numbers less than or equal to 2, expressed as $\{x \mid x \leq 2\}$. The graph of the solution set for $\{x \mid x \leq 2\}$ is pictured in Figure 3.4. The right-hand bracket at 2 indicates that 2 *is included* in the solution set.

Figure 3.4

It is convenient to express solution sets of inequalities using **interval notation**. The solution set $\{x \mid x > 6\}$ is written as $(6, \infty)$ using interval notation. In interval notation, parentheses are used to indicate exclusion of the endpoint. The $>$ and $<$ symbols in inequalities also indicate the exclusion of the endpoint. So when the inequality has a $>$ or $<$ symbol, the interval notation uses a parenthesis. This is consistent with the use of parentheses on the number line.

In this same example, $\{x \mid x > 6\}$, the solution set has no upper endpoint, so the infinity symbol, ∞, is used to indicate that the interval continues indefinitely. The solution set for $\{x \mid x < 3\}$ is written as $(-\infty, 3)$ in interval notation. Here the solution set has no lower endpoint, so a negative sign precedes the infinity symbol because the interval is extending indefinitely in the opposite direction. The infinity symbol always has a parenthesis in interval notation because there is no actual endpoint to include.

The solution set $\{x \mid x \geq 5\}$ is written as $[5, \infty)$ using interval notation. In interval notation square brackets are used to indicate inclusion of the endpoint. The $\geq$ and $\leq$ symbols in inequalities also indicate the inclusion of the endpoint. So when the inequality has a $\geq$ or $\leq$ symbol, the interval notation uses a square bracket. Again the use of a bracket in interval notation is consistent with the use of a bracket on the number line.

The examples in the table below contain some simple algebraic inequalities, their solution sets, graphs of the solution sets, and the solution sets written in interval notation. Look them over very carefully to be sure you understand the symbols.

Algebraic inequality	Solution set	Graph of solution set	Interval notation
$x < 2$	$\{x \mid x < 2\}$	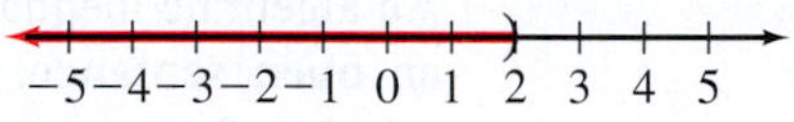	$(-\infty, 2)$
$x > -1$	$\{x \mid x > -1\}$	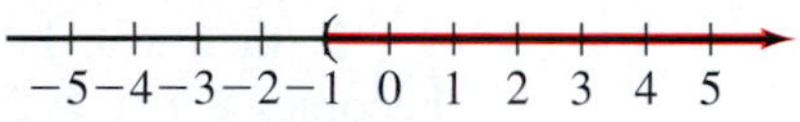	$(-1, \infty)$
$3 < x$	$\{x \mid x > 3\}$	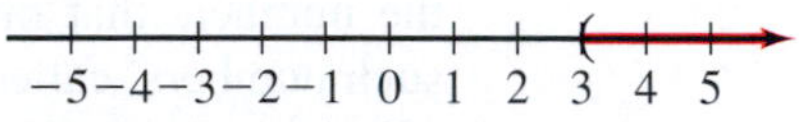	$(3, \infty)$
$x \geq 1$ ($\geq$ is read "greater than or equal to")	$\{x \mid x \geq 1\}$	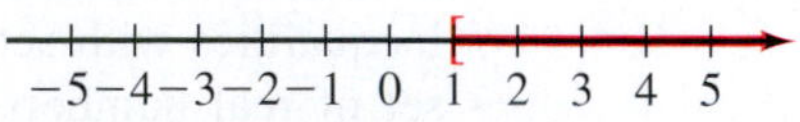	$[1, \infty)$
$x \leq 2$ ($\leq$ is read "less than or equal to")	$\{x \mid x \leq 2\}$	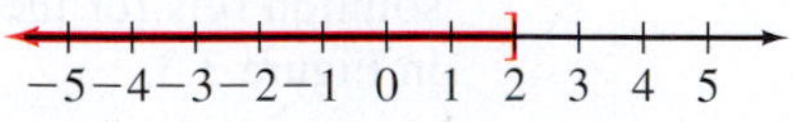	$(-\infty, 2]$
$1 \geq x$	$\{x \mid x \leq 1\}$	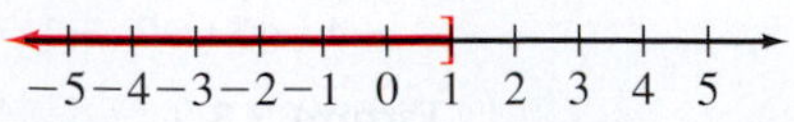	$(-\infty, 1]$

Figure 3.5

The general process for solving inequalities closely parallels that for solving equations. We continue to replace the given inequality with equivalent, but simpler inequalities. For example,

$$2x + 1 > 9 \qquad (1)$$

$$2x > 8 \qquad (2)$$

$$x > 4 \qquad (3)$$

are all equivalent inequalities; that is, they have the same solutions. Thus to solve inequality (1), we can solve inequality (3), which is obviously all numbers greater than 4. The exact procedure for simplifying inequalities is based primarily on two properties, and they become our topics of discussion at this time. The first of these is the **addition-subtraction property of inequality**.

Property 3.4 Addition-Subtraction Property of Inequality

For all real numbers a, b, and c,

1. $a > b$ if and only if $a + c > b + c$.
2. $a > b$ if and only if $a - c > b - c$.

Property 3.4 states that any number can be added to or subtracted from both sides of an inequality, and an equivalent inequality is produced. The property is stated in terms of $>$, but analogous properties exist for $<$, $\geq$, and $\leq$. Consider the use of this property in the next three examples.

Classroom Example
Solve $x - 5 < -3$ and graph the solutions.

EXAMPLE 1 Solve $x - 3 > -1$ and graph the solutions.

Solution

$$x - 3 > -1$$

$$x - 3 + 3 > -1 + 3 \qquad \text{Add 3 to both sides}$$

$$x > 2$$

The solution set is $\{x|x > 2\}$, and it can be graphed as shown in Figure 3.6. The solution, written in interval notation, is $(2, \infty)$.

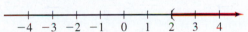

Figure 3.6

Classroom Example
Solve $x + 7 \geq 10$ and graph the solutions.

EXAMPLE 2 Solve $x + 4 \leq 5$ and graph the solutions.

Solution

$$x + 4 \leq 5$$

$$x + 4 - 4 \leq 5 - 4 \qquad \text{Subtract 4 from both sides}$$

$$x \leq 1$$

The solution set is $\{x|x \leq 1\}$, and it can be graphed as shown in Figure 3.7. The solution, written in interval notation, is $(-\infty, 1]$.

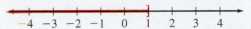

Figure 3.7

Classroom Example
Solve $3 < 8 + x$ and graph the solutions.

EXAMPLE 3 Solve $5 > 6 + x$ and graph the solutions.

Solution

$$5 > 6 + x$$

$$5 - 6 > 6 + x - 6 \quad \text{Subtract 6 from both sides}$$

$$-1 > x$$

Because $-1 > x$ is equivalent to $x < -1$, the solution set is $\{x \mid x < -1\}$, and it can be graphed as shown in Figure 3.8. The solution, written in interval notation, is $(-\infty, -1)$.

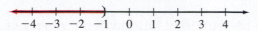

Figure 3.8

Now let's look at some numerical examples to see what happens when both sides of an inequality are multiplied or divided by some number.

$$4 > 3 \rightarrow 5(4) > 5(3) \rightarrow 20 > 15$$

$$-2 > -3 \rightarrow 4(-2) > 4(-3) \rightarrow -8 > -12$$

$$6 > 4 \rightarrow \frac{6}{2} > \frac{4}{2} \rightarrow 3 > 2$$

$$8 > -2 \rightarrow \frac{8}{4} > \frac{-2}{4} \rightarrow 2 > -\frac{1}{2}$$

Notice that multiplying or dividing both sides of an inequality by a positive number produces an inequality of the same sense. This means that if the original inequality is *greater than*, then the new inequality is *greater than*, and if the original is *less than*, then the resulting inequality is *less than*.

Now note what happens when we multiply or divide both sides by a negative number:

$$3 < 5 \rightarrow -2(3) > -2(5) \rightarrow -6 > -10$$

$$-4 < 1 \rightarrow -5(-4) > -5(1) \rightarrow 20 > -5$$

$$14 > 2 \rightarrow \frac{14}{-2} < \frac{2}{-2} \rightarrow -7 < -1$$

$$-3 > -6 \rightarrow \frac{-3}{-3} < \frac{-6}{-3} \rightarrow 1 < 2$$

Multiplying or dividing both sides of an inequality by a negative number *reverses the sense of the inequality*. Property 3.5 summarizes these ideas.

Property 3.5 Multiplication-Division Property of Inequality

(a) For all real numbers, a, b, and c, with $c > 0$,

1. $a > b$ if and only if $ac > bc$.

2. $a > b$ if and only if $\frac{a}{c} > \frac{b}{c}$.

(b) For all real numbers, a, b, and c, with $c < 0$,

1. $a > b$ if and only if $ac < bc$.

2. $a > b$ if and only if $\frac{a}{c} < \frac{b}{c}$.

Similar properties hold if each inequality is reversed or if $>$ is replaced with $\geq$, and $<$ is replaced with $\leq$. For example, if $a \leq b$ and $c < 0$, then $ac \geq bc$ and $\dfrac{a}{c} \geq \dfrac{b}{c}$.

Observe the use of Property 3.5 in the next three examples.

Classroom Example
Solve $3x < 9$.

EXAMPLE 4   Solve $2x > 4$.

Solution

$$2x > 4$$

$$\frac{2x}{2} > \frac{4}{2} \qquad \text{Divide both sides by 2}$$

$$x > 2$$

The solution set is $\{x \mid x > 2\}$ or $(2, \infty)$ in interval notation.

Classroom Example
Solve $\dfrac{5}{6}x \geq \dfrac{2}{3}$.

EXAMPLE 5   Solve $\dfrac{3}{4}x \leq \dfrac{1}{5}$.

Solution

$$\frac{3}{4}x \leq \frac{1}{5}$$

$$\frac{4}{3}\left(\frac{3}{4}x\right) \leq \frac{4}{3}\left(\frac{1}{5}\right) \qquad \text{Multiply both sides by } \frac{4}{3}$$

$$x \leq \frac{4}{15}$$

The solution set is $\left\{x \mid x \leq \dfrac{4}{15}\right\}$ or $\left(-\infty, \dfrac{4}{15}\right]$ in interval notation.

Classroom Example
Solve $-2x < 4$.

EXAMPLE 6   Solve $-3x > 9$.

Solution

$$-3x > 9$$

$$\frac{-3x}{-3} < \frac{9}{-3} \qquad \text{Divide both sides by } -3 \text{, which reverses the inequality}$$

$$x < -3$$

The solution set is $\{x \mid x < -3\}$ or $(-\infty, -3)$ in interval notation.

As we mentioned earlier, many of the same techniques used to solve equations may be used to solve inequalities. However, you must be extremely careful when you apply Property 3.5. Study the next examples and notice the similarities between solving equations and solving inequalities.

Classroom Example
Solve $8x - 2 < 14$.

EXAMPLE 7   Solve $4x - 3 > 9$.

Solution

$$4x - 3 > 9$$

$$4x - 3 + 3 > 9 + 3 \qquad \text{Add 3 to both sides}$$

$$4x > 12$$

$$\frac{4x}{4} > \frac{12}{4} \qquad \text{Divide both sides by 4}$$

$$x > 3$$

The solution set is $\{x \mid x > 3\}$ or $(3, \infty)$ in interval notation.

Classroom Example
Solve $-5x + 3 < 18$.

EXAMPLE 8   Solve $-3n + 5 < 11$.

Solution

$$-3n + 5 < 11$$

$$-3n + 5 - 5 < 11 - 5 \qquad \text{Subtract 5 from both sides}$$

$$-3n < 6$$

$$\frac{-3n}{-3} > \frac{6}{-3} \qquad \begin{array}{l}\text{Divide both sides by } -3, \text{ which reverses} \\ \text{the inequality}\end{array}$$

$$n > -2$$

The solution set is $\{n \mid n > -2\}$ or $(-2, \infty)$ in interval notation.

Checking the solutions for an inequality presents a problem. Obviously we cannot check all of the infinitely many solutions for a particular inequality. However, by checking at least one solution, especially when the multiplication-division property is used, we might catch the common mistake of forgetting to reverse the sense of the inequality. In Example 8 we are claiming that all numbers greater than -2 will satisfy the original inequality. Let's check one such number in the original inequality; we will check -1.

$$-3n + 5 < 11$$

$$-3(-1) + 5 \overset{?}{<} 11$$

$$3 + 5 \overset{?}{<} 11$$

$$8 < 11$$

Thus -1 satisfies the original inequality. If we had forgotten to reverse the sense of the inequality when we divided both sides by -3, our answer would have been $n < -2$, and the check would have detected the error.

Concept Quiz 3.5

For Problems 1–10, answer true or false.

1. Numerical statements of inequality are always true.

2. The algebraic statement $x + 4 > 6$ is called an open sentence.

3. The algebraic inequality $2x > 10$ has one solution.

4. The algebraic inequality $x < 3$ has an infinite number of solutions.

5. The set-builder notation $\{x \mid x < -5\}$ is read "the set of variables that are particular to $x < -5$."

6. When graphing the solution set of an inequality, a square bracket is used to include the endpoint.

7. The solution set of the inequality $x \geq 4$ is written $(4, \infty)$.

8. The solution set of the inequality $x < -5$ is written $(-\infty, -5)$.

9. When multiplying both sides of an inequality by a negative number, the sense of the inequality stays the same.

10. When adding a negative number to both sides of an inequality, the sense of the inequality stays the same.

Problem Set 3.5

For Problems 1–10, determine whether each numerical inequality is true or false. (Objective 1)

1. $2(3) - 4(5) < 5(3) - 2(-1) + 4$

2. $5 + 6(-3) - 8(-4) > 17$

3. $\frac{2}{3} - \frac{3}{4} + \frac{1}{6} > \frac{1}{5} + \frac{3}{4} - \frac{7}{10}$

4. $\frac{1}{2} + \frac{1}{3} < \frac{1}{3} + \frac{1}{4}$

5. $\left(-\frac{1}{2}\right)\left(\frac{4}{9}\right) > \left(\frac{3}{5}\right)\left(-\frac{1}{3}\right)$

6. $\left(\frac{5}{6}\right)\left(\frac{8}{12}\right) < \left(\frac{3}{7}\right)\left(\frac{14}{15}\right)$

7. $\frac{3}{4} + \frac{2}{3} \div \frac{1}{5} > \frac{2}{3} + \frac{1}{2} \div \frac{3}{4}$

8. $1.9 - 2.6 - 3.4 < 2.5 - 1.6 - 4.2$

9. $0.16 + 0.34 > 0.23 + 0.17$

10. $(0.6)(1.4) > (0.9)(1.2)$

For Problems 11–22, state the solution set and graph it on a number line. (Objectives 2 and 3)

11. $x > -2$

12. $x > -4$

13. $x \leq 3$

14. $x \leq 0$

15. $2 < x$

16. $-3 \leq x$

17. $-2 \geq x$

18. $1 > x$

19. $-x > 1$

20. $-x < 2$

21. $-2 < -x$

22. $-1 > -x$

For Problems 23–60, solve each inequality. (Objective 1)

23. $x + 6 < -14$

24. $x + 7 > -15$

25. $x - 4 \geq -13$

26. $x - 3 \leq -12$

27. $4x > 36$

28. $3x < 51$

29. $6x < 20$

30. $8x > 28$

31. $-5x > 40$

32. $-4x < 24$

33. $-7n \leq -56$

34. $-9n \geq -63$

35. $48 > -14n$

36. $36 < -8n$

37. $16 < 9 + n$

38. $19 > 27 + n$

39. $3x + 2 > 17$

40. $2x + 5 < 19$

41. $4x - 3 \leq 21$

42. $5x - 2 \geq 28$

43. $-2x - 1 \geq 41$

44. $-3x - 1 \leq 35$

45. $6x + 2 < 18$

46. $8x + 3 > 25$

47. $3 > 4x - 2$

48. $7 < 6x - 3$

49. $-2 < -3x + 1$

50. $-6 > -2x + 4$

51. $-38 \geq -9t - 2$

52. $36 \geq -7t + 1$

53. $5x - 4 - 3x > 24$

54. $7x - 8 - 5x < 38$

55. $4x + 2 - 6x < -1$

56. $6x + 3 - 8x > -3$

57. $-5 \geq 3t - 4 - 7t$

58. $6 \leq 4t - 7t - 10$

59. $-x - 4 - 3x > 5$

60. $-3 - x - 3x < 10$

Thoughts Into Words

61. Do the greater-than and less-than relationships possess the symmetric property? Explain your answer.

62. Is the solution set for $x < 3$ the same as for $3 > x$? Explain your answer.

63. How would you convince someone that it is necessary to reverse the inequality symbol when multiplying both sides of an inequality by a negative number?

Further Investigations

Solve each inequality.

64. $x + 3 < x - 4$

65. $x - 4 < x + 6$

66. $2x + 4 > 2x - 7$

67. $5x + 2 > 5x + 7$

68. $3x - 4 - 3x > 6$

69. $-2x + 7 + 2x > 1$

70. $-5 \leq -4x - 1 + 4x$

71. $-7 \geq 5x - 2 - 5x$

Answers to the Concept Quiz

1. False   **2.** True   **3.** False   **4.** True   **5.** False   **6.** True   **7.** False   **8.** True   **9.** False   **10.** True

---

3.6   Inequalities, Compound Inequalities, and Problem Solving

OBJECTIVES

1 Solve inequalities that involve the use of the distributive property

2 Solve inequalities that involve fractional forms

3 Determine the solution set for compound inequality statements

4 Solve word problems that translate into inequality statements

Inequalities

Let's begin this section by solving three inequalities with the same basic steps we used with equations. Again, be careful when applying the multiplication and division properties of inequality.

Classroom Example
Solve $6x + 9 \geq 2x - 7$.

EXAMPLE 1   Solve $5x + 8 \leq 3x - 10$.

Solution

$$5x + 8 \leq 3x - 10$$

$$5x + 8 - 3x \leq 3x - 10 - 3x \qquad \text{Subtract } 3x \text{ from both sides}$$

$$2x + 8 \leq -10$$

$$2x + 8 - 8 \leq -10 - 8 \qquad \text{Subtract 8 from both sides}$$

$$2x \leq -18$$

$$\frac{2x}{2} \leq \frac{-18}{2} \qquad \text{Divide both sides by 2}$$

$$x \leq -9$$

The solution set is $\{x \mid x \leq -9\}$ or $(-\infty, -9]$ in interval notation.

Classroom Example
Solve
$2(x - 5) + 7(x + 1) \leq 3(x - 2)$.

EXAMPLE 2   Solve $4(x + 3) + 3(x - 4) \geq 2(x - 1)$.

Solution

$$4(x + 3) + 3(x - 4) \geq 2(x - 1)$$

$$4x + 12 + 3x - 12 \geq 2x - 2 \qquad \text{Distributive property}$$

$$7x \geq 2x - 2 \qquad \text{Combine similar terms}$$

$$7x - 2x \geq 2x - 2 - 2x \qquad \text{Subtract } 2x \text{ from both sides}$$

$$5x \geq -2$$

$$\frac{5x}{5} \geq \frac{-2}{5} \qquad \text{Divide both sides by 5}$$

$$x \geq -\frac{2}{5}$$

The solution set is $\left\{ x \mid x \geq -\frac{2}{5} \right\}$ or $\left[-\frac{2}{5}, \infty \right)$ in interval notation.

Classroom Example
Solve $\frac{3}{8}n - \frac{1}{2}n > \frac{2}{3}$.

EXAMPLE 3   Solve $-\frac{3}{2}n + \frac{1}{6}n < \frac{3}{4}$.

Solution

$$-\frac{3}{2}n + \frac{1}{6}n < \frac{3}{4}$$

$$12\left(-\frac{3}{2}n + \frac{1}{6}n\right) < 12\left(\frac{3}{4}\right) \qquad \begin{array}{l}\text{Multiply both sides by 12, the LCD}\\\text{of all denominators}\end{array}$$

$$12\left(-\frac{3}{2}n\right) + 12\left(\frac{1}{6}n\right) < 12\left(\frac{3}{4}\right) \qquad \text{Distributive property}$$

$$-18n + 2n < 9$$

$$-16n < 9$$

$$\frac{-16n}{-16} > \frac{9}{-16} \qquad \begin{array}{l}\text{Divide both sides by } -16, \text{ which}\\\text{reverses the inequality}\end{array}$$

$$n > -\frac{9}{16}$$

The solution set is $\left\{ n \mid n > -\frac{9}{16} \right\}$ or $\left(-\frac{9}{16}, \infty\right)$ in interval notation.

In Example 3 we are claiming that all numbers greater than $-\frac{9}{16}$ will satisfy the original inequality. Let's check one number; we will check 0.

$$-\frac{3}{2}n + \frac{1}{6}n < \frac{3}{4}$$

$$-\frac{3}{2}(0) + \frac{1}{6}(0) \overset{?}{<} \frac{3}{4}$$

$$0 < \frac{3}{4}$$

The check resulted in a true statement, which means that 0 is in the solution set. Had we forgotten to reverse the inequality sign when we divided both sides by -16, then the solution set would have been $\left\{ n \mid n < -\frac{9}{16} \right\}$. Zero would not have been a member of that solution set, and we would have detected the error by the check.

Compound Inequalities

The words "and" and "or" are used in mathematics to form compound statements. We use "and" and "or" to join two inequalities to form a compound inequality.

Consider the compound inequality

$$x > 2 \qquad \text{and} \qquad x < 5$$

For the solution set, we must find values of x that make both inequalities true statements. The solution set of a compound inequality formed by the word "and" is the **intersection** of the solution sets of the two inequalities. The intersection of two sets, denoted by ∩, contains the elements that are common to both sets. For example, if $A = \{1, 2, 3, 4, 5, 6\}$ and $B = \{0, 2, 4, 6, 8, 10\}$, then $A \cap B = \{2, 4, 6\}$. So to find the solution set of the compound inequality $x > 2$ and $x < 5$, we find the solution set for each inequality and then determine the solutions that are common to both solution sets.

Classroom Example
Graph the solution set for the compound inequality $x > 3$ and $x < 7$, and write the solution set.

EXAMPLE 4

Graph the solution set for the compound inequality $x > 2$ and $x < 5$, and write the solution set.

Solution

$x > 2$ **(a)**

$x < 5$ **(b)**

$x > 2$ and $x < 5$ **(c)**

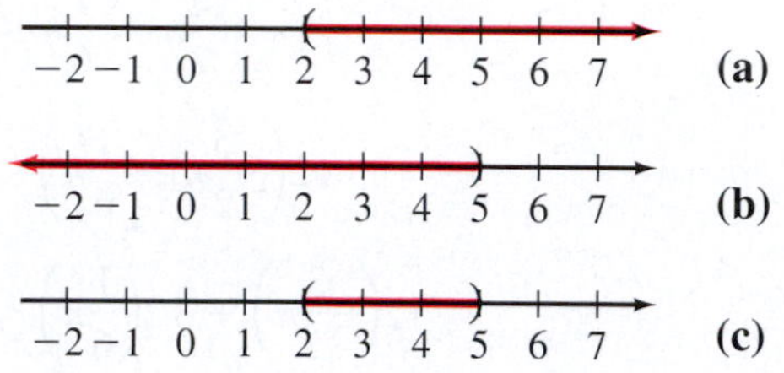

Figure 3.9

Thus all numbers greater than 2 and less than 5 are included in the solution set $\{x | 2 < x < 5\}$, and the graph is shown in Figure 3.9(c). In interval notation the solution set is $(2, 5)$.

Classroom Example
Graph the solution set for the compound inequality $x \geq -1$ and $x \geq -2$, and write the solution set.

EXAMPLE 5

Graph the solution set for the compound inequality $x \leq 1$ and $x \leq 4$, and write the solution set.

Solution

$x \leq 1$ **(a)**

$x \leq 4$ **(b)**

$x \leq 1$ and $x \leq 4$ **(c)**

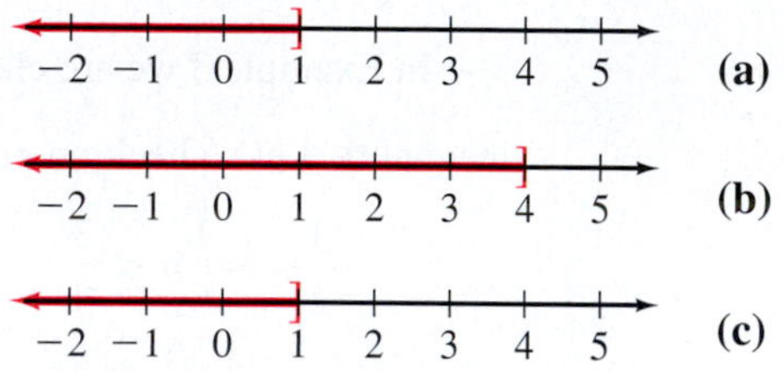

Figure 3.10

The intersection of the two solution sets is $x \leq 1$. The solution set $\{x | x \leq 1\}$ contains all the numbers that are less than or equal to 1, and the graph is shown in Figure 3.10(c). In interval notation the solution set is $(-\infty, 1]$.

The solution set of a compound inequality formed by the word "or" is the **union** of the solution sets of the two inequalities. The union of two sets, denoted by ∪, contains all the elements in both sets. For example, if $A = \{0, 1, 2\}$ and $B = \{1, 2, 3, 4\}$, then $A \cup B = \{0, 1, 2, 3, 4\}$. Note that even though 1 and 2 are in both set A and set B, there is no need to write them twice in $A \cup B$.

To find the solution set of the compound inequality

$x > 1$ or $x > 3$

we find the solution set for each inequality and then take all the values that satisfy either inequality or both.

EXAMPLE 6

Graph the solution set for $x > 1$ or $x > 3$ and write the solution set.

Solution

$x > 1$ **(a)**

$x > 3$ **(b)**

$x > 1$ or $x > 3$ **(c)**

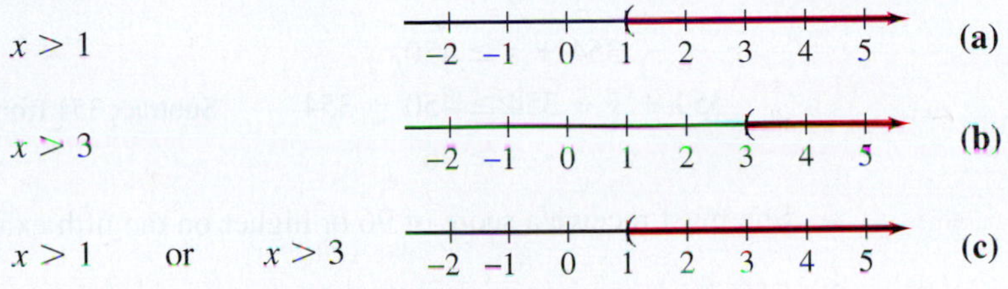

Figure 3.11

Thus all numbers greater than 1 are included in the solution set $\{x | x > 1\}$, and the graph is shown in Figure 3.11(c). The solution set is written as $(1, \infty)$ in interval notation.

EXAMPLE 7

Graph the solution set for $x \leq 0$ or $x \geq 2$ and write the solution set.

Solution

$x \leq 0$ **(a)**

$x \geq 2$ **(b)**

$x \leq 0$ or $x \geq 2$ **(c)**

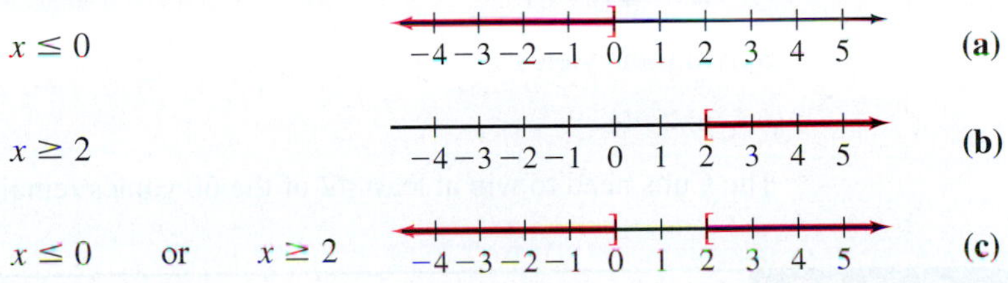

Figure 3.12

Thus all numbers less than or equal to 0 and all numbers greater than or equal to 2 are included in the solution set $\{x | x \leq 0 \text{ or } x \geq 2\}$, and the graph is shown in Figure 3.12(c). Since the solution set contains two intervals that are not continuous, a $\cup$ symbol is used in the interval notation. The solution set is written as $(-\infty, 0] \cup [2, \infty)$ in interval notation.

Back to Problem Solving

Let's consider some word problems that translate into inequality statements. We gave suggestions for solving word problems in Section 3.3; these suggestions still apply, except that here the situations described in the problems/examples will translate into inequalities instead of equations.

EXAMPLE 8

Ashley had scores of 95, 82, 93, and 84 on her first four exams of the semester. What score must she get on the fifth exam to have an average of 90 or higher for the five exams?

Solution

Let s represent the score needed on the fifth exam. Since the average is computed by adding all five scores and dividing by 5 (the number of scores), we have the following inequality to solve:

$$\frac{95 + 82 + 93 + 84 + s}{5} \geq 90$$

Solving this inequality, we obtain

$$\frac{354 + s}{5} \geq 90 \qquad \text{Simplify numerator of the left side}$$

$$5\left(\frac{354 + s}{5}\right) \geq 5(90) \qquad \text{Multiply both sides by 5}$$

$$354 + s \geq 450$$

$$354 + s - 354 \geq 450 - 354 \qquad \text{Subtract 354 from both sides}$$

$$s \geq 96$$

She must receive a score of 96 or higher on the fifth exam.

EXAMPLE 9

The Cubs have won 40 baseball games and have lost 62 games. They will play 60 more games. To win more than 50% of all their games, how many of the 60 games remaining must they win?

Solution

Let w represent the number of games the Cubs must win out of the 60 games remaining. Since they are playing a total of $40 + 62 + 60 = 162$ games, to win more than 50% of their games, they need to win more than 81 games. Thus we have the inequality

$$w + 40 > 81$$

Solving this yields

$$w > 41$$

The Cubs need to win at least 42 of the 60 games remaining.

Concept Quiz 3.6

For Problems 1–5, answer true or false.

1. The solution set of a compound inequality formed by the word "and" is an intersection of the solution sets of the two inequalities.

2. The solution set of a compound inequality formed by the words "and" or "or" is a union of the solution sets of the two inequalities.

3. The intersection of two sets contains the elements that are common to both sets.

4. The union of two sets contains all the elements in both sets.

5. The intersection of set A and set B is denoted by $A \cap B$.

For Problems 6–10, match the compound statement with the graph of its solution set (Figure 3.13).

6. $x > 4$ or $x < -1$ **A.**

7. $x > 4$ and $x > -1$ **B.**

8. $x > 4$ or $x > -1$ **C.**

9. $x \leq 4$ and $x \geq -1$ **D.**

10. $x > 4$ or $x \geq -1$ **E.**

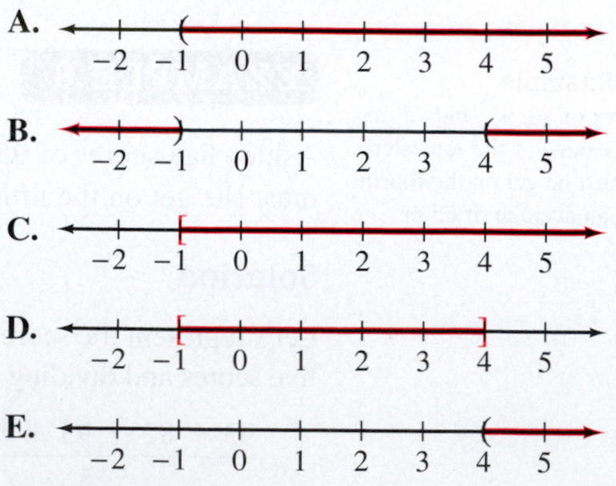

Figure 3.13

Problem Set 3.6

For Problems 1–50, solve each inequality. **(Objectives 1 and 2)**

1. $3x + 4 > x + 8$

2. $5x + 3 < 3x + 11$

3. $7x - 2 < 3x - 6$

4. $8x - 1 > 4x - 21$

5. $6x + 7 > 3x - 3$

6. $7x + 5 < 4x - 12$

7. $5n - 2 \leq 6n + 9$

8. $4n - 3 \geq 5n + 6$

9. $2t + 9 \geq 4t - 13$

10. $6t + 14 \leq 8t - 16$

11. $-3x - 4 < 2x + 7$

12. $-x - 2 > 3x - 7$

13. $-4x + 6 > -2x + 1$

14. $-6x + 8 < -4x + 5$

15. $5(x - 2) \leq 30$

16. $4(x + 1) \geq 16$

17. $2(n + 3) > 9$

18. $3(n - 2) < 7$

19. $-3(y - 1) < 12$

20. $-2(y + 4) > 18$

21. $-2(x + 6) > -17$

22. $-3(x - 5) < -14$

23. $3(x - 2) < 2(x + 1)$

24. $5(x + 3) > 4(x - 2)$

25. $4(x + 3) > 6(x - 5)$

26. $6(x - 1) < 8(x + 5)$

27. $3(x - 4) + 2(x + 3) < 24$

28. $2(x + 1) + 3(x + 2) > -12$

29. $5(n + 1) - 3(n - 1) > -9$

30. $4(n - 5) - 2(n - 1) < 13$

31. $\frac{1}{2}n - \frac{2}{3}n \geq -7$

32. $\frac{3}{4}n + \frac{1}{6}n \leq 1$

33. $\frac{3}{4}n - \frac{5}{6}n < \frac{3}{8}$

34. $\frac{2}{3}n - \frac{1}{2}n > \frac{1}{4}$

35. $\frac{3x}{5} - \frac{2}{3} > \frac{x}{10}$

36. $\frac{5x}{4} + \frac{3}{8} < \frac{7x}{12}$

37. $n \geq 3.4 + 0.15n$

38. $x \geq 2.1 + 0.3x$

39. $0.09t + 0.1(t + 200) > 77$

40. $0.07t + 0.08(t + 100) > 38$

41. $0.06x + 0.08(250 - x) \geq 19$

42. $0.08x + 0.09(2x) \leq 130$

43. $\frac{x - 1}{2} + \frac{x + 3}{5} > \frac{1}{10}$

44. $\frac{x + 3}{4} + \frac{x - 5}{7} < \frac{1}{28}$

45. $\frac{x + 2}{6} - \frac{x + 1}{5} < -2$

46. $\frac{x - 6}{8} - \frac{x + 2}{7} > -1$

47. $\frac{n + 3}{3} + \frac{n - 7}{2} > 3$

48. $\frac{n - 4}{4} + \frac{n - 2}{3} < 4$

49. $\frac{x - 3}{7} - \frac{x - 2}{4} \leq \frac{9}{14}$

50. $\frac{x - 1}{5} - \frac{x + 2}{6} \geq \frac{7}{15}$

For Problems 51–66, graph the solution set for each compound inequality. **(Objective 3)**

51. $x > -1$ and $x < 2$

52. $x > 1$ and $x < 4$

53. $x < -2$ or $x > 1$

54. $x < 0$ or $x > 3$

55. $x > -2$ and $x \leq 2$

56. $x \geq -1$ and $x < 3$

57. $x > -1$ and $x > 2$

58. $x < -2$ and $x < 3$

59. $x > -4$ or $x > 0$

60. $x < 2$ or $x < 4$

61. $x > 3$ and $x < -1$

62. $x < -3$ and $x > 6$

63. $x \leq 0$ or $x \geq 2$

64. $x \leq -2$ or $x \geq 1$

65. $x > -4$ or $x < 3$

66. $x > -1$ or $x < 2$

For Problems 67–78, solve each problem by setting up and solving an appropriate inequality. (Objective 4)

67. Five more than three times a number is greater than 26. Find all of the numbers that satisfy this relationship.

68. Fourteen increased by twice a number is less than or equal to three times the number. Find the numbers that satisfy this relationship.

69. Suppose that the perimeter of a rectangle is to be no greater than 70 inches, and the length of the rectangle must be 20 inches. Find the largest possible value for the width of the rectangle.

70. One side of a triangle is three times as long as another side. The third side is 27 centimeters long. If the perimeter of the triangle is to be no greater than 75 centimeters, find the greatest lengths that the other two sides can be.

71. Sue bowled 132 and 160 in her first two games. What must she bowl in the third game to have an average of at least 150 for the three games?

72. Mike has scores of 87, 81, and 74 on his first three algebra tests. What score must he get on the fourth test to have an average of 85 or higher for the four tests?

73. This semester Lance has scores of 96, 90, and 94 on his first three algebra exams. What must he average on the last two exams to have an average higher than 92 for all five exams?

74. The Mets have won 45 baseball games and lost 55 games. They have 62 more games to play. To win more than 50% of all their games, how many of the 62 games remaining must they win?

75. An Internet business has costs of $4000 plus $32 per sale. The business receives revenue of $48 per sale. What possible values for sales would ensure that the revenues exceed the costs?

76. The average height of the two forwards and the center of a basketball team is 6 feet, 8 inches. What must the average height of the two guards be so that the team's average height is at least 6 feet, 4 inches?

77. Scott shot rounds of 82, 84, 78, and 79 on the first four days of the golf tournament. What must he shoot on the fifth day of the tournament to average 80 or less for the 5 days?

78. Sydney earns $2300 a month. To qualify for a mortgage, her monthly payments must be less than 35% of her monthly income. Her monthly mortgage payments must be less than what amount in order to qualify for the mortgage?

Thoughts Into Words

79. Give an example of a compound statement using the word "and" outside the field of mathematics.

80. Give an example of a compound statement using the word "or" outside the field of mathematics.

81. Give a step-by-step description of how you would solve the inequality $3x - 2 > 4(x + 6)$.

Answers to the Concept Quiz

1. True   **2.** False   **3.** True   **4.** True   **5.** True   **6.** B   **7.** E   **8.** A   **9.** D   **10.** C

OBJECTIVE	SUMMARY	EXAMPLE
Solve first-degree equations.	Numerical equations can be true or false. Algebraic equations (open sentences) contain one or more variables. Solving an equation refers to the process of finding the number (or numbers) that makes an algebraic equation a true statement. A first-degree equation of one variable is an equation that contains only one variable, and this variable has an exponent of one. The properties found in this chapter provide the basis for solving equations. Be sure you are able to use these properties to solve the variety of equations presented.	
Solve equations using the addition-subtraction property of equality. (Section 3.1/Objective 1)	Any number can be added to or subtracted from both sides of an equation.	Solve $0.8 = x - 0.3$. **Solution** $0.8 = x - 0.3$ $0.8 + 0.3 = x - 0.3 + 0.3$ $1.1 = x$ The solution set is $\{1.1\}$.
Solve equations using the multiplication-division property of equality. (Section 3.1/Objective 2)	An equivalent equation is obtained whenever both sides of an equation are multiplied or divided by a nonzero real number.	Solve $-\dfrac{2}{3}x = 8$. **Solution** $-\dfrac{2}{3}x = 8$ $\left(-\dfrac{3}{2}\right)\left(-\dfrac{2}{3}x\right) = 8\left(-\dfrac{3}{2}\right)$ $x = -12$ The solution set is $\{-12\}$.
Solve equations using both the addition-subtraction property of equality and the multiplication-division property of equality. (Section 3.2/Objective 1)	To solve most equations, both properties must be applied.	Solve $5n - 2 = 8$. **Solution** $5n - 2 = 8$ $5n - 2 + 2 = 8 + 2$ $5n = 10$ $\dfrac{1}{5}(5n) = \dfrac{1}{5}(10)$ $n = 2$ The solution set is $\{2\}$.

(continued)

OBJECTIVE	SUMMARY	EXAMPLE
Solve equations that involve the use of the distributive property. (Section 3.4/Objective 1)	To solve equations in which the variable is part of an expression enclosed in parentheses, the distributive property is used. The distributive property removes the parentheses, and the resulting equation is solved in the usual way.	Solve $3(x - 4) = 2(x + 1)$. **Solution** $3(x - 4) = 2(x + 1)$ $3x - 12 = 2x + 2$ $x - 12 = 2$ $x = 14$ The solution set is $\{14\}$.
Solve equations that involve fractional forms. (Section 3.4/Objective 2)	When an equation contains several fractions, it is usually best to start by clearing the equation of all fractions. The fractions can be cleared by multiplying both sides of the equation by the least common denominator of all the denominators.	Solve $\dfrac{3n}{4} + \dfrac{n}{5} = \dfrac{7}{10}$. **Solution** $\dfrac{3n}{4} + \dfrac{n}{5} = \dfrac{7}{10}$ $20\left(\dfrac{3n}{4} + \dfrac{n}{5}\right) = 20\left(\dfrac{7}{10}\right)$ $20\left(\dfrac{3n}{4}\right) + 20\left(\dfrac{n}{5}\right) = 14$ $15n + 4n = 14$ $19n = 14$ $n = \dfrac{14}{19}$ The solution set is $\left\{\dfrac{14}{19}\right\}$.
Solve equations that are contradictions or identities. (Section 3.3/Objectives 2 and 3)	When an equation is not true for any value of x, then the equation is called "a contradiction." When an equation is true for any permissible value of x, then the equation is called "an identity."	Solve the following equations: **(a)** $2(x + 4) = 2x + 5$ **(b)** $4x - 8 = 2(2x - 4)$ **Solution** **(a)** $2(x + 4) = 2x + 5$ $2x + 8 = 2x + 5$ $2x - 2x + 8 = 2x - 2x + 5$ $8 = 5$ This is a false statement so there is no solution. The solution is $\varnothing$. **(b)** $4x - 8 = 2(2x - 4)$ $4x - 8 = 4x - 8$ $4x - 4x - 8 = 4x - 4x - 8$ $-8 = -8$ This is a true statement so any value of x is a solution. The solution set is $\{$All reals$\}$.

OBJECTIVE	SUMMARY	EXAMPLE
Show the solution set of an inequality in set-builder notation and by graphing. (Section 3.5/Objective 2)	The solution set of $x + 2 > 5$ is all numbers greater than 3. The solution set in set-builder notation is $\{x \mid x > 3\}$ and is read "the set of all x such that x is greater than 3." A number line is used to graph the solution. A parenthesis on the number line means that number is *not* included in the solution set. A bracket on the number line indicates that the number is included in the solution set.	Write the solution set of the inequalities in set-builder notation and graph the solution. **(a)** $x \leq 2$ **(b)** $x > -1$ **Solution** **(a)** $\{x \mid x \leq 2\}$ **(b)** $\{x \mid x > -1\}$
Solve first-degree inequalities. (Section 3.5/Objective 1)	Properties for solving inequalities are similar to the properties for solving equations—except for properties that involve multiplying or dividing by a negative number. When multiplying or dividing both sides of an inequality by a negative number, you must reverse the inequality symbol.	Solve $-4n - 3 > 7$. **Solution** $$-4n - 3 > 7$$ $$-4n > 10$$ $$\frac{-4n}{-4} < \frac{10}{-4}$$ $$n < \frac{-5}{2}$$ The solution set is $$\left\{ n \mid n < -\frac{5}{2} \right\} \text{ or } \left(-\infty, -\frac{5}{2} \right)$$
Solve inequalities that involve the use of the distributive property. (Section 3.6/Objective 1)	To solve inequalities when the variable is part of an expression enclosed in parentheses, use the distributive property. The distributive property removes the parentheses, and the resulting inequality is solved in the usual way.	Solve $15 < -2(x - 1) - 5$. **Solution** $$15 < -2x + 2 - 5$$ $$15 < -2x - 3$$ $$15 + 3 < -2x - 3 + 3$$ $$18 < -2x$$ $$\frac{18}{-2} > \frac{-2x}{-2}$$ $$-9 > x$$ The solution set is $\{x \mid x < -9\}$ or $(-\infty, -9)$.

(continued)

OBJECTIVE	SUMMARY	EXAMPLE
Solve inequalities that involve fractional forms. **(Section 3.6/Objective 2)**	When an inequality contains several fractions, it is usually best to clear the inequality of all fractions. The fractions can be cleared by multiplying both sides of the equation by the LCD of all the denominators.	Solve $\dfrac{3}{4}x < \dfrac{2}{3}$. **Solution** $$\frac{3}{4}x < \frac{2}{3}$$ $$\frac{4}{3}\left(\frac{3}{4}x\right) < \frac{4}{3}\left(\frac{2}{3}\right)$$ $$x < \frac{8}{9}$$ The solution set is $$\left\{x \mid x < \frac{8}{9}\right\} \text{ or } \left(-\infty, \frac{8}{9}\right).$$
Solve compound inequalities formed by the word "and." **(Section 3.6/Objective 3)**	The solution set of a compound inequality formed by the word "and" is the intersection of the solution sets of the two inequalities. To solve inequalities involving "and," we must satisfy all of the conditions. Thus the compound inequality $x > 1$ *and* $x < 3$ is satisfied by all numbers between 1 and 3.	Solve the compound inequality $x > -4$ and $x > 2$. **Solution** All of the conditions must be satisfied. Thus the compound inequality $x > -4$ and $x > 2$ is satisfied by all numbers greater than 2. The solution set is $\{x \mid x > 2\}$.
Solve compound inequalities formed by the word "or." **(Section 3.6/Objective 3)**	The solution set of a compound inequality formed by the word "or," is the union of the solution sets of the two inequalities. To solve inequalities involving "or" we must satisfy one or more of the conditions. Thus the compound inequality $x < 1$ *or* $x < 5$ is satisfied by all numbers less than 5.	Solve the compound inequality $x > -1$ or $x < 2$. **Solution** One or more of the conditions must be satisfied. Thus the compound inequality $x > -1$ or $x < 2$ is satisfied by all real numbers. The solution set is {All reals}.
Solve word problems. **(Section 3.2/Objective 2;** **Section 3.3/Objective 5;** **Section 3.4/Objective 3)**	Keep these suggestions in mind as you solve word problems: **1.** Read the problem carefully. **2.** Sketch any figure or diagram that might be helpful. **3.** Choose a meaningful variable. **4.** Look for a guideline. **5.** Form an equation. **6.** Solve the equation. **7.** Check your answer.	The difference of two numbers is 14. If 35 is the larger number, find the smaller number. **Solution** Let n represent the smaller number. **Guideline** Larger number $-$ smaller number $= 14$ $35 - n = 14$ $-n = -21$ $n = 21$ The smaller number is 21.

OBJECTIVE	SUMMARY	EXAMPLE
Solve word problems involving inequalities. (Section 3.6/Objective 4)	Follow the same steps as in the previous box but for step 5 substitute *Form an inequality* and for step 6 substitute *Solve the inequality*.	Martin must average at least 240 points for a series of three bowling games to get into the playoffs. If he has bowled games of 220 points and 210 points, what must his score be on the third game to get into the playoffs? **Solution** Let p represent the points for the third game. **Guideline** Average ≥ 240 $$\frac{220 + 210 + p}{3} \geq 240$$ $$430 + p \geq 720$$ $$p \geq 290$$ Martin must bowl 290 points or more.

Chapter 3   Review Problem Set

In Problems 1–20, solve each of the equations.

1. $9x - 2 = -29$

2. $-3 = -4y + 1$

3. $7 - 4x = 10$

4. $6y - 5 = 4y + 13$

5. $4n - 3 = 7n + 9$

6. $7(y - 4) = 4(y + 3)$

7. $2(x + 1) + 5(x - 3) = 11(x - 2)$

8. $-3(x + 6) = 5x - 3$

9. $\dfrac{2}{5}n - \dfrac{1}{2}n = \dfrac{7}{10}$

10. $\dfrac{3n}{4} + \dfrac{5n}{7} = \dfrac{1}{14}$

11. $\dfrac{x - 3}{6} + \dfrac{x + 5}{8} = \dfrac{11}{12}$

12. $\dfrac{n}{2} - \dfrac{n - 1}{4} = \dfrac{3}{8}$

13. $-2(x - 4) = -3(x + 8)$

14. $3x - 4x - 2 = 7x - 14 - 9x$

15. $5(n - 1) - 4(n + 2) = -3(n - 1) + 3n + 5$

16. $\dfrac{x - 3}{9} = \dfrac{x + 4}{8}$

17. $\dfrac{x - 1}{-3} = \dfrac{x + 2}{-4}$

18. $-(t - 3) - (2t + 1) = 3(t + 5) - 2(t + 1)$

19. $\dfrac{2x - 1}{3} = \dfrac{3x + 2}{2}$

20. $3(2t - 4) + 2(3t + 1) = -2(4t + 3) - (t - 1)$

For Problems 21–36, solve each inequality.

21. $3x - 2 > 10$

22. $-2x - 5 < 3$

23. $2x - 9 \geq x + 4$

24. $3x + 1 \leq 5x - 10$

25. $6(x - 3) > 4(x + 13)$

26. $2(x + 3) + 3(x - 6) < 14$

27. $\dfrac{2n}{5} - \dfrac{n}{4} < \dfrac{3}{10}$

28. $\dfrac{n + 4}{5} + \dfrac{n - 3}{6} > \dfrac{7}{15}$

29. $-16 < 8 + 2y - 3y$

30. $-24 > 5x - 4 - 7x$

31. $-3(n - 4) > 5(n + 2) + 3n$

32. $-4(n - 2) - (n - 1) < -4(n + 6)$

33. $\frac{3}{4}n - 6 \le \frac{2}{3}n + 4$

34. $\frac{1}{2}n - \frac{1}{3}n - 4 \ge \frac{3}{5}n + 2$

35. $-12 > -4(x - 1) + 2$

36. $36 < -3(x + 2) - 1$

For Problems 37–40, graph the solution set for each of the compound inequalities.

37. $x > -3$ and $x < 2$

38. $x < -1$ or $x > 4$

39. $x < 2$ or $x > 0$

40. $x > 1$ and $x > 0$

Set up an equation or an inequality to solve Problems 41–52.

41. Three-fourths of a number equals 18. Find the number.

42. Nineteen is 2 less than three times a certain number. Find the number.

43. The difference of two numbers is 21. If 12 is the smaller number, find the other number.

44. One subtracted from nine times a certain number is the same as 15 added to seven times the number. Find the number.

45. Monica has scores of 83, 89, 78, and 86 on her first four exams. What score must she receive on the fifth exam so that her average for all five exams is 85 or higher?

46. The sum of two numbers is 40. Six times the smaller number equals four times the larger. Find the numbers.

47. Find a number such that 2 less than two-thirds of the number is 1 more than one-half of the number.

48. Ameya's average score for her first three psychology exams is 84. What must she get on the fourth exam so that her average for the four exams is 85 or higher?

49. Miriam has 30 coins (nickels and dimes) that amount to $2.60. How many coins of each kind does she have?

50. Suppose that Russ has a bunch of nickels, dimes, and quarters amounting to $15.40. The number of dimes is 1 more than three times the number of nickels, and the number of quarters is twice the number of dimes. How many coins of each kind does he have?

51. The supplement of an angle is 14° more than three times the complement of the angle. Find the measure of the angle.

52. Pam rented a car from a rental agency that charges $25 a day and $0.20 per mile. She kept the car for 3 days and her bill was $215. How many miles did she drive during that 3-day period?

Chapter 3 Test

For Problems 1–12, solve each of the equations.

1. $7x - 3 = 11$

2. $-7 = -3x + 2$

3. $4n + 3 = 2n - 15$

4. $3n - 5 = 8n + 20$

5. $4(x - 2) = 5(x + 9)$

6. $9(x + 4) = 6(x - 3)$

7. $5(y - 2) + 2(y + 1) = 3(y - 6)$

8. $\dfrac{3}{5}x - \dfrac{2}{3} = \dfrac{1}{2}$

9. $\dfrac{x - 2}{4} = \dfrac{x + 3}{6}$

10. $\dfrac{x + 2}{3} + \dfrac{x - 1}{2} = 2$

11. $\dfrac{x - 3}{6} - \dfrac{x - 1}{8} = \dfrac{13}{24}$

12. $-5(n - 2) = -3(n + 7)$

For Problems 13–18, solve each of the inequalities.

13. $3x - 2 < 13$

14. $-2x + 5 \geq 3$

15. $3(x - 1) \leq 5(x + 3)$

16. $-4 > 7(x - 1) + 3$

17. $-2(x - 1) + 5(x - 2) < 5(x + 3)$

18. $\dfrac{1}{2}n + 2 \leq \dfrac{3}{4}n - 1$

For Problems 19 and 20, graph the solution set for each compound inequality.

19. $x \geq -2$ and $x \leq 4$

20. $x < 1$ or $x > 3$

For Problems 21–25, set up an equation or an inequality and solve each problem.

21. Jean-Paul received a cell phone bill for $98.24. Included in the $98.24 was a monthly-plan charge of $29.99 and a charge for 195 extra minutes. How much is Jean-Paul being charged for each extra minute?

22. Suppose that a triangular plot of ground is enclosed with 70 meters of fencing. The longest side of the lot is two times the length of the shortest side, and the third side is 10 meters longer than the shortest side. Find the length of each side of the plot.

23. Tina had scores of 86, 88, 89, and 91 on her first four history exams. What score must she get on the fifth exam to have an average of 90 or higher for the five exams?

24. Sean has 103 coins consisting of nickels, dimes, and quarters. The number of dimes is 1 less than twice the number of nickels, and the number of quarters is 2 more than three times the number of nickels. How many coins of each kind does he have?

25. In triangle ABC, the measure of angle C is one-half the measure of angle A, and the measure of angle B is 30° more than the measure of angle A. Find the measure of each angle of the triangle.

For Problems 1–8, simplify each numerical expressions.

1. $3(-4) - 2 + (-3)(-6) - 1$

2. $-(2)^7$

3. $6.2 - 7.1 - 3.4 + 1.9$

4. $-\dfrac{2}{3} + \dfrac{1}{2} - \dfrac{1}{4}$

5. $2 + 8 \div 2 \cdot 5$

6. $6(1 + 4) - (3 - 8)$

7. $(3 + 1)^2$

8. $5 - 3^2 - (-4)^2$

For Problems 9–12, evaluate each algebraic expression for the given value of the variables.

9. $-4x + 2y - xy$ for $x = -2$ and $y = 3$

10. $\dfrac{1}{5}x - \dfrac{2}{3}y$ for $x = -\dfrac{1}{2}$ and $y = \dfrac{1}{6}$

11. $0.2(x - y) - 0.3(x + y)$ for $x = 0.1$ and $y = -0.2$

12. $x^2 - y^2$ for $x = -3$ and $y = 5$

For Problems 13–18, simplify by combining similar terms. Apply the distributive property if necessary.

13. $\dfrac{3}{8}x + \dfrac{3}{7}y - \dfrac{5}{8}x + \dfrac{1}{7}y$

14. $-(x - 2) + 6(x + 4) - 2(x - 7)$

15. $-6x^2 + 5x + 2 + 9x^2 - 7x - 8$

16. $4y^2 + 3 - (2y^2 - 8y - 1)$

17. $5a^2 + a - 9 + a^2 - 3a$

18. $3(a + 2b) - 4(2a + c) + 5(3b - 2c)$

19. Find the greatest common factor of 48, 60, and 96.

20. Find the least common multiple of 9 and 12.

21. Express 300 as a product of primes.

22. Express 144 as a product of primes.

For Problems 23–30, perform the indicated operations.

23. $\dfrac{5}{12}\left(\dfrac{3}{8}\right)$

24. $\left(-\dfrac{3}{8}\right) \div \dfrac{1}{18}$

25. $\dfrac{3x^3}{5} \div \dfrac{x^2}{2y}$

26. $\dfrac{5}{x} - \dfrac{6}{y^2}$

27. $\dfrac{5}{3y} - \dfrac{2}{5y^2}$

28. $\dfrac{5x}{8}\left(-\dfrac{8x}{5}\right)$

29. $\left(\dfrac{5x^2y}{6xy^2}\right)\left(\dfrac{8x^2y^2}{30y}\right)$

30. $\left(\dfrac{ab^3}{3a^2}\right) \div \left(\dfrac{ab}{9b^2}\right)$

For Problems 31–36, solve each equation.

31. $3x + 2(x - 4) = 12$

32. $-3(x + 4) = -4(x - 1)$

33. $4(x + 7) = 4x - 9$

34. $\dfrac{x + 1}{4} - \dfrac{x - 2}{3} = -2$

35. $-2x + 3(x - 1) = x - 3$

36. $2(2x - 1) + 3(x - 3) = -4(x + 7)$

For Problems 37–40, solve the inequality.

37. $-5x + 2 > -18$

38. $x + 12 \le 4x - 15$

39. $2(x - 1) \ge 3(x - 6)$

40. $-2 < -(x - 1) - 4$

For Problems 41–46, graph the solution set for each of the compound inequalities.

41. $x \ge -1$ and $x < 3$

42. $x \ge 2$ and $x > 4$

43. $x \le -1$ or $x > 2$

44. $x < -2$ or $x < 3$

45. $x \le -1$ and $x \ge -4$

46. $x \ge -1$ or $x < 3$

For Problems 47–50, write an equation or inequality to represent each problem and solve the problem.

47. Find two consecutive odd integers where the smaller integer plus 5 times the larger integer equals 76.

48. On Friday and Saturday nights, the police made a total of 42 arrests at a DUI checkpoint. On Saturday night they made 6 more than three times the arrests of Friday night. Find the number of arrests for each night.

49. For a wedding reception, the caterer charges a $125 fee plus $35 per person for dinner. If Peter and Rynette must keep the cost of the caterer to less than $2500, how many people can attend the reception?

50. John has scored 87, 95, and 90 on three algebra tests. What must he score on the next test so that his average is 90 or higher.

4

Formulas and Problem Solving

The equation $s = 8 + 0.6s$ can be used to determine how much the owner of a pizza parlor must charge for a pizza if it costs $8 to make the pizza, and he wants to make a profit of 60% based on the selling price.

Kirk starts jogging at 5 miles per hour. One-half hour later, Consuela starts jogging on the same route at 7 miles per hour. How long will it take Consuela to catch Kirk? If we let t represent the time that Consuela jogs, then $t + \dfrac{1}{2}$ represents Kirk's time. We can use the equation $7t = 5\left(t + \dfrac{1}{2}\right)$ to determine that Consuela should catch Kirk in $1\dfrac{1}{4}$ hours.

We used the formula *distance equals rate times time*, which is usually expressed as $d = rt$, to set up the equation $7t = 5\left(t + \dfrac{1}{2}\right)$. Throughout this chapter, we will use a variety of formulas to solve problems that connect algebraic and geometric concepts.

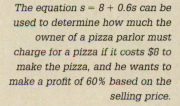 Video tutorials based on section learning objectives are available in a variety of delivery modes.

4.1   Ratio, Proportion, and Percent

OBJECTIVES

1 Solve proportions

2 Use a proportion to convert a fraction to a percent

3 Solve basic percent problems

4 Solve word problems using proportions

Ratio and Proportion

In Figure 4.1, as gear A revolves 4 times, gear B will revolve 3 times. We say that the gear ratio of A to B is 4 to 3, or the gear ratio of B to A is 3 to 4. Mathematically, a **ratio** is the comparison of two numbers by division. We can write the gear ratio of A to B as

$$4 \text{ to } 3 \quad \text{ or } \quad 4{:}3 \quad \text{ or } \quad \frac{4}{3}$$

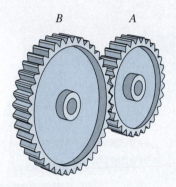

Figure 4.1

We express ratios as fractions in reduced form. For example, if there are 7500 women and 5000 men at a certain university, then the ratio of women to men is $\dfrac{7500}{5000} = \dfrac{3}{2}$.

A statement of equality between two ratios is called a **proportion**. For example,

$$\frac{2}{3} = \frac{8}{12}$$

is a proportion that states that the ratios $\dfrac{2}{3}$ and $\dfrac{8}{12}$ are equal. In the general proportion

$$\frac{a}{b} = \frac{c}{d}, \qquad b \neq 0 \text{ and } d \neq 0$$

if we multiply both sides of the equation by the common denominator, bd, we obtain

$$(bd)\left(\frac{a}{b}\right) = (bd)\left(\frac{c}{d}\right)$$

$$ad = bc$$

These products ad and bc are called *cross products* and are equal to each other. Let's state this as a property of proportions.

$$\frac{a}{b} = \frac{c}{d} \quad \text{if and only if } ad = bc, \text{ where } b \neq 0 \text{ and } d \neq 0$$

Classroom Example

Solve $\frac{n}{12} = \frac{2}{3}$.

EXAMPLE 1   Solve $\frac{x}{20} = \frac{3}{4}$.

Solution

$$\frac{x}{20} = \frac{3}{4}$$

$$4x = 60 \qquad \text{Cross products are equal}$$

$$x = 15$$

The solution set is $\{15\}$.

Classroom Example

Solve $\frac{x+1}{7} = \frac{x-4}{8}$.

EXAMPLE 2   Solve $\frac{x-3}{5} = \frac{x+2}{4}$.

Solution

$$\frac{x-3}{5} = \frac{x+2}{4}$$

$$4(x - 3) = 5(x + 2) \qquad \text{Cross products are equal}$$

$$4x - 12 = 5x + 10 \qquad \text{Distributive property}$$

$$-12 = x + 10 \qquad \text{Subtracted } 4x \text{ from both sides}$$

$$-22 = x \qquad \text{Subtracted 10 from both sides}$$

The solution set is $\{-22\}$.

If a variable appears in one or both of the denominators, then a proper restriction should be made to avoid division by zero, as the next example illustrates.

Classroom Example

Solve $\frac{9}{x+2} = \frac{3}{x-5}$.

EXAMPLE 3   Solve $\frac{7}{a-2} = \frac{4}{a+3}$.

Solution

$$\frac{7}{a-2} = \frac{4}{a+3}, \qquad a \neq 2 \text{ and } a \neq -3$$

$$7(a + 3) = 4(a - 2) \qquad \text{Cross products are equal}$$

$$7a + 21 = 4a - 8 \qquad \text{Distributive property}$$

$$3a + 21 = -8 \qquad \text{Subtracted } 4a \text{ from both sides}$$

$$3a = -29 \qquad \text{Subtracted 21 from both sides}$$

$$a = -\frac{29}{3} \qquad \text{Divided both sides by 3}$$

The solution set is $\left\{-\frac{29}{3}\right\}$.

EXAMPLE 4   Solve $\dfrac{x}{4} + 3 = \dfrac{x}{5}$.

Solution

This is *not* a proportion, so we can multiply both sides by 20 to clear the equation of all fractions.

$$\frac{x}{4} + 3 = \frac{x}{5}$$

$$20\left(\frac{x}{4} + 3\right) = 20\left(\frac{x}{5}\right) \qquad \text{Multiply both sides by 20}$$

$$20\left(\frac{x}{4}\right) + 20(3) = 20\left(\frac{x}{5}\right) \qquad \text{Distributive property}$$

$$5x + 60 = 4x$$

$$x + 60 = 0 \qquad \text{Subtracted } 4x \text{ from both sides}$$

$$x = -60 \qquad \text{Subtracted 60 from both sides}$$

The solution set is $\{-60\}$.

Remark: Example 4 demonstrates the importance of thinking first before pushing the pencil. Since the equation was not in the form of a proportion, we needed to revert to a previous technique for solving such equations.

Problem Solving Using Proportions

Some word problems can be conveniently set up and solved using the concepts of ratio and proportion. Consider the following examples.

EXAMPLE 5

On the map in Figure 4.2, 1 inch represents 20 miles. If two cities are $6\dfrac{1}{2}$ inches apart on the map, find the number of miles between the cities.

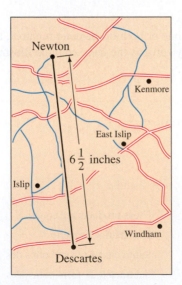

Figure 4.2

Solution

Let m represent the number of miles between the two cities. Now let's set up a proportion in which one ratio compares distances in inches on the map, and the other ratio compares *corresponding* distances in miles on land:

$$\frac{1}{6\frac{1}{2}} = \frac{20}{m}$$

To solve this equation, we equate the cross products:

$$m(1) = \left(6\frac{1}{2}\right)(20)$$

$$m = \left(\frac{13}{2}\right)(20) = 130$$

The distance between the two cities is 130 miles.

Classroom Example
A sum of $2600 is to be divided between two people in the ratio of 3 to 5. How much does each person receive?

EXAMPLE 6

A sum of $1750 is to be divided between two people in the ratio of 3 to 4. How much money does each person receive?

Solution

Let d represent the amount of money to be received by one person. Then $1750 - d$ represents the amount for the other person. We set up this proportion:

$$\frac{d}{1750 - d} = \frac{3}{4}$$

$$4d = 3(1750 - d)$$

$$4d = 5250 - 3d$$

$$7d = 5250$$

$$d = 750$$

If $d = 750$, then $1750 - d = 1000$; therefore, one person receives $750, and the other person receives $1000.

Percent

The word **percent** means "per one hundred," and we use the symbol % to express it. For example, we write 7 percent as 7%, which means $\frac{7}{100}$ or 0.07. In other words, percent is a special kind of ratio—namely, one in which the denominator is always 100. Proportions provide a convenient basis for changing common fractions to percents. Consider the next examples.

Classroom Example
Express $\frac{9}{25}$ as a percent.

EXAMPLE 7

Express $\frac{7}{20}$ as a percent.

Solution

We are asking "What number compares to 100 as 7 compares to 20?" Therefore, if we let n represent that number, we can set up the following proportion:

$$\frac{n}{100} = \frac{7}{20}$$

$$20n = 700$$

$$n = 35$$

Thus $\frac{7}{20} = \frac{35}{100} = 35\%$.

Classroom Example
Express $\frac{4}{9}$ as a percent.

EXAMPLE 8   Express $\frac{5}{6}$ as a percent.

Solution

$$\frac{n}{100} = \frac{5}{6}$$

$$6n = 500$$

$$n = \frac{500}{6} = \frac{250}{3} = 83\frac{1}{3}$$

Therefore, $\frac{5}{6} = 83\frac{1}{3}\%$.

Some Basic Percent Problems

What is 8% of 35? Fifteen percent of what number is 24? Twenty-one is what percent of 70? These are the three basic types of percent problems. We can solve each of these problems easily by translating into and solving a simple algebraic equation.

Classroom Example
What is 12% of 80?

EXAMPLE 9   What is 8% of 35?

Solution

Let n represent the number to be found. The word "is" refers to equality, and the word "of" means multiplication. Thus the question translates into

$$n = (8\%)(35)$$

which can be solved as follows:

$$n = (0.08)(35)$$
$$= 2.8$$

Therefore, 2.8 is 8% of 35.

Classroom Example
Six percent of what number is 9?

EXAMPLE 10   Fifteen percent of what number is 24?

Solution

Let n represent the number to be found.

$$(15\%)(n) = 24$$
$$0.15n = 24$$
$$15n = 2400 \qquad \text{Multiplied both sides by 100}$$
$$n = 160$$

Therefore, 15% of 160 is 24.

Classroom Example
Forty-two is what percent of 168?

EXAMPLE 11   Twenty-one is what percent of 70?

Solution

Let r represent the percent to be found.

$$21 = r(70)$$
$$\frac{21}{70} = r$$

$$\frac{3}{10} = r \qquad \text{Reduce!}$$

$$\frac{30}{100} = r \qquad \text{Changed } \frac{3}{10} \text{ to } \frac{30}{100}$$

$$30\% = r$$

Therefore, 21 is 30% of 70.

Classroom Example
Twenty-one is what percent of 12?

EXAMPLE 12   Seventy-two is what percent of 60?

Solution

Let r represent the percent to be found.

$$72 = r(60)$$

$$\frac{72}{60} = r$$

$$\frac{6}{5} = r$$

$$\frac{120}{100} = r \qquad \text{Changed } \frac{6}{5} \text{ to } \frac{120}{100}$$

$$120\% = r$$

Therefore, 72 is 120% of 60.

It is helpful to get into the habit of checking answers for *reasonableness*. We also suggest that you alert yourself to a potential computational error by estimating the answer before you actually do the problem. For example, prior to solving Example 12, you may have estimated as follows: Since 72 is greater than 60, you know that the answer has to be greater than 100%. Furthermore, 1.5 (or 150%) times 60 equals 90. Therefore, you can estimate the answer to be somewhere between 100% and 150%. That may seem like a rather rough estimate, but many times such an estimate will reveal a computational error.

Concept Quiz 4.1

For Problems 1–10, answer true or false.

1. A ratio is the comparison of two numbers by division.

2. The ratio of 7 to 3 can be written 3:7.

3. A proportion is a statement of equality between two ratios.

4. For the proportion $\frac{x}{3} = \frac{y}{5}$, the cross products would be $5x = 3y$.

5. The algebraic statement $\frac{w}{2} = \frac{w}{5} + 1$ is a proportion.

6. The word "percent" means parts per one thousand.

7. For the proportion $\frac{a+1}{a-2} = \frac{5}{7}x, a \neq -1$ and $a \neq 2$.

8. If the cross products of a proportion are $wx = yz$, then $\frac{x}{z} = \frac{y}{w}$.

9. One hundred twenty percent of 30 is 24.

10. Twelve is 30% of 40.

Problem Set 4.1

For Problems 1–36, solve each of the equations. **(Objective 1)**

1. $\dfrac{x}{6} = \dfrac{3}{2}$

2. $\dfrac{x}{9} = \dfrac{5}{3}$

3. $\dfrac{5}{12} = \dfrac{n}{24}$

4. $\dfrac{7}{8} = \dfrac{n}{16}$

5. $\dfrac{x}{3} = \dfrac{5}{2}$

6. $\dfrac{x}{7} = \dfrac{4}{3}$

7. $\dfrac{x-2}{4} = \dfrac{x+4}{3}$

8. $\dfrac{x-6}{7} = \dfrac{x+9}{8}$

9. $\dfrac{x+1}{6} = \dfrac{x+2}{4}$

10. $\dfrac{x-2}{6} = \dfrac{x-6}{8}$

11. $\dfrac{h}{2} - \dfrac{h}{3} = 1$

12. $\dfrac{h}{5} + \dfrac{h}{4} = 2$

13. $\dfrac{x+1}{3} - \dfrac{x+2}{2} = 4$

14. $\dfrac{x-2}{5} - \dfrac{x+3}{6} = -4$

15. $\dfrac{-4}{x+2} = \dfrac{-3}{x-7}$

16. $\dfrac{-9}{x+1} = \dfrac{-8}{x+5}$

17. $\dfrac{-1}{x-7} = \dfrac{5}{x-1}$

18. $\dfrac{3}{x-10} = \dfrac{-2}{x+6}$

19. $\dfrac{3}{2x-1} = \dfrac{2}{3x+2}$

20. $\dfrac{1}{4x+3} = \dfrac{2}{5x-3}$

21. $\dfrac{n+1}{n} = \dfrac{8}{7}$

22. $\dfrac{5}{6} = \dfrac{n}{n+1}$

23. $\dfrac{x-1}{2} - 1 = \dfrac{3}{4}$

24. $-2 + \dfrac{x+3}{4} = \dfrac{5}{6}$

25. $-3 - \dfrac{x+4}{5} = \dfrac{3}{2}$

26. $\dfrac{x-5}{3} + 2 = \dfrac{5}{9}$

27. $\dfrac{n}{150-n} = \dfrac{1}{2}$

28. $\dfrac{n}{200-n} = \dfrac{3}{5}$

29. $\dfrac{300-n}{n} = \dfrac{3}{2}$

30. $\dfrac{80-n}{n} = \dfrac{7}{9}$

31. $\dfrac{-1}{5x-1} = \dfrac{-2}{3x+7}$

32. $\dfrac{-3}{2x-5} = \dfrac{-4}{x-3}$

33. $\dfrac{2(x-1)}{3} = \dfrac{3(x+2)}{5}$

34. $\dfrac{4(x+3)}{7} = \dfrac{2(x-6)}{5}$

35. $\dfrac{3(2x-5)}{4} + 2 = \dfrac{4x-1}{2}$

36. $\dfrac{2(3x+1)}{3} - 1 = \dfrac{5(2x-7)}{6}$

For Problems 37–48, use proportions to change each common fraction to a percent. **(Objective 2)**

37. $\dfrac{11}{20}$

38. $\dfrac{17}{20}$

39. $\dfrac{3}{5}$

40. $\dfrac{7}{25}$

41. $\dfrac{1}{6}$

42. $\dfrac{5}{7}$

43. $\dfrac{3}{8}$

44. $\dfrac{1}{16}$

45. $\dfrac{3}{2}$

46. $\dfrac{5}{4}$

47. $\dfrac{12}{5}$

48. $\dfrac{13}{6}$

For Problems 49–60, answer the question by setting up and solving an appropriate equation. **(Objective 3)**

49. What is 7% of 38?

50. What is 35% of 52?

51. 15% of what number is 6.3?

52. 55% of what number is 38.5?

53. 76 is what percent of 95?

54. 72 is what percent of 120?

55. What is 120% of 50?

56. What is 160% of 70?

57. 46 is what percent of 40?

58. 26 is what percent of 20?

59. 160% of what number is 144?

60. 220% of what number is 66?

For Problems 61–77, solve each problem using a proportion. **(Objective 4)**

61. A blueprint has a scale in which 1 inch represents 6 feet. Find the dimensions of a rectangular room that measures $2\frac{1}{2}$ inches by $3\frac{1}{4}$ inches on the blueprint.

62. On a certain map, 1 inch represents 15 miles. If two cities are 7 inches apart on the map, find the number of miles between the cities.

63. Suppose that a car can travel 264 miles using 12 gallons of gasoline. How far will it go on 15 gallons?

64. Jesse used 10 gallons of gasoline to drive 170 miles. How much gasoline will he need to travel 238 miles?

65. If the ratio of the length of a rectangle to its width is $\frac{5}{2}$, and the width is 24 centimeters, find its length.

66. If the ratio of the width of a rectangle to its length is $\frac{4}{5}$, and the length is 45 centimeters, find the width.

67. A saltwater solution is made by dissolving 3 pounds of salt in 10 gallons of water. At this rate, how many pounds of salt are needed for 25 gallons of water? (See Figure 4.3.)

Figure 4.3

68. A home valued at $150,000 is assessed $2700 in real estate taxes. At the same rate, how much are the taxes on a home valued at $175,000?

69. If 20 pounds of fertilizer will cover 1500 square feet of lawn, how many pounds are needed for 2500 square feet?

70. It was reported that a flu epidemic is affecting six out of every ten college students in a certain part of the country. At this rate, how many students in that part of the country would be affected at a university of 15,000 students?

71. A preelection poll indicated that three out of every seven eligible voters were going to vote in an upcoming election. At this rate, how many people are expected to vote in a city of 210,000?

72. A board 28 feet long is cut into two pieces whose lengths are in the ratio of 2 to 5. Find the lengths of the two pieces.

73. In a nutrition plan the ratio of calories to grams of carbohydrates is 16 to 1. According to this ratio, how many grams of carbohydrates would be in a plan that has 2200 calories?

74. The ratio of male students to female students at a certain university is 5 to 4. If there is a total of 6975 students, find the number of male students and the number of female students.

75. An investment of $500 earns $45 in a year. At the same rate, how much additional money must be invested to raise the earnings to $72 per year?

76. A sum of $1250 is to be divided between two people in the ratio of 2 to 3. How much does each person receive?

77. An inheritance of $180,000 is to be divided between a child and the local cancer fund in the ratio of 5 to 1. How much money will the child receive?

Thoughts Into Words

78. Explain the difference between a ratio and a proportion.

79. What is wrong with the following procedure? Explain how it should be done.

$$\frac{x}{2} + 4 = \frac{x}{6}$$

$$6\left(\frac{x}{2} + 4\right) = 2(x)$$

$$3x + 24 = 2x$$

$$x = -24$$

80. Estimate an answer for each of the following problems. Also explain how you arrived at your estimate. Then work out the problem to see how well you estimated.

(a) The ratio of female students to male students at a small private college is 5 to 3. If there is a total of 1096 students, find the number of male students.

(b) If 15 pounds of fertilizer will cover 1200 square feet of lawn, how many pounds are needed for 3000 square feet?

(c) An investment of $5000 earns $300 interest in a year. At the same rate, how much money must be invested to earn $450?

(d) If the ratio of the length of a rectangle to its width is 5 to 3, and the length is 70 centimeters, find its width.

Further Investigations

Solve each of the following equations. Don't forget that division by zero is undefined.

81. $\dfrac{3}{x-2} = \dfrac{6}{2x-4}$

82. $\dfrac{8}{2x+1} = \dfrac{4}{x-3}$

83. $\dfrac{5}{x-3} = \dfrac{10}{x-6}$

84. $\dfrac{6}{x-1} = \dfrac{5}{x-1}$

85. $\dfrac{x-2}{2} = \dfrac{x}{2} - 1$

86. $\dfrac{x+3}{x} = 1 + \dfrac{3}{x}$

Answers to the Concept Quiz

1. True   **2.** False   **3.** True   **4.** True   **5.** False   **6.** False   **7.** False   **8.** True   **9.** False   **10.** True

---

4.2   More on Percents and Problem Solving

OBJECTIVES

1. Solve equations involving decimal numbers

2. Solve word problems involving discount

3. Solve word problems involving selling price

4. Use the simple interest formula to solve problems

We can solve the equation $x + 0.35 = 0.72$ by subtracting 0.35 from both sides of the equation. Another technique for solving equations that contain decimals is to clear the equation of all decimals by multiplying both sides by an appropriate power of 10. The following examples demonstrate both techniques in a variety of situations.

Classroom Example
Solve $0.3m = 81$.

EXAMPLE 1   Solve $0.5x = 14$.

Solution

$$0.5x = 14$$
$$5x = 140 \qquad \text{Multiplied both sides by 10}$$
$$x = 28 \qquad \text{Divided both sides by 5}$$

The solution set is $\{28\}$.

Classroom Example
Solve $d - 0.2d = 48$.

EXAMPLE 2   Solve $x + 0.04x = 5.2$.

Solution

$$x + 0.04x = 5.2$$
$$1.04x = 5.2 \qquad \text{Combined similar terms}$$
$$x = \frac{5.2}{1.04}$$
$$x = 5$$

The solution set is $\{5\}$.

Classroom Example
Solve $0.07x + 0.05x = 7.2$.

EXAMPLE 3   Solve $0.08y + 0.09y = 3.4$.

Solution

$$0.08y + 0.09y = 3.4$$
$$0.17y = 3.4 \qquad \text{Combined similar terms}$$
$$y = \frac{3.4}{0.17}$$
$$y = 20$$

The solution set is $\{20\}$.

Classroom Example
Solve
$0.09w = 240 - 0.05(w + 600)$.

EXAMPLE 4   Solve $0.10t = 560 - 0.12(t + 1000)$.

Solution

$$0.10t = 560 - 0.12(t + 1000)$$
$$10t = 56{,}000 - 12(t + 1000) \qquad \text{Multiplied both sides by 100}$$
$$10t = 56{,}000 - 12t - 12{,}000 \qquad \text{Distributive property}$$
$$22t = 44{,}000$$
$$t = 2000$$

The solution set is $\{2000\}$.

Problems Involving Percents

Many consumer problems can be solved with an equation approach. For example, we have this general guideline regarding discount sales:

> Original selling price − Discount = Discount sale price

Next we consider some examples using algebraic techniques along with this basic guideline.

Classroom Example
Dan bought a shirt at a 25% discount sale for $45. What was the original price of the shirt?

EXAMPLE 5

Amy bought a dress at a 30% discount sale for $35. What was the original price of the dress?

Solution

Let p represent the original price of the dress. We can use the basic discount guideline to set up an algebraic equation.

Original selling price − Discount = Discount sale price

$$(100\%)(p) \quad - \quad (30\%)(p) \quad = \quad \$35$$

Solving this equation, we obtain

$$(100\%)(p) - (30\%)(p) = 35$$
$$1.00p - 0.30p = 35 \qquad \text{Changed percents to decimals}$$
$$0.7p = 35$$
$$7p = 350$$
$$p = 50$$

The original price of the dress was $50.

Don't forget that if an item is on sale for 30% off, then you are going to pay 100% − 30% = 70% of the original price. So at a 30% discount sale, a $50 dress can be purchased for (70%)($50) = (0.70)($50) = $35. (Note that we just checked our answer for Example 5.)

Classroom Example
Find the cost of a $120 coat on sale for 15% off.

EXAMPLE 6

Find the cost of a $60 pair of jogging shoes on sale for 20% off (see Figure 4.4).

Figure 4.4

Solution

Let x represent the discount sale price. Since the shoes are on sale for 20% off, we must pay 80% of the original price.

$$x = (80\%)(60)$$
$$= (0.8)(60) = 48$$

The sale price is $48.

Here is another equation that we can use in consumer problems:

> Selling price = Cost + Profit

Profit (also called "markup, markon, margin, and margin of profit") may be stated in different ways. It may be stated as a percent of the selling price, a percent of the cost, or simply in terms of dollars and cents. Let's consider some problems where the profit is either a percent of the selling price or a percent of the cost.

Classroom Example
A retailer has some shoes that cost her $55 each. She wants to sell them at a profit of 40% of the cost. What selling price should be marked on the shoes?

EXAMPLE 7

A retailer has some shirts that cost him $20 each. He wants to sell them at a profit of 60% of the cost. What selling price should be marked on the shirts?

Solution

Let s represent the selling price. The basic relationship *selling price equals cost plus profit* can be used as a guideline.

$$\text{Selling price} = \text{Cost} + \text{Profit (\% of cost)}$$
$$\downarrow \qquad \quad \downarrow \qquad \quad \downarrow$$
$$s \qquad = \$20 + (60\%)(20)$$

Solving this equation, we obtain

$$s = 20 + (60\%)(20)$$
$$s = 20 + (0.6)(20) \qquad \text{Changed percent to decimal}$$
$$s = 20 + 12$$
$$s = 32$$

The selling price should be $32.

Classroom Example
Jorge bought an antique chair for $170 and later decided to resell it. He made a profit of 15% of the selling price. How much did he receive for the antique chair?

EXAMPLE 8

Kathrin bought a painting for $120 and later decided to resell it. She made a profit of 40% of the selling price. How much did she receive for the painting?

Solution

We can use the same basic relationship as a guideline, except this time the profit is a percent of the selling price. Let s represent the selling price.

Selling price = Cost + Profit (% of selling price)

$$s \quad = \quad 120 \quad + \quad (40\%)(s)$$

Solving this equation, we obtain

$$s = 120 + (40\%)(s)$$
$$s = 120 + 0.4s$$
$$0.6s = 120 \qquad \text{Subtracted } 0.4s \text{ from both sides}$$
$$s = \frac{120}{0.6} = 200$$

She received $200 for the painting.

Certain types of investment problems can be translated into algebraic equations. In some of these problems, we use the simple interest formula $i = Prt$, where i represents the amount of interest earned by investing P dollars at a yearly rate of r percent for t years.

Classroom Example
Isabel invested $5600 for 3 years and received $1092 in interest. Find the annual interest rate Isabel received on her investment.

EXAMPLE 9

John invested $9300 for 2 years and received $1395 in interest. Find the annual interest rate John received on his investment.

Solution

$$i = Prt$$
$$1395 = 9300r(2)$$
$$1395 = 18600r$$
$$\frac{1395}{18600} = r$$
$$0.075 = r$$

The annual interest rate is 7.5%.

EXAMPLE 10

How much principal must be invested to receive $1500 in interest when the investment is made for 3 years at an annual interest rate of 6.25%?

Solution

$$i = Prt$$
$$1500 = P(0.0625)(3)$$
$$1500 = P(0.1875)$$
$$\frac{1500}{0.1875} = P$$
$$8000 = P$$

The principal must be $8000.

EXAMPLE 11

How much monthly interest will be charged on a credit card bill with a balance of $754 when the credit card company charges an 18% annual interest rate?

Solution

$$i = Prt$$
$$i = 754(0.18)\left(\frac{1}{12}\right) \qquad \text{Remember, 1 month is } \frac{1}{12} \text{ of a year}$$
$$i = 11.31$$

The interest charge would be $11.31.

Concept Quiz 4.2

For Problems 1–10, answer true or false.

1. To clear the decimals from the equation $0.5x + 1.24 = 0.07x + 1.8$, you would multiply both sides of the equation by 10.

2. If an item is on sale for 35% off, then you are going to pay 65% of the original price.

3. Profit is always a percent of the selling price.

4. In the formula $i = Prt$, the r represents the interest return.

5. The basic relationship, *selling price equals cost plus profit*, can be used whether the profit is based on selling price or cost.

6. If a retailer buys a dozen golf balls for $28 and sells them for $36.40, she is making a 30% profit based on the cost.

7. The solution set for the equations $0.3x + 0.7(20 - x) = 8$ is {15}.

8. If a retailer buys a dozen golf balls for $28 and sells them for $40.00, she is making a 30% profit based on the selling price.

9. The cost of a $72 pair of shoes at a 20% discount sale is $54.

10. Five hundred dollars invested at a yearly rate of 7% simple interest earns $70 in 2 years.

Problem Set 4.2

For Problems 1–22, solve each of the equations. **(Objective 1)**

1. $x - 0.36 = 0.75$ **2.** $x - 0.15 = 0.42$

3. $x + 7.6 = 14.2$ **4.** $x + 11.8 = 17.1$

5. $0.62 - y = 0.14$ **6.** $7.4 - y = 2.2$

7. $0.7t = 56$ **8.** $1.3t = 39$

9. $x = 3.36 - 0.12x$ **10.** $x = 5.3 - 0.06x$

11. $s = 35 + 0.3s$ **12.** $s = 40 + 0.5s$

13. $s = 42 + 0.4s$ **14.** $s = 24 + 0.6s$

15. $0.07x + 0.08(x + 600) = 78$

16. $0.06x + 0.09(x + 200) = 63$

17. $0.09x + 0.1(2x) = 130.5$

18. $0.11x + 0.12(3x) = 188$

19. $0.08x + 0.11(500 - x) = 50.5$

20. $0.07x + 0.09(2000 - x) = 164$

21. $0.09x = 550 - 0.11(5400 - x)$

22. $0.08x = 580 - 0.1(6000 - x)$

For Problems 23–38, set up an equation and solve each problem. **(Objectives 2 and 3)**

23. Tom bought an electric drill at a 30% discount sale for $35. What was the original price of the drill?

24. Magda bought a dress for $140, which represents a 20% discount of the original price. What was the original price of the dress?

25. Find the cost of a $4800 wide-screen high-definition television that is on sale for 25% off.

26. Byron purchased a computer monitor at a 10% discount sale for $121.50. What was the original price of the monitor?

27. Suppose that Jack bought a $32 putter at the golf pro shop on sale for 35% off. How much did Jack pay for the putter?

28. Swati bought a 13-inch portable color TV for 20% off of the list price. The list price was $229.95. What did she pay for the TV?

29. Pierre bought a coat for $126 that was listed for $180. What rate of discount did he receive?

30. Phoebe paid $32 for a pair of sandals that was listed for $40. What rate of discount did she receive?

31. A retailer has some toe rings that cost him $5 each. He wants to sell them at a profit of 70% of the cost. What should the selling price be for the toe rings?

32. A retailer has some video games that cost her $25 each. She wants to sell them at a profit of 80% of the cost. What price should she charge for the video games?

33. The owner of a pizza parlor wants to make a profit of 55% of the cost for each pizza sold. If it costs $8 to make a pizza, at what price should it be sold?

34. Fresh produce in a food market usually has a high markup because of the loss to the retailer because of spoilage. If a head of lettuce costs a retailer $0.50, at what price should it be sold to realize a profit of 130% of the cost?

35. Jewelry has a very high markup rate. If a ring costs a jeweler $400, at what price should it be sold to gain a profit of 60% of the selling price?

36. If a box of candy costs a retailer $2.50 and he wants to make a profit of 50% based on the selling price, what price should he charge for the candy?

37. If the cost of a pair of shoes for a retailer is $32 and he sells them for $44.80, what is his rate of profit based on the cost?

38. A retailer has some skirts that cost her $24. If she sells them for $31.20, find her rate of profit based on the cost.

For Problems 39–46, use the formula $i = Prt$ to reach a solution. **(Objective 4)**

39. Find the annual interest rate if $560 in interest is earned when $3500 was invested for 2 years.

40. How much interest will be charged on a student loan if $8000 is borrowed for 9 months at a 19.2% annual interest rate?

41. How much principal, invested at 8% annual interest for 3 years, is needed to earn $1000?

42. How long will $2400 need to be invested at a 5.5% annual interest rate to earn $330?

43. What will be the interest earned on a $5000 certificate of deposit invested at 3.8% annual interest for 10 years?

44. One month a credit card company charged $38.15 in interest on a balance of $2725. What annual interest rate is the credit card company charging?

45. How much is a month's interest on a mortgage balance of $145,000 at a 6.5% annual interest rate?

46. For how many years must $2000 be invested at a 5.4% annual interest rate to earn $162?

Thoughts Into Words

47. What is wrong with the following procedure, and how should it be changed?

$$1.2x + 2 = 3.8$$
$$10(1.2x) + 2 = 10(3.8)$$
$$12x + 2 = 38$$
$$12x = 36$$
$$x = 3$$

48. From a consumer's viewpoint, would you prefer that a retailer figure profit based on the cost or on the selling price of an item? Explain your answer.

Further Investigations

49. A retailer buys an item for $40, resells it for $50, and claims that she is making only a 20% profit. Is her claim correct?

50. A store has a special discount sale of 40% off on all items. It also advertises an additional 10% off on items bought in quantities of a dozen or more. How much will it cost to buy a dozen items of some particular kind that regularly sell for $5 per item? (Be careful, a 40% discount followed by a 10% discount is not equal to a 50% discount.)

51. Is a 10% discount followed by a 40% discount the same as a 40% discount followed by a 10% discount? Justify your answer.

52. Some people use the following formula for determining the selling price of an item when the profit is based on a percent of the selling price:

$$\text{Selling price} = \frac{\text{Cost}}{100\% - \text{Percent of profit}}$$

Show how to develop this formula.

Solve each of the following equations and express the solutions in decimal form. Your calculator might be of some help.

53. $2.4x + 5.7 = 9.6$

54. $-3.2x - 1.6 = 5.8$

55. $0.08x + 0.09(800 - x) = 68.5$

56. $0.10x + 0.12(720 - x) = 80$

57. $7x - 0.39 = 0.03$

58. $9x - 0.37 = 0.35$

59. $0.2(t + 1.6) = 3.4$

60. $0.4(t - 3.8) = 2.2$

Answers to the Concept Quiz
1. False   **2.** True   **3.** False   **4.** False   **5.** True   **6.** True   **7.** True   **8.** True   **9.** False   **10.** True

4.3   Formulas

OBJECTIVES

1 Solve formulas for a specific variable when given the numerical values for the remaining variables

2 Solve formulas for a specific variable

3 Apply geometric formulas

4 Solve an equation for a specific variable

To find the distance traveled in 3 hours at a rate of 50 miles per hour, we multiply the rate by the time. Thus the distance is $50(3) = 150$ miles. We usually state the rule *distance equals rate times time* as a formula: $d = rt$. **Formulas** are simply rules we state in symbolic language and express as equations. Thus the formula $d = rt$ is an equation that involves three variables: d, r, and t.

As we work with formulas, it is often necessary to solve for a specific variable when we have numerical values for the remaining variables. Consider the following examples.

Classroom Example
Solve $d = rt$ for r, if $d = 210$ and $t = 3$.

EXAMPLE 1   Solve $d = rt$ for r if $d = 330$ and $t = 6$.

Solution

Substitute 330 for d and 6 for t in the given formula to obtain

$$330 = r(6)$$

Solving this equation yields

$$330 = 6r$$
$$55 = r$$

Classroom Example
Solve $C = \dfrac{5}{9}(F - 32)$ for F if $C = 25$.

EXAMPLE 2

Solve $C = \dfrac{5}{9}(F - 32)$ for F if $C = 10$. (This formula expresses the relationship between the Fahrenheit and Celsius temperature scales.)

Solution

Substitute 10 for C to obtain

$$10 = \frac{5}{9}(F - 32)$$

Solving this equation produces

$$\frac{9}{5}(10) = \frac{9}{5}\left(\frac{5}{9}\right)(F - 32) \qquad \text{Multiply both sides by } \frac{9}{5}$$
$$18 = F - 32$$
$$50 = F$$

Sometimes it may be convenient to change a formula's form by using the properties of equality. For example, the formula $d = rt$ can be changed as follows:

$$d = rt$$
$$\frac{d}{r} = \frac{rt}{r} \qquad \text{Divide both sides by } r$$
$$\frac{d}{r} = t$$

We say that the formula $d = rt$ has been *solved for the variable t*. The formula can also be *solved for r* as follows:

$$d = rt$$
$$\frac{d}{t} = \frac{rt}{t} \qquad \text{Divide both sides by } t$$
$$\frac{d}{t} = r$$

Geometric Formulas

There are several formulas in geometry that we use quite often. Let's briefly review them at this time; they will be used periodically throughout the remainder of the text. These formulas (along with some others) and Figures 4.5–4.15 are also listed in the inside front cover of this text.

Rectangle

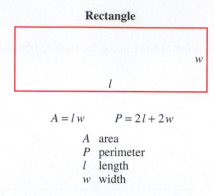

$$A = lw \qquad P = 2l + 2w$$

A area
P perimeter
l length
w width

Figure 4.5

Triangle

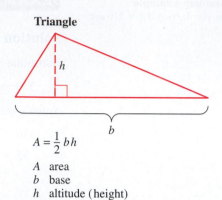

$$A = \frac{1}{2} bh$$

A area
b base
h altitude (height)

Figure 4.6

Trapezoid

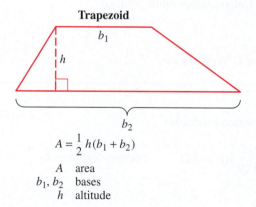

$$A = \frac{1}{2} h(b_1 + b_2)$$

A area
b_1, b_2 bases
h altitude

Figure 4.7

Parallelogram

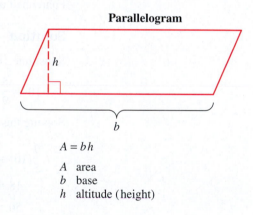

$$A = bh$$

A area
b base
h altitude (height)

Figure 4.8

Circle

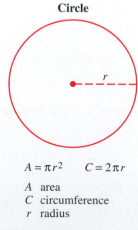

$$A = \pi r^2 \qquad C = 2\pi r$$

A area
C circumference
r radius

Figure 4.9

Sphere

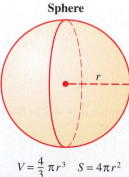

$$V = \frac{4}{3} \pi r^3 \qquad S = 4\pi r^2$$

S surface area
V volume
r radius

Figure 4.10

Rectangular Prism

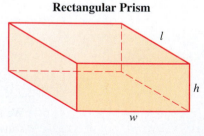

$$V = lwh \qquad S = 2hw + 2hl + 2lw$$

V volume
S total surface area
w width
l length
h altitude (height)

Figure 4.11

Prism

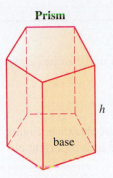

$V = Bh$

V volume
B area of base
h altitude (height)

Figure 4.12

Pyramid

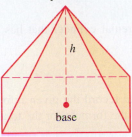

$V = \frac{1}{3}Bh$

V volume
B area of base
h altitude (height)

Figure 4.13

Right Circular Cylinder

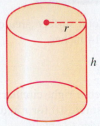

$V = \pi r^2 h$ $S = 2\pi r^2 + 2\pi rh$

V volume
S total surface area
r radius
h altitude (height)

Figure 4.14

Right Circular Cone

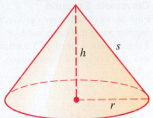

$V = \frac{1}{3}\pi r^2 h$ $S = \pi r^2 + \pi rs$

V volume
S total surface area
r radius
h altitude (height)
s slant height

Figure 4.15

Classroom Example
Solve $A = lw$ for w.

EXAMPLE 3   Solve $C = 2\pi r$ for r.

Solution

$$C = 2\pi r$$

$$\frac{C}{2\pi} = \frac{2\pi r}{2\pi} \qquad \text{Divide both sides by } 2\pi$$

$$\frac{C}{2\pi} = r$$

Classroom Example
Solve $A = \frac{1}{2}bh$ for b.

EXAMPLE 4   Solve $V = \frac{1}{3}Bh$ for h.

Solution

$$V = \frac{1}{3}Bh$$

$$3(V) = 3\left(\frac{1}{3}Bh\right) \qquad \text{Multiply both sides by 3}$$

$$3V = Bh$$

$$\frac{3V}{B} = \frac{Bh}{B} \qquad \text{Divide both sides by } B$$

$$\frac{3V}{B} = h$$

Classroom Example
Solve $A = \frac{1}{2}h(b_1 + b_2)$ for h.

EXAMPLE 5   Solve $P = 2l + 2w$ for w.

Solution

$$P = 2l + 2w$$

$$P - 2l = 2l + 2w - 2l \qquad \text{Subtract } 2l \text{ from both sides}$$

$$P - 2l = 2w$$

$$\frac{P - 2l}{2} = \frac{2w}{2} \qquad \text{Divide both sides by 2}$$

$$\frac{P - 2l}{2} = w$$

Classroom Example

Find the total surface area of a rectangular prism that has a width of 3 centimeters, a length of 5 centimeters, and a height of 8 centimeters.

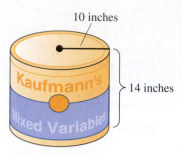

Figure 4.16

EXAMPLE 6

Find the total surface area of a right circular cylinder that has a radius of 10 inches and a height of 14 inches.

Solution

Let's sketch a right circular cylinder and record the given information as shown in Figure 4.16. Substitute 10 for r and 14 for h in the formula for the total surface area of a right circular cylinder to obtain

$$S = 2\pi r^2 + 2\pi rh$$
$$= 2\pi(10)^2 + 2\pi(10)(14)$$
$$= 200\pi + 280\pi$$
$$= 480\pi$$

The total surface area is 480π square inches.

In Example 6 we used the figure to record the given information, and it also served as a reminder of the geometric figure under consideration. Now let's consider an example where the figure helps us to analyze the problem.

Classroom Example

A painting is in a frame that has a width of 3 inches on all sides. The width of the painting is 11 inches, and the height is 16 inches. Find the area of the frame.

EXAMPLE 7

A sidewalk 3 feet wide surrounds a rectangular plot of ground that measures 75 feet by 100 feet. Find the area of the sidewalk.

Solution

We can make a sketch and record the given information as in Figure 4.17. The area of the sidewalk can be found by subtracting the area of the rectangular plot from the area of the plot

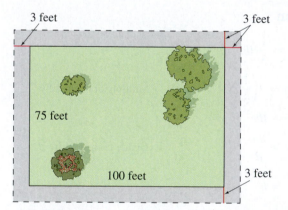

Figure 4.17

plus the sidewalk (the large dashed rectangle). The width of the large rectangle is $75 + 3 + 3 = 81$ feet, and its length is $100 + 3 + 3 = 106$ feet.

$$A = (81)(106) - (75)(100)$$
$$= 8586 - 7500 = 1086$$

The area of the sidewalk is 1086 square feet.

Changing Forms of Equations

In Chapter 8 you will be working with equations that contain two variables. At times you will need to solve for one variable in terms of the other variable—that is, change the form of the equation just like we have done with formulas. The next examples illustrate once again how we can use the properties of equality for such situations.

Classroom Example
Solve $4x + 3y = 9$ for x.

EXAMPLE 8   Solve $3x + y = 4$ for x.

Solution

$$3x + y = 4$$
$$3x + y - y = 4 - y \qquad \text{Subtract } y \text{ from both sides}$$
$$3x = 4 - y$$
$$\frac{3x}{3} = \frac{4 - y}{3} \qquad \text{Divide both sides by 3}$$
$$x = \frac{4 - y}{3}$$

Classroom Example
Solve $2x - 3y = 10$ for y.

EXAMPLE 9   Solve $4x - 5y = 7$ for y.

Solution

$$4x - 5y = 7$$
$$4x - 5y - 4x = 7 - 4x \qquad \text{Subtract } 4x \text{ from both sides}$$
$$-5y = 7 - 4x$$
$$\frac{-5y}{-5} = \frac{7 - 4x}{-5} \qquad \text{Divide both sides by } -5$$
$$y = \frac{7 - 4x}{-5}\left(\frac{-1}{-1}\right) \qquad \begin{array}{l}\text{Multiply numerator and denominator} \\ \text{of fraction on the right by } -1; \\ \text{we commonly do this so that the} \\ \text{denominator is positive}\end{array}$$
$$y = \frac{4x - 7}{5}$$

Classroom Example
Solve $y = mx + b$ for x.

EXAMPLE 10   Solve $y = mx + b$ for m.

Solution

$$y = mx + b$$
$$y - b = mx + b - b \qquad \text{Subtract } b \text{ from both sides}$$
$$y - b = mx$$
$$\frac{y - b}{x} = \frac{mx}{x} \qquad \text{Divide both sides by } x$$
$$\frac{y - b}{x} = m$$

Concept Quiz 4.3

For Problems 1–10, match the correct formula (next page) for each category below.

1. Volume of a right circular cone
2. Circumference of a circle
3. Volume of a rectangular prism
4. Area of a triangle
5. Area of a circle
6. Volume of a right circular cylinder
7. Volume of a prism
8. Surface area of a sphere
9. Area of a parallelogram
10. Volume of sphere

A. $A = \pi r^2$

B. $V = lwh$

C. $V = Bh$

D. $S = 4\pi r^2$

E. $V = \dfrac{1}{3}\pi r^2 h$

F. $A = bh$

G. $V = \dfrac{4}{3}\pi r^3$

H. $A = \dfrac{1}{2}bh$

I. $C = 2\pi r$

J. $V = \pi r^2 h$

Problem Set 4.3

For Problems 1–10, solve for the specified variable using the given facts. (Objective 1)

1. Solve $d = rt$ for t if $d = 336$ and $r = 48$.

2. Solve $d = rt$ for r if $d = 486$ and $t = 9$.

3. Solve $i = Prt$ for P if $i = 200$, $r = 0.08$, and $t = 5$.

4. Solve $i = Prt$ for t if $i = 540$, $P = 750$, and $r = 0.09$.

5. Solve $F = \dfrac{9}{5}C + 32$ for C if $F = 68$.

6. Solve $C = \dfrac{5}{9}(F - 32)$ for F if $C = 15$.

7. Solve $V = \dfrac{1}{3}Bh$ for B if $V = 112$ and $h = 7$.

8. Solve $V = \dfrac{1}{3}Bh$ for h if $V = 216$ and $B = 54$.

9. Solve $A = P + Prt$ for t if $A = 652$, $P = 400$, and $r = 0.07$.

10. Solve $A = P + Prt$ for P if $A = 1032$, $r = 0.06$, and $t = 12$.

For Problems 11–32, use the geometric formulas given in this section to help solve the problems. (Objective 3)

11. Find the perimeter of a rectangle that is 14 centimeters long and 9 centimeters wide.

12. If the perimeter of a rectangle is 80 centimeters and its length is 24 centimeters, find its width.

13. If the perimeter of a rectangle is 108 inches and its length is $3\dfrac{1}{4}$ feet, find its width in inches.

14. How many yards of fencing does it take to enclose a rectangular plot of ground that is 69 feet long and 42 feet wide?

15. A dirt path 4 feet wide surrounds a rectangular garden that is 38 feet long and 17 feet wide. Find the area of the dirt path.

16. Find the area of a cement walk 3 feet wide that surrounds a rectangular plot of ground 86 feet long and 42 feet wide.

17. Suppose that paint costs $8.00 per liter, and that 1 liter will cover 9 square meters of surface. We are going to paint (on one side only) 50 rectangular pieces of wood of the same size that have a length of 60 centimeters and a width of 30 centimeters. What will the cost of the paint be?

18. A lawn is in the shape of a triangle with one side 130 feet long and the altitude to that side 60 feet long. Will one bag of fertilizer that covers 4000 square feet be enough to fertilize the lawn?

19. Find the length of an altitude of a trapezoid with bases of 8 inches and 20 inches and an area of 98 square inches.

20. A flower garden is in the shape of a trapezoid with bases of 6 yards and 10 yards. The distance between the bases is 4 yards. Find the area of the garden.

21. In Figure 4.18 you'll notice that the diameter of a metal washer is 4 centimeters. The diameter of the hole is 2 centimeters. How many square centimeters of metal are there in 50 washers? Express the answer in terms of π.

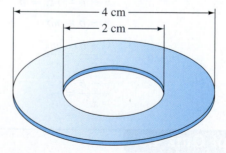

Figure 4.18

22. Find the area of a circular plot of ground that has a radius of length 14 meters. Use $3\dfrac{1}{7}$ as an approximation for π.

23. Find the area of a circular region that has a diameter of 1 yard. Express the answer in terms of π.

24. Find the area of a circular region if the circumference is 12π units. Express the answer in terms of π.

25. Find the total surface area and volume of a sphere that has a radius 9 inches long. Express the answers in terms of π.

26. A circular pool is 34 feet in diameter and has a flagstone walk around it that is 3 feet wide (see Figure 4.19). Find the area of the walk. Express the answer in terms of π.

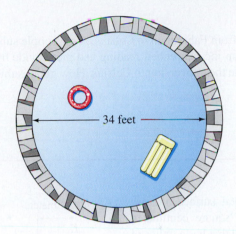

Figure 4.19

27. Find the volume and total surface area of a right circular cylinder that has a radius of 8 feet and a height of 18 feet. Express the answers in terms of π.

28. Find the total surface area and volume of a sphere that has a diameter 12 centimeters long. Express the answers in terms of π.

29. If the volume of a right circular cone is 324π cubic inches, and a radius of the base is 9 inches long, find the height of the cone.

30. Find the volume and total surface area of a tin can if the radius of the base is 3 centimeters, and the height of the can is 10 centimeters. Express the answers in terms of π.

31. If the total surface area of a right circular cone is 65π square feet, and a radius of the base is 5 feet long, find the slant height of the cone.

32. If the total surface area of a right circular cylinder is 104π square meters, and a radius of the base is 4 meters long, find the height of the cylinder.

For Problems 33–42, match the correct formula for each statement.

33. Area of a rectangle   A. $A = \pi r^2$

34. Circumference of a circle   B. $V = lwh$

35. Volume of a rectangular prism C. $P = 2l + 2w$

36. Area of a triangle   D. $V = \dfrac{4}{3}\pi r^3$

37. Area of a circle   E. $A = lw$

38. Volume of a right circular cylinder F. $A = bh$

39. Perimeter of a rectangle   G. $A = \dfrac{1}{2}h(b_1 + b_2)$

40. Volume of a sphere   H. $A = \dfrac{1}{2}bh$

41. Area of a parallelogram   I. $C = 2\pi r$

42. Area of a trapezoid   J. $V = \pi r^2 h$

For Problems 43–54, solve each formula for the indicated variable. (Before doing these problems, cover the right-hand column and see how many of these formulas you recognize!) (Objective 2)

43. $V = Bh$ for h

44. $A = lw$ for l

45. $V = \dfrac{1}{3}Bh$ for B

46. $A = \dfrac{1}{2}bh$ for h

47. $P = 2l + 2w$ for w

48. $V = \pi r^2 h$ for h

49. $V = \dfrac{1}{3}\pi r^2 h$ for h

50. $i = Prt$ for t

51. $F = \dfrac{9}{5}C + 32$ for C

52. $A = P + Prt$ for t

53. $A = 2\pi r^2 + 2\pi rh$ for h

54. $C = \dfrac{5}{9}(F - 32)$ for F

For Problems 55–70, solve each equation for the indicated variable. (Objective 4)

55. $3x + 7y = 9$ for x

56. $5x + 2y = 12$ for x

57. $9x - 6y = 13$ for y

58. $3x - 5y = 19$ for y

59. $-2x + 11y = 14$ for x

60. $-x + 14y = 17$ for x

61. $y = -3x - 4$ for x

62. $y = -7x + 10$ for x

63. $\dfrac{x - 2}{4} = \dfrac{y - 3}{6}$ for y

64. $\dfrac{x + 1}{3} = \dfrac{y - 5}{2}$ for y

65. $ax - by - c = 0$ for y

66. $ax + by = c$ for y

67. $\dfrac{x + 6}{2} = \dfrac{y + 4}{5}$ for x

68. $\dfrac{x - 3}{6} = \dfrac{y - 4}{8}$ for x

69. $m = \dfrac{y - b}{x}$ for y

70. $y = mx + b$ for x

Thoughts Into Words

71. Suppose that both the length and width of a rectangle are doubled. How does this affect the perimeter of the rectangle? Defend your answer.

72. Suppose that the length of a radius of a circle is doubled. How does this affect the area of the circle? Defend your answer.

73. To convert from Fahrenheit to Celsius, some people subtract 32 from the Fahrenheit reading and then divide by 2 to estimate the conversion to Celsius. How good is this estimate?

Further Investigations

For each of the following problems, use 3.14 as an approximation for π. Your calculator should be of some help with these problems.

74. Find the area of a circular plot of ground that has a radius 16.3 meters long. Express your answer to the nearest tenth of a square meter.

75. Find the area, to the nearest tenth of a square centimeter, of the ring in Figure 4.20.

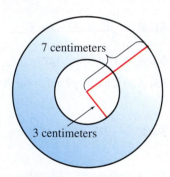

Figure 4.20

76. Find the area, to the nearest square inch, of each of these pizzas: 10-inch diameter, 12-inch diameter, 14-inch diameter.

77. Find the total surface area, to the nearest square centimeter, of the tin can shown in Figure 4.21.

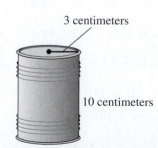

Figure 4.21

78. Find the total surface area, to the nearest square centimeter, of a baseball that has a radius of 4 centimeters (see Figure 4.22).

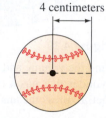

Figure 4.22

79. Find the volume, to the nearest cubic inch, of a softball that has a diameter of 5 inches (see Figure 4.23).

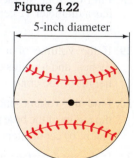

Figure 4.23

80. Find the volume, to the nearest cubic meter, of the rocket in Figure 4.24.

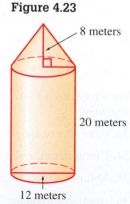

Figure 4.24

4.4

4.4   Problem Solving

OBJECTIVES

1 Apply problem solving techniques such as drawing diagrams, sketching figures, and using a guideline to solve word problems

2 Solve word problems involving simple interest

3 Solve word problems involving the perimeter of rectangles, triangles, or circles

4 Solve word problems involving distance, rate, and time

We begin this section by restating the suggestions for solving word problems that we offered in Section 3.3.

Suggestions for Solving Word Problems

1. Read the problem carefully, and make sure that you understand the meanings of all the words. Be especially alert for any technical terms used in the statement of the problem.

2. Read the problem a second time (perhaps even a third time) to get an overview of the situation being described and to determine the known facts as well as what is to be found.

3. Sketch any figure, diagram, or chart that might be helpful in analyzing the problem.

4. Choose a meaningful variable to represent an unknown quantity in the problem (perhaps t if time is the unknown quantity); represent any other unknowns in terms of that variable.

5. Look for a guideline that can be used to set up an equation. A guideline might be a formula such as *selling price equals cost plus profit*, or a relationship such as *interest earned from a 9% investment plus interest earned from a 10% investment equals total amount of interest earned*. A guideline may also be illustrated by a figure or diagram that you sketch for a particular problem.

6. Form an equation that contains the variable and that translates the conditions of the guideline from English into algebra.

7. Solve the equation, and use the solution to determine all the facts requested in the problem.

8. **Check all answers back into the original statement of the problem.**

Again we emphasize the importance of suggestion 5. Determining the guideline to follow when setting up the equation is key to analyzing a problem. Sometimes the guideline is a formula, such as one of the formulas we presented in the previous section and accompanying problem set. Let's consider an example of that type.

Classroom Example
How long will it take $750 to triple itself if it is invested at 4% simple interest?

EXAMPLE 1

How long will it take $500 to double itself if it is invested at 8% simple interest?

Solution

We can use the basic simple interest formula, $i = Prt$, where i represents interest, P is the principal (money invested), r is the rate (percent), and t is the time in years. For $500 to double itself means that we want $500 to earn another $500 in interest. Thus using $i = Prt$ as a guideline, we can proceed as follows:

$$i = Prt$$

$$500 = 500(8\%)(t)$$

Now let's solve this equation.

$$500 = 500(0.08)(t)$$

$$1 = 0.08t$$

$$100 = 8t$$

$$\frac{100}{8} = t$$

$$12\frac{1}{2} = t$$

It will take $12\frac{1}{2}$ years.

If the problem involves a geometric formula, then a sketch of the figure is helpful for recording the given information and analyzing the problem. The next example illustrates this idea.

Classroom Example
The length of a field is 15 meters less than three times its width, and the perimeter of the field is 450 meters. Find the length and width of the field.

EXAMPLE 2

The length of a football field is 40 feet more than twice its width, and the perimeter of the field is 1040 feet. Find the length and width of the field.

Solution

Since the length is stated in terms of the width, we can let w represent the width, and then $2w + 40$ represents the length, as shown in Figure 4.25. A guideline for this problem is the perimeter formula $P = 2l + 2w$. Thus the following equation can be set up and solved.

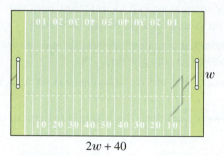

$$2w + 40$$

Figure 4.25

$$P = 2l + 2w$$

$$1040 = 2(2w + 40) + 2w$$

$$1040 = 4w + 80 + 2w$$

$$1040 = 6w + 80$$

$$960 = 6w$$

$$160 = w$$

If $w = 160$, then $2w + 40 = 2(160) + 40 = 360$. Thus, the football field is 360 feet long and 160 feet wide.

Sometimes the formulas we use when we are analyzing a problem are different than those we use as a guideline for setting up the equation. For example, uniform motion problems involve the formula $d = rt$, but the main guideline for setting up an equation for such problems is usually a statement about either *times*, *rates*, or *distances*. Let's consider an example.

EXAMPLE 3

Pablo leaves city A on a moped traveling toward city B at 18 miles per hour. At the same time, Cindy leaves city B on a moped traveling toward city A at 23 miles per hour. The distance between the two cities is 123 miles. How long will it take before Pablo and Cindy meet on their mopeds?

Solution

First, let's sketch a diagram as in Figure 4.26. Let t represent the time that Pablo travels. Then t also represents the time that Cindy travels.

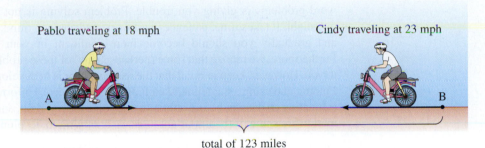

Pablo traveling at 18 mph   Cindy traveling at 23 mph

A   B

total of 123 miles

Figure 4.26

Distance Pablo travels + Distance Cindy travels = Total distance

$$18t \qquad + \qquad 23t \qquad = \qquad 123$$

Solving this equation yields

$$18t + 23t = 123$$
$$41t = 123$$
$$t = 3$$

They both travel for 3 hours.

Some people find it helpful to use a chart to organize the known and unknown facts in a uniform motion problem. We will illustrate with an example.

EXAMPLE 4

A car leaves a town traveling at 60 kilometers per hour. How long will it take a second car traveling at 75 kilometers per hour to catch the first car if the second car leaves 1 hour later?

Solution

Let t represent the time of the second car. Then $t + 1$ represents the time of the first car because it travels 1 hour longer. We can now record the information of the problem in a chart.

	Rate	Time	Distance
First car	60	$t + 1$	$60(t + 1)$
Second car	75	t	$75t$

$d = rt$

Because the second car is to overtake the first car, the distances must be equal.

Distance of second car = Distance of first car

$$75t = 60(t + 1)$$

Solving this equation yields

$$75t = 60(t + 1)$$
$$75t = 60t + 60$$
$$15t = 60$$
$$t = 4$$

The second car should overtake the first car in 4 hours. (Check the answer!)

We would like to offer one bit of advice at this time. Don't become discouraged if solving word problems is giving you trouble. Problem solving is not a skill that can be developed overnight. It takes time, patience, hard work, and an open mind. Keep giving it your best shot, and gradually you should become more confident in your approach to such problems. Furthermore, we realize that some (perhaps many) of these problems may not seem "practical" to you. However, keep in mind that the real goal here is to develop problem-solving techniques. Finding and using a guideline, sketching a figure to record information and help in the analysis, estimating an answer before attempting to solve the problem, and using a chart to record information are some of the important tools we are trying to develop.

Concept Quiz 4.4

Arrange the following steps for solving word problems in the correct order.

A. Declare a variable and represent any other unknown quantities in terms of that variable.

B. Check the answer back into the original statement of the problem.

C. Write an equation for the problem, and remember to look for a formula or guideline that could be used to write the equation.

D. Read the problem carefully, and be sure that you understand all the terms in the stated problem.

E. Sketch a diagram or figure that helps you analyze the problem.

F. Solve the equation, and determine the answer to the question asked in the problem.

Problem Set 4.4

For Problems 1–12, solve each of the equations. These equations are the types you will be using in Problems 13–40.

1. $950(0.12)t = 950$

2. $1200(0.09)t = 1200$

3. $w + \dfrac{1}{4}w - 1 = 19$

4. $w + \dfrac{2}{3}w + 1 = 41$

5. $500(0.08)t = 1000$

6. $800(0.11)t = 1600$

7. $s + (2s - 1) + (3s - 4) = 37$

8. $s + (3s - 2) + (4s - 4) = 42$

9. $\dfrac{5}{2}r + \dfrac{5}{2}(r + 6) = 135$

10. $\dfrac{10}{3}r + \dfrac{10}{3}(r - 3) = 90$

11. $24\left(t - \dfrac{2}{3}\right) = 18t + 8$

12. $16t + 8\left(\dfrac{9}{2} - t\right) = 60$

Solve each of the following problems. Keep in mind the suggestions we offered in this section. **(Objectives 2–4)**

13. How long will it take $4000 to double itself if it is invested at 8% simple interest?

14. How many years will it take $1000 to double itself if it is invested at 5% simple interest?

15. How long will it take $8000 to triple itself if it is invested at 6% simple interest?

16. How many years will it take $500 to earn $750 in interest if it is invested at 6% simple interest?

17. The length of a rectangle is three times its width. If the perimeter of the rectangle is 112 inches, find its length and width.

18. The width of a rectangle is one-half of its length. If the perimeter of the rectangle is 54 feet, find its length and width.

19. Suppose that the length of a rectangle is 2 centimeters less than three times its width. The perimeter of the rectangle is 92 centimeters. Find the length and width of the rectangle.

20. Suppose that the length of a certain rectangle is 1 meter more than five times its width. The perimeter of the rectangle is 98 meters. Find the length and width of the rectangle.

21. The width of a rectangle is 3 inches less than one-half of its length. If the perimeter of the rectangle is 42 inches, find the area of the rectangle.

22. The width of a rectangle is 1 foot more than one-third of its length. If the perimeter of the rectangle is 74 feet, find the area of the rectangle.

23. The perimeter of a triangle is 100 feet. The longest side is 3 feet less than twice the shortest side, and the third side is 7 feet longer than the shortest side. Find the lengths of the sides of the triangle.

24. A triangular plot of ground has a perimeter of 54 yards. The longest side is twice the shortest side, and the third side is 2 yards longer than the shortest side. Find the lengths of the sides of the triangle.

25. The second side of a triangle is 1 centimeter longer than three times the first side. The third side is 2 centimeters longer than the second side. If the perimeter is 46 centimeters, find the length of each side of the triangle.

26. The second side of a triangle is 3 meters shorter than twice the first side. The third side is 4 meters longer than the second side. If the perimeter is 58 meters, find the length of each side of the triangle.

27. The perimeter of an equilateral triangle is 4 centimeters more than the perimeter of a square, and the length of a side of the triangle is 4 centimeters more than the length of a side of the square. Find the length of a side of the equilateral triangle. (An equilateral triangle has three sides of the same length.)

28. Suppose that a square and an equilateral triangle have the same perimeter. Each side of the equilateral triangle is 6 centimeters longer than each side of the square. Find the length of each side of the square. (An equilateral triangle has three sides of the same length.)

29. Suppose that the length of a radius of a circle is the same as the length of a side of a square. If the circumference of the circle is 15.96 centimeters longer than the perimeter of the square, find the length of a radius of the circle. (Use 3.14 as an approximation for π.)

30. The circumference of a circle is 2.24 centimeters more than six times the length of a radius. Find the radius of the circle. (Use 3.14 as an approximation for π.)

31. Sandy leaves a town traveling in her car at a rate of 45 miles per hour. One hour later, Monica leaves the same town traveling the same route at a rate of 50 miles per hour. How long will it take Monica to overtake Sandy?

32. Two cars start from the same place traveling in opposite directions. One car travels 4 miles per hour faster than the other car. Find their speeds if after 5 hours they are 520 miles apart.

33. The distance between Jacksonville and Miami is 325 miles. A freight train leaves Jacksonville and travels toward Miami at 40 miles per hour. At the same time, a passenger train leaves Miami and travels toward Jacksonville at 90 miles per hour. How long will it take the two trains to meet?

34. Kirk starts jogging at 5 miles per hour. One-half hour later, Consuela starts jogging on the same route at 7 miles per hour. How long will it take Consuela to catch Kirk?

35. A car leaves a town traveling at 40 miles per hour. Two hours later a second car leaves the town traveling the same route and overtakes the first car in 5 hours and 20 minutes. How fast was the second car traveling?

36. Two airplanes leave St. Louis at the same time and fly in opposite directions (see Figure 4.27). If one travels at 500 kilometers per hour, and the other at 600 kilometers per hour, how long will it take for them to be 1925 kilometers apart?

Figure 4.27

37. Two trains leave at the same time, one traveling east and the other traveling west. At the end of $9\frac{1}{2}$ hours they are 1292 miles apart. If the rate of the train traveling east is 8 miles per hour faster than the rate of the other train, find their rates.

38. Dawn starts on a 58-mile trip on her moped at 20 miles per hour. After a while the motor stops, and she pedals the remainder of the trip at 12 miles per hour. The entire trip takes $3\frac{1}{2}$ hours. How far had Dawn traveled when the motor on the moped quit running?

39. Jeff leaves home on his bicycle and rides out into the country for 3 hours. On his return trip, along the same route, it takes him three-quarters of an hour longer. If his rate on the return trip was 2 miles per hour slower than on the trip out into the country, find the total roundtrip distance.

40. In $1\frac{1}{4}$ hours more time, Rita, riding her bicycle at 12 miles per hour, rode 2 miles farther than Sonya, who was riding her bicycle at 16 miles per hour. How long did each girl ride?

Thoughts Into Words

41. Suppose that your friend analyzes Problem 31 as follows: Sandy has traveled 45 miles before Monica starts. Since Monica travels 5 miles per hour faster than Sandy, it will take her $\frac{45}{5} = 9$ hours to catch Sandy. How would you react to this analysis of the problem?

42. Summarize the ideas about problem solving that you have acquired thus far in this course.

Answers to the Concept Quiz
D　　E　　A　　C　　F　　B

4.5	More about Problem Solving

OBJECTIVES

1 Solve word problems involving mixture

2 Solve word problems involving age

3 Solve word problems involving investments

We begin this section with an important but often overlooked facet of problem solving: the importance of looking back over your solution and considering some of the following questions.

1. Is your answer to the problem a reasonable answer? Does it agree with the answer you estimated before doing the problem?

2. Have you checked your answer by substituting it back into the conditions stated in the problem?

3. Do you now see another plan that you can use to solve the problem? Perhaps there is even another guideline that you can use.

4. Do you now see that this problem is closely related to another problem that you have solved previously?

5. Have you "tucked away for future reference" the technique you used to solve this problem?

Looking back over the solution of a newly solved problem can provide a foundation for solving problems in the future.

Now let's consider three examples of what we often refer to as mixture problems. No basic formula applies for all of these problems, but the suggestion that you *think in terms of a pure substance* is often helpful when setting up a guideline. For example, a phrase such as "30% solution of acid" means that 30% of the amount of solution is acid and the remaining 70% is water.

EXAMPLE 1

How many milliliters of pure acid must be added to 150 milliliters of a 30% solution of acid to obtain a 40% solution (see Figure 4.28)?

Remark: If a guideline is not apparent from reading the problem, it might help you to guess an answer and then check that guess. Suppose we guess that 30 milliliters of pure acid need to be added. To check, we must determine whether the final solution is 40% acid. Since we started with 0.30(150) = 45 milliliters of pure acid and added our guess of 30 milliliters, the final solution will have 45 + 30 = 75 milliliters of pure acid. The final amount of solution is 150 + 30 = 180 milliliters. Thus the final solution is $\dfrac{75}{180} = 41\dfrac{2}{3}\%$ pure acid.

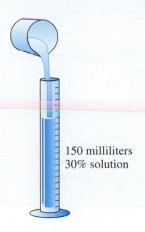

150 milliliters
30% solution

Figure 4.28

Solution

We hope that by guessing and checking the guess, you obtain the following guideline:

Amount of pure acid in original solution	+	Amount of pure acid to be added	=	Amount of pure acid in final solution

Let p represent the amount of pure acid to be added. Then using the guideline, we can form the following equation:

$$(30\%)(150) + p = 40\%(150 + p)$$

Now let's solve this equation to determine the amount of pure acid to be added.

$$(0.30)(150) + p = 0.40(150 + p)$$
$$45 + p = 60 + 0.4p$$
$$0.6p = 15$$
$$p = \frac{15}{0.6} = 25$$

We must add 25 milliliters of pure acid. (Perhaps you should check this answer.)

EXAMPLE 2

Suppose we have a supply of a 30% solution of alcohol and a 70% solution. How many quarts of each should be mixed to produce a 20-quart solution that is 40% alcohol?

Solution

We can use a guideline similar to the one we suggested in Example 1.

Pure alcohol in 30% solution	+	Pure alcohol in 70% solution	=	Pure alcohol in 40% solution

Let x represent the amount of 30% solution. Then $20 - x$ represents the amount of 70% solution. Now using the guideline, we translate into

$$(30\%)(x) + (70\%)(20 - x) = (40\%)(20)$$

Solving this equation, we obtain

$$0.30x + 0.70(20 - x) = 8$$
$$30x + 70(20 - x) = 800 \qquad \text{Multiplied both sides by 100}$$
$$30x + 1400 - 70x = 800$$
$$-40x = -600$$
$$x = 15$$

Therefore, $20 - x = 5$. We should mix 15 quarts of the 30% solution with 5 quarts of the 70% solution.

Classroom Example
A 20-gallon drink container is full and contains a 50% solution of fruit juice. How much needs to be drained out and replaced with pure fruit juice to obtain an 80% solution of fruit juice?

EXAMPLE 3

A 4-gallon radiator is full and contains a 40% solution of antifreeze. How much needs to be drained out and replaced with pure antifreeze to obtain a 70% solution?

Solution

This guideline can be used:

Pure antifreeze in the original solution	−	Pure antifreeze in the solution drained out	+	Pure antifreeze added	=	Pure antifreeze in the final solution

Let x represent the amount of pure antifreeze to be added. Then x also represents the amount of the 40% solution to be drained out. Thus the guideline translates into the following equation:

$$(40\%)(4) - (40\%)(x) + x = (70\%)(4)$$

Solving this equation, we obtain

$$0.4(4) - 0.4x + x = 0.7(4)$$
$$1.6 + 0.6x = 2.8$$
$$0.6x = 1.2$$
$$x = 2$$

Therefore, we must drain out 2 gallons of the 40% solution and then add 2 gallons of pure antifreeze. (Checking this answer is a worthwhile exercise for you!)

Classroom Example
A man invested a total of $12,000; part of it is invested at 5% and the remainder at 3%. His total yearly interest from the two investments is $500. How much did he invest at each rate?

EXAMPLE 4

A woman invests a total of $5000. Part of it is invested at 4% and the remainder at 6%. Her total yearly interest from the two investments is $260. How much did she invest at each rate?

Solution

Let x represent the amount invested at 6%. Then $5000 - x$ represents the amount invested at 4%. Use the following guideline:

Interest earned from 6% investment	+	Interest earned from 4% investment	=	Total interest earned
$(6\%)(x)$	+	$(4\%)(\$5000 - x)$	=	$260

Solving this equation yields

$$(6\%)(x) + (4\%)(5000 - x) = 260$$
$$0.06x + 0.04(5000 - x) = 260$$
$$6x + 4(5000 - x) = 26{,}000 \qquad \text{Multiplied both sides by 100}$$
$$6x + 20{,}000 - 4x = 26{,}000$$

$$2x + 20,000 = 26,000$$
$$2x = 6000$$
$$x = 3000$$

Therefore, $5000 - x = 2000$.

She invested $3000 at 6% and $2000 at 4%.

Classroom Example
An investor invests a certain amount of money at 4%. Then she finds a better deal and invests $2000 more than that amount at 6%. Her yearly income from the two investments is $1020. How much did she invest at each rate?

EXAMPLE 5

An investor invests a certain amount of money at 3%. Then he finds a better deal and invests $5000 more than that amount at 5%. His yearly income from the two investments is $650. How much did he invest at each rate?

Solution

Let x represent the amount invested at 3%. Then $x + 5000$ represents the amount invested at 5%.

$$(3\%)(x) + (5\%)(x + 5000) = 650$$
$$0.03x + 0.05(x + 5000) = 650$$
$$3x + 5(x + 5000) = 65,000 \qquad \text{Multiplied both sides by 100}$$
$$3x + 5x + 25,000 = 65,000$$
$$8x + 25,000 = 65,000$$
$$8x = 40,000$$
$$x = 5000$$

Therefore, $x + 5000 = 10,000$.

He invested $5000 at 3% and $10,000 at 5%.

Now let's consider a problem where the process of representing the various unknown quantities in terms of one variable is the key to solving the problem.

Classroom Example
Lauran is 2 years older than her sister Caitlin, and 2 years ago Caitlin was eight-ninths as old as Lauran. Find their present ages.

EXAMPLE 6

Jody is 6 years younger than her sister Cathy, and in 7 years Jody will be three-fourths as old as Cathy. Find their present ages.

Solution

By letting c represent Cathy's present age, we can represent all of the unknown quantities as follows:

$$c: \quad \text{Cathy's present age}$$
$$c - 6: \quad \text{Jody's present age}$$
$$c + 7: \quad \text{Cathy's age in 7 years}$$
$$c - 6 + 7 \text{ or } c + 1: \quad \text{Jody's age in 7 years}$$

The statement that Jody's age in 7 years will be three-fourths of Cathy's age at that time serves as the guideline. So we can set up and solve the following equation.

$$c + 1 = \frac{3}{4}(c + 7)$$
$$4c + 4 = 3(c + 7) \qquad \text{Multiplied both sides by 4}$$
$$4c + 4 = 3c + 21$$
$$c = 17$$

Therefore, Cathy's present age is 17, and Jody's present age is $17 - 6 = 11$.

Concept Quiz 4.5

For Problems 1–10, answer true or false.

1. The phrase "a 40% solution of alcohol" means that 40% of the amount of solution is alcohol.

2. The amount of pure acid in 300 ml of a 30% solution is 100 ml.

3. If we want to produce 10 quarts by mixing solution A and solution B, the amount of solution A needed could be represented by x and the amount of solution B would then be represented by $10 - x$.

4. The formula $d = rt$ is equivalent to $r = \dfrac{d}{t}$.

5. The formula $d = rt$ is equivalent to $t = \dfrac{r}{d}$.

6. If y represents John's current age, then his age four years ago would be represented by $y - 4$.

7. If Shane's current age is represented by x, then his age in 10 years would be represented by $10x$.

8. The solution set of $0.2x + 0.3(x - 2) = 0.9$ is $\{3\}$.

9. The solution set of $5x + 6\left(3\dfrac{1}{2} - x\right) = 19$ is $\{-1\}$.

10. The solution set of $0.04x - 0.5(x + 1.5) = -4.43$ is $\{8\}$.

Problem Set 4.5

For Problems 1–12, solve each equation. You will be using these types of equations in Problems 13–41.

1. $0.3x + 0.7(20 - x) = 0.4(20)$

2. $0.4x + 0.6(50 - x) = 0.5(50)$

3. $0.2(20) + x = 0.3(20 + x)$

4. $0.3(32) + x = 0.4(32 + x)$

5. $0.7(15) - x = 0.6(15 - x)$

6. $0.8(25) - x = 0.7(25 - x)$

7. $0.4(10) - 0.4x + x = 0.5(10)$

8. $0.2(15) - 0.2x + x = 0.4(15)$

9. $20x + 12\left(4\dfrac{1}{2} - x\right) = 70$

10. $30x + 14\left(3\dfrac{1}{2} - x\right) = 97$

11. $3t = \dfrac{11}{2}\left(t - \dfrac{3}{2}\right)$

12. $5t = \dfrac{7}{3}\left(t + \dfrac{1}{2}\right)$

Set up an equation and solve each of the following problems. **(Objectives 1–3)**

13. How many milliliters of pure acid must be added to 100 milliliters of a 10% acid solution to obtain a 20% solution?

14. How many liters of pure alcohol must be added to 20 liters of a 40% solution to obtain a 60% solution?

15. How many centiliters of distilled water must be added to 10 centiliters of a 50% acid solution to obtain a 20% acid solution?

16. How many milliliters of distilled water must be added to 50 milliliters of a 40% acid solution to reduce it to a 10% acid solution?

17. Suppose that we want to mix some 30% alcohol solution with some 50% alcohol solution to obtain 10 quarts of a 35% solution. How many quarts of each kind should we use?

18. We have a 20% alcohol solution and a 50% solution. How many pints must be used from each to obtain 8 pints of a 30% solution?

19. How much water needs to be removed from 20 gallons of a 30% salt solution to change it to a 40% salt solution?

20. How much water needs to be removed from 30 liters of a 20% salt solution to change it to a 50% salt solution?

21. Suppose that a 12-quart radiator contains a 20% solution of antifreeze. How much solution needs to be drained out and replaced with pure antifreeze to obtain a 40% solution of antifreeze?

22. A tank contains 50 gallons of a 40% solution of antifreeze. How much solution needs to be drained out and replaced with pure antifreeze to obtain a 50% solution?

23. How many gallons of a 15% salt solution must be mixed with 8 gallons of a 20% salt solution to obtain a 17% salt solution?

24. How many liters of a 10% salt solution must be mixed with 15 liters of a 40% salt solution to obtain a 20% salt solution?

25. Thirty ounces of a punch that contains 10% grapefruit juice is added to 50 ounces of a punch that contains 20% grapefruit juice. Find the percent of grapefruit juice in the resulting mixture.

26. Suppose that 20 gallons of a 20% salt solution is mixed with 30 gallons of a 25% salt solution. What is the percent of salt in the resulting solution?

27. Suppose that the perimeter of a square equals the perimeter of a rectangle. The width of the rectangle is 9 inches less than twice the side of the square, and the length of the rectangle is 3 inches less than twice the side of the square. Find the dimensions of the square and the rectangle.

28. The perimeter of a triangle is 40 centimeters. The longest side is 1 centimeter longer than twice the shortest side. The other side is 2 centimeters shorter than the longest side. Find the lengths of the three sides.

29. Andy starts walking from point A at 2 miles per hour. One-half hour later, Aaron starts walking from point A at $3\frac{1}{2}$ miles per hour and follows the same route. How long will it take Aaron to catch up with Andy?

30. Suppose that Karen, riding her bicycle at 15 miles per hour, rode 10 miles farther than Michelle, who was riding her bicycle at 14 miles per hour. Karen rode for 30 minutes longer than Michelle. How long did Michelle and Karen each ride their bicycles?

31. Pam is half as old as her brother Bill. Six years ago Bill was four times older than Pam. How old is each sibling now?

32. Suppose that the sum of the present ages of Tom and his father is 100 years. Ten years ago, Tom's father was three times as old as Tom was at that time. Find their present ages.

33. Annilee's present age is two-thirds of Jessie's present age. In 12 years, the sum of their ages will be 54 years. Find their present ages.

34. Suppose that Cory invested a certain amount of money at 3% interest, and he invested $750 more than that amount at 5%. His total yearly interest was $157.50. How much did he invest at each rate?

35. Nina received an inheritance of $12,000 from her grandmother. She invested part of it at 6% interest, and she invested the remainder at 8%. If the total yearly interest from both investments was $860, how much did she invest at each rate?

36. Udit received $1200 from his parents as a graduation present. He invested part of it at 4% interest, and he invested the remainder at 6%. If the total yearly interest amounted to $62, how much did he invest at each rate?

37. Sally invested a certain sum of money at 3%, twice that sum at 4%, and three times that sum at 5%. Her total yearly interest from all three investments was $130. How much did she invest at each rate?

38. If $2000 is invested at 8% interest, how much money must be invested at 11% interest so that the total return for both investments averages 10%?

39. Fawn invested a certain amount of money at 3% interest and she invested $1250 more than that amount at 5%. Her total yearly interest was $134.50. How much did she invest at each rate?

40. A sum of $2300 is invested, part of it at 5% interest and the remainder at 6%. If the interest earned by the 6% investment is $50 more than the interest earned by the 5% investment, find the amount invested at each rate.

41. If $3000 is invested at 3% interest, how much money must be invested at 6% so that the total return for both investments averages 5%?

42. How can $5400 be invested, part of it at 4% and the remainder at 5%, so that the two investments will produce the same amount of interest?

43. A sum of $6000 is invested, part of it at 5% interest and the remainder at 7%. If the interest earned by the 5% investment is $160 less than the interest earned by the 7% investment, find the amount invested at each rate.

Answers to the Concept Quiz

1. True   **2.** False   **3.** True   **4.** True   **5.** False   **6.** True   **7.** False   **8.** True   **9.** False   **10.** True

Chapter 4 Summary

OBJECTIVE	SUMMARY	EXAMPLE
Solve proportions. (Section 4.1/Objective 1)	A ratio is the comparison of two numbers by division. A statement of equality between two ratios is a proportion. In a proportion, the cross products are equal; that is to say: If $\dfrac{a}{b} = \dfrac{c}{d}$, then $ad = bc$, when $b \neq 0$ and $d \neq 0$.	Solve $\dfrac{5}{x-2} = \dfrac{6}{x+3}$. **Solution** $$\dfrac{5}{x-2} = \dfrac{6}{x+3}$$ $$6(x-2) = 5(x+3)$$ $$6x - 12 = 5x + 15$$ $$x - 12 = 15$$ $$x = 27$$ The solution set is $\{27\}$.
Solve word problems using proportions. (Section 4.1/Objective 4)	A variety of word problems can be set up and solved using proportions.	The scale on a blueprint shows that 1 inch represents 8 feet. If the length of the house on the blueprint is 4.5 inches, what is the length of the actual house? **Solution** Set up the proportion $\dfrac{1}{8} = \dfrac{4.5}{x}$ and solve for x. $$x = 4.5(8) = 36$$ The house is 36 feet long.
Use a proportion to convert a fraction to a percent. (Section 4.1/Objective 2)	The concept of *percent* means "per one hundred" and is therefore a ratio that has a denominator of 100. To convert a fraction to a percent, set up a proportion and solve.	Convert $\dfrac{3}{16}$ to a percent. **Solution** $$\dfrac{n}{100} = \dfrac{3}{16}$$ $$16n = 300$$ $$n = \dfrac{300}{16} = 18\dfrac{3}{4}$$ Therefore, $\dfrac{3}{16} = 18\dfrac{3}{4}\%$.
Solve basic percent problems. (Section 4.1/Objective 3)	There are three basic types of percent problems. Each of the types can be solved by translating the problem into a simple algebraic equation and solving.	**1.** What is 35% of 400? **Solution** $$n = 0.35(400) = 140$$ Therefore 140 is 35% of 400. **2.** Twelve percent of what number is 30? **Solution** $$0.12x = 30$$ $$x = \dfrac{30}{0.12} = 250$$ Therefore 12% of 250 is 30.

OBJECTIVE	SUMMARY	EXAMPLE
		3. Fifteen is what percent of 80? **Solution** $15 = x(80)$ $x = \dfrac{15}{80} = 0.1875$ Therefore, 15 is 18.75% of 80.
Solve equations involving decimal numbers. **(Section 4.2/Objective 1)**	To solve equations that contain decimals, you can either calculate with the decimals or clear the equation of all decimals by multiplying both sides of the equation by an appropriate power of 10.	Solve $0.04x + 0.05x = 1.8$. **Solution** Method 1 $0.04x + 0.05x = 1.8$ $0.09x = 1.8$ $x = \dfrac{1.8}{0.09} = 20$ The solution set is $\{20\}$. Method 2 $0.04x + 0.05x = 1.8$ $100(0.04x + 0.05x) = 100(1.8)$ $4x + 5x = 180$ $9x = 180$ $x = 20$ The solution set is $\{20\}$.
Solve word problems involving discount, selling price, cost, or profit. **(Section 4.2/Objectives 2 and 3)**	Many consumer problems involve the concept of *percent* and can be solved with an equation approach. The basic relationships *selling price equals cost plus profit* and *original selling price minus the discount equals the discount sale price* are frequently used to solve problems.	A computer store paid $20 each for some new games. What should the selling price be to make an 80% profit based on the cost? **Solution** Let s represent the selling price. $s = 20 + (80\%)(20)$ $s = 20 + 0.80(20)$ $s = 36$ Therefore the selling price should be $36.
Solve simple interest problems. **(Section 4.2/Objective 4)**	The simple interest formula $i = Prt$, where i represents the amount of interest earned by investing P dollars at a yearly interest rate of r percent for t years, is used to solve certain types of investment problems.	For how many years must $2000 be invested at 8% interest to earn $240? **Solution** $i = Prt$ $240 = 0.08(2000)(t)$ $240 = 160t$ $1.5 = t$ The $2000 must be invested for 1.5 years at 8% interest.
Solve formulas by evaluating. **(Section 4.3/Objective 2)**	Substitute the given values into the formula and solve for the remaining variable.	Solve $d = rt$ for r if $d = 140$ and $t = 2.5$. **Solution** $d = rt$ $140 = r(2.5)$ $56 = r$

(continued)

OBJECTIVE	SUMMARY	EXAMPLE
Apply geometric formulas. **(Section 4.3/Objective 3)**	Geometry formulas are used to solve problems involving geometric figures. The formulas we use quite often are presented in Section 4.3.	Find the width of a rectangle that has a length of 8 inches and a perimeter of 24 inches. **Solution** Use the formula $P = 2l + 2w$ $P = 2l + 2w$ $24 = 2(8) + 2w$ $24 = 16 + 2w$ $8 = 2w$ $4 = w$ Therefore the width is 4 inches.
Solve formulas for a specific variable. **(Section 4.3/Objective 2)**	We can solve a formula such as $P = 2l + 2w$ for l or for w by applying the properties of equality.	Solve $P = 2l + 2w$ for l. **Solution** $P = 2l + 2w$ $P - 2w = 2l$ $\dfrac{P - 2w}{2} = l$
Solve an equation for a specific variable. **(Section 4.3/Objective 4)**	In future chapters there will be equations with two variables, for example x and y. At times you will need to solve for one variable in terms of the other variable.	Solve $2x + 3y = 12$ for y. **Solution** $2x + 3y = 12$ $3y = -2x + 12$ $y = \dfrac{-2x + 12}{3}$
Solve word problems. **(Section 4.4/Objectives 2–4; Section 4.5/Objectives 1–3)**	Use these suggestions to help you solve word problems. 1. Read the problem carefully. 2. Sketch a figure, diagram, or chart to help organize the facts. 3. Choose a meaningful variable. Be sure to write down what the variable represents. 4. Look for a guideline. This is often the key component when solving a problem. Many times formulas are used as guidelines. 5. Use the guideline to set up an equation. 6. Solve the equation and answer the question in the problem. 7. Check your answer in the original statement of the problem.	Mary starts jogging at 4 miles per hour. One-half hour later, Tom starts jogging on the same route at 6 miles per hour. How long will it take Tom to catch up with Mary? **Solution** Let x represents Tom's time. Then because Mary left one-half hour before Tom, $x + \dfrac{1}{2}$ represents the time she is jogging. Note that the distance traveled by each person is the same. The distance can be found by applying the formula $d = rt$. Mary's distance = Tom's distance $4\left(x + \dfrac{1}{2}\right) = 6x$ $4x + 2 = 6x$ $2 = 2x$ $1 = x$ Therefore, it takes Tom 1 hour to catch up with Mary.

Chapter 4   Review Problem Set

In Problems 1–5, solve each of the equations.

1. $0.5x + 0.7x = 1.7$

2. $0.07t + 0.12(t - 3) = 0.59$

3. $0.1x + 0.12(1700 - x) = 188$

4. $x - 0.25x = 12$

5. $0.2(x - 3) = 14$

6. Solve $P = 2l + 2w$ for w if $P = 50$ and $l = 19$.

7. Solve $F = \dfrac{9}{5}C + 32$ for C if $F = 77$.

8. Solve $A = P + Prt$ for t.

9. Solve $2x - 3y = 13$ for x.

10. Find the area of a trapezoid that has one base 8 inches long, the other base 14 inches long, and the altitude between the two bases 7 inches.

11. If the area of a triangle is 27 square centimeters, and the length of one side is 9 centimeters, find the length of the altitude to that side.

12. If the total surface area of a right circular cylinder is 152π square feet, and the radius of the base is 4 feet long, find the height of the cylinder.

Set up an equation and solve each of the following problems.

13. Eighteen is what percent of 30?

14. The sum of two numbers is 96, and their ratio is 5 to 7. Find the numbers.

15. Fifteen percent of a certain number is 6. Find the number.

16. Suppose that the length of a certain rectangle is 5 meters longer than twice the width. The perimeter of the rectangle is 46 meters. Find the length and width of the rectangle.

17. Two airplanes leave Chicago at the same time and fly in opposite directions. If one travels at 350 miles per hour and the other at 400 miles per hour, how long will it take them to be 1125 miles apart?

18. How many liters of pure alcohol must be added to 10 liters of a 70% solution to obtain a 90% solution?

19. A copper wire 110 centimeters long was bent in the shape of a rectangle. The length of the rectangle was 10

centimeters longer than twice the width. Find the dimensions of the rectangle.

20. Seventy-eight yards of fencing were purchased to enclose a rectangular garden. The length of the garden is 1 yard shorter than three times its width. Find the length and width of the garden.

21. The ratio of the complement of an angle to the supplement of the angle is 7 to 16. Find the measure of the angle.

22. A car uses 18 gallons of gasoline for a 369-mile trip. At the same rate of consumption, how many gallons will it use on a 615-mile trip?

23. A sum of $2100 is invested, part of it at 3% interest and the remainder at 5%. If the interest earned by the 5% investment is $51 more than the interest from the 3% investment, find the amount invested at each rate.

24. A retailer has some sweaters that cost $28 each. At what price should the sweaters be sold to obtain a profit of 30% of the selling price?

25. Anastasia bought a dress on sale for $39, and the original price of the dress was $60. What percent discount did she receive?

26. One angle of a triangle has a measure of $47°$. Of the other two angles, one of them is $3°$ smaller than three times the other angle. Find the measures of the two remaining angles.

27. Connie rides out into the country on her bicycle at a rate of 10 miles per hour. An hour later Zak leaves from the same place that Connie did and rides his bicycle along the same route at 12 miles per hour. How long will it take Zak to catch Connie?

28. How many gallons of a 10% salt solution must be mixed with 12 gallons of a 15% salt solution to obtain a 12% salt solution?

29. Suppose that 20 ounces of a punch containing 20% orange juice is added to 30 ounces of a punch containing 30% orange juice. Find the percent of orange juice in the resulting mixture.

30. The sum of the present ages of Angie and her mother is 64 years. In 8 years, Angie will be three-fifths as old as her mother at that time. Find the present ages of Angie and her mother.

31. How much simple interest is due on a 2-year student loan when $3500 is borrowed at a 5.25% annual interest rate?

For Problems 1–10, solve each of the equations.

1. $\dfrac{x + 2}{4} = \dfrac{x - 3}{5}$

2. $\dfrac{-4}{2x - 1} = \dfrac{3}{3x + 5}$

3. $\dfrac{x - 1}{6} - \dfrac{x + 2}{5} = 2$

4. $\dfrac{x + 8}{7} - 2 = \dfrac{x - 4}{4}$

5. $\dfrac{n}{20 - n} = \dfrac{7}{3}$

6. $\dfrac{h}{4} + \dfrac{h}{6} = 1$

7. $0.05n + 0.06(400 - n) = 23$

8. $s = 35 + 0.5s$

9. $0.07n = 45.5 - 0.08(600 - n)$

10. $12t + 8\left(\dfrac{7}{2} - t\right) = 50$

11. Solve $F = \dfrac{9C + 160}{5}$ for C.

12. Solve $y = 2(x - 4)$ for x.

13. Solve $\dfrac{x + 3}{4} = \dfrac{y - 5}{9}$ for y.

For Problems 14–16, use the geometric formulas given in this chapter to help solve the problems.

14. Find the area of a circular region if the circumference is 16π centimeters. Express the answer in terms of π.

15. If the perimeter of a rectangle is 100 inches, and its length is 32 inches, find the area of the rectangle.

16. The area of a triangular plot of ground is 133 square yards. If the length of one side of the plot is 19 yards, find the length of the altitude to that side.

For Problems 17–25, set up an equation and solve each problem.

17. Express $\dfrac{5}{4}$ as a percent.

18. Thirty-five percent of what number is 24.5?

19. Cora bought a digital camera for $245, which represented a 30% discount of the original price. What was the original price of the digital camera?

20. A retailer has some lamps that cost her $40 each. She wants to sell them at a profit of 30% of the cost. What price should she charge for the lamps?

21. Hugh paid $48 for a pair of golf shoes that listed for $80. What rate of discount did he receive?

22. The election results in a certain precinct indicated that the ratio of female voters to male voters was 7 to 5. If a total of 1500 people voted, how many women voted?

23. A car leaves a city traveling at 50 miles per hour. One hour later a second car leaves the same city traveling the same route at 55 miles per hour. How long will it take the second car to overtake the first car?

24. How many centiliters of pure acid must be added to 6 centiliters of a 50% acid solution to obtain a 70% acid solution?

25. How long will it take $4000 to double itself if it is invested at 5% simple interest?

For Problems 1–10, simplify each algebraic expression by combining similar terms.

1. $7x - 9x - 14x$

2. $-10a - 4 + 13a + a - 2$

3. $5(x - 3) + 7(x + 6)$

4. $3(x - 1) - 4(2x - 1)$

5. $-3n - 2(n - 1) + 5(3n - 2) - n$

6. $6n + 3(4n - 2) - 2(2n - 3) - 5$

7. $\dfrac{1}{2}x - \dfrac{3}{4}x + \dfrac{2}{3}x - \dfrac{1}{6}x$

8. $\dfrac{1}{3}n - \dfrac{4}{15}n + \dfrac{5}{6}n - n$

9. $0.4x - 0.7x - 0.8x + x$

10. $0.5(x - 2) + 0.4(x + 3) - 0.2x$

For Problems 11–20, evaluate each algebraic expression for the given values of the variables.

11. $5x - 7y + 2xy$ for $x = -2$ and $y = 5$

12. $2ab - a + 6b$ for $a = 3$ and $b = -4$

13. $-3(x - 1) + 2(x + 6)$ for $x = -5$

14. $5(n + 3) - (n + 4) - n$ for $n = 7$

15. $\dfrac{3x - 2y}{2x - 3y}$ for $x = 3$ and $y = -6$

16. $\dfrac{3}{4}n - \dfrac{1}{3}n + \dfrac{5}{6}n$ for $n = -\dfrac{2}{3}$

17. $2a^2 - 4b^2$ for $a = 0.2$ and $b = -0.3$

18. $x^2 - 3xy - 2y^2$ for $x = \dfrac{1}{2}$ and $y = \dfrac{1}{4}$

19. $5x - 7y - 8x + 3y$ for $x = 9$ and $y = -8$

20. $\dfrac{3a - b - 4a + 3b}{a - 6b - 4b - 3a}$ for $a = -1$ and $b = 3$

For Problems 21–26, evaluate each of the expressions.

21. 3^4

22. -2^6

23. $\left(\dfrac{2}{3}\right)^3$

24. $\left(-\dfrac{1}{2}\right)^5$

25. $\left(\dfrac{1}{2} + \dfrac{1}{3}\right)^2$

26. $\left(\dfrac{3}{4} - \dfrac{7}{8}\right)^3$

For Problems 27–38, solve each equation.

27. $-5x + 2 = 22$

28. $3x - 4 = 7x + 4$

29. $3(4x - 1) = 6(2x - 1)$

30. $2(x - 1) - 3(x - 2) = 12$

31. $\dfrac{2}{5}x - \dfrac{1}{3} = \dfrac{1}{3}x + \dfrac{1}{2}$

32. $\dfrac{t - 2}{4} + \dfrac{t + 3}{3} = \dfrac{1}{6}$

33. $\dfrac{2n - 1}{5} - \dfrac{n + 2}{4} = 1$

34. $0.09x + 0.12(500 - x) = 54$

35. $-5(n + 1) - (n - 2) = 3(-2n - 1)$

36. $\dfrac{-2}{x - 1} = \dfrac{-3}{x + 4}$

37. $0.2x + 0.1(x - 4) = 0.7x - 1$

38. $-(t - 2) + (t - 4) = 2\left(t - \dfrac{1}{2}\right) - 3\left(t + \dfrac{1}{3}\right)$

For Problems 39–46, solve each of the inequalities.

39. $4x - 6 > 3x + 1$

40. $-3x - 6 < 12$

41. $-2(n - 1) \le 3(n - 2) + 1$

42. $\dfrac{2}{7}x - \dfrac{1}{4} \ge \dfrac{1}{4}x + \dfrac{1}{2}$

43. $0.08t + 0.1(300 - t) > 28$

44. $-4 > 5x - 2 - 3x$

45. $\dfrac{2}{3}n - 2 \ge \dfrac{1}{2}n + 1$

46. $-3 < -2(x - 1) - x$

For Problems 47–54, set up an equation or an inequality and solve each problem.

47. Erin's salary this year is $32,000. This represents $2000 more than twice her salary 5 years ago. Find her salary 5 years ago.

48. One of two supplementary angles is 45° smaller than four times the other angle. Find the measure of each angle.

49. Jaamal has 25 coins (nickels and dimes) that amount to $2.10. How many coins of each kind does he have?

50. Hana bowled 144 and 176 in her first two games. What must she bowl in the third game to have an average of at least 150 for the three games?

51. A board 30 feet long is cut into two pieces whose lengths are in the ratio of 2 to 3. Find the lengths of the two pieces.

52. A retailer has some shoes that cost him $32 per pair. He wants to sell them at a profit of 20% of the selling price. What price should he charge for the shoes?

53. Two cars start from the same place traveling in opposite directions. One car travels 5 miles per hour faster than the other car. Find their speeds if after 6 hours they are 570 miles apart.

54. How many liters of pure alcohol must be added to 15 liters of a 20% solution to obtain a 40% solution?

5 Coordinate Geometry and Linear Systems

Using the concept of slope, we can set up and solve the proportion $\dfrac{30}{100} = \dfrac{y}{5280}$ to determine the vertical change a street with a 30% grade has in 1 mile.

© Freddy Eliasson

René Descartes, a French mathematician of the 17th century, transformed geometric problems into an algebraic setting so that he could use the tools of algebra to solve those problems. This interface of algebraic and geometric ideas is the foundation of a branch of mathematics called analytic geometry; today more commonly called **coordinate geometry**.

We started to make this connection between algebra and geometry in Chapter 3 when graphing the solution sets of inequalities in one variable. For example, the solution set for $3x + 1 \geq 4$ is $\{x \mid x \geq 1\}$, and a geometric picture of this solution set is shown in Figure 5.1.

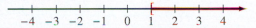

Figure 5.1

In this chapter we will associate pairs of real numbers with points in a geometric plane. This will provide the basis for obtaining pictures of algebraic equations and inequalities in two variables. Finally, we will work with systems of equations that will provide us with even more problem-solving power.

Video tutorials based on section learning objectives are available in a variety of delivery modes.

5.1   Cartesian Coordinate System

OBJECTIVES   **1** Plot points on a rectangular coordinate system

2 Solve equations for the specified variable

3 Draw graphs of equations by plotting points

In Section 2.3 we introduced the real number line (Figure 5.2), which is the result of setting up a one-to-one correspondence between the set of real numbers and the points on a line. Recall that the number associated with a point on the line is called the coordinate of that point.

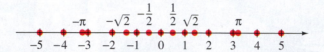

Figure 5.2

Now let's consider two lines (one vertical and one horizontal) that are perpendicular to each other at the point we associate with zero on both lines (Figure 5.3). We refer to these number lines as the horizontal and vertical **axes** and together as the **coordinate axes**; they partition the plane into four parts called **quadrants**. The quadrants are numbered counterclockwise from I to IV as indicated in Figure 5.3. The point of intersection of the two axes is called the **origin**.

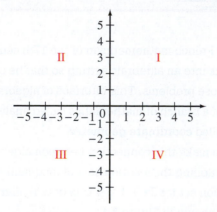

Figure 5.3

It is now possible to set up a one-to-one correspondence between ordered pairs of real numbers and the points in a plane. To each ordered pair of real numbers there corresponds a unique point in the plane, and to each point there corresponds a unique ordered pair of real numbers. We have indicated a part of this correspondence in Figure 5.4. The ordered pair (3, 1) corresponds to point *A* and denotes that the point *A* is located 3 units to the right of and 1 unit up from the origin. (The ordered pair (0, 0) corresponds to the origin.) The ordered pair (−2, 4) corresponds to point *B* and denotes that point *B* is located 2 units to the left of and 4 units up from the origin. Make sure that you agree with all of the other points plotted in Figure 5.4.

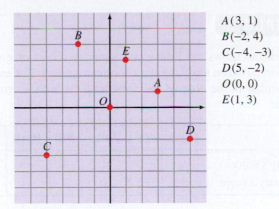

$A(3, 1)$
$B(-2, 4)$
$C(-4, -3)$
$D(5, -2)$
$O(0, 0)$
$E(1, 3)$

Figure 5.4

Remark: The notation $(-2, 4)$ was used earlier in this text to indicate an interval of the real number line. Now we are using the same notation to indicate an ordered pair of real numbers. This double meaning should not be confusing, because the context of the material will always indicate which meaning of the notation is being used. Throughout this chapter, we will be using the ordered-pair interpretation.

In general, we refer to the real numbers a and b in an ordered pair (a, b) associated with a point as the **coordinates of the point**. The first number, a, called the **abscissa**, is the directed distance of the point from the vertical axis measured parallel to the horizontal axis. The second number, b, called the **ordinate**, is the directed distance of the point from the horizontal axis measured parallel to the vertical axis (see Figure 5.5(a)). Thus in the first quadrant, all points have a positive abscissa and a positive ordinate. In the second quadrant, all points have a negative abscissa and a positive ordinate. We have indicated the signs in all four quadrants in Figure 5.5(b). This system of associating points with ordered pairs of real numbers is called the **Cartesian coordinate system** or the **rectangular coordinate system**.

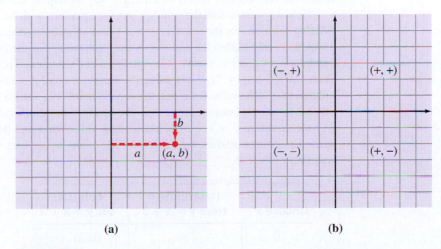

(a)   (b)

Figure 5.5

Plotting points on a rectangular coordinate system can be helpful when analyzing data to determine a trend or relationship. The following example shows the plot of some data.

Classroom Example
The chart below shows the changes in value of five stocks on Monday and then on Friday. Let Monday's value be the first number in the ordered pair, and let Friday's value be the second number in the ordered pair. List the ordered pairs. Plot the charted information on a rectangular coordinate system.

	A	B	C	D	E
Monday	+1	−2	−1	+3	+5
Friday	−3	+1	−4	−2	+3

EXAMPLE 1

The chart below shows the Friday and Saturday scores of golfers in terms of par. Plot the charted information on a rectangular coordinate system. For each golfer, let Friday's score be the first number in the ordered pair, and let Saturday's score be the second number in the ordered pair.

	Mark	Ty	Vinay	Bill	Herb	Rod
Friday's score	1	−2	−1	4	−3	0
Saturday's score	3	−2	0	7	−4	1

Solution

The ordered pairs are as follows:

Mark (1, 3)   Ty (−2, −2)

Vinay (−1, 0)   Bill (4, 7)

Herb (−3, −4)   Rod (0, 1)

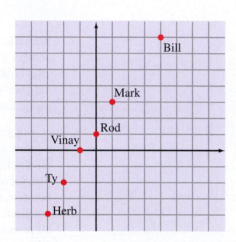

The points are plotted on the rectangular coordinate system in Figure 5.6. In the study of statistics, this graph of the charted data would be called a scatterplot. For this plot, the points appear to approximate a straight-line path, which suggests that there is a linear correlation between Friday's score and Saturday's score.

Figure 5.6

In Section 3.5, the real number line was used to display the solution set of an inequality. Recall that the solution set of an inequality such as $x > 3$ has an infinite number of solutions. Hence the real number line is an effective way to display the solution set of an inequality.

Now we want to find the solution sets for equations with two variables. Let's begin by considering solutions for the equation $y = x + 3$. A solution of an equation in two variables is a pair of numbers that makes the equation a true statement. For the equation $y = x + 3$, the pair of numbers $x = 4$ and $y = 7$ makes the equation a true statement. This pair of numbers can be written as an ordered pair (4, 7). When using the variables x and y, we agree that the first number of an ordered pair is the value for x, and the second number is a value for y. Likewise (2, 5) is a solution for $y = x + 3$ because $5 = 2 + 3$. We can find an infinite number of ordered pairs that satisfy the equation $y = x + 3$ by choosing a value for x and determining the corresponding value for y that satisfies the equation. The table below lists some of the solutions for the equation $y = x + 3$.

Choose x	Determine y from $y = x + 3$	Solution for $y = x + 3$
0	3	(0, 3)
1	4	(1, 4)
3	6	(3, 6)
5	8	(5, 8)
−1	2	(−1, 2)
−3	0	(−3, 0)
−5	−2	(−5, −2)

Because the number of solutions for the equation $y = x + 3$ is infinite, we do not have a convenient way to list the solution set. This is similar to inequalities for which the solution set is infinite. For inequalities we used a real number line to display the solution set. Now we will use the rectangular coordinate system to display the solution set of an equation in two variables.

On a rectangular coordinate system, where we label the horizontal axis as the x axis and the vertical axis as the y axis, we can locate the point associated with each ordered pair of numbers in the table. These points are shown in Figure 5.7(a). These points are only some of the infinite solutions of the equation $y = x + 3$. The straight line in Figure 5.7(b) that connects the points represents all the solutions of the equation; it is called the graph of the equation $y = x + 3$.

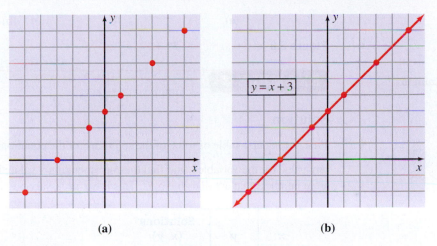

(a)   (b)

Figure 5.7

The next examples illustrate further the process of graphing equations.

Classroom Example
Graph $y = 3x + 2$.

EXAMPLE 2   Graph $y = 2x + 1$.

Solution

First, we set up a table of some of the solutions for the equation $y = 2x + 1$. You can choose any value for the variable x and find the corresponding y value. The values we chose for x are in the table and include positive integers, zero, and negative integers.

x	y	Solutions (x, y)
0	1	(0, 1)
1	3	(1, 3)
2	5	(2, 5)
−1	−1	(−1, −1)
−2	−3	(−2, −3)
−3	−5	(−3, −5)

From the table we plot the points associated with the ordered pairs as shown in Figure 5.8(a). Connecting these points with a straight line produces the graph of the equation $y = 2x + 1$ as shown in Figure 5.8(b).

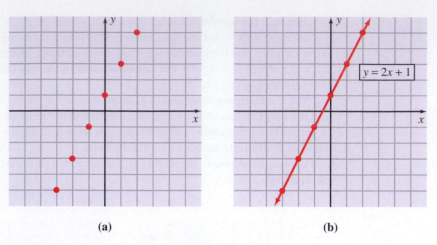

(a)   (b)

Figure 5.8

Classroom Example
Graph $y = -\dfrac{1}{4}x + 1$.

EXAMPLE 3   Graph $y = -\dfrac{1}{2}x + 3$.

Solution

First, we set up a table of some of the solutions for the equation $y = -\dfrac{1}{2}x + 3$. You can choose

any value for the variable x and find the corresponding y value; however the values we chose for
x are numbers that are divisible by 2. This is not necessary but does produce integer values for y.

x	y	Solutions (x, y)
0	3	(0, 3)
2	2	(2, 2)
4	1	(4, 1)
−2	4	(−2, 4)
−4	5	(−4, 5)

From the table we plot the points associated with the ordered pairs as shown in Figure 5.9(a).
Connecting these points with a straight line produces the graph of the equation $y = -\dfrac{1}{2}x + 3$
as shown in Figure 5.9(b).

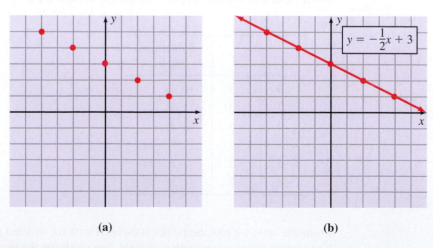

(a)   (b)

Figure 5.9

Classroom Example
Graph $2x + 3y = 6$.

EXAMPLE 4   Graph $6x + 3y = 9$.

Solution

First, let's change the form of the equation to make it easier to find solutions of the equation $6x + 3y = 9$. We can solve either for x in terms of y or for y in terms of x. Typically the equation is solved for y in terms of x so that's what we show here.

$$6x + 3y = 9$$
$$3y = -6x + 9$$
$$y = \frac{-6x}{3} + \frac{9}{3}$$
$$y = -2x + 3$$

Now we can set up a table of values. Plotting these points and connecting them produces Figure 5.10.

x	y
0	3
1	1
2	−1
−1	5

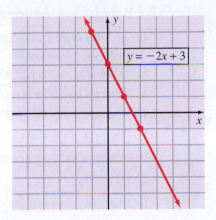

Figure 5.10

To graph an equation in two variables, x and y, keep these steps in mind:

1. Solve the equation for y in terms of x or for x in terms of y, if it is not already in such a form.
2. Set up a table of ordered pairs that satisfy the equation.
3. Plot the points associated with the ordered pairs.
4. Connect the points.

We conclude this section with two more examples that illustrate step 1.

Classroom Example
Solve $3x + 8y = 14$ for y.

EXAMPLE 5   Solve $4x + 9y = 12$ for y.

Solution

$$4x + 9y = 12$$
$$9y = -4x + 12 \qquad \text{Subtracted } 4x \text{ from both sides}$$
$$y = -\frac{4}{9}x + \frac{12}{9} \qquad \text{Divided both sides by 9}$$
$$y = -\frac{4}{9}x + \frac{4}{3} \qquad \text{Reduce the fraction}$$

EXAMPLE 6   Solve $2x - 6y = 3$ for y.

Solution

$$2x - 6y = 3$$

$$-6y = -2x + 3 \qquad \text{Subtracted } 2x \text{ from each side}$$

$$y = \frac{-2}{-6}x + \frac{3}{-6} \qquad \text{Divided both sides by } -6$$

$$y = \frac{1}{3}x - \frac{1}{2} \qquad \text{Reduced the fractions}$$

Concept Quiz 5.1

For Problems 1–5, answer true or false.

1. In a rectangular coordinate system, the coordinate axes partition the plane into four parts called quadrants.
2. Quadrants are named with Roman numerals and numbered clockwise.
3. The real numbers in an ordered pair are referred to as the coordinates of the point.
4. The equation $y = x + 3$ has an infinite number of ordered pairs that satisfy the equation.
5. The point of intersection of the coordinate axes is called the origin.

For Problems 6–10, match the points plotted in Figure 5.11 with their coordinates.

6. $(-3, 1)$
7. $(4, 0)$
8. $(3, -1)$
9. $(0, 4)$
10. $(-1, -3)$

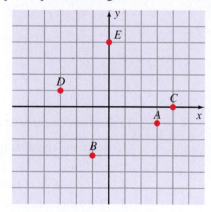

Figure 5.11

Problem Set 5.1

For Problems 1–10, plot the points on a rectangular coordinate system and determine if the points fall in a straight line. **(Objective 1)**

1. $(1, 4), (-2, 1), (-4, -1), (-3, 0)$
2. $(0, 2), (3, -1), (5, -3), (-3, 5)$
3. $(0, 3), (-2, 7), (3, -6)$
4. $(1, 4), (-1, 6), (6, -3)$

5. $(3, -1), (-3, -3), (0, -2)$
6. $(0, 1), (2, 0), (-2, 2)$
7. $(-4, 0), (-2, -1), (6, -5)$
8. $(1, 3), (-2, -6), (0, 0)$
9. $(-2, 4), (2, -4), (1, 0)$
10. $(2, 2), (-3, -6), (-1, -1)$

11. Maria, a biology student, designed an experiment to test the effects of changing the amount of light and the amount of water given to selected plants. In the experiment, the amounts of water and light given to a plant were randomly changed. The chart shows the amount of light and water above or below the normal amount given to the plant for six days. Plot the charted information on a rectangular coordinate system. Let the change in light be the first number in the ordered pair, and let the change in water be the second number in the ordered pair.

	Mon.	Tue.	Wed.	Thu.	Fri.	Sat.
Change in amount of light	1	−2	−1	4	−3	0
Change in amount of water	−3	4	−1	0	−5	1

12. Chase is studying the monthly percent changes in the stock price for two different companies. Using the data in the table below, plot the points for each month. Let the percent change for XM Inc. be the first number of the ordered pair, and let the percent change for Icom be the second number in the ordered pair.

	Jan.	Feb.	Mar.	Apr.	May	Jun.
XM Inc.	1	−2	−1	4	−3	0
Icom	−3	4	2	−5	−1	1

For Problems 13–20, (a) determine which of the ordered pairs satisfies the given equation, and then (b) graph the equation by plotting the points that satisfied the equation. **(Objective 1)**

13. $y = -x + 4$; $(1, 3), (0, 4), (2, -1), (-2, 6), (-1, 5)$

14. $y = -2x + 3$; $(1, 1), (1.5, 0), (3, -1), (0, 3), (-1, 5)$

15. $y = 2x - 3$; $(1, 2), (-3, 0), (0, -3), (2, 1), (-1, -5)$

16. $y = x + 4$; $(-1, 3), (2, -1), (-2, 2), (-1, 5), (0, 4)$

17. $y = \frac{2}{3}x$; $(1, 2), (-3, -2), (0, 0), (3, 6), (-6, -2)$

18. $y = \frac{1}{2}x$; $(1, 2), (2, 1), \left(0, \frac{1}{2}\right), (-2, 1) \left(-1, -\frac{1}{2}\right)$

19. $y = \frac{2}{5}x - \frac{1}{5}$; $\left(1, \frac{1}{5}\right), \left(-3, \frac{4}{5}\right), \left(-1, -\frac{3}{5}\right),$ $\left(2, \frac{2}{5}\right), (-2, -1)$

20. $y = -\frac{2}{3}x + \frac{1}{3}$; $\left(1, \frac{1}{3}\right), (0, 1), (5, -3), \left(\frac{1}{2}, 0\right), (-1, 0)$

For Problems 21–30, solve the given equation for the variable indicated. **(Objective 2)**

21. $3x + 7y = 13$ for y

22. $5x + 9y = 17$ for y

23. $x - 3y = 9$ for x

24. $2x - 7y = 5$ for x

25. $-x + 5y = 14$ for y

26. $-2x - y = 9$ for y

27. $-3x + y = 7$ for x

28. $-x - y = 9$ for x

29. $-2x + 3y = -5$ for y

30. $3x - 4y = -7$ for y

For Problems 31–50, graph each of the equations. **(Objective 3)**

31. $y = x + 1$

32. $y = x + 4$

33. $y = x - 2$

34. $y = -x - 1$

35. $y = 3x - 4$

36. $y = -2x - 1$

37. $y = -4x + 4$

38. $y = 2x - 6$

39. $y = \frac{1}{2}x + 3$

40. $y = \frac{1}{2}x - 2$

41. $x + 2y = 4$

42. $x + 3y = 6$

43. $2x - 5y = 10$

44. $5x - 2y = 10$

45. $y = -\frac{1}{3}x + 2$

46. $y = -\frac{2}{3}x - 3$

47. $y = x$

48. $y = -x$

49. $y = -3x + 2$

50. $3x - y = 4$

Thoughts Into Words

51. How would you convince someone that there are infinitely many ordered pairs of real numbers that satisfy the equation $x + y = 9$?

52. Explain why no points of the graph of the equation $y = x$ will be in the second quadrant.

Further Investigations

Not all graphs of equations are straight lines. Some equations produce graphs that are smooth curves. For example, the graph of $y = x^2$ will produce a smooth curve that is called a parabola. To graph $y = x^2$, let's proceed as in the previous problems, plot points associated with solutions to the equation $y = x^2$, and connect the points with a smooth curve.

x	$y = x^2$	Solutions (x, y)
0	0	$(0, 0)$
1	1	$(1, 1)$
2	4	$(2, 4)$
3	9	$(3, 9)$
-1	1	$(-1, 1)$
-2	4	$(-2, 4)$
-3	9	$(-3, 9)$

Then we plot the points associated with the solutions as shown in Figure 5.12(a). Finally, we connect the points with the smooth curve shown in Figure 5.12(b).

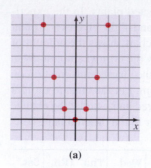
(a)

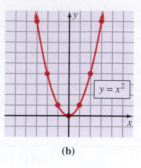
(b)

Figure 5.12

For Problems 53–62, graph each of the equations. Be sure to find a sufficient number of solutions so that the graph of the equation can be determined.

53. $y = x^2 + 2$ **54.** $y = x^2 - 3$

55. $y = -x^2$ **56.** $y = (x - 2)^2$

57. $y = x^3$ **58.** $y = x^3 - 4$

59. $y = (x - 3)^3$ **60.** $y = -x^3$

61. $y = x^4$ **62.** $y = -x^4$

Answers to the Concept Quiz

1. True   **2.** False   **3.** True   **4.** True   **5.** True   **6.** D   **7.** C   **8.** A   **9.** E   **10.** B

5.2   Graphing Linear Equations

OBJECTIVES

1 Find the x and y intercepts for the graphs of linear equations

2 Graph linear equations

3 Use linear equations to model problems

In the preceding section we graphed equations whose graphs were straight lines. In general, any equation of the form $Ax + By = C$, where A, B, and C are constants (A and B not both zero) and x and y are variables, is a *linear equation in two variables*, and its graph is a straight line.

We should clarify two points about this description of a linear equation in two variables. First, the choice of x and y for variables is arbitrary. We could use any two letters to represent the variables. An equation such as $3m + 2n = 7$ can be considered a linear equation in two variables. So that we are not constantly changing the labeling of the coordinate axes when graphing equations, however, it is much easier to use the same two variables in all equations. Thus we will go along with convention and use x and y as our variables. Second, the statement "any equation of the form $Ax + By = C$" technically means "any equation of the form $Ax + By = C$ or *equivalent* to the form." For example, the equation $y = x + 3$, which has a straight line graph, is equivalent to $-x + y = 3$.

All of the following are examples of linear equations in two variables.

$$y = x + 3 \qquad y = -3x + 2 \qquad y = \frac{2}{5}x + 1 \qquad y = 2x$$

$$3x - 2y = 6 \qquad x - 4y = 5 \qquad 5x - y = 10 \qquad y = \frac{2x + 4}{3}$$

The knowledge that any equation of the form $Ax + By = C$ produces a straight line graph, along with the fact that two points determine a straight line, makes graphing linear equations in two variables a simple process. We merely find two solutions, plot the corresponding points, and connect the points with a straight line. It is probably wise to find a third point as a check point. Let's consider an example.

Classroom Example
Graph $8x - 6y = 12$.

| **EXAMPLE 1** | Graph $2x - 3y = 6$. |

Solution

We recognize that equation $2x - 3y = 6$ is a linear equation in two variables, and therefore its graph will be a straight line. All that is necessary is to find two solutions and connect the points with a straight line. We will, however, also find a third solution to serve as a check point.

Let $x = 0$; then $2(0) - 3y = 6$

$$-3y = 6$$

$$y = -2 \qquad \text{Thus } (0, -2) \text{ is a solution}$$

Let $y = 0$; then $2x - 3(0) = 6$

$$2x = 6$$

$$x = 3 \qquad \text{Thus } (3, 0) \text{ is a solution}$$

Let $x = -3$; then $2(-3) - 3y = 6$

$$-6 - 3y = 6$$

$$-3y = 12$$

$$y = -4 \qquad \text{Thus } (-3, -4) \text{ is a solution}$$

We can plot the points associated with these three solutions and connect them with a straight line to produce the graph of $2x - 3y = 6$ in Figure 5.13.

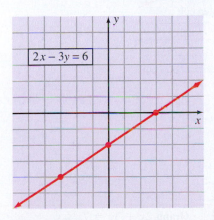

Figure 5.13

Let us briefly review our approach to Example 1. Notice that we did not begin the solution by solving either for y in terms of x or for x in terms of y. The reason for this is that we know the graph is a straight line and therefore there is no need for an extensive table of values. Thus there is no real benefit to changing the form of the original equation. The first two solutions indicate where the line intersects the coordinate axes. The ordinate of the point $(0, -2)$ is called the **y intercept**, and the abscissa of the point $(3, 0)$ is called the **x intercept** of this graph. That is, the graph of the equation $2x - 3y = 6$ has a y intercept of -2 and an x intercept of 3. In general, the intercepts are often easy to find. You can let $x = 0$ and solve for y to find the y intercept, and let $y = 0$ and solve for x to find the x intercept. The third solution, $(-3, -4)$, serves as a check point. If $(-3, -4)$ had not been on the line determined by the two intercepts, then we would have known that we had committed an error.

Classroom Example
Graph $2x + y = 3$.

EXAMPLE 2   Graph $x + 2y = 4$.

Solution

Without showing all of our work, we present the following table indicating the intercepts and a check point.

x	y	
0	2	Intercepts
4	0	
2	1	Check point

We plot the points $(0, 2)$, $(4, 0)$, and $(2, 1)$ and connect them with a straight line to produce the graph in Figure 5.14.

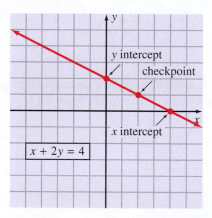

Figure 5.14

Classroom Example
Graph $4x + 5y = 3$.

EXAMPLE 3   Graph $2x + 3y = 7$.

Solution

The intercepts and a check point are given in the table. Finding intercepts may involve fractions, but the computation is usually easy. We plot the points from the table and show the graph of $2x + 3y = 7$ in Figure 5.15.

x	y	
0	$\frac{7}{3}$	Intercepts
$\frac{7}{2}$	0	
2	1	Check point

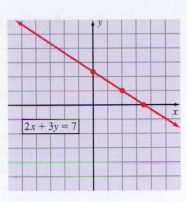

Figure 5.15

EXAMPLE 4   Graph $y = 2x$.

Solution

Notice that $(0, 0)$ is a solution; thus this line intersects both axes at the origin. Since both the x intercept and the y intercept are determined by the origin, $(0, 0)$, we need another point to graph the line. Then a third point should be found to serve as a check point. These results are summarized in the table. The graph of $y = 2x$ is shown in Figure 5.16.

x	y	
0	0	Intercept
2	4	Additional point
−1	−2	Check point

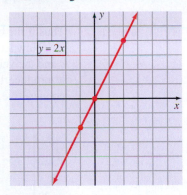

Figure 5.16

EXAMPLE 5   Graph $x = 3$.

Solution

Since we are considering linear equations in *two variables*, the equation $x = 3$ is equivalent to $x + 0(y) = 3$. Now we can see that any value of y can be used, but the x value must always be 3. Therefore, some of the solutions are $(3, 0)$, $(3, 1)$, $(3, 2)$, $(3, -1)$, and $(3, -2)$. The graph of all of the solutions is the vertical line indicated in Figure 5.17.

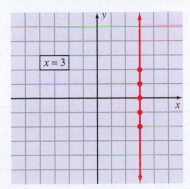

Figure 5.17

Applications of Linear Equations

Linear equations in two variables can be used to model many different types of real-world problems. For example, suppose that a retailer wants to sell some items at a profit of 30% of the cost of each item. If we let s represent the selling price and c the cost of each item, then we can use the equation

$$s = c + 0.3c = 1.3c$$

to determine the selling price of each item on the basis of the cost of the item. For example, if the cost of an item is \$4.50, then the retailer should sell it for $s = (1.3)(4.5) = \$5.85$.

By finding values that satisfy the equation $s = 1.3c$, we can create this table:

c	1	5	10	15	20
s	1.3	6.5	13	19.5	26

From the table we see that if the cost of an item is \$15, then the retailer should sell it for \$19.50 in order to make a profit of 30% of the cost. Furthermore, because this is a linear relationship, we can obtain exact values for costs that fall between the values given in the table. For example, because a c value of 12.5 is halfway between the c values of 10 and 15, the corresponding s value is halfway between the s values of 13 and 19.5. Therefore, a c value of 12.5 produces

$$s = 13 + \frac{1}{2}(19.5 - 13) = 16.25$$

Thus if the cost of an item is \$12.50, the retailer should sell it for \$16.25.

Now let's get a picture (graph) of this linear relationship. We can label the horizontal axis c and the vertical axis s, and we can use the origin along with one ordered pair from the table to produce the straight-line graph in Figure 5.18. (Because of the type of application, only nonnegative values for c and s are appropriate.)

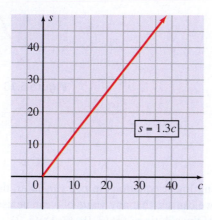

Figure 5.18

From the graph we can approximate s values on the basis of given c values. For example, if $c = 30$, then by reading up from 30 on the c axis to the line and then across to the s axis, we see that s is a little less than 40. (We get an exact s value of 39 by using the equation $s = 1.3c$.)

Many formulas that are used in various applications are linear equations in two variables. For example, the formula $C = \frac{5}{9}(F - 32)$, which converts temperatures from the Fahrenheit scale to the Celsius scale, is a linear relationship. Using this equation, we can determine that $14°F$ is equivalent to

$$C = \frac{5}{9}(14 - 32) = \frac{5}{9}(-18) = -10°C$$

Let's use the equation $C = \dfrac{5}{9}(F - 32)$ to create a table of values:

F	−22	−13	5	32	50	68	86
C	−30	−25	−15	0	10	20	30

Reading from the table we see, for example, that $-13°F = -25°C$ and $68°F = 20°C$.

To graph the equation $C = \dfrac{5}{9}(F - 32)$ we can label the horizontal axis F and the vertical axis C and plot two points that are given in the table. Figure 5.19 shows the graph of the equation.

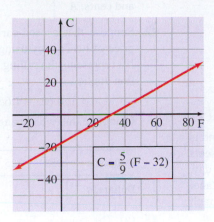

Figure 5.19

From the graph we can approximate C values on the basis of given F values. For example, if $F = 80°$, then by reading up from 80 on the F axis to the line and then across to the C axis, we see that C is approximately $25°$. Likewise, we can obtain approximate F values on the basis of given C values. For example, if $C = -25°$, then by reading across from -25 on the C axis to the line and then up to the F axis, we see that F is approximately $-15°$.

Concept Quiz 5.2

For Problems 1–10, answer true or false.

1. The graph of $y = x^2$ is a straight line.
2. Any equation of the form $Ax + By = C$, where A, B, and C (A and B not both zero) are constants and x and y are variables, has a graph that is a straight line.
3. The equations $2x + y = 4$ and $y = -2x + 4$ are equivalent.
4. The y intercept of the graph of $3x + 4y = -12$ is -4.
5. The x intercept of the graph of $3x + 4y = -12$ is -4.
6. Determining just two points is sufficient to graph a straight line.
7. The graph of $y = 4$ is a vertical line.
8. The graph of $x = 4$ is a vertical line.
9. The graph of $y = -1$ has a y intercept of -1.
10. The graph of every linear equation has a y intercept.

Problem Set 5.2

For Problems 1–36, graph each linear equation.

1. $x + y = 2$
2. $x + y = 4$
3. $x - y = 3$
4. $x - y = 1$
5. $x - y = -4$
6. $-x + y = 5$
7. $x + 2y = 2$
8. $x + 3y = 5$
9. $3x - y = 6$
10. $2x - y = -4$

11. $3x - 2y = 6$ **12.** $2x - 3y = 4$

13. $x - y = 0$ **14.** $x + y = 0$

15. $y = 3x$ **16.** $y = -2x$

17. $x = -2$ **18.** $y = 3$

19. $y = 0$ **20.** $x = 0$

21. $y = -2x - 1$ **22.** $y = 3x - 4$

23. $y = \dfrac{1}{2}x + 1$ **24.** $y = \dfrac{2}{3}x - 2$

25. $y = -\dfrac{1}{3}x - 2$ **26.** $y = -\dfrac{3}{4}x - 1$

27. $4x + 5y = -10$ **28.** $3x + 5y = -9$

29. $-2x + y = -4$ **30.** $-3x + y = -5$

31. $3x - 4y = 7$ **32.** $4x - 3y = 10$

33. $y + 4x = 0$ **34.** $y - 5x = 0$

35. $x = 2y$ **36.** $x = -3y$

37. Suppose that the daily profit from an ice cream stand is given by the equation $p = 2n - 4$, where n represents the number of gallons of ice cream mix used in a day, and p represents the number of dollars of profit. Label the horizontal axis n and the vertical axis p, and graph the equation $p = 2n - 4$ for nonnegative values of n.

38. The cost (c) of playing an online computer game for a time (t) in hours is given by the equation $c = 3t + 5$. Label the horizontal axis t and the vertical axis c, and graph the equation for nonnegative values of t.

39. The area of a sidewalk whose width is fixed at 3 feet can be given by the equation $A = 3l$, where A represents the area in square feet and l represents the length in feet. Label the horizontal axis l and the vertical axis A, and graph the equation $A = 3l$ for nonnegative values of l.

40. An online grocery store charges for delivery based on the equation $C = 0.30p$, where C represents the cost in dollars, and p represents the weight of the groceries in pounds. Label the horizontal axis p and the vertical axis C, and graph the equation $C = 0.30p$ for nonnegative values of p.

41. At \$0.06 per kilowatt-hour, the equation $A = 0.06t$ determines the amount, A, of an electric bill for t hours. Complete this table of values:

Hours, t	696	720	740	775	782
Dollars and cents, A					

42. Suppose that a used-car dealer determines the selling price of his cars by using a markup of 60% of the cost. If s represents the selling price, and c the cost, this equation applies:

$$s = c + 0.6c = 1.6c$$

Complete this table using the equation $s = 1.6c$.

Dollars, c	250	325	575	895	1095
Dollars, s					

43. (a) The equation $F = \dfrac{9}{5}C + 32$ converts temperatures from degrees Celsius to degrees Fahrenheit. Complete this table:

C	0	5	10	15	20	-5	-10	-15	-20	-25
F										

(b) Graph the equation $F = \dfrac{9}{5}C + 32$.

(c) Use your graph from part (b) to approximate values for F when C = 25°, 30°, -30°, and -40°.

(d) Check the accuracy of your readings from the graph in part (c) by using the equation $F = \dfrac{9}{5}C + 32$.

Thoughts Into Words

44. Your friend is having trouble understanding why the graph of the equation $y = 3$ is a horizontal line containing the point (0, 3). What might you do to help him?

45. How do we know that the graph of $y = -4x$ is a straight line that contains the origin?

46. Do all graphs of linear equations have x intercepts? Explain your answer.

47. How do we know that the graphs of $x - y = 4$ and $-x + y = -4$ are the same line?

Further Investigations

From our previous work with absolute value, we know that $|x + y| = 2$ is equivalent to $x + y = 2$ or $x + y = -2$. Therefore, the graph of $|x + y| = 2$ consists of the two lines $x + y = 2$ and $x + y = -2$. Graph each of the following equations.

48. $|x + y| = 1$ **49.** $|x - y| = 2$

50. $|2x + y| = 4$ **51.** $|3x - y| = 6$

5.3   Slope of a Line

OBJECTIVES

1 Find the slope of a line between two points

2 Given the equation of a line, find two points on the line, and use those points to determine the slope of the line

3 Graph lines, given a point and the slope

4 Solve word problems that involve slope

In Figure 5.20, note that the line associated with $4x - y = 4$ is *steeper* than the line associated with $2x - 3y = 6$. Mathematically, we use the concept of *slope* to discuss the steepness of lines. The **slope** of a line is the ratio of the vertical change to the horizontal change as we move from one point on a line to another point. We indicate this in Figure 5.21 with the points P_1 and P_2.

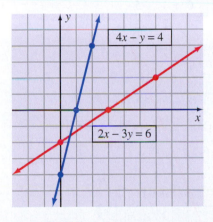

$4x - y = 4$

$2x - 3y = 6$

Figure 5.20

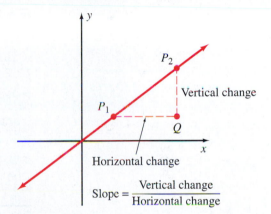

$$\text{Slope} = \frac{\text{Vertical change}}{\text{Horizontal change}}$$

Figure 5.21

We can give a precise definition for slope by considering the coordinates of the points P_1, P_2, and Q in Figure 5.22. Since P_1 and P_2 represent any two points on the line, we assign the coordinates (x_1, y_1) to P_1 and (x_2, y_2) to P_2. The point Q is the same distance from the y axis as P_2 and the same distance from the x axis as P_1. Thus we assign the coordinates (x_2, y_1) to Q (see Figure 5.22).

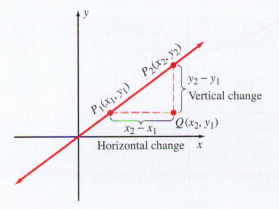

Figure 5.22

It should now be apparent that the vertical change is $y_2 - y_1$, and the horizontal change is $x_2 - x_1$. Thus we have the following definition for slope.

> ### Definition 5.1
>
> If points P_1 and P_2 with coordinates (x_1, y_1) and (x_2, y_2), respectively, are any two different points on a line, then the slope of the line (denoted by m) is
>
> $$m = \frac{y_2 - y_1}{x_2 - x_1}, \qquad x_1 \neq x_2$$

Using Definition 5.1, we can easily determine the slope of a line if we know the coordinates of two points on the line.

Classroom Example
Find the slope of the line determined by each of the following pairs of points:
(a) (3, 3) and (5, 6)
(b) (5, 3) and (3, −7)
(c) (−6, −1) and (−2, −1)

EXAMPLE 1

Find the slope of the line determined by each of the following pairs of points:

(a) $(2, 1)$ and $(4, 6)$ **(b)** $(3, 2)$ and $(-4, 5)$ **(c)** $(-4, -3)$ and $(-1, -3)$

Solution

(a) Let $(2, 1)$ be P_1 and $(4, 6)$ be P_2 as in Figure 5.23; then we have

$$m = \frac{y_2 - y_1}{x_2 - x_1} = \frac{6 - 1}{4 - 2} = \frac{5}{2}$$

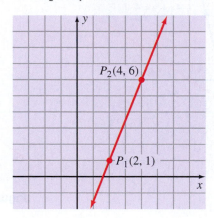

Figure 5.23

(b) Let $(3, 2)$ be P_1 and $(-4, 5)$ be P_2 as in Figure 5.24.

$$m = \frac{y_2 - y_1}{x_2 - x_1} = \frac{5 - 2}{-4 - 3} = \frac{3}{-7} = -\frac{3}{7}$$

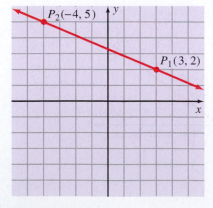

Figure 5.24

(c) Let $(-4, -3)$ be P_1 and $(-1, -3)$ be P_2 as in Figure 5.25.

$$m = \frac{y_2 - y_1}{x_2 - x_1} = \frac{-3 - (-3)}{-1 - (-4)} = \frac{0}{3} = 0$$

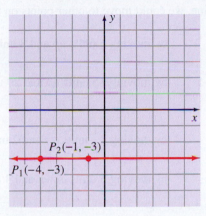

Figure 5.25

The designation of P_1 and P_2 in such problems is arbitrary and does not affect the value of the slope. For example, in part (a) of Example 1, if we let $(4, 6)$ be P_1 and $(2, 1)$ be P_2, then we obtain the same result for the slope as the following:

$$m = \frac{y_2 - y_1}{x_2 - x_1} = \frac{1 - 6}{2 - 4} = \frac{-5}{-2} = \frac{5}{2}$$

The parts of Example 1 illustrate the three basic possibilities for slope; that is, the slope of a line can be *positive, negative*, or *zero*. A line that has a positive slope rises as we move from left to right, as in part (a). A line that has a negative slope falls as we move from left to right, as in part (b). A horizontal line, as in part (c), has a slope of 0. Finally, we need to realize that **the concept of slope is undefined for vertical lines**. This is because, for any vertical line, the change in x as we move from one point to another is zero. Thus the ratio $\frac{y_2 - y_1}{x_2 - x_1}$ will have a denominator of zero and be undefined. So in Definition 5.1, the restriction $x_1 \neq x_2$ is made.

Classroom Example
Find the slope of the line determined by the equation $5x + 8y = 4$.

EXAMPLE 2   Find the slope of the line determined by the equation $3x + 4y = 12$.

Solution

Since we can use any two points on the line to determine the slope of the line, let's find the intercepts.

If $x = 0$, then $3(0) + 4y = 12$

$$4y = 12$$

$$y = 3 \qquad \text{Thus } (0, 3) \text{ is on the line}$$

If $y = 0$, then $3x + 4(0) = 12$

$$3x = 12$$

$$x = 4 \qquad \text{Thus } (4, 0) \text{ is on the line}$$

Using $(0, 3)$ as P_1 and $(4, 0)$ as P_2, we have

$$m = \frac{y_2 - y_1}{x_2 - x_1} = \frac{0 - 3}{4 - 0} = \frac{-3}{4} = -\frac{3}{4}$$

We need to emphasize one final idea pertaining to the concept of slope. The slope of a line is a **ratio** of vertical change to horizontal change. A slope of $\frac{3}{4}$ means that for every 3 units of vertical change, there is a corresponding 4 units of horizontal change. So starting at some point on the line, we could move to other points on the line as follows:

$$\frac{3}{4} = \frac{6}{8} \qquad \text{by moving 6 units } \textit{up} \text{ and 8 units to the } \textit{right}$$

$$\frac{3}{4} = \frac{15}{20} \qquad \text{by moving 15 units } \textit{up} \text{ and 20 units to the } \textit{right}$$

$$\frac{3}{4} = \frac{1\frac{1}{2}}{2} \qquad \text{by moving } 1\frac{1}{2} \text{ units } \textit{up} \text{ and 2 units to the } \textit{right}$$

$$\frac{3}{4} = \frac{-3}{-4} \qquad \text{by moving 3 units } \textit{down} \text{ and 4 units to the } \textit{left}$$

Likewise, a slope of $-\frac{5}{6}$ indicates that starting at some point on the line, we could move to other points on the line as follows:

$$-\frac{5}{6} = \frac{-5}{6} \qquad \text{by moving 5 units } \textit{down} \text{ and 6 units to the } \textit{right}$$

$$-\frac{5}{6} = \frac{5}{-6} \qquad \text{by moving 5 units } \textit{up} \text{ and 6 units to the } \textit{left}$$

$$-\frac{5}{6} = \frac{-10}{12} \qquad \text{by moving 10 units } \textit{down} \text{ and 12 units to the } \textit{right}$$

$$-\frac{5}{6} = \frac{15}{-18} \qquad \text{by moving 15 units } \textit{up} \text{ and 18 units to the } \textit{left}$$

Classroom Example
Graph the line that passes through the point $(-1, 3)$ and has a slope of $\frac{1}{4}$.

EXAMPLE 3

Graph the line that passes through the point $(0, -2)$ and has a slope of $\frac{1}{3}$.

Solution

To begin, plot the point $(0, -2)$. Because the slope $= \dfrac{\text{vertical change}}{\text{horizontal change}} = \dfrac{1}{3}$, we can locate another point on the line by starting from the point $(0, -2)$ and moving 1 unit up and 3 units to the right to obtain the point $(3, -1)$. Because two points determine a line, we can draw the line (Figure 5.26).

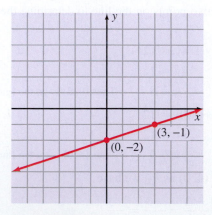

Figure 5.26

Remark: Because $m = \dfrac{1}{3} = \dfrac{-1}{-3}$, we can locate another point by moving 1 unit down and 3 units to the left from the point $(0, -2)$.

Classroom Example
Graph the line that passes through the point $(3, -2)$ and has a slope of -3.

EXAMPLE 4

Graph the line that passes through the point $(1, 3)$ and has a slope of -2.

Solution

To graph the line, plot the point $(1, 3)$. We know that $m = -2 = \dfrac{-2}{1}$. Furthermore, because

the slope $= \dfrac{\text{vertical change}}{\text{horizontal change}} = \dfrac{-2}{1}$, we can locate another point on the line by starting from the point $(1, 3)$ and moving 2 units down and 1 unit to the right to obtain the point $(2, 1)$. Because two points determine a line, we can draw the line (Figure 5.27).

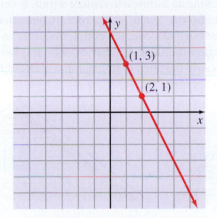

Figure 5.27

Remark: Because $m = -2 = \dfrac{-2}{1} = \dfrac{2}{-1}$ we can locate another point by moving 2 units up and 1 unit to the left from the point $(1, 3)$.

Applications of Slope

The concept of slope has many real-world applications even though the word "slope" is often not used. For example, the highway in Figure 5.28 is said to have a "grade" of 17%. This means that for every horizontal distance of 100 feet, the highway rises or drops 17 feet. In other words, the absolute value of the slope of the highway is $\dfrac{17}{100}$.

17 feet

100 feet

Figure 5.28

Classroom Example
A certain highway has a 4% grade.
How many feet does it rise in a
horizontal distance of 1 mile?

EXAMPLE 5

A certain highway has a 3% grade. How many feet does it rise in a horizontal distance of 1 mile?

Solution

A 3% grade means a slope of $\dfrac{3}{100}$. Therefore, if we let y represent the unknown vertical distance and use the fact that 1 mile = 5280 feet, we can set up and solve the following proportion:

$$\frac{3}{100} = \frac{y}{5280}$$

$$100y = 3(5280) = 15{,}840$$

$$y = 158.4$$

The highway rises 158.4 feet in a horizontal distance of 1 mile.

A roofer, when making an estimate to replace a roof, is concerned about not only the total area to be covered but also the "pitch" of the roof. (Contractors do not define pitch the same way that mathematicians define slope, but both terms refer to "steepness.") The two roofs in Figure 5.29 might require the same number of shingles, but the roof on the left will take longer to complete because the pitch is so great that scaffolding will be required.

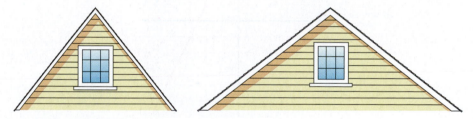

Figure 5.29

The concept of slope is also used in the construction of flights of stairs. The terms "rise" and "run" are commonly used, and the steepness (slope) of the stairs can be expressed as the ratio of rise to run. In Figure 5.30, the stairs on the left with the ratio of $\dfrac{10}{11}$ are steeper than the stairs on the right, which have a ratio of $\dfrac{7}{11}$.

Technically the concept of slope is involved in most situations where the idea of an incline is used. Hospital beds are constructed so that both the head-end and the foot-end can be raised or lowered; that is, the slope of either end of the bed can be changed. Likewise, treadmills are designed so that the incline (slope) of the platform can be raised or lowered as desired. Perhaps you can think of several other applications of the concept of slope.

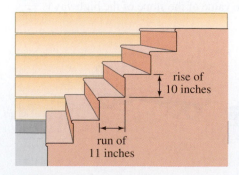
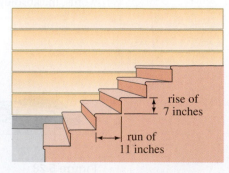

rise of
10 inches

run of
11 inches

rise of
7 inches

run of
11 inches

Figure 5.30

Concept Quiz 5.3

For Problems 1–10, answer true or false.

1. The concept of slope of a line pertains to the steepness of the line.
2. The slope of a line is the ratio of the horizontal change to the vertical change moving from one point to another point on the line.
3. A line that has a negative slope falls as we move from left to right.
4. The slope of a vertical line is 0.
5. The slope of a horizontal line is 0.
6. A line cannot have a slope of 0.
7. A slope of $\dfrac{-5}{2}$ is the same as a slope of $-\dfrac{5}{-2}$.
8. A slope of 5 means that for every unit of horizontal change there is a corresponding 5 units of vertical change.
9. The slope of the line determined by the equation $-x - 2y = 4$ is $-\dfrac{1}{2}$.
10. The slope of the line determined by the points $(-1, 4)$ and $(2, -1)$ is $\dfrac{5}{3}$.

Problem Set 5.3

For Problems 1–20, find the slope of the line determined by each pair of points. **(Objective 1)**

1. $(7, 5), (3, 2)$
2. $(9, 10), (6, 2)$
3. $(-1, 3), (-6, -4)$
4. $(-2, 5), (-7, -1)$
5. $(2, 8), (7, 2)$
6. $(3, 9), (8, 4)$
7. $(-2, 5), (1, -5)$
8. $(-3, 4), (2, -6)$
9. $(4, -1), (-4, -7)$
10. $(5, -3), (-5, -9)$
11. $(3, -4), (2, -4)$
12. $(-3, -6), (5, -6)$
13. $(-6, -1), (-2, -7)$
14. $(-8, -3), (-2, -11)$
15. $(-2, 4), (-2, -6)$
16. $(-4, -5), (-4, 9)$
17. $(-1, 10), (-9, 2)$
18. $(-2, 12), (-10, 2)$
19. $(a, b), (c, d)$
20. $(a, 0), (0, b)$

21. Find y if the line through the points $(7, 8)$ and $(2, y)$ has a slope of $\dfrac{4}{5}$.

22. Find y if the line through the points $(12, 14)$ and $(3, y)$ has a slope of $\dfrac{4}{3}$.

23. Find x if the line through the points $(-2, -4)$ and $(x, 2)$ has a slope of $-\dfrac{3}{2}$.

24. Find x if the line through the points $(6, -4)$ and $(x, 6)$ has a slope of $-\dfrac{5}{4}$.

For Problems 25–32, you are given one point on a line and the slope of the line. Find the coordinates of three other points on the line.

25. $(3, 2), m = \dfrac{2}{3}$
26. $(4, 1), m = \dfrac{5}{6}$
27. $(-2, -4), m = \dfrac{1}{2}$
28. $(-6, -2), m = \dfrac{2}{5}$
29. $(-3, 4), m = -\dfrac{3}{4}$
30. $(-2, 6), m = -\dfrac{3}{7}$
31. $(4, -5), m = -2$
32. $(6, -2), m = 4$

For Problems 33–40, sketch the line determined by each pair of points and decide whether the slope of the line is positive, negative, or zero.

33. $(2, 8), (7, 1)$
34. $(1, -2), (7, -8)$
35. $(-1, 3), (-6, -2)$
36. $(7, 3), (4, -6)$
37. $(-2, 4), (6, 4)$
38. $(-3, -4), (5, -4)$
39. $(-3, 5), (2, -7)$
40. $(-1, -1), (1, -9)$

For Problems 41–48, graph the line that passes through the given point and has the given slope. (Objective 3)

41. $(3, 1)$, $m = \dfrac{2}{3}$ **42.** $(-1, 0)$, $m = \dfrac{3}{4}$

43. $(-2, 3)$, $m = -1$ **44.** $(1, -4)$, $m = -3$

45. $(0, 5)$, $m = -\dfrac{1}{4}$ **46.** $(-3, 4)$, $m = -\dfrac{3}{2}$

47. $(2, -2)$, $m = \dfrac{3}{2}$ **48.** $(3, -4)$, $m = \dfrac{5}{2}$

For Problems 49–68, find the coordinates of two points on the given line, and then use those coordinates to find the slope of the line. (Objective 2)

49. $3x + 2y = 6$ **50.** $4x + 3y = 12$

51. $5x - 4y = 20$ **52.** $7x - 3y = 21$

53. $x + 5y = 6$ **54.** $2x + y = 4$

55. $2x - y = -7$ **56.** $x - 4y = -6$

57. $y = 3$ **58.** $x = 6$

59. $-2x + 5y = 9$ **60.** $-3x - 7y = 10$

61. $6x - 5y = -30$ **62.** $7x - 6y = -42$

63. $y = -3x - 1$ **64.** $y = -2x + 5$

65. $y = 4x$ **66.** $y = 6x$

67. $y = \dfrac{2}{3}x - \dfrac{1}{2}$ **68.** $y = -\dfrac{3}{4}x + \dfrac{1}{5}$

For Problems 69–73, solve word problems that involve slope. (Objective 4)

69. Suppose that a highway rises a distance of 135 feet in a horizontal distance of 2640 feet. Express the grade of the highway to the nearest tenth of a percent.

70. The grade of a highway up a hill is 27%. How much change in horizontal distance is there if the vertical height of the hill is 550 feet? Express the answer to the nearest foot.

71. If the ratio of rise to run is to be $\dfrac{3}{5}$ for some stairs, and the measure of the rise is 19 centimeters, find the measure of the run to the nearest centimeter.

72. If the ratio of rise to run is to be $\dfrac{2}{3}$ for some stairs, and the measure of the run is 28 centimeters, find the measure of the rise to the nearest centimeter.

73. A county ordinance requires a $2\dfrac{1}{4}$% "fall" for a sewage pipe from the house to the main pipe at the street. How much vertical drop must there be for a horizontal distance of 45 feet? Express the answer to the nearest tenth of a foot.

Thoughts Into Words

74. How would you explain the concept of slope to someone who was absent from class the day it was discussed?

75. If one line has a slope of $\dfrac{2}{3}$, and another line has a slope of 2, which line is steeper? Explain your answer.

76. Why do we say that the slope of a vertical line is undefined?

77. Suppose that a line has a slope of $\dfrac{3}{4}$ and contains the point $(5, 2)$. Are the points $(-3, -4)$ and $(14, 9)$ also on the line? Explain your answer.

Answers to the Concept Quiz

1. True   **2.** False   **3.** True   **4.** False   **5.** True   **6.** False   **7.** False   **8.** True   **9.** True   **10.** False

5.4   Writing Equations of Lines

OBJECTIVES

1 Find the equation of a line given
   a. a slope and a point
   b. two points on the line
   c. a point on the line and the equation of a line parallel or perpendicular to it

2 Become familiar with the point-slope form and the slope-intercept form of the equation of a straight line

3 Know the relationships for slopes of parallel and perpendicular lines

There are two basic types of problems in analytic or coordinate geometry:

1. Given an algebraic equation, find its geometric graph.

2. Given a set of conditions pertaining to a geometric figure, determine its algebraic equation.

We discussed problems of type 1 in the first two sections of this chapter. Now we want to consider a few problems of type 2 that deal with straight lines. In other words, given certain facts about a line, we need to be able to write its algebraic equation.

Classroom Example
Find the equation of the line that has a slope of $\frac{2}{3}$ and contains the point $(-3, 2)$.

EXAMPLE 1

Find the equation of the line that has a slope of $\frac{3}{4}$ and contains the point $(1, 2)$.

Solution

First, we draw the line as indicated in Figure 5.31. Since the slope is $\frac{3}{4}$, we can find a second point by moving 3 units up and 4 units to the right of the given point $(1, 2)$. (The point $(5, 5)$ merely helps to draw the line; it will not be used in analyzing the problem.) Now we choose a point (x, y) that represents any point on the line other than the given point $(1, 2)$. The slope determined by $(1, 2)$ and (x, y) is $\frac{3}{4}$.

Thus

$$\frac{y - 2}{x - 1} = \frac{3}{4}$$

$$3(x - 1) = 4(y - 2) \qquad \text{Cross multiply}$$

$$3x - 3 = 4y - 8$$

$$3x - 4y = -5$$

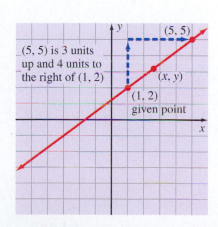

$(5, 5)$ is 3 units up and 4 units to the right of $(1, 2)$

(x, y)

$(1, 2)$ given point

Figure 5.31

EXAMPLE 2   Find the equation of the line that contains $(3, 4)$ and $(-2, 5)$.

Solution

First, we draw the line determined by the two given points in Figure 5.32. Since we know two points, we can find the slope.

$$m = \frac{y_2 - y_1}{x_2 - x_1} = \frac{5 - 4}{-2 - 3} = \frac{1}{-5} = -\frac{1}{5}$$

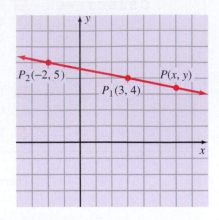

Figure 5.32

Now we can use the same approach as in Example 1. We form an equation using a variable point (x, y), one of the two given points (we choose $P_1(3, 4)$), and the slope of $-\frac{1}{5}$.

$$\frac{y - 4}{x - 3} = \frac{1}{-5} \qquad -\frac{1}{5} = \frac{1}{-5}$$
$$x - 3 = -5y + 20$$
$$x + 5y = 23$$

EXAMPLE 3

Find the equation of the line that has a slope of $\frac{1}{4}$ and a y intercept of 2.

Solution

A y intercept of 2 means that the point $(0, 2)$ is on the line. Since the slope is $\frac{1}{4}$, we can find another point by moving 1 unit up and 4 units to the right of $(0, 2)$. The line is drawn in Figure 5.33. We choose variable point (x, y) and proceed as in the preceding examples.

$$\frac{y - 2}{x - 0} = \frac{1}{4}$$
$$x = 4y - 8$$
$$x - 4y = -8$$

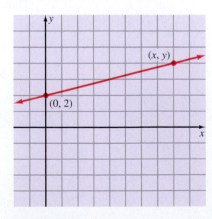

Figure 5.33

It may be helpful for you to pause for a moment and look back over Examples 1, 2, and 3. Notice that we used the same basic approach in all three examples; that is, we chose a variable point (x, y) and used it, along with another known point, to determine the equation of the line. You should also recognize that the approach we take in these examples can be generalized to produce some special forms of equations for straight lines.

Point-Slope Form

EXAMPLE 4

Find the equation of the line that has a slope of m and contains the point (x_1, y_1).

Solution

We choose (x, y) to represent any other point on the line in Figure 5.34. The slope of the line given by

$$m = \frac{y - y_1}{x - x_1}$$

enables us to obtain

$$y - y_1 = m(x - x_1)$$

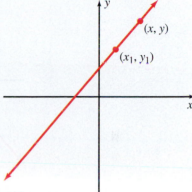

Figure 5.34

We refer to the equation

$$y - y_1 = m(x - x_1)$$

as the **point-slope form** of the equation of a straight line. Instead of the approach we used in Example 1, we could use the point-slope form to write the equation of a line with a given slope that contains a given point—as the next example illustrates.

EXAMPLE 5

Write the equation of the line that has a slope of $\frac{3}{5}$ and contains the point $(2, -4)$.

Solution

Substituting $\frac{3}{5}$ for m and $(2, -4)$ for (x_1, y_1) in the point-slope form, we obtain

$$y - y_1 = m(x - x_1)$$

$$y - (-4) = \frac{3}{5}(x - 2)$$

$$y + 4 = \frac{3}{5}(x - 2)$$

$5(y + 4) = 3(x - 2)$ Multiply both sides by 5

$5y + 20 = 3x - 6$

$26 = 3x - 5y$

Slope-Intercept Form

Now consider the equation of a line that has a slope of m and a y intercept of b (see Figure 5.35). A y intercept of b means that the line contains the point $(0, b)$; therefore, we can use the point-slope form.

$$y - y_1 = m(x - x_1)$$
$$y - b = m(x - 0) \qquad \text{$y_1 = b$ and $x_1 = 0$}$$
$$y - b = mx$$
$$y = mx + b$$

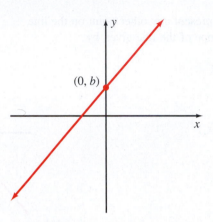

(0, b)

Figure 5.35

The equation

$$y = mx + b$$

is called the **slope-intercept form** of the equation of a straight line. We use it for three primary purposes, as the next three examples illustrate.

Classroom Example
Find the equation of the line that has a slope of $-\dfrac{2}{5}$ and a y intercept of -3.

EXAMPLE 6

Find the equation of the line that has a slope of $\dfrac{1}{4}$ and a y intercept of 2.

Solution

This is a restatement of Example 3, but this time we will use the slope-intercept form of a line ($y = mx + b$) to write its equation. From the statement of the problem we know that $m = \dfrac{1}{4}$ and $b = 2$. Thus substituting these values for m and b into $y = mx + b$, we obtain

$$y = mx + b$$
$$y = \frac{1}{4}x + 2$$
$$4y = x + 8$$
$$x - 4y = -8 \qquad \text{Same result as in Example 3}$$

Remark: It is acceptable to leave answers in slope-intercept form. We did not do that in Example 6 because we wanted to show that it was the same result as in Example 3.

EXAMPLE 7

Find the slope and y intercept of the line with the equation $2x + 3y = 4$.

Solution

We can solve the equation for y in terms of x, and then compare it to the slope-intercept form to determine its slope.

$$2x + 3y = 4$$
$$3y = -2x + 4$$
$$y = -\frac{2}{3}x + \frac{4}{3}$$

Compare this result to $y = mx + b$, and you see that the slope of the line is $-\frac{2}{3}$. Furthermore, the y intercept is $\frac{4}{3}$.

EXAMPLE 8

Graph the line determined by the equation $y = \frac{2}{3}x - 1$.

Solution

Comparing the given equation to the general slope-intercept form, we see that the slope of the line is $\frac{2}{3}$, and the y intercept is -1. Because the y intercept is -1, we can plot the point $(0, -1)$. Then because the slope is $\frac{2}{3}$, let's move 3 units to the right and 2 units up from $(0, -1)$ to locate the point $(3, 1)$. The two points $(0, -1)$ and $(3, 1)$ determine the line in Figure 5.36.

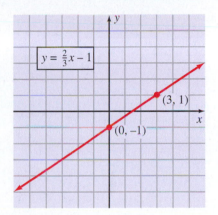

Figure 5.36

In general,

> If the equation of a nonvertical line is written in slope-intercept form, the coefficient of x is the slope of the line, and the constant term is the y intercept.

(Remember that the concept of slope is not defined for a vertical line.) Let's consider a few more examples.

EXAMPLE 9

Find the slope and y intercept of each of the following lines and graph the lines:

(a) $5x - 4y = 12$ (b) $-y = 3x - 4$ (c) $y = 2$

Solution

(a) We change $5x - 4y = 12$ to slope-intercept form to get

$$5x - 4y = 12$$
$$-4y = -5x + 12$$

$$4y = 5x - 12$$
$$y = \frac{5}{4}x - 3$$

The slope of the line is $\frac{5}{4}$ (the coefficient of x) and the y intercept is -3 (the constant term). To graph the line, we plot the y intercept, -3. Then because the slope is $\frac{5}{4}$, we can determine a second point, $(4, 2)$, by moving 5 units up and 4 units to the right from the y intercept. The graph is shown in Figure 5.37.

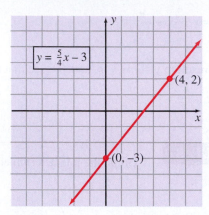

Figure 5.37

(b) We multiply both sides of the given equation by -1 to change it to slope-intercept form.

$$-y = 3x - 4$$
$$y = -3x + 4$$

The slope of the line is -3, and the y intercept is 4. To graph the line, we plot the y intercept, 4. Then because the slope is $-3 = \frac{-3}{1}$, we can find a second point, $(1, 1)$, by moving 3 units down and 1 unit to the right from the y intercept. The graph is shown in Figure 5.38.

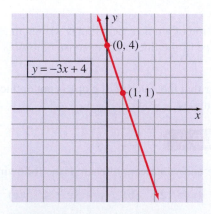

Figure 5.38

(c) We can write the equation $y = 2$ as

$$y = 0(x) + 2$$

The slope of the line is 0, and the y intercept is 2. To graph the line, we plot the y intercept, 2. Then because a line with a slope of 0 is horizontal, we draw a horizontal line through the y intercept. The graph is shown in Figure 5.39.

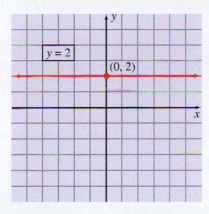

Figure 5.39

Parallel and Perpendicular Lines

We can use two important relationships between lines and their slopes to solve certain kinds of problems. It can be shown that nonvertical parallel lines have the same slope and that two nonvertical lines are perpendicular if the product of their slopes is -1. (Details for verifying these facts are left to another course.) In other words, if two lines have slopes m_1 and m_2, respectively, then

   1. The two lines are parallel if and only if $m_1 = m_2$;

   2. The two lines are perpendicular if and only if $(m_1)(m_2) = -1$.

The following examples demonstrate the use of these properties.

Classroom Example
(a) Verify that the graphs of
$2x - 5y = 7$ and $6x - 15y = 10$
are parallel lines.

(b) Verify that the graphs of
$3x + 7y = 4$ and $14x - 6y = 19$
are perpendicular lines.

EXAMPLE 10

  (a) Verify that the graphs of $2x + 3y = 7$ and $4x + 6y = 11$ are parallel lines.

  (b) Verify that the graphs of $8x - 12y = 3$ and $3x + 2y = 2$ are perpendicular lines.

Solution

  (a) Let's change each equation to slope-intercept form.

$$2x + 3y = 7 \quad \rightarrow \quad 3y = -2x + 7$$
$$y = -\frac{2}{3}x + \frac{7}{3}$$

$$4x + 6y = 11 \quad \rightarrow \quad 6y = -4x + 11$$
$$y = -\frac{4}{6}x + \frac{11}{6}$$
$$y = -\frac{2}{3}x + \frac{11}{6}$$

Both lines have a slope of $-\dfrac{2}{3}$, but they have different y intercepts. Therefore, the two lines are parallel.

(b) Solving each equation for y in terms of x, we obtain

$$8x - 12y = 3 \quad \rightarrow \quad -12y = -8x + 3$$

$$y = \frac{8}{12}x - \frac{3}{12}$$

$$y = \frac{2}{3}x - \frac{1}{4}$$

$$3x + 2y = 2 \quad \rightarrow \quad 2y = -3x + 2$$

$$y = -\frac{3}{2}x + 1$$

Because $\left(\frac{2}{3}\right)\left(-\frac{3}{2}\right) = -1$ (the product of the two slopes is -1), the lines are perpendicular.

Remark: The statement "the product of two slopes is -1" is the same as saying that the two slopes are negative reciprocals of each other; that is, $m_1 = -\dfrac{1}{m_2}$.

Classroom Example
Find the equation of the line that contains the point $(-2, 2)$ and is parallel to the line dertermined by $3x - y = 9$.

EXAMPLE 11

Find the equation of the line that contains the point $(1, 4)$ and is parallel to the line determined by $x + 2y = 5$.

Solution

First, let's draw a figure to help in our analysis of the problem (Figure 5.40). Because the line through $(1, 4)$ is to be parallel to the line determined by $x + 2y = 5$, it must have the same slope. Let's find the slope by changing $x + 2y = 5$ to the slope-intercept form.

$$x + 2y = 5$$

$$2y = -x + 5$$

$$y = -\frac{1}{2}x + \frac{5}{2}$$

The slope of both lines is $-\dfrac{1}{2}$. Now we can choose a variable point (x, y) on the line through $(1, 4)$ and proceed as we did in earlier examples.

$$\frac{y - 4}{x - 1} = \frac{1}{-2}$$

$$1(x - 1) = -2(y - 4)$$

$$x - 1 = -2y + 8$$

$$x + 2y = 9$$

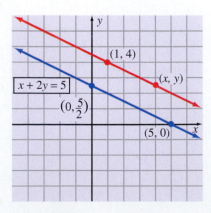

Figure 5.40

Classroom Example
Find the equation of the line that contains the point $(7, 5)$ and is perpendicular to the line determined by $4x - 3y = 6$.

EXAMPLE 12

Find the equation of the line that contains the point $(-1, -2)$ and is perpendicular to the line determined by $2x - y = 6$.

Solution

First, let's draw a figure to help in our analysis of the problem (Figure 5.41). Because the line through $(-1, -2)$ is to be perpendicular to the line determined by $2x - y = 6$, its slope must be the negative reciprocal of the slope of $2x - y = 6$. Let's find the slope of $2x - y = 6$ by changing it to the slope-intercept form.

$$2x - y = 6$$

$$-y = -2x + 6$$

$$y = 2x - 6 \quad \text{The slope is 2}$$

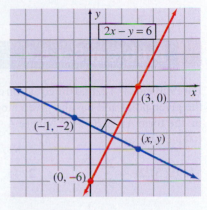

Figure 5.41

The slope of the desired line is $-\dfrac{1}{2}$ (the negative reciprocal of 2), and we can proceed as before by using a variable point (x, y).

$$\frac{y + 2}{x + 1} = \frac{1}{-2}$$

$$1(x + 1) = -2(y + 2)$$

$$x + 1 = -2y - 4$$

$$x + 2y = -5$$

We use two forms of equations of straight lines extensively. They are the **standard form** and the **slope-intercept form,** and we describe them as follows.

Standard Form $Ax + By = C$, where B and C are integers, and A is a nonnegative integer (A and B not both zero).

Slope-Intercept Form $y = mx + b$, where m is a real number representing the slope, and b is a real number representing the y intercept.

Concept Quiz 5.4

For Problems 1–10, answer true or false.

1. If two distinct lines have the same slope, then the lines are parallel.
2. If the slopes of two lines are reciprocals, then the lines are perpendicular.
3. In the standard form of the equation of a line $Ax + By = C$, A can be a rational number in fractional form.
4. In the slope-intercept form of an equation of a line $y = mx + b$, m is the slope.
5. In the standard form of the equation of a line $Ax + By = C$, A is the slope.
6. The slope of the line determined by the equation $3x - 2y = -4$ is $\dfrac{3}{2}$.

7. The concept of slope is not defined for the line $y = 2$.

8. The concept of slope is not defined for the line $x = 2$.

9. The lines determined by the equations $x - 3y = 4$ and $2x - 6y = 11$ are parallel lines.

10. The lines determined by the equations $x - 3y = 4$ and $x + 3y = 4$ are perpendicular lines.

Problem Set 5.4

For Problems 1–12, find the equation of the line that contains the given point and has the given slope. Express equations in the form $Ax + By = C$, where A, B, and C are integers. **(Objective 1a)**

1. $(2, 3)$, $m = \dfrac{2}{3}$

2. $(5, 2)$, $m = \dfrac{3}{7}$

3. $(-3, -5)$, $m = \dfrac{1}{2}$

4. $(5, -6)$, $m = \dfrac{3}{5}$

5. $(-4, 8)$, $m = -\dfrac{1}{3}$

6. $(-2, -4)$, $m = -\dfrac{5}{6}$

7. $(3, -7)$, $m = 0$

8. $(-3, 9)$, $m = 0$

9. $(0, 0)$, $m = -\dfrac{4}{9}$

10. $(0, 0)$, $m = \dfrac{5}{11}$

11. $(-6, -2)$, $m = 3$

12. $(2, -10)$, $m = -2$

For Problems 13–22, find the equation of the line that contains the two given points. Express equations in the form $Ax + By = C$, where A, B, and C are integers. **(Objective 1b)**

13. $(2, 3)$ and $(7, 10)$

14. $(1, 4)$ and $(9, 10)$

15. $(3, -2)$ and $(-1, 4)$

16. $(-2, 8)$ and $(4, -2)$

17. $(-1, -2)$ and $(-6, -7)$

18. $(-8, -7)$ and $(-3, -1)$

19. $(0, 0)$ and $(-3, -5)$

20. $(5, -8)$ and $(0, 0)$

21. $(0, 4)$ and $(7, 0)$

22. $(-2, 0)$ and $(0, -9)$

For Problems 23–32, find the equation of the line with the given slope and y intercept. Leave your answers in slope-intercept form. **(Objective 1a)**

23. $m = \dfrac{3}{5}$ and $b = 2$

24. $m = \dfrac{5}{9}$ and $b = 4$

25. $m = 2$ and $b = -1$

26. $m = 4$ and $b = -3$

27. $m = -\dfrac{1}{6}$ and $b = -4$

28. $m = -\dfrac{5}{7}$ and $b = -1$

29. $m = -1$ and $b = \dfrac{5}{2}$

30. $m = -2$ and $b = \dfrac{7}{3}$

31. $m = -\dfrac{5}{9}$ and $b = -\dfrac{1}{2}$

32. $m = -\dfrac{7}{12}$ and $b = -\dfrac{2}{3}$

For Problems 33–44, determine the slope and y intercept of the line represented by the given equation, and graph the line. **(Objective 2)**

33. $y = -2x - 5$

34. $y = \dfrac{2}{3}x + 4$

35. $3x - 5y = 15$

36. $7x + 5y = 35$

37. $-4x + 9y = 18$

38. $-6x + 7y = -14$

39. $-y = -\dfrac{3}{4}x + 4$

40. $5x - 2y = 0$

41. $-2x - 11y = 11$

42. $-y = \dfrac{2}{3}x + \dfrac{11}{2}$

43. $9x + 7y = 0$

44. $-5x - 13y = 26$

For Problems 45–60, write the equation of the line that satisfies the given conditions. Express final equations in standard form. **(Objectives 1a, 1c, and 3)**

45. x intercept of 2 and y intercept of -4

46. x intercept of -1 and y intercept of -3

47. x intercept of -3 and slope of $-\dfrac{5}{8}$

48. x intercept of 5 and slope of $-\dfrac{3}{10}$

49. Contains the point $(2, -4)$ and is parallel to the y axis

50. Contains the point $(-3, -7)$ and is parallel to the x axis

51. Contains the point $(5, 6)$ and is perpendicular to the y axis

52. Contains the point $(-4, 7)$ and is perpendicular to the x axis

53. Contains the point $(1, 3)$ and is parallel to the line $x + 5y = 9$

54. Contains the point $(-1, 4)$ and is parallel to the line $x - 2y = 6$

55. Contains the origin and is parallel to the line $4x - 7y = 3$

56. Contains the origin and is parallel to the line $-2x - 9y = 4$

57. Contains the point $(-1, 3)$ and is perpendicular to the line $2x - y = 4$

58. Contains the point $(-2, -3)$ and is perpendicular to the line $x + 4y = 6$

59. Contains the origin and is perpendicular to the line $-2x + 3y = 8$

60. Contains the origin and is perpendicular to the line $y = -5x$

Thoughts Into Words

61. Explain the importance of the slope-intercept form $(y = mx + b)$ of the equation of a line.

62. What does it mean to say that two points "determine" a line?

63. How would you describe coordinate geometry to a group of elementary algebra students?

64. How can you tell by inspection that $y = 2x - 4$ and $y = -3x - 1$ are not parallel lines?

Answers to the Concept Quiz
1. True   **2.** False   **3.** False   **4.** True   **5.** False   **6.** True   **7.** False   **8.** True   **9.** True   **10.** False

5.5 Systems of Two Linear Equations

OBJECTIVES

1. Solve linear systems of two equations by graphing

2. Solve linear systems of two equations by the substitution method

3. Recognize consistent and inconsistent systems of equations

4. Recognize dependent equations

5. Use a system of equations to solve word problems

Suppose we graph $x - 2y = 4$ and $x + 2y = 8$ on the same set of axes, as shown in Figure 5.42. The ordered pair $(6, 1)$, which is associated with the point of intersection of the two lines, satisfies both equations. In other words, we say that $(6, 1)$ is the solution for $x - 2y = 4$ and $x + 2y = 8$.

To check this, we can substitute 6 for x and 1 for y in both equations.

$x - 2y = 4$ becomes $6 - 2(1) = 4$ A true statement

$x + 2y = 8$ becomes $6 + 2(1) = 8$ A true statement

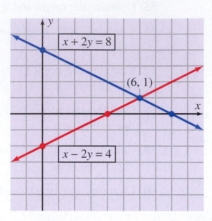

Figure 5.42

Thus we say that $\{(6, 1)\}$ is the solution set of the system

$$\begin{pmatrix} x - 2y = 4 \\ x + 2y = 8 \end{pmatrix}$$

Two or more linear equations in two variables considered together are called a **system of linear equations**. Here are three systems of linear equations:

$$\begin{pmatrix} x - 2y = 4 \\ x + 2y = 8 \end{pmatrix} \qquad \begin{pmatrix} 5x - 3y = 9 \\ 3x + 7y = 12 \end{pmatrix} \qquad \begin{pmatrix} 4x - y = 5 \\ 2x + y = 9 \\ 7x - 2y = 13 \end{pmatrix}$$

To **solve a system of linear equations** means to find all of the ordered pairs that are solutions of all of the equations in the system. There are several techniques for solving systems of linear equations. We will use three of them in this chapter: two methods are presented in this section, and a third method is presented in Section 5.6.

To solve a system of linear equations by **graphing**, we proceed as in the opening discussion of this section. We graph the equations on the same set of axes, and then the ordered pairs associated with any points of intersection are the solutions to the system. Let's consider another example.

Classroom Example
Solve the system $\begin{pmatrix} x + 2y = 5 \\ x - y = -4 \end{pmatrix}$.

EXAMPLE 1   Solve the system $\begin{pmatrix} x + y = 5 \\ x - 2y = -4 \end{pmatrix}$.

Solution

We can find the intercepts and a check point for each of the lines.

$x + y = 5$			**$x - 2y = -4$**		
x	**y**		**x**	**y**	
0	5		0	2	
5	0	Intercepts	−4	0	Intercepts
2	3	Check point	−2	1	Check point

Figure 5.43 shows the graphs of the two equations. It appears that $(2, 3)$ is the solution of the system. To check it we can substitute 2 for x and 3 for y in both equations.

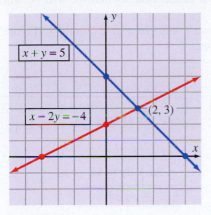

Figure 5.43

$x + y = 5$ becomes $2 + 3 = 5$ A true statement
$x - 2y = -4$ becomes $2 - 2(3) = -4$ A true statement

Therefore, $\{(2, 3)\}$ is the solution set.

It should be evident that solving systems of equations by graphing requires accurate graphs. In fact, unless the solutions are integers, it is really quite difficult to obtain exact solutions from a graph. For this reason, the systems to solve by graphing in this section have integer solutions. Furthermore, checking a solution takes on additional significance when the graphing approach is used. By checking, you can be absolutely sure that you are *reading* the correct solution from the graph.

Figure 5.44 shows the three possible cases for the graph of a system of two linear equations in two variables.

Case I   The graphs of the two equations are two lines intersecting in one point. There is one solution, and we call the system a **consistent system**.

Case II   The graphs of the two equations are parallel lines. There is no solution, and we call the system an **inconsistent system**.

Case III   The graphs of the two equations are the same line. There are infinitely many solutions to the system. Any pair of real numbers that satisfies one of the equations also satisfies the other equation, and we say the equations are **dependent**.

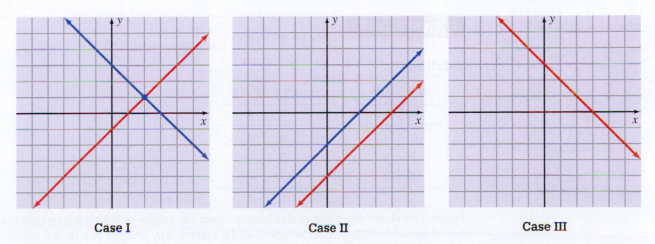

Case I   Case II   Case III

Figure 5.44

Thus as we solve a system of two linear equations in two variables, we know what to expect. The system will have no solutions, one ordered pair as a solution, or infinitely many ordered pairs as solutions. Most of the systems that we will be working with in this text have one solution.

An example of case I was given in Example 1 (Figure 5.43). The next two examples illustrate the other cases.

Classroom Example

Solve the system $\begin{pmatrix} 5x - y = 2 \\ 5x - y = 5 \end{pmatrix}$.

EXAMPLE 2   Solve the system $\begin{pmatrix} 2x + 3y = 6 \\ 2x + 3y = 12 \end{pmatrix}$.

Solution

2x + 3y = 6			**2x + 3y = 12**	
x	**y**		**x**	**y**
0	2		0	4
3	0		6	0
−3	4		3	2

Figure 5.45 shows the graph of the system. Since the lines are parallel, there is no solution to the system. The solution set is ∅.

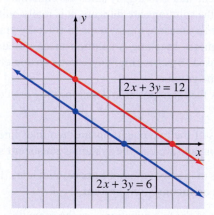

Figure 5.45

Classroom Example

Solve the system $\begin{pmatrix} 3x - 2y = 4 \\ 9x - 6y = 12 \end{pmatrix}$.

EXAMPLE 3   Solve the system $\begin{pmatrix} x + y = 3 \\ 2x + 2y = 6 \end{pmatrix}$.

Solution

x + y = 3			**2x + 2y = 6**	
x	**y**		**x**	**y**
0	3		0	3
3	0		3	0
1	2		1	2

Figure 5.46 shows the graph of this system. Since the graphs of both equations are the same line, there are infinitely many solutions to the system. Any ordered pair of real numbers that satisfies one equation also satisfies the other equation.

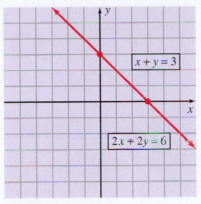

Figure 5.46

Substitution Method

As stated earlier, solving systems of equations by graphing requires accurate graphs. In fact, unless the solutions are integers, it is quite difficult to obtain exact solutions from a graph. Thus we will consider some other methods for solving systems of equations.

The **substitution method** works quite well with systems of two linear equations in two unknowns.

Step 1   Solve one of the equations for one variable in terms of the other variable, if neither equation is in such a form. (If possible, make a choice that will avoid fractions.)

Step 2   Substitute the expression obtained in step 1 into the other equation. This produces an equation in one variable.

Step 3   Solve the equation obtained in step 2.

Step 4   Use the solution obtained in step 3, along with the expression obtained in step 1, to determine the solution of the system.

Now let's look at some examples that illustrate the substitution method.

Classroom Example
Solve the system $\begin{pmatrix} 3x + y = 4 \\ y = x - 8 \end{pmatrix}$ by the substitution method.

EXAMPLE 4   Solve the system $\begin{pmatrix} x + y = 16 \\ y = x + 2 \end{pmatrix}$.

Solution

Because the second equation states that y equals $x + 2$, we can substitute $x + 2$ for y in the first equation.

$$x + y = 16 \xrightarrow{\text{Substitute } x + 2 \text{ for } y} x + (x + 2) = 16$$

Now we have an equation with one variable that we can solve in the usual way.

$$x + (x + 2) = 16$$
$$2x + 2 = 16$$
$$2x = 14$$
$$x = 7$$

Substituting 7 for x in one of the two original equations (let's use the second one) yields

$$y = 7 + 2 = 9$$

✓ Check

To check, we can substitute 7 for x and 9 for y in both of the original equations.

$$7 + 9 = 16 \quad \text{A true statement}$$
$$9 = 7 + 2 \quad \text{A true statement}$$

The solution set is $\{(7, 9)\}$.

Classroom Example
Solve the system $\begin{pmatrix} 2x - 9y = 14 \\ 3x - y = 6 \end{pmatrix}$
by the substitution method.

EXAMPLE 5   Solve the system $\begin{pmatrix} 3x - 7y = 2 \\ x + 4y = 1 \end{pmatrix}$.

Solution

Let's solve the second equation for x in terms of y.

$$x + 4y = 1$$
$$x = 1 - 4y$$

Now we can substitute $1 - 4y$ for x in the first equation.

$$3x - 7y = 2 \xrightarrow{\text{Substitute } 1 - 4y \text{ for } x} 3(1 - 4y) - 7y = 2$$

Let's solve this equation for y.

$$3(1 - 4y) - 7y = 2$$
$$3 - 12y - 7y = 2$$
$$-19y = -1$$
$$y = \frac{1}{19}$$

Finally, we can substitute $\frac{1}{19}$ for y in the equation $x = 1 - 4y$.

$$x = 1 - 4\left(\frac{1}{19}\right)$$
$$x = 1 - \frac{4}{19}$$
$$x = \frac{15}{19}$$

The solution set is $\left\{\left(\frac{15}{19}, \frac{1}{19}\right)\right\}$.

Classroom Example
Solve the system $\begin{pmatrix} 8x - 3y = 14 \\ 2x + 5y = -8 \end{pmatrix}$
by the substitution method.

EXAMPLE 6   Solve the system $\begin{pmatrix} 5x - 6y = -4 \\ 3x + 2y = -8 \end{pmatrix}$.

Solution

Note that solving either equation for either variable will produce a fractional form. Let's solve the second equation for y in terms of x.

$$3x + 2y = -8$$
$$2y = -8 - 3x$$
$$y = \frac{-8 - 3x}{2}$$

Now we can substitute $\frac{-8 - 3x}{2}$ for y in the first equation.

$$5x - 6y = -4 \xrightarrow{\text{Substitute } \frac{-8 - 3x}{2} \text{ for } y} 5x - 6\left(\frac{-8 - 3x}{2}\right) = -4$$

Solving the equation yields

$$5x - 6\left(\frac{-8 - 3x}{2}\right) = -4$$
$$5x - 3(-8 - 3x) = -4$$
$$5x + 24 + 9x = -4$$
$$14x = -28$$
$$x = -2$$

Substituting -2 for x in $y = \dfrac{-8 - 3x}{2}$ yields

$$y = \frac{-8 - 3(-2)}{2}$$

$$y = \frac{-8 + 6}{2}$$

$$y = \frac{-2}{2}$$

$$y = -1$$

The solution set is $\{(-2, -1)\}$.

Classroom Example

Solve the system $\begin{pmatrix} x + 3y = 9 \\ 2x + 6y = 11 \end{pmatrix}$

by the substitution method.

EXAMPLE 7 Solve the system $\begin{pmatrix} 2x + y = 4 \\ 4x + 2y = 7 \end{pmatrix}$.

Solution

Let's solve the first equation for y in terms of x.

$$2x + y = 4$$

$$y = 4 - 2x$$

Now we can substitute $4 - 2x$ for y in the second equation.

$$4x + 2y = 7 \quad \xrightarrow{\text{Substitute } 4 - 2x \text{ for } y} \quad 4x + 2(4 - 2x) = 7$$

Let's solve this equation for x.

$$4x + 2(4 - 2x) = 7$$

$$4x + 8 - 4x = 7$$

$$8 = 7$$

The statement $8 = 7$ is a contradiction, and therefore the original system is inconsistent; it has no solutions. The solution set is $\varnothing$.

Classroom Example

Solve the system $\begin{pmatrix} 8x - 2y = 6 \\ 4x = y + 3 \end{pmatrix}$

by the substitution method.

EXAMPLE 8 Solve the system $\begin{pmatrix} y = 2x + 1 \\ 4x - 2y = -2 \end{pmatrix}$.

Solution

Because the first equation states that y equals $2x + 1$, we can substitute $2x + 1$ for y in the second equation.

$$4x - 2y = -2 \quad \xrightarrow{\text{Substitute } 2x + 1 \text{ for } y} \quad 4x - 2(2x + 1) = -2$$

Let's solve this equation for x.

$$4x - 2(2x + 1) = -2$$

$$4x - 4x - 2 = -2$$

$$-2 = -2$$

We obtained a true statement, $-2 = -2$, which indicates that the system has an infinite number of solutions. Any ordered pair that satisfies one of the equations will also satisfy the other equation. Thus the solution set is any ordered pair on the line $y = 2x + 1$, and the solution set can be written as $\{(x, y) \mid y = 2x + 1\}$.

Problem Solving

Many word problems that we solved earlier in this text by using one variable and one equation can also be solved by using a system of two linear equations in two variables. In fact, in many of these problems you may find it much more natural to use two variables. Let's consider some examples.

Classroom Example
Sonora invested some money at 5% and $600 less than that amount at 3%. The yearly interest from the two investments was $190. How much did Sonora invest at each rate?

EXAMPLE 9

Anita invested some money at 8% and $400 more than that amount at 9%. The yearly interest from the two investments was $87. How much did Anita invest at each rate?

Solution

Let x represent the amount invested at 8%, and let y represent the amount invested at 9%. The problem translates into this system:

The amount invested at 9% was $400 more than at 8%

The yearly interest from the two investments was $87

$$\begin{pmatrix} y = x + 400 \\ 0.08x + 0.09y = 87 \end{pmatrix}$$

From the first equation, we can substitute $x + 400$ for y in the second equation and solve for x.

$$0.08x + 0.09(x + 400) = 87$$
$$0.08x + 0.09x + 36 = 87$$
$$0.17x = 51$$
$$x = 300$$

Therefore, Anita invested $300 at 8% and $300 + $400 = $700 at 9%.

Classroom Example
The proceeds of ticket sales at the children's theater were $6296. The price of a child's ticket was $12 and adult tickets were $20. If a total of 430 child and adult tickets were sold, how many of each were sold?

EXAMPLE 10

The proceeds from a concession stand that sold hamburgers and hot dogs at the baseball game were $575.50. The price of a hot dog was $2.50, and the price of a hamburger was $3.00. If a total of 213 hot dogs and hamburgers were sold, how many of each kind were sold?

Solution

Let x equal the number of hot dogs sold, and let y equal the number of hamburgers sold. The problem translates into this system:

The number sold

The proceeds from the sales

$$\begin{pmatrix} x + y = 213 \\ 2.50x + 3.00y = 575.50 \end{pmatrix}$$

Let's begin by solving the first equation for y.

$$x + y = 213$$
$$y = 213 - x$$

Now we will substitute $213 - x$ for y in the second equation and solve for x.

$$2.50x + 3.00(213 - x) = 575.50$$
$$2.50x + 639.00 - 3.00x = 575.50$$
$$-0.5x + 639.00 = 575.50$$
$$-0.5x = -63.50$$
$$x = 127$$

Therefore, there were 127 hot dogs sold and $213 - 127 = 86$ hamburgers sold.

Concept Quiz 5.5

For Problems 1–10, answer true or false.

1. To *solve a system of equations* means to find all the ordered pairs that satisfy all of the equations in the system.

2. A consistent system of linear equations will have more than one solution.

3. If the graph of a system of two linear equations results in two distinct parallel lines, then the system has no solution.

4. Every system of equations has a solution.

5. If the graphs of the two equations in a system are the same line, then the equations in the system are dependent.

6. To solve a system of two equations in variables x and y, it is sufficient to just find a value for x.

7. For the system $\begin{pmatrix} 2x + y = 4 \\ x + 5y = 10 \end{pmatrix}$, the ordered pair $(1, 2)$ is a solution.

8. Graphing a system of equations is the most accurate method to find the solution of the system.

9. The solution set of the system $\begin{pmatrix} x + 2y = 4 \\ 2x + 4y = 9 \end{pmatrix}$ is the null set.

10. The system $\begin{pmatrix} 2x + 2y = -4 \\ x = -y - 2 \end{pmatrix}$ has infinitely many solutions.

Problem Set 5.5

For Problems 1–20, use the graphing method to solve each system. **(Objective 1)**

1. $\begin{pmatrix} x + y = 1 \\ x - y = 3 \end{pmatrix}$

2. $\begin{pmatrix} x - y = 2 \\ x + y = -4 \end{pmatrix}$

3. $\begin{pmatrix} x + 2y = 4 \\ 2x - y = 3 \end{pmatrix}$

4. $\begin{pmatrix} 2x - y = -8 \\ x + y = 2 \end{pmatrix}$

5. $\begin{pmatrix} x + 3y = 6 \\ x + 3y = 3 \end{pmatrix}$

6. $\begin{pmatrix} y = -2x \\ y - 3x = 0 \end{pmatrix}$

7. $\begin{pmatrix} x + y = 0 \\ x - y = 0 \end{pmatrix}$

8. $\begin{pmatrix} 3x - y = 3 \\ 3x - y = -3 \end{pmatrix}$

9. $\begin{pmatrix} 3x - 2y = 5 \\ 2x + 5y = -3 \end{pmatrix}$

10. $\begin{pmatrix} 2x + 3y = 1 \\ 4x - 3y = -7 \end{pmatrix}$

11. $\begin{pmatrix} y = -2x + 3 \\ 6x + 3y = 9 \end{pmatrix}$

12. $\begin{pmatrix} y = 2x + 5 \\ x + 3y = -6 \end{pmatrix}$

13. $\begin{pmatrix} y = 5x - 2 \\ 4x + 3y = 13 \end{pmatrix}$

14. $\begin{pmatrix} y = x - 2 \\ 2x - 2y = 4 \end{pmatrix}$

15. $\begin{pmatrix} y = 4 - 2x \\ y = 7 - 3x \end{pmatrix}$

16. $\begin{pmatrix} y = 3x + 4 \\ y = 5x + 8 \end{pmatrix}$

17. $\begin{pmatrix} y = 2x \\ 3x - 2y = -2 \end{pmatrix}$

18. $\begin{pmatrix} y = 3x \\ 4x - 3y = 5 \end{pmatrix}$

19. $\begin{pmatrix} 7x - 2y = -8 \\ x = -2 \end{pmatrix}$

20. $\begin{pmatrix} 3x + 8y = -1 \\ y = -2 \end{pmatrix}$

For Problems 21–46, solve each system by using the substitution method. **(Objective 2)**

21. $\begin{pmatrix} x + y = 20 \\ x = y - 4 \end{pmatrix}$

22. $\begin{pmatrix} x + y = 23 \\ y = x - 5 \end{pmatrix}$

23. $\begin{pmatrix} y = -3x - 18 \\ 5x - 2y = -8 \end{pmatrix}$

24. $\begin{pmatrix} 4x - 3y = 33 \\ x = -4y - 25 \end{pmatrix}$

25. $\begin{pmatrix} x = -3y \\ 7x - 2y = -69 \end{pmatrix}$

26. $\begin{pmatrix} 9x - 2y = -38 \\ y = -5x \end{pmatrix}$

27. $\begin{pmatrix} x + 2y = 5 \\ 3x + 6y = -2 \end{pmatrix}$

28. $\begin{pmatrix} 4x + 2y = 6 \\ y = -2x + 3 \end{pmatrix}$

29. $\begin{pmatrix} 3x - 4y = 9 \\ x = 4y - 1 \end{pmatrix}$

30. $\begin{pmatrix} y = 3x - 5 \\ 2x + 3y = 6 \end{pmatrix}$

31. $\begin{pmatrix} y = \dfrac{2}{5}x - 1 \\ 3x + 5y = 4 \end{pmatrix}$

32. $\begin{pmatrix} y = \dfrac{3}{4}x - 5 \\ 5x - 4y = 9 \end{pmatrix}$

33. $\begin{pmatrix} 7x - 3y = -2 \\ x = \dfrac{3}{4}y + 1 \end{pmatrix}$

34. $\begin{pmatrix} 5x - y = 9 \\ x = \dfrac{1}{2}y - 3 \end{pmatrix}$

35. $\begin{pmatrix} 2x + y = 12 \\ 3x - y = 13 \end{pmatrix}$

36. $\begin{pmatrix} -x + 4y = -22 \\ x - 7y = 34 \end{pmatrix}$

37. $\begin{pmatrix} 4x + 3y = -40 \\ 5x - y = -12 \end{pmatrix}$

38. $\begin{pmatrix} x - 5y = 33 \\ -4x + 7y = -41 \end{pmatrix}$

39. $\begin{pmatrix} 3x + y = 2 \\ 11x - 3y = 5 \end{pmatrix}$

40. $\begin{pmatrix} 2x - y = 9 \\ 7x + 4y = 1 \end{pmatrix}$

41. $\begin{pmatrix} 4x - 8y = -12 \\ 3x - 6y = -9 \end{pmatrix}$

42. $\begin{pmatrix} 2x - 4y = -6 \\ 3x - 6y = 10 \end{pmatrix}$

43. $\begin{pmatrix} 4x - 5y = 3 \\ 8x + 15y = -24 \end{pmatrix}$

44. $\begin{pmatrix} 2x + 3y = 3 \\ 4x - 9y = -4 \end{pmatrix}$

45. $\begin{pmatrix} 6x - 3y = 4 \\ 5x + 2y = -1 \end{pmatrix}$

46. $\begin{pmatrix} 7x - 2y = 1 \\ 4x + 5y = 2 \end{pmatrix}$

For Problems 47–58, solve each problem by setting up and solving an appropriate system of linear equations. **(Objective 5)**

47. Doris invested some money at 7% and some money at 8%. She invested $6000 more at 8% than she did at 7%. Her total yearly interest from the two investments was $780. How much did Doris invest at each rate?

48. Suppose that Gus invested a total of $8000, part of it at 4% and the remainder at 6%. His yearly income from the two investments was $380. How much did he invest at each rate?

49. Find two numbers whose sum is 131 such that one number is 5 less than three times the other.

50. The difference of two numbers is 75. The larger number is 3 less than four times the smaller number. Find the numbers.

51. In a class of 50 students, the number of women is 2 more than five times the number of men. How many women are there in the class?

52. In a recent survey, 1000 registered voters were asked about their political preferences. The number of men in the survey was 5 less than one-half the number of women. Find the number of men in the survey.

53. The perimeter of a rectangle is 94 inches. The length of the rectangle is 7 inches more than the width. Find the dimensions of the rectangle.

54. Two angles are supplementary, and the measure of one of them is 20° less than three times the measure of the other angle. Find the measure of each angle.

55. A deposit slip listed $700 in cash to be deposited. There were 100 bills, some of them were five-dollar bills and the remainder were ten-dollar bills. How many bills of each denomination were deposited?

56. Cindy has 30 coins, consisting of dimes and quarters, which total $5.10. How many coins of each kind does she have?

57. The income from a student production was $27,500. The price of a student ticket was $8, and nonstudent tickets were sold at $15 each. Three thousand tickets were sold. How many tickets of each kind were sold?

58. Sue bought 3 packages of cookies and 2 sacks of potato chips for $7.35. Later she bought 2 packages of cookies and 5 sacks of potato chips for $9.63. Find the price of a package of cookies.

Thoughts Into Words

59. Discuss the strengths and weaknesses of solving a system of linear equations by graphing.

60. Determine a system of two linear equations for which the solution set is {(5, 7)}. Are there other systems that have the same solution set? If so, find at least one more system.

61. Give a general description of how to use the substitution method to solve a system of two linear equations in two variables.

62. Is it possible for a system of two linear equations in two variables to have exactly two solutions? Defend your answer.

63. Explain how you would use the substitution method to solve the system

$$\begin{pmatrix} 2x + 5y = 5 \\ 5x - y = 9 \end{pmatrix}$$

Answers to the Concept Quiz

1. True   **2.** False   **3.** True   **4.** False   **5.** True   **6.** False   **7.** False   **8.** False   **9.** True   **10.** True

5.6   Elimination-by-Addition Method

OBJECTIVES

1 Solve linear systems of equations by the elimination-by-addition method

2 Solve word problems using a system of two linear equations

We found in the previous section that the substitution method for solving a system of two equations and two unknowns works rather well. However, as the number of equations and unknowns increases, the substitution method becomes quite unwieldy. In this section we are going to introduce another method, called the **elimination-by-addition method**. We shall introduce it here, using systems of two linear equations in two unknowns. Later in the text, we shall extend its use to three linear equations in three unknowns.

The elimination-by-addition method involves replacing systems of equations with simpler equivalent systems until we obtain a system from which we can easily extract the solutions. **Equivalent systems of equations are systems that have exactly the same solution set.** We can apply the following operations or transformations to a system of equations to produce an equivalent system.

1. Any two equations of the system can be interchanged.

2. Both sides of any equation of the system can be multiplied by any nonzero real number.

3. Any equation of the system can be replaced by the sum of the equation and a nonzero multiple of another equation.

Now let's see how to apply these operations to solve a system of two linear equations in two unknowns.

Classroom Example
Solve the system $\begin{pmatrix} 6x + 9y = 9 \\ 4x - 9y = 21 \end{pmatrix}$ by the elimination-by-addition method.

EXAMPLE 1   Solve the system $\begin{pmatrix} 3x + 2y = 1 \\ 5x - 2y = 23 \end{pmatrix}$.   (1)
                                                                                      (2)

Solution

Let's replace equation (2) with an equation we form by multiplying equation (1) by 1 and then adding that result to equation (2).

$$\begin{pmatrix} 3x + 2y = 1 \\ 8x \qquad = 24 \end{pmatrix}$$ (3)
                                                                 (4)

From equation (4) we can easily obtain the value of x.

$$8x = 24$$
$$x = 3$$

Then we can substitute 3 for x in equation (3).

$$3x + 2y = 1$$
$$3(3) + 2y = 1$$
$$2y = -8$$
$$y = -4$$

The solution set is $\{(3, -4)\}$. Check it!

Classroom Example
Solve the system $\begin{pmatrix} 2x - 5y = 28 \\ 8x + y = 28 \end{pmatrix}$ using the elimination-by-addition method.

EXAMPLE 2   Solve the system $\begin{pmatrix} x + 5y = -2 \\ 3x - 4y = -25 \end{pmatrix}$.   (1)
                                                                                      (2)

Solution

Let's replace equation (2) with an equation we form by multiplying equation (1) by -3 and then adding that result to equation (2).

$$\begin{pmatrix} x + 5y = -2 \\ -19y = -19 \end{pmatrix} \qquad (3) \\ (4)$$

From equation (4) we can obtain the value y.

$$-19y = -19$$
$$y = 1$$

Now we can substitute 1 for y in equation (3)

$$x + 5y = -2$$
$$x + 5(1) = -2$$
$$x = -7$$

The solution set is $\{(-7, 1)\}$.

Note that our objective has been to produce an equivalent system of equations such that one of the variables can be *eliminated* from one equation. We accomplish this by multiplying one equation of the system by an appropriate number and then *adding* that result to the other equation. Thus the method is called *elimination by addition*. Let's look at another example.

Classroom Example
Solve the system $\begin{pmatrix} 3x - 5y = 31 \\ 7x + 2y = 4 \end{pmatrix}$
using the elimination-by-addition method.

EXAMPLE 3   Solve the system $\begin{pmatrix} 2x + 5y = 4 \\ 5x - 7y = -29 \end{pmatrix}$. $\qquad$ (1) (2)

Solution

Let's form an equivalent system whereby the second equation has no x term. First, we can multiply equation (2) by -2.

$$\begin{pmatrix} 2x + 5y = 4 \\ -10x + 14y = 58 \end{pmatrix} \qquad (3) \\ (4)$$

Now we can replace equation (4) with an equation that we form by multiplying equation (3) by 5 and then adding that result to equation (4).

$$\begin{pmatrix} 2x + 5y = 4 \\ 39y = 78 \end{pmatrix} \qquad (5) \\ (6)$$

From equation (6) we can find the value of y.

$$39y = 78$$
$$y = 2$$

Now we can substitute 2 for y in equation (5).

$$2x + 5y = 4$$
$$2x + 5(2) = 4$$
$$2x = -6$$
$$x = -3$$

The solution set is $\{(-3, 2)\}$.

Classroom Example
Solve the system $\begin{pmatrix} 5x - 6y = 3 \\ 8x + 4y = 10 \end{pmatrix}$
using the elimination-by-addition method.

EXAMPLE 4   Solve the system $\begin{pmatrix} 3x - 2y = 5 \\ 2x + 7y = 9 \end{pmatrix}$. $\qquad$ (1) (2)

Solution

We can start by multiplying equation (2) by -3.

$$\begin{pmatrix} 3x - 2y = 5 \\ -6x - 21y = -27 \end{pmatrix} \qquad (3) \\ (4)$$

Now we can replace equation (4) with an equation that we form by multiplying equation (3) by 2 and then adding that result to equation (4).

$$\begin{pmatrix} 3x - 2y = & 5 \\ -25y = & -17 \end{pmatrix}$$

$\qquad(5)$
$\qquad(6)$

From equation (6) we can find the value y.

$$-25y = -17$$
$$y = \frac{17}{25}$$

Now we can substitute $\dfrac{17}{25}$ for y in equation (5).

$$3x - 2y = 5$$
$$3x - 2\left(\frac{17}{25}\right) = 5$$
$$3x - \frac{34}{25} = 5$$
$$3x = 5 + \frac{34}{25}$$
$$3x = \frac{125}{25} + \frac{34}{25}$$
$$3x = \frac{159}{25}$$
$$x = \left(\frac{159}{25}\right)\left(\frac{1}{3}\right) = \frac{53}{25}$$

The solution set is $\left\{\left(\dfrac{53}{25}, \dfrac{17}{25}\right)\right\}$. (Perhaps you should check this result!)

Which Method to Use?

We can use both the elimination-by-addition and the substitution methods to obtain exact solutions for any system of two linear equations in two unknowns. Sometimes we need to decide which method to use on a particular system. As we have seen with the examples thus far in this section and with those in the previous section, many systems lend themselves to one or the other method by the original format of the equations. Let's emphasize that point with some more examples.

Classroom Example
Solve the system $\begin{pmatrix} 9x - 10y = 4 \\ 8x + 5y = 3 \end{pmatrix}$.

EXAMPLE 5   Solve the system $\begin{pmatrix} 4x - 3y = 4 \\ 10x + 9y = -1 \end{pmatrix}$.

$\qquad(1)$
$\qquad(2)$

Solution

Because changing the form of either equation in preparation for the substitution method would produce a fractional form, we are probably better off using the elimination-by-addition method. Let's replace equation (2) with an equation we form by multiplying equation (1) by 3 and then adding that result to equation (2).

$$\begin{pmatrix} 4x - 3y = 4 \\ 22x = 11 \end{pmatrix}$$

$\qquad(3)$
$\qquad(4)$

From equation (4) we can determine the value of x.

$$22x = 11$$
$$x = \frac{11}{22} = \frac{1}{2}$$

Now we can substitute $\dfrac{1}{2}$ for x in equation (3).

$$4x - 3y = 4$$

$$4\left(\frac{1}{2}\right) - 3y = 4$$

$$2 - 3y = 4$$

$$-3y = 2$$

$$y = -\frac{2}{3}$$

The solution set is $\left\{\left(\dfrac{1}{2}, -\dfrac{2}{3}\right)\right\}$

Classroom Example
Solve the system $\begin{pmatrix} x = -3y + 4 \\ 5x - 12y = 2 \end{pmatrix}$.

EXAMPLE 6   Solve the system $\begin{pmatrix} 6x + 5y = -3 \\ y = -2x - 7 \end{pmatrix}$. (1) (2)

Solution

Because the second equation is of the form y *equals*, let's use the substitution method. From the second equation we can substitute $-2x - 7$ for y in the first equation.

$$6x + 5y = -3 \quad\xrightarrow{\text{Substitute } -2x - 7 \text{ for } y}\quad 6x + 5(-2x - 7) = -3$$

Solving this equation yields

$$6x + 5(-2x - 7) = -3$$

$$6x - 10x - 35 = -3$$

$$-4x - 35 = -3$$

$$-4x = 32$$

$$x = -8$$

We substitute -8 for x in the second equation to get

$$y = -2(-8) - 7$$

$$y = 16 - 7 = 9$$

The solution set is $\{(-8, 9)\}$.

Sometimes we need to simplify the equations of a system before we can decide which method to use for solving the system. Let's consider an example of that type.

Classroom Example
Solve the system
$\begin{pmatrix} \dfrac{x - 2}{3} + \dfrac{y + 5}{4} = \dfrac{3}{2} \\ \dfrac{x + 1}{2} + \dfrac{y - 2}{5} = 2 \end{pmatrix}$.

EXAMPLE 7   Solve the system $\begin{pmatrix} \dfrac{x - 2}{4} + \dfrac{y + 1}{3} = 2 \\ \dfrac{x + 1}{7} + \dfrac{y - 3}{2} = \dfrac{1}{2} \end{pmatrix}$. (1) (2)

Solution

First, we need to simplify the two equations. Let's multiply both sides of equation (1) by 12 and simplify.

$$12\left(\frac{x - 2}{4} + \frac{y + 1}{3}\right) = 12(2)$$

$$3(x - 2) + 4(y + 1) = 24$$

$$3x - 6 + 4y + 4 = 24$$

$$3x + 4y - 2 = 24$$

$$3x + 4y = 26$$

Let's multiply both sides of equation (2) by 14.

$$14\left(\frac{x + 1}{7} + \frac{y - 3}{2}\right) = 14\left(\frac{1}{2}\right)$$

$$2(x + 1) + 7(y - 3) = 7$$

$$2x + 2 + 7y - 21 = 7$$

$$2x + 7y - 19 = 7$$

$$2x + 7y = 26$$

Now we have the following system to solve.

$$\begin{pmatrix} 3x + 4y = 26 \\ 2x + 7y = 26 \end{pmatrix} \qquad (3) \\ (4)$$

Probably the easiest approach is to use the elimination-by-addition method. We can start by multiplying equation (4) by -3.

$$\begin{pmatrix} 3x + 4y = 26 \\ -6x - 21y = -78 \end{pmatrix} \qquad (5) \\ (6)$$

Now we can replace equation (6) with an equation we form by multiplying equation (5) by 2 and then adding that result to equation (6).

$$\begin{pmatrix} 3x + 4y = 26 \\ -13y = -26 \end{pmatrix} \qquad (7) \\ (8)$$

From equation (8) we can find the value of y.

$$-13y = -26$$

$$y = 2$$

Now we can substitute 2 for y in equation (7).

$$3x + 4y = 26$$

$$3x + 4(2) = 26$$

$$3x = 18$$

$$x = 6$$

The solution set is $\{(6, 2)\}$.

Remark: Don't forget that to check a problem like Example 7, you must check the potential solutions back in the *original* equations.

In Section 5.5, we explained that you can tell whether a system of two linear equations in two unknowns has no solution, one solution, or infinitely many solutions, by graphing the equations of the system. That is, the two lines may be parallel (no solution), or they may intersect in one point (one solution), or they may coincide (infinitely many solutions).

From a practical viewpoint, the systems that have one solution deserve most of our attention. However, we do need to be able to deal with the other situations because they occur occasionally. The next two examples illustrate what happens when we encounter a *no solution* or *infinitely many solutions* situation when using the elimination-by-addition method.

Classroom Example
Solve the system $\begin{pmatrix} x - 5y = 4 \\ 2x - 10y = 1 \end{pmatrix}$.

EXAMPLE 8

Solve the system $\begin{pmatrix} 2x + y = 1 \\ 4x + 2y = 3 \end{pmatrix}$. $\qquad$ (1)
$\qquad$ (2)

Solution

Use the elimination-by-addition method. Let's replace equation (2) with an equation we form by multiplying equation (1) by -2 and then adding that result to form equation (2).

$$\begin{pmatrix} 2x + y = 1 \\ 0 + 0 = 1 \end{pmatrix} \qquad (3) \\ (4)$$

The false numerical statement $0 + 0 = 1$ implies that the system has no solution. Thus the solution set is $\varnothing$.

Classroom Example
Solve the system $\begin{pmatrix} 6x + 4y = 10 \\ 3x + 2y = 5 \end{pmatrix}$.

EXAMPLE 9

Solve the system $\begin{pmatrix} 5x + y = 2 \\ 10x + 2y = 4 \end{pmatrix}$. $\qquad$ (1)
$\qquad$ (2)

Solution

Use the elimination-by-addition method. Let's replace equation (2) with an equation we form by multiplying equation (1) by -2 and then adding that result to equation (2).

$$\begin{pmatrix} 5x + y = 2 \\ 0 + 0 = 0 \end{pmatrix} \qquad (3) \\ (4)$$

The *true numerical statement* $0 + 0 = 0$ implies that the system has *infinitely many solutions*. Any ordered pair that satisfies one of the equations will also satisfy the other equation. Thus the solution set can be expressed as

$$\{(x, y) | 5x + y = 2\}$$

Classroom Example
A 30% alcohol solution is to be mixed with a 55% alcohol solution to produce 25 liters of 40% alcohol solution. How many liters of each solution should be mixed?

EXAMPLE 10

A 25% chlorine solution is to be mixed with a 40% chlorine solution to produce 12 gallons of a 35% chlorine solution. How many gallons of each solution should be mixed?

Solution

Let x represent the gallons of 25% chlorine solution, and let y represent the gallons of 40% chlorine solution. Then one equation of the system will be $x + y = 12$. For the other equation we need to multiply the number of gallons of each solution by its percentage of chlorine. That gives the equation $0.25x + 0.40y = 0.35(12)$. So we need to solve the following system:

$$\begin{pmatrix} x + y = 12 \\ 0.25x + 0.40y = 4.2 \end{pmatrix} \qquad (1) \\ (2)$$

Use the elimination-by-addition method. Let's replace equation (2) with an equation we form by multiplying equation (1) by -0.25 and then adding that result to equation (2).

$$\begin{pmatrix} x + y = 12 \\ 0.15y = 1.2 \end{pmatrix} \qquad (3) \\ (4)$$

From equation (4) we can find the value of y.

$$0.15y = 1.2$$
$$y = 8$$

Now we can substitute 8 into equation (3).

$$x + 8 = 12$$
$$x = 4$$

Therefore, we need 4 gallons of the 25% chlorine solution and 8 gallons of the 40% solution.

Concept Quiz 5.6

For Problems 1–10, answer true or false.

1. Any two equations of a system can be interchanged to obtain an equivalent system.

2. Any equation of a system can be multiplied on both sides by zero to obtain an equivalent system.

3. The objective of the elimination-by-addition method is to produce an equivalent system with an equation in which one of the variables has been eliminated.

4. Either the substitution method or the elimination-by-addition method can be used for any linear system of equations.

5. If an equivalent system for an original system is $\begin{pmatrix} 3x - 5y = 7 \\ 0 + 0 = 0 \end{pmatrix}$, then the original system is inconsistent and has no solution.

6. The solution set of the system $\begin{pmatrix} 3x - 2y = -3 \\ 2x + 3y = 11 \end{pmatrix}$ is $\{(1, 3)\}$.

7. The solution set of the system $\begin{pmatrix} x - 5y = -17 \\ 3x + y = 4 \end{pmatrix}$ is $\{(-2, 3)\}$.

8. The system $\begin{pmatrix} 5x - 2y = 3 \\ 5x - 2y = 9 \end{pmatrix}$ has infinitely many solutions.

9. The system $\begin{pmatrix} 2x + 3y = 4 \\ 6x + 9y = 12 \end{pmatrix}$ has infinitely many solutions.

10. The system $\begin{pmatrix} x - 2y = 6 \\ 2x + y = 4 \end{pmatrix}$ has only one solution.

Problem Set 5.6

For Problems 1–16, use the elimination-by-addition method to solve each system. **(Objective 1)**

1. $\begin{pmatrix} 2x + 3y = -1 \\ 5x - 3y = 29 \end{pmatrix}$

2. $\begin{pmatrix} 3x - 4y = -30 \\ 7x + 4y = 10 \end{pmatrix}$

3. $\begin{pmatrix} 6x - 7y = 15 \\ 6x + 5y = -21 \end{pmatrix}$

4. $\begin{pmatrix} 5x + 2y = -4 \\ 5x - 3y = 6 \end{pmatrix}$

5. $\begin{pmatrix} x - 2y = -12 \\ 2x + 9y = 2 \end{pmatrix}$

6. $\begin{pmatrix} x - 4y = 29 \\ 3x + 2y = -11 \end{pmatrix}$

7. $\begin{pmatrix} 4x + 7y = -16 \\ 6x - y = -24 \end{pmatrix}$

8. $\begin{pmatrix} 6x + 7y = 17 \\ 3x + y = -4 \end{pmatrix}$

9. $\begin{pmatrix} 3x - 2y = 5 \\ 2x + 5y = -3 \end{pmatrix}$

10. $\begin{pmatrix} 4x + 3y = -4 \\ 3x - 7y = 34 \end{pmatrix}$

11. $\begin{pmatrix} 7x - 2y = 4 \\ 7x - 2y = 9 \end{pmatrix}$

12. $\begin{pmatrix} 5x - y = 6 \\ 10x - 2y = 12 \end{pmatrix}$

13. $\begin{pmatrix} 5x + 4y = 1 \\ 3x - 2y = -1 \end{pmatrix}$

14. $\begin{pmatrix} 2x - 7y = -2 \\ 3x + y = 1 \end{pmatrix}$

15. $\begin{pmatrix} 8x - 3y = 13 \\ 4x + 9y = 3 \end{pmatrix}$

16. $\begin{pmatrix} 10x - 8y = -11 \\ 8x + 4y = -1 \end{pmatrix}$

For Problems 17–44, solve each system by using either the substitution or the elimination-by-addition method, whichever seems more appropriate. **(Objective 1)**

17. $\begin{pmatrix} 5x + 3y = -7 \\ 7x - 3y = 55 \end{pmatrix}$

18. $\begin{pmatrix} 4x - 7y = 21 \\ -4x + 3y = -9 \end{pmatrix}$

19. $\begin{pmatrix} x = 5y + 7 \\ 4x + 9y = 28 \end{pmatrix}$

20. $\begin{pmatrix} 11x - 3y = -60 \\ y = -38 - 6x \end{pmatrix}$

21. $\begin{pmatrix} x = -6y + 79 \\ x = 4y - 41 \end{pmatrix}$

22. $\begin{pmatrix} y = 3x + 34 \\ y = -8x - 54 \end{pmatrix}$

23. $\begin{pmatrix} 4x - 3y = 2 \\ 5x - y = 3 \end{pmatrix}$

24. $\begin{pmatrix} 3x - y = 9 \\ 5x + 7y = 1 \end{pmatrix}$

25. $\begin{pmatrix} 5x - 2y = 1 \\ 10x - 4y = 7 \end{pmatrix}$

26. $\begin{pmatrix} 4x + 7y = 2 \\ 9x - 2y = 1 \end{pmatrix}$

27. $\begin{pmatrix} 3x - 2y = 7 \\ 5x + 7y = 1 \end{pmatrix}$

28. $\begin{pmatrix} 2x - 3y = 4 \\ y = \dfrac{2}{3}x - \dfrac{4}{3} \end{pmatrix}$

29. $\begin{pmatrix} -2x + 5y = -16 \\ x = \dfrac{3}{4}y + 1 \end{pmatrix}$

30. $\begin{pmatrix} y = \dfrac{2}{3}x - \dfrac{3}{4} \\ 2x + 3y = 11 \end{pmatrix}$

31. $\left(\begin{array}{l} y = \dfrac{2}{3}x - 4 \\ 5x - 3y = 9 \end{array} \right)$
32. $\left(\begin{array}{l} 5x - 3y = 7 \\ x = \dfrac{3y}{4} - \dfrac{1}{3} \end{array} \right)$

33. $\left(\begin{array}{l} \dfrac{x}{6} + \dfrac{y}{3} = 3 \\ \dfrac{5x}{2} - \dfrac{y}{6} = -17 \end{array} \right)$
34. $\left(\begin{array}{l} \dfrac{3x}{4} - \dfrac{2y}{3} = 31 \\ \dfrac{7x}{5} + \dfrac{y}{4} = 22 \end{array} \right)$

35. $\left(\begin{array}{l} -(x - 6) + 6(y + 1) = 58 \\ 3(x + 1) - 4(y - 2) = -15 \end{array} \right)$

36. $\left(\begin{array}{l} -2(x + 2) + 4(y - 3) = -34 \\ 3(x + 4) - 5(y + 2) = 23 \end{array} \right)$

37. $\left(\begin{array}{l} 5(x + 1) - (y + 3) = -6 \\ 2(x - 2) + 3(y - 1) = 0 \end{array} \right)$

38. $\left(\begin{array}{l} 2(x - 1) - 3(y + 2) = 30 \\ 3(x + 2) + 2(y - 1) = -4 \end{array} \right)$

39. $\left(\begin{array}{l} \dfrac{1}{2}x - \dfrac{1}{3}y = 12 \\ \dfrac{3}{4}x + \dfrac{2}{3}y = 4 \end{array} \right)$
40. $\left(\begin{array}{l} \dfrac{2}{3}x + \dfrac{1}{5}y = 0 \\ \dfrac{3}{2}x - \dfrac{3}{10}y = -15 \end{array} \right)$

41. $\left(\begin{array}{l} \dfrac{2x}{3} - \dfrac{y}{2} = -\dfrac{5}{4} \\ \dfrac{x}{4} + \dfrac{5y}{6} = \dfrac{17}{16} \end{array} \right)$
42. $\left(\begin{array}{l} \dfrac{x}{2} + \dfrac{y}{3} = \dfrac{5}{72} \\ \dfrac{x}{4} + \dfrac{5y}{2} = -\dfrac{17}{48} \end{array} \right)$

43. $\left(\begin{array}{l} \dfrac{3x + y}{2} + \dfrac{x - 2y}{5} = 8 \\ \dfrac{x - y}{3} - \dfrac{x + y}{6} = \dfrac{10}{3} \end{array} \right)$

44. $\left(\begin{array}{l} \dfrac{x - y}{4} - \dfrac{2x - y}{3} = -\dfrac{1}{4} \\ \dfrac{2x + y}{3} + \dfrac{x + y}{2} = \dfrac{17}{6} \end{array} \right)$

For Problems 45–55, solve each problem by setting up and solving an appropriate system of equations. **(Objective 2)**

45. A 10% salt solution is to be mixed with a 20% salt solution to produce 20 gallons of a 17.5% salt solution. How many gallons of the 10% solution and how many gallons of the 20% solution will be needed?

46. A small town library buys a total of 35 books that cost $1022. Some of the books cost $22 each, and the remainder cost $34 per book. How many books of each price did the library buy?

47. Suppose that on a particular day the cost of 3 tennis balls and 2 golf balls is $12. The cost of 6 tennis balls

and 3 golf balls is $21. Find the cost of 1 tennis ball and the cost of 1 golf ball.

48. For moving purposes, the Hendersons bought 25 cardboard boxes for $97.50. There were two kinds of boxes; the large ones cost $7.50 per box, and the small ones cost $3 per box. How many boxes of each kind did they buy?

49. A motel in a suburb of Chicago rents single rooms for $62 per day and double rooms for $82 per day. If a total of 55 rooms were rented for $4210, how many of each kind were rented?

50. Suppose that one solution is 50% alcohol and another solution is 80% alcohol. How many liters of each solution should be mixed to make 10.5 liters of a 70% alcohol solution?

51. If the numerator of a certain fraction is increased by 5, and the denominator is decreased by 1, the resulting fraction is $\dfrac{8}{3}$. However, if the numerator of the original fraction is doubled, and the denominator of the original fraction is increased by 7, the resulting fraction is $\dfrac{6}{11}$. Find the original fraction.

52. A man bought 2 pounds of coffee and 1 pound of butter for a total of $18.75. A month later the prices had not changed (this makes it a fictitious problem), and he bought 3 pounds of coffee and 2 pounds of butter for $29.50. Find the price per pound of both the coffee and the butter.

53. Suppose that we have a rectangular book cover. If the width is increased by 2 centimeters, and the length is decreased by 1 centimeter, then the area is increased by 28 square centimeters. However, if the width is decreased by 1 centimeter, and the length is increased by 2 centimeters, then the area is increased by 10 square centimeters. Find the dimensions of the book cover.

54. A blueprint indicates a master bedroom in the shape of a rectangle. If the width is increased by 2 feet, and the length remains the same, then the area is increased by 36 square feet. However, if the width is increased by 1 foot, and the length is increased by 2 feet, then the area is increased by 48 square feet. Find the dimensions of the room as indicated on the blueprint.

Thoughts Into Words

55. Explain how you would use the elimination-by-addition method to solve the system

$$\begin{pmatrix} 3x - 4y = -1 \\ 2x - 5y = 9 \end{pmatrix}$$

56. How do you decide whether to solve a system of linear equations in two variables by using the substitution method or the elimination-by-addition method?

Further Investigations

57. There is another way of telling whether a system of two linear equations in two unknowns is consistent, inconsistent, or dependent without taking the time to graph each equation. It can be shown that any system of the form

$$a_1 x + b_1 y = c_1$$
$$a_2 x + b_2 y = c_2$$

has one and only one solution if

$$\frac{a_1}{a_2} \neq \frac{b_1}{b_2}$$

that it has no solution if

$$\frac{a_1}{a_2} = \frac{b_1}{b_2} \neq \frac{c_1}{c_2}$$

and that it has infinitely many solutions if

$$\frac{a_1}{a_2} = \frac{b_1}{b_2} = \frac{c_1}{c_2}$$

Determine whether each of the following systems is consistent, inconsistent, or dependent.

(a) $\begin{pmatrix} 4x - 3y = 7 \\ 9x + 2y = 5 \end{pmatrix}$

(b) $\begin{pmatrix} 5x - y = 6 \\ 10x - 2y = 19 \end{pmatrix}$

(c) $\begin{pmatrix} 5x - 4y = 11 \\ 4x + 5y = 12 \end{pmatrix}$

(d) $\begin{pmatrix} x + 2y = 5 \\ x - 2y = 9 \end{pmatrix}$

(e) $\begin{pmatrix} x - 3y = 5 \\ 3x - 9y = 15 \end{pmatrix}$

(f) $\begin{pmatrix} 4x + 3y = 7 \\ 2x - y = 10 \end{pmatrix}$

(g) $\begin{pmatrix} 3x + 2y = 4 \\ y = -\dfrac{3}{2}x - 1 \end{pmatrix}$

(h) $\begin{pmatrix} y = \dfrac{4}{3}x - 2 \\ 4x - 3y = 6 \end{pmatrix}$

58. A system such as

$$\begin{pmatrix} \dfrac{3}{x} + \dfrac{2}{y} = 2 \\ \dfrac{2}{x} - \dfrac{3}{y} = \dfrac{1}{4} \end{pmatrix}$$

is not a system of linear equations but can be transformed into a linear system by changing variables. For example, when we substitute u for $\dfrac{1}{x}$ and v for $\dfrac{1}{y}$ in the above system we get

$$\begin{pmatrix} 3u + 2v = 2 \\ 2u - 3v = \dfrac{1}{4} \end{pmatrix}$$

We can solve this "new" system either by elimination by addition or by substitution (we will leave the details for you) to produce $u = \dfrac{1}{2}$ and $v = \dfrac{1}{4}$. Therefore, because $u = \dfrac{1}{x}$ and $v = \dfrac{1}{y}$, we have

$$\frac{1}{x} = \frac{1}{2} \qquad \text{and} \qquad \frac{1}{y} = \frac{1}{4}$$

Solving these equations yields

$$x = 2 \qquad \text{and} \qquad y = 4$$

The solution set of the original system is $\{(2, 4)\}$.

Solve each of the following systems.

(a) $\begin{pmatrix} \dfrac{1}{x} + \dfrac{2}{y} = \dfrac{7}{12} \\ \dfrac{3}{x} - \dfrac{2}{y} = \dfrac{5}{12} \end{pmatrix}$

(b) $\begin{pmatrix} \dfrac{2}{x} + \dfrac{3}{y} = \dfrac{19}{15} \\ -\dfrac{2}{x} + \dfrac{1}{y} = -\dfrac{7}{15} \end{pmatrix}$

(c) $\begin{pmatrix} \dfrac{3}{x} - \dfrac{2}{y} = \dfrac{13}{6} \\ \dfrac{2}{x} + \dfrac{3}{y} = 0 \end{pmatrix}$

(d) $\left(\begin{array}{l} \dfrac{4}{x} + \dfrac{1}{y} = 11 \\[2mm] \dfrac{3}{x} - \dfrac{5}{y} = -9 \end{array} \right)$

(f) $\left(\begin{array}{l} \dfrac{2}{x} - \dfrac{7}{y} = \dfrac{9}{10} \\[2mm] \dfrac{5}{x} + \dfrac{4}{y} = -\dfrac{41}{20} \end{array} \right)$

(e) $\left(\begin{array}{l} \dfrac{5}{x} - \dfrac{2}{y} = 23 \\[2mm] \dfrac{4}{x} + \dfrac{3}{y} = \dfrac{23}{2} \end{array} \right)$

59. Solve the following system for x and y:

$$\left(\begin{array}{l} a_1 x + b_1 y = c_1 \\ a_2 x + b_2 y = c_2 \end{array} \right)$$

Answers to the Concept Quiz

1. True   **2.** False   **3.** True   **4.** True   **5.** False   **6.** True   **7.** False   **8.** False   **9.** True   **10.** True

---

5.7   Graphing Linear Inequalities

OBJECTIVES

1 Graph linear inequalities

2 Graph systems of two linear inequalities

Linear inequalities in two variables are of the form $Ax + By > C$ or $Ax + By < C$, where A, B, and C are real numbers, and A and B not both zero. (Combined linear equality and inequality statements are of the form $Ax + By \geq C$ or $Ax + By \leq C$.) Graphing linear inequalities is almost as easy as graphing linear equations. The following discussion leads to a simple step-by-step process. Let's consider the next equation and related inequalities.

$$x - y = 2$$
$$x - y > 2$$
$$x - y < 2$$

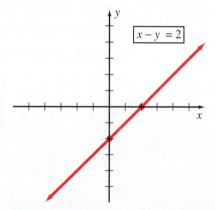

Figure 5.47

The graph of $x - y = 2$ is shown in Figure 5.47. The line divides the plane into two half-planes, one above the line and one below the line. In Figure 5.48(a) we have indicated coordinates for several points above the line. Note that for each point, the ordered pair of real numbers satisfies the inequality $x - y < 2$. This is true for *all points* in the half-plane above the line. Therefore, the graph of $x - y < 2$ is the half-plane above the line, indicated by the shaded region in Figure 5.48(b). We use a dashed line to indicate that points on the line do not satisfy $x - y < 2$.

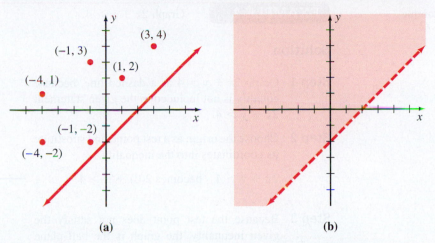

Figure 5.48

In Figure 5.49(a) the coordinates of several points below the line $x - y = 2$ are indicated. Note that for each point, the ordered pair of real numbers satisfies the inequality $x - y > 2$. This is true for *all points* in the half-plane below the line. Therefore, the graph of $x - y > 2$ is the half-plane below the line, as indicated by the shaded region in Figure 5.49(b).

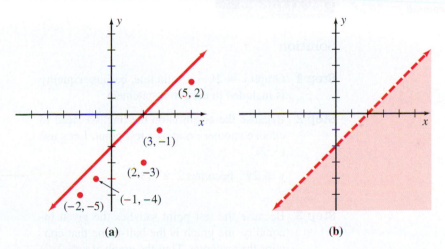

Figure 5.49

On the basis of this discussion, we suggest the following steps for graphing linear inequalities.

Step 1 Graph the corresponding equality. Use a solid line if equality is included in the given statement. Use a dashed line if equality is not included.

Step 2 Choose a "test point" that is not on the line and substitute its coordinates into the inequality statement. (The origin is a convenient point to use if it is not on the line.)

Step 3 The graph of the given inequality is:

   (a) the half-plane that contains the test point if the inequality is satisfied by the coordinates of the point, or

   (b) the half-plane that does not contain the test point if the inequality is not satisfied by the coordinates of the point.

Let's apply these steps to some examples.

EXAMPLE 1 Graph $2x + y > 4$.

Solution

Step 1 Graph $2x + y = 4$ as a dashed line, because equality is not included in the given statement $2x + y > 4$.

Step 2 Choose the origin as a test point, and substitute its coordinates into the inequality.

$2x + y > 4$ becomes $2(0) + 0 > 4$

Step 3 Because the test point does not satisfy the given inequality, the graph is the half-plane that does not contain the test point. Thus the graph of $2x + y > 4$ is the half-plane above the line, as indicated in Figure 5.50.

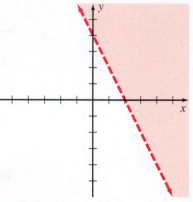

Figure 5.50

EXAMPLE 2 Graph $y \leq 2x$.

Solution

Step 1 Graph $y = 2x$ as a solid line, because equality is included in the given statement.

Step 2 Because the origin is on the line, we need to choose another point as a test point. Let's use $(3, 2)$.

$y \leq 2x$ becomes $2 \leq 2(3)$

Step 3 Because the test point satisfies the given inequality, the graph is the half-plane that contains the test point. Thus the graph of $y \leq 2x$ is the line, along with the half-plane below the line, as indicated in Figure 5.51.

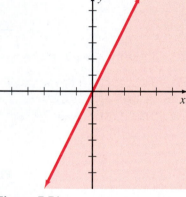

Figure 5.51

Systems of Linear Inequalities

It is now easy to use a graphing approach to solve a system of linear inequalities. For example, the solution set of a system of linear inequalities, such as

$$\begin{pmatrix} x + y < 1 \\ x - y > 1 \end{pmatrix}$$

is the intersection of the solution sets of the individual inequalities. In Figure 5.52(a) we indicated the solution set for $x + y < 1$, and in Figure 5.52(b) we indicated the solution set for $x - y > 1$. Then in Figure 5.52(c) we shaded the region that represents the intersection of the two shaded regions in parts (a) and (b); thus it is the solution of the given system. The shaded region in Figure 5.52(c) consists of all points that are below the line $x + y = 1$ *and also* are below the line $x - y = 1$.

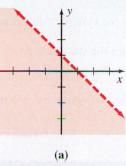
(a)

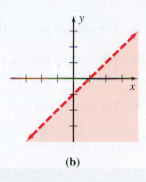
(b)

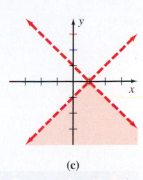
(c)

Figure 5.52

Let's solve another system of linear inequalities.

EXAMPLE 3   Solve the system $\begin{pmatrix} 2x + 3y \le 6 \\ x - 4y < 4 \end{pmatrix}$.

Solution

Let's first graph the individual inequalities. The solution set for $2x + 3y \le 6$ is shown in Figure 5.53(a), and the solution set for $x - 4y < 4$ is shown in Figure 5.53(b). (Note the solid line in part (a) and the dashed line in part (b).) Then in Figure 5.53(c) we shaded the intersection of the graphs in parts (a) and (b). Thus we represented the solution set for the given system by the shaded region in Figure 5.53(c). This region consists of all points that are on or below the line $2x + 3y = 6$ *and also* above the line $x - 4y = 4$.

Classroom Example
Solve this system by graphing:
$\begin{pmatrix} x - 2y < 3 \\ 3x + 4y \ge -8 \end{pmatrix}$

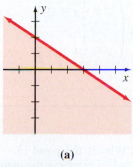
(a)

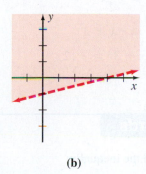
(b)

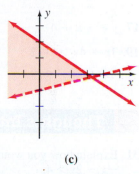
(c)

Figure 5.53

Remark: Remember that the shaded region in Figure 5.53(c) represents the solution set of the given system. Parts (a) and (b) were drawn only to help determine the final shaded region. With some practice, you may be able to go directly to part (c) without actually sketching the graphs of the individual inequalities.

Concept Quiz 5.7

For Problems 1–8, answer true or false.

1. The ordered pair $(2, -3)$ satisfies the linear inequality $2x + y > 1$.

2. A dashed line on the graph indicates that the points on the line do not satisfy the inequality.

3. Any point can be used as test point to determine the half-plane that is the solution of the inequality.

4. The solution of a system of inequalities is the intersection of the solution sets of the individual inequalities.

5. The ordered pair (1, 4) satisfies the system of linear inequalities $\begin{pmatrix} x + y > 2 \\ 2x + y < 3 \end{pmatrix}$.

6. The ordered pair (3, −2) satisfies the statement $5x - 2y \geq 19$.

7. The ordered pair (1, −3) satisfies the inequality $-2x - 3y < 4$.

8. The ordered pair (−4, −1) satisfies the system the system of inequalities $\begin{pmatrix} x + y < 5 \\ 2x - 3y > 6 \end{pmatrix}$.

Problem Set 5.7

For Problems 1–20, graph each inequality. (Objective 1)

1. $x + y > 1$

2. $2x + y > 4$

3. $3x + 2y < 6$

4. $x + 3y < 3$

5. $2x - y \geq 4$

6. $x - 2y \geq 2$

7. $4x - 3y \leq 12$

8. $3x - 4y \leq 12$

9. $y > -x$

10. $y < x$

11. $2x - y \geq 0$

12. $3x - y \leq 0$

13. $-x + 2y < -2$

14. $-2x + y > -2$

15. $y \leq \dfrac{1}{2}x - 2$

16. $y \geq -\dfrac{1}{2}x + 1$

17. $y \geq -x + 4$

18. $y \leq -x - 3$

19. $3x + 4y > -12$

20. $4x + 3y > -12$

For Problems 21–30, indicate the solution set for each system of linear inequalities by shading the appropriate region. (Objective 2)

21. $\begin{pmatrix} 2x + 3y > 6 \\ x - y < 2 \end{pmatrix}$

22. $\begin{pmatrix} x - 2y < 4 \\ 3x + y > 3 \end{pmatrix}$

23. $\begin{pmatrix} x - 3y \geq 3 \\ 3x + y \leq 3 \end{pmatrix}$

24. $\begin{pmatrix} 4x + 3y \leq 12 \\ 4x - y \geq 4 \end{pmatrix}$

25. $\begin{pmatrix} y \geq 2x \\ y < x \end{pmatrix}$

26. $\begin{pmatrix} y \leq -x \\ y > -3x \end{pmatrix}$

27. $\begin{pmatrix} y < -x + 1 \\ y > -x - 1 \end{pmatrix}$

28. $\begin{pmatrix} y > x - 2 \\ y < x + 3 \end{pmatrix}$

29. $\begin{pmatrix} y < \dfrac{1}{2}x + 2 \\ y < \dfrac{1}{2}x - 1 \end{pmatrix}$

30. $\begin{pmatrix} y > -\dfrac{1}{2}x - 2 \\ y > -\dfrac{1}{2}x + 1 \end{pmatrix}$

Thoughts Into Words

31. Explain how you would graph the inequality $-x - 2y > 4$.

32. Why is the point (3, −2) not a good test point to use when graphing the inequality $3x - 2y \leq 13$?

Further Investigations

For Problems 33–36, indicate the solution set for each system of linear inequalities by shading the appropriate region.

33. $\begin{pmatrix} y > x + 1 \\ y < x - 1 \end{pmatrix}$

34. $\begin{pmatrix} x \geq 0 \\ y \geq 0 \\ 3x + 4y \leq 12 \\ 2x + y \leq 4 \end{pmatrix}$

35. $\begin{pmatrix} x \geq 0 \\ y \geq 0 \\ 2x + y \leq 4 \\ 2x - 3y \leq 6 \end{pmatrix}$

36. $\begin{pmatrix} x \geq 0 \\ y \geq 0 \\ 3x + 5y \geq 15 \\ 5x + 3y \geq 15 \end{pmatrix}$

Answers to the Concept Quiz

1. False   2. True   3. False   4. True   5. False   6. True   7. False   8. False

Chapter 5 Summary

OBJECTIVE	SUMMARY	EXAMPLE
Plot points on a rectangular coordinate system. (Section 5.1/Objective 1)	The rectangular coordinate system involves a one-to-one correspondence between ordered pairs of real numbers and the points in a plane. For an ordered pair (x, y), x is the directed distance of the point from the vertical axis measured parallel to the horizontal axis, and y is the directed distance of the point from the horizontal axis measured parallel to the vertical axis.	Plot the point $(-3, 2)$. **Solution** From the origin move 3 units to the left along the x axis and then move up 2 units parallel to the y axis.
Draw graphs of equations by plotting points. (Section 5.1/Objective 3)	To graph an equation in two variables, x and y, keep these steps in mind: 1. Solve the equation for y in terms of x if it is not already in such a form. 2. Set up a table of ordered pairs that satisfies the equation. 3. Plot the ordered pairs. 4. Connect the points.	Graph $x + 2y = 6$. **Solution** Solving $x + 2y = 6$ for y results in the equation $y = -\frac{1}{2}x + 3$. $\begin{array}{c\|ccc} x & -2 & 0 & 2 \\ \hline y & 4 & 3 & 2 \end{array}$
Find the x and y intercepts for the graph of a linear equation. (Section 5.2/Objective 1)	The x intercept is the x coordinate of the point where the graph intersects the x axis. The y intercept is the y coordinate of the point where the graph intersects the y axis. To find the x intercept, substitute 0 for y in the equation and then solve for x. To find the y intercept, substitute 0 for x in the equation and then solve for y.	Find the x and y intercepts for the graph of a line with the equation $2x - y = -4$. **Solution** Let $y = 0$. $2x - 0 = -4$ $2x = -4$ $x = -2$ So the x intercept is -2. Let $x = 0$. $2(0) - y = -4$ $-y = -4$ $y = 4$ So the y intercept is 4.

(continued)

OBJECTIVE	SUMMARY	EXAMPLE
Graph linear equations. **(Section 5.2/Objective 2)**	To graph a linear equation, we can find two solutions, plot the corresponding points, and connect the points with a straight line. If the equation is in slope-intercept form, $y = mx + b$, one method for graphing is to use the y intercept and the slope to produce two solutions. If the equation is in standard form, $Ax + By = C$, one method for graphing is to use the x and y intercepts for the two solutions. It is advisable to find a third solution and plot its corresponding point to serve as a check.	Graph $y = 2x - 5$. **Solution** From the equation $y = 2x - 5$ we know the slope is 2 and the y intercept is -5. Plot the point $(0, -5)$. From the point $(0, -5)$ move up two units and over to the right 1 unit, because the slope is 2. Draw a straight line connecting the points.
Graph linear inequalities. **(Section 5.7/Objectives 1 and 2)**	To graph a linear inequality, first graph the line for the corresponding equality. Use a solid line if the equality is included in the given statement or a dashed line if the equality is not included. Then a test point is used to determine which half-plane is included in the solution set. See page 239 for the detailed steps. The solution set of a system of linear inequalities is the intersection of the solution sets of the individual inequalities.	Graph $2x + y \le -4$. **Solution** First graph $2x + y = -4$. Choose $(0, 0)$ as a test point. Substituting $(0, 0)$ into the inequality yields $0 \le -4$. Because the test point $(0, 0)$ makes the inequality a false statement, the half-plane not containing the point $(0, 0)$ is in the solution.
Find the slope of a line between two points. **(Section 5.3/Objective 1)**	If points P_1 and P_2 with coordinates (x_1, y_1) and (x_2, y_2), respectively, are any two points on a line, then the slope of the line (denoted by m) is given by $m = \dfrac{y_2 - y_1}{x_2 - x_1}$, $x_1 \ne x_2$. The slope of a line is a ratio of vertical change to horizontal change. The slope can be negative, positive, or zero. The concept of slope is not defined for vertical lines.	Find the slope of a line that passes through the points $(1, -4)$ and $(3, 5)$. **Solution** $m = \dfrac{5 - (-4)}{3 - 1} = \dfrac{9}{2}$

OBJECTIVE	SUMMARY	EXAMPLE
Graph lines, given a point and a slope. (Section 5.3/Objective 3)	To graph a line, given a point and the slope, first plot the point. Then from the point locate another point by moving the number of units of vertical change and horizontal change determined from the given slope.	Graph the line that passes through the point $(-4, 3)$ and has a slope of $-\dfrac{5}{2}$. **Solution** Plot the point $(-4, 3)$. Because $m = -\dfrac{5}{2} = \dfrac{-5}{2}$, from the point $(-4, 3)$ move down 5 units and 2 units to the right to locate another point. Draw the line through the points.
Apply the slope-intercept form of an equation. (Section 5.4/Objective 2)	The equation $y = mx + b$ is called the slope-intercept form of the equation of a straight line. If the equation of a nonvertical line is written in slope-intercept form, then the coefficient of x is the slope of the line, and the constant term is the y intercept.	Determine the slope and y intercept of the line with the equation $5x - 3y = 12$. **Solution** Solve for y to have the equation in slope-intercept form. $5x - 3y = 12$ $-3y = -5x + 12$ $y = \dfrac{5}{3}x - 4$ Therefore, the slope of the line is $\dfrac{5}{3}$ and the y intercept is -4.
Apply the concept of slope. (Section 5.3/Objective 4)	The concept of slope is used in situations involving an incline. The grade of a highway and the incline of a treadmill are a couple of examples that deal with slope.	The backyard of a house built on a hillside has a slope of $\dfrac{3}{2}$. In the backyard, how much does the vertical distance change for a 40-foot change in horizontal distance? **Solution** Let y represent the change in the vertical distance. Then solve the following proportion. $\dfrac{3}{2} = \dfrac{y}{40}$ $2y = 120$ $y = 60$ The change in vertical distance would be 60 feet.

(continued)

OBJECTIVE	SUMMARY	EXAMPLE
Find the equation of a line given a point and a slope. (Section 5.4/Objective 1a)	The point-slope form of the equation of a line, $y - y_1 = m(x - x_1)$, can be used to find the equation. Substitute the slope and the coordinates of a point into the form. Your answer can be left in standard form or in slope-intercept form, depending upon the instructions.	Find the equation of a line that has a slope of 3 and passes through the point $(-2, 5)$. Write the result in slope-intercept form. **Solution** $$y - y_1 = m(x - x_1)$$ $$y - 5 = 3[x - (-2)]$$ $$y - 5 = 3(x + 2)$$ $$y - 5 = 3x + 6$$ $$y = 3x + 11$$
Find the equation of a line given two points. (Section 5.4/Objective 1b)	To use the point-slope form we must first determine the slope. Then substitute the slope and the coordinates of a point into the point-slope form and simplify.	Find the equation of a line that passes through $(-2, 5)$ and $(1, 3)$. Write the result in standard form. **Solution** $$m = \frac{3 - 5}{1 - (-2)} = \frac{-2}{3}$$ Now substitute the slope and either point into the point-slope form. $$y - y_1 = m(x - x_1)$$ $$y - 3 = \frac{-2}{3}(x - 1)$$ $$3(y - 3) = -2(x - 1)$$ $$3y - 9 = -2x + 2$$ $$2x + 3y = 11$$
Find the equation of a line, given a point on the line and the equation of another line that is parallel or perpendicular to that line. (Section 5.4/Objective 1c)	If two lines have slopes m_1 and m_2, respectively, then 1. The two are parallel if and only if $m_1 = m_2$. 2. The lines are perpendicular if and only if $(m_1)(m_2) = -1$.	Find the equation of a line that passes through the point $(1, 4)$ and is perpendicular to the line $y = -\frac{2}{5}x + 3$. **Solution** The slope of the given line is $-\frac{2}{5}$, so the negative reciprocal is $\frac{5}{2}$. Use the point $(1, 4)$ and the slope $\frac{5}{2}$ in the point-slope form to obtain $y - 4 = \frac{5}{2}(x - 1)$. Simplifying gives the equation $5x - 2y = -3$.

OBJECTIVE	SUMMARY	EXAMPLE
Solve a linear system of two equations by graphing. (Section 5.5/Objective 1)	Solving a system of two linear equations by graphing produces one of the three following possibilities: 1. The graphs of the two equations are intersecting lines, which indicates one solution for the system, called a consistent system. 2. The graphs of the two equations are parallel lines, which indicates no solution for the system, called an inconsistent system. 3. The graphs of the two equations are the same line, which indicates infinitely many solutions for the system. We refer to the equations as a set of dependent equations.	Solve $\begin{pmatrix} 2x - y = 1 \\ x - y = -2 \end{pmatrix}$ by graphing. **Solution** Graph both lines on the same coordinate plane. The lines intersect at the point $(3, 5)$ which is the solution for the system.
Use the substitution method to solve a system of equations. (Section 5.5/Objective 2)	Here are the steps of the substitution method for solving a system of equations: **Step 1** Solve one of the equations for one variable in terms of the other variable. **Step 2** Substitute the expression obtained in step 1 into the other equation to produce an equation with one variable. **Step 3** Solve the equation obtained in step 2. **Step 4** Use the solution obtained in step 3, along with the expression obtained in step 1, to determine the solution of the system.	Solve $\begin{pmatrix} x - 2y = -9 \\ 3x + 2y = 5 \end{pmatrix}$. **Solution** Solve the first equation for x: $x = 2y - 9$ Substitute $2y - 9$ for x in the second equation and solve for y: $3(2y - 9) + 2y = 5$ $6y - 27 + 2y = 5$ $8y = 32$ $y = 4$ To find x, substitute 4 for y in the equation already solved for x: $x = 2(4) - 9$ $x = -1$ The solution is set $\{(-1, 4)\}$.

(continued)

OBJECTIVE	SUMMARY	EXAMPLE
Use the elimination-by-addition method to solve a system of equations. **(Section 5.6/Objective 1)**	The elimination-by-addition method involves replacing systems of equations with equivalent systems until we reach a system for which the solutions can be easily determined. We can perform the following operations: **1.** Any two equations of the system can be interchanged. **2.** Both sides of any equation of the system can be multiplied by any nonzero real number. **3.** Any equation of the system can be replaced by the sum of that equation and a nonzero multiple of another equation.	Solve $\begin{pmatrix} 2x + 5y = 11 \\ 3x - y = 8 \end{pmatrix}$. **Solution** Let's replace the first equation with an equation we form by multiplying the second equation by 5 and then adding the result to the first equation: $\begin{pmatrix} 17x = 51 \\ 3x - y = 8 \end{pmatrix}$ From the first equation we can determine that $x = 3$. To find y, substitute 3 for x in the first equation: $2(3) + 5y = 11$ $6 + 5y = 11$ $5y = 5$ $y = 1$ The solution set is $\{(3, 1)\}$.
Determine which method to use to solve a system of equations. **(Section 5.6/Objective 1)**	We have studied three methods for solving systems of equations. **1.** The graphing method works well if you want a visual representation of the problem. However, it may be impossible to get an exact solution from the graph. **2.** The substitution method works well if one of the equations is already solved for a variable. It gives exact solutions. **3.** The elimination-by-addition method gives exact solutions.	Decide which method, elimination-by-addition or substitution, would be the most fitting to solve the given system of equations: $\begin{pmatrix} 2x + y = 8 \\ y = 3x + 1 \end{pmatrix}$ **Solution** The substitution method would work well for this problem, because the second equation is already solved for y. Using the substitution method we can determine that the solution set is $\left\{ \left(\dfrac{7}{5}, \dfrac{26}{5} \right) \right\}$.
Solve word problems using a system of equations. **(Section 5.5/Objective 5; Section 5.6/Objective 2)**	Many word problems presented earlier in the text could be solved using a system of equations. These problems may seem easier to solve using a system of equations. Either the elimination-by-addition or the substitution method can be used to solve the resulting system of equations.	In a recent survey, 500 students were asked about their college majors. The number of women in the survey was 40 less than twice the number of men in the survey. Find the number of women in the survey. **Solution** Let x represent the number of women in the survey, and let y represent the number of men in the survey. Then with the information provided we can write the following system of equations: $\begin{pmatrix} x + y = 500 \\ x = 2y - 40 \end{pmatrix}$ Solving this system we can determine that the number of women in the survey is 320.

Chapter 5   Review Problem Set

For Problems 1–10, graph each of the equations.

1. $2x - 5y = 10$

2. $y = -\dfrac{1}{3}x + 1$

3. $y = -2x$

4. $3x + 4y = 12$

5. $2x - 3y = 0$

6. $2x + y = 2$

7. $x - y = 4$

8. $x + 2y = -2$

9. $y = \dfrac{2}{3}x - 1$

10. $y = 3x$

For Problems 11–16, determine the slope and y intercept and graph the line.

11. $2x - 5y = 10$

12. $y = -\dfrac{1}{3}x + 1$

13. $x + 2y = 2$

14. $3x + y = -2$

15. $2x - y = 4$

16. $3x - 4y = 12$

17. Find the slope of the line determined by the points $(3, -4)$ and $(-2, 5)$.

18. Find the slope of the line $5x - 6y = 30$.

19. Write the equation of the line that has a slope of $-\dfrac{5}{7}$ and contains the point $(2, -3)$.

20. Write the equation of the line that contains the points $(2, 5)$ and $(-1, -3)$.

21. Write the equation of the line that has a slope of $\dfrac{2}{9}$ and a y intercept of -1.

22. Write the equation of the line that contains the point $(2, 4)$ and is perpendicular to the x axis.

23. Solve the system $\begin{pmatrix} 2x + y = 4 \\ x - y = 5 \end{pmatrix}$ by using the graphing method.

For Problems 24–35, solve each system by using either the elimination-by-addition method or the substitution method.

24. $\begin{pmatrix} 2x - y = 1 \\ 3x - 2y = -5 \end{pmatrix}$

25. $\begin{pmatrix} 2x + 5y = 7 \\ x = -3y + 1 \end{pmatrix}$

26. $\begin{pmatrix} 3x + 2y = 7 \\ 4x - 5y = 3 \end{pmatrix}$

27. $\begin{pmatrix} 9x + 2y = 140 \\ x + 5y = 135 \end{pmatrix}$

28. $\begin{pmatrix} \dfrac{1}{2}x + \dfrac{1}{4}y = -5 \\ \dfrac{2}{3}x - \dfrac{1}{2}y = 0 \end{pmatrix}$

29. $\begin{pmatrix} x + y = 1000 \\ 0{,}07x + 0.09y = 82 \end{pmatrix}$

30. $\begin{pmatrix} y = 5x + 2 \\ 10x - 2y = 1 \end{pmatrix}$

31. $\begin{pmatrix} 5x - 7y = 9 \\ y = 3x - 2 \end{pmatrix}$

32. $\begin{pmatrix} 10t + u = 6u \\ t + u = 12 \end{pmatrix}$

33. $\begin{pmatrix} t = 2u \\ 10t + u - 36 = 10u + t \end{pmatrix}$

34. $\begin{pmatrix} u = 2t + 1 \\ 10t + u + 10u + t = 110 \end{pmatrix}$

35. $\begin{pmatrix} y = -\dfrac{2}{3}x \\ \dfrac{1}{3}x - y = -9 \end{pmatrix}$

For Problems 36–41, graph each of the inequalities.

36. $y > \dfrac{2}{3}x - 1$

37. $x - 2y \le 4$

38. $y \le -2x$

39. $3x + 2y > -6$

For Problems 40–41, graph the solution set of the system of inequalities.

40. $\begin{pmatrix} y > 2x - 4 \\ y \le -x + 3 \end{pmatrix}$

41. $\begin{pmatrix} 2x + y < 6 \\ x - 3y > 3 \end{pmatrix}$

Solve each of the following problems by setting up and solving a system of two linear equations in two variables.

42. The sum of two numbers is 113. The larger number is 1 less than twice the smaller number. Find the numbers.

43. Last year Mark invested a certain amount of money at 6% annual interest and $500 more than that amount at 8%. He received $390.00 in interest. How much did he invest at each rate?

44. Cindy has 43 coins consisting of nickels and dimes. The total value of the coins is $3.40. How many coins of each kind does she have?

45. The length of a rectangle is 1 inch more than three times the width. If the perimeter of the rectangle is 50 inches, find the length and width.

46. The width of a rectangle is 5 inches less than the length. If the perimeter of the rectangle is 38 inches, find the length and width.

47. Alex has 32 coins consisting of quarters and dimes. The total value of the coins is $4.85. How many coins of each kind does he have?

48. Two angles are complementary, and one of them is 6° less than twice the other one. Find the measure of each angle.

49. Two angles are supplementary, and the larger angle is 20° less than three times the smaller angle. Find the measure of each angle.

50. Four cheeseburgers and five milkshakes cost a total of $25.50. Two milkshakes cost $1.75 more than one cheeseburger. Find the cost of a cheeseburger and also find the cost of a milkshake.

51. Three bottles of orange juice and two bottles of water cost $6.75. On the other hand, two bottles of juice and three bottles of water cost $6.15. Find the cost per bottle of each.

For Problems 1–4, determine the slope and y intercept, and graph each equation.

1. $5x + 3y = 15$

2. $-2x + y = -4$

3. $y = -\dfrac{1}{2}x - 2$

4. $3x + y = 0$

5. Find x if the line through the points $(4, 7)$ and $(x, 13)$ has a slope of $\dfrac{3}{2}$.

6. Find y if the line through the points $(1, y)$ and $(6, 5)$ has a slope of $-\dfrac{3}{5}$.

7. If a line has a slope of $\dfrac{1}{4}$ and passes through the point $(3, 5)$, find the coordinates of two other points on the line.

8. If a line has a slope of -3 and passes through the point $(2, 1)$, find the coordinates of two other points on the line.

9. Suppose that a highway rises at a distance of 85 feet over a horizontal distance of 1850 feet. Express the grade of the highway to the nearest tenth of a percent.

10. Find the x intercept of the graph of $y = 4x + 8$.

For Problems 11–13, express each equation in $Ax + By = C$ form, where A, B, and C are integers.

11. Determine the equation of the line that has a slope of $-\dfrac{3}{5}$ and a y intercept of 4.

12. Determine the equation of the line containing the point $(4, -2)$ and having a slope of $\dfrac{4}{9}$.

13. Determine the equation of the line that contains the points $(4, 6)$ and $(-2, -3)$.

14. Solve the system $\begin{pmatrix} 3x - 2y = -4 \\ 2x + 3y = 19 \end{pmatrix}$ by graphing.

15. Solve the system $\begin{pmatrix} x - 3y = -9 \\ 4x + 7y = 40 \end{pmatrix}$ using the elimination-by-addition method.

16. Solve the system $\begin{pmatrix} 5x + y = -14 \\ 6x - 7y = -66 \end{pmatrix}$ using the substitution method.

17. Solve the system $\begin{pmatrix} 2x - 7y = 26 \\ 3x + 2y = -11 \end{pmatrix}$.

18. Solve the system $\begin{pmatrix} 8x + 5y = -6 \\ 4x - y = 18 \end{pmatrix}$.

For Problems 19–23, graph each equation or inequality.

19. $5x + 3y = 15$

20. $y = \dfrac{2}{3}x$

21. $y = 2x - 3$

22. $y \geq 2x - 4$

23. $x + 3y < -3$

For Problems 24 and 25, solve each problem by setting up and solving a system of two linear equations in two variables.

24. Three reams of paper and 4 notebooks cost $19.63. Four reams of paper and 1 notebook cost $16.25. Find the cost of each item.

25. The length of a rectangle is 1 inch less than twice the width of the rectangle. If the perimeter of the rectangle is 40 inches, find the length of the rectangle.

For Problems 1–3, simplify each numerical expression.

1. $3.9 - 4.6 - 1.2 + 0.4$

2. $\left(\dfrac{1}{2}\right)^3 + \left(\dfrac{1}{4}\right)^2 - \dfrac{1}{8}$

3. $(0.2)^3 - (0.4)^3 + (1.2)^2$

For Problems 4–6, evaluate each algebraic expression for the given values of the variables.

4. $\dfrac{1}{2}x - \dfrac{2}{5}y + xy$ for $x = -3$ and $y = \dfrac{1}{2}$

5. $1.1x - 2.3y + 2.5x + 1.6y$ for $x = 0.4$ and $y = 0.7$

6. $2(x + 6) - 3(x + 9) - 4(x - 5)$ for $x = 17$

For Problems 7–9, perform the indicated operations and express answers in simplest form.

7. $\dfrac{3}{x} + \dfrac{2}{y} - \dfrac{4}{xy}$

8. $\left(\dfrac{7xy}{9xy^2}\right)\left(\dfrac{12x}{14y}\right)$

9. $\left(\dfrac{2ab^2}{7a}\right) \div \left(\dfrac{4b}{21a}\right)$

For Problems 10–14, solve each equation.

10. $2(x - 3) - (x + 4) = 3(x + 10)$

11. $\dfrac{2}{3}x + \dfrac{1}{4} - \dfrac{1}{2}x = -1$

12. $\dfrac{x + 1}{4} - \dfrac{x - 2}{6} = \dfrac{3}{8}$

13. $\dfrac{x + 6}{7} = \dfrac{x - 2}{8}$

14. $0.06x + 0.07(1800 - x) = 120$

15. Solve $2x - 3y = 13$ for y.

16. Solve the system $\begin{pmatrix} 5x - 2y = -20 \\ 2x + 3y = 11 \end{pmatrix}$.

17. Solve the system $\begin{pmatrix} 4x - 9y = 47 \\ 7x + y = 32 \end{pmatrix}$.

18. Solve the inequality $3(x - 2) < 4(x + 6)$

19. Solve the inequality $\dfrac{1}{4}x - 2 \geq \dfrac{2}{3}x + 1$

20. Graph the equation $y = -2x - 3$.

21. Graph the equation $y = -2x^2 - 3$.

22. Graph the inequality $x - 2y > 4$.

For Problems 23–25, use an equation, an inequality, or a system of equations to help solve each problem.

23. Last week on an algebra test, the highest grade was 9 points less than three times the lowest grade. The sum of the two grades was 135. Find the lowest and highest grades on the test.

24. Suppose that Derwin shot rounds of 82, 84, 78, and 79 on the first four days of a golf tournament. What must he shoot on the fifth day of the tournament to average 80 or less for the five days?

25. A 10% salt solution is to be mixed with a 15% salt solution to produce 10 gallons of a 13% salt solution. How many gallons of the 10% salt solution and how many gallons of the 15% salt solution will be needed?

6

Exponents and Polynomials

The average distance between the sun and the earth is approximately 93,000,000 miles. Using scientific notation, 93,000,000 can be written as $(9.3)(10^7)$.

© noid

A rectangular dock that measures 12 feet by 16 feet is treated with a uniform strip of nonslip coating along both sides and both ends. How wide is the strip if one-half of the dock is treated? If we let *x* represent the width of the strip, then we can use the equation $(16 - 2x)(12 - 2x) = \frac{1}{2}(12)(16)$ to determine that the width of the strip is 2 feet.

The equation we used to solve this problem is called a **quadratic equation**. Quadratic equations belong to a larger classification called **polynomial equations**. To solve problems involving polynomial equations, we need to develop some basic skills that pertain to polynomials. That is to say, we need to be able to add, subtract, multiply, divide, and factor polynomials. Chapters 5 and 6 will help you develop those skills as you work through problems that involve quadratic equations.

Video tutorials based on section learning objectives are available in a variety of delivery modes.

6.1   Addition and Subtraction of Polynomials

OBJECTIVES

1. Know the definition of monomial, binomial, trinomial, and polynomial

2. Determine the degree of a polynomial

3. Add polynomials

4. Subtract polynomials using either a vertical or a horizontal format

In earlier chapters, we called algebraic expressions such as $4x$, $5y$, $-6ab$, $7x^2$, and $-9xy^2z^3$ "terms." Recall that a term is an indicated product that may contain any number of factors. The variables in a term are called "literal factors," and the numerical factor is called the "numerical coefficient" of the term. Thus in $-6ab$, a and b are literal factors, and the numerical coefficient is -6. Terms that have the same literal factors are called "similar" or "like" terms.

Terms that contain variables with only whole numbers as exponents are called **monomials**. The previously listed terms, $4x$, $5y$, $-6ab$, $7x^2$, and $-9xy^2z^3$ are all monomials. (We will work with some algebraic expressions later, such as $7x^{-1}y^{-1}$ and $4a^{-2}b^{-3}$, which are not monomials.) The **degree of a monomial** is the sum of the exponents of the literal factors. Here are some examples:

$4xy$ is of degree 2

$5x$ is of degree 1

$14a^2b$ is of degree 3

$-17xy^2z^3$ is of degree 6

$-9y^4$ is of degree 4

If the monomial contains only one variable, then the exponent of the variable is the degree of the monomial. Any nonzero constant term is said to be of degree zero.

A **polynomial** is a monomial or a finite sum (or difference) of monomials. The **degree of a polynomial** is the degree of the term with the highest degree in the polynomial. Some special classifications of polynomials are made according to the number of terms. We call a one-term polynomial a **monomial**, a two-term polynomial a **binomial**, and a three-term polynomial a **trinomial**. The following examples illustrate some of this terminology:

The polynomial $5x^3y^4$ is a monomial of degree 7

The polynomial $4x^2y - 3xy$ is a binomial of degree 3

The polynomial $5x^2 - 6x + 4$ is a trinomial of degree 2

The polynomial $9x^4 - 7x^3 + 6x^2 + x - 2$ is given no special name but is of degree 4

Adding Polynomials

In the preceding chapters, you have worked many problems involving the addition and subtraction of polynomials. For example, simplifying $4x^2 + 6x + 7x^2 - 2x$ to $11x^2 + 4x$ by combining similar terms can actually be considered the addition problem $(4x^2 + 6x) + (7x^2 - 2x)$. At this time we will simply review and extend some of those ideas.

Classroom Example
Add $4m^3 - 3m + 5$ and
$6m^3 + 7m - 2$.

EXAMPLE 1   Add $5x^2 + 7x - 2$ and $9x^2 - 12x + 13$.

Solution

We commonly use the horizontal format for a problem like this. Thus

$$(5x^2 + 7x - 2) + (9x^2 - 12x + 13) = (5x^2 + 9x^2) + (7x - 12x) + (-2 + 13)$$
$$= 14x^2 - 5x + 11$$

The commutative, associative, and distributive properties provide the basis for rearranging, regrouping, and combining similar terms.

Classroom Example
Add $9n + 4$, $2n - 5$, and $3n - 11$.

EXAMPLE 2   Add $5x - 1$, $3x + 4$, and $9x - 7$.

Solution

$$(5x - 1) + (3x + 4) + (9x - 7) = (5x + 3x + 9x) + [-1 + 4 + (-7)]$$
$$= 17x - 4$$

Classroom Example
Add $-y^3 - 4y + 1$, $3y^2 + 6y + 7$, and $-10y + 3$.

EXAMPLE 3   Add $-x^2 + 2x - 1$, $2x^3 - x + 4$, and $-5x + 6$.

Solution

$$(-x^2 + 2x - 1) + (2x^3 - x + 4) + (-5x + 6)$$
$$= (2x^3) + (-x^2) + (2x - x - 5x) + (-1 + 4 + 6)$$
$$= 2x^3 - x^2 - 4x + 9$$

Subtracting Polynomials

Recall from Chapter 2 that $a - b = a + (-b)$. We define subtraction as *adding the opposite*. This same idea extends to polynomials in general. The opposite of a polynomial is formed by taking the opposite of each term. For example, the opposite of $(2x^2 - 7x + 3)$ is $-2x^2 + 7x - 3$. Symbolically, we express this as

$$-(2x^2 - 7x + 3) = -2x^2 + 7x - 3$$

Now consider some subtraction problems.

Classroom Example
Subtract $5d^2 + 2d - 6$ from $9d^2 - 4d + 1$.

EXAMPLE 4   Subtract $2x^2 + 9x - 3$ from $5x^2 - 7x - 1$.

Solution

Use the horizontal format:

$$(5x^2 - 7x - 1) - (2x^2 + 9x - 3) = (5x^2 - 7x - 1) + (-2x^2 - 9x + 3)$$
$$= (5x^2 - 2x^2) + (-7x - 9x) + (-1 + 3)$$
$$= 3x^2 - 16x + 2$$

Classroom Example
Subtract $-3x^2 + 2x - 4$ from $6x^2 - 3$.

EXAMPLE 5   Subtract $-8y^2 - y + 5$ from $2y^2 + 9$.

Solution

$$(2y^2 + 9) - (-8y^2 - y + 5) = (2y^2 + 9) + (8y^2 + y - 5)$$
$$= (2y^2 + 8y^2) + (y) + (9 - 5)$$
$$= 10y^2 + y + 4$$

Later when dividing polynomials, you will need to use a vertical format to subtract polynomials. Let's consider two such examples.

Classroom Example
Subtract $4n^2 + 9n - 7$ from
$12n^2 - 5n + 2.$

EXAMPLE 6   Subtract $3x^2 + 5x - 2$ from $9x^2 - 7x - 1$.

Solution

$$\begin{array}{l} 9x^2 - 7x - 1 \\ \underline{3x^2 + 5x - 2} \end{array}$$

Notice which polynomial goes on the bottom and the alignment of similar terms in columns

Now we can mentally form the opposite of the bottom polynomial and add.

$$\begin{array}{l} 9x^2 - 7x - 1 \\ \underline{3x^2 + 5x - 2} \\ 6x^2 - 12x + 1 \end{array}$$

The opposite of $3x^2 + 5x - 2$ is $-3x^2 - 5x + 2$

Classroom Example
Subtract $23x^4 + 11x^3 + 2x$ from
$15x^3 - 5x^2 + 3x.$

EXAMPLE 7   Subtract $15y^3 + 5y^2 + 3$ from $13y^3 + 7y - 1$.

Solution

$$\begin{array}{l} 13y^3 + 7y - 1 \\ \underline{15y^3 + 5y^2 + 3} \\ -2y^3 - 5y^2 + 7y - 4 \end{array}$$

Similar terms are arranged in columns

We mentally formed the opposite of the bottom polynomial and added

We can use the distributive property along with the properties $a = 1(a)$ and $-a = -1(a)$ when adding and subtracting polynomials. The next examples illustrate this approach.

Classroom Example
Perform the indicated operations:
$(6a + 5) + (2a - 1) - (3a - 9)$

EXAMPLE 8   Perform the indicated operations:
$$(3x - 4) + (2x - 5) - (7x - 1)$$

Solution

$$\begin{aligned} &(3x - 4) + (2x - 5) - (7x - 1) \\ &= 1(3x - 4) + 1(2x - 5) - 1(7x - 1) \\ &= 1(3x) - 1(4) + 1(2x) - 1(5) - 1(7x) - 1(-1) \\ &= 3x - 4 + 2x - 5 - 7x + 1 \\ &= 3x + 2x - 7x - 4 - 5 + 1 \\ &= -2x - 8 \end{aligned}$$

Certainly we can do some of the steps mentally; Example 9 gives a possible format.

Classroom Example
Perform the indicated operations:
$(-3x^2 + 2x - 1) -$
$(7x^2 - 4x + 9) + (4x^2 + 9x - 10)$

EXAMPLE 9   Perform the indicated operations:
$$(-y^2 + 5y - 2) - (-2y^2 + 8y + 6) + (4y^2 - 2y - 5)$$

Solution

$$\begin{aligned} &(-y^2 + 5y - 2) - (-2y^2 + 8y + 6) + (4y^2 - 2y - 5) \\ &= -y^2 + 5y - 2 + 2y^2 - 8y - 6 + 4y^2 - 2y - 5 \\ &= -y^2 + 2y^2 + 4y^2 + 5y - 8y - 2y - 2 - 6 - 5 \\ &= 5y^2 - 5y - 13 \end{aligned}$$

When we use the horizontal format, as in Examples 8 and 9, we use parentheses to indicate a quantity. In Example 8 the quantities $(3x - 4)$ and $(2x - 5)$ are to be added; from this result we are to subtract the quantity $(7x - 1)$. Brackets, [ ], are also sometimes used as grouping symbols, especially if there is a need to indicate quantities within quantities. To remove the grouping symbols, perform the indicated operations, starting with the innermost set of symbols. Let's consider two examples of this type.

Classroom Example
Perform the indicated operations:
$8q - [3q + (q - 5)]$

EXAMPLE 10   Perform the indicated operations:
$$3x - [2x + (3x - 1)]$$

Solution

First we need to add the quantities $2x$ and $(3x - 1)$.

$$3x - [2x + (3x - 1)] = 3x - (2x + 3x - 1)$$
$$= 3x - (5x - 1)$$

Now we need to subtract the quantity $(5x - 1)$ from $3x$.

$$3x - (5x - 1) = 3x - 5x + 1$$
$$= -2x + 1$$

Classroom Example
Perform the indicated operations:
$17 - \{6m - [3 - (m + 2)] - 9m\}$

EXAMPLE 11   Perform the indicated operations:
$$8 - \{7x - [2 + (x - 1)] + 4x\}$$

Solution

Start with the innermost set of grouping symbols (the parentheses) and proceed as follows:

$$8 - \{7x - [2 + (x - 1)] + 4x\} = 8 - [7x - (x + 1) + 4x]$$
$$= 8 - (7x - x - 1 + 4x)$$
$$= 8 - (10x - 1)$$
$$= 8 - 10x + 1$$
$$= -10x + 9$$

For a final example in this section, we look at polynomials in a geometric setting.

Classroom Example
Suppose that a triangle and a square have the dimensions as shown below:

Find a polynomial that represents the sum of the areas of the two figures.

EXAMPLE 12

Suppose that a parallelogram and a rectangle have dimensions as indicated in Figure 6.1. Find a polynomial that represents the sum of the areas of the two figures.

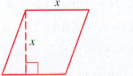

Figure 6.1

Solution

Using the area formulas $A = bh$ and $A = lw$ for parallelograms and rectangles, respectively, we can represent the sum of the areas of the two figures as follows:

Area of the parallelogram $x(x) = x^2$

Area of the rectangle $20(x) = 20x$

We can represent the total area by $x^2 + 20x$.

Concept Quiz 6.1

For Problems 1–5, answer true or false.

1. The degree of the monomial $4x^2y$ is 3.
2. The degree of the polynomial $2x^4 + 5x^3 + 7x^2 - 4x + 6$ is 10.
3. A three-term polynomial is called a binomial.
4. A polynomial is a monomial or a finite sum or difference of monomials.
5. Monomial terms must have whole number exponents for each variable.

For Problems 6–10, match the polynomial with its description.

6. $5xy^2$
7. $5xy^2 + 3x^2$
8. $5x^2y + 3xy^4$
9. $3x^5 + 2x^3 + 5x - 1$
10. $3x^2y^3$

A. Monomial of degree 5
B. Binomial of degree 5
C. Monomial of degree 3
D. Binomial of degree 3
E. Polynomial of degree 5

Problem Set 6.1

For Problems 1–8, determine the degree of each polynomial. (Objective 2)

1. $7x^2y + 6xy$
2. $4xy - 7x$
3. $5x^2 - 9$
4. $8x^2y^2 - 2xy^2 - x$
5. $5x^3 - x^2 - x + 3$
6. $8x^4 - 2x^2 + 6$
7. $5xy$
8. $-7x + 4$

For Problems 9–22, add the polynomials. (Objective 3)

9. $3x + 4$ and $5x + 7$
10. $3x - 5$ and $2x - 9$
11. $-5y - 3$ and $9y + 13$
12. $x^2 - 2x - 1$ and $-2x^2 + x + 4$
13. $-2x^2 + 7x - 9$ and $4x^2 - 9x - 14$
14. $3a^2 + 4a - 7$ and $-3a^2 - 7a + 10$
15. $5x - 2$, $3x - 7$, and $9x - 10$
16. $-x - 4$, $8x + 9$, and $-7x - 6$
17. $2x^2 - x + 4$, $-5x^2 - 7x - 2$, and $9x^2 + 3x - 6$
18. $-3x^2 + 2x - 6$, $6x^2 + 7x + 3$, and $-4x^2 - 9$
19. $-4n^2 - n - 1$ and $4n^2 + 6n - 5$
20. $-5n^2 + 7n - 9$ and $-5n - 4$
21. $2x^2 - 7x - 10$, $-6x - 2$ and $-9x^2 + 5$
22. $7x - 11$, $-x^2 - 5x + 9$ and $-4x + 5$

For Problems 23–34, subtract the polynomials using a horizontal format. (Objective 4)

23. $7x + 1$ from $12x + 6$
24. $10x + 3$ from $14x + 13$
25. $5x - 2$ from $3x - 7$
26. $7x - 2$ from $2x + 3$
27. $-x - 1$ from $-4x + 6$
28. $-3x + 2$ from $-x - 9$
29. $x^2 - 7x + 2$ from $3x^2 + 8x - 4$
30. $2x^2 + 6x - 1$ from $8x^2 - 2x + 6$
31. $-2n^2 - 3n + 4$ from $3n^2 - n + 7$
32. $3n^2 - 7n - 9$ from $-4n^2 + 6n + 10$
33. $-4x^3 - x^2 + 6x - 1$ from $-7x^3 + x^2 + 6x - 12$
34. $-4x^2 + 6x - 2$ from $-3x^3 + 2x^2 + 7x - 1$

For Problems 35–44, subtract the polynomials using a vertical format. (Objective 4)

35. $3x - 2$ from $12x - 4$
36. $-4x + 6$ from $7x - 3$
37. $-5a - 6$ from $-3a + 9$
38. $7a - 11$ from $-2a - 1$
39. $8x^2 - x + 6$ from $6x^2 - x + 11$
40. $3x^2 - 2$ from $-2x^2 + 6x - 4$
41. $-2x^3 - 6x^2 + 7x - 9$ from $4x^3 + 6x^2 + 7x - 14$
42. $4x^3 + x - 10$ from $3x^2 - 6$
43. $2x^2 - 6x - 14$ from $4x^3 - 6x^2 + 7x - 2$
44. $3x - 7$ from $7x^3 + 6x^2 - 5x - 4$

For Problems 45–64, perform the indicated operations.
(**Objectives 3 and 4**)

45. $(5x + 3) - (7x - 2) + (3x + 6)$

46. $(3x - 4) + (9x - 1) - (14x - 7)$

47. $(-x - 1) - (-2x + 6) + (-4x - 7)$

48. $(-3x + 6) + (-x - 8) - (-7x + 10)$

49. $(x^2 - 7x - 4) + (2x^2 - 8x - 9) - (4x^2 - 2x - 1)$

50. $(3x^2 + x - 6) - (8x^2 - 9x + 1) - (7x^2 + 2x - 6)$

51. $(-x^2 - 3x + 4) + (-2x^2 - x - 2) - (-4x^2 + 7x + 10)$

52. $(-3x^2 - 2) + (7x^2 - 8) - (9x^2 - 2x - 4)$

53. $(3a - 2b) - (7a + 4b) - (6a - 3b)$

54. $(5a + 7b) + (-8a - 2b) - (5a + 6b)$

55. $(n - 6) - (2n^2 - n + 4) + (n^2 - 7)$

56. $(3n + 4) - (n^2 - 9n + 10) - (-2n + 4)$

57. $7x + [3x - (2x - 1)]$

58. $-6x + [-2x - (5x + 2)]$

59. $-7n - [4n - (6n - 1)]$

60. $9n - [3n - (5n + 4)]$

61. $(5a - 1) - [3a + (4a - 7)]$

62. $(-3a + 4) - [-7a + (9a - 1)]$

63. $13x - \{5x - [4x - (x - 6)]\}$

64. $-10x - \{7x - [3x - (2x - 3)]\}$

65. Subtract $5x - 3$ from the sum of $4x - 2$ and $7x + 6$.

66. Subtract $7x + 5$ from the sum of $9x - 4$ and $-3x - 2$.

67. Subtract the sum of $-2n - 5$ and $-n + 7$ from $-8n + 9$.

68. Subtract the sum of $7n - 11$ and $-4n - 3$ from $13n - 4$.

69. Find a polynomial that represents the perimeter of the rectangle in Figure 6.2.

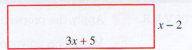

Figure 6.2

70. Find a polynomial that represents the area of the shaded region in Figure 6.3. The length of a radius of the larger circle is r units, and the length of a radius of the smaller circle is 4 units.

Figure 6.3

71. Find a polynomial that represents the sum of the areas of the rectangles and squares in Figure 6.4.

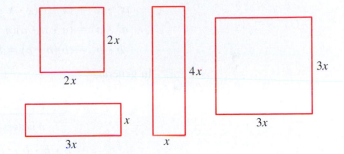

Figure 6.4

72. Find a polynomial that represents the total surface area of the rectangular solid in Figure 6.5.

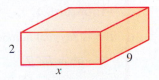

Figure 6.5

Thoughts Into Words

73. Explain how to subtract the polynomial $3x^2 + 6x - 2$ from $4x^2 + 7$.

74. Is the sum of two binomials always another binomial? Defend your answer.

75. Is the sum of two binomials ever a trinomial? Defend your answer.

6.2   Multiplying Monomials

OBJECTIVES

1. Apply the properties of exponents to multiply monomials

2. Multiply a polynomial by a monomial

3. Use products of monomials to represent the area or volume of geometric figures

In Section 2.4, we used exponents and some of the basic properties of real numbers to simplify algebraic expressions into a more compact form; for example,

$$(3x)(4xy) = 3 \cdot 4 \cdot x \cdot x \cdot y = 12x^2y$$

Actually we were **multiplying monomials**, and it is this topic that we will pursue now. We can make multiplying monomials easier by using some basic properties of exponents. These properties are the direct result of the definition of an exponent. The following examples lead to the first property:

$$x^2 \cdot x^3 = (x \cdot x)(x \cdot x \cdot x) = x^5$$
$$a^3 \cdot a^4 = (a \cdot a \cdot a)(a \cdot a \cdot a \cdot a) = a^7$$
$$b \cdot b^2 = (b)(b \cdot b) = b^3$$

In general,

$$b^n \cdot b^m = \underbrace{(b \cdot b \cdot b \cdot \cdots \cdot b)}_{n \text{ factors of } b}\underbrace{(b \cdot b \cdot b \cdot \cdots \cdot b)}_{m \text{ factors of } b}$$

$$= \underbrace{b \cdot b \cdot b \cdot \cdots \cdot b}_{(n + m) \text{ factors of } b}$$

$$= b^{n+m}$$

Property 6.1

If b is any real number, and n and m are positive integers, then

$$b^n \cdot b^m = b^{n+m}$$

Property 6.1 states that when multiplying powers with the same base, add exponents.

EXAMPLE 1   Multiply:

(a) $x^4 \cdot x^3$ (b) $a^8 \cdot a^7$

Solution

(a) $x^4 \cdot x^3 = x^{4+3} = x^7$ (b) $a^8 \cdot a^7 = a^{8+7} = a^{15}$

Another property of exponents is demonstrated by these examples:

$$(x^2)^3 = x^2 \cdot x^2 \cdot x^2 = x^{2+2+2} = x^6$$
$$(a^3)^2 = a^3 \cdot a^3 = a^{3+3} = a^6$$
$$(b^3)^4 = b^3 \cdot b^3 \cdot b^3 \cdot b^3 = b^{3+3+3+3} = b^{12}$$

In general,

$$(b^n)^m = \underbrace{b^n \cdot b^n \cdot b^n \cdot \cdots \cdot b^n}_{m \text{ factors of } b^n}$$

$$= b^{\overbrace{n+n+n+\cdots+n}^{m \text{ of these } n\text{'s}}}$$

$$= b^{mn}$$

Property 6.2

If b is any real number, and m and n are positive integers, then

$$(b^n)^m = b^{mn}$$

Property 6.2 states that when raising a power to a power, multiply exponents.

Classroom Example
Raise each to the indicated power:
(a) $(m^2)^6$ **(b)** $(n^7)^9$

EXAMPLE 2 Raise each to the indicated power:

(a) $(x^4)^3$ **(b)** $(a^5)^6$

Solution

(a) $(x^4)^3 = x^{3 \cdot 4} = x^{12}$ **(b)** $(a^5)^6 = a^{6 \cdot 5} = a^{30}$

The third property of exponents we will use in this section raises a monomial to a power.

$$(2x)^3 = (2x)(2x)(2x) = 2 \cdot 2 \cdot 2 \cdot x \cdot x \cdot x = 2^3 \cdot x^3$$
$$(3a^4)^2 = (3a^4)(3a^4) = 3 \cdot 3 \cdot a^4 \cdot a^4 = (3)^2(a^4)^2$$
$$(-2xy^5)^2 = (-2xy^5)(-2xy^5) = (-2)(-2)(x)(x)(y^5)(y^5) = (-2)^2(x)^2(y^5)^2$$

In general,

$$(ab)^n = \underbrace{ab \cdot ab \cdot ab \cdot \cdots \cdot ab}_{n \text{ factors of } ab}$$

$$= (\underbrace{a \cdot a \cdot a \cdot \cdots \cdot a}_{n \text{ factors of } a})(\underbrace{b \cdot b \cdot b \cdot \cdots \cdot b}_{n \text{ factors of } b})$$

$$= a^n b^n$$

Property 6.3

If a and b are real numbers, and n is a positive integer, then

$$(ab)^n = a^n b^n$$

Property 6.3 states that when raising a monomial to a power, raise each factor to that power.

Classroom Example
Raise each to the indicated power:
(a) $(3m^2n)^2$ **(b)** $(-5c^2 d^6)^3$

EXAMPLE 3 Raise each to the indicated power:

(a) $(2x^2y^3)^4$ **(b)** $(-3ab^5)^3$

Solution

(a) $(2x^2y^3)^4 = (2)^4(x^2)^4(y^3)^4 = 16x^8y^{12}$

(b) $(-3ab^5)^3 = (-3)^3(a^1)^3(b^5)^3 = -27a^3b^{15}$

Consider the following examples in which we use the properties of exponents to help simplify the process of multiplying monomials.

1. $(3x^3)(5x^4) = 3 \cdot 5 \cdot x^3 \cdot x^4$

$= 15x^7$ $x^3 \cdot x^4 = x^{3+4} = x^7$

2. $(-4a^2b^3)(6ab^2) = -4 \cdot 6 \cdot a^2 \cdot a \cdot b^3 \cdot b^2$

$= -24a^3b^5$

3. $(xy)(7xy^5) = 1 \cdot 7 \cdot x \cdot x \cdot y \cdot y^5$ The numerical coefficient of xy is 1

$= 7x^2y^6$

4. $\left(\dfrac{3}{4}x^2y^3\right)\left(\dfrac{1}{2}x^3y^5\right) = \dfrac{3}{4} \cdot \dfrac{1}{2} \cdot x^2 \cdot x^3 \cdot y^3 \cdot y^5$

$= \dfrac{3}{8}x^5y^8$

It is a simple process to raise a monomial to a power when using the properties of exponents. Study the next examples.

5. $(2x^3)^4 = (2)^4(x^3)^4$ by using $(ab)^n = a^nb^n$

$= (2)^4(x^{12})$ by using $(b^n)^m = b^{mn}$

$= 16x^{12}$

6. $(-2a^4)^5 = (-2)^5(a^4)^5$

$= -32a^{20}$

7. $\left(\dfrac{2}{5}x^2y^3\right)^3 = \left(\dfrac{2}{5}\right)^3(x^2)^3(y^3)^3$

$= \dfrac{8}{125}x^6y^9$

8. $(0.2a^6b^7)^2 = (0.2)^2(a^6)^2(b^7)^2$

$= 0.04a^{12}b^{14}$

Sometimes problems involve first raising monomials to a power and then multiplying the resulting monomials, as in the following examples.

9. $(3x^2)^3(2x^3)^2 = (3)^3(x^2)^3(2)^2(x^3)^2$

$= (27)(x^6)(4)(x^6)$

$= 108x^{12}$

10. $(-x^2y^3)^5(-2x^2y)^2 = (-1)^5(x^2)^5(y^3)^5(-2)^2(x^2)^2(y)^2$

$= (-1)(x^{10})(y^{15})(4)(x^4)(y^2)$

$= -4x^{14}y^{17}$

The distributive property along with the properties of exponents form a basis for finding the product of a monomial and a polynomial. The next examples illustrate these ideas.

11. $(3x)(2x^2 + 6x + 1) = (3x)(2x^2) + (3x)(6x) + (3x)(1)$

$= 6x^3 + 18x^2 + 3x$

12. $(5a^2)(a^3 - 2a^2 - 1) = (5a^2)(a^3) - (5a^2)(2a^2) - (5a^2)(1)$

$= 5a^5 - 10a^4 - 5a^2$

13. $(-2xy)(6x^2y - 3xy^2 - 4y^3) = (-2xy)(6x^2y) - (-2xy)(3xy^2) - (-2xy)(4y^3)$

$= -12x^3y^2 + 6x^2y^3 + 8xy^4$

Once you feel comfortable with this process, you may want to perform most of the work mentally and then simply write down the final result. See whether you understand the following examples.

14. $3x(2x + 3) = 6x^2 + 9x$

15. $-4x(2x^2 - 3x - 1) = -8x^3 + 12x^2 + 4x$

16. $ab(3a^2b - 2ab^2 - b^3) = 3a^3b^2 - 2a^2b^3 - ab^4$

We conclude this section by making a connection between algebra and geometry.

Classroom Example
Suppose that the dimensions of a right circular cylinder are represented by radius x and height $4x$. Express the volume and total surface area of the cylinder.

EXAMPLE 4

Suppose that the dimensions of a rectangular solid are represented by x, $2x$, and $3x$ as shown in Figure 6.6. Express the volume and total surface area of the figure.

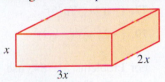

Figure 6.6

Solution

Using the formula $V = lwh$, we can express the volume of the rectangular solid as $(2x)(3x)(x)$, which equals $6x^3$. The total surface area can be described as follows:

Area of front and back rectangles: $2(x)(3x) = 6x^2$

Area of left side and right side: $2(2x)(x) = 4x^2$

Area of top and bottom: $2(2x)(3x) = 12x^2$

We can represent the total surface area by $6x^2 + 4x^2 + 12x^2$ or $22x^2$.

Concept Quiz 6.2

For Problems 1–10, answer true or false.

1. When multiplying factors with the same base, add the exponents.

2. $3^2 \cdot 3^2 = 9^4$

3. $2x^2 \cdot 3x^3 = 6x^6$

4. $(x^2)^3 = x^5$

5. $(-4x^3)^2 = -4x^6$

6. To simplify $(3x^2y)(2x^3y^2)^4$, use the order of operations to first raise $2x^3y^2$ to the fourth power, and then multiply the monomials.

7. $(3x^2y)^3 = 27x^6y^3$

8. $(-2xy)(3x^2y^3) = -6x^3y^4$

9. $(-x^2y)(xy^3)(xy) = x^4y^5$

10. $(2x^2y^3)^3(-xy^2) = -8x^7y^{11}$

Problem Set 6.2

For Problems 1–30, multiply using the properties of exponents to help with the manipulation. **(Objective 1)**

1. $(5x)(9x)$

2. $(7x)(8x)$

3. $(3x^2)(7x)$

4. $(9x)(4x^3)$

5. $(-3xy)(2xy)$

6. $(6xy)(-3xy)$

7. $(-2x^2y)(-7x)$

8. $(-5xy^2)(-4y)$

9. $(4a^2b^2)(-12ab)$

10. $(-3a^3b)(13ab^2)$

11. $(-xy)(-5x^3)$

12. $(-7y^2)(-x^2y)$

13. $(8ab^2c)(13a^2c)$

14. $(9abc^3)(14bc^2)$

15. $(5x^2)(2x)(3x^3)$

16. $(4x)(2x^2)(6x^4)$

17. $(4xy)(-2x)(7y^2)$

18. $(5y^2)(-3xy)(5x^2)$

19. $(-2ab)(-ab)(-3b)$

20. $(-7ab)(-4a)(-ab)$

21. $(6cd)(-3c^2d)(-4d)$

22. $(2c^3d)(-6d^3)(-5cd)$

23. $\left(\frac{2}{3}xy\right)\left(\frac{3}{5}x^2y^4\right)$

24. $\left(-\frac{5}{6}x\right)\left(\frac{8}{3}x^2y\right)$

25. $\left(-\frac{7}{12}a^2b\right)\left(\frac{8}{21}b^4\right)$

26. $\left(-\frac{9}{5}a^3b^4\right)\left(-\frac{15}{6}ab^2\right)$

27. $(0.4x^5)(0.7x^3)$

28. $(-1.2x^4)(0.3x^2)$

29. $(-4ab)(1.6a^3b)$

30. $(-6a^2b)(-1.4a^2b^4)$

For Problems 31–46, raise each monomial to the indicated power. Use the properties of exponents to help with the manipulation. **(Objective 1)**

31. $(2x^4)^2$

32. $(3x^3)^2$

33. $(-3a^2b^3)^2$

34. $(-8a^4b^5)^2$

35. $(3x^2)^3$

36. $(2x^4)^3$

37. $(-4x^4)^3$

38. $(-3x^3)^3$

39. $(9x^4y^5)^2$

40. $(8x^6y^4)^2$

41. $(2x^2y)^4$

42. $(2x^2y^3)^5$

43. $(-3a^3b^2)^4$

44. $(-2a^4b^2)^4$

45. $(-x^2y)^6$

46. $(-x^2y^3)^7$

For Problems 47–60, multiply by using the distributive property. **(Objective 2)**

47. $5x(3x + 2)$

48. $7x(2x + 5)$

49. $3x^2(6x - 2)$

50. $4x^2(7x - 2)$

51. $-4x(7x^2 - 4)$

52. $-6x(9x^2 - 5)$

53. $2x(x^2 - 4x + 6)$

54. $3x(2x^2 - x + 5)$

55. $-6a(3a^2 - 5a - 7)$

56. $-8a(4a^2 - 9a - 6)$

57. $7xy(4x^2 - x + 5)$

58. $5x^2y(3x^2 + 7x - 9)$

59. $-xy(9x^2 - 2x - 6)$

60. $xy^2(6x^2 - x - 1)$

For Problems 61–70, remove the parentheses by multiplying and then simplify by combining similar terms. **(Objective 2)** For example,
$$3(x - y) + 2(x - 3y) = 3x - 3y + 2x - 6y = 5x - 9y$$

61. $5(x + 2y) + 4(2x + 3y)$

62. $3(2x + 5y) + 2(4x + y)$

63. $4(x - 3y) - 3(2x - y)$

64. $2(5x - 3y) - 5(x + 4y)$

65. $2x(x^2 - 3x - 4) + x(2x^2 + 3x - 6)$

66. $3x(2x^2 - x + 5) - 2x(x^2 + 4x + 7)$

67. $3[2x - (x - 2)] - 4(x - 2)$

68. $2[3x - (2x + 1)] - 2(3x - 4)$

69. $-4(3x + 2) - 5[2x - (3x + 4)]$

70. $-5(2x - 1) - 3[x - (4x - 3)]$

For Problems 71–80, perform the indicated operations and simplify. **(Objective 1)**

71. $(3x)^2(2x^3)$

72. $(-2x)^3(4x^5)$

73. $(-3x)^3(-4x)^2$

74. $(3xy)^2(2x^2y)^4$

75. $(5x^2y)^2(xy^2)^3$

76. $(-x^2y)^3(6xy)^2$

77. $(-a^2bc^3)^3(a^3b)^2$

78. $(ab^2c^3)^4(-a^2b)^3$

79. $(-2x^2y^2)^4(-xy^3)^3$

80. $(-3xy)^3(-x^2y^3)^4$

For Problems 81–84, use the products of polynomials to represent the area or volume of the geometric figure. **(Objective 3)**

81. Express in simplified form the sum of the areas of the two rectangles shown in Figure 6.7.

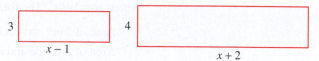

Figure 6.7

82. Express in simplified form the volume and the total surface area of the rectangular solid in Figure 6.8.

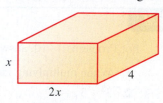

Figure 6.8

83. Represent the area of the shaded region in Figure 6.9. The length of a radius of the smaller circle is x, and the length of a radius of the larger circle is $2x$.

Figure 6.9

84. Represent the area of the shaded region in Figure 6.10.

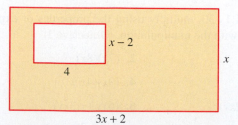

Figure 6.10

Thoughts Into Words

85. How would you explain to someone why the product of x^3 and x^4 is x^7 and not x^{12}?

86. Suppose your friend was absent from class the day that this section was discussed. How would you help her understand why the property $(b^n)^m = b^{mn}$ is true?

87. How can Figure 6.11 be used to geometrically demonstrate that $x(x + 2) = x^2 + 2x$?

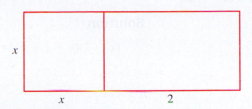

Figure 6.11

Further Investigations

For Problems 88–97, find each of the indicated products. Assume that the variables in the exponents represent positive integers; for example,

$$(x^{2n})(x^{4n}) = x^{2n + 4n} = x^{6n}$$

88. $(x^n)(x^{3n})$

89. $(x^{2n})(x^{5n})$

90. $(x^{2n-1})(x^{3n+2})$

91. $(x^{5n+2})(x^{n-1})$

92. $(x^3)(x^{4n-5})$

93. $(x^{6n-1})(x^4)$

94. $(2x^n)(3x^{2n})$

95. $(4x^{3n})(-5x^{7n})$

96. $(-6x^{2n+4})(5x^{3n-4})$

97. $(-3x^{5n-2})(-4x^{2n+2})$

Answers to the Concept Quiz

1. True   **2.** False   **3.** False   **4.** False   **5.** False   **6.** True   **7.** True   **8.** True   **9.** False   **10.** True

---

6.3   Multiplying Polynomials

OBJECTIVES

1. Use the distributive property to find the product of two binomials

2. Use the shortcut pattern to find the product of two binomials

3. Use a pattern to find the square of a binomial

4. Use a pattern to find the product of $(a + b)(a - b)$

In general, to go from multiplying a monomial times a polynomial to multiplying two polynomials requires the use of the distributive property. Consider some examples.

Classroom Example
Find the product of $(a + 6)$ and $(b + 3)$.

EXAMPLE 1   Find the product of $(x + 3)$ and $(y + 4)$.

Solution

$$(x + 3)(y + 4) = x(y + 4) + 3(y + 4)$$
$$= x(y) + x(4) + 3(y) + 3(4)$$
$$= xy + 4x + 3y + 12$$

Notice that each term of the first polynomial is multiplied times each term of the second polynomial.

Classroom Example
Find the product of $(m - 5)$ and $(m + 2n + 4)$.

EXAMPLE 2   Find the product of $(x - 2)$ and $(y + z + 5)$.

Solution

$$
\begin{aligned}
(x - 2)(y + z + 5) &= x(y + z + 5) - 2(y + z + 5) \\
&= x(y) + x(z) + x(5) - 2(y) - 2(z) - 2(5) \\
&= xy + xz + 5x - 2y - 2z - 10
\end{aligned}
$$

Usually, multiplying polynomials will produce similar terms that can be combined to simplify the resulting polynomial.

Classroom Example
Multiply $(x + 5)(x + 8)$.

EXAMPLE 3   Multiply $(x + 3)(x + 2)$.

Solution

$$
\begin{aligned}
(x + 3)(x + 2) &= x(x + 2) + 3(x + 2) \\
&= x^2 + 2x + 3x + 6 \\
&= x^2 + 5x + 6 \qquad \text{Combine like terms}
\end{aligned}
$$

Classroom Example
Multiply $(k + 6)(k - 2)$.

EXAMPLE 4   Multiply $(x - 4)(x + 9)$.

Solution

$$
\begin{aligned}
(x - 4)(x + 9) &= x(x + 9) - 4(x + 9) \\
&= x^2 + 9x - 4x - 36 \\
&= x^2 + 5x - 36 \qquad \text{Combine like terms}
\end{aligned}
$$

Classroom Example
Multiply $(a + 1)(2a^2 + a + 5)$.

EXAMPLE 5   Multiply $(x + 4)(x^2 + 3x + 2)$.

Solution

$$
\begin{aligned}
(x + 4)(x^2 + 3x + 2) &= x(x^2 + 3x + 2) + 4(x^2 + 3x + 2) \\
&= x^3 + 3x^2 + 2x + 4x^2 + 12x + 8 \\
&= x^3 + 7x^2 + 14x + 8
\end{aligned}
$$

Classroom Example
Multiply $(3m - n) \cdot (2m^2 + 4mn - 3n^2)$.

EXAMPLE 6   Multiply $(2x - y)(3x^2 - 2xy + 4y^2)$.

Solution

$$
\begin{aligned}
(2x - y)(3x^2 - 2xy + 4y^2) &= 2x(3x^2 - 2xy + 4y^2) - y(3x^2 - 2xy + 4y^2) \\
&= 6x^3 - 4x^2y + 8xy^2 - 3x^2y + 2xy^2 - 4y^3 \\
&= 6x^3 - 7x^2y + 10xy^2 - 4y^3
\end{aligned}
$$

Perhaps the most frequently used type of multiplication problem is the product of two binomials. It will be a big help later if you can become proficient at multiplying binomials without showing all of the intermediate steps. This is quite easy to do if you use a three-step shortcut pattern demonstrated by the following examples.

Classroom Example
Multiply $(x + 9)(x + 2)$.

EXAMPLE 7   Multiply $(x + 5)(x + 7)$.

Solution

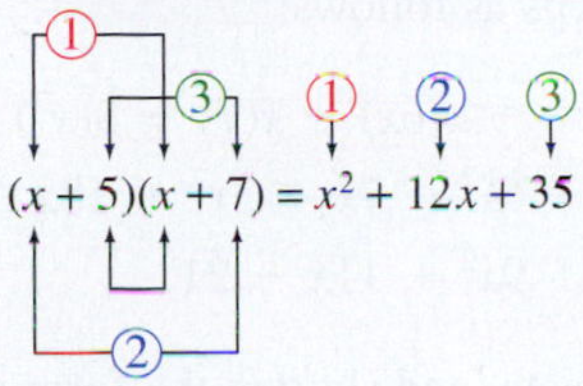

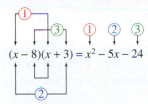

$(x + 5)(x + 7) = x^2 + 12x + 35$

Figure 6.12

Step 1   Multiply $x \cdot x$.
Step 2   Multiply $5 \cdot x$ and $7 \cdot x$ and combine them.
Step 3   Multiply $5 \cdot 7$.

Classroom Example
Multiply $(h + 4)(h - 7)$.

EXAMPLE 8   Multiply $(x - 8)(x + 3)$.

Solution

$(x - 8)(x + 3) = x^2 - 5x - 24$

Figure 6.13

Classroom Example
Multiply $(5n + 1)(4n - 3)$.

EXAMPLE 9   Multiply $(3x + 2)(2x - 5)$.

Solution

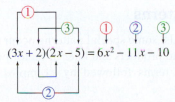

$(3x + 2)(2x - 5) = 6x^2 - 11x - 10$

Figure 6.14

The mnemonic device **FOIL** is often used to remember the pattern for multiplying binomials. The letters in FOIL represent First, Outside, Inside, and Last. If you look back at Examples 7 through 9, step 1 is to find the product of the first terms in each binomial; step 2 is to find the product of the outside terms and the inside terms; and step 3 is to find the product of the last terms in each binomial.

Now see whether you can use the pattern to find these products:

$(x + 3)(x + 7)$

$(3x + 1)(2x + 5)$

$(x - 2)(x - 3)$

$(4x + 5)(x - 2)$

Your answers should be as follows: $x^2 + 10x + 21$; $6x^2 + 17x + 5$; $x^2 - 5x + 6$; and $4x^2 - 3x - 10$.

Keep in mind that the shortcut pattern applies only to finding the product of two binomials. For other situations, such as finding the product of a binomial and a trinomial, we suggest showing the intermediate steps as follows:

$$(x + 3)(x^2 + 6x - 7) = x(x^2) + x(6x) - x(7) + 3(x^2) + 3(6x) - 3(7)$$
$$= x^3 + 6x^2 - 7x + 3x^2 + 18x - 21$$
$$= x^3 + 9x^2 + 11x - 21$$

Perhaps you could omit the first step, and shorten the form as follows:

$$(x - 4)(x^2 - 5x - 6) = x^3 - 5x^2 - 6x - 4x^2 + 20x + 24$$
$$= x^3 - 9x^2 + 14x + 24$$

Remember that you are multiplying each term of the first polynomial times each term of the second polynomial and combining similar terms.

Exponents are also used to indicate repeated multiplication of polynomials. For example, we can write $(x + 4)(x + 4)$ as $(x + 4)^2$. Thus to square a binomial we simply write it as the product of two equal binomials and apply the shortcut pattern.

$$(x + 4)^2 = (x + 4)(x + 4) = x^2 + 8x + 16$$
$$(x - 5)^2 = (x - 5)(x - 5) = x^2 - 10x + 25$$
$$(2x + 3)^2 = (2x + 3)(2x + 3) = 4x^2 + 12x + 9$$

When you square binomials, be careful not to forget the middle term. That is to say, $(x + 3)^2 \neq x^2 + 3^2$; instead, $(x + 3)^2 = (x + 3)(x + 3) = x^2 + 6x + 9$.

The next example suggests a format to use when cubing a binomial.

$$(x + 4)^3 = (x + 4)(x + 4)(x + 4)$$
$$= (x + 4)(x^2 + 8x + 16)$$
$$= x(x^2 + 8x + 16) + 4(x^2 + 8x + 16)$$
$$= x^3 + 8x^2 + 16x + 4x^2 + 32x + 64$$
$$= x^3 + 12x^2 + 48x + 64$$

Special Product Patterns

When we multiply binomials, some special patterns occur that you should recognize. We can use these patterns to find products and later to factor polynomials. We will state each of the patterns in general terms followed by examples to illustrate the use of each pattern.

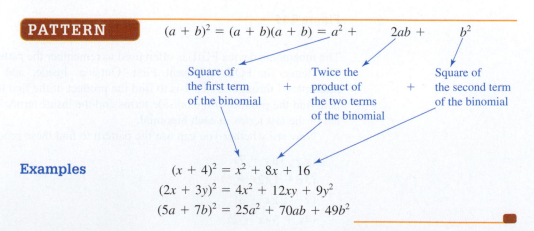

PATTERN $(a + b)^2 = (a + b)(a + b) = a^2 + 2ab + b^2$

| Square of the first term of the binomial | + | Twice the product of the two terms of the binomial | + | Square of the second term of the binomial |

Examples

$$(x + 4)^2 = x^2 + 8x + 16$$
$$(2x + 3y)^2 = 4x^2 + 12xy + 9y^2$$
$$(5a + 7b)^2 = 25a^2 + 70ab + 49b^2$$

PATTERN $(a - b)^2 = (a - b)(a - b) = a^2 \quad - \quad 2ab \quad + \quad b^2$

Square of   Twice the   Square of
the first term   −   product of   +   the second term
of the binomial   the two terms   of the binomial
                   of the binomial

Examples

$$(x - 8)^2 = x^2 - 16x + 64$$
$$(3x - 4y)^2 = 9x^2 - 24xy + 16y^2$$
$$(4a - 9b)^2 = 16a^2 - 72ab + 81b^2$$

PATTERN $(a + b)(a - b) \quad = \quad a^2 \quad - \quad b^2$

Square of   Square of
the first term   −   the second term
of the binomial   of the binomial

Examples

$$(x + 7)(x - 7) = x^2 - 49$$
$$(2x + y)(2x - y) = 4x^2 - y^2$$
$$(3a - 2b)(3a + 2b) = 9a^2 - 4b^2$$

As you might expect, there are geometric interpretations for many of the algebraic concepts presented in this section. We will give you the opportunity to make some of these connections between algebra and geometry in the next problem set. We conclude this section with a problem that allows us to use some algebra and geometry.

EXAMPLE 10

Classroom Example
A square piece of cardboard has a side 20 inches long. From each corner a square piece x inches on a side is cut out. The flaps are then turned up to form an open box. Find the polynomials that represent the volume and the exterior surface area of the box.

A rectangular piece of tin is 16 inches long and 12 inches wide as shown in Figure 6.15. From each corner a square piece x inches on a side is cut out. The flaps are then turned up to form an open box. Find polynomials that represent the volume and the exterior surface area of the box.

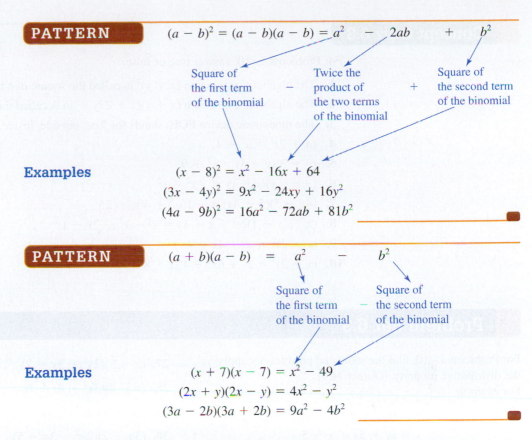

Figure 6.15

Solution

The length of the box is $16 - 2x$, the width is $12 - 2x$, and the height is x. From the volume formula $V = lwh$, the polynomial $(16 - 2x)(12 - 2x)(x)$, which simplifies to $4x^3 - 56x^2 + 192x$, represents the volume.

The outside surface area of the box is the area of the original piece of tin minus the four corners that were cut off. Therefore, the polynomial $16(12) - 4x^2$ or $192 - 4x^2$ represents the outside surface area of the box.

Concept Quiz 6.3

For Problems 1–10, answer true or false.

1. The algebraic expression $(x + y)^2$ is called the square of a binomial.
2. The algebraic expression $(x + y)(x + 2xy + y)$ is called the product of two binomials.
3. The mnemonic device FOIL stands for first, outside, inside, and last.
4. $(a + 2)^2 = a^2 + 4$
5. $(y + 3)(y - 3) = y^2 + 9$
6. $(-x + y)(x - y) = -x^2 - y^2$
7. $(4x - 5)(5 - 4x) = -16x^2 + 40x - 25$
8. $(x^2 - x - 1)(x^2 + x + 1) = x^4 - x^2 - 2x - 1$
9. $(x + 2)^2 = x^2 + 4x + 4$
10. $(x + 2)^3 = x^3 + 6x^2 + 12x + 8$

Problem Set 6.3

For Problems 1–10, find the indicated products by applying the distributive property. **(Objective 1)**
For example,

$$(x + 1)(y + 5) = x(y) + x(5) + 1(y) + 1(5)$$
$$= xy + 5x + y + 5$$

1. $(x + 2)(y + 3)$
2. $(x + 3)(y + 6)$
3. $(x - 4)(y + 1)$
4. $(x - 5)(y + 7)$
5. $(x - 5)(y - 6)$
6. $(x - 7)(y - 9)$
7. $(x + 2)(y + z + 1)$
8. $(x + 4)(y - z + 4)$
9. $(2x + 3)(3y + 1)$
10. $(3x - 2)(2y - 5)$

For Problems 11–36, find the indicated products by applying the distributive property and combining similar terms. **(Objective 1)**
Use the following format to show your work:

$$(x + 3)(x + 8) = x(x) + x(8) + 3(x) + 3(8)$$
$$= x^2 + 8x + 3x + 24$$
$$= x^2 + 11x + 24$$

11. $(x + 3)(x + 7)$
12. $(x + 4)(x + 2)$
13. $(x + 8)(x - 3)$
14. $(x + 9)(x - 6)$
15. $(x - 7)(x + 1)$
16. $(x - 10)(x + 8)$
17. $(n - 4)(n - 6)$
18. $(n - 3)(n - 7)$
19. $(3n + 1)(n + 6)$
20. $(4n + 3)(n + 6)$
21. $(5x - 2)(3x + 7)$
22. $(3x - 4)(7x + 1)$
23. $(x + 3)(x^2 + 4x + 9)$
24. $(x + 2)(x^2 + 6x + 2)$
25. $(x + 4)(x^2 - x - 6)$
26. $(x + 5)(x^2 - 2x - 7)$

27. $(x - 5)(2x^2 + 3x - 7)$
28. $(x - 4)(3x^2 + 4x - 6)$
29. $(2a - 1)(4a^2 - 5a + 9)$
30. $(3a - 2)(2a^2 - 3a - 5)$
31. $(3a + 5)(a^2 - a - 1)$
32. $(5a + 2)(a^2 + a - 3)$
33. $(x^2 + 2x + 3)(x^2 + 5x + 4)$
34. $(x^2 - 3x + 4)(x^2 + 5x - 2)$
35. $(x^2 - 6x - 7)(x^2 + 3x - 9)$
36. $(x^2 - 5x - 4)(x^2 + 7x - 8)$

For Problems 37–80, find the indicated products by using the shortcut pattern for multiplying binomials. **(Objective 2)**

37. $(x + 2)(x + 9)$
38. $(x + 3)(x + 8)$
39. $(x + 6)(x - 2)$
40. $(x + 8)(x - 6)$
41. $(x + 3)(x - 11)$
42. $(x + 4)(x - 10)$
43. $(n - 4)(n - 3)$
44. $(n - 5)(n - 9)$
45. $(n + 6)(n + 12)$
46. $(n + 8)(n + 13)$
47. $(y + 3)(y - 7)$
48. $(y + 2)(y - 12)$
49. $(y - 7)(y - 12)$
50. $(y - 4)(y - 13)$
51. $(x - 5)(x + 7)$
52. $(x - 1)(x + 9)$
53. $(x - 14)(x + 8)$
54. $(x - 15)(x + 6)$
55. $(a + 10)(a - 9)$
56. $(a + 7)(a - 6)$
57. $(2a + 1)(a + 6)$
58. $(3a + 2)(a + 4)$
59. $(5x - 2)(x + 7)$
60. $(2x - 3)(x + 8)$
61. $(3x - 7)(2x + 1)$
62. $(5x - 6)(4x + 3)$

63. $(4a + 3)(3a - 4)$

64. $(5a + 4)(4a - 5)$

65. $(6n - 5)(2n - 3)$

66. $(4n - 3)(6n - 7)$

67. $(7x - 4)(2x + 3)$

68. $(8x - 5)(3x + 7)$

69. $(5 - x)(9 - 2x)$

70. $(4 - 3x)(2 + x)$

71. $(-2x + 3)(4x - 5)$

72. $(-3x + 1)(9x - 2)$

73. $(-3x - 1)(3x - 4)$

74. $(-2x - 5)(4x + 1)$

75. $(8n + 3)(9n - 4)$

76. $(6n + 5)(9n - 7)$

77. $(3 - 2x)(9 - x)$

78. $(5 - 4x)(4 - 5x)$

79. $(-4x + 3)(-5x - 2)$

80. $(-2x + 7)(-7x - 3)$

81. John, who has not mastered squaring a binomial, turns in the homework shown below. Identify any mistakes in John's work, and then write your answer for the problem.

(a) $(x + 4)^2 = x^2 + 16$

(b) $(x - 5)^2 = x^2 - 5x + 25$

(c) $(x - 6)^2 = x^2 - 12x - 36$

(d) $(3x + 2)^2 = 3x^2 + 12x + 4$

(e) $(x + 1)^2 = x^2 + 2x + 2$

(f) $(x - 3)^2 = x^2 + 6x + 9$

For Problems 82–112, use one of the appropriate patterns $(a + b)^2 = a^2 + 2ab + b^2$, $(a - b)^2 = a^2 - 2ab + b^2$, or $(a + b)(a - b) = a^2 - b^2$ to find the indicated products.
(Objectives 3 and 4)

82. $(x + 3)^2$

83. $(x + 7)^2$

84. $(x + 9)^2$

85. $(5x - 2)(5x + 2)$

86. $(6x + 1)(6x - 1)$

87. $(x - 1)^2$

88. $(x - 4)^2$

89. $(3x + 7)^2$

90. $(2x + 9)^2$

91. $(2x - 3)^2$

92. $(4x - 5)^2$

93. $(2x + 3y)(2x - 3y)$

94. $(3a - b)(3a + b)$

95. $(1 - 5n)^2$

96. $(2 - 3n)^2$

97. $(3x + 4y)^2$

98. $(2x + 5y)^2$

99. $(3 + 4y)^2$

100. $(7 + 6y)^2$

101. $(1 + 7n)(1 - 7n)$

102. $(2 + 9n)(2 - 9n)$

103. $(4a - 7b)^2$

104. $(6a - b)^2$

105. $(x + 8y)^2$

106. $(x + 6y)^2$

107. $(5x - 11y)(5x + 11y)$

108. $(7x - 9y)(7x + 9y)$

109. $x(8x + 1)(8x - 1)$

110. $3x(5x + 7)(5x - 7)$

111. $-2x(4x + y)(4x - y)$

112. $-4x(2 - 3x)(2 + 3x)$

For Problems 113–120, find the indicated products. Don't forget that $(x + 2)^3$ means $(x + 2)(x + 2)(x + 2)$.

113. $(x + 2)^3$

114. $(x + 4)^3$

115. $(x - 3)^3$

116. $(x - 1)^3$

117. $(2n + 1)^3$

118. $(3n + 2)^3$

119. $(3n - 2)^3$

120. $(4n - 3)^3$

121. Explain how Figure 6.16 can be used to demonstrate geometrically that $(x + 3)(x + 5) = x^2 + 8x + 15$.

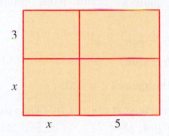

Figure 6.16

122. Explain how Figure 6.17 can be used to demonstrate geometrically that $(x + 5)(x - 3) = x^2 + 2x - 15$.

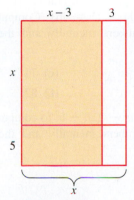

Figure 6.17

123. A square piece of cardboard is 14 inches long on each side. From each corner a square piece x inches on a side is cut out as shown in Figure 6.18. The flaps are then turned up to form an open box. Find polynomials that represent the volume and the exterior surface area of the box.

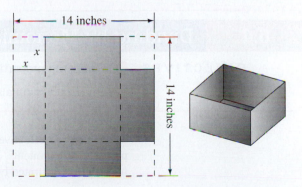

Figure 6.18

Thoughts Into Words

124. Describe the process of multiplying two polynomials.

125. Illustrate as many uses of the distributive property as you can.

126. Determine the number of terms in the product of $(x + y + z)$ and $(a + b + c)$ without doing the multiplication. Explain how you arrived at your answer.

Further Investigations

127. The following two patterns result from cubing binomials:

$$(a + b)^3 = a^3 + 3a^2b + 3ab^2 + b^3$$
$$(a - b)^3 = a^3 - 3a^2b + 3ab^2 - b^3$$

Use these patterns to redo Problems 113–120.

128. Find a pattern for the expansion of $(a + b)^4$. Then use the pattern to expand $(x + 2)^4$, $(x + 3)^4$, and $(2x + 1)^4$.

129. We can use some of the product patterns to do arithmetic computations mentally. For example, let's use the pattern $(a + b)^2 = a^2 + 2ab + b^2$ to mentally compute 31^2. Your thought process should be $31^2 = (30 + 1)^2 = 30^2 + 2(30)(1) + 1^2 = 961$. Compute each of the following numbers mentally and then check your answers.

  (a) 21^2 **(b)** 41^2 **(c)** 71^2

  (d) 32^2 **(e)** 52^2 **(f)** 82^2

130. Use the pattern $(a - b)^2 = a^2 - 2ab + b^2$ to compute each of the following numbers mentally and then check your answers.

  (a) 19^2 **(b)** 29^2 **(c)** 49^2

  (d) 79^2 **(e)** 38^2 **(f)** 58^2

131. Every whole number with a units digit of 5 can be represented by the expression $10x + 5$, where x is a whole number. For example, $35 = 10(3) + 5$ and $145 = 10(14) + 5$. Now observe the following pattern for squaring such a number:

$$(10x + 5)^2 = 100x^2 + 100x + 25$$
$$= 100x(x + 1) + 25$$

The pattern inside the dashed box can be stated as "add 25 to the product of x, $x + 1$, and 100." Thus to mentally compute 35^2 we can think $35^2 = 3(4)(100) + 25 = 1225$. Compute each of the following numbers mentally and then check your answers.

  (a) 15^2 **(b)** 25^2 **(c)** 45^2

  (d) 55^2 **(e)** 65^2 **(f)** 75^2

  (g) 85^2 **(h)** 95^2 **(i)** 105^2

Answers to the Concept Quiz

1. True   **2.** False   **3.** True   **4.** False   **5.** False   **6.** False   **7.** True   **8.** True   **9.** True   **10.** True

6.4   Dividing by Monomials

OBJECTIVES

  1   Apply the properties of exponents to divide monomials

  2   Divide polynomials by monomials

To develop an effective process for dividing by a monomial, we must rely on yet another property of exponents. This property is also a direct consequence of the definition of exponent and is illustrated by the following examples.

$$\frac{x^5}{x^2} = \frac{\cancel{x} \cdot \cancel{x} \cdot x \cdot x \cdot x}{\cancel{x} \cdot \cancel{x}} = x^3$$

$$\frac{a^4}{a^3} = \frac{\cancel{a} \cdot \cancel{a} \cdot \cancel{a} \cdot a}{\cancel{a} \cdot \cancel{a} \cdot \cancel{a}} = a$$

$$\frac{y^7}{y^3} = \frac{\cancel{y} \cdot \cancel{y} \cdot \cancel{y} \cdot y \cdot y \cdot y \cdot y}{\cancel{y} \cdot \cancel{y} \cdot \cancel{y}} = y^4$$

$$\frac{x^4}{x^4} = \frac{\cancel{x} \cdot \cancel{x} \cdot \cancel{x} \cdot \cancel{x}}{\cancel{x} \cdot \cancel{x} \cdot \cancel{x} \cdot \cancel{x}} = 1$$

$$\frac{y^3}{y^3} = \frac{\cancel{y} \cdot \cancel{y} \cdot \cancel{y}}{\cancel{y} \cdot \cancel{y} \cdot \cancel{y}} = 1$$

Property 6.4

If b is any nonzero real number, and n and m are positive integers, then

1. $\dfrac{b^n}{b^m} = b^{n-m}$ when $n > m$

2. $\dfrac{b^n}{b^m} = 1$ when $n = m$

(The situation $n < m$ will be discussed in a later section.)

Applying Property 6.4 to the previous examples yields these results:

$$\frac{x^5}{x^2} = x^{5-2} = x^3$$

$$\frac{a^4}{a^3} = a^{4-3} = a^1 \qquad \text{Usually written as } a$$

$$\frac{y^7}{y^3} = y^{7-3} = y^4$$

$$\frac{x^4}{x^4} = 1$$

$$\frac{y^3}{y^3} = 1$$

Property 6.4, along with our knowledge of dividing integers, provides the basis for dividing a monomial by another monomial. Consider the next examples:

$$\frac{16x^5}{2x^3} = 8x^{5-3} = 8x^2 \qquad\qquad \frac{-81a^{12}}{-9a^4} = 9a^{12-4} = 9a^8$$

$$\frac{-35x^9}{5x^4} = -7x^{9-4} = -7x^5 \qquad \frac{45x^4}{9x^4} = 5 \qquad \frac{x^4}{x^4} = 1$$

$$\frac{56y^6}{-7y^2} = -8y^{6-2} = -8y^4 \qquad \frac{54x^3y^7}{-6xy^5} = -9x^{3-1}y^{7-5} = -9x^2y^2$$

Recall that $\dfrac{a+b}{c} = \dfrac{a}{c} + \dfrac{b}{c}$; this property serves as the basis for dividing a polynomial by a monomial. Consider these examples:

$$\frac{25x^3 + 10x^2}{5x} = \frac{25x^3}{5x} + \frac{10x^2}{5x} = 5x^2 + 2x$$

$$\frac{-35x^8 - 28x^6}{7x^3} = \frac{-35x^8}{7x^3} - \frac{28x^6}{7x^3} = -5x^5 - 4x^3 \qquad \frac{a-b}{c} = \frac{a}{c} - \frac{b}{c}$$

> To divide a polynomial by a monomial, we simply divide each term of the polynomial by the monomial. Here are some additional examples.

$$\frac{12x^3y^2 - 14x^2y^5}{-2xy} = \frac{12x^3y^2}{-2xy} - \frac{14x^2y^5}{-2xy} = -6x^2y + 7xy^4$$

$$\frac{48ab^5 + 64a^2b}{-16ab} = \frac{48ab^5}{-16ab} + \frac{64a^2b}{-16ab} = -3b^4 - 4a$$

$$\frac{33x^6 - 24x^5 - 18x^4}{3x} = \frac{33x^6}{3x} - \frac{24x^5}{3x} - \frac{18x^4}{3x}$$

$$= 11x^5 - 8x^4 - 6x^3$$

As with many skills, once you feel comfortable with the process, you may want to perform some of the steps mentally. Your work could take on the following format:

$$\frac{24x^4y^5 - 56x^3y^9}{8x^2y^3} = 3x^2y^2 - 7xy^6$$

$$\frac{13a^2b - 12ab^2}{-ab} = -13a + 12b$$

Concept Quiz 6.4

For Problems 1–10, answer true or false.

1. When dividing factors with the same base, add the exponents.

2. $\dfrac{10a^6}{2a^2} = 8a^4$

3. $\dfrac{y^8}{y^4} = y^2$

4. $\dfrac{6x^5 + 3x}{3x} = 2x^4$

5. $\dfrac{x^3}{x^3} = 0$

6. $\dfrac{x^4}{-x^4} = -1$

7. $\dfrac{-x^6}{x^6} = -1$

8. $\dfrac{24x^6}{4x^3} = 6x^2$

9. $\dfrac{24x^6 - 18x^4 + 12x^2}{2x^2} = 12x^4 - 9x^2 + 6$

10. $\dfrac{-30x^5 + 20x^4 - 10x^3}{-5x^2} = 6x^3 - 4x^2 + 2x$

Problem Set 6.4

For Problems 1–24, divide the monomials. (Objective 1)

1. $\dfrac{x^{10}}{x^2}$

2. $\dfrac{x^{12}}{x^5}$

3. $\dfrac{4x^3}{2x}$

4. $\dfrac{8x^5}{4x^3}$

5. $\dfrac{-16n^6}{2n^2}$

6. $\dfrac{-54n^8}{6n^4}$

7. $\dfrac{72x^3}{-9x^3}$

8. $\dfrac{84x^5}{-7x^5}$

9. $\dfrac{65x^2y^3}{5xy}$

10. $\dfrac{70x^3y^4}{5x^2y}$

11. $\dfrac{-91a^4b^6}{-13a^3b^4}$

12. $\dfrac{-72a^5b^4}{-12ab^2}$

13. $\dfrac{18x^2y^6}{xy^2}$

14. $\dfrac{24x^3y^4}{x^2y^2}$

15. $\dfrac{32x^6y^2}{-x}$

16. $\dfrac{54x^5y^3}{-y^2}$

17. $\dfrac{-96x^5y^7}{12y^3}$

18. $\dfrac{-84x^4y^9}{14x^4}$

19. $\dfrac{-ab}{ab}$

20. $\dfrac{6ab}{-ab}$

21. $\dfrac{56a^2b^3c^5}{4abc}$

22. $\dfrac{60a^3b^2c}{15a^2c}$

23. $\dfrac{-80xy^2z^6}{-5xyz^2}$

24. $\dfrac{-90x^3y^2z^8}{-6xy^2z^4}$

For Problems 25–50, perform each division of polynomials by monomials. (Objective 2)

25. $\dfrac{8x^4 + 12x^5}{2x^2}$

26. $\dfrac{12x^3 + 16x^6}{4x}$

27. $\dfrac{9x^6 - 24x^4}{3x^3}$

28. $\dfrac{35x^8 - 45x^6}{5x^4}$

29. $\dfrac{-28n^5 + 36n^2}{4n^2}$

30. $\dfrac{-42n^6 + 54n^4}{6n^4}$

31. $\dfrac{35x^6 - 56x^5 - 84x^3}{7x^2}$

32. $\dfrac{27x^7 - 36x^5 - 45x^3}{3x}$

33. $\dfrac{-24n^8 + 48n^5 - 78n^3}{-6n^3}$

34. $\dfrac{-56n^9 + 84n^6 - 91n^2}{-7n^2}$

35. $\dfrac{-60a^7 - 96a^3}{-12a}$

36. $\dfrac{-65a^8 - 78a^4}{-13a^2}$

37. $\dfrac{27x^2y^4 - 45xy^4}{-9xy^3}$

38. $\dfrac{-40x^4y^7 + 64x^5y^8}{-8x^3y^4}$

39. $\dfrac{48a^2b^2 + 60a^3b^4}{-6ab}$

40. $\dfrac{45a^3b^4 - 63a^2b^6}{-9ab^2}$

41. $\dfrac{12a^2b^2c^2 - 52a^2b^3c^5}{-4a^2bc}$

42. $\dfrac{48a^3b^2c + 72a^2b^4c^5}{-12ab^2c}$

43. $\dfrac{9x^2y^3 - 12x^3y^4}{-xy}$

44. $\dfrac{-15x^3y + 27x^2y^4}{xy}$

45. $\dfrac{-42x^6 - 70x^4 + 98x^2}{14x^2}$

46. $\dfrac{-48x^8 - 80x^6 + 96x^4}{16x^4}$

47. $\dfrac{15a^3b - 35a^2b - 65ab^2}{-5ab}$

48. $\dfrac{-24a^4b^2 + 36a^3b - 48a^2b}{-6ab}$

49. $\dfrac{-xy + 5x^2y^3 - 7x^2y^6}{xy}$

50. $\dfrac{-9x^2y^3 - xy + 14xy^4}{-xy}$

Thoughts Into Words

51. How would you explain to someone why the quotient of x^8 and x^2 is x^6 and not x^4?

52. Your friend is having difficulty with problems such as $\dfrac{12x^2y}{xy}$ and $\dfrac{36x^3y^2}{-xy}$ for which there appears to be no numerical coefficient in the denominator. What can you tell him that might help?

6.5   Dividing by Binomials

OBJECTIVE **1** Divide polynomials by binomials

Perhaps the easiest way to explain the process of dividing a polynomial by a binomial is to work a few examples and describe the step-by-step procedure as we go along.

Classroom Example
Divide $x^2 - x - 20$ by $x + 4$.

EXAMPLE 1   Divide $x^2 + 5x + 6$ by $x + 2$.

Solution

Step 1 Use the conventional long division format from arithmetic, and arrange both the dividend and the divisor in descending powers of the variable.

$$x + 2 \overline{)x^2 + 5x + 6}$$

Step 2 Find the first term of the quotient by dividing the first term of the dividend by the first term of the divisor.

$$x + 2 \overline{)x^2 + 5x + 6} \qquad \frac{x^2}{x} = x$$

Step 3 Multiply the entire divisor by the term of the quotient found in step 2, and position this product to be subtracted from the dividend.

$$x + 2 \overline{)x^2 + 5x + 6} \qquad x(x + 2) =$$
$$\underline{}x^2 + 2x \qquad\qquad x^2 + 2x$$

Step 4 Subtract.
Remember to add the opposite!

$$x + 2 \overline{)x^2 + 5x + 6}$$
$$\underline{x^2 + 2x}$$
$$3x + 6$$

Step 5 Repeat the process beginning with step 2; use the polynomial that resulted from the subtraction in step 4 as a new dividend.

$$x + 3$$
$$x + 2 \overline{)x^2 + 5x + 6} \qquad \frac{3x}{x} = 3$$
$$\underline{x^2 + 2x}$$
$$3x + 6 \qquad 3(x + 2) =$$
$$\underline{3x + 6} \qquad\qquad 3x + 6$$

Thus $(x^2 + 5x + 6) \div (x + 2) = x + 3$, which can be checked by multiplying $(x + 2)$ and $(x + 3)$.

$$(x + 2)(x + 3) = x^2 + 5x + 6$$

A division problem such as $(x^2 + 5x + 6) \div (x + 2)$ can also be written as $\dfrac{x^2 + 5x + 6}{x + 2}$.

Using this format, we can express the final result for Example 1 as $\dfrac{x^2 + 5x + 6}{x + 2} = x + 3$.
(Technically, the restriction $x \neq -2$ should be made to avoid division by zero.)

In general, to check a division problem we can multiply the divisor times the quotient and add the remainder, which can be expressed as

Dividend = (Divisor)(Quotient) + Remainder

Sometimes the remainder is expressed as a fractional part of the divisor. The relationship then becomes

$$\frac{\text{Dividend}}{\text{Divisor}} = \text{Quotient} + \frac{\text{Remainder}}{\text{Divisor}}$$

Classroom Example
Divide $3v^2 - 19v - 14$ by $v - 7$.

EXAMPLE 2 Divide $2x^2 - 3x - 20$ by $x - 4$.

Solution

Step 1 $x - 4\overline{)2x^2 - 3x - 20}$

Step 2 $x - 4\overline{)2x^2 - 3x - 20}^{\,2x}$ $\dfrac{2x^2}{x} = 2x$

Step 3 $x - 4\overline{)2x^2 - 3x - 20}^{\,2x}$ $2x(x - 4) = 2x^2 - 8x$
$\quad\quad\quad\underline{2x^2 - 8x}$

Step 4 $x - 4\overline{)2x^2 - 3x - 20}^{\,2x}$
$\quad\quad\quad\underline{2x^2 - 8x}$
$\quad\quad\quad\quad\quad 5x - 20$

Step 5 $x - 4\overline{)2x^2 - 3x - 20}^{\,2x\ +\ 5}$ $\dfrac{5x}{x} = 5$
$\quad\quad\quad\underline{2x^2 - 8x}$
$\quad\quad\quad\quad\quad 5x - 20$ $5(x - 4) = 5x - 20$
$\quad\quad\quad\quad\quad\underline{5x - 20}$

✓ Check

$(x - 4)(2x + 5) = 2x^2 - 3x - 20$

Therefore, $\dfrac{2x^2 - 3x - 20}{x - 4} = 2x + 5$.

Now let's continue to think in terms of the step-by-step division process, but we will organize our work in the typical long division format.

Classroom Example
Divide $10b^2 + 3b - 4$ by $2b - 1$.

EXAMPLE 3 Divide $12x^2 + x - 6$ by $3x - 2$.

Solution

$3x - 2\overline{)12x^2\ +\ x\ -\ 6}^{\,4x\ +\ 3}$
$\quad\quad\quad\underline{12x^2 - 8x}$
$\quad\quad\quad\quad\quad\ 9x - 6$
$\quad\quad\quad\quad\quad\underline{9x - 6}$

✓ Check

$(3x - 2)(4x + 3) = 12x^2 + x - 6$

Therefore, $\dfrac{12x^2 + x - 6}{3x - 2} = 4x + 3$.

Each of the next three examples illustrates another aspect of the division process. Study them carefully; then you should be ready to work the exercises in the next problem set.

Classroom Example
Perform the division
$(3x^2 - 8x - 25) \div (x - 5)$

EXAMPLE 4   Perform the division $(7x^2 - 3x - 4) \div (x - 2)$.

Solution

$$
\begin{array}{r}
7x\ +\ 11 \\
x - 2 \overline{\smash{)}7x^2 -\ \ 3x\ -\ 4} \\
\underline{7x^2 - 14x} \\
11x\ -\ 4 \\
\underline{11x\ -\ 22} \\
18
\end{array}
$$
$\longleftarrow$ A remainder of 18

✓ Check

Just as in arithmetic, we check by *adding* the remainder to the product of the divisor and quotient.

$$(x - 2)(7x + 11) + 18 \overset{?}{=} 7x^2 - 3x - 4$$
$$7x^2 - 3x - 22 + 18 \overset{?}{=} 7x^2 - 3x - 4$$
$$7x^2 - 3x - 4 = 7x^2 - 3x - 4$$

Therefore, $\dfrac{7x^2 - 3x - 4}{x - 2} = 7x + 11 + \dfrac{18}{x - 2}$.

Classroom Example
Perform the division $\dfrac{y^3 + 1}{y + 1}$.

EXAMPLE 5   Perform the division $\dfrac{x^3 - 8}{x - 2}$.

Solution

$$
\begin{array}{r}
x^2 + 2x\ + 4 \\
x - 2 \overline{\smash{)}x^3 + 0x^2 + 0x\ -\ 8} \\
\underline{x^3 - 2x^2} \\
2x^2 + 0x\ -\ 8 \\
\underline{2x^2 - 4x} \\
4x\ -\ 8 \\
\underline{4x\ -\ 8}
\end{array}
$$
$\longleftarrow$ Notice the insertion of x^2 and x terms with zero coefficients

✓ Check

$$(x - 2)(x^2 + 2x + 4) \overset{?}{=} x^3 - 8$$
$$x^3 + 2x^2 + 4x - 2x^2 - 4x - 8 \overset{?}{=} x^3 - 8$$
$$x^3 - 8 = x^3 - 8$$

Therefore, $\dfrac{x^3 - 8}{x - 2} = x^2 + 2x + 4$.

Classroom Example
Perform the division
$\dfrac{z^3 - 6z^2 - 29z + 6}{z^2 + 3z}$

EXAMPLE 6   Perform the division $\dfrac{x^3 + 5x^2 - 3x - 4}{x^2 + 2x}$.

Solution

$$
\begin{array}{r}
x\ +\ 3 \\
x^2 + 2x \overline{\smash{)}x^3 + 5x^2 - 3x - 4} \\
\underline{x^3 + 2x^2} \\
3x^2 - 3x - 4 \\
\underline{3x^2 + 6x} \\
-9x\ -\ 4
\end{array}
$$
$\longleftarrow$ A remainder of $-9x - 4$

We stop the division process when the degree of the remainder is less than the degree of the divisor.

✓ **Check**

$$(x^2 + 2x)(x + 3) + (-9x - 4) \overset{?}{=} x^3 + 5x^2 - 3x - 4$$
$$x^3 + 3x^2 + 2x^2 + 6x - 9x - 4 \overset{?}{=} x^3 + 5x^2 - 3x - 4$$
$$x^3 + 5x^2 - 3x - 4 = x^3 + 5x^2 - 3x - 4$$

Therefore, $\dfrac{x^3 + 5x^2 - 3x - 4}{x^2 + 2x} = x + 3 + \dfrac{-9x - 4}{x^2 + 2x}$.

Concept Quiz 6.5

For Problems 1–10, answer true or false.

1. A division problem written as $(x^2 - x - 6) \div (x - 1)$ could also be written as $\dfrac{x^2 - x - 6}{x - 1}$.

2. The division of $\dfrac{x^2 + 7x + 12}{x + 3} = x + 4$ could be checked by multiplying $(x + 4)$ by $(x + 3)$.

3. For the division problem $(2x^2 + 5x + 9) \div (2x + 1)$ the remainder is 7. The remainder for the division problem can be expressed as $\dfrac{7}{2x + 1}$.

4. In general, to check a division problem we can multiply the divisor times the quotient and subtract the remainder.

5. If a term is inserted to act as a placeholder, then the coefficient of the term must be zero.

6. When performing division, the process ends when the degree of the remainder is less than the degree of the divisor.

7. The remainder is 0 when $x^3 - 1$ is divided by $x - 1$.

8. The remainder is 0 when $x^3 + 1$ is divided by $x + 1$.

9. The remainder is 0 when $x^3 - 1$ is divided by $x + 1$.

10. The remainder is 0 when $x^3 + 1$ is divided by $x - 1$.

Problem Set 6.5

For Problems 1–40, perform the divisions. **(Objective 1)**

1. $(x^2 + 16x + 48) \div (x + 4)$

2. $(x^2 + 15x + 54) \div (x + 6)$

3. $(x^2 - 5x - 14) \div (x - 7)$

4. $(x^2 + 8x - 65) \div (x - 5)$

5. $(x^2 + 11x + 28) \div (x + 3)$

6. $(x^2 + 11x + 15) \div (x + 2)$

7. $(x^2 - 4x - 39) \div (x - 8)$

8. $(x^2 - 9x - 30) \div (x - 12)$

9. $(5n^2 - n - 4) \div (n - 1)$

10. $(7n^2 - 61n - 90) \div (n - 10)$

11. $(8y^2 + 53y - 19) \div (y + 7)$

12. $(6y^2 + 47y - 72) \div (y + 9)$

13. $(20x^2 - 31x - 7) \div (5x + 1)$

14. $(27x^2 + 21x - 20) \div (3x + 4)$

15. $(6x^2 + 25x + 8) \div (2x + 7)$

16. $(12x^2 + 28x + 27) \div (6x + 5)$

17. $(2x^3 - x^2 - 2x - 8) \div (x - 2)$

18. $(3x^3 - 7x^2 - 26x + 24) \div (x - 4)$

19. $(5n^3 + 11n^2 - 15n - 9) \div (n + 3)$

20. $(6n^3 + 29n^2 - 6n - 5) \div (n + 5)$

21. $(n^3 - 40n + 24) \div (n - 6)$

22. $(n^3 - 67n - 24) \div (n + 8)$

23. $(x^3 - 27) \div (x - 3)$

24. $(x^3 + 8) \div (x + 2)$

25. $\dfrac{27x^3 - 64}{3x - 4}$

26. $\dfrac{8x^3 + 27}{2x + 3}$

27. $\dfrac{1 + 3n^2 - 2n}{n + 2}$

28. $\dfrac{x + 5 + 12x^2}{3x - 2}$

29. $\dfrac{9t^2 + 3t + 4}{-1 + 3t}$

30. $\dfrac{4n^2 + 6n - 1}{4 + 2n}$

31. $\dfrac{6n^3 - 5n^2 - 7n + 4}{2n - 1}$

32. $\dfrac{21n^3 + 23n^2 - 9n - 10}{3n + 2}$

33. $\dfrac{4x^3 + 23x^2 - 30x + 32}{x + 7}$

34. $\dfrac{5x^3 - 12x^2 + 13x - 14}{x - 1}$

35. $(x^3 + 2x^2 - 3x - 1) \div (x^2 - 2x)$

36. $(x^3 - 6x^2 - 2x + 1) \div (x^2 + 3x)$

37. $(2x^3 - 4x^2 + x - 5) \div (x^2 + 4x)$

38. $(2x^3 - x^2 - 3x + 5) \div (x^2 + x)$

39. $(x^4 - 16) \div (x + 2)$

40. $(x^4 - 81) \div (x - 3)$

Thoughts Into Words

41. Give a step-by-step description of how you would do the division problem $(2x^3 + 8x^2 - 29x - 30) \div (x + 6)$.

42. How do you know by inspection that the answer to the following division problem is incorrect?

$$(3x^3 - 7x^2 - 22x + 8) \div (x - 4) = 3x^2 + 5x + 1$$

Answers to the Concept Quiz
1. True   **2.** True   **3.** True   **4.** False   **5.** True   **6.** True   **7.** True   **8.** True   **9.** False   **10.** False

---

6.6   Integral Exponents and Scientific Notation

OBJECTIVES

1 Apply the properties of exponents including negative and zero exponents

2 Write numbers in scientific notation

3 Write numbers expressed in scientific notation in standard decimal notation

4 Use scientific notation to evaluate numerical expressions

Thus far in this text we have used only positive integers as exponents. The next definitions and properties serve as a basis for our work with exponents.

Definition 6.1

If n is a positive integer and b is any real number, then

$$b^n = \underbrace{bbb \cdots b}_{n \text{ factors of } b}$$

Property 6.5

If m and n are positive integers, and a and b are real numbers, except $b \neq 0$ whenever it appears in a denominator, then

1. $b^n \cdot b^m = b^{n+m}$

2. $(b^n)^m = b^{mn}$

3. $(ab)^n = a^n b^n$

4. $\left(\dfrac{a}{b}\right)^n = \dfrac{a^n}{b^n}$ Part 4 has not been stated previously

5. $\dfrac{b^n}{b^m} = b^{n-m}$ When $n > m$

     $\dfrac{b^n}{b^m} = 1$ When $n = m$

Property 6.5 pertains to the use of positive integers as exponents. Zero and the negative integers can also be used as exponents. First, let's consider the use of 0 as an exponent. We want to use 0 as an exponent in such a way that the basic properties of exponents will continue to hold. Consider the example $x^4 \cdot x^0$. If part 1 of Property 6.5 is to hold, then

$$x^4 \cdot x^0 = x^{4+0} = x^4$$

Note that x^0 acts like 1 because $x^4 \cdot x^0 = x^4$. This suggests the following definition.

Definition 6.2

If b is a nonzero real number, then

$$b^0 = 1$$

According to Definition 6.2 the following statements are all true.

$$4^0 = 1$$

$$(-628)^0 = 1$$

$$\left(\frac{4}{7}\right)^0 = 1$$

$$n^0 = 1, \qquad n \neq 0$$

$$(x^2 y^5)^0 = 1, \qquad x \neq 0 \text{ and } y \neq 0$$

A similar line of reasoning indicates how negative integers should be used as exponents. Consider the example $x^3 \cdot x^{-3}$. If part 1 of Property 6.5 is to hold, then

$$x^3 \cdot x^{-3} = x^{3+(-3)} = x^0 = 1$$

Thus x^{-3} must be the reciprocal of x^3 because their product is 1; that is,

$$x^{-3} = \frac{1}{x^3}$$

This process suggests the following definition.

> **Definition 6.3**
>
> If n is a positive integer, and b is a nonzero real number, then
>
> $$b^{-n} = \frac{1}{b^n}$$

According to Definition 6.3, the following statements are all true.

$$x^{-6} = \frac{1}{x^6}$$

$$2^{-3} = \frac{1}{2^3} = \frac{1}{8}$$

$$10^{-2} = \frac{1}{10^2} = \frac{1}{100} \quad \text{or} \quad 0.01$$

$$\frac{1}{x^{-4}} = \frac{1}{\dfrac{1}{x^4}} = x^4$$

$$\left(\frac{2}{3}\right)^{-2} = \frac{1}{\left(\dfrac{2}{3}\right)^2} = \frac{1}{\dfrac{4}{9}} = \frac{9}{4}$$

Remark: Note in the last example that $\left(\dfrac{2}{3}\right)^{-2} = \left(\dfrac{3}{2}\right)^{2}$. In other words, to raise a fraction to a negative power, we can invert the fraction and raise it to the corresponding positive power.

We can verify (we will not do so in this text) that all parts of Property 6.5 hold for *all integers*. In fact, we can replace part 5 with this statement.

> **Replacement for part 5 of Property 6.5**
>
> $$\frac{b^n}{b^m} = b^{n-m} \quad \text{for all integers } n \text{ and } m$$

The next examples illustrate the use of this new concept. In each example, we simplify the original expression and use only positive exponents in the final result.

$$\frac{x^2}{x^5} = x^{2-5} = x^{-3} = \frac{1}{x^3}$$

$$\frac{a^{-3}}{a^{-7}} = a^{-3-(-7)} = a^{-3+7} = a^4$$

$$\frac{y^{-5}}{y^{-2}} = y^{-5-(-2)} = y^{-5+2} = y^{-3} = \frac{1}{y^3}$$

$$\frac{x^{-6}}{x^{-6}} = x^{-6-(-6)} = x^{-6+6} = x^0 = 1$$

The properties of exponents provide a basis for simplifying certain types of numerical expressions, as the following examples illustrate.

$$2^{-4} \cdot 2^6 = 2^{-4+6} = 2^2 = 4$$

$$10^5 \cdot 10^{-6} = 10^{5+(-6)} = 10^{-1} = \frac{1}{10} \qquad \text{or} \qquad 0.1$$

$$\frac{10^2}{10^{-2}} = 10^{2-(-2)} = 10^{2+2} = 10^4 = 10,000$$

$$(2^{-3})^{-2} = 2^{-3(-2)} = 2^6 = 64$$

Having the use of all integers as exponents also expands the type of work that we can do with algebraic expressions. In each of the following examples we simplify a given expression and use only positive exponents in the final result.

$$x^8 x^{-2} = x^{8+(-2)} = x^6$$

$$a^{-4}a^{-3} = a^{-4+(-3)} = a^{-7} = \frac{1}{a^7}$$

$$(y^{-3})^4 = y^{-3(4)} = y^{-12} = \frac{1}{y^{12}}$$

$$(x^{-2}y^4)^{-3} = (x^{-2})^{-3}(y^4)^{-3} = x^6 y^{-12} = \frac{x^6}{y^{12}}$$

$$\left(\frac{x^{-1}}{y^2}\right)^{-2} = \frac{(x^{-1})^{-2}}{(y^2)^{-2}} = \frac{x^2}{y^{-4}} = x^2 y^4$$

$$(4x^{-2})(3x^{-1}) = 12x^{-2+(-1)} = 12x^{-3} = \frac{12}{x^3}$$

$$\left(\frac{12x^{-6}}{6x^{-2}}\right)^{-2} = (2x^{-6-(-2)})^{-2} = (2x^{-4})^{-2} \qquad \text{\color{blue}{Divide the coefficients}} \; \frac{12}{6} = 2$$

$$= (2)^{-2}(x^{-4})^{-2}$$

$$= \left(\frac{1}{2^2}\right)(x^8) = \frac{x^8}{4}$$

Scientific Notation

Many scientific applications of mathematics involve the use of very large and very small positive numbers. For example:

The speed of light is approximately 29,979,200,000 centimeters per second.

A light year (the distance light travels in 1 year) is approximately 5,865,696,000,000 miles.

A gigahertz equals 1,000,000,000 hertz.

The length of a typical virus cell equals 0.000000075 of a meter.

The length of a diameter of a water molecule is 0.0000000003 of a meter.

Working with numbers of this type in standard form is quite cumbersome. It is much more convenient to represent very small and very large numbers in **scientific notation**, sometimes called scientific form. A number is in scientific notation when it is written as the product of a number between 1 and 10 (including 1) and an integral power of 10. Symbolically, a number in scientific notation has the form $(N)(10^k)$, where $1 \le N < 10$, and k is an integer. For example, 621 can be written as $(6.21)(10^2)$, and 0.0023 can be written as $(2.3)(10^{-3})$.

To switch from ordinary notation to scientific notation, you can use the following procedure.

> Write the given number as the product of a number greater than or equal to 1 and less than 10, and an integral power of 10. To determine the exponent of 10, count the number of places that the decimal point moved when going from the original number to the number greater than or equal to 1 and less than 10. This exponent is (a) negative if the original number is less than 1, (b) positive if the original number is greater than 10, and (c) zero if the original number itself is between 1 and 10.

Thus we can write

$$0.000179 = (1.79)(10^{-4}) \qquad \text{According to part (a) of the procedure}$$
$$8175 = (8.175)(10^{3}) \qquad \text{According to part (b)}$$
$$3.14 = (3.14)(10^{0}) \qquad \text{According to part (c)}$$

We can express the applications given earlier in scientific notation as follows:

Speed of light: $29{,}979{,}200{,}000 = (2.99792)(10^{10})$ centimeters per second

Light year: $5{,}865{,}696{,}000{,}000 = (5.865696)(10^{12})$ miles

Gigahertz: $1{,}000{,}000{,}000 = (1)(10^{9})$ hertz

Length of a virus cell: $0.000000075 = (7.5)(10^{-8})$ meter

Length of the diameter of a water molecule $= 0.0000000003$
$$= (3)(10^{-10}) \text{ meter}$$

To switch from scientific notation to ordinary decimal notation you can use the following procedure.

> Move the decimal point the number of places indicated by the exponent of 10. The decimal point is moved to the right if the exponent is positive and to the left if it is negative.

Thus we can write

$$(4.71)(10^{4}) = 47{,}100 \qquad \text{Two zeros are needed for place value purposes}$$
$$(1.78)(10^{-2}) = 0.0178 \qquad \text{One zero is needed for place value purposes}$$

The use of scientific notation along with the properties of exponents can make some arithmetic problems much easier to evaluate. The next examples illustrate this point.

Classroom Example
Evaluate $(6000)(0.00072)$.

EXAMPLE 1   Evaluate $(4000)(0.000012)$.

Solution

$$
\begin{aligned}
(4000)(0.000012) &= (4)(10^{3})(1.2)(10^{-5}) \\
&= (4)(1.2)(10^{3})(10^{-5}) \\
&= (4.8)(10^{-2}) \\
&= 0.048
\end{aligned}
$$

Classroom Example
Evaluate $\dfrac{840{,}000}{0.024}$.

EXAMPLE 2   Evaluate $\dfrac{960{,}000}{0.032}$.

Solution

$$
\begin{aligned}
\frac{960{,}000}{0.032} &= \frac{(9.6)(10^{5})}{(3.2)(10^{-2})} \\
&= (3)(10^{7}) \qquad \frac{10^{5}}{10^{-2}} = 10^{5-(-2)} = 10^{7} \\
&= 30{,}000{,}000
\end{aligned}
$$

Classroom Example

Evaluate $\dfrac{(7000)(0.0000009)}{(0.0012)(30,000)}$.

EXAMPLE 3   Evaluate $\dfrac{(6000)(0.00008)}{(40,000)(0.006)}$.

Solution

$$\frac{(6000)(0.00008)}{(40,000)(0.006)} = \frac{(6)(10^3)(8)(10^{-5})}{(4)(10^4)(6)(10^{-3})}$$

$$= \frac{(48)(10^{-2})}{(24)(10^1)}$$

$$= (2)(10^{-3}) \qquad \frac{10^{-2}}{10^1} = 10^{-2-1} = 10^{-3}$$

$$= 0.002$$

Concept Quiz 6.6

For Problems 1–10, answer true or false.

1. Any nonzero number raised to the zero power is equal to one.

2. The algebraic expression x^{-2} is the reciprocal of x^2 for $x \neq 0$.

3. To raise a fraction to a negative exponent, we can invert the fraction and raise it to the corresponding positive exponent.

4. $\dfrac{1}{y^{-3}} = y^{-3}$

5. A number in scientific notation has the form $(N)(10^k)$ where $1 \leq N < 10$, and k is any real number.

6. A number is less than zero if the exponent is negative when the number is written in scientific notation.

7. $\dfrac{1}{x^{-2}} = x^2$

8. $\dfrac{10^{-2}}{10^{-4}} = 100$

9. $(3.11)(10^{-2}) = 311$

10. $(5.24)(10^{-1}) = 0.524$

Problem Set 6.6

For Problems 1–30, evaluate each numerical expression.
(Objective 1)

1. 3^{-2} 2. 2^{-5} 3. 4^{-3} 4. 5^{-2}

5. $\left(\dfrac{3}{2}\right)^{-1}$ 6. $\left(\dfrac{3}{4}\right)^{-2}$

7. $\dfrac{1}{2^{-4}}$ 8. $\dfrac{1}{3^{-1}}$

9. $\left(-\dfrac{4}{3}\right)^0$ 10. $\left(-\dfrac{1}{2}\right)^{-3}$

11. $\left(-\dfrac{2}{3}\right)^{-3}$ 12. $(-16)^0$

13. $(-2)^{-2}$ 14. $(-3)^{-2}$

15. $-(3^{-2})$

16. $-(2^{-2})$

17. $\dfrac{1}{\left(\dfrac{3}{4}\right)^{-3}}$

18. $\dfrac{1}{\left(\dfrac{3}{2}\right)^{-4}}$

19. $2^6 \cdot 2^{-9}$

20. $3^5 \cdot 3^{-2}$

21. $3^6 \cdot 3^{-3}$

22. $2^{-7} \cdot 2^2$

23. $\dfrac{10^2}{10^{-1}}$

24. $\dfrac{10^1}{10^{-3}}$

25. $\dfrac{10^{-1}}{10^2}$

26. $\dfrac{10^{-2}}{10^{-2}}$

27. $(2^{-1} \cdot 3^{-2})^{-1}$

28. $(3^{-1} \cdot 4^{-2})^{-1}$

29. $\left(\dfrac{4^{-1}}{3}\right)^{-2}$

30. $\left(\dfrac{3}{2^{-1}}\right)^{-3}$

81. $\left(\dfrac{2x^{-1}}{x^{-2}}\right)^{-3}$

82. $\left(\dfrac{3x^{-2}}{x^{-5}}\right)^{-1}$

83. $\left(\dfrac{18x^{-1}}{9x}\right)^{-2}$

84. $\left(\dfrac{35x^2}{7x^{-1}}\right)^{-1}$

For Problems 31–84, simplify each algebraic expression and express your answers using positive exponents only. **(Objective 1)**

31. $x^6 x^{-1}$

32. $x^{-2} x^7$

33. $n^{-4} n^2$

34. $n^{-8} n^3$

35. $a^{-2} a^{-3}$

36. $a^{-4} a^{-6}$

37. $(2x^3)(4x^{-2})$

38. $(5x^{-4})(6x^7)$

39. $(3x^{-6})(9x^2)$

40. $(8x^{-8})(4x^2)$

41. $(5y^{-1})(-3y^{-2})$

42. $(-7y^{-3})(9y^{-4})$

43. $(8x^{-4})(12x^4)$

44. $(-3x^{-2})(-6x^2)$

45. $\dfrac{x^7}{x^{-3}}$

46. $\dfrac{x^2}{x^{-4}}$

47. $\dfrac{n^{-1}}{n^3}$

48. $\dfrac{n^{-2}}{n^5}$

49. $\dfrac{4n^{-1}}{2n^{-3}}$

50. $\dfrac{12n^{-2}}{3n^{-5}}$

51. $\dfrac{-24x^{-6}}{8x^{-2}}$

52. $\dfrac{56x^{-5}}{-7x^{-1}}$

53. $\dfrac{-52y^{-2}}{-13y^{-2}}$

54. $\dfrac{-91y^{-3}}{-7y^{-3}}$

55. $(x^{-3})^{-2}$

56. $(x^{-1})^{-5}$

57. $(x^2)^{-2}$

58. $(x^3)^{-1}$

59. $(x^3 y^4)^{-1}$

60. $(x^4 y^{-2})^{-2}$

61. $(x^{-2} y^{-1})^3$

62. $(x^{-3} y^{-4})^2$

63. $(2n^{-2})^3$

64. $(3n^{-1})^4$

65. $(4n^3)^{-2}$

66. $(2n^2)^{-3}$

67. $(3a^{-2})^4$

68. $(5a^{-1})^2$

69. $(5x^{-1})^{-2}$

70. $(4x^{-2})^{-2}$

71. $(2x^{-2} y^{-1})^{-1}$

72. $(3x^2 y^{-3})^{-2}$

73. $\left(\dfrac{x^2}{y}\right)^{-1}$

74. $\left(\dfrac{y^2}{x^3}\right)^{-2}$

75. $\left(\dfrac{a^{-1}}{b^2}\right)^{-4}$

76. $\left(\dfrac{a^3}{b^{-2}}\right)^{-3}$

77. $\left(\dfrac{x^{-1}}{y^{-3}}\right)^{-2}$

78. $\left(\dfrac{x^{-3}}{y^{-4}}\right)^{-1}$

79. $\left(\dfrac{x^2}{x^3}\right)^{-1}$

80. $\left(\dfrac{x^4}{x}\right)^{-2}$

85. The U.S. Social Security Administration pays approximately \$492,000,000,000 in monthly benefits to all beneficiaries. Write this number in scientific notation.

86. In 2008 approximately 5,419,200,000 pennies were made. Write, in scientific notation, the number of pennies made.

87. The thickness of a dollar bill is 0.0043 inch. Write this number in scientific notation.

88. The average length of an Ebola virus cell is approximately 0.0000002 meter. Write this number in scientific notation.

89. The diameter of Jupiter is approximately 89,000 miles. Write this number in scientific notation.

90. Avogadro's Number, which has a value of approximately 602,200,000,000,000,000,000,000, refers to the calculated value of the number of molecules in a gram mole of any chemical substance. Write Avogadro's Number in scientific notation.

91. The cooling fan for an old computer hard drive has a thickness of 0.025 meter. Write this number in scientific notation.

92. A sheet of 20-weight bond paper has a thickness of 0.097 millimeter. Write the thickness in scientific notation.

93. According to the 2004 annual report for Coca-Cola, the Web site CokePLAY.com was launched in Korea, and it was visited by 11,000,000 people in its first six months. Write the number of hits for the Web site in scientific notation.

94. In 2005, Wal-Mart reported that they served 138,000,000 customers worldwide each week. Write, in scientific notation, the number of customers for each week.

For Problems 95–106, write each number in standard decimal form; for example, $(1.4)(10^3) = 1400$. **(Objective 3)**

95. $(8)(10^3)$

96. $(6)(10^2)$

97. $(5.21)(10^4)$

98. $(7.2)(10^3)$

99. $(1.14)(10^7)$

100. $(5.64)(10^8)$

101. $(7)(10^{-2})$

102. $(8.14)(10^{-1})$

103. $(9.87)(10^{-4})$

104. $(4.37)(10^{-5})$

105. $(8.64)(10^{-6})$

106. $(3.14)(10^{-7})$

For Problems 107–118, use scientific notation and the properties of exponents to evaluate each numerical expression. **(Objective 4)**

107. $(0.007)(120)$

108. $(0.0004)(13)$

109. $(5,000,000)(0.00009)$

110. $(800,000)(0.0000006)$

111. $\dfrac{6000}{0.0015}$

112. $\dfrac{480}{0.012}$

113. $\dfrac{0.00086}{4300}$

114. $\dfrac{0.0057}{30,000}$

115. $\dfrac{0.00039}{0.0013}$

116. $\dfrac{0.0000082}{0.00041}$

117. $\dfrac{(0.0008)(0.07)}{(20,000)(0.0004)}$

118. $\dfrac{(0.006)(600)}{(0.00004)(30)}$

For Problems 119–122, convert the numbers to scientific notation and compute the answer. **(Objectives 2 and 4)**

119. The U.S. Social Security Administration pays approximately $42,000,000,000 in benefits to all beneficiaries. If there are 48,000,000 beneficiaries, find the average dollar amount each beneficiary receives.

120. In 1998 approximately 10,200,000,000 pennies were made. If the population was 270,000,000, find the average number of pennies produced per person. Round the answer to the nearest whole number.

121. The thickness of a dollar bill is 0.0043 inch. How tall will a stack of 1,000,000, dollars be? Express the answer to the nearest foot.

122. The diameter of Jupiter is approximately 11 times larger than the diameter of Earth. Find the diameter of Earth given that Jupiter has a diameter of approximately 89,000 miles. Express the answer to the nearest mile.

Thoughts Into Words

123. Is the following simplification process correct?

$$(2^{-2})^{-1} = \left(\frac{1}{2^2}\right)^{-1} = \left(\frac{1}{4}\right)^{-1} = \frac{1}{\left(\frac{1}{4}\right)^1} = 4$$

Can you suggest a better way to do the problem?

124. Explain the importance of scientific notation.

Further Investigations

125. Use your calculator to do Problems 1–16. Be sure that your answers are equivalent to the answers you obtained without the calculator.

126. Use your calculator to evaluate $(140,000)^2$. Your answer should be displayed in scientific notation; the format of the display depends on the particular calculator. For example, it may look like $\boxed{1.96\ \ 10}$ or $\boxed{1.96\text{E} + 10}$. Thus in ordinary notation the answer is 19,600,000,000. Use your calculator to evaluate each expression. Express final answers in ordinary notation.

(a) $(9000)^3$

(b) $(4000)^3$

(c) $(150,000)^2$

(d) $(170,000)^2$

(e) $(0.012)^5$

(f) $(0.0015)^4$

(g) $(0.006)^3$

(h) $(0.02)^6$

127. Use your calculator to check your answers to Problems 107–118.

Chapter 6 Summary

OBJECTIVE	SUMMARY	EXAMPLE
Classify polynomials by size and degree. (Section 6.1/Objective 1) (Section 6.1/Objective 2)	Terms with variables that contain only whole number exponents are called **monomials**. A **polynomial** is a monomial or a finite sum of monomials. The degree of a monomial is the sum of the exponents of the literal factors. The degree of a polynomial is the degree of the term with the highest degree in the polynomial. A one-term polynomial is called a **monomial**, a two-term polynomial is called a **binomial**, and a three-term polynomial is called a **trinomial**.	The polynomial $6x^2y^3$ is a monomial of degree 5. The polynomial $3x^2y^2 - 7xy^2$ is a binomial of degree 4. The polynomial $8x^2 + 3x - 1$ is a trinomial of degree 2. The polynomial $x^4 + 6x^3 - 5x^2 + 12x - 2$ is a polynomial of degree 4.
Add and subtract polynomials. (Section 6.1/Objective 3) (Section 6.1/Objective 4)	Addition and subtraction of polynomials is based on using the distributive property and combining similar terms. A polynomial is subtracted by *adding the opposite*. The opposite of a polynomial is formed by taking the opposite of each term.	(a) Add $3x^2 - 2x + 1$ and $4x^2 + x - 5$. (b) Perform the subtraction $\quad (4x - 3) - (2x + 7)$ **Solution** (a) $(3x^2 - 2x + 1) + (4x^2 + x - 5)$ $\quad = (3x^2 + 4x^2) + (-2x + x) + (1 - 5)$ $\quad = (3 + 4)x^2 + (-2 + 1)x + (1 - 5)$ $\quad = 7x^2 - x - 4$ (b) $(4x - 3) - (2x + 7)$ $\quad = (4x - 3) + (-2x - 7)$ $\quad = (4x - 2x) + (-3 - 7)$ $\quad = 2x - 10$
Perform operations on polynomials involving both addition and subtraction. (Section 6.1/Objective 3) (Section 6.1/Objective 4)	We can use the distributive property along with the properties $a = 1(a)$ and $-a = -1(a)$ when adding and subtracting polynomials.	Perform the indicated operations: $3y + 4 - (6y - 5)$ **Solution** $3y + 4 - (6y - 5)$ $= 3y + 4 - 1(6y - 5)$ $= 3y + 4 - 1(6y) - 1(-5)$ $= 3y + 4 - 6y + 5 = -3y + 9$
Apply properties of exponents to multiply monomials. (Section 6.2/Objective 1)	When multiplying powers with the same base, add the exponents. When raising a power to a power, multiply the exponents.	Find the products: (a) $(3x^2)(4x^3)$ (b) $(n^2)^4$ **Solution** (a) $(3x^2)(4x^3)$ $\quad = 3 \cdot 4 \cdot x^{2+3}$ $\quad = 12x^5$ (b) $(n^2)^4$ $\quad = n^{2 \cdot 4}$ $\quad = n^8$
Apply properties of exponents to raise a monomial to a power. (Section 6.2/Objective 1)	When raising a monomial to a power, raise each factor to that power.	Simplify $(6ab^2)^3$. **Solution** $(6ab^2)^3$ $= 6^3 \cdot a^3 \cdot b^6$ $= 216a^3b^6$

OBJECTIVE	SUMMARY	EXAMPLE
Find the product of two polynomials. (Section 6.3/Objective 1)	The distributive property along with the properties of exponents form a basis for multiplying polynomials.	Multiply $(2x + 3)(x^2 - 4x + 5)$. **Solution** $(2x + 3)(x^2 - 4x + 5)$ $= 2x(x^2 - 4x + 5) + 3(x^2 - 4x + 5)$ $= 2x^3 - 8x^2 + 10x + 3x^2 - 12x + 15$ $= 2x^3 - 5x^2 - 2x + 15$
Use the shortcut pattern to find the product of two binomials. (Section 6.3/Objective 2)	The product of two binomials is perhaps the most frequently used type of multiplication problem. A three-step shortcut pattern, called FOIL, is often used to find the product of two binomials. FOIL stands for first, outside, inside, and last.	Multiply $(2x + 5)(x - 3)$. **Solution** $(2x + 5)(x - 3)$ $= 2x(x) + (2x)(-3) + (5)(x) + (5)(-3)$ $= 2x^2 - 6x + 5x - 15$ $= 2x^2 - x - 15$
Raise a binomial to a power. (Section 6.3/Objective 1)	When raising a binomial to a power, rewrite the problem as a product of binomials. Then you can apply the FOIL shortcut to multiply two binomials. Then apply the distributive property to find the product.	Find the product $(x + 4)^3$. **Solution** $(x + 4)^3 = (x + 4)(x + 4)(x + 4)$ $\quad = (x + 4)(x^2 + 8x + 16)$ $\quad = x(x^2 + 8x + 16) + 4(x^2 + 8x + 16)$ $\quad = x^3 + 8x^2 + 16x + 4x^2 + 32x + 64$ $\quad = x^3 + 12x^2 + 48x + 64$
Use a special product pattern to find products. (Section 6.3/Objective 3)	When multiplying binomials you should be able to recognize the following special patterns and use them to find the product. $(a + b)^2 = a^2 + 2ab + b^2$ $(a - b)^2 = a^2 - 2ab + b^2$ $(a + b)(a - b) = a^2 - b^2$	Find the product $(5x + 6)^2$. **Solution** $(5x + 6)^2 = (5x)^2 + 2(5x)(6) + (6)^2$ $\quad = 25x^2 + 60x + 36$
Apply polynomials to geometric problems. (Section 6.2/Objective 3)	Polynomials can be applied to represent perimeters, areas, and volumes of geometric figures.	A piece of aluminum that is 15 inches by 20 inches has a square piece x inches on a side cut out from two corners. Find the area of the aluminum piece after the corners are removed. **Solution** The area before the corners are removed is found by applying the formula $A = lw$. Therefore $A = 15(20) = 300$. Each of the two corners removed has an area of x^2. Therefore $2x^2$ must be subtracted from the area of the aluminum piece. The area of the aluminum piece after the corners are removed is $300 - 2x^2$.
Apply properties of exponents to divide monomials. (Section 6.4/Objective 1)	The following properties of exponents, along with our knowledge of dividing integers, serves as a basis for dividing monomials. $\dfrac{b^n}{b^m} = b^{n-m}$ when $n > m$ and $b \neq 0$ $\dfrac{b^n}{b^m} = 1$ when $n = m$ and $b \neq 0$	Divide $\dfrac{-32\,a^5 b^4}{8a^3 b}$. **Solution** $\dfrac{-32\,a^5 b^4}{8a^3 b} = -4a^2 b^3$

(continued)

OBJECTIVE	SUMMARY	EXAMPLE
Divide polynomials by monomials. (Section 6.4/Objective 2)	Dividing a polynomial by a monomial is based on the property $$\dfrac{a + b}{c} = \dfrac{a}{c} + \dfrac{b}{c}$$	Divide $\dfrac{8x^5 - 16x^4 + 4x^3}{4x}$. **Solution** $$\dfrac{8x^5 - 16x^4 + 4x^3}{4x} = \dfrac{8x^5}{4x} + \dfrac{-16x^4}{4x} + \dfrac{4x^3}{4x}$$ $$= 2x^4 - 4x^3 + x^2$$
Divide polynomials by binomials. (Section 6.5/Objective 1)	Use the conventional long division format from arithmetic. Arrange both the dividend and the divisor in descending powers of the variable. You may have to insert terms with zero coefficients if terms with some powers of the variable are missing.	Divide $(y^3 - 5y - 2) \div (y + 2)$. **Solution** $$\begin{array}{r} y^2 - 2y - 1 \\ y+2 \overline{)y^3 + 0y^2 - 5y - 2} \\ \underline{y^3 + 2y^2} \\ -2y^2 - 5y \\ \underline{-2y^2 - 4y} \\ -y - 2 \\ \underline{-y - 2} \end{array}$$
Apply the properties of exponents, including negative and zero exponents, to simplify expressions. (Section 6.6/Objective 1)	By definition, if b is a nonzero real number, then $b^0 = 1$. By definition, if n is a positive integer, and b is a nonzero real number, then $b^{-n} = \dfrac{1}{b^n}$. When simplifying expressions use only positive exponents in the final result.	Simplify $\dfrac{28n^{-6}}{4n^{-2}}$. Express the answer using positive exponents only. **Solution** $$\dfrac{28n^{-6}}{4n^{-2}} = 7n^{-6-(-2)} = 7n^{-6+2} = 7n^{-4} = \dfrac{7}{n^4}$$
Write numbers expressed in scientific notation in standard decimal notation. (Section 6.6/Objective 3)	To change from scientific notation to ordinary decimal notation, move the decimal point the number of places indicated by the exponent of 10. The decimal point is moved to the right if the exponent is positive and to the left if it is negative.	Write $(3.28)(10^{-4})$ in standard decimal notation. **Solution** $(3.28)(10^{-4}) = 0.000328$
Write numbers in scientific notation. (Section 6.6/Objective 2)	To represent a number in scientific notation, express it as a product of a number between 1 and 10 (including 1) and an integral power of ten.	Write 657,000,000 in scientific notation. **Solution** Count the number of decimal places from the existing decimal point until you have a number between 1 and 10. For the scientific notation raise 10 to that number: $657{,}000{,}000 = (6.57)(10^8)$
Use scientific notation to evaluate numerical expressions. (Section 6.6/Objective 4)	Scientific notation can be used to make arithmetic problems easier to evaluate.	Change each number to scientific notation and evaluate the expression. Express the answer in standard notation: $(61{,}000)(0.000005)$ **Solution** $(61{,}000)(0.000005) = (6.1)(10^4)(5)(10^{-6})$ $\phantom{(61{,}000)(0.000005)} = (6.1)(5)(10^4)(10^{-6})$ $\phantom{(61{,}000)(0.000005)} = (30.5)(10^{-2})$ $\phantom{(61{,}000)(0.000005)} = 0.305$

Chapter 6   Review Problem Set

For Problems 1–4, perform the additions and subtractions.

1. $(5x^2 - 6x + 4) + (3x^2 - 7x - 2)$

2. $(7y^2 + 9y - 3) - (4y^2 - 2y + 6)$

3. $(2x^2 + 3x - 4) + (4x^2 - 3x - 6) - (3x^2 - 2x - 1)$

4. $(-3x^2 - 2x + 4) - (x^2 - 5x - 6) - (4x^2 + 3x - 8)$

For Problems 5–12, remove parentheses and combine similar terms.

5. $5(2x - 1) + 7(x + 3) - 2(3x + 4)$

6. $3(2x^2 - 4x - 5) - 5(3x^2 - 4x + 1)$

7. $6(y^2 - 7y - 3) - 4(y^2 + 3y - 9)$

8. $3(a - 1) - 2(3a - 4) - 5(2a + 7)$

9. $-(a + 4) + 5(-a - 2) - 7(3a - 1)$

10. $-2(3n - 1) - 4(2n + 6) + 5(3n + 4)$

11. $3(n^2 - 2n - 4) - 4(2n^2 - n - 3)$

12. $-5(-n^2 + n - 1) + 3(4n^2 - 3n - 7)$

For Problems 13–20, find the indicated products.

13. $(5x^2)(7x^4)$

14. $(-6x^3)(9x^5)$

15. $(-4xy^2)(-6x^2y^3)$

16. $(2a^3b^4)(-3ab^5)$

17. $(2a^2b^3)^3$

18. $(-3xy^2)^2$

19. $5x(7x + 3)$

20. $(-3x^2)(8x - 1)$

For Problems 21–40, find the indicated products. Be sure to simplify your answers.

21. $(x + 9)(x + 8)$

22. $(3x + 7)(x + 1)$

23. $(x - 5)(x + 2)$

24. $(y - 4)(y - 9)$

25. $(2x - 1)(7x + 3)$

26. $(4a - 7)(5a + 8)$

27. $(3a - 5)^2$

28. $(x + 6)(2x^2 + 5x - 4)$

29. $(5n - 1)(6n + 5)$

30. $(3n + 4)(4n - 1)$

31. $(2n + 1)(2n - 1)$

32. $(4n - 5)(4n + 5)$

33. $(2a + 7)^2$

34. $(3a + 5)^2$

35. $(x - 2)(x^2 - x + 6)$

36. $(2x - 1)(x^2 + 4x + 7)$

37. $(a + 5)^3$

38. $(a - 6)^3$

39. $(x^2 - x - 1)(x^2 + 2x + 5)$

40. $(n^2 + 2n + 4)(n^2 - 7n - 1)$

For Problems 41–48, perform the divisions.

41. $\dfrac{36x^4y^5}{-3xy^2}$

42. $\dfrac{-56a^5b^7}{-8a^2b^3}$

43. $\dfrac{-18x^4y^3 - 54x^6y^2}{6x^2y^2}$

44. $\dfrac{-30a^5b^{10} + 39a^4b^8}{-3ab}$

45. $\dfrac{56x^4 - 40x^3 - 32x^2}{4x^2}$

46. $(x^2 + 9x - 1) \div (x + 5)$

47. $(21x^2 - 4x - 12) \div (3x + 2)$

48. $(2x^3 - 3x^2 + 2x - 4) \div (x - 2)$

For Problems 49–60, evaluate each expression.

49. $3^2 + 2^2$

50. $(3 + 2)^2$

51. 2^{-4}

52. $(-5)^0$

53. -5^0

54. $\dfrac{1}{3^{-2}}$

55. $\left(\dfrac{3}{4}\right)^{-2}$

56. $\dfrac{1}{\left(\dfrac{1}{4}\right)^{-1}}$

57. $\dfrac{1}{(-2)^{-3}}$

58. $2^{-1} + 3^{-2}$

59. $3^0 + 2^{-2}$

60. $(2 + 3)^{-2}$

For Problems 61–72, simplify each of the following, and express your answers using positive exponents only.

61. x^5x^{-8}

62. $(3x^5)(4x^{-2})$

63. $\dfrac{x^{-4}}{x^{-6}}$

64. $\dfrac{x^{-6}}{x^{-4}}$

65. $\dfrac{24a^5}{3a^{-1}}$

66. $\dfrac{48n^{-2}}{12n^{-1}}$

67. $(x^{-2}y)^{-1}$

68. $(a^2b^{-3})^{-2}$

69. $(2x)^{-1}$

70. $(3n^2)^{-2}$

71. $(2n^{-1})^{-3}$

72. $(4ab^{-1})(-3a^{-1}b^2)$

For Problems 73–76, write each expression in standard decimal form.

73. $(6.1)(10^2)$

74. $(5.6)(10^4)$

75. $(8)(10^{-2})$

76. $(9.2)(10^{-4})$

For Problems 77–80, write each number in scientific notation.

77. 9000

78. 47

79. 0.047

80. 0.00021

For Problems 81–84, use scientific notation and the properties of exponents to evaluate each expression.

81. $(0.00004)(12,000)$

82. $(0.0021)(2000)$

83. $\dfrac{0.0056}{0.0000028}$

84. $\dfrac{0.00078}{39,000}$

1. Find the sum of $-7x^2 + 6x - 2$ and $5x^2 - 8x + 7$.

2. Subtract $-x^2 + 9x - 14$ from $-4x^2 + 3x + 6$.

3. Remove parentheses and combine similar terms for the expression $3(2x - 1) - 6(3x - 2) - (x + 7)$.

4. Find the product $(-4xy^2)(7x^2y^3)$.

5. Find the product $(2x^2y)^2(3xy^3)$.

For Problems 6–12, find the indicated products and express answers in simplest form.

6. $(x - 9)(x + 2)$

7. $(n + 14)(n - 7)$

8. $(5a + 3)(8a + 7)$

9. $(3x - 7y)^2$

10. $(x + 3)(2x^2 - 4x - 7)$

11. $(9x - 5y)(9x + 5y)$

12. $(3x - 7)(5x - 11)$

13. Find the indicated quotient: $\dfrac{-96x^4y^5}{-12x^2y}$.

14. Find the indicated quotient: $\dfrac{56x^2y - 72xy^2}{-8xy}$.

15. Find the indicated quotient:

$(2x^3 + 5x^2 - 22x + 15) \div (2x - 3)$

16. Find the indicated quotient:

$(4x^3 + 23x^2 + 36) \div (x + 6)$

17. Evaluate $\left(\dfrac{2}{3}\right)^{-3}$.

18. Evaluate $4^{-2} + 4^{-1} + 4^0$.

19. Evaluate $\dfrac{1}{2^{-4}}$.

20. Find the product $(-6x^{-4})(4x^2)$, and express the answer using a positive exponent.

21. Simplify $\left(\dfrac{8x^{-1}}{2x^2}\right)^{-1}$, and express the answer using positive exponents.

22. Simplify $(x^{-3}y^5)^{-2}$, and express the answer using positive exponents.

23. Write 0.00027 in scientific notation.

24. Express $(9.2)(10^6)$ in standard decimal form.

25. Evaluate $(0.000002)(3000)$.

For Problems 1–10, evaluate each of the numerical expressions.

1. $5 + 3(2 - 7)^2 \div 3 \cdot 5$

2. $8 \div 2 \cdot (-1) + 3$

3. $7 - 2^2 \cdot 5 \div (-1)$

4. $4 + (-2) - 3(6)$

5. $(-3)^4$

6. -2^5

7. $\left(\dfrac{2}{3}\right)^{-1}$

8. $\dfrac{1}{4^{-2}}$

9. $\left(\dfrac{1}{2} - \dfrac{1}{3}\right)^{-2}$

10. $2^0 + 2^{-1} + 2^{-2}$

For Problems 11–16, evaluate each algebraic expression for the given values of the variables.

11. $\dfrac{2x + 3y}{x - y}$ for $x = \dfrac{1}{2}$ and $y = -\dfrac{1}{3}$

12. $\dfrac{2}{5}n - \dfrac{1}{3}n - n + \dfrac{1}{2}n$ for $n = -\dfrac{3}{4}$

13. $\dfrac{3a - 2b - 4a + 7b}{-a - 3a + b - 2b}$ for $a = -1$ and $b = -\dfrac{1}{3}$

14. $-2(x - 4) + 3(2x - 1) - (3x - 2)$ for $x = -2$

15. $(x^2 + 2x - 4) - (x^2 - x - 2) + (2x^2 - 3x - 1)$ for $x = -1$

16. $2(n^2 - 3n - 1) - (n^2 + n + 4) - 3(2n - 1)$ for $n = 3$

For Problems 17–29, find the indicated products.

17. $(3x^2y^3)(-5xy^4)$

18. $(-6ab^4)(-2b^3)$

19. $(-2x^2y^5)^3$

20. $-3xy(2x - 5y)$

21. $(5x - 2)(3x - 1)$

22. $(7x - 1)(3x + 4)$

23. $(-x - 2)(2x + 3)$

24. $(7 - 2y)(7 + 2y)$

25. $(x - 2)(3x^2 - x - 4)$

26. $(2x - 5)(x^2 + x - 4)$

27. $(2n + 3)^3$

28. $(1 - 2n)^3$

29. $(x^2 - 2x + 6)(2x^2 + 5x - 6)$

For Problems 30–34, perform the indicated divisions.

30. $\dfrac{-52x^3y^4}{13xy^2}$

31. $\dfrac{-126a^3b^5}{-9a^2b^3}$

32. $\dfrac{56xy^2 - 64x^3y - 72x^4y^4}{8xy}$

33. $(2x^3 + 2x^2 - 19x - 21) \div (x + 3)$

34. $(3x^3 + 17x^2 + 6x - 4) \div (3x - 1)$

For Problems 35–38, simplify each expression, and express your answers using positive exponents only.

35. $(-2x^3)(3x^{-4})$

36. $\dfrac{4x^{-2}}{2x^{-1}}$

37. $(3x^{-1}y^{-2})^{-1}$

38. $(xy^2z^{-1})^{-2}$

For Problems 39–41, use scientific notation and the properties of exponents to help evaluate each numerical expression.

39. $(0.00003)(4000)$

40. $(0.0002)(0.003)^2$

41. $\dfrac{0.00034}{0.0000017}$

For Problems 42–49, solve each of the equations.

42. $5x + 8 = 6x - 3$

43. $-2(4x - 1) = -5x + 3 - 2x$

44. $\dfrac{y}{2} - \dfrac{y}{3} = 8$

45. $6x + 8 - 4x = 10(3x + 2)$

46. $1.6 - 2.4x = 5x - 65$

47. $-3(x - 1) + 2(x + 3) = -4$

48. $\dfrac{3n + 1}{5} + \dfrac{n - 2}{3} = \dfrac{2}{15}$

49. $0.06x + 0.08(1500 - x) = 110$

For Problems 50–55, solve each of the inequalities.

50. $2x - 7 \leq -3(x + 4)$

51. $6x + 5 - 3x > 5$

52. $4(x - 5) + 2(3x + 6) < 0$

53. $-5x + 3 > -4x + 5$

54. $\dfrac{3x}{4} - \dfrac{x}{2} \leq \dfrac{5x}{6} - 1$

55. $0.08(700 - x) + 0.11x \geq 65$

For Problems 56–59, graph each equation.

56. $y = 2x + 3$ **57.** $y = -5x$

58. $x - 2y = 6$ **59.** $y = -\dfrac{1}{2}x + 2$

60. Write the equation of the line that has a slope of $-\dfrac{1}{4}$ and contains the point $(3, -4)$.

61. Find the slope of a line determined by the equation $-2x - 5y = 6$.

62. Write the equation of the line that is perpendicular to the line determined by the equation $7x - 2y = 5$ and contains the point $(-6, 2)$.

For Problems 63–69, set up an equation and solve each problem.

63. The sum of 4 and three times a certain number is the same as the sum of the number and 10. Find the number.

64. Fifteen percent of some number is 6. Find the number.

65. Lou has 18 coins consisting of dimes and quarters. If the total value of the coins is $3.30, how many coins of each denomination does he have?

66. A sum of $1500 is invested, part of it at 8% interest and the remainder at 9%. If the total interest amounts to $128, find the amount invested at each rate.

67. How many gallons of water must be added to 15 gallons of a 12% salt solution to change it to a 10% salt solution?

68. Two airplanes leave Atlanta at the same time and fly in opposite directions. If one travels at 400 miles per hour and the other at 450 miles per hour, how long will it take them to be 2975 miles apart?

69. The length of a rectangle is 1 meter more than twice its width. If the perimeter of the rectangle is 44 meters, find the length and width.

7 Factoring, Solving Equations, and Problem Solving

Algebraic equations can be used to solve a large variety of problems dealing with geometric relationships.

© CSLD

A flower garden is in the shape of a right triangle with one leg 7 meters longer than the other leg and the hypotenuse 1 meter longer than the longer leg. Find the lengths of all three sides of the right triangle. A popular geometric formula, called the Pythagorean theorem, serves as a guideline for setting up an equation to solve this problem. We can use the equation $x^2 + (x + 7)^2 = (x + 8)^2$ to determine that the sides of the right triangle are 5 meters, 12 meters, and 13 meters long.

The distributive property has allowed us to combine similar terms and multiply polynomials. In this chapter, we will see yet another use of the distributive property as we learn how to **factor polynomials**. Factoring polynomials will allow us to solve other kinds of equations, which will in turn help us to solve a greater variety of word problems.

Video tutorials based on section learning objectives are available in a variety of delivery modes.

<table>
<tr><td>**7.1**</td><td>**Factoring by Using the Distributive Property**</td></tr>
</table>

OBJECTIVES

1. Find the greatest common factor

2. Factor out the greatest common factor

3. Factor by grouping

4. Solve equations by factoring

In Chapter 1 we found the *greatest common factor* of two or more whole numbers by inspection or by using the prime factored form of the numbers. For example, by inspection we see that the greatest common factor of 8 and 12 is 4. This means that 4 is the largest whole number that is a factor of both 8 and 12. If it is difficult to determine the greatest common factor by inspection, then we can use the prime factorization technique as follows:

$$42 = 2 \cdot 3 \cdot 7$$
$$70 = 2 \cdot 5 \cdot 7$$

We see that $2 \cdot 7 = 14$ is the greatest common factor of 42 and 70.

It is meaningful to extend the concept of greatest common factor to monomials. Consider the next example.

Classroom Example
Find the greatest common factor of $9m^2$ and $12m^3$.

| **EXAMPLE 1** | Find the greatest common factor of $8x^2$ and $12x^3$. |

Solution

$$8x^2 = 2 \cdot 2 \cdot 2 \cdot x \cdot x$$
$$12x^3 = 2 \cdot 2 \cdot 3 \cdot x \cdot x \cdot x$$

Therefore, the greatest common factor is $2 \cdot 2 \cdot x \cdot x = 4x^2$.

By "the greatest common factor of two or more monomials," we mean the monomial with the largest numerical coefficient and highest power of the variables that is a factor of the given monomials.

Classroom Example
Find the greatest common factor of $20m^3n^2$, $30mn^3$, and $45mn$.

| **EXAMPLE 2** | Find the greatest common factor of $16x^2y$, $24x^3y^2$, and $32xy$. |

Solution

$$16x^2y = 2 \cdot 2 \cdot 2 \cdot 2 \cdot x \cdot x \cdot y$$
$$24x^3y^2 = 2 \cdot 2 \cdot 2 \cdot 3 \cdot x \cdot x \cdot x \cdot y \cdot y$$
$$32xy = 2 \cdot 2 \cdot 2 \cdot 2 \cdot 2 \cdot x \cdot y$$

Therefore, the greatest common factor is $2 \cdot 2 \cdot 2 \cdot x \cdot y = 8xy$.

We have used the distributive property to multiply a polynomial by a monomial; for example,

$$3x(x + 2) = 3x^2 + 6x$$

Suppose we start with $3x^2 + 6x$ and want to express it in factored form. We use the distributive property in the form $ab + ac = a(b + c)$.

$$3x^2 + 6x = 3x(x) + 3x(2) \qquad \text{3x is the greatest common factor of } 3x^2 \text{ and } 6x$$
$$= 3x(x + 2) \qquad \text{Use the distributive property}$$

The next four examples further illustrate this process of **factoring out the greatest common monomial factor**.

Classroom Example
Factor $18x^4 - 24x^2$.

EXAMPLE 3   Factor $12x^3 - 8x^2$.

Solution

$$12x^3 - 8x^2 = 4x^2(3x) - 4x^2(2)$$
$$= 4x^2(3x - 2) \qquad ab - ac = a(b - c)$$

Classroom Example
Factor $15ab^3 + 27a^2b$.

EXAMPLE 4   Factor $12x^2y + 18xy^2$.

Solution

$$12x^2y + 18xy^2 = 6xy(2x) + 6xy(3y)$$
$$= 6xy(2x + 3y)$$

Classroom Example
Factor $12k^2 - 33k^3 + 51k^5$.

EXAMPLE 5   Factor $24x^3 + 30x^4 - 42x^5$.

Solution

$$24x^3 + 30x^4 - 42x^5 = 6x^3(4) + 6x^3(5x) - 6x^3(7x^2)$$
$$= 6x^3(4 + 5x - 7x^2)$$

Classroom Example
Factor $5y^4 + 5y^3$.

EXAMPLE 6   Factor $9x^2 + 9x$.

Solution

$$9x^2 + 9x = 9x(x) + 9x(1)$$
$$= 9x(x + 1)$$

We want to emphasize the point made just before Example 3. It is important to realize that we are factoring out the *greatest* common monomial factor. We could factor an expression such as $9x^2 + 9x$ in Example 6 as $9(x^2 + x)$, $3(3x^2 + 3x)$, $3x(3x + 3)$, or even $\frac{1}{2}(18x^2 + 18x)$, but it is the form $9x(x + 1)$ that we want. We can accomplish this by factoring out the greatest common monomial factor; we sometimes refer to this process as **factoring completely**. A polynomial with integral coefficients is in completely factored form if these conditions are met:

    1. It is expressed as a product of polynomials with integral coefficients.

    2. No polynomial, other than a monomial, within the factored form can be further factored into polynomials with integral coefficients.

Thus $9(x^2 + x)$, $3(3x^2 + 3x)$, and $3x(3x + 3)$ are not completely factored because they violate condition 2. The form $\frac{1}{2}(18x^2 + 18x)$ violates both conditions 1 and 2.

Sometimes there may be a **common binomial factor** rather than a common monomial factor. For example, each of the two terms of $x(y + 2) + z(y + 2)$ has a binomial factor of $(y + 2)$. Thus we can factor $(y + 2)$ from each term and get

$$x(y + 2) + z(y + 2) = (y + 2)(x + z)$$

Consider a few more examples involving a common binomial factor.

$$a(b + c) - d(b + c) = (b + c)(a - d)$$
$$x(x + 2) + 3(x + 2) = (x + 2)(x + 3)$$
$$x(x + 5) - 4(x + 5) = (x + 5)(x - 4)$$

It may be that the original polynomial exhibits no apparent common monomial or binomial factor, which is the case with

$$ab + 3a + bc + 3c$$

However, by factoring a from the first two terms and c from the last two terms, we see that

$$ab + 3a + bc + 3c = a(b + 3) + c(b + 3)$$

Now a common binomial factor of $(b + 3)$ is obvious, and we can proceed as before.

$$a(b + 3) + c(b + 3) = (b + 3)(a + c)$$

This factoring process is called **factoring by grouping**. Let's consider two more examples of factoring by grouping.

$x^2 - x + 5x - 5 = x(x - 1) + 5(x - 1)$	Factor x from first two terms and 5 from last two terms
$ = (x - 1)(x + 5)$	Factor common binomial factor of $(x - 1)$ from both terms
$6x^2 - 4x - 3x + 2 = 2x(3x - 2) - 1(3x - 2)$	Factor $2x$ from first two terms and -1 from last two terms
$ = (3x - 2)(2x - 1)$	Factor common binomial factor of $(3x - 2)$ from both terms

Back to Solving Equations

Suppose we are told that the product of two numbers is 0. What do we know about the numbers? Do you agree with our conclusion that at least one of the numbers must be 0? The next property formalizes this idea.

Property 7.1

For all real numbers a and b,

$ab = 0$ if and only if $a = 0$ or $b = 0$

Property 7.1 provides us with another technique for solving equations.

Classroom Example
Solve $m^2 - 8m = 0$.

EXAMPLE 7   Solve $x^2 + 6x = 0$.

Solution

To solve equations by applying Property 7.1, one side of the equation must be a product, and the other side of the equation must be zero. This equation already has zero on the right-hand side of the equation, but the left-hand side of this equation is a sum. We will factor the left-hand side, $x^2 + 6x$, to change the sum into a product.

$x^2 + 6x = 0$		
$x(x + 6) = 0$		Factor
$x = 0$ or $x + 6 = 0$		$ab = 0$ if and only if $a = 0$ or $b = 0$
$x = 0$ or $x = -6$		

The solution set is $\{-6, 0\}$. (Be sure to check both values in the original equation.)

Classroom Example
Solve $x^2 = 17x$.

EXAMPLE 8   Solve $x^2 = 12x$.

Solution

In order to solve this equation by Property 7.1, we will first get zero on the right-hand side of the equation by adding $-12x$ to each side. Then we factor the expression on the left-hand side of the equation.

$$x^2 = 12x$$
$$x^2 - 12x = 0 \qquad\qquad\qquad\qquad \text{Added } -12x \text{ to both sides}$$
$$x(x - 12) = 0$$
$$x = 0 \quad \text{or} \quad x - 12 = 0 \qquad ab = 0 \text{ if and only if } a = 0 \text{ or } b = 0$$
$$x = 0 \quad \text{or} \qquad\quad x = 12$$

The solution set is $\{0, 12\}$.

Remark: Notice in Example 8 that we did not divide both sides of the original equation by x. This would cause us to lose the solution of 0.

Classroom Example
Solve $5d^2 - 7d = 0$.

EXAMPLE 9   Solve $4x^2 - 3x = 0$.

Solution

$$4x^2 - 3x = 0$$
$$x(4x - 3) = 0$$
$$x = 0 \quad \text{or} \quad 4x - 3 = 0 \qquad ab = 0 \text{ if and only if } a = 0 \text{ or } b = 0$$
$$x = 0 \quad \text{or} \qquad\quad 4x = 3$$
$$x = 0 \quad \text{or} \qquad\quad x = \frac{3}{4}$$

The solution set is $\left\{0, \dfrac{3}{4}\right\}$.

Classroom Example
Solve $w(w - 8) + 11(w - 8) = 0$.

EXAMPLE 10   Solve $x(x + 2) + 3(x + 2) = 0$.

Solution

In order to solve this equation by Property 7.1, we will factor the left-hand side of the equation. The greatest common factor of the terms is $(x + 2)$.

$$x(x + 2) + 3(x + 2) = 0$$
$$(x + 2)(x + 3) = 0$$
$$x + 2 = 0 \quad \text{or} \quad x + 3 = 0 \qquad ab = 0 \text{ if and only if } a = 0 \text{ or } b = 0$$
$$x = -2 \quad \text{or} \qquad x = -3$$

The solution set is $\{-3, -2\}$.

Each time we expand our equation-solving capabilities, we also acquire more techniques for solving problems. Let's solve a geometric problem with the ideas we learned in this section.

Classroom Example
The area of a square is numerically equal to three times its perimeter. Find the length of a side of the square.

Figure 7.1

EXAMPLE 11

The area of a square is numerically equal to twice its perimeter. Find the length of a side of the square.

Solution

Sketch a square and let s represent the length of each side (see Figure 7.1). Then the area is represented by s^2 and the perimeter by $4s$. Thus

$$s^2 = 2(4s)$$
$$s^2 = 8s$$
$$s^2 - 8s = 0$$
$$s(s - 8) = 0$$
$$s = 0 \quad \text{or} \quad s - 8 = 0$$
$$s = 0 \quad \text{or} \quad s = 8$$

Since 0 is not a reasonable answer to the problem, the solution is 8. (Be sure to check this solution in the original statement of the problem!)

Concept Quiz 7.1

For Problems 1–10, answer true or false.

1. The greatest common factor of $6x^2y^3 - 12x^3y^2 + 18x^4y$ is $2x^2y$.
2. If the factored form of a polynomial can be factored further, then it has not met the conditions to be considered "factored completely."
3. Common factors are always monomials.
4. If the product of x and y is zero, then x is zero or y is zero.
5. The factored form $3a(2a^2 + 4)$ is factored completely.
6. The solutions for the equation $x(x + 2) = 7$ are 7 and 5.
7. The solution set for $x^2 = 7x$ is $\{7\}$.
8. The solution set for $x(x - 2) - 3(x - 2) = 0$ is $\{2, 3\}$.
9. The solution set for $-3x = x^2$ is $\{-3, 0\}$.
10. The solution set for $x(x + 6) = 2(x + 6)$ is $\{-6\}$.

Problem Set 7.1

For Problems 1–10, find the greatest common factor of the given expressions. (Objective 1)

1. $24y$ and $30xy$
2. $32x$ and $40xy$
3. $60x^2y$ and $84xy^2$
4. $72x^3$ and $63x^2$
5. $42ab^3$ and $70a^2b^2$
6. $48a^2b^2$ and $96ab^4$
7. $6x^3$, $8x$, and $24x^2$
8. $72xy$, $36x^2y$, and $84xy^2$
9. $16a^2b^2$, $40a^2b^3$, and $56a^3b^4$
10. $70a^3b^3$, $42a^2b^4$, and $49ab^5$

For Problems 11–46, factor each polynomial completely. (Objective 2)

11. $8x + 12y$
12. $18x + 24y$
13. $14xy - 21y$
14. $24x - 40xy$
15. $18x^2 + 45x$
16. $12x + 28x^3$
17. $12xy^2 - 30x^2y$
18. $28x^2y^2 - 49x^2y$
19. $36a^2b - 60a^3b^4$
20. $65ab^3 - 45a^2b^2$
21. $16xy^3 + 25x^2y^2$
22. $12x^2y^2 + 29x^2y$
23. $64ab - 72cd$
24. $45xy - 72zw$

25. $9a^2b^4 - 27a^2b$ **26.** $7a^3b^5 - 42a^2b^6$

27. $52x^4y^2 + 60x^6y$ **28.** $70x^5y^3 - 42x^8y^2$

29. $40x^2y^2 + 8x^2y$ **30.** $84x^2y^3 + 12xy^3$

31. $12x + 15xy + 21x^2$

32. $30x^2y + 40xy + 55y$

33. $2x^3 - 3x^2 + 4x$

34. $x^4 + x^3 + x^2$

35. $44y^5 - 24y^3 - 20y^2$

36. $14a - 18a^3 - 26a^5$

37. $14a^2b^3 + 35ab^2 - 49a^3b$

38. $24a^3b^2 + 36a^2b^4 - 60a^4b^3$

39. $x(y + 1) + z(y + 1)$

40. $a(c + d) + 2(c + d)$

41. $a(b - 4) - c(b - 4)$

42. $x(y - 6) - 3(y - 6)$

43. $x(x + 3) + 6(x + 3)$

44. $x(x - 7) + 9(x - 7)$

45. $2x(x + 1) - 3(x + 1)$

46. $4x(x + 8) - 5(x + 8)$

For Problems 47–60, use the process of *factoring by grouping* to factor each polynomial. **(Objective 3)**

47. $5x + 5y + bx + by$

48. $7x + 7y + zx + zy$

49. $bx - by - cx + cy$

50. $2x - 2y - ax + ay$

51. $ac + bc + a + b$

52. $x + y + ax + ay$

53. $x^2 + 5x + 12x + 60$

54. $x^2 + 3x + 7x + 21$

55. $x^2 - 2x - 8x + 16$

56. $x^2 - 4x - 9x + 36$

57. $2x^2 + x - 10x - 5$

58. $3x^2 + 2x - 18x - 12$

59. $6n^2 - 3n - 8n + 4$

60. $20n^2 + 8n - 15n - 6$

For Problems 61–84, solve each equation. **(Objective 4)**

61. $x^2 - 8x = 0$ **62.** $x^2 - 12x = 0$

63. $x^2 + x = 0$ **64.** $x^2 + 7x = 0$

65. $n^2 = 5n$ **66.** $n^2 = -2n$

67. $2y^2 - 3y = 0$ **68.** $4y^2 - 7y = 0$

69. $7x^2 = -3x$ **70.** $5x^2 = -2x$

71. $3n^2 + 15n = 0$ **72.** $6n^2 - 24n = 0$

73. $4x^2 = 6x$ **74.** $12x^2 = 8x$

75. $7x - x^2 = 0$ **76.** $9x - x^2 = 0$

77. $13x = x^2$ **78.** $15x = -x^2$

79. $5x = -2x^2$ **80.** $7x = -5x^2$

81. $x(x + 5) - 4(x + 5) = 0$

82. $x(3x - 2) - 7(3x - 2) = 0$

83. $4(x - 6) - x(x - 6) = 0$

84. $x(x + 9) = 2(x + 9)$

For Problems 85–91, set up an equation and solve each problem. **(Objective 4)**

85. The square of a number equals nine times that number. Find the number.

86. Suppose that four times the square of a number equals 20 times that number. What is the number?

87. The area of a square is numerically equal to five times its perimeter. Find the length of a side of the square.

88. The area of a square is 14 times as large as the area of a triangle. One side of the triangle is 7 inches long, and the altitude to that side is the same length as a side of the square. Find the length of a side of the square. Also find the areas of both figures, and be sure that your answer checks.

89. Suppose that the area of a circle is numerically equal to the perimeter of a square, and that the length of a radius of the circle is equal to the length of a side of the square. Find the length of a side of the square. Express your answer in terms of π.

90. One side of a parallelogram, an altitude to that side, and one side of a rectangle all have the same measure. If an adjacent side of the rectangle is 20 centimeters long, and the area of the rectangle is twice the area of the parallelogram, find the areas of both figures.

91. The area of a rectangle is twice the area of a square. If the rectangle is 6 inches long, and the width of the rectangle is the same as the length of a side of the square, find the dimensions of both the rectangle and the square.

Thoughts Into Words

92. Suppose that your friend factors $24x^2y + 36xy$ like this:

$$24x^2y + 36xy = 4xy(6x + 9)$$
$$= (4xy)(3)(2x + 3)$$
$$= 12xy(2x + 3)$$

Is this correct? Would you make any suggestions for changing her method?

93. The following solution is given for the equation $x(x - 10) = 0$.

$$x(x - 10) = 0$$
$$x^2 - 10x = 0$$
$$x(x - 10) = 0$$
$$x = 0 \quad \text{or} \quad x - 10 = 0$$
$$x = 0 \quad \text{or} \quad x = 10$$

The solution set is $\{0, 10\}$. Is this a correct solution? Would you suggest any changes to the method?

Further Investigations

94. The total surface area of a right circular cylinder is given by the formula $A = 2\pi r^2 + 2\pi rh$, where r represents the radius of a base, and h represents the height of the cylinder. For computational purposes, it may be more convenient to change the form of the right side of the formula by factoring it.

$$A = 2\pi r^2 + 2\pi rh = 2\pi r(r + h)$$

Use $A = 2\pi r(r + h)$ to find the total surface area of each of the following cylinders. Use $\dfrac{22}{7}$ as an approximation for π.

(a) $r = 7$ centimeters and $h = 12$ centimeters

(b) $r = 14$ meters and $h = 20$ meters

(c) $r = 3$ feet and $h = 4$ feet

(d) $r = 5$ yards and $h = 9$ yards

95. The formula $A = P + Prt$ yields the total amount of money accumulated (A) when P dollars is invested at r percent simple interest for t years. For computational purposes, it may be convenient to change the right side of the formula by factoring.

$$A = P + Prt = P(1 + rt)$$

Use $A = P(1 + rt)$ to find the total amount of money accumulated for each of the following investments.

(a) $100 at 8% for 2 years

(b) $200 at 9% for 3 years

(c) $500 at 10% for 5 years

(d) $1000 at 10% for 10 years

For Problems 96–99, solve for the indicated variable.

96. $ax + bx = c$ for x

97. $b^2x^2 - cx = 0$ for x

98. $5ay^2 = by$ for y

99. $y + ay - by - c = 0$ for y

Answers to the Concept Quiz

1. False   **2.** True   **3.** False   **4.** True   **5.** False   **6.** False   **7.** False   **8.** True   **9.** True   **10.** False

---

7.2	Factoring the Difference of Two Squares

OBJECTIVES   **1** Factor the difference of two squares

2 Solve equations by factoring the difference of two squares

In Section 6.3 we noted some special multiplication patterns. One of these patterns was

$$(a - b)(a + b) = a^2 - b^2$$

We can view this same pattern as follows:

> **Difference of Two Squares**
>
> $a^2 - b^2 = (a - b)(a + b)$

To apply the pattern is a fairly simple process, as these next examples illustrate. The steps inside the box are often performed mentally.

$$
\begin{array}{lll}
x^2 - 36 = & (x)^2 - (6)^2 & = (x - 6)(x + 6) \\
4x^2 - 25 = & (2x)^2 - (5)^2 & = (2x - 5)(2x + 5) \\
9x^2 - 16y^2 = & (3x)^2 - (4y)^2 & = (3x - 4y)(3x + 4y) \\
64 - y^2 = & (8)^2 - (y)^2 & = (8 - y)(8 + y)
\end{array}
$$

Since multiplication is commutative, the order of writing the factors is not important. For example, $(x - 6)(x + 6)$ can also be written as $(x + 6)(x - 6)$.

You must be careful not to assume an analogous factoring pattern for the sum of two squares; it does not exist. For example, $x^2 + 4 \neq (x + 2)(x + 2)$ because $(x + 2)(x + 2) = x^2 + 4x + 4$. We say that the **sum of two squares is not factorable using integers**. The phrase "using integers" is necessary because $x^2 + 4$ could be written as $\frac{1}{2}(2x^2 + 8)$, but such factoring is of no help. Furthermore, we do not consider $(1)(x^2 + 4)$ as factoring $x^2 + 4$.

It is possible that both the technique of *factoring out a common monomial factor* and the pattern *difference of two squares* can be applied to the same polynomial. In general, it is best to look for a common monomial factor first.

Classroom Example
Factor $7x^2 - 28$.

EXAMPLE 1   Factor $2x^2 - 50$.

Solution

$$
\begin{array}{ll}
2x^2 - 50 = 2(x^2 - 25) & \text{Common factor of 2} \\
 = 2(x - 5)(x + 5) & \text{Difference of squares}
\end{array}
$$

In Example 1, by expressing $2x^2 - 50$ as $2(x - 5)(x + 5)$, we say that the algebraic expression has been **factored completely**. That means that the factors 2, $x - 5$, and $x + 5$ cannot be factored any further using integers.

Classroom Example
Factor completely $32m^3 - 50m$.

EXAMPLE 2   Factor completely $18y^3 - 8y$.

Solution

$$
\begin{array}{ll}
18y^3 - 8y = 2y(9y^2 - 4) & \text{Common factor of } 2y \\
 = 2y(3y - 2)(3y + 2) & \text{Difference of squares}
\end{array}
$$

Sometimes it is possible to apply the difference-of-squares pattern more than once. Consider the next example.

EXAMPLE 3   Factor completely $x^4 - 16$.

Solution

$$x^4 - 16 = (x^2 + 4)(x^2 - 4)$$
$$= (x^2 + 4)(x + 2)(x - 2)$$

The following examples should help you to summarize the factoring ideas presented thus far.

$$5x^2 + 20 = 5(x^2 + 4)$$
$$25 - y^2 = (5 - y)(5 + y)$$
$$3 - 3x^2 = 3(1 - x^2) = 3(1 + x)(1 - x)$$
$$36x^2 - 49y^2 = (6x - 7y)(6x + 7y)$$

$a^2 + 9$ is not factorable using integers

$9x + 17y$ is not factorable using integers

Solving Equations

Each time we learn a new factoring technique, we also develop more tools for solving equations. Let's consider how we can use the difference-of-squares factoring pattern to help solve certain kinds of equations.

EXAMPLE 4   Solve $x^2 = 25$.

Solution

$$x^2 = 25$$
$$x^2 - 25 = 0 \qquad \text{Added } -25 \text{ to both sides}$$
$$(x + 5)(x - 5) = 0$$
$$x + 5 = 0 \qquad \text{or} \qquad x - 5 = 0 \quad \text{Remember: } ab = 0 \text{ if and only if } a = 0 \text{ or } b = 0$$
$$x = -5 \qquad \text{or} \qquad x = 5$$

The solution set is $\{-5, 5\}$. Check these answers!

EXAMPLE 5   Solve $9x^2 = 25$.

Solution

$$9x^2 = 25$$
$$9x^2 - 25 = 0$$
$$(3x + 5)(3x - 5) = 0$$
$$3x + 5 = 0 \qquad \text{or} \qquad 3x - 5 = 0$$
$$3x = -5 \qquad \text{or} \qquad 3x = 5$$
$$x = -\frac{5}{3} \qquad \text{or} \qquad x = \frac{5}{3}$$

The solution set is $\left\{-\frac{5}{3}, \frac{5}{3}\right\}$.

Classroom Example
Solve $3x^2 = 27$.

EXAMPLE 6   Solve $5y^2 = 20$.

Solution

$$5y^2 = 20$$

$$\frac{5y^2}{5} = \frac{20}{5} \qquad \text{Divide both sides by 5}$$

$$y^2 = 4$$

$$y^2 - 4 = 0$$

$$(y + 2)(y - 2) = 0$$

$$y + 2 = 0 \qquad \text{or} \qquad y - 2 = 0$$

$$y = -2 \qquad \text{or} \qquad y = 2$$

The solution set is $\{-2, 2\}$. Check it!

Classroom Example
Solve $c^3 - 36c = 0$.

EXAMPLE 7   Solve $x^3 - 9x = 0$.

Solution

$$x^3 - 9x = 0$$

$$x(x^2 - 9) = 0$$

$$x(x - 3)(x + 3) = 0$$

$$x = 0 \qquad \text{or} \qquad x - 3 = 0 \qquad \text{or} \qquad x + 3 = 0$$

$$x = 0 \qquad \text{or} \qquad x = 3 \qquad \text{or} \qquad x = -3$$

The solution set is $\{-3, 0, 3\}$.

The more we know about solving equations, the more easily we can solve word problems.

Classroom Example
The combined area of two squares is 360 square inches. Each side of one square is three times as long as a side of the other square. Find the lengths of the sides of each square.

EXAMPLE 8

The combined area of two squares is 20 square centimeters. Each side of one square is twice as long as a side of the other square. Find the lengths of the sides of each square.

Solution

We can sketch two squares and label the sides of the smaller square s (see Figure 7.2). Then the sides of the larger square are $2s$. Since the sum of the areas of the two squares is 20 square centimeters, we can set up and solve the following equation:

$$s^2 + (2s)^2 = 20$$

$$s^2 + 4s^2 = 20$$

$$5s^2 = 20$$

$$s^2 = 4$$

$$s^2 - 4 = 0$$

$$(s + 2)(s - 2) = 0$$

$$s + 2 = 0 \qquad \text{or} \qquad s - 2 = 0$$

$$s = -2 \qquad \text{or} \qquad s = 2$$

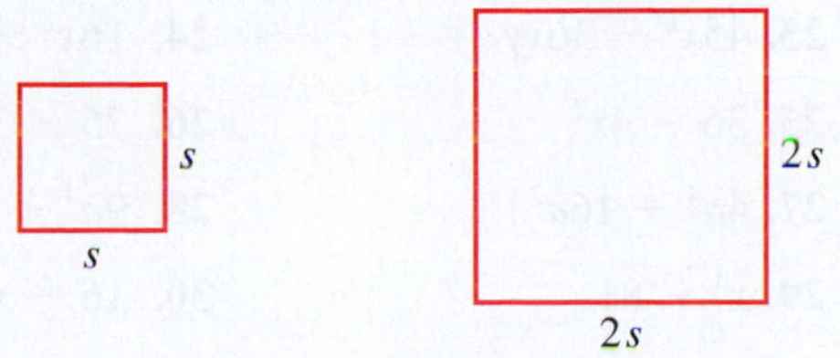

Figure 7.2

Since s represents the length of a side of a square, we must disregard the solution -2. Thus one square has sides of length 2 centimeters and the other square has sides of length $2(2) = 4$ centimeters.

Concept Quiz 7.2

For Problems 1–10, answer true or false.

1. A binomial that has two perfect square terms that are subtracted is called the difference of two squares.

2. The sum of two squares is factorable using integers.

3. When factoring it is usually best to look for a common factor first.

4. The polynomial $4x^2 + y^2$ factors into $(2x + y)(2x + y)$.

5. The completely factored form of $y^4 - 81$ is $(y^2 + 9)(y^2 - 9)$.

6. The solution set for $x^2 = -16$ is $\{-4\}$.

7. The solution set for $5x^3 - 5x = 0$ is $\{-1, 0, 1\}$.

8. The solution set for $x^4 - 9x^2 = 0$ is $\{-3, 0, 3\}$.

9. The completely factored form of $x^4 - 1$ is $(x + 1)(x - 1)(x^2 + 1)$.

10. The completely factored form of $2x^3y - 8xy$ is $2xy(x + 2)(x - 2)$.

Problem Set 7.2

For Problems 1–12, use the difference-of-squares pattern to factor each polynomial. **(Objective 1)**

1. $x^2 - 1$

2. $x^2 - 25$

3. $x^2 - 100$

4. $x^2 - 121$

5. $x^2 - 4y^2$

6. $x^2 - 36y^2$

7. $9x^2 - y^2$

8. $49y^2 - 64x^2$

9. $36a^2 - 25b^2$

10. $4a^2 - 81b^2$

11. $1 - 4n^2$

12. $4 - 9n^2$

For Problems 13–44, factor each polynomial completely. Indicate any that are not factorable using integers. Don't forget to look for a common monomial factor first. **(Objective 1)**

13. $5x^2 - 20$

14. $7x^2 - 7$

15. $8x^2 + 32$

16. $12x^2 + 60$

17. $2x^2 - 18y^2$

18. $8x^2 - 32y^2$

19. $x^3 - 25x$

20. $2x^3 - 2x$

21. $x^2 + 9y^2$

22. $18x - 42y$

23. $45x^2 - 36xy$

24. $16x^2 + 25y^2$

25. $36 - 4x^2$

26. $75 - 3x^2$

27. $4a^4 + 16a^2$

28. $9a^4 + 81a^2$

29. $x^4 - 81$

30. $16 - x^4$

31. $x^4 + x^2$

32. $x^5 + 2x^3$

33. $3x^3 + 48x$

34. $6x^3 + 24x$

35. $5x - 20x^3$

36. $4x - 36x^3$

37. $4x^2 - 64$

38. $9x^2 - 9$

39. $75x^3y - 12xy^3$

40. $32x^3y - 18xy^3$

41. $16x^4 - 81y^4$

42. $x^4 - 1$

43. $81 - x^4$

44. $81x^4 - 16y^4$

For Problems 45–68, solve each equation. **(Objective 2)**

45. $x^2 = 9$

46. $x^2 = 1$

47. $4 = n^2$

48. $144 = n^2$

49. $9x^2 = 16$

50. $4x^2 = 9$

51. $n^2 - 121 = 0$

52. $n^2 - 81 = 0$

53. $25x^2 = 4$

54. $49x^2 = 36$

55. $3x^2 = 75$

56. $7x^2 = 28$

57. $3x^3 - 48x = 0$

58. $x^3 - x = 0$

59. $n^3 = 16n$

60. $2n^3 = 8n$

61. $5 - 45x^2 = 0$

62. $3 - 12x^2 = 0$

63. $4x^3 - 400x = 0$

64. $2x^3 - 98x = 0$

65. $64x^2 = 81$

66. $81x^2 = 25$

67. $36x^3 = 9x$

68. $64x^3 = 4x$

For Problems 69–80, set up an equation and solve the problem. **(Objective 2)**

69. Forty-nine less than the square of a number equals zero. Find the number.

70. The cube of a number equals nine times the number. Find the number.

71. Suppose that five times the cube of a number equals 80 times the number. Find the number.

72. Ten times the square of a number equals 40. Find the number.

73. The sum of the areas of two squares is 234 square inches. Each side of the larger square is five times the length of a side of the smaller square. Find the length of a side of each square.

74. The difference of the areas of two squares is 75 square feet. Each side of the larger square is twice the length of a side of the smaller square. Find the length of a side of each square.

75. Suppose that the length of a certain rectangle is $2\frac{1}{2}$ times its width, and the area of that same rectangle is 160 square centimeters. Find the length and width of the rectangle.

76. Suppose that the width of a certain rectangle is three-fourths of its length, and the area of that same rectangle is 108 square meters. Find the length and width of the rectangle.

77. The sum of the areas of two circles is 80π square meters. Find the length of a radius of each circle if one of them is twice as long as the other.

78. The area of a triangle is 98 square feet. If one side of the triangle and the altitude to that side are of equal length, find the length.

79. The total surface area of a right circular cylinder is 100π square centimeters. If a radius of the base and the altitude of the cylinder are the same length, find the length of a radius.

80. The total surface area of a right circular cone is 192π square feet. If the slant height of the cone is equal in length to a diameter of the base, find the length of a radius.

Thoughts Into Words

81. How do we know that the equation $x^2 + 1 = 0$ has no solutions in the set of real numbers?

82. Why is the following factoring process incomplete?

$16x^2 - 64 = (4x + 8)(4x - 8)$

How could the factoring be done?

83. Consider the following solution:

$$4x^2 - 36 = 0$$
$$4(x^2 - 9) = 0$$
$$4(x + 3)(x - 3) = 0$$
$$4 = 0 \quad \text{or} \quad x + 3 = 0 \quad \text{or} \quad x - 3 = 0$$
$$4 = 0 \quad \text{or} \quad x = -3 \quad \text{or} \quad x = 3$$

The solution set is $\{-3, 3\}$. Is this a correct solution? Do you have any suggestions to offer the person who did this problem?

Further Investigations

The following patterns can be used to factor the sum and difference of two cubes.

$$a^3 + b^3 = (a + b)(a^2 - ab + b^2)$$
$$a^3 - b^3 = (a - b)(a^2 + ab + b^2)$$

Consider these examples.

$$x^3 + 8 = (x)^3 + (2)^3 = (x + 2)(x^2 - 2x + 4)$$
$$x^3 - 1 = (x)^3 - (1)^3 = (x - 1)(x^2 + x + 1)$$

Use the sum and difference-of-cubes patterns to factor each polynomial.

84. $x^3 + 1$

85. $x^3 - 8$

86. $n^3 - 27$

87. $n^3 + 64$

88. $8x^3 + 27y^3$

89. $27a^3 - 64b^3$

90. $1 - 8x^3$

91. $1 + 27a^3$

92. $x^3 + 8y^3$

93. $8x^3 - y^3$

94. $a^3b^3 - 1$

95. $27x^3 - 8y^3$

96. $8 + n^3$

97. $125x^3 + 8y^3$

98. $27n^3 - 125$

99. $64 + x^3$

Answers to the Concept Quiz

1. True   **2.** False   **3.** True   **4.** False   **5.** False   **6.** False   **7.** True   **8.** True   **9.** True   **10.** True

---

| **7.3** | Factoring Trinomials of the Form $x^2 + bx + c$ |

OBJECTIVES

1 Factor trinomials of the form $x^2 + bx + c$

2 Use factoring of trinomials to solve equations

3 Solve word problems involving consecutive numbers

4 Use the Pythagorean theorem to solve problems

One of the most common types of factoring used in algebra is the expression of a trinomial as the product of two binomials. In this section we will consider trinomials for which the coefficient of the squared term is 1; that is, trinomials of the form $x^2 + bx + c$.

Again, to develop a factoring technique, we first look at some multiplication ideas. Consider the product $(x + r)(x + s)$, and use the distributive property to show how each term of the resulting trinomial is formed.

$$(x + r)(x + s) = x(x) + x(s) + r(x) + r(s)$$

$$x^2 + (s + r)x + rs$$

Notice that the coefficient of the middle term is the *sum* of r and s, and the last term is the *product* of r and s. These two relationships are used in the next examples.

Classroom Example
Factor $x^2 + 12x + 27$.

EXAMPLE 1   Factor $x^2 + 7x + 12$.

Solution

We need to fill in the blanks with two numbers whose product is 12 and whose sum is 7.

$$x^2 + 7x + 12 = (x + \underline{\quad})(x + \underline{\quad})$$

To assist in finding the numbers, we can set up a table of the factors of 12.

Product	Sum
$1(12) = 12$	$1 + 12 = 13$
$2(6) = 12$	$2 + 6 = 8$
$3(4) = 12$	$3 + 4 = 7$

The bottom line contains the numbers that we need. Thus

$$x^2 + 7x + 12 = (x + 3)(x + 4)$$

Classroom Example
Factor $b^2 - 11b + 28$.

EXAMPLE 2   Factor $x^2 - 11x + 24$.

Solution

To factor $x^2 - 11x + 24$, we want to find two numbers whose product is 24 and whose sum is -11.

Product	Sum
$(-1)(-24) = 24$	$-1 + (-24) = -25$
$(-2)(-12) = 24$	$-2 + (-12) = -14$
$(-3)(-8) = 24$	$-3 + (-8) = -11$
$(-4)(-6) = 24$	$-4 + (-6) = -10$

The third line contains the numbers that we want. Thus

$$x^2 - 11x + 24 = (x - 3)(x - 8)$$

Classroom Example
Factor $x^2 + 2x - 24$.

EXAMPLE 3   Factor $x^2 + 3x - 10$.

Solution

To factor $x^2 + 3x - 10$, we want to find two numbers whose product is -10 and whose sum is 3.

Product	Sum
$1(-10) = -10$	$1 + (-10) = -9$
$-1(10) = -10$	$-1 + 10 = 9$
$2(-5) = -10$	$2 + (-5) = -3$
$-2(5) = -10$	$-2 + 5 = 3$

The bottom line is the key line. Thus

$$x^2 + 3x - 10 = (x + 5)(x - 2)$$

Classroom Example
Factor $n^2 - 4n - 32$.

EXAMPLE 4   Factor $x^2 - 2x - 8$.

Solution

We are looking for two numbers whose product is -8 and whose sum is -2.

Product	Sum
$1(-8) = -8$	$1 + (-8) = -7$
$-1(8) = -8$	$-1 + 8 = 7$
$2(-4) = -8$	$2 + (-4) = -2$
$-2(4) = -8$	$-2 + 4 = 2$

The third line has the information we want.

$$x^2 - 2x - 8 = (x - 4)(x + 2)$$

The tables in the last four examples illustrate one way of organizing your thoughts for such problems. We show complete tables; that is, for Example 4, we include the bottom line even though the desired numbers are obtained in the third line. If you use such tables, keep in mind that as soon as you get the desired numbers, the table need not be completed any further. Furthermore, you may be able to find the numbers without using a table. The key ideas are the product and sum relationships.

Classroom Example
Factor $x^2 + 13x - 14$.

EXAMPLE 5   Factor $x^2 - 13x + 12$.

Solution

Product	Sum
$(-1)(-12) = 12$	$(-1) + (-12) = -13$

We need not complete the table.

$$x^2 - 13x + 12 = (x - 1)(x - 12)$$

In the next example, we refer to the concept of absolute value. Recall that the absolute value of any nonzero real number is positive. For example,

$$|4| = 4 \qquad \text{and} \qquad |-4| = 4$$

Classroom Example
Factor $x^2 - 6x - 7$.

EXAMPLE 6   Factor $x^2 - x - 56$.

Solution

Notice that the coefficient of the middle term is -1. Therefore, we are looking for two numbers whose product is -56; because their sum is -1, the absolute value of the negative number must be one larger than the absolute value of the positive number. The numbers are -8 and 7, and we have

$$x^2 - x - 56 = (x - 8)(x + 7)$$

Classroom Example
Factor $m^2 + m + 3$.

EXAMPLE 7   Factor $x^2 + 10x + 12$.

Solution

Product	Sum
$1(12) = 12$	$1 + 12 = 13$
$2(6) = 12$	$2 + 6 = 8$
$3(4) = 12$	$3 + 4 = 7$

Since the table is complete and no two factors of 12 produce a sum of 10, we conclude that

$$x^2 + 10x + 12$$

is not factorable using integers.

In a problem such as Example 7, we need to be sure that we have tried all possibilities before we conclude that the trinomial is not factorable.

Back to Solving Equations

The property $ab = 0$ if and only if $a = 0$ or $b = 0$ continues to play an important role as we solve equations that involve the factoring ideas of this section. Consider the following examples.

Classroom Example
Solve $x^2 + 15x + 26 = 0$.

EXAMPLE 8   Solve $x^2 + 8x + 15 = 0$.

Solution

$$x^2 + 8x + 15 = 0$$
$$(x + 3)(x + 5) = 0 \qquad\qquad \text{Factor the left side}$$
$$x + 3 = 0 \qquad \text{or} \qquad x + 5 = 0 \qquad \text{Use } ab = 0 \text{ if and only if } a = 0 \text{ or } b = 0$$
$$x = -3 \qquad \text{or} \qquad x = -5$$

The solution set is $\{-5, -3\}$.

Classroom Example
Solve $x^2 - 8x - 9 = 0$.

EXAMPLE 9   Solve $x^2 + 5x - 6 = 0$.

Solution

$$x^2 + 5x - 6 = 0$$
$$(x + 6)(x - 1) = 0$$
$$x + 6 = 0 \qquad \text{or} \qquad x - 1 = 0$$
$$x = -6 \qquad \text{or} \qquad x = 1$$

The solution set is $\{-6, 1\}$.

Classroom Example
Solve $m^2 - 8m = 33$.

EXAMPLE 10   Solve $y^2 - 4y = 45$.

Solution

$$y^2 - 4y = 45$$
$$y^2 - 4y - 45 = 0$$
$$(y - 9)(y + 5) = 0$$
$$y - 9 = 0 \qquad \text{or} \qquad y + 5 = 0$$
$$y = 9 \qquad \text{or} \qquad y = -5$$

The solution set is $\{-5, 9\}$.

Don't forget that we can always check to be absolutely sure of our solutions. Let's check the solutions for Example 10. If $y = 9$, then $y^2 - 4y = 45$ becomes

$$9^2 - 4(9) \stackrel{?}{=} 45$$
$$81 - 36 \stackrel{?}{=} 45$$
$$45 = 45$$

If $y = -5$, then $y^2 - 4y = 45$ becomes

$$(-5)^2 - 4(-5) \stackrel{?}{=} 45$$
$$25 + 20 \stackrel{?}{=} 45$$
$$45 = 45$$

Back to Problem Solving

The more we know about factoring and solving equations, the more easily we can solve word problems.

Classroom Example
Find two consecutive odd integers whose product is 35.

EXAMPLE 11   Find two consecutive integers whose product is 72.

Solution

Let n represent one integer. Then $n + 1$ represents the next integer.

$$n(n + 1) = 72 \qquad \text{The product of the two integers is 72}$$
$$n^2 + n = 72$$
$$n^2 + n - 72 = 0$$
$$(n + 9)(n - 8) = 0$$
$$n + 9 = 0 \qquad \text{or} \qquad n - 8 = 0$$
$$n = -9 \qquad \text{or} \qquad n = 8$$

If $n = -9$, then $n + 1 = -9 + 1 = -8$. If $n = 8$, then $n + 1 = 8 + 1 = 9$. Thus the consecutive integers are -9 and -8 or 8 and 9.

Classroom Example
A triangular lot has a height that is 8 yards longer than the base. The area of the lot is 24 square yards. Find the base and height of the lot.

EXAMPLE 12

A rectangular plot is 6 meters longer than it is wide. The area of the plot is 16 square meters. Find the length and width of the plot.

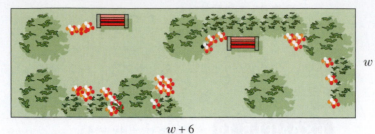

w

$w + 6$

Figure 7.3

Solution

We let w represent the width of the plot, and then $w + 6$ represents the length (see Figure 7.3). Using the area formula $A = lw$, we obtain

$$w(w + 6) = 16$$
$$w^2 + 6w = 16$$
$$w^2 + 6w - 16 = 0$$
$$(w + 8)(w - 2) = 0$$
$$w + 8 = 0 \quad \text{or} \quad w - 2 = 0$$
$$w = -8 \quad \text{or} \quad w = 2$$

The solution of -8 is not possible for the width of a rectangle, so the plot is 2 meters wide and its length $(w + 6)$ is 8 meters.

The Pythagorean theorem, an important theorem pertaining to right triangles, can also serve as a guideline for solving certain types of problems. The Pythagorean theorem states that **in any right triangle, the square of the longest side** (called the hypotenuse) **is equal to the sum of the squares of the other two sides** (called legs); see Figure 7.4. We can use this theorem to help solve problems.

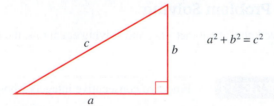

c

b

a

$a^2 + b^2 = c^2$

Figure 7.4

Classroom Example
Suppose that the lengths of the three sides of a right triangle are consecutive even integers. Find the lengths of the three sides.

EXAMPLE 13

Suppose that the lengths of the three sides of a right triangle are consecutive whole numbers. Find the lengths of the three sides.

Solution

Let s represent the length of the shortest leg. Then $s + 1$ represents the length of the other leg, and $s + 2$ represents the length of the hypotenuse. Using the Pythagorean theorem as a guideline, we obtain the following equation:

$$\underbrace{\text{Sum of squares of two legs}}_{s^2 + (s+1)^2} = \underbrace{\text{Square of hypotenuse}}_{(s+2)^2}$$

Solving this equation yields

$$s^2 + s^2 + 2s + 1 = s^2 + 4s + 4 \qquad \text{Remember } (a+b)^2 = a^2 + 2ab + b^2$$
$$2s^2 + 2s + 1 = s^2 + 4s + 4$$
$$s^2 + 2s + 1 = 4s + 4 \qquad \text{Add } -s^2 \text{ to both sides}$$
$$s^2 - 2s + 1 = 4$$
$$s^2 - 2s - 3 = 0$$
$$(s - 3)(s + 1) = 0$$
$$s - 3 = 0 \qquad \text{or} \qquad s + 1 = 0$$
$$s = 3 \qquad \text{or} \qquad s = -1$$

The solution of -1 is not possible for the length of a side, so the shortest side is of length 3. The other two sides ($s + 1$ and $s + 2$) have lengths of 4 and 5.

Concept Quiz 7.3

For Problems 1–10, answer true or false.

1. Any trinomial of the form $x^2 + bx + c$ can be factored (using integers) into the product of two binomials.
2. To factor $x^2 - 4x - 60$ we look for two numbers whose product is -60 and whose sum is -4.
3. A trinomial of the form $x^2 + bx + c$ will never have a common factor other than 1.
4. If n represents an odd integer, then $n + 1$ represents the next consecutive odd integer.
5. The Pythagorean theorem only applies to right triangles.
6. In a right triangle the longest side is called the hypotenuse.
7. The polynomial $x^2 + 25x + 72$ is not factorable.
8. The polynomial $x^2 + 27x + 72$ is not factorable.
9. The solution set of the equation $x^2 + 2x - 63 = 0$ is $\{-9, 7\}$.
10. The solution set of the equation $x^2 - 5x - 66 = 0$ is $\{-11, -6\}$.

Problem Set 7.3

For Problems 1–30, factor each trinomial completely. Indicate any that are not factorable using integers. (Objective 1)

1. $x^2 + 10x + 24$
2. $x^2 + 9x + 14$
3. $x^2 + 13x + 40$
4. $x^2 + 11x + 24$
5. $x^2 - 11x + 18$
6. $x^2 - 5x + 4$
7. $n^2 - 11n + 28$
8. $n^2 - 7n + 10$
9. $n^2 + 6n - 27$
10. $n^2 + 3n - 18$
11. $n^2 - 6n - 40$
12. $n^2 - 4n - 45$
13. $t^2 + 12t + 24$
14. $t^2 + 20t + 96$
15. $x^2 - 18x + 72$
16. $x^2 - 14x + 32$
17. $x^2 + 5x - 66$
18. $x^2 + 11x - 42$
19. $y^2 - y - 72$
20. $y^2 - y - 30$
21. $x^2 + 21x + 80$
22. $x^2 + 21x + 90$
23. $x^2 + 6x - 72$
24. $x^2 - 8x - 36$
25. $x^2 - 10x - 48$
26. $x^2 - 12x - 64$
27. $x^2 + 3xy - 10y^2$
28. $x^2 - 4xy - 12y^2$
29. $a^2 - 4ab - 32b^2$
30. $a^2 + 3ab - 54b^2$

For Problems 31–50, solve each equation. (Objective 2)

31. $x^2 + 10x + 21 = 0$
32. $x^2 + 9x + 20 = 0$
33. $x^2 - 9x + 18 = 0$
34. $x^2 - 9x + 8 = 0$
35. $x^2 - 3x - 10 = 0$
36. $x^2 - x - 12 = 0$

37. $n^2 + 5n - 36 = 0$

38. $n^2 + 3n - 18 = 0$

39. $n^2 - 6n - 40 = 0$

40. $n^2 - 8n - 48 = 0$

41. $t^2 + t - 56 = 0$

42. $t^2 + t - 72 = 0$

43. $x^2 - 16x + 28 = 0$

44. $x^2 - 18x + 45 = 0$

45. $x^2 + 11x = 12$

46. $x^2 + 8x = 20$

47. $x(x - 10) = -16$

48. $x(x - 12) = -35$

49. $-x^2 - 2x + 24 = 0$

50. $-x^2 + 6x + 16 = 0$

For Problems 51–68, set up an equation and solve each problem. **(Objectives 3 and 4)**

51. Find two consecutive integers whose product is 56.

52. Find two consecutive odd whole numbers whose product is 63.

53. Find two consecutive even whole numbers whose product is 168.

54. One number is 2 larger than another number. The sum of their squares is 100. Find the numbers.

55. Find four consecutive integers such that the product of the two larger integers is 22 less than twice the product of the two smaller integers.

56. Find three consecutive integers such that the product of the two smaller integers is 2 more than ten times the largest integer.

57. One number is 3 smaller than another number. The square of the larger number is 9 larger than ten times the smaller number. Find the numbers.

58. The area of the floor of a rectangular room is 84 square feet. The length of the room is 5 feet more than its width. Find the length and width of the room.

59. Suppose that the width of a certain rectangle is 3 inches less than its length. The area is numerically 6 less than twice the perimeter. Find the length and width of the rectangle.

60. The sum of the areas of a square and a rectangle is 64 square centimeters. The length of the rectangle is 4 centimeters more than a side of the square, and the width of the rectangle is 2 centimeters more than a side of the square. Find the dimensions of the square and the rectangle.

61. The perimeter of a rectangle is 30 centimeters, and the area is 54 square centimeters. Find the length and width of the rectangle. [*Hint*: Let w represent the width; then $15 - w$ represents the length.]

62. The perimeter of a rectangle is 44 inches, and its area is 120 square inches. Find the length and width of the rectangle.

63. An apple orchard contains 84 trees. The number of trees per row is five more than the number of rows. Find the number of rows.

64. A room contains 54 chairs. The number of rows is 3 less than the number of chairs per row. Find the number of rows.

65. Suppose that one leg of a right triangle is 7 feet shorter than the other leg. The hypotenuse is 2 feet longer than the longer leg. Find the lengths of all three sides of the right triangle.

66. Suppose that one leg of a right triangle is 7 meters longer than the other leg. The hypotenuse is 1 meter longer than the longer leg. Find the lengths of all three sides of the right triangle.

67. Suppose that the length of one leg of a right triangle is 2 inches less than the length of the other leg. If the length of the hypotenuse is 10 inches, find the length of each leg.

68. The length of one leg of a right triangle is 3 centimeters more than the length of the other leg. The length of the hypotenuse is 15 centimeters. Find the lengths of the two legs.

Thoughts Into Words

69. What does the expression "not factorable using integers" mean to you?

70. Discuss the role that factoring plays in solving equations.

71. Explain how you would solve the equation

$$(x - 3)(x + 4) = 0$$

and also how you would solve

$$(x - 3)(x + 4) = 8$$

Further Investigations

For Problems 72–75, factor each trinomial and assume that all variables appearing as exponents represent positive integers.

72. $x^{2a} + 10x^a + 24$ **73.** $x^{2a} + 13x^a + 40$

74. $x^{2a} - 2x^a - 8$ **75.** $x^{2a} + 6x^a - 27$

76. Suppose that we want to factor $n^2 + 26n + 168$ so that we can solve the equation $n^2 + 26n + 168 = 0$. We need to find two positive integers whose product is 168 and whose sum is 26. Since the constant term, 168, is rather large, let's look at it in prime factored form:

$$168 = 2 \cdot 2 \cdot 2 \cdot 3 \cdot 7$$

Now we can mentally form two numbers by using all of these factors in different combinations. Using two 2s and the 3 in one number and the other 2 and the 7 in another number produces $2 \cdot 2 \cdot 3 = 12$ and $2 \cdot 7 = 14$.

Therefore, we can solve the given equation as follows:

$$n^2 + 26n + 168 = 0$$
$$(n + 12)(n + 14) = 0$$

$n + 12 = 0$ or $n + 14 = 0$

       $n = -12$ or $n = -14$

The solution set is $\{-14, -12\}$.

Solve each of the following equations.

(a) $n^2 + 30n + 216 = 0$

(b) $n^2 + 35n + 294 = 0$

(c) $n^2 - 40n + 384 = 0$

(d) $n^2 - 40n + 375 = 0$

(e) $n^2 + 6n - 432 = 0$

(f) $n^2 - 16n - 512 = 0$

Answers to the Concept Quiz

1. False   **2.** True   **3.** True   **4.** False   **5.** True   **6.** True   **7.** True   **8.** False   **9.** True   **10.** False

---

7.4   Factoring Trinomials of the Form $ax^2 + bx + c$

OBJECTIVES

1   Factor trinomials where the leading coefficient is not 1

2   Solve equations that involve factoring

Now let's consider factoring trinomials where the coefficient of the squared term is not 1. We present here an informal trial and error technique that works quite well for certain types of trinomials. This technique simply relies on our knowledge of the multiplication of binomials.

Classroom Example
Factor $3x^2 + 10x + 8$.

> **EXAMPLE 1**   Factor $2x^2 + 7x + 3$.

Solution

By looking at the first term, $2x^2$, and the positive signs of the other two terms, we know that the binomials are of the form

$$(2x + \underline{\hspace{0.5cm}})(x + \underline{\hspace{0.5cm}})$$

Since the factors of the constant term, 3, are 1 and 3, we have only two possibilities to try:

$$(2x + 3)(x + 1) \qquad \text{or} \qquad (2x + 1)(x + 3)$$

By checking the middle term of both of these products, we find that the second one yields the correct middle term of $7x$. Therefore,

$$2x^2 + 7x + 3 = (2x + 1)(x + 3)$$

Classroom Example
Factor $15y^2 - 13y + 2$.

EXAMPLE 2   Factor $6x^2 - 17x + 5$.

Solution

First, we note that $6x^2$ can be written as $2x \cdot 3x$ or $6x \cdot x$. Second, since the middle term of the trinomial is negative, and the last term is positive, we know that the binomials are of the form

$$(2x - \underline{\hspace{1cm}})(3x - \underline{\hspace{1cm}}) \qquad \text{or} \qquad (6x - \underline{\hspace{1cm}})(x - \underline{\hspace{1cm}})$$

Since the factors of the constant term, 5, are 1 and 5, we have the following possibilities:

$$(2x - 5)(3x - 1) \qquad (2x - 1)(3x - 5)$$
$$(6x - 5)(x - 1) \qquad (6x - 1)(x - 5)$$

By checking the middle term for each of these products, we find that the product $(2x - 5)(3x - 1)$ produces the desired term of $-17x$. Therefore,

$$6x^2 - 17x + 5 = (2x - 5)(3x - 1)$$

Classroom Example
Factor $18n^2 + 12n - 16$.

EXAMPLE 3   Factor $8x^2 - 8x - 30$.

Solution

First, we note that the polynomial $8x^2 - 8x - 30$ has a common factor of 2. Factoring out the common factor gives us $2(4x^2 - 4x - 15)$. Now we need to factor $4x^2 - 4x - 15$.

We note that $4x^2$ can be written as $4x \cdot x$ or $2x \cdot 2x$. The last term, -15, can be written as $(1)(-15)$, $(-1)(15)$, $(3)(-5)$, or $(-3)(5)$. Thus we can generate the possibilities for the binomial factors as follows:

Using 1 and -15	**Using -1 and 15**
$(4x - 15)(x + 1)$	$(4x - 1)(x + 15)$
$(4x + 1)(x - 15)$	$(4x + 15)(x - 1)$
$(2x + 1)(2x - 15)$	$(2x - 1)(2x + 15)$

Using 3 and -5	**Using -3 and 5**
$(4x + 3)(x - 5)$	$(4x - 3)(x + 5)$
$(4x - 5)(x + 3)$	$(4x + 5)(x - 3)$
✓ $(2x - 5)(2x + 3)$	$(2x + 5)(2x - 3)$

By checking the middle term of each of these products, we find that the product indicated with a check mark produces the desired middle term of $-4x$. Therefore,

$$8x^2 - 8x - 30 = 2(4x^2 - 4x - 15) = 2(2x - 5)(2x + 3)$$

Let's pause for a moment and look back over Examples 1, 2, and 3. Obviously, Example 3 created the most difficulty because we had to consider so many possibilities. We have suggested one possible format for considering the possibilities, but as you practice such problems, you may develop a format that works better for you. Regardless of the format that you use, the key idea is to organize your work so that you consider all possibilities. Let's look at another example.

Classroom Example
Factor $9x^2 + 10x + 4$.

EXAMPLE 4   Factor $4x^2 + 6x + 9$.

Solution

First, we note that $4x^2$ can be written as $4x \cdot x$ or $2x \cdot 2x$. Second, since the middle term is positive and the last term is positive, we know that the binomials are of the form

$$(4x + \underline{\hspace{1cm}})(x + \underline{\hspace{1cm}}) \qquad \text{or} \qquad (2x + \underline{\hspace{1cm}})(2x + \underline{\hspace{1cm}})$$

Since 9 can be written as $9 \cdot 1$ or $3 \cdot 3$, we have only the five following possibilities to try:

$$(4x + 9)(x + 1) \qquad (4x + 1)(x + 9)$$
$$(4x + 3)(x + 3) \qquad (2x + 1)(2x + 9)$$
$$(2x + 3)(2x + 3)$$

When we try all of these possibilities, we find that none of them yields a middle term of $6x$. Therefore, $4x^2 + 6x + 9$ is *not factorable* using integers.

Remark: Example 4 illustrates the importance of organizing your work so that you try *all* possibilities before you conclude that a particular trinomial is not factorable.

Another Approach

There is another, more systematic technique that you may wish to use with some trinomials. It is an extension of the method we used in the previous section. Recall that at the beginning of Section 7.3 we looked at the following product:

$$(x + r)(x + s) = x(x) + x(s) + r(x) + r(s)$$
$$= x^2 + (s + r)x + rs$$

Sum of r and s Product of r and s

Now let's look at this product:

$$(px + r)(qx + s) = px(qx) + px(s) + r(qx) + r(s)$$
$$= (pq)x^2 + (ps + rq)x + rs$$

Notice that the product of the coefficient of the x^2 term, (pq), and the constant term, (rs), is $pqrs$. Likewise, the product of the two coefficients of x, (ps and rq), is also $pqrs$. Therefore, the two coefficients of x must have a sum of $ps + rq$ and a product of $pqrs$. To begin the factoring process, we will look for two factors of the product $pqrs$ whose sum is equal to the coefficient of the x term. This may seem a little confusing, but the next few examples will show how easy it is to apply.

Classroom Example
Factor $2y^2 + 19y + 24$.

EXAMPLE 5   Factor $3x^2 + 14x + 8$.

Solution

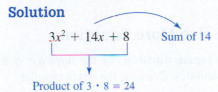

$$3x^2 + 14x + 8 \qquad \text{Sum of } 14$$

Product of $3 \cdot 8 = 24$

We need to find two integers whose product is 24 and whose sum is 14. Obviously, 2 and 12 satisfy these conditions. Therefore, we can express the middle term of the trinomial, $14x$, as $2x + 12x$ and proceed as follows:

$$3x^2 + 14x + 8 = 3x^2 + 2x + 12x + 8$$
$$= x(3x + 2) + 4(3x + 2) \qquad \text{Factor by grouping}$$
$$= (3x + 2)(x + 4)$$

Classroom Example
Factor $18x^2 - 13x + 2$.

EXAMPLE 6   Factor $16x^2 - 26x + 3$.

Solution

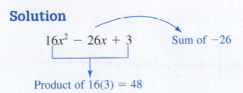

$$16x^2 - 26x + 3 \qquad \text{Sum of } -26$$

Product of $16(3) = 48$

We need two integers whose product is 48 and whose sum is -26. The integers -2 and -24 satisfy these conditions and allow us to express the middle term, $-26x$, as $-2x - 24x$. Then we can factor as follows:

$$16x^2 - 26x + 3 = 16x^2 - 2x - 24x + 3$$
$$= 2x(8x - 1) - 3(8x - 1) \qquad \text{Factor by grouping}$$
$$= (8x - 1)(2x - 3)$$

Classroom Example
Factor $8p^2 + 22p - 21$.

EXAMPLE 7   Factor $6x^2 - 5x - 6$.

Solution

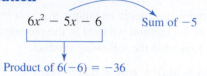

$6x^2 - 5x - 6$ Sum of -5

Product of $6(-6) = -36$

We need two integers whose product is -36 and whose sum is -5. Furthermore, since the sum is negative, the absolute value of the negative number must be greater than the absolute value of the positive number. A little searching will determine that the numbers are -9 and 4. Thus we can express the middle term of $-5x$ as $-9x + 4x$ and proceed as follows:

$$6x^2 - 5x - 6 = 6x^2 - 9x + 4x - 6$$
$$= 3x(2x - 3) + 2(2x - 3)$$
$$= (2x - 3)(3x + 2)$$

Now that we have shown you two possible techniques for factoring trinomials of the form $ax^2 + bx + c$, the ball is in your court. Practice may not make you perfect at factoring, but it will surely help. We are not promoting one technique over the other; that is an individual choice. Many people find the trial and error technique that we presented first to be very useful if the number of possibilities for the factors is fairly small. However, as the list of possibilities grows, the second technique does have the advantage of being systematic. So perhaps having both techniques at your fingertips is your best bet.

Now We Can Solve More Equations

The ability to factor certain trinomials of the form $ax^2 + bx + c$ provides us with greater equation-solving capabilities. Consider the next examples.

Classroom Example
Solve $2x^2 + 17x + 21 = 0$.

EXAMPLE 8   Solve $3x^2 + 17x + 10 = 0$.

Solution

$$3x^2 + 17x + 10 = 0$$
$$(x + 5)(3x + 2) = 0 \qquad \text{Factoring } 3x^2 + 17x + 10 \text{ as } (x + 5)(3x + 2)$$
$$\text{may require some extra work on scratch paper}$$

$$x + 5 = 0 \quad \text{or} \quad 3x + 2 = 0 \qquad ab = 0 \text{ if and only if } a = 0 \text{ or } b = 0$$
$$x = -5 \quad \text{or} \quad 3x = -2$$
$$x = -5 \quad \text{or} \quad x = -\frac{2}{3}$$

The solution set is $\left\{-5, -\dfrac{2}{3}\right\}$. Check it!

Classroom Example
Solve $40x^2 + 3x - 1 = 0$.

EXAMPLE 9   Solve $24x^2 + 2x - 15 = 0$.

Solution

$$24x^2 + 2x - 15 = 0$$

$$(4x - 3)(6x + 5) = 0$$

$$4x - 3 = 0 \qquad \text{or} \qquad 6x + 5 = 0$$

$$4x = 3 \qquad \text{or} \qquad 6x = -5$$

$$x = \frac{3}{4} \qquad \text{or} \qquad x = -\frac{5}{6}$$

The solution set is $\left\{-\dfrac{5}{6}, \dfrac{3}{4}\right\}$.

Concept Quiz 7.4

For Problems 1–10, answer true or false.

1. Any trinomial of the form $ax^2 + bx + c$ can be factored (using integers) into the product of two binomials.

2. To factor $2x^2 - x - 3$, we look for two numbers whose product is -3 and whose sum is -1.

3. A trinomial of the form $ax^2 + bx + c$ will never have a common factor other than 1.

4. The factored form $(x + 3)(2x + 4)$ is factored completely.

5. The difference-of-squares polynomial $9x^2 - 25$ could be written as the trinomial $9x^2 + 0x - 25$.

6. The polynomial $12x^2 + 11x - 12$ is not factorable.

7. The solution set of the equation $6x^2 + 13x - 5 = 0$ is $\left\{\dfrac{1}{3}, \dfrac{2}{5}\right\}$.

8. The solution set of the equation $18x^2 - 39x + 20 = 0$ is $\left\{\dfrac{5}{6}, \dfrac{4}{3}\right\}$.

9. The completely factored form of $-3x^3y - 3x^2y + 18xy$ is $-3xy(x - 2)(x + 3)$.

10. The completely factored form of $-x^3 - 7x^2 - 12x$ is $-x(x^2 + 7x + 12)$.

Problem Set 7.4

For Problems 1–50, factor each of the trinomials completely. Indicate any that are not factorable using integers. **(Objective 1)**

1. $3x^2 + 7x + 2$

2. $2x^2 + 9x + 4$

3. $6x^2 + 19x + 10$

4. $12x^2 + 19x + 4$

5. $4x^2 - 25x + 6$

6. $5x^2 - 22x + 8$

7. $12x^2 - 31x + 20$

8. $8x^2 - 30x + 7$

9. $5y^2 - 33y - 14$

10. $6y^2 - 4y - 16$

11. $4n^2 + 26n - 48$

12. $4n^2 + 17n - 15$

13. $2x^2 + x + 7$

14. $7x^2 + 19x + 10$

15. $18x^2 + 45x + 7$

16. $10x^2 + x - 5$

17. $7x^2 - 30x + 8$

18. $6x^2 - 17x + 12$

19. $8x^2 + 2x - 21$

20. $9x^2 + 15x - 14$

21. $9t^3 - 15t^2 - 14t$

22. $12t^3 - 20t^2 - 25t$

23. $12y^2 + 79y - 35$

24. $9y^2 + 52y - 12$

25. $6n^2 + 2n - 5$

26. $20n^2 - 27n + 9$

27. $14x^2 + 55x + 21$

28. $15x^2 + 34x + 15$

29. $20x^2 - 31x + 12$

30. $8t^2 - 3t - 4$

31. $16n^2 - 8n - 15$

32. $25n^2 - 20n - 12$

33. $24x^2 - 50x + 25$

34. $24x^2 - 41x + 12$

35. $2x^2 + 25x + 72$

36. $2x^2 + 23x + 56$

37. $21a^2 + a - 2$

38. $14a^2 + 5a - 24$

39. $12a^2 - 31a - 15$

40. $10a^2 - 39a - 4$

41. $12x^2 + 36x + 27$

42. $27x^2 - 36x + 12$

43. $6x^2 - 5xy + y^2$

44. $12x^2 + 13xy + 3y^2$

45. $20x^2 + 7xy - 6y^2$

46. $8x^2 - 6xy - 35y^2$

47. $5x^2 - 32x + 12$

48. $3x^2 - 35x + 50$

49. $8x^2 - 55x - 7$

50. $12x^2 - 67x - 30$

For Problems 51–80, solve each equation. **(Objective 2)**

51. $2x^2 + 13x + 6 = 0$

52. $3x^2 + 16x + 5 = 0$

53. $12x^2 + 11x + 2 = 0$

54. $15x^2 + 56x + 20 = 0$

55. $3x^2 - 25x + 8 = 0$

56. $4x^2 - 31x + 21 = 0$

57. $15n^2 - 41n + 14 = 0$

58. $6n^2 - 31n + 40 = 0$

59. $6t^2 + 37t - 35 = 0$

60. $2t^2 + 15t - 27 = 0$

61. $16y^2 - 18y - 9 = 0$

62. $9y^2 - 15y - 14 = 0$

63. $9x^2 - 6x - 8 = 0$

64. $12n^2 + 28n - 5 = 0$

65. $10x^2 - 29x + 10 = 0$

66. $4x^2 - 16x + 15 = 0$

67. $6x^2 + 19x = -10$

68. $12x^2 + 17x = -6$

69. $16x(x + 1) = 5$

70. $5x(5x + 2) = 8$

71. $35n^2 - 34n - 21 = 0$

72. $18n^2 - 3n - 28 = 0$

73. $4x^2 - 45x + 50 = 0$

74. $7x^2 - 65x + 18 = 0$

75. $7x^2 + 46x - 21 = 0$

76. $2x^2 + 7x - 30 = 0$

77. $12x^2 - 43x - 20 = 0$

78. $14x^2 - 13x - 12 = 0$

79. $18x^2 + 55x - 28 = 0$

80. $24x^2 + 17x - 20 = 0$

Thoughts Into Words

81. Explain your thought process when factoring $24x^2 - 17x - 20$.

82. Your friend factors $8x^2 - 32x + 32$ as follows:

$8x^2 - 32x + 32 = (4x - 8)(2x - 4)$

$\qquad\qquad\qquad = 4(x - 2)(2)(x - 2)$

$\qquad\qquad\qquad = 8(x - 2)(x - 2)$

Is she correct? Do you have any suggestions for her?

83. Your friend solves the equation $8x^2 - 32x + 32 = 0$ as follows:

$8x^2 - 32x + 32 = 0$

$(4x - 8)(2x - 4) = 0$

$4x - 8 = 0 \qquad$ or $\qquad 2x - 4 = 0$

$4x = 8 \qquad$ or $\qquad 2x = 4$

$x = 2 \qquad$ or $\qquad x = 2$

The solution set is $\{2\}$. Is she correct? Do you have any suggestions for her?

Further Investigations

84. Consider the following approach to factoring $20x^2 + 39x + 18$:

$20x^2 + 39x + 18 \qquad$ Sum of 39

Product of $20(18) = 360$

We need two integers whose sum is 39 and whose product is 360. To help find these integers, let's prime factor 360.

$360 = 2 \cdot 2 \cdot 2 \cdot 3 \cdot 3 \cdot 5$

Now by grouping these factors in various ways, we find that $2 \cdot 2 \cdot 2 \cdot 3 = 24$, $3 \cdot 5 = 15$, and $24 + 15 = 39$.

So the numbers are 15 and 24, and we can express the middle term of the given trinomial, $39x$, as $15x + 24x$. Therefore, we can complete the factoring as follows:

$$20x^2 + 39x + 18 = 20x^2 + 15x + 24x + 18$$
$$= 5x(4x + 3) + 6(4x + 3)$$
$$= (4x + 3)(5x + 6)$$

Factor each of the following trinomials.

(a) $20x^2 + 41x + 20$

(b) $24x^2 - 79x + 40$

(c) $30x^2 + 23x - 40$

(d) $36x^2 + 65x - 36$

Answers to the Concept Quiz

1. False   **2.** False   **3.** False   **4.** False   **5.** True   **6.** True   **7.** False   **8.** True   **9.** True   **10.** False

7.5   Factoring, Solving Equations, and Problem Solving

OBJECTIVES

1 Factor perfect-square trinomials

2 Recognize the different types of factoring patterns

3 Use factoring to solve equations

4 Solve word problems that involve factoring

Factoring

Before we summarize our work with factoring techniques, let's look at two more special factoring patterns. These patterns emerge when multiplying binomials. Consider the following examples.

$$(x + 5)^2 = (x + 5)(x + 5) = x^2 + 10x + 25$$
$$(2x + 3)^2 = (2x + 3)(2x + 3) = 4x^2 + 12x + 9$$
$$(4x + 7)^2 = (4x + 7)(4x + 7) = 16x^2 + 56x + 49$$

In general, $(a + b)^2 = (a + b)(a + b) = a^2 + 2ab + b^2$. Also,

$$(x - 6)^2 = (x - 6)(x - 6) = x^2 - 12x + 36$$
$$(3x - 4)^2 = (3x - 4)(3x - 4) = 9x^2 - 24x + 16$$
$$(5x - 2)^2 = (5x - 2)(5x - 2) = 25x^2 - 20x + 4$$

In general, $(a - b)^2 = (a - b)(a - b) = a^2 - 2ab + b^2$. Therefore, we have the following patterns.

Perfect Square Trinomials

$$a^2 + 2ab + b^2 = (a + b)^2$$
$$a^2 - 2ab + b^2 = (a - b)^2$$

Trinomials of the form $a^2 + 2ab + b^2$ or $a^2 - 2ab + b^2$ are called **perfect square trinomials**. They are easy to recognize because of the nature of their terms. For example, $9x^2 + 30x + 25$ is a perfect square trinomial for these reasons:

1. The first term is a square: $(3x)^2$.

2. The last term is a square: $(5)^2$.

3. The middle term is twice the product of the quantities being squared in the first and last terms: $2(3x)(5)$.

Likewise, $25x^2 - 40xy + 16y^2$ is a perfect square trinomial for these reasons:

1. The first term is a square: $(5x)^2$.
2. The last term is a square: $(4y)^2$.
3. The middle term is twice the product of the quantities being squared in the first and last terms: $2(5x)(4y)$.

Once we know that we have a perfect square trinomial, the factoring process follows immediately from the two basic patterns.

$$9x^2 + 30x + 25 = (3x + 5)^2$$
$$25x^2 - 40xy + 16y^2 = (5x - 4y)^2$$

Here are some additional examples of perfect square trinomials and their factored form.

$$
\begin{aligned}
x^2 - 16x + 64 &= (x)^2 - 2(x)(8) + (8)^2 &= (x - 8)^2 \\
16x^2 - 56x + 49 &= (4x)^2 - 2(4x)(7) + (7)^2 &= (4x - 7)^2 \\
25x^2 + 20xy + 4y^2 &= (5x)^2 + 2(5x)(2y) + (2y)^2 &= (5x + 2y)^2 \\
1 + 6y + 9y^2 &= (1)^2 + 2(1)(3y) + (3y)^2 &= (1 + 3y)^2 \\
4m^2 - 4mn + n^2 &= (2m)^2 - 2(2m)(n) + (n)^2 &= (2m - n)^2
\end{aligned}
$$

You may want to do this step mentally after you feel comfortable with the process

We have considered some basic factoring techniques in this chapter one at a time, but you must be able to apply them as needed in a variety of situations. So, let's first summarize the techniques and then consider some examples.

In this chapter we have discussed these techniques:

1. Factoring by using the distributive property to factor out the greatest common monomial or binomial factor
2. Factoring by grouping
3. Factoring by applying the difference-of-squares pattern
4. Factoring by applying the perfect-square-trinomial pattern
5. Factoring trinomials of the form $x^2 + bx + c$ into the product of two binomials
6. Factoring trinomials of the form $ax^2 + bx + c$ into the product of two binomials

As a general guideline, **always look for a greatest common monomial factor first**, and then proceed with the other factoring techniques.

In each of the following examples we have factored completely whenever possible. Study them carefully, and notice the factoring techniques we used.

1. $2x^2 + 12x + 10 = 2(x^2 + 6x + 5) = 2(x + 1)(x + 5)$
2. $4x^2 + 36 = 4(x^2 + 9)$ Remember that the sum of two squares is not factorable using integers unless there is a common factor
3. $4t^2 + 20t + 25 = (2t + 5)^2$ If you fail to recognize a perfect trinomial square, no harm is done; simply proceed to factor into the product of two binomials, and then you will recognize that the two binomials are the same

4. $x^2 - 3x - 8$ is not factorable using integers. This becomes obvious from the table.

Product	Sum
$1(-8) = -8$	$1 + (-8) = -7$
$-1(8) = -8$	$-1 + 8 = 7$
$2(-4) = -8$	$2 + (-4) = -2$
$-2(4) = -8$	$-2 + 4 = 2$

No two factors of -8 produce a sum of -3.

5. $6y^2 - 13y - 28 = (2y - 7)(3y + 4)$. We found the binomial factors as follows:

$(y + \underline{\quad})(6y - \underline{\quad})$ Factors of 28

or $1 \cdot 28$ or $28 \cdot 1$

$(y - \underline{\quad})(6y + \underline{\quad})$ $2 \cdot 14$ or $14 \cdot 2$

or $4 \cdot 7$ or $\boxed{7 \cdot 4}$

$(2y - \underline{\quad})(3y + \underline{\quad})$ ◀

or

$(2y + \underline{\quad})(3y - \underline{\quad})$

6. $32x^2 - 50y^2 = 2(16x^2 - 25y^2) = 2(4x + 5y)(4x - 5y)$

Solving Equations by Factoring

As stated in the preface, there is a common thread that runs throughout this text: namely, *learn a skill*; next, *use the skill to help solve equations*; and then *use equations to help solve application problems*. This thread becomes evident in this chapter. After presenting a factoring technique, we immediately solved some equations using this technique, and then considered some applications involving such equations. The following steps summarize the equation solving process in this chapter.

1. Organize all terms of the polynomial on the same side of the equation with zero on the other side.

2. Factor the polynomial. This will involve a variety of factoring techniques presented in this chapter.

3. Set each factor equal to zero and solve for the unknown.

4. Check your solutions back into the original equation.

Let's consider some examples.

Classroom Example
Solve $x^2 = 64x$.

EXAMPLE 1   Solve $x^2 = 25x$.

Solution

$$x^2 = 25x$$
$$x^2 - 25x = 0 \qquad \text{Added } -25x \text{ to both sides}$$
$$x(x - 25) = 0$$
$$x = 0 \quad \text{or} \quad x - 25 = 0$$
$$x = 0 \quad \text{or} \quad x = 25$$

The solution set is $\{0, 25\}$. Check it!

Classroom Example
Solve $m^3 - 144m = 0$.

EXAMPLE 2   Solve $x^3 - 36x = 0$.

Solution

$$x^3 - 36x = 0$$
$$x(x^2 - 36) = 0$$
$$x(x + 6)(x - 6) = 0$$

$x = 0$ or $x + 6 = 0$ or $x - 6 = 0$ If $abc = 0$, then $a = 0$, $b = 0$,

$x = 0$ or $x = -6$ or $x = 6$ or $c = 0$

The solution set is $\{-6, 0, 6\}$. Does it check?

Classroom Example
Solve $8x^2 + 10x - 7 = 0$.

EXAMPLE 3   Solve $10x^2 - 13x - 3 = 0$.

Solution

$$10x^2 - 13x - 3 = 0$$
$$(5x + 1)(2x - 3) = 0$$

$5x + 1 = 0$ or $2x - 3 = 0$

$5x = -1$ or $2x = 3$

$x = -\dfrac{1}{5}$ or $x = \dfrac{3}{2}$

The solution set is $\left\{-\dfrac{1}{5}, \dfrac{3}{2}\right\}$. Does it check?

Classroom Example
Solve $9x^2 - 24x + 16 = 0$.

EXAMPLE 4   Solve $4x^2 - 28x + 49 = 0$.

Solution

$$4x^2 - 28x + 49 = 0$$
$$(2x - 7)^2 = 0$$
$$(2x - 7)(2x - 7) = 0$$

$2x - 7 = 0$ or $2x - 7 = 0$

$2x = 7$ or $2x = 7$

$x = \dfrac{7}{2}$ or $x = \dfrac{7}{2}$

The solution set is $\left\{\dfrac{7}{2}\right\}$.

Pay special attention to the next example. We need to change the form of the original equation before we can apply the property $ab = 0$ if and only if $a = 0$ or $b = 0$. A necessary condition of this property is that an indicated product is set equal to zero.

Classroom Example
Solve $(x + 6)(x + 3) = 4$.

EXAMPLE 5   Solve $(x + 1)(x + 4) = 40$.

Solution

$$(x + 1)(x + 4) = 40$$
$$x^2 + 5x + 4 = 40$$
$$x^2 + 5x - 36 = 0$$
$$(x + 9)(x - 4) = 0$$

$$x + 9 = 0 \qquad \text{or} \qquad x - 4 = 0$$
$$x = -9 \qquad \text{or} \qquad x = 4$$

The solution set is $\{-9, 4\}$. Check it!

EXAMPLE 6 Solve $2n^2 + 16n - 40 = 0$.

Solution

$$2n^2 + 16n - 40 = 0$$
$$2(n^2 + 8n - 20) = 0$$
$$n^2 + 8n - 20 = 0 \qquad \text{Multiplied both sides by } \frac{1}{2}$$
$$(n + 10)(n - 2) = 0$$
$$n + 10 = 0 \qquad \text{or} \qquad n - 2 = 0$$
$$n = -10 \qquad \text{or} \qquad n = 2$$

The solution set is $\{-10, 2\}$. Does it check?

Problem Solving

Reminder: Throughout this book we highlight the need to *learn a skill, to use that skill to help solve equations*, and then *to use equations to help solve problems*. Our new factoring skills have provided more ways of solving equations, which in turn gives us more power to solve word problems. We conclude the chapter by solving a few more problems.

EXAMPLE 7

Find two numbers whose product is 65 if one of the numbers is 3 more than twice the other number.

Solution

Let n represent one of the numbers; then $2n + 3$ represents the other number. Since their product is 65, we can set up and solve the following equation:

$$n(2n + 3) = 65$$
$$2n^2 + 3n - 65 = 0$$
$$(2n + 13)(n - 5) = 0$$
$$2n + 13 = 0 \qquad \text{or} \qquad n - 5 = 0$$
$$2n = -13 \qquad \text{or} \qquad n = 5$$
$$n = -\frac{13}{2} \qquad \text{or} \qquad n = 5$$

If $n = -\dfrac{13}{2}$, then $2n + 3 = 2\left(-\dfrac{13}{2}\right) + 3 = -10$. However, if $n = 5$, then $2n + 3 = 2(5) + 3 = 13$. Thus the numbers are $-\dfrac{13}{2}$ and -10, or 5 and 13.

EXAMPLE 8

The area of a triangular sheet of paper is 14 square inches. One side of the triangle is 3 inches longer than the altitude to that side. Find the length of the one side and the length of the altitude to that side.

Solution

Let h represent the altitude to the side. Then $h + 3$ represents the side of the triangle (see Figure 7.5).

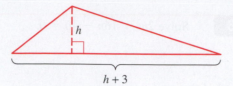

Figure 7.5

Since the formula for finding the area of a triangle is $A = \dfrac{1}{2}bh$, we have

$$\frac{1}{2}h(h + 3) = 14$$

$$h(h + 3) = 28 \qquad \text{\color{blue}Multiplied both sides by 2}$$

$$h^2 + 3h = 28$$

$$h^2 + 3h - 28 = 0$$

$$(h + 7)(h - 4) = 0$$

$$h + 7 = 0 \qquad \text{or} \qquad h - 4 = 0$$

$$h = -7 \qquad \text{or} \qquad h = 4$$

The solution of -7 is not reasonable. Thus the altitude is 4 inches, and the length of the side to which that altitude is drawn is 7 inches.

Classroom Example
A photograph measures 14 inches wide and 19 inches long. A strip of uniform width is to be cut off from both ends and both sides of the photograph in order to reduce the area of the photograph to 176 square inches. Find the width of the strip.

EXAMPLE 9

A strip with a uniform width is shaded along both sides and both ends of a rectangular poster with dimensions 12 inches by 16 inches. How wide is the strip if one-half of the area of the poster is shaded?

Solution

Let x represent the width of the shaded strip of the poster in Figure 7.6. The area of the strip is one-half of the area of the poster; therefore, it is $\dfrac{1}{2}(12)(16) = 96$ square inches. Furthermore, we can represent the area of the strip around the poster by the words *the area of the poster minus the area of the unshaded portion*.

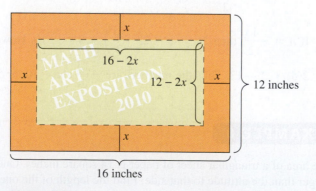

Figure 7.6

Thus we can set up and solve the following equation:

Area of poster $-$ Area of unshaded portion $=$ Area of strip

$$16(12) \quad - \quad (16 - 2x)(12 - 2x) \quad = \quad 96$$

$$192 - (192 - 56x + 4x^2) = 96$$
$$192 - 192 + 56x - 4x^2 = 96$$
$$-4x^2 + 56x - 96 = 0$$
$$x^2 - 14x + 24 = 0$$
$$(x - 12)(x - 2) = 0$$
$$x - 12 = 0 \quad \text{or} \quad x - 2 = 0$$
$$x = 12 \quad \text{or} \quad x = 2$$

Obviously, the strip cannot be 12 inches wide because the total width of the poster is 12 inches. Thus we must disregard the solution of 12 and conclude that the strip is 2 inches wide.

Concept Quiz 7.5

For Problems 1–7, match each factoring problem with the name of the type of pattern that would be used to factor the problem.

1. $x^2 + 2xy + y^2$ **A.** Trinomial with an x-squared coefficient of one

2. $x^2 - y^2$ **B.** Common binomial factor

3. $ax + ay + bx + by$ **C.** Difference of two squares

4. $x^2 + bx + c$ **D.** Common factor

5. $ax^2 + bx + c$ **E.** Factor by grouping

6. $ax^2 + ax + a$ **F.** Perfect-square trinomial

7. $(a + b)x + (a + b)y$ **G.** Trinomial with an x-squared coefficient of not one

Problem Set 7.5

For Problems 1–12, factor each of the perfect square trinomials. **(Objective 1)**

1. $x^2 + 4x + 4$ **2.** $x^2 + 18x + 81$

3. $x^2 - 10x + 25$ **4.** $x^2 - 24x + 144$

5. $9n^2 + 12n + 4$ **6.** $25n^2 + 30n + 9$

7. $16a^2 - 8a + 1$ **8.** $36a^2 - 84a + 49$

9. $4 + 36x + 81x^2$ **10.** $1 - 4x + 4x^2$

11. $16x^2 - 24xy + 9y^2$

12. $64x^2 + 16xy + y^2$

For Problems 13–40, factor each polynomial completely. Indicate any that are not factorable using integers. **(Objective 2)**

13. $2x^2 + 17x + 8$ **14.** $x^2 + 19x$

15. $2x^3 - 72x$ **16.** $30x^2 - x - 1$

17. $n^2 - 7n - 60$ **18.** $4n^3 - 100n$

19. $3a^2 - 7a - 4$ **20.** $a^2 + 7a - 30$

21. $8x^2 + 72$ **22.** $3y^3 - 36y^2 + 96y$

23. $9x^2 + 30x + 25$ **24.** $5x^2 - 5x - 6$

25. $15x^2 + 65x + 70$ **26.** $4x^2 - 20xy + 25y^2$

27. $24x^2 + 2x - 15$ **28.** $9x^2y - 27xy$

29. $xy + 5y - 8x - 40$

30. $xy - 3y + 9x - 27$

31. $20x^2 + 31xy - 7y^2$

32. $2x^2 - xy - 36y^2$ **33.** $24x^2 + 18x - 81$

34. $30x^2 + 55x - 50$ **35.** $12x^2 + 6x + 30$

36. $24x^2 - 8x + 32$ **37.** $5x^4 - 80$

38. $3x^5 - 3x$

39. $x^2 + 12xy + 36y^2$

40. $4x^2 - 28xy + 49y^2$

For Problems 41–70, solve each equation. (Objective 3)

41. $4x^2 - 20x = 0$ **42.** $-3x^2 - 24x = 0$

43. $x^2 - 9x - 36 = 0$ **44.** $x^2 + 8x - 20 = 0$

45. $-2x^3 + 8x = 0$ **46.** $4x^3 - 36x = 0$

47. $6n^2 - 29n - 22 = 0$ **48.** $30n^2 - n - 1 = 0$

49. $(3n - 1)(4n - 3) = 0$ **50.** $(2n - 3)(7n + 1) = 0$

51. $(n - 2)(n + 6) = -15$

52. $(n + 3)(n - 7) = -25$

53. $2x^2 = 12x$ **54.** $-3x^2 = 15x$

55. $t^3 - 2t^2 - 24t = 0$ **56.** $2t^3 - 16t^2 - 18t = 0$

57. $12 - 40x + 25x^2 = 0$

58. $12 - 7x - 12x^2 = 0$

59. $n^2 - 28n + 192 = 0$

60. $n^2 + 33n + 270 = 0$

61. $(3n + 1)(n + 2) = 12$

62. $(2n + 5)(n + 4) = -1$

63. $x^3 = 6x^2$

64. $x^3 = -4x^2$

65. $9x^2 - 24x + 16 = 0$

66. $25x^2 + 60x + 36 = 0$

67. $x^3 + 10x^2 + 25x = 0$

68. $x^3 - 18x^2 + 81x = 0$

69. $24x^2 + 17x - 20 = 0$

70. $24x^2 + 74x - 35 = 0$

For Problems 71–88, set up an equation and solve each problem. (Objective 4)

71. Find two numbers whose product is 15 such that one of the numbers is seven more than four times the other number.

72. Find two numbers whose product is 12 such that one of the numbers is four less than eight times the other number.

73. Find two numbers whose product is -1. One of the numbers is three more than twice the other number.

74. Suppose that the sum of the squares of three consecutive integers is 110. Find the integers.

75. A number is one more than twice another number. The sum of the squares of the two numbers is 97. Find the numbers.

76. A number is one less than three times another number. If the product of the two numbers is 102, find the numbers.

77. In an office building, a room contains 54 chairs. The number of chairs per row is three less than twice the number of rows. Find the number of rows and the number of chairs per row.

78. An apple orchard contains 85 trees. The number of trees in each row is three less than four times the number of rows. Find the number of rows and the number of trees per row.

79. Suppose that the combined area of two squares is 360 square feet. Each side of the larger square is three times as long as a side of the smaller square. How big is each square?

80. The area of a rectangular slab of sidewalk is 45 square feet. Its length is 3 feet more than four times its width. Find the length and width of the slab.

81. The length of a rectangular sheet of paper is 1 centimeter more than twice its width, and the area of the rectangle is 55 square centimeters. Find the length and width of the rectangle.

82. Suppose that the length of a certain rectangle is three times its width. If the length is increased by 2 inches, and the width increased by 1 inch, the newly formed rectangle has an area of 70 square inches. Find the length and width of the original rectangle.

83. The area of a triangle is 51 square inches. One side of the triangle is 1 inch less than three times the length of the altitude to that side. Find the length of that side and the length of the altitude to that side.

84. Suppose that a square and a rectangle have equal areas. Furthermore, suppose that the length of the rectangle is twice the length of a side of the square, and the width of the rectangle is 4 centimeters less than the length of a side of the square. Find the dimensions of both figures.

85. A strip of uniform width is to be cut off of both sides and both ends of a sheet of paper that is 8 inches by 11 inches, in order to reduce the size of the paper to an area of 40 square inches. Find the width of the strip.

86. The sum of the areas of two circles is 100π square centimeters. The length of a radius of the larger circle is 2 centimeters more than the length of a radius of the smaller circle. Find the length of a radius of each circle.

87. The sum of the areas of two circles is 180π square inches. The length of a radius of the smaller circle is 6 inches less than the length of a radius of the larger circle. Find the length of a radius of each circle.

88. A strip of uniform width is shaded along both sides and both ends of a rectangular poster that is 18 inches by 14 inches. How wide is the strip if the unshaded portion of the poster has an area of 165 square inches?

Thoughts Into Words

89. When factoring polynomials, why do you think that it is best to look for a greatest common monomial factor first?

90. Explain how you would solve $(4x - 3)(8x + 5) = 0$ and also how you would solve $(4x - 3)(8x + 5) = -9$.

91. Explain how you would solve $(x + 2)(x + 3) = (x + 2) \cdot (3x - 1)$. Do you see more than one approach to this problem?

Answers to the Concept Quiz

1. F or A   **2.** C   **3.** E   **4.** A   **5.** G   **6.** D   **7.** B

Chapter 7 Summary

OBJECTIVE	SUMMARY	EXAMPLE
Find the greatest common factor. **(Section 7.1/Objective 1)**	By "the greatest common factor of two or more monomials" we mean the monomial with the largest numerical coefficient and the highest power of the variables, which is a factor of each given monomial.	Find the greatest common factor of $12a^3b^3$, $18a^2b^2$, and $54ab^4$. **Solution** The largest numerical coefficient that is a factor of all three terms is 6. The highest exponent for a that is a factor of all three terms is 1. The highest exponent for b that is a factor of all three terms is 2. So the greatest common factor is $6ab^2$.
Factor out the greatest common factor. **(Section 7.1/Objective 2)**	The distributive property in the form $ab + ac = a(b + c)$ provides the basis for factoring out a greatest common monomial or binomial factor.	Factor $36x^2y + 18x - 27xy$. **Solution** $36x^2y + 18x - 27xy$ $= 9x(4xy) + 9x(2) - 9x(3y)$ $= 9x(4xy + 2 - 3y)$
Factor by grouping. **(Section 7.1/Objective 3)**	Rewriting an expression such as $ab + 3a + bc + 3c$ as $a(b + 3) + c(b + 3)$ and then factoring out the common binomial factor of $b + 3$ so that $a(b + 3) + c(b + 3)$ becomes $(b + 3) \cdot (a + c)$, is called factoring by grouping.	Factor $15xy + 6x + 10y^2 + 4y$. **Solution** $15xy + 6x + 10y^2 + 4y$ $= 3x(5y + 2) + 2y(5y + 2)$ $= (5y + 2)(3x + 2y)$
Factor the difference of two squares. **(Section 7.2/Objective 1)**	The factoring pattern, the difference of two squares, is $a^2 - b^2 = (a - b)(a + b)$. Be careful not to apply the pattern to the sum of two squares, such as $a^2 + b^2$. There is no factoring pattern for the sum of two squares.	Factor $4x^2 - 81y^2$. **Solution** $4x^2 - 81y^2 = (2x + 9y)(2x - 9y)$
Factor trinomials of the form $x^2 + bx + c$. **(Section 7.3/Objective 1)**	The following multiplication pattern provides a technique for factoring trinomials of the form $x^2 + bx + c$. $(x + r)(x + s) = x^2 + rx + sx + rs$ $\qquad\qquad\quad = x^2 + (r + s)x + rs$ For trinomials of the form $x^2 + bx + c$, we want two factors of c whose sum will be equal to b.	Factor $x^2 + 2x - 35$. **Solution** The factors of -35 that sum to 2 are 7 and -5. $x^2 + 2x - 35 = (x + 7)(x - 5)$

OBJECTIVE	SUMMARY	EXAMPLE
Factor trinomials in which the leading coefficient is not 1. **(Section 7.4/Objective 1)**	We presented two different techniques for factoring trinomials of the form $ax^2 + bx + c$. To review these techniques, turn to Section 7.4 and study the examples. The examples here show the two techniques.	Factor: **(a)** $3x^2 + 7x + 4$ **(b)** $2x^2 - 5x + 3$ **Solution** **(a)** To factor $3x^2 + 7x + 4$, we can look at the first term and the sign situation to determine that the factors are of the form $(3x + \underline{\quad})$ and $(x + \underline{\quad})$. By trial and error we can arrive at the correct factors: $$3x^2 + 7x + 4 = (3x + 4)(x + 1)$$ **(b)** $2x^2 - 5x + 3$ Sum of -5 Product of $2(3) = 6$ So we want two numbers whose sum is -5 and whose product is 6. The numbers are -2 and -3. Therefore, we rewrite the middle term as $-2x - 3x$. $$\begin{aligned} 2x^2 - 5x + 3 &= 2x^2 - 2x - 3x + 3 \\ &= 2x(x - 1) - 3(x - 1) \\ &= (x - 1)(2x - 3) \end{aligned}$$
Factor perfect-square trinomials. **(Section 7.5/Objective 1)**	Perfect-square trinomials are easy to recognize because of the nature of their terms. The first term and the last term will be the squares of a quantity. The middle term is twice the product of the quantities being squared in the first and last terms.	Factor $16x^2 + 56x + 49$. **Solution** $$\begin{aligned} & 16x^2 + 56x + 49 \\ &= (4x^2) + 2(4x)(7) + (7)^2 \\ &= (4x + 7)^2 \end{aligned}$$
Recognize the different types of factoring. **(Section 7.5/Objective 2)**	As a general guideline for factoring completely, always look for a greatest common factor *first*, and then proceed with one or more of the following techniques. **1.** Apply the difference-of-squares pattern. **2.** Apply the perfect-square pattern. **3.** Factor a trinomial of the form $x^2 + bx + c$ into the product of two binomials. **4.** Factor a trinomial of the form $ax^2 + bx + c$ into the product of two binomials.	Factor $3x^2 + 12xy + 12y^2$. **Solution** $$\begin{aligned} & 3x^2 + 12xy + 12y^2 \\ &= 3(x^2 + 4xy + 4y^2) \\ &= 3(x + 2y)^2 \end{aligned}$$
Use factoring to solve equations. **(Section 7.1/Objective 4; Section 7.2/Objective 2; Section 7.3/Objective 2; Section 7.4/Objective 2; Section 7.5/Objective 3)**	Property 7.1 states that for all real numbers a and b, $ab = 0$ if and only if $a = 0$ or $b = 0$. To solve equations by applying this property, first set the equation equal to zero. Proceed by factoring the other side of the equation. Then set each factor equal to 0 and solve the equations.	Solve $2x^2 = -10x$. **Solution** $2x^2 = -10x$ $2x^2 + 10x = 0$ $2x(x + 5) = 0$ $2x = 0$ or $x + 5 = 0$ $x = 0$ or $x = -5$ The solution is $\{-5, 0\}$.

(continued)

OBJECTIVE	SUMMARY	EXAMPLE
Solve word problems that involve factoring. **(Section 7.3/Objectives 3 and 4; Section 7.5/Objective 4)**	Knowledge of factoring has expanded the techniques available for solving word problems. This chapter introduced the Pythagorean theorem. The theorem pertains to right triangles and states that in any right triangle, the square of the longest side is equal to the sum of the squares of the other two sides. The formula for the theorem is written as $a^2 + b^2 = c^2$, where a and b are the legs of the triangle and c is the hypotenuse.	The length of one leg of a right triangle is 1 inch more than the length of the other leg. The length of the hypotenuse is 5 inches. Find the length of the two legs. **Solution** Let x represent the length of one leg of the triangle. Then $x + 1$ will represent the length of the other leg. We know the hypotenuse is equal to 5. Apply Pythagorean theorem. $x^2 + (x + 1)^2 = 5^2$ $x^2 + x^2 + 2x + 1 = 25$ $2x^2 + 2x - 24 = 0$ $2(x^2 + x - 12) = 0$ $x^2 + x - 12 = 0$ $(x + 4)(x - 3) = 0$ $x + 4 = 0$ or $x - 3 = 0$ $x = -4$ or $x = 3$ Because the length of a side of the triangle cannot be negative, the only viable answer is 3. Therefore, the length of one leg of the triangle is 3 inches, and the length of the other leg is 4 inches.

Chapter 7   Review Problem Set

For Problems 1–24, factor completely. Indicate any polynomials that are not factorable using integers.

1. $x^2 - 9x + 14$ **2.** $3x^2 + 21x$

3. $9x^2 - 4$ **4.** $4x^2 + 8x - 5$

5. $25x^2 - 60x + 36$ **6.** $n^3 + 13n^2 + 40n$

7. $y^2 + 11y - 12$ **8.** $3xy^2 + 6x^2y$

9. $x^4 - 1$ **10.** $18n^2 + 9n - 5$

11. $x^2 + 7x + 24$ **12.** $4x^2 - 3x - 7$

13. $3n^2 + 3n - 90$ **14.** $x^3 - xy^2$

15. $2x^2 + 3xy - 2y^2$ **16.** $4n^2 - 6n - 40$

17. $5x + 5y + ax + ay$

18. $21t^2 - 5t - 4$ **19.** $2x^3 - 2x$

20. $3x^3 - 108x$ **21.** $16x^2 + 40x + 25$

22. $xy - 3x - 2y + 6$

23. $15x^2 - 7xy - 2y^2$ **24.** $6n^4 - 5n^3 + n^2$

For Problems 25–44, solve each equation.

25. $x^2 + 4x - 12 = 0$ **26.** $x^2 = 11x$

27. $2x^2 + 3x - 20 = 0$

28. $9n^2 + 21n - 8 = 0$

29. $6n^2 = 24$

30. $16y^2 + 40y + 25 = 0$

31. $t^3 - t = 0$

32. $28x^2 + 71x + 18 = 0$

33. $x^2 + 3x - 28 = 0$

34. $(x - 2)(x + 2) = 21$

35. $5n^2 + 27n = 18$

36. $4n^2 + 10n = 14$

37. $2x^3 - 8x = 0$

38. $x^2 - 20x + 96 = 0$

39. $4t^2 + 17t - 15 = 0$

40. $3(x + 2) - x(x + 2) = 0$

41. $(2x - 5)(3x + 7) = 0$

42. $(x + 4)(x - 1) = 50$

43. $-7n - 2n^2 = -15$

44. $-23x + 6x^2 = -20$

Set up an equation and solve each of the following problems.

45. The larger of two numbers is one less than twice the smaller number. The difference of their squares is 33. Find the numbers.

46. The length of a rectangle is 2 centimeters less than five times the width of the rectangle. The area of the rectangle is 16 square centimeters. Find the length and width of the rectangle.

47. Suppose that the combined area of two squares is 104 square inches. Each side of the larger square is five times as long as a side of the smaller square. Find the size of each square.

48. The longer leg of a right triangle is one unit shorter than twice the length of the shorter leg. The hypotenuse is one unit longer than twice the length of the shorter leg. Find the lengths of the three sides of the triangle.

49. The product of two numbers is 26, and one of the numbers is one larger than six times the other number. Find the numbers.

50. Find three consecutive positive odd whole numbers such that the sum of the squares of the two smaller numbers is nine more than the square of the largest number.

51. The number of books per shelf in a bookcase is one less than nine times the number of shelves. If the bookcase contains 140 books, find the number of shelves.

52. The combined area of a square and a rectangle is 225 square yards. The length of the rectangle is eight times the width of the rectangle, and the length of a side of the square is the same as the width of the

rectangle. Find the dimensions of the square and the rectangle.

53. Suppose that we want to find two consecutive integers such that the sum of their squares is 613. What are they?

54. If numerically the volume of a cube equals the total surface area of the cube, find the length of an edge of the cube.

55. The combined area of two circles is 53π square meters. The length of a radius of the larger circle is 1 meter more than three times the length of a radius of the smaller circle. Find the length of a radius of each circle.

56. The product of two consecutive odd whole numbers is one less than five times their sum. Find the whole numbers.

57. Sandy has a photograph that is 14 centimeters long and 8 centimeters wide. She wants to reduce the length and width by the same amount so that the area is decreased by 40 square centimeters. By what amount should she reduce the length and width?

58. Suppose that a strip of uniform width is plowed along both sides and both ends of a garden that is 120 feet long and 90 feet wide (see Figure 7.7). How wide is the strip if the garden is half plowed?

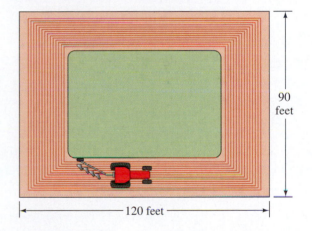

90 feet

120 feet

Figure 7.7

For Problems 1–10, factor each expression completely.

1. $x^2 + 3x - 10$

2. $x^2 - 5x - 24$

3. $2x^3 - 2x$

4. $x^2 + 21x + 108$

5. $18n^2 + 21n + 6$

6. $ax + ay + 2bx + 2by$

7. $4x^2 + 17x - 15$

8. $6x^2 + 24$

9. $30x^3 - 76x^2 + 48x$

10. $28 + 13x - 6x^2$

For Problems 11–21, solve each equation.

11. $7x^2 = 63$

12. $x^2 + 5x - 6 = 0$

13. $4n^2 = 32n$

14. $(3x - 2)(2x + 5) = 0$

15. $(x - 3)(x + 7) = -9$

16. $x^3 + 16x^2 + 48x = 0$

17. $9(x - 5) - x(x - 5) = 0$

18. $3t^2 + 35t = 12$

19. $8 - 10x - 3x^2 = 0$

20. $3x^3 = 75x$

21. $25n^2 - 70n + 49 = 0$

For Problems 22–25, set up an equation and solve each problem.

22. The length of a rectangle is 2 inches less than twice its width. If the area of the rectangle is 112 square inches, find the length of the rectangle.

23. The length of one leg of a right triangle is 4 centimeters more than the length of the other leg. The length of the hypotenuse is 8 centimeters more than the length of the shorter leg. Find the length of the shorter leg.

24. A room contains 112 chairs. The number of chairs per row is five less than three times the number of rows. Find the number of chairs per row.

25. If numerically the volume of a cube equals twice the total surface area, find the length of an edge of the cube.

For Problems 1–6, evaluate each algebraic expression for the given values of the variables. You may first want to simplify the expression or change its form by factoring.

1. $3x - 2xy - 7x + 5xy$ for $x = \dfrac{1}{2}$ and $y = 3$

2. $7(a - b) - 3(a - b) - (a - b)$ for $a = -3$ and $b = -5$

3. $ab + b^2$ for $a = 0.4$ and $b = 0.6$

4. $x^2 - y^2$ for $x = -6$ and $y = 4$

5. $x^2 + x - 72$ for $x = -10$

6. $3(x - 2) - (x + 3) - 5(x + 6)$ for $x = -6$

For Problems 7–14, evaluate each numerical expression.

7. 3^{-3}

8. $\left(\dfrac{2}{3}\right)^{-1}$

9. $\left(\dfrac{1}{2} + \dfrac{1}{3}\right)^0$

10. $\left(\dfrac{1}{3} + \dfrac{1}{4}\right)^{-1}$

11. -4^{-2}

12. -4^2

13. $\dfrac{1}{\left(\dfrac{2}{5}\right)^2}$

14. $(-3)^{-3}$

For Problems 15–26, perform the indicated operations and express answers in simplest form.

15. $\dfrac{7}{5x} + \dfrac{2}{x} - \dfrac{3}{2x}$

16. $\dfrac{4x}{5y} \div \dfrac{12x^2}{10y^2}$

17. $(-5x^2y)(7x^3y^4)$

18. $(9ab^3)^2$

19. $(-3n^2)(5n^2 + 6n - 2)$

20. $(5x - 1)(3x + 4)$

21. $(2x + 5)^2$

22. $(x + 2)(2x^2 - 3x - 1)$

23. $(x^2 - x - 1)(x^2 + 2x - 3)$

24. $(-2x - 1)(3x - 7)$

25. $\dfrac{24x^2y^3 - 48x^4y^5}{8xy^2}$

26. $(28x^2 - 19x - 20) \div (4x - 5)$

For Problems 27–36, factor each polynomial completely.

27. $3x^3 + 15x^2 + 27x$

28. $x^2 - 100$

29. $5x^2 - 22x + 8$

30. $8x^2 - 22x - 63$

31. $n^2 + 25n + 144$

32. $nx + ny - 2x - 2y$

33. $3x^3 - 3x$

34. $2x^3 - 6x^2 - 108x$

35. $36x^2 - 60x + 25$

36. $3x^2 - 5xy - 2y^2$

For Problems 37–46, solve each equation.

37. $3(x - 2) - 2(x + 6) = -2(x + 1)$

38. $x^2 = -11x$

39. $0.2x - 3(x - 0.4) = 1$

40. $5n^2 - 5 = 0$

41. $x^2 + 5x - 6 = 0$

42. $\dfrac{2x + 1}{2} + \dfrac{3x - 4}{3} = 1$

43. $2(x - 1) - x(x - 1) = 0$

44. $6x^2 + 19x - 7 = 0$

45. $(2x - 1)(x - 8) = 0$

46. $(x + 1)(x + 6) = 24$

For Problems 47–51, solve each inequality.

47. $-3x - 2 \geq 1$

48. $18 < 2(x - 4)$

49. $3(x - 2) - 2(x + 1) \leq -(x + 5)$

50. $\dfrac{2}{3}x - \dfrac{1}{4}x - 1 > 3$

51. $0.08x + 0.09(2x) \leq 130$

For Problems 52–59, graph each equation.

52. $y = -3x + 5$

53. $y = \dfrac{1}{4}x + 2$

54. $3x - y = 3$

55. $y = 2x^2 - 4$

56. Find the slope of the line determined by the equation $-5x + 6y = -10$.

57. Write the equation of the line that contains the points $(-2, 4)$ and $(1, -7)$.

58. Write the equation of the line that is parallel to the line $-3x + 8y = 51$ and contains the point $(2, 7)$.

59. Write the equation of the line that has an x intercept of -6 and a slope of $\dfrac{2}{9}$.

For Problems 60–63, solve each system of equations.

60. $\begin{pmatrix} 7x - 2y = -34 \\ x + 2y = -14 \end{pmatrix}$

61. $\begin{pmatrix} 5x + 3y = -9 \\ 3x - 5y = 15 \end{pmatrix}$

62. $\begin{pmatrix} 6x - 11y = -58 \\ 8x + y = 1 \end{pmatrix}$

63. $\begin{pmatrix} \dfrac{1}{2}x + \dfrac{2}{3}y = -1 \\ \dfrac{2}{5}x - \dfrac{1}{3}y = -6 \end{pmatrix}$

64. Is 91 a prime or composite number?

65. Find the greatest common factor of 18 and 48.

66. Find the least common multiple of 6, 8, and 9.

67. Express $\dfrac{7}{4}$ as a percent.

68. Express 0.0024 in scientific notation.

69. Express $(3.14)(10^3)$ in ordinary decimal notation.

70. Graph on a number line the solutions for the compound inequality $x < 0$ or $x > 3$.

71. Find the area of a circular region if the circumference is 8π centimeters. Express the answer in terms of π.

72. Thirty percent of what number is 5.4?

73. Graph the inequality $3x - 2y < -6$.

For Problems 74–89, use an equation, an inequality, or a system of equations to help solve each problem.

74. One leg of a right triangle is 2 inches longer than the other leg. The hypotenuse is 4 inches longer than the shorter leg. Find the lengths of the three sides of the right triangle.

75. How many milliliters of a 65% solution of hydrochloric acid must be added to 40 milliliters of a 30% solution of hydrochloric acid to obtain a 55% solution?

76. A landscaping border 28 feet long is bent into the shape of a rectangle. The length of the rectangle is 2 feet more than the width. Find the dimensions of the rectangle.

77. Two motorcyclists leave Daytona Beach at the same time and travel in opposite directions. If one travels at 55 miles per hour and the other travels at 65 miles per hour, how long will it take for them to be 300 miles apart?

78. Find the length of an altitude of a trapezoid with bases of 10 centimeters and 22 centimeters and an area of 120 square centimeters.

79. If a car uses 16 gallons of gasoline for a 352-mile trip, at the same rate of consumption, how many gallons will it use on a 594-mile trip?

80. If two angles are supplementary, and the larger angle is 20° less than three times the smaller angle, find the measure of each angle.

81. Find the measures of the three angles of a triangle if the largest angle is 10° more than twice the smallest, and the other angle is 10° larger than the smallest angle.

82. Zorka has 175 coins consisting of pennies, nickels, and dimes. The number of dimes is five more than twice the number of pennies, and the number of nickels is 10 more than the number of pennies. How many coins of each kind does she have?

83. Rashed has some dimes and quarters amounting to $7.65. The number of quarters is three less than twice the number of dimes. How many coins of each kind does he have?

84. Ashley has scores of 85, 87, 90, and 91 on her first four algebra tests. What score must she get on the fifth test to have an average of 90 or better for the five tests?

85. The ratio of girls to boys in a certain school is six to five. If there is a total of 1650 students in the school, find the number of girls and the number of boys.

86. If a ring costs a jeweler $750, at what price should it be sold for the jeweler to make a profit of 70% based on the selling price?

87. Suppose that the jeweler in Problem 86 would be satisfied with a 70% profit based on the cost of the ring. At what price should he sell the ring?

88. How many quarts of pure alcohol must be added to 6 quarts of a 30% solution to obtain a 40% solution?

89. Suppose that the cost of five tennis balls and four golf balls is $17. Furthermore, suppose that at the same prices, the cost of three tennis balls and seven golf balls is $20.55. Find the cost of one tennis ball and the cost of one golf ball.

8 A Transition from Elementary Algebra to Intermediate Algebra

A quadratic equation can be solved to determine the width of a uniform strip trimmed off both the sides and ends of a sheet of paper to obtain a specified area for the sheet of paper.

© photogolfer

Observe the following five rows of numbers.

Row 1						1		1		
Row 2					1		2		1	
Row 3				1		3		3		1
Row 4			1		4		6		4		1
Row 5		1		5		10		10		5		1

This configuration can be extended indefinitely. Do you see a pattern that will create row 6? Of what significance is this configuration of numbers? These questions are answered in Section 8.4.

As the title indicates, our primary objective in this chapter is to review briefly some concepts of elementary algebra and, in some instances, to extend the concepts into intermediate algebra territory. For example, in Sections 8.1 and 8.2, we review some basic techniques for solving equations and inequalities. Then, in Section 8.3, these techniques are extended to solve equations and inequalities that involve absolute value.

In Section 8.4, operations on polynomials are reviewed, and the multiplication of binomials is extended to binomial expansions in general. Likewise, in Section 8.5 the division of polynomials is reviewed and then extended to

Video tutorials based on section learning objectives are available in a variety of delivery modes.

synthetic division. Various techniques for factoring polynomials are reviewed in Section 8.6, and one or two new techniques are introduced.

This chapter should help you make a smooth transition from elementary algebra to intermediate algebra. We have indicated the new material in this chapter with a "New" symbol.

8.1	Equations: A Brief Review

OBJECTIVES

1. Apply properties of equality to solve linear equations

2. Solve proportions

3. Solve systems of two linear equations

4. Write equations to represent word problems and solve the equations

An **algebraic equation** such as $3x + 1 = 13$ is neither true nor false as it stands; for this reason, it is sometimes called an open sentence. Each time that a number is substituted for x (the variable), the algebraic equation $3x + 1 = 13$ becomes a **numerical statement** that is either true or false. For example, if $x = 2$, then $3x + 1 = 13$ becomes $3(2) + 1 = 13$, which is a false statement. If $x = 4$, then $3x + 1 = 13$ becomes $3(4) + 1 = 13$, which is a true statement. **Solving an algebraic equation** refers to the process of finding the number (or numbers) that make(s) the algebraic equation a true numerical statement. Such numbers are called the **solutions** or **roots** of the equation and are said to *satisfy* the equation. The set of all solutions of an equation is called its solution set. Thus the solution set of $3x + 1 = 13$ is {4}.

Equivalent equations are equations that have the same solution set. For example,

$$3x + 1 = 13 \qquad 3x = 12 \qquad \text{and} \qquad x = 4$$

are equivalent equations because {4} is the solution set of each. The general procedure for solving an equation is to continue replacing the given equation with equivalent but simpler equations until an equation of the form "variable = constant" or "constant = variable" is obtained. Thus in the previous example, $3x + 1 = 13$ simplifies to $3x = 12$, which simplifies to $x = 4$, from which the solution set {4} is obvious.

Techniques for solving equations revolve around the following basic properties of equality:

Property 8.1 Properties of Equality

For all real numbers, a, b, and c,

1. $a = a$ Reflexive property

2. If $a = b$, then $b = a$.   Symmetric property

3. If $a = b$ and $b = c$, then $a = c$.   Transitive property

4. If $a = b$, then a may be replaced by b, or b may be replaced by a, in any statement without changing the meaning of the statement.   Substitution property

5. $a = b$ if and only if $a + c = b + c$.   Addition property

6. $a = b$ if and only if $ac = bc$, where $c \neq 0$.   Multiplication property

---

In Chapter 3 we stated an *addition-subtraction property of equality* but pointed out that because subtraction is defined in terms of *adding the opposite*, only an addition property is technically necessary. Likewise, because division can be defined in terms of *multiplying by the reciprocal*, only a multiplication property is necessary.

Now let's use some examples to review the process of solving equations.

Classroom Example
Solve the equation $4 + 5x = -8$.

> ### EXAMPLE 1   Solve the equation $-3x + 1 = -8$.

Solution

$$-3x + 1 = -8$$
$$-3x = -9 \qquad \text{Added} \quad 1 \text{ to both sides}$$
$$x = 3 \qquad \text{Multiplied both sides by } -\frac{1}{3}$$

The solution set is $\{3\}$.

Don't forget that to be absolutely sure of a solution set, we must check the solution(s) back into the original equation. Thus for Example 1, substituting 3 for x in $-3x + 1 = -8$ produces $-3(3) + 1 = -8$, which is a true statement. Our solution set is indeed $\{3\}$. We will not use the space to show all checks, but remember their importance.

Classroom Example
Find the solution set of
$6y + 1 - 2y = y + 3 + 2y$.

> ### EXAMPLE 2   Find the solution set of $-3n + 6 + 5n = 9n - 4 - 6n$.

Solution

$$-3n + 6 + 5n = 9n - 4 - 6n$$
$$2n + 6 = 3n - 4 \qquad \text{Combined similar terms on both sides}$$
$$6 = n - 4 \qquad \text{Added } -2n \text{ to both sides}$$
$$10 = n \qquad \text{Added 4 to both sides}$$

The solution set is $\{10\}$.

Equations Containing Parentheses

If an equation contains parentheses, we may need to apply the distributive property, combine similar terms, and then apply the addition and multiplication properties.

Classroom Example
Solve $4(3x + 1) = -(7 - x)$.

> ### EXAMPLE 3   Solve $3(2x - 5) - 2(4x + 3) = -1$.

Solution

$$3(2x - 5) - 2(4x + 3) = -1$$
$$3(2x) - 3(5) - 2(4x) - 2(3) = -1 \qquad \text{Apply the distributive property twice}$$
$$6x - 15 - 8x - 6 = -1$$
$$-2x - 21 = -1 \qquad \text{Combine similar terms}$$
$$-2x = 20$$
$$x = -10$$

The solution set is $\{-10\}$. Perhaps you should check this solution!

Equations Containing Fractional Forms

If an equation contains fractional forms, then it is usually easier to multiply both sides by the least common denominator of all of the denominators.

> **EXAMPLE 4**   Solve $\dfrac{3x + 2}{4} + \dfrac{2x - 5}{6} = \dfrac{3}{8}$.

Solution

$$\frac{3x + 2}{4} + \frac{2x - 5}{6} = \frac{3}{8}$$

$$24\left(\frac{3x + 2}{4} + \frac{2x - 5}{6}\right) = 24\left(\frac{3}{8}\right) \qquad \text{24 is the LCD of 4, 6, and 8}$$

$$24\left(\frac{3x + 2}{4}\right) + 24\left(\frac{2x - 5}{6}\right) = 24\left(\frac{3}{8}\right) \qquad \text{Apply the distributive property}$$

$$6(3x + 2) + 4(2x - 5) = 9$$

$$18x + 12 + 8x - 20 = 9$$

$$26x - 8 = 9$$

$$26x = 17$$

$$x = \frac{17}{26}$$

The solution set is $\left\{\dfrac{17}{26}\right\}$.

If the equation contains some decimal fractions, then multiplying both sides of the equation by an appropriate power of 10 usually works quite well.

> **EXAMPLE 5**   Solve $0.04x + 0.06(1200 - x) = 66$.

Solution

$$0.04x + 0.06(1200 - x) = 66$$

$$100[0.04x + 0.06(1200 - x)] = 100(66)$$

$$100(0.04x) + 100[0.06(1200 - x)] = 100(66)$$

$$4x + 6(1200 - x) = 6600$$

$$4x + 7200 - 6x = 6600$$

$$-2x + 7200 = 6600$$

$$-2x = -600$$

$$x = 300$$

The solution set is $\{300\}$.

Proportions

Recall that a statement of equality between two ratios is a **proportion**. For example, $\dfrac{3}{4} = \dfrac{15}{20}$ is a proportion that states that the ratios $\dfrac{3}{4}$ and $\dfrac{15}{20}$ are equal. A general property of proportions states the following:

$$\frac{a}{b} = \frac{c}{d} \quad \text{if and only if } ad = bc, \text{ where } b \neq 0 \text{ and } d \neq 0$$

The products ad and bc are commonly called **cross products**. Thus the property states that the cross products in a proportion are equal. This becomes the basis of another equation-solving process. If a variable appears in one or both denominators, then restrictions need to be imposed to avoid division by zero.

Classroom Example
Solve $\dfrac{2x-5}{3} = \dfrac{x}{4}$.

EXAMPLE 6   Solve $\dfrac{3}{5x-2} = \dfrac{4}{7x+3}$.

Solution

$$\frac{3}{5x-2} = \frac{4}{7x+3}, \qquad x \neq \frac{2}{5} \text{ and } x \neq -\frac{3}{7}$$

$$3(7x+3) = 4(5x-2) \qquad \text{Cross products are equal}$$

$$21x + 9 = 20x - 8 \qquad \text{Apply the distributive property on both sides}$$

$$x = -17$$

The solution set is $\{-17\}$.

Equations That Are Algebraic Identities

An equation that is satisfied by all numbers for which both sides of the equation are defined is called an **algebraic identity**. For example,

$$5(x+1) = 5x + 5 \qquad\qquad x^2 - 9 = (x+3)(x-3)$$

$$\frac{1}{x} + \frac{2}{x} = \frac{3}{x} \qquad\qquad x^2 - x - 12 = (x-4)(x+3)$$

are all algebraic identities. In the third identity, x cannot equal zero, so the statement $\dfrac{1}{x} + \dfrac{2}{x} = \dfrac{3}{x}$ is true for all real numbers except 0. The other identities listed are true for all real numbers. Sometimes the original form of the equation does not explicitly indicate that it is an identity. Consider the following example.

Classroom Example
Solve $3x + 2(x-6) = 5x - 12$.

EXAMPLE 7   Solve $5(x+3) - 3(x-5) = 2(x+15)$.

Solution

$$5(x+3) - 3(x-5) = 2(x+15)$$

$$5x + 15 - 3x + 15 = 2x + 30$$

$$2x + 30 = 2x + 30$$

At this step it becomes obvious that we have an algebraic identity. No restrictions are necessary, so the solution set, written in set builder notation, is $\{x \mid x \text{ is a real number}\}$.

Equations That Are Contradictions

By inspection we can tell that the equation $x + 1 = x + 2$ has no solutions, because adding 1 to a number cannot produce the same result as adding 2 to that number. Thus the solution set is $\varnothing$ (the null set or empty set). Likewise, we can determine by inspection that the solution set for each of the following equations is $\varnothing$.

$$3x - 1 = 3x - 4 \qquad 5(x-1) = 5x - 1 \qquad \frac{1}{x} + \frac{2}{x} = \frac{5}{x}$$

Now suppose that the solution set is *not* obvious by inspection.

Classroom Example
Solve $10x + 8 = 2(5x - 1)$.

EXAMPLE 8   Solve $4(x - 2) - 2(x + 3) = 2(x + 6)$.

Solution

$$4(x - 2) - 2(x + 3) = 2(x + 6)$$
$$4x - 8 - 2x - 6 = 2x + 12$$
$$2x - 14 = 2x + 12$$

At this step we might recognize that the solution set is $\varnothing$, or we could continue by adding $-2x$ to both sides to produce

$$-14 = 12$$

Because we have logically arrived at a contradiction ($-14 = 12$), the solution set is $\varnothing$.

Problem Solving

Remember that one theme throughout this text is *learn a skill*, then *use the skill to help solve equations and inequalities*, and then *use equations and inequalities to help solve problems*. Being able to solve problems is the end result of this sequence; it is what we want to achieve. You may want to turn back to Section 3.3 and refresh your memory of the problem-solving suggestions given there.

Suggestion 5 is *look for a **guideline** that can be used to set up an equation*. Such a guideline may or may not be explicitly stated in the problem.

Classroom Example
A 6-ounce bottle is full and contains a 10% solution of hydrogen peroxide. How much needs to be drained out and replaced with pure hydrogen peroxide to obtain a 15% solution?

EXAMPLE 9

A 10-gallon container is full and contains a 40% solution of antifreeze. How much needs to be drained out and replaced with pure antifreeze to obtain a 70% solution?

Solution

We can use the following guideline for this problem.

Pure antifreeze in the original solution	$-$	Pure antifreeze in the solution drained out	$+$	Pure antifreeze added	$=$	Pure antifreeze in the final solution

Let x represent the amount of pure antifreeze to be added. Then x also represents the amount of the 40% solution to be drained out. Thus the guideline translates into the equation

$$40\%(10) - 40\%(x) + x = 70\%(10)$$

We can solve this equation as follows:

$$0.4(10) - 0.4x + x = 0.7(10)$$
$$4 + 0.6x = 7$$
$$0.6x = 3$$
$$6x = 30$$
$$x = 5$$

Therefore, we need to drain out 5 gallons of the 40% solution and replace it with 5 gallons of pure antifreeze. (Be sure you can check this answer!)

Recall that in Chapter 5 we found that sometimes it is easier to solve a problem using two equations and two unknowns than using one equation with one unknown. Also at that time, we used three different techniques for solving a system of two linear equations in two variables:

(1) a graphing approach, which was not very efficient when we wanted exact solutions, (2) a substitution method, and (3) an elimination-by-addition method. Now let's use a system of equations to help solve a problem. Furthermore, we will solve the system twice to review both the substitution method and the elimination-by-addition method.

Classroom Example
For a breakfast meeting, Maria bought 10 bagels and two gallons of coffee, which together cost $17.40. The next week she bought 8 bagels and one gallon of coffee for $11.37 from the same shop. Find the price of a bagel and the price of a gallon of coffee. Write a system of equations that represents the problem, and solve the system by the substitution method or the elimination-by-addition method.

EXAMPLE 10

Jose bought 1 pound of bananas and 3 pounds of tomatoes for $8.16. At the same prices, Jessica bought 4 pounds of bananas and 5 pounds of tomatoes for $15.21. Find the price per pound for bananas and for tomatoes.

Solution

Let b represent the price per pound for bananas, and let t represent the price per pound for tomatoes. The problem translates into the following system of equations.

$$\begin{pmatrix} b + 3t = 8.16 \\ 4b + 5t = 15.21 \end{pmatrix}$$

Let's solve this system using the substitution method. The first equation can be written as $b = 8.16 - 3t$. Now we can substitute $8.16 - 3t$ for b in the second equation.

$$4(8.16 - 3t) + 5t = 15.21$$
$$32.64 - 12t + 5t = 15.21$$
$$32.64 - 7t = 15.21$$
$$-7t = -17.43$$
$$t = 2.49$$

Finally, we can substitute 2.49 for t in $b = 8.16 - 3t$.

$$b = 8.16 - 3(2.49)$$
$$= 8.16 - 7.47$$
$$= 0.69$$

Therefore, the price for bananas is $0.69 per pound, and the price for tomatoes is $2.49 per pound.

For review purposes, let's solve the system of equations in Example 10 using the elimination-by-addition method.

$$\begin{pmatrix} b + 3t = 8.16 \\ 4b + 5t = 15.21 \end{pmatrix}$$

Multiply the first equation by -4, and add that result to the second equation to produce a new second equation.

$$\begin{pmatrix} b + 3t = 8.16 \\ -7t = -17.43 \end{pmatrix}$$

Now we can determine the value of t.

$$-7t = -17.43$$
$$t = 2.49$$

Substitute 2.49 for t in $b + 3t = 8.16$.

$$b + 3(2.49) = 8.16$$
$$b = 0.69$$

A t value of 2.49 and a b value of 0.69 agree with our previous work.

Concept Quiz 8.1

For Problems 1–5, solve the equations by inspection, and match the equation with its solution set.

1. $2x + 5 = 2x + 8$ **A.** $\{0\}$

2. $\dfrac{2}{x} = \dfrac{1}{7}$ **B.** $\{x \mid x \text{ is a real number}\}$

           C. $\{14\}$

3. $\dfrac{1}{x} + \dfrac{3}{x} = \dfrac{1}{3}$ **D.** $\varnothing$

4. $4x - 2 = 2(2x - 1)$ **E.** $\{12\}$

5. $5x + 7 = 3x + 7$

For Problems 6–10, answer true or false.

6. The solution set of the equation $\dfrac{x - 2}{4} - 3 = \dfrac{x + 1}{4}$ is the null set.

7. The solution set for $\dfrac{x + 2}{3} - 1 = \dfrac{x - 1}{3}$ is the set of real numbers.

8. The solution set of $3x(x - 1) = 6$ is $\{2, 5\}$.

9. The solution set of the system of equations $\begin{pmatrix} x + y = 4 \\ 2x + 2y = 8 \end{pmatrix}$ is the set of all ordered pairs of real numbers.

10. The solution set of the system of equations $\begin{pmatrix} 2x - 3y = 4 \\ 3x + y = 5 \end{pmatrix}$ is $\left\{ \dfrac{19}{11}, -\dfrac{2}{11} \right\}$.

Problem Set 8.1

For Problems 1–26, solve each equation.
(Objectives 1 and 2)

1. $5x - 4 = 16$

2. $-4x + 3 = -13$

3. $-6 = 7x + 1$

4. $8 = 6x - 4$

5. $-2x + 8 = -3x + 14$

6. $7 - 3x = 6 + 3x$

7. $4x - 6 - 5x = 3x + 1 - x$

8. $x - 2 - 3x = 2x + 1 - 5x$

9. $6(2x + 5) = 5(3x - 4)$

10. $-2(4x - 7) = -(9x + 4)$

11. $-2(3x - 1) - 3(2x + 5) = -5(2x - 8)$

12. $4(5x - 2) + (x - 4) = 6(3x - 10)$

13. $\dfrac{2}{3}x - \dfrac{1}{2} + \dfrac{1}{4}x = \dfrac{5}{8}$

14. $\dfrac{3}{4}x + \dfrac{2}{5} - x = \dfrac{3}{10}$

15. $\dfrac{2x - 1}{3} - \dfrac{3x + 2}{5} = -2$

16. $\dfrac{4x + 1}{6} + \dfrac{2x - 3}{5} = \dfrac{4}{15}$

17. $\dfrac{-3}{2x - 1} = \dfrac{2}{4x + 7}$

18. $\dfrac{4}{5x - 2} = \dfrac{6}{7x + 3}$

19. $\dfrac{x + 2}{3} + 1 = \dfrac{x - 2}{3}$

20. $3(x + 1) - 5(x - 2) = -2(x - 7)$

21. $0.09x + 0.11(x + 125) = 68.75$

22. $0.08(x + 200) = 0.07x + 20$

23. $\dfrac{x - 1}{5} - 2 = \dfrac{x - 11}{5}$

24. $3(x - 1) + 2(x + 4) = 5(x + 1)$

25. $0.05x - 4(x + 0.5) = 1.2$

26. $0.5(3x + 0.7) = 20.6$

For Problems 27–36, solve each system of equations. (Objective 3)

27. $\begin{pmatrix} x - 2y = 10 \\ 3x + 7y = -22 \end{pmatrix}$ 28. $\begin{pmatrix} 4x - 5y = -21 \\ 3x + y = 8 \end{pmatrix}$

29. $\begin{pmatrix} 5x - 2y = -3 \\ 3x - 4y = 15 \end{pmatrix}$ 30. $\begin{pmatrix} 7x - 5y = 6 \\ 3x + 2y = -14 \end{pmatrix}$

31. $\begin{pmatrix} 2x - y = 6 \\ 4x - 2y = -1 \end{pmatrix}$ 32. $\begin{pmatrix} y = 4x - 3 \\ y = 3x + 1 \end{pmatrix}$

33. $\begin{pmatrix} 3x + 4y = -18 \\ 5x - 7y = -30 \end{pmatrix}$ 34. $\begin{pmatrix} 4x - 6y = -4 \\ 2x - 3y = -2 \end{pmatrix}$

35. $\begin{pmatrix} \dfrac{1}{2}x - \dfrac{2}{3}y = -8 \\ \dfrac{3}{2}x + \dfrac{1}{3}y = -10 \end{pmatrix}$ 36. $\begin{pmatrix} \dfrac{3}{4}x + \dfrac{2}{5}y = 13 \\ \dfrac{1}{3}x - \dfrac{3}{10}y = 1 \end{pmatrix}$

For Problems 37–52, use an equation or a system of equations to help you solve each problem. (Objective 4)

37. Find three consecutive integers whose sum is −45.

38. Tina is paid time-and-a-half for each hour worked over 40 hours in a week. Last week she worked 45 hours and earned $437. What is her normal hourly rate?

39. There are 51 students in a certain class. The number of females is 5 less than three times the number of males. Find the number of females and the number of males in the class.

40. The sum of the present ages of Eric and his father is 58 years. In 10 years, his father will be twice as old as Eric will be at that time. Find their present ages.

41. Kaitlin went on a shopping trip, spending a total of $124 on a skirt, a sweater, and a pair of shoes. The cost of the sweater was $\dfrac{8}{7}$ of the cost of the skirt. The shoes cost $8 less than the skirt. Find the cost of each item.

42. The ratio of male students to female students at a certain university is 5 to 7. If there is a total of 16,200 students, find the number of male and the number of female students.

43. If each of two opposite sides of a square is increased by 3 centimeters, and each of the other two sides is decreased by 2 centimeters, the area is increased by 8 square centimeters. Find the length of a side of the square.

44. Desa invested a certain amount of money at 4% interest and $1500 more than that amount at 7%. Her total yearly interest was $435. How much did she invest at each rate?

45. Suppose that an item costs a retailer $50. How much more profit could be gained by fixing a 50% profit based on selling price rather than a 50% profit based on the cost?

46. Find three consecutive integers such that the sum of the smallest integer and the largest integer is equal to twice the middle integer.

47. The ratio of the weight of sodium to that of chlorine in common table salt is 5 to 3. Find the amount of each element in a salt block that weighs 200 pounds.

48. Sean bought 5 lemons and 3 limes for $3.90. At the same prices, Kim bought 4 lemons and 7 limes for $5.65. Find the price per lemon and the price per lime.

49. One leg of a right triangle is 7 meters longer than the other leg. If the length of the hypotenuse is 17 meters, find the length of each leg.

50. The perimeter of a rectangle is 44 inches, and its area is 112 square inches. Find the length and width of the rectangle.

51. Domenica and Javier start from the same location at the same time and ride their bicycles in opposite directions for 4 hours, at which time they are 140 miles apart. Domenica rides 3 miles per hour faster than Javier. Find the rate of each rider.

52. A container has 6 liters of a 40% alcohol solution in it. How much pure alcohol should be added to raise it to a 60% solution?

Thoughts Into Words

53. Explain how a trial-and-error approach could be used to solve Problem 49.

54. Now try a trial-and-error approach to solve Problem 41. What kind of difficulty are you having?

55. Suppose that your friend analyzes Problem 51 as follows: If they are 140 miles apart in 4 hours, then together they would need to average $\dfrac{140}{4} = 35$ miles per hour. Accordingly, we need two numbers whose sum is 35, and one number must be 3 larger than the other number. Thus Javier rides at 16 miles per hour and Domenica at 19 miles per hour. How would you react to this analysis of the problem?

8.2	Inequalities: A Brief Review

OBJECTIVES

1. Review of properties and techniques for solving inequalities

2. Express solution sets in interval notation

3. Review solving compound inequalities

4. Review solving word problems that involve inequalities

Just as we use the symbol $=$ to represent "is equal to," we also use the symbols $<$ and $>$ to represent "is less than" and "is greater than," respectively. Thus various **statements of inequality** can be made.

$a < b$ means a is less than b

$a \leq b$ means a is less than or equal to b

$a > b$ means a is greater than b

$a \geq b$ means a is greater than or equal to b

An **algebraic inequality** such as $x - 2 < 6$ is neither true nor false as it stands and is called an open sentence. For each numerical value substituted for x, the algebraic inequality $x - 2 < 6$ becomes a numerical statement of inequality that is true or false. For example, if $x = 10$, then $x - 2 < 6$ becomes $10 - 2 < 6$, which is false. If $x = 5$, then $x - 2 < 6$ becomes $5 - 2 < 6$, which is true. **Solving an algebraic inequality** is the process of finding the numbers that make it a true numerical statement. Such numbers are called the **solutions** of the inequality and are said to **satisfy** it.

The general process for solving inequalities closely parallels that for solving equations. We continue to replace the given inequality with equivalent but simpler inequalities until the solution set is obvious. The following property provides the basis for producing equivalent inequalities. (Because subtraction can be defined in terms of addition, and division can be defined in terms of multiplication, we state the property at this time in terms of only addition and multiplication.)

Property 8.2

1. For all real numbers a, b, and c,

   $a > b$ if and only if $a + c > b + c$.

2. For all real numbers, a, b, and c, with $c > 0$,

   $a > b$ if and only if $ac > bc$

3. For all real numbers, a, b, and c, with $c < 0$,

   $a > b$ if and only if $ac < bc$

Similar properties exist if $>$ is replaced by $<$, $\leq$, or $\geq$. Part 1 of Property 8.2 is commonly called the **addition property of inequality**. Parts 2 and 3 together make up the **multiplication property of inequality**. Pay special attention to part 3. If both sides of an inequality are multiplied by a negative number, the inequality symbol must be reversed. For example, if both sides of $-3 < 5$ are multiplied by -2, then the inequality $6 > -10$ is produced. In the following example, note the use of the distributive property, as well as both the addition and multiplication properties of inequality.

Classroom Example
Solve $2(2x + 3) \geq 2x - 6$.

EXAMPLE 1   Solve $3(2x - 1) < 8x - 7$.

Solution

$$3(2x - 1) < 8x - 7$$

$6x - 3 < 8x - 7$	Apply distributive property to left side
$-2x - 3 < -7$	Add $-8x$ to both sides
$-2x < -4$	Add 3 to both sides
$-\dfrac{1}{2}(-2x) > -\dfrac{1}{2}(-4)$	Multiply both sides by $-\dfrac{1}{2}$, which reverses the inequality
$x > 2$	

The solution set is $\{x \mid x > 2\}$.

A graph of the solution set $\{x \mid x > 2\}$ in Example 1 is shown in Figure 8.1. The parenthesis indicates that 2 does not belong to the solution set.

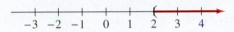

Figure 8.1

Checking the solutions of an inequality presents a problem. Obviously, we cannot check all of the infinitely many solutions for a particular inequality. However, by checking at least one solution, especially when the multiplication property has been used, we might catch a mistake of forgetting to change the type of inequality. In Example 1 we are claiming that all numbers greater than 2 will satisfy the original inequality. Let's check the number 3.

$$3(2x - 1) < 8x - 7$$
$$3[2(3) - 1] \overset{?}{<} 8(3) - 7$$
$$3(5) < 17$$

$15 < 17$	It checks!

Interval Notation

It is also convenient to express solution sets of inequalities by using **interval notation**. For example, the notation $(2, \infty)$ refers to the interval of all real numbers greater than 2. As on the graph in Figure 8.1, the left-hand parenthesis indicates that 2 is not to be included. The infinity symbol, ∞, along with the right-hand parenthesis, indicates that there is no right-hand endpoint. In the following table (which includes Figure 8.2) is a partial list of

Set	Graph	Interval notation
$\{x \mid x > a\}$	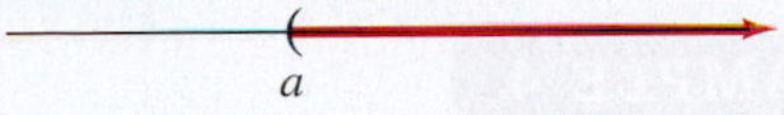	(a, ∞)
$\{x \mid x \geq a\}$		$[a, \infty)$
$\{x \mid x < b\}$		$(-\infty, b)$
$\{x \mid x \leq b\}$		$(-\infty, b]$

Figure 8.2

interval notations, along with the sets and graphs that they represent. Note the use of square brackets to *include* endpoints.

Classroom Example
Solve $2x + 4 - 5x \leq -2x + 7$, and express the solution set in interval notation.

EXAMPLE 2

Solve $-3x + 5x - 2 \geq 8x - 7 - 9x$, and express the solution set in interval notation.

Solution

$$-3x + 5x - 2 \geq 8x - 7 - 9x$$

$$2x - 2 \geq -x - 7 \qquad \text{Combine similar terms on both sides}$$

$$3x - 2 \geq -7 \qquad \text{Add } x \text{ to both sides}$$

$$3x \geq -5 \qquad \text{Add 2 to both sides}$$

$$\frac{1}{3}(3x) \geq \frac{1}{3}(-5) \qquad \text{Multiply both sides by } \frac{1}{3}$$

$$x \geq -\frac{5}{3}$$

The solution set is $\left[-\frac{5}{3}, \infty \right)$.

Classroom Example
Solve $5(2x + 1) < -3(-4x + 2)$, and express the solution set in interval notation.

EXAMPLE 3

Solve $4(x - 3) > 9(x + 1)$, and express the solution set in interval notation.

Solution

$$4(x - 3) > 9(x + 1)$$

$$4x - 12 > 9x + 9 \qquad \text{Apply the distributive property}$$

$$-5x - 12 > 9 \qquad \text{Add } -9x \text{ to both sides}$$

$$-5x > 21 \qquad \text{Add 12 to both sides}$$

$$-\frac{1}{5}(-5x) < -\frac{1}{5}(21) \qquad \text{Multiply both sides by } -\frac{1}{5}, \text{ which reverses the inequality}$$

$$x < -\frac{21}{5}$$

The solution set is $\left(-\infty, -\frac{21}{5} \right)$.

The next example will solve the inequality without indicating the justification for each step. Be sure that you can supply the reasons for the steps.

Classroom Example
Solve $6x - 4(x + 7) > 3(x - 9)$, and express the solution set in interval notation.

EXAMPLE 4

Solve $3(2x + 1) - 2(2x + 5) < 5(3x - 2)$, and express the solution set in interval notation.

Solution

$$3(2x + 1) - 2(2x + 5) < 5(3x - 2)$$

$$6x + 3 - 4x - 10 < 15x - 10$$

$$2x - 7 < 15x - 10$$

$$-13x - 7 < -10$$

$$-13x < -3$$

$$-\frac{1}{13}(-13x) > -\frac{1}{13}(-3)$$

$$x > \frac{3}{13}$$

The solution set is $\left(\frac{3}{13}, \infty\right)$.

Classroom Example
Solve $13 \leq 5(2x + 1) - 3(x + 2)$, and express the solution set in interval notation.

EXAMPLE 5

Solve $4 \leq 2(x - 3) - (x + 4)$, and express the solution set in interval notation.

Solution

$$4 \leq 2(x - 3) - (x + 4)$$

$$4 \leq 2x - 6 - x - 4$$

$$4 \leq x - 10$$

$$14 \leq x$$

$$x \geq 14$$

The solution set is $[14, \infty)$.

Remark: In the solution for Example 5, the solution set could be determined from the statement $14 \leq x$. However, you may find it easier to use the equivalent statement $x \geq 14$ to determine the solution set.

Classroom Example
Solve $\frac{y + 5}{3} - \frac{1}{6} \geq \frac{y - 4}{2}$, and express the solution set in interval notation.

EXAMPLE 6

Solve $\frac{x - 4}{6} - \frac{x - 2}{9} \leq \frac{5}{18}$, and express the solution set in interval notation.

Solution

$$\frac{x - 4}{6} - \frac{x - 2}{9} \leq \frac{5}{18}$$

$$18\left(\frac{x - 4}{6} - \frac{x - 2}{9}\right) \leq 18\left(\frac{5}{18}\right) \qquad \text{Multiply both sides by the LCD}$$

$$18\left(\frac{x - 4}{6}\right) - 18\left(\frac{x - 2}{9}\right) \leq 18\left(\frac{5}{18}\right) \qquad \text{Distributive property}$$

$$3(x - 4) - 2(x - 2) \leq 5$$

$$3x - 12 - 2x + 4 \leq 5$$

$$x - 8 \leq 5$$

$$x \leq 13$$

The solution set is $(-\infty, 13]$.

Classroom Example
Solve $0.05x + 0.04(1000 - x)$
≥ 46.50, and express the solution set
in interval notation.

EXAMPLE 7

Solve $0.08x + 0.09(x + 100) \geq 43$, and express the solution set in interval notation.

Solution

$$0.08x + 0.09(x + 100) \geq 43$$
$$100[0.08x + 0.09(x + 100)] \geq 100(43) \quad \text{Multiply both sides by 100}$$
$$100(0.08x) + 100[0.09(x + 100)] \geq 4300$$
$$8x + 9(x + 100) \geq 4300$$
$$8x + 9x + 900 \geq 4300$$
$$17x + 900 \geq 4300$$
$$17x \geq 3400$$
$$x \geq 200$$

The solution set is $[200, \infty)$.

Compound Statements

We use the words "and" and "or" in mathematics to form **compound statements**. The following are examples of compound numerical statements that use "and." We call such statements **conjunctions**. We agree to call a conjunction true only if all of its component parts are true. Statements 1 and 2 below are true, but statements 3, 4, and 5 are false.

   1. $3 + 4 = 7$ and $-4 < -3$ True

   2. $-3 < -2$ and $-6 > -10$ True

   3. $6 > 5$ and $-4 < -8$ False

   4. $4 < 2$ and $0 < 10$ False

   5. $-3 + 2 = 1$ and $5 + 4 = 8$ False

We call compound statements that use "or" **disjunctions**. The following are examples of disjunctions that involve numerical statements.

   6. $0.14 > 0.13$ or $0.235 < 0.237$ True

   7. $\dfrac{3}{4} > \dfrac{1}{2}$ or $-4 + (-3) = 10$ True

   8. $-\dfrac{2}{3} > \dfrac{1}{3}$ or $(0.4)(0.3) = 0.12$ True

   9. $\dfrac{2}{5} < -\dfrac{2}{5}$ or $7 + (-9) = 16$ False

A disjunction is true if at least one of its component parts is true. In other words, disjunctions are false only if all of the component parts are false. Thus statements 6, 7, and 8 are true, but statement 9 is false.

   Now let's consider finding solutions for some compound statements that involve algebraic inequalities. Keep in mind that our previous agreements for labeling conjunctions and disjunctions true or false form the basis for our reasoning.

Classroom Example
Graph the solution set for the conjunction $x \geq 0$ and $x < 5$. Express the solution set in interval notation and set-builder notation.

EXAMPLE 8

Graph the solution set for the conjunction $x > -1$ and $x < 3$. Also express the solution set in interval notation and in set-builder notation.

Solution

The key word is "and," so we need to satisfy both inequalities. Thus all numbers between -1 and 3 are solutions, which we indicate on a number line as in Figure 8.3.

Figure 8.3

Using interval notation, we can represent the interval enclosed in parentheses in Figure 8.3 by $(-1, 3)$. Using set-builder notation, we can express the same interval as $\{x \mid -1 < x < 3\}$, where the statement $-1 < x < 3$ is read "negative one is less than x, and x is less than three." In other words, x is between -1 and 3.

Example 8 represents another concept that pertains to sets. The set of all elements common to two sets is called the **intersection** of the two sets. Thus in Example 8 we found the intersection of the two sets $\{x \mid x > -1\}$ and $\{x \mid x < 3\}$ to be the set $\{x \mid -1 < x < 3\}$. In general, we define the intersection of two sets as follows:

Definition 8.1

The **intersection** of two sets A and B (written $A \cap B$) is the set of all elements that are in both set A and set B. Using set builder notation, we can write

$$A \cap B = \{x \mid x \in A \text{ and } x \in B\}$$

Classroom Example
Graph the solution set for the conjunction $3x + 4 \le 1$ *and* $5x + 2 \le -13$, and express it using interval notation.

EXAMPLE 9

Graph the solution set for the conjunction $3x + 1 > -5$ *and* $2x + 5 > 7$, and express it using interval notation.

Solution

First, let's simplify both inequalities.

$$3x + 1 > -5 \quad \text{and} \quad 2x + 5 > 7$$
$$3x > -6 \quad \text{and} \quad 2x > 2$$
$$x > -2 \quad \text{and} \quad x > 1$$

Because this is a conjunction, we must satisfy both inequalities. Thus all numbers greater than 1 are solutions, and the solution set is $(1, \infty)$. We show the graph of the solution set in Figure 8.4.

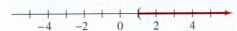

Figure 8.4

We can solve a conjunction such as $3x + 1 > -3$ and $3x + 1 < 7$, in which the same algebraic expression (in this case $3x + 1$) is contained in both inequalities, by using the **compact form** $-3 < 3x + 1 < 7$ as follows:

$$-3 < 3x + 1 < 7$$
$$-4 < 3x < 6 \qquad \text{Add } -1 \text{ to the left side, middle, and right side}$$
$$-\frac{4}{3} < x < 2 \qquad \text{Multiply through by } \frac{1}{3}$$

The solution set is $\left(-\frac{4}{3}, 2\right)$.

The word "and" ties the concept of a conjunction to the set concept of intersection. In a like manner, the word "or" links the idea of a disjunction to the set concept of **union**. We define the union of two sets as follows:

> **Definition 8.2**
>
> The **union** of two sets A and B (written $A \cup B$) is the set of all elements that are in set A or in set B, or in both. Using set builder notation, we can write
>
> $$A \cup B = \{x | x \in A \ \text{or} \ x \in B\}$$

Classroom Example
Graph the solution set for the disjunction $x \le 2$ or $x \ge 5$, and express it using interval notation.

EXAMPLE 10

Graph the solution set for the disjunction $x < -1$ *or* $x > 2$, and express it using interval notation.

Solution

The key word is "or," so all numbers that satisfy either inequality (or both) are solutions. Thus all numbers less than -1, along with all numbers greater than 2, are the solutions. The graph of the solution set is shown in Figure 8.5.

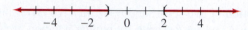

Figure 8.5

Using interval notation and the set concept of union, we can express the solution set as $(-\infty, -1) \cup (2, \infty)$.

Example 10 illustrates that in terms of set vocabulary, the solution set of a disjunction is the union of the solution sets of the component parts of the disjunction. Note that there is *no compact form* for writing $x < -1$ or $x > 2$.

Classroom Example
Graph the solution set for the disjunction $3x > x + 12$ or $7x - 4 \ge 10$, and express it using interval notation.

EXAMPLE 11

Graph the solution set for the disjunction $2x - 5 < -11$ *or* $5x + 1 \ge 6$, and express it using interval notation.

Solution

First, let's simplify both inequalities.

$$
\begin{array}{rcl}
2x - 5 < -11 & \text{or} & 5x + 1 \ge 6 \\
2x < -6 & \text{or} & 5x \ge 5 \\
x < -3 & \text{or} & x \ge 1
\end{array}
$$

This is a disjunction, and all numbers less than -3, along with all numbers greater than or equal to 1, will satisfy it. Thus the solution set is $(-\infty, -3) \cup [1, \infty)$. Its graph is shown in Figure 8.6.

Figure 8.6

In summary, to solve a compound sentence involving an inequality, proceed as follows:

1. Solve separately each inequality in the compound sentence.

2. If it is a *conjunction*, the solution set is the *intersection* of the solution sets of each inequality.

3. If it is a *disjunction*, the solution set is the *union* of the solution sets of each inequality.

The following agreements (Figure 8.7) on the use of interval notation should be added to the list on page 349.

Set	Graph	Interval notation
$\{x \mid a < x < b\}$	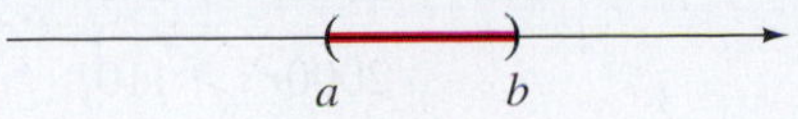	(a, b)
$\{x \mid a \leq x < b\}$	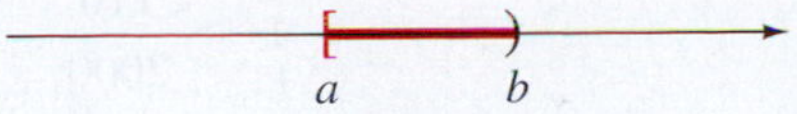	$[a, b)$
$\{x \mid a < x \leq b\}$	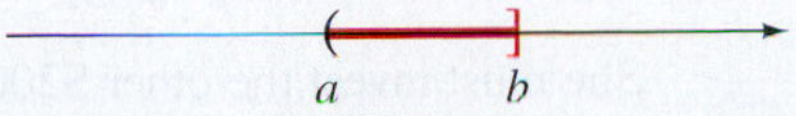	$(a, b]$
$\{x \mid a \leq x \leq b\}$	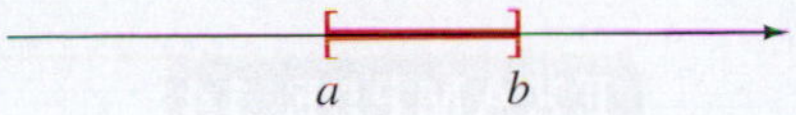	$[a, b]$
$\{x \mid x \text{ is a real number}\}$		$(-\infty, \infty)$

Figure 8.7

Problem Solving

We will conclude this section with some word problems that contain inequality statements.

Classroom Example
For the first four months of the year, Javier made $1200, $2000, $1800, and $3000 selling cars. What must his salary be in May to have an average monthly salary of more than $2200 for the five months?

EXAMPLE 12

Sari had scores of 94, 84, 86, and 88 on her first four exams of the semester. What score must she obtain on the fifth exam to have an average of 90 or better for the five exams?

Solution

Let s represent the score Sari needs on the fifth exam. Because the average is computed by adding all scores and dividing by the number of scores, we have the following inequality to solve.

$$\frac{94 + 84 + 86 + 88 + s}{5} \geq 90$$

Solving this inequality, we obtain

$$\frac{352 + s}{5} \geq 90$$

$$5\left(\frac{352 + s}{5}\right) \geq 5(90) \qquad \text{Multiply both sides by 5}$$

$$352 + s \geq 450$$

$$s \geq 98$$

Sari must receive a score of 98 or better.

Classroom Example
An investor has $3000 invested at 4% and $2000 invested at 6%. How much more money should be invested at 6% so that the two investments together yield more than $330 of yearly interest?

EXAMPLE 13

An investor has $4000 to invest. Suppose she invests $2000 at 8% interest. At what rate must she invest the other $2000 so that the two investments together yield more than $270 of yearly interest?

Solution

Let r represent the unknown rate of interest. We can use the following guideline to set up an inequality.

Interest from 8% investment	+	Interest from $r\%$ investment	>	$270
$(8\%)(\$2000)$	+	$r(\$2000)$		$270

Solving this inequality yields

$$160 + 2000r > 270$$

$$2000r > 110$$

$$r > \frac{110}{2000}$$

$$r > 0.055 \qquad \textcolor{blue}{\text{Change to a decimal}}$$

She must invest the other $2000 at a rate greater than 5.5%.

EXAMPLE 14

If the temperature for a 24-hour period ranged between 41°F and 59°F, inclusive (that is, $41 \leq F \leq 59$), what was the range in Celsius degrees?

Solution

Use the formula $F = \dfrac{9}{5}C + 32$, to solve the following compound inequality.

$$41 \leq \frac{9}{5}C + 32 \leq 59$$

Solving this yields

$$9 \leq \frac{9}{5}C \leq 27 \qquad \textcolor{blue}{\text{Add } -32}$$

$$\frac{5}{9}(9) \leq \frac{5}{9}\left(\frac{9}{5}C\right) \leq \frac{5}{9}(27) \qquad \textcolor{blue}{\text{Multiply by } \frac{5}{9}}$$

$$5 \leq C \leq 15$$

The range was between 5°C and 15°C, inclusive.

Concept Quiz 8.2

For Problems 1–5, match the inequality statements with the solution set expressed in interval notation.

1. $-2x > 6$

    A. $(3, \infty)$

2. $x + 3 > 6$

    B. $[-1, 3]$

3. $x + 1 \geq 0$ and $x \geq 3$

    C. $[3, \infty)$

4. $x + 1 \geq 0$ or $x \geq 3$

    D. $(-\infty, -3)$

5. $-2 \leq 2x \leq 6$

    E. $[-1, \infty)$

For Problems 6–10, answer true or false.

6. The solution set of the disjunction $x < 2$ or $x > 1$ is $(-\infty, \infty)$.

7. The solution set of the conjunction $x < 2$ and $x > 1$ is the null set.

8. The solution set of the inequality $-x - 2 > 4$ is $(-6, \infty)$.

9. The solution set of the conjunction $1 \leq -2x - 3 \leq 4$ is $\left[-\dfrac{7}{2}, -2\right]$.

10. The solution set of the disjunction $3x - 1 > 4$ or $3 > 2$ is $(-\infty, \infty)$.

Problem Set 8.2

For Problems 1–24, solve each inequality and express the solution set using interval notation. (Objective 1)

1. $6x - 2 > 4x - 14$

2. $9x + 5 < 6x - 10$

3. $2x - 7 < 6x + 13$

4. $2x - 3 > 7x + 22$

5. $4(x - 3) \leq -2(x + 1)$

6. $3(x - 1) \geq -(x + 4)$

7. $5(x - 4) - 6(x + 2) < 4$

8. $3(x + 2) - 4(x - 1) < 6$

9. $-3(3x + 2) - 2(4x + 1) \geq 0$

10. $-4(2x - 1) - 3(x + 2) \geq 0$

11. $-(x - 3) + 2(x - 1) < 3(x + 4)$

12. $3(x - 1) - (x - 2) > -2(x + 4)$

13. $7(x + 1) - 8(x - 2) < 0$

14. $5(x - 6) - 6(x + 2) < 0$

15. $\dfrac{x + 3}{8} - \dfrac{x + 5}{5} \geq \dfrac{3}{10}$

16. $\dfrac{x - 4}{6} - \dfrac{x - 2}{9} \leq \dfrac{5}{18}$

17. $\dfrac{4x - 3}{6} - \dfrac{2x - 1}{12} < -2$

18. $\dfrac{3x + 2}{9} - \dfrac{2x + 1}{3} > -1$

19. $0.06x + 0.08(250 - x) \geq 19$

20. $0.08x + 0.09(2x) \geq 130$

21. $0.09x + 0.1(x + 200) > 77$

22. $0.07x + 0.08(x + 100) > 38$

23. $x \geq 3.4 + 0.15x$

24. $x \geq 2.1 + 0.3x$

For Problems 25–30, solve each compound inequality and graph the solution sets. Express the solution sets in interval notation. (Objective 3)

25. $2x - 1 \geq 5$ and $x > 0$

26. $3x + 2 > 17$ and $x \geq 0$

27. $5x - 2 < 0$ and $3x - 1 > 0$

28. $x + 1 > 0$ and $3x - 4 < 0$

29. $3x + 2 < -1$ or $3x + 2 > 1$

30. $5x - 2 < -2$ or $5x - 2 > 2$

For Problems 31–36, solve each compound inequality using the compact form. Express the solution sets in interval notation. (Objective 3)

31. $-6 < 4x - 5 < 6$

32. $-2 < 3x + 4 < 2$

33. $-4 \leq \dfrac{x - 1}{3} \leq 4$

34. $-1 \leq \dfrac{x + 2}{4} \leq 1$

35. $-3 < 2 - x < 3$

36. $-4 < 3 - x < 4$

For Problems 37–44, solve each problem by setting up and solving an appropriate inequality. (Objective 4)

37. Mona invests $1000 at 8% yearly interest. How much does she have to invest at 9% so that the total yearly interest from the two investments exceeds $98?

38. Marsha bowled 142 and 170 in her first two games. What must she bowl in the third game to have an average of at least 160 for the three games?

39. Candace had scores of 95, 82, 93, and 84 on her first four exams of the semester. What score must she obtain on the fifth exam to have an average of 90 or better for the five exams?

40. Suppose that Derwin shot rounds of 82, 84, 78, and 79 on the first four days of a golf tournament. What must he shoot on the fifth day of the tournament to average 80 or less for the five days?

41. The temperatures for a 24-hour period ranged between $-4°F$ and $23°F$, inclusive. What was the range in Celsius degrees? (Use $F = \dfrac{9}{5}C + 32$.)

42. Oven temperatures for baking various foods usually range between $325°F$ and $425°F$, inclusive. Express this range in Celsius degrees. (Round answers to the nearest degree.)

43. A person's intelligence quotient (I) is found by dividing mental age (M), as indicated by standard tests, by chronological age (C), and then multiplying this ratio by 100. The formula $I = \dfrac{100M}{C}$ can be used. If the I range of a group of 11-year-olds is given by $80 \leq I \leq 140$, find the range of the mental age of this group.

44. Repeat Problem 43 for an I range of 70 to 125, inclusive, for a group of 9-year-olds.

Thoughts Into Words

45. Do the *less than* and *greater than* relations possess a symmetric property similar to the symmetric property of equality? Defend your answer.

46. Explain the difference between a conjunction and a disjunction. Give an example of each (outside the field of mathematics).

47. How do you know by inspection that the solution set of the inequality $x + 3 > x + 2$ is the entire set of real numbers?

48. Find the solution set for each of the following compound statements, and in each case explain your reasoning.

a. $x < 3$ and $5 > 2$

b. $x < 3$ or $5 > 2$

c. $x < 3$ and $6 < 4$

d. $x < 3$ or $6 < 4$

Answers to the Concept Quiz

1. D   **2.** A   **3.** C   **4.** E   **5.** B   **6.** True   **7.** False   **8.** False   **9.** True   **10.** True

---

8.3   Equations and Inequalities Involving Absolute Value

OBJECTIVES

1   Know the definition and properties of absolute value

2   Solve absolute value equations

3   Solve absolute value inequalities

Absolute Value

In Chapter 1 we used the concept of absolute value to describe precisely how to operate with positive and negative numbers. At that time we gave a geometric description of absolute value as the distance between a number and zero on the number line. For example, using vertical bars to denote absolute value, we can state that $|-3| = 3$ because the distance between -3 and 0 on the number line is 3 units. Likewise, $|2| = 2$ because the distance between 2 and 0 on the number line is 2 units. Using the distance interpretation, we can also state that $|0| = 0$ (Figure 8.8).

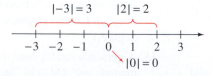

Figure 8.8

More formally, we define the concept of absolute value as follows:

Definition 8.3

For all real numbers a,

1. If $a \geq 0$, then $|a| = a$.

2. If $a < 0$, then $|a| = -a$.

Applying Definition 8.3, we obtain the following results:

$$|6| = 6$$ By applying part 1 of Definition 8.3

$$|0| = 0$$ By applying part 1 of Definition 8.3

$$|-7| = -(-7) = 7$$ By applying part 2 of Definition 8.3

Note the following ideas about absolute value:

1. The absolute value of a positive number is the number itself.

2. The absolute value of a negative number is its opposite.

3. The absolute value of any number except zero is always positive.

4. The absolute value of zero is zero.

5. A number and its opposite have the same absolute value.

We summarize these ideas in the following properties.

Properties of Absolute Value

The variables a and b represent any real number.

1. $|a| \geq 0$
2. $|a| = |-a|$
3. $|a - b| = |b - a|$ $a - b$ and $b - a$ are opposites of each other

---

Equations Involving Absolute Value

The interpretation of absolute value as distance on a number line provides a straightforward approach to solving a variety of equations and inequalities involving absolute value. First, let's consider some equations.

Classroom Example
Solve $|y| = 8$.

EXAMPLE 1   Solve $|x| = 2$.

Solution

Think in terms of *distance between the number and zero,* and you will see that x must be 2 or -2. That is, the equation $|x| = 2$ is equivalent to

$$x = -2 \quad \text{or} \quad x = 2$$

The solution set is $\{-2, 2\}$.

Classroom Example
Solve $|a - 4| = 7$.

EXAMPLE 2   Solve $|x + 2| = 5$.

Solution

The number, $x + 2$, must be -5 or 5. Thus $|x + 2| = 5$ is equivalent to

$$x + 2 = -5 \quad \text{or} \quad x + 2 = 5$$

Solving each equation of the disjunction yields

$$x + 2 = -5 \quad \text{or} \quad x + 2 = 5$$
$$x = -7 \quad \text{or} \quad x = 3$$

The solution set is $\{-7, 3\}$.

✓ **Check**

$$|x + 2| = 5 \qquad |x + 2| = 5$$
$$|-7 + 2| \overset{?}{=} 5 \qquad |3 + 2| \overset{?}{=} 5$$
$$|-5| \overset{?}{=} 5 \qquad |5| \overset{?}{=} 5$$
$$5 = 5 \qquad 5 = 5$$

The following general property should seem reasonable from the distance interpretation of absolute value.

> **Property 8.3**
>
> $|ax + b| = k$ is equivalent to $ax + b = -k$ or $ax + b = k$, where k is a positive number.

Example 3 demonstrates our format for solving equations of the form $|ax + b| = k$.

Classroom Example
Solve $|2b - 5| = 11$.

EXAMPLE 3   Solve $|5x + 3| = 7$.

Solution

$$|5x + 3| = 7$$
$$5x + 3 = -7 \quad \text{or} \quad 5x + 3 = 7$$
$$5x = -10 \quad \text{or} \quad 5x = 4$$
$$x = -2 \quad \text{or} \quad x = \frac{4}{5}$$

The solution set is $\left\{ -2, \dfrac{4}{5} \right\}$. Check these solutions!

Inequalities Involving Absolute Value

The *distance interpretation* for absolute value also provides a good basis for solving some inequalities that involve absolute value. Consider the following examples.

Classroom Example
Solve $|x| < 5$, and graph the solution.

EXAMPLE 4   Solve $|x| < 2$, and graph the solution set.

Solution

The number, x, must be *less than 2 units away from zero*. Thus $|x| < 2$ is equivalent to

$$x > -2 \quad \text{and} \quad x < 2$$

The solution set is $(-2, 2)$, and its graph is shown in Figure 8.9.

Figure 8.9

Classroom Example
Solve $|x - 4| < 2$, and graph the
solution.

EXAMPLE 5   Solve $|x + 3| < 1$, and graph the solutions.

Solution

Let's continue to think in terms of *distance* on a number line. The number, $x + 3$, must be *less than 1 unit away from zero*. Thus $|x + 3| < 1$ is equivalent to

$$x + 3 > -1 \quad \text{and} \quad x + 3 < 1$$

Solving this conjunction yields

$$
\begin{array}{ccc}
x + 3 > -1 & \text{and} & x + 3 < 1 \\
x > -4 & \text{and} & x < -2
\end{array}
$$

The solution set is $(-4, -2)$, and its graph is shown in Figure 8.10.

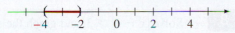

Figure 8.10

Take another look at Examples 4 and 5. The following general property should seem reasonable.

> **Property 8.4**
>
> $|ax + b| < k$ is equivalent to $ax + b > -k$ and $ax + b < k$, where k is a positive number.

Remember that we can write a conjunction such as $ax + b > -k$ and $ax + b < k$ in the compact form $-k < ax + b < k$. The compact form provides a very convenient format for solving inequalities such as $|3x - 1| < 8$, as Example 6 illustrates.

Classroom Example

Solve $|4x - 6| \leq 12$, and graph the
solution.

EXAMPLE 6   Solve $|3x - 1| < 8$, and graph the solutions.

Solution

$$
\begin{aligned}
|3x - 1| &< 8 \\
-8 < 3x - 1 &< 8 \\
-7 < 3x &< 9 \qquad &&\text{Add 1 to left side, middle, and right side} \\
\frac{1}{3}(-7) < \frac{1}{3}(3x) &< \frac{1}{3}(9) \qquad &&\text{Multiply through by } \frac{1}{3} \\
-\frac{7}{3} < x &< 3
\end{aligned}
$$

The solution set is $\left(-\dfrac{7}{3}, 3\right)$, and its graph is shown in Figure 8.11.

Figure 8.11

The distance interpretation also clarifies a property that pertains to *greater than* situations involving absolute value. Consider the following examples.

Classroom Example
Solve $|y| > 3$, and graph the solution.

EXAMPLE 7   Solve $|x| > 1$, and graph the solutions.

Solution

The number, x, must be *more than 1 unit away from zero.* Thus $|x| > 1$ is equivalent to

$$x < -1 \quad \text{or} \quad x > 1$$

The solution set is $(-\infty, -1) \cup (1, \infty)$, and its graph is shown in Figure 8.12.

Figure 8.12

Classroom Example
Solve $|a + 2| > 3$, and graph the solution.

EXAMPLE 8   Solve $|x - 1| > 3$, and graph the solutions.

Solution

The number, $x - 1$, must be *more than 3 units away from zero.* Thus $|x - 1| > 3$ is equivalent to

$$x - 1 < -3 \quad \text{or} \quad x - 1 > 3$$

Solving this disjunction yields

$$x - 1 < -3 \quad \text{or} \quad x - 1 > 3$$
$$x < -2 \quad \text{or} \quad x > 4$$

The solution set is $(-\infty, -2) \cup (4, \infty)$, and its graph is shown in Figure 8.13.

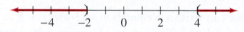

Figure 8.13

Examples 7 and 8 illustrate the following general property.

Property 8.5

$|ax + b| > k$ is equivalent to $ax + b < -k$ or $ax + b > k$, where k is a positive number.

Therefore, solving inequalities of the form $|ax + b| > k$ can take on the format shown in Example 9.

Classroom Example
Solve $\left|\dfrac{1}{2}x - 1\right| > 2$, and graph the solution.

EXAMPLE 9   Solve $|3x - 1| > 2$, and graph the solutions.

Solution

$$|3x - 1| > 2$$
$$3x - 1 < -2 \quad \text{or} \quad 3x - 1 > 2$$
$$3x < -1 \quad \text{or} \quad 3x > 3$$
$$x < -\frac{1}{3} \quad \text{or} \quad x > 1$$

The solution set is $\left(-\infty, -\dfrac{1}{3}\right) \cup (1, \infty)$, and its graph is shown in Figure 8.14.

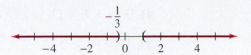

Figure 8.14

Properties 8.3, 8.4, and 8.5 provide the basis for solving a variety of equations and inequalities that involve absolute value. However, if at any time you become doubtful about what property applies, don't forget the distance interpretation. Furthermore, note that in each of the properties, k is a positive number. If k is a nonpositive number, we can determine the solution sets by inspection, as indicated by the following examples.

The solution set of $|x + 3| = 0$ is $\{-3\}$ because the number $x + 3$ has to be 0.

$|2x - 5| = -3$ has *no solutions* because the absolute value (distance) cannot be negative. [The solution set is $\varnothing$, the null set.]

$|x - 7| < -4$ has *no solutions* because we cannot obtain an absolute value less than -4. [The solution set is $\varnothing$.]

$|2x - 1| > -1$ is *satisfied by all real numbers* because the absolute value of $(2x - 1)$, regardless of what number is substituted for x, will always be greater than -1. [The solution set is the set of all real numbers, which we can express in interval notation as $(-\infty, \infty)$].

Concept Quiz 8.3

For Problems 1–10, answer true or false.

1. The absolute value of a negative number is the opposite of the number.
2. The absolute value of a number is always positive or zero.
3. The absolute value of a number is equal to the absolute value of its opposite.
4. The compound statement $x < 1$ or $x > 3$ can be written in compact form $3 < x < 1$.
5. The solution set for the equation $|x + 5| = 0$ is the null set, $\varnothing$.
6. The solution set for $|x - 2| \geq -6$ is all real numbers.
7. The solution set for $|x + 1| < -3$ is all real numbers.
8. The solution set for $|x - 4| \leq 0$ is $\{4\}$.
9. If a solution set in interval notation is $(-4, -2)$, then using set builder notation, it can be expressed as $\{x | -4 < x < -2\}$.
10. If a solution set in interval notation is $(-\infty, -2) \cup (4, \infty)$, then using set builder notation, it can be expressed as $\{x | x < -2 \text{ or } x > 4\}$.

Problem Set 8.3

For Problems 1–12, solve each equation. **(Objective 2)**

1. $|x - 1| = 8$
2. $|x + 2| = 9$
3. $|2x - 4| = 6$
4. $|3x - 4| = 14$
5. $|3x + 4| = 11$
6. $|5x - 7| = 14$
7. $|4 - 2x| = 6$
8. $|3 - 4x| = 8$
9. $\left| x - \dfrac{3}{4} \right| = \dfrac{2}{3}$
10. $\left| x + \dfrac{1}{2} \right| = \dfrac{3}{5}$
11. $|2x - 3| + 2 = 5$
12. $|3x - 1| - 1 = 9$

For Problems 13–26, solve each inequality and graph the solutions. **(Objective 3)**

13. $|x| < 5$
14. $|x| < 1$
15. $|x| \leq 2$
16. $|x| \leq 4$
17. $|x| > 2$
18. $|x| > 3$

19. $|x - 1| < 2$

20. $|x - 2| < 4$

21. $|x + 2| \leq 4$

22. $|x + 1| \leq 1$

23. $|x + 2| > 1$

24. $|x + 1| > 3$

25. $|x - 3| \geq 2$

26. $|x - 2| \geq 1$

For Problems 27–50, solve each inequality and express the solution set in interval notation. **(Objective 3)**

27. $|x - 2| > 6$

28. $|x - 3| > 9$

29. $|x + 3| < 5$

30. $|x + 1| < 8$

31. $|2x - 1| \leq 9$

32. $|3x + 1| \leq 13$

33. $|4x + 2| \geq 12$

34. $|5x - 2| \geq 10$

35. $|2 - x| > 4$

36. $|4 - x| > 3$

37. $|1 - 2x| < 2$

38. $|2 - 3x| < 5$

39. $|5x + 9| \leq 16$

40. $|7x - 6| \geq 22$

41. $|-2x + 7| \leq 13$

42. $|-3x - 4| \leq 15$

43. $\left| \dfrac{x - 3}{4} \right| < 2$

44. $\left| \dfrac{x + 2}{3} \right| < 1$

45. $\left| \dfrac{2x + 1}{2} \right| > 1$

46. $\left| \dfrac{3x - 1}{4} \right| > 3$

47. $|x + 7| - 3 \geq 4$

48. $|x - 2| + 4 \geq 10$

49. $|2x - 1| + 1 \leq 6$

50. $|4x + 3| - 2 \leq 5$

For Problems 51–60, solve each equation and inequality *by inspection.* **(Objectives 2 and 3)**

51. $|2x + 1| = -4$

52. $|5x - 1| = -2$

53. $|3x - 1| > -2$

54. $|4x + 3| < -4$

55. $|5x - 2| = 0$

56. $|3x - 1| = 0$

57. $|4x - 6| < -1$

58. $|x + 9| > -6$

59. $|x + 4| < 0$

60. $|x + 6| > 0$

Thoughts Into Words

61. Explain how you would solve the inequality $|2x + 5| > -3$

62. Why is 2 the only solution for $|x - 2| \leq 0$?

63. Explain how you would solve the equation $|2x - 3| = 0$

Further Investigations

For Problems 64–69, solve each equation.

64. $|3x + 1| = |2x + 3|$
    [*Hint:* $3x + 1 = 2x + 3$ or $3x + 1 = -(2x + 3)$]

65. $|-2x - 3| = |x + 1|$

66. $|2x - 1| = |x - 3|$

67. $|x - 2| = |x + 6|$

68. $|x + 1| = |x - 4|$

69. $|x + 1| = |x - 1|$

70. Use the definition of absolute value to help prove Property 8.3.

71. Use the definition of absolute value to help prove Property 8.4.

72. Use the definition of absolute value to help prove Property 8.5.

Answers to the Concept Quiz

1. True   **2.** True   **3.** True   **4.** False   **5.** False   **6.** True   **7.** False   **8.** True   **9.** True   **10.** True

8.4   Polynomials: A Brief Review and Binomial Expansions

OBJECTIVES

1. Review the addition and subtraction of polynomials
2. Review the properties of exponents
3. Review the multiplication of polynomials
4. Review the division of monomials
5. Write binomial expansions

Polynomials

Recall that algebraic expressions such as $5x$, $-6y^2$, $2x^{-1}y^{-2}$, $14a^2b$, $5x^{-4}$, and $-17ab^2c^3$ are called **terms**. Terms that contain variables with only nonnegative integers as exponents are called **monomials**. Of the previously listed terms, $5x$, $-6y^2$, $14a^2b$, and $-17ab^2c^3$ are monomials. The **degree** of a monomial is the sum of the exponents of the literal factors. For example, $7xy$ is of degree 2, whereas $14a^2b$ is of degree 3, and $-17ab^2c^3$ is of degree 6. If the monomial contains only one variable, then the exponent of that variable is the degree of the monomial. For example, $5x^3$ is of degree 3, and $-8y^4$ is of degree 4. Any nonzero constant term, such as 8, is of degree zero.

A **polynomial** is a monomial or a finite sum of monomials. Thus all of the following are polynomials.

$$4x^2 \qquad 3x^2 - 2x - 4 \qquad 7x^4 - 6x^3 + 5x^2 + 2x - 1$$

$$3x^2y - 2y \qquad \frac{1}{5}a^2 - \frac{2}{3}b^2 \qquad 14$$

In addition to calling a polynomial with one term a monomial, we also classify polynomials with two terms as **binomials** and those with three terms as **trinomials**. The **degree of a polynomial** is the degree of the term with the highest degree in the polynomial. The following examples illustrate some of this terminology.

The polynomial $4x^3y^4$ is a monomial in two variables of degree 7.

The polynomial $4x^2y - 2xy$ is a binomial in two variables of degree 3.

The polynomial $9x^2 - 7x - 1$ is a trinomial in one variable of degree 2.

Addition and Subtraction of Polynomials

Both adding polynomials and subtracting them rely on basically the same ideas. The commutative, associative, and distributive properties provide the basis for rearranging, regrouping, and combining similar terms. Consider the following addition problems.

$$(4x^2 + 5x + 1) + (7x^2 - 9x + 4) = (4x^2 + 7x^2) + (5x - 9x) + (1 + 4)$$
$$= 11x^2 - 4x + 5$$

$$(5x - 3) + (3x + 2) + (8x + 6) = (5x + 3x + 8x) + (-3 + 2 + 6)$$
$$= 16x + 5$$

The definition of subtraction as *adding the opposite* $[a - b = a + (-b)]$ extends to polynomials in general. The opposite of a polynomial can be formed by taking the opposite of each term. For example, the opposite of $3x^2 - 7x + 1$ is $-3x^2 + 7x - 1$. Symbolically, this is expressed as

$$-(3x^2 - 7x + 1) = -3x^2 + 7x - 1$$

You can also think in terms of the property $-x = -1(x)$ and the distributive property. Therefore,

$$-(3x^2 - 7x + 1) = -1(3x^2 - 7x + 1) = -3x^2 + 7x - 1$$

Now consider the following subtraction problems.

$$(7x^2 - 2x - 4) - (3x^2 + 7x - 1) = (7x^2 - 2x - 4) + (-3x^2 - 7x + 1)$$
$$= (7x^2 - 3x^2) + (-2x - 7x) + (-4 + 1)$$
$$= 4x^2 - 9x - 3$$

$$(4y^2 + 7) - (-3y^2 + y - 2) = (4y^2 + 7) + (3y^2 - y + 2)$$
$$= (4y^2 + 3y^2) + (-y) + (7 + 2)$$
$$= 7y^2 - y + 9$$

As we will see in Section 8.5, sometimes a vertical format is used, especially for subtraction of polynomials. Suppose, for example, that we want to subtract $4x^2 - 7xy + 5y^2$ from $3x^2 - 2xy + y^2$.

$$\begin{array}{l} 3x^2 - 2xy + y^2 \\ \underline{4x^2 - 7xy + 5y^2} \end{array}$$ Note which polynomial goes on the bottom and how the similar terms are aligned

Now we can *mentally form the opposite of the bottom polynomial* and add.

$$\begin{array}{l} 3x^2 - 2xy + y^2 \\ \underline{4x^2 - 7xy + 5y^2} \\ -x^2 + 5xy - 4y^2 \end{array}$$ The opposite of $4x^2 - 7xy + 5y^2$ is $-4x^2 + 7xy - 5y^2$

Products and Quotients of Monomials

Some basic properties of exponents play an important role in the multiplying and dividing of polynomials. These properties were introduced in Chapter 6, so at this time let's restate them and include (at the right) a "name tag." The name tags can be used for reference purposes, and they help reinforce the meaning of each specific part of the property.

> ### Property 8.6
>
> If m and n are integers and a and b are real numbers, with $b \neq 0$, whenever it appears in a denominator, then
>
> 1. $b^n \cdot b^m = b^{n+m}$ Product of two like bases with powers
> 2. $(b^n)^m = b^{mn}$ Power of a power
> 3. $(ab)^n = a^n b^n$ Power of a product
> 4. $\left(\dfrac{a}{b}\right)^n = \dfrac{a^n}{b^n}$ Power of a quotient
> 5. $\dfrac{b^n}{b^m} = b^{n-m}$ Quotient of two like bases with powers

Part 1 of Property 8.6, along with the commutative and associative properties of multiplication, form the basis for multiplying monomials. In the following examples, the steps enclosed in dashed boxes can be performed mentally whenever you feel comfortable with the process.

$$(-5a^3 b^4)(7a^2 b^5) = -5 \cdot 7 \cdot a^3 \cdot a^2 \cdot b^4 \cdot b^5$$
$$= -35a^{3+2} b^{4+5}$$
$$= -35a^5 b^9$$

$$(3x^2 y)(4x^3 y^2) = 3 \cdot 4 \cdot x^2 \cdot x^3 \cdot y \cdot y^2$$
$$= 12x^{2+3} y^{1+2}$$
$$= 12x^5 y^3$$

$$(-ab^2)(-5a^2 b) = (-1)(-5)(a)(a^2)(b^2)(b)$$
$$= 5a^{1+2} b^{2+1}$$
$$= 5a^3 b^3$$

$$(2x^2y^2)(3x^2y)(4y^3) = 2 \cdot 3 \cdot 4 \cdot x^2 \cdot x^2 \cdot y^2 \cdot y \cdot y^3$$
$$= 24x^{2+2}y^{2+1+3}$$
$$= 24x^4y^6$$

The following examples show how part 2 of Property 8.6 is used to find "a power of a power."

$$(x^4)^5 = x^{5(4)} = x^{20} \qquad (y^6)^3 = y^{3(6)} = y^{18} \qquad (2^3)^7 = 2^{7(3)} = 2^{21}$$

Parts 2 and 3 of Property 8.6 form the basis for raising a monomial to a power, as in the next examples.

$$(x^2y^3)^4 = (x^2)^4(y^3)^4 \qquad (3a^5)^3 = (3)^3(a^5)^3 \qquad (-2xy^4)^5 = (-2)^5(x)^5(y^4)^5$$
$$= x^8y^{12} \qquad\qquad = 27a^{15} \qquad\qquad = -32x^5y^{20}$$

Dividing Monomials

Part 5 of Property 8.6, along with our knowledge of dividing integers, provides the basis for dividing monomials. The following examples demonstrate the process.

$$\frac{24x^5}{3x^2} = 8x^{5-2} = 8x^3 \qquad\qquad \frac{-36a^{13}}{-12a^5} = 3a^{13-5} = 3a^8$$

$$\frac{-56x^9}{7x^4} = -8x^{9-4} = -8x^5 \qquad\qquad \frac{72b^5}{8b^5} = 9\left(\frac{b^5}{b^5} = 1\right)$$

$$\frac{48y^7}{-12y} = -4y^{7-1} = -4y^6 \qquad\qquad \frac{12x^4y^7}{2x^2y^4} = 6x^{4-2}y^{7-4} = 6x^2y^3$$

Multiplying Polynomials

The distributive property is usually stated as $a(b + c) = ab + ac$, but it can be extended as follows.

$$a(b + c + d) = ab + ac + ad$$
$$a(b + c + d + e) = ab + ac + ad + ae \qquad \text{etc.}$$

The commutative and associative properties, the properties of exponents, and the distributive property work together to form a basis for finding the product of a monomial and a polynomial. The following example illustrates this idea.

$$3x^2(2x^2 + 5x + 3) = 3x^2(2x^2) + 3x^2(5x) + 3x^2(3)$$
$$= 6x^4 + 15x^3 + 9x^2$$

Extending the method of finding the product of a monomial and a polynomial to finding the product of two polynomials is again based on the distributive property.

$$(x + 2)(y + 5) = x(y + 5) + 2(y + 5)$$
$$= x(y) + x(5) + 2(y) + 2(5)$$
$$= xy + 5x + 2y + 10$$

Note that we are multiplying each term of the first polynomial times each term of the second polynomial.

$$(x - 3)(y + z + 3) = x(y + z + 3) - 3(y + z + 3)$$
$$= xy + xz + 3x - 3y - 3z - 9$$

Frequently, multiplying polynomials produces similar terms that can be combined, which simplifies the resulting polynomial.

$$(x + 5)(x + 7) = x(x + 7) + 5(x + 7)$$
$$= x^2 + 7x + 5x + 35$$
$$= x^2 + 12x + 35$$

$$(x - 2)(x^2 - 3x + 4) = x(x^2 - 3x + 4) - 2(x^2 - 3x + 4)$$
$$= x^3 - 3x^2 + 4x - 2x^2 + 6x - 8$$
$$= x^3 - 5x^2 + 10x - 8$$

It helps to be able to find the product of two binomials without showing all of the intermediate steps. This is quite easy to do with the three-step shortcut pattern demonstrated (Figure 8.15) in the following example.

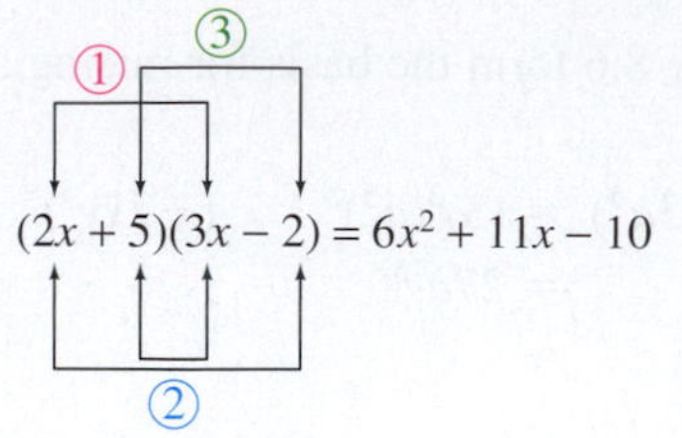

Step 1 Multiply $(2x)(3x)$.

Step 2 Multiply $(5)(3x)$ and $(2x)(-2)$ and combine.

Step 3 Multiply $(5)(-2)$.

$$(2x + 5)(3x - 2) = 6x^2 + 11x - 10$$

Figure 8.15

Now see whether you can use the pattern to find the following products.

$$(x + 2)(x + 6) = ?$$
$$(x - 3)(x + 5) = ?$$
$$(2x + 5)(3x + 7) = ?$$
$$(3x - 1)(4x - 3) = ?$$

Your answers should be $x^2 + 8x + 12$, $x^2 + 2x - 15$, $6x^2 + 29x + 35$, and $12x^2 - 13x + 3$. Keep in mind that this shortcut pattern applies only to finding the product of two binomials.

Remark: Shortcuts can be very helpful for certain manipulations in mathematics. But a word of caution: Do not lose the understanding of what you are doing. Make sure that you are able to do the manipulation without the shortcut.

Exponents can also be used to indicate repeated multiplication of polynomials. For example, $(3x - 4y)^2$ means $(3x - 4y)(3x - 4y)$, and $(x + 4)^3$ means $(x + 4)(x + 4)(x + 4)$. Therefore, raising a polynomial to a power is merely another multiplication problem.

$$(3x - 4y)^2 = (3x - 4y)(3x - 4y)$$
$$= 9x^2 - 24xy + 16y^2$$

[*Hint:* When squaring a binomial, be careful not to forget the middle term. That is, $(x + 5)^2 \neq x^2 + 25$; instead, $(x + 5)^2 = x^2 + 10x + 25$.]

$$(x + 4)^3 = (x + 4)(x + 4)(x + 4)$$
$$= (x + 4)(x^2 + 8x + 16)$$
$$= x(x^2 + 8x + 16) + 4(x^2 + 8x + 16)$$
$$= x^3 + 8x^2 + 16x + 4x^2 + 32x + 64$$
$$= x^3 + 12x^2 + 48x + 64$$

Special Patterns

When multiplying binomials, some special patterns occur that you should learn to recognize. These patterns can be used to find products, and some of them will be helpful later when you are factoring polynomials.

$$(a + b)^2 = a^2 + 2ab + b^2$$
$$(a - b)^2 = a^2 - 2ab + b^2$$
$$(a + b)(a - b) = a^2 - b^2$$
$$(a + b)^3 = a^3 + 3a^2b + 3ab^2 + b^3$$
$$(a - b)^3 = a^3 - 3a^2b + 3ab^2 - b^3$$

The three following examples illustrate the first three patterns, respectively.

$$(2x + 3)^2 = (2x)^2 + 2(2x)(3) + (3)^2$$
$$= 4x^2 + 12x + 9$$

$$(5x - 2)^2 = (5x)^2 - 2(5x)(2) + (2)^2$$
$$= 25x^2 - 20x + 4$$

$$(3x + 2y)(3x - 2y) = (3x)^2 - (2y)^2 = 9x^2 - 4y^2$$

In the first two examples, the resulting trinomial is called a **perfect-square trinomial**; it is the result of squaring a binomial. In the third example, the resulting binomial is called the **difference of two squares**. Later we will use both of these patterns when factoring polynomials.

The patterns for cubing of a binomial are helpful primarily when you are multiplying. These patterns can shorten the work of cubing a binomial, as the next two examples illustrate.

$$(3x + 2)^3 = (3x)^3 + 3(3x)^2(2) + 3(3x)(2)^2 + (2)^3$$
$$= 27x^3 + 54x^2 + 36x + 8$$

$$(5x - 2y)^3 = (5x)^3 - 3(5x)^2(2y) + 3(5x)(2y)^2 - (2y)^3$$
$$= 125x^3 - 150x^2y + 60xy^2 - 8y^3$$

Keep in mind that these multiplying patterns are useful shortcuts, but if you forget them, you can simply revert to applying the distributive property.

Binomial Expansion Pattern

It is possible to write the expansion of $(a + b)^n$, where n is *any* positive integer, without showing all of the intermediate steps of multiplying and combining similar terms. To do this, let's observe some patterns in the following examples; each one can be verified by direct multiplication.

$$(a + b)^1 = a + b$$
$$(a + b)^2 = a^2 + 2ab + b^2$$
$$(a + b)^3 = a^3 + 3a^2b + 3ab^2 + b^3$$
$$(a + b)^4 = a^4 + 4a^3b + 6a^2b^2 + 4ab^3 + b^4$$
$$(a + b)^5 = a^5 + 5a^4b + 10a^3b^2 + 10a^2b^3 + 5ab^4 + b^5$$

First, note the patterns of the exponents for a and b on a term-by-term basis. The exponents of a begin with the exponent of the binomial and decrease by 1, term by term, until the last term, which has $a^0 = 1$. The exponents of b begin with zero ($b^0 = 1$) and increase by 1, term by term, until the last term, which contains b to the power of the original binomial. In other words, the variables in the expansion of $(a + b)^n$ have the pattern

$$a^n, \quad a^{n-1}b, \quad a^{n-2}b^2, \quad \dots, \quad ab^{n-1}, \quad b^n$$

where for each term, the *sum* of the exponents of a and b is n.

Next, let's arrange the *coefficients* in a triangular formation; this yields an easy-to-remember pattern.

```
            1       1
        1       2       1
    1       3       3       1
  1     4       6       4       1
1     5      10      10       5      1
```

Row number n in the formation above contains the coefficients of the expansion of $(a + b)^n$. For example, the fifth row contains 1 5 10 10 5 1, and these numbers are the coefficients of the terms in the expansion of $(a + b)^5$. Furthermore, each can be formed from the previous row as follows:

1. Start and end each row with 1.

2. All other entries result from adding the two numbers in the row immediately above, one number to the left and one number to the right.

Thus from row 5, we can form row 6.

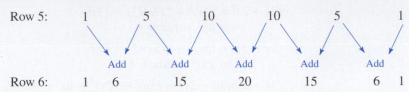

Row 5:   1   5   10   10   5   1

Add   Add   Add   Add   Add

Row 6:   1   6   15   20   15   6   1

Now we can use these seven coefficients and our discussion about the exponents to write out the expansion for $(a + b)^6$.

$$(a + b)^6 = a^6 + 6a^5b + 15a^4b^2 + 20a^3b^3 + 15a^2b^4 + 6ab^5 + b^6$$

Remark: The triangular formation of numbers that we have been discussing is often referred to as **Pascal's triangle**. This is in honor of Blaise Pascal, a 17th-century mathematician, to whom the discovery of this pattern is attributed.

Let's consider two more examples using Pascal's triangle and the exponent relationships.

Classroom Example
Expand $(x + y)^5$.

EXAMPLE 1   Expand $(a - b)^4$.

Solution

We can treat $a - b$ as $a + (-b)$ and use the fourth row of Pascal's triangle to obtain the coefficients.

$$[a + (-b)]^4 = a^4 + 4a^3(-b) + 6a^2(-b)^2 + 4a(-b)^3 + (-b)^4$$
$$= a^4 - 4a^3b + 6a^2b^2 - 4ab^3 + b^4$$

Classroom Example
Expand $(3x + 2y)^5$.

EXAMPLE 2   Expand $(2x + 3y)^5$.

Solution

Let $2x = a$ and $3y = b$. The coefficients come from the fifth row of Pascal's triangle.

$$(2x + 3y)^5 = (2x)^5 + 5(2x)^4(3y) + 10(2x)^3(3y)^2 + 10(2x)^2(3y)^3 + 5(2x)(3y)^4 + (3y)^5$$
$$= 32x^5 + 240x^4y + 720x^3y^2 + 1080x^2y^3 + 810xy^4 + 243y^5$$

Concept Quiz 8.4

For Problems 1–8, answer true or false.

1. The degree of the polynomial $3x^4 - 2x^3 + 7x^2 + 5x - 1$ is 4.

2. The polynomial $4x^2y^3$ is a binomial in two variables with degree 5.

3. $(-5xy^3)^2 = (-5)^2xy^6$.

4. $\dfrac{-15x^8}{3x^2} = -5x^4$.

5. The pattern for squaring a binomial is $(a + b)^2 = a^2 + 2ab + b^2$.

6. $4x^2 + 20x + 25$ is a perfect square trinomial.

7. The last term of the binomial expansion for $(2x + 3y)^4$ is $3y^4$.

8. Pascal's triangle is used to find the exponents in a binomial expansion.

Problem Set 8.4

For Problems 1–8, simplify the following using addition of polynomials. (Objective 1)

1. $8x - 2x + 7 + 3x - 10$

2. $15 - 2x - 4x + 3 + x$

3. $(4x^2 + 3x - 1) + (5x - 2)$

4. $(x^2 - 9) + (3x^2 - 5x - 1)$

5. $5(2x^2 - 3x + 7) - 3(-2x^2 + 8)$

6. $4(-3x^2 + 1) - (-2x^2 + 8x - 5)$

7. $(x^2 - 2xy - 2y^2) - (-x^2 + 3y^2)$

8. $3(x^2 - xy - 2y^2) - 4(x^2 - 5xy + 3y^2)$

For Problems 9–48, find each indicated product. (Objectives 2 and 3)

9. $\left(-\dfrac{1}{2}xy\right)\left(\dfrac{1}{3}x^2y^3\right)$

10. $\left(\dfrac{3}{4}x^4y^5\right)(-x^2y)$

11. $(3x)(-2x^2)(-5x^3)$

12. $(-2x)(-6x^3)(x^2)$

13. $(-6x^2)(3x^3)(x^4)$

14. $(-7x^2)(3x)(4x^3)$

15. $(x^2y)(-3xy^2)(x^3y^3)$

16. $(xy^2)(-5xy)(x^2y^4)$

17. $(-3ab^3)^4$

18. $(-2a^2b^4)^4$

19. $-(2ab)^4$

20. $-(3ab)^4$

21. $(4x + 5)(x + 7)$

22. $(6x + 5)(x + 3)$

23. $(3y - 1)(3y + 1)$

24. $(5y - 2)(5y + 2)$

25. $(7x - 2)(2x + 1)$

26. $(6x - 1)(3x + 2)$

27. $(1 + t)(5 - 2t)$

28. $(3 - t)(2 + 4t)$

29. $(3t + 7)^2$

30. $(4t + 6)^2$

31. $(2 - 5x)(2 + 5x)$

32. $(6 - 3x)(6 + 3x)$

33. $(x + 1)(x - 2)(x - 3)$

34. $(x - 1)(x + 4)(x - 6)$

35. $(x - 3)(x + 3)(x - 1)$

36. $(x - 5)(x + 5)(x - 8)$

37. $(x - 4)(x^2 + 5x - 4)$

38. $(x + 6)(2x^2 - x - 7)$

39. $(2x - 3)(x^2 + 6x + 10)$

40. $(3x + 4)(2x^2 - 2x - 6)$

41. $(4x - 1)(3x^2 - x + 6)$

42. $(5x - 2)(6x^2 + 2x - 1)$

43. $(x^2 + 2x + 1)(x^2 + 3x + 4)$

44. $(x^2 - x + 6)(x^2 - 5x - 8)$

45. $(4x - 1)^3$

46. $(3x - 2)^3$

47. $(5x + 2)^3$

48. $(4x - 5)^3$

For Problems 49–58, find the indicated products. Assume that all variables which appear as exponents represent positive integers. (Objective 3)

49. $(x^n - 4)(x^n + 4)$

50. $(x^{3a} - 1)(x^{3a} + 1)$

51. $(x^a + 6)(x^a - 2)$

52. $(x^a + 4)(x^a - 9)$

53. $(2x^n + 5)(3x^n - 7)$

54. $(3x^n + 5)(4x^n - 9)$

55. $(x^{2a} - 7)(x^{2a} - 3)$

56. $(x^{2a} + 6)(x^{2a} - 4)$

57. $(2x^n + 5)^2$

58. $(3x^n - 7)^2$

For Problems 59–68, find each quotient. (Objective 4)

59. $\dfrac{9x^4y^5}{3xy^2}$

60. $\dfrac{12x^2y^7}{6x^2y^3}$

61. $\dfrac{25x^5y^6}{-5x^2y^4}$

62. $\dfrac{56x^6y^4}{-7x^2y^3}$

63. $\dfrac{-54ab^2c^3}{-6abc}$

64. $\dfrac{-48a^3bc^5}{-6a^2c^4}$

65. $\dfrac{-18x^2y^2z^6}{xyz^2}$

66. $\dfrac{-32x^4y^5z^8}{x^2yz^3}$

67. $\dfrac{a^3b^4c^7}{-abc^5}$

68. $\dfrac{-a^4b^5c}{a^2b^4c}$

For Problems 69–80, use Pascal's triangle to help expand each of the following. (Objective 5)

69. $(a + b)^7$

70. $(a + b)^8$

71. $(x - y)^5$

72. $(x - y)^6$

73. $(x + 2y)^4$

74. $(2x + y)^5$

75. $(2a - b)^6$

76. $(3a - b)^4$

77. $(x^2 + y)^7$

78. $(x + 2y^2)^7$

79. $(2a - 3b)^5$

80. $(4a - 3b)^3$

Thoughts Into Words

81. Determine the number of terms in the product of $(x + y)$ and $(a + b + c + d)$ without doing the multiplication. Explain how you arrived at your answer.

82. How would you convince someone that $x^6 \div x^2$ is x^4 and not x^3?

Answers to the Concept Quiz

1. True **2.** False **3.** False **4.** False **5.** True **6.** True **7.** False **8.** False

8.5 Dividing Polynomials: Synthetic Division

OBJECTIVES

1 Review the division of a polynomial

2 Perform synthetic division

In the previous section, we reviewed the process of dividing monomials by monomials. In Section 2.2, we used $\dfrac{a}{b} + \dfrac{c}{b} = \dfrac{a + c}{b}$ and $\dfrac{a}{b} - \dfrac{c}{b} = \dfrac{a - c}{b}$ as the basis for adding and subtracting rational numbers and rational expressions. These same equalities, viewed as $\dfrac{a + c}{b} = \dfrac{a}{b} + \dfrac{c}{b}$ and $\dfrac{a - c}{b} = \dfrac{a}{b} - \dfrac{c}{b}$, along with our knowledge of dividing monomials, provide the basis for dividing polynomials by monomials. Consider the following examples:

$$\frac{18x^3 + 24x^2}{6x} = \frac{18x^3}{6x} + \frac{24x^2}{6x} = 3x^2 + 4x$$

$$\frac{35x^2y^3 - 55x^3y^4}{5xy^2} = \frac{35x^2y^3}{5xy^2} - \frac{55x^3y^4}{5xy^2} = 7xy - 11x^2y^2$$

To divide a polynomial by a monomial, we divide each term of the polynomial by the monomial. As with many skills, once you feel comfortable with the process, you may then want to perform some of the steps mentally. Your work could take on the following format:

$$\frac{40x^4y^5 + 72x^5y^7}{8x^2y} = 5x^2y^4 + 9x^3y^6 \qquad \frac{36a^3b^4 - 45a^4b^6}{-9a^2b^3} = -4ab + 5a^2b^3$$

In Section 6.5, we used some examples to introduce the process of dividing a polynomial by a binomial. The first example of that section gave a detailed step-by-step procedure for this division process. The following example uses some "think steps" to help you review that procedure.

Classroom Example
Divide $3x^2 - x - 10$ by $x - 2$.

EXAMPLE 1 Divide $(5x^2 + 6x - 8)$ by $(x + 2)$.

Solution

$$
\begin{array}{r}
5x - 4 \\
x + 2 \overline{)5x^2 + 6x - 8} \\
\underline{5x^2 + 10x} \\
-4x - 8 \\
\underline{-4x - 8} \\
0
\end{array}
$$

Think Steps

1. $\dfrac{5x^2}{x} = 5x$

2. $5x(x + 2) = 5x^2 + 10x$

3. $(5x^2 + 6x - 8) - (5x^2 + 10x) = -4x - 8$

4. $\dfrac{-4x}{x} = -4$

5. $-4(x + 2) = -4x - 8$

Recall that to check a division problem, we can multiply the divisor times the quotient and add the remainder. In other words,

Dividend = (Divisor)(Quotient) + (Remainder)

Sometimes the remainder is expressed as a fractional part of the divisor. The relationship then becomes

$$\frac{\text{Dividend}}{\text{Divisor}} = \text{Quotient} + \frac{\text{Remainder}}{\text{Divisor}}$$

Classroom Example
Divide $(5x^2 + 19x - 6)$ by $(x + 4)$.

EXAMPLE 2   Divide $(2x^2 - 3x + 1)$ by $(x - 5)$.

Solution

$$
\begin{array}{r}
2x + 7 \\
x - 5\overline{)2x^2 - 3x + 1} \\
\underline{2x^2 - 10x} \\
7x + 1 \\
\underline{7x - 35} \\
36 \quad \longleftarrow \quad \text{Remainder}
\end{array}
$$

Thus

$$\frac{2x^2 - 3x + 1}{x - 5} = 2x + 7 + \frac{36}{x - 5}, \qquad x \neq 5$$

Classroom Example
Divide $(2x^3 - 6x - 4)$ by $(x + 1)$.

EXAMPLE 3   Divide $(t^3 - 8)$ by $(t - 2)$.

Solution

$$
\begin{array}{r}
t^2 + 2t + 4 \\
t - 2\overline{)t^3 + 0t^2 + 0t - 8} \quad \longleftarrow \\
\underline{t^3 - 2t^2} \\
2t^2 + 0t - 8 \\
\underline{2t^2 - 4t} \\
4t - 8 \\
\underline{4t - 8} \\
0
\end{array}
$$

Note the insertion of a *t-squared* term and a *t* term with zero coefficients

Thus we can say that the quotient is $t^2 + 2t + 4$ and the remainder is 0.

Classroom Example
Find the quotient and remainder for $(x^3 + x^2 + 8x + 2) \div (x^2 + 3x)$.

EXAMPLE 4   Find the quotient and remainder for $(y^3 + 3y^2 - 2y - 1) \div (y^2 + 2y)$.

Solution

$$
\begin{array}{r}
y + 1 \\
y^2 + 2y\overline{)y^3 + 3y^2 - 2y - 1} \\
\underline{y^3 + 2y^2} \\
y^2 - 2y - 1 \\
\underline{y^2 + 2y} \\
-4y - 1 \quad \longleftarrow \quad \text{Remainder of } -4y - 1
\end{array}
$$

The division process is complete when the degree of the remainder is less than the degree of the divisor. Thus the quotient is $y + 1$ and the remainder is $-4y - 1$.

Synthetic Division

If the divisor is of the form $x - c$, where c is a constant, then the typical long-division algorithm can be simplified to a process called **synthetic division**. First, let's consider another division problem and use the regular-division algorithm. Then, in a step-by-step fashion, we will demonstrate some shortcuts that will lead us into the synthetic-division procedure. Consider the division problem $(2x^4 + x^3 - 17x^2 + 13x + 2) \div (x - 2)$.

$$
\begin{array}{r}
2x^3 + 5x^2 - 7x - 1 \\
x - 2 \overline{\smash{)}\ 2x^4 + x^3 - 17x^2 + 13x + 2} \\
\underline{2x^4 - 4x^3} \\
5x^3 - 17x^2 \\
\underline{5x^3 - 10x^2} \\
-7x^2 + 13x \\
\underline{-7x^2 + 14x} \\
-x + 2 \\
\underline{-x + 2}
\end{array}
$$

Because the dividend is written in descending powers of x, the quotient is produced in descending powers of x. In other words, the numerical coefficients are the *key issues,* so let's rewrite the problem in terms of its coefficients.

$$
\begin{array}{r}
2 \quad 5 \quad -7 \quad -1 \\
1 - 2 \overline{\smash{)}\ 2 \quad 1 \quad -17 \quad 13 \quad 2} \\
② \ -4 \\
5 \quad \boxed{-17} \\
⑤ \ -10 \\
-7 \quad \boxed{13} \\
\boxed{-7} \ 14 \\
-1 \quad ② \\
\boxed{-1} \ 2
\end{array}
$$

Now observe that the circled numbers are simply repetitions of the numbers directly above them in the format. Thus the circled numbers can be omitted, and the format will be as follows (disregard the arrows for the moment).

$$
\begin{array}{r}
2 \quad 5 \quad -7 \quad -1 \\
1 - 2 \overline{\smash{)}\ 2 \quad 1 \quad -17 \quad 13 \quad 2} \\
-4 \\
5 \\
-10 \\
-7 \\
14 \\
-1 \\
2
\end{array}
$$

Next, by moving some numbers up (indicated by the arrows) and by not writing the 1 that is the coefficient of x in the divisor, we obtain the following more compact form.

$$
\begin{array}{r}
2 \quad 5 \quad -7 \quad -1 \\
-2 \overline{\smash{)}\ 2 \quad 1 \quad -17 \quad 13 \quad 2} \\
-4 \quad -10 \quad 14 \quad 2 \\
\overline{5 \quad -7 \quad -1 \quad 0}
\end{array}
$$

(1)
(2)
(3)
(4)

Note that line 4 reveals all of the coefficients of the quotient (line 1) except for the first coefficient, 2. Thus we can omit line 1, begin line 4 with the first coefficient, and then use the following form.

$$
\begin{array}{r}
-2 \overline{\smash{)}\ 2 \quad 1 \quad -17 \quad 13 \quad 2} \\
-4 \quad -10 \quad 14 \quad 2 \\
\overline{2 \quad 5 \quad -7 \quad -1 \quad 0}
\end{array}
$$

(5)
(6)
(7)

Line 7 contains the coefficients of the quotient, where the zero indicates the remainder. Finally, by changing the constant in the divisor to 2 (instead of -2), which changes the signs of the numbers in line 6, we can *add* the corresponding entries in lines 5 and 6 rather than subtract. Thus the final synthetic-division form for this problem is

$$
\begin{array}{r}
2\,\overline{)\,2 \quad 1 \quad -17 \quad 13 \quad 2\,} \\
 4 \quad\;\; 10 \quad -14 \;\; -2 \\
\hline
2 \quad 5 \quad -7 \quad -1 \quad\;\; 0
\end{array}
$$

The first four entries of the bottom row are the coefficients and the constant term of the quotient $(2x^3 + 5x^2 - 7x - 1)$, and the last entry is the remainder (0).

Now we will consider another problem and indicate a step-by-step procedure for setting up and carrying out the synthetic-division process. Suppose that we want to do the division problem

$$x + 4\,\overline{)\,2x^3 + 5x^2 - 13x - 2\,}$$

Step 1 Write the coefficients of the dividend as follows:

$$\overline{)\,2 \quad 5 \quad -13 \quad -2\,}$$

Step 2 In the divisor, use -4 instead of 4 so that later we can add rather than subtract.

$$-4\,\overline{)\,2 \quad 5 \quad -13 \quad -2\,}$$

Step 3 Bring down the first coefficient of the dividend.

$$
\begin{array}{r}
-4\,\overline{)\,2 \quad 5 \quad -13 \quad -2\,} \\
\hline
2
\end{array}
$$

Step 4 Multiply that first coefficient times the divisor, which yields $2(-4) = -8$. Add this result to the second coefficient of the dividend.

$$
\begin{array}{r}
-4\,\overline{)\,2 \quad\;\; 5 \quad -13 \quad -2\,} \\
-8 \\
\hline
2 \quad -3
\end{array}
$$

Step 5 Multiply $(-3)(-4)$, which yields 12. Add this result to the third coefficient of the dividend.

$$
\begin{array}{r}
-4\,\overline{)\,2 \quad\;\; 5 \quad -13 \quad -2\,} \\
-8 \quad\;\; 12 \\
\hline
2 \quad -3 \quad -1
\end{array}
$$

Step 6 Multiply $(-1)(-4)$, which yields 4. Add this result to the last term of the dividend.

$$
\begin{array}{r}
-4\,\overline{)\,2 \quad\;\; 5 \quad -13 \quad -2\,} \\
-8 \quad\;\; 12 \quad\;\; 4 \\
\hline
2 \quad -3 \quad -1 \quad\;\; 2
\end{array}
$$

The last row indicates a quotient of $2x^2 - 3x - 1$ and a remainder of 2.

Now let's consider some examples in which we show only the final compact form of synthetic division.

Classroom Example
Find the quotient and remainder for $(x^3 - 5x^2 + 8x - 4) \div (x - 2)$.

EXAMPLE 5   Find the quotient and the remainder for

$$(x^3 + 8x^2 + 13x - 6) \div (x + 3).$$

Solution

$$
\begin{array}{r|rrrr}
-3) & 1 & 8 & 13 & -6 \\
   &   & -3 & -15 & 6 \\
\hline
   & 1 & 5 & -2 & 0
\end{array}
$$

Thus the quotient is $x^2 + 5x - 2$, and the remainder is zero.

Classroom Example
Find the quotient and remainder for
$(2x^4 + 14x^3 + 21x^2 - 17x - 22) \div (x + 4)$.

EXAMPLE 6   Find the quotient and the remainder for

$$(3x^4 + 5x^3 - 29x^2 - 45x + 14) \div (x - 3).$$

Solution

$$
\begin{array}{r|rrrrr}
3) & 3 & 5 & -29 & -45 & 14 \\
   &   & 9 & 42 & 39 & -18 \\
\hline
   & 3 & 14 & 13 & -6 & -4
\end{array}
$$

Thus the quotient is $3x^3 + 14x^2 + 13x - 6$, and the remainder is -4.

Classroom Example
Find the quotient and remainder for
$(3x^4 - 10x^2 - 4) \div (x + 2)$.

EXAMPLE 7   Find the quotient and the remainder for

$$(4x^4 - 2x^3 + 6x - 1) \div (x - 1).$$

Solution

$$
\begin{array}{r|rrrrr}
1) & 4 & -2 & 0 & 6 & -1 \\
   &   & 4 & 2 & 2 & 8 \\
\hline
   & 4 & 2 & 2 & 8 & 7
\end{array}
$$

Note that a zero has been inserted as the coefficient of the missing x^2 term

Thus the quotient is $4x^3 + 2x^2 + 2x + 8$, and the remainder is 7.

Classroom Example
Find the quotient and remainder for
$(x^3 - 1) \div (x - 1)$.

EXAMPLE 8   Find the quotient and the remainder for

$$(x^4 + 16) \div (x + 2).$$

Solution

$$
\begin{array}{r|rrrrr}
-2) & 1 & 0 & 0 & 0 & 16 \\
    &   & -2 & 4 & -8 & 16 \\
\hline
    & 1 & -2 & 4 & -8 & 32
\end{array}
$$

Note that zeros have been inserted as coefficients of the missing terms in the dividend

Thus the quotient is $x^3 - 2x^2 + 4x - 8$, and the remainder is 32.

Concept Quiz 8.5

For Problems 1–10, answer true or false.

1. To divide a polynomial by a monomial, we divide each term of the polynomial by the monomial.

2. To check a division problem, we can multiply the divisor by the quotient and add the remainder. The result should equal the dividend.

3. Synthetic division is used to simplify the process of dividing a polynomial by a monomial.

4. Synthetic division is used when the divisor is of the form $x - c$, where c is a constant.

5. The synthetic division process is used only when the remainder is zero.

6. If $2xy - x^2y^2 - 3xy^2$ is divided by $-xy$, the quotient is $-2 - xy + 3y$.

7. If $x^4 - 1$ is divided by $x + 1$, the remainder is 0.

8. If $x^4 + 1$ is divided by $x + 1$, the remainder is 0.

9. If $x^3 - 1$ is divided by $x + 1$, the remainder is 0.

10. If $x^3 + 1$ is divided by $x + 1$, the remainder is 0.

Problem Set 8.5

For Problems 1–10, perform the indicated divisions of polynomials by monomials. **(Objective 1)**

1. $\dfrac{9x^4 + 18x^3}{3x}$

2. $\dfrac{12x^3 - 24x^2}{6x^2}$

3. $\dfrac{-24x^6 + 36x^8}{4x^2}$

4. $\dfrac{-35x^5 - 42x^3}{-7x^2}$

5. $\dfrac{15a^3 - 25a^2 - 40a}{5a}$

6. $\dfrac{-16a^4 + 32a^3 - 56a^2}{-8a}$

7. $\dfrac{13x^3 - 17x^2 + 28x}{-x}$

8. $\dfrac{14xy - 16x^2y^2 - 20x^3y^4}{-xy}$

9. $\dfrac{-18x^2y^2 + 24x^3y^2 - 48x^2y^3}{6xy}$

10. $\dfrac{-27a^3b^4 - 36a^2b^3 + 72a^2b^5}{9a^2b^2}$

For Problems 11–22, find the quotient and remainder for each division problem. **(Objective 1)**

11. $(12x^2 + 7x - 10) \div (3x - 2)$

12. $(20x^2 - 39x + 18) \div (5x - 6)$

13. $(3t^3 + 7t^2 - 10t - 4) \div (3t + 1)$

14. $(4t^3 - 17t^2 + 7t + 10) \div (4t - 5)$

15. $(6x^2 + 19x + 11) \div (3x + 2)$

16. $(20x^2 + 3x - 1) \div (5x + 2)$

17. $(3x^3 + 2x^2 - 5x - 1) \div (x^2 + 2x)$

18. $(4x^3 - 5x^2 + 2x - 6) \div (x^2 - 3x)$

19. $(5y^3 - 6y^2 - 7y - 2) \div (y^2 - y)$

20. $(8y^3 - y^2 - y + 5) \div (y^2 + y)$

21. $(4a^3 - 2a^2 + 7a - 1) \div (a^2 - 2a + 3)$

22. $(5a^3 + 7a^2 - 2a - 9) \div (a^2 + 3a - 4)$

For Problems 23–46, use *synthetic division* to determine the quotient and remainder for each division problem. **(Objective 2)**

23. $(3x^2 + x - 4) \div (x - 1)$

24. $(2x^2 - 5x - 3) \div (x - 3)$

25. $(x^2 + 2x - 10) \div (x - 4)$

26. $(x^2 - 10x + 15) \div (x - 8)$

27. $(4x^2 + 5x - 4) \div (x + 2)$

28. $(5x^2 + 18x - 8) \div (x + 4)$

29. $(x^3 - 2x^2 - x + 2) \div (x - 2)$

30. $(x^3 - 5x^2 + 2x + 8) \div (x + 1)$

31. $(3x^4 - x^3 + 2x^2 - 7x - 1) \div (x + 1)$

32. $(2x^3 - 5x^2 - 4x + 6) \div (x - 2)$

33. $(x^3 - 7x - 6) \div (x + 2)$

34. $(x^3 + 6x^2 - 5x - 1) \div (x - 1)$

35. $(x^4 + 4x^3 - 7x - 1) \div (x - 3)$

36. $(2x^4 + 3x^2 + 3) \div (x + 2)$

37. $(x^3 + 6x^2 + 11x + 6) \div (x + 3)$

38. $(x^3 - 4x^2 - 11x + 30) \div (x - 5)$

39. $(x^5 - 1) \div (x - 1)$

40. $(x^5 - 1) \div (x + 1)$

41. $(x^5 + 1) \div (x - 1)$

42. $(x^5 + 1) \div (x + 1)$

43. $(2x^3 + 3x^2 - 2x + 3) \div \left(x + \dfrac{1}{2}\right)$

44. $(9x^3 - 6x^2 + 3x - 4) \div \left(x - \dfrac{1}{3}\right)$

45. $(4x^4 - 5x^2 + 1) \div \left(x - \dfrac{1}{2}\right)$

46. $(3x^4 - 2x^3 + 5x^2 - x - 1) \div \left(x + \dfrac{1}{3}\right)$

Thoughts Into Words

47. How do you know by inspection that a quotient of $3x^2 + 5x + 1$ and a remainder of 0 cannot be the correct answer for the division problem $(3x^3 - 7x^2 - 22x + 8) \div (x - 4)$?

48. Why is synthetic division restricted to situations where the divisor is of the format $x - c$?

Answers to the Concept Quiz

1. True   **2.** True   **3.** False   **4.** True   **5.** False   **6.** False   **7.** True   **8.** False   **9.** False   **10.** True

---

8.6 Factoring: A Brief Review and a Step Further

OBJECTIVES

1 Review factoring techniques

2 Factor the sum or difference of two cubes

3 Factor trinomials by a systematic technique

4 Review solving equations and word problems that involve factoring

---

Chapter 7 was organized as follows: First a factoring technique was introduced. Next some equations were solved using that factoring technique. Then this type of equation was used to solve some word problems. In this section we will briefly review and expand upon that material by first considering all of the factoring techniques, then solving a few equations, and finally solving some word problems.

Factoring: Use of the Distributive Property

In general, factoring is the reverse of multiplication. Previously, we have used the distributive property to find the product of a monomial and a polynomial, as in the next examples.

$$3(x + 2) = 3(x) + 3(2) = 3x + 6$$
$$5(2x - 1) = 5(2x) - 5(1) = 10x - 5$$
$$x(x^2 + 6x - 4) = x(x^2) + x(6x) - x(4) = x^3 + 6x^2 - 4x$$

We shall also use the distributive property (in the form $ab + ac = a(b + c)$) to reverse the process—that is, to factor a given polynomial. Consider the following examples. (The steps in the dashed boxes can be done mentally.)

$$3x + 6 = \boxed{3(x) + 3(2)} = 3(x + 2)$$
$$10x - 5 = \boxed{5(2x) - 5(1)} = 5(2x - 1)$$
$$x^3 + 6x^2 - 4x = \boxed{x(x^2) + x(6x) - x(4)} = x(x^2 + 6x - 4)$$

Note that in each example a given polynomial has been factored into the product of a monomial and a polynomial. Obviously, polynomials could be factored in a variety of ways. Consider some factorizations of $3x^2 + 12x$.

$$3x^2 + 12x = 3x(x + 4) \qquad \text{or} \qquad 3x^2 + 12x = 3(x^2 + 4x) \qquad \text{or}$$
$$3x^2 + 12x = x(3x + 12) \qquad \text{or} \qquad 3x^2 + 12x = \frac{1}{2}(6x^2 + 24x)$$

We are, however, primarily interested in the first of the previous factorization forms, which we refer to as the **completely factored form**. A polynomial with integral coefficients is in completely factored form if:

> **1.** It is expressed as a product of polynomials with *integral coefficients,* and
>
> **2.** No polynomial, other than a monomial, within the factored form can be further factored into polynomials with integral coefficients.

Do you see why only the first of the above factored forms of $3x^2 + 12x$ is said to be in completely factored form? In each of the other three forms, the polynomial inside the parentheses can be factored further. Moreover, in the last form, $\frac{1}{2}(6x^2 + 24x)$, the condition of using only integral coefficients is violated.

This factoring process, $ab + ac = a(b + c)$, is referred to as **factoring out the highest common factor**. The key idea is to recognize the largest monomial factor that is common to all terms. For example, we observe that each term of $2x^3 + 4x^2 + 6x$ has a factor of $2x$. Thus we write

$$2x^3 + 4x^2 + 6x = 2x(\underline{\qquad\qquad})$$

and insert within the parentheses the result of dividing $2x^3 + 4x^2 + 6x$ by $2x$.

$$2x^3 + 4x^2 + 6x = 2x(x^2 + 2x + 3)$$

The following examples further demonstrate this process of factoring out the highest common monomial factor.

$$12x^3 + 16x^2 = 4x^2(3x + 4) \qquad 6x^2y^3 + 27xy^4 = 3xy^3(2x + 9y)$$
$$8ab - 18b = 2b(4a - 9) \qquad 8y^3 + 4y^2 = 4y^2(2y + 1)$$
$$30x^3 + 42x^4 - 24x^5 = 6x^3(5 + 7x - 4x^2)$$

Note that in each example, the common monomial factor itself is not in a completely factored form. For example, $4x^2(3x + 4)$ is not written as $2 \cdot 2x \cdot x \cdot (3x + 4)$.

Sometimes there may be a common binomial factor rather than a common monomial factor. For example, each of the two terms of the expression $x(y + 2) + z(y + 2)$ has a binomial factor of $(y + 2)$. Thus we can factor $(y + 2)$ from each term, and our result is

$$x(y + 2) + z(y + 2) = (y + 2)(x + z)$$

It may be that the original polynomial exhibits no apparent common monomial or binomial factor, which is the case with $ab + 3a + bc + 3c$. However, by factoring a from the first two terms and c from the last two terms, we get

$$ab + 3a + bc + 3c = a(b + 3) + c(b + 3)$$

Now a common binomial factor of $(b + 3)$ is obvious, and we can proceed as before.

$$a(b + 3) + c(b + 3) = (b + 3)(a + c)$$

We refer to this factoring process as *factoring by grouping.* Let's consider a few more examples of this type.

$$ab^2 - 4b^2 + 3a - 12 = b^2(a - 4) + 3(a - 4) \qquad \text{Factor } b^2 \text{ from the first two terms and 3 from the last two terms}$$
$$= (a - 4)(b^2 + 3) \qquad \text{Factor the common binomial from both terms}$$

$$x^2 - x + 5x - 5 = x(x - 1) + 5(x - 1) \qquad \text{Factor } x \text{ from the first two terms and 5 from the last two terms}$$
$$= (x - 1)(x + 5) \qquad \text{Factor the common binomial from both terms}$$

$$x^2 + 2x - 3x - 6 = x(x + 2) - 3(x + 2) \qquad \text{Factor } x \text{ from the first two terms and } -3 \text{ from the last two terms}$$
$$= (x + 2)(x - 3) \qquad \text{Factor the common binomial from both terms}$$

It may be necessary to rearrange some terms before applying the distributive property. Terms that contain common factors need to be grouped together, and this may be done in more than one way. The next example illustrates this idea.

$$4a^2 - bc^2 - a^2b + 4c^2 = 4a^2 - a^2b + 4c^2 - bc^2$$
$$= a^2(4 - b) + c^2(4 - b)$$
$$= (4 - b)(a^2 + c^2) \qquad \text{or}$$

$$4a^2 - bc^2 - a^2b + 4c^2 = 4a^2 + 4c^2 - bc^2 - a^2b$$
$$= 4(a^2 + c^2) - b(c^2 + a^2)$$
$$= 4(a^2 + c^2) - b(a^2 + c^2)$$
$$= (a^2 + c^2)(4 - b)$$

Factoring: Difference of Two Squares

In Section 6.3, we examined some special multiplication patterns. One of these patterns was

$$(a + b)(a - b) = a^2 - b^2$$

This same pattern, viewed as a factoring pattern, is referred to as the difference of two squares.

> **Difference of Two Squares**
>
> $$a^2 - b^2 = (a + b)(a - b)$$

Applying the pattern is fairly simple, as these next examples demonstrate. Again, the steps in dashed boxes are usually performed mentally.

$$x^2 - 16 = (x)^2 - (4)^2 = (x + 4)(x - 4)$$
$$4x^2 - 25 = (2x)^2 - (5)^2 = (2x + 5)(2x - 5)$$
$$16x^2 - 9y^2 = (4x)^2 - (3y)^2 = (4x + 3y)(4x - 3y)$$
$$1 - a^2 = (1)^2 - (a)^2 = (1 + a)(1 - a)$$

Multiplication is commutative, so the order in which we write the factors is not important. For example, $(x + 4)(x - 4)$ can also be written as $(x - 4)(x + 4)$.

You must be careful not to assume an analogous factoring pattern for the *sum* of two squares; *it does not exist*. For example, $x^2 + 4 \neq (x + 2)(x + 2)$ because $(x + 2)(x + 2) = x^2 + 4x + 4$. We say that a polynomial such as $x^2 + 4$ is a **prime polynomial** or that it is *not factorable using integers*.

Sometimes the difference-of-two-squares pattern can be applied more than once, as the next examples illustrate.

$$x^4 - y^4 = (x^2 + y^2)(x^2 - y^2) = (x^2 + y^2)(x + y)(x - y)$$

$$16x^4 - 81y^4 = (4x^2 + 9y^2)(4x^2 - 9y^2) = (4x^2 + 9y^2)(2x + 3y)(2x - 3y)$$

It may also be that the squares are other than simple monomial squares, as in the next three examples.

$$(x + 3)^2 - y^2 = [(x + 3) + y][(x + 3) - y] = (x + 3 + y)(x + 3 - y)$$

$$4x^2 - (2y + 1)^2 = [2x + (2y + 1)][2x - (2y + 1)]$$
$$= (2x + 2y + 1)(2x - 2y - 1)$$

$$(x - 1)^2 - (x + 4)^2 = [(x - 1) + (x + 4)][(x - 1) - (x + 4)]$$
$$= (x - 1 + x + 4)(x - 1 - x - 4)$$
$$= (2x + 3)(-5)$$

It is possible to apply both the technique of *factoring out a common monomial factor* and the pattern of the *difference of two squares* to the same problem. *In general, it is best to look first for a common monomial factor.* Consider the following examples.

$$2x^2 - 50 = 2(x^2 - 25)$$
$$= 2(x + 5)(x - 5)$$
$$48y^3 - 27y = 3y(16y^2 - 9)$$
$$= 3y(4y + 3)(4y - 3)$$
$$9x^2 - 36 = 9(x^2 - 4)$$
$$= 9(x + 2)(x - 2)$$

Word of Caution The polynomial $9x^2 - 36$ can be factored as follows:

$$9x^2 - 36 = (3x + 6)(3x - 6)$$
$$= 3(x + 2)(3)(x - 2)$$
$$= 9(x + 2)(x - 2)$$

However, when one is taking this approach, there seems to be a tendency to stop at the step $(3x + 6)(3x - 6)$. Therefore, remember the suggestion to *look first for a common monomial factor.*

Factoring: Sum or Difference of Two Cubes

As we pointed out before, there exists no sum-of-squares pattern analogous to the difference-of-squares factoring pattern. That is, a polynomial such as $x^2 + 9$ is not factorable using integers. However, patterns do exist for both *the sum and the difference of two cubes*. These patterns are as follows:

Sum and Difference of Two Cubes

$$a^3 + b^3 = (a + b)(a^2 - ab + b^2)$$

$$a^3 - b^3 = (a - b)(a^2 + ab + b^2)$$

Note how we apply these patterns in the next four examples.

$$x^3 + 27 = (x)^3 + (3)^3 = (x + 3)(x^2 - 3x + 9)$$
$$8a^3 + 125b^3 = (2a)^3 + (5b)^3 = (2a + 5b)(4a^2 - 10ab + 25b^2)$$
$$x^3 - 1 = (x)^3 - (1)^3 = (x - 1)(x^2 + x + 1)$$
$$27y^3 - 64x^3 = (3y)^3 - (4x)^3 = (3y - 4x)(9y^2 + 12xy + 16x^2)$$

Factoring: Trinomials of the Form $x^2 + bx + c$

To factor trinomials of the form $x^2 + bx + c$ (that is, trinomials for which the coefficient of the squared term is 1), we can use the result of the following multiplication problem.

$$(x + a)(x + b) = x(x + b) + a(x + b)$$
$$= x(x) + x(b) + a(x) + a(b)$$
$$= x^2 + (b + a)x + ab$$
$$= x^2 + (a + b)x + ab$$

Note that the coefficient of x is the *sum of a and b*, and the last term is the *product of a and b*. Let's consider some examples to review the use of these ideas.

Classroom Example
Factor $x^2 + 11x + 24$.

EXAMPLE 1   Factor $x^2 + 8x + 12$.

Solution

We need to complete the following with two integers whose product is 12 and whose sum is 8.

$$x^2 + 8x + 12 = (x + \underline{\hspace{1cm}})(x + \underline{\hspace{1cm}})$$

The possible pairs of factors of 12 are 1(12), 2(6), and 3(4). Because $6 + 2 = 8$, we can complete the factoring as follows:

$$x^2 + 8x + 12 = (x + 6)(x + 2)$$

To *check* our answer, we find the product of $(x + 6)$ and $(x + 2)$.

Classroom Example
Factor $x^2 - 9x + 18$.

EXAMPLE 2   Factor $x^2 - 10x + 24$.

Solution

We need two integers whose product is 24 and whose sum is -10. Let's use a small table to organize our thinking.

Product	Sum
$(-1)(-24) = 24$	$-1 + (-24) = -25$
$(-2)(-12) = 24$	$-2 + (-12) = -14$
$(-3)(-8) = 24$	$-3 + (-8) = -11$
$(-4)(-6) = 24$	$-4 + (-6) = -10$

The bottom line contains the numbers that we need. Thus

$$x^2 - 10x + 24 = (x - 4)(x - 6)$$

Classroom Example
Factor $x^2 - 4x - 12$.

EXAMPLE 3   Factor $x^2 + 7x - 30$.

Solution

We need two integers whose product is -30 and whose sum is 7.

Product	Sum
$(-1)(30) = -30$	$-1 + 30 = 29$
$1(-30) = -30$	$1 + (-30) = -29$
$2(-15) = -30$	$2 + (-15) = -13$
$-2(15) = -30$	$-2 + 15 = 13$
$-3(10) = -30$	$-3 + 10 = 7$

No need to search any further

The numbers that we need are -3 and 10, and we can complete the factoring.

$$x^2 + 7x - 30 = (x + 10)(x - 3)$$

Classroom Example
Factor $x^2 + 8x + 10$.

EXAMPLE 4   Factor $x^2 + 7x + 16$.

Solution

We need two integers whose product is 16 and whose sum is 7.

Product	Sum
$1(16) = 16$	$1 + 16 = 17$
$2(8) = 16$	$2 + 8 = 10$
$4(4) = 16$	$4 + 4 = 8$

We have exhausted all possible pairs of factors of 16, and no two factors have a sum of 7, so we conclude that $x^2 + 7x + 16$ *is not factorable using integers.*

The tables in Examples 2, 3, and 4 were used to illustrate one way of organizing your thoughts for such problems. Normally you would probably factor such problems mentally without taking the time to formulate a table. Note, however, that in Example 4 the table helped us to be absolutely sure that we tried all the possibilities. Whether or not you use the table, keep in mind that the key ideas are the product and sum relationships.

Classroom Example
Factor $x^2 - 7x - 120$.

EXAMPLE 5   Factor $t^2 + 2t - 168$.

Solution

We need two integers whose product is -168 and whose sum is 2. Because the absolute value of the constant term is rather large, it might help to look at it in prime factored form.

$$168 = 2 \cdot 2 \cdot 2 \cdot 3 \cdot 7$$

Now we can mentally form two numbers by using all of these factors in different combinations. Using two 2s and a 3 in one number and the other 2 and the 7 in the second number produces $2 \cdot 2 \cdot 3 = 12$ and $2 \cdot 7 = 14$. The coefficient of the middle term of the trinomial is 2, so we know that we must use 14 and -12. Thus we obtain

$$t^2 + 2t - 168 = (t + 14)(t - 12)$$

Factoring: Trinomials of the Form $ax^2 + bx + c$

Now let's consider factoring trinomials where the coefficient of the squared term is not 1. In Section 7.4, we used an informal trial-and-error process for such trinomials. This technique is based on our knowledge of multiplication of binomials and works quite well for certain trinomials. Let's review the process with an example.

Classroom Example
Factor $3x^2 - 13x - 10$.

EXAMPLE 6   Factor $5x^2 - 18x - 8$.

Solution

The first term, $5x^2$, can be written as $x \cdot 5x$. The last term, -8, can be written as $(-2)(4)$, $(2)(-4)$, $(-1)(8)$, or $(1)(-8)$. Therefore, we have the following possibilities to try.

$(x - 2)(5x + 4)$	$(x + 4)(5x - 2)$
$(x + 2)(5x - 4)$	$(x - 4)(5x + 2)$
$(x - 1)(5x + 8)$	$(x + 8)(5x - 1)$
$(x + 1)(5x - 8)$	$(x - 8)(5x + 1)$

By checking the middle terms, we find that $(x - 4)(5x + 2)$ yields the desired middle term of $-18x$. Thus

$$5x^2 - 18x - 8 = (x - 4)(5x + 2)$$

Certainly, as the number of possibilities increases, this trial-and-error technique for factoring becomes more tedious. The key idea is to organize your work so that all possibilities are considered. We have suggested one possible format in the previous examples. However, as you practice such problems, you may devise a format that works better for you. Whatever works best for you is the right approach.

There is another, more systematic technique that you may wish to use with some trinomials. It is an extension of the technique we used earlier with trinomials where the coefficient of the squared term was 1. To see the basis of this technique, consider the following general product:

$$\begin{aligned}(px + r)(qx + s) &= px(qx) + px(s) + r(qx) + r(s) \\ &= (pq)x^2 + ps(x) + rq(x) + rs \\ &= (pq)x^2 + (ps + rq)x + rs\end{aligned}$$

Note that the product of the coefficient of x^2 and the constant term is $pqrs$. Likewise, the product of the two coefficients of x (ps and rq) is also $pqrs$. Therefore, the coefficient of x must be a sum of the form $ps + rq$, such that the product of the coefficient of x^2 and the constant term is $pqrs$. Now let's see how this works in some specific examples.

Classroom Example
Factor $8x^2 + 22x + 15$.

EXAMPLE 7   Factor $6x^2 + 17x + 5$.

Solution

$$6x^2 + 17x + 5 \qquad \text{Sum of 17}$$

Product of $6 \cdot 5 = 30$

We need two integers whose sum is 17 and whose product is 30. The integers 2 and 15 satisfy these conditions. Therefore the middle term, $17x$, of the given trinomial can be expressed as $2x + 15x$, and we can proceed as follows:

$$\begin{aligned}6x^2 + 17x + 5 &= 6x^2 + 2x + 15x + 5 \\ &= 2x(3x + 1) + 5(3x + 1) \qquad \text{Factor by grouping} \\ &= (3x + 1)(2x + 5)\end{aligned}$$

Classroom Example
Factor $8x^2 - 10x - 3$.

EXAMPLE 8   Factor $5x^2 - 18x - 8$.

Solution

$$5x^2 - 18x - 8 \qquad \text{Sum of } -18$$

Product of $5(-8) = -40$

We need two integers whose sum is -18 and whose product is -40. The integers -20 and 2 satisfy these conditions. Therefore the middle term, $-18x$, of the trinomial can be written as $-20x + 2x$, and we can factor as follows:

$$\begin{aligned}5x^2 - 18x - 8 &= 5x^2 - 20x + 2x - 8 \\ &= 5x(x - 4) + 2(x - 4) \qquad \text{Factor by grouping} \\ &= (x - 4)(5x + 2)\end{aligned}$$

Classroom Example
Factor $12x^2 + 8x - 15$.

EXAMPLE 9   Factor $24x^2 + 2x - 15$.

Solution

$$24x^2 + 2x - 15 \qquad \text{Sum of 2}$$

Product of $24(-15) = -360$

We need two integers whose sum is 2 and whose product is -360. To help find these integers, let's factor 360 into primes.

$$360 = 2 \cdot 2 \cdot 2 \cdot 3 \cdot 3 \cdot 5$$

Now by grouping these factors in various ways, we find that $2 \cdot 2 \cdot 5 = 20$ and $2 \cdot 3 \cdot 3 = 18$, so we can use the integers 20 and -18 to produce a sum of 2 and a product of -360. Therefore, the middle term, $2x$, of the trinomial can be expressed as $20x - 18x$, and we can proceed as follows:

$$\begin{aligned}
24x^2 + 2x - 15 &= 24x^2 + 20x - 18x - 15 \\
&= 4x(6x + 5) - 3(6x + 5) \\
&= (6x + 5)(4x - 3)
\end{aligned}$$

Factoring: Perfect-Square Trinomials

In Section 6.3 we used the following two patterns to square binomials.

$$(a + b)^2 = a^2 + 2ab + b^2 \qquad \text{and} \qquad (a - b)^2 = a^2 - 2ab + b^2$$

These patterns can also be used for factoring purposes.

$$a^2 + 2ab + b^2 = (a + b)^2 \qquad \text{and} \qquad a^2 - 2ab + b^2 = (a - b)^2$$

The trinomials on the left sides are called **perfect-square trinomials**; they are the result of squaring a binomial. We can always factor perfect-square trinomials using the usual techniques for factoring trinomials. However, they are easily recognized by the nature of their terms. For example, $4x^2 + 12x + 9$ is a perfect-square trinomial because

1. The first term is a perfect square. $(2x)^2$
2. The last term is a perfect square. $(3)^2$
3. The middle term is twice the product of the quantities $2(2x)(3)$
   being squared in the first and last terms.

Likewise, $9x^2 - 30x + 25$ is a perfect-square trinomial because

1. The first term is a perfect square. $(3x)^2$
2. The last term is a perfect square. $(5)^2$
3. The middle term is the negative of twice the product of $-2(3x)(5)$
   the quantities being squared in the first and last terms.

Once we know that we have a perfect-square trinomial, then the factors follow immediately from the two basic patterns. Thus

$$4x^2 + 12x + 9 = (2x + 3)^2 \qquad 9x^2 - 30x + 25 = (3x - 5)^2$$

Here are some additional examples of perfect-square trinomials and their factored forms.

$$\begin{aligned}
x^2 + 14x + 49 &= (x)^2 + 2(x)(7) + (7)^2 & = (x + 7)^2 \\
n^2 - 16n + 64 &= (n)^2 - 2(n)(8) + (8)^2 & = (n - 8)^2 \\
36a^2 + 60ab + 25b^2 &= (6a)^2 + 2(6a)(5b) + (5b)^2 & = (6a + 5b)^2 \\
16x^2 - 8xy + y^2 &= (4x)^2 - 2(4x)(y) + (y)^2 & = (4x - y)^2
\end{aligned}$$

Perhaps you will want to do this step mentally after you feel comfortable with the process

Solving Equations

One reason why factoring is an important algebraic skill is that it extends our techniques for solving equations. Each factoring technique provides us with more power to solve equations. Let's review this process with two examples.

Classroom Example
Solve $y^3 - 2y^2 - 15y = 0$.

EXAMPLE 10   Solve $x^3 - 49x = 0$.

Solution

$$x^3 - 49x = 0$$
$$x(x^2 - 49) = 0$$
$$x(x + 7)(x - 7) = 0$$
$$x = 0 \quad \text{or} \quad x + 7 = 0 \quad \text{or} \quad x - 7 = 0$$
$$x = 0 \quad \text{or} \quad x = -7 \quad \text{or} \quad x = 7$$

The solution set is $\{-7, 0, 7\}$.

Classroom Example
Solve $4x(x - 1) = 3$.

EXAMPLE 11   Solve $9a(a + 1) = 4$.

Solution

$$9a(a + 1) = 4$$
$$9a^2 + 9a = 4$$
$$9a^2 + 9a - 4 = 0$$
$$(3a + 4)(3a - 1) = 0$$
$$3a + 4 = 0 \quad \text{or} \quad 3a - 1 = 0$$
$$3a = -4 \quad \text{or} \quad 3a = 1$$
$$a = -\frac{4}{3} \quad \text{or} \quad a = \frac{1}{3}$$

The solution set is $\left\{-\frac{4}{3}, \frac{1}{3}\right\}$.

Problem Solving

Finally, one of the end results of being able to factor and solve equations is that we can use these skills to help solve problems. Let's conclude this section with two problem-solving situations.

Classroom Example
A spreadsheet for recording the grades of students has 720 cells. The number of rows (where student's name are entered) is 6 more than the number of columns (where the score for each assignment is entered). How many students are entered in the spreadsheet, and how many assignments are recorded?

EXAMPLE 12

A room contains 78 chairs. The number of chairs per row is 1 more than twice the number of rows. Find the number of rows and the number of chairs per row.

Solution

Let r represent the number of rows. Then $2r + 1$ represents the number of chairs per row.

$$r(2r + 1) = 78 \qquad \text{The number of rows times the number of chairs per}$$
$$2r^2 + r = 78 \qquad \text{row yields the total number of chairs}$$
$$2r^2 + r - 78 = 0$$
$$(2r + 13)(r - 6) = 0$$

$$2r + 13 = 0 \quad \text{or} \quad r - 6 = 0$$
$$2r = -13 \quad \text{or} \quad r = 6$$
$$r = -\frac{13}{2} \quad \text{or} \quad r = 6$$

The solution $-\dfrac{13}{2}$ must be disregarded, so there are 6 rows and $2r + 1$ or $2(6) + 1 = 13$ chairs per row.

EXAMPLE 13

Suppose that the volume of a right circular cylinder is numerically equal to the total surface area of the cylinder. If the height of the cylinder is equal to the length of a radius of the base, find the height.

Solution

Because $r = h$, the formula for volume $V = \pi r^2 h$ becomes $V = \pi r^3$ and the formula for the total surface area $S = 2\pi r^2 + 2\pi rh$ becomes $S = 2\pi r^2 + 2\pi r^2$, or $S = 4\pi r^2$. Therefore, we can set up and solve the following equation.

$$\pi r^3 = 4\pi r^2$$
$$\pi r^3 - 4\pi r^2 = 0$$
$$\pi r^2(r - 4) = 0$$
$$\pi r^2 = 0 \quad \text{or} \quad r - 4 = 0$$
$$r = 0 \quad \text{or} \quad r = 4$$

0 is not a reasonable answer, so the height must be 4 units.

Concept Quiz 8.6

For Problems 1–10, answer true or false.

1. The factored form $4xy^2(2x + 5y)$ is factored completely.
2. The polynomial $ax + ay - bx - by$ can be factored by grouping, but its equivalent $ax - bx + ay - by$ cannot be factored by grouping.
3. If a polynomial is not factorable using integers, it is referred to as a prime polynomial.
4. The polynomial $x^3 + 64$ can be factored into $(x + 3)(x + 3)(x + 3)$.
5. The polynomial $x^2 + 10x + 24$ is not factorable using integers.
6. The polynomial $6x^2 - 10x - 3$ is not factorable using integers.
7. $9x^2 + 4 = (3x + 2)(3x + 2)$
8. $15x^2 - 17x - 4 = (5x + 1)(3x - 4)$
9. The solution set for $4x^2 + x - 18 = 0$ is $\left\{-\dfrac{9}{4}, 2\right\}$.
10. The solution set for $x(2x - 1) = 1$ is $\left\{\dfrac{1}{2}, 1\right\}$.

Problem Set 8.6

For Problems 1–50, factor completely each of the polynomials. Indicate any that are not factorable using integers.
(Objectives 1 and 2)

1. $6xy - 8xy^2$
2. $4a^2b^2 + 12ab^3$
3. $x(z + 3) + y(z + 3)$
4. $5(x + y) + a(x + y)$
5. $3x + 3y + ax + ay$
6. $ac + bc + a + b$
7. $ax - ay - bx + by$
8. $2a^2 - 3bc - 2ab + 3ac$

9. $9x^2 - 25$

10. $4x^2 + 9$

11. $1 - 81n^2$

12. $9x^2y^2 - 64$

13. $(x + 4)^2 - y^2$

14. $x^2 - (y - 1)^2$

15. $9s^2 - (2t - 1)^2$

16. $4a^2 - (3b + 1)^2$

17. $x^2 - 5x - 14$

18. $a^2 + 5a - 24$

19. $15 - 2x - x^2$

20. $40 - 6x - x^2$

21. $x^2 + 7x - 36$

22. $x^2 - 4xy - 5y^2$

23. $3x^2 - 11x + 10$

24. $2x^2 - 7x - 30$

25. $10x^2 - 33x - 7$

26. $8y^2 + 22y - 21$

27. $4x^2 + 16$

28. $n^3 - 49n$

29. $x^3 - 9x$

30. $12n^2 + 59n + 72$

31. $9a^2 - 42a + 49$

32. $1 - 16x^4$

33. $2n^3 + 6n^2 + 10n$

34. $x^2 - (y - 7)^2$

35. $10x^2 + 39x - 27$

36. $3x^2 + x - 5$

37. $36a^2 - 12a + 1$

38. $18n^3 + 39n^2 - 15n$

39. $8x^2 + 2xy - y^2$

40. $12x^2 + 7xy - 10y^2$

41. $2n^2 - n - 5$

42. $25t^2 - 100$

43. $2n^3 + 14n^2 - 20n$

44. $25n^2 + 64$

For Problems 45–54, factor completely each of the sums or differences of cubes. **(Objective 2)**

45. $x^3 - 8$

46. $x^3 + 64$

47. $64x^3 + 27y^3$

48. $27x^3 - 8y^3$

49. $4x^3 + 32$

50. $2x^3 - 54$

51. $a^3 + 1000$

52. $1 - 8b^3$

53. $x^4y - xy^4$

54. $8x^3 + 8$

For Problems 55–88, solve each equation. **(Objective 4)**

55. $x^2 + 4x + 3 = 0$

56. $x^2 + 7x + 10 = 0$

57. $x^2 + 18x + 72 = 0$

58. $n^2 + 20n + 91 = 0$

59. $n^2 - 13n + 36 = 0$

60. $n^2 - 10n + 16 = 0$

61. $x^2 + 4x - 12 = 0$

62. $x^2 + 7x - 30 = 0$

63. $w^2 - 4w = 5$

64. $s^2 - 4s = 21$

65. $n^2 + 25n + 156 = 0$

66. $n(n - 24) = -128$

67. $3t^2 + 14t - 5 = 0$

68. $4t^2 - 19t - 30 = 0$

69. $6x^2 + 25x + 14 = 0$

70. $25x^2 + 30x + 8 = 0$

71. $3t(t - 4) = 0$

72. $4x^2 + 12x + 9 = 0$

73. $-6n^2 + 13n - 2 = 0$

74. $(x + 1)^2 - 4 = 0$

75. $2n^3 = 72n$

76. $a(a - 1) = 2$

77. $(x - 5)(x + 3) = 9$

78. $3w^3 - 24w^2 + 36w = 0$

79. $9x^2 - 6x + 1 = 0$

80. $16t^2 - 72t + 81 = 0$

81. $n^2 + 7n - 44 = 0$

82. $2x^3 = 50x$

83. $3x^2 = 75$

84. $x^2 + x - 2 = 0$

85. $2x^3 + 3x^2 - 2x = 0$

86. $3x^3 = 48x$

87. $20x^3 + 25x^2 - 105x = 0$ **88.** $12x^3 + 12x^2 - 9x = 0$

For Problems 89–98, set up an equation and solve each problem. **(Objective 4)**

89. Suppose that the volume of a sphere is numerically equal to twice the surface area of the sphere. Find the length of a radius of the sphere.

90. Suppose that a radius of a sphere is equal in length to a radius of a circle. If the volume of the sphere is numerically equal to four times the area of the circle, find the length of a radius for both the sphere and the circle.

91. Find two integers whose product is 104 such that one of the integers is 3 less than twice the other integer.

92. The perimeter of a rectangle is 32 inches, and the area is 60 square inches. Find the length and width of the rectangle.

93. The lengths of the three sides of a right triangle are represented by consecutive even whole numbers. Find the lengths of the three sides.

94. The area of a triangular sheet of paper is 28 square inches. One side of the triangle is 2 inches more than three times the length of the altitude to that side. Find the length of that side and the altitude to that side.

95. The total surface area of a right circular cylinder is 54π square inches. If the altitude of the cylinder is twice the length of a radius, find the altitude of the cylinder.

96. The Ortegas have an apple orchard that contains 90 trees. The number of trees in each row is 3 more than twice the number of rows. Find the number of rows and the number of trees per row.

97. The combined area of a square and a rectangle is 64 square centimeters. The width of the rectangle is 2 centimeters more than the length of a side of the square, and the length of the rectangle is 2 centimeters more than its width. Find the dimensions of the square and the rectangle.

98. The cube of a number equals nine times the same number. Find the number.

Thoughts Into Words

99. Suppose that your friend factors $36x^2y + 48xy^2$ as follows:

$$36x^2y + 48xy^2 = (4xy)(9x + 12y)$$
$$= (4xy)(3)(3x + 4y)$$
$$= 12xy(3x + 4y)$$

Is this a correct approach? Would you have any suggestion to offer your friend?

100. Your classmate solves the equation $3ax + bx = 0$ for x as follows:

$$3ax + bx = 0$$
$$3ax = -bx$$
$$x = \frac{-bx}{3a}$$

How should he know that the solution is incorrect? How would you help him obtain the correct solution?

101. Consider the following solution:

$$6x^2 - 24 = 0$$
$$6(x^2 - 4) = 0$$
$$6(x + 2)(x - 2) = 0$$

$$6 = 0 \quad \text{or} \quad x + 2 = 0 \quad \text{or} \quad x - 2 = 0$$
$$6 = 0 \quad \text{or} \quad x = -2 \quad \text{or} \quad x = 2$$

The solution set is $\{-2, 2\}$.

Is this a correct solution? Would you have any suggestion to offer the person who used this approach?

102. Explain how you would solve the equation $(x + 6) \cdot (x - 4) = 0$ and also how you would solve $(x + 6) \cdot (x - 4) = -16$.

103. Explain how you would solve the equation $3(x - 1) \cdot (x + 2) = 0$ and also how you would solve the equation $x(x - 1)(x + 2) = 0$.

104. Consider the following two solutions for the equation $(x + 3)(x - 4) = (x + 3)(2x - 1)$.

Solution A

$$(x + 3)(x - 4) = (x + 3)(2x - 1)$$
$$(x + 3)(x - 4) - (x + 3)(2x - 1) = 0$$
$$(x + 3)[x - 4 - (2x - 1)] = 0$$
$$(x + 3)(x - 4 - 2x + 1) = 0$$
$$(x + 3)(-x - 3) = 0$$

$$x + 3 = 0 \quad \text{or} \quad -x - 3 = 0$$
$$x = -3 \quad \text{or} \quad -x = 3$$
$$x = -3 \quad \text{or} \quad x = -3$$

The solution set is $\{-3\}$.

Solution B

$$(x + 3)(x - 4) = (x + 3)(2x - 1)$$
$$x^2 - x - 12 = 2x^2 + 5x - 3$$
$$0 = x^2 + 6x + 9$$
$$0 = (x + 3)^2$$
$$x + 3 = 0$$
$$x = -3$$

The solution set is $\{-3\}$.

Are both approaches correct? Which approach would you use, and why?

Answers to the Concept Quiz
1. True   **2.** False   **3.** True   **4.** False   **5.** False   **6.** True   **7.** False   **8.** True   **9.** True   **10.** False

Chapter 8   Summary

OBJECTIVE	SUMMARY	EXAMPLE
Solve first-degree equations. (Section 8.1/Objective 1)	"Solving an algebraic equation" refers to the process of finding the number (or numbers) that make(s) the algebraic equation a true numerical statement. Two properties of equality play an important role in solving equations. **Addition Property of Equality** $a = b$ if and only if $a + c = b + c$ **Multiplication Property of Equality** For $c \neq 0$, $a = b$ if and only if $ac = bc$	Solve $3(2x - 1) = 2x + 6 - 5x$. **Solution** $3(2x - 1) = 2x + 6 - 5x$ $6x - 3 = -3x + 6$ $9x - 3 = 6$ $9x = 9$ $x = 1$ The solution set is $\{1\}$.
Solve equations that involve fractional forms. (Section 8.1/Objective 1)	When an equation contains several fractions, it is usually best to start by clearing the equation of all fractions. The fractions can be cleared by multiplying both sides of the equation by the least common denominator of all the denominators.	Solve $\dfrac{3n}{4} + \dfrac{n}{5} = \dfrac{7}{10}$. **Solution** The LCD is 20. $$20\left(\frac{3n}{4} + \frac{n}{5}\right) = 20\left(\frac{7}{10}\right)$$ $$20\left(\frac{3n}{4}\right) + 20\left(\frac{n}{5}\right) = 20\left(\frac{7}{10}\right)$$ $$5(3n) + 4(n) = 2(7)$$ $$15n + 4n = 14$$ $$19n = 14$$ $$n = \frac{14}{19}$$ The solution set is $\left\{\dfrac{14}{19}\right\}$.
Solve proportions. (Section 8.1/Objective 2)	A statement of equality between two ratios is a proportion. A general property of proportions is as follows: $\dfrac{a}{b} = \dfrac{c}{d}$ if and only if $ad = bc$, where $b \neq 0$ and $d \neq 0$ The products ad and bc are called cross products.	Solve $\dfrac{x + 4}{5} = \dfrac{x}{3}$. **Solution** $\dfrac{x + 4}{5} = \dfrac{x}{3}$ $3(x + 4) = 5x$ $3x + 12 = 5x$ $12 = 2x$ $6 = x$ The solution set is $\{6\}$.

OBJECTIVE	SUMMARY	EXAMPLE
Solve equations that are contradictions or identities. (Section 8.1/Objective 1)	When an equation is not true for any value of x, then the equation is called a "**contradiction**." When an equation is true for any permissible value of x, then the equation is called an "**identity**."	Solve the equations. **(a)** $2(x + 4) = 2x + 5$ **(b)** $4x - 8 = 2(2x - 4)$ **Solution** **(a)** $2(x + 4) = 2x + 5$ $\quad 2x + 8 = 2x + 5$ $\quad 2x - 2x + 8 = 2x - 2x + 5$ $\quad\quad\quad\quad 8 = 5$ This is a false statement so there is no solution; the solution set is $\varnothing$. **(b)** $4x - 8 = 2(2x - 4)$ $\quad 4x - 8 = 4x - 8$ $\quad 4x - 4x - 8 = 4x - 4x - 8$ $\quad\quad\quad\quad -8 = -8$ This is a true statement so any value of x is a solution. The solution set is (All reals).
Solve systems of linear equations by substitution. (Section 8.1/Objective 3)	We can describe the **substitution method** of solving a system of equations as follows: **Step 1** Solve one of the equations for one variable in terms of the other variable if neither equation is in such a form. (If possible, make a choice that will avoid fractions.) **Step 2** Substitute the expression obtained in step 1 into the other equation to produce an equation with one variable. **Step 3** Solve the equation obtained in step 2. **Step 4** Use the solution obtained in step 3, along with the expression obtained in step 1, to determine the solution of the system.	Solve the system $\begin{pmatrix} 3x + y = -9 \\ 2x + 3y = 8 \end{pmatrix}$. **Solution** Solving the first equation for y gives the equation $y = -3x - 9$. In the second equation, substitute $-3x - 9$ for y and solve. $2x + 3(-3x - 9) = 8$ $\quad 2x - 9x - 27 = 8$ $\quad\quad\quad -7x = 35$ $\quad\quad\quad\quad x = -5$ Now to find the value of y, substitute -5 for x in the equation $y = -3x - 9$. $y = -3(-5) - 9 = 6$ The solution set of the system is $\{(-5, 6)\}$.
Solve systems of equations by the elimination-by-addition method. (Section 8.1/Objective 3)	The **elimination-by-addition method** involves the replacement of a system of equations with equivalent systems until a system is obtained whereby the solutions can be easily determined. The following operations or transformations can be performed on a system to produce an equivalent system. 1. Any two equations of the system can be interchanged. 2. Both sides of any equation of the system can be multiplied by any nonzero real number. 3. Any equation of the system can be replaced by the *sum* of that equation and a nonzero multiple of another equation.	Solve the system $\begin{pmatrix} 2x - 5y = 31 \\ 4x + 3y = 23 \end{pmatrix}$. **Solution** Let's multiply the first equation by -2, and add the result to the second equation to eliminate the x variable. Then the equivalent system is $\begin{pmatrix} 2x - 5y = 31 \\ 13y = -39 \end{pmatrix}$ Now solving the second equation for y, we obtain $y = -3$. Substitute -3 for y in either of the original equations and solve for x. $2x - 5(-3) = 31$ $\quad 2x + 15 = 31$ $\quad\quad\quad 2x = 16$ $\quad\quad\quad\quad x = 8$ The solution set of the system is $\{(8, -3)\}$

(continued)

OBJECTIVE	SUMMARY	EXAMPLE
Solve word problems. (Section 8.1/Objective 4)	Keep these suggestions in mind as you solve word problems: **1.** Read the problem carefully. **2.** Sketch any figure or diagram that might be helpful. **3.** Choose a meaningful variable. **4.** Look for a guideline. **5.** Form an equation. **6.** Solve the equation. **7.** Check your answer.	The length of a rectangle is 4 feet less than twice the width. The perimeter of the rectangle is 34 feet. Find the length and width. **Solution** Let w represent the width and then $2w - 4$ represents the length. Use the formula $P = 2w + 2l$. $34 = 2w + 2(2w - 4)$ $34 = 2w + 4w - 8$ $42 = 6w$ $7 = w$ So the width is 7 feet, and the length is $2(7) - 4 = 10$ feet.
Write solution sets in interval notation. (Section 8.2/Objective 2)	The solution set for an algebraic inequality can be written in interval notation. See Figure 8.2 on pages 349 for examples of various algebraic inequalities and how their solution sets would be written in interval notation.	Express the solution set for $x \leq 4$ in interval notation. **Solution** For the solution set we want all numbers less than or equal to 4. In interval notation the solution set is written as $(-\infty, 4]$.
Solve first-degree inequalities. (Section 8.2/Objective 1)	Properties for solving inequalities are similar to the properties for solving equations, except for properties that involve multiplying or dividing by a negative number. When multiplying or dividing both sides of an inequality by a negative number, you must reverse the inequality symbol.	Solve $-4n - 3 > 7$. **Solution** $-4n - 3 > 7$ $-4n > 10$ $\dfrac{-4n}{-4} < \dfrac{10}{-4}$ $n < -\dfrac{5}{2}$ The solution set is $\left(-\infty, -\dfrac{5}{2}\right)$.
Solve compound inequalities formed by the word "and." (Section 8.2/Objective 3)	Inequalities connected with the word "and" form a compound statement called a conjunction. The solution set of a compound inequality formed by the word "and" is the **intersection** of the solution sets of the two inequalities. To solve inequalities involving "and," we must satisfy all of the conditions. Thus the compound inequality $x > 1$ and $x < 3$ is satisfied by all numbers between 1 and 3.	Solve the compound inequality $x > -4$ and $x > 2$, and graph the solution set. **Solution** All of the conditions must be satisfied. Thus the compound inequaltiy $x > -4$ and $x > 2$ is satisfied by all numbers greater than 2. The solution set is $\{x \mid x > 2\}$.
Solve compound inequalities formed by the word "or." (Section 8.2/Objective 3)	Inequalities connected with the word "or" form a compound statement called a disjunction. The solution set of a compound inequality formed by the word "or" is the **union** of the solution sets of the two inequalities. To solve inequalities involving "or," we must satisfy one or more of the conditions. Thus the compound inequality $x < 1$ or $x < 5$ is satisfied by all numbers less than 5.	Solve the compound inequality $x > -1$, or $x < 2$ and graph the solution set. **Solution** One or more of the conditions must be satisfied. Thus the compound inequality $x > -1$ or $x < 2$ is satisfied by all real numbers. The solution set is $\{$All reals$\}$.

OBJECTIVE	SUMMARY	EXAMPLE
Solve word problems involving inequalities. (Section 8.2/Objective 4)	Keep these suggestions in mind as you solve word problems: 1. Read the problem carefully. 2. Sketch any figure or diagram that might be helpful. 3. Choose a meaningful variable. 4. Look for a guideline. 5. Form an inequality. 6. Solve the inequality. 7. Check your answer.	Martin must average at least 240 points for a series of three bowling games to get into the playoffs. If he has bowled games of 220 points and 210 points, what must his score be on the third game to get into the playoffs? **Solution** Let p represent the points for the third game. **Guideline** Average ≥ 240 $$\frac{220 + 210 + p}{3} \geq 240$$ $$430 + p \geq 720$$ $$p \geq 290$$ Martin must bowl 290 points or more.
Solve absolute value equations. (Section 8.3/Objective 2)	Property 8.3 states that $|ax + b| = k$ is equivalent to $ax + b = k$ or $ax + b = -k$, where k is a positive number. This property is applied to solve absolute value equations.	Solve $|2x - 5| = 9$. **Solution** $|2x - 5| = 9$ $2x - 5 = 9$ or $2x - 5 = -9$ $2x = 14$ or $2x = -4$ $x = 7$ or $x = -2$ The solution set is $\{-2, 7\}$.
Solve absolute value inequalities. (Section 8.3/Objective 3)	Property 8.4 states that $|ax + b| < k$ is equivalent to $ax + b > -k$ and $ax + b < k$, where k is a positive number. This could be written in the compact form $$-k < ax + b < k$$ Property 8.5 states that $|ax + b| > k$ is equivalent to $ax + b < -k$ and $ax + b > k$, where k is a positive number. This disjunction can not be written in a compact form.	Solve $|x + 5| > 8$. **Solution** $|x + 5| > 8$ $x + 5 < -8$ or $x + 5 > 8$ $x < -13$ or $x > 3$ The solution set is $(-\infty, -13) \cup (3, \infty)$.
Add and subtract polynomial expressions. (Section 8.4/Objective 1)	Similar terms, or like terms, have the same literal factors. The commutative, associative, and distributive properties provide the basis for rearranging, regrouping, and combining similar terms.	Perform the indicated operations: $4x - [9x^2 - 2(7x - 3x^2)]$ **Solution** $4x - [9x^2 - 2(7x - 3x^2)]$ $= 4x - (9x^2 - 14x + 6x^2)$ $= 4x - (15x^2 - 14x)$ $= 4x - 15x^2 + 14x$ $= -15x^2 + 18x$

(continued)

OBJECTIVE	SUMMARY	EXAMPLE
Multiply monomials and raise a monomial to an exponent. (Section 8.4/Objective 2)	The following properties provide the basis for multiplying monomials. 1. $b^n \cdot b^m = b^{n+m}$ 2. $(b^n)^m = b^{mn}$ 3. $(ab)^n = a^n b^n$	Simplify each of the following: (a) $(-5a^4 b)(2a^2 b^3)$ (b) $(-3x^3 y)^2$ **Solution** (a) $(-5a^4 b)(2a^2 b^3) = -10a^6 b^4$ (b) $(-3x^3 y)^2 = (-3)^2 (x^3)^2 (y)^2$ $\quad\quad = 9x^6 y^2$
Multiply polynomials. (Section 8.4/Objective 3)	To multiply two polynomials, every term of the first polynomial is multiplied by each term of the second polynomial. Multiplying polynomials often produces similar terms that can be combined to simplify the resulting polynomial.	Find the indicated product: $(3x + 4)(x^2 + 6x - 5)$ **Solution** $(3x + 4)(x^2 + 6x - 5)$ $= 3x(x^2 + 6x - 5) + 4(x^2 + 6x - 5)$ $= 3x^3 + 18x^2 - 15x + 4x^2 + 24x - 20$ $= 3x^3 + 22x^2 + 9x - 20$
Find the square of a binomial using a shortcut pattern. (Section 8.4/Objective 3)	The patterns for squaring a binomial are $(a + b)^2 = a^2 + 2ab + b^2$ and $(a - b)^2 = a^2 - 2ab + b^2$	Expand $(4x - 3)^2$. **Solution** $(4x - 3)^2 = (4x)^2 + 2(4x)(-3) + (-3)^2$ $\quad\quad = 16x^2 - 24x + 9$
Use a pattern to find the product of $(a + b)(a - b)$. (Section 8.4/Objective 3)	The pattern is $(a + b)(a - b) = a^2 - b^2$.	Find the product: $(x - 3y)(x + 3y)$ **Solution** $(x - 3y)(x + 3y) = (x)^2 - (3y)^2$ $\quad\quad = x^2 - 9y^2$
Find the cube of a binomial. (Section 8.4/Objective 3)	The patterns for cubing a binomial are $(a + b)^3 = a^2 + 3a^2 b + 3ab^2 + b^3$ and $(a - b)^3 = a^2 - 3a^2 b + 3ab^2 - b^3$	Expand $(2a + 5)^3$. **Solution** $(2a + 5)^3 = (2a)^3 + 3(2a)^2 (5)$ $\quad\quad\quad + 3(2a)(5)^2 + (5)^3$ $= 8a^3 + 60a^2 + 150a + 125$
Divide monomials. (Section 8.4/Objective 4)	The following properties provide the basis for dividing monomials. 1. $\dfrac{b^n}{b^m} = b^{n-m}$ if $n > m$ 2. $\dfrac{b^n}{b^m} = 1$ if $n = m$	Find the quotient: $\dfrac{8x^5 y^4}{-8xy^2}$ **Solution** $\dfrac{8x^5 y^4}{-8xy^2} = -x^4 y^2$

OBJECTIVE	SUMMARY	EXAMPLE
Expand a binomial using a pattern. **(Section 8.4/Objective 5)**	The expansion of a binomial such as $(a + b)^7$ can be accomplished by using the coefficients from the seventh row of Pascal's triangle and using the exponents for a and b from the following pattern: $a^n, \quad a^n b, \quad a^{n-2} b^2, \quad \ldots, \quad ab^{n-1}, \quad b^n$	Expand $(a + b)^4$. **Solution** The coefficients from Pascal's triangle are 1, 4, 6, 4, and 1. Using the pattern for the exponents, we arrive at $a^4 + 4a^3 b + 6a^2 b^2 + 4ab^3 + b^4$
Divide polynomials. **(Section 8.5/Objective 1)**	1. To divide a polynomial by a monomial, divide each term of the polynomial by the monomial. 2. The procedure for dividing a polynomial by a polynomial resembles the long-division process.	Divide $(2x^2 + 11x + 19)$ by $(x + 3)$. **Solution** $$\begin{array}{r} 2x + 5 \\ x + 3 \overline{\smash{)}2x^2 + 11x + 19} \\ \underline{2x^2 + 6x} \\ 5x + 19 \\ \underline{5x + 15} \\ 4 \end{array}$$ The quotient is $2x + 5$ and the remainder is 4.
Use synthetic division to divide polynomials. **(Section 8.5/Objective 2)**	Synthetic division is a shortcut to the long-division process, when the divisor is of the form $x - k$.	Divide $(x^4 - 3x^2 + 5x + 6)$ by $(x + 2)$. **Solution** $$\begin{array}{r|rrrrr} -2 & 1 & 0 & -3 & 5 & 6 \\ & & -2 & 4 & -2 & -6 \\ \hline & 1 & -2 & 1 & 3 & 0 \end{array}$$ The quotient is $x^3 - 2x^2 + x + 3$.
Factor out the highest common monomial factor. **(Section 8.6/Objective 1)**	The distributive property in the form $ab + ac = a(b + c)$ is the basis for factoring out the highest common monomial factor.	Factor $-4x^3 y^4 - 2x^4 y^3 - 6x^5 y^2$. **Solution** $-4x^3 y^4 - 2x^4 y^3 - 6x^5 y^2$ $= -2x^3 y^2 (2y^2 + xy + 3x^2)$
Factor by grouping. **(Section 8.6/Objective 1)**	It may be that the polynomial exhibits no common monomial or binomial factor. However, by factoring common factors from groups of terms, a common factor may be evident.	Factor $2xz + 6x + yz + 3y$. **Solution** $2xz + 6x + yz + 3y$ $= 2x(z + 3) + y(z + 3)$ $= (z + 3)(2x + y)$
Factor the difference of two squares. **(Section 8.6/Objective 1)**	The factoring pattern $a^2 - b^2 = (a + b)(a - b)$ is called the difference of two squares.	Factor $36a^2 - 25b^2$. **Solution** $36a^2 - 25b^2 = (6a - 5b)(6a + 5b)$
Factor the sum or difference of two cubes. **(Section 8.6/Objective 2)**	The factoring patterns $a^3 + b^3 = (a + b)(a^2 - ab + b^2)$ and $a^3 - b^3 = (a - b)(a^2 + ab + b^2)$ are called the sum of two cubes or the difference of two cubes.	Factor $8x^3 + 27y^3$. **Solution** $8x^3 + 27y^3$ $= (2x + 3y)(4x^2 - 6xy + 9y^2)$

(continued)

OBJECTIVE	SUMMARY	EXAMPLE
Factor trinomials of the form $x^2 + bx + c$. (Section 8.6/Objective 3)	Expressing a trinominal (for which the coefficient of the squared term is 1) as a product of two binomials is based on the relationship $(x + a)(x + b) = x^2 + (a + b)x + ab$ The coefficient of the middle term is the sum of a and b, and the last term is the product of a and b.	Factor $x^2 - 2x - 35$. **Solution** $x^2 - 2x - 35 = (x - 7)(x + 5)$
Factor trinomials of the form $ax^2 + bx + c$. (Section 8.6/Objective 3)	Two methods were presented for factoring trinomials of the form $ax^2 + bx + c$. One technique is to try the various possibilities of factors and check by multiplying. This method is referred to as trial and error. The other method is a structured technique and is shown in Section 8.6, Examples 7 and 8.	Factor $4x^2 + 16x + 15$. **Solution** Multiply 4 times 15 to get 60. The factors of 60 that add to 16 are 6 and 10. Rewrite the problem and factor by grouping. $4x^2 + 16x + 15$ $= 4x^2 + 10x + 6x + 15$ $= 2x(2x + 5) + 3(2x + 5)$ $= (2x + 5)(2x + 3)$
Factor perfect square trinomials. (Section 8.6/Objective 3)	A perfect-square trinomial is the result of squaring a binomial. There are two basic perfect-square trinomial factoring patterns. $a^2 + 2ab + b^2 = (a + b)^2$ and $a^2 - 2ab + b^2 = (a - b)^2$	Factor $16x^2 + 40x + 25$. **Solution** $16x^2 + 40x + 25 = (4x + 5)^2$
Solve equations. (Section 8.6/Objective 4)	The factoring techniques in this chapter, along with the property $ab = 0$, provide the basis for some additional equation-solving skills.	Solve $x^2 - 11x + 28 = 0$. **Solution** $x^2 - 11x + 28 = 0$ $(x - 7)(x - 4) = 0$ $x - 7 = 0$ or $x - 4 = 0$ $x = 7$ or $x = 4$ The solution set is $\{4, 7\}$.
Solve word problems. (Section 8.6/Objective 4)	The ability to solve more types of equations increases our capabilities to solve more word problems.	Suppose that the area of a square is numerically equal to three times its perimeter. Find the length of a side of the square. **Solution** Let x represent the length of a side of the square. The area is x^2 and the perimeter is $4x$. Knowing that the area is numerically equal to three times the perimeter, the following equation can be formed: $x^2 = 3(4x)$ By solving this equation, we can determine that the length of a side of the square is 12 units.

Chapter 8   Review Problem Set

For Problems 1–24, solve each of the equations.

1. $5(x - 6) = 3(x + 2)$

2. $2(2x + 1) - (x - 4) = 4(x + 5)$

3. $-(2n - 1) + 3(n + 2) = 7$

4. $2(3n - 4) + 3(2n - 3) = -2(n + 5)$

5. $\dfrac{3t - 2}{4} = \dfrac{2t + 1}{3}$

6. $\dfrac{x + 6}{5} + \dfrac{x - 1}{4} = 2$

7. $1 - \dfrac{2x - 1}{6} = \dfrac{3x}{8}$

8. $\dfrac{2x + 1}{3} + \dfrac{3x - 1}{5} = \dfrac{1}{10}$

9. $\dfrac{3n - 1}{2} - \dfrac{2n + 3}{7} = 1$

10. $|3x - 1| = 11$

11. $0.06x + 0.08(x + 100) = 15$

12. $0.4(t - 6) = 0.3(2t + 5)$

13. $0.1(n + 300) = 0.09n + 32$

14. $0.2(x - 0.5) - 0.3(x + 1) = 0.4$

15. $4x^2 - 36 = 0$ **16.** $x^2 + 5x - 6 = 0$

17. $49n^2 - 28n + 4 = 0$

18. $(3x - 1)(5x + 2) = 0$

19. $(3x - 4)^2 - 25 = 0$

20. $6a^3 = 54a$ **21.** $7n(7n + 2) = 8$

22. $30w^2 - w - 20 = 0$

23. $3t^3 - 27t^2 + 24t = 0$ **24.** $-4n^2 - 39n + 10 = 0$

For Problems 25–29, solve each equation for x.

25. $ax - b = b + 2$

26. $ax = bx + c$

27. $m(x + a) = p(x + b)$

28. $5x - 7y = 11$

29. $\dfrac{x - a}{b} = \dfrac{y + 1}{c}$

For Problems 30–39, solve each inequality and express the solutions using interval notation.

30. $5x - 2 \geq 4x - 7$ **31.** $3 - 2x < -5$

32. $2(3x - 1) - 3(x - 3) > 0$

33. $3(x + 4) \leq 5(x - 1)$

34. $\dfrac{5}{6}n - \dfrac{1}{3}n < \dfrac{1}{6}$

35. $\dfrac{n - 4}{5} + \dfrac{n - 3}{6} > \dfrac{7}{15}$

36. $s \geq 4.5 + 0.25s$

37. $0.07x + 0.09(500 - x) \geq 43$

38. $|2x - 1| < 11$

39. $|3x + 1| > 10$

For Problems 40–43, graph the solutions of each compound inequality.

40. $x > -1$ and $x < 1$ **41.** $x > 2$ or $x \leq -3$

42. $x > 2$ and $x > 3$ **43.** $x < 2$ or $x > -1$

For Problems 44–65, perform the indicated operations and simplify each of the following.

44. $(3x - 2) + (4x - 6) + (-2x + 5)$

45. $(8x^2 + 9x - 3) - (5x^2 - 3x - 1)$

46. $(6x^2 - 2x - 1) + (4x^2 + 2x + 5) - (-2x^2 + x - 1)$

47. $(-5x^2y^3)(4x^3y^4)$ **48.** $(-2a^2)(3ab^2)(a^2b^3)$

49. $5a^2(3a^2 - 2a - 1)$ **50.** $(4x - 3y)(6x + 5y)$

51. $(x + 4)(3x^2 - 5x - 1)$ **52.** $(4x^2y^3)^4$

53. $(3x - 2y)^2$ **54.** $(-2x^2y^3z)^3$

55. $\dfrac{-39x^3y^4}{3xy^3}$

56. $[3x - (2x - 3y + 1)] - [2y - (x - 1)]$

57. $(x^2 - 2x - 5)(x^2 + 3x - 7)$

58. $(7 - 3x)(3 + 5x)$ **59.** $-(3ab)(2a^2b^3)^2$

60. $\left(\dfrac{1}{2}ab\right)(8a^3b^2)(-2a^3)$ **61.** $(7x - 9)(x + 4)$

62. $(3x + 2)(2x^2 - 5x + 1)$ **63.** $(3x^{n+1})(2x^{3n-1})$

64. $(2x + 5y)^2$ **65.** $(x - 2)^3$

For Problems 66–69, use synthetic division to find the quotient and remainder for each of the following.

66. $(3x^3 - 10x^2 + 2x + 41) \div (x - 2)$

67. $(5x^3 + 8x^2 + x + 6) \div (x + 1)$

68. $(x^4 - x^3 - 19x^2 - 22x - 3) \div (x + 3)$

69. $(2x^4 - 5x^3 - 16x^2 + 14x + 8) \div (x - 4)$

For Problems 70–91, factor each polynomial completely. Indicate any that are not factorable using integers.

70. $x^2 + 3x - 28$ **71.** $2t^2 - 18$

72. $4n^2 + 9$ **73.** $12n^2 - 7n + 1$

74. $x^6 - x^2$ **75.** $x^3 - 6x^2 - 72x$

76. $6a^3b + 4a^2b^2 - 2a^2bc$ **77.** $x^2 - (y - 1)^2$

78. $8x^2 + 12$ **79.** $12x^2 + x - 35$

80. $16n^2 - 40n + 25$ **81.** $4n^2 - 8n$

82. $3w^3 + 18w^2 - 24w$ **83.** $20x^2 + 3xy - 2y^2$

84. $16a^2 - 64a$ **85.** $3x^3 - 15x^2 - 18x$

86. $n^2 - 8n - 128$ **87.** $t^4 - 22t^2 - 75$

88. $35x^2 - 11x - 6$ **89.** $15 - 14x + 3x^2$

90. $64n^3 - 27$ **91.** $16x^3 + 250$

Solve each of Problems 92–107 by setting up and solving an appropriate equation, inequality, or system of equations.

92. The width of a rectangle is 2 meters more than one-third of the length. The perimeter of the rectangle is 44 meters. Find the length and width of the rectangle.

93. A total of $5000 was invested, part of it at 7% interest and the remainder at 8%. If the total yearly interest from both investments amounted to $380, how much was invested at each rate?

94. Susan's average score for her first three psychology exams is 84. What must she get on the fourth exam so that her average for the four exams is 85 or better?

95. Find three consecutive integers such that the sum of one-half of the smallest and one-third of the largest is one less than the other integer.

96. Pat is paid time-and-a-half for each hour he works over 36 hours in a week. Last week he worked 42 hours for a total of $472.50. What is his normal hourly rate?

97. Marcela has a collection of nickels, dimes, and quarters worth $24.75. The number of dimes is 10 more than twice the number of nickels, and the number of quarters is 25 more than the number of dimes. How many coins of each kind does she have?

98. If the complement of an angle is one-tenth of the supplement of the angle, find the measure of the angle.

99. A retailer has some sweaters that cost her $38 each. She wants to sell them at a profit of 20% of her cost. What price should she charge for the sweaters?

100. Nora scored 16, 22, 18, and 14 points for each of the first four basketball games. How many points does she need to score in the fifth game so that her average for the first five games is at least 20 points per game?

101. Gladys leaves a town driving at a rate of 40 miles per hour. Two hours later, Reena leaves from the same place traveling the same route. She catches Gladys in 5 hours and 20 minutes. How fast was Reena traveling?

102. In $1\frac{1}{4}$ hours more time, Rita, riding her bicycle at 12 miles per hour, rode 2 miles farther than Sonya, who was riding her bicycle at 16 miles per hour. How long did each girl ride?

103. How many cups of orange juice must be added to 50 cups of a punch that is 10% orange juice to obtain a punch that is 20% orange juice?

104. Two cars leave an intersection at the same time, one traveling north and the other traveling east. Some time later, they are 20 miles apart, and the car going east has traveled 4 miles farther than the other car. How far has each car traveled?

105. The perimeter of a rectangle is 32 meters, and its area is 48 square meters. Find the length and width of the rectangle.

106. A room contains 144 chairs. The number of chairs per row is two less than twice the number of rows. Find the number of rows and the number of chairs per row.

107. The area of a triangle is 39 square feet. The length of one side is 1 foot more than twice the altitude to that side. Find the length of that side and the altitude to the side.

For Problems 1–4, perform the indicated operations and simplify each expression.

1. $(-3x - 1) + (9x - 2) - (4x + 8)$

2. $(5x - 7)(4x + 9)$

3. $(x + 6)(2x^2 - x - 5)$

4. $(x - 4y)^3$

5. Find the quotient and remainder for the division problem $(6x^3 - 19x^2 + 3x + 20) \div (3x - 5)$.

6. Find the quotient and remainder for the division problem $(3x^4 + 8x^3 - 5x^2 - 12x - 15) \div (x + 3)$.

7. Factor $x^2 - xy + 4x - 4y$ completely.

8. Factor $12x^2 - 3$ completely.

For Problems 9–18, solve each equation.

9. $3(2x - 1) - 2(x + 5) = -(x - 3)$

10. $\dfrac{3t - 2}{4} = \dfrac{5t + 1}{5}$

11. $|4x - 3| = 9$

12. $\dfrac{1 - 3x}{4} + \dfrac{2x + 3}{3} = 1$

13. $0.05x + 0.06(1500 - x) = 83.5$

14. $4n^2 = n$

15. $4x^2 - 12x + 9 = 0$

16. $3x^3 + 21x^2 - 54x = 0$

17. $12 + 13x - 35x^2 = 0$

18. $n(3n - 5) = 2$

For Problems 19–21, solve each inequality and use interval notation to express the solutions.

19. $|6x - 4| < 10$

20. $\dfrac{x - 2}{6} - \dfrac{x + 3}{9} > -\dfrac{1}{2}$

21. $2(x - 1) - 3(3x + 1) \geq -6(x - 5)$

For Problems 22–25, use an equation, an inequality, or a system of equations to help solve each problem.

22. How many cups of grapefruit juice must be added to 30 cups of a punch that is 8% grapefruit juice to obtain a punch that is 10% grapefruit juice?

23. Rex has scores of 85, 92, 87, 88, and 91 on the first five exams. What score must he make on the sixth exam to have an average of 90 or better for all six exams?

24. If the complement of an angle is $\dfrac{2}{11}$ of the supplement of the angle, find the measure of the angle.

25. The combined area of a square and a rectangle is 57 square feet. The width of the rectangle is 3 feet more than the length of a side of the square, and the length of the rectangle is 5 feet more than the length of a side of the square. Find the length of the rectangle.

9

Rational Expressions

Carpenters often work together to complete construction jobs. Rational numbers can be used to express the rate at which a carpenter works.

© Todd Taulman

It takes Pat 12 hours to complete a task. After he had been working on this task for 3 hours, he was joined by his brother, Liam, and together they finished the job in 5 hours. How long would it take Liam to do the job by himself? We can use the *fractional equation* $\frac{5}{12} + \frac{5}{h} = \frac{3}{4}$ to determine that Liam could do the entire job by himself in 15 hours.

Rational expressions are to algebra what rational numbers are to arithmetic. Most of the work we will do with rational expressions in this chapter parallels the work you have previously done with arithmetic fractions. The same basic properties we use to explain reducing, adding, subtracting, multiplying, and dividing arithmetic fractions will serve as a basis for our work with rational expressions. The techniques of factoring that we studied in Chapter 7 will also play an important role in our discussions. At the end of this chapter, we will work with some fractional equations that contain rational expressions.

Video tutorials based on section learning objectives are available in a variety of delivery modes.

9.1 Simplifying Rational Expressions

OBJECTIVES

1 Reduce rational numbers

2 Simplify rational expressions

We reviewed the basic operations with rational numbers in an informal setting in Chapter 2. In this review, we relied primarily on your knowledge of arithmetic. At this time, we want to become a little more formal with our review so that we can use the work with rational numbers as a basis for operating with rational expressions. We will define a rational expression shortly.

You will recall that any number that can be written in the form $\frac{a}{b}$, where a and b are integers and $b \neq 0$, is called a rational number. The following are examples of rational numbers.

$$\frac{1}{2} \qquad \frac{3}{4} \qquad \frac{15}{7} \qquad \frac{-5}{6} \qquad \frac{7}{-8} \qquad \frac{-12}{-17}$$

Numbers such as 6, -4, 0, $4\frac{1}{2}$, 0.7, and 0.21 are also rational, because we can express them as the indicated quotient of two integers. For example,

$$6 = \frac{6}{1} = \frac{12}{2} = \frac{18}{3} \quad \text{and so on} \qquad\qquad 4\frac{1}{2} = \frac{9}{2}$$

$$-4 = \frac{4}{-1} = \frac{-4}{1} = \frac{8}{-2} \quad \text{and so on} \qquad\qquad 0.7 = \frac{7}{10}$$

$$0 = \frac{0}{1} = \frac{0}{2} = \frac{0}{3} \quad \text{and so on} \qquad\qquad 0.21 = \frac{21}{100}$$

Because a rational number is the quotient of two integers, our previous work with division of integers can help us understand the various forms of rational numbers. If the signs of the numerator and denominator are different, then the rational number is negative. If the signs of the numerator and denominator are the same, then the rational number is positive. The next examples and Property 9.1 show the equivalent forms of rational numbers. Generally, it is preferred to express the denominator of a rational number as a positive integer.

$$\frac{8}{-2} = \frac{-8}{2} = -\frac{8}{2} = -4 \qquad \frac{12}{3} = \frac{-12}{-3} = 4$$

Observe the following general properties.

Property 9.1

1. $\dfrac{-a}{b} = \dfrac{a}{-b} = -\dfrac{a}{b}$ where $b \neq 0$

2. $\dfrac{-a}{-b} = \dfrac{a}{b}$ where $b \neq 0$

Therefore, a rational number such as $\dfrac{-2}{5}$ can also be written as $\dfrac{2}{-5}$ or $-\dfrac{2}{5}$.

We use the following property, often referred to as the **fundamental principle of fractions,** to reduce fractions to lowest terms or express fractions in simplest or reduced form.

> **Property 9.2 Fundamental Principle of Fractions**
>
> If b and k are nonzero integers and a is any integer, then
>
> $$\frac{a \cdot k}{b \cdot k} = \frac{a}{b}$$

Let's apply Properties 9.1 and 9.2 to the following examples.

Classroom Example
Reduce $\dfrac{14}{21}$ to lowest terms.

EXAMPLE 1 Reduce $\dfrac{18}{24}$ to lowest terms.

Solution

$$\frac{18}{24} = \frac{3 \cdot 6}{4 \cdot 6} = \frac{3}{4}$$

Classroom Example
Change $\dfrac{32}{56}$ to simplest form.

EXAMPLE 2 Change $\dfrac{40}{48}$ to simplest form.

Solution

$$\frac{\overset{5}{\cancel{40}}}{\underset{6}{\cancel{48}}} = \frac{5}{6}$$ A common factor of 8 was divided out of both numerator and denominator

Classroom Example
Express $\dfrac{-28}{44}$ in reduced form.

EXAMPLE 3 Express $\dfrac{-36}{63}$ in reduced form.

Solution

$$\frac{-36}{63} = -\frac{36}{63} = -\frac{4 \cdot 9}{7 \cdot 9} = -\frac{4}{7}$$

Classroom Example
Reduce $\dfrac{36}{-84}$ to simplest form.

EXAMPLE 4 Reduce $\dfrac{72}{-90}$ to simplest form.

Solution

$$\frac{72}{-90} = -\frac{72}{90} = -\frac{2 \cdot 2 \cdot 2 \cdot 3 \cdot 3}{2 \cdot 3 \cdot 3 \cdot 5} = -\frac{4}{5}$$

Note the different terminology used in Examples 1–4. Regardless of the terminology, keep in mind that the number is not being changed, but the form of the numeral representing the number is being changed. In Example 1, $\dfrac{18}{24}$ and $\dfrac{3}{4}$ are equivalent fractions; they name the same number. Also note the use of prime factors in Example 4.

Simplifying Rational Expressions

A **rational expression** is the indicated quotient of two polynomials. The following are examples of rational expressions.

$$\frac{3x^2}{5} \qquad \frac{x-2}{x+3} \qquad \frac{x^2+5x-1}{x^2-9} \qquad \frac{xy^2+x^2y}{xy} \qquad \frac{a^3-3a^2-5a-1}{a^4+a^3+6}$$

Because we must avoid division by zero, no values that create a denominator of zero can be assigned to variables. Thus the rational expression $\dfrac{x-2}{x+3}$ is meaningful for all values of x except $x = -3$. Rather than making restrictions for each individual expression, we will merely assume that all denominators represent nonzero real numbers.

Property 9.2 $\left(\dfrac{a \cdot k}{b \cdot k} = \dfrac{a}{b}\right)$ serves as the basis for simplifying rational expressions, as the next examples illustrate.

Classroom Example

Simplify $\dfrac{18mn}{45m}$.

EXAMPLE 5

Simplify $\dfrac{15xy}{25y}$.

Solution

$$\frac{15xy}{25y} = \frac{3 \cdot \cancel{5} \cdot x \cdot \cancel{y}}{\cancel{5} \cdot 5 \cdot \cancel{y}} = \frac{3x}{5}$$

Classroom Example

Simplify $\dfrac{-12}{36ab^2}$.

EXAMPLE 6

Simplify $\dfrac{-9}{18x^2y}$.

Solution

$$\frac{-9}{18x^2y} = -\frac{\overset{1}{\cancel{9}}}{\underset{2}{\cancel{18}}x^2y} = -\frac{1}{2x^2y}$$ A common factor of 9 was divided out of numerator and denominator

Classroom Example

Simplify $\dfrac{-42x^3y^2}{-54x^2y^2}$.

EXAMPLE 7

Simplify $\dfrac{-28a^2b^2}{-63a^2b^3}$.

Solution

$$\frac{-28a^2b^2}{-63a^2b^3} = \frac{4 \cdot \cancel{7} \cdot \cancel{a^2} \cdot \cancel{b^2}}{9 \cdot \cancel{7} \cdot \cancel{a^2} \cdot \underset{b}{\cancel{b^3}}} = \frac{4}{9b}$$

The factoring techniques from Chapter 7 can be used to factor numerators and/or denominators so that we can apply the property $\dfrac{a \cdot k}{b \cdot k} = \dfrac{a}{b}$. Examples 8–12 should clarify this process.

Classroom Example

Simplify $\dfrac{x^2 - 7x}{x^2 - 49}$.

EXAMPLE 8

Simplify $\dfrac{x^2 + 4x}{x^2 - 16}$.

Solution

$$\frac{x^2 + 4x}{x^2 - 16} = \frac{x(x + 4)}{(x - 4)(x + 4)} = \frac{x}{x - 4}$$

Classroom Example

Simplify $\dfrac{9x^2 + 6x + 1}{3x + 1}$.

EXAMPLE 9

Simplify $\dfrac{4a^2 + 12a + 9}{2a + 3}$.

Solution

$$\frac{4a^2 + 12a + 9}{2a + 3} = \frac{(2a + 3)(2a + 3)}{1(2a + 3)} = \frac{2a + 3}{1} = 2a + 3$$

Classroom Example
Simplify $\dfrac{7n^2 + 23n + 6}{21n^2 - n - 2}$.

EXAMPLE 10   Simplify $\dfrac{5n^2 + 6n - 8}{10n^2 - 3n - 4}$.

Solution

$$\frac{5n^2 + 6n - 8}{10n^2 - 3n - 4} = \frac{(5n - 4)(n + 2)}{(5n - 4)(2n + 1)} = \frac{n + 2}{2n + 1}$$

Classroom Example
Simplify $\dfrac{3x^3y - 12xy}{x^2y - xy - 6y}$.

EXAMPLE 11   Simplify $\dfrac{6x^3y - 6xy}{x^3 + 5x^2 + 4x}$.

Solution

$$\frac{6x^3y - 6xy}{x^3 + 5x^2 + 4x} = \frac{6xy(x^2 - 1)}{x(x^2 + 5x + 4)} = \frac{6xy(x + 1)(x - 1)}{x(x + 1)(x + 4)} = \frac{6y(x - 1)}{x + 4}$$

Note that in Example 11 we left the numerator of the final fraction in factored form. This is often done if expressions other than monomials are involved. Either $\dfrac{6y(x - 1)}{x + 4}$ or $\dfrac{6xy - 6y}{x + 4}$ is an acceptable answer.

Remember that the quotient of any nonzero real number and its opposite is -1. For example, $\dfrac{6}{-6} = -1$ and $\dfrac{-8}{8} = -1$. Likewise, the indicated quotient of any polynomial and its opposite is equal to -1; that is,

$$\frac{a}{-a} = -1 \qquad \text{because } a \text{ and } -a \text{ are opposites}$$

$$\frac{a - b}{b - a} = -1 \quad \text{because } a - b \text{ and } b - a \text{ are opposites}$$

$$\frac{x^2 - 4}{4 - x^2} = -1 \quad \text{because } x^2 - 4 \text{ and } 4 - x^2 \text{ are opposites}$$

Example 12 shows how we use this idea when simplifying rational expressions.

Classroom Example
Simplify $\dfrac{6a^2 - 17a + 5}{15a - 6a^2}$.

EXAMPLE 12   Simplify $\dfrac{6a^2 - 7a + 2}{10a - 15a^2}$.

Solution

$$\frac{6a^2 - 7a + 2}{10a - 15a^2} = \frac{(2a - 1)(3a - 2)}{5a(2 - 3a)} \qquad \frac{3a - 2}{2 - 3a} = -1$$

$$= (-1)\left(\frac{2a - 1}{5a}\right)$$

$$= -\frac{2a - 1}{5a} \qquad \text{or} \qquad \frac{1 - 2a}{5a}$$

Concept Quiz 9.1

For Problems 1–10, answer true or false.

1. When a rational number is being reduced, the form of the numeral is being changed but not the number it represents.

2. A rational number is the ratio of two integers where the denominator is not zero.

3. -3 is a rational number.

4. The rational expression $\dfrac{x + 2}{x + 3}$ is meaningful for all values of x except when $x = -2$ and $x = 3$.

5. The binomials $x - y$ and $y - x$ are opposites.

6. The binomials $x + 3$ and $x - 3$ are opposites.

7. The rational expression $\dfrac{2 - x}{x + 2}$ reduces to -1.

8. The rational expression $\dfrac{x - y}{y - x}$ reduces to -1.

9. $\dfrac{x^2 + 5x - 14}{x^2 + 2x + 1} = \dfrac{5x - 14}{2x + 1}$

10. The rational expression $\dfrac{2x - x^2}{x^2 - 4}$ reduces to $\dfrac{x}{x + 2}$.

Problem Set 9.1

For Problems 1–8, express each rational number in reduced form. **(Objective 1)**

1. $\dfrac{27}{36}$

2. $\dfrac{14}{21}$

3. $\dfrac{45}{54}$

4. $\dfrac{-14}{42}$

5. $\dfrac{24}{-60}$

6. $\dfrac{45}{-75}$

7. $\dfrac{-16}{-56}$

8. $\dfrac{-30}{-42}$

For Problems 9–50, simplify each rational expression. **(Objective 2)**

9. $\dfrac{12xy}{42y}$

10. $\dfrac{21xy}{35x}$

11. $\dfrac{18a^2}{45ab}$

12. $\dfrac{48ab}{84b^2}$

13. $\dfrac{-14y^3}{56xy^2}$

14. $\dfrac{-14x^2y^3}{63xy^2}$

15. $\dfrac{54c^2d}{-78cd^2}$

16. $\dfrac{60x^3z}{-64xyz^2}$

17. $\dfrac{-40x^3y}{-24xy^4}$

18. $\dfrac{-30x^2y^2z^2}{-35xz^3}$

19. $\dfrac{x^2 - 4}{x^2 + 2x}$

20. $\dfrac{xy + y^2}{x^2 - y^2}$

21. $\dfrac{18x + 12}{12x - 6}$

22. $\dfrac{20x + 50}{15x - 30}$

23. $\dfrac{a^2 + 7a + 10}{a^2 - 7a - 18}$

24. $\dfrac{a^2 + 4a - 32}{3a^2 + 26a + 16}$

25. $\dfrac{2n^2 + n - 21}{10n^2 + 33n - 7}$

26. $\dfrac{4n^2 - 15n - 4}{7n^2 - 30n + 8}$

27. $\dfrac{5x^2 + 7}{10x}$

28. $\dfrac{12x^2 + 11x - 15}{20x^2 - 23x + 6}$

29. $\dfrac{6x^2 + x - 15}{8x^2 - 10x - 3}$

30. $\dfrac{4x^2 + 8x}{x^3 + 8}$

31. $\dfrac{3x^2 - 12x}{x^3 - 64}$

32. $\dfrac{x^2 - 14x + 49}{6x^2 - 37x - 35}$

33. $\dfrac{3x^2 + 17x - 6}{9x^2 - 6x + 1}$

34. $\dfrac{9y^2 - 1}{3y^2 + 11y - 4}$

35. $\dfrac{2x^3 + 3x^2 - 14x}{x^2y + 7xy - 18y}$

36. $\dfrac{3x^3 + 12x}{9x^2 + 18x}$

37. $\dfrac{5y^2 + 22y + 8}{25y^2 - 4}$

38. $\dfrac{16x^3y + 24x^2y^2 - 16xy^3}{24x^2y + 12xy^2 - 12y^3}$

39. $\dfrac{15x^3 - 15x^2}{5x^3 + 5x}$

40. $\dfrac{5n^2 + 18n - 8}{3n^2 + 13n + 4}$

41. $\dfrac{4x^2y + 8xy^2 - 12y^3}{18x^3y - 12x^2y^2 - 6xy^3}$

42. $\dfrac{3 + x - 2x^2}{2 + x - x^2}$

43. $\dfrac{3n^2 + 16n - 12}{7n^2 + 44n + 12}$

44. $\dfrac{x^4 - 2x^2 - 15}{2x^4 + 9x^2 + 9}$

45. $\dfrac{8 + 18x - 5x^2}{10 + 31x + 15x^2}$

46. $\dfrac{6x^4 - 11x^2 + 4}{2x^4 + 17x^2 - 9}$

47. $\dfrac{27x^4 - x}{6x^3 + 10x^2 - 4x}$

48. $\dfrac{64x^4 + 27x}{12x^3 - 27x^2 - 27x}$

49. $\dfrac{-40x^3 + 24x^2 + 16x}{20x^3 + 28x^2 + 8x}$ **50.** $\dfrac{-6x^3 - 21x^2 + 12x}{-18x^3 - 42x^2 + 120x}$

For Problems 51–58, simplify each rational expression. You will need to use factoring by grouping. **(Objective 2)**

51. $\dfrac{xy + ay + bx + ab}{xy + ay + cx + ac}$ **52.** $\dfrac{xy + 2y + 3x + 6}{xy + 2y + 4x + 8}$

53. $\dfrac{ax - 3x + 2ay - 6y}{2ax - 6x + ay - 3y}$ **54.** $\dfrac{x^2 - 2x + ax - 2a}{x^2 - 2x + 3ax - 6a}$

55. $\dfrac{5x^2 + 5x + 3x + 3}{5x^2 + 3x - 30x - 18}$ **56.** $\dfrac{x^2 + 3x + 4x + 12}{2x^2 + 6x - x - 3}$

57. $\dfrac{2st - 30 - 12s + 5t}{3st - 6 - 18s + t}$ **58.** $\dfrac{nr - 6 - 3n + 2r}{nr + 10 + 2r + 5n}$

For Problems 59–68, simplify each rational expression. You may want to refer to Example 12 of this section. **(Objective 2)**

59. $\dfrac{5x - 7}{7 - 5x}$ **60.** $\dfrac{4a - 9}{9 - 4a}$

61. $\dfrac{n^2 - 49}{7 - n}$ **62.** $\dfrac{9 - y}{y^2 - 81}$

63. $\dfrac{2y - 2xy}{x^2y - y}$ **64.** $\dfrac{3x - x^2}{x^2 - 9}$

65. $\dfrac{2x^3 - 8x}{4x - x^3}$ **66.** $\dfrac{x^2 - (y - 1)^2}{(y - 1)^2 - x^2}$

67. $\dfrac{n^2 - 5n - 24}{40 + 3n - n^2}$ **68.** $\dfrac{x^2 + 2x - 24}{20 - x - x^2}$

Thoughts Into Words

69. Compare the concept of a rational number in arithmetic to the concept of a rational expression in algebra.

70. What role does factoring play in the simplifying of rational expressions?

71. Why is the rational expression $\dfrac{x + 3}{x^2 - 4}$ undefined for $x = 2$ and $x = -2$ but defined for $x = -3$?

72. How would you convince someone that $\dfrac{x - 4}{4 - x} = -1$ for all real numbers except 4?

Graphing Calculator Activities

This is the first of many appearances of a group of problems called Graphing Calculator Activities. These problems are specifically designed for those of you who have access to a graphing calculator or a computer with an appropriate software package. Within the framework of these problems, you will be given the opportunity to reinforce concepts we discussed in the text; lay groundwork for concepts we will introduce later in the text; predict shapes and locations of graphs on the basis of your previous graphing experiences; solve problems that are unreasonable (or perhaps impossible) to solve without a graphing utility; and, in general, become familiar with the capabilities and limitations of your graphing utility.

This first set of activities is designed to help you get started with your graphing utility by setting different boundaries for the viewing rectangle; you will notice the effect on the graphs produced. These boundaries are usually set by using a menu displayed by a key marked either WINDOW or RANGE. You may need to consult the user's manual for specific key-punching instructions.

73. Graph the equation $y = \dfrac{1}{x}$ using the following boundaries.

  a. $-15 \leq x \leq 15$ and $-10 \leq y \leq 10$

  b. $-10 \leq x \leq 10$ and $-10 \leq y \leq 10$

  c. $-5 \leq x \leq 5$ and $-5 \leq y \leq 5$

74. Graph the equation $y = \dfrac{-2}{x^2}$ using the following boundaries.

  a. $-15 \leq x \leq 15$ and $-10 \leq y \leq 10$

  b. $-5 \leq x \leq 5$ and $-10 \leq y \leq 10$

  c. $-5 \leq x \leq 5$ and $-10 \leq y \leq 1$

75. Graph the two equations $y = \pm\sqrt{x}$ on the same set of axes using the following boundaries. Let $Y_1 = \sqrt{x}$ and $Y_2 = -\sqrt{x}$.

  a. $-15 \leq x \leq 15$ and $-10 \leq y \leq 10$

  b. $-1 \leq x \leq 15$ and $-10 \leq y \leq 10$

  c. $-1 \leq x \leq 15$ and $-5 \leq y \leq 5$

76. Graph $y = \dfrac{1}{x}, y = \dfrac{5}{x}, y = \dfrac{10}{x}$, and $y = \dfrac{20}{x}$ on the same set of axes. (Choose your own boundaries.) What effect does increasing the constant seem to have on the graph?

77. Graph $y = \dfrac{10}{x}$ and $y = \dfrac{-10}{x}$ on the same set of axes. What relationship exists between the two graphs?

78. Graph $y = \dfrac{10}{x^2}$ and $y = \dfrac{-10}{x^2}$ on the same set of axes. What relationship exists between the two graphs?

79. Use a graphing calculator to give visual support for your answers for Problems 21–30.

80. Use a graphing calculator to give visual support for your answers for Problems 59–62.

9.2	Multiplying and Dividing Rational Expressions

OBJECTIVES

1 Multiply rational numbers

2 Multiply rational expressions

3 Divide rational numbers

4 Divide rational expressions

5 Simplify problems that involve both multiplication and division of rational expressions

We define multiplication of rational numbers in common fraction form as follows:

> **Definition 9.1 Multiplication of Fractions**
>
> If a, b, c, and d are integers, and b and d are not equal to zero, then
>
> $$\frac{a}{b} \cdot \frac{c}{d} = \frac{a \cdot c}{b \cdot d} = \frac{ac}{bd}$$

To multiply rational numbers in common fraction form, we *multiply numerators and multiply denominators,* as the following examples demonstrate. (The steps in the dashed boxes are usually done mentally.)

$$\frac{2}{3} \cdot \frac{4}{5} = \frac{2 \cdot 4}{3 \cdot 5} = \frac{8}{15}$$

$$\frac{-3}{4} \cdot \frac{5}{7} = \frac{-3 \cdot 5}{4 \cdot 7} = \frac{-15}{28} = -\frac{15}{28}$$

$$-\frac{5}{6} \cdot \frac{13}{3} = \frac{-5}{6} \cdot \frac{13}{3} = \frac{-5 \cdot 13}{6 \cdot 3} = \frac{-65}{18} = -\frac{65}{18}$$

We also agree, when multiplying rational numbers, to express the final product in reduced form. The following examples show some different formats used to multiply and simplify rational numbers.

$$\frac{3}{4} \cdot \frac{4}{7} = \frac{3 \cdot 4}{4 \cdot 7} = \frac{3}{7}$$

$$\overset{1}{\underset{1}{\cancel{\frac{8}{9}}}} \cdot \overset{3}{\underset{4}{\cancel{\frac{27}{32}}}} = \frac{3}{4} \qquad \text{A common factor of 9 was divided out of 9 and 27, and a common factor of 8 was divided out of 8 and 32}$$

$$\left(-\frac{28}{25}\right)\left(-\frac{65}{78}\right) = \frac{2 \cdot 2 \cdot 7 \cdot \cancel{5} \cdot \cancel{13}}{\cancel{5} \cdot 5 \cdot 2 \cdot 3 \cdot \cancel{13}} = \frac{14}{15}$$

We should recognize that a negative times a negative is positive; also, note the use of prime factors to help us recognize common factors

Multiplying Rational Expressions

Multiplication of rational expressions follows the same basic pattern as multiplication of rational numbers in common fraction form. That is to say, we multiply numerators and multiply denominators and express the final product in simplified or reduced form. Let's consider some examples.

$$\frac{3x}{4y} \cdot \frac{8y^2}{9x} = \frac{\cancel{3} \cdot \overset{2}{\cancel{8}} \cdot \cancel{x} \cdot \overset{y}{\cancel{y^2}}}{\underset{3}{\cancel{4} \cdot \cancel{9}} \cdot \cancel{x} \cdot \cancel{y}} = \frac{2y}{3}$$

Note that we use the commutative property of multiplication to rearrange the factors in a form that allows us to identify common factors of the numerator and denominator

$$\frac{-4a}{6a^2b^2} \cdot \frac{9ab}{12a^2} = -\frac{\overset{3}{\cancel{4} \cdot \cancel{9}} \cdot \cancel{a^2} \cdot \cancel{b}}{\underset{2}{\cancel{6}} \cdot \underset{3}{\cancel{12}} \cdot \underset{a^2}{\cancel{a^4}} \cdot \underset{b}{\cancel{b^2}}} = -\frac{1}{2a^2b}$$

$$\frac{12x^2y}{-18xy} \cdot \frac{-24xy^2}{56y^3} = \frac{\overset{2}{\cancel{12}} \cdot \overset{3}{\cancel{24}} \cdot \overset{x^2}{\cancel{x^3}} \cdot \cancel{y^3}}{\underset{3}{\cancel{18}} \cdot \underset{7}{\cancel{56}} \cdot \cancel{x} \cdot \underset{y}{\cancel{y^4}}} = \frac{2x^2}{7y}$$

You should recognize that the first fraction is negative, and the second fraction is negative. Thus the product is positive.

If the rational expressions contain polynomials (other than monomials) that are factorable, then our work may take on the following format.

Classroom Example
Multiply and simplify

$$\frac{m}{n^2 - 9} \cdot \frac{n - 3}{m^3}$$

EXAMPLE 1   Multiply and simplify $\dfrac{y}{x^2 - 4} \cdot \dfrac{x + 2}{y^2}$.

Solution

$$\frac{y}{x^2 - 4} \cdot \frac{x + 2}{y^2} = \frac{\cancel{y}(\cancel{x + 2})}{\underset{y}{\cancel{y^2}}(\cancel{x + 2})(x - 2)} = \frac{1}{y(x - 2)}$$

In Example 1, note that we combined the steps of multiplying numerators and denominators and factoring the polynomials. Also note that we left the final answer in factored form. Either $\dfrac{1}{y(x - 2)}$ or $\dfrac{1}{xy - 2y}$ would be an acceptable answer.

Classroom Example
Multiply and simplify:

$$\frac{m^2 + m}{m + 4} \cdot \frac{m^2 - 4m + 3}{m^4 - m^2}$$

EXAMPLE 2   Multiply and simplify $\dfrac{x^2 - x}{x + 5} \cdot \dfrac{x^2 + 5x + 4}{x^4 - x^2}$.

Solution

$$\frac{x^2 - x}{x + 5} \cdot \frac{x^2 + 5x + 4}{x^4 - x^2} = \frac{x(x - 1)}{x + 5} \cdot \frac{(x + 1)(x + 4)}{x^2(x - 1)(x + 1)}$$

$$= \frac{\cancel{x}(\cancel{x - 1})(\cancel{x + 1})(x + 4)}{(x + 5)(\underset{x}{\cancel{x^2}})(\cancel{x - 1})(\cancel{x + 1})} = \frac{x + 4}{x(x + 5)}$$

Classroom Example
Multiply and simplify:

$$\frac{8x^2 + 10x - 3}{6x^2 + 7x - 3} \cdot \frac{3x^2 + 20x - 7}{8x^2 + 18x - 5}$$

EXAMPLE 3

Multiply and simplify $\dfrac{6n^2 + 7n - 5}{n^2 + 2n - 24} \cdot \dfrac{4n^2 + 21n - 18}{12n^2 + 11n - 15}$.

Solution

$$\frac{6n^2 + 7n - 5}{n^2 + 2n - 24} \cdot \frac{4n^2 + 21n - 18}{12n^2 + 11n - 15}$$

$$= \frac{(3n + 5)(2n - 1)(4n - 3)(n + 6)}{(n + 6)(n - 4)(3n + 5)(4n - 3)} = \frac{2n - 1}{n - 4}$$

Dividing Rational Numbers

We define division of rational numbers in common fraction form as follows:

> **Definition 9.2 Division of Fractions**
>
> If a, b, c, and d are integers, and b, c, and d are not equal to zero, then
>
> $$\frac{a}{b} \div \frac{c}{d} = \frac{a}{b} \cdot \frac{d}{c} = \frac{ad}{bc}$$

Definition 9.2 states that to divide two rational numbers in fraction form, we **invert the divisor and multiply**. We call the numbers $\dfrac{c}{d}$ and $\dfrac{d}{c}$ "reciprocals" or "multiplicative inverses" of each other, because their product is 1. Thus we can describe division by saying "to divide by a fraction, multiply by its reciprocal." The following examples demonstrate the use of Definition 9.2.

$$\frac{7}{8} \div \frac{5}{6} = \frac{7}{\underset{4}{8}} \cdot \frac{\overset{3}{6}}{5} = \frac{21}{20}, \qquad \frac{-5}{9} \div \frac{15}{18} = -\frac{5}{9} \cdot \frac{\overset{2}{18}}{\underset{3}{15}} = -\frac{2}{3}$$

$$\frac{14}{-19} \div \frac{21}{-38} = \left(-\frac{14}{19}\right) \div \left(-\frac{21}{38}\right) = \left(-\frac{\overset{2}{14}}{19}\right)\left(-\frac{\overset{2}{38}}{\underset{3}{21}}\right) = \frac{4}{3}$$

Dividing Rational Expressions

We define division of algebraic rational expressions in the same way that we define division of rational numbers. That is, the quotient of two rational expressions is the product we obtain when we multiply the first expression by the reciprocal of the second. Consider the following examples.

Classroom Example
Divide and simplify:

$$\frac{18mn^3}{32m^3n^2} \div \frac{9m^3n^2}{12m^2n^2}$$

EXAMPLE 4

Divide and simplify $\dfrac{16x^2y}{24xy^3} \div \dfrac{9xy}{8x^2y^2}$.

Solution

$$\frac{16x^2y}{24xy^3} \div \frac{9xy}{8x^2y^2} = \frac{16x^2y}{24xy^3} \cdot \frac{8x^2y^2}{9xy} = \frac{16 \cdot 8 \cdot \overset{x^2}{x^4} \cdot y^3}{\underset{3}{24} \cdot 9 \cdot x^2 \cdot \underset{y}{y^4}} = \frac{16x^2}{27y}$$

Classroom Example
Divide and simplify:

$$\frac{4x^2 + 36}{8x^2 + 4x} \div \frac{x^4 - 81}{2x^2 - 5x - 3}$$

EXAMPLE 5

Divide and simplify $\dfrac{3a^2 + 12}{3a^2 - 15a} \div \dfrac{a^4 - 16}{a^2 - 3a - 10}$.

Solution

$$\frac{3a^2 + 12}{3a^2 - 15a} \div \frac{a^4 - 16}{a^2 - 3a - 10} = \frac{3a^2 + 12}{3a^2 - 15a} \cdot \frac{a^2 - 3a - 10}{a^4 - 16}$$

$$= \frac{3(a^2 + 4)}{3a(a - 5)} \cdot \frac{(a - 5)(a + 2)}{(a^2 + 4)(a + 2)(a - 2)}$$

$$= \frac{\overset{1}{\cancel{3}}(\cancel{a^2 + 4})(\cancel{a - 5})(\cancel{a + 2})}{\underset{1}{\cancel{3}}a(\cancel{a - 5})(\cancel{a^2 + 4})(\cancel{a + 2})(a - 2)}$$

$$= \frac{1}{a(a - 2)}$$

Classroom Example
Divide and simplify:

$$\frac{35x^3 - 8x^2 - 3x}{45x^2 - x - 2} \div (7x - 3)$$

EXAMPLE 6

Divide and simplify $\dfrac{28t^3 - 51t^2 - 27t}{49t^2 + 42t + 9} \div (4t - 9)$.

Solution

$$\frac{28t^3 - 51t^2 - 27t}{49t^2 + 42t + 9} \div \frac{4t - 9}{1} = \frac{28t^3 - 51t^2 - 27t}{49t^2 + 42t + 9} \cdot \frac{1}{4t - 9}$$

$$= \frac{t(7t + 3)(4t - 9)}{(7t + 3)(7t + 3)} \cdot \frac{1}{(4t - 9)}$$

$$= \frac{t(\cancel{7t + 3})(\cancel{4t - 9})}{(\cancel{7t + 3})(7t + 3)(\cancel{4t - 9})}$$

$$= \frac{t}{7t + 3}$$

In a problem such as Example 6, it may be helpful to write the divisor with a denominator of 1. Thus we write $4t - 9$ as $\dfrac{4t - 9}{1}$; its reciprocal is obviously $\dfrac{1}{4t - 9}$.

Let's consider one final example that involves both multiplication and division.

Classroom Example
Perform the indicated operations and simplify:

$$\frac{5x^2 + 13x - 6}{2xy^2 - 3y^2} \cdot \frac{2x^2 + 5x - 12}{x^3 + 3x^2}$$

$$\div \frac{2x^2 + 13x + 20}{x^2y}$$

EXAMPLE 7

Perform the indicated operations and simplify:

$$\frac{x^2 + 5x}{3x^2 - 4x - 20} \cdot \frac{x^2y + y}{2x^2 + 11x + 5} \div \frac{xy^2}{6x^2 - 17x - 10}$$

Solution

$$\frac{x^2 + 5x}{3x^2 - 4x - 20} \cdot \frac{x^2y + y}{2x^2 + 11x + 5} \div \frac{xy^2}{6x^2 - 17x - 10}$$

$$= \frac{x^2 + 5x}{3x^2 - 4x - 20} \cdot \frac{x^2y + y}{2x^2 + 11x + 5} \cdot \frac{6x^2 - 17x - 10}{xy^2}$$

$$= \frac{x(x+5)}{(3x-10)(x+2)} \cdot \frac{y(x^2+1)}{(2x+1)(x+5)} \cdot \frac{(2x+1)(3x-10)}{xy^2}$$

$$= \frac{x(x+5)(y)(x^2+1)(2x+1)(3x-10)}{(3x-10)(x+2)(2x+1)(x+5)(x)(y^2)} = \frac{x^2+1}{y(x+2)}$$

Concept Quiz 9.2

For Problems 1–10, answer true or false.

1. To multiply two rational numbers in fraction form, we need to change to equivalent fractions with a common denominator.

2. When multiplying rational expressions that contain polynomials, the polynomials are factored so that common factors can be divided out.

3. In the division problem $\dfrac{2x^2y}{3z} \div \dfrac{4x^3}{5y^2}$, the fraction $\dfrac{4x^3}{5y^2}$ is the divisor.

4. The numbers $-\dfrac{2}{3}$ and $\dfrac{3}{2}$ are multiplicative inverses.

5. To divide two numbers in fraction form, we invert the divisor and multiply.

6. If $x \neq 0$, then $\left(\dfrac{4xy}{x}\right)\left(\dfrac{3y}{2x}\right) = \dfrac{6y^2}{x}$.

7. $\dfrac{3}{4} \div \dfrac{4}{3} = 1$.

8. If $x \neq 0$ and $y \neq 0$, then $\dfrac{5x^2y}{2y} \div \dfrac{10x^2}{3y} = \dfrac{3}{4}$.

9. If $x \neq 0$ and $y \neq 0$, then $\dfrac{1}{x} \div \dfrac{1}{y} = xy$.

10. If $x \neq y$, then $\dfrac{1}{x-y} \div \dfrac{1}{y-x} = -1$.

Problem Set 9.2

For Problems 1–12, perform the indicated operations involving rational numbers. Express final answers in reduced form. (Objectives 1 and 3)

1. $\dfrac{7}{12} \cdot \dfrac{6}{35}$

2. $\dfrac{5}{8} \cdot \dfrac{12}{20}$

3. $\dfrac{-4}{9} \cdot \dfrac{18}{30}$

4. $\dfrac{-6}{9} \cdot \dfrac{36}{48}$

5. $\dfrac{3}{-8} \cdot \dfrac{-6}{12}$

6. $\dfrac{-12}{16} \cdot \dfrac{18}{-32}$

7. $\left(-\dfrac{5}{7}\right) \div \dfrac{6}{7}$

8. $\left(-\dfrac{5}{9}\right) \div \dfrac{10}{3}$

9. $\dfrac{-9}{5} \div \dfrac{27}{10}$

10. $\dfrac{4}{7} \div \dfrac{16}{-21}$

11. $\dfrac{4}{9} \cdot \dfrac{6}{11} \div \dfrac{4}{15}$

12. $\dfrac{2}{3} \cdot \dfrac{6}{7} \div \dfrac{8}{3}$

For Problems 13–50, perform the indicated operations involving rational expressions. Express final answers in simplest form. (Objectives 2, 4, and 5)

13. $\dfrac{6xy}{9y^4} \cdot \dfrac{30x^3y}{-48x}$

14. $\dfrac{-14xy^4}{18y^2} \cdot \dfrac{24x^2y^3}{35y^2}$

15. $\dfrac{5a^2b^2}{11ab} \cdot \dfrac{22a^3}{15ab^2}$

16. $\dfrac{10a^2}{5b^2} \cdot \dfrac{15b^3}{2a^4}$

17. $\dfrac{5xy}{8y^2} \cdot \dfrac{18x^2y}{15}$

18. $\dfrac{4x^2}{5y^2} \cdot \dfrac{15xy}{24x^2y^2}$

19. $\dfrac{5x^4}{12x^2y^3} \div \dfrac{9}{5xy}$

20. $\dfrac{7x^2y}{9xy^3} \div \dfrac{3x^4}{2x^2y^2}$

21. $\dfrac{9a^2c}{12bc^2} \div \dfrac{21ab}{14c^3}$

22. $\dfrac{3ab^3}{4c} \div \dfrac{21ac}{12bc^3}$

23. $\dfrac{9x^2y^3}{14x} \cdot \dfrac{21y}{15xy^2} \cdot \dfrac{10x}{12y^3}$

24. $\dfrac{5xy}{7a} \cdot \dfrac{14a^2}{15x} \cdot \dfrac{3a}{8y}$

25. $\dfrac{3x+6}{5y} \cdot \dfrac{x^2+4}{x^2+10x+16}$

26. $\dfrac{5xy}{x+6} \cdot \dfrac{x^2-36}{x^2-6x}$

27. $\dfrac{5a^2+20a}{a^3-2a^2} \cdot \dfrac{a^2-a-12}{a^2-16}$

28. $\dfrac{2a^2+6}{a^2-a} \cdot \dfrac{a^3-a^2}{8a-4}$

29. $\dfrac{3n^2+15n-18}{3n^2+10n-48} \cdot \dfrac{6n^2-n-40}{4n^2+6n-10}$

30. $\dfrac{6n^2+11n-10}{3n^2+19n-14} \cdot \dfrac{2n^2+6n-56}{2n^2-3n-20}$

31. $\dfrac{9y^2}{x^2+12x+36} \div \dfrac{12y}{x^2+6x}$

32. $\dfrac{7xy}{x^2-4x+4} \div \dfrac{14y}{x^2-4}$

33. $\dfrac{x^2-4xy+4y^2}{7xy^2} \div \dfrac{4x^2-3xy-10y^2}{20x^2y+25xy^2}$

34. $\dfrac{x^2+5xy-6y^2}{xy^2-y^3} \cdot \dfrac{2x^2+15xy+18y^2}{xy+4y^2}$

35. $\dfrac{5-14n-3n^2}{1-2n-3n^2} \cdot \dfrac{9+7n-2n^2}{27-15n+2n^2}$

36. $\dfrac{6-n-2n^2}{12-11n+2n^2} \cdot \dfrac{24-26n+5n^2}{2+3n+n^2}$

37. $\dfrac{3x^4+2x^2-1}{3x^4+14x^2-5} \cdot \dfrac{x^4-2x^2-35}{x^4-17x^2+70}$

38. $\dfrac{2x^4+x^2-3}{2x^4+5x^2+2} \cdot \dfrac{3x^4+10x^2+8}{3x^4+x^2-4}$

39. $\dfrac{3x^2-20x+25}{2x^2-7x-15} \div \dfrac{9x^2-3x-20}{12x^2+28x+15}$

40. $\dfrac{21t^2+t-2}{2t^2-17t-9} \div \dfrac{12t^2-5t-3}{8t^2-2t-3}$

41. $\dfrac{10t^3+25t}{20t+10} \cdot \dfrac{2t^2-t-1}{t^5-t}$

42. $\dfrac{t^4-81}{t^2-6t+9} \cdot \dfrac{6t^2-11t-21}{5t^2+8t-21}$

43. $\dfrac{4t^2+t-5}{t^3-t^2} \cdot \dfrac{t^4+6t^3}{16t^2+40t+25}$

44. $\dfrac{9n^2-12n+4}{n^2-4n-32} \cdot \dfrac{n^2+4n}{3n^3-2n^2}$

45. $\dfrac{nr+3n+2r+6}{nr+3n-3r-9} \cdot \dfrac{n^2-9}{n^3-4n}$

46. $\dfrac{xy+xc+ay+ac}{xy-2xc+ay-2ac} \cdot \dfrac{2x^3-8x}{12x^3+20x^2-8x}$

47. $\dfrac{x^2-x}{4y} \cdot \dfrac{10xy^2}{2x-2} \div \dfrac{3x^2+3x}{15x^2y^2}$

48. $\dfrac{4xy^2}{7x} \cdot \dfrac{14x^3y}{12y} \div \dfrac{7y}{9x^3}$

49. $\dfrac{a^2-4ab+4b^2}{6a^2-4ab} \cdot \dfrac{3a^2+5ab-2b^2}{6a^2+ab-b^2} \div \dfrac{a^2-4b^2}{8a+4b}$

50. $\dfrac{2x^2+3x}{2x^3-10x^2} \cdot \dfrac{x^2-8x+15}{3x^3-27x} \div \dfrac{14x+21}{x^2-6x-27}$

Thoughts Into Words

51. Explain in your own words how to divide two rational expressions.

52. Suppose that your friend missed class the day the material in this section was discussed. How could you draw on her background in arithmetic to explain to her how to multiply and divide rational expressions?

53. Give a step-by-step description of how to do the following multiplication problem.

$$\dfrac{x^2+5x+6}{x^2-2x-8} \cdot \dfrac{x^2-16}{16-x^2}$$

Graphing Calculator Activities

54. Use your graphing calculator to check Examples 3–7.

55. Use your graphing calculator to check your answers for Problems 27–34.

56. Use your graphing calculator to graph $y = \dfrac{1}{x}$ again.

Then predict the graphs of $y = \dfrac{4}{x}$, $y = \dfrac{2}{x}$, and $y = \dfrac{8}{x}$. Finally, using a graphing calculator, graph all four equations on the same set of axes to check your predictions.

57. Draw rough sketches of the graphs of $y = \dfrac{1}{x^2}$, $y = \dfrac{4}{x^2}$, and $y = \dfrac{6}{x^2}$. Then check your sketches by using your graphing calculator to graph all three equations on the same set of axes.

58. Use your graphing calculator to graph $y = \dfrac{1}{x^2 + 1}$.

Now predict the graphs of $y = \dfrac{3}{x^2 + 1}$, $y = \dfrac{5}{x^2 + 1}$, and $y = \dfrac{8}{x^2 + 1}$. Finally, check your predictions by using your graphing calculator to graph all four equations on the same set of axes.

Answers to the Concept Quiz

1. False **2.** True **3.** True **4.** False **5.** True **6.** True **7.** False **8.** False **9.** False **10.** True

---

9.3 Adding and Subtracting Rational Expressions

OBJECTIVES

1 Add and subtract rational numbers

2 Add and subtract rational expressions

We can define addition and subtraction of rational numbers as follows:

> **Definition 9.3 Addition and Subtraction of Fractions**
>
> If a, b, and c are integers, and b is not zero, then
>
> $$\frac{a}{b} + \frac{c}{b} = \frac{a + c}{b} \qquad \text{Addition}$$
>
> $$\frac{a}{b} - \frac{c}{b} = \frac{a - c}{b} \qquad \text{Subtraction}$$

We can add or subtract rational numbers with a common denominator by adding or subtracting the numerators and placing the result over the common denominator. The following examples illustrate Definition 9.3.

$$\frac{2}{9} + \frac{3}{9} = \frac{2 + 3}{9} = \frac{5}{9}$$

$$\frac{7}{8} - \frac{3}{8} = \frac{7 - 3}{8} = \frac{4}{8} = \frac{1}{2} \qquad \text{Don't forget to reduce!}$$

$$\frac{4}{6} + \frac{-5}{6} = \frac{4 + (-5)}{6} = \frac{-1}{6} = -\frac{1}{6}$$

$$\frac{7}{10} + \frac{4}{-10} = \frac{7}{10} + \frac{-4}{10} = \frac{7 + (-4)}{10} = \frac{3}{10}$$

We use this same *common denominator* approach when adding or subtracting rational expressions, as in these next examples.

$$\frac{3}{x} + \frac{9}{x} = \frac{3+9}{x} = \frac{12}{x}$$

$$\frac{8}{x-2} - \frac{3}{x-2} = \frac{8-3}{x-2} = \frac{5}{x-2}$$

$$\frac{9}{4y} + \frac{5}{4y} = \frac{9+5}{4y} = \frac{14}{4y} = \frac{7}{2y} \qquad \text{Don't forget to simplify the final answer!}$$

$$\frac{n^2}{n-1} - \frac{1}{n-1} = \frac{n^2-1}{n-1} = \frac{(n+1)(n-1)}{n-1} = n+1$$

$$\frac{6a^2}{2a+1} + \frac{13a+5}{2a+1} = \frac{6a^2+13a+5}{2a+1} = \frac{(2a+1)(3a+5)}{2a+1} = 3a+5$$

In each of the previous examples that involve rational expressions, we should technically restrict the variables to exclude division by zero. For example, $\dfrac{3}{x} + \dfrac{9}{x} = \dfrac{12}{x}$ is true for all real number values for x, except $x = 0$. Likewise, $\dfrac{8}{x-2} - \dfrac{3}{x-2} = \dfrac{5}{x-2}$ as long as x does not equal 2. Rather than taking the time and space to write down restrictions for each problem, we will merely assume that such restrictions exist.

If rational numbers that do not have a common denominator are to be added or subtracted, then we apply the fundamental principle of fractions $\left(\dfrac{a}{b} = \dfrac{ak}{bk}\right)$ *to obtain equivalent fractions with a common denominator.* Equivalent fractions are fractions such as $\dfrac{1}{2}$ and $\dfrac{2}{4}$ that name the same number. Consider the following example.

$$\frac{1}{2} + \frac{1}{3} = \frac{3}{6} + \frac{2}{6} = \frac{3+2}{6} = \frac{5}{6}$$

$$\left(\begin{array}{c} \frac{1}{2} \text{ and } \frac{3}{6} \\ \text{are equivalent} \\ \text{fractions} \end{array}\right) \quad \left(\begin{array}{c} \frac{1}{3} \text{ and } \frac{2}{6} \\ \text{are equivalent} \\ \text{fractions} \end{array}\right)$$

Note that we chose 6 as our common denominator, and 6 is the *least common multiple* of the original denominators 2 and 3. The least common multiple of a set of whole numbers is the smallest nonzero whole number divisible by each of the numbers. In general, we use the least common multiple of the denominators of the fractions to be added or subtracted as a *least common denominator* (LCD).

A least common denominator may be found by inspection or by using the prime-factored forms of the numbers. Let's consider some examples and use each of these techniques.

Classroom Example
Subtract $\dfrac{7}{9} - \dfrac{1}{6}$.

EXAMPLE 1   Subtract $\dfrac{5}{6} - \dfrac{3}{8}$.

Solution

By inspection, we can see that the LCD is 24. Thus both fractions can be changed to equivalent fractions, each with a denominator of 24.

$$\frac{5}{6} - \frac{3}{8} = \left(\frac{5}{6}\right)\left(\frac{4}{4}\right) - \left(\frac{3}{8}\right)\left(\frac{3}{3}\right) = \frac{20}{24} - \frac{9}{24} = \frac{11}{24}$$

Form of 1   Form of 1

In Example 1, note that the fundamental principle of fractions, $\dfrac{a}{b} = \dfrac{a \cdot k}{b \cdot k}$, can be written as $\dfrac{a}{b} = \left(\dfrac{a}{b}\right)\left(\dfrac{k}{k}\right)$. This latter form emphasizes the fact that 1 is the multiplication identity element.

Classroom Example
Perform the indicated operations:
$$\dfrac{1}{4} + \dfrac{3}{7} - \dfrac{5}{28}$$

EXAMPLE 2

Perform the indicated operations: $\dfrac{3}{5} + \dfrac{1}{6} - \dfrac{13}{15}$.

Solution

Again by inspection, we can determine that the LCD is 30. Thus we can proceed as follows:

$$\frac{3}{5} + \frac{1}{6} - \frac{13}{15} = \left(\frac{3}{5}\right)\left(\frac{6}{6}\right) + \left(\frac{1}{6}\right)\left(\frac{5}{5}\right) - \left(\frac{13}{15}\right)\left(\frac{2}{2}\right)$$

$$= \frac{18}{30} + \frac{5}{30} - \frac{26}{30} = \frac{18 + 5 - 26}{30}$$

$$= \frac{-3}{30} = -\frac{1}{10} \qquad \text{Don't forget to reduce!}$$

Classroom Example
Add $\dfrac{4}{9} + \dfrac{7}{15}$.

EXAMPLE 3

Add $\dfrac{7}{18} + \dfrac{11}{24}$.

Solution

Let's use the prime-factored forms of the denominators to help find the LCD.

$$18 = 2 \cdot 3 \cdot 3 \qquad 24 = 2 \cdot 2 \cdot 2 \cdot 3$$

The LCD must contain three factors of 2 because 24 contains three 2s. The LCD must also contain two factors of 3 because 18 has two 3s. Thus the LCD $= 2 \cdot 2 \cdot 2 \cdot 3 \cdot 3 = 72$. Now we can proceed as usual.

$$\frac{7}{18} + \frac{11}{24} = \left(\frac{7}{18}\right)\left(\frac{4}{4}\right) + \left(\frac{11}{24}\right)\left(\frac{3}{3}\right) = \frac{28}{72} + \frac{33}{72} = \frac{61}{72}$$

To add and subtract rational expressions with different denominators, follow the same basic routine that you follow when you add or subtract rational numbers with different denominators. Study the following examples carefully and note the similarity to our previous work with rational numbers.

Classroom Example
Add $\dfrac{2x + 3}{5} + \dfrac{x + 4}{2}$.

EXAMPLE 4

Add $\dfrac{x + 2}{4} + \dfrac{3x + 1}{3}$.

Solution

By inspection, we see that the LCD is 12.

$$\frac{x + 2}{4} + \frac{3x + 1}{3} = \left(\frac{x + 2}{4}\right)\left(\frac{3}{3}\right) + \left(\frac{3x + 1}{3}\right)\left(\frac{4}{4}\right)$$

$$= \frac{3(x + 2)}{12} + \frac{4(3x + 1)}{12}$$

$$= \frac{3(x + 2) + 4(3x + 1)}{12}$$

$$= \frac{3x + 6 + 12x + 4}{12}$$

$$= \frac{15x + 10}{12}$$

Note the final result in Example 4. The numerator, $15x + 10$, could be factored as $5(3x + 2)$. However, because this produces no common factors with the denominator, the fraction cannot be simplified. Thus the final answer can be left as $\dfrac{15x + 10}{12}$. It would also be acceptable to express it as $\dfrac{5(3x + 2)}{12}$.

Classroom Example
Subtract $\dfrac{x - 3}{3} - \dfrac{x + 12}{12}$.

EXAMPLE 5 Subtract $\dfrac{a - 2}{2} - \dfrac{a - 6}{6}$.

Solution

By inspection, we see that the LCD is 6.

$$\frac{a - 2}{2} - \frac{a - 6}{6} = \left(\frac{a - 2}{2}\right)\left(\frac{3}{3}\right) - \frac{a - 6}{6}$$

$$= \frac{3(a - 2)}{6} - \frac{a - 6}{6}$$

$$= \frac{3(a - 2) - (a - 6)}{6} \qquad \text{Be careful with this sign as you move to the next step!}$$

$$= \frac{3a - 6 - a + 6}{6}$$

$$= \frac{2a}{6} = \frac{a}{3} \qquad \text{Don't forget to simplify}$$

Classroom Example
Perform the indicated operations:
$\dfrac{x + 2}{12} - \dfrac{x - 4}{6} + \dfrac{3x - 5}{20}$

EXAMPLE 6 Perform the indicated operations: $\dfrac{x + 3}{10} + \dfrac{2x + 1}{15} - \dfrac{x - 2}{18}$.

Solution

If you cannot determine the LCD by inspection, then use the prime-factored forms of the denominators.

$$10 = 2 \cdot 5 \qquad 15 = 3 \cdot 5 \qquad 18 = 2 \cdot 3 \cdot 3$$

The LCD must contain one factor of 2, two factors of 3, and one factor of 5. Thus the LCD is $2 \cdot 3 \cdot 3 \cdot 5 = 90$.

$$\frac{x + 3}{10} + \frac{2x + 1}{15} - \frac{x - 2}{18} = \left(\frac{x + 3}{10}\right)\left(\frac{9}{9}\right) + \left(\frac{2x + 1}{15}\right)\left(\frac{6}{6}\right) - \left(\frac{x - 2}{18}\right)\left(\frac{5}{5}\right)$$

$$= \frac{9(x + 3)}{90} + \frac{6(2x + 1)}{90} - \frac{5(x - 2)}{90}$$

$$= \frac{9(x + 3) + 6(2x + 1) - 5(x - 2)}{90}$$

$$= \frac{9x + 27 + 12x + 6 - 5x + 10}{90}$$

$$= \frac{16x + 43}{90}$$

A denominator that contains variables does not create any serious difficulties; our approach remains basically the same.

EXAMPLE 7   Add $\dfrac{3}{2x} + \dfrac{5}{3y}$.

Solution

Using an LCD of $6xy$, we can proceed as follows:

$$\frac{3}{2x} + \frac{5}{3y} = \left(\frac{3}{2x}\right)\left(\frac{3y}{3y}\right) + \left(\frac{5}{3y}\right)\left(\frac{2x}{2x}\right)$$

$$= \frac{9y}{6xy} + \frac{10x}{6xy}$$

$$= \frac{9y + 10x}{6xy}$$

EXAMPLE 8   Subtract $\dfrac{7}{12ab} - \dfrac{11}{15a^2}$.

Solution

We can prime factor the numerical coefficients of the denominators to help find the LCD.

$$\left.\begin{array}{l} 12ab = 2 \cdot 2 \cdot 3 \cdot a \cdot b \\ 15a^2 = 3 \cdot 5 \cdot a^2 \end{array}\right\} \longrightarrow \quad \text{LCD} = 2 \cdot 2 \cdot 3 \cdot 5 \cdot a^2 \cdot b = 60a^2b$$

$$\frac{7}{12ab} - \frac{11}{15a^2} = \left(\frac{7}{12ab}\right)\left(\frac{5a}{5a}\right) - \left(\frac{11}{15a^2}\right)\left(\frac{4b}{4b}\right)$$

$$= \frac{35a}{60a^2b} - \frac{44b}{60a^2b}$$

$$= \frac{35a - 44b}{60a^2b}$$

EXAMPLE 9   Add $\dfrac{x}{x-3} + \dfrac{4}{x}$.

Solution

By inspection, the LCD is $x(x - 3)$.

$$\frac{x}{x-3} + \frac{4}{x} = \left(\frac{x}{x-3}\right)\left(\frac{x}{x}\right) + \left(\frac{4}{x}\right)\left(\frac{x-3}{x-3}\right)$$

$$= \frac{x^2}{x(x-3)} + \frac{4(x-3)}{x(x-3)}$$

$$= \frac{x^2 + 4(x-3)}{x(x-3)}$$

$$= \frac{x^2 + 4x - 12}{x(x-3)} \quad \text{or} \quad \frac{(x+6)(x-2)}{x(x-3)}$$

EXAMPLE 10   Subtract $\dfrac{2x}{x+1} - 3$.

Solution

$$\frac{2x}{x+1} - 3 = \frac{2x}{x+1} - 3\left(\frac{x+1}{x+1}\right)$$

$$= \frac{2x}{x + 1} - \frac{3(x + 1)}{x + 1}$$

$$= \frac{2x - 3(x + 1)}{x + 1}$$

$$= \frac{2x - 3x - 3}{x + 1}$$

$$= \frac{-x - 3}{x + 1}$$

Concept Quiz 9.3

For Problems 1–10, answer true or false.

1. The addition problem $\dfrac{2x}{x + 4} + \dfrac{1}{x + 4}$ is equal to $\dfrac{2x + 1}{x + 4}$ for all values of x except $x = -\dfrac{1}{2}$ and $x = -4$.

2. Any common denominator can be used to add rational expressions, but typically we use the least common denominator.

3. The fractions $\dfrac{2x^2}{3y}$ and $\dfrac{10x^2z}{15yz}$ are equivalent fractions.

4. The least common multiple of the denominators is always the lowest common denominator.

5. To simplify the expression $\dfrac{5}{2x - 1} + \dfrac{3}{1 - 2x}$, we could use $2x - 1$ for the common denominator.

6. If $x \neq \dfrac{1}{2}$, then $\dfrac{5}{2x - 1} + \dfrac{3}{1 - 2x} = \dfrac{2}{2x - 1}$.

7. $\dfrac{3}{-4} - \dfrac{-2}{3} = \dfrac{17}{12}$

8. $\dfrac{4x - 1}{5} + \dfrac{2x + 1}{6} = \dfrac{x}{5}$

9. $\dfrac{x}{4} - \dfrac{3x}{2} + \dfrac{5x}{3} = \dfrac{5x}{12}$

10. If $x \neq 0$, then $\dfrac{2}{3x} - \dfrac{3}{2x} - 1 = \dfrac{-5 - 6x}{6x}$.

Problem Set 9.3

For Problems 1–12, perform the indicated operations involving rational numbers. Be sure to express your answers in reduced form. **(Objective 1)**

1. $\dfrac{1}{4} + \dfrac{5}{6}$

2. $\dfrac{3}{5} + \dfrac{1}{6}$

3. $\dfrac{7}{8} - \dfrac{3}{5}$

4. $\dfrac{7}{9} - \dfrac{1}{6}$

5. $\dfrac{6}{5} + \dfrac{1}{-4}$

6. $\dfrac{7}{8} + \dfrac{5}{-12}$

7. $\dfrac{8}{15} + \dfrac{3}{25}$

8. $\dfrac{5}{9} - \dfrac{11}{12}$

9. $\dfrac{1}{5} + \dfrac{5}{6} - \dfrac{7}{15}$

10. $\dfrac{2}{3} - \dfrac{7}{8} + \dfrac{1}{4}$

11. $\dfrac{1}{3} - \dfrac{1}{4} - \dfrac{3}{14}$

12. $\dfrac{5}{6} - \dfrac{7}{9} - \dfrac{3}{10}$

For Problems 13–66, add or subtract the rational expressions as indicated. Be sure to express your answers in simplest form. **(Objective 2)**

13. $\dfrac{2x}{x - 1} + \dfrac{4}{x - 1}$

14. $\dfrac{3x}{2x + 1} - \dfrac{5}{2x + 1}$

15. $\dfrac{4a}{a + 2} + \dfrac{8}{a + 2}$

16. $\dfrac{6a}{a - 3} - \dfrac{18}{a - 3}$

17. $\dfrac{3(y-2)}{7y} + \dfrac{4(y-1)}{7y}$

18. $\dfrac{2x-1}{4x^2} + \dfrac{3(x-2)}{4x^2}$

19. $\dfrac{x-1}{2} + \dfrac{x+3}{3}$

20. $\dfrac{x-2}{4} + \dfrac{x+6}{5}$

21. $\dfrac{2a-1}{4} + \dfrac{3a+2}{6}$

22. $\dfrac{a-4}{6} + \dfrac{4a-1}{8}$

23. $\dfrac{n+2}{6} - \dfrac{n-4}{9}$

24. $\dfrac{2n+1}{9} - \dfrac{n+3}{12}$

25. $\dfrac{3x-1}{3} - \dfrac{5x+2}{5}$

26. $\dfrac{4x-3}{6} - \dfrac{8x-2}{12}$

27. $\dfrac{x-2}{5} - \dfrac{x+3}{6} + \dfrac{x+1}{15}$

28. $\dfrac{x+1}{4} + \dfrac{x-3}{6} - \dfrac{x-2}{8}$

29. $\dfrac{3}{8x} + \dfrac{7}{10x}$

30. $\dfrac{5}{6x} - \dfrac{3}{10x}$

31. $\dfrac{5}{7x} - \dfrac{11}{4y}$

32. $\dfrac{5}{12x} - \dfrac{9}{8y}$

33. $\dfrac{4}{3x} + \dfrac{5}{4y} - 1$

34. $\dfrac{7}{3x} - \dfrac{8}{7y} - 2$

35. $\dfrac{7}{10x^2} + \dfrac{11}{15x}$

36. $\dfrac{7}{12a^2} - \dfrac{5}{16a}$

37. $\dfrac{10}{7n} - \dfrac{12}{4n^2}$

38. $\dfrac{6}{8n^2} - \dfrac{3}{5n}$

39. $\dfrac{3}{n^2} - \dfrac{2}{5n} + \dfrac{4}{3}$

40. $\dfrac{1}{n^2} + \dfrac{3}{4n} - \dfrac{5}{6}$

41. $\dfrac{3}{x} - \dfrac{5}{3x^2} - \dfrac{7}{6x}$

42. $\dfrac{7}{3x^2} - \dfrac{9}{4x} - \dfrac{5}{2x}$

43. $\dfrac{6}{5t^2} - \dfrac{4}{7t^3} + \dfrac{9}{5t^3}$

44. $\dfrac{5}{7t} + \dfrac{3}{4t^2} + \dfrac{1}{14t}$

45. $\dfrac{5b}{24a^2} - \dfrac{11a}{32b}$

46. $\dfrac{9}{14x^2y} - \dfrac{4x}{7y^2}$

47. $\dfrac{7}{9xy^3} - \dfrac{4}{3x} + \dfrac{5}{2y^2}$

48. $\dfrac{7}{16a^2b} + \dfrac{3a}{20b^2}$

49. $\dfrac{2x}{x-1} + \dfrac{3}{x}$

50. $\dfrac{3x}{x-4} - \dfrac{2}{x}$

51. $\dfrac{a-2}{a} - \dfrac{3}{a+4}$

52. $\dfrac{a+1}{a} - \dfrac{2}{a+1}$

53. $\dfrac{-3}{4n+5} - \dfrac{8}{3n+5}$

54. $\dfrac{-2}{n-6} - \dfrac{6}{2n+3}$

55. $\dfrac{-1}{x+4} + \dfrac{4}{7x-1}$

56. $\dfrac{-3}{4x+3} + \dfrac{5}{2x-5}$

57. $\dfrac{7}{3x-5} - \dfrac{5}{2x+7}$

58. $\dfrac{5}{x-1} - \dfrac{3}{2x-3}$

59. $\dfrac{5}{3x-2} + \dfrac{6}{4x+5}$

60. $\dfrac{3}{2x+1} + \dfrac{2}{3x+4}$

61. $\dfrac{3x}{2x+5} + 1$

62. $2 + \dfrac{4x}{3x-1}$

63. $\dfrac{4x}{x-5} - 3$

64. $\dfrac{7x}{x+4} - 2$

65. $-1 - \dfrac{3}{2x+1}$

66. $-2 - \dfrac{5}{4x-3}$

67. Recall that the indicated quotient of a polynomial and its opposite is -1. For example, $\dfrac{x-2}{2-x}$ simplifies to -1. Keep this idea in mind as you add or subtract the following rational expressions.

(a) $\dfrac{1}{x-1} - \dfrac{x}{x-1}$ **(b)** $\dfrac{3}{2x-3} - \dfrac{2x}{2x-3}$

(c) $\dfrac{4}{x-4} - \dfrac{x}{x-4} + 1$ **(d)** $-1 + \dfrac{2}{x-2} - \dfrac{x}{x-2}$

68. Consider the addition problem $\dfrac{8}{x-2} + \dfrac{5}{2-x}$. Note that the denominators are opposites of each other. If the property $\dfrac{a}{-b} = -\dfrac{a}{b}$ is applied to the second fraction, we have $\dfrac{5}{2-x} = -\dfrac{5}{x-2}$. Thus we proceed as follows:

$$\dfrac{8}{x-2} + \dfrac{5}{2-x} = \dfrac{8}{x-2} - \dfrac{5}{x-2} = \dfrac{8-5}{x-2} = \dfrac{3}{x-2}$$

Use this approach to do the following problems.

(a) $\dfrac{7}{x-1} + \dfrac{2}{1-x}$ **(b)** $\dfrac{5}{2x-1} + \dfrac{8}{1-2x}$

(c) $\dfrac{4}{a-3} - \dfrac{1}{3-a}$ **(d)** $\dfrac{10}{a-9} - \dfrac{5}{9-a}$

(e) $\dfrac{x^2}{x-1} - \dfrac{2x-3}{1-x}$ **(f)** $\dfrac{x^2}{x-4} - \dfrac{3x-28}{4-x}$

Thoughts Into Words

69. What is the difference between the concept of least common multiple and the concept of least common denominator?

70. A classmate tells you that she finds the least common multiple of two counting numbers by listing the multiples of each number and then choosing the smallest number that appears in both lists. Is this a correct procedure? What is the weakness of this procedure?

71. For which of the real numbers does $\dfrac{x}{x-3} + \dfrac{4}{x}$ equal $\dfrac{(x+6)(x-2)}{x(x-3)}$? Explain your answer.

72. Suppose that your friend does an addition problem as follows:

$$\frac{5}{8} + \frac{7}{12} = \frac{5(12) + 8(7)}{8(12)} = \frac{60 + 56}{96} = \frac{116}{96} = \frac{29}{24}$$

Is this answer correct? If not, what advice would you offer your friend?

Graphing Calculator Activities

73. Use your graphing calculator to check your answers for Problems 19–28.

74. There is another way to use the graphing calculator to check some real numbers in two one-variable algebraic expressions that we claim are equal. In Example 4 we claim that

$$\frac{x+2}{4} + \frac{3x+1}{3} = \frac{15x+10}{12}$$

for all real numbers. Let's check this claim for $x = 3$, $x = 7$, and $x = -5$. Enter $Y_1 = \dfrac{x+2}{4} + \dfrac{3x+1}{3}$ and

$Y_2 = \dfrac{15x+10}{12}$ as though we were going to graph them (Figure 9.1).

Figure 9.1

Return to the home screen and store $x = 3$ and evaluate Y_1 and Y_2; then store $x = 7$ and evaluate Y_1 and Y_2; finally, store $x = -5$ and evaluate Y_1 and Y_2. Part of this procedure is shown in Figure 9.2.

```
3→X
                           3
Y₁
               4.583333333
Y₂
               4.583333333
7→X
                           7
Y₁
               9.583333333
Y₂
               9.583333333
```

Figure 9.2

Use this technique to check at least three values of x for Problems 57–66.

75. Once again, let's start with the graph of $y = \dfrac{1}{x}$. Now draw rough sketches of $y = \dfrac{1}{x-2}$, $y = \dfrac{1}{x-4}$, and $y = \dfrac{1}{x+2}$. Finally, use your graphing calculator to graph all four equations on the same set of axes.

76. Use your graphing calculator to graph $y = \dfrac{1}{x^3}$. Then predict the graphs of $y = \dfrac{1}{(x-1)^3}$, $y = \dfrac{1}{(x-4)^3}$, and $y = \dfrac{1}{(x+2)^3}$. Finally, use your graphing calculator to graph all four equations on the same set of axes.

77. Use your graphing calculator to obtain the graph of $y = \dfrac{1}{x^2 + 1}$. Then predict the graphs of $y = \dfrac{1}{(x-2)^2 + 1}$, $y = \dfrac{1}{(x-4)^2 + 1}$, and $y = \dfrac{1}{(x+3)^2 + 1}$. Finally, use your graphing calculator to check your predictions.

9.4 More on Rational Expressions and Complex Fractions

OBJECTIVES
1. Add and subtract rational expressions
2. Simplify complex fractions

In this section, we expand our work with adding and subtracting rational expressions, and we discuss the process of simplifying complex fractions. Before we begin, however, this seems like an appropriate time to offer a bit of advice regarding your study of algebra. Success in algebra depends on having a good understanding of the concepts and being able to perform the various computations. As for the computational work, you should adopt a carefully organized format that shows as many steps as you need in order to minimize the chances of making careless errors. Don't be eager to find shortcuts for certain computations before you have a thorough understanding of the steps involved in the process. This advice is especially appropriate at the beginning of this section.

Study Examples 1–4 very carefully. Note that the same basic procedure is followed in solving each problem:

Step 1 Factor the denominators.
Step 2 Find the LCD.
Step 3 Change each fraction to an equivalent fraction that has the LCD as its denominator.
Step 4 Combine the numerators and place over the LCD.
Step 5 Simplify by performing the addition or subtraction.
Step 6 Look for ways to reduce the resulting fraction.

Classroom Example
Add $\dfrac{8}{a^2 - 2a} + \dfrac{4}{a}$.

EXAMPLE 1   Add $\dfrac{8}{x^2 - 4x} + \dfrac{2}{x}$.

Solution

$$\frac{8}{x^2 - 4x} + \frac{2}{x} = \frac{8}{x(x-4)} + \frac{2}{x}$$ Factor the denominators

The LCD is $x(x-4)$. Find the LCD

$$= \frac{8}{x(x-4)} + \left(\frac{2}{x}\right)\left(\frac{x-4}{x-4}\right)$$ Change each fraction to an equivalent fraction that has the LCD as its denominator

$$= \frac{8 + 2(x-4)}{x(x-4)}$$ Combine the numerators and place over the LCD

$$= \frac{8 + 2x - 8}{x(x-4)}$$ Simplify by performing the addition or subtraction

$$= \frac{2x}{x(x-4)}$$

$$= \frac{2}{x-4}$$ Reduce

Classroom Example
Subtract $\dfrac{x}{x^2-9} - \dfrac{7}{x+3}$.

EXAMPLE 2   Subtract $\dfrac{a}{a^2-4} - \dfrac{3}{a+2}$.

Solution

$$\frac{a}{a^2-4} - \frac{3}{a+2} = \frac{a}{(a+2)(a-2)} - \frac{3}{a+2} \qquad \text{Factor the denominators}$$

The LCD is $(a+2)(a-2)$. $\qquad\qquad\qquad$ Find the LCD

$$= \frac{a}{(a+2)(a-2)} - \left(\frac{3}{a+2}\right)\left(\frac{a-2}{a-2}\right) \qquad \begin{array}{l}\text{Change each fraction to an equivalent fraction}\\\text{that has the LCD as its denominator}\end{array}$$

$$= \frac{a-3(a-2)}{(a+2)(a-2)} \qquad\qquad\qquad \text{Combine numerators and place over the LCD}$$

$$= \frac{a-3a+6}{(a+2)(a-2)} \qquad\qquad\qquad \text{Simplify by performing the addition or subtraction}$$

$$= \frac{-2a+6}{(a+2)(a-2)} \quad \text{or} \quad \frac{-2(a-3)}{(a+2)(a-2)}$$

Classroom Example
Add:
$$\frac{2x}{x^2+5x+6} + \frac{5}{x^2-5x-14}$$

EXAMPLE 3   Add $\dfrac{3n}{n^2+6n+5} + \dfrac{4}{n^2-7n-8}$.

Solution

$$\frac{3n}{n^2+6n+5} + \frac{4}{n^2-7n-8}$$

$$= \frac{3n}{(n+5)(n+1)} + \frac{4}{(n-8)(n+1)} \qquad \text{Factor the denominators}$$

The LCD is $(n+5)(n+1)(n-8)$. $\qquad$ Find the LCD

$$= \left(\frac{3n}{(n+5)(n+1)}\right)\left(\frac{n-8}{n-8}\right)$$

$$+ \left(\frac{4}{(n-8)(n+1)}\right)\left(\frac{n+5}{n+5}\right) \qquad \begin{array}{l}\text{Change each fraction to an equivalent fraction}\\\text{that has the LCD as its denominator}\end{array}$$

$$= \frac{3n(n-8) + 4(n+5)}{(n+5)(n+1)(n-8)} \qquad \text{Combine numerators and place over the LCD}$$

$$= \frac{3n^2 - 24n + 4n + 20}{(n+5)(n+1)(n-8)} \qquad \text{Simplify by performing the addition or subtraction}$$

$$= \frac{3n^2 - 20n + 20}{(n+5)(n+1)(n-8)}$$

Classroom Example
Perform the indicated operations:
$$\frac{4x^2}{x^4-16} + \frac{x}{x^2-4} - \frac{1}{x-2}$$

EXAMPLE 4   Perform the indicated operations:
$$\frac{2x^2}{x^4-1} + \frac{x}{x^2-1} - \frac{1}{x-1}$$

Solution

$$\frac{2x^2}{x^4-1} + \frac{x}{x^2-1} - \frac{1}{x-1}$$

$$= \frac{2x^2}{(x^2+1)(x+1)(x-1)} + \frac{x}{(x+1)(x-1)} - \frac{1}{x-1} \qquad \text{Factor the denominators}$$

The LCD is $(x^2+1)(x+1)(x-1)$. $\qquad\qquad\qquad$ Find the LCD

$$= \frac{2x^2}{(x^2 + 1)(x + 1)(x - 1)}$$

$$+ \left(\frac{x}{(x + 1)(x - 1)}\right)\left(\frac{x^2 + 1}{x^2 + 1}\right)$$

$$- \left(\frac{1}{x - 1}\right)\frac{(x^2 + 1)(x + 1)}{(x^2 + 1)(x + 1)}$$

Change each fraction to an equivalent fraction that has the LCD as its denominator

$$= \frac{2x^2 + x(x^2 + 1) - (x^2 + 1)(x + 1)}{(x^2 + 1)(x + 1)(x - 1)}$$

Combine numerators and place over the LCD

$$= \frac{2x^2 + x^3 + x - x^3 - x^2 - x - 1}{(x^2 + 1)(x + 1)(x - 1)}$$

Simplify by performing the addition or subtraction

$$= \frac{x^2 - 1}{(x^2 + 1)(x + 1)(x - 1)}$$

$$= \frac{(x + 1)(x - 1)}{(x^2 + 1)(x + 1)(x - 1)}$$

$$= \frac{1}{x^2 + 1}$$

Reduce

Simplifying Complex Fractions

Complex fractions are fractional forms that contain rational numbers or rational expressions in the numerators and/or denominators. The following are examples of complex fractions.

$$\frac{\dfrac{4}{x}}{\dfrac{2}{xy}} \qquad \frac{\dfrac{1}{2} + \dfrac{3}{4}}{\dfrac{5}{6} - \dfrac{3}{8}} \qquad \frac{\dfrac{3}{x} + \dfrac{2}{y}}{\dfrac{5}{x} - \dfrac{6}{y^2}} \qquad \frac{\dfrac{1}{x} + \dfrac{1}{y}}{2} \qquad \frac{-3}{\dfrac{2}{x} - \dfrac{3}{y}}$$

It is often necessary to **simplify** a complex fraction. We will take each of these five examples and examine some techniques for simplifying complex fractions.

Classroom Example

Simplify $\dfrac{\dfrac{6}{m}}{\dfrac{3}{m^2 n}}$.

EXAMPLE 5

Simplify $\dfrac{\dfrac{4}{x}}{\dfrac{2}{xy}}$.

Solution

This type of problem is a simple division problem.

$$\frac{\dfrac{4}{x}}{\dfrac{2}{xy}} = \frac{4}{x} \div \frac{2}{xy}$$

$$= \frac{\overset{2}{\cancel{4}}}{\cancel{x}} \cdot \frac{\cancel{xy}}{\cancel{2}} = 2y$$

Classroom Example

Simplify $\dfrac{\dfrac{3}{4} - \dfrac{1}{3}}{\dfrac{5}{6} + \dfrac{2}{9}}$.

EXAMPLE 6

Simplify $\dfrac{\dfrac{1}{2} + \dfrac{3}{4}}{\dfrac{5}{6} - \dfrac{3}{8}}$.

Let's look at two possible ways to simplify such a problem.

Solution A

Here we will simplify the numerator by performing the addition and simplify the denominator by performing the subtraction. Then the problem is a simple division problem as in Example 5.

$$\dfrac{\dfrac{1}{2} + \dfrac{3}{4}}{\dfrac{5}{6} - \dfrac{3}{8}} = \dfrac{\dfrac{2}{4} + \dfrac{3}{4}}{\dfrac{20}{24} - \dfrac{9}{24}}$$

$$= \dfrac{\dfrac{5}{4}}{\dfrac{11}{24}} = \dfrac{5}{\cancel{4}} \cdot \dfrac{\overset{6}{\cancel{24}}}{11}$$

$$= \dfrac{30}{11}$$

Solution B

Here we find the LCD of all four denominators (2, 4, 6, and 8). The LCD is 24. Use this LCD to multiply the entire complex fraction by a form of 1, specifically $\dfrac{24}{24}$.

$$\dfrac{\dfrac{1}{2} + \dfrac{3}{4}}{\dfrac{5}{6} - \dfrac{3}{8}} = \left(\dfrac{24}{24}\right)\left(\dfrac{\dfrac{1}{2} + \dfrac{3}{4}}{\dfrac{5}{6} - \dfrac{3}{8}}\right)$$

$$= \dfrac{24\left(\dfrac{1}{2} + \dfrac{3}{4}\right)}{24\left(\dfrac{5}{6} - \dfrac{3}{8}\right)}$$

$$= \dfrac{24\left(\dfrac{1}{2}\right) + 24\left(\dfrac{3}{4}\right)}{24\left(\dfrac{5}{6}\right) - 24\left(\dfrac{3}{8}\right)}$$

$$= \dfrac{12 + 18}{20 - 9} = \dfrac{30}{11}$$

Classroom Example

Simplify $\dfrac{\dfrac{1}{x} + \dfrac{3}{y}}{\dfrac{4}{x} - \dfrac{2}{y^2}}$.

EXAMPLE 7

Simplify $\dfrac{\dfrac{3}{x} + \dfrac{2}{y}}{\dfrac{5}{x} - \dfrac{6}{y^2}}$.

Solution A

Simplify the numerator and the denominator. Then the problem becomes a division problem.

$$\frac{\dfrac{3}{x} + \dfrac{2}{y}}{\dfrac{5}{x} - \dfrac{6}{y^2}} = \frac{\left(\dfrac{3}{x}\right)\left(\dfrac{y}{y}\right) + \left(\dfrac{2}{y}\right)\left(\dfrac{x}{x}\right)}{\left(\dfrac{5}{x}\right)\left(\dfrac{y^2}{y^2}\right) - \left(\dfrac{6}{y^2}\right)\left(\dfrac{x}{x}\right)}$$

$$= \frac{\dfrac{3y}{xy} + \dfrac{2x}{xy}}{\dfrac{5y^2}{xy^2} - \dfrac{6x}{xy^2}}$$

$$= \frac{\dfrac{3y + 2x}{xy}}{\dfrac{5y^2 - 6x}{xy^2}}$$

$$= \frac{3y + 2x}{xy} \div \frac{5y^2 - 6x}{xy^2}$$

$$= \frac{3y + 2x}{\cancel{xy}} \cdot \frac{\overset{y}{\cancel{xy^2}}}{5y^2 - 6x}$$

$$= \frac{y(3y + 2x)}{5y^2 - 6x}$$

Solution B

Here we find the LCD of all four denominators (x, y, x, and y^2). The LCD is xy^2. Use this LCD to multiply the entire complex fraction by a form of 1, specifically $\dfrac{xy^2}{xy^2}$.

$$\frac{\dfrac{3}{x} + \dfrac{2}{y}}{\dfrac{5}{x} - \dfrac{6}{y^2}} = \left(\frac{xy^2}{xy^2}\right)\left(\frac{\dfrac{3}{x} + \dfrac{2}{y}}{\dfrac{5}{x} - \dfrac{6}{y^2}}\right)$$

$$= \frac{xy^2\left(\dfrac{3}{x} + \dfrac{2}{y}\right)}{xy^2\left(\dfrac{5}{x} - \dfrac{6}{y^2}\right)}$$

$$= \frac{xy^2\left(\dfrac{3}{x}\right) + xy^2\left(\dfrac{2}{y}\right)}{xy^2\left(\dfrac{5}{x}\right) - xy^2\left(\dfrac{6}{y^2}\right)}$$

$$= \frac{3y^2 + 2xy}{5y^2 - 6x} \quad \text{or} \quad \frac{y(3y + 2x)}{5y^2 - 6x}$$

Certainly either approach (Solution A or Solution B) will work with problems such as Examples 6 and 7. Examine Solution B in both examples carefully. This approach works effectively with complex fractions where the LCD of all the denominators is easy to find. (Don't be misled by the length of Solution B for Example 6; we were especially careful to show every step.)

Classroom Example
Simplify $\dfrac{\dfrac{1}{m} - \dfrac{1}{n}}{3}$.

EXAMPLE 8   Simplify $\dfrac{\dfrac{1}{x} + \dfrac{1}{y}}{2}$.

Solution

The number 2 can be written as $\dfrac{2}{1}$; thus the LCD of all three denominators (x, y, and 1) is xy.

Therefore, let's multiply the entire complex fraction by a form of 1, specifically $\dfrac{xy}{xy}$.

$$\left(\dfrac{\dfrac{1}{x} + \dfrac{1}{y}}{\dfrac{2}{1}}\right)\left(\dfrac{xy}{xy}\right) = \dfrac{xy\left(\dfrac{1}{x}\right) + xy\left(\dfrac{1}{y}\right)}{2xy}$$

$$= \dfrac{y + x}{2xy}$$

Classroom Example
Simplify $\dfrac{-5}{\dfrac{4}{x} - \dfrac{8}{y}}$.

EXAMPLE 9   Simplify $\dfrac{-3}{\dfrac{2}{x} - \dfrac{3}{y}}$.

Solution

$$\left(\dfrac{\dfrac{-3}{1}}{\dfrac{2}{x} - \dfrac{3}{y}}\right)\left(\dfrac{xy}{xy}\right) = \dfrac{-3(xy)}{xy\left(\dfrac{2}{x}\right) - xy\left(\dfrac{3}{y}\right)}$$

$$= \dfrac{-3xy}{2y - 3x}$$

Let's conclude this section with an example that has a complex fraction as part of an algebraic expression.

Classroom Example
Simplify $1 + \dfrac{x}{1 + \dfrac{1}{x}}$.

EXAMPLE 10   Simplify $1 - \dfrac{n}{1 - \dfrac{1}{n}}$.

Solution

First simplify the complex fraction $\dfrac{n}{1 - \dfrac{1}{n}}$ by multiplying by $\dfrac{n}{n}$.

$$\left(\dfrac{n}{1 - \dfrac{1}{n}}\right)\left(\dfrac{n}{n}\right) = \dfrac{n^2}{n - 1}$$

Now we can perform the subtraction.

$$1 - \dfrac{n^2}{n - 1} = \left(\dfrac{n - 1}{n - 1}\right)\left(\dfrac{1}{1}\right) - \dfrac{n^2}{n - 1}$$

$$= \dfrac{n - 1}{n - 1} - \dfrac{n^2}{n - 1}$$

$$= \dfrac{n - 1 - n^2}{n - 1} \quad \text{or} \quad \dfrac{-n^2 + n - 1}{n - 1}$$

Concept Quiz 9.4

For Problems 1–7, answer true or false.

1. A complex fraction can be described as a fraction within a fraction.

2. Division can simplify the complex fraction $\dfrac{\dfrac{2y}{x}}{\dfrac{6}{x^2}}$.

3. The complex fraction $\dfrac{\dfrac{3}{x-2}+\dfrac{2}{x+2}}{\dfrac{7x}{(x+2)(x-2)}}$ simplifies to $\dfrac{5x+2}{7x}$ for all values of x except $x = 0$.

4. The complex fraction $\dfrac{\dfrac{1}{3}-\dfrac{5}{6}}{\dfrac{1}{6}+\dfrac{5}{9}}$ simplifies to $-\dfrac{9}{13}$.

5. One method for simplifying a complex fraction is to multiply the entire fraction by a form of 1.

6. The complex fraction $\dfrac{\dfrac{3}{4}-\dfrac{1}{2}}{\dfrac{2}{3}}$ simplifies to $\dfrac{3}{8}$.

7. The complex fraction $\dfrac{\dfrac{7}{8}-\dfrac{1}{18}}{\dfrac{5}{6}+\dfrac{4}{15}}$ simplifies to $\dfrac{59}{33}$.

8. Arrange in order the following steps for adding rational expressions:
  A. Combine numerators and place over the LCD.
  B. Find the LCD.
  C. Reduce.
  D. Factor the denominators.
  E. Simplify by performing addition or subtraction.
  F. Change each fraction to an equivalent fraction that has the LCD as its denominator.

Problem Set 9.4

For Problems 1–40, perform the indicated operations, and express your answers in simplest form. **(Objective 1)**

1. $\dfrac{2x}{x^2+4x}+\dfrac{5}{x}$

2. $\dfrac{3x}{x^2-6x}+\dfrac{4}{x}$

3. $\dfrac{4}{x^2+7x}-\dfrac{1}{x}$

4. $\dfrac{-10}{x^2-9x}-\dfrac{2}{x}$

5. $\dfrac{x}{x^2-1}+\dfrac{5}{x+1}$

6. $\dfrac{2x}{x^2-16}+\dfrac{7}{x-4}$

7. $\dfrac{6a+4}{a^2-1}-\dfrac{5}{a-1}$

8. $\dfrac{4a-4}{a^2-4}-\dfrac{3}{a+2}$

9. $\dfrac{2n}{n^2-25}-\dfrac{3}{4n+20}$

10. $\dfrac{3n}{n^2-36}-\dfrac{2}{5n+30}$

11. $\dfrac{5}{x}-\dfrac{5x-30}{x^2+6x}+\dfrac{x}{x+6}$

12. $\dfrac{3}{x+1}+\dfrac{x+5}{x^2-1}-\dfrac{3}{x-1}$

13. $\dfrac{3}{x^2+9x+14}+\dfrac{5}{2x^2+15x+7}$

14. $\dfrac{6}{x^2+11x+24}+\dfrac{4}{3x^2+13x+12}$

15. $\dfrac{1}{a^2 - 3a - 10} - \dfrac{4}{a^2 + 4a - 45}$

16. $\dfrac{6}{a^2 - 3a - 54} - \dfrac{10}{a^2 + 5a - 6}$

17. $\dfrac{3a}{8a^2 - 2a - 3} + \dfrac{1}{4a^2 + 13a - 12}$

18. $\dfrac{2a}{6a^2 + 13a - 5} + \dfrac{a}{2a^2 + a - 10}$

19. $\dfrac{5}{x^2 + 3} - \dfrac{2}{x^2 + 4x - 21}$

20. $\dfrac{7}{x^2 + 1} - \dfrac{3}{x^2 + 7x - 60}$

21. $\dfrac{3x}{x^2 - 6x + 9} - \dfrac{2}{x - 3}$

22. $\dfrac{3}{x + 4} + \dfrac{2x}{x^2 + 8x + 16}$

23. $\dfrac{5}{x^2 - 1} + \dfrac{9}{x^2 + 2x + 1}$ **24.** $\dfrac{6}{x^2 - 9} - \dfrac{9}{x^2 - 6x + 9}$

25. $\dfrac{2}{y^2 + 6y - 16} - \dfrac{4}{y + 8} - \dfrac{3}{y - 2}$

26. $\dfrac{7}{y - 6} - \dfrac{10}{y + 12} + \dfrac{4}{y^2 + 6y - 72}$

27. $x - \dfrac{x^2}{x - 2} + \dfrac{3}{x^2 - 4}$

28. $x + \dfrac{5}{x^2 - 25} - \dfrac{x^2}{x + 5}$

29. $\dfrac{x + 3}{x + 10} + \dfrac{4x - 3}{x^2 + 8x - 20} + \dfrac{x - 1}{x - 2}$

30. $\dfrac{2x - 1}{x + 3} + \dfrac{x + 4}{x - 6} + \dfrac{3x - 1}{x^2 - 3x - 18}$

31. $\dfrac{n}{n - 6} + \dfrac{n + 3}{n + 8} + \dfrac{12n + 26}{n^2 + 2n - 48}$

32. $\dfrac{n - 1}{n + 4} + \dfrac{n}{n + 6} + \dfrac{2n + 18}{n^2 + 10n + 24}$

33. $\dfrac{4x - 3}{2x^2 + x - 1} - \dfrac{2x + 7}{3x^2 + x - 2} - \dfrac{3}{3x - 2}$

34. $\dfrac{2x + 5}{x^2 + 3x - 18} - \dfrac{3x - 1}{x^2 + 4x - 12} + \dfrac{5}{x - 2}$

35. $\dfrac{n}{n^2 + 1} + \dfrac{n^2 + 3n}{n^4 - 1} - \dfrac{1}{n - 1}$

36. $\dfrac{2n^2}{n^4 - 16} - \dfrac{n}{n^2 - 4} + \dfrac{1}{n + 2}$

37. $\dfrac{15x^2 - 10}{5x^2 - 7x + 2} - \dfrac{3x + 4}{x - 1} - \dfrac{2}{5x - 2}$

38. $\dfrac{32x + 9}{12x^2 + x - 6} - \dfrac{3}{4x + 3} - \dfrac{x + 5}{3x - 2}$

39. $\dfrac{t + 3}{3t - 1} + \dfrac{8t^2 + 8t + 2}{3t^2 - 7t + 2} - \dfrac{2t + 3}{t - 2}$

40. $\dfrac{t - 3}{2t + 1} + \dfrac{2t^2 + 19t - 46}{2t^2 - 9t - 5} - \dfrac{t + 4}{t - 5}$

For Problems 41–64, simplify each complex fraction. **(Objective 2)**

41. $\dfrac{\dfrac{1}{2} - \dfrac{1}{4}}{\dfrac{5}{8} + \dfrac{3}{4}}$ **42.** $\dfrac{\dfrac{3}{8} + \dfrac{3}{4}}{\dfrac{5}{8} - \dfrac{7}{12}}$

43. $\dfrac{\dfrac{3}{28} - \dfrac{5}{14}}{\dfrac{5}{7} + \dfrac{1}{4}}$ **44.** $\dfrac{\dfrac{5}{9} + \dfrac{7}{36}}{\dfrac{3}{18} - \dfrac{5}{12}}$

45. $\dfrac{\dfrac{5}{6y}}{\dfrac{10}{3xy}}$ **46.** $\dfrac{\dfrac{9}{8xy^2}}{\dfrac{5}{4x^2}}$

47. $\dfrac{\dfrac{3}{x} - \dfrac{2}{y}}{\dfrac{4}{y} - \dfrac{7}{xy}}$ **48.** $\dfrac{\dfrac{9}{x} + \dfrac{7}{x^2}}{\dfrac{5}{y} + \dfrac{3}{y^2}}$

49. $\dfrac{\dfrac{6}{a} - \dfrac{5}{b^2}}{\dfrac{12}{a^2} + \dfrac{2}{b}}$ **50.** $\dfrac{\dfrac{4}{ab} - \dfrac{3}{b^2}}{\dfrac{1}{a} + \dfrac{3}{b}}$

51. $\dfrac{\dfrac{2}{x} - 3}{\dfrac{3}{y} + 4}$ **52.** $\dfrac{1 + \dfrac{3}{x}}{1 - \dfrac{6}{x}}$

53. $\dfrac{3 + \dfrac{2}{n + 4}}{5 - \dfrac{1}{n + 4}}$ **54.** $\dfrac{4 + \dfrac{6}{n - 1}}{7 - \dfrac{4}{n - 1}}$

55. $\dfrac{5 - \dfrac{2}{n-3}}{4 - \dfrac{1}{n-3}}$

56. $\dfrac{\dfrac{3}{n-5} - 2}{1 - \dfrac{4}{n-5}}$

61. $\dfrac{\dfrac{3a}{2 - \dfrac{1}{a}} - 1}$

62. $\dfrac{\dfrac{a}{\dfrac{1}{a} + 4} + 1}$

57. $\dfrac{\dfrac{-1}{y-2} + \dfrac{5}{x}}{\dfrac{3}{x} - \dfrac{4}{xy - 2x}}$

58. $\dfrac{\dfrac{-2}{x} - \dfrac{4}{x+2}}{\dfrac{3}{x^2 + 2x} + \dfrac{3}{x}}$

63. $2 - \dfrac{x}{3 - \dfrac{2}{x}}$

64. $1 + \dfrac{x}{1 + \dfrac{1}{x}}$

59. $\dfrac{\dfrac{2}{x-3} - \dfrac{3}{x+3}}{\dfrac{5}{x^2 - 9} - \dfrac{2}{x-3}}$

60. $\dfrac{\dfrac{2}{x-y} + \dfrac{3}{x+y}}{\dfrac{5}{x+y} - \dfrac{1}{x^2 - y^2}}$

Thoughts Into Words

65. Which of the two techniques presented in the text would you use to simplify $\dfrac{\dfrac{1}{4} + \dfrac{1}{3}}{\dfrac{3}{4} - \dfrac{1}{6}}$? Which technique would you use to simplify $\dfrac{\dfrac{3}{8} - \dfrac{5}{7}}{\dfrac{7}{9} + \dfrac{6}{25}}$? Explain your choice for each problem.

66. Give a step-by-step description of how to do the following addition problem.

$$\dfrac{3x + 4}{8} + \dfrac{5x - 2}{12}$$

Graphing Calculator Activities

67. Before doing this problem, refer to Problem 74 of Problem Set 9.3. Now check your answers for Problems 13–22 of this problem set, using both a graphical approach and the approach described in Problem 74 of Problem Set 9.3.

68. Again, let's start with the graph of $y = \dfrac{1}{x}$. Draw some rough sketches of the graphs of the following equations: $y = -\dfrac{1}{x}, y = -\dfrac{3}{x},$ and $y = -\dfrac{5}{x}$. Then use your graphing calculator to graph all four equations on the same set of axes.

69. Let's start with the graph of $y = \dfrac{1}{x^2}$. Draw rough sketches of the graphs of $y = -\dfrac{1}{x^2}, y = -\dfrac{4}{x^2},$ and $y = -\dfrac{6}{x^2}$. Then use your graphing calculator to graph all four equations on the same set of axes.

70. Use your graphing calculator to graph $y = \dfrac{1}{(x-1)^2}$.

Then draw rough sketches of $y = -\dfrac{1}{(x-1)^2}$, $y = \dfrac{3}{(x-1)^2},$ and $y = -\dfrac{3}{(x-1)^2}$. Finally, use your graphing calculator to graph all four equations on the same set of axes.

Answers to the Concept Quiz

1. True   **2.** True   **3.** False   **4.** True   **5.** True   **6.** True   **7.** False   **8.** D, B, F, A, E, C

9.5   Equations Containing Rational Expressions

OBJECTIVES

1 Solve rational equations

2 Solve proportions

3 Solve word problems involving ratios

The fractional equations used in this text are of two basic types. One type has only constants as denominators, and the other type contains variables in the denominators.

In Chapter 3, we considered fractional equations that involve only constants in the denominators. Let's briefly review our approach to solving such equations, because we will be using that same basic technique to solve any type of fractional equation.

Classroom Example
Solve $\dfrac{x+5}{2} + \dfrac{x-3}{6} = \dfrac{2}{3}$.

EXAMPLE 1   Solve $\dfrac{x-2}{3} + \dfrac{x+1}{4} = \dfrac{1}{6}$.

Solution

$$\frac{x-2}{3} + \frac{x+1}{4} = \frac{1}{6}$$

$$12\left(\frac{x-2}{3} + \frac{x+1}{4}\right) = 12\left(\frac{1}{6}\right) \qquad \text{Multiply both sides by 12, which is the LCD of all of the denominators}$$

$$4(x-2) + 3(x+1) = 2$$

$$4x - 8 + 3x + 3 = 2$$

$$7x - 5 = 2$$

$$7x = 7$$

$$x = 1$$

The solution set is $\{1\}$. Check it!

If an equation contains a variable (or variables) in one or more denominators, then we proceed in essentially the same way as in Example 1 *except that we must avoid any value of the variable that makes a denominator zero.* Consider the following examples.

Classroom Example
Solve $\dfrac{3}{n} + \dfrac{1}{4} = \dfrac{5}{n}$.

EXAMPLE 2   Solve $\dfrac{5}{n} + \dfrac{1}{2} = \dfrac{9}{n}$.

Solution

First, we need to realize that n cannot equal zero. (Let's indicate this restriction so that it is not forgotten!) Then we can proceed.

$$\frac{5}{n} + \frac{1}{2} = \frac{9}{n}, \qquad n \neq 0$$

$$2n\left(\frac{5}{n} + \frac{1}{2}\right) = 2n\left(\frac{9}{n}\right) \qquad \text{Multiply both sides by the LCD, which is } 2n$$

$$10 + n = 18$$

$$n = 8$$

The solution set is $\{8\}$. Check it!

EXAMPLE 3

Solve $\dfrac{35 - x}{x} = 7 + \dfrac{3}{x}$.

Solution

$$\frac{35 - x}{x} = 7 + \frac{3}{x}, \qquad x \neq 0$$

$$x\left(\frac{35 - x}{x}\right) = x\left(7 + \frac{3}{x}\right) \qquad \text{Multiply both sides by } x$$

$$35 - x = 7x + 3$$

$$32 = 8x$$

$$4 = x$$

The solution set is $\{4\}$.

EXAMPLE 4

Solve $\dfrac{3}{a - 2} = \dfrac{4}{a + 1}$.

Solution

$$\frac{3}{a - 2} = \frac{4}{a + 1}, \qquad a \neq 2 \text{ and } a \neq -1$$

$$(a - 2)(a + 1)\left(\frac{3}{a - 2}\right) = (a - 2)(a + 1)\left(\frac{4}{a + 1}\right) \qquad \begin{array}{l}\text{Multiply both sides by}\\ (a - 2)(a + 1)\end{array}$$

$$3(a + 1) = 4(a - 2)$$

$$3a + 3 = 4a - 8$$

$$11 = a$$

The solution set is $\{11\}$.

Keep in mind that listing the restrictions at the beginning of a problem does not replace checking the potential solutions. In Example 4, the answer 11 needs to be checked in the original equation.

EXAMPLE 5

Solve $\dfrac{a}{a - 2} + \dfrac{2}{3} = \dfrac{2}{a - 2}$.

Solution

$$\frac{a}{a - 2} + \frac{2}{3} = \frac{2}{a - 2}, \qquad a \neq 2$$

$$3(a - 2)\left(\frac{a}{a - 2} + \frac{2}{3}\right) = 3(a - 2)\left(\frac{2}{a - 2}\right) \qquad \begin{array}{l}\text{Multiply both sides}\\ \text{by } 3(a - 2)\end{array}$$

$$3a + 2(a - 2) = 6$$

$$3a + 2a - 4 = 6$$

$$5a = 10$$

$$a = 2$$

Because our initial restriction was $a \neq 2$, we conclude that this equation has no solution. Thus the solution set is $\varnothing$.

Solving Proportions

A **ratio** is the comparison of two numbers by division. We often use the fractional form to express ratios. For example, we can write the ratio of a to b as $\dfrac{a}{b}$. A statement of equality between two ratios is called a **proportion**. Thus if $\dfrac{a}{b}$ and $\dfrac{c}{d}$ are two equal ratios, we can form the proportion $\dfrac{a}{b} = \dfrac{c}{d}$ ($b \neq 0$ and $d \neq 0$). We deduce an important property of proportions as follows:

$$\frac{a}{b} = \frac{c}{d}, \qquad b \neq 0 \text{ and } d \neq 0$$

$$bd\left(\frac{a}{b}\right) = bd\left(\frac{c}{d}\right) \qquad \text{Multiply both sides by } bd$$

$$ad = bc$$

Cross-Multiplication Property of Proportions

If $\dfrac{a}{b} = \dfrac{c}{d}$ ($b \neq 0$ and $d \neq 0$), then $ad = bc$.

We can treat some fractional equations as proportions and solve them by using the cross-multiplication idea, as in the next examples.

Classroom Example
Solve $\dfrac{4}{x+3} = \dfrac{9}{x-4}$.

EXAMPLE 6

Solve $\dfrac{5}{x+6} = \dfrac{7}{x-5}$.

Solution

$$\frac{5}{x+6} = \frac{7}{x-5}, \qquad x \neq -6 \text{ and } x \neq 5$$

$$5(x-5) = 7(x+6) \qquad \text{Apply the cross-multiplication property}$$

$$5x - 25 = 7x + 42$$

$$-67 = 2x$$

$$-\frac{67}{2} = x$$

The solution set is $\left\{-\dfrac{67}{2}\right\}$.

Classroom Example
Solve $\dfrac{x}{9} = \dfrac{3}{x-6}$.

EXAMPLE 7

Solve $\dfrac{x}{7} = \dfrac{4}{x+3}$.

Solution

$$\frac{x}{7} = \frac{4}{x+3}, \qquad x \neq -3$$

$$x(x+3) = 7(4) \qquad \text{Cross-multiplication property}$$

$$x^2 + 3x = 28$$

$$x^2 + 3x - 28 = 0$$

$$(x+7)(x-4) = 0$$

$$x + 7 = 0 \qquad \text{or} \qquad x - 4 = 0$$

$$x = -7 \qquad \text{or} \qquad x = 4$$

The solution set is $\{-7, 4\}$. Check these solutions in the original equation.

Solving Word Problems Involving Ratios

We can conveniently set up some problems and solve them using the concepts of ratio and proportion. Let's conclude this section with two such examples.

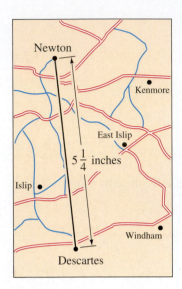

Figure 9.3

EXAMPLE 8

On a certain map, $1\frac{1}{2}$ inches represents 25 miles. If two cities are $5\frac{1}{4}$ inches apart on the map, find the number of miles between the cities (see Figure 9.3).

Solution

Let m represent the number of miles between the two cities. To set up the proportion, we will use a ratio of inches on the map to miles. Be sure to keep the ratio "inches on the map to miles" the same for both sides of the proportion.

$$\frac{1\frac{1}{2}}{25} = \frac{5\frac{1}{4}}{m}, \qquad m \neq 0$$

$$\frac{\frac{3}{2}}{25} = \frac{\frac{21}{4}}{m}$$

$$\frac{3}{2}m = 25\left(\frac{21}{4}\right) \qquad \text{Cross-multiplication property}$$

$$\frac{2}{3}\left(\frac{3}{2}m\right) = \frac{2}{3}(25)\left(\frac{\overset{7}{\cancel{21}}}{\underset{2}{\cancel{4}}}\right) \qquad \text{Multiply both sides by } \frac{2}{3}$$

$$m = \frac{175}{2} = 87\frac{1}{2}$$

The distance between the two cities is $87\frac{1}{2}$ miles.

EXAMPLE 9

A sum of $3500 is to be divided between two people in the ratio of 2 to 3. How much does each person receive?

Solution

Let d represent the amount of money that one person receives. Then $3500 - d$ represents the amount for the other person.

$$\frac{d}{3500 - d} = \frac{2}{3}, \qquad d \neq 3500$$

$$3d = 2(3500 - d)$$

$$3d = 7000 - 2d$$

$$5d = 7000$$

$$d = 1400$$

If $d = 1400$, then $3500 - d$ equals 2100. Therefore, one person receives $1400, and the other person receives $2100.

Concept Quiz 9.5

For Problems 1–3, answer true or false.

1. In solving rational equations, any value of the variable that makes a denominator zero cannot be a solution of the equation.

2. One method of solving rational equations is to multiply both sides of the equation by the lowest common denominator of the fractions in the equation.

3. In solving a rational equation that is a proportion, cross products can be set equal to each other.

For Problems 4–8, match each equation with its solution set.

Equations

4. $\dfrac{3}{x + 1} = \dfrac{3}{x - 1}$

5. $\dfrac{x}{5} = \dfrac{3x}{15}$

6. $\dfrac{2x + 1}{7} = \dfrac{3x}{7}$

7. $\dfrac{-x + 9}{x - 4} = \dfrac{5}{x - 4}$

8. $\dfrac{4}{x + 2} = \dfrac{4}{2x - 1}$

Solution Sets

A. {All real numbers}

B. $\varnothing$

C. {3}

D. {1}

9. Identify the following equations as a proportion or not a proportion.

(a) $\dfrac{2x}{x + 1} + x = \dfrac{7}{x + 1}$ (b) $\dfrac{x - 8}{2x + 5} = \dfrac{7}{9}$ (c) $5 + \dfrac{2x}{x + 6} = \dfrac{x - 3}{x + 4}$

10. Select all the equations that could represent the following problem: John bought three bottles of energy drink for $5.07. If the price remains the same, what will eight bottles of the energy drink cost?

(a) $\dfrac{3}{5.07} = \dfrac{x}{8}$ (b) $\dfrac{5.07}{8} = \dfrac{x}{3}$ (c) $\dfrac{3}{8} = \dfrac{5.07}{x}$ (d) $\dfrac{5.07}{3} = \dfrac{x}{8}$

Problem Set 9.5

For Problems 1–44, solve each equation. **(Objectives 1 and 2)**

1. $\dfrac{x + 1}{4} + \dfrac{x - 2}{6} = \dfrac{3}{4}$

2. $\dfrac{x + 2}{5} + \dfrac{x - 1}{6} = \dfrac{3}{5}$

3. $\dfrac{x + 3}{2} - \dfrac{x - 4}{7} = 1$

4. $\dfrac{x + 4}{3} - \dfrac{x - 5}{9} = 1$

5. $\dfrac{5}{n} + \dfrac{1}{3} = \dfrac{7}{n}$

6. $\dfrac{3}{n} + \dfrac{1}{6} = \dfrac{11}{3n}$

7. $\dfrac{7}{2x} + \dfrac{3}{5} = \dfrac{2}{3x}$

8. $\dfrac{9}{4x} + \dfrac{1}{3} = \dfrac{5}{2x}$

9. $\dfrac{3}{4x} + \dfrac{5}{6} = \dfrac{4}{3x}$

10. $\dfrac{5}{7x} - \dfrac{5}{6} = \dfrac{1}{6x}$

11. $\dfrac{47 - n}{n} = 8 + \dfrac{2}{n}$

12. $\dfrac{45 - n}{n} = 6 + \dfrac{3}{n}$

13. $\dfrac{n}{65 - n} = 8 + \dfrac{2}{65 - n}$

14. $\dfrac{n}{70 - n} = 7 + \dfrac{6}{70 - n}$

15. $n + \dfrac{1}{n} = \dfrac{17}{4}$

16. $n + \dfrac{1}{n} = \dfrac{37}{6}$

17. $n - \dfrac{2}{n} = \dfrac{23}{5}$

18. $n - \dfrac{3}{n} = \dfrac{26}{3}$

19. $\dfrac{5}{7x - 3} = \dfrac{3}{4x - 5}$

20. $\dfrac{3}{2x - 1} = \dfrac{5}{3x + 2}$

21. $\dfrac{-2}{x - 5} = \dfrac{1}{x + 9}$

22. $\dfrac{5}{2a - 1} = \dfrac{-6}{3a + 2}$

23. $\dfrac{x}{x + 1} - 2 = \dfrac{3}{x - 3}$

24. $\dfrac{x}{x - 2} + 1 = \dfrac{8}{x - 1}$

25. $\dfrac{a}{a + 5} - 2 = \dfrac{3a}{a + 5}$

26. $\dfrac{a}{a - 3} - \dfrac{3}{2} = \dfrac{3}{a - 3}$

27. $\dfrac{5}{x + 6} = \dfrac{6}{x - 3}$

28. $\dfrac{3}{x - 1} = \dfrac{4}{x + 2}$

29. $\dfrac{3x - 7}{10} = \dfrac{2}{x}$

30. $\dfrac{x}{-4} = \dfrac{3}{12x - 25}$

31. $\dfrac{x}{x - 6} - 3 = \dfrac{6}{x - 6}$

32. $\dfrac{x}{x + 1} + 3 = \dfrac{4}{x + 1}$

33. $\dfrac{3s}{s + 2} + 1 = \dfrac{35}{2(3s + 1)}$

34. $\dfrac{s}{2s - 1} - 3 = \dfrac{-32}{3(s + 5)}$

35. $2 - \dfrac{3x}{x-4} = \dfrac{14}{x+7}$ **36.** $-1 + \dfrac{2x}{x+3} = \dfrac{-4}{x+4}$

37. $\dfrac{n+6}{27} = \dfrac{1}{n}$ **38.** $\dfrac{n}{5} = \dfrac{10}{n-5}$

39. $\dfrac{3n}{n-1} - \dfrac{1}{3} = \dfrac{-40}{3n-18}$ **40.** $\dfrac{n}{n+1} + \dfrac{1}{2} = \dfrac{-2}{n+2}$

41. $\dfrac{-3}{4x+5} = \dfrac{2}{5x-7}$ **42.** $\dfrac{7}{x+4} = \dfrac{3}{x-8}$

43. $\dfrac{2x}{x-2} + \dfrac{15}{x^2 - 7x + 10} = \dfrac{3}{x-5}$

44. $\dfrac{x}{x-4} - \dfrac{2}{x+3} = \dfrac{20}{x^2 - x - 12}$

For Problems 45–56, set up an algebraic equation and solve each problem. **(Objective 3)**

45. A sum of $1750 is to be divided between two people in the ratio of 3 to 4. How much does each person receive?

46. A blueprint has a scale in which 1 inch represents 5 feet. Find the dimensions of a rectangular room that measures $3\dfrac{1}{2}$ inches by $5\dfrac{3}{4}$ inches on the blueprint.

47. One angle of a triangle has a measure of $60°$, and the measures of the other two angles are in the ratio of 2 to 3. Find the measures of the other two angles.

48. The ratio of the complement of an angle to its supplement is 1 to 4. Find the measure of the angle.

49. If a home valued at $150,000 is assessed $2500 in real estate taxes, then what are the taxes on a home valued at $210,000 if assessed at the same rate?

50. The ratio of male students to female students at a certain university is 5 to 7. If there is a total of 16,200 students, find the number of male students and the number of female students.

51. Suppose that, together, Laura and Tammy sold $120.75 worth of candy for the annual school fair. If the ratio of Tammy's sales to Laura's sales was 4 to 3, how much did each sell?

52. The total value of a house and a lot is $168,000. If the ratio of the value of the house to the value of the lot is 7 to 1, find the value of the house.

53. A 20-foot board is to be cut into two pieces whose lengths are in the ratio of 7 to 3. Find the lengths of the two pieces.

54. An inheritance of $300,000 is to be divided between a son and the local heart fund in the ratio of 3 to 1. How much money will the son receive?

55. Suppose that in a certain precinct, 1150 people voted in the last presidential election. If the ratio of female voters to male voters was 3 to 2, how many females and how many males voted?

56. The perimeter of a rectangle is 114 centimeters. If the ratio of its width to its length is 7 to 12, find the dimensions of the rectangle.

Thoughts Into Words

57. How could you do Problem 53 without using algebra?

58. How can you tell by inspection that the equation
$$\dfrac{x}{x+2} = \dfrac{-2}{x+2}$$ has no solution?

59. How would you help someone solve the equation
$$\dfrac{3}{x} - \dfrac{4}{x} = \dfrac{-1}{x}?$$

Graphing Calculator Activities

60. Use your graphing calculator and supply a partial check for Problems 25–34.

61. Use your graphing calculator to help solve each of the following equations. Be sure to check your answers.

a. $\dfrac{1}{6} + \dfrac{1}{x} = \dfrac{5}{18}$ **b.** $\dfrac{1}{50} + \dfrac{1}{40} = \dfrac{1}{x}$

c. $\dfrac{2050}{x+358} = \dfrac{260}{x}$ **d.** $\dfrac{280}{x} = \dfrac{300}{x+2} + 20$

e. $\dfrac{x}{2x-8} + \dfrac{16}{x^2 - 16} = \dfrac{1}{2}$

f. $\dfrac{3}{x-5} - \dfrac{2}{2x+1} = \dfrac{x+3}{2x^2 - 9x - 5}$

g. $2 + \dfrac{4}{x-2} = \dfrac{8}{x^2 - 2x}$

OBJECTIVES
1. Solve rational equations with denominators that require factoring
2. Solve formulas that involve fractional forms
3. Solve rate-time word problems

Let's begin this section by considering a few more fractional equations. We will continue to solve them using the same basic techniques as in the previous section. That is, we will multiply both sides of the equation by the least common denominator of all of the denominators in the equation, with the necessary restrictions to avoid division by zero. Some of the denominators in these problems will require factoring before we can determine a least common denominator.

Classroom Example
Solve $\dfrac{x}{3x + 9} + \dfrac{9}{x^2 - 9} = \dfrac{1}{3}$.

EXAMPLE 1 Solve $\dfrac{x}{2x - 8} + \dfrac{16}{x^2 - 16} = \dfrac{1}{2}$.

Solution

$$\frac{x}{2x - 8} + \frac{16}{x^2 - 16} = \frac{1}{2}$$

$$\frac{x}{2(x - 4)} + \frac{16}{(x + 4)(x - 4)} = \frac{1}{2}, \qquad x \neq 4 \text{ and } x \neq -4$$

$$2(x - 4)(x + 4)\left(\frac{x}{2(x - 4)} + \frac{16}{(x + 4)(x - 4)}\right) = 2(x + 4)(x - 4)\left(\frac{1}{2}\right) \qquad \begin{array}{l}\text{Multiply both}\\\text{sides by the LCD,}\\2(x - 4)\ (x + 4)\end{array}$$

$$x(x + 4) + 2(16) = (x + 4)(x - 4)$$

$$x^2 + 4x + 32 = x^2 - 16$$

$$4x = -48$$

$$x = -12$$

The solution set is $\{-12\}$. Perhaps you should check it!

In Example 1, note that the restrictions were not indicated until the denominators were expressed in factored form. It is usually easier to determine the necessary restrictions at this step.

Classroom Example
Solve
$$\frac{4}{x + 5} + \frac{3}{3x - 2} = \frac{x + 12}{3x^2 + 13x - 10}$$

EXAMPLE 2 Solve $\dfrac{3}{n - 5} - \dfrac{2}{2n + 1} = \dfrac{n + 3}{2n^2 - 9n - 5}$.

Solution

$$\frac{3}{n - 5} - \frac{2}{2n + 1} = \frac{n + 3}{2n^2 - 9n - 5}$$

$$\frac{3}{n - 5} - \frac{2}{2n + 1} = \frac{n + 3}{(2n + 1)(n - 5)}, \qquad n \neq -\frac{1}{2} \text{ and } n \neq 5$$

$$(2n + 1)(n - 5)\left(\frac{3}{n - 5} - \frac{2}{2n + 1}\right) = (2n + 1)(n - 5)\left(\frac{n + 3}{(2n + 1)(n - 5)}\right) \qquad \begin{array}{l}\text{Multiply}\\\text{both sides}\\\text{by the LCD,}\\(2n + 1) \cdot\\(n - 5)\end{array}$$

$$3(2n + 1) - 2(n - 5) = n + 3$$

$$6n + 3 - 2n + 10 = n + 3$$

$$4n + 13 = n + 3$$
$$3n = -10$$
$$n = -\frac{10}{3}$$

The solution set is $\left\{-\dfrac{10}{3}\right\}$.

Classroom Example
Solve $3 - \dfrac{9}{x + 3} = \dfrac{27}{x^2 + 3x}$.

EXAMPLE 3   Solve $2 + \dfrac{4}{x - 2} = \dfrac{8}{x^2 - 2x}$.

Solution

$$2 + \frac{4}{x - 2} = \frac{8}{x^2 - 2x}$$

$$2 + \frac{4}{x - 2} = \frac{8}{x(x - 2)}, \qquad x \neq 0 \text{ and } x \neq 2$$

$$x(x - 2)\left(2 + \frac{4}{x - 2}\right) = x(x - 2)\left(\frac{8}{x(x - 2)}\right) \qquad \text{Multiply both sides by the LCD, } x(x - 2)$$

$$2x(x - 2) + 4x = 8$$
$$2x^2 - 4x + 4x = 8$$
$$2x^2 = 8$$
$$x^2 = 4$$
$$x^2 - 4 = 0$$
$$(x + 2)(x - 2) = 0$$
$$x + 2 = 0 \qquad \text{or} \qquad x - 2 = 0$$
$$x = -2 \qquad \text{or} \qquad x = 2$$

Because our initial restriction indicated that $x \neq 2$, the only solution is -2. Thus the solution set is $\{-2\}$.

Solving Formulas That Involve Fractional Forms

In Section 4.3, we discussed using the properties of equality to change the form of various formulas. For example, we considered the simple interest formula $A = P + Prt$ and changed its form by solving for P as follows:

$$A = P + Prt$$
$$A = P(1 + rt)$$

$$\frac{A}{1 + rt} = P \qquad \text{Multiply both sides by } \frac{1}{1 + rt}$$

If the formula is in the form of a fractional equation, then the techniques of these last two sections are applicable. Consider the following example.

Classroom Example
Solve the future value formula for r:
$$A = P\left(1 + \frac{r}{n}\right)$$

EXAMPLE 4

If the original cost of some business property is C dollars, and it is depreciated linearly over N years, then its value V at the end of T years is given by

$$V = C\left(1 - \frac{T}{N}\right)$$

Solve this formula for N in terms of V, C, and T.

Solution

$$V = C\left(1 - \frac{T}{N}\right)$$

$$V = C - \frac{CT}{N}$$

$$N(V) = N\left(C - \frac{CT}{N}\right) \qquad \text{Multiply both sides by } N$$

$$NV = NC - CT$$

$$NV - NC = -CT$$

$$N(V - C) = -CT$$

$$N = \frac{-CT}{V - C}$$

$$N = -\frac{CT}{V - C}$$

Solving Rate-Time Word Problems

In Section 4.4 we solved some uniform motion problems. The formula $d = rt$ was used in the analysis of the problems, and we used guidelines that involve distance relationships. Now let's consider some uniform motion problems for which guidelines involving either times or rates are appropriate. These problems will generate fractional equations to solve.

Classroom Example
An airplane travels 2852 miles in the same time that a car travels 299 miles. If the rate of the plane is 555 miles per hour greater than the rate of the car, find the rate of each.

EXAMPLE 5

An airplane travels 2050 miles in the same time that a car travels 260 miles. If the rate of the plane is 358 miles per hour greater than the rate of the car, find the rate of each.

Solution

Let r represent the rate of the car. Then $r + 358$ represents the rate of the plane. The fact that the times are equal can be a guideline. Remember from the basic formula, $d = rt$, that $t = \dfrac{d}{r}$.

Time of plane   Equals   Time of car

$$\frac{\text{Distance of plane}}{\text{Rate of plane}} = \frac{\text{Distance of car}}{\text{Rate of car}}$$

$$\frac{2050}{r + 358} = \frac{260}{r}$$

$$2050r = 260(r + 358)$$

$$2050r = 260r + 93{,}080$$

$$1790r = 93{,}080$$

$$r = 52$$

If $r = 52$, then $r + 358$ equals 410. Thus the rate of the car is 52 miles per hour, and the rate of the plane is 410 miles per hour.

Classroom Example
It takes a freight train 1 hour longer to travel 180 miles than it takes an express train to travel 195 miles. The rate of the express train is 20 miles per hour greater than the rate of the freight train. Find the times and rates of both trains.

EXAMPLE 6

It takes a freight train 2 hours longer to travel 300 miles than it takes an express train to travel 280 miles. The rate of the express train is 20 miles per hour greater than the rate of the freight train. Find the times and rates of both trains.

Solution

Let t represent the time of the express train. Then $t + 2$ represents the time of the freight train. Let's record the information of this problem in a table.

	Distance	Time	Rate $= \dfrac{\text{distance}}{\text{time}}$
Express train	280	t	$\dfrac{280}{t}$
Freight train	300	$t + 2$	$\dfrac{300}{t + 2}$

The fact that the rate of the express train is 20 miles per hour greater than the rate of the freight train can be a guideline.

$$\underset{\text{Rate of express}}{\underbrace{\dfrac{280}{t}}} \quad \underset{\text{Equals}}{=} \quad \underset{\text{Rate of freight train plus 20}}{\underbrace{\dfrac{300}{t + 2} + 20}}$$

$$t(t + 2)\left(\dfrac{280}{t}\right) = t(t + 2)\left(\dfrac{300}{t + 2} + 20\right) \qquad \text{Multiply both sides by } t(t + 2)$$

$$280(t + 2) = 300t + 20t(t + 2)$$
$$280t + 560 = 300t + 20t^2 + 40t$$
$$280t + 560 = 340t + 20t^2$$
$$0 = 20t^2 + 60t - 560$$
$$0 = t^2 + 3t - 28$$
$$0 = (t + 7)(t - 4)$$
$$t + 7 = 0 \qquad \text{or} \qquad t - 4 = 0$$
$$t = -7 \qquad \text{or} \qquad t = 4$$

The negative solution must be discarded, so the time of the express train (t) is 4 hours, and the time of the freight train ($t + 2$) is 6 hours. The rate of the express train $\left(\dfrac{280}{t}\right)$ is $\dfrac{280}{4} = $ 70 miles per hour, and the rate of the freight train $\left(\dfrac{300}{t + 2}\right)$ is $\dfrac{300}{6} = 50$ miles per hour.

Remark: Note that to solve Example 5 we went directly to a guideline without the use of a table, but for Example 6 we used a table. Remember that this is a personal preference; we are merely acquainting you with a variety of techniques.

Uniform motion problems are a special case of a larger group of problems we refer to as **rate-time problems**. For example, if a certain machine can produce 150 items in 10 minutes, then we say that the machine is producing at a rate of $\dfrac{150}{10} = 15$ items per minute. Likewise, if a person can do a certain job in 3 hours, then, assuming a constant rate of work, we say that the person is working at a rate of $\frac{1}{3}$ of the job per hour. In general, if Q is the quantity of something done in t units of time, then the rate, r, is given by $r = \frac{Q}{t}$. We state the rate in terms of *so much quantity per unit of time.* (In uniform motion problems the "quantity" is distance.) Let's consider some examples of rate-time problems.

EXAMPLE 7

If Jim can mow a lawn in 50 minutes, and his son, Todd, can mow the same lawn in 40 minutes, how long will it take them to mow the lawn if they work together?

Solution

Jim's rate is $\frac{1}{50}$ of the lawn per minute, and Todd's rate is $\frac{1}{40}$ of the lawn per minute. If we let m represent the number of minutes that they work together, then $\frac{1}{m}$ represents their rate when working together. Therefore, because the sum of the individual rates must equal the rate working together, we can set up and solve the following equation.

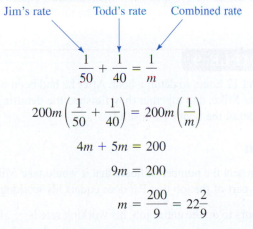

$$\frac{1}{50} + \frac{1}{40} = \frac{1}{m}$$

$$200m\left(\frac{1}{50} + \frac{1}{40}\right) = 200m\left(\frac{1}{m}\right)$$

$$4m + 5m = 200$$

$$9m = 200$$

$$m = \frac{200}{9} = 22\frac{2}{9}$$

It should take them $22\frac{2}{9}$ minutes.

EXAMPLE 8

Working together, Lucia and Kate can type a term paper in $3\frac{3}{5}$ hours. Lucia can type the paper by herself in 6 hours. How long would it take Kate to type the paper by herself?

Solution

Their rate working together is $\dfrac{1}{3\frac{3}{5}} = \dfrac{1}{\frac{18}{5}} = \dfrac{5}{18}$ of the job per hour, and Lucia's rate is $\frac{1}{6}$ of the job per hour. If we let h represent the number of hours that it would take Kate to do the job by herself, then her rate is $\frac{1}{h}$ of the job per hour. Thus we have

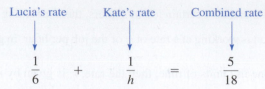

Solving this equation yields

$$18h\left(\frac{1}{6} + \frac{1}{h}\right) = 18h\left(\frac{5}{18}\right)$$

$$3h + 18 = 5h$$

$$18 = 2h$$

$$9 = h$$

It would take Kate 9 hours to type the paper by herself.

Our final example of this section illustrates another approach that some people find meaningful for rate-time problems. For this approach, think in terms of fractional parts of the job. For example, if a person can do a certain job in 5 hours, then at the end of 2 hours, he or she has done $\frac{2}{5}$ of the job. (Again, assume a constant rate of work.) At the end of 4 hours, he or she has finished $\frac{4}{5}$ of the job; and, in general, at the end of h hours, he or she has done $\frac{h}{5}$ of the job.

Just as for the motion problems in which distance equals rate times the time, here the fractional part done equals the working rate times the time. Let's see how this works in a problem.

Classroom Example
It takes Wayne 9 hours to tile a backsplash. After he had been working for 2 hours, he was joined by Greg, and together they finished the task in 4 hours. How long would it take Greg to do the job by himself?

EXAMPLE 9

It takes Pat 12 hours to detail a boat. After he had been working for 3 hours, he was joined by his brother Mike, and together they finished the detailing in 5 hours. How long would it take Mike to detail the boat by himself?

Solution

Let h represent the number of hours that it would take Mike to do the detailing by himself. The fractional part of the job that Pat does equals his working rate times his time. Because it takes Pat 12 hours to do the entire job, his working rate is $\frac{1}{12}$. He works for 8 hours (3 hours before Mike and then 5 hours with Mike). Therefore, Pat's part of the job is $\frac{1}{12}(8) = \frac{8}{12}$. The fractional part of the job that Mike does equals his working rate times his time. Because h represents Mike's time to do the entire job, his working rate is $\frac{1}{h}$; he works for 5 hours. Therefore, Mike's part of the job is $\frac{1}{h}(5) = \frac{5}{h}$. Adding the two fractional parts together results in 1 entire job being done. Let's also show this information in chart form and set up our guideline. Then we can set up and solve the equation.

	Time to do entire job	Working rate	Time working	Fractional part of the job done
Pat	12	$\frac{1}{12}$	8	$\frac{8}{12}$
Mike	h	$\frac{1}{h}$	5	$\frac{5}{h}$

Fractional part of
the job that Pat does

Fractional part of
the job that Mike does

$$\frac{8}{12} + \frac{5}{h} = 1$$

$$12h\left(\frac{8}{12} + \frac{5}{h}\right) = 12h(1)$$

$$12h\left(\frac{8}{12}\right) + 12h\left(\frac{5}{h}\right) = 12h$$

$$8h + 60 = 12h$$

$$60 = 4h$$

$$15 = h$$

It would take Mike 15 hours to detail the boat by himself.

Concept Quiz 9.6

For Problems 1–10, answer true or false.

1. Assuming uniform motion, the rate at which a car travels is equal to the time traveled divided by the distance traveled.

2. If a worker can lay 640 square feet of tile in 8 hours, we can say his rate of work is 80 square feet per hour.

3. If a person can complete two jobs in 5 hours, then the person is working at the rate of $\frac{5}{2}$ of the job per hour.

4. In a time-rate problem involving two workers, the sum of their individual rates must equal the rate working together.

5. If a person works at the rate of $\frac{2}{15}$ of the job per hour, then at the end of 3 hours the job would be $\frac{6}{15}$ completed.

6. If a person can do a job in 7 hours, then at the end of 5 hours he or she will have completed $\frac{5}{7}$ of the job.

7. If a person can do a job in h hours, then at the end of 3 hours he or she will have completed $\frac{h}{3}$ of the job.

8. The equation $A = P + Prt$ cannot be solved for P, because P occurs in two different terms.

9. If Zorka can complete a certain task in 5 hours, and Mitzie can complete the same task in 9 hours, then working together they should be able to complete the task in 7 hours.

10. Uniform motion problems are one type of rate-time problem.

Problem Set 9.6

For Problems 1–30, solve each equation. **(Objective 1)**

1. $\dfrac{x}{4x - 4} + \dfrac{5}{x^2 - 1} = \dfrac{1}{4}$

2. $\dfrac{x}{3x - 6} + \dfrac{4}{x^2 - 4} = \dfrac{1}{3}$

3. $3 + \dfrac{6}{t - 3} = \dfrac{6}{t^2 - 3t}$

4. $2 + \dfrac{4}{t - 1} = \dfrac{4}{t^2 - t}$

5. $\dfrac{3}{n - 5} + \dfrac{4}{n + 7} = \dfrac{2n + 11}{n^2 + 2n - 35}$

6. $\dfrac{2}{n + 3} + \dfrac{3}{n - 4} = \dfrac{2n - 1}{n^2 - n - 12}$

7. $\dfrac{5x}{2x + 6} - \dfrac{4}{x^2 - 9} = \dfrac{5}{2}$

8. $\dfrac{3x}{5x + 5} - \dfrac{2}{x^2 - 1} = \dfrac{3}{5}$

9. $1 + \dfrac{1}{n - 1} = \dfrac{1}{n^2 - n}$

10. $3 + \dfrac{9}{n - 3} = \dfrac{27}{n^2 - 3n}$

11. $\dfrac{2}{n-2} - \dfrac{n}{n+5} = \dfrac{10n+15}{n^2+3n-10}$

12. $\dfrac{n}{n+3} + \dfrac{1}{n-4} = \dfrac{11-n}{n^2-n-12}$

13. $\dfrac{2}{2x-3} - \dfrac{2}{10x^2-13x-3} = \dfrac{x}{5x+1}$

14. $\dfrac{1}{3x+4} + \dfrac{6}{6x^2+5x-4} = \dfrac{x}{2x-1}$

15. $\dfrac{2x}{x+3} - \dfrac{3}{x-6} = \dfrac{29}{x^2-3x-18}$

16. $\dfrac{x}{x-4} - \dfrac{2}{x+8} = \dfrac{63}{x^2+4x-32}$

17. $\dfrac{a}{a-5} + \dfrac{2}{a-6} = \dfrac{2}{a^2-11a+30}$

18. $\dfrac{a}{a+2} + \dfrac{3}{a+4} = \dfrac{14}{a^2+6a+8}$

19. $\dfrac{-1}{2x-5} + \dfrac{2x-4}{4x^2-25} = \dfrac{5}{6x+15}$

20. $\dfrac{-2}{3x+2} + \dfrac{x-1}{9x^2-4} = \dfrac{3}{12x-8}$

21. $\dfrac{7y+2}{12y^2+11y-15} - \dfrac{1}{3y+5} = \dfrac{2}{4y-3}$

22. $\dfrac{5y-4}{6y^2+y-12} - \dfrac{2}{2y+3} = \dfrac{5}{3y-4}$

23. $\dfrac{2n}{6n^2+7n-3} - \dfrac{n-3}{3n^2+11n-4} = \dfrac{5}{2n^2+11n+12}$

24. $\dfrac{x+1}{2x^2+7x-4} - \dfrac{x}{2x^2-7x+3} = \dfrac{1}{x^2+x-12}$

25. $\dfrac{1}{2x^2-x-1} + \dfrac{3}{2x^2+x} = \dfrac{2}{x^2-1}$

26. $\dfrac{2}{n^2+4n} + \dfrac{3}{n^2-3n-28} = \dfrac{5}{n^2-6n-7}$

27. $\dfrac{x+1}{x^3-9x} - \dfrac{1}{2x^2+x-21} = \dfrac{1}{2x^2+13x+21}$

28. $\dfrac{x}{2x^2+5x} - \dfrac{x}{2x^2+7x+5} = \dfrac{2}{x^2+x}$

29. $\dfrac{4t}{4t^2-t-3} + \dfrac{2-3t}{3t^2-t-2} = \dfrac{1}{12t^2+17t+6}$

30. $\dfrac{2t}{2t^2+9t+10} + \dfrac{1-3t}{3t^2+4t-4} = \dfrac{4}{6t^2+11t-10}$

For Problems 31–44, solve each equation for the indicated variable. **(Objective 2)**

31. $y = \dfrac{5}{6}x + \dfrac{2}{9}$ for x

32. $y = \dfrac{3}{4}x - \dfrac{2}{3}$ for x

33. $\dfrac{-2}{x-4} = \dfrac{5}{y-1}$ for y

34. $\dfrac{7}{y-3} = \dfrac{3}{x+1}$ for y

35. $I = \dfrac{100M}{C}$ for M

36. $V = C\left(1 - \dfrac{T}{N}\right)$ for T

37. $\dfrac{R}{S} = \dfrac{T}{S+T}$ for R

38. $\dfrac{1}{R} = \dfrac{1}{S} + \dfrac{1}{T}$ for R

39. $\dfrac{y-1}{x-3} = \dfrac{b-1}{a-3}$ for y

40. $y = -\dfrac{a}{b}x + \dfrac{c}{d}$ for x

41. $\dfrac{x}{a} + \dfrac{y}{b} = 1$ for y

42. $\dfrac{y-b}{x} = m$ for y

43. $\dfrac{y-1}{x+6} = \dfrac{-2}{3}$ for y

44. $\dfrac{y+5}{x-2} = \dfrac{3}{7}$ for y

Set up an equation and solve each of the following problems. **(Objective 3)**

45. Kent drives his Mazda 270 miles in the same time that it takes Dave to drive his Nissan 250 miles. If Kent averages 4 miles per hour faster than Dave, find their rates.

46. Suppose that Wendy rides her bicycle 30 miles in the same time that it takes Kim to ride her bicycle 20 miles. If Wendy rides 5 miles per hour faster than Kim, find the rate of each.

47. An inlet pipe can fill a tank (see Figure 9.4) in 10 minutes. A drain can empty the tank in 12 minutes. If the tank is empty, and both the pipe and drain are open, how long will it take before the tank overflows?

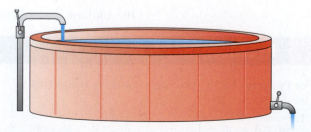

Figure 9.4

48. Barry can do a certain job in 3 hours, whereas it takes Sanchez 5 hours to do the same job. How long would it take them to do the job working together?

49. Connie can type 600 words in 5 minutes less than it takes Katie to type 600 words. If Connie types at a rate of 20 words per minute faster than Katie types, find the typing rate of each woman.

50. Ryan can mow a lawn in 1 hour, and his son, Malik, can mow the same lawn in 50 minutes. One day Malik started mowing the lawn by himself and worked for 30 minutes. Then Ryan joined him and they finished the lawn. How long did it take them to finish mowing the lawn after Ryan started to help?

51. Plane A can travel 1400 miles in 1 hour less time than it takes plane B to travel 2000 miles. The rate of plane B is 50 miles per hour greater than the rate of plane A. Find the times and rates of both planes.

52. To travel 60 miles, it takes Sue, riding a moped, 2 hours less time than it takes Doreen to travel 50 miles riding a bicycle. Sue travels 10 miles per hour faster than Doreen. Find the times and rates of both girls.

53. It takes Amy twice as long to clean the office as it does Nancy. How long would it take each girl to clean the office by herself if they can clean the office together in 40 minutes?

54. If two inlet pipes are both open, they can fill a pool in 1 hour and 12 minutes. One of the pipes can fill the pool by itself in 2 hours. How long would it take the other pipe to fill the pool by itself?

55. Rod agreed to mow a vacant lot for $12. It took him an hour longer than he had anticipated, so he earned $1 per hour less than he had originally calculated. How long had he anticipated that it would take him to mow the lot?

56. Last week Al bought some golf balls for $20. The next day they were on sale for $0.50 per ball less, and he bought $22.50 worth of balls. If he purchased 5 more balls on the second day than on the first day, how many did he buy each day and at what price per ball?

57. Debbie rode her bicycle out into the country for a distance of 24 miles. On the way back, she took a much shorter route of 12 miles and made the return trip in one-half hour less time. If her rate out into the country was 4 miles per hour greater than her rate on the return trip, find both rates.

58. Felipe jogs for 10 miles and then walks another 10 miles. He jogs $2\frac{1}{2}$ miles per hour faster than he walks, and the entire distance of 20 miles takes 6 hours. Find the rate at which he walks and the rate at which he jogs.

Thoughts Into Words

59. Why is it important to consider more than one way to do a problem?

60. Write a paragraph or two summarizing the new ideas about problem solving you have acquired thus far in this course.

Graphing Calculator Activities

In Section 4.5 we solved mixture-of-solution problems. You can use the graphing calculator in a more general approach to problems of this type: How much pure alcohol should be added to 6 liters of a 40% alcohol solution to raise it to a 60% alcohol solution?

We let x represent the amount of pure alcohol to be added to the solution. For this more general approach we want to write a rational expression that represents the concentration of pure alcohol in the final solution. The amount of pure alcohol we are starting with is 40% of the 6 liters, which equals $0.40(6) = 2.4$ liters. Because we are adding x liters of pure alcohol to the solution, the expression $2.4 + x$ represents the amount of pure alcohol in the final solution. The final amount of solution is $6 + x$. The rational expression $\dfrac{2.4 + x}{6 + x}$ represents the concentration of pure alcohol in the final solution. Let's graph the equation $y = \dfrac{2.4 + x}{6 + x}$ as shown in Figure 9.5.

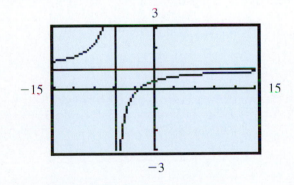

Figure 9.5

The y axis is the concentration of alcohol, so that will be a number between 0.40 and 1.0. The x axis is the amount of alcohol to be added, so x will be a nonnegative number. Therefore let's change the viewing window so that $0 \le x \le 15$

and $0 \leq y \leq 2$ to obtain Figure 9.6. Now we can use the graph to answer a variety of questions about this problem.

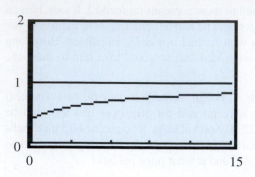

Figure 9.6

1. How much pure alcohol needs to be added to raise the 40% solution to a 60% alcohol solution? (*Answer:* Using the trace feature of the graphing utility, we find that $y = 0.6$ when $x = 3$. Therefore, 3 liters of pure alcohol need to be added.)

2. How much pure alcohol needs to be added to raise the 40% solution to a 70% alcohol solution? (*Answer:* Using the trace feature, we find that $y = 0.7$ when $x = 6$. Therefore, 6 liters of pure alcohol need to be added.)

3. What percent of alcohol do we have if we add 9 liters of pure alcohol to the 6 liters of a 40% solution?

(*Answer:* Using the trace feature, we find that $y = 0.76$ when $x = 9$. Therefore, adding 9 liters of pure alcohol will give us a 76% alcohol solution.)

Now use this approach with your graphing utility to solve the following problems.

61. Suppose that x ounces of pure acid have been added to 14 ounces of a 15% acid solution.

   a. Set up the rational expression that represents the concentration of pure acid in the final solution.

   b. Graph the rational equation that displays the level of concentration.

   c. How many ounces of pure acid need to be added to the 14 ounces of a 15% acid solution to raise it to a 40.5% acid solution? Check your answer.

   d. How many ounces of pure acid need to be added to the 14 ounces of 15% acid solution to raise it to a 50% acid solution? Check your answer.

   e. What percent of acid do we obtain if we add 12 ounces of pure acid to the 14 ounces of 15% acid solution? Check your answer.

62. Solve the following problem both algebraically and graphically: One solution contains 50% alcohol, and another solution contains 80% alcohol. How many liters of each solution should be mixed to produce 10.5 liters of a 70% alcohol solution? Check your answer.

Answers to the Concept Quiz

1. False   **2.** True   **3.** False   **4.** True   **5.** True   **6.** True   **7.** False   **8.** False   **9.** False   **10.** True

Chapter 9 Summary

OBJECTIVE	SUMMARY	EXAMPLE
Reduce rational numbers and rational expressions. (Section 9.1/Objectives 1 and 2)	Any number that can be written in the form $\frac{a}{b}$, where a and b are integers and $b \neq 0$, is a rational number. A rational expression is defined as the indicated quotient of two polynomials. The fundamental principle of fractions, $\frac{a \cdot k}{b \cdot k} = \frac{a}{b}$, is used when reducing rational numbers or rational expressions.	Simplify $\frac{x^2 - 2x - 15}{x^2 + x - 6}$. **Solution** $$\frac{x^2 - 2x - 15}{x^2 + x - 6}$$ $$= \frac{(x + 3)(x - 5)}{(x + 3)(x - 2)} = \frac{x - 5}{x - 2}$$
Multiply rational numbers and rational expressions. (Section 9.2/Objectives 1 and 2)	Multiplication of rational expressions is based on the following definition: $\frac{a}{b} \cdot \frac{c}{d} = \frac{ac}{bd}$, where $b \neq 0$ and $d \neq 0$	Find the product: $$\frac{3y^2 + 12y}{y^3 - 2y^2} \cdot \frac{y^2 - 3y + 2}{y^2 + 7y + 12}$$ **Solution** $$\frac{3y^2 + 12y}{y^3 - 2y^2} \cdot \frac{y^2 - 3y + 2}{y^2 + 7y + 12}$$ $$= \frac{3y(y + 4)}{y^2(y - 2)} \cdot \frac{(y - 2)(y - 1)}{(y + 3)(y + 4)}$$ $$= \frac{3y(y + 4)}{y^2(y - 2)} \cdot \frac{(y - 2)(y - 1)}{(y + 3)(y + 4)}$$ $$= \frac{3(y - 1)}{y(y + 3)}$$
Divide rational numbers and rational expressions. (Section 9.2/Objectives 3 and 4)	Division of rational expressions is based on the following definition: $\frac{a}{b} \div \frac{c}{d} = \frac{a}{b} \cdot \frac{d}{c} = \frac{ad}{bc}$, where $b \neq 0$, $c \neq 0$, and $d \neq 0$	Find the quotient: $$\frac{6xy}{x^2 - 6x + 9} \div \frac{18x}{x^2 - 9}$$ **Solution** $$\frac{6xy}{x^2 - 6x + 9} \div \frac{18x}{x^2 - 9}$$ $$= \frac{6xy}{x^2 - 6x + 9} \cdot \frac{x^2 - 9}{18x}$$ $$= \frac{6xy}{(x - 3)(x - 3)} \cdot \frac{(x + 3)(x - 3)}{18x}$$ $$= \frac{6xy}{(x - 3)(x - 3)} \cdot \frac{(x + 3)(x - 3)}{18x}$$ $$= \frac{y(x + 3)}{3(x - 3)}$$

(continued)

OBJECTIVE	SUMMARY	EXAMPLE
Simplify problems that involve both multiplication and division of rational expressions. **(Section 9.2/Objective 5)**	Perform the multiplications and divisions from left to right according to the order of operations. You can change division to multiplication by multiplying by the reciprocal and then finding the product.	Perform the indicated operations: $$\dfrac{6xy^3}{5x} \div \dfrac{3xy}{10} \cdot \dfrac{y}{7x^2}$$ **Solution** $$\dfrac{6xy^3}{5x} \div \dfrac{3xy}{10} \cdot \dfrac{y}{7x^2}$$ $$= \dfrac{6xy^3}{5x} \cdot \dfrac{10}{3xy} \cdot \dfrac{y}{7x^2}$$ $$= \dfrac{\overset{2}{6}xy^3}{5x} \cdot \dfrac{\overset{2}{10}}{3xy} \cdot \dfrac{y}{7x^2}$$ $$= \dfrac{4y^3}{7x^3}$$
Add and subtract rational numbers or rational expressions. **(Section 9.3/Objectives 1 and 2; Section 9.4/Objective 1)**	Addition and subtraction of rational expressions are based on the following definitions. $$\dfrac{a}{b} + \dfrac{c}{b} = \dfrac{a+c}{b} \quad \text{Addition}$$ $$\dfrac{a}{b} - \dfrac{c}{b} = \dfrac{a-c}{b} \quad \text{Subtraction}$$ The following basic procedure is used to add or subtract rational expressions. **1.** Factor the denominators. **2.** Find the LCD. **3.** Change each fraction to an equivalent fraction that has the LCD as the denominator. **4.** Combine the numerators and place over the LCD. **5.** Simplify by performing the addition or subtraction in the numerator. **6.** If possible, reduce the resulting fraction.	Subtract $$\dfrac{2}{x^2 - 2x - 3} - \dfrac{5}{x^2 + 5x + 4}$$ **Solution** $$\dfrac{2}{x^2 - 2x - 3} - \dfrac{5}{x^2 + 5x + 4}$$ $$= \dfrac{2}{(x-3)(x+1)} - \dfrac{5}{(x+1)(x+4)}$$ The LCD is $(x-3)(x+1)(x+4)$. $$= \dfrac{2(x+4)}{(x-3)(x+1)(x+4)}$$ $$- \dfrac{5(x-3)}{(x+1)(x+4)(x-3)}$$ $$= \dfrac{2(x+4) - 5(x-3)}{(x-3)(x+1)(x+4)}$$ $$= \dfrac{2x + 8 - 5x + 15}{(x-3)(x+1)(x+4)}$$ $$= \dfrac{-3x + 23}{(x-3)(x+1)(x+4)}$$

OBJECTIVE	SUMMARY	EXAMPLE
Simplify complex fractions. (Section 9.4/Objective 2)	Fractions that contain rational numbers or rational expressions in the numerators or denominators are called complex fractions. In Section 9.4 two methods were shown for simplifying complex fractions.	Simplify $\dfrac{\dfrac{2}{x} - \dfrac{3}{y}}{\dfrac{4}{x^2} + \dfrac{5}{y}}$. **Solution** $$\dfrac{\dfrac{2}{x} - \dfrac{3}{y}}{\dfrac{4}{x^2} + \dfrac{5}{y}}$$ Multiply the numerator and denominator by $x^2 y$: $$\dfrac{x^2 y\left(\dfrac{2}{x} - \dfrac{3}{y}\right)}{x^2 y\left(\dfrac{4}{x^2} + \dfrac{5}{y}\right)}$$ $$= \dfrac{x^2 y\left(\dfrac{2}{x}\right) + x^2 y\left(-\dfrac{3}{y}\right)}{x^2 y\left(\dfrac{4}{x^2}\right) + x^2 y\left(\dfrac{5}{y}\right)}$$ $$= \dfrac{2xy - 3x^2}{4y + 5x^2}$$
Solve rational equations. (Section 9.5/Objective 1)	To solve a rational equation, it is often easiest to begin by multiplying both sides of the equation by the LCD of all the denominators in the equation. Recall that any value of the variable that makes the denominators zero cannot be a solution to the equation.	Solve $\dfrac{2}{3x} + \dfrac{5}{12} = \dfrac{1}{4x}$. **Solution** $$\dfrac{2}{3x} + \dfrac{5}{12} = \dfrac{1}{4x}, \qquad x \neq 0$$ Multiply both sides by $12x$: $$12x\left(\dfrac{2}{3x} + \dfrac{5}{12}\right) = 12x\left(\dfrac{1}{4x}\right)$$ $$12x\left(\dfrac{2}{3x}\right) + 12x\left(\dfrac{5}{12}\right) = 12x\left(\dfrac{1}{4x}\right)$$ $$8 + 5x = 3$$ $$5x = -5$$ $$x = -1$$ The solution set is $\{-1\}$.

(continued)

OBJECTIVE	SUMMARY	EXAMPLE
Solve proportions. (Section 9.5/Objective 2)	A ratio is the comparison of two numbers by division. A proportion is a statement of equality between two ratios. Proportions can be solved using the cross-multiplication property of proportions.	Solve $\dfrac{5}{2x-1} = \dfrac{3}{x+4}$. **Solution** $\dfrac{5}{2x-1} = \dfrac{3}{x+4}$, $\qquad x \neq -4, x \neq \dfrac{1}{2}$ $3(2x-1) = 5(x+4)$ $6x - 3 = 5x + 20$ $x = 23$ The solution set is $\{23\}$.
Solve rational equations where the denominators require factoring. (Section 9.6/Objective 1)	It may be necessary to factor the denominators in a rational equation in order to determine the LCD of all the denominators.	Solve $\dfrac{7x}{3x+12} - \dfrac{2}{x^2-16} = \dfrac{7}{3}$ **Solution** $\dfrac{7x}{3x+12} - \dfrac{2}{x^2-16} = \dfrac{7}{3}$, $\quad x \neq -4, x \neq 4$ $\dfrac{7x}{3(x+4)} - \dfrac{2}{(x-4)(x+4)} = \dfrac{7}{3}$ Multiply both sides by $3(x+4)(x-4)$: $7x(x-4) - 2(3) = 7(x+4)(x-4)$ $7x^2 - 28x - 6 = 7x^2 - 112$ $-28x = -106$ $x = \dfrac{-106}{-28} = \dfrac{53}{14}$ The solution set is $\left\{ \dfrac{53}{14} \right\}$.
Solve formulas that involve fractional forms. (Section 9.6/Objective 2)	The techniques that are used for solving rational equations can also be used to change the form of formulas.	Solve $\dfrac{x}{2a} - \dfrac{y}{2b} = 1$ for y. **Solution** $\dfrac{x}{2a} - \dfrac{y}{2b} = 1$ Multiply both sides by $2ab$: $2ab\left(\dfrac{x}{2a} - \dfrac{y}{2b} \right) = 2ab(1)$ $bx - ay = 2ab$ $-ay = 2ab - bx$ $y = \dfrac{2ab - bx}{-a}$ $y = \dfrac{-2ab + bx}{a}$

OBJECTIVE	SUMMARY	EXAMPLE
Solve word problems involving ratios. (Section 9.5/Objective 3)	Many real-world situations can be solved by using ratios and setting up a proportion to be solved.	At a law firm, the ratio of female attorneys to male attorneys is 1 to 4. If the firm has a total of 125 attorneys, find the number of female attorneys. **Solution** Let x represent the number of female attorneys. Then $125 - x$ represents the number of male attorneys. The following proportion can be set up: $$\frac{x}{125 - x} = \frac{1}{4}$$ Solve by cross-multiplication: $$\frac{x}{125 - x} = \frac{1}{4}$$ $$4x = 1(125 - x)$$ $$4x = 125 - x$$ $$5x = 125$$ $$x = 25$$ There are 25 female attorneys.
Solve rate-time word problems. (Section 9.6/Objective 3)	Uniform motion problems are a special case of rate-time problems. In general, if Q is the quantity of some job done in t time units, then the rate, r, is given by $r = \dfrac{Q}{t}$.	At a veterinarian clinic, it takes Laurie twice as long to feed the animals as it does Janet. How long would it take each person to feed the animals by herself if they can feed the animals together in 60 minutes? **Solution** Let t represent the time it takes Janet to feed the animals. Then $2t$ represents the time it would take Laurie to feed the animals. Laurie's rate plus Janet's rate equals the rate working together. $$\frac{1}{2t} + \frac{1}{t} = \frac{1}{60}$$ Multiply both sides by $60t$: $$60t\left(\frac{1}{2t} + \frac{1}{t}\right) = 60t\left(\frac{1}{60}\right)$$ $$30 + 60 = t$$ $$90 = t$$ It would take Janet 90 minutes working alone to feed the animals, and it would take Laurie 180 minutes working alone to feed the animals.

Chapter 9 Review Problem Set

For Problems 1–6, simplify each rational expression.

1. $\dfrac{26x^2y^3}{39x^4y^2}$

2. $\dfrac{a^2 - 9}{a^2 + 3a}$

3. $\dfrac{n^2 - 3n - 10}{n^2 + n - 2}$

4. $\dfrac{x^4 - 1}{x^3 - x}$

5. $\dfrac{8x^3 - 2x^2 - 3x}{12x^2 - 9x}$

6. $\dfrac{x^4 - 7x^2 - 30}{2x^4 + 7x^2 + 3}$

For Problems 7–10, simplify each complex fraction.

7. $\dfrac{\dfrac{5}{8} - \dfrac{1}{2}}{\dfrac{1}{6} + \dfrac{3}{4}}$

8. $\dfrac{\dfrac{3}{2x} + \dfrac{5}{3y}}{\dfrac{4}{x} - \dfrac{3}{4y}}$

9. $\dfrac{\dfrac{3}{x - 2} - \dfrac{4}{x^2 - 4}}{\dfrac{2}{x + 2} + \dfrac{1}{x - 2}}$

10. $1 - \dfrac{1}{2 - \dfrac{1}{x}}$

For Problems 11–24, perform the indicated operations, and express your answers in simplest form.

11. $\dfrac{6xy^2}{7y^3} \div \dfrac{15x^2y}{5x^2}$

12. $\dfrac{9ab}{3a + 6} \cdot \dfrac{a^2 - 4a - 12}{a^2 - 6a}$

13. $\dfrac{n^2 + 10n + 25}{n^2 - n} \cdot \dfrac{5n^3 - 3n^2}{5n^2 + 22n - 15}$

14. $\dfrac{x^2 - 2xy - 3y^2}{x^2 + 9y^2} \div \dfrac{2x^2 + xy - y^2}{2x^2 - xy}$

15. $\dfrac{2x + 1}{5} + \dfrac{3x - 2}{4}$

16. $\dfrac{3}{2n} + \dfrac{5}{3n} - \dfrac{1}{9}$

17. $\dfrac{3x}{x + 7} - \dfrac{2}{x}$

18. $\dfrac{10}{x^2 - 5x} + \dfrac{2}{x}$

19. $\dfrac{3}{n^2 - 5n - 36} + \dfrac{2}{n^2 + 3n - 4}$

20. $\dfrac{3}{2y + 3} + \dfrac{5y - 2}{2y^2 - 9y - 18} - \dfrac{1}{y - 6}$

21. $\dfrac{2x^2 y}{3x} \cdot \dfrac{xy^2}{6} \div \dfrac{x}{9y}$

22. $\dfrac{10x^4 y^3}{8x^2 y} \div \dfrac{5}{xy^2} \cdot \dfrac{3y}{x}$

23. $\dfrac{8x}{2x - 6} \div \dfrac{2x - 1}{x^2 - 9} \cdot \dfrac{2x^2 + x - 1}{x^2 + 7x + 12}$

24. $\dfrac{2 - x}{6} \cdot \dfrac{x + 1}{x^2 - 4} \div \dfrac{x^2 + 2x + 1}{10}$

For Problems 25–36, solve each equation.

25. $\dfrac{4x + 5}{3} + \dfrac{2x - 1}{5} = 2$

26. $\dfrac{3}{4x} + \dfrac{4}{5} = \dfrac{9}{10x}$

27. $\dfrac{a}{a - 2} - \dfrac{3}{2} = \dfrac{2}{a - 2}$

28. $\dfrac{4}{5y - 3} = \dfrac{2}{3y + 7}$

29. $n + \dfrac{1}{n} = \dfrac{53}{14}$

30. $\dfrac{1}{2x - 7} + \dfrac{x - 5}{4x^2 - 49} = \dfrac{4}{6x - 21}$

31. $\dfrac{x}{2x + 1} - 1 = \dfrac{-4}{7(x - 2)}$

32. $\dfrac{2x}{-5} = \dfrac{3}{4x - 13}$

33. $\dfrac{2n}{2n^2 + 11n - 21} - \dfrac{n}{n^2 + 5n - 14} = \dfrac{3}{n^2 + 5n - 14}$

34. $\dfrac{2}{t^2 - t - 6} + \dfrac{t + 1}{t^2 + t - 12} = \dfrac{t}{t^2 + 6t + 8}$

35. Solve $\dfrac{y - 6}{x + 1} = \dfrac{3}{4}$ for y.

36. Solve $\dfrac{x}{a} - \dfrac{y}{b} = 1$ for y.

For Problems 37–43, set up an equation, and solve the problem.

37. A sum of $1400 is to be divided between two people in the ratio of $\dfrac{3}{5}$. How much does each person receive?

38. At a restaurant the tips are split between the busboy and the waiter in the ratio of 2 to 7. Find the amount each received in tips if there was a total of $162 in tips.

39. Working together, Dan and Julio can mow a lawn in 12 minutes. Julio can mow the lawn by himself in 10 minutes less time than it takes Dan by himself. How long does it take each of them to mow the lawn alone?

40. Suppose that car A can travel 250 miles in 3 hours less time than it takes car B to travel 440 miles. The rate of car B is 5 miles per hour faster than that of car A. Find the rates of both cars.

41. Mark can overhaul an engine in 20 hours, and Phil can do the same job by himself in 30 hours. If they both work together for a time and then Mark finishes the job by himself in 5 hours, how long did they work together?

42. Kelly contracted to paint a house for $640. It took him 20 hours longer than he had anticipated, so he earned $1.60 per hour less than he had calculated. How long had he anticipated that it would take him to paint the house?

43. Nasser rode his bicycle 66 miles in $4\dfrac{1}{2}$ hours. For the first 40 miles he averaged a certain rate, and then for the last 26 miles he reduced his rate by 3 miles per hour. Find his rate for the last 26 miles.

For Problems 1–4, simplify each rational expression.

1. $\dfrac{39x^2y^3}{72x^3y}$

2. $\dfrac{3x^2 + 17x - 6}{x^3 - 36x}$

3. $\dfrac{6n^2 - 5n - 6}{3n^2 + 14n + 8}$

4. $\dfrac{2x - 2x^2}{x^2 - 1}$

For Problems 5–13, perform the indicated operations, and express your answers in simplest form.

5. $\dfrac{5x^2y}{8x} \cdot \dfrac{12y^2}{20xy}$

6. $\dfrac{5a + 5b}{20a + 10b} \cdot \dfrac{a^2 - ab}{2a^2 + 2ab}$

7. $\dfrac{3x^2 + 10x - 8}{5x^2 + 19x - 4} \div \dfrac{3x^2 - 23x + 14}{x^2 - 3x - 28}$

8. $\dfrac{3x - 1}{4} + \dfrac{2x + 5}{6}$

9. $\dfrac{5x - 6}{3} - \dfrac{x - 12}{6}$

10. $\dfrac{3}{5n} + \dfrac{2}{3} - \dfrac{7}{3n}$

11. $\dfrac{3x}{x - 6} + \dfrac{2}{x}$

12. $\dfrac{9}{x^2 - x} - \dfrac{2}{x}$

13. $\dfrac{3}{2n^2 + n - 10} + \dfrac{5}{n^2 + 5n - 14}$

14. Simplify the complex fraction $\dfrac{\dfrac{3}{2x} - \dfrac{1}{6}}{\dfrac{2}{3x} + \dfrac{3}{4}}$.

15. Solve $3x - 5y = 2$ for y.

16. Solve $\dfrac{x + 2}{y - 4} = \dfrac{3}{4}$ for y.

For Problems 17–22, solve each equation.

17. $\dfrac{x - 1}{2} - \dfrac{x + 2}{5} = -\dfrac{3}{5}$

18. $\dfrac{5}{4x} + \dfrac{3}{2} = \dfrac{7}{5x}$

19. $\dfrac{-3}{4n - 1} = \dfrac{-2}{3n + 11}$

20. $n - \dfrac{5}{n} = 4$

21. $\dfrac{6}{x - 4} - \dfrac{4}{x + 3} = \dfrac{8}{x - 4}$

22. $\dfrac{1}{3x - 1} + \dfrac{x - 2}{9x^2 - 1} = \dfrac{7}{6x - 2}$

For Problems 23–25, set up an equation and then solve the problem.

23. The denominator of a rational number is 9 less than three times the numerator. The number in simplest form is $\dfrac{3}{8}$. Find the number.

24. It takes Jodi three times as long to deliver papers as it does Jannie. Together they can deliver the papers in 15 minutes. How long would it take Jodi by herself?

25. René can ride her bike 60 miles in 1 hour less time than it takes Sue to ride 60 miles. René's rate is 3 miles per hour faster than Sue's rate. Find René's rate.

Chapters 1–9 Cumulative Review Problem Set

1. Simplify the numerical expression $16 \div 4(2) + 8$.

2. Simplify the numerical expression
   $$(-2)^2 + (-2)^3 - 3^2$$

3. Evaluate $-2xy + 5y^2$ for $x = -3$ and $y = 4$.

4. Evaluate $3(n - 2) + 4(n - 4) - 8(n - 3)$
   for $n = -\dfrac{1}{2}$.

For Problems 5–14, perform the indicated operations and then simplify.

5. $(6a^2 + 3a - 4) + (8a + 6) + (a^2 - 1)$

6. $(x^2 + 5x + 2) - (3x^2 - 4x + 6)$

7. $(2x^2y)(-xy^4)$ 8. $(4xy^3)^2$

9. $(-3a^3)^2(4ab^2)$ 10. $(4a^2b)(-3a^3b^2)(2ab)$

11. $-3x^2(6x^2 - x + 4)$ 12. $(5x + 3y)(2x - y)$

13. $(x + 4y)^2$ 14. $(a + 3b)(a^2 - 4ab + b^2)$

For Problems 15–20, factor each polynomial completely.

15. $x^2 - 5x + 6$ 16. $6x^2 - 5x - 4$

17. $2x^2 - 8x + 6$ 18. $3x^2 + 18x - 48$

19. $9m^2 - 16n^2$ 20. $27a^3 + 8$

21. Simplify $\dfrac{-28x^2y^5}{4x^4y}$.   22. Simplify $\dfrac{4x - x^2}{x - 4}$.

For Problems 23–28, perform the indicated operations and express the answer in simplest form.

23. $\dfrac{6xy}{2x + 4} \cdot \dfrac{x^2 - 3x - 10}{3xy - 3y}$

24. $\dfrac{x^2 - 3x - 4}{x^2 - 1} \div \dfrac{x^2 - x - 12}{x^2 + 6x - 7}$

25. $\dfrac{7n - 3}{5} - \dfrac{n + 4}{2}$ 26. $\dfrac{3}{x^2 + x - 6} + \dfrac{5}{x^2 - 9}$

27. $\dfrac{\dfrac{2}{x} + \dfrac{3}{y}}{6}$ 28. $\dfrac{\dfrac{1}{n^2} - \dfrac{1}{m^2}}{\dfrac{1}{m} + \dfrac{1}{n}}$

For Problems 29–38, solve the equation.

29. $8n - 3(n + 2) = 2n + 12$

30. $0.2(y - 6) = 0.02y + 3.12$

31. $\dfrac{x + 1}{4} + \dfrac{3x + 2}{2} = 5$

32. $\dfrac{5}{8}(x + 2) - \dfrac{1}{2}x = 2$

33. $|3x - 2| = 8$

34. $|x + 8| - 4 = 16$

35. $x^2 + 7x - 8 = 0$

36. $2x^2 + 13x + 15 = 0$

37. $n - \dfrac{3}{n} = \dfrac{26}{3}$

38. $\dfrac{3}{n - 7} + \dfrac{4}{n + 2} = \dfrac{27}{n^2 - 5n - 14}$

39. Solve the formula $A = P + Prt$ for P.

40. Solve the formula $V = \dfrac{1}{3}BH$ for B.

For Problems 41–45, solve the inequality, and express the solution in interval notation.

41. $-3x + 2(x - 4) \geq -10$

42. $-10 < 3x + 2 < 8$

43. $|4x + 3| < 15$

44. $|2x + 6| \geq 20$

45. $|x + 4| - 6 > 0$

46. The owner of a local café wants to make a profit of 80% of the cost for each Caesar salad sold. If it costs $3.20 to make a Caesar salad, at what price should each salad be sold?

47. Find the discount sale price of a $920 television that is on sale for 25% off.

48. Suppose that the length of a rectangle is 8 inches less than twice the width. The perimeter of the rectangle is 122 inches. Find the length and width of the rectangle.

49. Two planes leave Kansas City at the same time and fly in opposite directions. If one travels at 450 miles per hour and the other travels at 400 miles per hour, how long will it take for them to be 3400 miles apart?

50. A sum of $68,000 is to be divided between two partners in the ratio of $\dfrac{1}{4}$. How much does each person receive?

51. Victor can rake the lawn in 20 minutes, and his sister Lucia can rake the same lawn in 30 minutes. How long will it take them to rake the lawn if they work together?

52. One leg of a right triangle is 7 inches less than the other leg. The hypotenuse is 1 inch longer than the longer of the two legs. Find the length of the three sides of the right triangle.

53. How long will it take $1500 to double itself at 6% simple interest?

54. A collection of 40 coins consisting of dimes and quarters has a value of $5.95. Find the number of each kind of coin.

10

Exponents and Radicals

By knowing the time it takes for the pendulum to swing from one side to the other side and back, the formula $T = 2\pi\sqrt{\dfrac{L}{32}}$ can be used to find the length of the pendulum.

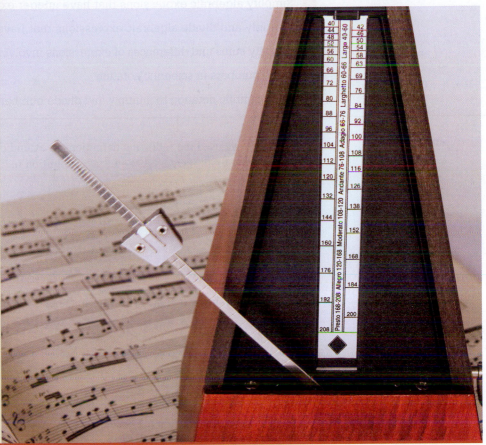

© Adam Fraise

How long will it take a pendulum that is 1.5 feet long to swing from one side to the other side and back? The formula $T = 2\pi\sqrt{\dfrac{L}{32}}$ can be used to determine that it will take approximately 1.4 seconds.

It is not uncommon in mathematics to find two separately developed concepts that are closely related to each other. In this chapter, we will first develop the concepts of exponent and root individually and then show how they merge to become even more functional as a unified idea.

Video tutorials based on section learning objectives are available in a variety of delivery modes.

10.1   Integral Exponents and Scientific Notation Revisited

OBJECTIVES

1. Simplify numerical expressions that have integer exponents
2. Simplify algebraic expressions that have integer exponents
3. Multiply and divide algebraic expressions that have integer exponents
4. Simplify sums and differences of expressions involving integer exponents
5. Write numbers in scientific notation
6. Use scientific notation to multiply and divide numbers

In Section 6.6 we used the following definitions to extend our work with exponents from the positive integers to all integers. Let's restate Definition 6.2:

> **Definition 6.2 Exponent of Zero**
>
> If b is a nonzero real number, then
>
> $$b^0 = 1$$

According to Definition 6.2, the following statements are all true.

$$5^0 = 1 \qquad\qquad (-413)^0 = 1$$

$$\left(\frac{3}{11}\right)^0 = 1 \qquad\qquad n^0 = 1, \quad n \neq 0$$

$$(x^3 y^4)^0 = 1, \qquad x \neq 0, y \neq 0$$

Restatement of Definition 6.3:

> **Definition 6.3 Negative Exponent**
>
> If n is a positive integer, and b is a nonzero real number, then
>
> $$b^{-n} = \frac{1}{b^n}$$

According to Definition 6.3, the following statements are all true.

$$x^{-5} = \frac{1}{x^5} \qquad\qquad 2^{-4} = \frac{1}{2^4} = \frac{1}{16}$$

$$10^{-2} = \frac{1}{10^2} = \frac{1}{100} \quad \text{or} \quad 0.01 \qquad\qquad \frac{2}{x^{-3}} = \frac{2}{\dfrac{1}{x^3}} = (2)\left(\frac{x^3}{1}\right) = 2x^3$$

$$\left(\frac{3}{4}\right)^{-2} = \frac{1}{\left(\dfrac{3}{4}\right)^2} = \frac{1}{\dfrac{9}{16}} = \frac{16}{9}$$

The following properties of exponents were stated both in Chapter 6 and in Chapter 8, but we will restate them here for your convenience.

Restatement of Property 8.6:

Property 8.6 Properties for Integer Exponents

If m and n are integers, and a and b are real numbers (and $b \neq 0$ whenever it appears in a denominator), then

1. $b^n \cdot b^m = b^{n+m}$ Product of two powers

2. $(b^n)^m = b^{mn}$ Power of a power

3. $(ab)^n = a^n b^n$ Power of a product

4. $\left(\dfrac{a}{b}\right)^n = \dfrac{a^n}{b^n}$ Power of a quotient

5. $\dfrac{b^n}{b^m} = b^{n-m}$ Quotient of two powers

Having the use of all integers as exponents enables us to work with a large variety of numerical and algebraic expressions. Let's consider some examples that illustrate the use of the various parts of Property 8.6.

Classroom Example
Simplify each of the following numerical expressions:

(a) $10^{-5} \cdot 10^2$

(b) $(3^{-2})^{-2}$

(c) $(3^{-1} \cdot 5^2)^{-1}$

(d) $\left(\dfrac{3^{-3}}{5^{-2}}\right)^{-1}$

(e) $\dfrac{10^{-6}}{10^{-9}}$

EXAMPLE 1   Simplify each of the following numerical expressions:

(a) $10^{-3} \cdot 10^2$ (b) $(2^{-3})^{-2}$ (c) $(2^{-1} \cdot 3^2)^{-1}$

(d) $\left(\dfrac{2^{-3}}{3^{-2}}\right)^{-1}$ (e) $\dfrac{10^{-2}}{10^{-4}}$

Solution

(a) $10^{-3} \cdot 10^2 = 10^{-3+2}$ Product of two powers

$= 10^{-1}$

$= \dfrac{1}{10^1} = \dfrac{1}{10}$

(b) $(2^{-3})^{-2} = 2^{(-2)(-3)}$ Power of a power

$= 2^6 = 64$

(c) $(2^{-1} \cdot 3^2)^{-1} = (2^{-1})^{-1}(3^2)^{-1}$ Power of a product

$= 2^1 \cdot 3^{-2}$

$= \dfrac{2^1}{3^2} = \dfrac{2}{9}$

(d) $\left(\dfrac{2^{-3}}{3^{-2}}\right)^{-1} = \dfrac{(2^{-3})^{-1}}{(3^{-2})^{-1}}$ Power of a quotient

$= \dfrac{2^3}{3^2} = \dfrac{8}{9}$

(e) $\dfrac{10^{-2}}{10^{-4}} = 10^{-2-(-4)}$ Quotient of two powers

$= 10^2 = 100$

EXAMPLE 2

Simplify each of the following; express final results without using zero or negative integers as exponents:

(a) $x^2 \cdot x^{-5}$ **(b)** $(x^{-2})^4$ **(c)** $(x^2y^{-3})^{-4}$

(d) $\left(\dfrac{a^3}{b^{-5}}\right)^{-2}$ **(e)** $\dfrac{x^{-4}}{x^{-2}}$

Solution

(a) $x^2 \cdot x^{-5} = x^{2+(-5)}$ Product of two powers

$\qquad\qquad\quad = x^{-3}$

$\qquad\qquad\quad = \dfrac{1}{x^3}$

(b) $(x^{-2})^4 = x^{4(-2)}$ Power of a power

$\qquad\qquad = x^{-8}$

$\qquad\qquad = \dfrac{1}{x^8}$

(c) $(x^2y^{-3})^{-4} = (x^2)^{-4}(y^{-3})^{-4}$ Power of a product

$\qquad\qquad\qquad = x^{-4(2)}y^{-4(-3)}$

$\qquad\qquad\qquad = x^{-8}y^{12}$

$\qquad\qquad\qquad = \dfrac{y^{12}}{x^8}$

(d) $\left(\dfrac{a^3}{b^{-5}}\right)^{-2} = \dfrac{(a^3)^{-2}}{(b^{-5})^{-2}}$ Power of a quotient

$\qquad\qquad\quad = \dfrac{a^{-6}}{b^{10}}$

$\qquad\qquad\quad = \dfrac{1}{a^6b^{10}}$

(e) $\dfrac{x^{-4}}{x^{-2}} = x^{-4-(-2)}$ Quotient of two powers

$\qquad\quad = x^{-2}$

$\qquad\quad = \dfrac{1}{x^2}$

EXAMPLE 3

Find the indicated products and quotients; express your results using positive integral exponents only:

(a) $(3x^2y^{-4})(4x^{-3}y)$ **(b)** $\dfrac{12a^3b^2}{-3a^{-1}b^5}$ **(c)** $\left(\dfrac{15x^{-1}y^2}{5xy^{-4}}\right)^{-1}$

Solution

(a) $(3x^2y^{-4})(4x^{-3}y) = 12x^{2+(-3)}y^{-4+1}$

$\qquad\qquad\qquad\qquad = 12x^{-1}y^{-3}$

$\qquad\qquad\qquad\qquad = \dfrac{12}{xy^3}$

(b) $\dfrac{12a^3b^2}{-3a^{-1}b^5} = -4a^{3-(-1)}b^{2-5}$

$= -4a^4b^{-3}$

$= -\dfrac{4a^4}{b^3}$

(c) $\left(\dfrac{15x^{-1}y^2}{5xy^{-4}}\right)^{-1} = \left(3x^{-1-1}y^{2-(-4)}\right)^{-1}$ — Note that we are first simplifying inside the parentheses

$= (3x^{-2}y^6)^{-1}$

$= 3^{-1}x^2y^{-6}$

$= \dfrac{x^2}{3y^6}$

The next three examples of this section show the simplification of numerical and algebraic expressions that involve sums and differences. In such cases, we use Definition 6.2 to change from negative to positive exponents so that we can proceed in the usual way.

Classroom Example
Simplify $3^{-2} + 5^{-1}$.

EXAMPLE 4 Simplify $2^{-3} + 3^{-1}$.

Solution

$2^{-3} + 3^{-1} = \dfrac{1}{2^3} + \dfrac{1}{3^1}$

$= \dfrac{1}{8} + \dfrac{1}{3}$

$= \dfrac{3}{24} + \dfrac{8}{24}$ — Use 24 as the LCD

$= \dfrac{11}{24}$

Classroom Example
Simplify $(6^{-1} - 2^{-3})^{-1}$.

EXAMPLE 5 Simplify $(4^{-1} - 3^{-2})^{-1}$.

Solution

$(4^{-1} - 3^{-2})^{-1} = \left(\dfrac{1}{4^1} - \dfrac{1}{3^2}\right)^{-1}$ — Apply $b^{-n} = \dfrac{1}{b^n}$ to 4^{-1} and to 3^{-2}

$= \left(\dfrac{1}{4} - \dfrac{1}{9}\right)^{-1}$

$= \left(\dfrac{9}{36} - \dfrac{4}{36}\right)^{-1}$ — Use 36 as the LCD

$= \left(\dfrac{5}{36}\right)^{-1}$

$= \dfrac{1}{\left(\dfrac{5}{36}\right)^1}$ — Apply $b^{-n} = \dfrac{1}{b^n}$

$= \dfrac{1}{\dfrac{5}{36}} = \dfrac{36}{5}$

EXAMPLE 6

Express $a^{-1} + b^{-2}$ as a single fraction involving positive exponents only.

Solution

$$a^{-1} + b^{-2} = \frac{1}{a^1} + \frac{1}{b^2} \qquad \text{Use } ab^2 \text{ as the common denominator}$$

$$= \left(\frac{1}{a}\right)\left(\frac{b^2}{b^2}\right) + \left(\frac{1}{b^2}\right)\left(\frac{a}{a}\right) \qquad \text{Change to equivalent fractions with } ab^2 \text{ as the common denominator}$$

$$= \frac{b^2}{ab^2} + \frac{a}{ab^2}$$

$$= \frac{b^2 + a}{ab^2}$$

Scientific Notation

Many applications of mathematics involve the use of very large or very small numbers. Working with numbers of this type in standard decimal form is quite cumbersome. It is much more convenient to represent very small and very large numbers in *scientific notation*. Although negative numbers can be written in scientific form, we will restrict our discussion to positive numbers. The expression $(N)(10^k)$, where N is a number greater than or equal to 1 and less than 10, written in decimal form, and k is any integer, is commonly called **scientific notation** or the scientific form of a number. Consider the following examples, which show a comparison between ordinary decimal notation and scientific notation.

Ordinary decimal notation	Scientific notation
2.14	$(2.14)(10^0)$
31.78	$(3.178)(10^1)$
412.9	$(4.129)(10^2)$
8,000,000	$(8)(10^6)$
0.14	$(1.4)(10^{-1})$
0.0379	$(3.79)(10^{-2})$
0.00000049	$(4.9)(10^{-7})$

To switch from ordinary notation to scientific notation, you can use the following procedure.

Write the given number as the product of a number greater than or equal to 1 and less than 10, and a power of 10. The exponent of 10 is determined by counting the number of places that the decimal point was moved when going from the original number to the number greater than or equal to 1 and less than 10. This exponent is (a) negative if the original number is less than 1, (b) positive if the original number is greater than 10, and (c) 0 if the original number itself is between 1 and 10.

Thus we can write

$$0.00467 = (4.67)(10^{-3})$$
$$87,000 = (8.7)(10^4)$$
$$3.1416 = (3.1416)(10^0)$$

To switch from scientific notation to ordinary decimal notation, you can use the following procedure.

> Move the decimal point the number of places indicated by the exponent of 10. The decimal point is moved to the right if the exponent is positive and to the left if the exponent is negative.

Thus we can write

$$(4.78)(10^4) = 47,800$$
$$(8.4)(10^{-3}) = 0.0084$$

Scientific notation can frequently be used to simplify numerical calculations. We merely change the numbers to scientific notation and use the appropriate properties of exponents. Consider the following examples.

Classroom Example
Perform the indicated operations:
(a) $(0.00051)(4000)$
(b) $\dfrac{8,600,000}{0.00043}$
(c) $\dfrac{(0.000052)(0.032)}{(0.000016)(0.00104)}$
(d) $\sqrt{0.000025}$

EXAMPLE 7

Convert each number to scientific notation and perform the indicated operations. Express the result in ordinary decimal notation:

(a) $(0.00024)(20,000)$

(b) $\dfrac{7,800,000}{0.0039}$

(c) $\dfrac{(0.00069)(0.0034)}{(0.0000017)(0.023)}$

(d) $\sqrt{0.000004}$

Solution

(a) $(0.00024)(20,000) = (2.4)(10^{-4})(2)(10^4)$

$$= (2.4)(2)(10^{-4})(10^4)$$
$$= (4.8)(10^0)$$
$$= (4.8)(1)$$
$$= 4.8$$

(b) $\dfrac{7,800,000}{0.0039} = \dfrac{(7.8)(10^6)}{(3.9)(10^{-3})}$

$$= (2)(10^9)$$
$$= 2,000,000,000$$

(c) $\dfrac{(0.00069)(0.0034)}{(0.0000017)(0.023)} = \dfrac{(6.9)(10^{-4})(3.4)(10^{-3})}{(1.7)(10^{-6})(2.3)(10^{-2})}$

$$= \dfrac{\overset{3}{(6.9)}\overset{2}{(3.4)}(10^{-7})}{\underset{}{(1.7)}\underset{}{(2.3)}(10^{-8})}$$

$$= (6)(10^1)$$
$$= 60$$

(d) $\sqrt{0.000004} = \sqrt{(4)(10^{-6})}$

$$= ((4)(10^{-6}))^{\frac{1}{2}}$$
$$= (4)^{\frac{1}{2}}(10^{-6})^{\frac{1}{2}}$$
$$= (2)(10^{-3})$$
$$= 0.002$$

Classroom Example
The speed of light is approximately
$(1.86)(10^5)$ miles per second. When
Saturn is $(8.9)(10^8)$ miles away from
the sun, how long does it take light
from the sun to reach Saturn?

EXAMPLE 8

The speed of light is approximately $(1.86)(10^5)$ miles per second. When Earth is $(9.3)(10^7)$ miles away from the sun, how long does it take light from the sun to reach Earth?

Solution

We will use the formula $t = \dfrac{d}{r}$.

$$t = \frac{(9.3)(10^7)}{(1.86)(10^5)}$$

$$t = \frac{(9.3)}{(1.86)}(10^2) \qquad \text{Subtract exponents}$$

$$t = (5)(10^2) = 500 \text{ seconds}$$

At this distance it takes light about 500 seconds to travel from the sun to Earth. To find the answer in minutes, divide 500 seconds by 60 seconds/minute. That gives a result of approximately 8.33 minutes.

Many calculators are equipped to display numbers in scientific notation. The display panel shows the number between 1 and 10 and the appropriate exponent of 10. For example, evaluating $(3,800,000^2)$ yields

> 1.444E13

Thus $(3,800,000^2) = (1.444)(10^{13}) = 14,440,000,000,000$.
Similarly, the answer for (0.000168^2) is displayed as

> 2.8224E-8

Thus $(0.000168^2) = (2.8224)(10^{-8}) = 0.000000028224$.
Calculators vary as to the number of digits displayed in the number between 1 and 10 when scientific notation is used. For example, we used two different calculators to estimate (6729^6) and obtained the following results.

> 9.2833E22

> 9.283316768E22

Obviously, you need to know the capabilities of your calculator when working with problems in scientific notation. Many calculators also allow the entry of a number in scientific notation. Such calculators are equipped with an enter-the-exponent key (often labeled as $\boxed{\text{EE}}$ or $\boxed{\text{EEX}}$). Thus a number such as $(3.14)(10^8)$ might be entered as follows:

Enter	Press	Display		Enter	Press	Display
3.14	$\boxed{\text{EE}}$	3.14E	or	3.14	$\boxed{\text{EE}}$	3.14^{00}
8		3.14E8		8		3.14^{08}

A $\boxed{\text{MODE}}$ key is often used on calculators to let you choose normal decimal notation, scientific notation, or engineering notation. (The abbreviations Norm, Sci, and Eng are commonly used.) If the calculator is in scientific mode, then a number can be entered and changed to scientific form by pressing the $\boxed{\text{ENTER}}$ key. For example, when we enter 589 and press the $\boxed{\text{ENTER}}$ key, the display will show 5.89E2. Likewise, when the calculator is in scientific mode, the answers to computational problems are given in scientific form. For example, the answer for $(76)(533)$ is given as 4.0508E4.

It should be evident from this brief discussion that even when you are using a calculator, you need to have a thorough understanding of scientific notation.

Concept Quiz 10.1

For Problems 1–10, answer true or false.

1. $\left(\dfrac{2}{5}\right)^{-2} = \left(\dfrac{5}{2}\right)^2$

2. $(3)^0 (3)^2 = 9^2$

3. $(2)^{-4}(2)^4 = 2$

4. $(2^{-2} \cdot 2^{-3})^{-1} = \dfrac{1}{16}$

5. $\left(\dfrac{3^{-2}}{3^{-1}}\right)^2 = \dfrac{1}{9}$

6. $\dfrac{1}{\left(\dfrac{2}{3}\right)^{-3}} = \dfrac{8}{27}$

7. $\dfrac{x^{-6}}{x^{-3}} = x^2$

8. $x^{-1} - x^{-2} = \dfrac{x-1}{x^2}$

9. A positive number written in scientific notation has the form $(N)(10^k)$, where $1 \le N < 10$ and k is an integer.

10. When the scientific notation form of a number has a negative exponent, then the number is less than zero.

Problem Set 10.1

For Problems 1–42, simplify each numerical expression.
(Objective 1)

1. 3^{-3}

2. 2^{-4}

3. -10^{-2}

4. 10^{-3}

5. $\dfrac{1}{3^{-4}}$

6. $\dfrac{1}{2^{-6}}$

7. $-\left(\dfrac{1}{3}\right)^{-3}$

8. $\left(\dfrac{1}{2}\right)^{-3}$

9. $\left(-\dfrac{1}{2}\right)^{-3}$

10. $\left(\dfrac{2}{7}\right)^{-2}$

11. $\left(-\dfrac{3}{4}\right)^0$

12. $\dfrac{1}{\left(\dfrac{4}{5}\right)^{-2}}$

13. $\dfrac{1}{\left(\dfrac{3}{7}\right)^{-2}}$

14. $-\left(\dfrac{5}{6}\right)^0$

15. $2^7 \cdot 2^{-3}$

16. $3^{-4} \cdot 3^6$

17. $10^{-5} \cdot 10^2$

18. $10^4 \cdot 10^{-6}$

19. $10^{-1} \cdot 10^{-2}$

20. $10^{-2} \cdot 10^{-2}$

21. $(3^{-1})^{-3}$

22. $(2^{-2})^{-4}$

23. $(5^3)^{-1}$

24. $(3^{-1})^3$

25. $(2^3 \cdot 3^{-2})^{-1}$

26. $(2^{-2} \cdot 3^{-1})^{-3}$

27. $(4^2 \cdot 5^{-1})^2$

28. $(2^{-3} \cdot 4^{-1})^{-1}$

29. $\left(\dfrac{2^{-1}}{5^{-2}}\right)^{-1}$

30. $\left(\dfrac{2^{-4}}{3^{-2}}\right)^{-2}$

31. $\left(\dfrac{2^{-1}}{3^{-2}}\right)^2$

32. $\left(\dfrac{3^2}{5^{-1}}\right)^{-1}$

33. $\dfrac{3^3}{3^{-1}}$

34. $\dfrac{2^{-2}}{2^3}$

35. $\dfrac{10^{-2}}{10^2}$

36. $\dfrac{10^{-2}}{10^{-5}}$

37. $2^{-2} + 3^{-2}$

38. $2^{-4} + 5^{-1}$

39. $\left(\dfrac{1}{3}\right)^{-1} - \left(\dfrac{2}{5}\right)^{-1}$

40. $\left(\dfrac{3}{2}\right)^{-1} - \left(\dfrac{1}{4}\right)^{-1}$

41. $(2^{-3} + 3^{-2})^{-1}$

42. $(5^{-1} - 2^{-3})^{-1}$

For Problems 43–62, simplify each expression. Express final results without using zero or negative integers as exponents. (Objective 2)

43. $x^2 \cdot x^{-8}$

44. $x^{-3} \cdot x^{-4}$

45. $a^3 \cdot a^{-5} \cdot a^{-1}$

46. $b^{-2} \cdot b^3 \cdot b^{-6}$

47. $(a^{-4})^2$

48. $(b^4)^{-3}$

49. $(x^2y^{-6})^{-1}$

50. $(x^5y^{-1})^{-3}$

51. $(ab^3c^{-2})^{-4}$

52. $(a^3b^{-3}c^{-2})^{-5}$

53. $(2x^3y^{-4})^{-3}$

54. $(4x^5y^{-2})^{-2}$

55. $\left(\dfrac{x^{-1}}{y^{-4}}\right)^{-3}$

56. $\left(\dfrac{y^3}{x^{-4}}\right)^{-2}$

57. $\left(\dfrac{3a^{-2}}{2b^{-1}}\right)^{-2}$

58. $\left(\dfrac{2xy^2}{5a^{-1}b^{-2}}\right)^{-1}$

59. $\dfrac{x^{-6}}{x^{-4}}$

60. $\dfrac{a^{-2}}{a^2}$

61. $\dfrac{a^3b^{-2}}{a^{-2}b^{-4}}$

62. $\dfrac{x^{-3}y^{-4}}{x^2y^{-1}}$

For Problems 63–74, find the indicated products and quotients. Express final results using positive integral exponents only. (Objective 3)

63. $(2xy^{-1})(3x^{-2}y^4)$

64. $(-4x^{-1}y^2)(6x^3y^{-4})$

65. $(-7a^2b^{-5})(-a^{-2}b^7)$

66. $(-9a^{-3}b^{-6})(-12a^{-1}b^4)$

67. $\dfrac{28x^{-2}y^{-3}}{4x^{-3}y^{-1}}$

68. $\dfrac{63x^2y^{-4}}{7xy^{-4}}$

69. $\dfrac{-72a^2b^{-4}}{6a^3b^{-7}}$

70. $\dfrac{108a^{-5}b^{-4}}{9a^{-2}b}$

71. $\left(\dfrac{35x^{-1}y^{-2}}{7x^4y^3}\right)^{-1}$

72. $\left(\dfrac{-48ab^2}{-6a^3b^5}\right)^{-2}$

73. $\left(\dfrac{-36a^{-1}b^{-6}}{4a^{-1}b^4}\right)^{-2}$

74. $\left(\dfrac{8xy^3}{-4x^4y}\right)^{-3}$

For Problems 75–84, express each of the following as a single fraction involving positive exponents only. (Objective 4)

75. $x^{-2} + x^{-3}$

76. $x^{-1} + x^{-5}$

77. $x^{-3} - y^{-1}$

78. $2x^{-1} - 3y^{-2}$

79. $3a^{-2} + 4b^{-1}$

80. $a^{-1} + a^{-1}b^{-3}$

81. $x^{-1}y^{-2} - xy^{-1}$

82. $x^2y^{-2} - x^{-1}y^{-3}$

83. $2x^{-1} - 3x^{-2}$

84. $5x^{-2}y + 6x^{-1}y^{-2}$

For Problems 85–92, write each of the following in scientific notation. (Objective 5)

For example, $27,800 = (2.78)(10^4)$

85. 40,000,000

86. 500,000,000

87. 376.4

88. 9126.21

89. 0.347

90. 0.2165

91. 0.0214

92. 0.0037

For Problems 93–100, write each of the following in ordinary decimal notation. (Objective 5)

For example, $(3.18)(10^2) = 318$

93. $(3.14)(10^{10})$

94. $(2.04)(10^{12})$

95. $(4.3)(10^{-1})$

96. $(5.2)(10^{-2})$

97. $(9.14)(10^{-4})$

98. $(8.76)(10^{-5})$

99. $(5.123)(10^{-8})$

100. $(6)(10^{-9})$

For Problems 101–104, use scientific notation and the properties of exponents to help you perform the following operations. (Objective 6)

101. $\dfrac{(60,000)(0.006)}{(0.0009)(400)}$

102. $\dfrac{(0.00063)(960,000)}{(3,200)(0.0000021)}$

103. $\dfrac{(0.0045)(60,000)}{(1800)(0.00015)}$

104. $\dfrac{(0.00016)(300)(0.028)}{0.064}$

Thoughts Into Words

105. Is the following simplification process correct?

$$(3^{-2})^{-1} = \left(\dfrac{1}{3^2}\right)^{-1} = \left(\dfrac{1}{9}\right)^{-1} = \dfrac{1}{\left(\dfrac{1}{9}\right)^1} = 9$$

Could you suggest a better way to do the problem?

106. Explain how to simplify $(2^{-1} \cdot 3^{-2})^{-1}$ and also how to simplify $(2^{-1} + 3^{-2})^{-1}$.

Further Investigations

107. Use a calculator to check your answers for Problems 1–42.

108. Use a calculator to simplify each of the following numerical expressions. Express your answers to the nearest hundredth.

(a) $(2^{-3} + 3^{-3})^{-2}$

(b) $(4^{-3} - 2^{-1})^{-2}$

(c) $(5^{-3} - 3^{-5})^{-1}$

(d) $(6^{-2} + 7^{-4})^{-2}$

(e) $(7^{-3} - 2^{-4})^{-2}$

(f) $(3^{-4} + 2^{-3})^{-3}$

Answers to the Concept Quiz

1. True   **2.** False   **3.** False   **4.** False   **5.** True   **6.** True   **7.** False   **8.** True   **9.** True   **10.** False

10.2   Roots and Radicals

OBJECTIVES

1 Evaluate roots of numbers

2 Express a radical in simplest radical form

3 Rationalizing the denominator to simplify radicals

4 Applications of radicals

To **square a number** means to raise it to the second power—that is, to use the number as a factor twice.

$$4^2 = 4 \cdot 4 = 16 \qquad \text{Read "four squared equals sixteen"}$$

$$10^2 = 10 \cdot 10 = 100$$

$$\left(\frac{1}{2}\right)^2 = \frac{1}{2} \cdot \frac{1}{2} = \frac{1}{4}$$

$$(-3)^2 = (-3)(-3) = 9$$

A **square root of a number** is one of its two equal factors. Thus 4 is a square root of 16 because $4 \cdot 4 = 16$. Likewise, -4 is also a square root of 16 because $(-4)(-4) = 16$. In general, a is a square root of b if $a^2 = b$. The following generalizations are a direct consequence of the previous statement.

1. Every positive real number has two square roots; one is positive and the other is negative. They are opposites of each other.

2. Negative real numbers have no real number square roots because any real number, except zero, is positive when squared.

3. The square root of 0 is 0.

The symbol $\sqrt{}$, called a **radical sign**, is used to designate the nonnegative or principal square root. The number under the radical sign is called the **radicand**. The entire expression, such as $\sqrt{16}$, is called a **radical**.

$$\sqrt{16} = 4 \qquad \sqrt{16} \text{ indicates the nonnegative or principal square root of 16}$$

$$-\sqrt{16} = -4 \qquad -\sqrt{16} \text{ indicates the negative square root of 16}$$

$$\sqrt{0} = 0 \qquad \text{Zero has only one square root. Technically, we could write } -\sqrt{0} = -0 = 0$$

$\sqrt{-4}$ is not a real number

$-\sqrt{-4}$ is not a real number

In general, the following definition is useful.

> **Definition 10.1 Principal Square Root**
>
> If $a \geq 0$ and $b \geq 0$, then $\sqrt{b} = a$ if and only if $a^2 = b$; a is called the **principal square root of b**.

To **cube a number** means to raise it to the third power—that is, to use the number as a factor three times.

$$2^3 = 2 \cdot 2 \cdot 2 = 8 \qquad \text{Read "two cubed equals eight"}$$
$$4^3 = 4 \cdot 4 \cdot 4 = 64$$
$$\left(\frac{2}{3}\right)^3 = \frac{2}{3} \cdot \frac{2}{3} \cdot \frac{2}{3} = \frac{8}{27}$$
$$(-2)^3 = (-2)(-2)(-2) = -8$$

A **cube root of a number** is one of its three equal factors. Thus 2 is a cube root of 8 because $2 \cdot 2 \cdot 2 = 8$. (In fact, 2 is the only real number that is a cube root of 8.) Furthermore, -2 is a cube root of -8 because $(-2)(-2)(-2) = -8$. (In fact, -2 is the only real number that is a cube root of -8.)

In general, a is a cube root of b if $a^3 = b$. The following generalizations are a direct consequence of the previous statement.

1. Every positive real number has one positive real number cube root.

2. Every negative real number has one negative real number cube root.

3. The cube root of 0 is 0.

Remark: Technically, every nonzero real number has three cube roots, but only one of them is a real number. The other two roots are classified as imaginary numbers. We are restricting our work at this time to the set of real numbers.

The symbol $\sqrt[3]{}$ designates the cube root of a number. Thus we can write

$$\sqrt[3]{8} = 2 \qquad\qquad \sqrt[3]{\frac{1}{27}} = \frac{1}{3}$$

$$\sqrt[3]{-8} = -2 \qquad\qquad \sqrt[3]{-\frac{1}{27}} = -\frac{1}{3}$$

In general, the following definition is useful.

> **Definition 10.2 Cube Root of a Number**
>
> $\sqrt[3]{b} = a$ if and only if $a^3 = b$.

In Definition 10.2, if b is a positive number, then a, the cube root, is a positive number; whereas if b is a negative number, then a, the cube root, is a negative number. The number a is called the principal cube root of b or simply the cube root of b.

The concept of root can be extended to fourth roots, fifth roots, sixth roots, and, in general, nth roots.

> **Definition 10.3 nth Root of a Number**
>
> The nth root of b is a if and only if $a^n = b$.

We can make the following generalizations.

If n is an even positive integer, then the following statements are true.

1. Every positive real number has exactly two real *n*th roots—one positive and one negative. For example, the real fourth roots of 16 are 2 and -2.

2. Negative real numbers do not have real *n*th roots. For example, there are no real fourth roots of -16.

If n is an odd positive integer greater than 1, then the following statements are true.

1. Every real number has exactly one real *n*th root.

2. The real *n*th root of a positive number is positive. For example, the fifth root of 32 is 2.

3. The real *n*th root of a negative number is negative. For example, the fifth root of -32 is -2.

The symbol $\sqrt[n]{}$ designates the principal *n*th root. To complete our terminology, the *n* in the radical $\sqrt[n]{b}$ is called the **index** of the radical. If $n = 2$, we commonly write $\sqrt{b}$ instead of $\sqrt[2]{b}$.

The following chart can help summarize this information with respect to $\sqrt[n]{b}$, where *n* is a positive integer greater than 1.

	If *b* is positive	If *b* is zero	If *b* is negative
n is even	$\sqrt[n]{b}$ is a positive real number	$\sqrt[n]{b} = 0$	$\sqrt[n]{b}$ is not a real number
n is odd	$\sqrt[n]{b}$ is a positive real number	$\sqrt[n]{b} = 0$	$\sqrt[n]{b}$ is a negative real number

Consider the following examples.

$\sqrt[4]{81} = 3$ because $3^4 = 81$

$\sqrt[5]{32} = 2$ because $2^5 = 32$

$\sqrt[5]{-32} = -2$ because $(-2)^5 = -32$

$\sqrt[4]{-16}$ is not a real number   because any real number, except zero, is positive when raised to the fourth power

The following property is a direct consequence of Definition 10.3.

> **Property 10.1**
>
> 1. $\left(\sqrt[n]{b}\right)^n = b$ *n* is any positive integer greater than 1
>
> 2. $\sqrt[n]{b^n} = b$ *n* is any positive integer greater than 1 if $b \geq 0$; *n* is an odd positive integer greater than 1 if $b < 0$

Because the radical expressions in parts (1) and (2) of Property 10.1 are both equal to *b*, by the transitive property they are equal to each other. Hence $\sqrt[n]{b^n} = \left(\sqrt[n]{b}\right)^n$. The arithmetic is usually easier to simplify when we use the form $\left(\sqrt[n]{b}\right)^n$. The following examples demonstrate the use of Property 10.1.

$$\sqrt{144^2} = \left(\sqrt{144}\right)^2 = 12^2 = 144$$

$$\sqrt[3]{64^3} = \left(\sqrt[3]{64}\right)^3 = 4^3 = 64$$

$$\sqrt[3]{(-8)^3} = \left(\sqrt[3]{-8}\right)^3 = (-2)^3 = -8$$

$$\sqrt[4]{16^4} = \left(\sqrt[4]{16}\right)^4 = 2^4 = 16$$

Let's use some examples to lead into the next very useful property of radicals.

$$\sqrt{4 \cdot 9} = \sqrt{36} = 6 \qquad \text{and} \qquad \sqrt{4} \cdot \sqrt{9} = 2 \cdot 3 = 6$$

$$\sqrt{16 \cdot 25} = \sqrt{400} = 20 \qquad \text{and} \qquad \sqrt{16} \cdot \sqrt{25} = 4 \cdot 5 = 20$$

$$\sqrt[3]{8 \cdot 27} = \sqrt[3]{216} = 6 \qquad \text{and} \qquad \sqrt[3]{8} \cdot \sqrt[3]{27} = 2 \cdot 3 = 6$$

$$\sqrt[3]{(-8)(27)} = \sqrt[3]{-216} = -6 \qquad \text{and} \qquad \sqrt[3]{-8} \cdot \sqrt[3]{27} = (-2)(3) = -6$$

In general, we can state the following property.

> ## Property 10.2
>
> $\sqrt[n]{bc} = \sqrt[n]{b}\sqrt[n]{c}$, when $\sqrt[n]{b}$ and $\sqrt[n]{c}$ are real numbers.

Property 10.2 states that the nth root of a product is equal to the product of the nth roots.

Expressing a Radical in Simplest Radical Form

The definition of nth root, along with Property 10.2, provides the basis for changing radicals to simplest radical form. The concept of **simplest radical form** takes on additional meaning as we encounter more complicated expressions, but for now it simply means that the radicand is not to contain any perfect powers of the index. Let's consider some examples to clarify this idea.

Classroom Example
Express each of the following in simplest radical form:

(a) $\sqrt{12}$

(b) $\sqrt{32}$

(c) $\sqrt[3]{48}$

(d) $\sqrt[3]{40}$

EXAMPLE 1   Express each of the following in simplest radical form:

(a) $\sqrt{8}$ (b) $\sqrt{45}$ (c) $\sqrt[3]{24}$ (d) $\sqrt[3]{54}$

Solution

(a) $\sqrt{8} = \sqrt{4 \cdot 2} = \sqrt{4}\sqrt{2} = 2\sqrt{2}$ 4 is a perfect square

(b) $\sqrt{45} = \sqrt{9 \cdot 5} = \sqrt{9}\sqrt{5} = 3\sqrt{5}$ 9 is a perfect square

(c) $\sqrt[3]{24} = \sqrt[3]{8 \cdot 3} = \sqrt[3]{8}\sqrt[3]{3} = 2\sqrt[3]{3}$ 8 is a perfect cube

(d) $\sqrt[3]{54} = \sqrt[3]{27 \cdot 2} = \sqrt[3]{27}\sqrt[3]{2} = 3\sqrt[3]{2}$ 27 is a perfect cube

The first step in each example is to express the radicand of the given radical as the product of two factors, one of which must be a perfect nth power other than 1. Also, observe the radicands of the final radicals. In each case, the radicand cannot have a factor that is a perfect nth power other than 1. We say that the final radicals $2\sqrt{2}, 3\sqrt{5}, 2\sqrt[3]{3}$, and $3\sqrt[3]{2}$ are in *simplest radical form*.

You may vary the steps somewhat in changing to simplest radical form, but the final result should be the same. Consider some different approaches to changing $\sqrt{72}$ to simplest form:

$$\sqrt{72} = \sqrt{9}\sqrt{8} = 3\sqrt{8} = 3\sqrt{4}\sqrt{2} = 3 \cdot 2\sqrt{2} = 6\sqrt{2} \qquad \text{or}$$

$$\sqrt{72} = \sqrt{4}\sqrt{18} = 2\sqrt{18} = 2\sqrt{9}\sqrt{2} = 2 \cdot 3\sqrt{2} = 6\sqrt{2} \qquad \text{or}$$

$$\sqrt{72} = \sqrt{36}\sqrt{2} = 6\sqrt{2}$$

Another variation of the technique for changing radicals to simplest form is to prime factor the radicand and then to look for perfect nth powers in exponential form. The following example illustrates the use of this technique.

EXAMPLE 2   Express each of the following in simplest radical form:

(a) $\sqrt{50}$ (b) $3\sqrt{80}$ (c) $\sqrt[3]{108}$

Solution

(a) $\sqrt{50} = \sqrt{2 \cdot 5 \cdot 5} = \sqrt{5^2}\sqrt{2} = 5\sqrt{2}$

(b) $3\sqrt{80} = 3\sqrt{2 \cdot 2 \cdot 2 \cdot 2 \cdot 5} = 3\sqrt{2^4}\sqrt{5} = 3 \cdot 2^2\sqrt{5} = 12\sqrt{5}$

(c) $\sqrt[3]{108} = \sqrt[3]{2 \cdot 2 \cdot 3 \cdot 3 \cdot 3} = \sqrt[3]{3^3}\sqrt[3]{4} = 3\sqrt[3]{4}$

Another property of nth roots is demonstrated by the following examples.

$$\sqrt{\frac{36}{9}} = \sqrt{4} = 2 \qquad \text{and} \qquad \frac{\sqrt{36}}{\sqrt{9}} = \frac{6}{3} = 2$$

$$\sqrt[3]{\frac{64}{8}} = \sqrt[3]{8} = 2 \qquad \text{and} \qquad \frac{\sqrt[3]{64}}{\sqrt[3]{8}} = \frac{4}{2} = 2$$

$$\sqrt[3]{\frac{-8}{64}} = \sqrt[3]{-\frac{1}{8}} = -\frac{1}{2} \qquad \text{and} \qquad \frac{\sqrt[3]{-8}}{\sqrt[3]{64}} = \frac{-2}{4} = -\frac{1}{2}$$

In general, we can state the following property.

Property 10.3

$$\sqrt[n]{\frac{b}{c}} = \frac{\sqrt[n]{b}}{\sqrt[n]{c}}, \text{ when } \sqrt[n]{b} \text{ and } \sqrt[n]{c} \text{ are real numbers, and } c \neq 0.$$

Property 10.3 states that the nth root of a quotient is equal to the quotient of the nth roots.

To evaluate radicals such as $\sqrt{\dfrac{4}{25}}$ and $\sqrt[3]{\dfrac{27}{8}}$, for which the numerator and denominator of the fractional radicand are perfect nth powers, you may use Property 10.3 or merely rely on the definition of nth root.

$$\sqrt{\frac{4}{25}} = \frac{\sqrt{4}}{\sqrt{25}} = \frac{2}{5} \quad \text{or} \quad \sqrt{\frac{4}{25}} = \frac{2}{5} \quad \text{because} \quad \frac{2}{5} \cdot \frac{2}{5} = \frac{4}{25}$$

$$\uparrow \text{Property 10.3} \qquad\qquad\qquad \uparrow \text{Definition of } n\text{th root}$$

$$\sqrt[3]{\frac{27}{8}} = \frac{\sqrt[3]{27}}{\sqrt[3]{8}} = \frac{3}{2} \quad \text{or} \quad \sqrt[3]{\frac{27}{8}} = \frac{3}{2} \quad \text{because} \quad \frac{3}{2} \cdot \frac{3}{2} \cdot \frac{3}{2} = \frac{27}{8}$$

Radicals such as $\sqrt{\dfrac{28}{9}}$ and $\sqrt[3]{\dfrac{24}{27}}$, in which only the denominators of the radicand are perfect nth powers, can be simplified as follows:

$$\sqrt{\frac{28}{9}} = \frac{\sqrt{28}}{\sqrt{9}} = \frac{\sqrt{28}}{3} = \frac{\sqrt{4}\sqrt{7}}{3} = \frac{2\sqrt{7}}{3}$$

$$\sqrt[3]{\frac{24}{27}} = \frac{\sqrt[3]{24}}{\sqrt[3]{27}} = \frac{\sqrt[3]{24}}{3} = \frac{\sqrt[3]{8}\sqrt[3]{3}}{3} = \frac{2\sqrt[3]{3}}{3}$$

Before we consider more examples, let's summarize some ideas that pertain to the simplifying of radicals. A radical is said to be in *simplest radical form* if the following conditions are satisfied.

---

1. No fraction appears with a radical sign. $\sqrt{\dfrac{3}{4}}$ violates this condition

2. No radical appears in the denominator. $\dfrac{\sqrt{2}}{\sqrt{3}}$ violates this condition

3. No radicand, when expressed in prime-factored form, contains a factor raised to a power equal to or greater than the index. $\sqrt{2^3 \cdot 5}$ violates this condition

---

Rationalizing the Denominator to Simplify Radicals

Now let's consider an example in which neither the numerator nor the denominator of the radicand is a perfect *n*th power.

Classroom Example

Simplify $\sqrt{\dfrac{5}{7}}$.

EXAMPLE 3

Simplify $\sqrt{\dfrac{2}{3}}$.

Solution

$$\sqrt{\frac{2}{3}} = \frac{\sqrt{2}}{\sqrt{3}} = \frac{\sqrt{2}}{\sqrt{3}} \cdot \frac{\sqrt{3}}{\sqrt{3}} = \frac{\sqrt{6}}{3}$$

↑ Form of 1

We refer to the process we used to simplify the radical in Example 3 as **rationalizing the denominator**. Note that the denominator becomes a rational number. The process of rationalizing the denominator can often be accomplished in more than one way, as we will see in the next example.

Classroom Example

Simplify $\dfrac{\sqrt{7}}{\sqrt{12}}$.

EXAMPLE 4

Simplify $\dfrac{\sqrt{5}}{\sqrt{8}}$.

Solution A

$$\frac{\sqrt{5}}{\sqrt{8}} = \frac{\sqrt{5}}{\sqrt{8}} \cdot \frac{\sqrt{8}}{\sqrt{8}} = \frac{\sqrt{40}}{8} = \frac{\sqrt{4}\sqrt{10}}{8} = \frac{2\sqrt{10}}{8} = \frac{\sqrt{10}}{4}$$

Solution B

$$\frac{\sqrt{5}}{\sqrt{8}} = \frac{\sqrt{5}}{\sqrt{8}} \cdot \frac{\sqrt{2}}{\sqrt{2}} = \frac{\sqrt{10}}{\sqrt{16}} = \frac{\sqrt{10}}{4}$$

Solution C

$$\frac{\sqrt{5}}{\sqrt{8}} = \frac{\sqrt{5}}{\sqrt{4}\sqrt{2}} = \frac{\sqrt{5}}{2\sqrt{2}} = \frac{\sqrt{5}}{2\sqrt{2}} \cdot \frac{\sqrt{2}}{\sqrt{2}} = \frac{\sqrt{10}}{2\sqrt{4}} = \frac{\sqrt{10}}{2(2)} = \frac{\sqrt{10}}{4}$$

The three approaches to Example 4 again illustrate the need to think first and only then push the pencil. You may find one approach easier than another. To conclude this section, study the following examples and check the final radicals against the three conditions previously listed for simplest radical form.

Classroom Example
Simplify each of the following:

(a) $\dfrac{3\sqrt{5}}{4\sqrt{3}}$ (b) $\dfrac{6\sqrt{11}}{5\sqrt{27}}$

(c) $\sqrt[3]{\dfrac{7}{25}}$ (d) $\dfrac{\sqrt[3]{3}}{\sqrt[3]{36}}$

EXAMPLE 5   Simplify each of the following:

(a) $\dfrac{3\sqrt{2}}{5\sqrt{3}}$ (b) $\dfrac{3\sqrt{7}}{2\sqrt{18}}$ (c) $\sqrt[3]{\dfrac{5}{9}}$ (d) $\dfrac{\sqrt[3]{5}}{\sqrt[3]{16}}$

Solution

(a) $\dfrac{3\sqrt{2}}{5\sqrt{3}} = \dfrac{3\sqrt{2}}{5\sqrt{3}} \cdot \dfrac{\sqrt{3}}{\sqrt{3}} = \dfrac{3\sqrt{6}}{5\sqrt{9}} = \dfrac{3\sqrt{6}}{15} = \dfrac{\sqrt{6}}{5}$

↑ Form of 1

(b) $\dfrac{3\sqrt{7}}{2\sqrt{18}} = \dfrac{3\sqrt{7}}{2\sqrt{18}} \cdot \dfrac{\sqrt{2}}{\sqrt{2}} = \dfrac{3\sqrt{14}}{2\sqrt{36}} = \dfrac{3\sqrt{14}}{12} = \dfrac{\sqrt{14}}{4}$

↑ Form of 1

(c) $\sqrt[3]{\dfrac{5}{9}} = \dfrac{\sqrt[3]{5}}{\sqrt[3]{9}} = \dfrac{\sqrt[3]{5}}{\sqrt[3]{9}} \cdot \dfrac{\sqrt[3]{3}}{\sqrt[3]{3}} = \dfrac{\sqrt[3]{15}}{\sqrt[3]{27}} = \dfrac{\sqrt[3]{15}}{3}$

↑ Form of 1

(d) $\dfrac{\sqrt[3]{5}}{\sqrt[3]{16}} = \dfrac{\sqrt[3]{5}}{\sqrt[3]{16}} \cdot \dfrac{\sqrt[3]{4}}{\sqrt[3]{4}} = \dfrac{\sqrt[3]{20}}{\sqrt[3]{64}} = \dfrac{\sqrt[3]{20}}{4}$

↑ Form of 1

Applications of Radicals

Many real-world applications involve radical expressions. For example, police often use the formula $S = \sqrt{30Df}$ to estimate the speed of a car on the basis of the length of the skid marks at the scene of an accident. In this formula, S represents the speed of the car in miles per hour, D represents the length of the skid marks in feet, and f represents a coefficient of friction. For a particular situation, the coefficient of friction is a constant that depends on the type and condition of the road surface.

Classroom Example
Using 0.46 as a coefficient of friction, determine how fast a car was traveling if it skidded 275 feet.

EXAMPLE 6

Using 0.35 as a coefficient of friction, determine how fast a car was traveling if it skidded 325 feet.

Solution

Substitute 0.35 for f and 325 for D in the formula.

$$S = \sqrt{30Df} = \sqrt{30(325)(0.35)} = 58 \quad \text{to the nearest whole number}$$

The car was traveling at approximately 58 miles per hour.

The **period** of a pendulum is the time it takes to swing from one side to the other side and back. The formula

$$T = 2\pi\sqrt{\dfrac{L}{32}}$$

where T represents the time in seconds and L the length in feet, can be used to determine the period of a pendulum (see Figure 10.1).

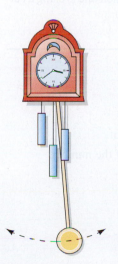

Figure 10.1

EXAMPLE 7

Find, to the nearest tenth of a second, the period of a pendulum of length 3.5 feet.

Solution

Let's use 3.14 as an approximation for π and substitute 3.5 for L in the formula.

$$T = 2\pi\sqrt{\frac{L}{32}} = 2(3.14)\sqrt{\frac{3.5}{32}} = 2.1 \quad \text{to the nearest tenth}$$

The period is approximately 2.1 seconds.

Radical expressions are also used in some geometric applications. For example, the area of a triangle can be found by using a formula that involves a square root. If a, b, and c represent the lengths of the three sides of a triangle, the formula $K = \sqrt{s(s - a)(s - b)(s - c)}$, known as Heron's formula, can be used to determine the area (K) of the triangle. The letter s represents the semiperimeter of the triangle; that is, $s = \dfrac{a + b + c}{2}$.

EXAMPLE 8

Find the area of a triangular piece of sheet metal that has sides of lengths 17 inches, 19 inches, and 26 inches.

Solution

First, let's find the value of s, the semiperimeter of the triangle.

$$s = \frac{17 + 19 + 26}{2} = 31$$

Now we can use Heron's formula.

$$K = \sqrt{s(s - a)(s - b)(s - c)} = \sqrt{31(31 - 17)(31 - 19)(31 - 26)}$$
$$= \sqrt{31(14)(12)(5)}$$
$$= \sqrt{20{,}640}$$
$$= 161.4 \quad \text{to the nearest tenth}$$

Thus the area of the piece of sheet metal is approximately 161.4 square inches.

Remark: Note that in Examples 6–8, we did not simplify the radicals. When one is using a calculator to approximate the square roots, there is no need to simplify first.

Concept Quiz 10.2

For Problems 1–10, answer true or false.

1. The cube root of a number is one of its three equal factors.

2. Every positive real number has one positive real number square root.

3. The principal square root of a number is the positive square root of the number.

4. The symbol $\sqrt{}$ is called a radical.

5. The square root of 0 is not a real number.

6. The number under the radical sign is called the radicand.

7. Every positive real number has two square roots.

8. The n in the radical $\sqrt[n]{a}$ is called the index of the radical.

9. If n is an odd integer greater than 1, and b is a negative real number, then $\sqrt[n]{b}$ is a negative real number.

10. $\dfrac{3\sqrt{24}}{8}$ is in simplest radical form.

Problem Set 10.2

For Problems 1–20, evaluate each of the following. For example, $\sqrt{25} = 5$. **(Objective 1)**

1. $\sqrt{64}$ **2.** $\sqrt{49}$

3. $-\sqrt{100}$ **4.** $-\sqrt{81}$

5. $\sqrt[3]{27}$ **6.** $\sqrt[3]{216}$

7. $\sqrt[3]{-64}$ **8.** $\sqrt[3]{-125}$

9. $\sqrt[4]{81}$ **10.** $-\sqrt[4]{16}$

11. $\sqrt{\dfrac{16}{25}}$ **12.** $\sqrt{\dfrac{25}{64}}$

13. $-\sqrt{\dfrac{36}{49}}$ **14.** $\sqrt{\dfrac{16}{64}}$

15. $\sqrt{\dfrac{9}{36}}$ **16.** $\sqrt{\dfrac{144}{36}}$

17. $\sqrt[3]{\dfrac{27}{64}}$ **18.** $\sqrt[3]{-\dfrac{8}{27}}$

19. $\sqrt[3]{8^3}$ **20.** $\sqrt[4]{16^4}$

For Problems 21–74, change each radical to simplest radical form. **(Objectives 2 and 3)**

21. $\sqrt{27}$ **22.** $\sqrt{48}$

23. $\sqrt{32}$ **24.** $\sqrt{98}$

25. $\sqrt{80}$ **26.** $\sqrt{125}$

27. $\sqrt{160}$ **28.** $\sqrt{112}$

29. $4\sqrt{18}$ **30.** $5\sqrt{32}$

31. $-6\sqrt{20}$ **32.** $-4\sqrt{54}$

33. $\dfrac{2}{5}\sqrt{75}$ **34.** $\dfrac{1}{3}\sqrt{90}$

35. $\dfrac{3}{2}\sqrt{24}$ **36.** $\dfrac{3}{4}\sqrt{45}$

37. $-\dfrac{5}{6}\sqrt{28}$ **38.** $-\dfrac{2}{3}\sqrt{96}$

39. $\sqrt{\dfrac{19}{4}}$ **40.** $\sqrt{\dfrac{22}{9}}$

41. $\sqrt{\dfrac{27}{16}}$ **42.** $\sqrt{\dfrac{8}{25}}$

43. $\sqrt{\dfrac{75}{81}}$ **44.** $\sqrt{\dfrac{24}{49}}$

45. $\sqrt{\dfrac{2}{7}}$ **46.** $\sqrt{\dfrac{3}{8}}$

47. $\sqrt{\dfrac{2}{3}}$ **48.** $\sqrt{\dfrac{7}{12}}$

49. $\dfrac{\sqrt{5}}{\sqrt{12}}$ **50.** $\dfrac{\sqrt{3}}{\sqrt{7}}$

51. $\dfrac{\sqrt{11}}{\sqrt{24}}$ **52.** $\dfrac{\sqrt{5}}{\sqrt{48}}$

53. $\dfrac{\sqrt{18}}{\sqrt{27}}$ **54.** $\dfrac{\sqrt{10}}{\sqrt{20}}$

55. $\dfrac{\sqrt{35}}{\sqrt{7}}$ **56.** $\dfrac{\sqrt{42}}{\sqrt{6}}$

57. $\dfrac{2\sqrt{3}}{\sqrt{7}}$ **58.** $\dfrac{3\sqrt{2}}{\sqrt{6}}$

59. $-\dfrac{4\sqrt{12}}{\sqrt{5}}$ **60.** $\dfrac{-6\sqrt{5}}{\sqrt{18}}$

61. $\dfrac{3\sqrt{2}}{4\sqrt{3}}$ **62.** $\dfrac{6\sqrt{5}}{5\sqrt{12}}$

63. $\dfrac{-8\sqrt{18}}{10\sqrt{50}}$ **64.** $\dfrac{4\sqrt{45}}{-6\sqrt{20}}$

65. $\sqrt[3]{16}$ **66.** $\sqrt[3]{40}$

67. $2\sqrt[3]{81}$ **68.** $-3\sqrt[3]{54}$

69. $\dfrac{2}{\sqrt[3]{9}}$ **70.** $\dfrac{3}{\sqrt[3]{3}}$

71. $\dfrac{\sqrt[3]{27}}{\sqrt[3]{4}}$ **72.** $\dfrac{\sqrt[3]{8}}{\sqrt[3]{16}}$

73. $\dfrac{\sqrt[3]{6}}{\sqrt[3]{4}}$ **74.** $\dfrac{\sqrt[3]{4}}{\sqrt[3]{2}}$

For Problems 75–80, use radicals to solve the problems. **(Objective 4)**

75. Use a coefficient of friction of 0.4 in the formula from Example 6 and find the speeds of cars that left skid marks of lengths 150 feet, 200 feet, and 350 feet. Express your answers to the nearest mile per hour.

76. Use the formula from Example 7, and find the periods of pendulums of lengths 2 feet, 3 feet, and 4.5 feet. Express your answers to the nearest tenth of a second.

77. Find, to the nearest square centimeter, the area of a triangle that measures 14 centimeters by 16 centimeters by 18 centimeters.

78. Find, to the nearest square yard, the area of a triangular plot of ground that measures 45 yards by 60 yards by 75 yards.

79. Find the area of an equilateral triangle, each of whose sides is 18 inches long. Express the area to the nearest square inch.

80. Find, to the nearest square inch, the area of the quadrilateral in Figure 10.2.

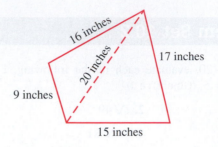

Figure 10.2

Thoughts Into Words

81. Why is $\sqrt{-9}$ not a real number?

82. Why is it that we say 25 has two square roots (5 and −5), but we write $\sqrt{25} = 5$?

83. How is the multiplication property of 1 used when simplifying radicals?

84. How could you find a whole number approximation for $\sqrt{2750}$ if you did not have a calculator or table available?

Further Investigations

85. Use your calculator to find a rational approximation, to the nearest thousandth, for (a) through (i).

(a) $\sqrt{2}$ (b) $\sqrt{75}$

(c) $\sqrt{156}$ (d) $\sqrt{691}$

(e) $\sqrt{3249}$ (f) $\sqrt{45{,}123}$

(g) $\sqrt{0.14}$ (h) $\sqrt{0.023}$

(i) $\sqrt{0.8649}$

86. Sometimes a fairly good estimate can be made of a radical expression by using whole number approximations. For example, $5\sqrt{35} + 7\sqrt{50}$ is approximately $5(6) + 7(7) = 79$. Using a calculator, we find that

$5\sqrt{35} + 7\sqrt{50} = 79.1$, to the nearest tenth. In this case our whole number estimate is very good. For (a) through (f), first make a whole number estimate, and then use your calculator to see how well you estimated.

(a) $3\sqrt{10} - 4\sqrt{24} + 6\sqrt{65}$

(b) $9\sqrt{27} + 5\sqrt{37} - 3\sqrt{80}$

(c) $12\sqrt{5} + 13\sqrt{18} + 9\sqrt{47}$

(d) $3\sqrt{98} - 4\sqrt{83} - 7\sqrt{120}$

(e) $4\sqrt{170} + 2\sqrt{198} + 5\sqrt{227}$

(f) $-3\sqrt{256} - 6\sqrt{287} + 11\sqrt{321}$

Answers to the Concept Quiz

1. True   **2.** True   **3.** True   **4.** False   **5.** False   **6.** True   **7.** True   **8.** True   **9.** True   **10.** False

10.3   Simplifying and Combining Radicals

OBJECTIVES

1 Simplify expressions by combining radicals

2 Simplify radicals that contain variables

Recall our use of the distributive property as the basis for combining similar terms. For example,

$$3x + 2x = (3 + 2)x = 5x$$

$$8y - 5y = (8 - 5)y = 3y$$

$$\frac{2}{3}a^2 + \frac{3}{4}a^2 = \left(\frac{2}{3} + \frac{3}{4}\right)a^2 = \left(\frac{8}{12} + \frac{9}{12}\right)a^2 = \frac{17}{12}a^2$$

In a like manner, expressions that contain radicals can often be simplified by using the distributive property, as follows:

$$3\sqrt{2} + 5\sqrt{2} = (3 + 5)\sqrt{2} = 8\sqrt{2}$$

$$7\sqrt[3]{5} - 3\sqrt[3]{5} = (7 - 3)\sqrt[3]{5} = 4\sqrt[3]{5}$$

$$4\sqrt{7} + 5\sqrt{7} + 6\sqrt{11} - 2\sqrt{11} = (4 + 5)\sqrt{7} + (6 - 2)\sqrt{11} = 9\sqrt{7} + 4\sqrt{11}$$

Note that *in order to be added or subtracted, radicals must have the same index and the same radicand.* Thus we cannot simplify an expression such as $5\sqrt{2} + 7\sqrt{11}$.

Simplifying by combining radicals sometimes requires that you first express the given radicals in simplest form and then apply the distributive property. The following examples illustrate this idea.

Classroom Example
Simplify $2\sqrt{12} + 5\sqrt{48} - 7\sqrt{3}$.

EXAMPLE 1   Simplify $3\sqrt{8} + 2\sqrt{18} - 4\sqrt{2}$.

Solution

$$3\sqrt{8} + 2\sqrt{18} - 4\sqrt{2} = 3\sqrt{4}\sqrt{2} + 2\sqrt{9}\sqrt{2} - 4\sqrt{2}$$

$$= 3 \cdot 2 \cdot \sqrt{2} + 2 \cdot 3 \cdot \sqrt{2} - 4\sqrt{2}$$

$$= 6\sqrt{2} + 6\sqrt{2} - 4\sqrt{2}$$

$$= (6 + 6 - 4)\sqrt{2} = 8\sqrt{2}$$

Classroom Example
Simplify $\frac{3}{4}\sqrt{98} - \frac{2}{5}\sqrt{50}$.

EXAMPLE 2   Simplify $\frac{1}{4}\sqrt{45} + \frac{1}{3}\sqrt{20}$.

Solution

$$\frac{1}{4}\sqrt{45} + \frac{1}{3}\sqrt{20} = \frac{1}{4}\sqrt{9}\sqrt{5} + \frac{1}{3}\sqrt{4}\sqrt{5}$$

$$= \frac{1}{4} \cdot 3 \cdot \sqrt{5} + \frac{1}{3} \cdot 2 \cdot \sqrt{5}$$

$$= \frac{3}{4}\sqrt{5} + \frac{2}{3}\sqrt{5} = \left(\frac{3}{4} + \frac{2}{3}\right)\sqrt{5}$$

$$= \left(\frac{9}{12} + \frac{8}{12}\right)\sqrt{5} = \frac{17}{12}\sqrt{5}$$

Classroom Example
Simplify $3\sqrt[3]{4} + 5\sqrt[3]{32} - 2\sqrt[3]{108}$.

EXAMPLE 3   Simplify $5\sqrt[3]{2} - 2\sqrt[3]{16} - 6\sqrt[3]{54}$.

Solution

$$5\sqrt[3]{2} - 2\sqrt[3]{16} - 6\sqrt[3]{54} = 5\sqrt[3]{2} - 2\sqrt[3]{8}\sqrt[3]{2} - 6\sqrt[3]{27}\sqrt[3]{2}$$

$$= 5\sqrt[3]{2} - 2 \cdot 2 \cdot \sqrt[3]{2} - 6 \cdot 3 \cdot \sqrt[3]{2}$$

$$= 5\sqrt[3]{2} - 4\sqrt[3]{2} - 18\sqrt[3]{2}$$

$$= (5 - 4 - 18)\sqrt[3]{2}$$

$$= -17\sqrt[3]{2}$$

Simplifying Radicals That Contain Variables

Before we discuss the process of simplifying radicals that contain variables, there is one technicality that we should call to your attention. Let's look at some examples to clarify the point. Consider the radical $\sqrt{x^2}$.

Let $x = 3$; then $\sqrt{x^2} = \sqrt{3^2} = \sqrt{9} = 3$.

Let $x = -3$; then $\sqrt{x^2} = \sqrt{(-3)^2} = \sqrt{9} = 3$.

Thus if $x \geq 0$, then $\sqrt{x^2} = x$, *but* if $x < 0$, then $\sqrt{x^2} = -x$. Using the concept of absolute value, we can state that for all real numbers, $\sqrt{x^2} = |x|$.

Now consider the radical $\sqrt{x^3}$. Because x^3 is negative when x is negative, we need to restrict x to the nonnegative real numbers when working with $\sqrt{x^3}$. Thus we can write, "if $x \geq 0$, then $\sqrt{x^3} = \sqrt{x^2}\sqrt{x} = x\sqrt{x}$," and no absolute-value sign is necessary. Finally, let's consider the radical $\sqrt[3]{x^3}$.

Let $x = 2$; then $\sqrt[3]{x^3} = \sqrt[3]{2^3} = \sqrt[3]{8} = 2$.

Let $x = -2$; then $\sqrt[3]{x^3} = \sqrt[3]{(-2)^3} = \sqrt[3]{-8} = -2$.

Thus it is correct to write, "$\sqrt[3]{x^3} = x$ for all real numbers," and again no absolute-value sign is necessary.

The previous discussion indicates that, technically, every radical expression involving variables in the radicand needs to be analyzed individually in terms of any necessary restrictions imposed on the variables. To help you gain experience with this skill, examples and problems are discussed under *Further Investigations* in the problem set. For now, however, to avoid considering such restrictions on a problem-to-problem basis, we shall merely assume that all variables represent positive real numbers. Let's consider the process of simplifying radicals that contain variables in the radicand. Study the following examples, and note that the same basic approach we used in Section 10.2 is applied here.

Classroom Example
Simplify each of the following:
(a) $\sqrt{125m^3}$
(b) $\sqrt{28m^5n^9}$
(c) $\sqrt{147x^6y^7}$
(d) $\sqrt[3]{128m^{10}n^5}$

EXAMPLE 4   Simplify each of the following:

(a) $\sqrt{8x^3}$ **(b)** $\sqrt{45x^3y^7}$ **(c)** $\sqrt{180a^4b^3}$ **(d)** $\sqrt[3]{40x^4y^8}$

Solution

(a) $\sqrt{8x^3} = \sqrt{4x^2}\sqrt{2x} = 2x\sqrt{2x}$ $4x^2$ is a perfect square

(b) $\sqrt{45x^3y^7} = \sqrt{9x^2y^6}\sqrt{5xy} = 3xy^3\sqrt{5xy}$ $9x^2y^6$ is a perfect square

(c) If the numerical coefficient of the radicand is quite large, you may want to look at it in the prime-factored form.

$$\sqrt{180a^4b^3} = \sqrt{2 \cdot 2 \cdot 3 \cdot 3 \cdot 5 \cdot a^4 \cdot b^3}$$

$$= \sqrt{36 \cdot 5 \cdot a^4 \cdot b^3}$$

$$= \sqrt{36a^4b^2}\sqrt{5b}$$

$$= 6a^2b\sqrt{5b}$$

(d) $\sqrt[3]{40x^4y^8} = \sqrt[3]{8x^3y^6}\ \sqrt[3]{5xy^2} = 2xy^2\sqrt[3]{5xy^2}$ $8x^3y^6$ is a perfect cube

Before we consider more examples, let's restate (in such a way that includes radicands containing variables) the conditions necessary for a radical to be in simplest radical form.

1. A radicand contains no polynomial factor raised to a power equal to or greater than the index of the radical.	$\sqrt{x^3}$ violates this condition
2. No fraction appears within a radical sign.	$\sqrt{\dfrac{2x}{3y}}$ violates this condition
3. No radical appears in the denominator.	$\dfrac{3}{\sqrt[3]{4x}}$ violates this condition

Classroom Example
Express each of the following in simplest radical form:

(a) $\sqrt{\dfrac{5m}{7n}}$ (b) $\dfrac{\sqrt{3}}{\sqrt{8x^5}}$

(c) $\dfrac{\sqrt{125m^7}}{\sqrt{48n^4}}$

(d) $\dfrac{7}{\sqrt[3]{9y}}$

(e) $\dfrac{\sqrt[3]{81a^7}}{\sqrt[3]{32b^2}}$

EXAMPLE 5   Express each of the following in simplest radical form:

(a) $\sqrt{\dfrac{2x}{3y}}$ (b) $\dfrac{\sqrt{5}}{\sqrt{12a^3}}$ (c) $\dfrac{\sqrt{8x^2}}{\sqrt{27y^5}}$ (d) $\dfrac{3}{\sqrt[3]{4x}}$ (e) $\dfrac{\sqrt[3]{16x^2}}{\sqrt[3]{9y^5}}$

Solution

(a) $\sqrt{\dfrac{2x}{3y}} = \dfrac{\sqrt{2x}}{\sqrt{3y}} = \dfrac{\sqrt{2x}}{\sqrt{3y}} \cdot \dfrac{\sqrt{3y}}{\sqrt{3y}} = \dfrac{\sqrt{6xy}}{3y}$

Form of 1

(b) $\dfrac{\sqrt{5}}{\sqrt{12a^3}} = \dfrac{\sqrt{5}}{\sqrt{12a^3}} \cdot \dfrac{\sqrt{3a}}{\sqrt{3a}} = \dfrac{\sqrt{15a}}{\sqrt{36a^4}} = \dfrac{\sqrt{15a}}{6a^2}$

Form of 1

(c) $\dfrac{\sqrt{8x^2}}{\sqrt{27y^5}} = \dfrac{\sqrt{4x^2}\sqrt{2}}{\sqrt{9y^4}\sqrt{3y}} = \dfrac{2x\sqrt{2}}{3y^2\sqrt{3y}} = \dfrac{2x\sqrt{2}}{3y^2\sqrt{3y}} \cdot \dfrac{\sqrt{3y}}{\sqrt{3y}} = \dfrac{2x\sqrt{6y}}{(3y^2)(3y)} = \dfrac{2x\sqrt{6y}}{9y^3}$

(d) $\dfrac{3}{\sqrt[3]{4x}} = \dfrac{3}{\sqrt[3]{4x}} \cdot \dfrac{\sqrt[3]{2x^2}}{\sqrt[3]{2x^2}} = \dfrac{3\sqrt[3]{2x^2}}{\sqrt[3]{8x^3}} = \dfrac{3\sqrt[3]{2x^2}}{2x}$

(e) $\dfrac{\sqrt[3]{16x^2}}{\sqrt[3]{9y^5}} = \dfrac{\sqrt[3]{16x^2}}{\sqrt[3]{9y^5}} \cdot \dfrac{\sqrt[3]{3y}}{\sqrt[3]{3y}} = \dfrac{\sqrt[3]{48x^2y}}{\sqrt[3]{27y^6}} = \dfrac{\sqrt[3]{8}\sqrt[3]{6x^2y}}{3y^2} = \dfrac{2\sqrt[3]{6x^2y}}{3y^2}$

Note that in part (c) we did some simplifying first before rationalizing the denominator, whereas in part (b) we proceeded immediately to rationalize the denominator. This is an individual choice, and you should probably do it both ways a few times to decide which you prefer.

Concept Quiz 10.3

For Problems 1–10, answer true or false.

1. In order to be combined when adding, radicals must have the same index and the same radicand.

2. If $x \ge 0$, then $\sqrt{x^2} = x$.

3. For all real numbers, $\sqrt{x^2} = x$.

4. For all real numbers, $\sqrt[3]{x^3} = x$.

5. A radical is not in simplest radical form if it has a fraction within the radical sign.

6. If a radical contains a factor raised to a power that is equal to the index of the radical, then the radical is not in simplest radical form.

7. The radical $\dfrac{1}{\sqrt{x}}$ is in simplest radical form.

8. $3\sqrt{2} + 4\sqrt{3} = 7\sqrt{5}$.

9. If $x > 0$, then $\sqrt{45x^3} = 3x^2\sqrt{5x}$.

10. If $x > 0$, then $\dfrac{4\sqrt{x^5}}{3\sqrt{4x^2}} = \dfrac{2x\sqrt{x}}{3}$.

Problem Set 10.3

For Problems 1–20, use the distributive property to help simplify each of the following. **(Objective 1)**
For example,

$$3\sqrt{8} - \sqrt{32} = 3\sqrt{4}\sqrt{2} - \sqrt{16}\sqrt{2}$$
$$= 3(2)\sqrt{2} - 4\sqrt{2}$$
$$= 6\sqrt{2} - 4\sqrt{2}$$
$$= (6 - 4)\sqrt{2} = 2\sqrt{2}$$

1. $5\sqrt{18} - 2\sqrt{2}$

2. $7\sqrt{12} + 4\sqrt{3}$

3. $7\sqrt{12} + 10\sqrt{48}$

4. $6\sqrt{8} - 5\sqrt{18}$

5. $-2\sqrt{50} - 5\sqrt{32}$

6. $-2\sqrt{20} - 7\sqrt{45}$

7. $3\sqrt{20} - \sqrt{5} - 2\sqrt{45}$

8. $6\sqrt{12} + \sqrt{3} - 2\sqrt{48}$

9. $-9\sqrt{24} + 3\sqrt{54} - 12\sqrt{6}$

10. $13\sqrt{28} - 2\sqrt{63} - 7\sqrt{7}$

11. $\dfrac{3}{4}\sqrt{7} - \dfrac{2}{3}\sqrt{28}$

12. $\dfrac{3}{5}\sqrt{5} - \dfrac{1}{4}\sqrt{80}$

13. $\dfrac{3}{5}\sqrt{40} + \dfrac{5}{6}\sqrt{90}$

14. $\dfrac{3}{8}\sqrt{96} - \dfrac{2}{3}\sqrt{54}$

15. $\dfrac{3\sqrt{18}}{5} - \dfrac{5\sqrt{72}}{6} + \dfrac{3\sqrt{98}}{4}$

16. $\dfrac{-2\sqrt{20}}{3} + \dfrac{3\sqrt{45}}{4} - \dfrac{5\sqrt{80}}{6}$

17. $5\sqrt[3]{3} + 2\sqrt[3]{24} - 6\sqrt[3]{81}$

18. $-3\sqrt[3]{2} - 2\sqrt[3]{16} + \sqrt[3]{54}$

19. $-\sqrt[3]{16} + 7\sqrt[3]{54} - 9\sqrt[3]{2}$

20. $4\sqrt[3]{24} - 6\sqrt[3]{3} + 13\sqrt[3]{81}$

For Problems 21–64, express each of the following in simplest radical form. All variables represent positive real numbers. **(Objective 2)**

21. $\sqrt{32x}$

22. $\sqrt{50y}$

23. $\sqrt{75x^2}$

24. $\sqrt{108y^2}$

25. $\sqrt{20x^2y}$

26. $\sqrt{80xy^2}$

27. $\sqrt{64x^3y^7}$

28. $\sqrt{36x^5y^6}$

29. $\sqrt{54a^4b^3}$

30. $\sqrt{96a^7b^8}$

31. $\sqrt{63x^6y^8}$

32. $\sqrt{28x^4y^{12}}$

33. $2\sqrt{40a^3}$

34. $4\sqrt{90a^5}$

35. $\dfrac{2}{3}\sqrt{96xy^3}$

36. $\dfrac{4}{5}\sqrt{125x^4y}$

37. $\sqrt{\dfrac{2x}{5y}}$

38. $\sqrt{\dfrac{3x}{2y}}$

39. $\sqrt{\dfrac{5}{12x^4}}$

40. $\sqrt{\dfrac{7}{8x^2}}$

41. $\dfrac{5}{\sqrt{18y}}$

42. $\dfrac{3}{\sqrt{12x}}$

43. $\dfrac{\sqrt{7x}}{\sqrt{8y^5}}$

44. $\dfrac{\sqrt{5y}}{\sqrt{18x^3}}$

45. $\dfrac{\sqrt{18y^3}}{\sqrt{16x}}$

46. $\dfrac{\sqrt{2x^3}}{\sqrt{9y}}$

47. $\dfrac{\sqrt{24a^2b^3}}{\sqrt{7ab^6}}$

48. $\dfrac{\sqrt{12a^2b}}{\sqrt{5a^3b^3}}$

49. $\sqrt[3]{24y}$

50. $\sqrt[3]{16x^2}$

51. $\sqrt[3]{16x^4}$

52. $\sqrt[3]{54x^3}$

53. $\sqrt[3]{56x^6y^8}$

54. $\sqrt[3]{81x^5y^6}$

55. $\sqrt[3]{\dfrac{7}{9x^2}}$

56. $\sqrt[3]{\dfrac{5}{2x}}$

57. $\dfrac{\sqrt[3]{3y}}{\sqrt[3]{16x^4}}$

58. $\dfrac{\sqrt[3]{2y}}{\sqrt[3]{3x}}$

59. $\dfrac{\sqrt[3]{12xy}}{\sqrt[3]{3x^2y^5}}$

60. $\dfrac{5}{\sqrt[3]{9xy^2}}$

61. $\sqrt{8x + 12y}$ [*Hint:* $\sqrt{8x + 12y} = \sqrt{4(2x + 3y)}$]

62. $\sqrt{4x + 4y}$

63. $\sqrt{16x + 48y}$

64. $\sqrt{27x + 18y}$

For Problems 65–74, use the distributive property to help simplify each of the following. All variables represent positive real numbers. **(Objective 2)**

65. $-3\sqrt{4x} + 5\sqrt{9x} + 6\sqrt{16x}$

66. $-2\sqrt{25x} - 4\sqrt{36x} + 7\sqrt{64x}$

67. $2\sqrt{18x} - 3\sqrt{8x} - 6\sqrt{50x}$

68. $4\sqrt{20x} + 5\sqrt{45x} - 10\sqrt{80x}$

69. $5\sqrt{27n} - \sqrt{12n} - 6\sqrt{3n}$

70. $4\sqrt{8n} + 3\sqrt{18n} - 2\sqrt{72n}$

71. $7\sqrt{4ab} - \sqrt{16ab} - 10\sqrt{25ab}$

72. $4\sqrt{ab} - 9\sqrt{36ab} + 6\sqrt{49ab}$

73. $-3\sqrt{2x^3} + 4\sqrt{8x^3} - 3\sqrt{32x^3}$

74. $2\sqrt{40x^5} - 3\sqrt{90x^5} + 5\sqrt{160x^5}$

Thoughts Into Words

75. Is the expression $3\sqrt{2} + \sqrt{50}$ in simplest radical form? Defend your answer.

76. Your friend simplified $\dfrac{\sqrt{6}}{\sqrt{8}}$ as follows:

$$\frac{\sqrt{6}}{\sqrt{8}} \cdot \frac{\sqrt{8}}{\sqrt{8}} = \frac{\sqrt{48}}{8} = \frac{\sqrt{16}\sqrt{3}}{8} = \frac{4\sqrt{3}}{8} = \frac{\sqrt{3}}{2}$$

Is this a correct procedure? Can you show her a better way to do this problem?

77. Does $\sqrt{x + y}$ equal $\sqrt{x} + \sqrt{y}$? Defend your answer.

Further Investigations

78. Use your calculator and evaluate each expression in Problems 1–16. Then evaluate the simplified expression that you obtained when doing these problems. Your two results for each problem should be the same.

Consider these problems, in which the variables could represent any real number. However, we would still have the restriction that the radical would represent a real number. In other words, the radicand must be nonnegative.

$\sqrt{98x^2} = \sqrt{49x^2}\sqrt{2} = 7|x|\sqrt{2}$ An absolute-value sign is necessary to ensure that the principal root is nonnegative

$\sqrt{24x^4} = \sqrt{4x^4}\sqrt{6} = 2x^2\sqrt{6}$ Because x^2 is nonnegative, there is no need for an absolute-value sign to ensure that the principal root is nonnegative

$\sqrt{25x^3} = \sqrt{25x^2}\sqrt{x} = 5x\sqrt{x}$ Because the radicand is defined to be nonnegative, x must be nonnegative, and there is no need for an absolute-value sign to ensure that the principal root is nonnegative

$\sqrt{18b^5} = \sqrt{9b^4}\sqrt{2b} = 3b^2\sqrt{2b}$ An absolute-value sign is not necessary to ensure that the principal root is nonnegative

$\sqrt{12y^6} = \sqrt{4y^6}\sqrt{3} = 2|y^3|\sqrt{3}$ An absolute-value sign is necessary to ensure that the principal root is nonnegative

79. Do the following problems, in which the variable could be any real number as long as the radical represents a real number. Use absolute-value signs in the answers as necessary.

(a) $\sqrt{125x^2}$

(b) $\sqrt{16x^4}$

(c) $\sqrt{8b^3}$

(d) $\sqrt{3y^5}$

(e) $\sqrt{288x^6}$

(f) $\sqrt{28m^8}$

(g) $\sqrt{128c^{10}}$

(h) $\sqrt{18d^7}$

(i) $\sqrt{49x^2}$

(j) $\sqrt{80n^{20}}$

(k) $\sqrt{81h^3}$

10.4   Products and Quotients of Radicals

OBJECTIVES

1   Multiply two radicals

2   Use the distributive property to multiply radical expressions

3   Rationalize binomial denominators

As we have seen, Property 10.2 $\left(\sqrt[n]{bc} = \sqrt[n]{b}\sqrt[n]{c}\right)$ is used to express one radical as the product of two radicals and also to express the product of two radicals as one radical. In fact, we have used the property for both purposes within the framework of simplifying radicals. For example,

$$\frac{\sqrt{3}}{\sqrt{32}} = \frac{\sqrt{3}}{\sqrt{16}\sqrt{2}} = \frac{\sqrt{3}}{4\sqrt{2}} = \frac{\sqrt{3}}{4\sqrt{2}} \cdot \frac{\sqrt{2}}{\sqrt{2}} = \frac{\sqrt{6}}{8}$$

$$\uparrow \qquad\qquad \uparrow \qquad\qquad\qquad\qquad\qquad \uparrow \qquad\qquad \uparrow$$

$$\sqrt[n]{bc} = \sqrt[n]{b}\sqrt[n]{c} \qquad\qquad\qquad\qquad \sqrt[n]{b}\sqrt[n]{c} = \sqrt[n]{bc}$$

The following examples demonstrate the use of Property 10.2 to multiply radicals and to express the product in simplest form.

Classroom Example
Multiply and simplify where possible:

(a) $(4\sqrt{2})(6\sqrt{7})$

(b) $(5\sqrt{18})(2\sqrt{2})$

(c) $(2\sqrt{5})(5\sqrt{10})$

(d) $(8\sqrt[3]{9})(5\sqrt[3]{6})$

EXAMPLE 1   Multiply and simplify where possible:

(a) $(2\sqrt{3})(3\sqrt{5})$ **(b)** $(3\sqrt{8})(5\sqrt{2})$

(c) $(7\sqrt{6})(3\sqrt{8})$ **(d)** $(2\sqrt[3]{6})(5\sqrt[3]{4})$

Solution

(a) $(2\sqrt{3})(3\sqrt{5}) = 2 \cdot 3 \cdot \sqrt{3} \cdot \sqrt{5} = 6\sqrt{15}$

(b) $(3\sqrt{8})(5\sqrt{2}) = 3 \cdot 5 \cdot \sqrt{8} \cdot \sqrt{2} = 15\sqrt{16} = 15 \cdot 4 = 60$

(c) $(7\sqrt{6})(3\sqrt{8}) = 7 \cdot 3 \cdot \sqrt{6} \cdot \sqrt{8} = 21\sqrt{48} = 21\sqrt{16}\sqrt{3}$

$$= 21 \cdot 4 \cdot \sqrt{3} = 84\sqrt{3}$$

(d) $(2\sqrt[3]{6})(5\sqrt[3]{4}) = 2 \cdot 5 \cdot \sqrt[3]{6} \cdot \sqrt[3]{4} = 10\sqrt[3]{24}$

$$= 10\sqrt[3]{8}\sqrt[3]{3}$$

$$= 10 \cdot 2 \cdot \sqrt[3]{3}$$

$$= 20\sqrt[3]{3}$$

Using the Distributive Property to Multiply Radical Expressions

Recall the use of the distributive property when finding the product of a monomial and a polynomial. For example, $3x^2(2x + 7) = 3x^2(2x) + 3x^2(7) = 6x^3 + 21x^2$. In a similar manner, the distributive property and Property 10.2 provide the basis for finding certain special products that involve radicals. The following examples illustrate this idea.

Classroom Example
Multiply and simplify where possible:

(a) $\sqrt{2}(\sqrt{10} + \sqrt{8})$

(b) $3\sqrt{5}(\sqrt{10} + \sqrt{15})$

(c) $\sqrt{7a}(\sqrt{14a} - \sqrt{28ab})$

(d) $(3\sqrt[3]{12})(4\sqrt[3]{9})$

EXAMPLE 2 Multiply and simplify where possible:

(a) $\sqrt{3}(\sqrt{6} + \sqrt{12})$

(b) $2\sqrt{2}(4\sqrt{3} - 5\sqrt{6})$

(c) $\sqrt{6x}(\sqrt{8x} + \sqrt{12xy})$

(d) $\sqrt[3]{2}(5\sqrt[3]{4} - 3\sqrt[3]{16})$

Solution

(a) $\sqrt{3}(\sqrt{6} + \sqrt{12}) = \sqrt{3}\sqrt{6} + \sqrt{3}\sqrt{12}$

$= \sqrt{18} + \sqrt{36}$

$= \sqrt{9}\sqrt{2} + 6$

$= 3\sqrt{2} + 6$

(b) $2\sqrt{2}(4\sqrt{3} - 5\sqrt{6}) = (2\sqrt{2})(4\sqrt{3}) - (2\sqrt{2})(5\sqrt{6})$

$= 8\sqrt{6} - 10\sqrt{12}$

$= 8\sqrt{6} - 10\sqrt{4}\sqrt{3}$

$= 8\sqrt{6} - 20\sqrt{3}$

(c) $\sqrt{6x}(\sqrt{8x} + \sqrt{12xy}) = (\sqrt{6x})(\sqrt{8x}) + (\sqrt{6x})(\sqrt{12xy})$

$= \sqrt{48x^2} + \sqrt{72x^2y}$

$= \sqrt{16x^2}\sqrt{3} + \sqrt{36x^2}\sqrt{2y}$

$= 4x\sqrt{3} + 6x\sqrt{2y}$

(d) $\sqrt[3]{2}(5\sqrt[3]{4} - 3\sqrt[3]{16}) = (\sqrt[3]{2})(5\sqrt[3]{4}) - (\sqrt[3]{2})(3\sqrt[3]{16})$

$= 5\sqrt[3]{8} - 3\sqrt[3]{32}$

$= 5 \cdot 2 - 3\sqrt[3]{8}\sqrt[3]{4}$

$= 10 - 6\sqrt[3]{4}$

The distributive property also plays a central role in determining the product of two binomials. For example, $(x + 2)(x + 3) = x(x + 3) + 2(x + 3) = x^2 + 3x + 2x + 6 = x^2 + 5x + 6$. Finding the product of two binomial expressions that involve radicals can be handled in a similar fashion, as in the next examples.

Classroom Example
Find the following products and simplify:

(a) $(\sqrt{2} - \sqrt{7})(\sqrt{5} + \sqrt{3})$

(b) $(5\sqrt{6} + \sqrt{3})(4\sqrt{5} - 6\sqrt{3})$

(c) $(\sqrt{10} + \sqrt{3})(\sqrt{10} - \sqrt{3})$

(d) $(\sqrt{m} - \sqrt{n})(\sqrt{m} + \sqrt{n})$

EXAMPLE 3 Find the following products and simplify:

(a) $(\sqrt{3} + \sqrt{5})(\sqrt{2} + \sqrt{6})$

(b) $(2\sqrt{2} - \sqrt{7})(3\sqrt{2} + 5\sqrt{7})$

(c) $(\sqrt{8} + \sqrt{6})(\sqrt{8} - \sqrt{6})$

(d) $(\sqrt{x} + \sqrt{y})(\sqrt{x} - \sqrt{y})$

Solution

(a) $(\sqrt{3} + \sqrt{5})(\sqrt{2} + \sqrt{6}) = \sqrt{3}(\sqrt{2} + \sqrt{6}) + \sqrt{5}(\sqrt{2} + \sqrt{6})$

$= \sqrt{3}\sqrt{2} + \sqrt{3}\sqrt{6} + \sqrt{5}\sqrt{2} + \sqrt{5}\sqrt{6}$

$= \sqrt{6} + \sqrt{18} + \sqrt{10} + \sqrt{30}$

$= \sqrt{6} + 3\sqrt{2} + \sqrt{10} + \sqrt{30}$

(b) $(2\sqrt{2} - \sqrt{7})(3\sqrt{2} + 5\sqrt{7}) = 2\sqrt{2}(3\sqrt{2} + 5\sqrt{7})$

$- \sqrt{7}(3\sqrt{2} + 5\sqrt{7})$

$= (2\sqrt{2})(3\sqrt{2}) + (2\sqrt{2})(5\sqrt{7})$

$- (\sqrt{7})(3\sqrt{2}) - (\sqrt{7})(5\sqrt{7})$

$= 12 + 10\sqrt{14} - 3\sqrt{14} - 35$

$= -23 + 7\sqrt{14}$

(c) $(\sqrt{8} + \sqrt{6})(\sqrt{8} - \sqrt{6}) = \sqrt{8}(\sqrt{8} - \sqrt{6}) + \sqrt{6}(\sqrt{8} - \sqrt{6})$

$$= \sqrt{8}\sqrt{8} - \sqrt{8}\sqrt{6} + \sqrt{6}\sqrt{8} - \sqrt{6}\sqrt{6}$$

$$= 8 - \sqrt{48} + \sqrt{48} - 6$$

$$= 2$$

(d) $(\sqrt{x} + \sqrt{y})(\sqrt{x} - \sqrt{y}) = \sqrt{x}(\sqrt{x} - \sqrt{y}) + \sqrt{y}(\sqrt{x} - \sqrt{y})$

$$= \sqrt{x}\sqrt{x} - \sqrt{x}\sqrt{y} + \sqrt{y}\sqrt{x} - \sqrt{y}\sqrt{y}$$

$$= x - \sqrt{xy} + \sqrt{xy} - y$$

$$= x - y$$

Rationalizing Binomial Denominators

Note parts (c) and (d) of Example 3; they fit the special-product pattern $(a + b)(a - b) = a^2 - b^2$. Furthermore, in each case the final product is in rational form. The factors $a + b$ and $a - b$ are called **conjugates**. This suggests a way of rationalizing the denominator in an expression that contains a binomial denominator with radicals. We will multiply by the conjugate of the binomial denominator. Consider the following example.

Classroom Example
Simplify $\dfrac{2}{\sqrt{7} - \sqrt{3}}$ by rationalizing the denominator.

EXAMPLE 4   Simplify $\dfrac{4}{\sqrt{5} + \sqrt{2}}$ by rationalizing the denominator.

Solution

$$\frac{4}{\sqrt{5} + \sqrt{2}} = \frac{4}{\sqrt{5} + \sqrt{2}} \cdot \left(\frac{\sqrt{5} - \sqrt{2}}{\sqrt{5} - \sqrt{2}}\right) \quad \text{Form of 1}$$

$$= \frac{4(\sqrt{5} - \sqrt{2})}{(\sqrt{5} + \sqrt{2})(\sqrt{5} - \sqrt{2})} = \frac{4(\sqrt{5} - \sqrt{2})}{5 - 2}$$

$$= \frac{4(\sqrt{5} - \sqrt{2})}{3} \quad \text{or} \quad \frac{4\sqrt{5} - 4\sqrt{2}}{3}$$

Either answer
is acceptable

The next examples further illustrate the process of rationalizing and simplifying expressions that contain binomial denominators.

Classroom Example
For each of the following, rationalize the denominator and simplify:

(a) $\dfrac{\sqrt{6}}{\sqrt{2} - 8}$

(b) $\dfrac{5}{2\sqrt{5} - 3\sqrt{2}}$

(c) $\dfrac{\sqrt{y} - 6}{\sqrt{y} + 2}$

(d) $\dfrac{8\sqrt{m} + 5\sqrt{n}}{\sqrt{m} - \sqrt{n}}$

EXAMPLE 5

For each of the following, rationalize the denominator and simplify:

(a) $\dfrac{\sqrt{3}}{\sqrt{6} - 9}$ **(b)** $\dfrac{7}{3\sqrt{5} + 2\sqrt{3}}$ **(c)** $\dfrac{\sqrt{x} + 2}{\sqrt{x} - 3}$ **(d)** $\dfrac{2\sqrt{x} - 3\sqrt{y}}{\sqrt{x} + \sqrt{y}}$

Solution

(a) $\dfrac{\sqrt{3}}{\sqrt{6} - 9} = \dfrac{\sqrt{3}}{\sqrt{6} - 9} \cdot \dfrac{\sqrt{6} + 9}{\sqrt{6} + 9}$

$$= \frac{\sqrt{3}(\sqrt{6} + 9)}{(\sqrt{6} - 9)(\sqrt{6} + 9)}$$

$$= \frac{\sqrt{18} + 9\sqrt{3}}{6 - 81}$$

$$= \frac{3\sqrt{2} + 9\sqrt{3}}{-75}$$

$$= \frac{3(\sqrt{2} + 3\sqrt{3})}{(-3)(25)}$$

$$= -\frac{\sqrt{2} + 3\sqrt{3}}{25} \quad \text{or} \quad \frac{-\sqrt{2} - 3\sqrt{3}}{25}$$

(b) $\dfrac{7}{3\sqrt{5} + 2\sqrt{3}} = \dfrac{7}{3\sqrt{5} + 2\sqrt{3}} \cdot \dfrac{3\sqrt{5} - 2\sqrt{3}}{3\sqrt{5} - 2\sqrt{3}}$

$$= \frac{7(3\sqrt{5} - 2\sqrt{3})}{(3\sqrt{5} + 2\sqrt{3})(3\sqrt{5} - 2\sqrt{3})}$$

$$= \frac{7(3\sqrt{5} - 2\sqrt{3})}{45 - 12}$$

$$= \frac{7(3\sqrt{5} - 2\sqrt{3})}{33} \quad \text{or} \quad \frac{21\sqrt{5} - 14\sqrt{3}}{33}$$

(c) $\dfrac{\sqrt{x} + 2}{\sqrt{x} - 3} = \dfrac{\sqrt{x} + 2}{\sqrt{x} - 3} \cdot \dfrac{\sqrt{x} + 3}{\sqrt{x} + 3} = \dfrac{(\sqrt{x} + 2)(\sqrt{x} + 3)}{(\sqrt{x} - 3)(\sqrt{x} + 3)}$

$$= \frac{x + 3\sqrt{x} + 2\sqrt{x} + 6}{x - 9}$$

$$= \frac{x + 5\sqrt{x} + 6}{x - 9}$$

(d) $\dfrac{2\sqrt{x} - 3\sqrt{y}}{\sqrt{x} + \sqrt{y}} = \dfrac{2\sqrt{x} - 3\sqrt{y}}{\sqrt{x} + \sqrt{y}} \cdot \dfrac{\sqrt{x} - \sqrt{y}}{\sqrt{x} - \sqrt{y}}$

$$= \frac{(2\sqrt{x} - 3\sqrt{y})(\sqrt{x} - \sqrt{y})}{(\sqrt{x} + \sqrt{y})(\sqrt{x} - \sqrt{y})}$$

$$= \frac{2x - 2\sqrt{xy} - 3\sqrt{xy} + 3y}{x - y}$$

$$= \frac{2x - 5\sqrt{xy} + 3y}{x - y}$$

Concept Quiz 10.4

For Problems 1–10, answer true or false.

1. The property $\sqrt[n]{x}\sqrt[n]{y} = \sqrt[n]{xy}$ can be used to express the product of two radicals as one radical.

2. The product of two radicals always results in an expression that has a radical even after simplifying.

3. The conjugate of $5 + \sqrt{3}$ is $-5 - \sqrt{3}$.

4. The product of $2 - \sqrt{7}$ and $2 + \sqrt{7}$ is a rational number.

5. To rationalize the denominator for the expression $\dfrac{2\sqrt{5}}{4 - \sqrt{5}}$, we would multiply by $\dfrac{\sqrt{5}}{\sqrt{5}}$.

6. To rationalize the denominator for the expression $\dfrac{\sqrt{x} + 8}{\sqrt{x} - 4}$, we would multiply the numerator and denominator by $\sqrt{x} - 4$.

7. $\dfrac{\sqrt{8} + \sqrt{12}}{\sqrt{2}} = 2 + \sqrt{6}$

8. $\dfrac{\sqrt{2}}{\sqrt{8} + \sqrt{12}} = \dfrac{1}{2 + \sqrt{6}}$

9. The product of $5 + \sqrt{3}$ and $-5 - \sqrt{3}$ is -28.

10. The product of $\sqrt{5} - 1$ and $\sqrt{5} + 1$ is 24.

Problem Set 10.4

For Problems 1–14, multiply and simplify where possible. **(Objective 1)**

1. $\sqrt{6}\sqrt{12}$
2. $\sqrt{8}\sqrt{6}$
3. $(3\sqrt{3})(2\sqrt{6})$
4. $(5\sqrt{2})(3\sqrt{12})$
5. $(4\sqrt{2})(-6\sqrt{5})$
6. $(-7\sqrt{3})(2\sqrt{5})$
7. $(-3\sqrt{3})(-4\sqrt{8})$
8. $(-5\sqrt{8})(-6\sqrt{7})$
9. $(5\sqrt{6})(4\sqrt{6})$
10. $(3\sqrt{7})(2\sqrt{7})$
11. $(2\sqrt[3]{4})(6\sqrt[3]{2})$
12. $(4\sqrt[3]{3})(5\sqrt[3]{9})$
13. $(4\sqrt[3]{6})(7\sqrt[3]{4})$
14. $(9\sqrt[3]{6})(2\sqrt[3]{9})$

For Problems 15–52, find the following products and express answers in simplest radical form. All variables represent nonnegative real numbers. **(Objective 2)**

15. $\sqrt{2}(\sqrt{3} + \sqrt{5})$
16. $\sqrt{3}(\sqrt{7} + \sqrt{10})$
17. $3\sqrt{5}(2\sqrt{2} - \sqrt{7})$
18. $5\sqrt{6}(2\sqrt{5} - 3\sqrt{11})$
19. $2\sqrt{6}(3\sqrt{8} - 5\sqrt{12})$
20. $4\sqrt{2}(3\sqrt{12} + 7\sqrt{6})$
21. $-4\sqrt{5}(2\sqrt{5} + 4\sqrt{12})$
22. $-5\sqrt{3}(3\sqrt{12} - 9\sqrt{8})$
23. $3\sqrt{x}(5\sqrt{2} + \sqrt{y})$
24. $\sqrt{2x}(3\sqrt{y} - 7\sqrt{5})$
25. $\sqrt{xy}(5\sqrt{xy} - 6\sqrt{x})$
26. $4\sqrt{x}(2\sqrt{xy} + 2\sqrt{x})$
27. $\sqrt{5y}(\sqrt{8x} + \sqrt{12y^2})$
28. $\sqrt{2x}(\sqrt{12xy} - \sqrt{8y})$
29. $5\sqrt{3}(2\sqrt{8} - 3\sqrt{18})$
30. $2\sqrt{2}(3\sqrt{12} - \sqrt{27})$
31. $(\sqrt{3} + 4)(\sqrt{3} - 7)$
32. $(\sqrt{2} + 6)(\sqrt{2} - 2)$
33. $(\sqrt{5} - 6)(\sqrt{5} - 3)$
34. $(\sqrt{7} - 2)(\sqrt{7} - 8)$
35. $(3\sqrt{5} - 2\sqrt{3})(2\sqrt{7} + \sqrt{2})$

36. $(\sqrt{2} + \sqrt{3})(\sqrt{5} - \sqrt{7})$
37. $(2\sqrt{6} + 3\sqrt{5})(\sqrt{8} - 3\sqrt{12})$
38. $(5\sqrt{2} - 4\sqrt{6})(2\sqrt{8} + \sqrt{6})$
39. $(2\sqrt{6} + 5\sqrt{5})(3\sqrt{6} - \sqrt{5})$
40. $(7\sqrt{3} - \sqrt{7})(2\sqrt{3} + 4\sqrt{7})$
41. $(3\sqrt{2} - 5\sqrt{3})(6\sqrt{2} - 7\sqrt{3})$
42. $(\sqrt{8} - 3\sqrt{10})(2\sqrt{8} - 6\sqrt{10})$
43. $(\sqrt{6} + 4)(\sqrt{6} - 4)$
44. $(\sqrt{7} - 2)(\sqrt{7} + 2)$
45. $(\sqrt{2} + \sqrt{10})(\sqrt{2} - \sqrt{10})$
46. $(2\sqrt{3} + \sqrt{11})(2\sqrt{3} - \sqrt{11})$
47. $(\sqrt{2x} + \sqrt{3y})(\sqrt{2x} - \sqrt{3y})$
48. $(2\sqrt{x} - 5\sqrt{y})(2\sqrt{x} + 5\sqrt{y})$
49. $2\sqrt[3]{3}(5\sqrt[3]{4} + \sqrt[3]{6})$
50. $2\sqrt[3]{2}(3\sqrt[3]{6} - 4\sqrt[3]{5})$
51. $3\sqrt[3]{4}(2\sqrt[3]{2} - 6\sqrt[3]{4})$
52. $3\sqrt[3]{3}(4\sqrt[3]{9} + 5\sqrt[3]{7})$

For Problems 53–76, rationalize the denominator and simplify. All variables represent positive real numbers. **(Objective 3)**

53. $\dfrac{2}{\sqrt{7} + 1}$
54. $\dfrac{6}{\sqrt{5} + 2}$
55. $\dfrac{3}{\sqrt{2} - 5}$
56. $\dfrac{-4}{\sqrt{6} - 3}$

57. $\dfrac{1}{\sqrt{2} + \sqrt{7}}$ **58.** $\dfrac{3}{\sqrt{3} + \sqrt{10}}$ **67.** $\dfrac{2}{\sqrt{x} + 4}$ **68.** $\dfrac{3}{\sqrt{x} + 7}$

59. $\dfrac{\sqrt{2}}{\sqrt{10} - \sqrt{3}}$ **60.** $\dfrac{\sqrt{3}}{\sqrt{7} - \sqrt{2}}$ **69.** $\dfrac{\sqrt{x}}{\sqrt{x} - 5}$ **70.** $\dfrac{\sqrt{x}}{\sqrt{x} - 1}$

61. $\dfrac{\sqrt{3}}{2\sqrt{5} + 4}$ **62.** $\dfrac{\sqrt{7}}{3\sqrt{2} - 5}$ **71.** $\dfrac{\sqrt{x} - 2}{\sqrt{x} + 6}$ **72.** $\dfrac{\sqrt{x} + 1}{\sqrt{x} - 10}$

63. $\dfrac{6}{3\sqrt{7} - 2\sqrt{6}}$ **64.** $\dfrac{5}{2\sqrt{5} + 3\sqrt{7}}$ **73.** $\dfrac{\sqrt{x}}{\sqrt{x} + 2\sqrt{y}}$ **74.** $\dfrac{\sqrt{y}}{2\sqrt{x} - \sqrt{y}}$

65. $\dfrac{\sqrt{6}}{3\sqrt{2} + 2\sqrt{3}}$ **66.** $\dfrac{3\sqrt{6}}{5\sqrt{3} - 4\sqrt{2}}$ **75.** $\dfrac{3\sqrt{y}}{2\sqrt{x} - 3\sqrt{y}}$ **76.** $\dfrac{2\sqrt{x}}{3\sqrt{x} + 5\sqrt{y}}$

Thoughts Into Words

77. How would you help someone rationalize the denominator and simplify $\dfrac{4}{\sqrt{8} + \sqrt{12}}$?

78. Discuss how the distributive property has been used thus far in this chapter.

79. How would you simplify the expression $\dfrac{\sqrt{8} + \sqrt{12}}{\sqrt{2}}$?

Further Investigations

80. Use your calculator to evaluate each expression in Problems 53–66. Then evaluate the results you obtained when you did the problems.

Answers to the Concept Quiz

1. True   **2.** False   **3.** False   **4.** True   **5.** False   **6.** False   **7.** True   **8.** False   **9.** False   **10.** False

10.5   Radical Equations

OBJECTIVES

1   Solve radical equations

2   Solve radical equations for real-world problems

We often refer to equations that contain radicals with variables in a radicand as **radical equations**. In this section we discuss techniques for solving such equations that contain one or more radicals. To solve radical equations, we need the following property of equality.

> **Property 10.4**
>
> Let a and b be real numbers and n be a positive integer.
>
> If $a = b$, then $a^n = b^n$

Property 10.4 states that we can raise both sides of an equation to a positive integral power. However, raising both sides of an equation to a positive integral power sometimes produces results that do not satisfy the original equation. Let's consider two examples to illustrate this point.

Classroom Example
Solve $\sqrt{2x - 1} = 3$.

EXAMPLE 1   Solve $\sqrt{2x - 5} = 7$.

Solution

$$\sqrt{2x - 5} = 7$$
$$(\sqrt{2x - 5})^2 = 7^2 \qquad \text{Square both sides}$$
$$2x - 5 = 49$$
$$2x = 54$$
$$x = 27$$

✔ **Check**

$$\sqrt{2x - 5} = 7$$
$$\sqrt{2(27) - 5} \overset{?}{=} 7$$
$$\sqrt{49} \overset{?}{=} 7$$
$$7 = 7$$

The solution set for $\sqrt{2x - 5} = 7$ is $\{27\}$.

Classroom Example
Solve $\sqrt{5y + 9} = -8$.

EXAMPLE 2   Solve $\sqrt{3a + 4} = -4$.

Solution

$$\sqrt{3a + 4} = -4$$
$$(\sqrt{3a + 4})^2 = (-4)^2 \qquad \text{Square both sides}$$
$$3a + 4 = 16$$
$$3a = 12$$
$$a = 4$$

✔ **Check**

$$\sqrt{3a + 4} = -4$$
$$\sqrt{3(4) + 4} \overset{?}{=} -4$$
$$\sqrt{16} \overset{?}{=} -4$$
$$4 \neq -4$$

Because 4 does not check, the original equation has no real number solution. Thus the solution set is $\varnothing$.

In general, raising both sides of an equation to a positive integral power produces an equation that has all of the solutions of the original equation, but it may also have some extra solutions that do not satisfy the original equation. Such extra solutions are called **extraneous solutions**. Therefore, when using Property 10.4, you *must* check each potential solution in the original equation.

Let's consider some examples to illustrate different situations that arise when we are solving radical equations.

Classroom Example
Solve $\sqrt{2x + 6} = x + 3$.

EXAMPLE 3   Solve $\sqrt{2t - 4} = t - 2$.

Solution

$$\sqrt{2t - 4} = t - 2$$

$$(\sqrt{2t - 4})^2 = (t - 2)^2 \qquad \text{Square both sides}$$

$$2t - 4 = t^2 - 4t + 4$$

$$0 = t^2 - 6t + 8$$

$$0 = (t - 2)(t - 4) \qquad \text{Factor the right side}$$

$$t - 2 = 0 \quad \text{or} \quad t - 4 = 0 \qquad \text{Apply: } ab = 0 \text{ if and only}$$
$$\qquad \qquad \qquad \qquad \qquad \qquad \text{if } a = 0 \text{ or } b = 0$$

$$t = 2 \quad \text{or} \qquad t = 4$$

✔ **Check**

$$\sqrt{2t - 4} = t - 2 \qquad\qquad\qquad \sqrt{2t - 4} = t - 2$$

$$\sqrt{2(2) - 4} \stackrel{?}{=} 2 - 2 \quad \text{when } t = 2 \quad \text{or} \quad \sqrt{2(4) - 4} \stackrel{?}{=} 4 - 2 \quad \text{when } t = 4$$

$$\sqrt{0} \stackrel{?}{=} 0 \qquad\qquad\qquad\qquad\qquad \sqrt{4} \stackrel{?}{=} 2$$

$$0 = 0 \qquad\qquad\qquad\qquad\qquad\qquad 2 = 2$$

The solution set is $\{2, 4\}$.

Classroom Example
Solve $\sqrt{m} + 2 = m$.

EXAMPLE 4   Solve $\sqrt{y} + 6 = y$.

Solution

$$\sqrt{y} + 6 = y$$

$$\sqrt{y} = y - 6$$

$$(\sqrt{y})^2 = (y - 6)^2 \qquad \text{Square both sides}$$

$$y = y^2 - 12y + 36$$

$$0 = y^2 - 13y + 36$$

$$0 = (y - 4)(y - 9) \qquad \text{Factor the right side}$$

$$y - 4 = 0 \quad \text{or} \quad y - 9 = 0 \qquad \text{Apply: } ab = 0 \text{ if and}$$
$$\qquad \qquad \qquad \qquad \qquad \qquad \text{only if } a = 0 \text{ or } b = 0$$

$$y = 4 \quad \text{or} \qquad y = 9$$

✔ **Check**

$$\sqrt{y} + 6 = y \qquad\qquad\qquad \sqrt{y} + 6 = y$$

$$\sqrt{4} + 6 \stackrel{?}{=} 4 \quad \text{when } y = 4 \quad \text{or} \quad \sqrt{9} + 6 \stackrel{?}{=} 9 \quad \text{when } y = 9$$

$$2 + 6 \stackrel{?}{=} 4 \qquad\qquad\qquad\qquad\qquad 3 + 6 \stackrel{?}{=} 9$$

$$8 \neq 4 \qquad\qquad\qquad\qquad\qquad\qquad 9 = 9$$

The only solution is 9; the solution set is $\{9\}$.

In Example 4 above, note that we changed the form of the original equation $\sqrt{y} + 6 = y$ to $\sqrt{y} = y - 6$ before we squared both sides. Note that squaring both sides of $\sqrt{y} + 6 = y$ produces $y + 12\sqrt{y} + 36 = y^2$, which is a much more complex equation that still contains a radical. Here again, it pays to think ahead before carrying out all the steps. Now let's consider an example involving a cube root.

Classroom Example
Solve $\sqrt[3]{x^2 + 2} = 3$.

EXAMPLE 5   Solve $\sqrt[3]{n^2 - 1} = 2$.

Solution

$$\sqrt[3]{n^2 - 1} = 2$$

$$\left(\sqrt[3]{n^2 - 1}\right)^3 = 2^3 \qquad \text{Cube both sides}$$

$$n^2 - 1 = 8$$

$$n^2 - 9 = 0$$

$$(n + 3)(n - 3) = 0$$

$$n + 3 = 0 \qquad \text{or} \qquad n - 3 = 0$$

$$n = -3 \qquad \text{or} \qquad n = 3$$

✔ **Check**

$$\sqrt[3]{n^2 - 1} = 2 \qquad\qquad\qquad \sqrt[3]{n^2 - 1} = 2$$

$$\sqrt[3]{(-3)^2 - 1} \stackrel{?}{=} 2 \quad \text{when } n = -3 \quad \text{or} \quad \sqrt[3]{3^2 - 1} \stackrel{?}{=} 2 \quad \text{when } n = 3$$

$$\sqrt[3]{8} \stackrel{?}{=} 2 \qquad\qquad\qquad\qquad \sqrt[3]{8} \stackrel{?}{=} 2$$

$$2 = 2 \qquad\qquad\qquad\qquad\qquad 2 = 2$$

The solution set is $\{-3, 3\}$.

It may be necessary to square both sides of an equation, simplify the resulting equation, and then square both sides again. The next example illustrates this type of problem.

Classroom Example
Solve $\sqrt{x + 4} = 1 + \sqrt{x - 1}$.

EXAMPLE 6   Solve $\sqrt{x + 2} = 7 - \sqrt{x + 9}$.

Solution

$$\sqrt{x + 2} = 7 - \sqrt{x + 9}$$

$$\left(\sqrt{x + 2}\right)^2 = \left(7 - \sqrt{x + 9}\right)^2 \qquad \text{Square both sides}$$

$$x + 2 = 49 - 14\sqrt{x + 9} + x + 9$$

$$x + 2 = x + 58 - 14\sqrt{x + 9}$$

$$-56 = -14\sqrt{x + 9}$$

$$4 = \sqrt{x + 9}$$

$$(4)^2 = \left(\sqrt{x + 9}\right)^2 \qquad \text{Square both sides}$$

$$16 = x + 9$$

$$7 = x$$

✔ **Check**

$$\sqrt{x + 2} = 7 - \sqrt{x + 9}$$

$$\sqrt{7 + 2} \stackrel{?}{=} 7 - \sqrt{7 + 9} \quad \text{when } x = 7$$

$$\sqrt{9} \stackrel{?}{=} 7 - \sqrt{16}$$

$$3 \stackrel{?}{=} 7 - 4$$

$$3 = 3$$

The solution set is $\{7\}$.

Solving Radical Equations for Real-World Problems

In Section 10.2 we used the formula $S = \sqrt{30Df}$ to approximate how fast a car was traveling on the basis of the length of skid marks. (Remember that S represents the speed of the car in miles per hour, D represents the length of the skid marks in feet, and f represents a coefficient of friction.) This same formula can be used to estimate the length of skid marks that are produced by cars traveling at different rates on various types of road surfaces. To use the formula for this purpose, let's change the form of the equation by solving for D.

$$\sqrt{30Df} = S$$
$$30Df = S^2 \qquad \text{The result of squaring both sides of the original equation}$$
$$D = \frac{S^2}{30f} \qquad \text{D, S, and f are positive numbers, so this final equation and the original one are equivalent}$$

Classroom Example
Suppose that for a particular road surface, the coefficient of friction is 0.27. How far will a car skid when the brakes are applied at 65 miles per hour?

EXAMPLE 7

Suppose that for a particular road surface, the coefficient of friction is 0.35. How far will a car skid when the brakes are applied at 60 miles per hour?

Solution

We can substitute 0.35 for f and 60 for S in the formula $D = \frac{S^2}{30f}$.

$$D = \frac{60^2}{30(0.35)} = 343 \quad \text{to the nearest whole number}$$

The car will skid approximately 343 feet.

Remark: Pause for a moment and think about the result in Example 7. The coefficient of friction 0.35 refers to a wet concrete road surface. Note that a car traveling at 60 miles per hour on such a surface will skid more than the length of a football field.

Concept Quiz 10.5

For Problems 1–10, answer true or false.

1. To solve a radical equation, we can raise each side of the equation to a positive integer power.
2. Solving the equation that results from squaring each side of an original equation may not give all the solutions of the original equation.
3. The equation $\sqrt[3]{x-1} = -2$ has a solution.
4. Potential solutions that do not satisfy the original equation are called extraneous solutions.
5. The equation $\sqrt{x+1} = -2$ has no real number solutions.
6. The solution set for $\sqrt{x+2} = x$ is $\{1, 4\}$.
7. The solution set for $\sqrt{x+1} + \sqrt{x-2} = -3$ is the null set.
8. The solution set for $\sqrt[3]{x+2} = -2$ is the null set.
9. The solution set for the equation $\sqrt{x^2 - 2x + 1} = x - 3$ is $\{2\}$.
10. The solution set for the equation $\sqrt{5x+1} + \sqrt{x+4} = 3$ is $\{0\}$.

Problem Set 10.5

For Problems 1–56, solve each equation. Don't forget to check each of your potential solutions. **(Objective 1)**

1. $\sqrt{5x} = 10$

2. $\sqrt{3x} = 9$

3. $\sqrt{2x + 4} = 0$

4. $\sqrt{4x + 5} = 0$

5. $2\sqrt{n} = 5$

6. $5\sqrt{n} = 3$

7. $3\sqrt{n} - 2 = 0$

8. $2\sqrt{n} - 7 = 0$

9. $\sqrt{3y + 1} = 4$

10. $\sqrt{2y - 3} = 5$

11. $\sqrt{4y - 3} - 6 = 0$

12. $\sqrt{3y + 5} - 2 = 0$

13. $\sqrt{3x - 1} + 1 = 4$

14. $\sqrt{4x - 1} - 3 = 2$

15. $\sqrt{2n + 3} - 2 = -1$

16. $\sqrt{5n + 1} - 6 = -4$

17. $\sqrt{2x - 5} = -1$

18. $\sqrt{4x - 3} = -4$

19. $\sqrt{5x + 2} = \sqrt{6x + 1}$

20. $\sqrt{4x + 2} = \sqrt{3x + 4}$

21. $\sqrt{3x + 1} = \sqrt{7x - 5}$

22. $\sqrt{6x + 5} = \sqrt{2x + 10}$

23. $\sqrt{3x - 2} - \sqrt{x + 4} = 0$

24. $\sqrt{7x - 6} - \sqrt{5x + 2} = 0$

25. $5\sqrt{t - 1} = 6$

26. $4\sqrt{t + 3} = 6$

27. $\sqrt{x^2 + 7} = 4$

28. $\sqrt{x^2 + 3} - 2 = 0$

29. $\sqrt{x^2 + 13x + 37} = 1$

30. $\sqrt{x^2 + 5x - 20} = 2$

31. $\sqrt{x^2 - x + 1} = x + 1$

32. $\sqrt{n^2 - 2n - 4} = n$

33. $\sqrt{x^2 + 3x + 7} = x + 2$

34. $\sqrt{x^2 + 2x + 1} = x + 3$

35. $\sqrt{-4x + 17} = x - 3$

36. $\sqrt{2x - 1} = x - 2$

37. $\sqrt{n + 4} = n + 4$

38. $\sqrt{n + 6} = n + 6$

39. $\sqrt{3y} = y - 6$

40. $2\sqrt{n} = n - 3$

41. $4\sqrt{x + 5} = x$

42. $\sqrt{-x - 6} = x$

43. $\sqrt[3]{x - 2} = 3$

44. $\sqrt[3]{x + 1} = 4$

45. $\sqrt[3]{2x + 3} = -3$

46. $\sqrt[3]{3x - 1} = -4$

47. $\sqrt[3]{2x + 5} = \sqrt[3]{4 - x}$

48. $\sqrt[3]{3x - 1} = \sqrt[3]{2 - 5x}$

49. $\sqrt{x + 19} - \sqrt{x + 28} = -1$

50. $\sqrt{x + 4} = \sqrt{x - 1} + 1$

51. $\sqrt{3x + 1} + \sqrt{2x + 4} = 3$

52. $\sqrt{2x - 1} - \sqrt{x + 3} = 1$

53. $\sqrt{n - 4} + \sqrt{n + 4} = 2\sqrt{n - 1}$

54. $\sqrt{n - 3} + \sqrt{n + 5} = 2\sqrt{n}$

55. $\sqrt{t + 3} - \sqrt{t - 2} = \sqrt{7 - t}$

56. $\sqrt{t + 7} - 2\sqrt{t - 8} = \sqrt{t - 5}$

For Problems 57–59, use the appropriate formula to solve the problems. **(Objective 2)**

57. Use the formula given in Example 7 with a coefficient of friction of 0.95. How far will a car skid at 40 miles per hour? at 55 miles per hour? at 65 miles per hour? Express the answers to the nearest foot.

58. Solve the formula $T = 2\pi\sqrt{\dfrac{L}{32}}$ for L. (Remember that in this formula, which was used in Section 10.2, T represents the period of a pendulum expressed in seconds, and L represents the length of the pendulum in feet.)

59. In Problem 58, you should have obtained the equation $L = \dfrac{8T^2}{\pi^2}$. What is the length of a pendulum that has a period of 2 seconds? of 2.5 seconds? of 3 seconds? Express your answers to the nearest tenth of a foot.

Thoughts Into Words

60. Your friend makes an effort to solve the equation $3 + 2\sqrt{x} = x$ as follows:

$$(3 + 2\sqrt{x})^2 = x^2$$

$$9 + 12\sqrt{x} + 4x = x^2$$

At this step he stops and doesn't know how to proceed. What help would you give him?

61. Explain why possible solutions for radical equations *must* be checked.

62. Explain the concept of extraneous solutions.

Answers to the Concept Quiz

1. True   **2.** False   **3.** True   **4.** True   **5.** True   **6.** False   **7.** True   **8.** False   **9.** False   **10.** True

10.6 Merging Exponents and Roots

OBJECTIVES

1 Evaluate a number raised to a rational exponent

2 Write an expression with rational exponents as a radical

3 Write radical expressions as expressions with rational exponents

4 Simplify algebraic expressions that have rational exponents

5 Multiply and divide radicals with different indexes

Recall that the basic properties of positive integral exponents led to a definition for the use of negative integers as exponents. In this section, the properties of integral exponents are used to form definitions for the use of rational numbers as exponents. These definitions will tie together the concepts of exponent and root.

Let's consider the following comparisons.

From our study of radicals, we know that

If $(b^n)^m = b^{mn}$ is to hold when n equals a rational number of the form $\dfrac{1}{p}$, where p is a positive integer greater than 1, then

$$(\sqrt{5})^2 = 5 \qquad\qquad (5^{\frac{1}{2}})^2 = 5^{2\left(\frac{1}{2}\right)} = 5^1 = 5$$

$$(\sqrt[3]{8})^3 = 8 \qquad\qquad (8^{\frac{1}{3}})^3 = 8^{3\left(\frac{1}{3}\right)} = 8^1 = 8$$

$$(\sqrt[4]{21})^4 = 21 \qquad\qquad (21^{\frac{1}{4}})^4 = 21^{4\left(\frac{1}{4}\right)} = 21^1 = 21$$

It would seem reasonable to make the following definition.

Definition 10.4

If b is a real number, n is a positive integer greater than 1, and $\sqrt[n]{b}$ exists, then

$$b^{\frac{1}{n}} = \sqrt[n]{b}$$

Definition 10.4 states that $b^{\frac{1}{n}}$ means the nth root of b. We shall assume that b and n are chosen so that $\sqrt[n]{b}$ exists. For example, $(-25)^{\frac{1}{2}}$ is not meaningful at this time because $\sqrt{-25}$ is not a real number. Consider the following examples, which demonstrate the use of Definition 10.4.

$$25^{\frac{1}{2}} = \sqrt{25} = 5 \qquad\qquad 16^{\frac{1}{4}} = \sqrt[4]{16} = 2$$

$$8^{\frac{1}{3}} = \sqrt[3]{8} = 2 \qquad\qquad \left(\frac{36}{49}\right)^{\frac{1}{2}} = \sqrt{\frac{36}{49}} = \frac{6}{7}$$

$$(-27)^{\frac{1}{3}} = \sqrt[3]{-27} = -3$$

The following definition provides the basis for the use of *all* rational numbers as exponents.

Definition 10.5

If $\dfrac{m}{n}$ is a rational number, where n is a positive integer greater than 1, and b is a real number such that $\sqrt[n]{b}$ exists, then

$$b^{\frac{m}{n}} = \sqrt[n]{b^m} = (\sqrt[n]{b})^m$$

In Definition 10.5, note that the denominator of the exponent is the index of the radical and that the numerator of the exponent is either the exponent of the radicand or the exponent of the root.

Whether we use the form $\sqrt[n]{b^m}$ or the form $\left(\sqrt[n]{b}\right)^m$ for computational purposes depends somewhat on the magnitude of the problem. Let's use both forms on two problems to illustrate this point.

$$8^{\frac{2}{3}} = \sqrt[3]{8^2} \qquad \text{or} \qquad 8^{\frac{2}{3}} = \left(\sqrt[3]{8}\right)^2$$
$$= \sqrt[3]{64} \qquad\qquad\qquad = 2^2$$
$$= 4 \qquad\qquad\qquad\quad = 4$$

$$27^{\frac{2}{3}} = \sqrt[3]{27^2} \qquad \text{or} \qquad 27^{\frac{2}{3}} = \left(\sqrt[3]{27}\right)^2$$
$$= \sqrt[3]{729} \qquad\qquad\qquad = 3^2$$
$$= 9 \qquad\qquad\qquad\quad = 9$$

To compute $8^{\frac{2}{3}}$, either form seems to work about as well as the other one. However, to compute $27^{\frac{2}{3}}$, it should be obvious that $\left(\sqrt[3]{27}\right)^2$ is much easier to handle than $\sqrt[3]{27^2}$.

Classroom Example
Simplify each of the following numerical expressions:
(a) $49^{\frac{3}{2}}$ (b) $81^{\frac{3}{4}}$
(c) $8^{-\frac{5}{3}}$ (d) $(-27)^{\frac{4}{3}}$
(e) $-64^{\frac{1}{3}}$

EXAMPLE 1   Simplify each of the following numerical expressions:

(a) $25^{\frac{3}{2}}$ (b) $16^{\frac{3}{4}}$ (c) $(32)^{-\frac{2}{5}}$ (d) $(-64)^{\frac{2}{3}}$ (e) $-8^{\frac{1}{3}}$

Solution

(a) $25^{\frac{3}{2}} = \left(\sqrt{25}\right)^3 = 5^3 = 125$

(b) $16^{\frac{3}{4}} = \left(\sqrt[4]{16}\right)^3 = 2^3 = 8$

(c) $(32)^{-\frac{2}{5}} = \dfrac{1}{(32)^{\frac{2}{5}}} = \dfrac{1}{\left(\sqrt[5]{32}\right)^2} = \dfrac{1}{2^2} = \dfrac{1}{4}$

(d) $(-64)^{\frac{2}{3}} = \left(\sqrt[3]{-64}\right)^2 = (-4)^2 = 16$

(e) $-8^{\frac{1}{3}} = -\sqrt[3]{8} = -2$

The basic laws of exponents that we stated in Property 8.6 are true for all rational exponents. Therefore, from now on we will use Property 8.6 for rational as well as integral exponents.

Some problems can be handled better in exponential form and others in radical form. Thus we must be able to switch forms with a certain amount of ease. Let's consider some examples that require a switch from one form to the other.

Classroom Example
Write each of the following expressions in radical form:
(a) $m^{\frac{3}{5}}$ (b) $6a^{\frac{4}{7}}$
(c) $m^{\frac{1}{3}}n^{\frac{2}{3}}$ (d) $(a+b)^{\frac{3}{4}}$

EXAMPLE 2   Write each of the following expressions in radical form:

(a) $x^{\frac{3}{4}}$ (b) $3y^{\frac{2}{5}}$ (c) $x^{\frac{1}{4}}y^{\frac{3}{4}}$ (d) $(x+y)^{\frac{2}{3}}$

Solution

(a) $x^{\frac{3}{4}} = \sqrt[4]{x^3}$

(b) $3y^{\frac{2}{5}} = 3\sqrt[5]{y^2}$

(c) $x^{\frac{1}{4}}y^{\frac{3}{4}} = (xy^3)^{\frac{1}{4}} = \sqrt[4]{xy^3}$

(d) $(x+y)^{\frac{2}{3}} = \sqrt[3]{(x+y)^2}$

EXAMPLE 3   Write each of the following using positive rational exponents:

(a) $\sqrt{xy}$ (b) $\sqrt[3]{a^3b}$ (c) $4\sqrt[3]{x^2}$ (d) $\sqrt[5]{(x+y)^4}$

Classroom Example
Write each of the following using positive rational exponents:

(a) $\sqrt{ab}$ (b) $\sqrt[3]{m^2n}$

(c) $5\sqrt[4]{x^3}$ (d) $\sqrt[6]{(m+n)^5}$

Solution

(a) $\sqrt{xy} = (xy)^{\frac{1}{2}} = x^{\frac{1}{2}}y^{\frac{1}{2}}$

(b) $\sqrt[4]{a^3b} = (a^3b)^{\frac{1}{4}} = a^{\frac{3}{4}}b^{\frac{1}{4}}$

(c) $4\sqrt[3]{x^2} = 4x^{\frac{2}{3}}$

(d) $\sqrt[5]{(x+y)^4} = (x+y)^{\frac{4}{5}}$

The properties of exponents provide the basis for simplifying algebraic expressions that contain rational exponents, as these next examples illustrate.

Classroom Example
Simplify each of the following. Express final results using positive exponents only:

(a) $\left(5x^{\frac{1}{3}}\right)\left(2x^{\frac{3}{5}}\right)$ (b) $\left(3m^{\frac{1}{4}}n^{\frac{1}{6}}\right)^3$

(c) $\dfrac{18y^{\frac{1}{6}}}{9y^{\frac{1}{4}}}$ (d) $\left(\dfrac{5x^{\frac{1}{7}}}{8y^{\frac{3}{5}}}\right)^3$

EXAMPLE 4

Simplify each of the following. Express final results using positive exponents only:

(a) $\left(3x^{\frac{1}{2}}\right)\left(4x^{\frac{2}{3}}\right)$ (b) $\left(5a^{\frac{1}{3}}b^{\frac{1}{2}}\right)^2$ (c) $\dfrac{12y^{\frac{1}{3}}}{6y^{\frac{1}{2}}}$ (d) $\left(\dfrac{3x^{\frac{2}{5}}}{2y^{\frac{2}{3}}}\right)^4$

Solution

(a) $\left(3x^{\frac{1}{2}}\right)\left(4x^{\frac{2}{3}}\right) = 3 \cdot 4 \cdot x^{\frac{1}{2}} \cdot x^{\frac{2}{3}}$

$= 12x^{\frac{1}{2}+\frac{2}{3}}$ $b^n \cdot b^m = b^{n+m}$

$= 12x^{\frac{3}{6}+\frac{4}{6}}$ Use 6 as LCD

$= 12x^{\frac{7}{6}}$

(b) $\left(5a^{\frac{1}{3}}b^{\frac{1}{2}}\right)^2 = 5^2 \cdot \left(a^{\frac{1}{3}}\right)^2 \cdot \left(b^{\frac{1}{2}}\right)^2$ $(ab)^n = a^nb^n$

$= 25a^{\frac{2}{3}}b$ $(b^n)^m = b^{mn}$

(c) $\dfrac{12y^{\frac{1}{3}}}{6y^{\frac{1}{2}}} = 2y^{\frac{1}{3}-\frac{1}{2}}$ $\dfrac{b^n}{b^m} = b^{n-m}$

$= 2y^{\frac{2}{6}-\frac{3}{6}}$

$= 2y^{-\frac{1}{6}}$

$= \dfrac{2}{y^{\frac{1}{6}}}$

(d) $\left(\dfrac{3x^{\frac{2}{5}}}{2y^{\frac{2}{3}}}\right)^4 = \dfrac{\left(3x^{\frac{2}{5}}\right)^4}{\left(2y^{\frac{2}{3}}\right)^4}$ $\left(\dfrac{a}{b}\right)^n = \dfrac{a^n}{b^n}$

$= \dfrac{3^4 \cdot \left(x^{\frac{2}{5}}\right)^4}{2^4 \cdot \left(y^{\frac{2}{3}}\right)^4}$ $(ab)^n = a^nb^n$

$= \dfrac{81x^{\frac{8}{5}}}{16y^{\frac{8}{3}}}$ $(b^n)^m = b^{mn}$

Multiplying and Dividing Radicals with Different Indexes

The link between exponents and roots also provides a basis for multiplying and dividing some radicals even if they have different indexes. The general procedure is as follows:

1. Change from radical form to exponential form.

2. Apply the properties of exponents.

3. Then change back to radical form.

The three parts of Example 5 illustrate this process.

EXAMPLE 5

Perform the indicated operations and express the answers in simplest radical form:

(a) $\sqrt{2}\sqrt[3]{2}$ (b) $\dfrac{\sqrt{5}}{\sqrt[3]{5}}$ (c) $\dfrac{\sqrt{4}}{\sqrt[3]{2}}$

Solution

(a) $\sqrt{2}\sqrt[3]{2} = 2^{\frac{1}{2}} \cdot 2^{\frac{1}{3}}$

$= 2^{\frac{1}{2}+\frac{1}{3}}$

$= 2^{\frac{3}{6}+\frac{2}{6}}$ Use 6 as LCD

$= 2^{\frac{5}{6}}$

$= \sqrt[6]{2^5} = \sqrt[6]{32}$

(b) $\dfrac{\sqrt{5}}{\sqrt[3]{5}} = \dfrac{5^{\frac{1}{2}}}{5^{\frac{1}{3}}}$

$= 5^{\frac{1}{2}-\frac{1}{3}}$

$= 5^{\frac{3}{6}-\frac{2}{6}}$ Use 6 as LCD

$= 5^{\frac{1}{6}} = \sqrt[6]{5}$

(c) $\dfrac{\sqrt{4}}{\sqrt[3]{2}} = \dfrac{4^{\frac{1}{2}}}{2^{\frac{1}{3}}}$

$= \dfrac{(2^2)^{\frac{1}{2}}}{2^{\frac{1}{3}}}$

$= \dfrac{2^1}{2^{\frac{1}{3}}}$

$= 2^{1-\frac{1}{3}}$

$= 2^{\frac{2}{3}} = \sqrt[3]{2^2} = \sqrt[3]{4}$

Concept Quiz 10.6

For Problems 1–10, answer true or false.

1. Assuming the nth root of x exists, $\sqrt[n]{x}$ can be written as $x^{\frac{1}{n}}$.

2. An exponent of $\dfrac{1}{3}$ means that we need to find the cube root of the number.

3. To evaluate $16^{\frac{2}{3}}$ we would find the square root of 16 and then cube the result.

4. When an expression with a rational exponent is written as a radical expression, the denominator of the rational exponent is the index of the radical.

5. The expression $\sqrt[n]{x^m}$ is equivalent to $\left(\sqrt[n]{x}\right)^m$.

6. $-16^{-3} = \dfrac{1}{64}$

7. $\dfrac{\sqrt{7}}{\sqrt[3]{7}} = \sqrt[6]{7}$

8. $(16)^{-\frac{3}{4}} = \dfrac{1}{8}$

9. $\dfrac{\sqrt[3]{16}}{\sqrt{2}} = 2\sqrt{2}$

10. $\sqrt[3]{64^2} = 16$

Problem Set 10.6

For Problems 1–30, evaluate each numerical expression.
(Objective 1)

1. $81^{\frac{1}{2}}$

2. $64^{\frac{1}{2}}$

3. $27^{\frac{1}{3}}$

4. $(-32)^{\frac{1}{5}}$

5. $(-8)^{\frac{1}{3}}$

6. $\left(-\dfrac{27}{8}\right)^{\frac{1}{3}}$

7. $-25^{\frac{1}{2}}$

8. $-64^{\frac{1}{3}}$

9. $36^{-\frac{1}{2}}$

10. $81^{-\frac{1}{2}}$

11. $\left(\frac{1}{27}\right)^{-\frac{1}{3}}$

12. $\left(-\frac{8}{27}\right)^{-\frac{1}{3}}$

13. $4^{\frac{3}{2}}$

14. $64^{\frac{1}{3}}$

15. $27^{\frac{4}{3}}$

16. $4^{\frac{7}{2}}$

17. $(-1)^{\frac{7}{3}}$

18. $(-8)^{\frac{4}{3}}$

19. $-4^{\frac{5}{2}}$

20. $-16^{\frac{3}{2}}$

21. $\left(\frac{27}{8}\right)^{\frac{4}{3}}$

22. $\left(\frac{8}{125}\right)^{\frac{2}{3}}$

23. $\left(\frac{1}{8}\right)^{-\frac{2}{3}}$

24. $\left(-\frac{1}{27}\right)^{-\frac{2}{3}}$

25. $64^{-\frac{7}{6}}$

26. $32^{-\frac{4}{5}}$

27. $-25^{\frac{3}{2}}$

28. $-16^{\frac{3}{4}}$

29. $125^{\frac{4}{3}}$

30. $81^{\frac{5}{4}}$

For Problems 31–44, write each of the following in radical form. (Objective 2)
For example,

$$3x^{\frac{2}{3}} = 3\sqrt[3]{x^2}$$

31. $x^{\frac{4}{3}}$

32. $x^{\frac{2}{5}}$

33. $3x^{\frac{1}{2}}$

34. $5x^{\frac{1}{4}}$

35. $(2y)^{\frac{1}{3}}$

36. $(3xy)^{\frac{1}{2}}$

37. $(2x - 3y)^{\frac{1}{2}}$

38. $(5x + y)^{\frac{1}{3}}$

39. $(2a - 3b)^{\frac{2}{3}}$

40. $(5a + 7b)^{\frac{3}{5}}$

41. $x^{\frac{2}{3}}y^{\frac{1}{3}}$

42. $x^{\frac{3}{7}}y^{\frac{5}{7}}$

43. $-3x^{\frac{1}{5}}y^{\frac{2}{5}}$

44. $-4x^{\frac{3}{4}}y^{\frac{1}{4}}$

For Problems 45–58, write each of the following using positive rational exponents. (Objective 3)
For example,

$$\sqrt{ab} = (ab)^{\frac{1}{2}} = a^{\frac{1}{2}}b^{\frac{1}{2}}$$

45. $\sqrt{5y}$

46. $\sqrt{2xy}$

47. $3\sqrt{y}$

48. $5\sqrt{ab}$

49. $\sqrt[3]{xy^2}$

50. $\sqrt[5]{x^2y^4}$

51. $\sqrt[4]{a^2b^3}$

52. $\sqrt[6]{ab^5}$

53. $\sqrt[5]{(2x - y)^3}$

54. $\sqrt[7]{(3x - y)^4}$

55. $5x\sqrt{y}$

56. $4y\sqrt[3]{x}$

57. $-\sqrt[3]{x + y}$

58. $-\sqrt[5]{(x - y)^2}$

For Problems 59–80, simplify each of the following. Express final results using positive exponents only. (Objective 4)
For example,

$$\left(2x^{\frac{1}{2}}\right)\left(3x^{\frac{1}{3}}\right) = 6x^{\frac{5}{6}}$$

59. $\left(2x^{\frac{2}{5}}\right)\left(6x^{\frac{1}{4}}\right)$

60. $\left(3x^{\frac{1}{4}}\right)\left(5x^{\frac{1}{3}}\right)$

61. $\left(y^{\frac{2}{3}}\right)\left(y^{-\frac{1}{4}}\right)$

62. $\left(y^{\frac{3}{4}}\right)\left(y^{-\frac{1}{2}}\right)$

63. $\left(x^{\frac{2}{5}}\right)\left(4x^{-\frac{1}{2}}\right)$

64. $\left(2x^{\frac{1}{3}}\right)\left(x^{-\frac{1}{2}}\right)$

65. $\left(4x^{\frac{1}{2}}y\right)^2$

66. $\left(3x^{\frac{1}{4}}y^{\frac{1}{5}}\right)^3$

67. $(8x^6y^3)^{\frac{1}{3}}$

68. $(9x^2y^4)^{\frac{1}{2}}$

69. $\dfrac{24x^{\frac{3}{5}}}{6x^{\frac{1}{5}}}$

70. $\dfrac{18x^{\frac{1}{2}}}{9x^{\frac{1}{3}}}$

71. $\dfrac{48b^{\frac{1}{3}}}{12b^{\frac{3}{4}}}$

72. $\dfrac{56a^{\frac{1}{6}}}{8a^{\frac{1}{4}}}$

73. $\left(\dfrac{6x^{\frac{2}{5}}}{7y^{\frac{2}{3}}}\right)^2$

74. $\left(\dfrac{2x^{\frac{1}{3}}}{3y^{\frac{1}{4}}}\right)^4$

75. $\left(\dfrac{x^2}{y^3}\right)^{-\frac{1}{2}}$

76. $\left(\dfrac{a^3}{b^{-2}}\right)^{-\frac{1}{3}}$

77. $\left(\dfrac{18x^{\frac{1}{3}}}{9x^{\frac{1}{4}}}\right)^2$

78. $\left(\dfrac{72x^{\frac{3}{4}}}{6x^{\frac{1}{2}}}\right)^2$

79. $\left(\dfrac{60a^{\frac{1}{5}}}{15a^{\frac{3}{4}}}\right)^2$

80. $\left(\dfrac{64a^{\frac{1}{3}}}{16a^{\frac{5}{9}}}\right)^3$

For Problems 81–90, perform the indicated operations and express answers in simplest radical form (see Example 5). (Objective 5)

81. $\sqrt[3]{3}\sqrt{3}$

82. $\sqrt{2}\sqrt[4]{2}$

83. $\sqrt[4]{6}\sqrt{6}$

84. $\sqrt[3]{5}\sqrt{5}$

85. $\dfrac{\sqrt[3]{3}}{\sqrt[4]{3}}$

86. $\dfrac{\sqrt{2}}{\sqrt[3]{2}}$

87. $\dfrac{\sqrt[3]{8}}{\sqrt[4]{4}}$

88. $\dfrac{\sqrt{9}}{\sqrt[3]{3}}$

89. $\dfrac{\sqrt[4]{27}}{\sqrt{3}}$

90. $\dfrac{\sqrt[3]{16}}{\sqrt[6]{4}}$

Thoughts Into Words

91. Your friend keeps getting an error message when evaluating $-4^{\frac{5}{2}}$ on his calculator. What error is he probably making?

92. Explain how you would evaluate $27^{\frac{2}{3}}$ without a calculator.

Further Investigations

93. Use your calculator to evaluate each of the following.

(a) $\sqrt[3]{1728}$ (b) $\sqrt[3]{5832}$

(c) $\sqrt[4]{2401}$ (d) $\sqrt[4]{65{,}536}$

(e) $\sqrt[5]{161{,}051}$ (f) $\sqrt[5]{6{,}436{,}343}$

94. Definition 5.7 states that

$$b^{\frac{m}{n}} = \sqrt[n]{b^m} = \left(\sqrt[n]{b}\right)^m$$

Use your calculator to verify each of the following.

(a) $\sqrt[3]{27^2} = \left(\sqrt[3]{27}\right)^2$ (b) $\sqrt[3]{8^5} = \left(\sqrt[3]{8}\right)^5$

(c) $\sqrt[4]{16^3} = \left(\sqrt[4]{16}\right)^3$ (d) $\sqrt[3]{16^2} = \left(\sqrt[3]{16}\right)^2$

(e) $\sqrt[5]{9^4} = \left(\sqrt[5]{9}\right)^4$ (f) $\sqrt[3]{12^4} = \left(\sqrt[3]{12}\right)^4$

95. Use your calculator to evaluate each of the following.

(a) $16^{\frac{5}{2}}$ (b) $25^{\frac{7}{2}}$ (c) $16^{\frac{9}{4}}$

(d) $27^{\frac{5}{3}}$ (e) $343^{\frac{2}{3}}$ (f) $512^{\frac{4}{3}}$

96. Use your calculator to estimate each of the following to the nearest one-thousandth.

(a) $7^{\frac{4}{3}}$ (b) $10^{\frac{4}{5}}$

(c) $12^{\frac{3}{5}}$ (d) $19^{\frac{2}{5}}$

(e) $7^{\frac{3}{4}}$ (f) $10^{\frac{5}{4}}$

97. (a) Because $\dfrac{4}{5} = 0.8$, we can evaluate $10^{\frac{4}{5}}$ by evaluating $10^{0.8}$, which involves a shorter sequence of calculator steps. Evaluate parts (b), (c), (d), (e), and (f) of Problem 96 and take advantage of the decimal exponents.

(b) What problem is created when we try to evaluate $7^{\frac{4}{3}}$ by changing the exponent to decimal form?

OBJECTIVE	SUMMARY	EXAMPLE
Simplify numerical expressions that have integer exponents. (Section 10.1/Objective 1)	The concept of exponent is expanded to include negative exponents and exponents of zero. If b is a nonzero number, then $b^0 = 1$. If n is a positive integer, and b is a nonzero number, then $b^{-n} = \dfrac{1}{b^n}$.	Simplify $\left(\dfrac{2}{5}\right)^{-2}$. **Solution** $$\left(\dfrac{2}{5}\right)^{-2} = \dfrac{2^{-2}}{5^{-2}} = \dfrac{5^2}{2^2} = \dfrac{25}{4}$$
Simplify algebraic expressions that have integer exponents. (Section 10.1/Objective 2)	The properties for integer exponents listed on page 459 form the basis for manipulating with integer exponents. These properties, along with Definition 6.3; that is, $b^{-n} = \dfrac{1}{b^n}$, enable us to simplify algebraic expressions and express the results with positive exponents.	Simplify $(2x^{-3}y)^{-2}$ and express the final result using positive exponents. **Solution** $(2x^{-3}y)^{-2} = 2^{-2}x^6y^{-2}$ $$= \dfrac{x^6}{2^2y^2} = \dfrac{x^6}{4y^2}$$
Multiply and divide algebraic expressions that have integer exponents. (Section 10.1/Objective 3)	The previous remark also applies to simplifying multiplication and division problems that involve integer exponents.	Simplify $(-3x^5y^{-2})(4x^{-1}y^{-1})$ and express the final result using positive exponents. **Solution** $(-3x^5y^{-2})(4x^{-1}y^{-1}) = -12x^4y^{-3}$ $$= -\dfrac{12x^4}{y^3}$$
Simplify sums and differences of expressions involving integer exponents. (Section 10.1/Objective 4)	Find the sum or difference of expressions involving integer exponents, and change all expressions having negative or zero exponents to equivalent expressions with positive exponents only. To find the sum or difference, it may be necessary to find a common denominator.	Simplify $5x^{-2} + 6y^{-1}$ and express the result as a single fraction involving positive exponents only. **Solution** $5x^{-2} + 6y^{-1} = \dfrac{5}{x^2} + \dfrac{6}{y}$ $$= \dfrac{5}{x^2} \cdot \dfrac{y}{y} + \dfrac{6}{y} \cdot \dfrac{x^2}{x^2}$$ $$= \dfrac{5y + 6x^2}{x^2y}$$
Write numbers in scientific notation. (Section 10.1/Objective 5)	Scientific notation is often used to write numbers that are very small or very large in magnitude. The scientific form of a number is expressed as $(N)(10^k)$, where the absolute value of N is a number greater than or equal to 1 and less than 10, written in decimal form, and k is an integer.	Write each of the following in scientific notation: **(a)** 0.000000843 **(b)** 456,000,000,000 **Solution** **(a)** $0.000000843 = (8.43)(10^{-7})$ **(b)** $456,000,000,000 = (4.56)(10^{11})$
Convert numbers from scientific notation to ordinary decimal notation. (Section 10.1/Objective 5)	To switch from scientific notation to ordinary notation, move the decimal point the number of places indicated by the exponent of 10. The decimal point is moved to the right if the exponent is positive and to the left if the exponent is negative.	Write each of the following in ordinary decimal notation: **(a)** $(8.5)(10^{-5})$ **(b)** $(3.4)(10^6)$ **Solution** **(a)** $(8.5)(10^{-5}) = 0.000085$ **(b)** $(3.4)(10^6) = 3,400,000$

(continued)

OBJECTIVE	SUMMARY	EXAMPLE
Perform calculations with numbers using scientific notation. **(Section 10.1/Objective 6)**	Scientific notation can often be used to simplify numerical calculations.	Use scientific notation and the properties of exponents to simplify $\dfrac{0.0000084}{0.002}$. **Solution** Change the numbers to scientific notation and use the appropriate properties of exponents. Express the result in standard decimal notation. $\dfrac{0.0000084}{0.002} = \dfrac{(8.4)(10^{-6})}{(2)(10^{-3})}$ $= (4.2)(10^{-3}) = 0.0042$
Express a radical in simplest radical form. **(Section 10.2/Objective 2)**	The principal nth root of b is designated by $\sqrt[n]{b}$, where n is the index and b is the radicand. A radical expression is in simplest form if: 1. A radicand contains no polynomial factor raised to a power equal to or greater than the index of the radical; 2. No fraction appears within a radical sign; and 3. No radical appears in the denominator. The following properties are used to express radicals in simplest form: $\sqrt[n]{bc} = \sqrt[n]{b}\sqrt[n]{c}$ and $\sqrt[n]{\dfrac{b}{c}} = \dfrac{\sqrt[n]{b}}{\sqrt[n]{c}}$	Simplify $\sqrt{150a^3b^2}$. Assume all variables represent nonnegative values. **Solution** $\sqrt{150a^3b^2} = \sqrt{25a^2b^2}\sqrt{6b}$ $= 5ab\sqrt{6b}$
Rationalize the denominator to simplify radicals. **(Section 10.2/Objective 3)**	If a radical appears in the denominator, it will be necessary to rationalize the denominator for the expression to be in simplest form.	Simplify $\dfrac{2\sqrt{18}}{\sqrt{5}}$. **Solution** $\dfrac{2\sqrt{18}}{\sqrt{5}} = \dfrac{2\sqrt{9}\sqrt{2}}{\sqrt{5}}$ $= \dfrac{2(3)\sqrt{2}}{\sqrt{5}} = \dfrac{6\sqrt{2}}{\sqrt{5}}$ $= \dfrac{6\sqrt{2}}{\sqrt{5}} \cdot \dfrac{\sqrt{5}}{\sqrt{5}} = \dfrac{6\sqrt{10}}{\sqrt{25}}$ $= \dfrac{6\sqrt{10}}{5}$
Simplify expressions by combining radicals. **(Section 10.3/Objective 1)**	Simplifying by combining radicals sometimes requires that we first express the given radicals in simplest form.	Simplify $\sqrt{24} - \sqrt{54} + 8\sqrt{6}$. **Solution** $\sqrt{24} - \sqrt{54} + 8\sqrt{6}$ $= \sqrt{4}\sqrt{6} - \sqrt{9}\sqrt{6} + 8\sqrt{6}$ $= 2\sqrt{6} - 3\sqrt{6} + 8\sqrt{6}$ $= (2 - 3 + 8)(\sqrt{6})$ $= 7\sqrt{6}$

OBJECTIVE	SUMMARY	EXAMPLE
Multiply two radicals. (Section 10.4/Objective 1)	The property $\sqrt[n]{b}\sqrt[n]{c} = \sqrt[n]{bc}$ is used to find the product of two radicals.	Multiply $\sqrt[3]{4x^2y}\sqrt[3]{6x^2y^2}$. **Solution** $$\sqrt[3]{4x^2y}\sqrt[3]{6x^2y^2} = \sqrt[3]{24x^4y^3}$$ $$= \sqrt[3]{8x^3y^3}\sqrt[3]{3x}$$ $$= 2xy\sqrt[3]{3x}$$
Use the distributive property to multiply radical expressions. (Section 10.4/Objective 2)	The distributive property and the property $\sqrt[n]{b}\sqrt[n]{c} = \sqrt[n]{bc}$ are used to find products of radical expressions.	Multiply $\sqrt{2x}(\sqrt{6x} + \sqrt{18xy})$ and simplify where possible. **Solution** $$\sqrt{2x}(\sqrt{6x} + \sqrt{18xy})$$ $$= \sqrt{12x^2} + \sqrt{36x^2y}$$ $$= \sqrt{4x^2}\sqrt{3} + \sqrt{36x^2}\sqrt{y}$$ $$= 2x\sqrt{3} + 6x\sqrt{y}$$
Rationalize binomial denominators. (Section 10.4/Objective 3)	The factors $(a - b)$ and $(a + b)$ are called conjugates. To rationalize a binomial denominator involving radicals, multiply the numerator or denominator by the conjugate of the denominator.	Simplify $\dfrac{3}{\sqrt{7} - \sqrt{5}}$ by rationalizing the denominator. **Solution** $$\frac{3}{\sqrt{7} - \sqrt{5}}$$ $$= \frac{3}{(\sqrt{7} - \sqrt{5})} \cdot \frac{(\sqrt{7} + \sqrt{5})}{(\sqrt{7} + \sqrt{5})}$$ $$= \frac{3(\sqrt{7} + \sqrt{5})}{\sqrt{49} - \sqrt{25}} = \frac{3(\sqrt{7} + \sqrt{5})}{7 - 5}$$ $$= \frac{3(\sqrt{7} + \sqrt{5})}{2}$$
Solve radical equations. (Section 10.5/Objective 1)	Equations with variables in a radicand are called radical equations. Radical equations are solved by raising each side of the equation to the appropriate power. However, raising both sides of the equation to a power may produce extraneous roots. Therefore, you must check each potential solution.	Solve $\sqrt{x} + 20 = x$. **Solution** $$\sqrt{x} + 20 = x$$ $$\sqrt{x} = x - 20 \quad \text{Isolate the radical}$$ $$(\sqrt{x})^2 = (x - 20)^2$$ $$x = x^2 - 40x + 400$$ $$0 = x^2 - 41x + 400$$ $$0 = (x - 25)(x - 16)$$ $$x = 25 \quad \text{or} \quad x = 16$$ ✔ **Check** $$\sqrt{x} + 20 = x$$ **If $x = 25$** **If $x = 16$** $$\sqrt{25} + 20 \overset{?}{=} 25 \qquad \sqrt{16} + 20 \overset{?}{=} 16$$ $$25 = 25 \qquad\qquad 24 \neq 16$$ The solution set is $\{25\}$.

(continued)

OBJECTIVE	SUMMARY	EXAMPLE
Solve radical equations for real-world problems. **(Section 10.5/Objective 2)**	Various formulas involve radical equations. These formulas are solved in the same manner as radical equations.	Use the formula $\sqrt{30Df} = S$ (given in Section 10.5) to determine the coefficient of friction, to the nearest hundredth, if a car traveling at 50 miles per hour skidded 300 feet. **Solution** Solve $\sqrt{30Df} = S$ for f. $$(\sqrt{30Df})^2 = S^2$$ $$30Df = S^2$$ $$f = \frac{S^2}{30D}$$ Substituting the values for S and D gives $$f = \frac{50^2}{30(300)}$$ $$= 0.28 \quad \text{to the nearest hundredth}$$
Evaluate a number raised to a rational exponent. **(Section 10.6/Objective 1)**	To simplify a number raised to a rational exponent, we apply either the property $b^{\frac{1}{n}} = \sqrt[n]{b}$ or the property $b^{\frac{m}{n}} = \sqrt[n]{b^m} = (\sqrt[n]{b})^m$. When simplifying $b^{\frac{m}{n}}$, the arithmetic computations are usually easiest using the form $(\sqrt[n]{b})^m$, where the nth root is taken first, and that result is raised to the m power.	Simplify $16^{\frac{3}{2}}$. **Solution** $$16^{\frac{3}{2}} = (16^{\frac{1}{2}})^3$$ $$= 4^3$$ $$= 64$$
Write an expression with rational exponents as a radical. **(Section 10.6/Objective 2)**	If $\dfrac{m}{n}$ is a rational number, n is a positive integer greater than 1, and b is a real number such that $\sqrt[n]{b}$ exists, then $b^{\frac{m}{n}} = \sqrt[n]{b^m} = (\sqrt[n]{b})^m$.	Write $x^{\frac{3}{5}}$ in radical form. **Solution** $$x^{\frac{3}{5}} = \sqrt[5]{x^3}$$
Write radical expressions as expressions with rational exponents. **(Section 10.6/Objective 3)**	The index of the radical will be the denominator of the rational exponent.	Write $\sqrt[4]{x^3y}$ using positive rational exponents. **Solution** $$\sqrt[4]{x^3y} = x^{\frac{3}{4}}y^{\frac{1}{4}}$$
Simplify algebraic expressions that have rational exponents. **(Section 10.6/Objective 4)**	Properties of exponents are used to simplify products and quotients involving rational exponents.	Simplify $\left(4x^{\frac{1}{3}}\right)\left(-3x^{-\frac{3}{4}}\right)$ and express the result with positive exponents only. **Solution** $$\left(4x^{\frac{1}{3}}\right)\left(-3x^{-\frac{3}{4}}\right) = -12x^{\frac{1}{3}-\frac{3}{4}}$$ $$= -12x^{-\frac{5}{12}}$$ $$= \frac{-12}{x^{\frac{5}{12}}}$$
Multiply and divide radicals with different indexes. **(Section 10.6/Objective 5)**	The link between rational exponents and roots provides the basis for multiplying and dividing radicals with different indexes.	Multiply $\sqrt[3]{y^2}\sqrt{y}$ and express in simplest radical form. **Solution** $$\sqrt[3]{y^2}\sqrt{y} = y^{\frac{2}{3}}y^{\frac{1}{2}}$$ $$= y^{\frac{2}{3}+\frac{1}{2}} = y^{\frac{7}{6}}$$ $$= \sqrt[6]{y^7} = y\sqrt[6]{y}$$

Chapter 10   Review Problem Set

For Problems 1–6, evaluate the numerical expression.

1. 4^{-3}

2. $\left(\dfrac{2}{3}\right)^{-2}$

3. $(3^2 \cdot 3^{-3})^{-1}$

4. $(4^{-2} \cdot 4^2)^{-1}$

5. $\left(\dfrac{3^{-1}}{3^2}\right)^{-1}$

6. $\left(\dfrac{5^2}{5^{-1}}\right)^{-1}$

For Problems 7–18, simplify and express the final result using positive exponents.

7. $(x^{-3}y^4)^{-2}$

8. $\left(\dfrac{2a^{-1}}{3b^4}\right)^{-3}$

9. $\left(\dfrac{4a^{-2}}{3b^{-2}}\right)^{-2}$

10. $(5x^3y^{-2})^{-3}$

11. $\left(\dfrac{6x^{-2}}{2x^4}\right)^{-2}$

12. $\left(\dfrac{8y^2}{2y^{-1}}\right)^{-1}$

13. $(-5x^{-3})(2x^6)$

14. $(a^{-4}b^3)(3ab^2)$

15. $\dfrac{a^{-1}b^{-2}}{a^4b^{-5}}$

16. $\dfrac{x^3y^5}{x^{-1}y^6}$

17. $\dfrac{-12x^3}{6x^5}$

18. $\dfrac{10a^2b^3}{-5ab^4}$

For Problems 19–22, express as a single fraction involving positive exponents only.

19. $x^{-2} + y^{-1}$

20. $a^{-2} - 2a^{-1}b^{-1}$

21. $2x^{-1} + 3y^{-2}$

22. $(2x)^{-1} + 3y^{-2}$

For Problems 23–34, express the radical in simplest radical form. Assume the variables represent positive real numbers.

23. $\sqrt{54}$

24. $\sqrt{48x^3y}$

25. $\sqrt[3]{56}$

26. $\sqrt[3]{108x^4y^8}$

27. $\dfrac{3}{4}\sqrt{150}$

28. $\dfrac{2}{3}\sqrt{45xy^3}$

29. $\dfrac{4\sqrt{3}}{\sqrt{6}}$

30. $\sqrt{\dfrac{5}{12x^3}}$

31. $\dfrac{\sqrt[3]{2}}{\sqrt[3]{9}}$

32. $\sqrt{\dfrac{9}{5}}$

33. $\sqrt{\dfrac{3x^3}{7}}$

34. $\dfrac{\sqrt{8x^2}}{\sqrt{2x}}$

For Problems 35–38, use the distributive property to help simplify the expression.

35. $3\sqrt{45} - 2\sqrt{20} - \sqrt{80}$

36. $4\sqrt[3]{24} + 3\sqrt[3]{3} - 2\sqrt[3]{81}$

37. $3\sqrt{24} - \dfrac{2\sqrt{54}}{5} + \dfrac{\sqrt{96}}{4}$

38. $-2\sqrt{12x} + 3\sqrt{27x} - 5\sqrt{48x}$

For Problems 39–48, multiply and simplify. Assume the variables represent nonnegative real numbers.

39. $(3\sqrt{8})(4\sqrt{5})$

40. $(5\sqrt[3]{2})(6\sqrt[3]{4})$

41. $(\sqrt{6xy})(\sqrt{10x})$

42. $(-3\sqrt{6xy^3})(\sqrt{6y})$

43. $3\sqrt{2}(4\sqrt{6} - 2\sqrt{7})$

44. $(\sqrt{x} + 3)(\sqrt{x} - 5)$

45. $(2\sqrt{5} - \sqrt{3})(2\sqrt{5} + \sqrt{3})$

46. $(3\sqrt{2} + \sqrt{6})(5\sqrt{2} - 3\sqrt{6})$

47. $(2\sqrt{a} + \sqrt{b})(3\sqrt{a} - 4\sqrt{b})$

48. $(4\sqrt{8} - \sqrt{2})(\sqrt{8} + 3\sqrt{2})$

For Problems 49–52, rationalize the denominator and simplify.

49. $\dfrac{4}{\sqrt{7} - 1}$

50. $\dfrac{\sqrt{3}}{\sqrt{8} + \sqrt{5}}$

51. $\dfrac{3}{2\sqrt{3} + 3\sqrt{5}}$

52. $\dfrac{3\sqrt{2}}{2\sqrt{6} - \sqrt{10}}$

For Problems 53–60, solve the equation.

53. $\sqrt{7x - 3} = 4$

54. $\sqrt{2y + 1} = \sqrt{5y - 11}$

55. $\sqrt{2x} = x - 4$

56. $\sqrt{n^2 - 4n - 4} = n$

57. $\sqrt[3]{2x - 1} = 3$

58. $\sqrt[3]{t^2 + 9t - 1} = 3$

59. $\sqrt{x^2 + 3x - 6} = x$

60. $\sqrt{x + 1} - \sqrt{2x} = -1$

61. The formula $S = \sqrt{30Df}$ is used to approximate the speed S, where D represents the length of the skid marks in feet and f represents the coefficient of friction for the road surface. Suppose that the coefficient of friction is 0.38. How far will a car skid, to the nearest foot, when the brakes are applied at 75 miles per hour?

62. The formula $T = 2\pi\sqrt{\dfrac{L}{32}}$ is used for pendulum motion, where T represents the period of the pendulum in seconds, and L represents the length of the pendulum in feet. Find the length of a pendulum, to the nearest tenth of a foot, if the period is 2.4 seconds.

For Problems 63–70, simplify.

63. $4^{\frac{5}{2}}$

64. $(-1)^{\frac{2}{3}}$

65. $\left(\frac{8}{27}\right)^{\frac{2}{3}}$

66. $-16^{\frac{3}{2}}$

67. $(27)^{-\frac{2}{3}}$

68. $(32)^{-\frac{2}{5}}$

69. $9^{\frac{3}{2}}$

70. $16^{\frac{3}{4}}$

For Problems 71–74, write the expression in radical form.

71. $x^{\frac{1}{3}}y^{\frac{2}{3}}$

72. $a^{\frac{3}{4}}$

73. $4y^{\frac{1}{2}}$

74. $(x + 5y)^{\frac{2}{3}}$

For Problems 75–78, write the expression using positive rational exponents.

75. $\sqrt[5]{x^3y}$

76. $\sqrt[3]{4a^2}$

77. $6\sqrt[4]{y^2}$

78. $\sqrt[3]{(3a + b)^5}$

For Problems 79–84, simplify and express the final result using positive exponents.

79. $\left(4x^{\frac{1}{2}}\right)\left(5x^{\frac{1}{5}}\right)$

80. $\frac{42a^{\frac{3}{4}}}{6a^{\frac{1}{3}}}$

81. $\left(\frac{x^3}{y^4}\right)^{-\frac{1}{3}}$

82. $\left(-3a^{\frac{1}{4}}\right)\left(2a^{-\frac{1}{2}}\right)$

83. $\left(x^{\frac{4}{5}}\right)^{-\frac{1}{2}}$

84. $\frac{-24y^{\frac{2}{3}}}{4y^{\frac{1}{4}}}$

For Problems 85–88, perform the indicated operation and express the answer in simplest radical form.

85. $\sqrt[4]{3}\sqrt{3}$

86. $\sqrt[3]{9}\sqrt{3}$

87. $\frac{\sqrt[3]{5}}{\sqrt[4]{5}}$

88. $\frac{\sqrt[3]{16}}{\sqrt{2}}$

For Problems 89–92, write the number in scientific notation.

89. 540,000,000

90. 84,000

91. 0.000000032

92. 0.000768

For Problems 93–96, write the number in ordinary decimal notation.

93. $(1.4)(10^{-6})$

94. $(6.38)(10^{-4})$

95. $(4.12)(10^7)$

96. $(1.25)(10^5)$

For Problems 97–104, use scientific notation and the properties of exponents to help perform the calculations.

97. $(0.00002)(0.0003)$

98. $(120,000)(300,000)$

99. $(0.000015)(400,000)$

100. $\frac{0.000045}{0.0003}$

101. $\frac{(0.00042)(0.0004)}{0.006}$

102. $\sqrt{0.000004}$

103. $\sqrt[3]{0.000000008}$

104. $4,000,000^{\frac{3}{2}}$

For Problems 1–4, simplify each of the numerical expressions.

1. $(4)^{-\frac{5}{2}}$

2. $-16^{\frac{5}{4}}$

3. $\left(\dfrac{2}{3}\right)^{-4}$

4. $\left(\dfrac{2^{-1}}{2^{-2}}\right)^{-2}$

For Problems 5–9, express each radical expression in simplest radical form. Assume the variables represent positive real numbers.

5. $\sqrt{63}$

6. $\sqrt[3]{108}$

7. $\sqrt{52x^4y^3}$

8. $\dfrac{5\sqrt{18}}{3\sqrt{12}}$

9. $\sqrt{\dfrac{7}{24x^3}}$

10. Multiply and simplify: $(4\sqrt{6})(3\sqrt{12})$

11. Multiply and simplify: $(3\sqrt{2} + \sqrt{3})(\sqrt{2} - 2\sqrt{3})$

12. Simplify by combining similar radicals:

$$2\sqrt{50} - 4\sqrt{18} - 9\sqrt{32}$$

13. Rationalize the denominator and simplify:

$$\dfrac{3\sqrt{2}}{4\sqrt{3} - \sqrt{8}}$$

14. Simplify and express the answer using positive exponents: $\left(\dfrac{2x^{-1}}{3y}\right)^{-2}$

15. Simplify and express the answer using positive exponents: $\dfrac{-84a^{\frac{1}{2}}}{7a^{\frac{4}{5}}}$

16. Express $x^{-1} + y^{-3}$ as a single fraction involving positive exponents.

17. Multiply and express the answer using positive exponents: $(3x^{-\frac{1}{2}})(-4x^{\frac{3}{4}})$

18. Multiply and simplify:

$$(3\sqrt{5} - 2\sqrt{3})(3\sqrt{5} + 2\sqrt{3})$$

For Problems 19 and 20, use scientific notation and the properties of exponents to help with the calculations.

19. $\dfrac{(0.00004)(300)}{0.00002}$

20. $\sqrt{0.000009}$

For Problems 21–25, solve each equation.

21. $\sqrt{3x + 1} = 3$

22. $\sqrt[3]{3x + 2} = 2$

23. $\sqrt{x} = x - 2$

24. $\sqrt{5x - 2} = \sqrt{3x + 8}$

25. $\sqrt{x^2 - 10x + 28} = 2$

1. Evaluate each of the following numerical expressions.

  a. $\left(\dfrac{3}{4} - \dfrac{1}{3}\right)^{-1}$ **b.** $(2^{-1} + 3^{-2})^{-2}$

  c. $\sqrt[3]{-\dfrac{1}{8}}$ **d.** $(-27)^{\frac{4}{3}}$

For Problems 2–5, perform the indicated operations and express results using positive exponents only.

2. $\dfrac{-64x^{-1}y^3}{16x^3y^{-2}}$ **3.** $(-4x^{-2}y^{-3})(3x^2y^{-1})$

4. $(3x - 1)(2x^2 + 6x - 4)$

5. $(4x^3 + 17x^2 - 44x - 12) \div (x + 6)$

6. Solve the system $\begin{pmatrix} 5x - 3y = -31 \\ 4x + 7y = 41 \end{pmatrix}$.

7. Express each of the following in simplest radical form.

  a. $3\sqrt{56}$ **b.** $2\sqrt[3]{56}$

  c. $\dfrac{3\sqrt{2}}{4\sqrt{6}}$ **d.** $\sqrt{\dfrac{6}{8}}$

8. Write the equation of the line that is parallel to the line $7x + 3y = 9$ and contains the point $(-4, -6)$.

9. Twenty-five percent of what number is 18?

10. Evaluate $-4(2a - b) + 6(b - 2a) - (3a + 4b)$ for $a = -15$ and $b = 14$.

11. Evaluate $\dfrac{56x^{-1}y^{-3}}{8x^{-2}y^{-4}}$ for $x = -\dfrac{1}{2}$ and $y = \dfrac{2}{3}$.

12. Prime-factor each of the following composite numbers.

  a. 52   **b.** 80

  c. 91   **d.** 78

13. Simplify the complex fraction $\dfrac{\dfrac{5}{3x} - \dfrac{2}{y}}{\dfrac{3}{x} + \dfrac{5}{3y}}$.

For Problems 14–16, perform the indicated operations and express your answers in simplest form.

14. $\dfrac{3x - 7}{5} - \dfrac{2x + 1}{4}$ **15.** $\dfrac{5x^2y}{7xy} \div \dfrac{15y}{14x}$

16. $\left(\dfrac{x^2 + 5x}{3x^2 + 2x - 8}\right)\left(\dfrac{2x^2 - 8}{x^3 + 3x^2 - 10x}\right)$

For Problems 17–21, solve each equation.

17. $x(x - 4) - 3(x - 4) = 0$

18. $(x + 2)(x - 5) = -6$

19. $(3x - 5)(2x + 7) = 0$

20. $|2x - 5| = -4$ **21.** $\sqrt{5 + 2x} = 1 + \sqrt{2x}$

For Problems 22–25, use an equation or a system of equations to help solve each problem.

22. The area of a triangle is 51 square inches. The length of one side of the triangle is 1 inch less than three times the length of the altitude to that side. Find the length of that side and the length of the altitude to that side.

23. Brad is 6 years older than Pedro. Five years ago Pedro's age was three-fourths of Brad's age at that time. Find the present ages of Brad and Pedro.

24. Karla sold an autographed sports card for $97.50. This selling price represented a 30% profit for her, on the basis of what she originally paid for the card. Find Karla's original cost for the autographed sports card.

25. A rectangular piece of cardboard is 4 inches longer than it is wide. From each of its corners, a square piece 2 inches on a side is cut out. The flaps are then turned up to form an open box, which has a volume of 42 cubic inches. Find the length and width of the original piece of cardboard.

11

Quadratic Equations and Inequalities

The Pythagorean theorem is applied throughout the construction industry when right angles are involved.

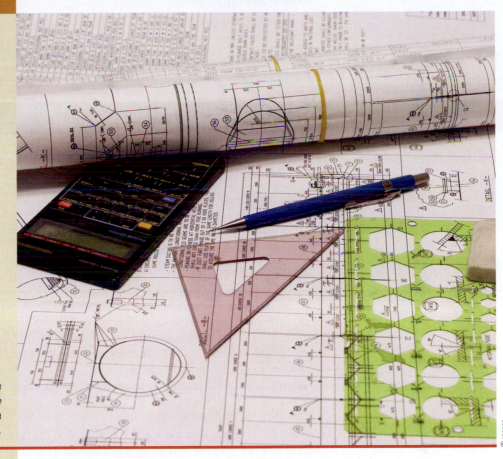

© ragsac

A crime scene investigator must record the dimensions of a rectangular bedroom. Because the access to one wall is blocked, the investigator can only measure the other wall and the diagonal of the rectangular room. The investigator determines that one wall measures 12 feet and the diagonal of the room measures 15 feet. By applying the Pythagorean theorem and solving the resulting quadratic equation, $a^2 + 12^2 = 15^2$, the investigator can determine that the room measures 12 feet by 9 feet.

Solving equations is one of the central themes of this text. Let's pause for a moment and reflect on the different types of equations that we have solved in the last five chapters.

As the chart on the next page shows, we have solved second-degree equations in one variable, but only those for which the polynomial is factorable. In this chapter we will expand our work to include more general types of second-degree equations, as well as inequalities in one variable.

Video tutorials based on section learning objectives are available in a variety of delivery modes.

Type of equation	Example	Solution set
First-degree equations in one variable	$3(x + 4) = -2x + 4x - 10$	$\{-22\}$
Second-degree equations in one variable *that are factorable*	$x^2 - x - 6 = 0$	$\{-2, 3\}$
Fractional equations	$\dfrac{2}{x^2 - 9} + \dfrac{3}{x + 3} = \dfrac{4}{x - 3}$	$\{-19\}$
Radical equations	$\sqrt{3x - 2} = 5$	$\{9\}$

11.1 Complex Numbers

OBJECTIVES

1. Know about the set of complex numbers
2. Add and subtract complex numbers
3. Simplify radicals involving negative numbers
4. Perform operations on radicals involving negative numbers
5. Multiply complex numbers
6. Divide complex numbers

Because the square of any real number is nonnegative, a simple equation such as $x^2 = -4$ has no solutions in the set of real numbers. To handle this situation, we can expand the set of real numbers into a larger set called the *complex numbers*. In this section we will instruct you on how to manipulate complex numbers.

To provide a solution for the equation $x^2 + 1 = 0$, we use the number i, such that

$$i^2 = -1$$

The number i is not a real number and is often called the **imaginary unit**, but the number i^2 is the real number -1. The imaginary unit i is used to define a complex number as follows:

> **Definition 11.1**
>
> A **complex number** is any number that can be expressed in the form
>
> $$a + bi$$
>
> where a and b are real numbers.

The form $a + bi$ is called the **standard form** of a complex number. The real number a is called the **real part** of the complex number, and b is called the **imaginary part**. (Note that b is a real number even though it is called the imaginary part.) The following list exemplifies this terminology.

1. The number $7 + 5i$ is a complex number that has a real part of 7 and an imaginary part of 5.

2. The number $\dfrac{2}{3} + i\sqrt{2}$ is a complex number that has a real part of $\dfrac{2}{3}$ and an imaginary part of $\sqrt{2}$. (It is easy to mistake $\sqrt{2}i$ for $\sqrt{2i}$. Thus we commonly write $i\sqrt{2}$ instead of $\sqrt{2}i$ to avoid any difficulties with the radical sign.)

3. The number $-4 - 3i$ can be written in the standard form $-4 + (-3i)$ and therefore is a complex number that has a real part of -4 and an imaginary part of -3. [The form $-4 - 3i$ is often used, but we know that it means $-4 + (-3i)$.]

4. The number $-9i$ can be written as $0 + (-9i)$; thus it is a complex number that has a real part of 0 and an imaginary part of -9. (Complex numbers, such as $-9i$, for which $a = 0$ and $b \neq 0$ are called *pure imaginary numbers*.)

5. The real number 4 can be written as $4 + 0i$ and is thus a complex number that has a real part of 4 and an imaginary part of 0.

Look at item 5 in this list. We see that the set of real numbers is a subset of the set of complex numbers. The following diagram indicates the organizational format of the complex numbers.

Complex numbers $a + bi$ where a and b are real numbers

Real numbers
$a + bi$ where $b = 0$

Imaginary numbers
$a + bi$ where $b \neq 0$

Pure imaginary numbers
$a + bi$ where $a = 0$ and $b \neq 0$

Two complex numbers $a + bi$ and $c + di$ are said to be **equal** if and only if $a = c$ and $b = d$.

Adding and Subtracting Complex Numbers

To *add complex numbers*, we simply add their real parts and add their imaginary parts. Thus

$$(a + bi) + (c + di) = (a + c) + (b + d)i$$

The following example shows addition of two complex numbers.

EXAMPLE 1   Add the complex numbers:

(a) $(4 + 3i) + (5 + 9i)$ (b) $(-6 + 4i) + (8 - 7i)$

(c) $\left(\dfrac{1}{2} + \dfrac{3}{4}i\right) + \left(\dfrac{2}{3} + \dfrac{1}{5}i\right)$

Solution

(a) $(4 + 3i) + (5 + 9i) = (4 + 5) + (3 + 9)i = 9 + 12i$

(b) $(-6 + 4i) + (8 - 7i) = (-6 + 8) + (4 - 7)i$
$$= 2 - 3i$$

(c) $\left(\dfrac{1}{2} + \dfrac{3}{4}i\right) + \left(\dfrac{2}{3} + \dfrac{1}{5}i\right) = \left(\dfrac{1}{2} + \dfrac{2}{3}\right) + \left(\dfrac{3}{4} + \dfrac{1}{5}\right)i$

$$= \left(\dfrac{3}{6} + \dfrac{4}{6}\right) + \left(\dfrac{15}{20} + \dfrac{4}{20}\right)i$$

$$= \dfrac{7}{6} + \dfrac{19}{20}i$$

The set of complex numbers is closed with respect to addition; that is, the sum of two complex numbers is a complex number. Furthermore, the commutative and associative properties of addition hold for all complex numbers. The addition identity element is $0 + 0i$ (or simply the real number 0). The additive inverse of $a + bi$ is $-a - bi$, because

$$(a + bi) + (-a - bi) = 0$$

To *subtract complex numbers*, $c + di$ from $a + bi$, add the additive inverse of $c + di$. Thus

$$(a + bi) - (c + di) = (a + bi) + (-c - di)$$
$$= (a - c) + (b - d)i$$

In other words, we subtract the real parts and subtract the imaginary parts, as in the next examples.

1. $(9 + 8i) - (5 + 3i) = (9 - 5) + (8 - 3)i$
$$= 4 + 5i$$

2. $(3 - 2i) - (4 - 10i) = (3 - 4) + (-2 - (-10))i$
$$= -1 + 8i$$

Simplifying Radicals Involving Negative Numbers

Because $i^2 = -1$, i is a square root of -1, so we let $i = \sqrt{-1}$. It should be evident that $-i$ is also a square root of -1, because

$$(-i)^2 = (-i)(-i) = i^2 = -1$$

Thus, in the set of complex numbers, -1 has two square roots, i and $-i$. We express these symbolically as

$$\sqrt{-1} = i \quad \text{and} \quad -\sqrt{-1} = -i$$

Let us extend our definition so that in the set of complex numbers every negative real number has two square roots. We simply define $\sqrt{-b}$, where b is a positive real number, to be the number whose square is $-b$. Thus

$$\left(\sqrt{-b}\right)^2 = -b \quad \text{for } b > 0$$

Furthermore, because $\left(i\sqrt{b}\right)\left(i\sqrt{b}\right) = i^2(b) = -1(b) = -b$, we see that

$$\sqrt{-b} = i\sqrt{b}$$

In other words, a square root of any negative real number can be represented as the product of a real number and the imaginary unit i. Consider the following examples.

Classroom Example
Simplify each of the following:

(a) $\sqrt{-9}$

(b) $\sqrt{-19}$

(c) $\sqrt{-32}$

EXAMPLE 2   Simplify each of the following:

(a) $\sqrt{-4}$ **(b)** $\sqrt{-17}$ **(c)** $\sqrt{-24}$

Solution

(a) $\sqrt{-4} = i\sqrt{4} = 2i$

(b) $\sqrt{-17} = i\sqrt{17}$

(c) $\sqrt{-24} = i\sqrt{24} = i\sqrt{4}\sqrt{6} = 2i\sqrt{6}$ Note that we simplified the radical $\sqrt{24}$ to $2\sqrt{6}$

We should also observe that $-\sqrt{-b}$ (where $b > 0$) is a square root of $-b$ because

$$\left(-\sqrt{-b}\right)^2 = \left(-i\sqrt{b}\right)^2 = i^2(b) = -1(b) = -b$$

Thus in the set of complex numbers, $-b$ (where $b > 0$) has two square roots, $i\sqrt{b}$ and $-i\sqrt{b}$. We express these symbolically as

$$\sqrt{-b} = i\sqrt{b} \quad \text{and} \quad -\sqrt{-b} = -i\sqrt{b}$$

Performing Operations on Radicals Involving Negative Numbers

We must be very careful with the use of the symbol $\sqrt{-b}$, where $b > 0$. Some real number properties that involve the square root symbol do not hold if the square root symbol does not represent a real number. For example, $\sqrt{a}\sqrt{b} = \sqrt{ab}$ does not hold if a and b are both negative numbers.

Correct $\quad \sqrt{-4}\sqrt{-9} = (2i)(3i) = 6i^2 = 6(-1) = -6$

Incorrect $\quad \sqrt{-4}\sqrt{-9} = \sqrt{(-4)(-9)} = \sqrt{36} = 6$

To avoid difficulty with this idea, you should rewrite all expressions of the form $\sqrt{-b}$ (where $b > 0$) in the form $i\sqrt{b}$ before doing any computations. The following example further demonstrates this point.

Classroom Example
Simplify each of the following:

(a) $\sqrt{-10}\sqrt{-5}$

(b) $\sqrt{-5}\sqrt{-20}$

(c) $\dfrac{\sqrt{-27}}{\sqrt{-3}}$

(d) $\dfrac{\sqrt{-39}}{\sqrt{13}}$

EXAMPLE 3   Simplify each of the following:

(a) $\sqrt{-6}\sqrt{-8}$ (b) $\sqrt{-2}\sqrt{-8}$ (c) $\dfrac{\sqrt{-75}}{\sqrt{-3}}$ (d) $\dfrac{\sqrt{-48}}{\sqrt{12}}$

Solution

(a) $\sqrt{-6}\sqrt{-8} = (i\sqrt{6})(i\sqrt{8}) = i^2\sqrt{48} = (-1)\sqrt{16}\sqrt{3} = -4\sqrt{3}$

(b) $\sqrt{-2}\sqrt{-8} = (i\sqrt{2})(i\sqrt{8}) = i^2\sqrt{16} = (-1)(4) = -4$

(c) $\dfrac{\sqrt{-75}}{\sqrt{-3}} = \dfrac{i\sqrt{75}}{i\sqrt{3}} = \dfrac{\sqrt{75}}{\sqrt{3}} = \sqrt{\dfrac{75}{3}} = \sqrt{25} = 5$

(d) $\dfrac{\sqrt{-48}}{\sqrt{12}} = \dfrac{i\sqrt{48}}{\sqrt{12}} = i\sqrt{\dfrac{48}{12}} = i\sqrt{4} = 2i$

Multiplying Complex Numbers

Complex numbers have a binomial form, so we find the product of two complex numbers in the same way that we find the product of two binomials. Then, by replacing i^2 with -1, we are able to simplify and express the final result in standard form. Consider the following example.

Classroom Example
Simplify each of the following:

(a) $(5 + 2i)(1 + 7i)$

(b) $(3 - 7i)(-2 + 3i)$

(c) $(4 - 7i)^2$

(d) $(4 - 6i)(4 + 6i)$

EXAMPLE 4   Find the product of each of the following:

(a) $(2 + 3i)(4 + 5i)$ (b) $(-3 + 6i)(2 - 4i)$

(c) $(1 - 7i)^2$ (d) $(2 + 3i)(2 - 3i)$

Solution

(a) $(2 + 3i)(4 + 5i) = 2(4 + 5i) + 3i(4 + 5i)$

$\qquad\qquad\qquad\quad = 8 + 10i + 12i + 15i^2$

$\qquad\qquad\qquad\quad = 8 + 22i + 15i^2$

$\qquad\qquad\qquad\quad = 8 + 22i + 15(-1) = -7 + 22i$

(b) $(-3 + 6i)(2 - 4i) = -3(2 - 4i) + 6i(2 - 4i)$

$\qquad\qquad\qquad\qquad = -6 + 12i + 12i - 24i^2$

$\qquad\qquad\qquad\qquad = -6 + 24i - 24(-1)$

$\qquad\qquad\qquad\qquad = -6 + 24i + 24 = 18 + 24i$

(c) $(1 - 7i)^2 = (1 - 7i)(1 - 7i)$

$\qquad\qquad\quad = 1(1 - 7i) - 7i(1 - 7i)$

$\qquad\qquad\quad = 1 - 7i - 7i + 49i^2$

$$= 1 - 14i + 49(-1)$$
$$= 1 - 14i - 49$$
$$= -48 - 14i$$

(d) $(2 + 3i)(2 - 3i) = 2(2 - 3i) + 3i(2 - 3i)$
$$= 4 - 6i + 6i - 9i^2$$
$$= 4 - 9(-1)$$
$$= 4 + 9$$
$$= 13$$

Example 4(d) illustrates an important situation: The complex numbers $2 + 3i$ and $2 - 3i$ are conjugates of each other. In general, we say that two complex numbers $a + bi$ and $a - bi$ are called **conjugates** of each other. *The product of a complex number and its conjugate is always a real number*, which can be shown as follows:

$$(a + bi)(a - bi) = a(a - bi) + bi(a - bi)$$
$$= a^2 - abi + abi - b^2i^2$$
$$= a^2 - b^2(-1)$$
$$= a^2 + b^2$$

Dividing Complex Numbers

We use conjugates to simplify expressions such as $\dfrac{3i}{5 + 2i}$ that indicate the quotient of two complex numbers. To eliminate i in the denominator and change the indicated quotient to the standard form of a complex number, we can multiply both the numerator and the denominator by the conjugate of the denominator as follows:

$$\frac{3i}{5 + 2i} = \frac{3i(5 - 2i)}{(5 + 2i)(5 - 2i)}$$
$$= \frac{15i - 6i^2}{25 - 4i^2}$$
$$= \frac{15i - 6(-1)}{25 - 4(-1)}$$
$$= \frac{15i + 6}{29}$$
$$= \frac{6}{29} + \frac{15}{29}i$$

The following example further clarifies the process of dividing complex numbers.

Classroom Example
Find the quotient of each of the following:

(a) $\dfrac{2 + 4i}{2 - 5i}$

(b) $\dfrac{9 - 2i}{4i}$

EXAMPLE 5   Find the quotient of each of the following:

(a) $\dfrac{2 - 3i}{4 - 7i}$ **(b)** $\dfrac{4 - 5i}{2i}$

Solution

(a) $\dfrac{2 - 3i}{4 - 7i} = \dfrac{(2 - 3i)(4 + 7i)}{(4 - 7i)(4 + 7i)}$ $4 + 7i$ is the conjugate of $4 - 7i$

$$= \frac{8 + 14i - 12i - 21i^2}{16 - 49i^2}$$

$$= \frac{8 + 2i - 21(-1)}{16 - 49(-1)}$$

$$= \frac{8 + 2i + 21}{16 + 49}$$

$$= \frac{29 + 2i}{65}$$

$$= \frac{29}{65} + \frac{2}{65}i$$

(b) $\dfrac{4 - 5i}{2i} = \dfrac{(4 - 5i)(-2i)}{(2i)(-2i)}$ $-2i$ is the conjugate of $2i$

$$= \frac{-8i + 10i^2}{-4i^2}$$

$$= \frac{-8i + 10(-1)}{-4(-1)}$$

$$= \frac{-8i - 10}{4}$$

$$= -\frac{5}{2} - 2i$$

In Example 5(b), in which the denominator is a pure imaginary number, we can change to standard form by choosing a multiplier other than the conjugate. Consider the following alternative approach for Example 5(b).

$$\frac{4 - 5i}{2i} = \frac{(4 - 5i)(i)}{(2i)(i)}$$

$$= \frac{4i - 5i^2}{2i^2}$$

$$= \frac{4i - 5(-1)}{2(-1)}$$

$$= \frac{4i + 5}{-2}$$

$$= -\frac{5}{2} - 2i$$

Concept Quiz 11.1

For Problems 1–10, answer true or false.

1. The number i is a real number and is called the imaginary unit.
2. The number $4 + 2i$ is a complex number that has a real part of 4.
3. The number $-3 - 5i$ is a complex number that has an imaginary part of 5.
4. Complex numbers that have a real part of 0 are called pure imaginary numbers.
5. The set of real numbers is a subset of the set of complex numbers.
6. Any real number x can be written as the complex number $x + 0i$.
7. By definition, i^2 is equal to -1.
8. The complex numbers $-2 + 5i$ and $2 - 5i$ are conjugates.
9. The product of two complex numbers is never a real number.
10. In the set of complex numbers, -16 has two square roots.

Problem Set 11.1

For Problems 1–8, label each statement true or false. **(Objective 1)**

1. Every complex number is a real number.

2. Every real number is a complex number.

3. The real part of the complex number $6i$ is 0.

4. Every complex number is a pure imaginary number.

5. The sum of two complex numbers is always a complex number.

6. The imaginary part of the complex number 7 is 0.

7. The sum of two complex numbers is sometimes a real number.

8. The sum of two pure imaginary numbers is always a pure imaginary number.

For Problems 9–26, add or subtract as indicated. **(Objective 2)**

9. $(6 + 3i) + (4 + 5i)$ 10. $(5 + 2i) + (7 + 10i)$

11. $(-8 + 4i) + (2 + 6i)$ 12. $(5 - 8i) + (-7 + 2i)$

13. $(3 + 2i) - (5 + 7i)$ 14. $(1 + 3i) - (4 + 9i)$

15. $(-7 + 3i) - (5 - 2i)$ 16. $(-8 + 4i) - (9 - 4i)$

17. $(-3 - 10i) + (2 - 13i)$ 18. $(-4 - 12i) + (-3 + 16i)$

19. $(4 - 8i) - (8 - 3i)$ 20. $(12 - 9i) - (14 - 6i)$

21. $(-1 - i) - (-2 - 4i)$ 22. $(-2 - 3i) - (-4 - 14i)$

23. $\left(\dfrac{3}{2} + \dfrac{1}{3}i\right) + \left(\dfrac{1}{6} - \dfrac{3}{4}i\right)$ 24. $\left(\dfrac{2}{3} - \dfrac{1}{5}i\right) + \left(\dfrac{3}{5} - \dfrac{3}{4}i\right)$

25. $\left(-\dfrac{5}{9} + \dfrac{3}{5}i\right) - \left(\dfrac{4}{3} - \dfrac{1}{6}i\right)$ 26. $\left(\dfrac{3}{8} - \dfrac{5}{2}i\right) - \left(\dfrac{5}{6} + \dfrac{1}{7}i\right)$

For Problems 27–42, write each of the following in terms of i and simplify. **(Objective 3)**
For example,

$$\sqrt{-20} = i\sqrt{20} = i\sqrt{4}\sqrt{5} = 2i\sqrt{5}$$

27. $\sqrt{-81}$ 28. $\sqrt{-49}$

29. $\sqrt{-14}$ 30. $\sqrt{-33}$

31. $\sqrt{-\dfrac{16}{25}}$ 32. $\sqrt{-\dfrac{64}{36}}$

33. $\sqrt{-18}$ 34. $\sqrt{-84}$

35. $\sqrt{-75}$ 36. $\sqrt{-63}$

37. $3\sqrt{-28}$ 38. $5\sqrt{-72}$

39. $-2\sqrt{-80}$ 40. $-6\sqrt{-27}$

41. $12\sqrt{-90}$ 42. $9\sqrt{-40}$

For Problems 43–60, write each of the following in terms of i, perform the indicated operations, and simplify. **(Objective 4)**
For example,

$$\sqrt{-3}\sqrt{-8} = (i\sqrt{3})(i\sqrt{8})$$
$$= i^2\sqrt{24}$$
$$= (-1)\sqrt{4}\sqrt{6}$$
$$= -2\sqrt{6}$$

43. $\sqrt{-4}\sqrt{-16}$ 44. $\sqrt{-81}\sqrt{-25}$

45. $\sqrt{-3}\sqrt{-5}$ 46. $\sqrt{-7}\sqrt{-10}$

47. $\sqrt{-9}\sqrt{-6}$ 48. $\sqrt{-8}\sqrt{-16}$

49. $\sqrt{-15}\sqrt{-5}$ 50. $\sqrt{-2}\sqrt{-20}$

51. $\sqrt{-2}\sqrt{-27}$ 52. $\sqrt{-3}\sqrt{-15}$

53. $\sqrt{6}\sqrt{-8}$ 54. $\sqrt{-75}\sqrt{3}$

55. $\dfrac{\sqrt{-25}}{\sqrt{-4}}$ 56. $\dfrac{\sqrt{-81}}{\sqrt{-9}}$

57. $\dfrac{\sqrt{-56}}{\sqrt{-7}}$ 58. $\dfrac{\sqrt{-72}}{\sqrt{-6}}$

59. $\dfrac{\sqrt{-24}}{\sqrt{6}}$ 60. $\dfrac{\sqrt{-96}}{\sqrt{2}}$

For Problems 61–84, find each of the products and express the answers in the standard form of a complex number. **(Objective 5)**

61. $(5i)(4i)$ 62. $(-6i)(9i)$

63. $(7i)(-6i)$ 64. $(-5i)(-12i)$

65. $(3i)(2 - 5i)$ 66. $(7i)(-9 + 3i)$

67. $(-6i)(-2 - 7i)$ 68. $(-9i)(-4 - 5i)$

69. $(3 + 2i)(5 + 4i)$ 70. $(4 + 3i)(6 + i)$

71. $(6 - 2i)(7 - i)$ 72. $(8 - 4i)(7 - 2i)$

73. $(-3 - 2i)(5 + 6i)$ 74. $(-5 - 3i)(2 - 4i)$

75. $(9 + 6i)(-1 - i)$ 76. $(10 + 2i)(-2 - i)$

77. $(4 + 5i)^2$ 78. $(5 - 3i)^2$

79. $(-2 - 4i)^2$ 80. $(-3 - 6i)^2$

81. $(6 + 7i)(6 - 7i)$ 82. $(5 - 7i)(5 + 7i)$

83. $(-1 + 2i)(-1 - 2i)$ 84. $(-2 - 4i)(-2 + 4i)$

For Problems 85–100, find each of the following quotients, and express the answers in the standard form of a complex number. (Objective 6)

85. $\dfrac{3i}{2+4i}$

86. $\dfrac{4i}{5+2i}$

87. $\dfrac{-2i}{3-5i}$

88. $\dfrac{-5i}{2-4i}$

89. $\dfrac{-2+6i}{3i}$

90. $\dfrac{-4-7i}{6i}$

91. $\dfrac{2}{7i}$

92. $\dfrac{3}{10i}$

93. $\dfrac{2+6i}{1+7i}$

94. $\dfrac{5+i}{2+9i}$

95. $\dfrac{3+6i}{4-5i}$

96. $\dfrac{7-3i}{4-3i}$

97. $\dfrac{-2+7i}{-1+i}$

98. $\dfrac{-3+8i}{-2+i}$

99. $\dfrac{-1-3i}{-2-10i}$

100. $\dfrac{-3-4i}{-4-11i}$

101. Some of the solution sets for quadratic equations in the next sections in this chapter will contain complex numbers such as $(-4+\sqrt{-12})/2$ and $(-4-\sqrt{-12})/2$. We can simplify the first number as follows.

$$\frac{-4+\sqrt{-12}}{2} = \frac{-4+i\sqrt{12}}{2}$$

$$= \frac{-4+2i\sqrt{3}}{2} = \frac{2(-2+i\sqrt{3})}{2}$$

$$= -2+i\sqrt{3}$$

Simplify each of the following complex numbers. (Objective 3)

(a) $\dfrac{-4-\sqrt{-12}}{2}$

(b) $\dfrac{6+\sqrt{-24}}{4}$

(c) $\dfrac{-1-\sqrt{-18}}{2}$

(d) $\dfrac{-6+\sqrt{-27}}{3}$

(e) $\dfrac{10+\sqrt{-45}}{4}$

(f) $\dfrac{4-\sqrt{-48}}{2}$

Thoughts Into Words

102. Why is the set of real numbers a subset of the set of complex numbers?

103. Can the sum of two nonreal complex numbers be a real number? Defend your answer.

104. Can the product of two nonreal complex numbers be a real number? Defend your answer.

Answers to the Concept Quiz

1. False **2.** True **3.** False **4.** True **5.** True **6.** True **7.** True **8.** False **9.** False **10.** True

---

11.2	Quadratic Equations

OBJECTIVES

1. Solve quadratic equations by factoring
2. Solve quadratic equations of the form $x^2 = a$
3. Solve problems pertaining to right triangles and 30°–60° triangles

A second-degree equation in one variable contains the variable with an exponent of 2, but no higher power. Such equations are also called *quadratic equations*. The following are examples of quadratic equations.

$$x^2 = 36 \qquad y^2 + 4y = 0 \qquad x^2 + 5x - 2 = 0$$

$$3n^2 + 2n - 1 = 0 \qquad 5x^2 + x + 2 = 3x^2 - 2x - 1$$

A **quadratic equation** in the variable x can also be defined as any equation that can be written in the form

$$ax^2 + bx + c = 0$$

where a, b, and c are real numbers and $a \neq 0$. The form $ax^2 + bx + c = 0$ is called the *standard form* of a quadratic equation.

In previous chapters you solved quadratic equations (the term *quadratic* was not used at that time) by factoring and applying the property, $ab = 0$ if and only if $a = 0$ or $b = 0$. Let's review a few such examples.

Classroom Example
Solve $4x^2 + 11x - 3 = 0$.

EXAMPLE 1   Solve $3n^2 + 14n - 5 = 0$.

Solution

$$3n^2 + 14n - 5 = 0$$
$$(3n - 1)(n + 5) = 0 \qquad \text{Factor the left side}$$
$$3n - 1 = 0 \quad \text{or} \quad n + 5 = 0 \qquad \text{Apply: } ab = 0 \text{ if and only if } a = 0 \text{ or } b = 0$$
$$3n = 1 \quad \text{or} \quad n = -5$$
$$n = \frac{1}{3} \quad \text{or} \quad n = -5$$

The solution set is $\left\{-5, \dfrac{1}{3}\right\}$.

Classroom Example
Solve $2\sqrt{y} = y - 3$.

EXAMPLE 2   Solve $2\sqrt{x} = x - 8$.

Solution

$$2\sqrt{x} = x - 8$$
$$(2\sqrt{x})^2 = (x - 8)^2 \qquad \text{Square both sides}$$
$$4x = x^2 - 16x + 64$$
$$0 = x^2 - 20x + 64$$
$$0 = (x - 16)(x - 4) \qquad \text{Factor the right side}$$
$$x - 16 = 0 \quad \text{or} \quad x - 4 = 0 \qquad \text{Apply: } ab = 0 \text{ if and only if } a = 0 \text{ or } b = 0$$
$$x = 16 \quad \text{or} \quad x = 4$$

✓ **Check**

If $x = 16$	If $x = 4$
$2\sqrt{x} = x - 8$	$2\sqrt{x} = x - 8$
$2\sqrt{16} \overset{?}{=} 16 - 8$	$2\sqrt{4} \overset{?}{=} 4 - 8$
$2(4) \overset{?}{=} 8$	$2(2) \overset{?}{=} -4$
$8 = 8$	$4 \neq -4$

The solution set is $\{16\}$.

We should make two comments about Example 2. First, remember that applying the property if $a = b$, then $a^n = b^n$ might produce extraneous solutions. Therefore, we *must* check all potential solutions. Second, the equation $2\sqrt{x} = x - 8$ is said to be of *quadratic form* because it can be written as $2x^{\frac{1}{2}} = \left(x^{\frac{1}{2}}\right)^2 - 8$. More will be said about the phrase *quadratic form* later.

Solving Quadratic Equations of the Form $x^2 = a$

Let's consider quadratic equations of the form $x^2 = a$, where x is the variable and a is any real number. We can solve $x^2 = a$ as follows:

$$x^2 = a$$
$$x^2 - a = 0$$
$$x^2 - (\sqrt{a})^2 = 0 \qquad\qquad a = (\sqrt{a})^2$$
$$(x - \sqrt{a})(x + \sqrt{a}) = 0 \qquad\qquad \text{Factor the left side}$$
$$x - \sqrt{a} = 0 \qquad \text{or} \qquad x + \sqrt{a} = 0 \qquad \text{Apply: } ab = 0 \text{ if and only if } a = 0 \text{ or } b = 0$$
$$x = \sqrt{a} \qquad \text{or} \qquad x = -\sqrt{a}$$

The solutions are $\sqrt{a}$ and $-\sqrt{a}$. We can state this result as a general property and use it to solve certain types of quadratic equations.

> ### Property 11.1
>
> For any real number a,
>
> $$x^2 = a \quad \text{if and only if } x = \sqrt{a} \text{ or } x = -\sqrt{a}$$
>
> (The statement $x = \sqrt{a}$ or $x = -\sqrt{a}$ can be written as $x = \pm\sqrt{a}$.)

Property 11.1, along with our knowledge of square roots, makes it very easy to solve quadratic equations of the form $x^2 = a$.

Classroom Example
Solve $m^2 = 48$.

EXAMPLE 3   Solve $x^2 = 45$.

Solution

$$x^2 = 45$$
$$x = \pm\sqrt{45}$$
$$x = \pm 3\sqrt{5} \qquad\qquad\qquad \sqrt{45} = \sqrt{9}\sqrt{5} = 3\sqrt{5}$$

The solution set is $\{\pm 3\sqrt{5}\}$.

Classroom Example
Solve $n^2 = -25$.

EXAMPLE 4   Solve $x^2 = -9$.

Solution

$$x^2 = -9$$
$$x = \pm\sqrt{-9}$$
$$x = \pm 3i \qquad\qquad\qquad \sqrt{-9} = i\sqrt{9} = 3i$$

Thus the solution set is $\{\pm 3i\}$.

Classroom Example
Solve $5x^2 = 16$.

EXAMPLE 5   Solve $7n^2 = 12$.

Solution

$$7n^2 = 12$$
$$n^2 = \frac{12}{7}$$

$$n = \pm\sqrt{\frac{12}{7}}$$

$$n = \pm\frac{2\sqrt{21}}{7} \qquad \sqrt{\frac{12}{7}} = \frac{\sqrt{12}}{\sqrt{7}} \cdot \frac{\sqrt{7}}{\sqrt{7}} = \frac{\sqrt{84}}{7} = \frac{\sqrt{4}\sqrt{21}}{7} = \frac{2\sqrt{21}}{7}$$

The solution set is $\left\{\pm\frac{2\sqrt{21}}{7}\right\}$.

Classroom Example
Solve $(4x - 3)^2 = 49$.

EXAMPLE 6   Solve $(3n + 1)^2 = 25$.

Solution

$$(3n + 1)^2 = 25$$
$$(3n + 1) = \pm\sqrt{25}$$
$$3n + 1 = \pm 5$$
$$3n + 1 = 5 \qquad \text{or} \qquad 3n + 1 = -5$$
$$3n = 4 \qquad \text{or} \qquad 3n = -6$$
$$n = \frac{4}{3} \qquad \text{or} \qquad n = -2$$

The solution set is $\left\{-2, \frac{4}{3}\right\}$.

Classroom Example
Solve $(x + 4)^2 = -18$.

EXAMPLE 7   Solve $(x - 3)^2 = -10$.

Solution

$$(x - 3)^2 = -10$$
$$x - 3 = \pm\sqrt{-10}$$
$$x - 3 = \pm i\sqrt{10}$$
$$x = 3 \pm i\sqrt{10}$$

Thus the solution set is $\{3 \pm i\sqrt{10}\}$.

Remark: Take another look at the equations in Examples 4 and 7. We should immediately realize that the solution sets will consist only of nonreal complex numbers, because any nonzero real number squared is positive.

Sometimes it may be necessary to change the form before we can apply Property 11.1. Let's consider one example to illustrate this idea.

Classroom Example
Solve $2(5x - 1)^2 + 9 = 53$.

EXAMPLE 8   Solve $3(2x - 3)^2 + 8 = 44$.

Solution

$$3(2x - 3)^2 + 8 = 44$$
$$3(2x - 3)^2 = 36$$
$$(2x - 3)^2 = 12$$
$$2x - 3 = \pm\sqrt{12}$$
$$2x - 3 = \pm 2\sqrt{3}$$

$$2x = 3 \pm 2\sqrt{3}$$

$$x = \frac{3 \pm 2\sqrt{3}}{2}$$

The solution set is $\left\{\dfrac{3 \pm 2\sqrt{3}}{2}\right\}$.

Solving Problems Pertaining to Right Triangles and 30°–60° Triangles

Our work with radicals, Property 11.1, and the Pythagorean theorem form a basis for solving a variety of problems that pertain to right triangles.

EXAMPLE 9

A 50-foot rope hangs from the top of a flagpole. When pulled taut to its full length, the rope reaches a point on the ground 18 feet from the base of the pole. Find the height of the pole to the nearest tenth of a foot.

Solution

Let's make a sketch (Figure 11.1) and record the given information. Use the Pythagorean theorem to solve for p as follows:

$$p^2 + 18^2 = 50^2$$
$$p^2 + 324 = 2500$$
$$p^2 = 2176$$
$$p = \sqrt{2176} = 46.6 \quad \text{to the nearest tenth}$$

The height of the flagpole is approximately 46.6 feet.

There are two special kinds of right triangles that we use extensively in later mathematics courses. The first is the **isosceles right triangle**, which is a right triangle that has both legs of the same length. Let's consider a problem that involves an isosceles right triangle.

EXAMPLE 10

Find the length of each leg of an isosceles right triangle that has a hypotenuse of length 5 meters.

Solution

Let's sketch an isosceles right triangle and let x represent the length of each leg (Figure 11.2). Then we can apply the Pythagorean theorem.

$$x^2 + x^2 = 5^2$$
$$2x^2 = 25$$
$$x^2 = \frac{25}{2}$$
$$x = \pm\sqrt{\frac{25}{2}} = \pm\frac{5}{\sqrt{2}} = \pm\frac{5\sqrt{2}}{2}$$

Each leg is $\dfrac{5\sqrt{2}}{2}$ meters long.

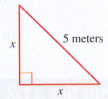

Figure 11.2

Remark: In Example 9 we made no attempt to express $\sqrt{2176}$ in simplest radical form because the answer was to be given as a rational approximation to the nearest tenth. However,

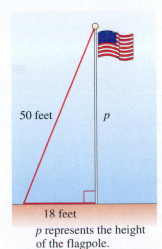

50 feet p

18 feet

p represents the height of the flagpole.

Figure 11.1

in Example 10 we left the final answer in radical form and therefore expressed it in simplest radical form.

The second special kind of right triangle that we use frequently is one that contains acute angles of 30° and 60°. In such a right triangle, which we refer to as a **30°–60° right triangle**, the side opposite the 30° angle is equal in length to one-half of the length of the hypotenuse. This relationship, along with the Pythagorean theorem, provides us with another problem-solving technique.

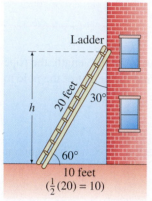

Figure 11.3

EXAMPLE 11

Suppose that a 20-foot ladder is leaning against a building and makes an angle of 60° with the ground. How far up the building does the top of the ladder reach? Express your answer to the nearest tenth of a foot.

Solution

Figure 11.3 depicts this situation. The side opposite the 30° angle equals one-half of the hypotenuse, so it is of length $\frac{1}{2}(20) = 10$ feet. Now we can apply the Pythagorean theorem.

$$h^2 + 10^2 = 20^2$$
$$h^2 + 100 = 400$$
$$h^2 = 300$$
$$h = \sqrt{300} = 17.3 \quad \text{to the nearest tenth}$$

The ladder touches the building at a point approximately 17.3 feet from the ground.

Concept Quiz 11.2

For Problems 1–10, answer true or false.

1. The quadratic equation $-3x^2 + 5x - 8 = 0$ is in standard form.
2. The solution set of the equation $(x + 1)^2 = -25$ will consist only of nonreal complex numbers.
3. An isosceles right triangle is a right triangle that has a hypotenuse of the same length as one of the legs.
4. In a 30°–60° right triangle, the hypotenuse is equal in length to twice the length of the side opposite the 30° angle.
5. The equation $2x^2 + x^3 - x + 4 = 0$ is a quadratic equation.
6. The solution set for $4x^2 = 8x$ is $\{2\}$.
7. The solution set for $3x^2 = 8x$ is $\left\{0, \dfrac{8}{3}\right\}$.
8. The solution set for $x^2 - 8x - 48 = 0$ is $\{-12, 4\}$.
9. If the length of each leg of an isosceles right triangle is 4 inches, then the hypotenuse is of length $4\sqrt{2}$ inches.
10. If the length of the leg opposite the 30° angle in a right triangle is 6 centimeters, then the length of the other leg is 12 centimeters.

Problem Set 11.2

For Problems 1–20, solve each of the quadratic equations by factoring and applying the property $ab = 0$ if and only if $a = 0$ or $b = 0$. If necessary, return to Chapter 7 and review the factoring techniques presented there. **(Objective 1)**

1. $x^2 - 9x = 0$
2. $x^2 + 5x = 0$
3. $x^2 = -3x$
4. $x^2 = 15x$
5. $3y^2 + 12y = 0$
6. $6y^2 - 24y = 0$

7. $5n^2 - 9n = 0$

8. $4n^2 + 13n = 0$

9. $x^2 + x - 30 = 0$

10. $x^2 - 8x - 48 = 0$

11. $x^2 - 19x + 84 = 0$

12. $x^2 - 21x + 104 = 0$

13. $2x^2 + 19x + 24 = 0$

14. $4x^2 + 29x + 30 = 0$

15. $15x^2 + 29x - 14 = 0$

16. $24x^2 + x - 10 = 0$

17. $25x^2 - 30x + 9 = 0$

18. $16x^2 - 8x + 1 = 0$

19. $6x^2 - 5x - 21 = 0$

20. $12x^2 - 4x - 5 = 0$

For Problems 21–26, solve each radical equation. Don't forget, you *must* check potential solutions. **(Objective 1)**

21. $3\sqrt{x} = x + 2$

22. $3\sqrt{2x} = x + 4$

23. $\sqrt{2x} = x - 4$

24. $\sqrt{x} = x - 2$

25. $\sqrt{3x + 6} = x$

26. $\sqrt{5x + 10} = x$

For Problems 27–62, use Property 11.1 to help solve each quadratic equation. **(Objective 2)**

27. $x^2 = 1$

28. $x^2 = 81$

29. $x^2 = -36$

30. $x^2 = -49$

31. $x^2 = 14$

32. $x^2 = 22$

33. $n^2 - 28 = 0$

34. $n^2 - 54 = 0$

35. $3t^2 = 54$

36. $4t^2 = 108$

37. $2t^2 = 7$

38. $3t^2 = 8$

39. $15y^2 = 20$

40. $14y^2 = 80$

41. $10x^2 + 48 = 0$

42. $12x^2 + 50 = 0$

43. $24x^2 = 36$

44. $12x^2 = 49$

45. $(x - 2)^2 = 9$

46. $(x + 1)^2 = 16$

47. $(x + 3)^2 = 25$

48. $(x - 2)^2 = 49$

49. $(x + 6)^2 = -4$

50. $(3x + 1)^2 = 9$

51. $(2x - 3)^2 = 1$

52. $(2x + 5)^2 = -4$

53. $(n - 4)^2 = 5$

54. $(n - 7)^2 = 6$

55. $(t + 5)^2 = 12$

56. $(t - 1)^2 = 18$

57. $(3y - 2)^2 = -27$

58. $(4y + 5)^2 = 80$

59. $3(x + 7)^2 + 4 = 79$

60. $2(x + 6)^2 - 9 = 63$

61. $2(5x - 2)^2 + 5 = 25$

62. $3(4x - 1)^2 + 1 = -17$

For Problems 63–68, a and b represent the lengths of the legs of a right triangle, and c represents the length of the hypotenuse. Express answers in simplest radical form. **(Objective 3)**

63. Find c if $a = 4$ centimeters and $b = 6$ centimeters.

64. Find c if $a = 3$ meters and $b = 7$ meters.

65. Find a if $c = 12$ inches and $b = 8$ inches.

66. Find a if $c = 8$ feet and $b = 6$ feet.

67. Find b if $c = 17$ yards and $a = 15$ yards.

68. Find b if $c = 14$ meters and $a = 12$ meters.

For Problems 69–72, use the isosceles right triangle in Figure 11.4. Express your answers in simplest radical form. **(Objective 3)**

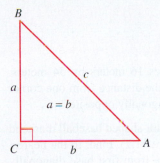

Figure 11.4

69. If $b = 6$ inches, find c.

70. If $a = 7$ centimeters, find c.

71. If $c = 8$ meters, find a and b.

72. If $c = 9$ feet, find a and b.

For Problems 73–78, use the triangle in Figure 11.5. Express your answers in simplest radical form. **(Objective 3)**

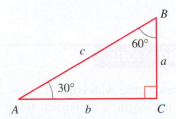

Figure 11.5

73. If $a = 3$ inches, find b and c.

74. If $a = 6$ feet, find b and c.

75. If $c = 14$ centimeters, find a and b.

76. If $c = 9$ centimeters, find a and b.

77. If $b = 10$ feet, find a and c.

78. If $b = 8$ meters, find a and c.

79. A 24-foot ladder resting against a house reaches a windowsill 16 feet above the ground. How far is the foot of the ladder from the foundation of the house? Express your answer to the nearest tenth of a foot.

80. A 62-foot guy-wire makes an angle of 60° with the ground and is attached to a telephone pole (see Figure 11.6). Find the distance from the base of the pole to the point on the pole where the wire is attached. Express your answer to the nearest tenth of a foot.

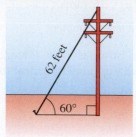

Figure 11.6

81. A rectangular plot measures 16 meters by 34 meters. Find, to the nearest meter, the distance from one corner of the plot to the corner diagonally opposite.

82. Consecutive bases of a square-shaped baseball diamond are 90 feet apart (see Figure 11.7). Find, to the nearest tenth of a foot, the distance from first base diagonally across the diamond to third base.

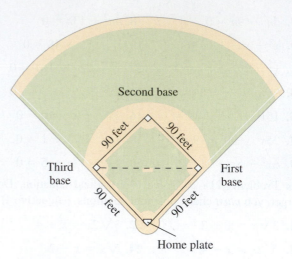

Figure 11.7

83. A diagonal of a square parking lot is 75 meters. Find, to the nearest meter, the length of a side of the lot.

Thoughts Into Words

84. Explain why the equation $(x + 2)^2 + 5 = 1$ has no real number solutions.

85. Suppose that your friend solved the equation $(x + 3)^2 = 25$ as follows:

$$(x + 3)^2 = 25$$
$$x^2 + 6x + 9 = 25$$
$$x^2 + 6x - 16 = 0$$
$$(x + 8)(x - 2) = 0$$
$$x + 8 = 0 \quad \text{or} \quad x - 2 = 0$$
$$x = -8 \quad \text{or} \quad x = 2$$

Is this a correct approach to the problem? Would you offer any suggestion about an easier approach to the problem?

Further Investigations

86. Suppose that we are given a cube with edges 12 centimeters in length. Find the length of a diagonal from a lower corner to the diagonally opposite upper corner. Express your answer to the nearest tenth of a centimeter.

87. Suppose that we are given a rectangular box with a length of 8 centimeters, a width of 6 centimeters, and a height of 4 centimeters. Find the length of a diagonal from a lower corner to the upper corner diagonally opposite. Express your answer to the nearest tenth of a centimeter.

88. The converse of the Pythagorean theorem is also true. It states, "If the measures a, b, and c of the sides of a triangle are such that $a^2 + b^2 = c^2$, then the triangle is a right triangle with a and b the measures of the legs and c the measure of the hypotenuse." Use the converse of the Pythagorean theorem to determine which of the triangles with sides of the following measures are right triangles.

(a) 9, 40, 41   (b) 20, 48, 52

(c) 19, 21, 26   (d) 32, 37, 49

(e) 65, 156, 169   (f) 21, 72, 75

89. Find the length of the hypotenuse (h) of an isosceles right triangle if each leg is s units long. Then use this relationship to redo Problems 69–72.

90. Suppose that the side opposite the 30° angle in a 30°–60° right triangle is s units long. Express the length of the hypotenuse and the length of the other leg in terms of s. Then use these relationships and redo Problems 73–78.

Answers to the Concept Quiz

1. True   2. True   3. False   4. True   5. False   6. False   7. True   8. False   9. True   10. False

11.3   Completing the Square

O B J E C T I V E   **1** Solve quadratic equations by completing the square

Thus far we have solved quadratic equations by factoring and applying the property, $ab = 0$ if and only if $a = 0$ or $b = 0$, or by applying the property, $x^2 = a$ if and only if $x = \pm\sqrt{a}$. In this section we examine another method called *completing the square*, which will give us the power to solve any quadratic equation.

A factoring technique we studied in Chapter 7 relied on recognizing *perfect-square trinomials*. In each of the following, the perfect-square trinomial on the right side is the result of squaring the binomial on the left side.

$$(x + 4)^2 = x^2 + 8x + 16 \qquad (x - 6)^2 = x^2 - 12x + 36$$
$$(x + 7)^2 = x^2 + 14x + 49 \qquad (x - 9)^2 = x^2 - 18x + 81$$
$$(x + a)^2 = x^2 + 2ax + a^2$$

Note that in each of the square trinomials, the constant term is equal to the square of one-half of the coefficient of the x term. This relationship enables us to form a perfect-square trinomial by adding a proper constant term. To find the constant term, take one-half of the coefficient of the x term and then square the result. For example, suppose that we want to form a perfect-square trinomial from $x^2 + 10x$. The coefficient of the x term is 10. Because $\dfrac{1}{2}(10) = 5$, and $5^2 = 25$, the constant term should be 25. The perfect-square trinomial that can be formed is $x^2 + 10x + 25$. This perfect-square trinomial can be factored and expressed as $(x + 5)^2$. Let's use the previous ideas to help solve some quadratic equations.

Classroom Example
Solve $x^2 + 8x - 5 = 0$.

EXAMPLE 1   Solve $x^2 + 10x - 2 = 0$.

Solution

$$x^2 + 10x - 2 = 0$$
$$x^2 + 10x = 2 \qquad \text{Isolate the } x^2 \text{ and } x \text{ terms}$$
$$\frac{1}{2}(10) = 5 \quad \text{and} \quad 5^2 = 25 \qquad \text{Take } \frac{1}{2} \text{ of the coefficient of the } x \text{ term and then square the result}$$
$$x^2 + 10x + 25 = 2 + 25 \qquad \text{Add 25 to } both \text{ sides of the equation}$$
$$(x + 5)^2 = 27 \qquad \text{Factor the perfect-square trinomial}$$
$$x + 5 = \pm\sqrt{27} \qquad \text{Now solve by applying Property 11.1}$$
$$x + 5 = \pm3\sqrt{3}$$
$$x = -5 \pm 3\sqrt{3}$$

The solution set is $\{-5 \pm 3\sqrt{3}\}$.

The method of completing the square to solve a quadratic equation is merely what the name implies. A perfect-square trinomial is formed, then the equation can be changed to the necessary form for applying the property "$x^2 = a$ if and only if $x = \pm\sqrt{a}$." Let's consider another example.

Classroom Example
Solve $x^2 + 10x + 33 = 0$.

EXAMPLE 2   Solve $x^2 + 4x + 7 = 0$.

Solution

$$x^2 + 4x + 7 = 0$$
$$x^2 + 4x = -7 \qquad \text{Isolate the } x^2 \text{ and } x \text{ terms}$$

$$x^2 + 4x + 4 = -7 + 4$$

$$(x + 2)^2 = -3$$

$$x + 2 = \pm\sqrt{-3}$$

$$x + 2 = \pm i\sqrt{3}$$

$$x = -2 \pm i\sqrt{3}$$

The solution set is $\{-2 \pm i\sqrt{3}\}$.

$\frac{1}{2}(4) = 2$ and $2^2 = 4$

Factor the perfect-square trinomial

Now solve by applying Property 11.1

Let's pause for a moment and give a little visual support for our answer in Example 2. Figure 11.8 shows the graph of $y = x^2 + 4x + 7$. Because it does not intersect the x axis, the equation $x^2 + 4x + 7 = 0$ has no real number solutions. This supports our solution set of two nonreal complex numbers. To be absolutely sure that we have the correct complex numbers, we should substitute them back into the original equation.

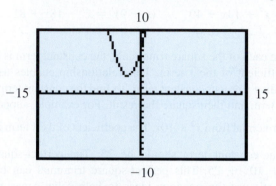

Figure 11.8

Classroom Example
Solve $m^2 - 3m - 5 = 0$.

EXAMPLE 3   Solve $x^2 - 3x + 1 = 0$.

Solution

$$x^2 - 3x + 1 = 0$$

$$x^2 - 3x = -1$$

$$x^2 - 3x + \frac{9}{4} = -1 + \frac{9}{4} \qquad\qquad \frac{1}{2}(3) = \frac{3}{2} \text{ and } \left(\frac{3}{2}\right)^2 = \frac{9}{4}$$

$$\left(x - \frac{3}{2}\right)^2 = \frac{5}{4}$$

$$x - \frac{3}{2} = \pm\sqrt{\frac{5}{4}}$$

$$x - \frac{3}{2} = \pm\frac{\sqrt{5}}{2}$$

$$x = \frac{3}{2} \pm \frac{\sqrt{5}}{2}$$

$$x = \frac{3 \pm \sqrt{5}}{2}$$

The solution set is $\left\{\dfrac{3 \pm \sqrt{5}}{2}\right\}$.

In Example 3 note that because the coefficient of the x term is odd, we are forced into the realm of fractions. Using common fractions rather than decimals enables us to apply our previous work with radicals.

The relationship for a perfect-square trinomial that states that the constant term is equal to the square of one-half of the coefficient of the x term holds only if the coefficient of x^2 is 1. Thus we must make an adjustment when solving quadratic equations that have a coefficient of x^2 other than 1. We will need to apply the multiplication property of equality so that the coefficient of the x^2 term becomes 1. The next example shows how to make this adjustment.

Classroom Example
Solve $3y^2 - 24y + 26 = 0$.

EXAMPLE 4   Solve $2x^2 + 12x - 5 = 0$.

Solution

$$2x^2 + 12x - 5 = 0$$

$$2x^2 + 12x = 5$$

$$x^2 + 6x = \frac{5}{2} \qquad \text{Multiply both sides by } \frac{1}{2}$$

$$x^2 + 6x + 9 = \frac{5}{2} + 9 \qquad \frac{1}{2}(6) = 3, \text{ and } 3^2 = 9$$

$$x^2 + 6x + 9 = \frac{23}{2}$$

$$(x + 3)^2 = \frac{23}{2}$$

$$x + 3 = \pm\sqrt{\frac{23}{2}}$$

$$x + 3 = \pm\frac{\sqrt{46}}{2} \qquad \sqrt{\frac{23}{2}} = \frac{\sqrt{23}}{\sqrt{2}} \cdot \frac{\sqrt{2}}{\sqrt{2}} = \frac{\sqrt{46}}{2}$$

$$x = -3 \pm \frac{\sqrt{46}}{2}$$

$$x = \frac{-6}{2} \pm \frac{\sqrt{46}}{2} \qquad \text{Common denominator of 2}$$

$$x = \frac{-6 \pm \sqrt{46}}{2}$$

The solution set is $\left\{ \dfrac{-6 \pm \sqrt{46}}{2} \right\}$.

As we mentioned earlier, we can use the method of completing the square to solve *any* quadratic equation. To illustrate, let's use it to solve an equation that could also be solved by factoring.

Classroom Example
Solve $t^2 - 10t + 21 = 0$ by completing the square.

EXAMPLE 5   Solve $x^2 - 2x - 8 = 0$ by completing the square.

Solution

$$x^2 - 2x - 8 = 0$$

$$x^2 - 2x = 8$$

$$x^2 - 2x + 1 = 8 + 1 \qquad \frac{1}{2}(-2) = -1 \text{ and } (-1)^2 = 1$$

$$(x - 1)^2 = 9$$

$$x - 1 = \pm 3$$

$$x - 1 = 3 \quad \text{or} \quad x - 1 = -3$$

$$x = 4 \quad \text{or} \quad x = -2$$

The solution set is $\{-2, 4\}$.

Solving the equation in Example 5 by factoring would be easier than completing the square. Remember, however, that the method of completing the square will work with any quadratic equation.

Concept Quiz 11.3

For Problems 1–10, answer true or false.

1. In a perfect-square trinomial of the form $x^2 + bx + c$, the constant term is equal to one-half the coefficient of the x term.

2. The method of completing the square will solve any quadratic equation.

3. Every quadratic equation solved by completing the square will have real number solutions.

4. The completing-the-square method cannot be used if the quadratic equation could be solved by factoring.

5. To use the completing-the-square method for solving the equation $3x^2 + 2x = 5$, we would first divide both sides of the equation by 3.

6. The equation $x^2 + 2x = 0$ cannot be solved by using the method of completing the square.

7. To solve the equation $x^2 - 5x = 1$ by completing the square, we would start by adding $\dfrac{25}{4}$ to both sides of the equation.

8. To solve the equation $x^2 - 2x = 14$ by completing the square, we must first change the form of the equation to $x^2 - 2x - 14 = 0$.

9. The solution set of the equation $x^2 - 2x = 14$ is $\{1 \pm \sqrt{15}\}$.

10. The solution set of the equation $x^2 - 5x - 1 = 0$ is $\left\{\dfrac{5 \pm \sqrt{29}}{2}\right\}$.

Problem Set 11.3

For Problems 1–14, solve each quadratic equation by using (a) the factoring method and (b) the method of completing the square. **(Objective 1)**

1. $x^2 - 4x - 60 = 0$
2. $x^2 + 6x - 16 = 0$
3. $x^2 - 14x = -40$
4. $x^2 - 18x = -72$
5. $x^2 - 5x - 50 = 0$
6. $x^2 + 3x - 18 = 0$
7. $x(x + 7) = 8$
8. $x(x - 1) = 30$
9. $2n^2 - n - 15 = 0$
10. $3n^2 + n - 14 = 0$
11. $3n^2 + 7n - 6 = 0$
12. $2n^2 + 7n - 4 = 0$
13. $n(n + 6) = 160$
14. $n(n - 6) = 216$

For Problems 15–38, use the method of completing the square to solve each quadratic equation. **(Objective 1)**

15. $x^2 + 4x - 2 = 0$
16. $x^2 + 2x - 1 = 0$
17. $x^2 + 6x - 3 = 0$
18. $x^2 + 8x - 4 = 0$
19. $y^2 - 10y = 1$
20. $y^2 - 6y = -10$
21. $n^2 - 8n + 17 = 0$
22. $n^2 - 4n + 2 = 0$
23. $n(n + 12) = -9$
24. $n(n + 14) = -4$
25. $n^2 + 2n + 6 = 0$
26. $n^2 + n - 1 = 0$
27. $x^2 + 3x - 2 = 0$
28. $x^2 + 5x - 3 = 0$
29. $x^2 + 5x + 1 = 0$
30. $x^2 + 7x + 2 = 0$
31. $y^2 - 7y + 3 = 0$
32. $y^2 - 9y + 30 = 0$
33. $2x^2 + 4x - 3 = 0$
34. $2t^2 - 4t + 1 = 0$
35. $3n^2 - 6n + 5 = 0$
36. $3x^2 + 12x - 2 = 0$
37. $3x^2 + 5x - 1 = 0$
38. $2x^2 + 7x - 3 = 0$

For Problems 39–60, solve each quadratic equation using the method that seems most appropriate.

39. $x^2 + 8x - 48 = 0$
40. $x^2 + 5x - 14 = 0$
41. $2n^2 - 8n = -3$
42. $3x^2 + 6x = 1$
43. $(3x - 1)(2x + 9) = 0$
44. $(5x + 2)(x - 4) = 0$
45. $(x + 2)(x - 7) = 10$
46. $(x - 3)(x + 5) = -7$
47. $(x - 3)^2 = 12$
48. $x^2 = 16x$

49. $3n^2 - 6n + 4 = 0$ **50.** $2n^2 - 2n - 1 = 0$

51. $n(n + 8) = 240$ **52.** $t(t - 26) = -160$

53. $3x^2 + 5x = -2$ **54.** $2x^2 - 7x = -5$

55. $4x^2 - 8x + 3 = 0$ **56.** $9x^2 + 18x + 5 = 0$

57. $x^2 + 12x = 4$ **58.** $x^2 + 6x = -11$

59. $4(2x + 1)^2 - 1 = 11$ **60.** $5(x + 2)^2 + 1 = 16$

61. Use the method of completing the square to solve $ax^2 + bx + c = 0$ for x, where a, b, and c are real numbers and $a \neq 0$.

Thoughts Into Words

62. Explain the process of completing the square to solve a quadratic equation.

63. Give a step-by-step description of how to solve $3x^2 + 9x - 4 = 0$ by completing the square.

Further Investigations

Solve Problems 64–67 for the indicated variable. Assume that all letters represent positive numbers.

64. $\dfrac{x^2}{a^2} - \dfrac{y^2}{b^2} = 1$ for y

65. $\dfrac{x^2}{a^2} + \dfrac{y^2}{b^2} = 1$ for x

66. $s = \dfrac{1}{2}gt^2$ for t

67. $A = \pi r^2$ for r

Solve each of the following equations for x.

68. $x^2 + 8ax + 15a^2 = 0$

69. $x^2 - 5ax + 6a^2 = 0$

70. $10x^2 - 31ax - 14a^2 = 0$

71. $6x^2 + ax - 2a^2 = 0$

72. $4x^2 + 4bx + b^2 = 0$

73. $9x^2 - 12bx + 4b^2 = 0$

Answers to the Concept Quiz

1. False **2.** True **3.** False **4.** False **5.** True **6.** False **7.** True **8.** False **9.** True **10.** True

---

11.4   Quadratic Formula

OBJECTIVES

1 Use the quadratic formula to solve quadratic equations

2 Determine the nature of roots to quadratic equations

As we saw in the last section, the method of completing the square can be used to solve any quadratic equation. Thus if we apply the method of completing the square to the equation $ax^2 + bx + c = 0$, where a, b, and c are real numbers and $a \neq 0$, we can produce a formula for solving quadratic equations. This formula can then be used to solve any quadratic equation. Let's solve $ax^2 + bx + c = 0$ by completing the square.

$$ax^2 + bx + c = 0$$

$$ax^2 + bx = -c \qquad \text{Isolate the } x^2 \text{ and } x \text{ terms}$$

$$x^2 + \frac{b}{a}x = -\frac{c}{a} \qquad \text{Multiply both sides by } \frac{1}{a}$$

$$x^2 + \frac{b}{a}x + \frac{b^2}{4a^2} = -\frac{c}{a} + \frac{b^2}{4a^2} \qquad \frac{1}{2}\left(\frac{b}{a}\right) = \frac{b}{2a} \text{ and } \left(\frac{b}{2a}\right)^2 = \frac{b^2}{4a^2}$$

$$x^2 + \frac{b}{a}x + \frac{b^2}{4a^2} = -\frac{4ac}{4a^2} + \frac{b^2}{4a^2}$$

Complete the square by adding $\frac{b^2}{4a^2}$ to both sides

Common denominator of $4a^2$ on right side

$$x^2 + \frac{b}{a}x + \frac{b^2}{4a^2} = \frac{b^2}{4a^2} - \frac{4ac}{4a^2}$$

Commutative property

$$\left(x + \frac{b}{2a}\right)^2 = \frac{b^2 - 4ac}{4a^2}$$

The right side is combined into a single fraction

$$x + \frac{b}{2a} = \pm\sqrt{\frac{b^2 - 4ac}{4a^2}}$$

$$x + \frac{b}{2a} = \pm\frac{\sqrt{b^2 - 4ac}}{\sqrt{4a^2}}$$

$$x + \frac{b}{2a} = \pm\frac{\sqrt{b^2 - 4ac}}{2a}$$

$\sqrt{4a^2} = |2a|$ but $2a$ can be used because of the use of $\pm$

$$x + \frac{b}{2a} = \frac{\sqrt{b^2 - 4ac}}{2a} \qquad \text{or} \qquad x + \frac{b}{2a} = -\frac{\sqrt{b^2 - 4ac}}{2a}$$

$$x = -\frac{b}{2a} + \frac{\sqrt{b^2 - 4ac}}{2a} \qquad \text{or} \qquad x = -\frac{b}{2a} - \frac{\sqrt{b^2 - 4ac}}{2a}$$

$$x = \frac{-b + \sqrt{b^2 - 4ac}}{2a} \qquad \text{or} \qquad x = \frac{-b - \sqrt{b^2 - 4ac}}{2a}$$

The quadratic formula is usually stated as follows:

> ### Quadratic Formula
>
> $$x = \frac{-b \pm \sqrt{b^2 - 4ac}}{2a}, \qquad a \neq 0$$

We can use the quadratic formula to solve *any* quadratic equation by expressing the equation in the standard form $ax^2 + bx + c = 0$ and substituting the values for a, b, and c into the formula. Let's consider some examples, and let's use a graphical approach first to predict approximate solutions whenever possible.

Classroom Example
Solve $n^2 - 5n - 9 = 0$.

EXAMPLE 1   Solve $x^2 + 5x + 2 = 0$.

Solution

Figure 11.9 shows the graph of the equation.

$$y = x^2 + 5x + 2$$

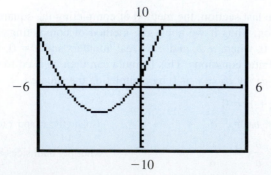

One x intercept appears to be between -1 and 0 and the other between -5 and -4

Figure 11.9

The given equation is in standard form with $a = 1$, $b = 5$, and $c = 2$. Let's substitute these values into the formula and simplify.

$$x = \frac{-b \pm \sqrt{b^2 - 4ac}}{2a}$$

$$x = \frac{-5 \pm \sqrt{5^2 - 4(1)(2)}}{2(1)}$$

$$= \frac{-5 \pm \sqrt{25 - 8}}{2}$$

$$= \frac{-5 \pm \sqrt{17}}{2}$$

The solution set is $\left\{ \dfrac{-5 \pm \sqrt{17}}{2} \right\}$.

Let's compare the solutions, $\dfrac{-5 + \sqrt{17}}{2}$ and $\dfrac{-5 - \sqrt{17}}{2}$, with our estimates from the graphical approach. The decimal approximations of the solutions are -0.44 and -4.56, rounded to the hundredths place. These results agree with our graphical-approach prediction of a solution between -1 and 0 and another solution between -5 and -4.

Classroom Example
Solve $a^2 + 8a + 5 = 0$.

EXAMPLE 2   Solve $x^2 - 2x - 4 = 0$.

Solution

The graph of the equation $y = x^2 - 2x - 4$ is shown in Figure 11.10.

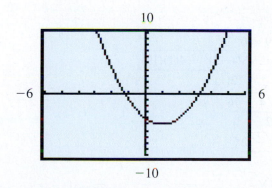

One x intercept appears to be between -2 and -1 and the other between 3 and 4

Figure 11.10

We need to think of $x^2 - 2x - 4 = 0$ as $x^2 + (-2)x + (-4) = 0$ to determine the values $a = 1$, $b = -2$, and $c = -4$. Let's substitute these values into the quadratic formula and simplify.

$$x = \frac{-b \pm \sqrt{b^2 - 4ac}}{2a}$$

$$x = \frac{-(-2) \pm \sqrt{(-2)^2 - 4(1)(-4)}}{2(1)}$$

$$= \frac{2 \pm \sqrt{4 + 16}}{2}$$

$$= \frac{2 \pm \sqrt{20}}{2}$$

$$= \frac{2 \pm 2\sqrt{5}}{2}$$

$$= \frac{2(1 \pm \sqrt{5})}{2}$$

The solution set is $\{1 \pm \sqrt{5}\}$.

The decimal approximations for the solutions $1 - \sqrt{5}$ and $1 + \sqrt{5}$ are -1.24 and 3.24, rounded to the hundredths place. These results agree with our graphical-approach prediction of a solution between -2 and -1 and another solution between 3 and 4.

Classroom Example
Solve $x^2 + 2x + 5 = 0$.

EXAMPLE 3   Solve $x^2 - 2x + 19 = 0$.

Solution

Figure 11.11 shows the graph of the equation $y = x^2 - 2x + 19$.

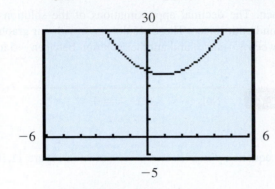

There are no x intercepts; the solutions are nonreal complex numbers

Figure 11.11

$$x^2 - 2x + 19 = 0$$

$$x = \frac{-(-2) \pm \sqrt{(-2)^2 - 4(1)(19)}}{2(1)}$$

$$= \frac{2 \pm \sqrt{4 - 76}}{2}$$

$$= \frac{2 \pm \sqrt{-72}}{2}$$

$$= \frac{2 \pm 6i\sqrt{2}}{2} \qquad \sqrt{-72} = i\sqrt{72} = i\sqrt{36}\sqrt{2} = 6i\sqrt{2}$$

$$= \frac{2(1 \pm 3i\sqrt{2})}{2}$$

$$= 1 \pm 3i\sqrt{2}$$

The solution set is $\{1 \pm 3i\sqrt{2}\}$.

Classroom Example
Solve $2b^2 + 6b - 5 = 0$.

EXAMPLE 4   Solve $2x^2 + 4x - 3 = 0$.

Solution

Let's begin by estimating the solutions using a graphical approach. The graph of the equation $y = 2x^2 + 4x - 3$ is shown in Figure 11.12.

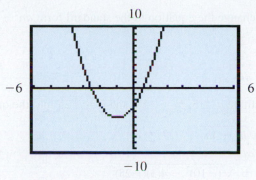

One x intercept is between -3 and -2, and the other is between 0 and 1

Figure 11.12

$$x = \frac{-b \pm \sqrt{b^2 - 4ac}}{2a}$$

$$x = \frac{-4 \pm \sqrt{4^2 - 4(2)(-3)}}{2(2)}$$

$$x = \frac{-4 \pm \sqrt{16 + 24}}{4}$$

$$x = \frac{-4 \pm \sqrt{40}}{4}$$

$$x = \frac{-4 \pm 2\sqrt{10}}{4}$$

$$x = \frac{-2 \pm \sqrt{10}}{2}$$

The solution set is $\left\{ \dfrac{-2 \pm \sqrt{10}}{2} \right\}$.

The decimal approximations for the solutions $\dfrac{-2 - \sqrt{10}}{2}$ and $\dfrac{-2 + \sqrt{10}}{2}$ are -2.58 and 0.58 rounded to the hundredths place. These results agree with our graphical-approach prediction of a solution between -3 and -2 and another solution between 0 and 1.

Classroom Example
Solve $x(5x - 7) = 6$.

EXAMPLE 5   Solve $n(3n - 10) = 25$.

Solution

Figure 11.13 shows the graph of the equation $y = x(3x - 10) - 25$.

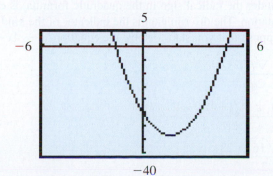

One x intercept is between -2 and -1, and the other appears to be 5

Figure 11.13

First, we need to change the equation to the standard form $an^2 + bn + c = 0$.

$$n(3n - 10) = 25$$
$$3n^2 - 10n = 25$$
$$3n^2 - 10n - 25 = 0$$

Now we can substitute $a = 3$, $b = -10$, and $c = -25$ into the quadratic formula.

$$n = \frac{-b \pm \sqrt{b^2 - 4ac}}{2a}$$

$$n = \frac{-(-10) \pm \sqrt{(-10)^2 - 4(3)(-25)}}{2(3)}$$

$$n = \frac{10 \pm \sqrt{100 + 300}}{2(3)}$$

$$n = \frac{10 \pm \sqrt{400}}{6}$$

$$n = \frac{10 \pm 20}{6}$$

$$n = \frac{10 + 20}{6} \quad \text{or} \quad n = \frac{10 - 20}{6}$$

$$n = 5 \quad \text{or} \quad n = -\frac{5}{3}$$

The solution set is $\left\{-\dfrac{5}{3}, 5\right\}$.

The solutions are $-1\dfrac{2}{3}$ and 5. These results agree with our graphical approach of a solution between -2 and -1 and another solution of 5.

In Example 5, note that we used the variable n. The quadratic formula is usually stated in terms of x, but it certainly can be applied to quadratic equations in other variables. Also note in Example 5 that the polynomial $3n^2 - 10n - 25$ can be factored as $(3n + 5)(n - 5)$. Therefore, we could also solve the equation $3n^2 - 10n - 25 = 0$ by using the factoring approach. Section 11.5 will offer some guidance about which approach to use for a particular equation.

Determining the Nature of Roots of Quadratic Equations

The quadratic formula makes it easy to determine the nature of the roots of a quadratic equation without completely solving the equation. The value of the expression

$$b^2 - 4ac$$

which appears under the radical sign in the quadratic formula, is called the **discriminant** of the quadratic equation. The discriminant is the indicator of the kind of roots the equation has. For example, suppose that you start to solve the equation $x^2 - 4x + 7 = 0$ as follows:

$$x = \frac{-b \pm \sqrt{b^2 - 4ac}}{2a}$$

$$x = \frac{-(-4) \pm \sqrt{(-4)^2 - 4(1)(7)}}{2(1)}$$

$$x = \frac{4 \pm \sqrt{16 - 28}}{2}$$

$$x = \frac{4 \pm \sqrt{-12}}{2}$$

At this stage you should be able to look ahead and realize that you will obtain two nonreal complex solutions for the equation. (Note, by the way, that these solutions are complex conjugates.) In other words, the discriminant (-12) indicates what type of roots you will obtain.

We make the following general statements relative to the roots of a quadratic equation of the form $ax^2 + bx + c = 0$.

---

1. If $b^2 - 4ac < 0$, then the equation has two nonreal complex solutions.

2. If $b^2 - 4ac = 0$, then the equation has one real solution.

3. If $b^2 - 4ac > 0$, then the equation has two real solutions.

The following examples illustrate each of these situations. (You may want to solve the equations completely to verify the conclusions.)

Equation	Discriminant	Nature of roots
$x^2 - 3x + 7 = 0$	$\begin{aligned} b^2 - 4ac &= (-3)^2 - 4(1)(7) \\ &= 9 - 28 \\ &= -19 \end{aligned}$	Two nonreal complex solutions
$9x^2 - 12x + 4 = 0$	$\begin{aligned} b^2 - 4ac &= (-12)^2 - 4(9)(4) \\ &= 144 - 144 \\ &= 0 \end{aligned}$	One real solution
$2x^2 + 5x - 3 = 0$	$\begin{aligned} b^2 - 4ac &= (5)^2 - 4(2)(-3) \\ &= 25 + 24 \\ &= 49 \end{aligned}$	Two real solutions

Remark: A clarification is called for at this time. Previously we made the statement that if $b^2 - 4ac = 0$, then the equation has one real solution. Technically, such an equation has two solutions, but they are equal. For example, each factor of $(x - 7)(x - 7) = 0$ produces a solution, but both solutions are the number 7. We sometimes refer to this as one real solution with a *multiplicity of two*. Using the idea of multiplicity of roots, we can say that every quadratic equation has two roots.

Classroom Example
Use the discriminant to determine whether the equation $3x^2 - 7x + 2 = 0$ has two nonreal complex solutions, one real solution with a multiplicity of 2, or two real solutions.

EXAMPLE 6

Use the discriminant to determine if the equation $5x^2 + 2x + 7 = 0$ has two nonreal complex solutions, one real solution with a multiplicity of two, or two real solutions.

Solution

For the equation $5x^2 + 2x + 7 = 0$, $a = 5$, $b = 2$, and $c = 7$.

$$\begin{aligned} b^2 - 4ac &= (2)^2 - 4(5)(7) \\ &= 4 - 140 \\ &= -136 \end{aligned}$$

Because the discriminant is negative, the solutions will be two nonreal complex numbers. ▪

Most students become very adept at applying the quadratic formula to solve quadratic equations but make errors when reducing the answers. The next example shows two different methods for simplifying the answers.

Classroom Example
Solve $7m^2 + 4m - 2 = 0$.

EXAMPLE 7   Solve $3x^2 - 8x + 2 = 0$.

Solution

Here $a = 3$, $b = -8$, and $c = 2$. Let's substitute these values into the quadratic formula and simplify.

$$x = \frac{-b \pm \sqrt{b^2 - 4ac}}{2a}$$

$$x = \frac{-(-8) \pm \sqrt{(-8)^2 - 4(3)(2)}}{2(3)}$$

$$x = \frac{8 \pm \sqrt{64 - 24}}{6}$$

$$x = \frac{8 \pm \sqrt{40}}{6} = \frac{8 \pm 2\sqrt{10}}{6} \qquad \sqrt{40} = \sqrt{4}\sqrt{10} = 2\sqrt{10}$$

Now to simplify, one method is to factor 2 out of the numerator and reduce.

$$x = \frac{8 \pm 2\sqrt{10}}{6} = \frac{2(4 \pm \sqrt{10})}{6} = \frac{\cancel{2}(4 \pm \sqrt{10})}{\underset{3}{\cancel{6}}} = \frac{4 \pm \sqrt{10}}{3}$$

Another method for simplifying the answer is to write the result as two separate fractions and reduce each fraction.

$$x = \frac{8 \pm 2\sqrt{10}}{6} = \frac{8}{6} \pm \frac{2\sqrt{10}}{6} = \frac{4}{3} \pm \frac{\sqrt{10}}{3} = \frac{4 \pm \sqrt{10}}{3}$$

Be very careful when simplifying your result because that is a common source of incorrect answers.

Concept Quiz 11.4

For Problems 1–10, answer true or false.

1. The quadratic formula can be used to solve any quadratic equation.

2. The number $\sqrt{b^2 - 4ac}$ is called the discriminant of the quadratic equation.

3. Every quadratic equation will have two solutions.

4. The quadratic formula cannot be used if the quadratic equation can be solved by factoring.

5. To use the quadratic formula for solving the equation $3x^2 + 2x - 5 = 0$, you must first divide both sides of the equation by 3.

6. The equation $9x^2 + 30x + 25 = 0$ has one real solution with a multiplicity of 2.

7. The equation $2x^2 + 3x + 4 = 0$ has two nonreal complex solutions.

8. The equation $x^2 + 9 = 0$ has two real solutions.

9. Because the quadratic formula has a denominator, it could be simplified and written as $x = -b \pm \dfrac{\sqrt{b^2 - 4ac}}{2a}$.

10. Rachel reduced the result $x = \dfrac{6 \pm 5\sqrt{7}}{2}$ to obtain $x = 3 \pm \dfrac{5\sqrt{7}}{2}$. Her result is correct.

Problem Set 11.4

For Problems 1–10, simplify and reduce each expression.

1. $\dfrac{2 \pm \sqrt{20}}{4}$

2. $\dfrac{4 \pm \sqrt{20}}{6}$

3. $\dfrac{-6 \pm \sqrt{27}}{3}$

4. $\dfrac{-9 \pm \sqrt{54}}{3}$

5. $\dfrac{6 \pm \sqrt{18}}{9}$

6. $\dfrac{12 \pm \sqrt{32}}{8}$

7. $\dfrac{-10 \pm \sqrt{75}}{10}$

8. $\dfrac{-4 \pm \sqrt{8}}{4}$

9. $\dfrac{-6 \pm \sqrt{48}}{4}$

10. $\dfrac{-8 \pm \sqrt{72}}{4}$

For Problems 11–50, use the quadratic formula to solve each of the quadratic equations. **(Objective 1)**

11. $x^2 + 2x - 1 = 0$

12. $x^2 + 4x - 1 = 0$

13. $n^2 + 5n - 3 = 0$

14. $n^2 + 3n - 2 = 0$

15. $a^2 - 8a = 4$

16. $a^2 - 6a = 2$

17. $n^2 + 5n + 8 = 0$

18. $2n^2 - 3n + 5 = 0$

19. $x^2 - 18x + 80 = 0$

20. $x^2 + 19x + 70 = 0$

21. $-y^2 = -9y + 5$

22. $-y^2 + 7y = 4$

23. $2x^2 + x - 4 = 0$

24. $2x^2 + 5x - 2 = 0$

25. $4x^2 + 2x + 1 = 0$

26. $3x^2 - 2x + 5 = 0$

27. $3a^2 - 8a + 2 = 0$

28. $2a^2 - 6a + 1 = 0$

29. $-2n^2 + 3n + 5 = 0$

30. $-3n^2 - 11n + 4 = 0$

31. $3x^2 + 19x + 20 = 0$

32. $2x^2 - 17x + 30 = 0$

33. $36n^2 - 60n + 25 = 0$

34. $9n^2 + 42n + 49 = 0$

35. $4x^2 - 2x = 3$

36. $6x^2 - 4x = 3$

37. $5x^2 - 13x = 0$

38. $7x^2 + 12x = 0$

39. $3x^2 = 5$

40. $4x^2 = 3$

41. $6t^2 + t - 3 = 0$

42. $2t^2 + 6t - 3 = 0$

43. $n^2 + 32n + 252 = 0$

44. $n^2 - 4n - 192 = 0$

45. $12x^2 - 73x + 110 = 0$

46. $6x^2 + 11x - 255 = 0$

47. $-2x^2 + 4x - 3 = 0$

48. $-2x^2 + 6x - 5 = 0$

49. $-6x^2 + 2x + 1 = 0$

50. $-2x^2 + 4x + 1 = 0$

For each quadratic equation in Problems 51–60, first use the discriminant to determine whether the equation has two nonreal complex solutions, one real solution with a multiplicity of two, or two real solutions. Then solve the equation. **(Objective 2)**

51. $x^2 + 4x - 21 = 0$

52. $x^2 - 3x - 54 = 0$

53. $9x^2 - 6x + 1 = 0$

54. $4x^2 + 20x + 25 = 0$

55. $x^2 - 7x + 13 = 0$

56. $2x^2 - x + 5 = 0$

57. $15x^2 + 17x - 4 = 0$

58. $8x^2 + 18x - 5 = 0$

59. $3x^2 + 4x = 2$

60. $2x^2 - 6x = -1$

Thoughts Into Words

61. Your friend states that the equation $-2x^2 + 4x - 1 = 0$ must be changed to $2x^2 - 4x + 1 = 0$ (by multiplying both sides by -1) before the quadratic formula can be applied. Is she right about this? If not, how would you convince her she is wrong?

62. Another of your friends claims that the quadratic formula can be used to solve the equation $x^2 - 9 = 0$. How would you react to this claim?

63. Why must we change the equation $3x^2 - 2x = 4$ to $3x^2 - 2x - 4 = 0$ before applying the quadratic formula?

Further Investigations

The solution set for $x^2 - 4x - 37 = 0$ is $\{2 \pm \sqrt{41}\}$. With a calculator, we found a rational approximation, to the nearest one-thousandth, for each of these solutions.

$$2 - \sqrt{41} = -4.403 \quad \text{and} \quad 2 + \sqrt{41} = 8.403$$

Thus the solution set is $\{-4.403, 8.403\}$, with the answers rounded to the nearest one-thousandth.

Solve each of the equations in Problems 64–73, expressing solutions to the nearest one-thousandth.

64. $x^2 - 6x - 10 = 0$ **65.** $x^2 - 16x - 24 = 0$

66. $x^2 + 6x - 44 = 0$ **67.** $x^2 + 10x - 46 = 0$

68. $x^2 + 8x + 2 = 0$ **69.** $x^2 + 9x + 3 = 0$

70. $4x^2 - 6x + 1 = 0$ **71.** $5x^2 - 9x + 1 = 0$

72. $2x^2 - 11x - 5 = 0$ **73.** $3x^2 - 12x - 10 = 0$

For Problems 74–76, use the discriminant to help solve each problem.

74. Determine k so that the solutions of $x^2 - 2x + k = 0$ are complex but nonreal.

75. Determine k so that $4x^2 - kx + 1 = 0$ has two equal real solutions.

76. Determine k so that $3x^2 - kx - 2 = 0$ has real solutions.

Graphing Calculator Activities

77. Use your graphing calculator to verify your answers for Problems 1–10.

78. Use your graphing calculator to give visual support for your answers for Problems 17, 18, 25, 26, 47, and 48.

79. Use your graphing calculator to verify your solutions, to the nearest tenth, for Problems 13, 14, 15, 16, 21, 22, 23, 24, 35, 36, 41, 42, 49, and 50.

Answers to the Concept Quiz

1. True   **2.** False   **3.** True   **4.** False   **5.** False   **6.** True   **7.** True   **8.** False   **9.** False   **10.** True

11.5   More Quadratic Equations and Applications

OBJECTIVES

1. Solve quadratic equations selecting the most appropriate method

2. Solve equations that are quadratic in form

3. Solve word problems involving quadratic equations

4. Solve word problems involving interest compounded annually

Which method should be used to solve a particular quadratic equation? There is no hard and fast answer to that question; it depends on the type of equation and on your personal preference. In the following examples we will state reasons for choosing a specific technique. However, keep in mind that usually this is a decision you must make as the need arises. That's why you need to be familiar with the strengths and weaknesses of each method.

Classroom Example
Solve $3x^2 - x - 5 = 0$.

EXAMPLE 1   Solve $2x^2 - 3x - 1 = 0$.

Solution

Because of the leading coefficient of 2 and the constant term of -1, there are very few factoring possibilities to consider. Therefore, with such problems, first try the factoring approach. Unfortunately, this particular polynomial is not factorable using integers. Let's use the quadratic formula to solve the equation.

$$x = \frac{-b \pm \sqrt{b^2 - 4ac}}{2a}$$

$$x = \frac{-(-3) \pm \sqrt{(-3)^2 - 4(2)(-1)}}{2(2)}$$

$$x = \frac{3 \pm \sqrt{9 + 8}}{4}$$

$$x = \frac{3 \pm \sqrt{17}}{4}$$

The solution set is $\left\{ \dfrac{3 \pm \sqrt{17}}{4} \right\}$.

Classroom Example

Solve $\dfrac{2}{x} + \dfrac{6}{x + 3} = 1$.

EXAMPLE 2   Solve $\dfrac{3}{n} + \dfrac{10}{n + 6} = 1$.

Solution

$$\frac{3}{n} + \frac{10}{n + 6} = 1, \qquad n \neq 0 \text{ and } n \neq -6$$

$$n(n + 6)\left(\frac{3}{n} + \frac{10}{n + 6} \right) = 1(n)(n + 6) \qquad \text{Multiply both sides by } n(n + 6),$$
$$\text{which is the LCD}$$

$$3(n + 6) + 10n = n(n + 6)$$

$$3n + 18 + 10n = n^2 + 6n$$

$$13n + 18 = n^2 + 6n$$

$$0 = n^2 - 7n - 18$$

This equation is an easy one to consider for possible factoring, and it factors as follows:

$$0 = (n - 9)(n + 2)$$

$$n - 9 = 0 \qquad \text{or} \qquad n + 2 = 0$$

$$n = 9 \qquad \text{or} \qquad n = -2$$

The solution set is $\{-2, 9\}$.

We should make a comment about Example 2. Note the indication of the initial restrictions $n \neq 0$ and $n \neq -6$. Remember that we need to do this when solving fractional equations.

Classroom Example

Solve $m^2 + 20m + 96 = 0$.

EXAMPLE 3   Solve $x^2 + 22x + 112 = 0$.

Solution

The size of the constant term makes the factoring approach a little cumbersome for this problem. Furthermore, because the leading coefficient is 1, and the coefficient of the x term is even, the method of completing the square will work effectively.

$$x^2 + 22x + 112 = 0$$

$$x^2 + 22x = -112$$

$$x^2 + 22x + 121 = -112 + 121$$

$$(x + 11)^2 = 9$$

$$x + 11 = \pm\sqrt{9}$$

$$x + 11 = \pm 3$$

$$x + 11 = 3 \qquad \text{or} \qquad x + 11 = -3$$

$$x = -8 \qquad \text{or} \qquad x = -14$$

The solution set is $\{-14, -8\}$.

Classroom Example
Solve $x^4 + 2x^2 - 360 = 0$.

EXAMPLE 4   Solve $x^4 - 4x^2 - 96 = 0$.

Solution

An equation such as $x^4 - 4x^2 - 96 = 0$ is not a quadratic equation, but we can solve it using the techniques that we use on quadratic equations. That is, we can factor the polynomial and apply the property "$ab = 0$ if and only if $a = 0$ or $b = 0$" as follows:

$$x^4 - 4x^2 - 96 = 0$$
$$(x^2 - 12)(x^2 + 8) = 0$$

$x^2 - 12 = 0$	or	$x^2 + 8 = 0$
$x^2 = 12$	or	$x^2 = -8$
$x = \pm\sqrt{12}$	or	$x = \pm\sqrt{-8}$
$x = \pm 2\sqrt{3}$	or	$x = \pm 2i\sqrt{2}$

The solution set is $\{\pm 2\sqrt{3},\ \pm 2i\sqrt{2}\}$.

Remark: Another approach to Example 4 would be to substitute y for x^2 and y^2 for x^4. The equation $x^4 - 4x^2 - 96 = 0$ becomes the quadratic equation $y^2 - 4y - 96 = 0$. Thus we say that $x^4 - 4x^2 - 96 = 0$ is of *quadratic form*. Then we could solve the quadratic equation $y^2 - 4y - 96 = 0$ and use the equation $y = x^2$ to determine the solutions for x.

Solving Word Problems Involving Quadratic Equations

Before we conclude this section with some word problems that can be solved using quadratic equations, let's restate the suggestions we made in an earlier chapter for solving word problems.

Suggestions for Solving Word Problems

1. Read the problem carefully, and make certain that you understand the meanings of all the words. Be especially alert for any technical terms used in the statement of the problem.

2. Read the problem a second time (perhaps even a third time) to get an overview of the situation being described and to determine the known facts, as well as what is to be found.

3. Sketch any figure, diagram, or chart that might be helpful in analyzing the problem.

4. Choose a meaningful variable to represent the unknown quantity in the problem (perhaps l, if the length of a rectangle is the unknown quantity), and represent any other unknowns in terms of that variable.

5. Look for a guideline that you can use to set up an equation. A guideline might be a formula such as $A = lw$ or a relationship such as "the fractional part of a job done by Bill plus the fractional part of the job done by Mary equals the total job."

6. Form an equation that contains the variable and that translates the conditions of the guideline from English to algebra.

7. Solve the equation and use the solutions to determine all facts requested in the problem.

8. **Check all answers back into the original statement of the problem.**

Keep these suggestions in mind as we now consider some word problems.

Classroom Example
A margin of 1 inch surrounds the front of a card, which leaves 39 square inches for graphics. If the height of the card is three times the width, what are the dimensions of the card?

EXAMPLE 5

A page for a magazine contains 70 square inches of type. The height of a page is twice the width. If the margin around the type is to be 2 inches uniformly, what are the dimensions of a page?

Solution

Let x represent the width of a page. Then $2x$ represents the height of a page. Now let's draw and label a model of a page (Figure 11.14).

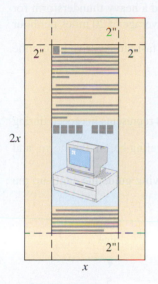

Figure 11.14

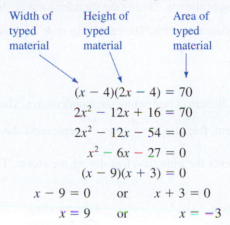

$$\begin{array}{ccc}
\text{Width of} & \text{Height of} & \text{Area of} \\
\text{typed} & \text{typed} & \text{typed} \\
\text{material} & \text{material} & \text{material}
\end{array}$$

$$(x - 4)(2x - 4) = 70$$
$$2x^2 - 12x + 16 = 70$$
$$2x^2 - 12x - 54 = 0$$
$$x^2 - 6x - 27 = 0$$
$$(x - 9)(x + 3) = 0$$
$$x - 9 = 0 \quad \text{or} \quad x + 3 = 0$$
$$x = 9 \quad \text{or} \quad x = -3$$

Disregard the negative solution; the page must be 9 inches wide, and its height is $2(9) = 18$ inches.

Let's use our knowledge of quadratic equations to analyze some applications of the business world. For example, if P dollars is invested at r rate of interest compounded annually for t years, then the amount of money, A, accumulated at the end of t years is given by the formula

$$A = P(1 + r)^t$$

This compound interest formula serves as a guideline for the next problem.

Classroom Example
Suppose that $2500 is invested at a certain rate of interest compounded annually for 2 years. If the accumulated value at the end of 2 years is $2704, find the rate of interest.

EXAMPLE 6

Suppose that $2000 is invested at a certain rate of interest compounded annually for 2 years. If the accumulated value at the end of 2 years is $2205, find the rate of interest.

Solution

Let r represent the rate of interest. Substitute the known values into the compound interest formula to yield

$$A = P(1 + r)^t$$
$$2205 = 2000(1 + r)^2$$

Solving this equation, we obtain

$$\frac{2205}{2000} = (1 + r)^2$$
$$1.1025 = (1 + r)^2$$
$$\pm\sqrt{1.1025} = 1 + r$$
$$\pm 1.05 = 1 + r$$
$$1 + r = 1.05 \quad \text{or} \quad 1 + r = -1.05$$

$$r = -1 + 1.05 \qquad \text{or} \qquad r = -1 - 1.05$$
$$r = 0.05 \qquad \text{or} \qquad r = -2.05$$

We must disregard the negative solution, so that $r = 0.05$ is the only solution. Change 0.05 to a percent, and the rate of interest is 5%.

Classroom Example
After hiking 9 miles of a 10-mile hike, Sam hurt his foot. For the remainder of the hike, his rate was two miles per hour slower than before he hurt his foot. The entire hike took $2\frac{3}{4}$ hours. How fast did he hike before hurting his foot?

EXAMPLE 7

On a 130-mile trip from Orlando to Sarasota, Roberto encountered a heavy thunderstorm for the last 40 miles of the trip. During the thunderstorm he drove an average of 20 miles per hour slower than before the storm. The entire trip took $2\frac{1}{2}$ hours. How fast did he drive before the storm?

Solution

Let x represent Roberto's rate before the thunderstorm. Then $x - 20$ represents his speed during the thunderstorm. Because $t = \dfrac{d}{r}$, then $\dfrac{90}{x}$ represents the time traveling before the storm, and $\dfrac{40}{x - 20}$ represents the time traveling during the storm. The following guideline sums up the situation.

Time traveling before the storm	+	Time traveling after the storm	=	Total time
$\dfrac{90}{x}$	+	$\dfrac{40}{x - 20}$	=	$\dfrac{5}{2}$

Solving this equation, we obtain

$$2x(x - 20)\left(\frac{90}{x} + \frac{40}{x - 20}\right) = 2x(x - 20)\left(\frac{5}{2}\right)$$

$$2x(x - 20)\left(\frac{90}{x}\right) + 2x(x - 20)\left(\frac{40}{x - 20}\right) = 2x(x - 20)\left(\frac{5}{2}\right)$$

$$180(x - 20) + 2x(40) = 5x(x - 20)$$

$$180x - 3600 + 80x = 5x^2 - 100x$$

$$0 = 5x^2 - 360x + 3600$$

$$0 = 5(x^2 - 72x + 720)$$

$$0 = 5(x - 60)(x - 12)$$

$$x - 60 = 0 \qquad \text{or} \qquad x - 12 = 0$$

$$x = 60 \qquad \text{or} \qquad x = 12$$

We discard the solution of 12 because it would be impossible to drive 20 miles per hour slower than 12 miles per hour; thus Roberto's rate before the thunderstorm was 60 miles per hour.

Classroom Example
James bought a shipment of monitors for \$6000. When he had sold all but 10 monitors at a profit of \$100 per monitor, he had regained the entire cost of the shipment. How many monitors were sold and at what price per monitor?

EXAMPLE 8

A computer installer agreed to do an installation for \$150. It took him 2 hours longer than he expected, and therefore he earned \$2.50 per hour less than he anticipated. How long did he expect the installation would take?

Solution

Let x represent the number of hours he expected the installation to take. Then $x + 2$ represents the number of hours the installation actually took. The rate of pay is represented by the pay divided by the number of hours. The following guideline is used to write the equation.

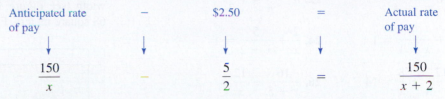

Anticipated rate of pay	−	$2.50	=	Actual rate of pay
$\dfrac{150}{x}$	−	$\dfrac{5}{2}$	=	$\dfrac{150}{x+2}$

Solving this equation, we obtain

$$2x(x+2)\left(\frac{150}{x}-\frac{5}{2}\right)=2x(x+2)\left(\frac{150}{x+2}\right)$$

$$2(x+2)(150)-x(x+2)(5)=2x(150)$$

$$300(x+2)-5x(x+2)=300x$$

$$300x+600-5x^2-10x=300x$$

$$-5x^2-10x+600=0$$

$$-5(x^2+2x-120)=0$$

$$-5(x+12)(x-10)=0$$

$$x=-12 \quad \text{or} \quad x=10$$

Disregard the negative answer. Therefore he anticipated that the installation would take 10 hours.

This next problem set contains a large variety of word problems. Not only are there some business applications similar to those we discussed in this section, but there are also more problems of the types we discussed in Chapters 7 and 9. Try to give them your best shot without referring to the examples in earlier chapters.

Concept Quiz 11.5

For Problems 1–5, choose the method that you think is most appropriate for solving the given equation.

1. $2x^2+6x-3=0$ A. Factoring

2. $(x+1)^2=36$ B. Square-root property (Property 11.1)

3. $x^2-3x+2=0$ C. Completing the square

4. $x^2+6x=19$ D. Quadratic formula

5. $4x^2+2x-5=0$

6. $4x^2=3$

7. $x^2-4x-12=0$

Problem Set 11.5

For Problems 1–20, solve each quadratic equation using the method that seems most appropriate to you. **(Objective 1)**

1. $x^2-4x-6=0$

2. $x^2-8x-4=0$

3. $3x^2+23x-36=0$

4. $n^2+22n+105=0$

5. $x^2-18x=9$

6. $x^2+20x=25$

7. $2x^2-3x+4=0$

8. $3y^2-2y+1=0$

9. $135+24n+n^2=0$

10. $28-x-2x^2=0$

11. $(x-2)(x+9)=-10$

12. $(x+3)(2x+1)=-3$

13. $2x^2-4x+7=0$

14. $3x^2-2x+8=0$

15. $x^2-18x+15=0$

16. $x^2-16x+14=0$

17. $20y^2+17y-10=0$

18. $12x^2+23x-9=0$

19. $4t^2+4t-1=0$

20. $5t^2+5t-1=0$

For Problems 21–40, solve each equation. **(Objective 2)**

21. $n + \dfrac{3}{n} = \dfrac{19}{4}$

22. $n - \dfrac{2}{n} = -\dfrac{7}{3}$

23. $\dfrac{3}{x} + \dfrac{7}{x-1} = 1$

24. $\dfrac{2}{x} + \dfrac{5}{x+2} = 1$

25. $\dfrac{12}{x-3} + \dfrac{8}{x} = 14$

26. $\dfrac{16}{x+5} - \dfrac{12}{x} = -2$

27. $\dfrac{3}{x-1} - \dfrac{2}{x} = \dfrac{5}{2}$

28. $\dfrac{4}{x+1} + \dfrac{2}{x} = \dfrac{5}{3}$

29. $\dfrac{6}{x} + \dfrac{40}{x+5} = 7$

30. $\dfrac{12}{t} + \dfrac{18}{t+8} = \dfrac{9}{2}$

31. $\dfrac{5}{n-3} - \dfrac{3}{n+3} = 1$

32. $\dfrac{3}{t+2} + \dfrac{4}{t-2} = 2$

33. $x^4 - 18x^2 + 72 = 0$

34. $x^4 - 21x^2 + 54 = 0$

35. $3x^4 - 35x^2 + 72 = 0$

36. $5x^4 - 32x^2 + 48 = 0$

37. $3x^4 + 17x^2 + 20 = 0$

38. $4x^4 + 11x^2 - 45 = 0$

39. $6x^4 - 29x^2 + 28 = 0$

40. $6x^4 - 31x^2 + 18 = 0$

For Problems 41–70, set up an equation and solve each problem. **(Objectives 3 and 4)**

41. Find two consecutive whole numbers such that the sum of their squares is 145.

42. Find two consecutive odd whole numbers such that the sum of their squares is 74.

43. Two positive integers differ by 3, and their product is 108. Find the numbers.

44. Suppose that the sum of two numbers is 20, and the sum of their squares is 232. Find the numbers.

45. Find two numbers such that their sum is 10 and their product is 22.

46. Find two numbers such that their sum is 6 and their product is 7.

47. Suppose that the sum of two whole numbers is 9, and the sum of their reciprocals is $\dfrac{1}{2}$. Find the numbers.

48. The difference between two whole numbers is 8, and the difference between their reciprocals is $\dfrac{1}{6}$. Find the two numbers.

49. The sum of the lengths of the two legs of a right triangle is 21 inches. If the length of the hypotenuse is 15 inches, find the length of each leg.

50. The length of a rectangular floor is 1 meter less than twice its width. If a diagonal of the rectangle is 17 meters, find the length and width of the floor.

51. A rectangular plot of ground measuring 12 meters by 20 meters is surrounded by a sidewalk of a uniform width (see Figure 11.15). The area of the sidewalk is 68 square meters. Find the width of the walk.

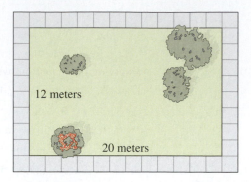

Figure 11.15

52. A 5-inch by 7-inch picture is surrounded by a frame of uniform width. The area of the picture and frame together is 80 square inches. Find the width of the frame.

53. The perimeter of a rectangle is 44 inches, and its area is 112 square inches. Find the length and width of the rectangle.

54. A rectangular piece of cardboard is 2 units longer than it is wide. From each of its corners a square piece 2 units on a side is cut out. The flaps are then turned up to form an open box that has a volume of 70 cubic units. Find the length and width of the original piece of cardboard.

55. Charlotte's time to travel 250 miles is 1 hour more than Lorraine's time to travel 180 miles. Charlotte drove 5 miles per hour faster than Lorraine. How fast did each one travel?

56. Larry's time to travel 156 miles is 1 hour more than Terrell's time to travel 108 miles. Terrell drove 2 miles per hour faster than Larry. How fast did each one travel?

57. On a 570-mile trip, Andy averaged 5 miles per hour faster for the last 240 miles than he did for the first 330 miles. The entire trip took 10 hours. How fast did he travel for the first 330 miles?

58. On a 135-mile bicycle excursion, Maria averaged 5 miles per hour faster for the first 60 miles than she did for the last 75 miles. The entire trip took 8 hours. Find her rate for the first 60 miles.

59. It takes Terry 2 hours longer to do a certain job than it takes Tom. They worked together for 3 hours; then Tom left and Terry finished the job in 1 hour. How long would it take each of them to do the job alone?

60. Suppose that Arlene can mow the entire lawn in 40 minutes less time with the power mower than she can with the push mower. One day the power mower broke down after she had been mowing for 30 minutes. She finished the lawn with the push mower in 20 minutes. How long does it take Arlene to mow the entire lawn with the power mower?

61. A student did a word processing job for $24. It took him 1 hour longer than he expected, and therefore he earned $4 per hour less than he anticipated. How long did he expect that it would take to do the job?

62. A group of students agreed that each would chip in the same amount to pay for a party that would cost $100. Then they found 5 more students interested in the party and in sharing the expenses. This decreased the amount each had to pay by $1. How many students were involved in the party and how much did each student have to pay?

63. A group of students agreed that each would contribute the same amount to buy their favorite teacher an $80 birthday gift. At the last minute, 2 of the students decided not to chip in. This increased the amount that the remaining students had to pay by $2 per student. How many students actually contributed to the gift?

64. The formula $D = \dfrac{n(n-3)}{2}$ yields the number of diagonals, D, in a polygon of n sides. Find the number of sides of a polygon that has 54 diagonals.

65. The formula $S = \dfrac{n(n+1)}{2}$ yields the sum, S, of the first n natural numbers $1, 2, 3, 4, \ldots$. How many consecutive natural numbers starting with 1 will give a sum of 1275?

66. At a point 16 yards from the base of a tower, the distance to the top of the tower is 4 yards more than the height of the tower (see Figure 11.16). Find the height of the tower.

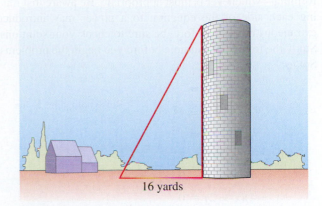

Figure 11.16

16 yards

67. Suppose that $500 is invested at a certain rate of interest compounded annually for 2 years. If the accumulated value at the end of 2 years is $594.05, find the rate of interest.

68. Suppose that $10,000 is invested at a certain rate of interest compounded annually for 2 years. If the accumulated value at the end of 2 years is $12,544, find the rate of interest.

69. What rate of interest compounded annually is needed for an investment of $8000 to accumulate to $8988.80 at the end of two years?

70. What rate of interest compounded annually is needed for an investment of $6500 to accumulate to $7166.25 at the end of two years?

Thoughts Into Words

71. How would you solve the equation $x^2 - 4x = 252$? Explain your choice of the method that you would use.

72. Explain how you would solve $(x-2)(x-7) = 0$ and also how you would solve $(x-2)(x-7) = 4$.

73. One of our problem-solving suggestions is to look for a guideline that can be used to help determine an equation. What does this suggestion mean to you?

74. Can a quadratic equation with integral coefficients have exactly one nonreal complex solution? Explain your answer.

Further Investigations

For Problems 75–81, solve each equation.

75. $x - 9\sqrt{x} + 18 = 0$ [*Hint:* Let $y = \sqrt{x}$.]

76. $x - 4\sqrt{x} + 3 = 0$

77. $x + \sqrt{x} - 2 = 0$

78. $x^{\frac{2}{3}} + x^{\frac{1}{3}} - 6 = 0$ [*Hint:* Let $y = x^{\frac{1}{3}}$.]

79. $6x^{\frac{2}{3}} - 5x^{\frac{1}{3}} - 6 = 0$ 80. $x^{-2} + 4x^{-1} - 12 = 0$

81. $12x^{-2} - 17x^{-1} - 5 = 0$

The following equations are also quadratic in form. To solve, begin by raising each side of the equation to the appropriate power so that the exponent will become an integer. Then, to solve the resulting quadratic equation, you may use the square-root property, factoring, or the quadratic formula—whichever is most appropriate. Be aware that raising each side of the equation to a power may introduce extraneous roots; therefore, be sure to check your solutions. Study the following example before you begin the problems. Solve

$$(x + 3)^{\frac{2}{3}} = 1$$
$$[(x + 3)^{\frac{2}{3}}]^3 = 1^3 \qquad \text{Raise both sides to the third power}$$
$$(x + 3)^2 = 1$$
$$x^2 + 6x + 9 = 1$$
$$x^2 + 6x + 8 = 0$$
$$(x + 4)(x + 2) = 0$$
$$x + 4 = 0 \qquad \text{or} \qquad x + 2 = 0$$
$$x = -4 \qquad \text{or} \qquad x = -2$$

Both solutions do check. The solution set is $\{-4, -2\}$.

For problems 82–90, solve each equation.

82. $(5x + 6)^{\frac{1}{2}} = x$

83. $(3x + 4)^{\frac{1}{2}} = x$

84. $x^{\frac{2}{3}} = 2$

85. $x^{\frac{2}{5}} = 2$

86. $(2x + 6)^{\frac{1}{2}} = x$

87. $(2x - 4)^{\frac{2}{3}} = 1$

88. $(4x + 5)^{\frac{2}{3}} = 2$

89. $(6x + 7)^{\frac{1}{2}} = x + 2$

90. $(5x + 21)^{\frac{1}{2}} = x + 3$

Graphing Calculator Activities

91. Use your graphing calculator to give visual support for your answers for Problems 33–40.

92. In the text we have used graphs to *predict* solutions of equations and to give *visual support* for solutions that we obtained algebraically. Now use your graphing calculator to *find*, to the nearest tenth, the real number solutions for each of the following equations.

a. $x^2 - 8x + 14 = 0$ **b.** $x^2 + 2x - 2 = 0$

c. $x^2 + 4x - 16 = 0$ **d.** $x^2 + 2x - 111 = 0$

e. $x^4 + x^2 - 12 = 0$ **f.** $x^4 - 19x^2 - 20 = 0$

g. $2x^2 - x - 2 = 0$ **h.** $3x^2 - 2x - 2 = 0$

Answers to the Concept Quiz

Answers for Problems 1–5 may vary. **1.** D   **2.** B   **3.** A   **4.** C   **5.** D   **6.** B   **7.** A

11.6   Quadratic and Other Nonlinear Inequalities

OBJECTIVES

1. Solve quadratic and polynomial inequalities

2. Solve rational inequalities

We refer to the equation $ax^2 + bx + c = 0$ as the standard form of a quadratic equation in one variable. Similarly, the following forms express **quadratic inequalities** in one variable:

$$ax^2 + bx + c > 0 \qquad ax^2 + bx + c < 0$$
$$ax^2 + bx + c \geq 0 \qquad ax^2 + bx + c \leq 0$$

We can use the number line very effectively to help solve quadratic inequalities for which the quadratic polynomial is factorable. Let's consider some examples to illustrate the procedure.

Classroom Example
Solve and graph the solutions for
$x^2 + 4x - 21 \geq 0$.

EXAMPLE 1   Solve and graph the solutions for $x^2 + 2x - 8 > 0$.

Solution

First, let's factor the polynomial:

$$x^2 + 2x - 8 > 0$$

$$(x + 4)(x - 2) > 0$$

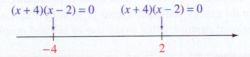

Figure 11.17

On a number line (Figure 11.17), we indicate that at $x = 2$ and $x = -4$, the product $(x + 4) \cdot (x - 2)$ equals zero. The numbers -4 and 2 divide the number line into three intervals: (1) the numbers less than -4, (2) the numbers between -4 and 2, and (3) the numbers greater than 2. We can choose a *test number* from each of these intervals and see how it affects the signs of the factors $x + 4$ and $x - 2$ and, consequently, the sign of the product of these factors. For example, if $x < -4$ (try $x = -5$), then $x + 4$ is negative and $x - 2$ is negative, so their product is positive. If $-4 < x < 2$ (try $x = 0$), then $x + 4$ is positive and $x - 2$ is negative, so their product is negative. If $x > 2$ (try $x = 3$), then $x + 4$ is positive and $x - 2$ is positive, so their product is positive. This information can be conveniently arranged using a number line, as shown in Figure 11.18. Note the open circles at -4 and 2 to indicate that they are not included in the solution set.

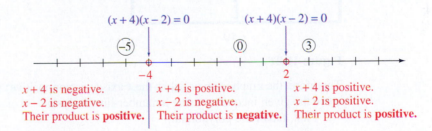

Figure 11.18

Thus the given inequality, $x^2 + 2x - 8 > 0$, is satisfied by numbers less than -4 along with numbers greater than 2. Using interval notation, the solution set is $(-\infty, -4) \cup (2, \infty)$. These solutions can be shown on a number line (Figure 11.19).

Figure 11.19

How can we give graphical support for a problem such as Example 1? Suppose that we graph $y = x^2 + 2x - 8$, as in Figure 11.20. Because $y = x^2 + 2x - 8$, and we want $x^2 + 2x - 8$ to be greater than zero, let's consider the part of the graph above the x axis—in other words, where y is positive. The x intercepts appear to be -4 and 2. Therefore, y is positive when $x < -4$ and also when $x > 2$. Thus the solution set given in Example 1 appears to be correct.

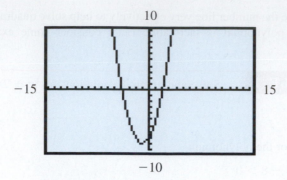

Figure 11.20

We refer to numbers such as -4 and 2 in the preceding example (where the given polynomial or algebraic expression equals zero or is undefined) as **critical numbers**. Let's consider some additional examples that make use of critical numbers and test numbers.

Classroom Example
Solve and graph the solutions for
$x^2 + 3x - 10 < 0$.

EXAMPLE 2   Solve and graph the solutions for $x^2 + 2x - 3 \le 0$.

Solution

For this problem, let's use a graphical approach to *predict* the solution set. Let's graph $y = x^2 + 2x - 3$, as shown in Figure 11.21. The x intercepts appear to be -3 and 1.

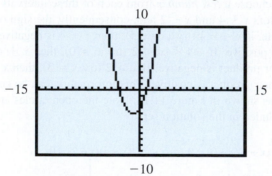

Figure 11.21

Therefore, the graph is on or below the x axis (y is nonpositive) when $-3 \le x \le 1$. Now let's solve the given inequality using a number-line analysis.

First, factor the polynomial:

$$x^2 + 2x - 3 \le 0$$
$$(x + 3)(x - 1) \le 0$$

Second, locate the values for which $(x + 3)(x - 1)$ equals zero. We put dots at -3 and 1 to remind ourselves that these two numbers are to be included in the solution set because the given statement includes equality. Now let's choose a test number from each of the three intervals, and record the sign behavior of the factors $x + 3$ and $x - 1$ (Figure 11.22).

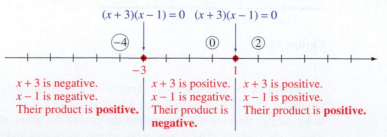

Figure 11.22

Therefore, the solution set is $[-3, 1]$, and it can be graphed as in Figure 11.23.

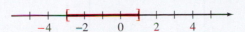

Figure 11.23

Solving Inequalities of Quotients

Examples 1 and 2 have indicated a systematic approach for solving quadratic inequalities when the polynomial is factorable. This same type of number line analysis can also be used to solve indicated quotients such as $\dfrac{x + 1}{x - 5} > 0$.

Classroom Example
Solve and graph the solutions for
$\dfrac{x - 2}{x + 6} \geq 0$.

EXAMPLE 3   Solve and graph the solutions for $\dfrac{x + 1}{x - 5} > 0$.

Solution

First, indicate that at $x = -1$ the given quotient equals zero, and at $x = 5$ the quotient is undefined. Second, choose test numbers from each of the three intervals, and record the sign behavior of $x + 1$ and $x - 5$ as in Figure 11.24.

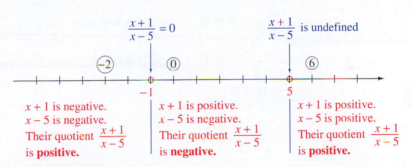

Figure 11.24

Therefore, the solution set is $(-\infty, -1) \cup (5, \infty)$, and its graph is shown in Figure 11.25.

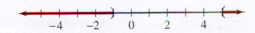

Figure 11.25

Classroom Example
Solve $\dfrac{m + 1}{m + 3} \leq 0$.

EXAMPLE 4   Solve $\dfrac{x + 2}{x + 4} \leq 0$.

Solution

The indicated quotient equals zero at $x = -2$ and is undefined at $x = -4$. (Note that -2 is to be included in the solution set, but -4 is not to be included.) Now let's choose some test numbers and record the sign behavior of $x + 2$ and $x + 4$ as in Figure 11.26.

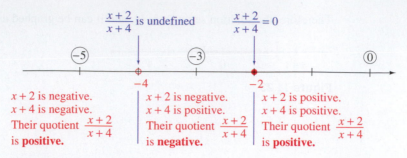

Figure 11.26

Therefore, the solution set is $(-4, -2]$.

The final example illustrates that sometimes we need to change the form of the given inequality before we use the number line analysis.

Classroom Example
Solve $\dfrac{x}{x+4} \leq 2$.

EXAMPLE 5   Solve $\dfrac{x}{x+2} \geq 3$.

Solution

First, let's change the form of the given inequality as follows:

$$\frac{x}{x+2} \geq 3$$

$$\frac{x}{x+2} - 3 \geq 0 \qquad \text{Add } -3 \text{ to both sides}$$

$$\frac{x - 3(x+2)}{x+2} \geq 0 \qquad \text{Express the left side over a common denominator}$$

$$\frac{x - 3x - 6}{x+2} \geq 0$$

$$\frac{-2x - 6}{x+2} \geq 0$$

Now we can proceed as we did with the previous examples. If $x = -3$, then $\dfrac{-2x-6}{x+2}$ equals zero; and if $x = -2$, then $\dfrac{-2x-6}{x+2}$ is undefined. Then, choosing test numbers, we can record the sign behavior of $-2x - 6$ and $x + 2$ as in Figure 11.27.

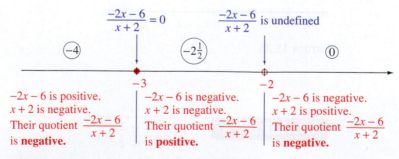

Figure 11.27

Therefore, the solution set is $[-3, -2)$. Perhaps you should check a few numbers from this solution set back into the original inequality!

Concept Quiz 11.6

For Problems 1–10, answer true or false.

1. When solving the inequality $(x + 3)(x - 2) > 0$, we are looking for values of x that make the product of $x + 3$ and $x - 2$ a positive number.

2. The solution set of the inequality $x^2 + 4 > 0$ is all real numbers.

3. The solution set of the inequality $x^2 \leq 0$ is the null set.

4. The critical numbers for the inequality $(x + 4)(x - 1) \leq 0$ are -4 and -1.

5. The number 2 is included in the solution set of the inequality $\dfrac{x + 4}{x - 2} \geq 0$.

6. The solution set of $(x - 2)^2 \geq 0$ is the set of all real numbers.

7. The solution set of $\dfrac{x + 2}{x - 3} \leq 0$ is $(-2, 3)$.

8. The solution set of $\dfrac{x - 1}{x} > 2$ is $(-1, 0)$.

9. The solution set of the inequality $(x - 2)^2(x + 1)^2 < 0$ is $\varnothing$.

10. The solution set of the inequality $(x - 4)(x + 3)^2 \leq 0$ is $(-\infty, 4]$.

Problem Set 11.6

For Problems 1–12, solve each inequality and graph its solution set on a number line. **(Objective 1)**

1. $(x + 2)(x - 1) > 0$
2. $(x - 2)(x + 3) > 0$

3. $(x + 1)(x + 4) < 0$
4. $(x - 3)(x - 1) < 0$

5. $(2x - 1)(3x + 7) \geq 0$
6. $(3x + 2)(2x - 3) \geq 0$

7. $(x + 2)(4x - 3) \leq 0$
8. $(x - 1)(2x - 7) \leq 0$

9. $(x + 1)(x - 1)(x - 3) > 0$

10. $(x + 2)(x + 1)(x - 2) > 0$

11. $x(x + 2)(x - 4) \leq 0$
12. $x(x + 3)(x - 3) \leq 0$

For Problems 13–38, solve each inequality. **(Objective 1)**

13. $x^2 + 2x - 35 < 0$
14. $x^2 + 3x - 54 < 0$

15. $x^2 - 11x + 28 > 0$
16. $x^2 + 11x + 18 > 0$

17. $3x^2 + 13x - 10 \leq 0$
18. $4x^2 - x - 14 \leq 0$

19. $8x^2 + 22x + 5 \geq 0$
20. $12x^2 - 20x + 3 \geq 0$

21. $x(5x - 36) > 32$
22. $x(7x + 40) < 12$

23. $x^2 - 14x + 49 \geq 0$
24. $(x + 9)^2 \geq 0$

25. $4x^2 + 20x + 25 \leq 0$
26. $9x^2 - 6x + 1 \leq 0$

27. $(x + 1)(x - 3)^2 > 0$
28. $(x - 4)^2(x - 1) \leq 0$

29. $4 - x^2 < 0$
30. $2x^2 - 18 \geq 0$

31. $4(x^2 - 36) < 0$
32. $-4(x^2 - 36) \geq 0$

33. $5x^2 + 20 > 0$
34. $-3x^2 - 27 \geq 0$

35. $x^2 - 2x \geq 0$
36. $2x^2 + 6x < 0$

37. $3x^3 + 12x^2 > 0$
38. $2x^3 + 4x^2 \leq 0$

For Problems 39–56, solve each inequality. **(Objective 2)**

39. $\dfrac{x + 1}{x - 2} > 0$
40. $\dfrac{x - 1}{x + 2} > 0$

41. $\dfrac{x - 3}{x + 2} < 0$
42. $\dfrac{x + 2}{x - 4} < 0$

43. $\dfrac{2x - 1}{x} \geq 0$
44. $\dfrac{x}{3x + 7} \geq 0$

45. $\dfrac{-x + 2}{x - 1} \leq 0$
46. $\dfrac{3 - x}{x + 4} \leq 0$

47. $\dfrac{2x}{x + 3} > 4$
48. $\dfrac{x}{x - 1} > 2$

49. $\dfrac{x - 1}{x - 5} \leq 2$
50. $\dfrac{x + 2}{x + 4} \leq 3$

51. $\dfrac{x + 2}{x - 3} > -2$
52. $\dfrac{x - 1}{x - 2} < -1$

53. $\dfrac{3x + 2}{x + 4} \leq 2$
54. $\dfrac{2x - 1}{x + 2} \geq -1$

55. $\dfrac{x + 1}{x - 2} < 1$
56. $\dfrac{x + 3}{x - 4} \geq 1$

Thoughts Into Words

57. Explain how to solve the inequality $(x + 1)(x - 2) \cdot (x - 3) > 0$.

58. Explain how to solve the inequality $(x - 2)^2 > 0$ by inspection.

59. Your friend looks at the inequality $1 + \dfrac{1}{x} > 2$ and, without any computation, states that the solution set is all real numbers between 0 and 1. How can she do that?

60. Why is the solution set for $(x - 2)^2 \geq 0$ the set of all real numbers?

61. Why is the solution set for $(x - 2)^2 \leq 0$ the set $\{2\}$?

Further Investigations

62. The product $(x - 2)(x + 3)$ is positive if both factors are negative *or* if both factors are positive. Therefore, we can solve $(x - 2)(x + 3) > 0$ as follows:

$(x - 2 < 0 \text{ and } x + 3 < 0)$ or $(x - 2 > 0 \text{ and } x + 3 > 0)$

$(x < 2 \text{ and } x < -3)$ or $(x > 2 \text{ and } x > -3)$

$x < -3$ or $x > 2$

The solution set is $(-\infty, -3) \cup (2, \infty)$. Use this type of analysis to solve each of the following.

(a) $(x - 2)(x + 7) > 0$ **(b)** $(x - 3)(x + 9) \geq 0$

(c) $(x + 1)(x - 6) \leq 0$ **(d)** $(x + 4)(x - 8) < 0$

(e) $\dfrac{x + 4}{x - 7} > 0$ **(f)** $\dfrac{x - 5}{x + 8} \leq 0$

Graphing Calculator Activities

63. Use your graphing calculator to give visual support for our answers for Examples 3, 4, and 5.

64. Use your graphing calculator to give visual support for your answers for Problems 30–38.

65. Use your graphing calculator to help determine the solution sets for the following inequalities. Express the solution sets in interval notation.

a. $x^2 + x - 2 \geq 0$ **b.** $(3 - x)(x + 4) < 0$

c. $x^3 - 3x^2 - x + 3 > 0$ **d.** $-x^3 + 7x - 6 \leq 0$

e. $x^3 - 21x - 20 < 0$

Answers to the Concept Quiz

1. True   **2.** True   **3.** False   **4.** False   **5.** False   **6.** True   **7.** False   **8.** True   **9.** True   **10.** True

Chapter 11   Summary

OBJECTIVE	SUMMARY	EXAMPLE
Know the set of complex numbers. (Section 11.1/ Objective 1)	A number of the form $a + bi$, where a and b are real numbers and i is the imaginary unit defined by $i = \sqrt{-1}$, is a complex number. Two complex numbers are said to be equal if and only if $a = c$ and $b = d$.	
Add and subtract complex numbers. (Section 11.1/ Objective 2)	We describe the addition and subtraction of complex numbers as follows: $(a + bi) + (c + di) = (a + c) + (b + d)i$ and $(a + bi) - (c + di) = (a - c) + (b - d)i$	Add the complex numbers: $(3 - 6i) + (-7 - 3i)$ **Solution** $(3 - 6i) + (-7 - 3i)$ $= (3 - 7) + (-6 - 3)i$ $= -4 - 9i$
Simplify radicals involving negative numbers. (Section 11.1/Objective 3)	We can represent a square root of any negative real number as the product of a real number and the imaginary unit i. That is, $\sqrt{-b} = i\sqrt{b}$, where b is a positive real number.	Write $\sqrt{-48}$ in terms of i and simplify. **Solution** $\sqrt{-48} = \sqrt{-1}\sqrt{48}$ $= i\sqrt{16}\sqrt{3}$ $= 4i\sqrt{3}$
Perform operations on radicals involving negative numbers. (Section 11.1/Objective 4)	Before performing any operations, represent the square root of any negative real number as the product of a real number and the imaginary unit i.	Perform the indicated operation and simplify: $\dfrac{\sqrt{-28}}{\sqrt{-4}}$ **Solution** $\dfrac{\sqrt{-28}}{\sqrt{-4}} = \dfrac{i\sqrt{28}}{i\sqrt{4}} = \dfrac{\sqrt{28}}{\sqrt{4}} = \sqrt{\dfrac{28}{4}} = \sqrt{7}$
Multiply complex numbers. (Section 11.1/Objective 5)	The product of two complex numbers follows the same pattern as the product of two binomials. The conjugate of $a + bi$ is $a - bi$. The product of a complex number and its conjugate is a real number. When simplifying, replace any i^2 with -1.	Find the product $(2 + 3i)(4 - 5i)$ and express the answer in standard form of a complex number. **Solution** $(2 + 3i)(4 - 5i) = 8 + 2i - 15i^2$ $= 8 + 2i - 15(-1)$ $= 23 + 2i$
Divide complex numbers. (Section 11.1/Objective 6)	To simplify expressions that indicate the quotient of complex numbers, such as $\dfrac{4 + 3i}{5 - 2i}$, multiply the numerator and denominator by the conjugate of the denominator.	Find the quotient $\dfrac{2 + 3i}{4 - i}$ and express the answer in standard form of a complex number. **Solution** Multiply the numerator and denominator by $4 + i$, the conjugate of the denominator. $\dfrac{2 + 3i}{4 - i} = \dfrac{(2 + 3i)}{(4 - i)} \cdot \dfrac{(4 + i)}{(4 + i)}$ $= \dfrac{8 + 14i + 3i^2}{16 - i^2}$ $= \dfrac{8 + 14i + 3(-1)}{16 - (-1)}$ $= \dfrac{5 + 14i}{17} = \dfrac{5}{17} + \dfrac{14}{17}i$

(continued)

OBJECTIVE	SUMMARY	EXAMPLE
Solve quadratic equations by factoring. (Section 11.2/Objective 1)	The standard form for a quadratic equation in one variable is $ax^2 + bx + c = 0$, where a, b, and c are real numbers and $a \neq 0$. Some quadratic equations can be solved by factoring and applying the property, $ab = 0$ if and only if $a = 0$ or $b = 0$.	Solve $2x^2 + x - 3 = 0$. **Solution** $$2x^2 + x - 3 = 0$$ $$(2x + 3)(x - 1) = 0$$ $2x + 3 = 0 \qquad$ or $\qquad x - 1 = 0$ $$x = -\frac{3}{2} \quad \text{or} \quad x = 1$$ The solution set is $\left\{ -\frac{3}{2}, 1 \right\}$.
Solve quadratic equations of the form $x^2 = a$. (Section 11.2/Objective 2)	We can solve some quadratic equations by applying the property, $x^2 = a$ if and only if $x = \pm\sqrt{a}$.	Solve $3(x + 7)^2 = 24$. **Solution** $$3(x + 7)^2 = 24$$ First divide both sides of the equation by 3: $$(x + 7)^2 = 8$$ $$x + 7 = \pm\sqrt{8}$$ $$x + 7 = \pm 2\sqrt{2}$$ $$x = -7 \pm 2\sqrt{2}$$ The solution set is $\{-7 \pm 2\sqrt{2}\}$.
Solve quadratic equations by completing the square. (Section 11.3/Objective 1)	To solve a quadratic equation by completing the square, first put the equation in the form $x^2 + bx = k$. Then (1) take one-half of b, square that result, and add to each side of the equation; (2) factor the left side; and (3) apply the property, $x^2 = a$ if and only if $x = \pm\sqrt{a}$.	Solve $x^2 + 12x - 2 = 0$. **Solution** $$x^2 + 12x - 2 = 0$$ $$x^2 + 12x = 2$$ $$x^2 + 12x + 36 = 2 + 36$$ $$(x + 6)^2 = 38$$ $$x + 6 = \pm\sqrt{38}$$ $$x = -6 \pm \sqrt{38}$$ The solution set is $\{-6 \pm \sqrt{38}\}$.
Use the quadratic formula to solve quadratic equations. (Section 11.4/Objective 1)	Any quadratic equation of the form $ax^2 + bx + c = 0$, where $a \neq 0$, can be solved by the quadratic formula, which is usually stated as $$x = \frac{-b \pm \sqrt{b^2 - 4ac}}{2a}$$	Solve $3x^2 - 5x - 6 = 0$. **Solution** $$3x^2 - 5x - 6 = 0$$ $a = 3, b = -5,$ and $c = -6$ $$x = \frac{-(-5) \pm \sqrt{(-5)^2 - 4(3)(-6)}}{2(3)}$$ $$x = \frac{5 \pm \sqrt{97}}{6}$$ The solution set is $\left\{ \dfrac{5 \pm \sqrt{97}}{6} \right\}$.

OBJECTIVE	SUMMARY	EXAMPLE
Determine the nature of roots to quadratic equations. (Section 11.4/Objective 2)	The discriminant $b^2 - 4ac$ can be used to determine the nature of the roots of a quadratic equation. 1. If $b^2 - 4ac$ is less than zero, then the equation has two nonreal complex solutions. 2. If $b^2 - 4ac$ is equal to zero, then the equation has two equal real solutions. 3. If $b^2 - 4ac$ is greater than zero, then the equation has two unequal real solutions.	Use the discriminant to determine the nature of the solutions for the equation $2x^2 + 3x + 5 = 0$. **Solution** $2x^2 + 3x + 5 = 0$ For $a = 2$, $b = 3$, and $c = 5$, $b^2 - 4ac = (3)^2 - 4(2)(5) = -31$. Because the discriminant is less than zero, the equation has two nonreal complex solutions.
Solve quadratic equations by selecting the most appropriate method. (Section 11.5/Objective 1)	There are three major methods for solving a quadratic equation. 1. Factoring 2. Completing the square 3. Quadratic formula Consider which method is most appropriate before you begin solving the equation.	Solve $x^2 - 4x + 9 = 0$. **Solution** This equation does not factor. This equation can easily be solved by completing the square, because $a = 1$ and b is an even number. $x^2 - 4x + 9 = 0$ $x^2 - 4x = -9$ $x^2 - 4x + 4 = -9 + 4$ $(x + 4)^2 = -5$ $x + 4 = \pm\sqrt{-5}$ $x = -4 \pm i\sqrt{5}$ The solution set is $\{-4 \pm i\sqrt{5}\}$.
Solve problems pertaining to right triangles and $30°-60°$ triangles. (Section 11.2/Objective 3)	There are two special kinds of right triangles that are used in later mathematics courses. The **isosceles right triangle** is a right triangle that has both legs of the same length. In a $\mathbf{30°-60°}$ **right triangle**, the side opposite the 30° angle is equal in length to one-half the length of the hypotenuse.	Find the length of each leg of an isosceles right triangle that has a hypotenuse of length 6 inches. **Solution** Let x represent the length of each leg: $x^2 + x^2 = 6^2$ $2x^2 = 36$ $x^2 = 18$ $x = \pm\sqrt{18} = \pm 3\sqrt{2}$ Disregard the negative solution. The length of each leg is $3\sqrt{2}$ inches.
Solve word problems involving quadratic equations. (Section 11.5/Objective 3)	Keep the following suggestions in mind as you solve word problems. 1. Read the problem carefully. 2. Sketch any figure, diagram, or chart that might help you organize and analyze the problem. 3. Choose a meaningful variable. 4. Look for a guideline that can be used to set up an equation.	Find two consecutive odd whole numbers such that the sum of their squares is 290. **Solution** Let x represent the first whole number. Then $x + 2$ would represent the next consecutive odd whole number. $x^2 + (x + 2)^2 = 290$ $x^2 + x^2 + 4x + 4 = 290$ $2x^2 + 4x - 286 = 0$

(continued)

OBJECTIVE	SUMMARY	EXAMPLE
	5. Form an equation that translates the guideline from English into algebra. **6.** Solve the equation and answer the question posed in the problem. **7.** Check all answers back into the original statement of the problem.	$2(x^2 + 2x - 143) = 0$ $2(x + 13)(x - 11) = 0$ $x = -13 \quad$ or $\quad x = 11$ Disregard the solution of -13 because it is not a whole number. The whole numbers are 11 and 13.
Solve quadratic or polynomial inequalities. (Section 11.6/Objective 1)	To solve quadratic inequalities that are factorable polynomials, the critical numbers are found by factoring the polynomial. The critical numbers partition the number line into regions. A test point from each region is used to determine if the values in that region make the inequality a true statement. The answer is usually expressed in interval notation.	Solve $x^2 + x - 6 \leq 0$. **Solution** Solve the equation $x^2 + x - 6 = 0$ to find the critical numbers. $$x^2 + x - 6 = 0$$ $$(x + 3)(x - 2) = 0$$ $$x = -3 \quad \text{or} \quad x = 2$$ The critical numbers are -3 and 2. Choose a test point from each of the intervals $(-\infty, -3)$, $(-3, 2)$, and $(2, \infty)$. Evaluating the inequality $x^2 + x - 6 \leq 0$ for each of the test points shows that $(-3, 2)$ is the only interval of values that makes the inequality a true statement. Because the inequality includes the endpoints of the interval, the solution is $[-3, 2]$.
Solve rational inequalities. (Section 11.6/Objective 2)	To solve inequalities involving rational expressions, use the same basic approach as for solving quadratic equations. Be careful to avoid any values that make the denominator zero.	Solve $\dfrac{x + 1}{2x - 3} \geq 0$. **Solution** Set the numerator equal to zero and then set the denominator equal to zero to find the critical numbers. $$x + 1 = 0 \quad \text{and} \quad 2x - 3 = 0$$ $$x = -1 \quad \text{and} \quad x = \frac{3}{2}$$ The critical numbers are -1 and $\dfrac{3}{2}$. Evaluate the inequality with a test point from each of the intervals $(-\infty, -1)$, $\left(-1, \dfrac{3}{2}\right)$, and $\left(\dfrac{3}{2}, \infty\right)$; this shows that the values in the intervals $(-\infty, -1)$ and $\left(\dfrac{3}{2}, \infty\right)$ make the inequality a true statement. Because the inequality includes the "equal to" statement, the solution should include -1 but not $\dfrac{3}{2}$, because $\dfrac{3}{2}$ would make the quotient undefined. The solution set is $(-\infty, -1] \cup \left(\dfrac{3}{2}, \infty\right)$.

Chapter 11   Review Problem Set

For Problems 1–4, perform the indicated operations and express the answers in the standard form of a complex number.

1. $(-7 + 3i) + (9 - 5i)$ **2.** $(4 - 10i) - (7 - 9i)$

3. $(6 - 3i) - (-2 + 5i)$ **4.** $(-4 + i) + (2 + 3i)$

For Problems 5–8, write each expression in terms of i and simplify.

5. $\sqrt{-8}$ **6.** $\sqrt{-25}$

7. $3\sqrt{-16}$ **8.** $2\sqrt{-18}$

For Problems 9–18, perform the indicated operation and simplify.

9. $\sqrt{-2}\sqrt{-6}$ **10.** $\sqrt{-2}\sqrt{18}$

11. $\dfrac{\sqrt{-42}}{\sqrt{-6}}$ **12.** $\dfrac{\sqrt{-6}}{\sqrt{2}}$

13. $5i(3 - 6i)$ **14.** $(5 - 7i)(6 + 8i)$

15. $(-2 - 3i)(4 - 8i)$ **16.** $(4 - 3i)(4 + 3i)$

17. $\dfrac{4 + 3i}{6 - 2i}$ **18.** $\dfrac{-1 - i}{-2 + 5i}$

For Problems 19 and 20, perform the indicated operations and express the answer in the standard form of a complex number.

19. $\dfrac{3 + 4i}{2i}$ **20.** $\dfrac{-6 + 5i}{-i}$

For Problems 21–24, solve each of the quadratic equations by factoring.

21. $x^2 + 8x = 0$ **22.** $x^2 = 6x$

23. $x^2 - 3x - 28 = 0$ **24.** $2x^2 + x - 3 = 0$

For Problems 25–28, use Property 11.1 to help solve each quadratic equation.

25. $2x^2 = 90$ **26.** $(y - 3)^2 = -18$

27. $(2x + 3)^2 = 24$ **28.** $a^2 - 27 = 0$

For Problems 29–32, use the method of completing the square to solve the quadratic equation.

29. $y^2 + 18y - 10 = 0$ **30.** $n^2 + 6n + 20 = 0$

31. $x^2 - 10x + 1 = 0$ **32.** $x^2 + 5x - 2 = 0$

For Problems 33–36, use the quadratic formula to solve the equation.

33. $x^2 + 6x + 4 = 0$ **34.** $x^2 + 4x + 6 = 0$

35. $3x^2 - 2x + 4 = 0$ **36.** $5x^2 - x - 3 = 0$

For Problems 37–40, find the discriminant of each equation and determine whether the equation has (1) two nonreal complex solutions, (2) one real solution with a multiplicity of 2, or (3) two real solutions. Do not solve the equations.

37. $4x^2 - 20x + 25 = 0$ **38.** $5x^2 - 7x + 31 = 0$

39. $7x^2 - 2x - 14 = 0$ **40.** $5x^2 - 2x = 4$

For Problems 41–59, solve each equation.

41. $x^2 - 17x = 0$ **42.** $(x - 2)^2 = 36$

43. $(2x - 1)^2 = -64$ **44.** $x^2 - 4x - 21 = 0$

45. $x^2 + 2x - 9 = 0$ **46.** $x^2 - 6x = -34$

47. $4\sqrt{x} = x - 5$ **48.** $3n^2 + 10n - 8 = 0$

49. $n^2 - 10n = 200$ **50.** $3a^2 + a - 5 = 0$

51. $x^2 - x + 3 = 0$ **52.** $2x^2 - 5x + 6 = 0$

53. $2a^2 + 4a - 5 = 0$ **54.** $t(t + 5) = 36$

55. $x^2 + 4x + 9 = 0$ **56.** $(x - 4)(x - 2) = 80$

57. $\dfrac{3}{x} + \dfrac{2}{x + 3} = 1$ **58.** $2x^4 - 23x^2 + 56 = 0$

59. $\dfrac{3}{n - 2} = \dfrac{n + 5}{4}$

For Problems 60–70, set up an equation and solve each problem.

60. The wing of an airplane is in the shape of a $30°-60°$ right triangle. If the side opposite the 30° angle measures 20 feet, find the measure of the other two sides of the wing. Round the answers to the nearest tenth of a foot.

61. An agency is using photo surveillance of a rectangular plot of ground that measures 40 meters by 25 meters. If during the surveillance, someone is observed moving from one corner of the plot to the corner diagonally opposite, how far has the observed person moved? Round the answer to the nearest tenth of a meter.

62. One leg of an isosceles right triangle measures 4 inches. Find the length of the hypotenuse of the triangle. Express the answer in radical form.

63. Find two numbers whose sum is 6 and whose product is 2.

64. A landscaper agreed to design and plant a flower bed for $40. It took him three hours less than he anticipated, and therefore he earned $3 per hour more than he anticipated. How long did he expect it would take to design and plant the flower bed?

65. Andre traveled 270 miles in 1 hour more than it took Sandy to travel 260 miles. Sandy drove 7 miles per hour faster than Andre. How fast did each one travel?

66. The area of a square is numerically equal to twice its perimeter. Find the length of a side of the square.

67. Find two consecutive even whole numbers such that the sum of their squares is 164.

68. The perimeter of a rectangle is 38 inches, and its area is 84 square inches. Find the length and width of the rectangle.

69. It takes Billy 2 hours longer to do a certain job than it takes Reena. They worked together for 2 hours; then Reena left, and Billy finished the job in 1 hour. How long would it take each of them to do the job alone?

70. A company has a rectangular parking lot 40 meters wide and 60 meters long. The company plans to increase the area of the lot by 1100 square meters by adding a strip of equal width to one side and one end. Find the width of the strip to be added.

For Problems 71–78, solve each inequality and express the solution set using interval notation.

71. $x^2 + 3x - 10 > 0$ **72.** $2x^2 + x - 21 \leq 0$

73. $4x^2 - 1 \leq 0$ **74.** $x^2 - 7x + 10 > 0$

75. $\dfrac{x - 4}{x + 6} \geq 0$ **76.** $\dfrac{2x - 1}{x + 1} > 4$

77. $\dfrac{3x + 1}{x - 4} < 2$ **78.** $\dfrac{3x + 1}{x - 1} \leq 0$

1. Find the product $(3 - 4i)(5 + 6i)$, and express the result in the standard form of a complex number.

2. Find the quotient $\dfrac{2 - 3i}{3 + 4i}$, and express the result in the standard form of a complex number.

For Problems 3–15, solve each equation.

3. $x^2 = 7x$

4. $(x - 3)^2 = 16$

5. $x^2 + 3x - 18 = 0$

6. $x^2 - 2x - 1 = 0$

7. $5x^2 - 2x + 1 = 0$

8. $x^2 + 30x = -224$

9. $(3x - 1)^2 + 36 = 0$

10. $(5x - 6)(4x + 7) = 0$

11. $(2x + 1)(3x - 2) = 55$

12. $n(3n - 2) = 40$

13. $x^4 + 12x^2 - 64 = 0$

14. $\dfrac{3}{x} + \dfrac{2}{x + 1} = 4$

15. $3x^2 - 2x - 3 = 0$

16. Does the equation $4x^2 + 20x + 25 = 0$ have (a) two nonreal complex solutions, (b) two equal real solutions, or (c) two unequal real solutions?

17. Does the equation $4x^2 - 3x = -5$ have (a) two nonreal complex solutions, (b) two equal real solutions, or (c) two unequal real solutions?

For Problems 18–20, solve each inequality and express the solution set using interval notation.

18. $x^2 - 3x - 54 \le 0$

19. $\dfrac{3x - 1}{x + 2} > 0$

20. $\dfrac{x - 2}{x + 6} \ge 3$

For Problems 21–25, set up an equation and solve each problem.

21. A 24-foot ladder leans against a building and makes an angle of 60° with the ground. How far up on the building does the top of the ladder reach? Express your answer to the nearest tenth of a foot.

22. A rectangular plot of ground measures 16 meters by 24 meters. Find, to the nearest meter, the distance from one corner of the plot to the diagonally opposite corner.

23. Amy agreed to clean her brother's room for $36. It took her 1 hour longer than she expected, and therefore she earned $3 per hour less than she anticipated. How long did she expect it would take to clean the room?

24. The perimeter of a rectangle is 41 inches, and its area is 91 square inches. Find the length of its shortest side.

25. The sum of two numbers is 6, and their product is 4. Find the larger of the two numbers.

For Problems 1–5, evaluate each algebraic expression for the given values of the variables.

1. $\dfrac{4a^2b^3}{12a^3b}$ for $a = 5$ and $b = -8$

2. $\dfrac{\dfrac{1}{x} + \dfrac{1}{y}}{\dfrac{1}{x} - \dfrac{1}{y}}$ for $x = 4$ and $y = 7$

3. $\dfrac{3}{n} + \dfrac{5}{2n} - \dfrac{4}{3n}$ for $n = 25$

4. $\dfrac{4}{x-1} - \dfrac{2}{x+2}$ for $x = \dfrac{1}{2}$

5. $2\sqrt{2x+y} - 5\sqrt{3x-y}$ for $x = 5$ and $y = 6$

For Problems 6–17, perform the indicated operations and express the answers in simplified form.

6. $(3a^2b)(-2ab)(4ab^3)$

7. $(x+3)(2x^2 - x - 4)$

8. $\dfrac{6xy^2}{14y} \cdot \dfrac{7x^2y}{8x}$

9. $\dfrac{a^2 + 6a - 40}{a^2 - 4a} \div \dfrac{2a^2 + 19a - 10}{a^3 + a^2}$

10. $\dfrac{3x+4}{6} - \dfrac{5x-1}{9}$

11. $\dfrac{4}{x^2 + 3x} + \dfrac{5}{x}$

12. $\dfrac{3n^2 + n}{n^2 + 10n + 16} \cdot \dfrac{2n^2 - 8}{3n^3 - 5n^2 - 2n}$

13. $\dfrac{3}{5x^2 + 3x - 2} - \dfrac{2}{5x^2 - 22x + 8}$

14. $\dfrac{y^3 - 7y^2 + 16y - 12}{y - 2}$

15. $(4x^3 - 17x^2 + 7x + 10) \div (4x - 5)$

16. $(3\sqrt{2} + 2\sqrt{5})(5\sqrt{2} - \sqrt{5})$

17. $(\sqrt{x} - 3\sqrt{y})(2\sqrt{x} + 4\sqrt{y})$

For Problems 18–25, evaluate each of the numerical expressions.

18. $-\sqrt{\dfrac{9}{64}}$

19. $\sqrt[3]{-\dfrac{8}{27}}$

20. $\sqrt[3]{0.008}$

21. $32^{-\frac{1}{5}}$

22. $3^0 + 3^{-1} + 3^{-2}$

23. $-9^{\frac{3}{2}}$

24. $\left(\dfrac{3}{4}\right)^{-2}$

25. $\dfrac{1}{\left(\dfrac{2}{3}\right)^{-3}}$

For Problems 26–31, factor each of the algebraic expressions completely.

26. $3x^4 + 81x$

27. $6x^2 + 19x - 20$

28. $12 + 13x - 14x^2$

29. $9x^4 + 68x^2 - 32$

30. $2ax - ay - 2bx + by$

31. $27x^3 - 8y^3$

For Problems 32–55, solve each of the equations.

32. $3(x-2) - 2(3x+5) = 4(x-1)$

33. $0.06n + 0.08(n+50) = 25$

34. $4\sqrt{x} + 5 = x$

35. $\sqrt[3]{n^2 - 1} = -1$

36. $6x^2 - 24 = 0$

37. $a^2 + 14a + 49 = 0$

38. $3n^2 + 14n - 24 = 0$

39. $\dfrac{2}{5x-2} = \dfrac{4}{6x+1}$

40. $\sqrt{2x-1} - \sqrt{x+2} = 0$

41. $5x - 4 = \sqrt{5x-4}$

42. $|3x - 1| = 11$

43. $(3x - 2)(4x - 1) = 0$

44. $(2x + 1)(x - 2) = 7$

45. $\dfrac{5}{6x} - \dfrac{2}{3} = \dfrac{7}{10x}$

46. $\dfrac{3}{y+4} + \dfrac{2y-1}{y^2 - 16} = \dfrac{-2}{y-4}$

47. $6x^4 - 23x^2 - 4 = 0$

48. $3n^3 + 3n = 0$

49. $n^2 - 13n - 114 = 0$

50. $12x^2 + x - 6 = 0$

51. $x^2 - 2x + 26 = 0$

52. $(x + 2)(x - 6) = -15$

53. $(3x - 1)(x + 4) = 0$

54. $x^2 + 4x + 20 = 0$

55. $2x^2 - x - 4 = 0$

For Problems 56–65, solve each inequality and express the solution set using interval notation.

56. $6 - 2x \geq 10$

57. $4(2x - 1) < 3(x + 5)$

58. $\dfrac{n + 1}{4} + \dfrac{n - 2}{12} > \dfrac{1}{6}$

59. $|2x - 1| < 5$

60. $|3x + 2| > 11$

61. $\dfrac{1}{2}(3x - 1) - \dfrac{2}{3}(x + 4) \leq \dfrac{3}{4}(x - 1)$

62. $x^2 - 2x - 8 \leq 0$

63. $3x^2 + 14x - 5 > 0$

64. $\dfrac{x + 2}{x - 7} \geq 0$

65. $\dfrac{2x - 1}{x + 3} < 1$

66. Find the slope of the line determined by the points $(2, 7)$ and $(-1, -6)$.

67. Find the slope of the line determined by the equation $-3x + 5y = -2$.

68. Write the equation of the line that has a slope of $-\dfrac{3}{11}$ and contains the point $(4, -2)$.

69. Write the equation of the line that is perpendicular to the line $-2x + 5y = -12$ and contains the point $(1, 2)$.

For Problems 70–72, solve each system.

70. $\begin{pmatrix} 4x - 9y = 14 \\ x + 5y = -11 \end{pmatrix}$

71. $\begin{pmatrix} \dfrac{1}{2}x - \dfrac{1}{3}y = 7 \\ \dfrac{3}{4}x + \dfrac{2}{3}y = 0 \end{pmatrix}$

72. $\begin{pmatrix} 7x - 2y = 16 \\ 3x + 5y = 42 \end{pmatrix}$

73. Find the product $(-6x^{-3}y^{-4})(5xy^6)$, and express the result using positive exponents only.

74. Express each of the following in simplest radical form.

 a. $4\sqrt{54}$

 b. $2\sqrt[3]{54}$

 c. $\dfrac{3\sqrt{6}}{4\sqrt{10}}$

 d. $\sqrt{24x^3y^5}$

75. Express each of the following in scientific notation.

 a. 7652

 b. 0.000026

 c. 1.414

 d. 1000

For Problems 76–89, solve each problem by setting up and solving an appropriate equation, inequality, or system of equations.

76. How many liters of a 60% acid solution must be added to 14 liters of a 10% acid solution to produce a 25% acid solution?

77. A sum of $2250 is to be divided between two people in the ratio of 2 to 3. How much does each person receive?

78. The length of a picture without its border is 7 inches less than twice its width. If the border is 1 inch wide and its area is 62 square inches, what are the dimensions of the picture alone?

79. Working together, Lolita and Doug can paint a shed in 3 hours and 20 minutes. If Doug can paint the shed by himself in 10 hours, how long would it take Lolita to paint the shed by herself?

80. Angie bought some golf balls for $14. If each ball had cost $0.25 less, she could have purchased one more ball for the same amount of money. How many golf balls did Angie buy?

81. A jogger who can run an 8-minute mile starts a half-mile ahead of a jogger who can run a 6-minute mile. How long will it take the faster jogger to catch the slower jogger?

82. Suppose that $1000 is invested at a certain rate of interest compounded annually for 2 years. If the accumulated value at the end of 2 years is $1149.90, find the rate of interest.

83. A theater contains 120 chairs arranged in rows. The number of chairs per row is one less than twice the number of rows. Find the number of chairs per row.

84. Bjorn bought a number of shares of stock for $2800. A month later the value of the stock had increased $6 per share, and he sold all but 60 shares and regained his original investment of $2800. How many shares did he sell?

85. One angle of a triangle has a measure of $40°$, and the measures of the other two angles are in a ratio of 3 to 4. Find the measures of the other two angles.

86. The supplement of an angle is $10°$ more than five times its complement. Find the measure of the angle.

87. Megan has $5000 to invest. If she invests $3000 at 7% interest, at what rate must she invest the other $2000 so that the two investments together yield more than $380 in yearly interest?

88. Wilma went to the local market and bought 3 lemons and 5 oranges for $2.40. The same day and for the same prices, Fran bought 4 lemons and 7 oranges for $3.31. What was the price per lemon and the price per orange?

89. Larry has an oil painting that he bought for $700 and now wants to sell. How much more profit could he gain by fixing a 30% profit based on the selling price rather than a 30% profit based on the cost?

12

Coordinate Geometry: Lines, Parabolas, Circles, Ellipses, and Hyperbolas

René Descartes, a philosopher and mathematician, developed a system for locating a point on a plane. This system is our current rectangular coordinate grid used for graphing and is named the Cartesian coordinate system.

The graph in Figure 12.1 shows the number of gallons of lemonade sold each day at Marci's Little Lemonade Stand, depending on how high the temperature was each day. The owner of the stand would like to determine an algebraic equation that shows the relationship between the daily temperature and the number of gallons of lemonade sold each day.

The owner of Marci's Little Lemonade Stand decides that the relationship is linear and knows that two pairs of values can determine the equation for the relationship. For example, 20 gallons were sold when the temperature was 87°, and 60 gallons were sold when the temperature was 95°. Therefore it can be determined that the equation $y = 5x - 415$ describes the relationship in which x is the daily high temperature, and y is the number of gallons sold. So from the geometric information in Figure 12.1, we are able to determine an algebraic equation for the relationship.

It was René Descartes who connected algebraic and geometric ideas to found the branch of mathematics called analytic geometry—today more commonly called coordinate geometry. Basically, there are two kinds of problems in coordinate geometry: Given an algebraic equation, find its geometric graph;

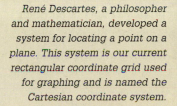 Video tutorials based on section learning objectives are available in a variety of delivery modes.

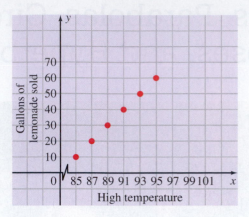

Figure 12.1

and given a set of conditions pertaining to a geometric graph, find its algebraic equation. We discuss problems of both types in this chapter, but our strong emphasis will be on developing graphing techniques.

With the graphing techniques developed in this chapter we will graph parabolas, circles, ellipses, and hyperbolas; we often refer to these curves as **conic sections**. Conic sections can be formed when a plane intersects a conical surface as shown in Figure 12.2. A flashlight produces a "cone of light" that can be cut by the plane of a wall to illustrate the conic sections. Try shining a flashlight against a wall at different angles to produce a circle, an ellipse, a parabola, and one branch of a hyperbola. (You may find it difficult to distinguish between a parabola and a branch of a hyperbola.)

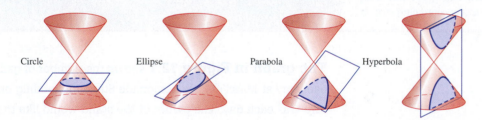

Circle   Ellipse   Parabola   Hyperbola

Figure 12.2

12.1   Distance, Slope, and Graphing Techniques

OBJECTIVES

1 Find the distance between two points

2 Determine the slope of a line

3 Determine the type of symmetries for the graph of an equation

4 Graph equations using intercepts, symmetries, and plotting points

We introduced the rectangular coordinate system in Chapter 5. Most of our work at that time concentrated on graphing linear equations (straight line graphs), including a graphical approach to solving systems of two linear equations. Then in Chapter 9 we began to use graphs occasionally to give some visual support for an algebraic computation. These graphs

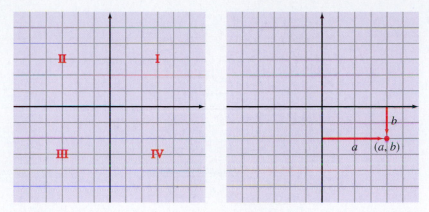

Figure 12.3   **Figure 12.4**

were calculator-generated, and no attempt was made to introduce any graphing techniques for sketching graphs. Now in this chapter, we will begin to develop some specific techniques to aid in the free-hand sketching of graphs. Let's begin by briefly reviewing some basic ideas pertaining to the rectangular coordinate system.

Consider two number lines, one vertical and one horizontal, perpendicular to each other at the point we associate with zero on both lines (Figure 12.3). We refer to these number lines as the **horizontal and vertical axes** or, together, as the **coordinate axes.** They partition the plane into four regions called **quadrants.** The quadrants are numbered counterclockwise from I through IV, as indicated in Figure 12.3. The point of intersection of the two axes is called the **origin.**

In general we refer to the real numbers a and b in an ordered pair (a, b), associated with a point, as the **coordinates of the point.** The first number, a, called the **abscissa,** is the directed distance of the point from the vertical axis measured parallel to the horizontal axis. The second number, b, called the **ordinate,** is the directed distance of the point from the horizontal axis measured parallel to the vertical axis (Figure 12.4). This system of associating points in a plane with pairs of real numbers is called the **rectangular coordinate system** or the **Cartesian coordinate system.**

Distance between Two Points

As we work with the rectangular coordinate system, it is sometimes necessary to express the length of certain line segments. In other words, we need to be able to find the distance between two points. Let's first consider two specific examples and then develop the general distance formula.

EXAMPLE 1

Find the distance between the points $A(2, 2)$ and $B(5, 2)$ and also between the points $C(-2, 5)$ and $D(-2, -4)$.

Solution

Let's plot the points and draw $\overline{AB}$ as in Figure 12.5. Because $\overline{AB}$ is parallel to the x axis, its length can be expressed as $|5 - 2|$ or $|2 - 5|$. (The absolute-value symbol is used to ensure a nonnegative value.) Thus the length of $\overline{AB}$ is 3 units. Likewise, the length of $\overline{CD}$ is $|5 - (-4)| = |-4 - 5| = 9$ units.

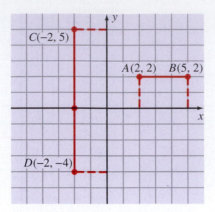

Figure 12.5

EXAMPLE 2 Find the distance between the points $A(2, 3)$ and $B(5, 7)$.

Solution

Let's plot the points and form a right triangle as indicated in Figure 12.6. Note that the coordinates of point C are $(5, 3)$. Because $\overline{AC}$ is parallel to the horizontal axis, its length is easily determined to be 3 units. Likewise, $\overline{CB}$ is parallel to the vertical axis, and its length is 4 units. Let d represent the length of $\overline{AB}$, and apply the Pythagorean theorem to obtain

$$d^2 = 3^2 + 4^2$$
$$d^2 = 9 + 16$$
$$d^2 = 25$$
$$d = \pm\sqrt{25} = \pm 5$$

"Distance between" is a nonnegative value, so the length of $\overline{AB}$ is 5 units.

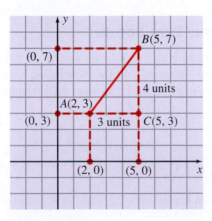

Figure 12.6

The approach we used in Example 2 becomes the basis for a general distance formula for finding the distance between any two points in a coordinate plane:

1. Let $P_1(x_1, y_1)$ and $P_2(x_2, y_2)$ represent any two points in a coordinate plane.
2. Form a right triangle as indicated in Figure 12.7. The coordinates of the vertex of the right angle, point R, are (x_2, y_1).

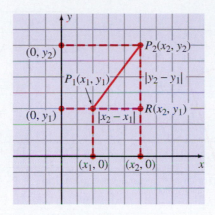

Figure 12.7

The length of $\overline{P_1R}$ is $|x_2 - x_1|$, and the length of $\overline{RP_2}$ is $|y_2 - y_1|$. (The absolute-value symbol is used to ensure a nonnegative value.) Let d represent the length of $\overline{P_1P_2}$ and apply the Pythagorean theorem to obtain

$$d^2 = |x_2 - x_1|^2 + |y_2 - y_1|^2$$

Because $|a|^2 = a^2$, the **distance formula** can be stated as

$$d = \sqrt{(x_2 - x_1)^2 + (y_2 - y_1)^2}$$

It makes no difference which point you call P_1 or P_2 when using the distance formula. If you forget the formula, don't panic. Just form a right triangle and apply the Pythagorean theorem as we did in Example 2. Let's consider an example that demonstrates the use of the distance formula.

Classroom Example
Find the distance between $(-1, 3)$ and $(-6, 8)$.

EXAMPLE 3   Find the distance between $(-1, 4)$ and $(1, 2)$.

Solution

Let $(-1, 4)$ be P_1 and $(1, 2)$ be P_2. Using the distance formula, we obtain

$$\begin{aligned} d &= \sqrt{[1 - (-1)]^2 + (2 - 4)^2} \\ &= \sqrt{2^2 + (-2)^2} \\ &= \sqrt{4 + 4} \\ &= \sqrt{8} = 2\sqrt{2} \qquad \text{Express the answer in simplest radical form} \end{aligned}$$

The distance between the two points is $2\sqrt{2}$ units.

In Example 3, we did not sketch a figure because of the simplicity of the problem. However, sometimes it is helpful to use a figure to organize the given information and aid in the analysis of the problem, as we see in the next example.

Classroom Example
Verify that the points $(2, -1)$, $(6, 5)$, and $(-4, 3)$ are vertices of an isosceles triangle.

EXAMPLE 4

Verify that the points $(-3, 6)$, $(3, 4)$, and $(1, -2)$ are vertices of an isosceles triangle. (An isosceles triangle has two sides of the same length.)

Solution

Let's plot the points and draw the triangle (Figure 12.8). Use the distance formula to find the lengths d_1, d_2, and d_3, as follows:

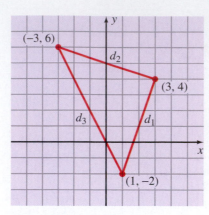

$$d_1 = \sqrt{(3 - 1)^2 + [4 - (-2)]^2}$$
$$= \sqrt{2^2 + 6^2} = \sqrt{40} = 2\sqrt{10}$$

$$d_2 = \sqrt{(-3 - 3)^2 + (6 - 4)^2}$$
$$= \sqrt{(-6)^2 + 2^2} = \sqrt{40} = 2\sqrt{10}$$

$$d_3 = \sqrt{(-3 - 1)^2 + [6 - (-2)]^2}$$
$$= \sqrt{(-4)^2 + 8^2} = \sqrt{80} = 4\sqrt{5}$$

Figure 12.8

Because $d_1 = d_2$, we know that it is an isosceles triangle.

Finding the Slope of a Line

In Chapter 5 we introduced the concept of slope of a line; in this section we will review the basic concepts of slope. In coordinate geometry, the concept of **slope** is used to describe the "steepness" of lines. The slope of a line is the ratio of the vertical change to the horizontal change as we move from one point on a line to another point.

A precise definition for slope can be given by considering the coordinates of the points P_1, P_2, and R as indicated in Figure 12.9. The horizontal change as we move from P_1 to P_2 is $x_2 - x_1$, and the vertical change is $y_2 - y_1$. Thus the following definition for slope is given.

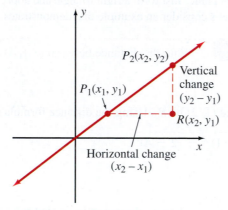

Figure 12.9

Definition 12.1 Slope of a Line

If points P_1 and P_2 with coordinates (x_1, y_1) and (x_2, y_2), respectively, are any two different points on a line, then the slope of the line (denoted by m) is

$$m = \frac{y_2 - y_1}{x_2 - x_1}, \qquad x_2 \neq x_1$$

Because $\dfrac{y_2 - y_1}{x_2 - x_1} = \dfrac{y_1 - y_2}{x_1 - x_2}$, how we designate P_1 and P_2 is not important. Let's use Definition 12.1 to find the slopes of some lines.

Classroom Example
Find the slope of the line determined by each of the following
pairs of points, and graph the lines:

(a) $(-3, -4)$ and $(2, 3)$
(b) $(-2, 4)$ and $(3, -5)$
(c) $(4, 2)$ and $(-3, 2)$

EXAMPLE 5

Find the slope of the line determined by each of the following pairs of points, and graph the lines:

(a) $(-1, 1)$ and $(3, 2)$ **(b)** $(4, -2)$ and $(-1, 5)$

(c) $(2, -3)$ and $(-3, -3)$

Solution

(a) Let $(-1, 1)$ be P_1 and $(3, 2)$ be P_2 (Figure 12.10).

$$m = \frac{y_2 - y_1}{x_2 - x_1} = \frac{2 - 1}{3 - (-1)} = \frac{1}{4}$$

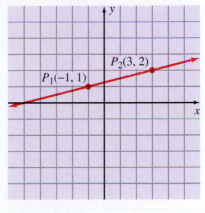

Figure 12.10

(b) Let $(4, -2)$ be P_1 and $(-1, 5)$ be P_2 (Figure 12.11).

$$m = \frac{y_2 - y_1}{x_2 - x_1} = \frac{5 - (-2)}{-1 - 4} = \frac{7}{-5} = -\frac{7}{5}$$

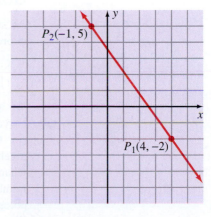

Figure 12.11

(c) Let $(2, -3)$ be P_1 and $(-3, -3)$ be P_2 (Figure 12.12).

$$m = \frac{y_2 - y_1}{x_2 - x_1}$$

$$= \frac{-3 - (-3)}{-3 - 2}$$

$$= \frac{0}{-5} = 0$$

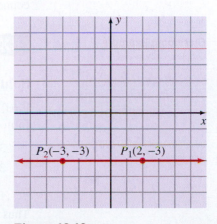

Figure 12.12

The three parts of Example 5 represent the three basic possibilities for slope; that is, the slope of a line can be positive, negative, or zero. A line that has a positive slope rises as we move from left to right, as in Figure 12.10. A line that has a negative slope falls as we move from left to right, as in Figure 12.11. A horizontal line, as in Figure 12.12, has a slope of zero. Finally, we need to realize that *the concept of slope is undefined for vertical lines.* This is due to the fact that for any vertical line, the horizontal change as we move from one point on the line to another is zero. Thus the ratio $\dfrac{y_2 - y_1}{x_2 - x_1}$ will have a denominator of zero and be undefined. Accordingly, the restriction $x_2 \neq x_1$ is imposed in Definition 12.1.

Suppose that we are given $3x - 2y = 6$ as the equation of a line. The slope of that line can be found by determining the coordinates of two points on the line and then using the definition of slope. If we let $y = 0$, then $x = 2$, and the point $(2, 0)$ is on the line. If we let $x = 0$, then $y = -3$ and the point $(0, -3)$ is on the line. Applying the definition of slope we obtain $\dfrac{0 - (-3)}{2 - 0} = \dfrac{3}{2}$.

Recall also from Chapter 5, we can change the given equation to slope-intercept form $(y = mx + b)$ and determine the slope from that equation.

$$3x - 2y = 6$$
$$-2y = -3x + 6$$
$$y = \frac{3}{2}x - 3$$

The slope is $\dfrac{3}{2}$.

Graphing Techniques

As stated in the introductory remarks for this chapter, there are two kinds of problems in coordinate geometry:

1. Given an algebraic equation, determine its geometric graph.

2. Given a set of conditions pertaining to a geometric figure, determine its algebraic equation.

We will work with both kinds of problems in this chapter, with an emphasis in this section on curve-sketching techniques.

One very important graphing technique is to be able to recognize that a certain kind of algebraic equation produces a certain kind of geometric graph. For example, from our work in Chapter 5, we know that any equation of the form $Ax + By = C$, where A, B, and C are real numbers (A and B not both zero) and x and y are variables, is a **linear equation** in two variables, and its graph is a straight line. Because two points determine a straight line, graphing linear equations is a simple process. We find two solutions, plot the corresponding points, and connect the points with a straight line. Let's consider an example.

Classroom Example
Graph $x - 2y = -4$

EXAMPLE 6   Graph $3y = 6 - 2x$.

Solution

First, we need to realize that $3y = 6 - 2x$ is equivalent to $2x + 3y = 6$ and therefore fits the form of a linear equation. Now we can determine two points. Let $x = 0$ in the original equation.

$$3y = 6 - 2(0)$$
$$3y = 6$$
$$y = 2$$

Thus the point $(0, 2)$ is on the line. Then let $y = 0$.

$$3(0) = 6 - 2x$$
$$2x = 6$$
$$x = 3$$

Thus the point $(3, 0)$ is on the line. Plotting the two points $(0, 2)$ and $(3, 0)$ and connecting them with a straight line produces Figure 12.13.

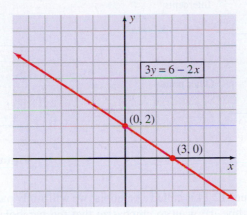

Figure 12.13

The points $(3, 0)$ and $(0, 2)$ in Figure 12.13 are special points. They are the points of the graph that are on the coordinate axes. That is, they yield the *x* intercept and the *y* intercept of the graph. Let's define in general the *intercepts* of a graph.

> The *x* coordinates of the points that a graph has in common with the *x* axis are called the **x intercepts** of the graph. (To compute the *x* intercepts, let $y = 0$ and solve for x.)
>
> The *y* coordinates of the points that a graph has in common with the *y* axis are called the **y intercepts** of the graph. (To compute the *y* intercepts, let $x = 0$ and solve for y.)

Each of the following examples, along with the follow-up discussion, will introduce another aspect of curve sketching. Then toward the end of the section, we will summarize these techniques for you.

Classroom Example
Graph $y = x^2 + 3$.

EXAMPLE 7   Graph $y = x^2 - 4$.

Solution

Let's begin by finding the intercepts. If $x = 0$, then

$$y = 0^2 - 4 = -4$$

The point $(0, -4)$ is on the graph. If $y = 0$, then

$$0 = x^2 - 4$$
$$0 = (x + 2)(x - 2)$$
$$x + 2 = 0 \quad \text{or} \quad x - 2 = 0$$
$$x = -2 \quad \text{or} \quad x = 2$$

The points $(-2, 0)$ and $(2, 0)$ are on the graph. The given equation is in a convenient form for setting up a table of values.

Plotting these points and connecting them with a smooth curve produces Figure 12.14.

x	y	
0	−4	
−2	0	Intercepts
2	0	
1	−3	
−1	−3	
3	5	Other points
−3	5	

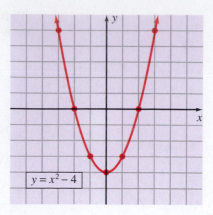

$y = x^2 - 4$

Figure 12.14

The curve in Figure 12.14 is called a parabola; we will study parabolas in more detail in a later chapter. At this time we want to emphasize that the parabola in Figure 12.14 is said to be *symmetric with respect to the y axis*. In other words, the y axis is a line of symmetry. Each half of the curve is a mirror image of the other half through the y axis. Note, in the table of values, that for each ordered pair (x, y), the ordered pair $(-x, y)$ is also a solution. A general test for y-axis symmetry can be stated as follows:

y-Axis Symmetry

The graph of an equation is symmetric with respect to the y axis if replacing x with −x results in an equivalent equation.

The equation $y = x^2 - 4$ exhibits symmetry with respect to the y axis because replacing x with −x produces $y = (-x)^2 - 4 = x^2 - 4$. Let's test some equations for such symmetry. We will replace x with −x and check for an equivalent equation.

Equation	Test for symmetry with respect to the y axis	Equivalent equation	Symmetric with respect to the y axis
$y = -x^2 + 2$	$y = -(-x)^2 + 2 = -x^2 + 2$	Yes	Yes
$y = 2x^2 + 5$	$y = 2(-x)^2 + 5 = 2x^2 + 5$	Yes	Yes
$y = x^4 + x^2$	$y = (-x)^4 + (-x)^2 = x^4 + x^2$	Yes	Yes
$y = x^3 + x^2$	$y = (-x)^3 + (-x)^2$ $= -x^3 + x^2$	No	No
$y = x^2 + 4x + 2$	$y = (-x)^2 + 4(-x) + 2$ $= x^2 - 4x + 2$	No	No

Some equations yield graphs that have x-axis symmetry. In the next example we will see the graph of a parabola that is symmetric with respect to the x axis.

Classroom Example

Graph $x = \frac{1}{2}y^2$.

EXAMPLE 8   Graph $x = y^2$.

Solution

First, we see that $(0, 0)$ is on the graph and determines both intercepts. Second, the given equation is in a convenient form for setting up a table of values.

Plotting these points and connecting them with a smooth curve produces Figure 12.15.

x	y	
0	0	Intercepts
1	1	
1	−1	Other points
4	2	
4	−2	

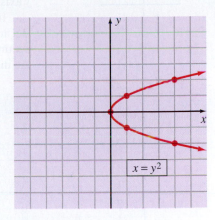

Figure 12.15

The parabola in Figure 12.15 is said to be *symmetric with respect to the x axis*. Each half of the curve is a mirror image of the other half through the *x* axis. Also note in the table of values, that for each ordered pair (x, y), the ordered pair $(x, -y)$ is a solution. A general test for *x*-axis symmetry can be stated as follows:

---

x-Axis Symmetry

The graph of an equation is symmetric with respect to the *x* axis if replacing *y* with $-y$ results in an equivalent equation.

The equation $x = y^2$ exhibits *x*-axis symmetry because replacing *y* with $-y$ produces $x = (-y)^2 = y^2$. Let's test some equations for *x*-axis symmetry. We will replace *y* with $-y$ and check for an equivalent equation.

Equation	Test for symmetry with respect to the x axis	Equivalent equation	Symmetric with respect to the x axis
$x = y^2 + 5$	$x = (-y)^2 + 5 = y^2 + 5$	Yes	Yes
$x = -3y^2$	$x = -3(-y)^2 = -3y^2$	Yes	Yes
$x = y^3 + 2$	$x = (-y)^3 + 2 = -y^3 + 2$	No	No
$x = y^2 - 5y + 6$	$x = (-y)^2 - 5(-y) + 6$	No	No
	$= y^2 + 5y + 6$		

Classroom Example
Graph $y = \dfrac{2}{x}$.

In addition to *y*-axis and *x*-axis symmetry, some equations yield graphs that have symmetry with respect to the origin. In the next example we will see a graph that is symmetric with respect to the origin.

EXAMPLE 9   Graph $y = \dfrac{1}{x}$.

Solution

First, let's find the intercepts. Let $x = 0$; then $y = \dfrac{1}{x}$ becomes $y = \dfrac{1}{0}$, and $\dfrac{1}{0}$ is undefined. Thus there is no *y* intercept. Let $y = 0$; then $y = \dfrac{1}{x}$ becomes $0 = \dfrac{1}{x}$, and there are no values of *x*

x	y
$\frac{1}{2}$	2
1	1
2	$\frac{1}{2}$
3	$\frac{1}{3}$
$-\frac{1}{2}$	-2
-1	-1
-2	$-\frac{1}{2}$
-3	$-\frac{1}{3}$

that will satisfy this equation. In other words, this graph has no points on either the x axis or the y axis. Second, let's set up a table of values and keep in mind that neither x nor y can equal zero.

In Figure 12.16(a) we plotted the points associated with the solutions from the table. Because the graph does not intersect either axis, it must consist of two branches. Thus connecting the points in the first quadrant with a smooth curve and then connecting the points in the third quadrant with a smooth curve, we obtain the graph shown in Figure 12.16(b).

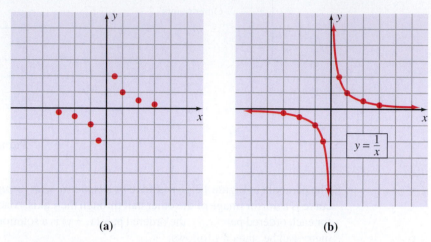

(a)   (b)

Figure 12.16

The curve in Figure 12.16 is said to be *symmetric with respect to the origin*. Each half of the curve is a mirror image of the other half through the origin. Note in the table of values, that for each ordered pair (x, y), the ordered pair $(-x, -y)$ is also a solution. A general test for origin symmetry can be stated as follows:

Origin Symmetry

The graph of an equation is symmetric with respect to the origin if replacing x with $-x$ and y with $-y$ results in an equivalent equation.

The equation $y = \dfrac{1}{x}$ exhibits symmetry with respect to the origin because replacing y with $-y$ and x with $-x$ produces $-y = \dfrac{1}{-x}$, which is equivalent to $y = \dfrac{1}{x}$. Let's test some equations for symmetry with respect to the origin. We will replace y with $-y$, replace x with $-x$, and then check for an equivalent equation.

Equation	Test for symmetry with respect to the origin	Equivalent equation	Symmetric with respect to the origin
$y = x^3$	$(-y) = (-x)^3$ $-y = -x^3$ $y = x^3$	Yes	Yes
$x^2 + y^2 = 4$	$(-x)^2 + (-y)^2 = 4$ $x^2 + y^2 = 4$	Yes	Yes
$y = x^2 - 3x + 4$	$(-y) = (-x)^2 - 3(-x) + 4$ $-y = x^2 + 3x + 4$ $y = -x^2 - 3x - 4$	No	No

Let's pause for a moment and pull together the graphing techniques that we have introduced thus far. The following list is a set of graphing suggestions. The order of the suggestions indicates the order in which we usually attack a new graphing problem.

1. Determine what type of symmetry the equation exhibits.

2. Find the intercepts.

3. Solve the equation for y in terms of x or for x in terms of y if it is not already in such a form.

4. Set up a table of ordered pairs that satisfy the equation. The type of symmetry will affect your choice of values in the table. (We will illustrate this in a moment.)

5. Plot the points associated with the ordered pairs from the table, and connect them with a smooth curve. Then, if appropriate, reflect this part of the curve according to the symmetry shown by the equation.

Classroom Example
Graph $x^2 y = 3$.

EXAMPLE 10   Graph $x^2 y = -2$.

Solution

Because replacing x with $-x$ produces $(-x)^2 y = -2$ or, equivalently, $x^2 y = -2$, the equation exhibits y-axis symmetry. There are no intercepts because neither x nor y can equal 0. Solving the equation for y produces $y = \dfrac{-2}{x^2}$. The equation exhibits y-axis symmetry, so let's use only positive values for x and then reflect the curve across the y axis.

Let's plot the points determined by the table, connect them with a smooth curve, and reflect this portion of the curve across the y axis. Figure 12.17 is the result of this process.

x	y
1	-2
2	$-\dfrac{1}{2}$
3	$-\dfrac{2}{9}$
4	$-\dfrac{1}{8}$
$\dfrac{1}{2}$	-8

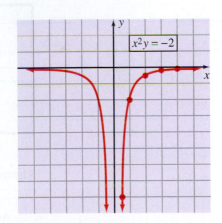

Figure 12.17

Classroom Example
Graph $x = \dfrac{1}{2} y^3$.

EXAMPLE 11   Graph $x = y^3$.

Solution

Because replacing x with $-x$ and y with $-y$ produces $-x = (-y)^3 = -y^3$, which is equivalent to $x = y^3$, the given equation exhibits origin symmetry. If $x = 0$, then $y = 0$, so the origin is a point of the graph. The given equation is in an easy form for deriving a table of values.

Let's plot the points determined by the table, connect them with a smooth curve, and reflect this portion of the curve through the origin to produce Figure 12.18.

x	y
0	0
1	1
8	2
$\frac{1}{8}$	$\frac{1}{2}$

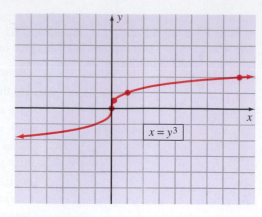

Figure 12.18

Classroom Example
Use a graphing utility to obtain a graph of the equation $x = \frac{1}{2}y^3$.

EXAMPLE 12   Use a graphing utility to obtain a graph of the equation $x = y^3$.

Solution

First, we may need to solve the equation for y in terms of x. (We say we "may need to" because some graphing utilities are capable of graphing two-variable equations without solving for y in terms of x.)

$$y = \sqrt[3]{x} = x^{1/3}$$

Now we can enter the expression $x^{1/3}$ for Y_1 and obtain the graph shown in Figure 12.19.

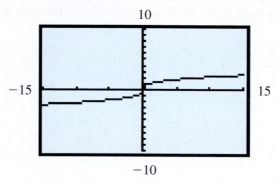

Figure 12.19

As indicated in Figure 12.19, the **viewing rectangle** of a graphing utility is a portion of the xy plane shown on the display of the utility. In this display, the boundaries were set so that $-15 \leq x \leq 15$ and $-10 \leq y \leq 10$. These boundaries were set automatically; however, boundaries can be reassigned as necessary, which is an important feature of graphing utilities.

Concept Quiz 12.1

For Problems 1–10, answer true or false.

1. The distance between P_1 and P_2 is the opposite of the distance between P_2 and P_1.

2. The slope of a line is the ratio of the vertical change to the horizontal change when moving from one point to another point on the line.

3. A slope of zero means there is no change in the vertical direction when moving from one point to another point on the line.

4. The equation $y = \sqrt{x}$ is a nonlinear equation.

5. When replacing y with $-y$ in an equation results in an equivalent equation, then the graph of the equation is symmetric with respect to the x axis.

6. If a parabola is symmetric with respect to the x axis, then each half of the curve is a mirror image of the other half through the x axis.

7. If the graph of an equation is symmetric with respect to the x axis, then it cannot be symmetric with respect to the y axis.

8. If the point $(-2, 5)$ is on a graph that is symmetric with respect to the y axis, then the point $(-2, -5)$ is also on the graph.

9. If for each ordered pair (x, y) that is a solution of the equation, the ordered pair $(-x, -y)$ is also a solution, then the graph of the equation is symmetric with respect to the origin.

10. The graph of a straight line is symmetric with respect to the origin only if the line passes through the origin.

Problem Set 12.1

For Problems 1–12, find the distance between each pair of points. Express answers in simplest radical form. **(Objective 1)**

1. $(-2, -1), (7, 11)$

2. $(2, 1), (10, 7)$

3. $(1, -1), (3, -4)$

4. $(-1, 3), (2, -2)$

5. $(6, -4), (9, -7)$

6. $(-5, 2), (-1, 6)$

7. $(-3, 3), (0, -3)$

8. $(-2, -4), (4, 0)$

9. $(1, -6), (-5, -6)$

10. $(-2, 3), (-2, -7)$

11. $(1, 7), (4, -2)$

12. $(6, 4), (-4, -8)$

13. Verify that the points $(-3, 1), (5, 7)$, and $(8, 3)$ are vertices of a right triangle. [*Hint*: If $a^2 + b^2 = c^2$, then it is a right triangle with the right angle opposite side c.]

14. Verify that the points $(0, 3), (2, -3)$, and $(-4, -5)$ are vertices of an isosceles triangle.

15. Verify that the points $(7, 12)$ and $(11, 18)$ divide the line segment joining $(3, 6)$ and $(15, 24)$ into three segments of equal length.

16. Verify that $(3, 1)$ is the midpoint of the line segment joining $(-2, 6)$ and $(8, -4)$.

For Problems 17–28, graph the line determined by the two points, and find the slope of the line. **(Objective 2)**

17. $(1, 2), (4, 6)$

18. $(3, 1), (-2, -2)$

19. $(-4, 5), (-1, -2)$

20. $(-2, 5), (3, -1)$

21. $(2, 6), (6, -2)$

22. $(-2, -1), (2, -5)$

23. $(-6, 1), (-1, 4)$

24. $(-3, 3), (2, 3)$

25. $(-2, -4), (2, -4)$

26. $(1, -5), (1, -1)$

27. $(0, -2), (4, 0)$

28. $(-4, 0), (0, -6)$

29. Find x if the line through $(-2, 4)$ and $(x, 6)$ has a slope of $\dfrac{2}{9}$.

30. Find y if the line through $(1, y)$ and $(4, 2)$ has a slope of $\dfrac{5}{3}$.

31. Find x if the line through $(x, 4)$ and $(2, -5)$ has a slope of $-\dfrac{9}{4}$.

32. Find y if the line through $(5, 2)$ and $(-3, y)$ has a slope of $-\dfrac{7}{8}$.

For each of the points in Problems 33–37, determine the points that are symmetric with respect to (a) the x axis, (b) the y axis, and (c) the origin. **(Objective 3)**

33. $(-3, 1)$

34. $(-2, -4)$

35. $(7, -2)$

36. $(0, -4)$

37. $(5, 0)$

For Problems 38–57, determine the type(s) of symmetry (symmetry with respect to the x axis, y axis, and/or origin) exhibited by the graph of each of the following equations. Do not sketch the graph. **(Objective 3)**

38. $x^2 + 2y = 4$

39. $-3x + 2y^2 = -4$

40. $x = -y^2 + 5$

41. $y = 4x^2 + 13$

42. $xy = -6$

43. $2x^2y^2 = 5$

44. $2x^2 + 3y^2 = 9$

45. $x^2 - 2x - y^2 = 4$

46. $y = x^2 - 6x - 4$

47. $y = 2x^2 - 7x - 3$

48. $y = x$

49. $y = 2x$

50. $y = x^4 + 4$

51. $y = x^4 - x^2 + 2$

52. $x^2 + y^2 = 13$

53. $x^2 - y^2 = -6$

54. $y = -4x^2 - 2$

55. $x = -y^2 + 9$

56. $x^2 + y^2 - 4x - 12 = 0$

57. $2x^2 + 3y^2 + 8y + 2 = 0$

For Problems 58–85, graph each of the equations. **(Objective 4)**

58. $y = x + 1$

59. $y = x - 4$

60. $y = 3x - 6$

61. $y = 2x + 4$

62. $y = -2x + 1$

63. $y = -3x - 1$

64. $2x + y = 6$

65. $2x - y = 4$

66. $y = x^2 - 1$

67. $y = x^2 + 2$

68. $y = -x^3$

69. $y = x^3$

70. $y = \dfrac{2}{x^2}$

71. $y = \dfrac{-1}{x^2}$

72. $y = 2x^2$

73. $y = -3x^2$

74. $xy = -3$

75. $xy = 2$

76. $x^2 y = 4$

77. $xy^2 = -4$

78. $y^3 = x^2$

79. $y^2 = x^3$

80. $y = \dfrac{-2}{x^2 + 1}$

81. $y = \dfrac{4}{x^2 + 1}$

82. $x = -y^3$

83. $y = x^4$

84. $y = -x^4$

85. $x = -y^3 + 2$

Thoughts Into Words

86. How would you explain the concept of slope to someone who was absent from class the day it was discussed?

87. If one line has a slope of $\dfrac{2}{5}$, and another line has a slope of $\dfrac{3}{7}$, which line is steeper? Explain your answer.

88. Suppose that a line has a slope of $\dfrac{2}{3}$ and contains the point $(4, 7)$. Are the points $(7, 9)$ and $(1, 3)$ also on the line? Explain your answer.

89. How would you convince someone that there are infinitely many ordered pairs of real numbers that satisfy $x + y = 7$?

90. Is a graph symmetric with respect to the origin if it is symmetric with respect to both axes? Defend your answer.

91. What is the graph of $x = 0$? What is the graph of $y = 0$? Explain your answers.

92. Is a graph symmetric with respect to both axes if it is symmetric with respect to the origin? Defend your answer.

Further Investigations

93. Sometimes it is necessary to find the coordinate of a point on a number line that is located somewhere between two given points. For example, suppose that we want to find the coordinate (x) of the point located two-thirds of the distance from 2 to 8. Because the total distance from 2 to 8 is $8 - 2 = 6$ units, we can start at 2 and move $\dfrac{2}{3}(6) = 4$ units toward 8. Thus $x = 2 + \dfrac{2}{3}(6) = 2 + 4 = 6$.

For each of the following, find the coordinate of the indicated point on a number line.

(a) Two-thirds of the distance from 1 to 10
(b) Three-fourths of the distance from -2 to 14
(c) One-third of the distance from -3 to 7
(d) Two-fifths of the distance from -5 to 6
(e) Three-fifths of the distance from -1 to -11
(f) Five-sixths of the distance from 3 to -7

94. Now suppose that we want to find the coordinates of point P, which is located two-thirds of the distance from $A(1, 2)$ to $B(7, 5)$ in a coordinate plane. We have plotted the given points A and B in Figure 12.20 to help with the analysis of this problem. Point D is

two-thirds of the distance from A to C because parallel lines cut off proportional segments on every transversal that intersects the lines.

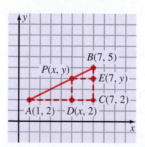

Figure 12.20

Thus $\overline{AC}$ can be treated as a segment of a number line, as shown in Figure 12.21. Therefore,

$$x = 1 + \dfrac{2}{3}(7 - 1) = 1 + \dfrac{2}{3}(6) = 5$$

Figure 12.21

Similarly, $\overline{CB}$ can be treated as a segment of a number line, as shown in Figure 12.22. Therefore,

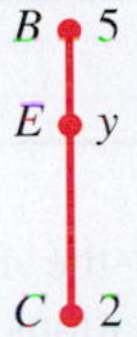

$$y = 2 + \frac{2}{3}(5 - 2) = 2 + \frac{2}{3}(3) = 4$$

The coordinates of point P are $(5, 4)$.

Figure 12.22

For each of the following, find the coordinates of the indicated point in the xy plane.

(a) One-third of the distance from $(2, 3)$ to $(5, 9)$
(b) Two-thirds of the distance from $(1, 4)$ to $(7, 13)$
(c) Two-fifths of the distance from $(-2, 1)$ to $(8, 11)$
(d) Three-fifths of the distance from $(2, -3)$ to $(-3, 8)$
(e) Five-eighths of the distance from $(-1, -2)$ to $(4, -10)$
(f) Seven-eighths of the distance from $(-2, 3)$ to $(-1, -9)$

95. Suppose we want to find the coordinates of the midpoint of a line segment. Let $P(x, y)$ represent the midpoint of the line segment from $A(x_1, y_1)$ to $B(x_2, y_2)$. Using the method from Problem 93, the formula for the x coordinate of the midpoint is $x = x_1 + \frac{1}{2}(x_2 - x_1)$.

This formula can be simplified algebraically to produce a simpler formula.

$$x = x_1 + \frac{1}{2}(x_2 - x_1)$$

$$x = x_1 + \frac{1}{2}x_2 - \frac{1}{2}x_1$$

$$x = \frac{1}{2}x_1 + \frac{1}{2}x_2$$

$$x = \frac{x_1 + x_2}{2}$$

Hence the x coordinate of the midpoint can be interpreted as the average of the x coordinates of the endpoints of the line segment. A similar argument for the y coordinate of the midpoint gives the following formula:

$$y = \frac{y_1 + y_2}{2}$$

For each of the pairs of points, use the formula to find the midpoint of the line segment between the points.

(a) $(3, 1)$ and $(7, 5)$
(b) $(-2, 8)$ and $(6, 4)$
(c) $(-3, 2)$ and $(5, 8)$
(d) $(4, 10)$ and $(9, 25)$
(e) $(-4, -1)$ and $(-10, 5)$
(f) $(5, 8)$ and $(-1, 7)$

Graphing Calculator Activities

For Problems 96–102, answer the "How does . . . ." question and then use your calculator to graph both equations on the same set of axes to check your answer.

96. How does the graph of $y = -\dfrac{5}{x}$ compare to the graph of $y = \dfrac{5}{x}$?

97. How does the graph of $-y = -\dfrac{5}{x}$ compare to the graph of $y = \dfrac{5}{x}$?

98. How does the graph of $y = -\dfrac{2}{x^2}$ compare to the graph of $y = \dfrac{2}{x^2}$?

99. How does the graph of $y = -2x$ compare to the graph of $y = 2x$?

100. How does the graph of $y = \dfrac{1}{x} + 2$ compare to the graph of $y = \dfrac{1}{x}$?

101. How does the graph of $y = \dfrac{1}{x^2} - 3$ compare to the graph of $y = \dfrac{1}{x^2}$?

102. How does the graph of $y = \dfrac{1}{x - 2}$ compare to the graph of $y = \dfrac{1}{x}$?

Answers to the Concept Quiz

1. False   **2.** True   **3.** True   **4.** True   **5.** True   **6.** True   **7.** False   **8.** False   **9.** True   **10.** True

12.2 Graphing Parabolas

OBJECTIVE   **1** Graph parabolas

In Chapter 11, we used a graphing calculator to graph some equations of the form $y = ax^2 + bx + c$. We did this to predict approximate solutions of quadratic equations and also to give visual support for solutions that we obtained algebraically. In each case, the graph was a curve called a **parabola**. In general, the graph of any equation of the form $y = ax^2 + bx + c$, where a, b, and c are real numbers, and $a \neq 0$, is a parabola.

At this time, we want to develop a very easy and systematic way of graphing parabolas without the use of a graphing calculator. As we work with parabolas, we will use the vocabulary indicated in Figure 12.23.

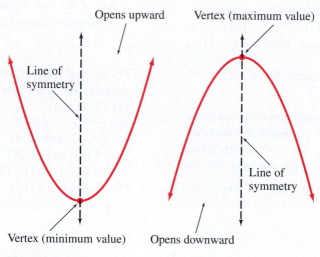

Figure 12.23

Let's begin by using the concepts of intercepts and symmetry to help us sketch the graph of the equation $y = x^2$.

EXAMPLE 1   Graph $y = x^2$.

Solution

If we replace x with $-x$, the given equation becomes $y = (-x)^2 = x^2$; therefore, we have y-axis symmetry. The origin, $(0, 0)$, is a point of the graph. We can recognize from the equation that 0 is the minimum value of y; hence the point $(0, 0)$ is the vertex of the parabola. Now we can set up a table of values that uses nonnegative values for x. Plot the points determined by the table, connect them with a smooth curve, and reflect that portion of the curve across the y axis to produce Figure 12.24.

x	y
0	0
$\dfrac{1}{2}$	$\dfrac{1}{4}$
1	1
2	4
3	9

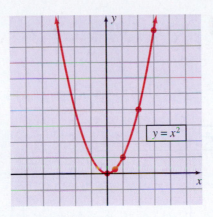

Figure 12.24

To graph parabolas, we need to be able to:

1. Find the vertex.
2. Determine whether the parabola opens upward or downward.
3. Locate two points on opposite sides of the line of symmetry.
4. Compare the parabola to the basic parabola $y = x^2$.

To graph parabolas produced by the various types of equations such as $y = x^2 + k$, $y = ax^2$, $y = (x - h)^2$, and $y = a(x - h)^2 + k$, we can compare these equations to that of the basic parabola, $y = x^2$. First, let's consider some equations of the form $y = x^2 + k$, where k is a constant.

Classroom Example
Graph $y = x^2 + 3$.

EXAMPLE 2   Graph $y = x^2 + 1$.

Solution

Let's set up a table of values to compare y values for $y = x^2 + 1$ to corresponding y values for $y = x^2$.

x	$y = x^2$	$y = x^2 + 1$
0	0	1
1	1	2
2	4	5
−1	1	2
−2	4	5

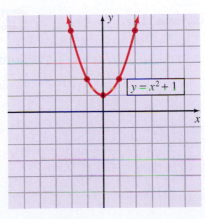

Figure 12.25

It should be evident that y values for $y = x^2 + 1$ are 1 *greater than* the corresponding y values for $y = x^2$. For example, if $x = 2$, then $y = 4$ for the equation $y = x^2$; but if $x = 2$, then $y = 5$ for the equation $y = x^2 + 1$. Thus the graph of $y = x^2 + 1$ is the same as the graph of $y = x^2$, but moved up 1 unit (Figure 12.25). The vertex will move from $(0, 0)$ to $(0, 1)$.

EXAMPLE 3   Graph $y = x^2 - 2$.

Solution

The y values for $y = x^2 - 2$ are *2 less than* the corresponding y values for $y = x^2$, as indicated in the following table.

x	$y = x^2$	$y = x^2 - 2$
0	0	−2
1	1	−1
2	4	2
−1	1	−1
−2	4	2

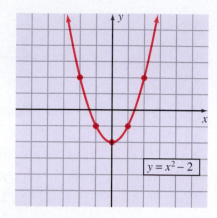

Figure 12.26

Thus the graph of $y = x^2 - 2$ is the same as the graph of $y = x^2$ but moved down 2 units (Figure 12.26). The vertex will move from $(0, 0)$ to $(0, -2)$.

In general, the graph of a quadratic equation of the form $y = x^2 + k$ is the same as the graph of $y = x^2$ but moved up or down $|k|$ units, depending on whether k is positive or negative.

Now, let's consider some quadratic equations of the form $y = ax^2$, where a is a nonzero constant.

EXAMPLE 4   Graph $y = 2x^2$.

Solution

Again, let's use a table to make some comparisons of y values.

x	$y = x^2$	$y = 2x^2$
0	0	0
1	1	2
2	4	8
−1	1	2
−2	4	8

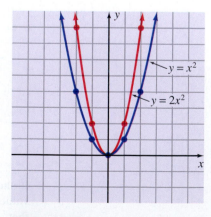

Figure 12.27

Obviously, the y values for $y = 2x^2$ are *twice* the corresponding y values for $y = x^2$. Thus the parabola associated with $y = 2x^2$ has the same vertex (the origin) as the graph of $y = x^2$, but it is narrower (Figure 12.27).

EXAMPLE 5   Graph $y = \frac{1}{2}x^2$.

Solution

The following table indicates some comparisons of y values.

x	$y = x^2$	$y = \frac{1}{2}x^2$
0	0	0
1	1	$\frac{1}{2}$
2	4	2
-1	1	$\frac{1}{2}$
-2	4	2

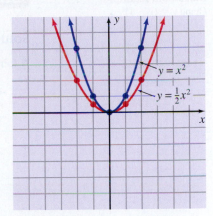

Figure 12.28

The y values for $y = \frac{1}{2}x^2$ are one-half of the corresponding y values for $y = x^2$. Therefore, the graph of $y = \frac{1}{2}x^2$ has the same vertex (the origin) as the graph of $y = x^2$, however it is wider (Figure 12.28).

EXAMPLE 6   Graph $y = -x^2$.

Solution

x	$y = x^2$	$y = -x^2$
0	0	0
1	1	-1
2	4	-4
-1	1	-1
-2	4	-4

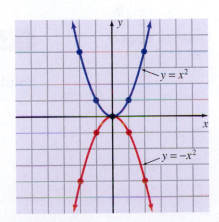

Figure 12.29

The y values for $y = -x^2$ are the opposites of the corresponding y values for $y = x^2$. Thus the graph of $y = -x^2$ has the same vertex (the origin) as the graph of $y = x^2$, but it is a reflection across the x axis of the basic parabola (Figure 12.29).

In general, the graph of a quadratic equation of the form $y = ax^2$ has its vertex at the origin and opens upward if a is positive and downward if a is negative. The parabola is narrower than the basic parabola if $|a| > 1$ and wider if $|a| < 1$.

Let's continue our investigation of quadratic equations by considering those of the form $y = (x - h)^2$, where h is a nonzero constant.

EXAMPLE 7   Graph $y = (x - 2)^2$.

Solution

A fairly extensive table of values reveals a pattern.

x	$y = x^2$	$y = (x - 2)^2$
-2	4	16
-1	1	9
0	0	4
1	1	1
2	4	0
3	9	1
4	16	4
5	25	9

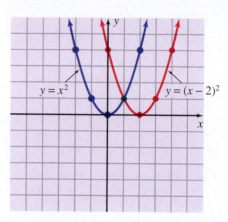

Figure 12.30

Note that $y = (x - 2)^2$ and $y = x^2$ take on the same y values, but for different values of x. More specifically, if $y = x^2$ achieves a certain y value at x equals a constant, then $y = (x - 2)^2$ achieves the same y value at x equals the *constant plus two*. In other words, the graph of $y = (x - 2)^2$ is the same as the graph of $y = x^2$ but moved 2 units to the right (Figure 12.30). The vertex will move from $(0, 0)$ to $(2, 0)$.

EXAMPLE 8   Graph $y = (x + 3)^2$.

Solution

x	$y = x^2$	$y = (x + 3)^2$
-3	9	0
-2	4	1
-1	1	4
0	0	9
1	1	16
2	4	25
3	9	36

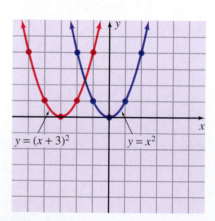

Figure 12.31

If $y = x^2$ achieves a certain y value at x equals a constant, then $y = (x + 3)^2$ achieves that same y value at x equals that *constant minus three*. Therefore, the graph of $y = (x + 3)^2$ is the same as the graph of $y = x^2$ but moved 3 units to the left (Figure 12.31). The vertex will move from $(0, 0)$ to $(-3, 0)$.

In general, the graph of a quadratic equation of the form $y = (x - h)^2$ is the same as the graph of $y = x^2$ but moved to the right h units if h is positive or moved to the left $|h|$ units if h is negative.

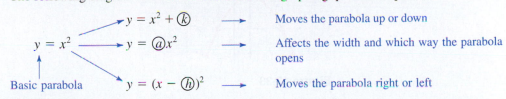

$$y = (x - 4)^2 \qquad\longrightarrow\qquad \text{Moved to the right 4 units}$$

$$y = (x + 2)^2 = [x - (-2)]^2 \qquad\longrightarrow\qquad \text{Moved to the left 2 units}$$

The following diagram summarizes our work with graphing quadratic equations.

$$y = x^2 + \textcircled{k} \qquad\longrightarrow\qquad \text{Moves the parabola up or down}$$

$$y = x^2 \longrightarrow y = \textcircled{a}x^2 \qquad\longrightarrow\qquad \text{Affects the width and which way the parabola opens}$$

Basic parabola $\qquad y = (x - \textcircled{h})^2 \qquad\longrightarrow\qquad$ Moves the parabola right or left

Equations of the form $y = x^2 + k$ and $y = ax^2$ are symmetric about the y axis. The next two examples of this section show how we can combine these ideas to graph a quadratic equation of the form $y = a(x - h)^2 + k$.

Classroom Example
Graph $y = 3(x + 2)^2 - 2$.

EXAMPLE 9   Graph $y = 2(x - 3)^2 + 1$.

Solution

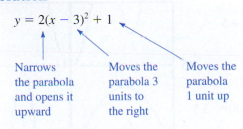

$$y = 2(x - 3)^2 + 1$$

Narrows the parabola and opens it upward

Moves the parabola 3 units to the right

Moves the parabola 1 unit up

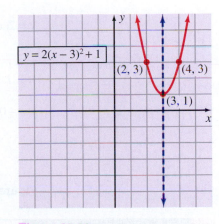

$y = 2(x - 3)^2 + 1$

$(2, 3) \qquad (4, 3)$

$(3, 1)$

Figure 12.32

The vertex will be located at the point $(3, 1)$. In addition to the vertex, two points are located to determine the parabola. The parabola is drawn in Figure 12.32.

Classroom Example
Graph $y = -\dfrac{1}{3}(x - 1)^2 + 3$.

EXAMPLE 10   Graph $y = -\dfrac{1}{2}(x + 1)^2 - 2$.

Solution

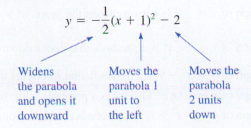

$$y = -\frac{1}{2}(x + 1)^2 - 2$$

Widens the parabola and opens it downward

Moves the parabola 1 unit to the left

Moves the parabola 2 units down

The parabola is drawn in Figure 12.33.

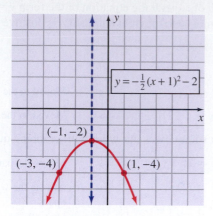

Figure 12.33

Finally, we can use a graphing utility to demonstrate some of the ideas of this section. Let's graph $y = x^2$, $y = -3(x - 7)^2 - 1$, $y = 2(x + 9)^2 + 5$, and $y = -0.2(x + 8)^2 - 3.5$ on the same set of axes, as shown in Figure 12.34. Certainly, Figure 12.34 is consistent with the ideas we presented in this section.

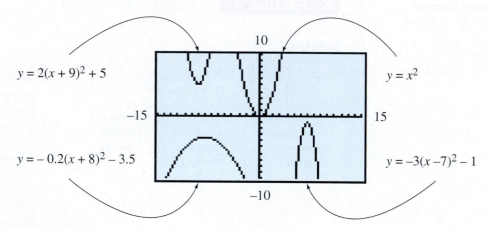

Figure 12.34

Concept Quiz 12.2

For Problems 1–10, answer true or false.

1. The graph of $y = (x - 3)^2$ is the same as the graph of $y = x^2$ but moved 3 units to the right.
2. The graph of $y = x^2 - 4$ is the same as the graph of $y = x^2$ but moved 4 units to the right.
3. The graph of $y = x^2 + 1$ is the same as the graph of $y = x^2$ but moved 1 unit up.
4. The graph of $y = -x^2$ is the same as the graph of $y = x^2$ but is reflected across the y axis.
5. The vertex of the parabola given by the equation $y = (x + 2)^2 - 5$ is located at $(-2, -5)$.
6. The graph of $y = \dfrac{1}{3}x^2$ is narrower than the graph of $y = x^2$.
7. The graph of $y = -\dfrac{2}{3}x^2$ is a parabola that opens downward.
8. The graph of $y = x^2 - 9$ is a parabola that intersects the x axis at $(9, 0)$ and $(-9, 0)$.
9. The graph of $y = -(x - 3)^2 - 7$ is a parabola with the vertex at $(3, -7)$.
10. The graph of $y = -x^2 + 6$ is a parabola that does not intersect the x axis.

Problem Set 12.2

For Problems 1–30, graph each parabola. **(Objective 1)**

1. $y = x^2 + 2$

2. $y = x^2 + 3$

3. $y = x^2 - 1$

4. $y = x^2 - 5$

5. $y = 4x^2$

6. $y = 3x^2$

7. $y = -3x^2$

8. $y = -4x^2$

9. $y = \dfrac{1}{3}x^2$

10. $y = \dfrac{1}{4}x^2$

11. $y = -\dfrac{1}{2}x^2$

12. $y = -\dfrac{2}{3}x^2$

13. $y = (x - 1)^2$

14. $y = (x - 3)^2$

15. $y = (x + 4)^2$

16. $y = (x + 2)^2$

17. $y = 3x^2 + 2$

18. $y = 2x^2 + 3$

19. $y = -2x^2 - 2$

20. $y = \dfrac{1}{2}x^2 - 2$

21. $y = (x - 1)^2 - 2$

22. $y = (x - 2)^2 + 3$

23. $y = (x + 2)^2 + 1$

24. $y = (x + 1)^2 - 4$

25. $y = 3(x - 2)^2 - 4$

26. $y = 2(x + 3)^2 - 1$

27. $y = -(x + 4)^2 + 1$

28. $y = -(x - 1)^2 + 1$

29. $y = -\dfrac{1}{2}(x + 1)^2 - 2$

30. $y = -3(x - 4)^2 - 2$

Thoughts Into Words

31. Write a few paragraphs that summarize the ideas we presented in this section for someone who was absent from class that day.

32. How would you convince someone that $y = (x + 3)^2$ is the basic parabola moved 3 units to the left but that $y = (x - 3)^2$ is the basic parabola moved 3 units to the right?

33. How does the graph of $-y = x^2$ compare to the graph of $y = x^2$? Explain your answer.

34. How does the graph of $y = 4x^2$ compare to the graph of $y = 2x^2$? Explain your answer.

Graphing Calculator Activities

35. Use a graphing calculator to check your graphs for Problems 21–30.

36. (a) Graph $y = x^2$, $y = 2x^2$, $y = 3x^2$, and $y = 4x^2$ on the same set of axes.

(b) Graph $y = x^2$, $y = \dfrac{3}{4}x^2$, $y = \dfrac{1}{2}x^2$, and $y = \dfrac{1}{5}x^2$ on the same set of axes.

(c) Graph $y = x^2$, $y = -x^2$, $y = -3x^2$, and $y = -\dfrac{1}{4}x^2$ on the same set of axes.

37. (a) Graph $y = x^2$, $y = (x - 2)^2$, $y = (x - 3)^2$, and $y = (x - 5)^2$ on the same set of axes.

(b) Graph $y = x^2$, $y = (x + 1)^2$, $y = (x + 3)^2$, and $y = (x + 6)^2$ on the same set of axes.

38. (a) Graph $y = x^2$, $y = (x - 2)^2 + 3$, $y = (x + 4)^2 - 2$, and $y = (x - 6)^2 - 4$ on the same set of axes.

(b) Graph $y = x^2$, $y = 2(x + 1)^2 + 4$, $y = 3(x - 1)^2 - 3$, and $y = \dfrac{1}{2}(x - 5)^2 + 2$ on the same set of axes.

(c) Graph $y = x^2$, $y = -(x - 4)^2 - 3$, $y = -2(x + 3)^2 - 1$, and $y = -\dfrac{1}{2}(x - 2)^2 + 6$ on the same set of axes.

39. (a) Graph $y = x^2 - 12x + 41$ and $y = x^2 + 12x + 41$ on the same set of axes. What relationship seems to exist between the two graphs?

(b) Graph $y = x^2 - 8x + 22$ and $y = -x^2 + 8x - 22$ on the same set of axes. What relationship seems to exist between the two graphs?

(c) Graph $y = x^2 + 10x + 29$ and $y = -x^2 + 10x - 29$ on the same set of axes. What relationship seems to exist between the two graphs?

(d) Summarize your findings for parts (a) through (c).

12.3 More Parabolas and Some Circles

OBJECTIVES

1 Graph quadratic equations of the form $y = ax^2 + bx + c$

2 Write the equation of a circle in standard form

3 Graph a circle

We are now ready to graph quadratic equations of the form $y = ax^2 + bx + c$, where a, b, and c are real numbers, and $a \neq 0$. The general approach is one of changing the form of the equation by completing the square.

$$y = ax^2 + bx + c \longrightarrow y = a(x - h)^2 + k$$

Then we can proceed to graph the parabolas as we did in the previous section. Let's consider some examples.

Classroom Example
Graph $y = x^2 + 4x + 5$.

EXAMPLE 1   Graph $y = x^2 + 6x + 8$.

Solution

$y = x^2 + 6x + 8$

$y = (x^2 + 6x + \underline{\quad}) - (\underline{\quad}) + 8$

$y = (x^2 + 6x + 9) - (9) + 8$

$y = (x + 3)^2 - 1$

Complete the square

$\frac{1}{2}(6) = 3$ and $3^2 = 9$. Add 9 and also subtract 9 to compensate for the 9 that was added

The graph of $y = (x + 3)^2 - 1$ is the basic parabola moved 3 units to the left and 1 unit down (Figure 12.35).

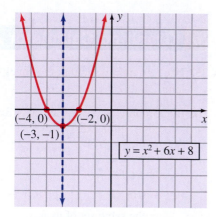

Figure 12.35

Classroom Example
Graph $y = x^2 + x - 1$.

EXAMPLE 2   Graph $y = x^2 - 3x - 1$.

Solution

$y = x^2 - 3x - 1$

$y = (x^2 - 3x + \underline{\quad}) - (\underline{\quad}) - 1$

$y = \left(x^2 - 3x + \frac{9}{4}\right) - \frac{9}{4} - 1$

$y = \left(x - \frac{3}{2}\right)^2 - \frac{13}{4}$

Complete the square

$\frac{1}{2}(-3) = -\frac{3}{2}$ and $\left(-\frac{3}{2}\right)^2 = \frac{9}{4}$.

Add and subtract $\frac{9}{4}$

The graph of $y = \left(x - \dfrac{3}{2}\right)^2 - \dfrac{13}{4}$ is the basic

parabola moved $1\dfrac{1}{2}$ units to the right and $3\dfrac{1}{4}$ units

down (Figure 12.36).

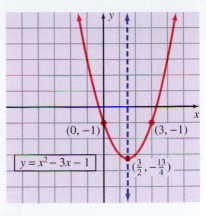

Figure 12.36

If the coefficient of x^2 is not 1, then a slight adjustment has to be made before we apply the process of completing the square. The next two examples illustrate this situation.

Classroom Example
Graph $y = 3x^2 - 12x + 14$.

EXAMPLE 3   Graph $y = 2x^2 + 8x + 9$.

Solution

$$y = 2x^2 + 8x + 9$$ Factor a 2 from the x-variable terms

$$y = 2(x^2 + 4x) + 9$$ Complete the square; note that the

$$y = 2(x^2 + 4x + \underline{\quad}) - (2)(\underline{\quad}) + 9$$ number being subtracted will be multiplied by a factor of 2

$$y = 2(x^2 + 4x + 4) - 2(4) + 9$$ $\dfrac{1}{2}(4) = 2$, and $2^2 = 4$

$$y = 2(x^2 + 4x + 4) - 8 + 9$$

$$y = 2(x + 2)^2 + 1$$

See Figure 12.37 for the graph of
$y = 2(x + 2)^2 + 1$.

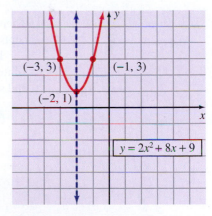

Figure 12.37

Classroom Example
Graph $y = -2x^2 - 4x - 5$.

EXAMPLE 4   Graph $y = -3x^2 + 6x - 5$.

Solution

$$y = -3x^2 + 6x - 5$$

$$y = -3(x^2 - 2x) - 5$$ Factor -3 from the x-variable terms

$$y = -3(x^2 - 2x + \underline{\quad}) - (-3)(\underline{\quad}) - 5$$ Complete the square; note that the number being subtracted will be multiplied by a factor of -3

$$y = -3(x^2 - 2x + 1) - (-3)(1) - 5$$ $\dfrac{1}{2}(-2) = -1$ and $(-1)^2 = 1$

$$y = -3(x^2 - 2x + 1) + 3 - 5$$
$$y = -3(x - 1)^2 - 2$$

The graph of $y = -3(x - 1)^2 - 2$ is shown in Figure 12.38.

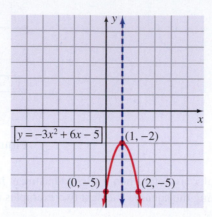

$y = -3x^2 + 6x - 5$ $(1, -2)$ $(0, -5)$ $(2, -5)$

Figure 12.38

Keep in mind that when working with parabolas, your knowledge of quadratic equations can be used to find the intercepts of the graphs. For example, consider the equation $y = x^2 + 6x + 8$. To find the x intercepts we can let $y = 0$ and solve the resulting quadratic equation.

$$x^2 + 6x + 8 = 0$$
$$(x + 2)(x + 4) = 0$$
$$x + 2 = 0 \quad \text{or} \quad x + 4 = 0$$
$$x = -2 \quad \text{or} \quad x = -4$$

Therefore, the parabola intersects the x axis at $(-2, 0)$ and $(-4, 0)$. In other words, the x intercepts are -2 and -4. (Now go back and look at Figure 12.35.)

Let's consider another example. Take a look at Figure 12.36. It appears that one x intercept is between -1 and 0, and the other x intercept is between 3 and 4. If we let $y = 0$ and solve the resulting equation we obtain the following results.

$$x^2 - 3x - 1 = 0$$
$$x = \frac{-(-3) \pm \sqrt{(-3)^2 - 4(1)(-1)}}{2(1)}$$
$$x = \frac{3 \pm \sqrt{9 + 4}}{2}$$
$$x = \frac{3 \pm \sqrt{13}}{2}$$

Thus the x intercepts are $\dfrac{3 - \sqrt{13}}{2}$ and $\dfrac{3 + \sqrt{13}}{2}$. Using 3.6 as an approximation for $\sqrt{13}$, the x intercepts are approximately $\dfrac{3 - 3.6}{2} = \dfrac{-0.6}{2} = -0.3$ and $\dfrac{3 + 3.6}{2} = \dfrac{6.6}{2} = 3.3$.

Now take a look at Figure 12.37. The parabola $y = 2x^2 + 8x + 9$ does not intersect the x axis. This indicates that the solutions of $2x^2 + 8x + 9 = 0$ should be nonreal complex numbers. Let's check this out.

$$2x^2 + 8x + 9 = 0$$
$$x = \frac{-8 \pm \sqrt{8^2 - 4(2)(9)}}{2(2)}$$

$$x = \frac{-8 \pm \sqrt{64 - 72}}{4}$$

$$x = \frac{-8 \pm \sqrt{-8}}{4}$$

$$x = \frac{-8 \pm 2i\sqrt{2}}{4}$$

$$x = \frac{-4 \pm i\sqrt{2}}{2}$$

The solutions of $2x^2 + 8x + 9 = 0$ are indeed nonreal complex numbers, and this agrees with the graph of $y = 2x^2 + 8x + 9$ in Figure 12.37 that indicates no real numbers x intercepts.

Circles

The distance formula, $d = \sqrt{(x_2 - x_1)^2 + (y_2 - y_1)^2}$ (developed in Section 12.1), when it applies to the definition of a circle, produces what is known as the **standard equation of a circle**. We start with a precise definition of a circle.

> ### Definition 12.2
>
> A **circle** is the set of all points in a plane equidistant from a given fixed point called the **center**. A line segment determined by the center and any point on the circle is called a **radius**.

Let's consider a circle that has a radius of length r and a center at (h, k) on a coordinate system (Figure 12.39).

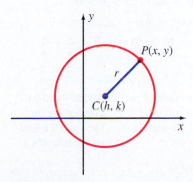

Figure 12.39

By using the distance formula, we can express the length of a radius (denoted by r) for any point $P(x, y)$ on the circle, as

$$r = \sqrt{(x - h)^2 + (y - k)^2}$$

Thus squaring both sides of the equation, we obtain the **standard form of the equation of a circle**:

$$(x - h)^2 + (y - k)^2 = r^2$$

We can use the standard form of the equation of a circle to solve two basic kinds of circle problems:

1. Given the coordinates of the center and the length of a radius of a circle, find its equation.

2. Given the equation of a circle, find its center and the length of a radius.

Let's look at some examples of such problems.

Classroom Example
Write the equation of a circle that has its center at $(-2, 3)$ and a radius of length 4 units.

EXAMPLE 5

Write the equation of a circle that has its center at $(3, -5)$ and a radius of length 6 units.

Solution

Let's substitute 3 for h, -5 for k, and 6 for r into the standard form $(x - h)^2 + (y - k)^2 = r^2$, which becomes $(x - 3)^2 + (y + 5)^2 = 6^2$ that we can simplify as follows:

$$(x - 3)^2 + (y + 5)^2 = 6^2$$
$$x^2 - 6x + 9 + y^2 + 10y + 25 = 36$$
$$x^2 + y^2 - 6x + 10y - 2 = 0$$

Note in Example 5 that we simplified the equation to the form $x^2 + y^2 + Dx + Ey + F = 0$, where D, E, and F are integers. This is another form that we commonly use when working with circles.

Classroom Example
Graph $x^2 + y^2 - 6x + 2y + 1 = 0$.

EXAMPLE 6   Graph $x^2 + y^2 + 4x - 6y + 9 = 0$.

Solution

This equation is of the form $x^2 + y^2 + Dx + Ey + F = 0$, so its graph is a circle. We can change the given equation into the form $(x - h)^2 + (y - k)^2 = r^2$ by completing the square on x and on y as follows:

$$x^2 + y^2 + 4x - 6y + 9 = 0$$
$$(x^2 + 4x + \underline{\quad}) + (y^2 - 6y + \underline{\quad}) = -9$$
$$(x^2 + 4x + 4) + (y^2 - 6y + 9) = -9 + 4 + 9$$

Added 4 to complete the square on x

Added 9 to complete the square on y

Added 4 and 9 to compensate for the 4 and 9 added on the left side

$$(x + 2)^2 + (y - 3)^2 = 4$$
$$[x - (-2)]^2 + (y - 3)^2 = 2^2$$

$\qquad h \qquad\qquad k \qquad\quad r$

The center of the circle is at $(-2, 3)$ and the length of a radius is 2 (Figure 12.40).

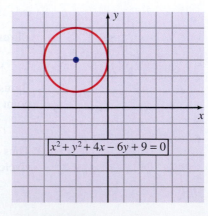

$x^2 + y^2 + 4x - 6y + 9 = 0$

Figure 12.40

As demonstrated by Examples 5 and 6, both forms, $(x - h)^2 + (y - k)^2 = r^2$ and $x^2 + y^2 + Dx + Ey + F = 0$, play an important role when we are solving problems that deal with circles.

Finally, we need to recognize that the standard form of a circle that has its center at the origin is $x^2 + y^2 = r^2$. This is merely the result of letting $h = 0$ and $k = 0$ in the general standard form.

$$(x - h)^2 + (y - k)^2 = r^2$$
$$(x - 0)^2 + (y - 0)^2 = r^2$$
$$x^2 + y^2 = r^2$$

Thus by inspection we can recognize that $x^2 + y^2 = 9$ is a circle with its center at the origin; the length of a radius is 3 units. Likewise, the equation of a circle that has its center at the origin and a radius of length 6 units is $x^2 + y^2 = 36$.

When using a graphing utility to graph a circle, we need to solve the equation for y in terms of x. This will produce two equations that can be graphed on the same set of axes. Furthermore, as with any graph, it may be necessary to change the boundaries on x or y (or both) to obtain a complete graph. If the circle appears oblong, you may want to use a zoom square option so that the graph will appear as a circle. Let's consider an example.

Classroom Example
Graph $x^2 + 20x + y^2 + 80 = 0$.

EXAMPLE 7   Use a graphing utility to graph $x^2 - 40x + y^2 + 351 = 0$.

Solution

First, we need to solve for y in terms of x.

$$x^2 - 40x + y^2 + 351 = 0$$
$$y^2 = -x^2 + 40x - 351$$
$$y = \pm\sqrt{-x^2 + 40x - 351}$$

Now we can make the following assignments.

$$Y_1 = \sqrt{-x^2 + 40x - 351}$$
$$Y_2 = -Y_1$$

(Note that we assigned Y_2 in terms of Y_1. By doing this we avoid repetitive key strokes and thus reduce the chance for errors. You may need to consult your user's manual for instructions on how to keystroke $-Y_1$.) Figure 12.41 shows the graph.

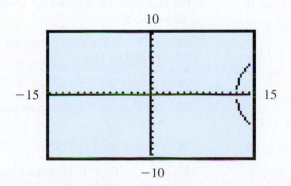

Figure 12.41

Because we know from the original equation that this graph should be a circle, we need to make some adjustments on the boundaries in order to get a complete graph. This can be done by completing the square on the original equation to change its form to $(x - 20)^2 + y^2 = 49$

or simply by a trial-and-error process. By changing the boundaries on x and y such that $-15 \leq x \leq 30$ and $-15 \leq y \leq 15$, we obtain Figure 12.42.

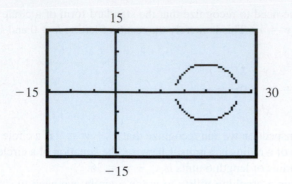

Figure 12.42

Concept Quiz 12.3

For Problems 1–10, answer true or false.

1. Equations of the form $y = ax^2 + bx + c$ can be changed by completing the square to the form $y = a(x - h)^2 + k$.

2. A circle is the set of points in a plane that are equidistant from a given fixed point.

3. A line segment determined by the center and any point on the circle is called the diameter.

4. The circle $(x + 2)^2 + (y - 5)^2 = 20$ has its center at $(2, -5)$.

5. The circle $(x - 4)^2 + (y + 3)^2 = 10$ has a radius of length 10.

6. The circle $x^2 + y^2 = 16$ has its center at the origin.

7. The graph of $y = -x^2 + 4x - 1$ does not intersect the x axis.

8. The only x intercept of the graph of $y = x^2 - 4x + 4$ is 2.

9. The origin is a point on the circle $x^2 + 4x + y^2 - 2y = 0$.

10. The vertex of the parabola $y = -2x^2 - 8x + 7$ is at $(-2, 15)$.

Problem Set 12.3

For Problems 1–22, graph each parabola. **(Objective 1)**

1. $y = x^2 - 6x + 13$
2. $y = x^2 - 4x + 7$
3. $y = x^2 + 2x + 6$
4. $y = x^2 + 8x + 14$
5. $y = x^2 - 5x + 3$
6. $y = x^2 + 3x + 1$
7. $y = x^2 + 7x + 14$
8. $y = x^2 - x - 1$
9. $y = 3x^2 - 6x + 5$
10. $y = 2x^2 + 4x + 7$
11. $y = 4x^2 - 24x + 32$
12. $y = 3x^2 + 24x + 49$
13. $y = -2x^2 - 4x - 5$
14. $y = -2x^2 + 8x - 5$
15. $y = -x^2 + 8x - 21$
16. $y = -x^2 - 6x - 7$
17. $y = 2x^2 - x + 2$
18. $y = 2x^2 + 3x + 1$
19. $y = 3x^2 + 2x + 1$
20. $y = 3x^2 - x - 1$
21. $y = -3x^2 - 7x - 2$
22. $y = -2x^2 + x - 2$

For Problems 23–34, find the x intercepts of each parabola. **(Objective 1)**

23. $y = 4x^2 - 24x + 32$ (Compare your answer to the graph of problem 11.)
24. $y = x^2 + 2x - 3$
25. $y = x^2 - 5x + 3$ (Compare your answer to the graph of problem 5.)
26. $y = x^2 + 8x + 14$
27. $y = x^2 - 6x + 13$ (Compare your answer to the graph of problem 1.)
28. $y = 2x^2 + 4x + 7$
29. $y = -3x^2 - 7x - 2$ (Compare your answer to the graph of problem 21.)
30. $y = -2x^2 - 7x - 3$

31. $y = -x^2 + 8x - 21$ (Compare your answer to the graph of problem 15.)

32. $y = -2x^2 - 4x - 5$

33. $y = 4x^2 - 12x + 9$ **34.** $y = 25x^2 + 20x + 4$

For Problems 35–46, find the center and the length of a radius of each circle. (Objective 2)

35. $x^2 + y^2 - 2x - 6y - 6 = 0$

36. $x^2 + y^2 + 4x - 12y + 39 = 0$

37. $x^2 + y^2 + 6x + 10y + 18 = 0$

38. $x^2 + y^2 - 10x + 2y + 1 = 0$

39. $x^2 + y^2 = 10$

40. $x^2 + y^2 + 4x + 14y + 50 = 0$

41. $x^2 + y^2 - 16x + 6y + 71 = 0$

42. $x^2 + y^2 = 12$

43. $x^2 + y^2 + 6x - 8y = 0$

44. $x^2 + y^2 - 16x + 30y = 0$

45. $4x^2 + 4y^2 + 4x - 32y + 33 = 0$

46. $9x^2 + 9y^2 - 6x - 12y - 40 = 0$

For Problems 47–56, graph each circle. (Objective 3)

47. $x^2 + y^2 = 25$ **48.** $x^2 + y^2 = 36$

49. $(x - 1)^2 + (y + 2)^2 = 9$ **50.** $(x + 3)^2 + (y - 2)^2 = 1$

51. $x^2 + y^2 + 6x - 2y + 6 = 0$

52. $x^2 + y^2 - 4x - 6y - 12 = 0$

53. $x^2 + y^2 + 4y - 5 = 0$ **54.** $x^2 + y^2 - 4x + 3 = 0$

55. $x^2 + y^2 + 4x + 4y - 8 = 0$

56. $x^2 + y^2 - 6x + 6y + 2 = 0$

For Problems 57–66, write the equation of each circle. Express the final equation in the form $x^2 + y^2 + Dx + Ey + F = 0$. (Objective 2)

57. Center at $(3, 5)$ and $r = 5$

58. Center at $(2, 6)$ and $r = 7$

59. Center at $(-4, 1)$ and $r = 8$

60. Center at $(-3, 7)$ and $r = 6$

61. Center at $(-2, -6)$ and $r = 3\sqrt{2}$

62. Center at $(-4, -5)$ and $r = 2\sqrt{3}$

63. Center at $(0, 0)$ and $r = 2\sqrt{5}$

64. Center at $(0, 0)$ and $r = \sqrt{7}$

65. Center at $(5, -8)$ and $r = 4\sqrt{6}$

66. Center at $(4, -10)$ and $r = 8\sqrt{2}$

67. Find the equation of the circle that passes through the origin and has its center at $(0, 4)$.

68. Find the equation of the circle that passes through the origin and has its center at $(-6, 0)$.

69. Find the equation of the circle that passes through the origin and has its center at $(-4, 3)$.

70. Find the equation of the circle that passes through the origin and has its center at $(8, -15)$.

Thoughts Into Words

71. What is the graph of $x^2 + y^2 = -4$? Explain your answer.

72. On which axis does the center of the circle $x^2 + y^2 - 8y + 7 = 0$ lie? Defend your answer.

73. Give a step-by-step description of how you would help someone graph the parabola $y = 2x^2 - 12x + 9$.

Further Investigations

74. The points (x, y) and (y, x) are mirror images of each other across the line $y = x$. Therefore, by interchanging x and y in the equation $y = ax^2 + bx + c$, we obtain the equation of its mirror image across the line $y = x$; namely, $x = ay^2 + by + c$. Thus to graph $x = y^2 + 2$, we can first graph $y = x^2 + 2$ and then reflect it across the line $y = x$, as indicated in Figure 12.43.

Graph each of the following parabolas.

(a) $x = y^2$

(b) $x = -y^2$

(c) $x = y^2 - 1$

(d) $x = -y^2 + 3$

(e) $x = -2y^2$

(f) $x = 3y^2$

(g) $x = y^2 + 4y + 7$

(h) $x = y^2 - 2y - 3$

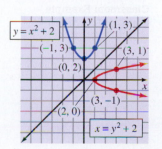

Figure 12.43

75. By expanding $(x - h)^2 + (y - k)^2 = r^2$, we obtain $x^2 - 2hx + h^2 + y^2 - 2ky + k^2 - r^2 = 0$. When we compare this result to the form $x^2 + y^2 + Dx + Ey + F = 0$, we see that $D = -2h$, $E = -2k$, and $F = h^2 + k^2 - r^2$. Therefore, the center and length of a radius of a circle can be found by using $h = \dfrac{D}{-2}$, $k = \dfrac{E}{-2}$, and $r = \sqrt{h^2 + k^2 - F}$. Use these relationships to find the center and the length of a radius of each of the following circles.

(a) $x^2 + y^2 - 2x - 8y + 8 = 0$
(b) $x^2 + y^2 + 4x - 14y + 49 = 0$
(c) $x^2 + y^2 + 12x + 8y - 12 = 0$
(d) $x^2 + y^2 - 16x + 20y + 115 = 0$
(e) $x^2 + y^2 - 12y - 45 = 0$
(f) $x^2 + y^2 + 14x = 0$

Graphing Calculator Activities

76. Use a graphing calculator to check your graphs for Problems 1–22.

77. Use a graphing calculator to graph the circles in Problems 39–41. Be sure that your graphs are consistent with the center and the length of a radius that you found when you did the problems.

78. Graph each of the following parabolas and circles. Be sure to set your boundaries so that you get a complete graph.
(a) $x^2 + 24x + y^2 + 135 = 0$
(b) $y = x^2 - 4x + 18$
(c) $x^2 + y^2 - 18y + 56 = 0$
(d) $x^2 + y^2 + 24x + 28y + 336 = 0$
(e) $y = -3x^2 - 24x - 58$
(f) $y = x^2 - 10x + 3$

Answers to the Concept Quiz

1. True   **2.** True   **3.** False   **4.** False   **5.** False   **6.** True   **7.** False   **8.** True   **9.** True   **10.** True

---

12.4   Graphing Ellipses

OBJECTIVES

1 Graph ellipses with centers at the origin

2 Graph ellipses with centers not at the origin

In the previous section, we found that the graph of the equation $x^2 + y^2 = 36$ is a circle of radius 6 units with its center at the origin. More generally, it is true that any equation of the form $Ax^2 + By^2 = C$, where $A = B$ and where A, B, and C are nonzero constants that have the same sign, is a circle with the center at the origin. For example, $3x^2 + 3y^2 = 12$ is equivalent to $x^2 + y^2 = 4$ (divide both sides of the given equation by 3), and thus it is a circle of radius 2 units with its center at the origin.

The general equation $Ax^2 + By^2 = C$ can be used to describe other geometric figures by changing the restrictions on A and B. For example, if A, B, and C are of the same sign, but $A \neq B$, then the graph of the equation $Ax^2 + By^2 = C$ is an **ellipse**. Let's consider two examples.

Classroom Example
Graph $4x^2 + 16y^2 = 64$.

> **EXAMPLE 1**   Graph $4x^2 + 25y^2 = 100$.

Solution

Let's find the x and y intercepts. Let $x = 0$; then

$$4(0)^2 + 25y^2 = 100$$
$$25y^2 = 100$$
$$y^2 = 4$$
$$y = \pm 2$$

Thus the points $(0, 2)$ and $(0, -2)$ are on the graph.
Let $y = 0$; then

$$4x^2 + 25(0)^2 = 100$$
$$4x^2 = 100$$
$$x^2 = 25$$
$$x = \pm 5$$

Thus the points $(5, 0)$ and $(-5, 0)$ are also on the
graph. We know that this figure is an ellipse, so we
plot the four points and we get a pretty good sketch of
the figure (Figure 12.44).

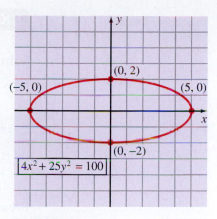

Figure 12.44

In Figure 12.44, the line segment with endpoints at $(-5, 0)$ and $(5, 0)$ is called the **major axis**
of the ellipse. The shorter line segment with endpoints at $(0, -2)$ and $(0, 2)$ is called the **minor
axis**. Establishing the endpoints of the major and minor axes provides a basis for sketching an
ellipse. The point of intersection of the major and minor axes is called the **center** of the ellipse.

Classroom Example
Graph $16x^2 + 9y^2 = 144$.

EXAMPLE 2   Graph $9x^2 + 4y^2 = 36$.

Solution

Again, let's find the x and y intercepts. Let $x = 0$; then

$$9(0)^2 + 4y^2 = 36$$
$$4y^2 = 36$$
$$y^2 = 9$$
$$y = \pm 3$$

Thus the points $(0, 3)$ and $(0, -3)$ are on the graph.
Let $y = 0$; then

$$9x^2 + 4(0)^2 = 36$$
$$9x^2 = 36$$
$$x^2 = 4$$
$$x = \pm 2$$

Thus the points $(2, 0)$ and $(-2, 0)$ are also on the
graph. The ellipse is sketched in Figure 12.45.

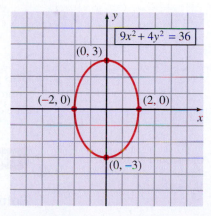

Figure 12.45

In Figure 12.45, the major axis has endpoints at $(0, -3)$ and $(0, 3)$, and the minor axis
has endpoints at $(-2, 0)$ and $(2, 0)$. The ellipses in Figures 12.44 and 12.45 are symmetric
about the x axis and about the y axis. In other words, both the x axis and the y axis serve as
axes of symmetry.

Graphing Ellipses with Centers Not at the Origin

Now we turn to some ellipses that have centers not at the origin but have major and minor axes
parallel to the x axis and the y axis. We can graph such ellipses in much the same way that we
handled circles in Section 12.2. Let's consider two examples to illustrate the procedure.

EXAMPLE 3   Graph $4x^2 + 24x + 9y^2 - 36y + 36 = 0$.

Solution

Let's complete the square on x and y as follows:

$$4x^2 + 24x + 9y^2 - 36y + 36 = 0$$
$$4(x^2 + 6x + \underline{\quad}) + 9(y^2 - 4y + \underline{\quad}) = -36$$
$$4(x^2 + 6x + 9) + 9(y^2 - 4y + 4) = -36 + 36 + 36$$
$$4(x + 3)^2 + 9(y - 2)^2 = 36$$
$$4[x - (-3)]^2 + 9(y - 2)^2 = 36$$

Because 4, 9, and 36 are of the same sign and $4 \neq 9$, the graph is an ellipse. The center of the ellipse is at $(-3, 2)$. We can find the endpoints of the major and minor axes as follows: Use the equation $4(x + 3)^2 + 9(y - 2)^2 = 36$ and let $y = 2$ (the y coordinate of the center).

$$4(x + 3)^2 + 9(2 - 2)^2 = 36$$
$$4(x + 3)^2 = 36$$
$$(x + 3)^2 = 9$$
$$x + 3 = \pm 3$$
$$x + 3 = 3 \quad \text{or} \quad x + 3 = -3$$
$$x = 0 \quad \text{or} \quad x = -6$$

This gives the points $(0, 2)$ and $(-6, 2)$. These are the coordinates of the endpoints of the major axis. Now let $x = -3$ (the x coordinate of the center).

$$4(-3 + 3)^2 + 9(y - 2)^2 = 36$$
$$9(y - 2)^2 = 36$$
$$(y - 2)^2 = 4$$
$$y - 2 = \pm 2$$
$$y - 2 = 2 \quad \text{or} \quad y - 2 = -2$$
$$y = 4 \quad \text{or} \quad y = 0$$

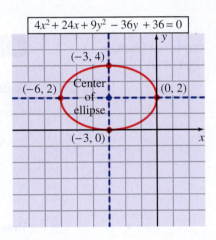

This gives the points $(-3, 4)$ and $(-3, 0)$. These are the coordinates of the endpoints of the minor axis. The ellipse is shown in Figure 12.46.

Figure 12.46

EXAMPLE 4   Graph $4x^2 - 16x + y^2 + 6y + 9 = 0$.

Solution

First let's complete the square on x and on y.

$$4x^2 - 16x + y^2 + 6y + 9 = 0$$
$$4(x^2 - 4x + \underline{\quad}) + (y^2 + 6y + \underline{\quad}) = -9$$
$$4(x^2 - 4x + 4) + (y^2 + 6y + 9) = -9 + 16 + 9$$
$$4(x - 2)^2 + (y + 3)^2 = 16$$

The center of the ellipse is at $(2, -3)$. Now let $x = 2$ (the x coordinate of the center).

$$4(2 - 2)^2 + (y + 3)^2 = 16$$
$$(y + 3)^2 = 16$$
$$y + 3 = \pm 4$$
$$y + 3 = -4 \quad \text{or} \quad y + 3 = 4$$
$$y = -7 \quad \text{or} \quad y = 1$$

This gives the points $(2, -7)$ and $(2, 1)$. These are the coordinates of the endpoints of the major axis. Now let $y = -3$ (the y coordinate of the center).

$$4(x - 2)^2 + (-3 + 3)^2 = 16$$
$$4(x - 2)^2 = 16$$
$$(x - 2)^2 = 4$$
$$x - 2 = \pm 2$$
$$x - 2 = -2 \quad \text{or} \quad x - 2 = 2$$
$$x = 0 \quad \text{or} \quad x = 4$$

This gives the points $(0, -3)$ and $(4, -3)$. These are the coordinates of the endpoints of the minor axis. The ellipse is shown in Figure 12.47.

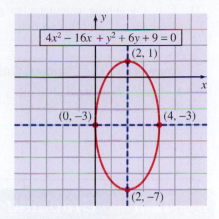

Figure 12.47

Concept Quiz 12.4

For Problems 1–10, answer true or false.

1. The length of the major axis of an ellipse is always greater than the length of the minor axis.

2. The major axis of an ellipse is always parallel to the x axis.

3. The axes of symmetry for an ellipse pass through the center of the ellipse.

4. The ellipse $9(x - 1)^2 + 4(y + 5)^2 = 36$ has its center at $(1, 5)$.

5. The x and y intercepts of the graph of an ellipse centered at the origin and symmetric to both axes are the endpoints of its axes.

6. The endpoints of the major axis of the ellipse $9x^2 + 4y^2 = 36$ are at $(-2, 0)$ and $(2, 0)$.

7. The endpoints of the minor axis of the ellipse $x^2 + 5y^2 = 15$ are at $\left(0, -\sqrt{3}\right)$ and $\left(0, \sqrt{3}\right)$.

8. The endpoints of the major axis of the ellipse $3(x - 2)^2 + 5(y + 3)^2 = 12$ are at $(0, -3)$ and $(4, -3)$.

9. The center of the ellipse $7x^2 - 14x + 8y^2 - 32y - 17 = 0$ is at $(1, 2)$.

10. The center of the ellipse $2x^2 + 12x + y^2 + 2 = 0$ is on the y axis.

Problem Set 12.4

For Problems 1–16, graph each ellipse. **(Objectives 1 and 2)**

1. $x^2 + 4y^2 = 36$

2. $x^2 + 4y^2 = 16$

3. $9x^2 + y^2 = 36$

4. $16x^2 + 9y^2 = 144$

5. $4x^2 + 3y^2 = 12$

6. $5x^2 + 4y^2 = 20$

7. $16x^2 + y^2 = 16$

8. $9x^2 + 2y^2 = 18$

9. $25x^2 + 2y^2 = 50$

10. $12x^2 + y^2 = 36$

11. $4x^2 + 8x + 16y^2 - 64y + 4 = 0$

12. $9x^2 - 36x + 4y^2 - 24y + 36 = 0$

13. $x^2 + 8x + 9y^2 + 36y + 16 = 0$

14. $4x^2 - 24x + y^2 + 4y + 24 = 0$

15. $4x^2 + 9y^2 - 54y + 45 = 0$

16. $x^2 + 2x + 4y^2 - 15 = 0$

Thoughts Into Words

17. Is the graph of $x^2 + y^2 = 4$ the same as the graph of $y^2 + x^2 = 4$? Explain your answer.

18. Is the graph of $x^2 + y^2 = 0$ a circle? If so, what is the length of a radius?

19. Is the graph of $4x^2 + 9y^2 = 36$ the same as the graph of $9x^2 + 4y^2 = 36$? Explain your answer.

20. What is the graph of $x^2 + 2y^2 = -16$? Explain your answer.

Graphing Calculator Activities

21. Use a graphing calculator to graph the ellipses in Examples 1–4 of this section.

22. Use a graphing calculator to check your graphs for Problems 11–16.

Answers to the Concept Quiz

1. True   **2.** False   **3.** True   **4.** False   **5.** True   **6.** False   **7.** True   **8.** True   **9.** True   **10.** False

---

12.5   Graphing Hyperbolas

OBJECTIVES

1.   Graph hyperbolas symmetric to both axes

2.   Graph hyperbolas not symmetric to both axes

The graph of an equation of the form $Ax^2 + By^2 = C$, where A, B, and C are nonzero real numbers and A and B are of unlike signs, is a **hyperbola**. Let's use some examples to illustrate a procedure for graphing hyperbolas.

Classroom Example
Graph $x^2 - y^2 = 16$.

> **EXAMPLE 1**   Graph $x^2 - y^2 = 9$.

Solution

If we let $y = 0$, we obtain

$$x^2 - 0^2 = 0$$
$$x^2 = 9$$
$$x = \pm 3$$

Thus the points $(3, 0)$ and $(-3, 0)$ are on the graph. If we let $x = 0$, we obtain

$$0^2 - y^2 = 9$$
$$-y^2 = 9$$
$$y^2 = -9$$

Because $y^2 = -9$ has no real number solutions, there are no points of the y axis on this graph. That is, the graph does not intersect the y axis. Now let's solve the given equation for y so that we have a more convenient form for finding other solutions.

$$x^2 - y^2 = 9$$
$$-y^2 = 9 - x^2$$
$$y^2 = x^2 - 9$$
$$y = \pm\sqrt{x^2 - 9}$$

The radicand, $x^2 - 9$, must be nonnegative, so the values we choose for x must be greater than or equal to 3 or less than or equal to -3. With this in mind, we can form the following table of values.

x	y	
3	0	Intercepts
−3	0	
4	$\pm\sqrt{7}$	
−4	$\pm\sqrt{7}$	Other points
5	± 4	
−5	± 4	

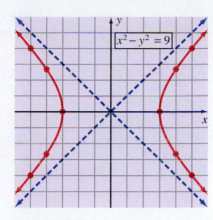

$x^2 - y^2 = 9$

Plot these points and draw the hyperbola as in Figure 12.48. (This graph is symmetric about both axes.)

Figure 12.48

Note the blue lines in Figure 12.48; they are called **asymptotes**. Each branch of the hyperbola approaches one of these lines but does not intersect it. Therefore, the ability to sketch the asymptotes of a hyperbola is very helpful when we are graphing the hyperbola. Fortunately, the equations of the asymptotes are easy to determine. They can be found by replacing the constant term in the given equation of the hyperbola with 0 and solving for y. (The reason why this works will become evident in a later course.) Thus for the hyperbola in Example 3, we obtain

$$x^2 - y^2 = 0$$
$$y^2 = x^2$$
$$y = \pm x$$

Thus the two lines $y = x$ and $y = -x$ are the asymptotes indicated by the blue lines in Figure 12.48.

Classroom Example
Graph $y^2 - 6x^2 = 9$.

EXAMPLE 2   Graph $y^2 - 5x^2 = 4$.

Solution

If we let $x = 0$, we obtain

$$y^2 - 5(0)^2 = 4$$
$$y^2 = 4$$
$$y = \pm 2$$

The points $(0, 2)$ and $(0, -2)$ are on the graph. If we let $y = 0$, we obtain

$$0^2 - 5x^2 = 4$$
$$-5x^2 = 4$$
$$x^2 = -\frac{4}{5}$$

Because $x^2 = -\dfrac{4}{5}$ has no real number solutions, we know that this hyperbola does not intersect the x axis. Solving the given equation for y yields

$$y^2 - 5x^2 = 4$$
$$y^2 = 5x^2 + 4$$
$$y = \pm\sqrt{5x^2 + 4}$$

The table below shows some additional solutions for the equation. The equations of the asymptotes are determined as follows:

$$y^2 - 5x^2 = 0$$
$$y^2 = 5x^2$$
$$y = \pm\sqrt{5}x$$

x	y	
0	2	
0	−2	Intercepts
1	±3	
−1	±3	Other points
2	± 2√6	
−2	± 2√6	

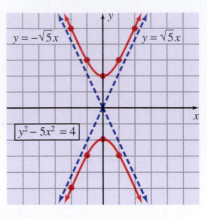

Figure 12.49

Sketch the asymptotes and plot the points determined by the table of values to determine the hyperbola in Figure 12.49. (Note that this hyperbola is also symmetric about the x axis and the y axis.)

Classroom Example
Graph $9x^2 - 16y^2 = 144$.

EXAMPLE 3   Graph $4x^2 - 9y^2 = 36$.

Solution

If we let $x = 0$, we obtain

$$4(0)^2 - 9y^2 = 36$$
$$-9y^2 = 36$$
$$y^2 = -4$$

Because $y^2 = -4$ has no real number solutions, we know that this hyperbola does not intersect the y axis. If we let $y = 0$, we obtain

$$4x^2 - 9(0)^2 = 36$$
$$4x^2 = 36$$
$$x^2 = 9$$
$$x = \pm 3$$

Thus the points $(3, 0)$ and $(-3, 0)$ are on the graph. Now let's solve the equation for y in terms of x and set up a table of values.

$$4x^2 - 9y^2 = 36$$
$$-9y^2 = 36 - 4x^2$$
$$9y^2 = 4x^2 - 36$$
$$y^2 = \frac{4x^2 - 36}{9}$$
$$y = \pm\frac{\sqrt{4x^2 - 36}}{3}$$

x	y	
3	0	
−3	0	Intercepts
4	$\pm\dfrac{2\sqrt{7}}{3}$	
−4	$\pm\dfrac{2\sqrt{7}}{3}$	Other points
5	$\pm\dfrac{8}{3}$	
−5	$\pm\dfrac{8}{3}$	

The equations of the asymptotes are found as follows:

$$4x^2 - 9y^2 = 0$$
$$-9y^2 = -4x^2$$
$$9y^2 = 4x^2$$
$$y^2 = \frac{4x^2}{9}$$
$$y = \pm\frac{2}{3}x$$

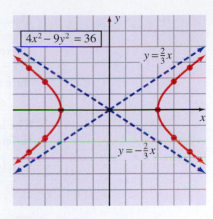

Figure 12.50

Sketch the asymptotes and plot the points determined by the table to determine the hyperbola as shown in Figure 12.50.

Graphing Hyperbolas Not Symmetric to Both Axes

Now let's consider hyperbolas that are not symmetric with respect to the origin but are symmetric with respect to lines parallel to one of the axes—that is, vertical and horizontal lines. Again, let's use examples to illustrate a procedure for graphing such hyperbolas.

Classroom Example
Graph $25x^2 + 50x - 4y^2 + 24y - 111 = 0$.

EXAMPLE 4   Graph $4x^2 - 8x - y^2 - 4y - 16 = 0$.

Solution

Completing the square on x and y, we obtain

$$4x^2 - 8x - y^2 - 4y - 16 = 0$$
$$4(x^2 - 2x + \underline{\quad}) - (y^2 + 4y + \underline{\quad}) = 16$$
$$4(x^2 - 2x + 1) - (y^2 + 4y + 4) = 16 + 4 - 4$$
$$4(x - 1)^2 - (y + 2)^2 = 16$$
$$4(x - 1)^2 - 1(y - (-2))^2 = 16$$

Because 4 and -1 are of opposite signs, the graph is a hyperbola. The center of the hyperbola is at $(1, -2)$.

Now using the equation $4(x - 1)^2 - (y + 2)^2 = 16$, we can proceed as follows: Let $y = -2$; then

$$4(x - 1)^2 - (-2 + 2)^2 = 16$$
$$4(x - 1)^2 = 16$$
$$(x - 1)^2 = 4$$
$$x - 1 = \pm 2$$
$$x - 1 = 2 \quad \text{or} \quad x - 1 = -2$$
$$x = 3 \quad \text{or} \quad x = -1$$

Thus the hyperbola intersects the horizontal line $y = -2$ at $(3, -2)$ and at $(-1, -2)$. Let $x = 1$; then

$$4(1 - 1)^2 - (y + 2)^2 = 16$$
$$-(y + 2)^2 = 16$$
$$(y + 2)^2 = -16$$

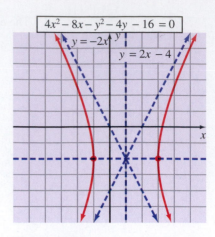

Because $(y + 2)^2 = -16$ has no real number solutions, we know that the hyperbola does not intersect the vertical line $x = 1$. We replace the constant term of $4(x - 1)^2 - (y + 2)^2 = 16$ with 0 and solve for y to produce the equations of the asymptotes as follows:

$$4(x - 1)^2 - (y + 2)^2 = 0$$

The left side can be factored using the pattern of the difference of squares.

$$[2(x - 1) + (y + 2)][2(x - 1) - (y + 2)] = 0$$
$$(2x - 2 + y + 2)(2x - 2 - y - 2) = 0$$
$$(2x + y)(2x - y - 4) = 0$$

$$2x + y = 0 \qquad \text{or} \qquad 2x - y - 4 = 0$$
$$y = -2x \qquad \text{or} \qquad 2x - 4 = y$$

Figure 12.51

Thus the equations of the asymptotes are $y = -2x$ and $y = 2x - 4$. Sketching the asymptotes and plotting the two points $(3, -2)$ and $(-1, -2)$, we can draw the hyperbola as shown in Figure 12.51.

Classroom Example
Graph $4y^2 - 16y - x^2 - 4x + 8 = 0$.

EXAMPLE 5   Graph $y^2 - 4y - 4x^2 - 24x - 36 = 0$.

Solution

First let's complete the square on x and on y.

$$y^2 - 4y - 4x^2 - 24x - 36 = 0$$
$$(y^2 - 4y + \underline{}) - 4(x^2 + 6x + \underline{}) = 36$$
$$(y^2 - 4y + 4) - 4(x^2 + 6x + 9) = 36 + 4 - 36$$
$$(y - 2)^2 - 4(x + 3)^2 = 4$$

The center of the hyperbola is at $(-3, 2)$. Now let $y = 2$.

$$(2 - 2)^2 - 4(x + 3)^2 = 4$$
$$-4(x + 3)^2 = 4$$
$$(x + 3)^2 = -1$$

Because $(x + 3)^2 = -1$ has no real number solutions, the graph does not intersect the line $y = 2$. Now let $x = -3$.

$$(y - 2)^2 - 4(-3 + 3)^2 = 4$$
$$(y - 2)^2 = 4$$
$$y - 2 = \pm 2$$

$$y - 2 = -2 \qquad \text{or} \qquad y - 2 = 2$$
$$y = 0 \qquad \text{or} \qquad y = 4$$

Therefore, the hyperbola intersects the line $x = -3$ at $(-3, 0)$ and $(-3, 4)$. Now, to find the equations of the asymptotes, let's replace the constant term of $(y - 2)^2 - 4(x + 3)^2 = 4$ with 0 and solve for y.

$$(y - 2)^2 - 4(x + 3)^2 = 0$$
$$[(y - 2) + 2(x + 3)][(y - 2) - 2(x + 3)] = 0$$
$$(y - 2 + 2x + 6)(y - 2 - 2x - 6) = 0$$
$$(y + 2x + 4)(y - 2x - 8) = 0$$

$$y + 2x + 4 = 0 \qquad \text{or} \quad y - 2x - 8 = 0$$
$$y = -2x - 4 \quad \text{or} \qquad\qquad y = 2x + 8$$

Therefore, the equations of the asymptotes are $y = -2x - 4$ and $y = 2x + 8$. Drawing the asymptotes and plotting the points $(-3, 0)$ and $(-3, 4)$, we can graph the hyperbola as shown in Figure 12.52.

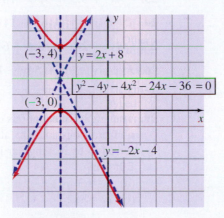

Figure 12.52

As a way of summarizing our work with conic sections, let's focus our attention on the continuity pattern used in this chapter. In Sections 12.2 and 12.3, we studied parabolas by considering variations of the basic quadratic equation $y = ax^2 + bx + c$. Also in Section 12.3, we used the definition of a circle to generate a standard form for the equation of a circle. Then in Sections 12.4 and 12.5, we discussed ellipses and hyperbolas, not from a definition viewpoint, but by considering variations of the equations $Ax^2 + By^2 = C$ and $A(x - h)^2 + B(y - k)^2 = C$. In a subsequent mathematics course, parabolas, ellipses, and hyperbolas will be developed from a definition viewpoint. That is, first each concept will be defined, and then the definition will be used to generate a standard form of its equation.

Concept Quiz 12.5

For Problems 1–10, answer true or false.

1. The graph of an equation of the form $Ax^2 + By^2 = C$, where A, B, and C are nonzero real numbers, is a hyperbola if A and B are of like sign.

2. The graph of a hyperbola always has two branches.

3. Each branch of the graph of a hyperbola intersects one of the asymptotes.

4. To find the equations for the asymptotes, we replace the constant term in the equation of the hyperbola with zero and solve for y.

5. The hyperbola $9(x + 1)^2 - 4(y - 3)^2 = 36$ has its center at $(-1, 3)$.

6. The asymptotes of the graph of a hyperbola intersect at the center of the hyperbola.

7. The equations of the asymptotes for the hyperbola $9x^2 - 4y^2 = 36$ are $y = -\dfrac{2}{3}x$ and $y = \dfrac{2}{3}x$.

8. The center of the hyperbola $(y - 2)^2 - 9(x - 6)^2 = 18$ is at $(2, 6)$.

9. The equations of the asymptotes for the hyperbola $(y - 2)^2 - 9(x - 6)^2 = 18$ are $y = 3x - 16$ and $y = -3x + 20$.

10. The center of the hyperbola $x^2 + 6x - y^2 + 4y - 15 = 0$ is at $(-3, 2)$.

Problem Set 12.5

For Problems 1–6, find the intercepts and the equations for the asymptotes.

1. $x^2 - 9y^2 = 16$ **2.** $16x^2 - y^2 = 25$

3. $y^2 - 9x^2 = 36$ **4.** $4x^2 - 9y^2 = 16$

5. $25x^2 - 9y^2 = 4$ **6.** $y^2 - x^2 = 16$

For Problems 7–18, graph each hyperbola. **(Objective 1)**

7. $x^2 - y^2 = 1$ **8.** $x^2 - y^2 = 4$

9. $y^2 - 4x^2 = 9$ **10.** $4y^2 - x^2 = 16$

11. $5x^2 - 2y^2 = 20$ **12.** $9x^2 - 4y^2 = 9$

13. $y^2 - 16x^2 = 4$ **14.** $y^2 - 9x^2 = 16$

15. $-4x^2 + y^2 = -4$ **16.** $-9x^2 + y^2 = -36$

17. $25y^2 - 3x^2 = 75$ **18.** $16y^2 - 5x^2 = 80$

For Problems 19–22, find the equations for the asymptotes.

19. $x^2 + 4x - y^2 - 6y - 30 = 0$

20. $y^2 - 8y - x^2 - 4x + 3 = 0$

21. $9x^2 - 18x - 4y^2 - 24y - 63 = 0$

22. $4x^2 + 24x - y^2 + 4y + 28 = 0$

For Problems 23–28, graph each hyperbola. **(Objective 2)**

23. $-4x^2 + 32x + 9y^2 - 18y - 91 = 0$

24. $x^2 - 4x - y^2 + 6y - 14 = 0$

25. $-4x^2 + 24x + 16y^2 + 64y - 36 = 0$

26. $x^2 + 4x - 9y^2 + 54y - 113 = 0$

27. $4x^2 - 24x - 9y^2 = 0$ **28.** $16y^2 + 64y - x^2 = 0$

29. The graphs of equations of the form $xy = k$, where k is a nonzero constant, are also hyperbolas, sometimes referred to as rectangular hyperbolas. Graph each of the following.

  (a) $xy = 3$ **(b)** $xy = 5$

  (c) $xy = -2$ **(d)** $xy = -4$

30. What is the graph of $xy = 0$? Defend your answer.

31. We have graphed various equations of the form $Ax^2 + By^2 = C$, where C is a nonzero constant. Now graph each of the following.

  (a) $x^2 + y^2 = 0$ **(b)** $2x^2 + 3y^2 = 0$

  (c) $x^2 - y^2 = 0$ **(d)** $4y^2 - x^2 = 0$

Thoughts Into Words

32. Explain the concept of an asymptote.

33. Explain how asymptotes can be used to help graph hyperbolas.

34. Are the graphs of $x^2 - y^2 = 0$ and $y^2 - x^2 = 0$ identical? Are the graphs of $x^2 - y^2 = 4$ and $y^2 - x^2 = 4$ identical? Explain your answers.

Graphing Calculator Activities

35. To graph the hyperbola in Example 1 of this section, we can make the following assignments for the graphing calculator.

$$Y_1 = \sqrt{x^2 - 9} \qquad Y_2 = -Y_1$$
$$Y_3 = x \qquad Y_4 = -Y_3$$

See if your graph agrees with Figure 12.48. Also graph the asymptotes and hyperbolas for Examples 2 and 3.

36. Use a graphing calculator to check your graphs for Problems 7–12.

37. Use a graphing calculator to check your graphs for Problems 23–28.

38. For each of the following equations, (1) predict the type and location of the graph, and (2) use your graphing calculator to check your predictions.

  (a) $x^2 + y^2 = 100$ **(b)** $x^2 - y^2 = 100$

  (c) $y^2 - x^2 = 100$ **(d)** $y = -x^2 + 9$

  (e) $2x^2 + y^2 = 14$ **(f)** $x^2 + 2y^2 = 14$

  (g) $x^2 + 2x + y^2 - 4 = 0$ **(h)** $x^2 + y^2 - 4y - 2 = 0$

  (i) $y = x^2 + 16$ **(j)** $y^2 = x^2 + 16$

  (k) $9x^2 - 4y^2 = 72$ **(l)** $4x^2 - 9y^2 = 72$

  (m) $y^2 = -x^2 - 4x + 6$

Chapter 12 Summary

OBJECTIVE	SUMMARY	EXAMPLE
Find the distance between two points. **(Section 12.1/Objective 1)**	The distance between any two points (x_1, y_1) and (x_2, y_2) is given by the distance formula $d = \sqrt{(x_2 - x_1)^2 + (y_2 - y_1)^2}$.	Find the distance between $(1, -5)$ and $(4, 2)$. **Solution** $d = \sqrt{(x_2 - x_1)^2 + (y_2 - y_1)^2}$ $d = \sqrt{(4 - 1)^2 + [2 - (-5)]^2}$ $d = \sqrt{(3)^2 + (7)^2}$ $d = \sqrt{9 + 49} = \sqrt{58}$
Determine the slope of a line. **(Section 12.1/Objective 2)**	The slope (denoted by m) of a line determined by the points (x_1, y_1) and (x_2, y_2) is given by the slope formula $m = \dfrac{y_2 - y_1}{x_2 - x_1}$ where $x_2 \neq x_1$.	Find the slope of a line that contains the points $(-1, 2)$ and $(7, 8)$. **Solution** Use the slope formula: $m = \dfrac{8 - 2}{7 - (-1)} = \dfrac{6}{8} = \dfrac{3}{4}$ Thus the slope of the line is $\dfrac{3}{4}$.
Determine if the graph of an equation is symmetric to the x axis, the y axis, or the origin. **(Section 12.1/Objective 3)**	The graph of an equation is symmetric with respect to the y axis if replacing x with $-x$ results in an equivalent equation. The graph of an equation is symmetric with respect to the x axis if replacing y with $-y$ results in an equivalent equation. The graph of an equation is symmetric with respect to origin if replacing x with $-x$ and y with $-y$ results in an equivalent equation.	Determine the type of symmetry exhibited by the graph of the following equation: $x = y^2 + 4$ **Solution** Replacing x with $-x$ gives $-x = y^2 + 4$. This is not an equivalent equation, so the graph will not exhibit y-axis symmetry. Replacing y with $-y$ gives $x = (-y)^2 + 4 = y^2 + 4$. This is an equivalent equation, so the graph will exhibit x-axis symmetry. Replacing x with $-x$ and y with $-y$ gives: $(-x) = (-y)^2 + 4$ $-x = y^2 + 4$ This is not an equivalent equation, so the graph will not exhibit symmetry with respect to the origin.

(continued)

OBJECTIVE	SUMMARY	EXAMPLE
Graph equations using intercepts, symmetries, and plotting points. **(Section 12.1/Objective 4)**	The following suggestions are offered for graphing an equation in two variables. 1. Determine what type of symmetry the equation exhibits. 2. Find the intercepts. 3. Solve the equation for y or x if it is not already in such a form. 4. Set up a table of ordered pairs that satisfies the equation. The type of symmetry will affect your choice of values in the table. 5. Plot the points associated with the ordered pairs and connect them with a smooth curve. Then, if appropriate, reflect this part of the curve according to the symmetry shown by the equation.	Graph $x - y^2 + 4 = 0$. **Solution** Replacing y with $-y$ gives an equivalent equation, so the graph will be symmetric with respect to the x axis. To find the x intercept, let $y = 0$ and solve for x. This gives an x intercept of -4. To find the y intercept, let $x = 0$ and solve for y. This gives y intercepts of 2 and -2. Solving the equation for x gives the equation $x = y^2 - 4$. Choose values for y to obtain the table of points. <table><tr><th>x</th><th>y</th></tr><tr><td>−3</td><td>1</td></tr><tr><td>0</td><td>2</td></tr><tr><td>5</td><td>3</td></tr></table> 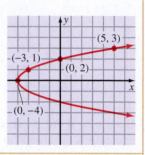
Graph parabolas. **(Section 12.2/Objective 1)**	To graph parabolas, you need to be able to: 1. Find the vertex. 2. Determine whether the parabola opens upward or downward. 3. Locate two points on opposite sides of the line of symmetry. 4. Compare the parabola to the basic parabola $y = x^2$. The following diagram summarizes the graphing of parabolas. 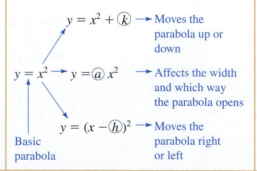	Graph $y = -(x + 2)^2 + 5$. **Solution** The vertex is located at $(-2, 5)$. The parabola opens downward. The points $(-4, 1)$ and $(0, 1)$ are on the parabola. 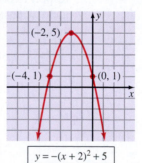

OBJECTIVE	SUMMARY	EXAMPLE
Graph quadratic equations of the form $y = ax^2 + bx + c$. **(Section 12.3/Objective 1)**	The general approach to graph equations of the form $y = ax^2 + bx + c$ is to change the form of the equation by completing the square.	Graph $y = x^2 - 6x + 11$. **Solution** $y = x^2 - 6x + 11$ $y = x^2 - 6x + 9 - 9 + 11$ $y = (x - 3)^2 + 2$ The vertex is located at $(3, 2)$. The parabola opens upward. The points $(1, 6)$ and $(5, 6)$ are on the parabola. 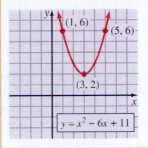
Write the equation of a circle in standard form. **(Section 12.3/Objective 2)**	The standard form of the equation of a circle is $(x - h)^2 + (y - k)^2 = r^2$. We can use the standard form of the equation of a circle to solve two basic kinds of circle problems: 1. Given the coordinates of the center and the length of a radius of a circle, find its equation. 2. Given the equation of a circle, find its center and the length of a radius.	Write the equation of a circle that has its center at $(-7, 3)$ and a radius of length 4 units. **Solution** Substitute -7 for h, 3 for k, and 4 for r in $(x - h)^2 + (y - k)^2 = r^2$. Then $[x - (-7)]^2 + (y - 3)^2 = 4^2$ $(x + 7)^2 + (y - 3)^2 = 16$ or $x^2 + y^2 + 14x - 6y + 42 = 0$
Graph a circle. **(Section 12.3/Objective 3)**	To graph a circle, put the equation in standard form $(x - h)^2 + (y - k)^2 = r^2$. It may be necessary to use completing the square to change the equation to standard form. The center will be at (h, k) and the length of a radius is r.	Graph $x^2 - 4x + y^2 + 2y - 4 = 0$. **Solution** Use completing the square to change the form of the equation. $x^2 - 4x + y^2 + 2y - 4 = 0$ $(x - 2)^2 + (y + 1)^2 = 9$ The center of the circle is at $(2, -1)$ and the length of a radius is 3. 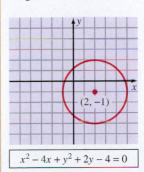

(continued)

OBJECTIVE	SUMMARY	EXAMPLE
Graph ellipses with centers at the origin. **(Section 12.4/Objective 1)**	The graph of the equation $Ax^2 + By^2 = C$, where A, B, and C are nonzero real numbers of the same sign and $A \neq B$, is an ellipse with the center at $(0, 0)$. The intercepts are the endpoints of the axes of the ellipse. The longer axis is called the major axis, and the shorter axis is called the minor axis. The center is at the point of intersection of the major and minor axes.	Graph $4x^2 + y^2 = 16$. **Solution** The coordinates of the intercepts are $(0, 4)$, $(0, -4)$, $(2, 0)$, and $(-2, 0)$.
Graph ellipses with centers not at the origin. **(Section 12.4/Objective 2)**	The standard form of the equation of an ellipse whose center is not at the origin is $$A(x - h)^2 + B(y - k)^2 = C$$ where A, B, and C are nonzero real numbers of the same sign and $A \neq B$. The center will be at (h, k). Completing the square is often used to change the equation of an ellipse to standard form.	Graph $9x^2 + 36x + 4y^2 - 24y + 36 = 0$ **Solution** Use completing the square to change the equation to the equivalent form $$9(x + 2)^2 + 4(y - 3)^2 = 36$$ The center is at $(-2, 3)$. Substitute -2 for x to obtain $(-2, 6)$ and $(-2, 0)$ as the endpoints of the major axis. Substitute 3 for y to obtain $(0, 3)$ and $(-4, 3)$ as the endpoints of the minor axis.

OBJECTIVE	SUMMARY	EXAMPLE
Graph hyperbolas symmetric to both axes. (Section 12.5/Objective 1)	The graph of the equation $Ax^2 + By^2 = C$, where A, B, and C are nonzero real numbers, and A and B, are of unlike signs, is a hyperbola that is symmetric to both axes. Each branch of the hyperbola approaches a line called the asymptote. The equation for the asymptotes can be found by replacing the constant term with zero and solving for y.	Graph $9x^2 - y^2 = 9$. **Solution** If $y = 0$, then $x = \pm 1$. Hence the points $(1, 0)$ and $(-1, 0)$ are on the graph. To find the asymptotes, replace the constant term with zero and solve for y. $9x^2 - y^2 = 0$ $9x^2 = y^2$ $y = \pm 3x$ So the equations of the asymptotes are $y = 3x$ and $y = -3x$.
Graph hyperbolas not symmetric to both axes. (Section 12.5/Objective 2)	The graph of the equation $A(x - h)^2 + B(y - k)^2 = C$, where A, B, and C are nonzero real numbers, and A and B are of unlike signs, is a hyperbola that is not symmetric with respect to both axes.	Graph $4y^2 + 40y - x^2 + 4x + 92 = 0$ **Solution** Complete the square to change the equation to the equivalent form $4(y + 5)^2 - (x - 2)^2 = 4$ If $x = 2$, then $y = -4$ or $y = -6$. Hence the points $(2, -4)$ and $(2, -6)$ are on the graph. To find the asymptotes, replace the constant term with zero and solve for y. The equations of the asymptotes are $y = \frac{1}{2}x - 6 \quad \text{and} \quad y = -\frac{1}{2}x - 4$
Graph conic sections.	Part of the challenge of graphing conic sections is to know the characteristics of the equations that produce each type of conic section.	

Chapter 12   Review Problem Set

1. Find the slope of the line determined by each pair of points.

   a. $(3, 4), (-2, -2)$ **b.** $(-2, 3), (4, -1)$

2. Find the slope of each of the following lines.

   a. $4x + y = 7$ **b.** $2x - 7y = 3$

3. Find the lengths of the sides of a triangle whose vertices are at $(2, 3)$, $(5, -1)$, and $(-4, -5)$.

For Problems 4–8, write the equation of the line that satisfies the stated conditions. Express final equations in standard form. (These are review problems from Chapter 5.)

4. Containing the points $(-1, 2)$ and $(3, -5)$

5. Having a slope of $-\dfrac{3}{7}$ and a y intercept of 4

6. Containing the point $(-1, -6)$ and having a slope of $\dfrac{2}{3}$

7. Containing the point $(2, 5)$ and parallel to the line $x - 2y = 4$

8. Containing the point $(-2, -6)$ and perpendicular to the line $3x + 2y = 12$

For Problems 9–13, determine the type(s) of symmetry (with respect to the x axis, y axis, and/or origin) exhibited by the graph of each equation. Do not sketch the graphs.

9. $x^2 = y + 1$

10. $y^2 = x^3$

11. $y = -2x$

12. $y = \dfrac{8}{x^2 + 6}$

13. $x^2 - 2y^2 = -3$

For Problems 14–19, graph each equation.

14. $2x - y = 6$

15. $y = -2x^2 - 1$

16. $y = -2x - 1$

17. $y = -4x$

18. $xy^2 = -1$

19. $y = \dfrac{-3}{x^2 + 1}$

20. A certain highway has a 6% grade. How many feet does it rise in a horizontal distance of 1 mile?

21. If the ratio of rise to run is to be $\dfrac{2}{3}$ for the steps of a staircase, and the run is 12 inches, find the rise.

For Problems 22–29, find the vertex of each parabola.

22. $y = x^2 + 6$

23. $y = -x^2 - 8$

24. $y = (x + 3)^2 - 1$

25. $y = (x + 1)^2$

26. $y = x^2 - 14x + 54$

27. $y = -x^2 + 12x - 44$

28. $y = 3x^2 + 24x + 39$

29. $y = 2x^2 + 8x + 5$

For Problems 30–33, write the equation of the circle satisfying the given conditions. Express your answers in the form $x^2 + y^2 + Dx + Ey + F = 0$.

30. Center at $(0, 0)$ and $r = 6$

31. Center at $(2, -6)$ and $r = 5$

32. Center at $(-4, -8)$ and $r = 2\sqrt{3}$

33. Center at $(0, 5)$ and passes through the origin.

For Problems 34–37, find the center and the length of a radius for each circle and then graph the circle.

34. $x^2 + 14x + y^2 - 8y + 16 = 0$

35. $x^2 + 16x + y^2 + 39 = 0$

36. $x^2 - 12x + y^2 + 16y = 0$

37. $x^2 + y^2 = 24$

For Problems 38–45, find the length of the major axis and the length of the minor axis of each ellipse.

38. $16x^2 + y^2 = 64$ 39. $16x^2 + 9y^2 = 144$

40. $4x^2 + 25y^2 = 100$ 41. $2x^2 + 7y^2 = 28$

42. $x^2 - 4x + 9y^2 + 54y + 76 = 0$

43. $x^2 + 6x + 4y^2 - 16y + 9 = 0$

44. $9x^2 + 72x + 4y^2 - 8y + 112 = 0$

45. $16x^2 - 32x + 9y^2 - 54y - 47 = 0$

For Problems 46–51, find the equations of the asymptotes of each hyperbola.

46. $x^2 - 9y^2 = 25$ 47. $4x^2 - y^2 = 16$

48. $9y^2 - 25x^2 = 36$ 49. $16y^2 - 4x^2 = 17$

50. $25x^2 + 100x - 4y^2 + 24y - 36 = 0$

51. $36y^2 - 288y - x^2 + 2x + 539 = 0$

For Problems 52–65, graph each equation.

52. $9x^2 + y^2 = 81$ 53. $9x^2 - y^2 = 81$

54. $y = -2x^2 + 3$ 55. $y = 4x^2 - 16x + 19$

56. $x^2 + 4x + y^2 + 8y + 11 = 0$

57. $4x^2 - 8x + y^2 + 8y + 4 = 0$

58. $y^2 + 6y - 4x^2 - 24x - 63 = 0$

59. $y = -2x^2 - 4x - 3$ 60. $x^2 - y^2 = -9$

61. $4x^2 + 16y^2 + 96y = 0$ 62. $(x - 3)^2 + (y + 1)^2 = 4$

63. $(x + 1)^2 + (y - 2)^2 = 4$

64. $x^2 + y^2 - 6x - 2y + 4 = 0$

65. $x^2 + y^2 - 2y - 8 = 0$

1. Find the slope of the line determined by the points $(-2, 4)$ and $(3, -2)$.

2. Find the slope of the line determined by the equation $3x - 7y = 12$.

3. Find the length of the line segment whose endpoints are $(4, 2)$ and $(-3, -1)$. Express the answer in simplest radical form.

4. Find the x intercepts of the parabola $y = 2x^2 + 9x - 5$.

5. Find the x intercepts of the parabola $y = -6x^2 - x + 15$.

6. What type of symmetry does the graph of $y = x^2 - 4$ possess?

7. What type of symmetry does the graph of $x^2 - 2x + y^2 = 10$ possess?

8. The grade of a highway up a hill is 25%. How much change in horizontal distance is there if the vertical height of the hill is 120 feet?

9. If the ratio of rise to run is to be $\dfrac{3}{4}$ for the steps of a staircase, and the rise is 32 centimeters, find the run to the nearest centimeter.

10. Find the vertex of the parabola $y = x^2 - 6x + 9$.

11. Find the vertex of the parabola $y = 4x^2 + 32x + 62$.

12. Write the equation of the circle with its center at $(2, 8)$ and a radius of length 3 units.

13. Find the center and the length of a radius of the circle $x^2 - 12x + y^2 + 8y + 3 = 0$.

14. Find the length of the major axis of the ellipse $9x^2 + 2y^2 = 32$.

15. Find the length of the minor axis of the ellipse $8x^2 - 32x + 5y^2 + 30y + 45 = 0$.

16. Find the equations of the asymptotes for the hyperbola $y^2 - 16x^2 = 36$.

For Problems 17–25, graph each equation.

17. $\dfrac{1}{3}x + \dfrac{1}{2}y = 2$

18. $y = \dfrac{-x - 1}{4}$

19. $x^2 - 4y^2 = -16$

20. $y = x^2 + 4x$

21. $x^2 + 2x + y^2 + 8y + 8 = 0$

22. $2x^2 + 3y^2 = 12$

23. $y = 2x^2 + 12x + 22$

24. $9x^2 - y^2 = 9$

25. $3x^2 - 12x + 5y^2 + 10y - 10 = 0$

1. Evaluate each of the following numerical expressions.

  a. $\sqrt[3]{-\dfrac{8}{27}}$ **b.** $-16^{\frac{5}{4}}$

  c. $\left(\dfrac{2^{-1}}{2^{-2}}\right)^{-2}$ **d.** $\left(\dfrac{1}{3}-\dfrac{1}{5}\right)^{-1}$

2. Express each of the following in simplest radical form.

  a. $\sqrt{96}$ **b.** $\sqrt[3]{40}$

  c. $\dfrac{4\sqrt{2}}{6\sqrt{3}}$ **d.** $\dfrac{2}{\sqrt{7}-\sqrt{2}}$

For Problems 3–7, perform the indicated operations and express the answers in simplifed form.

3. $(4xy^2)(-3x^2y^3)(2x)$

4. $(2x+1)(3x^2-4x-2)$

5. $(x^4+5x^3+5x^2-7x-12)\div(x+3)$

6. $\dfrac{2x-1}{4}-\dfrac{3x+2}{3}$

7. $(2\sqrt{3}-\sqrt{2})(3\sqrt{3}+2\sqrt{2})$

8. Identify the type of symmetry that each of the following equations exhibits.

  a. $xy=-6$

  b. $x^2+3x+y^2-4=0$

  c. $y=x^2-14$

  d. $x^2-2y^2-3y-6=0$

9. Evaluate $(3x^2-2x-1)-(3x^2-3x+2)$ for $x=-18$.

10. Evaluate $\dfrac{36x^{-2}y^4}{9x^{-3}y^3}$ for $x=6$ and $y=-8$.

11. Find the least common multiple of 6, 9, and 12.

12. Find the slope of the line determined by the equation $x-4y=-6$.

13. Find the equation of the line that contains the two points $(2,7)$ and $(-1,3)$.

For Problems 14–18, solve each equation.

14. $(x+2)(x-6)=-16$

15. $\dfrac{2}{3}(x-1)-\dfrac{3}{4}(2x+3)=\dfrac{3}{2}$

16. $0.06x+0.07(7000-x)=460$

17. $|3x-5|=10$

18. $3x^2-x-1=0$

For Problems 19–21, solve each inequality and express the solutions using interval notation.

19. $|2x-3|\le 5$

20. $\dfrac{x-2}{x-3}>0$

21. $x^2-5x-6<0$

For Problems 22–25, use an equation or a system of equations to help solve each problem.

22. One of two complementary angles is 6° larger than one-half of the other angle. Find the measure of each angle.

23. Abby has 37 coins, consisting only of dimes and quarters, worth \$7.45. How many dimes and how many quarters does she have?

24. Juan started walking at 4 miles per hour. An hour and a half later, Cathy starts jogging along the same route at 6 miles per hour. How long will it take Cathy to catch up with Juan?

25. Sue bought 3 packages of cookies and 2 sacks of potato chips for \$11.35. Later, at the same prices, she bought 2 packages of cookies and 5 sacks of potato chips for \$18.53. Find the price of a package of cookies.

For Problems 26–31, graph the following equations. Label the x and y intercepts on the graph.

26. $2x-y=4$ **27.** $x-3y=6$

28. $y=\dfrac{1}{2}x+3$ **29.** $y=-3x+1$

30. $y=-4$ **31.** $x=2$

For Problems 32 and 33, find the distance between the two points. Express the answer in simplest radical form.

32. $(3,-1)$ and $(-4,6)$

33. $(8,0)$ and $(3,4)$

For Problems 34–37, write the equation of a line that satisfies the given conditions. Express the answer in standard form.

34. x intercept of 2 and slope of $\dfrac{3}{5}$

35. Contains the points $(-1,4)$ and $(0,3)$

36. Contains the point $(-3, 5)$ and is parallel to the line $4x + 2y = -5$

37. Contains the point $(1, -2)$ and is perpendicular to the line $x - 3y = 3$

38. Find the center and the length of a radius of the circle $x^2 + 8x + y^2 + 4y - 5 = 0$

39. Find the coordinates of the vertex of the parabola $y = 3x^2 - 6x + 8$

40. Find the equations of the asymptotes for the hyperbola $16x^2 - 9y^2 = 25$

41. Graph the ellipse $25x^2 + 4y^2 = 100$.

42. Graph the hyperbola $9x^2 - 4y^2 = 36$.

43. Graph the parabola $y = -2(x - 1)^2 + 4$.

13

Functions

The price of goods may be decided by using a function to describe the relationship between the price and the demand. Such a function gives us a means of studying the demand when the price is varied.

© Matt Antonio

A golf pro-shop operator finds that she can sell 30 sets of golf clubs at $500 per set in a year. Furthermore, she predicts that for each $25 decrease in price, 3 additional sets of golf clubs could be sold. At what price should she sell the clubs to maximize gross income? We can use the quadratic function $f(x) = (30 + 3x)(500 - 25x)$ to determine that the clubs should be sold at $375 per set.

One of the fundamental concepts of mathematics is the concept of a function. Functions are used to unify mathematics and also to apply mathematics to many real-world problems. Functions provide a means of studying quantities that vary with one another—that is, change in one quantity causes a corresponding change in the other.

In this chapter we will (1) introduce the basic ideas that pertain to the function concept, (2) review and extend some concepts from Chapter 12, and (3) discuss some applications of functions.

Video tutorials based on section learning objectives are available in a variety of delivery modes.

13.1 Relations and Functions

OBJECTIVES

1. Determine if a relation is a function
2. Use function notation when evaluating a function
3. Specify the domain and range of a function
4. Find the difference quotient of a given function

Mathematically, a function is a special kind of *relation*, so we will begin our discussion with a simple definition of a relation.

Definition 13.1

A **relation** is a set of ordered pairs.

Thus a set of ordered pairs such as $\{(1, 2), (3, 7), (8, 14)\}$ is a relation. The set of all first components of the ordered pairs is the **domain** of the relation, and the set of all second components is the **range** of the relation. The relation $\{(1, 2), (3, 7), (8, 14)\}$ has a domain of $\{1, 3, 8\}$ and a range of $\{2, 7, 14\}$.

The ordered pairs we refer to in Definition 13.1 may be generated by various means, such as a graph or a chart. However, one of the most common ways of generating ordered pairs is by using equations. Because the solution set of an equation in two variables is a set of ordered pairs, such an equation describes a relation. Each of the following equations describes a relation between the variables x and y. We have listed some of the infinitely many ordered pairs (x, y) of each relation.

1. $x^2 + y^2 = 4$ $\left(1, \sqrt{3}\right), \left(1, -\sqrt{3}\right), (0, 2), (0, -2)$
2. $y^2 = x^3$ $(0, 0), (1, 1), (1, -1), (4, 8), (4, -8)$
3. $y = x + 2$ $(0, 2), (1, 3), (2, 4), (-1, 1), (5, 7)$
4. $y = \dfrac{1}{x - 1}$ $(0, -1), (2, 1), \left(3, \dfrac{1}{2}\right), \left(-1, -\dfrac{1}{2}\right), \left(-2, -\dfrac{1}{3}\right)$
5. $y = x^2$ $(0, 0), (1, 1), (2, 4), (-1, 1), (-2, 4)$

Now we direct your attention to the ordered pairs associated with equations 3, 4, and 5. Note that in each case, no two ordered pairs have the same first component. Such a set of ordered pairs is called a *function*.

Definition 13.2

A **function** is a relation in which no two ordered pairs have the same first component.

Stated another way, Definition 13.2 declares that a function is a relation wherein each member of the domain is assigned *one and only one* member of the range. The following table lists the five equations from above and determines if the generated ordered pairs fit the definition of a function.

Equation	Ordered pairs	Function
1. $x^2 + y^2 = 4$	$(1, \sqrt{3}), (1, -\sqrt{3}), (0, 2), (0, -2)$ [*Note*: The ordered pairs $(1, \sqrt{3})$ and $(1, -\sqrt{3})$ have the same first component and different second components.]	No
2. $y^2 = x^3$	$(0, 0), (1, 1), (1, -1), (4, 8), (4, -8)$ [*Note*: The ordered pairs $(1, 1)$ and $(1, -1)$ have the same first component and different second components.]	No
3. $y = x + 2$	$(0, 2), (1, 3), (2, 4), (-1, 1), (5, 7)$	Yes
4. $y = \dfrac{1}{x - 1}$	$(0, -1), (2, 1), \left(3, \dfrac{1}{2}\right), \left(-1, -\dfrac{1}{2}\right), \left(-2, -\dfrac{1}{3}\right)$	Yes
5. $y = x^2$	$(0, 0), (1, 1), (2, 4), (-1, 1), (-2, 4)$	Yes

Classroom Example
Do the following sets of ordered pairs determine a function? Specify the domain and range:
(a) $\{(0, 3), (1, 5), (4, -3), (6, 1)\}$
(b) $\{(6, 3), (5, 2), (-4, 2), (-6, 0)\}$
(c) $\{(-1, 1), (3, 2), (-1, 2), (4, 1)\}$

EXAMPLE 1

Determine if the following sets of ordered pairs determine a function. Specify the domain and range:

(a) $\{(1, 3), (2, 5), (3, 7), (4, 8)\}$

(b) $\{(2, 1), (2, 3), (2, 5), (2, 7)\}$

(c) $\{(0, -2), (2, -2), (4, 6), (6, 6)\}$

Solution

(a) $D = \{1, 2, 3, 4\}$ $R = \{3, 5, 7, 8\}$. Yes, the set of ordered pairs does determine a function. No first component is ever repeated. Therefore every first component has one and only one second component.

(b) $D = \{2\}$ $R = \{1, 3, 5, 7\}$. No, the set of ordered pairs does not determine a function. The ordered pairs $(2, 1)$ and $(2, 3)$ have the same first component and different second components.

(c) $D = \{0, 2, 4, 6\}$ $R = \{-2, 6\}$. Yes, the set of ordered pairs does determine a function. No first component is ever repeated. Therefore every first component has one and only one second component.

Classroom Example
Determine whether the relation $\{(x, y) | y^3 = x\}$ is a function and specify its domain and range.

EXAMPLE 2

Determine whether the relation $\{(x, y) | y^2 = x\}$ is a function and specify its domain and range.

Solution

Because $y^2 = x$ is equivalent to $y = \pm\sqrt{x}$, to each value of x where $x > 0$, there are assigned *two* values for y. Therefore, this relation is not a function. The expression $\sqrt{x}$ requires that x be nonnegative; therefore, the domain (D) is

$$D = \{x | x \geq 0\}$$

To each nonnegative real number, the relation assigns two real numbers, $\sqrt{x}$ and $-\sqrt{x}$. Thus the range (R) is

$$R = \{y | y \text{ is a real number}\}$$

Using Function Notation When Evaluating a Function

The three sets of ordered pairs listed in the table that generated functions could be named as shown below.

$$f = \{(x, y)|y = x + 2\} \qquad g = \left\{(x, y)\Big|y = \frac{1}{x - 1}\right\} \qquad h = \{(x, y)|y = x^2\}$$

For the first set of ordered pairs, the notation could be read "the function f is the set of ordered pairs (x, y) such that y is equal to $x + 2$."

Note that we named the previous functions f, g, and h. It is customary to name functions by means of a single letter, and the letters f, g, and h are often used. We would suggest more meaningful choices when functions are used to portray real-world situations. For example, if a problem involves a profit function, then naming the function p or even P would seem natural.

The symbol for a function can be used along with a variable that represents an element in the domain to represent the associated element in the range. For example, suppose that we have a function f specified in terms of the variable x. The symbol $f(x)$, which is read "f of x" or "the value of f at x," represents the element in the range associated with the element x from the domain. The function $f = \{(x, y)|y = x + 2\}$ can be written as $f = \{(x, f(x))| f(x) = x + 2\}$ and is usually shortened to read "f is the function determined by the equation $f(x) = x + 2$."

Remark: Be careful with the notation $f(x)$. As we stated above, it means the value of the function f at x. It does not mean f times x.

This **function notation** is very convenient for computing and expressing various values of the function. For example, the value of the function $f(x) = 3x - 5$ at $x = 1$ is

$$f(1) = 3(1) - 5 = -2$$

Likewise, the functional values for $x = 2$, $x = -1$, and $x = 5$ are

$$f(2) = 3(2) - 5 = 1$$
$$f(-1) = 3(-1) - 5 = -8$$
$$f(5) = 3(5) - 5 = 10$$

Thus this function f contains the ordered pairs $(1, -2)$, $(2, 1)$, $(-1, -8)$, $(5, 10)$, and in general all ordered pairs having the form $(x, f(x))$, where $f(x) = 3x - 5$ and x is any real number.

It may be helpful for you to picture the concept of a function in terms of a *function machine*, as in Figure 13.1. Each time that a value of x is put into the machine, the equation $f(x) = x + 2$ is used to generate one and only one value for $f(x)$ to be ejected from the machine. For example, if 3 is put into this machine, then $f(3) = 3 + 2 = 5$, and 5 is ejected. Thus the ordered pair $(3, 5)$ is one element of the function. Now let's look at some examples that involve evaluating functions.

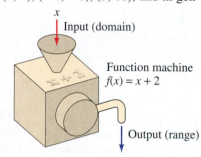

Figure 13.1

Classroom Example
Consider the function $f(x) = x^2 - 3$.
Evaluate $f(-1), f(0)$, and $f(5)$.

EXAMPLE 3   Evaluate $f(-2), f(0)$, and $f(4)$ for the function $f(x) = x^2$.

Solution

$$f(-2) = (-2)^2 = 4$$
$$f(0) = (0)^2 = 0$$
$$f(4) = (4)^2 = 16$$

Classroom Example
If $f(x) = -3x + 2$ and
$g(x) = x^2 + 4x - 7$, find $f(2)$,
$f(-3)$, $f(m)$, $f(2n)$, $g(-2)$,
$g(3)$, $g(a)$, and $g(a + 2)$.

EXAMPLE 4

If $f(x) = -2x + 7$ and $g(x) = x^2 - 5x + 6$, find $f(3)$, $f(-4)$, $f(b)$, $f(3c)$, $g(2)$, $g(-1)$, $g(a)$, and $g(a + 4)$.

Solution

$$f(x) = -2x + 7 \qquad\qquad g(x) = x^2 - 5x + 6$$
$$f(3) = -2(3) + 7 = 1 \qquad\qquad g(2) = (2)^2 - 5(2) + 6 = 0$$
$$f(-4) = -2(-4) + 7 = 15 \qquad g(-1) = (-1)^2 - 5(-1) + 6 = 12$$
$$f(b) = -2(b) + 7 = -2b + 7 \qquad g(a) = (a)^2 - 5(a) + 6 = a^2 - 5a + 6$$
$$f(3c) = -2(3c) + 7 = -6c + 7 \qquad g(a + 4) = (a + 4)^2 - 5(a + 4) + 6$$
$$= a^2 + 8a + 16 - 5a - 20 + 6$$
$$= a^2 + 3a + 2$$

In Example 4, note that we are working with two different functions in the same problem. Thus different names, f and g, are used.

Specifying the Domain and Range of a Function

For our purposes in this text, if the domain of a function is not specifically indicated or determined by a real-world application, then we assume the domain to be all *real number* replacements for the variable, which represents an element in the domain that will produce *real number* functional values. Consider the following examples.

Classroom Example
Specify the domain and range for
the function $f(x) = \sqrt{x - 2}$.

EXAMPLE 5   Specify the domain and range for the function $f(x) = x^2$.

Solution

Because any real number can be squared, the domain (D) is any real number.

$$D = \{x \mid x \text{ is a real number}\}$$

For the range, squaring a real number always produces a nonnegative value. Thus the range (R) is

$$R = \{f(x) \mid f(x) \geq 0\}$$

Classroom Example
Specify the domain for each of the
following:
(a) $f(x) = \dfrac{1}{x + 3}$
(b) $f(m) = \dfrac{1}{m^2 - 16}$
(c) $f(t) = \sqrt{t + 2}$

EXAMPLE 6   Specify the domain for each of the following:

(a) $f(x) = \dfrac{1}{x - 1}$ **(b)** $f(t) = \dfrac{1}{t^2 - 4}$ **(c)** $f(s) = \sqrt{s - 3}$

Solution

(a) We can replace x with any real number except 1, because 1 makes the denominator zero. Thus the domain is given by

$$D = \{x \mid x \neq 1\} \qquad \text{or} \qquad D: (-\infty, 1) \cup (1, \infty)$$

Here you may consider set builder notation to be easier than interval notation for expressing the domain.

(b) We need to eliminate any value of t that will make the denominator zero, so let's solve the equation $t^2 - 4 = 0$.

$$t^2 - 4 = 0$$
$$t^2 = 4$$
$$t = \pm 2$$

The domain is the set

$$D = \{t \mid t \neq -2 \text{ and } t \neq 2\} \qquad \text{or} \qquad D: (-\infty, -2) \cup (-2, 2) \cup (2, \infty)$$

When the domain is all real numbers except a few numbers, set builder notation is the more compact notation.

(c) The radicand, $s - 3$, must be nonnegative.

$$s - 3 \geq 0$$
$$s \geq 3$$

The domain is the set

$$D = \{s \mid s \geq 3\} \qquad \text{or} \qquad D: [3, \infty)$$

Finding the Difference Quotient of a Given Function

The quotient $\dfrac{f(a + h) - f(a)}{h}$ is often called a **difference quotient**, and we use it extensively with functions when studying the limit concept in calculus. The next two examples show how we found the difference quotient for two specific functions.

Classroom Example
If $f(x) = 2x + 3$, find

$$\frac{f(a + h) - f(a)}{h}$$

EXAMPLE 7   If $f(x) = 3x - 5$, find $\dfrac{f(a + h) - f(a)}{h}$.

Solution

$$f(a + h) = 3(a + h) - 5$$
$$= 3a + 3h - 5$$

and

$$f(a) = 3a - 5$$

Therefore,

$$f(a + h) - f(a) = (3a + 3h - 5) - (3a - 5)$$
$$= 3a + 3h - 5 - 3a + 5$$
$$= 3h$$

and

$$\frac{f(a + h) - f(a)}{h} = \frac{3h}{h} = 3$$

Classroom Example
If $f(x) = x^2 - 3x + 5$, find

$$\frac{f(a + h) - f(a)}{h}$$

EXAMPLE 8   If $f(x) = x^2 + 2x - 3$, find $\dfrac{f(a + h) - f(a)}{h}$.

Solution

$$f(a + h) = (a + h)^2 + 2(a + h) - 3$$
$$= a^2 + 2ah + h^2 + 2a + 2h - 3$$

and

$$f(a) = a^2 + 2a - 3$$

Therefore,

$$f(a + h) - f(a) = (a^2 + 2ah + h^2 + 2a + 2h - 3) - (a^2 + 2a - 3)$$
$$= a^2 + 2ah + h^2 + 2a + 2h - 3 - a^2 - 2a + 3$$
$$= 2ah + h^2 + 2h$$

and

$$\frac{f(a + h) - f(a)}{h} = \frac{2ah + h^2 + 2h}{h}$$
$$= \frac{h(2a + h + 2)}{h}$$
$$= 2a + h + 2$$

Functions and functional notation provide the basis for describing many real-world relationships. The next example illustrates this point.

EXAMPLE 9

Suppose a factory determines that the overhead for producing a quantity of a certain item is $500 and that the cost for producing each item is $25. Express the total expenses as a function of the number of items produced, and compute the expenses for producing 12, 25, 50, 75, and 100 items.

Solution

Let n represent the number of items produced. Then $25n + 500$ represents the total expenses. Let's use E to represent the expense function, so that we have

$$E(n) = 25n + 500 \quad \text{where } n \text{ is a whole number}$$

from which we obtain

$$E(12) = 25(12) + 500 = 800$$
$$E(25) = 25(25) + 500 = 1125$$
$$E(50) = 25(50) + 500 = 1750$$
$$E(75) = 25(75) + 500 = 2375$$
$$E(100) = 25(100) + 500 = 3000$$

Thus the total expenses for producing 12, 25, 50, 75, and 100 items are $800, $1125, $1750, $2375, and $3000, respectively.

Concept Quiz 13.1

For Problems 1–10, answer true or false.

1. A function is a special type of relation.
2. The relation {(John, Mary), (Mike, Ada), (Kyle, Jenn), (Mike, Sydney)} is a function.
3. Given $f(x) = 3x + 4$, the notation $f(7)$ means to find the value of f when $x = 7$.
4. The set of all first components of the ordered pairs of a relation is called the range.
5. The domain of a function can never be the set of all real numbers.

6. The domain of the function $f(x) = \dfrac{x}{x - 3}$ is the set of all real numbers.

7. The range of the function $f(x) = x + 1$ is the set of all real numbers.

8. If $f(x) = -x^2 - 1$, then $f(2) = -5$.

9. The range of the function $f(x) = \sqrt{x} - 1$ is the set of all real numbers greater than or equal to 1.

10. If $f(x) = -x^2 + 3x$, then $f(2a) = 4a^2 + 6a$.

Problem Set 13.1

For Problems 1–10, specify the domain and the range for each relation. Also state whether or not the relation is a function. **(Objectives 1 and 3)**

1. $\{(1, 5), (2, 8), (3, 11), (4, 14)\}$

2. $\{(0, 0), (2, 10), (4, 20), (6, 30), (8, 40)\}$

3. $\{(0, 5), (0, -5), (1, 2\sqrt{6}), (1, -2\sqrt{6})\}$

4. $\{(1, 1), (1, 2), (1, -1), (1, -2), (1, 3)\}$

5. $\{(1, 2), (2, 5), (3, 10), (4, 17), (5, 26)\}$

6. $\{(-1, 5), (0, 1), (1, -3), (2, -7)\}$

7. $\{(x, y)\,|\, 5x - 2y = 6\}$ **8.** $\{(x, y)\,|\, y = -3x\}$

9. $\{(x, y)\,|\, x^2 = y^3\}$ **10.** $\{(x, y)\,|\, x^2 - y^2 = 16\}$

For Problems 11–36, specify the domain for each of the functions. **(Objective 3)**

11. $f(x) = 7x - 2$ **12.** $f(x) = x^2 + 1$

13. $f(x) = \dfrac{1}{x - 1}$ **14.** $f(x) = \dfrac{-3}{x + 4}$

15. $g(x) = \dfrac{3x}{4x - 3}$ **16.** $g(x) = \dfrac{5x}{2x + 7}$

17. $h(x) = \dfrac{2}{(x + 1)(x - 4)}$

18. $h(x) = \dfrac{-3}{(x - 6)(2x + 1)}$

19. $f(x) = \dfrac{14}{x^2 + 3x - 40}$ **20.** $f(x) = \dfrac{7}{x^2 - 8x - 20}$

21. $f(x) = \dfrac{-4}{x^2 + 6x}$ **22.** $f(x) = \dfrac{9}{x^2 - 12x}$

23. $f(t) = \dfrac{4}{t^2 + 9}$ **24.** $f(t) = \dfrac{8}{t^2 + 1}$

25. $f(t) = \dfrac{3t}{t^2 - 4}$ **26.** $f(t) = \dfrac{-2t}{t^2 - 25}$

27. $h(x) = \sqrt{x + 4}$ **28.** $h(x) = \sqrt{5x - 3}$

29. $f(s) = \sqrt{4s - 5}$ **30.** $f(s) = \sqrt{s - 2} + 5$

31. $f(x) = \sqrt{x^2 - 16}$ **32.** $f(x) = \sqrt{x^2 - 49}$

33. $f(x) = \sqrt{x^2 - 3x - 18}$

34. $f(x) = \sqrt{x^2 + 4x - 32}$

35. $f(x) = \sqrt{1 - x^2}$ **36.** $f(x) = \sqrt{9 - x^2}$

37. If $f(x) = 5x - 2$, find $f(0)$, $f(2)$, $f(-1)$, and $f(-4)$.

38. If $f(x) = -3x - 4$, find $f(-2)$, $f(-1)$, $f(3)$, and $f(5)$.

39. If $f(x) = \dfrac{1}{2}x - \dfrac{3}{4}$, find $f(-2), f(0), f\left(\dfrac{1}{2}\right), f\left(\dfrac{2}{3}\right)$

40. If $g(x) = x^2 + 3x - 1$, find $g(1), g(-1), g(3)$, and $g(-4)$.

41. If $g(x) = 2x^2 - 5x - 7$, find $g(-1)$, $g(2)$, $g(-3)$, and $g(4)$.

42. If $h(x) = -x^2 - 3$, find $h(1), h(-1), h(-3)$, and $h(5)$.

43. If $h(x) = -2x^2 - x + 4$, find $h(-2)$, $h(-3)$, $h(4)$, and $h(5)$.

44. If $f(x) = \sqrt{x - 1}$, find $f(1)$, $f(5)$, $f(13)$, and $f(26)$.

45. If $f(x) = \sqrt{2x + 1}$, find $f(3)$, $f(4)$, $f(10)$, and $f(12)$.

46. If $f(x) = \dfrac{3}{x - 2}$, find $f(3)$, $f(0)$, $f(-1)$, and $f(-5)$.

47. If $f(x) = \dfrac{-4}{x + 3}$, find $f(1)$, $f(-1)$, $f(3)$, and $f(-6)$.

48. If $f(x) = -2x + 7$, find $f(a)$, $f(a + 2)$, and $f(a + h)$.

49. If $f(x) = x^2 - 7x$, find $f(a)$, $f(a - 3)$, and $f(a + h)$.

50. If $f(x) = x^2 - 4x + 10$, find $f(-a)$, $f(a - 4)$, and $f(a + h)$.

51. If $f(x) = 2x^2 - x - 1$, find $f(-a)$, $f(a + 1)$, and $f(a + h)$.

52. If $f(x) = -x^2 + 3x + 5$, find $f(-a)$, $f(a + 6)$, and $f(-a + 1)$.

53. If $f(x) = -x^2 - 2x - 7$, find $f(-a)$, $f(-a - 2)$, and $f(a + 7)$.

54. If $f(x) = 2x^2 - 7$ and $g(x) = x^2 + x - 1$, find $f(-2)$, $f(3)$, $g(-4)$, and $g(5)$.

55. If $f(x) = 5x^2 - 2x + 3$ and $g(x) = -x^2 + 4x - 5$, find $f(-2)$, $f(3)$, $g(-4)$, and $g(6)$.

56. If $f(x) = |3x - 2|$ and $g(x) = |x| + 2$, find $f(1)$, $f(-1)$, $g(2)$, and $g(-3)$.

57. If $f(x) = 3|x| - 1$ and $g(x) = -|x| + 1$, find $f(-2)$, $f(3)$, $g(-4)$, and $g(5)$.

For Problems 58–65, find $\dfrac{f(a + h) - f(a)}{h}$ for each of the given functions. (Objective 4)

58. $f(x) = 5x - 4$

59. $f(x) = -3x + 6$

60. $f(x) = x^2 + 5$

61. $f(x) = -x^2 - 1$

62. $f(x) = x^2 - 3x + 7$

63. $f(x) = 2x^2 - x + 8$

64. $f(x) = -3x^2 + 4x - 1$

65. $f(x) = -4x^2 - 7x - 9$

66. Suppose that the cost function for producing a certain item is given by $C(n) = 3n + 5$, where n represents the number of items produced. Compute $C(150)$, $C(500)$, $C(750)$, and $C(1500)$.

67. The height of a projectile fired vertically into the air (neglecting air resistance) at an initial velocity of 64 feet per second is a function of the time (t) and is given by the equation

$$h(t) = 64t - 16t^2$$

Compute $h(1)$, $h(2)$, $h(3)$, and $h(4)$.

68. In Company S, the profit function for selling n items is given by $P(n) = -n^2 + 500n - 61{,}500$. Compute $P(200)$, $P(230)$, $P(250)$, and $P(260)$.

69. A car rental agency charges \$50 per day plus \$0.32 a mile. Therefore, the daily charge for renting a car is a function of the number of miles traveled (m) and can be expressed as $C(m) = 50 + 0.32m$. Compute $C(75)$, $C(150)$, $C(225)$, and $C(650)$.

70. The equation $A(r) = \pi r^2$ expresses the area of a circular region as a function of the length of a radius (r). Use 3.14 as an approximation for π, and compute $A(2)$, $A(3)$, $A(12)$, and $A(17)$.

71. The equation $I(r) = 500r$ expresses the amount of simple interest earned by an investment of \$500 for 1 year as a function of the rate of interest (r). Compute $I(0.04)$, $I(0.06)$, $I(0.075)$, and $I(0.09)$.

Thoughts Into Words

72. Are all functions also relations? Are all relations also functions? Defend your answers.

73. What does it mean to say that the domain of a function may be restricted if the function represents a real-world situation? Give two or three examples of such situations.

74. Does $f(a + b) = f(a) + f(b)$ for all functions? Defend your answer.

75. Are there any functions for which $f(a + b) = f(a) + f(b)$? Defend your answer.

Answers to the Concept Quiz

1. True **2.** False **3.** True **4.** False **5.** False **6.** False **7.** True **8.** True **9.** False **10.** False

13.2   Functions: Their Graphs and Applications

OBJECTIVES

1. Graph linear functions

2. Applications of linear functions

3. Find the vertex of the graph of a quadratic function

4. Graph quadratic functions

5. Use quadratic functions to solve problems

In Section 5.1, we made statements such as "The graph of the solution set of the equation, $y = x - 1$ (or simply, the graph of the equation $y = x - 1$), is a line that contains the points $(0, -1)$ and $(1, 0)$." Because the equation $y = x - 1$ (which can be written as $f(x) = x - 1$) can be used to specify a function, that line we previously referred to is also called the *graph of the function specified by the equation* or simply the *graph of the function*. Generally speaking, the graph of any equation that determines a function is also called the **graph of the function**. Thus the graphing techniques we discussed earlier will continue to play an important role as we graph functions.

As we use the function concept in our study of mathematics, it is helpful to classify certain types of functions and become familiar with their equations, characteristics, and graphs. In this section we will discuss two special types of functions: **linear** and **quadratic functions**. These functions are merely an outgrowth of our earlier study of linear and quadratic equations.

Graphing Linear Functions

Any function that can be written in the form

$$f(x) = ax + b$$

where a and b are real numbers, is called a **linear function**. The following equations are examples of linear functions.

$$f(x) = -3x + 6 \qquad f(x) = 2x + 4 \qquad f(x) = -\frac{1}{2}x - \frac{3}{4}$$

Graphing linear functions is quite easy because the graph of every linear function is a straight line. Therefore, all we need to do is determine two points of the graph and draw the line determined by those two points. You may want to use a third point as a check point.

Classroom Example
Graph the function $f(x) = 2x - 3$.

EXAMPLE 1   Graph the function $f(x) = -3x + 6$.

Solution

Because $f(0) = 6$, the point $(0, 6)$ is on the graph. Likewise, because $f(1) = 3$, the point $(1, 3)$ is on the graph. Plot these two points, and draw the line determined by the two points to produce Figure 13.2.

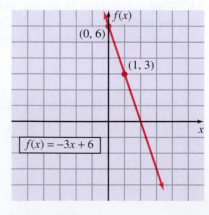

Figure 13.2

Remark: Note in Figure 13.2 that we labeled the vertical axis $f(x)$. We could also label it y, because $f(x) = -3x + 6$ and $y = -3x + 6$ mean the same thing. We will continue to use the label $f(x)$ in this chapter to help you adjust to the function notation.

EXAMPLE 2   Graph the function $f(x) = x$.

Solution

The equation $f(x) = x$ can be written as $f(x) = 1x + 0$; thus it is a linear function. Because $f(0) = 0$ and $f(2) = 2$, the points $(0, 0)$ and $(2, 2)$ determine the line in Figure 13.3. The function $f(x) = x$ is often called the **identity function**.

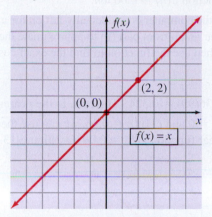

Figure 13.3

As we use function notation to graph functions, it is often helpful to think of the ordinate of every point on the graph as the value of the function at a specific value of x. Geometrically, this functional value is the directed distance of the point from the x axis, as illustrated in Figure 13.4 with the function $f(x) = 2x - 4$. For example, consider the graph of the function $f(x) = 2$. The function $f(x) = 2$ means that every functional value is 2, or, geometrically, that every point on the graph is 2 units above the x axis. Thus the graph is the horizontal line shown in Figure 13.5.

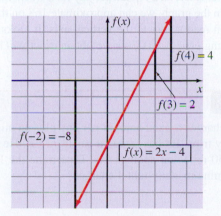

Figure 13.4

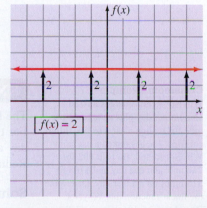

Figure 13.5

Any linear function of the form $f(x) = ax + b$, where $a = 0$, is called a **constant function**, and its graph is a horizontal line.

Applications of Linear Functions

We worked with some applications of linear equations in Section 5.2. Let's consider some additional applications that use the concept of a linear function to connect mathematics to the real world.

Classroom Example
The cost for burning a 75-watt bulb is given by the function $c(h) = 0.0045h$, where h represents the number of hours that the bulb is burning.
(a) How much does it cost to burn a 75-watt bulb for 4 hours per night for two weeks?
(b) Graph the function
$c(h) = 0.0045h$
(c) What is the approximate cost of allowing the bulb to burn continuously for two weeks?

EXAMPLE 3

The cost for operating a desktop computer is given by the function $c(h) = 0.0036h$, where h represents the number of hours that the computer is on.

(a) How much does it cost to operate the computer for 3 hours per night for a 30-day month?

(b) Graph the function $c(h) = 0.0036h$.

(c) Suppose that the computer is accidentally left on for a week while the owner is on vacation. Use the graph from part (b) to approximate the cost of operating the computer for a week. Then use the function to find the exact cost.

Solution

(a) $c(90) = 0.0036(90) = 0.324$. The cost, to the nearest cent, is $0.32.

(b) Because $c(0) = 0$ and $c(100) = 0.36$, we can use the points $(0, 0)$ and $(100, 0.36)$ to graph the linear function $c(h) = 0.0036h$ (Figure 13.6).

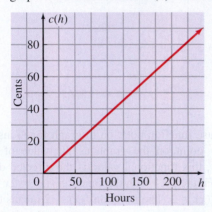

Figure 13.6

(c) If the computer is left on 24 hours per day for a week, then it runs for $24(7) = 168$ hours. Reading from the graph, we can approximate 168 on the horizontal axis, read up to the line, and then read across to the vertical axis. It looks as if it will cost approximately 60 cents. Using $c(h) = 0.0036h$, we obtain exactly $c(168) = 0.0036(168) = 0.6048$.

Classroom Example
WhyNot Rental charges a fixed amount per day plus an amount per mile for renting a car. For two different trips, Joe has rented a car from WhyNot Rental. He paid $92 for 212 miles on one day and $168 for 516 miles on another day. Determine the linear function that WhyNot Rental uses to determine its daily rental charges.

EXAMPLE 4

The EZ Car Rental charges a fixed amount per day plus an amount per mile for renting a car. For two different day trips, Ed has rented a car from EZ. He paid $70 for 100 miles on one day and $120 for 350 miles on another day. Determine the linear function that the EZ Car Rental uses to determine its daily rental charges.

Solution

The linear function $f(x) = ax + b$, where x represents the number of miles, models this situation. Ed's two day trips can be represented by the ordered pairs $(100, 70)$ and $(350, 120)$. From these two ordered pairs we can determine a, which is the slope of the line.

$$a = \frac{120 - 70}{350 - 100} = \frac{50}{250} = \frac{1}{5} = 0.2$$

Thus $f(x) = ax + b$ becomes $f(x) = 0.2x + b$. Now either ordered pair can be used to determine the value of b. Using $(100, 70)$, we have $f(100) = 70$; therefore,

$$f(100) = 0.2(100) + b = 70$$
$$b = 50$$

The linear function is $f(x) = 0.2x + 50$. In other words, the EZ Car Rental charges a daily fee of \$50 plus \$0.20 per mile.

Classroom Example

Suppose Joe also has access to MayBe Rental, which charges a daily fee of \$45 plus \$0.15 per mile. Should Joe use WhyNot Rental (from the previous example) or MayBe Rental?

EXAMPLE 5

Suppose that Ed (Example 4) also has access to the A-OK Car Rental agency, which charges a daily fee of \$25 plus \$0.30 per mile. Should Ed use the EZ Car Rental from Example 4 or A-OK Car Rental?

Solution

The linear function $g(x) = 0.3x + 25$, where x represents the number of miles, can be used to determine the daily charges of A-OK Car Rental. Let's graph this function and $f(x) = 0.2x + 50$ from Example 4 on the same set of axes (Figure 13.7).

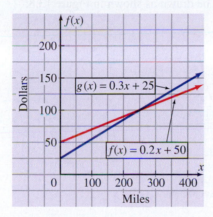

Figure 13.7

Now we see that the two functions have equal values at the point of intersection of the two lines. To find the coordinates of this point, we can set $0.3x + 25$ equal to $0.2x + 50$ and solve for x.

$$0.3x + 25 = 0.2x + 50$$
$$0.1x = 25$$
$$x = 250$$

If $x = 250$, then $0.3(250) + 25 = 100$, and the point of intersection is $(250, 100)$. Look again at the lines in Figure 13.7; we see that Ed should use A-OK Car Rental for day trips of less than 250 miles, but he should use EZ Car Rental for day trips of more than 250 miles.

Graphing Quadratic Functions

Any function that can be written in the form

$$f(x) = ax^2 + bx + c$$

where a, b, and c are real numbers with $a \neq 0$, is called a **quadratic function**. The following equations are examples of quadratic functions.

$$f(x) = 3x^2 \qquad f(x) = -2x^2 + 5x \qquad f(x) = 4x^2 - 7x + 1$$

The techniques discussed in Chapter 12 relative to graphing quadratic equations of the form $y = ax^2 + bx + c$ provide the basis for graphing quadratic functions. Let's review some work we did in Chapter 12 with an example.

Classroom Example
Graph the function:

$f(x) = 3x^2 + 12x + 11$

EXAMPLE 6   Graph the function $f(x) = 2x^2 - 4x + 5$.

Solution

$$f(x) = 2x^2 - 4x + 5$$
$$= 2(x^2 - 2x + \underline{\quad}) + 5 \qquad \text{Recall the process of completing the square!}$$
$$= 2(x^2 - 2x + 1) + 5 - 2$$
$$= 2(x - 1)^2 + 3$$

From this form we can obtain the following information about the parabola.

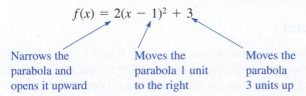

$$f(x) = 2(x - 1)^2 + 3$$

| Narrows the parabola and opens it upward | Moves the parabola 1 unit to the right | Moves the parabola 3 units up |

Thus the parabola can be drawn as shown in Figure 13.8.

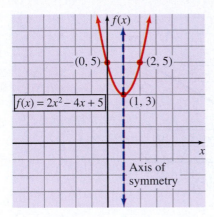

$(0, 5)$ $(2, 5)$

$f(x) = 2x^2 - 4x + 5$ $(1, 3)$

Axis of symmetry

Figure 13.8

In general, if we complete the square on

$$f(x) = ax^2 + bx + c$$

we obtain

$$f(x) = a\left(x^2 + \frac{b}{a}x + \underline{\quad}\right) + c$$

$$= a\left(x^2 + \frac{b}{a}x + \frac{b^2}{4a^2}\right) + c - \frac{b^2}{4a}$$

$$= a\left(x + \frac{b}{2a}\right)^2 + \frac{4ac - b^2}{4a}$$

Therefore, the parabola associated with $f(x) = ax^2 + bx + c$ has its vertex at $\left(-\dfrac{b}{2a}, \dfrac{4ac - b^2}{4a}\right)$, and the equation of its axis of symmetry is $x = -\dfrac{b}{2a}$. These facts are illustrated in Figure 13.9.

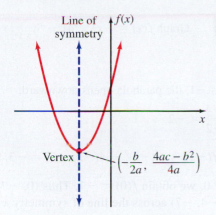

Figure 13.9

By using the information from Figure 13.9, we now have another way of graphing quadratic functions of the form $f(x) = ax^2 + bx + c$, shown by the following steps.

1. Determine whether the parabola opens upward (if $a > 0$) or downward (if $a < 0$).

2. Find $-\dfrac{b}{2a}$, which is the x coordinate of the vertex.

3. Find $f\left(-\dfrac{b}{2a}\right)$, which is the y coordinate of the vertex. $\left(\text{You could also find the } y\right.$
   coordinate by evaluating $\dfrac{4ac - b^2}{4a}$. $\Big)$

4. Locate another point on the parabola, and also locate its image across the line of symmetry, $x = -\dfrac{b}{2a}$.

The three points in steps 2, 3, and 4 should determine the general shape of the parabola. Let's use these steps in the following two examples.

Classroom Example
Graph $f(x) = 2x^2 - 8x + 9$.

EXAMPLE 7   Graph $f(x) = 3x^2 - 6x + 5$.

Solution

Step 1 Because $a = 3$, the parabola opens upward.

Step 2 $-\dfrac{b}{2a} = -\dfrac{-6}{6} = 1$

Step 3 $f\left(-\dfrac{b}{2a}\right) = f(1) = 3 - 6 + 5 = 2.$ Thus
the vertex is at $(1, 2)$.

Step 4 Letting $x = 2$, we obtain $f(2) = 12 - 12 + 5 = 5$. Thus $(2, 5)$ is on the graph and so is its reflection $(0, 5)$ across the line of symmetry $x = 1$.

The three points $(1, 2)$, $(2, 5)$, and $(0, 5)$ are used to graph the parabola in Figure 13.10.

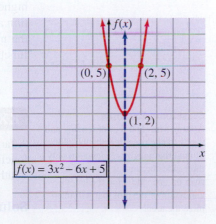

Figure 13.10

Classroom Example
Graph $f(x) = -2x^2 - 8x - 5$.

EXAMPLE 8   Graph $f(x) = -x^2 - 4x - 7$.

Solution

Step 1 Because $a = -1$, the parabola opens downward.

Step 2 $-\dfrac{b}{2a} = -\dfrac{-4}{-2} = -2$.

Step 3 $f\left(-\dfrac{b}{2a}\right) = f(-2) = -(-2)^2 - 4(-2) - 7 = -3$. So the vertex is at $(-2, -3)$.

Step 4 Letting $x = 0$, we obtain $f(0) = -7$. Thus $(0, -7)$ is on the graph and so is its reflection $(-4, -7)$ across the line of symmetry $x = -2$.

The three points $(-2, -3)$, $(0, -7)$, and $(-4, -7)$ are used to draw the parabola in Figure 13.11.

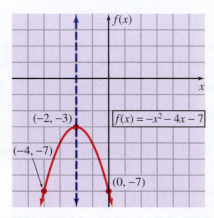

Figure 13.11

In summary, to graph a quadratic function, we have two methods.

1. We can express the function in the form $f(x) = a(x - h)^2 + k$, and use the values of a, h, and k to determine the parabola.

2. We can express the function in the form $f(x) = ax^2 + bx + c$, and use the approach demonstrated in Examples 7 and 8.

Problem Solving Using Quadratic Functions

As we have seen, the vertex of the graph of a quadratic function is either the lowest or the highest point on the graph. Thus the terms "minimum value" or "maximum value" of a function are often used in applications of the parabola. The x value of the vertex indicates where the minimum or maximum occurs, and $f(x)$ yields the minimum or maximum value of the function. Let's consider some examples that illustrate these ideas.

Classroom Example
A farmer has 560 feet of fencing and wants to enclose a rectangular plot of land that requires fencing on only three sides because it is bound by a river on one side. Find the length and width of the plot that will maximize the area.

EXAMPLE 9

A farmer has 120 rods of fencing and wants to enclose a rectangular plot of land that requires fencing on only three sides, because it is bound by a river on one side. Find the length and width of the plot that will maximize the area.

Solution

Let x represent the width; then $120 - 2x$ represents the length, as indicated in Figure 13.12.

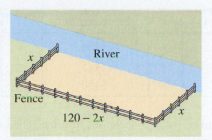

Figure 13.12

The function $A(x) = x(120 - 2x)$ represents the area of the plot in terms of the width x. Because

$$A(x) = x(120 - 2x)$$
$$= 120x - 2x^2$$
$$= -2x^2 + 120x$$

we have a quadratic function with $a = -2$, $b = 120$, and $c = 0$. Therefore, the x value where the maximum value of the function is obtained is

$$-\frac{b}{2a} = -\frac{120}{2(-2)} = 30$$

If $x = 30$, then $120 - 2x = 120 - 2(30) = 60$. Thus the farmer should make the plot 30 rods wide and 60 rods long to maximize the area at $(30)(60) = 1800$ square rods.

Classroom Example
Find two numbers whose sum is 20, such that the sum of their squares is a minimum.

EXAMPLE 10

Find two numbers whose sum is 30, such that the sum of their squares is a minimum.

Solution

Let x represent one of the numbers; then $30 - x$ represents the other number. By expressing the sum of the squares as a function of x, we obtain

$$f(x) = x^2 + (30 - x)^2$$

which can be simplified to

$$f(x) = x^2 + 900 - 60x + x^2$$
$$= 2x^2 - 60x + 900$$

This is a quadratic function with $a = 2$, $b = -60$, and $c = 900$. Therefore, the x value where the minimum occurs is

$$-\frac{b}{2a} = -\frac{-60}{4} = 15$$

If $x = 15$, then $30 - x = 30 - (15) = 15$. Thus the two numbers should both be 15.

Classroom Example
A travel agent can sell 42 tickets for a 3-day cruise at $450 each. For each $20 decrease in price, the number of tickets sold increases by four. At what price should the tickets be sold to maximize gross income?

EXAMPLE 11

A golf pro-shop operator finds that she can sell 30 sets of golf clubs at $500 per set in a year. Furthermore, she predicts that for each $25 decrease in price, three more sets of golf clubs could be sold. At what price should she sell the clubs to maximize gross income?

Solution

Sometimes, when we are analyzing such a problem, it helps to set up a table. This table is based on the condition that three additional sets can be sold for a $25 decrease in price.

Number of sets	×	Price per set	=	Income
30	×	$500	=	$15,000
33	×	$475	=	$15,675
36	×	$450	=	$16,200

Let x represent the number of \$25 decreases in price. Then we can express the income as a function of x as follows:

$$f(x) = (30 + 3x)(500 - 25x)$$

Number   Price
of sets   per set

When we simplify, we obtain

$$f(x) = 15{,}000 - 750x + 1500x - 75x^2$$
$$= -75x^2 + 750x + 15{,}000$$

Completing the square yields

$$f(x) = -75x^2 + 750x + 15{,}000$$
$$= -75(x^2 - 10x + \underline{}) + 15{,}000$$
$$= -75(x^2 - 10x + 25) + 15{,}000 + 1875$$
$$= -75(x - 5)^2 + 16{,}875$$

From this form we know that the vertex of the parabola is at (5, 16,875). Thus 5 decreases of \$25 each—that is, a \$125 reduction in price—will give a maximum income of \$16,875. The golf clubs should be sold at \$375 per set.

What we know about parabolas and the process of completing the square can be helpful when we are using a graphing utility to graph a quadratic function. Consider the following example.

Classroom Example
Use a graphing utility to obtain the graph of the quadratic function $f(x) = -x^2 - 13x - 41$.

EXAMPLE 12

Use a graphing utility to obtain the graph of the quadratic function

$$f(x) = -x^2 + 37x - 311$$

Solution

First, we know that the parabola opens downward and that its width is the same as that of the basic parabola $f(x) = x^2$. Then we can start the process of completing the square to determine an approximate location of the vertex.

$$f(x) = -x^2 + 37x - 311$$
$$= -(x^2 - 37x + \underline{}) - 311$$
$$= -\left[x^2 - 37x + \left(\frac{37}{2}\right)^2\right] - 311 + \left(\frac{37}{2}\right)^2$$
$$= -[(x^2 - 37x + (18.5)^2] - 311 + 342.25$$
$$= -(x - 18.5)^2 + 31.25$$

Thus the vertex is near $x = 18$ and $y = 31$. Setting the boundaries of the viewing rectangle so that $-2 \le x \le 25$ and $-10 \le y \le 35$, we obtain the graph shown in Figure 13.13.

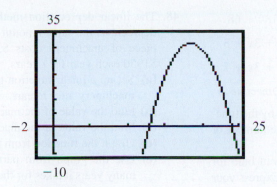

Figure 13.13

Remark: The graph in Figure 13.13 is sufficient for most purposes, because it shows the vertex and the x intercepts of the parabola. Certainly other boundaries could be used that would also give this information.

Concept Quiz 13.2

For Problems 1–10, answer true or false.

1. The function $f(x) = 3x^2 + 4$ is a linear function.

2. The graph of a linear function of the form $f(x) = b$ is a horizontal line.

3. The graph of a quadratic function is a parabola.

4. The vertex of the graph of a quadratic function is either the lowest or highest point on the graph.

5. The axis of symmetry for a parabola passes through the vertex of the parabola.

6. The parabola for the quadratic function $f(x) = -2x^2 + 3x + 7$ opens upward.

7. The linear function $f(x) = 1$ is called the identity function.

8. The parabola $f(x) = x^2 + 6x + 5$ is symmetric with respect to the line $x = 3$.

9. The vertex of the parabola $f(x) = -2x^2 + 4x + 2$ is at $(1, 4)$.

10. The graph of the function $f(x) = -3$ is symmetric with respect to the $f(x)$ axis.

Problem Set 13.2

For Problems 1–30, graph each of the following linear and quadratic functions. **(Objectives 1 and 4)**

1. $f(x) = 2x - 4$

2. $f(x) = 3x + 3$

3. $f(x) = -2x^2$

4. $f(x) = -4x^2$

5. $f(x) = -3x$

6. $f(x) = -4x$

7. $f(x) = -(x + 1)^2 - 2$

8. $f(x) = -(x - 2)^2 + 4$

9. $f(x) = -x + 3$

10. $f(x) = -2x - 4$

11. $f(x) = x^2 + 2x - 2$

12. $f(x) = x^2 - 4x - 1$

13. $f(x) = -x^2 + 6x - 8$

14. $f(x) = -x^2 - 8x - 15$

15. $f(x) = -3$

16. $f(x) = 1$

17. $f(x) = 2x^2 - 20x + 52$

18. $f(x) = 2x^2 + 12x + 14$

19. $f(x) = -3x^2 + 6x$

20. $f(x) = -4x^2 - 8x$

21. $f(x) = x^2 - x + 2$

22. $f(x) = x^2 + 3x + 2$

23. $f(x) = 2x^2 + 10x + 11$

24. $f(x) = 2x^2 - 10x + 15$

25. $f(x) = -2x^2 - 1$

26. $f(x) = -3x^2 + 2$

27. $f(x) = -3x^2 + 12x - 7$

28. $f(x) = -3x^2 - 18x - 23$

29. $f(x) = -2x^2 + 14x - 25$

30. $f(x) = -2x^2 - 10x - 14$

For Problems 31–40, find the x intercepts and the vertex of each parabola. **(Objectives 3)**

31. $f(x) = x^2 - 8x + 15$

32. $f(x) = x^2 - 16x + 63$

33. $f(x) = 2x^2 - 28x + 96$

34. $f(x) = 3x^2 - 60x + 297$

35. $f(x) = x^2 - 14x + 44$ **36.** $f(x) = x^2 - 18x + 68$

37. $f(x) = -x^2 + 9x - 21$ **38.** $f(x) = -2x^2 + 3x + 3$

39. $f(x) = -4x^2 + 4x + 4$ **40.** $f(x) = -2x^2 + 3x + 7$

To solve Problems 41–48, use linear functions. **(Objective 2)**

41. The cost for burning a 75-watt bulb is given by the function $c(h) = 0.0045h$, where h represents the number of hours that the bulb burns.

  (a) How much does it cost to burn a 75-watt bulb for 3 hours per night for a 31-day month? Express your answer to the nearest cent.

  (b) Graph the function $c(h) = 0.0045h$.

  (c) Use the graph in part (b) to approximate the cost of burning a 75-watt bulb for 225 hours.

  (d) Use $c(h) = 0.0045h$ to find the exact cost, to the nearest cent, of burning a 75-watt bulb for 225 hours.

42. The Rent-Me Car Rental charges $35 per day plus $0.32 per mile to rent a car. Determine a linear function that can be used to calculate daily car rentals. Then use that function to determine the cost of renting a car for a day and driving 150 miles; 230 miles; 360 miles; 430 miles.

43. The ABC Car Rental uses the function $f(x) = 100$ for any daily use of a car up to and including 200 miles. For driving more than 200 miles per day, ABC uses the function $g(x) = 100 + 0.25(x - 200)$ to determine the charges. How much would ABC charge for daily driving of 150 miles? of 230 miles? of 360 miles? of 430 miles?

44. Suppose that a car rental agency charges a fixed amount per day plus an amount per mile for renting a car. Heidi rented a car one day and paid $80 for 200 miles. On another day she rented a car from the same agency and paid $117.50 for 350 miles. Find the linear function that the agency could use to determine its daily rental charges.

45. A retailer has a number of items that she wants to sell and make a profit of 40% of the cost of each item. The function $s(c) = c + 0.4c = 1.4c$, where c represents the cost of an item, can be used to determine the selling price. Find the selling price of items that cost $1.50, $3.25, $14.80, $21, and $24.20.

46. Zack wants to sell five items that cost him $1.20, $2.30, $6.50, $12, and $15.60. He wants to make a profit of 60% of the cost. Create a function that you can use to determine the selling price of each item, and then use the function to calculate each selling price.

47. "All Items 20% Off Marked Price" is a sign at a local golf course. Create a function and then use it to determine how much one has to pay for each of the following marked items: a $9.50 hat, a $15 umbrella, a $75 pair of golf shoes, a $12.50 golf glove, a $750 set of golf clubs.

48. The linear depreciation method assumes that an item depreciates the same amount each year. Suppose a new piece of machinery costs $32,500 and it depreciates $1950 each year for t years.

  (a) Set up a linear function that yields the value of the machinery after t years.

  (b) Find the value of the machinery after 5 years.

  (c) Find the value of the machinery after 8 years.

  (d) Graph the function from part (a).

  (e) Use the graph from part (d) to approximate how many years it takes for the value of the machinery to become zero.

  (f) Use the function to determine how long it takes for the value of the machinery to become zero.

To solve Problems 49–56, use quadratic functions. **(Objective 5)**

49. Suppose that the cost function for a particular item is given by the equation $C(x) = 2x^2 - 320x + 12,920$, where x represents the number of items. How many items should be produced to minimize the cost?

50. Suppose that the equation $p(x) = -2x^2 + 280x - 1000$, where x represents the number of items sold, describes the profit function for a certain business. How many items should be sold to maximize the profit?

51. Find two numbers whose sum is 30, such that the sum of the square of one number plus ten times the other number is a minimum.

52. The height of a projectile fired vertically into the air (neglecting air resistance) at an initial velocity of 96 feet per second is a function of the time and is given by the equation $f(x) = 96x - 16x^2$, where x represents the time. Find the highest point reached by the projectile.

53. Two hundred forty meters of fencing is available to enclose a rectangular playground. What should be the dimensions of the playground to maximize the area?

54. Find two numbers whose sum is 50 and whose product is a maximum.

55. A movie rental company has 1000 subscribers, and each pays $15 per month. On the basis of a survey, company managers feel that for each decrease of $0.25 on the monthly rate, they could obtain 20 additional subscribers. At what rate will maximum revenue be obtained and how many subscribers will it take at that rate?

56. A restaurant advertises that it will provide beer, pizza, and wings for $50 per person at a Super Bowl party. It must have a guarantee of 30 people, and for each person in excess of 30, the restaurant will reduce the price per person for all attending by $0.50. How many people will it take to maximize the restaurant's revenue?

Thoughts Into Words

57. Give a step-by-step description of how you would use the ideas of this section to graph $f(x) = -4x^2 + 16x - 13$.

58. Is $f(x) = (3x - 2) - (2x + 1)$ a linear function? Explain your answer.

59. Suppose that Bianca walks at a constant rate of 3 miles per hour. Explain what it means that the distance Bianca walks is a linear function of the time that she walks.

Graphing Calculator Activities

60. Use a graphing calculator to check your graphs for Problems 17–30.

61. Graph each of the following parabolas, and keep in mind that you may need to change the dimensions of the viewing window to obtain a good picture.
(a) $f(x) = x^2 - 2x + 12$ **(b)** $f(x) = -x^2 - 4x - 16$
(c) $f(x) = x^2 + 12x + 44$ **(d)** $f(x) = x^2 - 30x + 229$
(e) $f(x) = -2x^2 + 8x - 19$

62. Graph each of the following parabolas, and use the TRACE feature to find whole number estimates of the vertex. Then either complete the square or use

$$\left(-\frac{b}{2a}, \frac{4ac - b^2}{4a}\right) \text{ to find the vertex.}$$

(a) $f(x) = x^2 - 6x + 3$ **(b)** $f(x) = x^2 - 18x + 66$
(c) $f(x) = -x^2 + 8x - 3$
(d) $f(x) = -x^2 + 24x - 129$
(e) $f(x) = 14x^2 - 7x + 1$
(f) $f(x) = -0.5x^2 + 5x - 8.5$

63. (a) Graph $f(x) = |x|$, $f(x) = 2|x|$, $f(x) = 4|x|$, and $f(x) = \frac{1}{2}|x|$ on the same set of axes.

(b) Graph $f(x) = |x|$, $f(x) = -|x|$, $f(x) = -3|x|$, and $f(x) = -\frac{1}{2}|x|$ on the same set of axes.

(c) Use your results from parts (a) and (b) to make a conjecture about the graphs of $f(x) = a|x|$, where a is a nonzero real number.

(d) Graph $f(x) = |x|$, $f(x) = |x| + 3$, $f(x) = |x| - 4$, and $f(x) = |x| + 1$ on the same set of axes. Make a conjecture about the graphs of $f(x) = |x| + k$, where k is a nonzero real number.

(e) Graph $f(x) = |x|$, $f(x) = |x - 3|$, $f(x) = |x - 1|$, and $f(x) = |x + 4|$ on the same set of axes. Make a conjecture about the graphs of $f(x) = |x - h|$, where h is a nonzero real number.

(f) On the basis of your results from parts (a) through (e), sketch each of the following graphs. Then use a graphing calculator to check your sketches.
(1) $f(x) = |x - 2| + 3$ **(2)** $f(x) = |x + 1| - 4$
(3) $f(x) = 2|x - 4| - 1$ **(4)** $f(x) = -3|x + 2| + 4$
(5) $f(x) = \frac{1}{2}|x - 3| - 2$

Answers to the Concept Quiz
1. False **2.** True **3.** True **4.** True **5.** True **6.** False **7.** False **8.** False **9.** True **10.** True

---

13.3   Graphing Made Easy via Transformations

OBJECTIVES

1 Know the basic graphs of the following functions: $f(x) = x^2$, $f(x) = x^3$, $f(x) = 1/x$, $f(x) = \sqrt{x}$, and $f(x) = |x|$

2 Graph functions by using translations

3 Graph functions by using reflections

4 Graph functions by using vertical stretching or shrinking

5 Graph functions by using successive transformations

Within mathematics there are several basic functions that we encounter throughout our work. Many functions that you have to graph will be shifts, reflections, stretching, and shrinking of these basic graphs. The objective of this section is to be able to graph functions by making transformations to the basic graphs.

The five basic functions you will encounter in this section are as follows:

$$f(x) = x^2 \qquad f(x) = x^3 \qquad f(x) = \frac{1}{x} \qquad f(x) = \sqrt{x} \qquad \text{and} \qquad f(x) = |x|$$

Figures 13.14–13.16 show the graphs of the functions $f(x) = x^2$, $f(x) = x^3$, and $f(x) = \frac{1}{x}$, respectively.

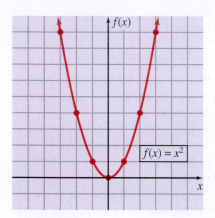

Figure 13.14

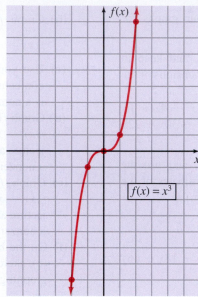

Figure 13.15

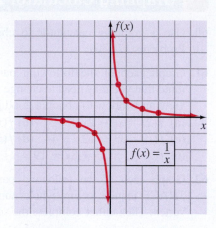

Figure 13.16

To graph a new function—that is, one you are not familiar with—use some of the graphing suggestions we offered in Chapter 12. We restate those suggestions in terms of function vocabulary and notation. Pay special attention to suggestions 2 and 3, where we have restated the concepts of intercepts and symmetry using function notation.

1. Determine the domain of the function.

2. Determine any types of symmetry that the equation possesses. If $f(-x) = f(x)$, then the function exhibits y-axis symmetry. If $f(-x) = -f(x)$, then the function exhibits origin symmetry. (Note that the definition of a function rules out the possibility that the graph of a function has x-axis symmetry.)

3. Find the y intercept (we are labeling the y axis with $f(x)$) by evaluating $f(0)$. Find the x intercept by finding the value(s) of x such that $f(x) = 0$.

4. Set up a table of ordered pairs that satisfy the equation. The type of symmetry and the domain will affect your choice of values of x in the table.

5. Plot the points associated with the ordered pairs and connect them with a smooth curve. Then, if appropriate, reflect this part of the curve according to any symmetries the graph exhibits.

Let's consider these suggestions as we determine the graphs of $f(x) = \sqrt{x}$ and $f(x) = |x|$.

To graph $f(x) = \sqrt{x}$, let's first determine the domain. The radicand must be nonnegative, so the domain is the set of nonnegative real numbers. Because $x \geq 0$, $f(-x)$ is not a real number; thus there is no symmetry for this graph. We see that $f(0) = 0$, so both intercepts are 0. That is, the origin $(0, 0)$ is a point of the graph. Now let's set up a table of values, keeping in mind that $x \geq 0$. Plotting these points and connecting them with a smooth curve produces Figure 13.17.

x	f(x)
0	0
1	1
4	2
9	3

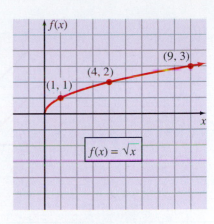

Figure 13.17

Sometimes a new function is defined in terms of old functions. In such cases, the definition plays an important role in the study of the new function. Consider the following example.

To graph $f(x) = |x|$, it is important to consider the definition of absolute value. The concept of absolute value is defined for all real numbers as

$$|x| = x \quad \text{if } x \geq 0$$
$$|x| = -x \quad \text{if } x < 0$$

Therefore, we express the absolute-value function as

$$f(x) = |x| = \begin{cases} x & \text{if } x \geq 0 \\ -x & \text{if } x < 0 \end{cases}$$

The graph of $f(x) = x$ for $x \geq 0$ is the ray in the first quadrant, and the graph of $f(x) = -x$ for $x < 0$ is the half-line in the second quadrant in Figure 13.18.

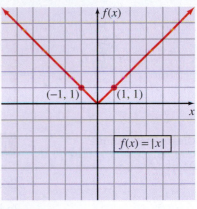

Figure 13.18

Remark: Note that the equation $f(x) = |x|$ does exhibit y-axis symmetry because $f(-x) = |-x| = |x|$. Even though we did not use the symmetry idea to sketch the curve, you should recognize that the symmetry does exist.

Graphing Functions by Using Translations

From our previous work in Chapter 12, we know that the graph of $f(x) = x^2 + 3$ is the graph of $f(x) = x^2$ moved up 3 units. Likewise, the graph of $f(x) = x^2 - 2$ is the graph of $f(x) = x^2$ moved down 2 units. Now we will describe in general the concept of *vertical translation*.

Classroom Example
Graph (a) $f(x) = x^2 + 3$ and
(b) $f(x) = x^2 - 1$.

Vertical Translation

The graph of $y = f(x) + k$ is the graph of $y = f(x)$ shifted k units upward if $k > 0$ or shifted $|k|$ units downward if $k < 0$.

EXAMPLE 1   Graph $f(x) = |x| + 2$ and $f(x) = |x| - 3$.

Solution

In Figure 13.19, we obtain the graph of $f(x) = |x| + 2$ by shifting the graph of $f(x) = |x|$ upward 2 units, and we obtain the graph of $f(x) = |x| - 3$ by shifting the graph of $f(x) = |x|$ downward 3 units. Remember that we can write $f(x) = |x| - 3$ as $f(x) = |x| + (-3)$.

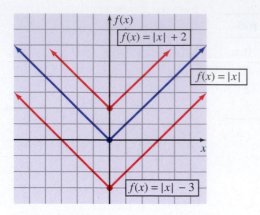

Figure 13.19

We also graphed horizontal translations of the basic parabola in Chapter 12. For example, the graph of $f(x) = (x - 4)^2$ is the graph of $f(x) = x^2$ shifted 4 units to the right, and the graph of $f(x) = (x + 5)^2$ is the graph of $f(x) = x^2$ shifted 5 units to the left. We describe the general concept of a *horizontal translation* as follows:

> ### Horizontal Translation
>
> The graph of $y = f(x - h)$ is the graph of $y = f(x)$ shifted h units to the right if $h > 0$ or shifted $|h|$ units to the left if $h < 0$.

EXAMPLE 2   Graph $f(x) = (x - 3)^3$ and $f(x) = (x + 2)^3$.

Solution

In Figure 13.20, we obtain the graph of $f(x) = (x - 3)^3$ by shifting the graph of $f(x) = x^3$ to the right 3 units. Likewise, we obtain the graph of $f(x) = (x + 2)^3$ by shifting the graph of $f(x) = x^3$ to the left 2 units.

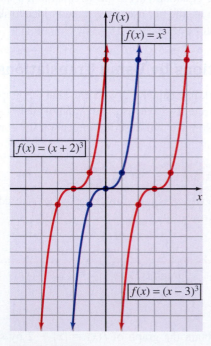

Figure 13.20

Graphing Functions by Using Reflections

From our work in Chapter 12, we know that the graph of $f(x) = -x^2$ is the graph of $f(x) = x^2$ reflected through the x axis. We describe the general concept of an *x-axis reflection* as follows:

> ### x-Axis Reflection
>
> The graph of $y = -f(x)$ is the graph of $y = f(x)$ reflected through the x axis.

Classroom Example
Graph $f(x) = -|x|$.

EXAMPLE 3   Graph $f(x) = -\sqrt{x}$.

Solution

In Figure 13.21, we obtain the graph of $f(x) = -\sqrt{x}$ by reflecting the graph of $f(x) = \sqrt{x}$ through the x axis. Reflections are sometimes referred to as mirror images. Thus in Figure 13.21, if we think of the x axis as a mirror, the graphs of $f(x) = \sqrt{x}$ and $f(x) = -\sqrt{x}$ are mirror images of each other.

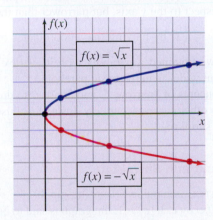

Figure 13.21

In Chapter 12, we did not consider a y-axis reflection of the basic parabola $f(x) = x^2$ because it is symmetric with respect to the y axis. In other words, a y-axis reflection of $f(x) = x^2$ produces the same figure in the same location. At this time we will describe the general concept of a y-axis reflection.

> ### y-Axis Reflection
>
> The graph of $y = f(-x)$ is the graph of $y = f(x)$ reflected through the y axis.

Now suppose that we want to do a y-axis reflection of $f(x) = \sqrt{x}$. Because $f(x) = \sqrt{x}$ is defined for $x \geq 0$, the y-axis reflection $f(x) = \sqrt{-x}$ is defined for $-x \geq 0$, which is equivalent to $x \leq 0$. Figure 13.22 shows the y-axis reflection of $f(x) = \sqrt{x}$.

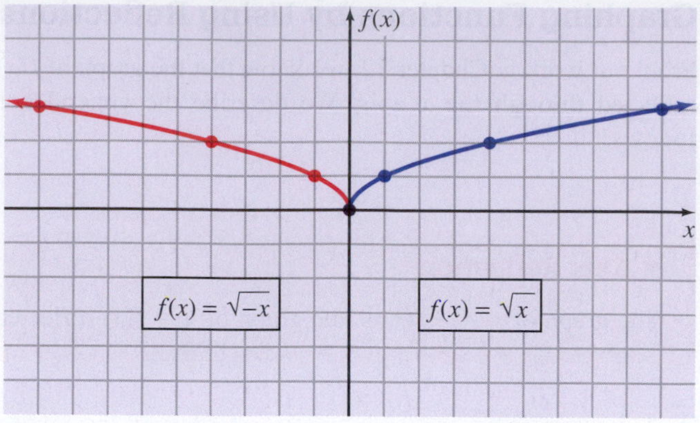

Figure 13.22

Graphing Functions by Using Vertical Stretching and Shrinking

Translations and reflections are called **rigid transformations** because the basic shape of the curve being transformed is not changed. In other words, only the positions of the graphs have changed. Now we want to consider some transformations that distort the shape of the original figure somewhat.

In Chapter 12, we graphed the equation $y = 2x^2$ by doubling the y coordinates of the ordered pairs that satisfy the equation $y = x^2$. We obtained a parabola with its vertex at the origin, symmetric with respect to the y axis, but narrower than the basic parabola. Likewise, we graphed the equation $y = \frac{1}{2}x^2$ by halving the y coordinates of the ordered pairs that satisfy $y = x^2$. In this case, we obtained a parabola with its vertex at the origin, symmetric with respect to the y axis, but wider than the basic parabola.

We can use the concepts of "narrower" and "wider" to describe parabolas, but they cannot be used to describe other curves accurately. Instead, we use the more general concepts of vertical stretching and shrinking.

Vertical Stretching and Shrinking

The graph of $y = cf(x)$ is obtained from the graph of $y = f(x)$ by multiplying the y coordinates of $y = f(x)$ by c. If $|c| > 1$, the graph is said to be *stretched* by a factor of $|c|$, and if $0 < |c| < 1$, the graph is said to be *shrunk* by a factor of $|c|$.

Classroom Example
Graph (a) $f(x) = 3|x|$ and (b) $f(x) = \frac{1}{3}|x|$.

EXAMPLE 4 Graph $f(x) = 2\sqrt{x}$ and $f(x) = \frac{1}{2}\sqrt{x}$.

Solution

In Figure 13.23, the graph of $f(x) = 2\sqrt{x}$ is obtained by doubling the y coordinates of points on the graph of $f(x) = \sqrt{x}$. Likewise, in Figure 13.23, the graph of $f(x) = \frac{1}{2}\sqrt{x}$ is obtained by halving the y coordinates of points on the graph of $f(x) = \sqrt{x}$.

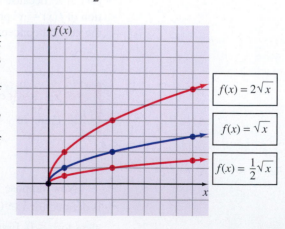

Figure 13.23

Graphing Functions by Using Successive Transformations

Some curves result from performing more than one transformation on a basic curve. Let's consider the graph of a function that involves a stretching, a reflection, a horizontal translation, and a vertical translation of the basic absolute-value function.

EXAMPLE 5   Graph $f(x) = -2|x - 3| + 1$.

Solution

This is the basic absolute-value curve stretched by a factor of two, reflected through the x axis, shifted 3 units to the right, and shifted 1 unit upward. To sketch the graph, we locate the point $(3, 1)$ and then determine a point on each of the rays. The graph is shown in Figure 13.24.

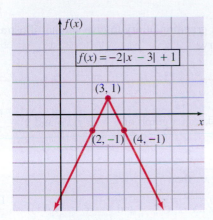

Figure 13.24

Remark: Note in Example 5 that we did not sketch the original basic curve $f(x) = |x|$ or any of the intermediate transformations. However, it is helpful to mentally picture each transformation. This locates the point $(3, 1)$ and establishes the fact that the two rays point downward. Then a point on each ray determines the final graph.

You also need to realize that changing the order of doing the transformations may produce an incorrect graph. In Example 5, performing the translations first, followed by the stretching and x-axis reflection, would produce an incorrect graph that has its vertex at $(3, -1)$ instead of $(3, 1)$. Unless parentheses indicate otherwise, stretchings, shrinkings, and x-axis reflections should be performed before translations.

EXAMPLE 6   Graph the function $f(x) = \dfrac{1}{x + 2} + 3$.

Solution

This is the basic curve $f(x) = \dfrac{1}{x}$ moved 2 units to the left and 3 units upward. Remember that the x axis is a horizontal asymptote, and the y axis a vertical asymptote for the curve $f(x) = \dfrac{1}{x}$. Thus for this curve, the vertical asymptote is shifted 2 units to the left, and its equation is $x = -2$. Likewise, the horizontal asymptote is shifted 3 units upward, and its equation is $y = 3$. Therefore, in Figure 13.25 we have drawn the asymptotes as dashed lines and then located a few points to help determine each branch of the curve.

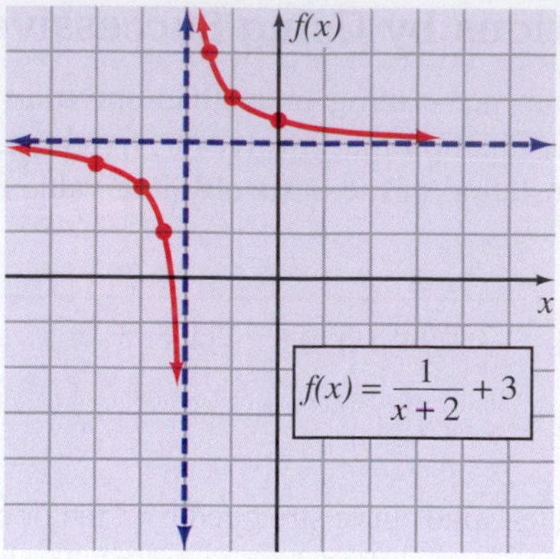

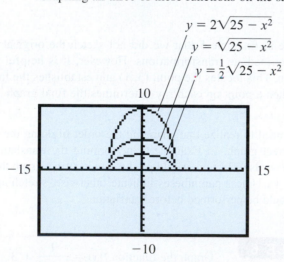

Figure 13.25

Finally, let's use a graphing utility to give another illustration of the concepts of stretching and shrinking a curve.

Classroom Example
If $f(x) = \sqrt{36 - x^2}$, sketch a graph of $y = 3\,(f(x))$ and $y = \frac{1}{3}f(x)$.

EXAMPLE 7

If $f(x) = \sqrt{25 - x^2}$, sketch a graph of $y = 2(f(x))$ and $y = \frac{1}{2}(f(x))$.

Solution

If $y = f(x) = \sqrt{25 - x^2}$, then

$$y = 2(f(x)) = 2\sqrt{25 - x^2} \qquad \text{and} \qquad y = \frac{1}{2}(f(x)) = \frac{1}{2}\sqrt{25 - x^2}$$

Graphing all three of these functions on the same set of axes produces Figure 13.26.

Figure 13.26

Concept Quiz 13.3

For Problems 1–5, match the function with the description of its graph relative to the graph of $f(x) = \sqrt{x}$.

1. $f(x) = \sqrt{x} + 3$ **A.** Stretched by a factor of three

2. $f(x) = -\sqrt{x}$ **B.** Reflected across the y axis

3. $f(x) = \sqrt{x + 3}$ **C.** Shifted up three units

4. $f(x) = \sqrt{-x}$ **D.** Reflected across the x axis

5. $f(x) = 3\sqrt{x}$ **E.** Shifted three units to the left

For Problems 6–10, answer true or false.

6. The graph of $f(x) = \sqrt{x} - 1$ is the graph of $f(x) = \sqrt{x}$ shifted 1 unit downward.

7. The graph of $f(x) = |x + 2|$ is symmetric with respect to the line $x = -2$.

8. The graph of $f(x) = x^3 - 3$ is the graph of $f(x) = x^3$ shifted 3 units downward.

9. The graph of $f(x) = \dfrac{1}{x + 2}$ intersects the line $x = -2$ at the point $(-2, 0)$.

10. The graph of $f(x) = \dfrac{1}{x} + 2$ is the graph of $f(x) = \dfrac{1}{x}$ shifted 2 units upward.

Problem Set 13.3

For Problems 1–6, match the function with its graph (Figures 13.27–13.32). **(Objective 1)**

1. $f(x) = x^2$ **2.** $f(x) = x^3$

3. $f(x) = \dfrac{1}{x}$ **4.** $f(x) = \sqrt{x}$

5. $f(x) = |x|$ **6.** $f(x) = x$

A.
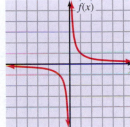

Figure 13.27

B.
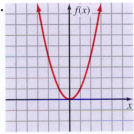

Figure 13.28

C.

Figure 13.29

D.

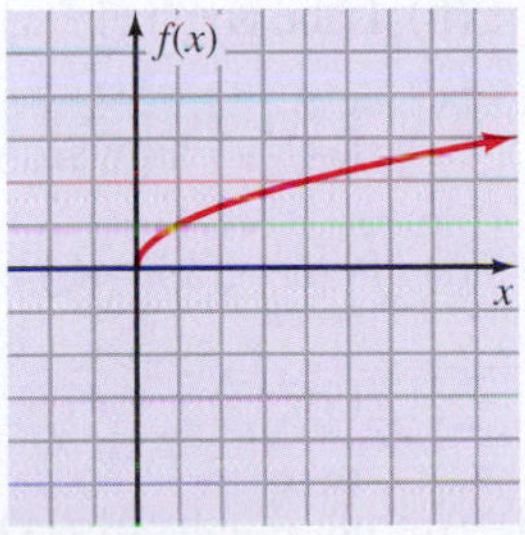

Figure 13.30

E.
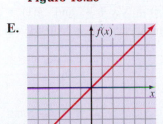

Figure 13.31

F.
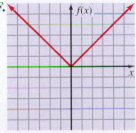

Figure 13.32

For Problems 7–50, graph each of the functions. **(Objectives 2–5)**

7. $f(x) = x^3 - 2$ **8.** $f(x) = |x| + 3$

9. $f(x) = \sqrt{x} + 3$ **10.** $f(x) = \dfrac{1}{x} - 2$

11. $f(x) = \sqrt{x + 3}$ **12.** $f(x) = \dfrac{1}{x - 2}$

13. $f(x) = |x + 1|$ **14.** $f(x) = (x - 2)^2$

15. $f(x) = -x^3$ **16.** $f(x) = -x^2$

17. $f(x) = \sqrt{-x}$ **18.** $f(x) = -\sqrt{x}$

19. $f(x) = \dfrac{1}{2}|x|$ **20.** $f(x) = \dfrac{1}{4}x^2$

21. $f(x) = 2x^2$ **22.** $f(x) = 2\sqrt{x}$

23. $f(x) = -(x - 4)^2 + 2$ **24.** $f(x) = -2(x + 3)^2 - 4$

25. $f(x) = |x - 1| + 2$ **26.** $f(x) = -|x + 2|$

27. $f(x) = -2\sqrt{x}$ **28.** $f(x) = 2\sqrt{x - 1}$

29. $f(x) = \sqrt{x + 2} - 3$ **30.** $f(x) = -\sqrt{x + 2} + 2$

31. $f(x) = \dfrac{2}{x - 1} + 3$ **32.** $f(x) = \dfrac{3}{x + 3} - 4$

33. $f(x) = \sqrt{2 - x}$ **34.** $f(x) = \sqrt{-1 - x}$

35. $f(x) = -3(x - 2)^2 - 1$ **36.** $f(x) = (x + 5)^2 - 2$

37. $f(x) = 3(x - 2)^3 - 1$ **38.** $f(x) = -2(x + 1)^3 + 2$

39. $f(x) = 2x^3 + 3$ **40.** $f(x) = -2x^3 - 1$

41. $f(x) = -2\sqrt{x + 3} + 4$ **42.** $f(x) = -3\sqrt{x - 1} + 2$

43. $f(x) = \dfrac{-2}{x + 2} + 2$ **44.** $f(x) = \dfrac{-1}{x - 1} - 1$

45. $f(x) = \dfrac{x - 1}{x}$ **46.** $f(x) = \dfrac{x + 2}{x}$

47. $f(x) = -3|x + 4| + 3$ **48.** $f(x) = -2|x - 3| - 4$

49. $f(x) = 4|x| + 2$ **50.** $f(x) = -3|x| - 4$

51. Suppose that the graph of $y = f(x)$ with a domain of $-2 \leq x \leq 2$ is shown in Figure 13.33.

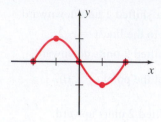

Figure 13.33

Sketch the graph of each of the following transformations of $y = f(x)$.

(a) $y = f(x) + 3$ **(b)** $y = f(x - 2)$
(c) $y = -f(x)$ **(d)** $y = f(x + 3) - 4$

52. Use the definition of absolute value to help you sketch the following graphs.

(a) $f(x) = x + |x|$ **(b)** $f(x) = x - |x|$

(c) $f(x) = |x| - x$ **(d)** $f(x) = \dfrac{x}{|x|}$

(e) $f(x) = \dfrac{|x|}{x}$

Thoughts Into Words

53. Is the graph of $f(x) = x^2 + 2x + 4$ a y-axis reflection of $f(x) = x^2 - 2x + 4$? Defend your answer.

54. Is the graph of $f(x) = x^2 - 4x - 7$ an x-axis reflection of $f(x) = x^2 + 4x + 7$? Defend your answer.

55. Your friend claims that the graph of $f(x) = \dfrac{2x + 1}{x}$ is the graph of $f(x) = \dfrac{1}{x}$ shifted 2 units upward. How could you verify whether she is correct?

Graphing Calculator Activities

56. Use a graphing calculator to check your graphs for Problems 39–50.

57. Use a graphing calculator to check your graphs for Problem 52.

58. For each of the following, answer the question on the basis of your knowledge of transformations, and then use a graphing calculator to check your answer.
(a) Is the graph of $f(x) = 2x^2 + 8x + 13$ a y-axis reflection of $f(x) = 2x^2 - 8x + 13$?
(b) Is the graph of $f(x) = 3x^2 - 12x + 16$ an x-axis reflection of $f(x) = -3x^2 + 12x - 16$?
(c) Is the graph of $f(x) = \sqrt{4 - x}$ a y-axis reflection of $f(x) = \sqrt{x + 4}$?
(d) Is the graph of $f(x) = \sqrt{3 - x}$ a y-axis reflection of $f(x) = \sqrt{x - 3}$?
(e) Is the graph of $f(x) = -x^3 + x + 1$ a y-axis reflection of $f(x) = x^3 - x + 1$?

(f) Is the graph of $f(x) = -(x - 2)^3$ an x-axis reflection of $f(x) = (x - 2)^3$?
(g) Is the graph of $f(x) = -x^3 - x^2 - x + 1$ an x-axis reflection of $f(x) = x^3 + x^2 + x - 1$?
(h) Is the graph of $f(x) = \dfrac{3x + 1}{x}$ a vertical translation of $f(x) = \dfrac{1}{x}$ upward 3 units?
(i) Is the graph of $f(x) = 2 + \dfrac{1}{x}$ a y-axis reflection of $f(x) = \dfrac{2x - 1}{x}$?

59. Are the graphs of $f(x) = 2\sqrt{x}$ and $g(x) = \sqrt{2x}$ identical? Defend your answer.

60. Are the graphs of $f(x) = \sqrt{x + 4}$ and $g(x) = \sqrt{-x + 4}$ y-axis reflections of each other? Defend your answer.

Answers to the Concept Quiz

1. C   **2.** D   **3.** E   **4.** B   **5.** A   **6.** False   **7.** True   **8.** True   **9.** False   **10.** True

| 13.4 | Composition of Functions |

OBJECTIVES

1. Find the composition of two functions and determine the domain

2. Determine functional values for composite functions

The basic operations of addition, subtraction, multiplication, and division can be performed on functions. However, there is an additional operation, called *composition*, that we will use in the next chapter. Let's start with the definition and an illustration of this operation.

> **Definition 13.3**
>
> The **composition** of functions f and g is defined by
>
> $(f \circ g)(x) = f(g(x))$
>
> for all x in the domain of g such that $g(x)$ is in the domain of f.

The left side, $(f \circ g)(x)$, of the equation in Definition 13.3 can be read "the composition of f and g," and the right side, $f(g(x))$, can be read "f of g of x." It may also be helpful for you to picture Definition 13.3 as two function machines hooked together to produce another function (often called a **composite function**) as illustrated in Figure 13.34. Note that what comes out of the function g is substituted into the function f. Thus composition is sometimes called the *substitution of functions*.

Figure 13.34 also vividly illustrates the fact that $f \circ g$ is defined for all x in the domain of g such that $g(x)$ is in the domain of f. In other words, what comes out of g must be capable of being fed into f. Let's consider some examples.

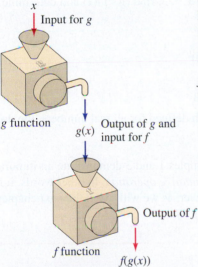

Figure 13.34

Classroom Example
If $f(x) = 2x^2$ and $g(x) = x + 1$, find $(f \circ g)(x)$ and determine its domain.

EXAMPLE 1

If $f(x) = x^2$ and $g(x) = x - 3$, find $(f \circ g)(x)$ and determine its domain.

Solution

Applying Definition 13.3, we obtain

$(f \circ g)(x) = f(g(x))$
$= f(x - 3)$
$= (x - 3)^2$

Because g and f are both defined for all real numbers, so is $f \circ g$.

Classroom Example

If $f(x) = \dfrac{1}{x}$ and $g(x) = x + 2$, find $(f \circ g)(x)$ and determine its domain.

EXAMPLE 2

If $f(x) = \sqrt{x}$ and $g(x) = x - 4$, find $(f \circ g)(x)$ and determine its domain.

Solution

Applying Definition 13.3, we obtain

$$(f \circ g)(x) = f(g(x))$$
$$= f(x - 4)$$
$$= \sqrt{x - 4}$$

The domain of g is all real numbers, but the domain of f is only the nonnegative real numbers. Thus $g(x)$, which is $x - 4$, has to be nonnegative. Therefore,

$$x - 4 \geq 0$$
$$x \geq 4$$

and the domain of $f \circ g$ is $D = \{x | x \geq 4\}$ or $D: [4, \infty)$.

Definition 13.3, with f and g interchanged, defines the composition of g and f as $(g \circ f)(x) = g(f(x))$.

Classroom Example

If $f(x) = 2x^2$ and $g(x) = x + 5$, find $(g \circ f)(x)$ and determine its domain.

EXAMPLE 3

If $f(x) = x^2$ and $g(x) = x - 3$, find $(g \circ f)(x)$ and determine its domain.

Solution

$$(g \circ f)(x) = g(f(x))$$
$$= g(x^2)$$
$$= x^2 - 3$$

Because f and g are both defined for all real numbers, the domain of $g \circ f$ is the set of all real numbers.

The results of Examples 1 and 3 demonstrate an important idea: that the composition of functions is *not a commutative operation*. In other words, it is not true that $f \circ g = g \circ f$ for all functions f and g. However, as we will see in the next chapter, there is a special class of functions for which $f \circ g = g \circ f$.

Classroom Example

If $f(x) = \dfrac{3}{x + 2}$ and $g(x) = \dfrac{2}{x}$, find $(f \circ g)(x)$ and $(g \circ f)(x)$. Determine the domain for each composite function.

EXAMPLE 4

If $f(x) = \dfrac{2}{x - 1}$ and $g(x) = \dfrac{1}{x}$, find $(f \circ g)(x)$ and $(g \circ f)(x)$. Determine the domain for each composite function.

Solution

$$(f \circ g)(x) = f(g(x))$$

$$= f\left(\frac{1}{x}\right)$$

$$= \frac{2}{\dfrac{1}{x} - 1} = \frac{2}{\dfrac{1 - x}{x}}$$

$$= \frac{2x}{1 - x}$$

The domain of g is all real numbers except 0, and the domain of f is all real numbers except 1. Because $g(x)$, which is $\dfrac{1}{x}$, cannot equal 1,

$$\frac{1}{x} \neq 1$$

$$x \neq 1$$

Therefore, the domain of $f \circ g$ is $D = \{x \mid x \neq 0 \text{ and } x \neq 1\}$ or $D: (-\infty, 0) \cup (0, 1) \cup (1, \infty)$.

$$(g \circ f)(x) = g(f(x))$$

$$= g\left(\frac{2}{x-1}\right)$$

$$= \frac{1}{\dfrac{2}{x-1}}$$

$$= \frac{x-1}{2}$$

The domain of f is all real numbers except 1, and the domain of g is all real numbers except 0. Because $f(x)$, which is $\dfrac{2}{x-1}$, will never equal 0, the domain of $g \circ f$ is $D = \{x \mid x \neq 1\}$ or $D: (-\infty, 1) \cup (1, \infty)$.

Determining the Function Values for Composite Functions

Classroom Example
If $f(x) = 4x - 1$ and $g(x) = \sqrt{x+3}$, determine each of the following.
(a) $(f \circ g)(x)$
(b) $(g \circ f)(x)$
(c) $(f \circ g)(6)$
(d) $(g \circ f)\left(\dfrac{1}{2}\right)$

EXAMPLE 5

If $f(x) = 2x + 3$ and $g(x) = \sqrt{x-1}$, determine each of the following.

   (a) $(f \circ g)(x)$ **(b)** $(g \circ f)(x)$ **(c)** $(f \circ g)(5)$ **(d)** $(g \circ f)(7)$

Solution

  (a) $(f \circ g)(x) = f(g(x))$

$$= f(\sqrt{x-1})$$

$$= 2\sqrt{x-1} + 3 \qquad D = \{x \mid x \geq 1\} \text{ or } D: [1, \infty)$$

  (b) $(g \circ f)(x) = g(f(x))$

$$= g(2x + 3)$$

$$= \sqrt{2x + 3 - 1}$$

$$= \sqrt{2x + 2} \qquad D = \{x \mid x \geq -1\} \text{ or } D: [-1, \infty)$$

  (c) $(f \circ g)(5) = 2\sqrt{5-1} + 3 = 7$

  (d) $(g \circ f)(7) = \sqrt{2(7) + 2} = 4$

Graphing a Composite Function Using a Graphing Utility

A graphing utility can be used to find the graph of a composite function without actually forming the function algebraically. Let's see how this works.

EXAMPLE 6

If $f(x) = x^3$ and $g(x) = x - 4$, use a graphing utility to obtain the graph of $y = (f \circ g)(x)$ and the graph of $y = (g \circ f)(x)$.

Solution

To find the graph of $y = (f \circ g)(x)$, we can make the following assignments:

$Y_1 = x - 4$

$Y_2 = (Y_1)^3$

(Note that we have substituted Y_1 for x in $f(x)$ and assigned this expression to Y_2, much the same way that we would algebraically.) Now, by showing only the graph of Y_2, we obtain Figure 13.35.

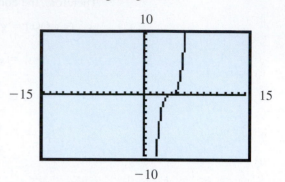

Figure 13.35

To find the graph of $y = (g \circ f)(x)$, we can make the following assignments:

$Y_1 = x^3$

$Y_2 = Y_1 - 4$

The graph of $y = (g \circ f)(x)$ is the graph of Y_2, as shown in Figure 13.36.

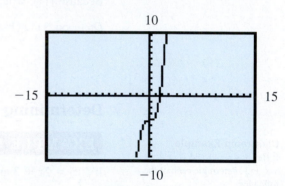

Figure 13.36

Take another look at Figures 13.35 and 13.36. Note that in Figure 13.35 the graph of $y = (f \circ g)(x)$ is the basic cubic curve $f(x) = x^3$ shifted 4 units to the right. Likewise, in Figure 13.36 the graph of $y = (g \circ f)(x)$ is the basic cubic curve shifted 4 units downward. These are examples of a more general concept of using composite functions to represent various geometric transformations.

Concept Quiz 13.4

For Problems 1–10, answer true or false.

1. The composition of functions is a commutative operation.
2. To find $(h \circ k)(x)$, the function k will be substituted into the function h.
3. The notation $(f \circ g)(x)$ is read "the substitution of g and f."
4. The domain for $(f \circ g)(x)$ is always the same as the domain for g.
5. The notation $f(g(x))$ is read "f of g of x."
6. If $f(x) = x + 2$ and $g(x) = 3x - 1$, then $f(g(2)) = 7$.
7. If $f(x) = x + 2$ and $g(x) = 3x - 1$, then $g(f(2)) = 7$.
8. If $f(x) = \sqrt{x - 1}$ and $g(x) = 2x - 3$, then $f(g(1))$ is undefined.
9. If $f(x) = \sqrt{x - 1}$ and $g(x) = 2x - 3$, then $g(f(1))$ is undefined.
10. If $f(x) = -x^2 - x + 2$ and $g(x) = x + 1$, then $f(g(x)) = x^2 + 2x + 1$.

Problem Set 13.4

For Problems 1–18, determine $(f \circ g)(x)$ and $(g \circ f)(x)$ for each pair of functions. Also specify the domain of $(f \circ g)(x)$ and $(g \circ f)(x)$. **(Objective 1)**

1. $f(x) = 3x$ and $g(x) = 5x - 1$

2. $f(x) = 4x - 3$ and $g(x) = -2x$

3. $f(x) = -2x + 1$ and $g(x) = 7x + 4$

4. $f(x) = 6x - 5$ and $g(x) = -x + 6$

5. $f(x) = 3x + 2$ and $g(x) = x^2 + 3$

6. $f(x) = -2x + 4$ and $g(x) = 2x^2 - 1$

7. $f(x) = 2x^2 - x + 2$ and $g(x) = -x + 3$

8. $f(x) = 3x^2 - 2x - 4$ and $g(x) = -2x + 1$

9. $f(x) = \dfrac{3}{x}$ and $g(x) = 4x - 9$

10. $f(x) = -\dfrac{2}{x}$ and $g(x) = -3x + 6$

11. $f(x) = \sqrt{x + 1}$ and $g(x) = 5x + 3$

12. $f(x) = 7x - 2$ and $g(x) = \sqrt{2x - 1}$

13. $f(x) = \dfrac{1}{x}$ and $g(x) = \dfrac{1}{x - 4}$

14. $f(x) = \dfrac{2}{x + 3}$ and $g(x) = -\dfrac{3}{x}$

15. $f(x) = \sqrt{x}$ and $g(x) = \dfrac{4}{x}$

16. $f(x) = \dfrac{2}{x}$ and $g(x) = |x|$

17. $f(x) = \dfrac{3}{2x}$ and $g(x) = \dfrac{1}{x + 1}$

18. $f(x) = \dfrac{4}{x - 2}$ and $g(x) = \dfrac{3}{4x}$

For Problems 19–26, show that $(f \circ g)(x) = x$ and $(g \circ f)(x) = x$ for each pair of functions. **(Objective 1)**

19. $f(x) = 3x$ and $g(x) = \dfrac{1}{3}x$

20. $f(x) = -2x$ and $g(x) = -\dfrac{1}{2}x$

21. $f(x) = 4x + 2$ and $g(x) = \dfrac{x - 2}{4}$

22. $f(x) = 3x - 7$ and $g(x) = \dfrac{x + 7}{3}$

23. $f(x) = \dfrac{1}{2}x + \dfrac{3}{4}$ and $g(x) = \dfrac{4x - 3}{2}$

24. $f(x) = \dfrac{2}{3}x - \dfrac{1}{5}$ and $g(x) = \dfrac{3}{2}x + \dfrac{3}{10}$

25. $f(x) = -\dfrac{1}{4}x - \dfrac{1}{2}$ and $g(x) = -4x - 2$

26. $f(x) = -\dfrac{3}{4}x + \dfrac{1}{3}$ and $g(x) = -\dfrac{4}{3}x + \dfrac{4}{9}$

For Problems 27–38, determine the indicated functional values. **(Objective 2)**

27. If $f(x) = 9x - 2$ and $g(x) = -4x + 6$, find $(f \circ g)(-2)$ and $(g \circ f)(4)$.

28. If $f(x) = -2x - 6$ and $g(x) = 3x + 10$, find $(f \circ g)(5)$ and $(g \circ f)(-3)$.

29. If $f(x) = 4x^2 - 1$ and $g(x) = 4x + 5$, find $(f \circ g)(1)$ and $(g \circ f)(4)$.

30. If $f(x) = -5x + 2$ and $g(x) = -3x^2 + 4$, find $(f \circ g)(-2)$ and $(g \circ f)(-1)$.

31. If $f(x) = \dfrac{1}{x}$ and $g(x) = \dfrac{2}{x - 1}$, find $(f \circ g)(2)$ and $(g \circ f)(-1)$.

32. If $f(x) = \dfrac{2}{x - 1}$ and $g(x) = -\dfrac{3}{x}$, find $(f \circ g)(1)$ and $(g \circ f)(-1)$.

33. If $f(x) = \dfrac{1}{x - 2}$ and $g(x) = \dfrac{4}{x - 1}$, find $(f \circ g)(3)$ and $(g \circ f)(2)$.

34. If $f(x) = \sqrt{x + 6}$ and $g(x) = 3x - 1$, find $(f \circ g)(-2)$ and $(g \circ f)(-2)$.

35. If $f(x) = \sqrt{3x - 2}$ and $g(x) = -x + 4$, find $(f \circ g)(1)$ and $(g \circ f)(6)$.

36. If $f(x) = -5x + 1$ and $g(x) = \sqrt{4x + 1}$, find $(f \circ g)(6)$ and $(g \circ f)(-1)$.

37. If $f(x) = |4x - 5|$ and $g(x) = x^3$, find $(f \circ g)(-2)$ and $(g \circ f)(2)$.

38. If $f(x) = -x^3$ and $g(x) = |2x + 4|$, find $(f \circ g)(-1)$ and $(g \circ f)(-3)$.

Thoughts Into Words

39. Explain why the composition of functions is not a commutative operation.

40. How would you explain the concept of composition of functions to a friend who missed class the day it was discussed?

Graphing Calculator Activities

41. For each of the following, use your graphing calculator to find the graph of $y = (f \circ g)(x)$ and $y = (g \circ f)(x)$. Then algebraically find $(f \circ g)(x)$ and $(g \circ f)(x)$ to see whether your results agree.

(a) $f(x) = x^2$ and $g(x) = x - 3$
(b) $f(x) = x^3$ and $g(x) = x + 4$
(c) $f(x) = x - 2$ and $g(x) = -x^3$
(d) $f(x) = x + 6$ and $g(x) = \sqrt{x}$
(e) $f(x) = \sqrt{x}$ and $g(x) = x - 5$

Answers to the Concept Quiz
1. False **2.** True **3.** False **4.** False **5.** True **6.** True **7.** False **8.** True **9.** False **10.** False

---

13.5   Direct and Inverse Variations

OBJECTIVES

1. Translate sentences into equations of variation

2. Determine the constant of variation

3. Solve variation word problems

"The distance a car travels at a fixed rate varies directly as the time." "At a constant temperature, the volume of an enclosed gas varies inversely as the pressure." Such statements illustrate two basic types of functional relationships, called **direct** and **inverse variation**, that are widely used, especially in the physical sciences. These relationships can be expressed by equations that specify functions. The purpose of this section is to investigate these special functions.

The statement "y varies directly as x" means

$$y = kx$$

where k is a nonzero constant called the **constant of variation**. The phrase "y is directly proportional to x" is also used to indicate direct variation; k is then referred to as the **constant of proportionality**.

Remark: Note that the equation $y = kx$ defines a function and could be written as $f(x) = kx$ by using function notation. However, in this section it is more convenient to avoid the function notation and use variables that are meaningful in terms of the physical entities involved in the problem.

Statements that indicate direct variation may also involve powers of x. For example, "y varies directly as the square of x" can be written as

$$y = kx^2$$

In general, "y varies directly as the nth power of x $(n > 0)$" means

$$y = kx^n$$

The three types of problems that deal with direct variation are

1. Translate an English statement into an equation that expresses the direct variation

2. Find the constant of variation from given values of the variables

3. Find additional values of the variables once the constant of variation has been determined

Let's consider an example of each of these types of problems.

Classroom Example
Translate the statement "the distance traveled varies directly as the time traveled" into an equation, and use k as the constant of variation.

EXAMPLE 1

Translate the statement, "the tension on a spring varies directly as the distance it is stretched," into an equation, and use k as the constant of variation.

Solution

If we let t represent the tension and d the distance, the equation becomes

$$t = kd$$

Classroom Example
If A varies directly as the square root of s and if $A = 60$ when $s = 36$, find the constant of variation.

EXAMPLE 2

If A varies directly as the square of s, and $A = 28$ when $s = 2$, find the constant of variation.

Solution

Because A varies directly as the square of s, we have

$$A = ks^2$$

Substituting $A = 28$ and $s = 2$, we obtain

$$28 = k(2)^2$$

Solving this equation for k yields

$$28 = 4k$$
$$7 = k$$

The constant of variation is 7.

Classroom Example
If r is directly proportional to t and if $r = 40$ when $t = 48$, find the value of r when $t = 84$.

EXAMPLE 3

If y is directly proportional to x, and if $y = 6$ when $x = 9$, find the value of y when $x = 24$.

Solution

The statement "y is directly proportional to x" translates into

$$y = kx$$

If we let $y = 6$ and $x = 9$, the constant of variation becomes

$$6 = k(9)$$
$$6 = 9k$$
$$\frac{6}{9} = k$$
$$\frac{2}{3} = k$$

Thus the specific equation is $y = \frac{2}{3}x$. Now, letting $x = 24$, we obtain

$$y = \frac{2}{3}(24) = 16$$

The required value of y is 16.

Inverse Variation

We define the second basic type of variation, called **inverse variation**, as follows: The statement "y varies inversely as x" means

$$y = \frac{k}{x}$$

where k is a nonzero constant; again we refer to it as the constant of variation. The phrase "y is inversely proportional to x" is also used to express inverse variation. As with direct variation, statements that indicate inverse variation may involve powers of x. For example, "y varies inversely as the square of x" can be written as

$$y = \frac{k}{x^2}$$

In general, "y varies inversely as the nth power of x $(n > 0)$" means

$$y = \frac{k}{x^n}$$

The following examples illustrate the three basic kinds of problems we run across that involve inverse variation.

Classroom Example
Translate the statement "the volume of a gas varies inversely as the pressure" into an equation that uses k as the constant of variation.

EXAMPLE 4

Translate the statement "the length of a rectangle of a fixed area varies inversely as the width" into an equation that uses k as the constant of variation.

Solution

Let l represent the length and w the width, and the equation is

$$l = \frac{k}{w}$$

Classroom Example
If m is inversely proportional to n and if $m = 6$ when $n = 15$, find the constant of variation.

EXAMPLE 5

If y is inversely proportional to x, and $y = 4$ when $x = 12$, find the constant of variation.

Solution

Because y is inversely proportional to x, we have

$$y = \frac{k}{x}$$

Substituting $y = 4$ and $x = 12$, we obtain

$$4 = \frac{k}{12}$$

Solving this equation for k by multiplying both sides of the equation by 12 yields

$$k = 48$$

The constant of variation is 48.

Classroom Example
Suppose that the time a car travels a fixed distance varies inversely with the speed. If it takes 4 hours at 70 miles per hour to travel that distance, how long would it take at 56 miles per hour?

EXAMPLE 6

Suppose the number of days it takes to complete a construction job varies inversely as the number of people assigned to the job. If it takes 7 people 8 days to do the job, how long would it take 14 people to complete the job?

Solution

Let d represent the number of days and p the number of people. The phrase "number of days . . . varies inversely as the number of people" translates into

$$d = \frac{k}{p}$$

Let $d = 8$ when $p = 7$, and the constant of variation becomes

$$8 = \frac{k}{7}$$

$$k = 56$$

Thus the specific equation is

$$d = \frac{56}{p}$$

Now let $p = 14$ to obtain

$$d = \frac{56}{14}$$

$$d = 4$$

It should take 14 people 4 days to complete the job.

The terms *direct* and *inverse*, as applied to variation, refer to the relative behavior of the variables involved in the equation. That is, in *direct* variation $(y = kx)$, an assignment of *increasing* absolute values for x produces increasing absolute values for y, whereas in *inverse* variation $\left(y = \dfrac{k}{x}\right)$, an assignment of increasing absolute values for x produces *decreasing* absolute values for y.

Solving Variation Problems with More Than Two Variables

Variation may involve more than two variables. The following table illustrates some variation statements and their equivalent algebraic equations that use k as the constant of variation.

Statements 1, 2, and 3 illustrate the concept of **joint variation**. Statements 4 and 5 show that both direct and inverse variation may occur in the same problem. Statement 6 combines joint variation with inverse variation.

The two final examples of this section illustrate some of these variation situations.

Variation statement	Algebraic equation
1. y varies jointly as x and z	$y = kxz$
2. y varies jointly as x, z, and w	$y = kxzw$
3. V varies jointly as h and the square of r	$V = khr^2$
4. h varies directly as V and inversely as w	$h = \dfrac{kV}{w}$
5. y is directly proportional to x and inversely proportional to the square of z	$y = \dfrac{kx}{z^2}$
6. y varies jointly as w and z and inversely as x	$y = \dfrac{kwz}{x}$

EXAMPLE 7

Suppose that *y* varies jointly as *x* and *z* and inversely as *w*. If *y* = 154 when *x* = 6, *z* = 11, and *w* = 3, find the constant of variation.

Solution

The statement "*y* varies jointly as *x* and *z* and inversely as *w*" translates into

$$y = \frac{kxz}{w}$$

Substitute *y* = 154, *x* = 6, *z* = 11, and *w* = 3 to obtain

$$154 = \frac{k(6)(11)}{3}$$

$$154 = 22k$$

$$7 = k$$

The constant of variation is 7.

EXAMPLE 8

The length of a rectangular box with a fixed height varies directly as the volume and inversely as the width. If the length is 12 centimeters when the volume is 960 cubic centimeters and the width is 8 centimeters, find the length when the volume is 700 centimeters and the width is 5 centimeters.

Solution

Use *l* for length, *V* for volume, and *w* for width, and the phrase "length varies directly as the volume and inversely as the width" translates into

$$l = \frac{kV}{w}$$

Substitute *l* = 12, *V* = 960, and *w* = 8, and the constant of variation becomes

$$12 = \frac{k(960)}{8} = 120k$$

$$\frac{1}{10} = k$$

Thus the specific equation is

$$l = \frac{\frac{1}{10}V}{w} = \frac{V}{10w}$$

Now let *V* = 700 and *w* = 5 to obtain

$$l = \frac{700}{10(5)} = \frac{700}{50} = 14$$

The length is 14 centimeters.

Concept Quiz 13.5

For Problems 1–5, answer true or false.

1. In the equation *y* = *kx*, the *k* is a quantity that varies directly as *y*.

2. The equation *y* = *kx* defines a function and could be written in functional notation as *f*(*x*) = *kx*.

3. Variation that involves more than two variables is called proportional variation.

4. Every equation of variation will have a constant of variation.

5. In joint variation, both direct and inverse variation may occur in the same problem.

For Problems 6–10, match the statement of variation with its equation.

6. y varies directly as x

7. y varies inversely as x

8. y varies directly as the square of x

9. y varies directly as the square root of x

10. y varies jointly as x and z

    A. $y = \dfrac{k}{x}$

    B. $y = kxz$

    C. $y = kx^2$

    D. $y = kx$

    E. $y = k\sqrt{x}$

Problem Set 13.5

For Problems 1–10, translate each statement of variation into an equation, and use k as the constant of variation. **(Objective 1)**

1. y varies inversely as the square of x.

2. y varies directly as the cube of x.

3. C varies directly as g and inversely as the cube of t.

4. V varies jointly as l and w.

5. The volume (V) of a sphere is directly proportional to the cube of its radius (r).

6. At a constant temperature, the volume (V) of a gas varies inversely as the pressure (P).

7. The surface area (S) of a cube varies directly as the square of the length of an edge (e).

8. The intensity of illumination (I) received from a source of light is inversely proportional to the square of the distance (d) from the source.

9. The volume (V) of a cone varies jointly as its height and the square of its radius.

10. The volume (V) of a gas varies directly as the absolute temperature (T) and inversely as the pressure (P).

For Problems 11–24, find the constant of variation for each of the stated conditions. **(Objective 2)**

11. y varies directly as x, and $y = 8$ when $x = 12$.

12. y varies directly as x, and $y = 60$ when $x = 24$.

13. y varies directly as the square of x, and $y = -144$ when $x = 6$.

14. y varies directly as the cube of x, and $y = 48$ when $x = -2$.

15. V varies jointly as B and h, and $V = 96$ when $B = 24$ and $h = 12$.

16. A varies jointly as b and h, and $A = 72$ when $b = 16$ and $h = 9$.

17. y varies inversely as x, and $y = -4$ when $x = \dfrac{1}{2}$.

18. y varies inversely as x, and $y = -6$ when $x = \dfrac{4}{3}$.

19. r varies inversely as the square of t, and $r = \dfrac{1}{8}$ when $t = 4$.

20. r varies inversely as the cube of t, and $r = \dfrac{1}{16}$ when $t = 4$.

21. y varies directly as x and inversely as z, and $y = 45$ when $x = 18$ and $z = 2$.

22. y varies directly as x and inversely as z, and $y = 24$ when $x = 36$ and $z = 18$.

23. y is directly proportional to x and inversely proportional to the square of z, and $y = 81$ when $x = 36$ and $z = 2$.

24. y is directly proportional to the square of x and inversely proportional to the cube of z, and $y = 4\dfrac{1}{2}$ when $x = 6$ and $z = 4$.

Solve each of the following problems. **(Objective 3)**

25. If y is directly proportional to x, and $y = 36$ when $x = 48$, find the value of y when $x = 12$.

26. If y is directly proportional to x, and $y = 42$ when $x = 28$, find the value of y when $x = 38$.

27. If y is inversely proportional to x, and $y = \dfrac{1}{9}$ when $x = 12$, find the value of y when $x = 8$.

28. If y is inversely proportional to x, and $y = \dfrac{1}{35}$ when $x = 14$, find the value of y when $x = 16$.

29. If A varies jointly as b and h, and $A = 60$ when $b = 12$ and $h = 10$, find A when $b = 16$ and $h = 14$.

30. If V varies jointly as B and h, and $V = 51$ when $B = 17$ and $h = 9$, find V when $B = 19$ and $h = 12$.

31. The volume of a gas at a constant temperature varies inversely as the pressure. What is the volume of a gas under pressure of 25 pounds if the gas occupies 15 cubic centimeters under a pressure of 20 pounds?

32. The time required for a car to travel a certain distance varies inversely as the rate at which it travels. If it takes 4 hours at 50 miles per hour to travel the distance, how long will it take at 40 miles per hour?

33. The volume (V) of a gas varies directly as the temperature (T) and inversely as the pressure (P). If $V = 48$ when $T = 320$ and $P = 20$, find V when $T = 280$ and $P = 30$.

34. The distance that a freely falling body falls varies directly as the square of the time it falls. If a body falls 144 feet in 3 seconds, how far will it fall in 5 seconds?

35. The period (the time required for one complete oscillation) of a simple pendulum varies directly as the square root of its length. If a pendulum 12 feet long has a period of 4 seconds, find the period of a pendulum 3 feet long.

36. The simple interest earned by a certain amount of money varies jointly as the rate of interest and the time (in years) that the money is invested. If the money is invested at 8% for 2 years, $80 is earned. How much is earned if the money is invested at 6% for 3 years?

37. The electrical resistance of a wire varies directly as its length and inversely as the square of its diameter. If the resistance of 200 meters of wire that has a diameter of $\dfrac{1}{2}$ centimeter is 1.5 ohms, find the resistance of 400 meters of wire with a diameter of $\dfrac{1}{4}$ centimeter.

38. The volume of a cylinder varies jointly as its altitude and the square of the radius of its base. If the volume of a cylinder is 1386 cubic centimeters when the radius of the base is 7 centimeters and its altitude is 9 centimeters, find the volume of a cylinder that has a base of radius 14 centimeters. The altitude of the cylinder is 5 centimeters.

39. The simple interest earned by a certain amount of money varies jointly as the rate of interest and the time (in years) that the money is invested.
   (a) If some money invested at 7% for 2 years earns $245, how much would the same amount earn at 5% for 1 year?
   (b) If some money invested at 4% for 3 years earns $273, how much would the same amount earn at 6% for 2 years?
   (c) If some money invested at 6% for 4 years earns $840, how much would the same amount earn at 8% for 2 years?

40. The period (the time required for one complete oscillation) of a simple pendulum varies directly as the square root of its length. If a pendulum 9 inches long has a period of 2.4 seconds, find the period of a pendulum 12 inches long. Express your answer to the nearest tenth of a second.

41. The volume of a cylinder varies jointly as its altitude and the square of the radius of its base. If a cylinder that has a base with a radius of 5 meters and has an altitude of 7 meters has a volume of 549.5 cubic meters, find the volume of a cylinder that has a base with a radius of 9 meters and has an altitude of 14 meters.

42. If y is directly proportional to x and inversely proportional to the square of z, and if $y = 0.336$ when $x = 6$ and $z = 5$, find the constant of variation.

43. If y is inversely proportional to the square root of x, and if $y = 0.08$ when $x = 225$, find y when $x = 625$.

Thoughts Into Words

44. How would you explain the difference between direct variation and inverse variation?

45. Suppose that y varies directly as the square of x. Does doubling the value of x also double the value of y? Explain your answer.

46. Suppose that y varies inversely as x. Does doubling the value of x also double the value of y? Explain your answer.

OBJECTIVE	SUMMARY	EXAMPLE
Determine if a relation is a function. (Section 13.1/Objective 1)	A relation is a set of ordered pairs; a function is a relation in which no two ordered pairs have the same first component. The domain of a relation (or function) is the set of all first components, and the range is the set of all second components.	Specify the domain and range of the relation and state whether or not it is a function. $\{(1, 8), (2, 7), (5, 6), (3, 8)\}$ **Solution** $D = \{1, 2, 3, 5\}, R = \{6, 7, 8\}$ It is a function.
Use function notation when evaluating a function. (Section 13.1/Objective 2)	Single letters such as f, g, and h are commonly used to name functions. The symbol $f(x)$ represents the element in the range associated with x from the domain.	If $f(x) = 2x^2 + 3x - 5$, find $f(4)$. **Solution** Substitute 4 for x in the equation. $f(4) = 2(4)^2 + 3(4) - 5$ $f(4) = 32 + 12 - 5$ $f(4) = 39$
Specify the domain and range of a function. (Section 13.1/Objective 3)	The domain of a function is the set of all real number replacements for the variable that will produce real number functional values. Replacement values that make a denominator zero or a radical expression undefined are excluded from the domain. The range is the set of functional values assigned by the function for all the members in the domain.	Specify the domain and range for $$f(x) = \frac{1}{x - 3}.$$ **Solution** The values that make the denominator zero must be excluded from the domain. To find those values, set the denominator equal to zero and solve. $$x - 3 = 0$$ $$x = 3$$ The domain is the set $$\left\{x \mid x \neq 3\right\} \quad \text{or} \quad (-\infty, 3) \cup (3, \infty)$$ The functional values will be all real numbers except zero. No member of the domain will produce a functional value of zero. The range is the set $\{f(x) \mid f(x) \neq 0\}$.
Find the difference quotient of a given function. (Section 13.1/Objective 4)	The quotient $\dfrac{f(a + h) - f(a)}{h}$ is called the difference quotient.	If $f(x) = 5x + 7$, find the difference quotient. **Solution** $$\frac{f(a + h) - f(a)}{h}$$ $$= \frac{5(a + h) + 7 - (5a + 7)}{h}$$ $$= \frac{5a + 5h + 7 - 5a - 7}{h}$$ $$= \frac{5h}{h} = 5$$

(continued)

OBJECTIVE	SUMMARY	EXAMPLE
Graph linear functions. (Section 13.2/Objective 1)	Any function that can be written in the form $f(x) = ax + b$, where a and b are real numbers, is a linear function. The graph of a linear function is a straight line.	Graph $f(x) = 3x + 1$. **Solution** Because $f(0) = 1$, the point $(0, 1)$ is on the graph. Also $f(1) = 4$, so the point $(1, 4)$ is on the graph.
Apply linear functions. (Section 13.2/Objective 2)	Linear functions and their graphs can be an aid in problem solving.	The FixItFast computer repair company uses the equation $C(m) = 2m + 15$, where m is the number of minutes for the service call, to determine the charge for a service call. Graph the function and use the graph to approximate the charge for a 25-minute service call. Then use the function to find the exact charge for a 25-minute service call. **Solution** Compare your approximation to the exact charge, $$C(25) = 2(25) + 15$$ $$= 65$$

OBJECTIVE	SUMMARY	EXAMPLE
Graph quadratic functions. (Section 13.2/Objective 4)	Any function that can be written in the form $f(x) = ax^2 + bx + c$, where a, b, and c are real numbers and $a \neq 0$, is a quadratic function. The graph of any quadratic function is a parabola, which can be drawn using either of the following methods. 1. Express the function in the form $f(x) = a(x - h)^2 + k$, and use the values of a, h, and k to determine the parabola. 2. Express the function in the form $f(x) = ax^2 + bx + c$, and use the facts that the vertex is at $$\left(-\frac{b}{2a}, f\left(-\frac{b}{2a}\right)\right)$$ and the axis of symmetry is $$x = -\frac{b}{2a}$$	Graph $f(x) = 2x^2 + 8x + 7$. **Solution** $f(x) = 2x^2 + 8x + 7$ $\quad = 2(x^2 + 4x) + 7$ $\quad = 2(x^2 + 4x + 4) - 8 + 7$ $\quad = 2(x + 2)^2 - 1$ $f(x) = 2(x + 2)^2 - 1$
Use quadratic functions to solve problems. (Section 13.2/Objective 5)	We can solve some applications that involve maximum and minimum values with our knowledge of parabolas, which are generated by quadratic functions.	Suppose the cost function for producing a particular item is given by the equation $C(x) = 3x^2 - 270x + 15{,}800$, where x represents the number of items produced. How many items should be produced to minimize the cost? **Solution** The function represents a parabola. The minimum will occur at the vertex, so we want to find the x coordinate of the vertex. $$x = -\frac{b}{2a}$$ $$x = -\frac{-270}{2(3)}$$ $$\quad = 45$$ Therefore, 45 items should be produced to minimize the cost.

(continued)

OBJECTIVE	SUMMARY
Know the five basic graphs shown here. In order to shift and reflect these graphs, it is necessary to know their basic shapes. **(Section 13.3/Objective 1)**	$f(x) = x^2$ 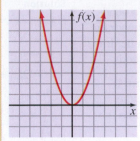 $f(x) = \dfrac{1}{x}$ 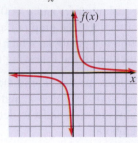 $f(x) = x^3$  $f(x) = \sqrt{x}$ 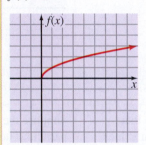 $f(x) = |x|$ 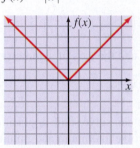

OBJECTIVE	SUMMARY	EXAMPLE
Graph functions by using translations. **(Section 13.3/Objective 2)**	**Vertical translation** The graph of $y = f(x) + k$ is the graph of $y = f(x)$ shifted k units upward if k is positive and $\|k\|$ units downward if k is negative. **Horizontal translation** The graph of $y = f(x - h)$ is the graph of $y = f(x)$ shifted h units to the right if h is positive and $\|h\|$ units to the left if h is negative.	Graph $f(x) = \|x + 4\|$. **Solution** To fit the form, change the equation to the equivalent form $f(x) = \|x - (-4)\|$ Because h is negative, the graph of $f(x) = \|x\|$ is shifted 4 units to the left. 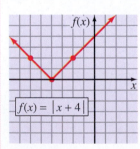 $f(x) = \|x + 4\|$

OBJECTIVE	SUMMARY	EXAMPLE
Graph functions by using reflections. **(Section 13.3/Objective 3)**	**x-axis reflection** The graph of $y = -f(x)$ is the graph of $y = f(x)$ reflected through the x axis. **y-axis reflection** The graph of $y = f(-x)$ is the graph of $y = f(x)$ reflected through the y axis.	Graph $f(x) = \sqrt{-x}$. **Solution** The graph of $f(x) = \sqrt{-x}$ is the graph of $f(x) = \sqrt{x}$ reflected through the y axis.
Graph functions by using vertical stretching or shrinking. **(Section 13.3/Objective 4)**	Vertical stretching and shrinking: The graph of $y = cf(x)$ is obtained from the graph of $y = f(x)$ by multiplying the y coordinates of $y = f(x)$ by c. If $|c| > 1$ then the graph is said to be stretched by a factor of $|c|$, and if $0 < |c| < 1$ then the graph is said to be shrunk by a factor of $|c|$.	Graph $f(x) = \dfrac{1}{4}x^2$. **Solution** The graph of $f(x) = \dfrac{1}{4}x^2$ is the graph of $f(x) = x^2$ shrunk by a factor of $\dfrac{1}{4}$.
Graph functions by using successive transformations. **(Section 13.3/Objective 5)**	Some curves are the result of performing more than one transformation on a basic curve. Unless parentheses indicate otherwise, stretchings, shrinkings, and x-axis reflections should be performed before translations.	Graph $f(x) = -2(x + 1)^2 + 3$. **Solution** $f(x) = -2(x + 1)^2 + 3$ Narrows the parabola and opens it downward Moves the parabola 1 unit to the left Moves the parabola 3 units up

(continued)

OBJECTIVE	SUMMARY	EXAMPLE
Find the composition of two functions and determine the domain. (Section 13.4/Objective 1)	The composition of two functions f and g is defined by $(f \circ g)(x) = f(g(x))$ for all x in the domain of g such that $g(x)$ is in the domain of f. Remember that the composition of functions is not a commutative operation.	If $f(x) = x + 5$ and $g(x) = x^2 + 4x - 6$, find $(g \circ f)(x)$. **Solution** In the function g, substitute $f(x)$ for x: $(g \circ f)(x)$ $\quad = (f(x))^2 + 4(f(x)) - 6$ $\quad = (x + 5)^2 + 4(x + 5) - 6$ $\quad = x^2 + 10x + 25 + 4x + 20 - 6$ $\quad = x^2 + 14x + 39$
Determine function values for composite functions. (Section 13.4/Objective 2)	Composite functions can be evaluated for values of x in the domain of the composite function.	If $f(x) = 3x - 1$ and $g(x) = -x^2 + 9$, find $(f \circ g)(-4)$. **Solution** First, form the composite function $(f \circ g)(x)$: $(f \circ g)(x) = 3(-x^2 + 9) - 1$ $\qquad\qquad = -3x^2 + 26$ To find $(f \circ g)(-4)$, substitute -4 for x in the composite function: $(f \circ g)(-4) = -3(-4)^2 + 26 = -22$
Translate sentences into equations of variation. (Section 13.5/Objective 1)	The equation $y = kx$ (k is a nonzero constant) defines a function called direct variation. The equation $y = \dfrac{k}{x}$ defines a function called inverse variation.	Translate the statement "The area of a circle varies directly as the square of the length of a radius." **Solution** Let A represent the area and let r represent the length of a radius. Then the equation of variation is $A = kr^2$.
Determine the constant of variation. (Section 13.5/Objective 2)	The constant of variation is a nonzero constant usually represented by k.	Roxanne's pay varies directly with the number of hours she works. If Roxanne is paid $292.50 for working 30 hours, find the constant of variation. **Solution** Because the pay varies directly with the number of hours worked, we have $p = kh$. Substituting $p = 292.50$ and $h = 30$, we obtain $292.50 = k(30)$. Solving this equation for k, yields $k = 9.75$. The constant of variation is 9.75.

OBJECTIVE	SUMMARY	EXAMPLE
Solve variation word problems. (Section 13.5/Objective 3)	To solve word problems involving variation, 1. Translate the problem to an equation of variation 2. Determine the value of k, the constant of variation 3. Substitute k and the given values of the variable(s) to solve the problem	The cost of electricity varies directly with the number of kilowatt hours used. If it costs $127.20 for 1200 kilowatt hours, what will 1500 kilowatt hours cost? **Solution** Let C represent the cost and w represent the number of kilowatt hours. The equation of variation is $C = kw$. Substitute 127.20 for C and 1200 for w and then solve for k. $127.20 = k(1200)$ $k = \dfrac{127.20}{1200} = 0.106$ Now find the cost when $w = 1500$: $C = 0.106(1500) = 159.$ Therefore, 1500 kilowatt hours cost $159.00.

Chapter 13   Review Problem Set

For Problems 1–4, determine if the following relations determine a function. Specify the domain and range.

1. $\{(9, 2), (8, 3), (7, 4), (6, 5)\}$

2. $\{(1, 1), (2, 3), (1, 5), (2, 7)\}$

3. $\{(0, -6), (0, -5), (0, -4), (0, -3)\}$

4. $\{(0, 8), (1, 8), (2, 8), (3, 8)\}$

5. If $f(x) = x^2 - 2x - 1$, find $f(2)$, $f(-3)$, and $f(a)$.

6. If $g(x) = \dfrac{2x - 4}{x + 2}$, find $f(2)$, $f(-1)$, and $f(3a)$.

For Problems 7–10, specify the domain of each function.

7. $f(x) = x^2 - 9$

8. $f(x) = \dfrac{4}{x - 5}$

9. $f(x) = \dfrac{3}{x^2 + 4x}$

10. $f(x) = \sqrt{x^2 - 25}$

11. If $f(x) = 6x + 8$, find $\dfrac{f(a + h) - f(a)}{h}$.

12. If $f(x) = 2x^2 + x - 7$, find $\dfrac{f(a + h) - f(a)}{h}$.

13. The cost for burning a 100-watt bulb is given by the function $c(h) = 0.006h$, where h represents the number of hours that the bulb burns. How much, to the nearest cent, does it cost to burn a 100-watt bulb for 4 hours per night for a 30-day month?

14. "All Items 30% Off Marked Price" is a sign in a local department store. Form a function and then use it to determine how much one must pay for each of the following marked items: a $65 pair of shoes, a $48 pair of slacks, a $15.50 belt.

For Problems 15–18, graph each of the functions.

15. $f(x) = \dfrac{3}{4}x + 2$

16. $f(x) = -3x + 2$

17. $f(x) = 4$

18. $f(x) = 2x$

19. A college math placement test has 50 questions. The score is computed by awarding 4 points for each correct answer and subtracting 2 points for each incorrect answer. Determine the linear function that would be used to compute the score. Use the function to determine the score when a student gets 35 questions correct.

20. An outpatient operating room charges each patient a fixed amount per surgery plus an amount per minute for use. One patient was charged $250 for a 30-minute surgery and another patient was charged $450 for a 90-minute surgery. Determine the linear function that would be used to compute the charge. Use the function to determine the charge when a patient has a 45-minute surgery.

For Problems 21–23, graph each of the functions.

21. $f(x) = x^2 + 2x + 2$ **22.** $f(x) = -\frac{1}{2}x^2$

23. $f(x) = -3x^2 + 6x - 2$

24. Find the coordinates of the vertex and the equation of the line of symmetry for each of the following parabolas.
   (a) $f(x) = x^2 + 10x - 3$

   (b) $f(x) = -2x^2 - 14x + 9$

25. Find two numbers whose sum is 40 and whose product is a maximum.

26. Find two numbers whose sum is 50 such that the square of one number plus six times the other number is a minimum.

27. A gardener has 60 yards of fencing and wants to enclose a rectangular garden that requires fencing only on three sides. Find the length and width of the plot that will maximize the area.

28. Suppose that 50 students are able to raise $250 for a gift when each one contributes $5. Furthermore, they figure that for each additional student they can find to contribute, the cost per student will decrease by a nickel. How many additional students will they need to maximize the amount of money they will have for a gift?

For Problems 29–40, graph the functions.

29. $f(x) = x^3 - 2$ **30.** $f(x) = |x| + 4$

31. $f(x) = (x + 1)^2$ **32.** $f(x) = \frac{1}{x - 4}$

33. $f(x) = \sqrt{-x}$ **34.** $f(x) = -|x|$

35. $f(x) = \frac{1}{3}|x|$ **36.** $f(x) = 2\sqrt{x}$

37. $f(x) = -\sqrt{x + 1} - 2$ **38.** $f(x) = \sqrt{x - 2} - 3$

39. $f(x) = -|x - 2|$ **40.** $f(x) = -(x + 3)^2 - 2$

For Problems 41–43, determine $(f \circ g)(x)$ and $(g \circ f)(x)$ for each pair of function.

41. If $f(x) = 2x - 3$ and $g(x) = 3x - 4$

42. $f(x) = x - 4$ and $g(x) = x^2 - 2x + 3$

43. $f(x) = x^2 - 5$ and $g(x) = -2x + 5$

44. If $f(x) = x^2 + 3x - 1$ and $g(x) = 4x - 7$, find $(f \circ g)(-2)$.

45. If $f(x) = \frac{1}{2}x - 6$ and $g(x) = 2x + 10$, find $(g \circ f)(-4)$.

46. Andrew's paycheck varies directly with the number of hours he works. If he is paid $475 when he works 38 hours, find his pay when he works 30 hours.

47. The surface area of a cube varies directly as the square of the length of an edge. If the surface area of a cube that has edges 8 inches long is 384 square inches, find the surface area of a cube that has edges 10 inches long.

48. The time it takes to fill an aquarium varies inversely with the square of the hose diameter. If it takes 40 minutes to fill the aquarium when the hose diameter is $\frac{1}{4}$ inch, find the time it takes to fill the aquarium when the hose diameter is $\frac{1}{2}$ inch.

49. The weight of a body above the surface of the earth varies inversely as the square of its distance from the center of the earth. Assume that the radius of the earth is 4000 miles. How much would a man weigh 1000 miles above the earth's surface if he weighs 200 pounds on the surface?

50. If y varies directly as x and inversely as z and if $y = 21$ when $x = 14$ and $z = 6$, find the constant of variation.

51. If y varies jointly as x and the square root of z and if $y = 60$ when $x = 2$ and $z = 9$, find y when $x = 3$ and $z = 16$.

1. For the function $f(x) = \dfrac{-3}{2x^2 + 7x - 4}$, determine the domain.

2. Determine the domain of the function $f(x) = \sqrt{5 - 3x}$.

3. If $f(x) = -\dfrac{1}{2}x + \dfrac{1}{3}$, find $f(-3)$.

4. If $f(x) = -x^2 - 6x + 3$, find $f(-2)$.

5. Find the vertex of the parabola
   $f(x) = -2x^2 - 24x - 69$.

6. If $f(x) = 3x^2 + 2x - 5$, find $\dfrac{f(a + h) - f(a)}{h}$.

7. If $f(x) = -3x + 4$ and $g(x) = 7x + 2$, find $(f \circ g)(x)$.

8. If $f(x) = 2x + 5$ and $g(x) = 2x^2 - x + 3$, find $(g \circ f)(x)$.

9. If $f(x) = \dfrac{3}{x - 2}$ and $g(x) = \dfrac{2}{x}$, find $(f \circ g)(x)$.

10. A movie rental club charges a fixed amount per month plus an amount for each movie rented. One month Carlos rented 6 movies and paid $14.00. The next month he rented 3 movies and paid $9.50. Find the linear function that the movie rental club uses to determine the amount due each month.

11. Find two numbers whose sum is 60, such that the sum of the square of one number plus twelve times the other number is a minimum.

12. If y varies inversely as x, and if $y = \dfrac{1}{2}$ when $x = -8$, find the constant of variation.

13. If y varies jointly as x and z, and if $y = 18$ when $x = 8$ and $z = 9$, find y when $x = 5$ and $z = 12$.

14. The simple interest earned by a certain amount of money varies jointly as the rate of interest and the time (in years) that the money is invested. If $140 is earned for a certain amount of money invested at 7% for 5 years, how much is earned if the same amount is invested at 8% for 3 years?

For Problems 15–17, use the concepts of translation and/or reflection to describe how the second curve can be obtained from the first curve.

15. $f(x) = x^3$, $f(x) = (x - 6)^3 - 4$

16. $f(x) = |x|$, $f(x) = -|x| + 8$

17. $f(x) = \sqrt{x}$, $f(x) = -\sqrt{x + 5} + 7$

For Problems 18–25, graph each function.

18. $f(x) = -x - 1$

19. $f(x) = -2x^2 - 12x - 14$

20. $f(x) = 2\sqrt{x} - 2$

21. $f(x) = 3|x - 2| - 1$

22. $f(x) = -\dfrac{1}{x} + 3$

23. $f(x) = \sqrt{-x + 2}$

24. $f(x) = -x^2 + 4$

25. $f(x) = (x - 2)^3 - 3$

1. Find the greatest common factor of 48, 60, and 84.

2. Identify the type of symmetry that each of the following equations exhibits.

   a. $2x^2 + y^2 - 3x - 4 = 0$

   b. $f(x) = 3x^2 + 4$

   c. $xy = 18$

   d. $f(x) = x$

3. Find the x intercepts of the graph of the function $f(x) = 2x^2 + 17x - 9$.

4. Find the vertex of the parabola $f(x) = 2x^2 + 4x$.

5. Evaluate $\dfrac{48x^{-3}y^{-2}}{12x^{-4}y^{-3}}$ for $x = -2$ and $y = -4$.

6. Express each of the following in simplest radical form. All variables represent positive real numbers.

   a. $\sqrt{32x^3y^4}$

   b. $\sqrt[3]{32x^3y^4}$

   c. $\dfrac{2\sqrt{6}}{3\sqrt{12}}$

   d. $\dfrac{\sqrt{2}}{3\sqrt{2} + \sqrt{3}}$

7. Evaluate $\dfrac{3x^2 + 5x + 2}{3x^2 - x - 2}$ for $x = 21$.

For Problems 8–11, perform the indicated operations and express the answers in simplest form.

8. $(3x + 4)(2x^2 - x - 5)$

9. $\dfrac{3}{4x} - \dfrac{2}{5x} + \dfrac{7}{10x}$

10. $(2x^4 - 13x^3 + 19x^2 - 25x + 25) \div (x - 5)$

11. $\dfrac{x^3 - 8}{x - 2}$

12. Express $(1.414)(10^{-3})$ in ordinary decimal notation.

13. 41.4 is what percent of 36?

14. Find the product $(6 + 5i)(-4 - 2i)$ and express the result in the standard form of a complex number.

15. Evaluate each of the following numerical expressions.

   a. $16^{\frac{3}{4}}$

   b. $\left(\dfrac{3}{2} - \dfrac{1}{4}\right)^{-1}$

   c. $-4^{-\frac{3}{2}}$

   d. $\sqrt[3]{\dfrac{64}{27}}$

16. Find the slope of the line determined by the equation $-2x - 3y = 7$.

17. Find the equation of the line that has a slope of $-\dfrac{3}{4}$ and a y intercept of 5.

For Problems 18–21, solve each equation.

18. $3(2x - 1) - (x + 2) = 2(-x + 3)$

19. $\dfrac{3}{2x - 1} = \dfrac{4}{3x - 2}$

20. $3x^2 - 2x + 1 = 0$

21. $\sqrt{x + 1} + \sqrt{x - 4} = 5$

For Problems 22–25, graph each equation.

22. $4x^2 - y^2 = 16$

23. $f(x) = -2x^2 + 8x - 5$

24. $xy^2 = -4$

25. $f(x) = -x - 3$

14

Exponential and Logarithmic Functions

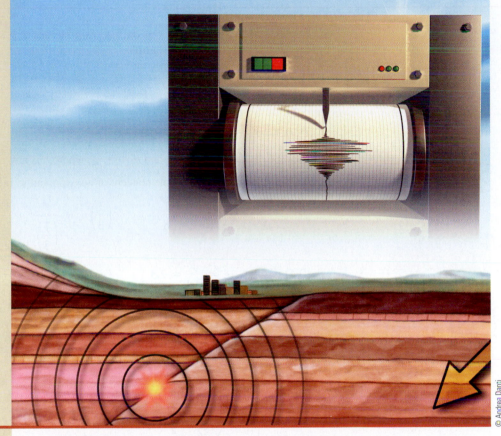

© Andrea Danti

The Richter number for reporting the intensity of an earthquake is calculated from a logarithm. Logarithmic scales are commonly used in science and mathematics to transform very large numbers to a smaller scale.

How long will it take $100 to triple itself if it is invested at 8% interest compounded continuously? We can use the formula $A = Pe^{rt}$ to generate the equation $300 = 100e^{0.08t}$, which can be solved for t by using logarithms. It will take approximately 13.7 years for the money to triple itself.

This chapter will expand the meaning of an exponent and introduce the concept of a logarithm. We will (1) work with some exponential functions, (2) work with some logarithmic functions, and (3) use the concepts of exponent and logarithm to expand our capabilities for solving problems. Your calculator will be a valuable tool throughout this chapter.

Video tutorials based on section learning objectives are available in a variety of delivery modes.

14.1   Exponents and Exponential Functions

OBJECTIVES
1. Solve exponential equations
2. Graph exponential functions

In Chapter 2, the expression b^n was defined as n factors of b, where n is any positive integer, and b is any real number. For example,

$$4^3 = 4 \cdot 4 \cdot 4 = 64$$

$$\left(\frac{1}{2}\right)^4 = \left(\frac{1}{2}\right)\left(\frac{1}{2}\right)\left(\frac{1}{2}\right)\left(\frac{1}{2}\right) = \frac{1}{16}$$

$$(-0.3)^2 = (-0.3)(-0.3) = 0.09$$

In Chapter 6, by defining $b^0 = 1$ and $b^{-n} = \frac{1}{b^n}$, where n is any positive integer and b is any nonzero real number, we extended the concept of an exponent to include all integers. For example,

$$2^{-3} = \frac{1}{2^3} = \frac{1}{2 \cdot 2 \cdot 2} = \frac{1}{8} \qquad \left(\frac{1}{3}\right)^{-2} = \frac{1}{\left(\frac{1}{3}\right)^2} = \frac{1}{\frac{1}{9}} = 9$$

$$(0.4)^{-1} = \frac{1}{(0.4)^1} = \frac{1}{0.4} = 2.5 \qquad (-0.98)^0 = 1$$

In Chapter 10, we also provided for the use of all rational numbers as exponents by defining $b^{\frac{m}{n}} = \sqrt[n]{b^m}$, where n is a positive integer greater than 1, and b is a real number such that $\sqrt[n]{b}$ exists. For example,

$$8^{\frac{2}{3}} = \sqrt[3]{8^2} = \sqrt[3]{64} = 4$$

$$16^{\frac{1}{4}} = \sqrt[4]{16^1} = 2$$

$$32^{-\frac{1}{5}} = \frac{1}{32^{\frac{1}{5}}} = \frac{1}{\sqrt[5]{32}} = \frac{1}{2}$$

To extend the concept of an exponent formally to include the use of irrational numbers requires some ideas from calculus and is therefore beyond the scope of this text. However, here's a glance at the general idea involved. Consider the number $2^{\sqrt{3}}$. By using the nonterminating and nonrepeating decimal representation 1.73205 . . . for $\sqrt{3}$, form the sequence of numbers $2^1, 2^{1.7}, 2^{1.73}, 2^{1.732}, 2^{1.7320}, 2^{1.73205}. \ldots$. It would seem reasonable that each successive power gets closer to $2^{\sqrt{3}}$. This is precisely what happens if b^n, where n is irrational, is properly defined by using the concept of a limit.

From now on, then, we can use any real number as an exponent, and the basic properties stated in Chapter 6 can be extended to include all real numbers as exponents. Let's restate those properties at this time with the restriction that the bases a and b are to be positive numbers (to avoid expressions such as $(-4)^{\frac{1}{2}}$, which do not represent real numbers).

Property 14.1

If a and b are positive real numbers, and m and n are any real numbers, then

1. $b^n \cdot b^m = b^{n+m}$ Product of two powers
2. $(b^n)^m = b^{mn}$ Power of a power
3. $(ab)^n = a^n b^n$ Power of a product
4. $\left(\frac{a}{b}\right)^n = \frac{a^n}{b^n}$ Power of a quotient
5. $\frac{b^n}{b^m} = b^{n-m}$ Quotient of two powers

Another property that can be used to solve certain types of equations involving exponents can be stated as follows:

> **Property 14.2**
>
> If $b > 0$, $b \neq 1$, and m and n are real numbers, then
>
> $$b^n = b^m \quad \text{if and only if } n = m$$

The following examples illustrate the use of Property 14.2.

Classroom Example
Solve $5^x = 125$.

EXAMPLE 1   Solve $2^x = 32$.

Solution

$$2^x = 32$$
$$2^x = 2^5 \qquad 32 = 2^5$$
$$x = 5 \qquad \text{Property 14.2}$$

The solution set is $\{5\}$.

Classroom Example
Solve $5^{2x} = \dfrac{1}{25}$.

EXAMPLE 2   Solve $3^{2x} = \dfrac{1}{9}$.

Solution

$$3^{2x} = \frac{1}{9} = \frac{1}{3^2}$$
$$3^{2x} = 3^{-2}$$
$$2x = -2 \qquad \text{Property 14.2}$$
$$x = -1$$

The solution set is $\{-1\}$.

Classroom Example
Solve $\left(\dfrac{1}{2}\right)^{x+1} = \dfrac{1}{32}$.

EXAMPLE 3   Solve $\left(\dfrac{1}{5}\right)^{x-2} = \dfrac{1}{125}$.

Solution

$$\left(\frac{1}{5}\right)^{x-2} = \frac{1}{125} = \left(\frac{1}{5}\right)^3$$
$$x - 2 = 3 \qquad \text{Property 14.2}$$
$$x = 5$$

The solution set is $\{5\}$.

Classroom Example
Solve $16^x = 64$.

EXAMPLE 4   Solve $8^x = 32$.

Solution

$$8^x = 32$$
$$\left(2^3\right)^x = 2^5 \qquad 8 = 2^3$$
$$2^{3x} = 2^5$$
$$3x = 5 \qquad \text{Property 14.2}$$
$$x = \frac{5}{3}$$

The solution set is $\left\{\dfrac{5}{3}\right\}$.

EXAMPLE 5   Solve $(3^{x+1})(9^{x-2}) = 27$.

Solution

$$(3^{x+1})(9^{x-2}) = 27$$

$$(3^{x+1})(3^2)^{x-2} = 3^3$$

$$(3^{x+1})(3^{2x-4}) = 3^3$$

$$3^{3x-3} = 3^3$$

$$3x - 3 = 3 \qquad \text{Property 14.2}$$

$$3x = 6$$

$$x = 2$$

The solution set is {2}.

Graphing Exponential Functions

If b is any positive number, then the expression b^x designates exactly one real number for every real value of x. Thus the equation $f(x) = b^x$ defines a function whose domain is the set of real numbers. Furthermore, if we impose the additional restriction $b \neq 1$, then any equation of the form $f(x) = b^x$ describes a one-to-one function and is called an **exponential function**. This leads to the following definition.

> **Definition 14.1**
>
> If $b > 0$ and $b \neq 1$, then the function f defined by
>
> $$f(x) = b^x$$
>
> where x is any real number, is called the **exponential function with base b**.

Now let's consider graphing some exponential functions.

EXAMPLE 6   Graph the function $f(x) = 2^x$.

Solution

First, let's set up a table of values:

x	$f(x) = 2^x$
-2	$\dfrac{1}{4}$
-1	$\dfrac{1}{2}$
0	1
1	2
2	4
3	8

Plot these points and connect them with a smooth curve to produce Figure 14.1.

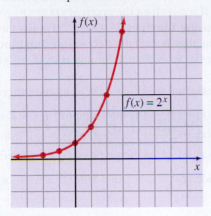

Figure 14.1

EXAMPLE 7   Graph $f(x) = \left(\dfrac{1}{2}\right)^x$.

Solution

Again, let's set up a table of values. Plot these points and connect them with a smooth curve to produce Figure 14.2.

x	$f(x) = \left(\dfrac{1}{2}\right)^x$
-2	4
-1	2
0	1
1	$\dfrac{1}{2}$
2	$\dfrac{1}{4}$
3	$\dfrac{1}{8}$

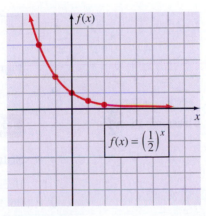

Figure 14.2

In the tables for Examples 6 and 7 we chose integral values for x to keep the computation simple. However, with the use of a calculator, we could easily acquire function values by using nonintegral exponents. Consider the following additional values for each of the tables.

$f(x) = 2^x$	
$f(0.5) \approx 1.41$	$f(-0.5) \approx 0.71$
$f(1.7) \approx 3.25$	$f(-2.6) \approx 0.16$

$f(x) = \left(\dfrac{1}{2}\right)^x$	
$f(0.7) \approx 0.62$	$f(-0.8) \approx 1.74$
$f(2.3) \approx 0.20$	$f(-2.1) \approx 4.29$

Use your calculator to check these results. Also, it would be worthwhile for you to go back and see that the points determined do fit the graphs in Figures 14.1 and 14.2.

The graphs in Figures 14.1 and 14.2 illustrate a general behavior pattern of exponential functions. That is, if $b > 1$, then the graph of $f(x) = b^x$ *goes up to the right* and the function is called an **increasing function**. If $0 < b < 1$, then the graph of $f(x) = b^x$ *goes down to the right* and the function is called a **decreasing function**. These facts are illustrated in Figure 14.3. Note that because $b^0 = 1$ for any $b > 0$, all graphs of $f(x) = b^x$ contain the point $(0, 1)$.

As you graph exponential functions, don't forget your previous graphing experiences.

1. The graph of $f(x) = 2^x - 4$ is the graph of $f(x) = 2^x$ moved down 4 units.

2. The graph of $f(x) = 2^{x+3}$ is the graph of $f(x) = 2^x$ moved 3 units to the left.

3. The graph of $f(x) = -2^x$ is the graph of $f(x) = 2^x$ reflected across the x axis.

We used a graphing calculator to graph these four functions on the same set of axes, as shown in Figure 14.4.

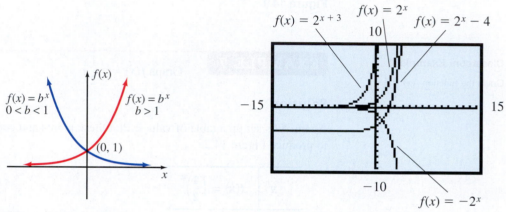

Figure 14.3   **Figure 14.4**

EXAMPLE 8   Graph $f(x) = 2^{-x^2}$.

Solution

Because $f(-x) = 2^{-(-x)^2} = 2^{-x^2} = f(x)$, we know that this curve is symmetric with respect to the y axis. Therefore, let's set up a table of values using nonnegative values for x. Plot these points, connect them with a smooth curve, and reflect this portion of the curve across the y axis to produce the graph in Figure 14.5.

x	2^{-x^2}
0	1
$\frac{1}{2}$	0.84
1	0.5
$\frac{3}{2}$	0.21
2	0.06

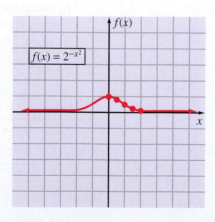

Figure 14.5

Finally, let's consider a problem in which a graphing utility gives us an approximate solution.

EXAMPLE 9

Use a graphing utility to obtain a graph of $f(x) = 50(2^x)$ and find an approximate value for x when $f(x) = 15,000$.

Solution

First, we must find an appropriate viewing rectangle. Because $50(2^{10}) = 51,200$, let's set the boundaries so that $0 \leq x \leq 10$ and $0 \leq y \leq 50,000$ with a scale of 10,000 on the y axis. (Certainly other boundaries could be used, but these will give us a graph that we can work with for this problem.) The graph of $f(x) = 50(2^x)$ is shown in Figure 14.6. Now we can use the TRACE and ZOOM features of the graphing utility to find that $x \approx 8.2$ at $y = 15,000$.

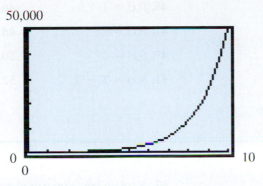

Figure 14.6

Remark: In Example 9, we used a graphical approach to solve the equation $50(2^x) = 15,000$. In Section 14.6, we will use an algebraic approach for solving that same kind of equation.

Concept Quiz 14.1

For Problems 1–10, answer true or false.

1. If $2^{x+1} = 2^{3x}$, then $x + 1 = 3x$.
2. The numerical expression 9^x is equivalent to 3^{2x}.
3. For the exponential function $f(x) = b^x$, the base b can be any positive number.
4. All the graphs of $f(x) = b^x$ for all positive values of b pass through the point $(0, 1)$.
5. The graphs of $f(x) = 3^x$ and $f(x) = \left(\dfrac{1}{3}\right)^x$ are reflections of each other across the y axis.
6. The solution set for $(8^x)(4^{x-2}) = 16$ is $\{1\}$.
7. The graph of $f(x) = 2^x - 4$ is the graph of $f(x) = 2^x$ shifted 4 units to the left.
8. The graph of $f(x) = 3^{-x}$ is the graph of $f(x) = 3^x$ reflected across the y axis.
9. The function $f(x) = 1^x$ is a constant function.
10. The function $f(x) = -3^x$ is a decreasing function.

Problem Set 14.1

For Problems 1–32, solve each of the equations. (Objective 1)

1. $2^x = 64$
2. $3^x = 81$
3. $3^{2x} = 27$
4. $2^{2x} = 16$
5. $\left(\dfrac{1}{2}\right)^x = \dfrac{1}{128}$
6. $\left(\dfrac{1}{4}\right)^x = \dfrac{1}{256}$
7. $3^{-x} = \dfrac{1}{243}$
8. $3^{x+1} = 9$

9. $6^{3x-1} = 36$

10. $2^{2x+3} = 32$

11. $5^{x+2} = 125$

12. $4^{x-3} = 16$

13. $\left(\dfrac{3}{4}\right)^n = \dfrac{64}{27}$

14. $\left(\dfrac{2}{3}\right)^n = \dfrac{9}{4}$

15. $16^x = 64$

16. $4^x = 8$

17. $\left(\dfrac{1}{2}\right)^{2x} = 64$

18. $\left(\dfrac{1}{3}\right)^{5x} = 243$

19. $6^{2x} + 3 = 39$

20. $5^{2x} - 2 = 123$

21. $27^{4x} = 9^{x+1}$

22. $32^x = 16^{1-x}$

23. $9^{4x-2} = \dfrac{1}{81}$

24. $8^{3x+2} = \dfrac{1}{16}$

25. $10^x = 0.1$

26. $10^x = 0.0001$

27. $(2^{x+1})(2^x) = 64$

28. $(2^{2x-1})(2^{x+2}) = 32$

29. $(27)(3^x) = 9^x$

30. $(3^x)(3^{5x}) = 81$

31. $(4^x)(16^{3x-1}) = 8$

32. $(8^{2x})(4^{2x-1}) = 16$

For Problems 33–52, graph each of the exponential functions.

33. $f(x) = 3^x$

34. $f(x) = 4^x$

35. $f(x) = \left(\dfrac{1}{3}\right)^x$

36. $f(x) = \left(\dfrac{1}{4}\right)^x$

37. $f(x) = \left(\dfrac{3}{2}\right)^x$

38. $f(x) = \left(\dfrac{2}{3}\right)^x$

39. $f(x) = 2^x - 3$

40. $f(x) = 2^x + 1$

41. $f(x) = 2^{x+2}$

42. $f(x) = 2^{x-1}$

43. $f(x) = -2^x$

44. $f(x) = -3^x$

45. $f(x) = 2^{-x-2}$

46. $f(x) = 2^{-x+1}$

47. $f(x) = 2^{x^2}$

48. $f(x) = 2^x + 2^{-x}$

49. $f(x) = 2^{|x|}$

50. $f(x) = 3^{1-x^2}$

51. $f(x) = 2^x - 2^{-x}$

52. $f(x) = 2^{-|x|}$

Thoughts Into Words

53. Explain how you would solve the equation $(2^{x+1})(8^{2x-3}) = 64$.

54. Why is the base of an exponential function restricted to positive numbers not including 1?

55. Explain how you would graph the function $f(x) = -\left(\dfrac{1}{3}\right)^x$.

Graphing Calculator Activities

56. Use a graphing calculator to check your graphs for Problems 33–52.

57. Graph $f(x) = 2^x$. Where should the graphs of $f(x) = 2^{x-5}, f(x) = 2^{x-7}$, and $f(x) = 2^{x+5}$ be located? Graph all three functions on the same set of axes with $f(x) = 2^x$.

58. Graph $f(x) = 3^x$. Where should the graphs of $f(x) = 3^x + 2$, $f(x) = 3^x - 3$, and $f(x) = 3^x - 7$ be located? Graph all three functions on the same set of axes with $f(x) = 3^x$.

59. Graph $f(x) = \left(\dfrac{1}{2}\right)^x$. Where should you locate the graphs of $f(x) = -\left(\dfrac{1}{2}\right)^x$, $f(x) = \left(\dfrac{1}{2}\right)^{-x}$, and $f(x) = -\left(\dfrac{1}{2}\right)^{-x}$? Graph all three functions on the same set of axes with $f(x) = \left(\dfrac{1}{2}\right)^x$.

60. Graph $f(x) = (1.5)^x$, $f(x) = (5.5)^x$, $f(x) = (0.3)^x$, and $f(x) = (0.7)^x$ on the same set of axes. Are these graphs consistent with Figure 14.3?

61. What is the solution for $3^x = 5$? Do you agree that it is between 1 and 2 because $3^1 = 3$ and $3^2 = 9$? Now graph $f(x) = 3^x - 5$ and use the ZOOM and TRACE features of your graphing calculator to find an approximation, to the nearest hundredth, for the x intercept. You should get an answer of 1.46. Do you see that this is an approximation for the solution of $3^x = 5$? Try it; raise 3 to the 1.46 power.

   Find an approximate solution, to the nearest hundredth, for each of the following equations by graphing the appropriate function and finding the x intercept.

   a. $2^x = 19$ b. $3^x = 50$ c. $4^x = 47$

   d. $5^x = 120$ e. $2^x = 1500$ f. $3^{x-1} = 34$

Answers to the Concept Quiz

1. True   **2.** True   **3.** False   **4.** True   **5.** True   **6.** False   **7.** False   **8.** True   **9.** True   **10.** True

14.2 Applications of Exponential Functions

OBJECTIVES

1 Solve exponential growth and compound interest problems

2 Solve exponential decay and half-life problems

3 Solve growth problems involving the number e

4 Graph exponential functions involving a base of e

Equations that describe exponential functions can represent many real-world situations that exhibit growth or decay. For example, suppose that an economist predicts an annual inflation rate of 5% for the next 10 years. This means an item that presently costs $8 will cost $8(105\%) = 8(1.05) = \$8.40$ a year from now. The same item will cost $(105\%)[8(105\%)] = 8(1.05)^2 = \$8.82$ in 2 years. In general, the equation

$$P = P_0(1.05)^t$$

yields the predicted price P of an item in t years at the annual inflation rate of 5%, where that item presently costs P_0. By using this equation, we can look at some future prices based on the prediction of a 5% inflation rate.

A $1.29 jar of mustard will cost $\$1.29(1.05)^3 = \$1.49$ in 3 years.

A $3.29 bag of potato chips will cost $\$3.29(1.05)^5 = \$4.20$ in 5 years.

A $7.69 can of coffee will cost $\$7.69(1.05)^7 = \$10.82$ in 7 years.

Compound Interest

Compound interest provides another illustration of exponential growth. Suppose that $500 (called the **principal**) is invested at an interest rate of 4% **compounded annually**. The interest earned the first year is $\$500(0.04) = \$20$, and this amount is added to the original $500 to form a new principal of $520 for the second year. The interest earned during the second year is $\$520(0.04) = \$20.80$, and this amount is added to $520, to form a new principal of $540.80 for the third year. Each year a new principal is formed by reinvesting the interest earned that year.

In general, suppose that a sum of money P (called the principal) is invested at an interest rate of r percent compounded annually. The interest earned the first year is Pr, and the new principal for the second year is $P + Pr$ or $P(1 + r)$. Note that the new principal for the second year can be found by multiplying the original principal P by $(1 + r)$. In like fashion, the new principal for the third year can be found by multiplying the previous principal $P(1 + r)$ by $(1 + r)$, thus obtaining $P(1 + r)^2$. If this process is continued, then after t years the total amount of money accumulated, A, is given by

$$A = P(1 + r)^t$$

Consider the following examples of investments made at a certain rate of interest compounded annually.

1. $750 invested for 5 years at 4% compounded annually produces

$$A = \$750(1.04)^5 = \$912.49$$

2. $1000 invested for 10 years at 7% compounded annually produces

$$A = \$1000(1.07)^{10} = \$1967.15$$

3. $5000 invested for 20 years at 6% compounded annually produces

$$A = \$5000(1.06)^{20} = \$16,035.68$$

If we invest money at a certain rate of interest to be compounded more than once a year, then we can adjust the basic formula, $A = P(1 + r)^t$, according to the number of compounding periods in the year. For example, for **compounding semiannually**, the formula becomes

$$A = P\left(1 + \frac{r}{2}\right)^{2t}$$ and for **compounding quarterly**, the formula becomes $A = P\left(1 + \frac{r}{4}\right)^{4t}$.

In general, we have the following formula, where n represents the number of compounding periods in a year.

$$A = P\left(1 + \frac{r}{n}\right)^{nt}$$

The following examples should clarify the use of this formula.

Classroom Example
Find the amount produced by the
given conditions:
(a) $800 invested for 6 years at
3.5% compounded semiannually
(b) $2000 invested for 15 years at
5.25% compounded quarterly
(c) $7000 invested for 25 years at
4% compounded monthly

EXAMPLE 1   Find the amount produced by the given conditions:

(a) $750 invested for 5 years at 4% compounded semiannually

(b) $1000 invested for 10 years at 7% compounded quarterly

(c) $5000 invested for 20 years at 6% compounded monthly

Solution

(a) $750 invested for 5 years at 4% compounded semiannually produces

$$A = \$750\left(1 + \frac{0.04}{2}\right)^{2(5)} = \$750(1.02)^{10} = \$914.25$$

(b) $1000 invested for 10 years at 7% compounded quarterly produces

$$A = \$1000\left(1 + \frac{0.07}{4}\right)^{4(10)} = \$1000(1.0175)^{40} = \$2001.60$$

(c) $5000 invested for 20 years at 6% compounded monthly produces

$$A = \$5000\left(1 + \frac{0.06}{12}\right)^{12(20)} = \$5000(1.005)^{240} = \$16,551.02$$

You may find it interesting to compare these results with those obtained earlier for compounding annually.

Exponential Decay

Suppose it is estimated that the value of a car depreciates 15% per year for the first 5 years. Thus a car that costs $19,500 will be worth $19,500(100\% - 15\%) = 19,500(85\%) = 19,500(0.85) = \$16,575$ in 1 year. In 2 years the value of the car will have depreciated to $19,500(0.85)^2 = \$14,089$ (nearest dollar). The equation

$$V = V_0(0.85)^t$$

yields the value V of a car in t years at the annual depreciation rate of 15%, where the car initially cost V_0. By using this equation, we can estimate some car values, to the nearest dollar, as follows:

A $17,000 car will be worth $\$17,000(0.85)^5 = \$7543$ in 5 years.

A $25,000 car will be worth $\$25,000(0.85)^4 = \$13,050$ in 4 years.

A $40,000 car will be worth $\$40,000(0.85)^3 = \$24,565$ in 3 years.

Another example of exponential decay is associated with radioactive substances. We can describe the rate of decay exponentially, on the basis of the half-life of a substance.

The *half-life* of a radioactive substance is the amount of time that it takes for one-half of an initial amount of the substance to disappear as the result of decay. For example, suppose that we have 200 grams of a certain substance that has a half-life of 5 days. After 5 days, $200\left(\dfrac{1}{2}\right) = 100$ grams remain. After 10 days, $200\left(\dfrac{1}{2}\right)^2 = 50$ grams remain. After 15 days, $200\left(\dfrac{1}{2}\right)^3 = 25$ grams remain. In general, after t days, $200\left(\dfrac{1}{2}\right)^{\frac{t}{5}}$ grams remain.

This discussion leads us to the following half-life formula. Suppose there is an initial amount, Q_0, of a radioactive substance with a half-life of h. The amount of substance remaining, Q, after a time period of t, is given by the formula

$$Q = Q_0\left(\frac{1}{2}\right)^{\frac{t}{h}}$$

The units of measure for t and h must be the same.

Classroom Example
A radioactive substance has a half-life of 4 years. If there are 600 milligrams of the substance initially, how many milligrams remain after 1 year? After 20 years?

EXAMPLE 2

Barium-140 has a half-life of 13 days. If there are 500 milligrams of barium initially, how many milligrams remain after 26 days? After 100 days?

Solution

When we use $Q_0 = 500$ and $h = 13$, the half-life formula becomes

$$Q = 500\left(\frac{1}{2}\right)^{\frac{t}{13}}$$

If $t = 26$, then

$$Q = 500\left(\frac{1}{2}\right)^{\frac{26}{13}}$$

$$= 500\left(\frac{1}{2}\right)^2$$

$$= 500\left(\frac{1}{4}\right) = 125$$

Thus 125 milligrams remain after 26 days.

If $t = 100$, then

$$Q = 500\left(\frac{1}{2}\right)^{\frac{100}{13}}$$

$$= 500(0.5)^{\frac{100}{13}}$$

$$= 2.4 \quad \text{to the nearest tenth of a milligram}$$

Thus approximately 2.4 milligrams remain after 100 days.

Remark: The solution to Example 2 clearly demonstrates one facet of the role of the calculator in the application of mathematics. We solved the first part of the problem easily without the calculator, but the calculator certainly was helpful for the second part of the problem.

Number *e*

An interesting situation occurs if we consider the compound interest formula for $P = \$1$, $r = 100\%$, and $t = 1$ year. The formula becomes $A = 1\left(1 + \dfrac{1}{n}\right)^n$. The accompanying table shows some values, rounded to eight decimal places, of $\left(1 + \dfrac{1}{n}\right)^n$ for different values of n.

n	$\left(1 + \dfrac{1}{n}\right)^n$
1	2.00000000
10	2.59374246
100	2.70481383
1000	2.71692393
10,000	2.71814593
100,000	2.71826824
1,000,000	2.71828047
10,000,000	2.71828169
100,000,000	2.71828181
1,000,000,000	2.71828183

The table suggests that as n increases, the value of $\left(1 + \dfrac{1}{n}\right)^n$ gets closer and closer to some fixed number. This does happen, and the fixed number is called e. To five decimal places, $e = 2.71828$.

The function defined by the equation $f(x) = e^x$ is the **natural exponential function**. It has a great many real-world applications; some we will look at in a moment. First, however, let's get a picture of the natural exponential function. Because $2 < e < 3$, the graph of $f(x) = e^x$ must fall between the graphs of $f(x) = 2^x$ and $f(x) = 3^x$. To be more specific, let's use our calculator to determine a table of values. Use the $\boxed{e^x}$ key, and round the results to the nearest tenth to obtain the following table. Plot the points determined by this table, and connect them with a smooth curve to produce Figure 14.7.

x	$f(x) = e^x$
0	1.0
1	2.7
2	7.4
−1	0.4
−2	0.1

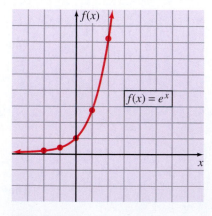

Figure 14.7

Let's return to the concept of compound interest. If the number of compounding periods in a year is increased indefinitely, we arrive at the concept of **compounding continuously**. Mathematically, this can be accomplished by applying the limit concept to the expression

$P\left(1 + \dfrac{r}{n}\right)^{nt}$. We will not show the details here, but the following result is obtained. The formula

$$A = Pe^{rt}$$

yields the accumulated value, A, of a sum of money, P, that has been invested for t years at a rate of r percent compounded continuously. The following examples show the use of the formula.

Classroom Example
Find the amount produced by the given conditions:
(a) $2000 invested for 4 years at 4% compounded continuously
(b) $5000 invested for 8 years at 3% compounded continuously
(c) $8000 invested for 12 years at 2% compounded continuously

EXAMPLE 3   Find the amount produced by the given conditions:

(a) $750 invested for 5 years at 4% compounded continuously

(b) $1000 invested for 10 years at 7% compounded continuously

(c) $5000 invested for 20 years at 6% compounded continuously

Solution

(a) $750 invested for 5 years at 4% compounded continuously produces

$$A = \$750e^{(0.04)(5)} = 750e^{0.20} = \$916.05$$

(b) $1000 invested for 10 years at 7% compounded continuously produces

$$A = \$1000e^{(0.07)(10)} = 1000e^{0.7} = \$2013.75$$

(c) $5000 invested for 20 years at 6% compounded continuously produces

$$A = \$5000e^{(0.06)(20)} = 5000e^{1.2} = \$16,600.58$$

Again you may find it interesting to compare these results with those you obtained earlier when using a different number of compounding periods.

The ideas behind compounding continuously carry over to other growth situations. The law of exponential growth,

$$Q(t) = Q_0e^{kt}$$

is used as a mathematical model for numerous growth-and-decay applications. In this equation, $Q(t)$ represents the quantity of a given substance at any time t; Q_0 is the initial amount of the substance (when $t = 0$); and k is a constant that depends on the particular application. If $k < 0$, then $Q(t)$ decreases as t increases, and we refer to the model as the **law of decay**. Let's consider some growth-and-decay applications.

Classroom Example
Suppose that in a certain culture, the equation $Q(t) = 2100e^{0.25t}$ expresses the number of bacteria present as a function of the time t, where t is expressed in days. Find (a) the initial number of bacteria and (b) the number of bacteria after 16 days.

EXAMPLE 4

Suppose that in a certain culture, the equation $Q(t) = 15,000e^{0.3t}$ expresses the number of bacteria present as a function of the time t, where t is expressed in hours. Find (a) the initial number of bacteria, and (b) the number of bacteria after 3 hours.

Solution

(a) The initial number of bacteria is produced when $t = 0$.

$$Q(0) = 15,000e^{0.3(0)}$$
$$= 15,000e^{0}$$
$$= 15,000 \qquad e^{0} = 1$$

(b) $Q(3) = 15{,}000e^{0.3(3)}$

$$= 15{,}000e^{0.9}$$

$$= 36{,}894 \quad \text{to the nearest whole number}$$

Therefore, there should be approximately 36,894 bacteria present after 3 hours.

Classroom Example
Suppose the number of bacteria present in a certain culture after t hours is given by the equation $Q(t) = Q_0e^{0.04t}$, where Q_0 represents the initial number of bacteria. If 9000 bacteria are present after 25 hours, how many bacteria were present initially?

EXAMPLE 5

Suppose the number of bacteria present in a certain culture after t minutes is given by the equation $Q(t) = Q_0e^{0.05t}$, where Q_0 represents the initial number of bacteria. If 5000 bacteria are present after 20 minutes, how many bacteria were present initially?

Solution

If 5000 bacteria are present after 20 minutes, then $Q(20) = 5000$.

$$5000 = Q_0e^{0.05(20)}$$

$$5000 = Q_0e^1$$

$$\frac{5000}{e} = Q_0$$

$$1839 = Q_0 \quad \text{to the nearest whole number}$$

Therefore, there were approximately 1839 bacteria present initially.

Classroom Example
The number of grams Q of a certain radioactive substance present after t minutes is given by $Q(t) = 500e^{-0.2t}$. How many grams remain after 20 minutes?

EXAMPLE 6

The number of grams Q of a certain radioactive substance present after t seconds is given by $Q(t) = 200e^{-0.3t}$. How many grams remain after 7 seconds?

Solution

We need to evaluate $Q(7)$.

$$Q(7) = 200e^{-0.3(7)}$$

$$= 200e^{-2.1}$$

$$= 24 \quad \text{to the nearest whole number}$$

Thus approximately 24 grams remain after 7 seconds.

Finally, let's consider two examples where we use a graphing utility to produce the graph.

Classroom Example
Suppose that $4000 was invested at 5% interest compounded continuously. How long would it take for the money to double?

EXAMPLE 7

Suppose that $1000 was invested at 6.5% interest compounded continuously. How long would it take for the money to double?

Solution

Substitute $1000 for P and 0.065 for r in the formula $A = Pe^{rt}$ to produce $A = 1000e^{0.065t}$. If we let $y = A$ and $x = t$, we can graph the equation $y = 1000e^{0.065x}$. By letting $x = 20$, we obtain $y = 1000e^{0.065(20)} = 1000e^{1.3} \approx 3670$. Therefore, let's set the boundaries of the viewing rectangle so that $0 \le x \le 20$ and $0 \le y \le 3700$ with a y scale of 1000. Then we obtain the graph in Figure 14.8. Now we want to find the value of x so that $y = 2000$. (The money is to double.) Using the ZOOM and TRACE features of the graphing utility, we can determine that an x value of approximately 10.7 will produce a y value of 2000. Thus it will take approximately 10.7 years for the $1000 investment to double.

3700

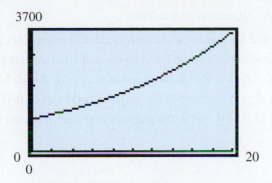

0
0   20
0

Figure 14.8

EXAMPLE 8   Graph the function $y = \dfrac{1}{\sqrt{2\pi}}e^{-x^2/2}$ and find its maximum value.

Solution

If $x = 0$, then $y = \dfrac{1}{\sqrt{2\pi}}e^0 = \dfrac{1}{\sqrt{2\pi}} \approx 0.4$. Let's set the boundaries of the viewing rectangle so that $-5 \le x \le 5$ and $0 \le y \le 1$ with a y scale of 0.1; the graph of the function is shown in Figure 14.9. From the graph, we see that the maximum value of the function occurs at $x = 0$, which we have already determined to be approximately 0.4.

1

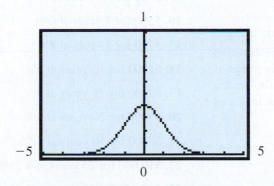

−5   5
0

Figure 14.9

Remark: The curve in Figure 14.9 is called the standard normal distribution curve. You may want to ask your instructor to explain what it means to assign grades on the basis of the normal distribution curve.

Concept Quiz 14.2

For Problems 1–5, match each type of problem with its formula.

1. Compound continuously

   A. $A = P\left(1 + \dfrac{r}{n}\right)^{nt}$

2. Exponential growth or decay

   B. $Q = Q_0\left(\dfrac{1}{2}\right)^{\frac{t}{h}}$

3. Interest compounded annually

   C. $A = P(1 + r)^t$

4. Interest compounded n times a year

   D. $Q(t) = Q_0 e^{kt}$

5. Half-life

   E. $A = Pe^{rt}$

For Problems 6 – 10, answer true or false.

6. $500 invested for 2 years at 7% compounded semiannually produces $573.76.

7. $500 invested for 2 years at 7% compounded continuously produces $571.14.

8. The graph of $f(x) = e^{x-5}$ is the graph of $f(x) = e^x$ shifted 5 units to the right.

9. The graph of $f(x) = e^x - 5$ is the graph of $f(x) = e^x$ shifted 5 units downward.

10. The graph of $f(x) = -e^x$ is the graph of $f(x) = e^x$ reflected across the x axis.

Problem Set 14.2

1. Assuming that the rate of inflation is 4% per year, the equation $P = P_0(1.04)^t$ yields the predicted price P, in t years, of an item that presently costs P_0. Find the predicted price of each of the following items for the indicated years ahead. **(Objective 1)**
  (a) $1.38 can of soup in 3 years
  (b) $3.43 container of cocoa mix in 5 years
  (c) $1.99 jar of coffee creamer in 4 years
  (d) $1.54 can of beans and bacon in 10 years
  (e) $18,000 car in 5 years (nearest dollar)
  (f) $180,000 house in 8 years (nearest dollar)
  (g) $500 TV set in 7 years (nearest dollar)

2. Suppose it is estimated that the value of a car depreciates 30% per year for the first 5 years. The equation $A = P_0(0.7)^t$ yields the value (A) of a car after t years if the original price is P_0. Find the value (to the nearest dollar) of each of the following cars after the indicated time. **(Objective 2)**
  (a) $16,500 car after 4 years
  (b) $22,000 car after 2 years
  (c) $27,000 car after 5 years
  (d) $40,000 car after 3 years

For Problems 3 – 14, use the formula $A = P\left(1 + \dfrac{r}{n}\right)^{nt}$ to find the total amount of money accumulated at the end of the indicated time period for each of the following investments. **(Objective 1)**

3. $200 for 6 years at 6% compounded annually

4. $250 for 5 years at 7% compounded annually

5. $500 for 7 years at 4% compounded semiannually

6. $750 for 8 years at 4% compounded semiannually

7. $800 for 9 years at 5% compounded quarterly

8. $1200 for 10 years at 4% compounded quarterly

9. $1500 for 5 years at 8% compounded monthly

10. $2000 for 10 years at 3% compounded monthly

11. $5000 for 15 years at 4.5% compounded annually

12. $7500 for 20 years at 6.5% compounded semiannually

13. $8000 for 10 years at 5.5% compounded quarterly

14. $10,000 for 25 years at 4.25% compounded monthly

For Problems 15 – 23, use the formula $A = Pe^{rt}$ to find the total amount of money accumulated at the end of the indicated time period by compounding continuously. **(Objective 3)**

15. $400 for 5 years at 7%

16. $500 for 7 years at 6%

17. $750 for 8 years at 8%

18. $1000 for 10 years at 5%

19. $2000 for 15 years at 7%

20. $5000 for 20 years at 8%

21. $7500 for 10 years at 6.5%

22. $10,000 for 25 years at 4.25%

23. $15,000 for 10 years at 5.75%

24. Complete the following chart, which illustrates what happens to $1000 invested at various rates of interest for different lengths of time but always compounded continuously. Round your answers to the nearest dollar.

$1000 compounded continuously			
8%	10%	12%	14%
5 years			
10 years			
15 years			
20 years			
25 years			

25. Complete the following chart, which illustrates what happens to $1000 invested at 6% for different lengths of time and different numbers of compounding periods. Round all of your answers to the nearest dollar.

$1000 at 6%			
Compounded annually			
1 year	5 years	10 years	20 years
Compounded semiannually			
1 year	5 years	10 years	20 years
Compounded quarterly			
1 year	5 years	10 years	20 years
Compounded monthly			
1 year	5 years	10 years	20 years
Compounded continuously			
1 year	5 years	10 years	20 years

26. Complete the following chart, which illustrates what happens to $1000 in 10 years for different rates of interest and different numbers of compounding periods. Round your answers to the nearest dollar.

$1000 for 10 years			
Compounded annually			
4%	5%	6%	7%
Compounded semiannually			
4%	5%	6%	7%
Compounded quarterly			
4%	5%	6%	7%
Compounded monthly			
4%	5%	6%	7%
Compounded continuously			
4%	5%	6%	7%

27. Suppose that Nora invested $500 at 8.25% compounded annually for 5 years, and Patti invested $500 at 8% compounded quarterly for 5 years. At the end of 5 years, who will have the most money and by how much?

28. Two years ago Daniel invested some money at 8% interest compounded annually. Today it is worth $758.16. How much did he invest two years ago?

29. What rate of interest (to the nearest hundredth of a percent) is needed so that an investment of $2500 will yield $3000 in 2 years if the money is compounded annually?

30. Suppose that a certain radioactive substance has a half-life of 20 years. If there are presently 2500 milligrams of the substance, how much, to the nearest milligram, will remain after 40 years? After 50 years?

31. Strontium-90 has a half-life of 29 years. If there are 400 grams of strontium initially, how much, to the nearest gram, will remain after 87 years? After 100 years?

32. The half-life of radium is approximately 1600 years. If the present amount of radium in a certain location is 500 grams, how much will remain after 800 years? Express your answer to the nearest gram.

For Problems 33–38, graph each of the exponential functions. **(Objective 4)**

33. $f(x) = e^x + 1$

34. $f(x) = e^x - 2$

35. $f(x) = 2e^x$

36. $f(x) = -e^x$

37. $f(x) = e^{2x}$

38. $f(x) = e^{-x}$

For Problems 39–44, express your answers to the nearest whole number. **(Objective 3)**

39. Suppose that in a certain culture, the equation $Q(t) = 1000e^{0.4t}$ expresses the number of bacteria present as a function of the time t, where t is expressed in hours. How many bacteria are present at the end of 2 hours? 3 hours? 5 hours?

40. The number of bacteria present at a given time under certain conditions is given by the equation $Q = 5000e^{0.05t}$, where t is expressed in minutes. How many bacteria are present at the end of 10 minutes? 30 minutes? 1 hour?

41. The number of bacteria present in a certain culture after t hours is given by the equation $Q = Q_0e^{0.3t}$, where Q_0 represents the initial number of bacteria. If 6640 bacteria are present after 4 hours, how many bacteria were present initially?

42. The number of grams Q of a certain radioactive substance present after t seconds is given by the equation $Q = 1500e^{-0.4t}$. How many grams remain after 5 seconds? 10 seconds? 20 seconds?

43. Suppose that the present population of a city is 75,000. Using the equation $P(t) = 75,000e^{0.01t}$ to estimate future growth, estimate the population
   (a) 10 years from now
   (b) 15 years from now
   (c) 25 years from now

44. Suppose that the present population of a city is 150,000. Use the equation $P(t) = 150,000e^{0.032t}$ to estimate future growth. Estimate the population
   (a) 10 years from now
   (b) 20 years from now
   (c) 30 years from now

45. The atmospheric pressure, measured in pounds per square inch, is a function of the altitude above sea level. The equation $P(a) = 14.7e^{-0.21a}$, where a is the altitude measured in miles, can be used to approximate atmospheric

pressure. Find the atmospheric pressure at each of the following locations. Express each answer to the nearest tenth of a pound per square inch.
(a) Mount McKinley in Alaska—altitude of 3.85 miles

(b) Denver, Colorado—the "mile-high" city
(c) Asheville, North Carolina—altitude of 1985 feet
   (5280 feet = 1 mile)
(d) Phoenix, Arizona—altitude of 1090 feet

Thoughts Into Words

46. Explain the difference between simple interest and compound interest.

47. Would it be better to invest $5000 at 6.25% interest compounded annually for 5 years or to invest $5000

at 6% interest compounded continuously for 5 years? Defend your answer.

Graphing Calculator Activities

48. Use a graphing calculator to check your graphs for Problems 33–38.

49. Graph $f(x) = 2^x$, $f(x) = e^x$, and $f(x) = 3^x$ on the same set of axes. Are these graphs consistent with the discussion prior to Figure 14.5?

50. Graph $f(x) = e^x$. Where should the graphs of $f(x) = e^{x-4}$, $f(x) = e^{x-6}$, and $f(x) = e^{x+5}$ be located? Graph all three functions on the same set of axes with $f(x) = e^x$.

51. Graph $f(x) = e^x$. Now predict the graphs for $f(x) = -e^x$, $f(x) = e^{-x}$, and $f(x) = -e^{-x}$. Graph all three functions on the same set of axes with $f(x) = e^x$.

52. How do you think the graphs of $f(x) = e^x$, $f(x) = e^{2x}$, and $f(x) = 2e^x$ will compare? Graph them on the same set of axes to see whether you were correct.

Answers to the Concept Quiz

1. E   **2.** D   **3.** C   **4.** A   **5.** B   **6.** True   **7.** False   **8.** True   **9.** True   **10.** True

---

14.3   Inverse Functions

OBJECTIVES

1 Determine if a function is one-to-one

2 Verify that two functions are inverse functions

3 Find inverse functions

4 Find intervals where a function is increasing or decreasing

The definition of a function is the basis for the **vertical line** test for functions. If each vertical line intersects a graph in no more than one point, then the graph represents a function. There is also a useful distinction between two basic types of functions. Consider the graphs of the two functions in Figure 14.10(a), $f(x) = 2x - 1$, and Figure 14.10(b), $g(x) = x^2$. In Figure 14.10(a), any *horizontal line* will intersect the graph in no more than one point. Therefore, every value of $f(x)$ has only one value of x associated with it. Any function that has this property of having exactly one value of x associated with each value of $f(x)$ is called a **one-to-one function.** Thus $g(x) = x^2$ is not a one-to-one function, because the horizontal line in Figure 14.10(b) intersects the parabola in two points.

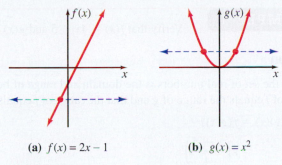

(a) $f(x) = 2x - 1$ **(b)** $g(x) = x^2$

Figure 14.10

The statement that for a function f to be a one-to-one function, *every value of $f(x)$ has only one value of x associated with it*, can be equivalently stated as "if $f(x_1) = f(x_2)$ for x_1 and x_2 in the domain of f, then $x_1 = x_2$." Let's use this last "if-then" statement to verify that $f(x) = 2x - 1$ is a one-to-one function. We start with the assumption that $f(x_1) - f(x_2)$.

$$2x_1 - 1 = 2x_2 - 1$$
$$2x_1 = 2x_2$$
$$x_1 = x_2$$

Thus $f(x) = 2x - 1$ is a one-to-one function.

To show that $g(x) = x^2$ is not a one-to-one function, we simply need to find two distinct real numbers in the domain of f that produce the same functional value. For example, $g(-2) = (-2)^2 = 4$, and $g(2) = 2^2 = 4$. Thus $g(x) = x^2$ is not a one-to-one function.

Now let's consider a one-to-one function f that assigns to each x in its domain D the value $f(x)$ in its range R (Figure 14.11a). We can define a new function g that goes from R to D; it assigns $f(x)$ in R back to x in D, as indicated in Figure 14.11(b).

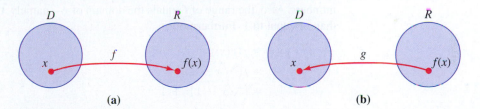

(a)   **(b)**

Figure 14.11

The functions f and g are called **inverse functions** of one another. The following definition precisely states this concept.

Definition 14.2

Let f be a one-to-one function with a domain of X and a range of Y. A function g with a domain of Y and a range of X is called the **inverse function** of f if

$$(f \circ g)(x) = x \quad \text{for every } x \text{ in } Y$$

and

$$(g \circ f)(x) = x \quad \text{for every } x \text{ in } X$$

In Definition 14.2, note that for f and g to be inverses of each other, the domain of f must equal the range of g, and the range of f must equal the domain of g. Furthermore, g must reverse the correspondences given by f, and f must reverse the correspondences given by g. In other words, inverse functions *undo* each other. Let's use Definition 14.2 to verify that two specific functions are inverses of each other.

Classroom Example

Verify that $f(x) = \dfrac{3}{2}x + 7$ and

$g(x) = \dfrac{2x - 14}{3}$ are inverse

functions.

EXAMPLE 1

Verify that $f(x) = 4x - 5$ and $g(x) = \dfrac{x + 5}{4}$ are inverse functions.

Solution

Because the set of real numbers is the domain and range of both functions, we know that the domain of f equals the range of g and that the range of f equals the domain of g. Furthermore,

$$(f \circ g)(x) = f(g(x))$$

$$= f\left(\frac{x + 5}{4}\right)$$

$$= 4\left(\frac{x + 5}{4}\right) - 5 = x$$

and

$$(g \circ f)(x) = g(f(x))$$

$$= g(4x - 5)$$

$$= \frac{4x - 5 + 5}{4} = x$$

Therefore, f and g are inverses of each other.

Classroom Example
Verify that $f(x) = (x + 3)^2$ for

$x \geq -3$ and $g(x) = \sqrt{x} - 3$ for

$x \geq 0$ are inverse functions.

EXAMPLE 2

Verify that $f(x) = x^2 + 1$ for $x \geq 0$ and $g(x) = \sqrt{x - 1}$ for $x \geq 1$ are inverse functions.

Solution

First, note that the domain of f equals the range of g—namely, the set of nonnegative real numbers. Also, the range of f equals the domain of g—namely, the set of real numbers greater than or equal to 1. Furthermore,

$$(f \circ g)(x) = f(g(x))$$

$$= f(\sqrt{x - 1})$$

$$= (\sqrt{x - 1})^2 + 1$$

$$= x - 1 + 1 = x$$

and

$$(g \circ f)(x) = g(f(x))$$

$$= g(x^2 + 1)$$

$$= \sqrt{x^2 + 1 - 1} = \sqrt{x^2} = x \qquad \sqrt{x^2} = x \text{ because } x \geq 1$$

Therefore, f and g are inverses of each other.

The inverse of a function f is commonly denoted by f^{-1}, read "f inverse or the inverse of f." Do not confuse the -1 in f^{-1} with a negative exponent. The symbol f^{-1} *does not* mean $1/f^1$ but rather refers to the inverse function of function f.

Remember that a function can also be thought of as a set of ordered pairs no two of which have the same first element. Along those lines, a one-to-one function further requires that no two of the ordered pairs have the same second element. Then, if the components of each ordered pair of a given one-to-one function are interchanged, the resulting function and the given function are inverses of each other. Thus if

$$f = \{(1, 4), (2, 7), (5, 9)\}$$

then

$$f^{-1} = \{(4, 1), (7, 2), (9, 5)\}$$

Graphically, two functions that are inverses of each other are **mirror images with reference to the line $y = x$.** This is due to the fact that ordered pairs (a, b) and (b, a) are reflections of each other with respect to the line $y = x$, as illustrated in Figure 14.12. (You will verify this in the next set of exercises.) Therefore, if the graph of a function f is known, as in Figure 14.13(a), then the graph of f^{-1} can be determined by reflecting f across the line $y = x$ (Figure 14.13b).

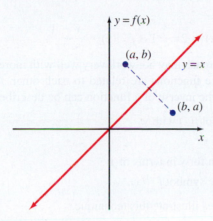

Figure 14.12

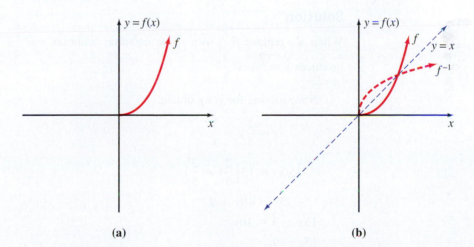

(a)   (b)

Figure 14.13

Finding Inverse Functions

The idea of inverse functions *undoing each other* provides the basis for an informal approach to finding the inverse of a function. Consider the function

$$f(x) = 2x + 1$$

To each x, this function assigns twice x plus 1. To undo this function, we can subtract 1 and divide by 2. Hence the inverse is

$$f^{-1}(x) = \frac{x - 1}{2}$$

Now let's verify that f and f^{-1} are indeed inverses of each other.

$$(f \circ f^{-1})(x) = f(f^{-1}(x))$$

$$= f\left(\frac{x-1}{2}\right)$$

$$= 2\left(\frac{x-1}{2}\right) + 1$$

$$= x - 1 + 1 = x$$

$$(f^{-1} \circ f)(x) = f^{-1}(f(x))$$

$$= f^{-1}(2x + 1)$$

$$= \frac{2x + 1 - 1}{2}$$

$$= \frac{2x}{2} = x$$

Thus the inverse of $f(x) = 2x + 1$ is $f^{-1}(x) = \dfrac{x-1}{2}$.

This informal approach may not work very well with more complex functions, but it does emphasize how inverse functions are related to each other. A more formal and systematic technique for finding the inverse of a function can be described as follows:

1. Replace the symbol $f(x)$ with y.
2. Interchange x and y.
3. Solve the equation for y in terms of x.
4. Replace y with the symbol $f^{-1}(x)$.

The following examples illustrate this technique.

Classroom Example
Find the inverse of $f(x) = \dfrac{3}{4}x + \dfrac{3}{5}$.

EXAMPLE 3   Find the inverse of $f(x) = \dfrac{2}{3}x + \dfrac{3}{5}$.

Solution

When we replace $f(x)$ with y, the equation becomes $y = \dfrac{2}{3}x + \dfrac{3}{5}$. Interchanging x and y produces $x = \dfrac{2}{3}y + \dfrac{3}{5}$.

Now, solving for y, we obtain

$$x = \frac{2}{3}y + \frac{3}{5}$$

$$15(x) = 15\left(\frac{2}{3}y + \frac{3}{5}\right)$$

$$15x = 10y + 9$$

$$15x - 9 = 10y$$

$$\frac{15x - 9}{10} = y$$

Finally, by replacing y with $f^{-1}(x)$, we can express the inverse function as

$$f^{-1}(x) = \frac{15x - 9}{10}$$

The domain of f is equal to the range of f^{-1} (both are the set of real numbers), and the range of f equals the domain of f^{-1} (both are the set of real numbers). Furthermore, we could show that $(f \circ f^{-1})(x) = x$ and $(f^{-1} \circ f)(x) = x$. We leave this for you to complete.

Does the function $f(x) = x^2 - 2$ have an inverse? Sometimes a graph of the function helps answer such a question. In Figure 14.14(a), it should be evident that f is not a one-to-one function and therefore cannot have an inverse. However, it should also be apparent from the graph that if we restrict the domain of f to the nonnegative real numbers, then f is a one-to-one function and should have an inverse (Figure 14.14b). The next example illustrates how to find the inverse function.

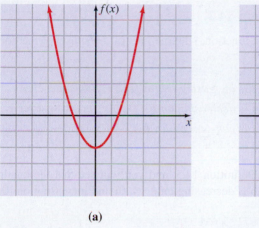

(a)

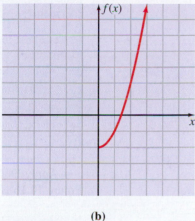

(b)

Figure 14.14

Classroom Example
Find the inverse of $f(x) = x^2 + 1$, where $x \geq 0$.

EXAMPLE 4   Find the inverse of $f(x) = x^2 - 2$, where $x \geq 0$.

Solution

When we replace $f(x)$ with y, the equation becomes

$$y = x^2 - 2, \qquad x \geq 0$$

Interchanging x and y produces

$$x = y^2 - 2, \qquad y \geq 0$$

Now let's solve for y; keep in mind that y is to be nonnegative.

$$x = y^2 - 2$$
$$x + 2 = y^2$$
$$\sqrt{x + 2} = y, \qquad x \geq -2$$

Finally, by replacing y with $f^{-1}(x)$, we can express the inverse function as

$$f^{-1}(x) = \sqrt{x + 2}, \qquad x \geq -2$$

The domain of f equals the range of f^{-1} (both are the nonnegative real numbers), and the range of f equals the domain of f^{-1} (both are the real numbers greater than or equal to -2). It can also be shown that $(f \circ f^{-1})(x) = x$ and $(f^{-1} \circ f)(x) = x$. Again, we leave this for you to complete.

Increasing and Decreasing Functions

In Section 14.1, we used exponential functions as examples of increasing and decreasing functions. In reality, one function can be both increasing and decreasing over certain intervals. For example, in Figure 14.15, the function f is said to be *increasing* on the intervals $(-\infty, x_1]$ and

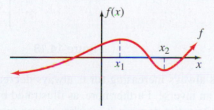

Figure 14.15

$[x_2, \infty)$ and is said to be *decreasing* on the interval $[x_1, x_2]$. More specifically, increasing and decreasing functions are defined as follows:

Definition 14.3

Let f be a function, with the interval I a subset of the domain of f. Let x_1 and x_2 be in I. Then

    1. f is *increasing* on I if $f(x_1) < f(x_2)$ whenever $x_1 < x_2$.

    2. f is *decreasing* on I if $f(x_1) > f(x_2)$ whenever $x_1 < x_2$.

    3. f is *constant* on I if $f(x_1) = f(x_2)$ for every x_1 and x_2.

Apply Definition 14.3, and you will see that the quadratic function $f(x) = x^2$ shown in Figure 14.16 is decreasing on $(-\infty, 0]$ and increasing on $[0, \infty)$. Likewise,

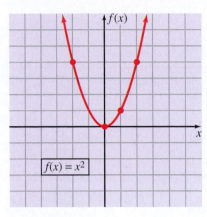

Figure 14.16

the linear function $f(x) = 2x$ in Figure 14.17 is increasing throughout its domain of real numbers, so we say that it is increasing on $(-\infty, \infty)$. The function $f(x) = -2x$ in Figure 14.18 is decreasing on $(-\infty, \infty)$. For our purposes in this text, we will rely on our knowledge of the graphs of the functions to determine where functions are increasing and decreasing. More formal techniques for determining where functions increase and decrease will be developed in the calculus.

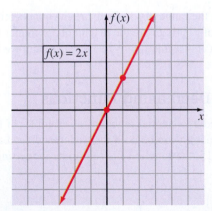

Figure 14.17

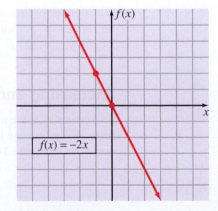

Figure 14.18

A function that is always increasing (or is always decreasing) over its entire domain is one to one and so has an inverse. Furthermore, as illustrated by Example 4, even if a function is not one to one over its entire domain, it may be so over some subset of the domain. It then has an inverse over this restricted domain.

As functions become more complex, a graphing utility can be used to help with the problems we have discussed in this section. For example, suppose that we want to know whether the function $f(x) = \dfrac{3x + 1}{x - 4}$ is a one-to-one function and therefore has an inverse. Using a graphing utility, we can quickly get a sketch of the graph (see Figure 14.19). Then, by applying the horizontal-line test to the graph, we can be fairly certain that the function is one-to-one.

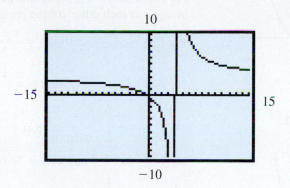

Figure 14.19

A graphing utility can also be used to help determine intervals on which a function is increasing or decreasing. For example, to determine such intervals for the function $f(x) = \sqrt{x^2 + 4}$, let's use a graphing utility to get a sketch of the curve (Figure 14.20). From this graph, we see that the function is decreasing on the interval $(-\infty, 0]$ and is increasing on the interval $[0, \infty)$.

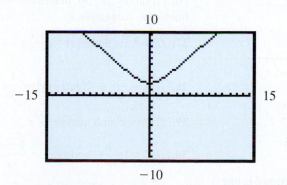

Figure 14.20

Concept Quiz 14.3

For Problems 1–7, answer true or false.

1. If a horizontal line intersects the graph of a function in exactly two points, then the function is said to be one to one.

2. The notation f^{-1} refers to the inverse of function f.

3. The graphs of two functions that are inverses of each other are mirror images with reference to the y axis.

4. If $g = \{(1, 3), (5, 9)\}$, then $g^{-1} = \{(3, 1), (9, 5)\}$.

5. Given that f and g are inverse functions, then the range of f is the domain of g.

6. A linear function whose graph has a negative slope is an increasing function.

7. A function that is increasing over its entire domain is a one-to-one function.

Problem Set 14.3

For Problems 1–6, determine whether the graph represents a one-to-one function. **(Objective 1)**

1.

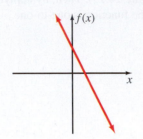

Figure 14.21

2.

Figure 14.22

3.

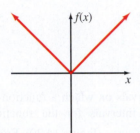

Figure 14.23

4.

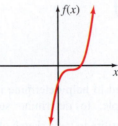

Figure 14.24

5.

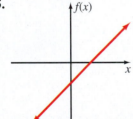

Figure 14.25

6.

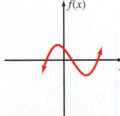

Figure 14.26

For Problems 7–14, determine whether the function f is one to one. **(Objective 1)**

7. $f(x) = 5x + 4$

8. $f(x) = -3x + 4$

9. $f(x) = x^3$

10. $f(x) = x^5 + 1$

11. $f(x) = |x| + 1$

12. $f(x) = -|x| - 2$

13. $f(x) = -x^4$

14. $f(x) = x^4 + 1$

For Problems 15–18, (a) list the domain and range of the function, (b) form the inverse function f^{-1}, and (c) list the domain and range of f^{-1}. **(Objective 3)**

15. $f = \{(1, 5), (2, 9), (5, 21)\}$

16. $f = \{(1, 1), (4, 2), (9, 3), (16, 4)\}$

17. $f = \{(0, 0), (2, 8), (-1, -1), (-2, -8)\}$

18. $f = \{(-1, 1), (-2, 4), (-3, 9), (-4, 16)\}$

For Problems 19–26, verify that the two given functions are inverses of each other. **(Objective 2)**

19. $f(x) = 5x - 9$ and $g(x) = \dfrac{x + 9}{5}$

20. $f(x) = -3x + 4$ and $g(x) = \dfrac{4 - x}{3}$

21. $f(x) = -\dfrac{1}{2}x + \dfrac{5}{6}$ and $g(x) = -2x + \dfrac{5}{3}$

22. $f(x) = x^3 + 1$ and $g(x) = \sqrt[3]{x - 1}$

23. $f(x) = \dfrac{1}{x - 1}$ for $x > 1$; $g(x) = \dfrac{x + 1}{x}$ for $x > 0$

24. $f(x) = x^2 + 2$ for $x \geq 0$; $g(x) = \sqrt{x - 2}$ for $x \geq 2$

25. $f(x) = \sqrt{2x - 4}$ for $x \geq 2$; $g(x) = \dfrac{x^2 + 4}{2}$ for $x \geq 0$

26. $f(x) = x^2 - 4$ for $x \geq 0$; $g(x) = \sqrt{x + 4}$ for $x \geq -4$

For Problems 27–36, determine whether f and g are inverse functions. **(Objective 2)**

27. $f(x) = 3x$ and $g(x) = -\dfrac{1}{3}x$

28. $f(x) = \dfrac{3}{4}x - 2$ and $g(x) = \dfrac{4}{3}x + \dfrac{8}{3}$

29. $f(x) = x^3$ and $g(x) = \sqrt[3]{x}$

30. $f(x) = \dfrac{1}{x + 1}$ and $g(x) = \dfrac{1 - x}{x}$

31. $f(x) = x$ and $g(x) = \dfrac{1}{x}$

32. $f(x) = \dfrac{3}{5}x + \dfrac{1}{3}$ and $g(x) = \dfrac{5}{3}x - 3$

33. $f(x) = x^2 - 3$ for $x \geq 0$, and
$g(x) = \sqrt{x + 3}$ for $x \geq -3$

34. $f(x) = |x - 1|$ for $x \geq 1$, and
$g(x) = |x + 1|$ for $x \geq 0$

35. $f(x) = \sqrt{x + 1}$ and $g(x) = x^2 - 1$ for $x \geq 0$

36. $f(x) = \sqrt{2x - 2}$ and $g(x) = \dfrac{1}{2}x^2 + 1$

For Problems 37–50, (a) find f^{-1}, and (b) verify that $(f \circ f^{-1})(x) = x$ and $(f^{-1} \circ f)(x) = x$. (Objective 3)

37. $f(x) = x - 4$

38. $f(x) = 2x - 1$

39. $f(x) = -3x - 4$

40. $f(x) = -5x + 6$

41. $f(x) = \dfrac{3}{4}x - \dfrac{5}{6}$

42. $f(x) = \dfrac{2}{3}x - \dfrac{1}{4}$

43. $f(x) = -\dfrac{2}{3}x$

44. $f(x) = \dfrac{4}{3}x$

45. $f(x) = \sqrt{x}$ for $x \geq 0$

46. $f(x) = \dfrac{1}{x}$ for $x \neq 0$

47. $f(x) = x^2 + 4$ for $x \geq 0$

48. $f(x) = x^2 + 1$ for $x \leq 0$

49. $f(x) = 1 + \dfrac{1}{x}$ for $x > 0$

50. $f(x) = \dfrac{x}{x + 1}$ for $x > -1$

For Problems 51–58, (a) find f^{-1} and (b) graph f and f^{-1} on the same set of axes. (Objective 3)

51. $f(x) = 3x$

52. $f(x) = -x$

53. $f(x) = 2x + 1$

54. $f(x) = -3x - 3$

55. $f(x) = \dfrac{2}{x - 1}$ for $x > 1$

56. $f(x) = \dfrac{-1}{x - 2}$ for $x > 2$

57. $f(x) = x^2 - 4$ for $x \geq 0$

58. $f(x) = \sqrt{x - 3}$ for $x \geq 3$

For Problems 59–66, find the intervals on which the given function is increasing and the intervals on which it is decreasing. (Objective 4)

59. $f(x) = x^2 + 1$

60. $f(x) = x^3$

61. $f(x) = -3x + 1$

62. $f(x) = (x - 3)^2 + 1$

63. $f(x) = -(x + 2)^2 - 1$

64. $f(x) = x^2 - 2x + 6$

65. $f(x) = -2x^2 - 16x - 35$

66. $f(x) = x^2 + 3x - 1$

Thoughts Into Words

67. Does the function $f(x) = 4$ have an inverse? Explain your answer.

68. Explain why every nonconstant linear function has an inverse.

69. Are the functions $f(x) = x^4$ and $g(x) = \sqrt[4]{x}$ inverses of each other? Explain your answer.

70. What does it mean to say that 2 and -2 are additive inverses of each other? What does it mean to say that 2 and $\dfrac{1}{2}$ are multiplicative inverses of each other? What does it mean to say that the functions $f(x) = x - 2$ and $f(x) = x + 2$ are inverses of each other? Do you think that the concept of "inverse" is being used in a consistent manner? Explain your answer.

Further Investigations

71. The function notation and the operation of composition can be used to find inverses. Say we want to find the inverse of $f(x) = 5x + 3$. We know that $f(f^{-1}(x))$ must produce x. Therefore,

$$f(f^{-1}(x)) = 5[f^{-1}(x)] + 3 = x$$

$$5[f^{-1}(x)] = x - 3$$

$$f^{-1}(x) = \dfrac{x - 3}{5}$$

Use this approach to find the inverse of each of the following functions.

a. $f(x) = 3x - 9$ **b.** $f(x) = -2x + 6$
c. $f(x) = -x + 1$ **d.** $f(x) = 2x$
e. $f(x) = -5x$ **f.** $f(x) = x^2 + 6$ for $x \geq 0$

72. If $f(x) = 2x + 3$ and $g(x) = 3x - 5$, find

a. $(f \circ g)^{-1}(x)$ **b.** $(f^{-1} \circ g^{-1})(x)$

c. $(g^{-1} \circ f^{-1})(x)$

73. For Problems 37–44, graph the given function, the inverse function that you found, and $f(x) = x$ on the same set of axes. In each case, the given function and its inverse should produce graphs that are reflections of each other through the line $f(x) = x$.

74. There is another way in which we can use the graphing calculator to help show that two functions are inverses of each other. Suppose we want to show that $f(x) = x^2 - 2$ for $x \geq 0$ and $g(x) = \sqrt{x + 2}$ for $x \geq -2$ are inverses of each other. Let's make the following assignments for our graphing calculator.

$$
\begin{aligned}
f: \quad & Y_1 = x^2 - 2 \\
g: \quad & Y_2 = \sqrt{x + 2} \\
f \circ g: \quad & Y_3 = (Y_2)^2 - 2 \\
g \circ f: \quad & Y_4 = \sqrt{Y_1 + 2}
\end{aligned}
$$

Now we can proceed as follows:

1. Graph $Y_1 = x^2 - 2$, and note that for $x > 0$, the range is greater than or equal to -2.

2. Graph $Y_2 = \sqrt{x + 2}$, and note that for $x \geq -2$, the range is greater than or equal to 0.

3. Graph $Y_3 = (Y_2)^2 - 2$ for $x \geq -2$, and observe the line $y = x$ for $x \geq -2$.

4. Graph $Y_4 = \sqrt{Y_1 + 2}$ for $x \geq 0$, and observe the line $y = x$ for $x \geq 0$.

Use this approach to check your answers for Problems 45–50.

75. Use the technique demonstrated in Problem 74 to show that

$$
f(x) = \frac{x}{\sqrt{x^2 + 1}}
$$

and

$$
g(x) = \frac{x}{\sqrt{1 - x^2}} \quad \text{for } -1 < x < 1
$$

are inverses of each other.

Answers to the Concept Quiz

1. False   **2.** True   **3.** False   **4.** True   **5.** True   **6.** False   **7.** True

14.4   Logarithms

OBJECTIVES

1 Switch equations between exponential and logarithmic form

2 Evaluate a logarithmic expression

3 Solve logarithmic equations by switching to exponential form

4 Apply the properties of logarithms

5 Solve logarithmic equations

In Sections 14.1 and 14.2, (1) we learned about exponential expressions of the form b^n, where b is any positive real number and n is any real number, (2) we used exponential expressions of the form b^n to define exponential functions, and (3) we used exponential functions to help solve problems. In the next three sections we will follow the same basic pattern with respect to a new concept, that of a *logarithm*. Let's begin with the following definition.

> ### Definition 14.4
>
> If r is any positive real number, then the unique exponent t such that $b^t = r$ is called the **logarithm of r with base b** and is denoted by $\log_b r$.

According to Definition 14.4, the logarithm of 8 base 2 is the exponent t such that $2^t = 8$; thus we can write $\log_2 8 = 3$. Likewise, we can write $\log_{10} 100 = 2$ because $10^2 = 100$. In general, we can remember Definition 14.4 in terms of the statement

$$\log_b r = t \quad \text{is equivalent to} \quad b^t = r$$

Thus we can easily switch back and forth between exponential and logarithmic forms of equations, as the next examples illustrate.

$$\log_3 81 = 4 \quad \text{is equivalent to} \quad 3^4 = 81$$
$$\log_{10} 100 = 2 \quad \text{is equivalent to} \quad 10^2 = 100$$
$$\log_{10} 0.001 = -3 \quad \text{is equivalent to} \quad 10^{-3} = 0.001$$
$$\log_2 128 = 7 \quad \text{is equivalent to} \quad 2^7 = 128$$
$$\log_m n = p \quad \text{is equivalent to} \quad m^p = n$$
$$2^4 = 16 \quad \text{is equivalent to} \quad \log_2 16 = 4$$
$$5^2 = 25 \quad \text{is equivalent to} \quad \log_5 25 = 2$$
$$\left(\frac{1}{2}\right)^4 = \frac{1}{16} \quad \text{is equivalent to} \quad \log_{\frac{1}{2}}\left(\frac{1}{16}\right) = 4$$
$$10^{-2} = 0.01 \quad \text{is equivalent to} \quad \log_{10} 0.01 = -2$$
$$a^b = c \quad \text{is equivalent to} \quad \log_a c = b$$

Classroom Example
Change the form of the equation $\log_b 6 = x$ to exponential form.

EXAMPLE 1   Change the form of the equation $\log_b 9 = x$ to exponential form.

Solution

$\log_b 9 = x$ is equivalent to $b^x = 9$

Classroom Example
Change the form of the equation $9^2 = 81$ to logarithmic form.

EXAMPLE 2   Change the form of the equation $7^2 = 49$ to logarithmic form.

Solution

$7^2 = 49$ is equivalent to $\log_7 49 = 2$

We can conveniently calculate some logarithms by changing to exponential form, as in the next examples.

Classroom Example
Evaluate $\log_5 125$.

EXAMPLE 3   Evaluate $\log_4 64$.

Solution

Let $\log_4 64 = x$. Then by switching to exponential form, we have $4^x = 64$, which we can solve as we did back in Section 14.1.

$$4^x = 64$$
$$4^x = 4^3$$
$$x = 3$$

Therefore, we can write $\log_4 64 = 3$.

Classroom Example
Evaluate $\log_{10} 0.0001$.

EXAMPLE 4 Evaluate $\log_{10} 0.1$.

Solution

Let $\log_{10} 0.1 = x$. Then by switching to exponential form, we have $10^x = 0.1$, which we can solve as follows:

$$10^x = 0.1$$

$$10^x = \frac{1}{10}$$

$$10^x = 10^{-1}$$

$$x = -1$$

Thus we obtain $\log_{10} 0.1 = -1$. ∎

The link between logarithms and exponents also provides the basis for solving some equations that involve logarithms, as the next two examples illustrate.

Classroom Example
Solve $\log_{25} x = \frac{3}{2}$.

EXAMPLE 5 Solve $\log_8 x = \frac{2}{3}$.

Solution

$$\log_8 x = \frac{2}{3}$$

$$8^{\frac{2}{3}} = x \quad \text{By switching to exponential form}$$

$$\sqrt[3]{8^2} = x$$

$$(\sqrt[3]{8})^2 = x$$

$$4 = x$$

The solution set is $\{4\}$. ∎

Classroom Example
Solve $\log_b 144 = 2$.

EXAMPLE 6 Solve $\log_b 1000 = 3$.

Solution

$$\log_b 1000 = 3$$

$$b^3 = 1000$$

$$b = 10$$

The solution set is $\{10\}$. ∎

Properties of Logarithms

There are some properties of logarithms that are a direct consequence of Definition 14.4 and our knowledge of exponents. For example, writing the exponential equations $b^1 = b$ and $b^0 = 1$ in logarithmic form yields the following property.

Property 14.3

For $b > 0$ and $b \neq 1$,

1. $\log_b b = 1$

2. $\log_b 1 = 0$

Thus we can write

$$\log_{10} 10 = 1$$
$$\log_2 2 = 1$$
$$\log_{10} 1 = 0$$
$$\log_5 1 = 0$$

By Definition 14.4, $\log_b r$ is the exponent t such that $b^t = r$. Therefore, raising b to the $\log_b r$ power must produce r. We state this fact in Property 14.4.

Property 14.4

For $b > 0$, $b \neq 1$, and $r > 0$

$$b^{\log_b r} = r$$

The following examples illustrate Property 14.4.

$$10^{\log_{10} 19} = 19$$
$$2^{\log_2 14} = 14$$
$$e^{\log_e 5} = 5$$

Because a logarithm is by definition an exponent, it would seem reasonable to predict that there are some properties of logarithms that correspond to the basic exponential properties. This is an accurate prediction; these properties provide a basis for computational work with logarithms. Let's state the first of these properties and show how it can be verified by using our knowledge of exponents.

Property 14.5

For positive real numbers b, r, and s, where $b \neq 1$,

$$\log_b rs = \log_b r + \log_b s$$

To verify Property 14.5 we can proceed as follows: Let $m = \log_b r$ and $n = \log_b s$. Change each of these equations to exponential form.

$$m = \log_b r \quad \text{becomes} \quad r = b^m$$
$$n = \log_b s \quad \text{becomes} \quad s = b^n$$

Thus the product rs becomes

$$rs = b^m \cdot b^n = b^{m+n}$$

Now by changing $rs = b^{m+n}$ back to logarithmic form, we obtain

$$\log_b rs = m + n$$

Replacing m with $\log_b r$ and n with $\log_b s$ yields

$$\log_b rs = \log_b r + \log_b s$$

The following three examples demonstrate a use of Property 14.5.

Classroom Example
If $\log_3 6 = 1.6309$ and $\log_3 2 = 0.6309$, evaluate $\log_3 12$.

EXAMPLE 7   If $\log_2 5 = 2.3219$ and $\log_2 3 = 1.5850$, evaluate $\log_2 15$.

Solution

Because $15 = 5 \cdot 3$, we can apply Property 14.5.

$$\log_2 15 = \log_2(5 \cdot 3)$$
$$= \log_2 5 + \log_2 3$$
$$= 2.3219 + 1.5850$$
$$= 3.9069$$

EXAMPLE 8

If $\log_{10} 178 = 2.2504$ and $\log_{10} 89 = 1.9494$, evaluate $\log_{10}(178 \cdot 89)$.

Solution

$$\begin{aligned}
\log_{10}(178 \cdot 89) &= \log_{10} 178 + \log_{10} 89 \\
&= 2.2504 + 1.9494 \\
&= 4.1998
\end{aligned}$$

EXAMPLE 9   If $\log_3 8 = 1.8928$, evaluate $\log_3 72$.

Solution

$$\begin{aligned}
\log_3 72 &= \log_3(9 \cdot 8) \\
&= \log_3 9 + \log_3 8 \\
&= 2 + 1.8928 \qquad \log_3 9 = 2 \text{ because } 3^2 = 9 \\
&= 3.8928
\end{aligned}$$

Because $\dfrac{b^m}{b^n} = b^{m-n}$, we would expect a corresponding property pertaining to logarithms. There is such a property, Property 14.6.

> **Property 14.6**
>
> For positive numbers b, r, and s, where $b \neq 1$,
>
> $$\log_b\left(\frac{r}{s}\right) = \log_b r - \log_b s$$

This property can be verified by using an approach similar to the one we used to verify Property 14.5. We leave it for you to do in an exercise in the next problem set.

We can use Property 14.6 to change a division problem into a subtraction problem, as in the next two examples.

EXAMPLE 10   If $\log_5 36 = 2.2266$ and $\log_5 4 = 0.8614$, evaluate $\log_5 9$.

Solution

Because $9 = \dfrac{36}{4}$, we can use Property 14.6.

$$\begin{aligned}
\log_5 9 &= \log_5\left(\frac{36}{4}\right) \\
&= \log_5 36 - \log_5 4 \\
&= 2.2266 - 0.8614 \\
&= 1.3652
\end{aligned}$$

EXAMPLE 11

Evaluate $\log_{10}\left(\dfrac{379}{86}\right)$ given that $\log_{10}379 = 2.5786$ and $\log_{10}86 = 1.9345$.

Solution

$$\log_{10}\left(\frac{379}{86}\right) = \log_{10}379 - \log_{10}86$$

$$= 2.5786 - 1.9345$$

$$= 0.6441$$

The next property of logarithms provides the basis for evaluating expressions such as $3^{\sqrt{2}}$, $\left(\sqrt{5}\right)^{\frac{2}{3}}$, and $(0.076)^{\frac{2}{3}}$. We cite the property, consider a basis for its justification, and offer illustrations of its use.

Property 14.7

If r is a positive real number, b is a positive real number other than 1, and p is any real number, then

$$\log_b r^p = p(\log_b r)$$

As you might expect, the exponential property $(b^n)^m = b^{nm}$ plays an important role in the verification of Property 14.7. This is an exercise for you in the next problem set. Let's look at some uses of Property 14.7.

EXAMPLE 12

Evaluate $\log_2 22^{\frac{1}{3}}$ given that $\log_2 22 = 4.4594$.

Solution

$$\log_2 22^{\frac{1}{3}} = \frac{1}{3}\log_2 22 \qquad \text{Property 14.7}$$

$$= \frac{1}{3}(4.4594)$$

$$= 1.4865$$

Together, the properties of logarithms enable us to change the forms of various logarithmic expressions. For example, an expression such as $\log_b\sqrt{\dfrac{xy}{z}}$ can be rewritten in terms of sums and differences of simpler logarithmic quantities as follows:

$$\log_b\sqrt{\frac{xy}{z}} = \log_b\left(\frac{xy}{z}\right)^{\frac{1}{2}}$$

$$= \frac{1}{2}\log_b\left(\frac{xy}{z}\right) \qquad \text{Property 14.7}$$

$$= \frac{1}{2}(\log_b xy - \log_b z) \qquad \text{Property 14.6}$$

$$= \frac{1}{2}(\log_b x + \log_b y - \log_b z) \qquad \text{Property 14.5}$$

EXAMPLE 13

Write each expression as the sums or differences of simpler logarithmic quantities. Assume that all variables represent positive real numbers:

(a) $\log_b xy^2$ (b) $\log_b \dfrac{\sqrt{x}}{y}$ (c) $\log_b \dfrac{x^3}{yz}$

Solution

(a) $\log_b xy^2 = \log_b x + \log_b y^2 = \log_b x + 2\log_b y$

(b) $\log_b \dfrac{\sqrt{x}}{y} = \log_b \sqrt{x} - \log_b y = \dfrac{1}{2}\log_b x - \log_b y$

(c) $\log_b \dfrac{x^3}{yz} = \log_b x^3 - \log_b yz = 3\log_b x - (\log_b y + \log_b z)$

$$= 3\log_b x - \log_b y - \log_b z$$

Sometimes we need to change from an indicated sum or difference of logarithmic quantities to an indicated product or quotient. This is especially helpful when solving certain kinds of equations that involve logarithms. Note in these next two examples how we can use the properties, along with the process of changing from logarithmic form to exponential form, to solve some equations.

EXAMPLE 14   Solve $\log_{10} x + \log_{10}(x + 9) = 1$.

Solution

$$\log_{10} x + \log_{10}(x + 9) = 1$$
$$\log_{10}[x(x + 9)] = 1 \qquad \text{Property 14.5}$$
$$10^1 = x(x + 9) \qquad \text{Change to exponential form}$$
$$10 = x^2 + 9x$$
$$0 = x^2 + 9x - 10$$
$$0 = (x + 10)(x - 1)$$

$x + 10 = 0 \qquad$ or $\qquad x - 1 = 0$
$x = -10 \qquad$ or $\qquad x = 1$

Because logarithms are defined only for positive numbers, x and $x + 9$ have to be positive. Therefore, the solution of -10 must be discarded. The solution set is $\{1\}$.

EXAMPLE 15   Solve $\log_5(x + 4) - \log_5 x = 2$.

Solution

$$\log_5(x + 4) - \log_5 x = 2$$
$$\log_5\left(\dfrac{x + 4}{x}\right) = 2 \qquad \text{Property 14.6}$$
$$5^2 = \dfrac{x + 4}{x} \qquad \text{Change to exponential form}$$
$$25 = \dfrac{x + 4}{x}$$
$$25x = x + 4$$

$$24x = 4$$
$$x = \frac{4}{24} = \frac{1}{6}$$

The solution set is $\left\{\frac{1}{6}\right\}$.

Because logarithms are defined only for positive numbers, we should realize that some logarithmic equations may not have any solutions. (The solution set is the null set.) It is also possible that a logarithmic equation has a negative solution, as the next example illustrates.

Classroom Example
Solve $\log_5 6 + \log_5(x + 7) = 2$.

EXAMPLE 16   Solve $\log_2 3 + \log_2(x + 4) = 3$.

Solution

$$\log_2 3 + \log_2(x + 4) = 3$$
$$\log_2 3(x + 4) = 3 \qquad \text{Property 14.5}$$
$$3(x + 4) = 2^3 \qquad \text{Change to exponential form}$$
$$3x + 12 = 8$$
$$3x = -4$$
$$x = -\frac{4}{3}$$

The only restriction is that $x + 4 > 0$ or $x > -4$. Therefore, the solution set is $\left\{-\frac{4}{3}\right\}$. Perhaps you should check this answer.

Concept Quiz 14.4

For Problems 1–10, answer true or false.

1. The $\log_m n = q$ is equivalent to $m^q = n$.

2. The $\log_7 7$ equals 0.

3. A logarithm is by definition an exponent.

4. The $\log_5 9^2$ is equivalent to $2\log_5 9$.

5. For the expression $\log_3 9$, the base of the logarithm is 9.

6. The expression $\log_2 x - \log_2 y + \log_2 z$ is equivalent to $\log_2 xyz$.

7. $\log_4 4 + \log_4 1 = 1$

8. $\log_2 8 - \log_3 9 + \log_4\left(\frac{1}{16}\right) = -1$

9. The solution set for $\log_{10} x + \log_{10}(x - 3) = 1$ is $\{10\}$.

10. The solution set for $\log_6 x + \log_6(x + 5) = 2$ is $\{4\}$.

Problem Set 14.4

For Problems 1–10, write each of the following in logarithmic form. For example, $2^3 = 8$ becomes $\log_2 8 = 3$ in logarithmic form. **(Objective 1)**

1. $2^7 = 128$

2. $3^3 = 27$

3. $5^3 = 125$

4. $2^6 = 64$

5. $10^3 = 1000$

6. $10^1 = 10$

7. $2^{-2} = \frac{1}{4}$

8. $3^{-4} = \frac{1}{81}$

9. $10^{-1} = 0.1$

10. $10^{-2} = 0.01$

For Problems 11–20, write each of the following in exponential form. For example, $\log_2 8 = 3$ becomes $2^3 = 8$ in exponential form. **(Objective 1)**

11. $\log_3 81 = 4$ **12.** $\log_2 256 = 8$

13. $\log_4 64 = 3$ **14.** $\log_5 25 = 2$

15. $\log_{10} 10,000 = 4$ **16.** $\log_{10} 100,000 = 5$

17. $\log_2\left(\dfrac{1}{16}\right) = -4$ **18.** $\log_5\left(\dfrac{1}{125}\right) = -3$

19. $\log_{10} 0.001 = -3$ **20.** $\log_{10} 0.000001 = -6$

For Problems 21–40, evaluate each expression. **(Objective 2)**

21. $\log_2 16$ **22.** $\log_3 9$

23. $\log_3 81$ **24.** $\log_2 512$

25. $\log_6 216$ **26.** $\log_4 256$

27. $\log_7 \sqrt{7}$ **28.** $\log_2 \sqrt[3]{2}$

29. $\log_{10} 1$ **30.** $\log_{10} 10$

31. $\log_{10} 0.1$ **32.** $\log_{10} 0.0001$

33. $10^{\log_{10} 5}$ **34.** $10^{\log_{10} 14}$

35. $\log_2\left(\dfrac{1}{32}\right)$ **36.** $\log_5\left(\dfrac{1}{25}\right)$

37. $\log_5(\log_2 32)$ **38.** $\log_2(\log_4 16)$

39. $\log_{10}(\log_7 7)$ **40.** $\log_2(\log_5 5)$

For Problems 41–50, solve each equation. **(Objective 3)**

41. $\log_7 x = 2$ **42.** $\log_2 x = 5$

43. $\log_8 x = \dfrac{4}{3}$ **44.** $\log_{16} x = \dfrac{3}{2}$

45. $\log_9 x = \dfrac{3}{2}$ **46.** $\log_8 x = -\dfrac{2}{3}$

47. $\log_4 x = -\dfrac{3}{2}$ **48.** $\log_9 x = -\dfrac{5}{2}$

49. $\log_x 2 = \dfrac{1}{2}$ **50.** $\log_x 3 = \dfrac{1}{2}$

For Problems 51–59, you are given $\log_2 5 = 2.3219$ and $\log_2 7 = 2.8074$. Evaluate each expression using Properties 14.5–14.7. **(Objective 4)**

51. $\log_2 35$ **52.** $\log_2\left(\dfrac{7}{5}\right)$

53. $\log_2 125$ **54.** $\log_2 49$

55. $\log_2 \sqrt{7}$ **56.** $\log_2 \sqrt[3]{5}$

57. $\log_2 175$ **58.** $\log_2 56$

59. $\log_2 80$

For Problems 60–68, you are given $\log_8 5 = 0.7740$ and $\log_8 11 = 1.1531$. Evaluate each expression using Properties 14.5–14.7. **(Objective 4)**

60. $\log_8 55$ **61.** $\log_8\left(\dfrac{5}{11}\right)$

62. $\log_8 25$ **63.** $\log_8 \sqrt{11}$

64. $\log_8 (5)^{\frac{2}{3}}$ **65.** $\log_8 88$

66. $\log_8 320$ **67.** $\log_8\left(\dfrac{25}{11}\right)$

68. $\log_8\left(\dfrac{121}{25}\right)$

For Problems 69–80, express each of the following as the sum or difference of simpler logarithmic quantities. Assume all variables represent positive real numbers. **(Objective 4)** For example,

$$\log_b \frac{x^3}{y^2} = \log_b x^3 - \log_b y^2$$
$$= 3\log_b x - 2\log_b y$$

69. $\log_b xyz$ **70.** $\log_b 5x$

71. $\log_b\left(\dfrac{y}{z}\right)$ **72.** $\log_b\left(\dfrac{x^2}{y}\right)$

73. $\log_b y^3 z^4$ **74.** $\log_b x^2 y^3$

75. $\log_b\left(\dfrac{x^{\frac{1}{2}} y^{\frac{1}{3}}}{z^4}\right)$ **76.** $\log_b x^{\frac{2}{3}} y^{\frac{3}{4}}$

77. $\log_b \sqrt[3]{x^2 z}$ **78.** $\log_b \sqrt{xy}$

79. $\log_b\left(x\sqrt{\dfrac{x}{y}}\right)$ **80.** $\log_b \sqrt{\dfrac{x}{y}}$

For Problems 81–97, solve each of the equations. **(Objective 5)**

81. $\log_3 x + \log_3 4 = 2$

82. $\log_7 5 + \log_7 x = 1$

83. $\log_{10} x + \log_{10}(x - 21) = 2$

84. $\log_{10} x + \log_{10}(x - 3) = 1$

85. $\log_2 x + \log_2(x - 3) = 2$

86. $\log_3 x + \log_3(x - 2) = 1$

87. $\log_{10}(2x - 1) - \log_{10}(x - 2) = 1$

88. $\log_{10}(9x - 2) = 1 + \log_{10}(x - 4)$

89. $\log_5(3x - 2) = 1 + \log_5(x - 4)$

90. $\log_6 x + \log_6(x + 5) = 2$

91. $\log_8(x + 7) + \log_8 x = 1$

92. $\log_6(x + 1) + \log_6(x - 4) = 2$

93. $\log_2 5 + \log_2(x + 6) = 3$

94. $\log_2(x - 1) - \log_2(x + 3) = 2$

95. $\log_5 x = \log_5(x + 2) + 1$

96. $\log_3(x + 3) + \log_3(x + 5) = 1$

97. $\log_2(x + 2) = 1 - \log_2(x + 3)$

98. Verify Property 14.6.

99. Verify Property 14.7.

Thoughts Into Words

100. Explain, without using Property 14.4, why $4^{\log_4 9}$ equals 9.

101. How would you explain the concept of a logarithm to someone who had just completed an elementary algebra course?

102. In the next section we will show that the logarithmic function $f(x) = \log_2 x$ is the inverse of the exponential function $f(x) = 2^x$. From that information, how could you sketch a graph of $f(x) = \log_2 x$?

Answers to the Concept Quiz

1. True   **2.** False   **3.** True   **4.** True   **5.** False   **6.** False   **7.** True   **8.** True   **9.** False   **10.** True

---

14.5   Logarithmic Functions

OBJECTIVES

1 Graph logarithmic functions

2 Evaluate common and natural logarithms using a calculator

3 Solve common and natural logarithmic equations using a calculator

We can now use the concept of a logarithm to define a new function.

> **Definition 14.5**
>
> If $b > 0$ and $b \neq 1$, then the function f defined by
>
> $$f(x) = \log_b x$$
>
> where x is any positive real number, is called the **logarithmic function with base b**.

We can obtain the graph of a specific logarithmic function in various ways. For example, we can change the equation $y = \log_2 x$ to the exponential equation $2^y = x$, where we can determine a table of values. The next set of exercises asks you to graph some logarithmic functions with this approach.

Classroom Example
Graph $f(x) = \log_3 x$.

We can obtain the graph of a logarithmic function by setting up a table of values directly from the logarithmic equation. Example 1 illustrates this approach.

EXAMPLE 1   Graph $f(x) = \log_2 x$.

Solution

Let's choose some values for x for which the corresponding values for $\log_2 x$ are easily determined. (Remember that logarithms are defined only for the positive real numbers.)

x	f(x)
$\frac{1}{8}$	-3
$\frac{1}{4}$	-2
$\frac{1}{2}$	-1
1	0
2	1
4	2
8	3

$\log_2 \dfrac{1}{8} = -3$ because $2^{-3} = \dfrac{1}{2^3} = \dfrac{1}{8}$

$\log_2 1 = 0$ because $2^0 = 1$

$f(x) = \log_2 x$

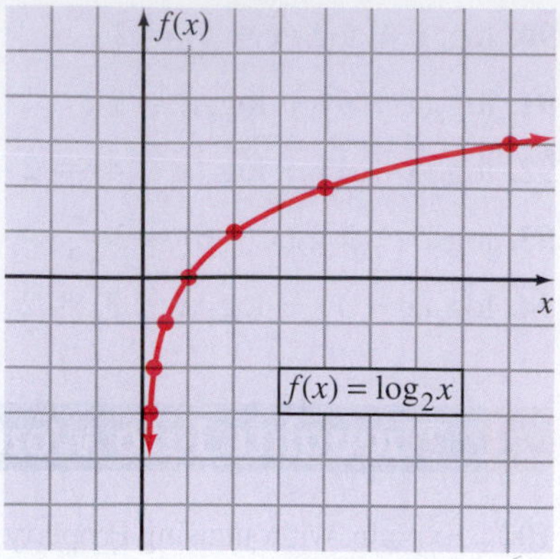

Figure 14.27

Plot these points and connect them with a smooth curve to produce Figure 14.27.

Suppose that we consider the following two functions f and g.

$f(x) = b^x$ Domain: all real numbers

                      Range: positive real numbers

$g(x) = \log_b x$ Domain: positive real numbers

                      Range: all real numbers

Furthermore, suppose that we consider the composition of f and g and the composition of g and f.

$$(f \circ g)(x) = f(g(x)) = f(\log_b x) = b^{\log_b x} = x$$
$$(g \circ f)(x) = g(f(x)) = g(b^x) = \log_b b^x = x \log_b b = x(1) = x$$

Therefore, because the domain of f is the range of g, the range of f is the domain of g, $f(g(x)) = x$, and $g(f(x)) = x$, the two functions f and g are inverses of each other.

Remember that the graphs of a function and its inverse are reflections of each other through the line $y = x$. Thus the graph of a logarithmic function can also be determined by reflecting the graph of its inverse exponential function through the line $y = x$. We see this in Figure 14.28, where the graph of $y = 2^x$ has been reflected across the line $y = x$ to produce the graph of $y = \log_2 x$.

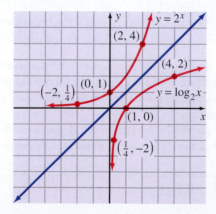

Figure 14.28

The general behavior patterns of exponential functions were illustrated by two graphs back in Figure 14.3. We can now reflect each of those graphs through the line

$y = x$ and observe the general behavior patterns of logarithmic functions, as shown in Figure 14.29.

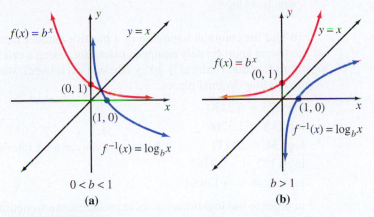

Figure 14.29

Finally, when graphing logarithmic functions, don't forget about variations of the basic curves.

1. The graph of $f(x) = 3 + \log_2 x$ is the graph of $f(x) = \log_2 x$ moved up 3 units. We commonly write $3 + \log_2 x$ because $\log_2 x + 3$ is apt to be confused with $\log_2(x + 3)$.

2. The graph of $f(x) = \log_2(x - 4)$ is the graph of $f(x) = \log_2 x$ moved 4 units to the right.

3. The graph of $f(x) = -\log_2 x$ is the graph of $f(x) = \log_2 x$ reflected across the x axis.

Common Logarithms—Base 10

The properties of logarithms we discussed in Section 14.4 are true for any valid base. Because the Hindu-Arabic numeration system that we use is a base-10 system, logarithms to base 10 have historically been used for computational purposes. Base-10 logarithms are called **common logarithms**.

Originally, common logarithms were developed to assist in complicated numerical calculations that involved products, quotients, and powers of real numbers. Today they are seldom used for that purpose, because the calculator and computer can much more effectively handle the messy computational problems. However, common logarithms do still occur in applications, so they are deserving of our attention.

As we know from earlier work, the definition of a logarithm provides the basis for evaluating $\log_{10} x$ for values of x that are integral powers of 10. Consider the following examples.

$\log_{10} 1000 = 3 \qquad$ because $10^3 = 1000$

$\log_{10} 100 = 2 \qquad$ because $10^2 = 100$

$\log_{10} 10 = 1 \qquad$ because $10^1 = 10$

$\log_{10} 1 = 0 \qquad$ because $10^0 = 1$

$\log_{10} 0.1 = -1 \qquad$ because $10^{-1} = \dfrac{1}{10} = 0.1$

$\log_{10} 0.01 = -2 \qquad$ because $10^{-2} = \dfrac{1}{10^2} = 0.01$

$\log_{10} 0.001 = -3 \qquad$ because $10^{-3} = \dfrac{1}{10^3} = 0.001$

When working with base-10 logarithms, it is customary to omit writing the numeral 10 to designate the base. Thus the expression $\log_{10} x$ is written as $\log x$, and a statement such as

$\log_{10} 1000 = 3$ becomes $\log 1000 = 3$. We will follow this practice from now on in this chapter, but don't forget that the base is understood to be 10.

$$\log_{10} x = \log x$$

To find the common logarithm of a positive number that is not an integral power of 10, we can use an appropriately equipped calculator. Using a calculator equipped with a common logarithm function (ordinarily a key labeled $\boxed{\log}$ is used), we obtained the following results, rounded to four decimal places.

$\log 1.75 = 0.2430$
$\log 23.8 = 1.3766$
$\log 134 = 2.1271$ Be sure that you can use a calculator and obtain these results
$\log 0.192 = -0.7167$
$\log 0.0246 = -1.6091$

In order to use logarithms to solve problems, we sometimes need to be able to determine a number when the logarithm of the number is known. That is, we may need to determine x when $\log x$ is known. Let's consider an example.

Classroom Example
Find x if $\log x = 0.6805$.

EXAMPLE 2   Find x if $\log x = 0.2430$.

Solution

If $\log x = 0.2430$, then by changing to exponential form, we have $10^{0.2430} = x$. Therefore, using the $\boxed{10^x}$ key, we can find x.

$$x = 10^{0.2430} \approx 1.749846689$$

Therefore, $x = 1.7498$, rounded to five significant digits.

Be sure that you can use your calculator to obtain the following results. We have rounded the values for x to five significant digits.

If $\log x = 0.7629$, then $x = 10^{0.7629} = 5.7930$.
If $\log x = 1.4825$, then $x = 10^{1.4825} = 30.374$.
If $\log x = 4.0214$, then $x = 10^{4.0214} = 10,505$.
If $\log x = -1.5162$, then $x = 10^{-1.5162} = 0.030465$.
If $\log x = -3.8921$, then $x = 10^{-3.8921} = 0.00012820$.

The **common logarithmic function** is defined by the equation $f(x) = \log x$. It should now be a simple matter to set up a table of values and sketch the function. We will have you do this in the next set of exercises. Remember that $f(x) = 10^x$ and $g(x) = \log x$ are inverses of each other. Therefore, we could also get the graph of $g(x) = \log x$ by reflecting the exponential curve $f(x) = 10^x$ across the line $y = x$.

Natural Logarithms—Base *e*

In many practical applications of logarithms, the number e (remember that $e \approx 2.71828$) is used as a base. Logarithms with a base of e are called **natural logarithms**, and the symbol $\ln x$ is commonly used instead of $\log_e x$.

$$\log_e x = \ln x$$

Natural logarithms can be found with an appropriately equipped calculator. Using a calculator with a natural logarithm function (ordinarily a key labeled $\boxed{\ln}$), we can obtain the following results, rounded to four decimal places.

$$\ln 3.21 = 1.1663$$
$$\ln 47.28 = 3.8561$$
$$\ln 842 = 6.7358$$
$$\ln 0.21 = -1.5606$$
$$\ln 0.0046 = -5.3817$$
$$\ln 10 = 2.3026$$

Be sure that you can use your calculator to obtain these results. Keep in mind the significance of a statement such as $\ln 3.21 = 1.1663$. By changing to exponential form, we are claiming that e raised to the 1.1663 power is approximately 3.21. Using a calculator, we obtain $e^{1.1663} = 3.210093293$.

Let's do a few more problems and find x when given $\ln x$. Be sure that you agree with these results.

If $\ln x = 2.4156$, then $x = e^{2.4156} = 11.196$.

If $\ln x = 0.9847$, then $x = e^{0.9847} = 2.6770$.

If $\ln x = 4.1482$, then $x = e^{4.1482} = 63.320$.

If $\ln x = -1.7654$, then $x = e^{-1.7654} = 0.17112$.

The **natural logarithmic function** is defined by the equation $f(x) = \ln x$. It is the inverse of the natural exponential function $f(x) = e^x$. Thus one way to graph $f(x) = \ln x$ is to reflect the graph of $f(x) = e^x$ across the line $y = x$. We will have you do this in the next set of problems.

Concept Quiz 14.5

For Problems 1–10, answer true or false.

1. The domain for the logarithmic function $f(x) = \log_b x$ is all real numbers.
2. Every logarithmic function has an inverse function that is an exponential function.
3. The base for common logarithms is 2.
4. Logarithms with a base of e are called empirical logarithms.
5. The symbol $\ln x$ is usually written instead of $\log_e x$.
6. The graph of $f(x) = \log_4 x$ is the reflection with reference to the line $y = x$ of the graph $f(x) = 4^x$.
7. If $\ln x = 1$, then $x = e$.
8. If $\log x = 1$, then $x = 10$.
9. The graph of $f(x) = \ln x$ is the graph of $g(x) = e^x$ reflected across the x axis.
10. The domain of the function $f(x) = \ln x$ is the set of real numbers.

Problem Set 14.5

For Problems 1–10, use a calculator to find each common logarithm. Express answers to four decimal places. **(Objective 2)**

1. $\log 7.24$
2. $\log 2.05$
3. $\log 52.23$
4. $\log 825.8$
5. $\log 3214.1$
6. $\log 14{,}189$
7. $\log 0.729$
8. $\log 0.04376$
9. $\log 0.00034$
10. $\log 0.000069$

For Problems 11–20, use your calculator to find x when given $\log x$. Express answers to five significant digits. **(Objective 3)**

11. $\log x = 2.6143$ **12.** $\log x = 1.5263$

13. $\log x = 4.9547$ **14.** $\log x = 3.9335$

15. $\log x = 1.9006$ **16.** $\log x = 0.5517$

17. $\log x = -1.3148$ **18.** $\log x = -0.1452$

19. $\log x = -2.1928$ **20.** $\log x = -2.6542$

For Problems 21–30, use your calculator to find each natural logarithm. Express answers to four decimal places. **(Objective 2)**

21. $\ln 5$ **22.** $\ln 18$

23. $\ln 32.6$ **24.** $\ln 79.5$

25. $\ln 430$ **26.** $\ln 371.8$

27. $\ln 0.46$ **28.** $\ln 0.524$

29. $\ln 0.0314$ **30.** $\ln 0.008142$

For Problems 31–40, use your calculator to find x when given $\ln x$. Express answers to five significant digits. **(Objective 3)**

31. $\ln x = 0.4721$ **32.** $\ln x = 0.9413$

33. $\ln x = 1.1425$ **34.** $\ln x = 2.7619$

35. $\ln x = 4.6873$ **36.** $\ln x = 3.0259$

37. $\ln x = -0.7284$ **38.** $\ln x = -1.6246$

39. $\ln x = -3.3244$ **40.** $\ln x = -2.3745$

For Problems 41–46, follow the suggested procedure to graph logarithmic functions. **(Objective 1)**

41. (a) Complete the following table and then graph $f(x) = \log x$. (Express the values for $\log x$ to the nearest tenth.)

x	0.1	0.5	1	2	4	8	10
$\log x$							

(b) Complete the following table and express values for 10^x to the nearest tenth.

x	−1	−0.3	0	0.3	0.6	0.9	1
10^x							

Then graph $f(x) = 10^x$ and reflect it across the line $y = x$ to produce the graph for $f(x) = \log x$.

42. (a) Complete the following table and then graph $f(x) = \ln x$. (Express the values for $\ln x$ to the nearest tenth.)

x	0.1	0.5	1	2	4	8	10
$\ln x$							

(b) Complete the following table and express values for e^x to the nearest tenth.

x	−2.3	−0.7	0	0.7	1.4	2.1	2.3
e^x							

Then graph $f(x) = e^x$ and reflect it across the line $y = x$ to produce the graph for $f(x) = \ln x$.

43. Graph $y = \log_{\frac{1}{2}} x$ by graphing $\left(\dfrac{1}{2}\right)^y = x$.

44. Graph $y = \log_2 x$ by graphing $2^y = x$.

45. Graph $f(x) = \log_3 x$ by reflecting the graph of $g(x) = 3^x$ across the line $y = x$.

46. Graph $f(x) = \log_4 x$ by reflecting the graph of $g(x) = 4^x$ across the line $y = x$.

For Problems 47–53, graph each of the functions. Remember that the graph of $f(x) = \log_2 x$ is given in Figure 14.27. **(Objective 1)**

47. $f(x) = 3 + \log_2 x$ **48.** $f(x) = -2 + \log_2 x$

49. $f(x) = \log_2(x + 3)$ **50.** $f(x) = \log_2(x - 2)$

51. $f(x) = \log_2 2x$ **52.** $f(x) = -\log_2 x$

53. $f(x) = 2\log_2 x$

For Problems 54–61, perform the following calculations and express answers to the nearest hundredth. (These calculations are in preparation for our work in the next section.)

54. $\dfrac{\log 7}{\log 3}$ **55.** $\dfrac{\ln 2}{\ln 7}$

56. $\dfrac{2\ln 3}{\ln 8}$ **57.** $\dfrac{\ln 5}{2\ln 3}$

58. $\dfrac{\ln 3}{0.04}$ **59.** $\dfrac{\ln 2}{0.03}$

60. $\dfrac{\log 2}{5\log 1.02}$ **61.** $\dfrac{\log 5}{3\log 1.07}$

Thoughts Into Words

62. Why is the number 1 excluded from being a base of a logarithmic function?

63. How do we know that $\log_2 6$ is between 2 and 3?

Graphing Calculator Activities

64. Graph $f(x) = x$, $f(x) = e^x$, and $f(x) = \ln x$ on the same set of axes.

65. Graph $f(x) = x$, $f(x) = 10^x$, and $f(x) = \log x$ on the same set of axes.

66. Graph $f(x) = \ln x$. How should the graphs of $f(x) = 2 \ln x$, $f(x) = 4 \ln x$, and $f(x) = 6 \ln x$ compare to the graph of $f(x) = \ln x$? Graph the three functions on the same set of axes with $f(x) = \ln x$.

67. Graph $f(x) = \log x$. Now predict the graphs for $f(x) = 2 + \log x$, $f(x) = -2 + \log x$, and $f(x) = -6 + \log x$. Graph the three functions on the same set of axes with $f(x) = \log x$.

68. Graph $f(x) = \ln x$. Now predict the graphs for $f(x) = \ln(x - 2)$, $f(x) = \ln(x - 6)$, and $f(x) = \ln(x + 4)$. Graph the three functions on the same set of axes with $f(x) = \ln x$.

69. For each of the following, (a) predict the general shape and location of the graph, and (b) use your graphing calculator to graph the function and thus check your prediction.
  (a) $f(x) = \log x + \ln x$
  (b) $f(x) = \log x - \ln x$
  (c) $f(x) = \ln x - \log x$
  (d) $f(x) = \ln x^2$

Answers to the Concept Quiz

1. False **2.** True **3.** False **4.** False **5.** True **6.** True **7.** True **8.** True **9.** False **10.** False

---

14.6 Exponential Equations, Logarithmic Equations, and Problem Solving

OBJECTIVES

1. Solve exponential equations
2. Solve logarithmic equations
3. Use logarithms to solve problems
4. Solve problems involving Richter numbers
5. Evaluate logarithms using the change-of-base formula

In Section 14.1 we solved exponential equations such as $3^x = 81$ when we expressed both sides of the equation as a power of 3 and then applied the property "if $b^n = b^m$, then $n = m$." However, if we try to use this same approach with an equation such as $3^x = 5$, we face the difficulty of expressing 5 as a power of 3. We can solve this type of problem by using the properties of logarithms and the following property of equality.

Property 14.8

If $x > 0$, $y > 0$, and $b \neq 1$, then

$$x = y \quad \text{if and only if} \quad \log_b x = \log_b y$$

Property 14.8 is stated in terms of any valid base b; however, for most applications we use either common logarithms (base 10) or natural logarithms (base e). Let's consider some examples.

EXAMPLE 1   Solve $3^x = 5$ to the nearest hundredth.

Solution

By using common logarithms, we can proceed as follows:

$$3^x = 5$$
$$\log 3^x = \log 5$$
$$x \log 3 = \log 5 \qquad\qquad \log r^p = p \log r$$
$$x = \frac{\log 5}{\log 3}$$
$$x = 1.46 \quad \text{to the nearest hundredth}$$

✔ **Check**

Because $3^{1.46} \approx 4.972754647$, we say that, to the nearest hundredth, the solution set for $3^x = 5$ is $\{1.46\}$.

EXAMPLE 2   Solve $e^{x+1} = 5$ to the nearest hundredth.

Solution

The base e is used in the exponential expression, so let's use natural logarithms to help solve this equation.

$$e^{x+1} = 5$$
$$\ln e^{x+1} = \ln 5 \qquad\qquad \text{Property 14.8}$$
$$(x + 1)\ln e = \ln 5 \qquad\qquad \ln r^p = p \ln r$$
$$(x + 1)(1) = \ln 5 \qquad\qquad \ln e = 1$$
$$x = \ln 5 - 1$$
$$x \approx 0.609437912$$
$$x = 0.61 \quad \text{to the nearest hundredth}$$

The solution set is $\{0.61\}$. Check it!

Logarithmic Equations

In Example 14 of Section 14.4 we solved the logarithmic equation

$$\log_{10} x + \log_{10}(x + 9) = 1$$

by simplifying the left side of the equation to $\log_{10}[x(x + 9)]$ and then changing the equation to exponential form to complete the solution. Now, using Property 14.8, we can solve such a logarithmic equation another way and also expand our equation-solving capabilities. Let's consider some examples.

EXAMPLE 3   Solve $\log x + \log(x - 15) = 2$.

Solution

Because $\log 100 = 2$, the given equation becomes

$$\log x + \log(x - 15) = \log 100$$

Now simplify the left side, apply Property 14.8, and proceed as follows:

$$\log(x)(x - 15) = \log 100$$
$$x(x - 15) = 100$$
$$x^2 - 15x - 100 = 0$$
$$(x - 20)(x + 5) = 0$$
$$x - 20 = 0 \quad \text{or} \quad x + 5 = 0$$
$$x = 20 \quad \text{or} \quad x = -5$$

The domain of a logarithmic function must contain only positive numbers, so x and $x - 15$ must be positive in this problem. Therefore, we discard the solution -5; the solution set is $\{20\}$.

Classroom Example
Solve $\ln(x + 17) = \ln(x - 3) + \ln 5$.

EXAMPLE 4   Solve $\ln(x + 2) = \ln(x - 4) + \ln 3$.

Solution

$$\ln(x + 2) = \ln(x - 4) + \ln 3$$
$$\ln(x + 2) = \ln[3(x - 4)]$$
$$x + 2 = 3(x - 4)$$
$$x + 2 = 3x - 12$$
$$14 = 2x$$
$$7 = x$$

The solution set is $\{7\}$.

Using Logarithms to Solve Problems

In Section 14.2 we used the compound interest formula

$$A = P\left(1 + \frac{r}{n}\right)^{nt}$$

to determine the amount of money (A) accumulated at the end of t years if P dollars is invested at r rate of interest compounded n times per year. Now let's use this formula to solve other types of problems that deal with compound interest.

Classroom Example
How long will $6000 take to double itself if it is invested at 3% interest compounded monthly?

EXAMPLE 5

How long will $500 take to double itself if it is invested at 12% interest compounded quarterly?

Solution

To "double itself" means that the $500 will grow into $1000. Thus

$$1000 = 500\left(1 + \frac{0.12}{4}\right)^{4t}$$
$$1000 = 500(1 + 0.03)^{4t}$$
$$1000 = 500(1.03)^{4t}$$

Multiply both sides of $1000 = 500(1.03)^{4t}$ by $\dfrac{1}{500}$ to yield

$$2 = (1.03)^{4t}$$

Therefore,

$$\log 2 = \log(1.03)^{4t}$$ Property 14.8

$$\log 2 = 4t \log 1.03$$ $\log r^p = p \log r$

Solve for t to obtain

$$\log 2 = 4t \log 1.03$$

$$\frac{\log 2}{\log 1.03} = 4t$$

$$\frac{\log 2}{4 \log 1.03} = t$$ Multiply both sides by $\frac{1}{4}$

$$t \approx 5.862443063$$

$$t = 5.9 \quad \text{to the nearest tenth}$$

Therefore, we are claiming that $500 invested at 12% interest compounded quarterly will double itself in approximately 5.9 years.

✔ Check

$500 invested at 12% compounded quarterly for 5.9 years will produce

$$A = \$500\left(1 + \frac{0.12}{4}\right)^{4(5.9)}$$

$$= \$500(1.03)^{23.6}$$

$$= \$1004.45$$

In Section 14.2, we also used the formula $A = Pe^{rt}$ when money was to be compounded continuously. At this time, with the help of natural logarithms, we can extend our use of this formula.

Classroom Example
How long will it take $300 to triple itself if it is invested at 4% interest compounded continuously?

EXAMPLE 6

How long will it take $100 to triple itself if it is invested at 8% interest that is compounded continuously?

Solution

To "triple itself" means that the $100 will grow into $300. Thus using the formula for interest that is compounded continuously, we can proceed as follows:

$$A = Pe^{rt}$$

$$\$300 = \$100e^{(0.08)t}$$

$$3 = e^{0.08t}$$

$$\ln 3 = \ln e^{0.08t}$$ Property 14.8

$$\ln 3 = 0.08t \ln e$$ $\ln r^p = p \ln r$

$$\ln 3 = 0.08t$$ $\ln e = 1$

$$\frac{\ln 3}{0.08} = t$$

$$t \approx 13.73265361$$

$$t = 13.7 \quad \text{to the nearest tenth}$$

Therefore, $100 invested at 8% interest compounded continuously will triple itself in approximately 13.7 years.

✔ Check

$100 invested at 8% compounded continuously for 13.7 years produces

$$A = Pe^{rt}$$
$$= \$100e^{0.08(13.7)}$$
$$= \$100e^{1.096}$$
$$\approx \$299.22$$

Classroom Example
For a certain virus, the equation $Q(t) = Q_0e^{-0.2t}$, where t is the time in hours and Q_0 is the initial number of virus particles, yields the number of virus particles as a function of time. How long will it take 5000 particles to be reduced to 500 particles?

EXAMPLE 7

For a certain virus, the equation $Q(t) = Q_0e^{0.35t}$, where t is the time in hours and Q_0 is the initial number of virus particles, yields the number of virus particles as a function of time. How long will it take 200 particles to increase to 2000 particles?

Solution

Substitute 200 for Q_0 and 2000 for $Q(t)$ in the equation and proceed to solve for t.

$$2000 = 200e^{0.35t}$$
$$10 = e^{0.35t}$$
$$\ln 10 = \ln e^{0.35t} \qquad \text{Take the natural logarithm of each side}$$
$$\ln 10 = 0.35t \ln e \qquad \text{Property 14.7}$$
$$\ln 10 = 0.35t \qquad \text{Property 14.3}$$
$$t = \frac{\ln 10}{0.35}$$
$$t \approx 6.578814551$$
$$t = 6.6 \quad \text{to the nearest tenth}$$

Therefore, it takes approximately 6.6 hours for 200 virus particles to grow to 2000 particles.

Richter Numbers

Seismologists use the Richter scale to measure and report the magnitude of earthquakes. The equation

$$R = \log \frac{I}{I_0} \qquad \text{R is called a Richter number}$$

compares the intensity I of an earthquake to a minimum or reference intensity I_0. The reference intensity is the smallest earth movement that can be recorded on a seismograph. Suppose that the intensity of an earthquake was determined to be 50,000 times the reference intensity. In this case, $I = 50,000I_0$, and the Richter number is calculated as follows:

$$R = \log \frac{50,000I_0}{I_0}$$
$$R = \log 50,000$$
$$R \approx 4.698970004$$

Thus a Richter number of 4.7 would be reported. Let's consider two more examples that involve Richter numbers.

EXAMPLE 8

An earthquake in San Francisco in 1989 was reported to have a Richter number of 6.9. How did its intensity compare to the reference intensity?

Solution

$$6.9 = \log \frac{I}{I_0}$$

$$10^{6.9} = \frac{I}{I_0}$$

$$I = (10^{6.9})(I_0)$$

$$I \approx 7943282 I_0$$

Thus its intensity was a little less than 8 million times the reference intensity.

EXAMPLE 9

An earthquake in Iran in 1990 had a Richter number of 7.7. Compare the intensity level of that earthquake to the one in San Francisco (Example 8).

Solution

From Example 8 we have $I = (10^{6.9})(I_0)$ for the earthquake in San Francisco. Using a Richter number of 7.7, we obtain $I = (10^{7.7})(I_0)$ for the earthquake in Iran. Therefore, by comparison,

$$\frac{(10^{7.7})(I_0)}{(10^{6.9})(I_0)} = 10^{7.7-6.9} = 10^{0.8} \approx 6.3$$

The earthquake in Iran was about 6 times as intense as the one in San Francisco.

Using the Change of Base for Logarithms with Bases Other Than 10 or *e*

Now let's use either common or natural logarithms to evaluate logarithms that have bases other than 10 or *e*. Consider the following example.

EXAMPLE 10   Evaluate $\log_3 41$.

Solution

Let $x = \log_3 41$. Change to exponential form to obtain

$$3^x = 41$$

Now we can apply Property 14.8 and proceed as follows:

$$\log 3^x = \log 41$$

$$x \log 3 = \log 41$$

$$x = \frac{\log 41}{\log 3}$$

$$x \approx \frac{1.6128}{0.4771} = 3.3804 \quad \text{rounded to four decimal places}$$

Using the method of Example 10 to evaluate $\log_a r$ produces the following formula, which we often refer to as the **change-of-base** formula for logarithms.

> **Property 14.9**
>
> If a, b, and r are positive numbers with $a \neq 1$ and $b \neq 1$, then
>
> $$\log_a r = \frac{\log_b r}{\log_b a}$$

By using Property 14.9, we can easily determine a relationship between logarithms of different bases. For example, suppose that in Property 14.9 we let $a = 10$ and $b = e$. Then

$$\log_a r = \frac{\log_b r}{\log_b a}$$

becomes

$$\log_{10} r = \frac{\log_e r}{\log_e 10}$$

which can be written as

$$\log_e r = (\log_e 10)(\log_{10} r)$$

Because $\log_e 10 = 2.3026$, rounded to four decimal places, we have

$$\log_e r = (2.3026)(\log_{10} r)$$

Thus the natural logarithm of any positive number is approximately equal to the common logarithm of the number times 2.3026.

Now we can use a graphing utility to graph logarithmic functions such as $f(x) = \log_2 x$. Using the change-of-base formula, we can express this function as $f(x) = \dfrac{\log x}{\log 2}$ or as $f(x) = \dfrac{\ln x}{\ln 2}$. The graph of $f(x) = \log_2 x$ is shown in Figure 14.30.

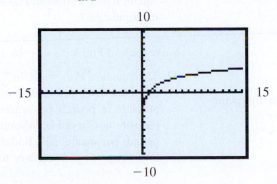

Figure 14.30

Concept Quiz 14.6

For Problems 1–8, answer true or false.

1. The equation $5^{2x+1} = 70$ can be solved by taking the logarithm of both sides of the equation.
2. All solutions of the equation $\log_b x + \log_b(x + 2) = 3$ must be positive numbers.
3. The Richter number compares the intensity of an earthquake to the logarithm of 100,000.
4. If the difference in Richter numbers for two earthquakes is 2, then it can be said that one earthquake was 2 times more intense than the other earthquake.
5. The expression $\log_3 7$ is equivalent to $\dfrac{\ln 7}{\ln 3}$.
6. The expression $\log_3 7$ is equivalent to $\dfrac{\log 7}{\log 3}$.
7. The solution set for $\ln(3x - 4) - \ln(x + 1) = \ln 2$ is $\{8\}$.
8. The value of $\log_5 47$ can be determined by solving the equation $5^x = 47$ for x.

Problem Set 14.6

For Problems 1–14, solve each exponential equation and express solutions to the nearest hundredth. **(Objective 1)**

1. $3^x = 32$ **2.** $2^x = 40$

3. $4^x = 21$ **4.** $5^x = 73$

5. $3^{x-2} = 11$ **6.** $2^{x+1} = 7$

7. $5^{3x+1} = 9$ **8.** $7^{2x-1} = 35$

9. $e^x = 5.4$ **10.** $e^x = 45$

11. $e^{x-2} = 13.1$ **12.** $e^{x-1} = 8.2$

13. $3e^x = 35.1$ **14.** $4e^x - 2 = 26$

For Problems 15–22, solve each logarithmic equation. **(Objective 2)**

15. $\log x + \log(x + 21) = 2$

16. $\log x + \log(x + 3) = 1$

17. $\log(3x - 1) = 1 + \log(5x - 2)$

18. $\log(2x - 1) - \log(x - 3) = 1$

19. $\log(x + 2) - \log(2x + 1) = \log x$

20. $\log(x + 1) - \log(x + 2) = \log\dfrac{1}{x}$

21. $\ln(2t + 5) = \ln 3 + \ln(t - 1)$

22. $\ln(3t - 4) - \ln(t + 1) = \ln 2$

For Problems 23–32, approximate each of the following logarithms to three decimal places. **(Objective 5)**

23. $\log_2 23$ **24.** $\log_3 32$

25. $\log_6 0.214$ **26.** $\log_5 1.4$

27. $\log_7 421$ **28.** $\log_8 514$

29. $\log_9 0.0017$ **30.** $\log_4 0.00013$

31. $\log_3 720$ **32.** $\log_2 896$

For Problems 33–41, solve each problem and express answers to the nearest tenth. **(Objective 3)**

33. How long will it take $7500 to be worth $10,000 if it is invested at 4% interest compounded quarterly?

34. How long will it take $1000 to double itself if it is invested at 5% interest compounded semiannually?

35. How long will it take $2000 to double itself if it is invested at 6% interest compounded continuously?

36. How long will it take $5000 to triple itself if it is invested at 4% interest compounded continuously?

37. For a certain strain of bacteria, the number present after t hours is given by the equation $Q = Q_0 e^{0.34t}$, where Q_0 represents the initial number of bacteria. How long will it take 400 bacteria to increase to 4000 bacteria?

38. A piece of machinery valued at $30,000 depreciates at a rate of 10% yearly. How long will it take until it has a value of $15,000?

39. The number of grams of a certain radioactive substance present after t hours is given by the equation $Q = Q_0 e^{-0.45t}$, where Q_0 represents the initial number of grams. How long will it take 2500 grams to be reduced to 1250 grams?

40. For a certain culture the equation $Q(t) = Q_0 e^{0.4t}$, where Q_0 is an initial number of bacteria and t is time measured in hours, yields the number of bacteria as a function of time. How long will it take 500 bacteria to increase to 2000?

41. Suppose that the equation $P(t) = P_0 e^{0.02t}$, where P_0 represents an initial population, and t is the time in years, is used to predict population growth. How long does this equation predict it would take a city of 50,000 to double its population?

Solve each of Problems 42–46. **(Objective 4)**

42. The equation $P(a) = 14.7e^{-0.21a}$, where a is the altitude above sea level measured in miles, yields the atmospheric pressure in pounds per square inch. If the atmospheric pressure at Cheyenne, Wyoming is approximately 11.53 pounds per square inch, find that city's altitude above sea level. Express your answer to the nearest hundred feet.

43. An earthquake in Los Angeles in 1971 had an intensity of approximately five million times the reference intensity. What was the Richter number associated with that earthquake?

44. An earthquake in San Francisco in 1906 was reported to have a Richter number of 8.3. How did its intensity compare to the reference intensity?

45. Calculate how many times more intense an earthquake with a Richter number of 7.3 is than an earthquake with a Richter number of 6.4.

46. Calculate how many times more intense an earthquake with a Richter number of 8.9 is than an earthquake with a Richter number of 6.2.

Thoughts Into Words

47. Explain how to determine $\log_7 46$ without using Property 14.9.

48. Explain the concept of a Richter number.

49. Explain how you would solve the equation $2^x = 64$ and also how you would solve the equation $2^x = 53$.

50. How do logarithms with a base of 9 compare to logarithms with a base of 3? Explain how you reached this conclusion.

Graphing Calculator Activities

51. Graph $f(x) = x$, $f(x) = 2^x$, and $f(x) = \log_2 x$ on the same set of axes.

52. Graph $f(x) = x$, $f(x) = (0.5)^x$, and $f(x) = \log_{0.5} x$ on the same set of axes.

53. Graph $f(x) = \log_2 x$. Now predict the graphs for $f(x) = \log_3 x$, $f(x) = \log_4 x$, and $f(x) = \log_8 x$. Graph these

three functions on the same set of axes with $f(x) = \log_2 x$.

54. Graph $f(x) = \log_5 x$. Now predict the graphs for $f(x) = 2\log_5 x$, $f(x) = -4\log_5 x$, and $f(x) = \log_5(x + 4)$. Graph these three functions on the same set of axes with $f(x) = \log_5 x$.

Answers to the Concept Quiz

1. True   **2.** True   **3.** False   **4.** False   **5.** True   **6.** True   **7.** False   **8.** True

OBJECTIVE	SUMMARY	EXAMPLE
Graph exponential functions. (Section 14.1/Objective 2)	A function defined by an equation of the form $f(x) = b^x$, where $b > 0$ and $b \neq 1$, is called an exponential function with base b.	Graph $f(x) = 4^x$. **Solution** Set up a table of values. <table><tr><td>x</td><td>$f(x)$</td></tr><tr><td>-2</td><td>$\frac{1}{16}$</td></tr><tr><td>-1</td><td>$\frac{1}{4}$</td></tr><tr><td>0</td><td>1</td></tr><tr><td>1</td><td>4</td></tr><tr><td>2</td><td>16</td></tr></table>
Solve exponential equations. (Section 14.1/Objective 1)	Some exponential equations can be solved by applying Property 14.2: If $b > 0$, $b \neq 1$, and m and n are real numbers, then $b^n = b^m$ if and only if $n = m$. This property can be applied only for equations that can be written in the form in which the bases are equal.	Solve $4^{3x} = 32$. **Solution** Rewrite each side of the equation as a power of 2. Then apply Property 14.2 to solve the equation. $$4^{3x} = 32$$ $$(2^2)^{3x} = 2^5$$ $$2^{6x} = 2^5$$ $$6x = 5$$ $$x = \frac{5}{6}$$ The solution set is $\left\{\frac{5}{6}\right\}$

OBJECTIVE	SUMMARY	EXAMPLE
Determine if a function is a one-to-one function. (Section 14.3/Objective 1)	The horizontal line test is used to determine if a graph is the graph of a one-to-one function. If the graph is the graph of a one-to-one function, then a horizontal line will intersect the graph in only one point.	Identify the graph as the graph of a one-to-one function or the graph of a function that is not one to one. **Solution** The graph is the graph of a one-to-one function, because a horizontal line intersects the graph in only one point.
Verify that two functions are inverse functions. (Section 14.3/Objective 2)	For two functions, f and g, to be inverse functions, the domain of f must equal the range of g, and the range of f must equal the domain of g. Also, g must reverse the correspondences given by f, and f must reverse the correspondences given by g.	Verify that $f(x) = \sqrt{x + 3}$ for $x \geq -3$ and $g(x) = x^2 - 3$ for $x \geq 0$. **Solution** f $D: x \geq -3$ $R: f(x) \geq 0$ g $D: x \geq 0$ $R: f(x) \geq -3$ $(f \circ g)(x) = \sqrt{(x^2 - 3) + 3} = x$ $(g \circ f)(x) = \left(\sqrt{x + 3}\right)^2 - 3 = x$ Hence, the functions are inverses.
Find the inverse of a function. (Section 14.3/Objective 3)	A technique for finding the inverse of a function is as follows: 1. Let $y = f(x)$. 2. Interchange x and y. 3. Solve the equation for y in terms of x. 4. $f^{-1}(x)$ is determined by the final equation. Graphically, two functions that are inverses of each other have graphs that are mirror images with reference to the line $y = x$. We can show that two functions f and f^{-1} are inverses of each other by verifying that 1. $(f^{-1} \circ f)(x) = x$ for all x in the domain of f 2. $(f \circ f^{-1})(x) = x$ for all x in the domain of f^{-1}	Find the inverse of the function $$f(x) = \frac{2}{5}x - 7$$ **Solution** 1. Let $y = \frac{2}{5}x - 7$. 2. Interchange x and y: $$x = \frac{2}{5}y - 7$$ 3. Solve for y: $x = \frac{2}{5}y - 7$ $5x = 2y - 35$ (Multiplied both sides by 5) $5x + 35 = 2y$ $y = \frac{5x + 35}{2}$ 4. $f^{-1}(x) = \frac{5x + 35}{2}$

(continued)

OBJECTIVE	SUMMARY	EXAMPLE
Find intervals where a function is increasing or decreasing. (Section 14.3/Objective 4)	Let f be function with the interval I a subset of the domain of f. Let x_1 and x_2 be in I. 1. f is increasing on I if $f(x_1) < f(x_2)$ whenever $x_1 < x_2$. 2. f is decreasing on I if $f(x_1) > f(x_2)$ whenever $x_1 < x_2$. 3. f is constant on I if $f(x_1) = f(x_2)$ for every x_1 and x_2 in I.	From the graph, determine the intervals where f is increasing or decreasing. **Solution** f is decreasing on $(-\infty, 3]$. f is increasing on $[3, \infty)$.
Switch equations between exponential and logarithmic form. (Section 14.4/Objective 1)	If r is any positive real number, then the unique exponent t such that $b^t = r$ is called the logarithm of r with base b and is denoted by $\log_b r$. The solutions to many problems require the ability to convert between the exponential and logarithmic form of equations. $\log_b r = t$ is equivalent to $b^t = r$	1. Write $6^2 = 36$ in logarithmic form. 2. Write $\log_3 81 = 4$ in exponential form. **Solution** 1. $6^2 = 36$ is equivalent to $\log_6 36 = 2$. 2. $\log_3 81 = 4$ is equivalent to $3^4 = 81$.
Evaluate a logarithmic expression. (Section 14.4/Objective 2)	Some logarithmic expressions can be evaluated by switching to exponential form and then applying Property 14.2.	Evaluate $\log_2\left(\dfrac{1}{8}\right)$. **Solution** Let $\log_2\left(\dfrac{1}{8}\right) = x$. Switching to exponential form gives $2^x = \dfrac{1}{8}$. $2^x = \dfrac{1}{8}$ $2^x = 2^{-3}$ $x = -3$ Therefore, $\log_2\left(\dfrac{1}{8}\right) = -3$.
Evaluate a common logarithm. (Section 14.5/Objective 2)	Base-10 logarithms are called common logarithms. When working with base-10 logarithms, it is customary to omit writing the numeral 10 to designate the base; hence $\log_{10} x$ is written as $\log x$. A calculator with a logarithm function key, typically labeled log, is used to evaluate a common logarithm.	Use a calculator to find $\log 245$. Express the answer to four decimal places. **Solution** Follow the instructions for your calculator: $\log 245 = 2.3892$

OBJECTIVE	SUMMARY	EXAMPLE
Evaluate a natural logarithm. **(Section 14.5/Objective 2)**	In many practical applications of logarithms, the number e (remember that $e \approx 2.71828$) is used as a base. Logarithms with a base of e are called natural logarithms and are written as $\ln x$.	Use a calculator to find $\ln 486$. Express the answer to four decimal places. **Solution** Follow the instructions for your calculator: $\ln 486 = 6.1862$
Evaluate logarithms using the change-of-base formula. **(Section 14.6/Objective 5)**	To evaluate logarithms other than base-10 or base-e, a change-of-base formula, $$\log_a r = \frac{\log_b r}{\log_b a}$$ is used. When applying the formula, you can choose base-10 or base-e for b.	Find $\log_8 724$. Round the answer to the nearest hundredth. **Solution** $$\log_8 724 = \frac{\log 724}{\log 8}$$ $$\approx 3.17$$
Use the properties of logarithms. **(Section 14.4/Objective 4)**	The following properties of logarithms are derived from the definition of a logarithm and the properties of exponents. For positive real numbers b, r, and s, where $b \neq 1$, we have: 1. $\log_b b = 1$ 2. $\log_b 1 = 0$ 3. $b^{\log_b r} = r$ 4. $\log_b rs = \log_b r + \log_b s$ 5. $\log_b\left(\dfrac{r}{s}\right) = \log_b r - \log_b s$ 6. $\log_b r^p = p\log_b r$ where p is any real number	Express $\log_b \dfrac{\sqrt{x}}{y^2 z}$ as indicated sums or differences of simpler logarithmic quantities. **Solution** $$\log_b \frac{\sqrt{x}}{y^2 z} = \log_b \sqrt{x} - \log_b(y^2 z)$$ $$= \frac{1}{2}\log_b x - \left[\log_b y^2 + \log_b z\right]$$ $$= \frac{1}{2}\log_b x - \left[2\log_b y + \log_b z\right]$$ $$= \frac{1}{2}\log_b x - 2\log_b y - \log_b z$$
Solve logarithmic equations by switching to exponential form. **(Section 14.4/Objectives 3 and 5)**	This technique is used when a single logarithmic expression is equal to a constant, such as $\log_4(3x + 5) = 2$. You may have to apply properties of logarithms to rewrite a sum or difference of logarithms as a single logarithm.	Solve $\log_3 x + \log_3(x + 8) = 2$. **Solution** Rewrite the left-hand side of the equation as a single logarithm. $\log_3 x(x + 8) = 2$ Switch to exponential form and solve the equation: $$x(x + 8) = 3^2$$ $$x^2 + 8x = 9$$ $$x^2 + 8x - 9 = 0$$ $$(x + 9)(x - 1) = 0$$ $$x = -9 \quad \text{or} \quad x = 1$$ *Always check your answer.* Because -9 would make the expression $\log_3 x$ undefined, the solution set is $\{1\}$.

(continued)

OBJECTIVE	SUMMARY	EXAMPLE
Graph logarithmic functions. **(Section 14.5/Objective 1)**	A function defined by an equation of the form $f(x) = \log_b x$, where $b > 0$ and $b \neq 1$, is called a logarithmic function. The equation $y = \log_b x$ is equivalent to $x = b^y$. The two functions $f(x) = b^x$ and $g(x) = \log_b x$ are inverses of each other.	Graph $f(x) = \log_3 x$. **Solution** Change $y = \log_3 x$ to $3^y = x$ and determine a table of values. <table><tr><th>x</th><th>y</th></tr><tr><td>$\frac{1}{9}$</td><td>-2</td></tr><tr><td>$\frac{1}{3}$</td><td>-1</td></tr><tr><td>1</td><td>0</td></tr><tr><td>3</td><td>1</td></tr><tr><td>9</td><td>2</td></tr></table>
Solve logarithmic equations of the form $\log_b x = \log_b y$ **(Section 14.6/Objective 2)**	The following property of equality is used for solving some logarithmic equations: If $x > 0$, and $y > 0$, and $b \neq 1$, then $x = y$ if and only if $\log_b x = \log_b y$. One application of this property is for solving equations where a single logarithm equals another single logarithm.	Solve $\ln(9x + 1) - \ln(x + 4) = \ln 4$. **Solution** Write the left-hand side of the equation as a single logarithm. $$\ln \frac{9x + 1}{x + 4} = \ln 4$$ Now apply the property of equality: if $\log_b x = \log_b y$, then $x = y$: $$\frac{9x + 1}{x + 4} = 4$$ $$9x + 1 = 4(x + 4)$$ $$9x + 1 = 4x + 16$$ $$5x = 15$$ $$x = 3$$ The answer does check so the solution set is {3}.

OBJECTIVE	SUMMARY	EXAMPLE
Solve exponential equations by taking the logarithm of each side. (Section 14.6/Objective 1)	Some exponential equations can be solved by applying the property of equality in the form "If $x = y$, then $\log_b x = \log_b y$, where $x > 0$, and $y > 0$, and $b \neq 1$." If the base of the exponential equation is e then use natural logarithms, and if the base is 10 then use common logarithms.	Solve $6^{3x-2} = 45$. Express the solution to the nearest hundredth. **Solution** Take the logarithm of both sides of the equation: $$\log 6^{3x-2} = \log 45$$ $$(3x - 2)\log 6 = \log 45$$ $$3x(\log 6) - 2\log 6 = \log 45$$ $$3x(\log 6) = \log 45 + 2\log 6$$ $$x = \frac{\log 45 + 2\log 6}{3\log 6}$$ $$x \approx 1.37$$ The solution set is {1.37}.
Solve exponential growth problems. (Section 14.2/Objective 1)	In general, the equation $P = P_0(1.06)^t$ yields the predicted price P of an item in t years at the annual inflation rate of 6%, when that item presently costs P_0.	Assuming that the rate of inflation is 3% per year, find the price of a $20.00 haircut in 5 years. **Solution** Use the formula $P = P_0(1.03)^t$ and substitute $20.00 for P_0 and 5 for t. $P = 20.00(1.03)^5 \approx 23.19$ The haircut would cost about $23.19.
Solve compound interest problems. (Section 14.2/Objective 1)	A general formula for any principal, P, invested for any number of years (t) at a rate of r percent compounded n times per year is $$A = P\left(1 + \frac{r}{n}\right)^{nt}$$ when A represents the total amount of money accumulated at the end of t years.	Find the total amount of money accumulated at the end of 8 years when $4000 is invested at 6% compounded quarterly. **Solution** $$A = 4000\left(1 + \frac{0.06}{4}\right)^{4(8)}$$ $$A = 4000(1.015)^{32}$$ $$A = 6441.30$$
Solve continuous compounding problems. (Section 14.2/Objective 3)	The value of $\left(1 + \frac{1}{n}\right)^n$, as n gets infinitely large, approaches the number e, where $e = 2.71828$ to five decimal places. The formula $A = Pe^{rt}$ yields the accumulated value, A, of a sum of money, P, that has been invested for t years at a rate of r percent compounded continuously.	If $8300 is invested at 6.5% interest compounded continuously, how much will accumulate in 10 years? **Solution** $$A = 8300e^{0.065(10)} = 8300e^{0.65}$$ $$= 15,898.99$$ The accumulated amount will be $15,898.99.

(continued)

OBJECTIVE	SUMMARY	EXAMPLE
Solve exponential decay half-life problems. (Section 14.2/Objective 2)	For radioactive substances, the rate of decay is exponential and can be described in terms of the half-life of a substance, which is the amount of time that it takes for one-half of an initial amount of the substance to disappear as the result of decay. Suppose there is an initial amount, Q_0, of a radioactive substance with a half-life of h. The amount of substance remaining, Q, after a time period of t is given by the formula $Q = Q_0\left(\dfrac{1}{2}\right)^{\frac{t}{h}}$ The units of measure for t and h must be the same.	Molybdenum-99 has a half-life of 66 hours. If there are 200 milligrams of molybdenum-99 intially, how many milligrams remain after 16 hours? **Solution** $Q = 200\left(\dfrac{1}{2}\right)^{\frac{16}{66}}$ $Q = 169.1$ to the nearest tenth of a milligram Thus approximately 169.1 milligrams remain after 16 hours.
Solve growth problems involving the number e. (Section 14.2/Objective 3)	The law of exponential growth $Q(t) = Q_0 e^{kt}$ is used as a mathematical model for growth-and-decay problems. In this equation, $Q(t)$ represents the quantity of a given substance at any time t; Q_0 is the initial amount of the substance (when $t = 0$); and k is a constant that depends on the particular application. If $k < 0$, then $Q(t)$ decreases as t increases, and we refer to the model as the law of decay.	The number of bacteria present in a certain culture after t hours is given by the equation $Q(t) = Q_0 e^{0.42t}$, where Q_0 represents the initial number of bacteria. How long will it take 100 bacteria to increase to 500 bacteria? **Solution** Substitute 100 for Q_0 and 500 for $Q(t)$ into $Q(t) = Q_0 e^{0.42t}$. $500 = 100 e^{0.42t}$ $5 = e^{0.42t}$ $\ln 5 = \ln e^{0.42t}$ $\ln 5 = 0.42t$ $t = \dfrac{\ln 5}{0.42} = 3.83$ to the nearest hundredth It will take approximately 3.83 hours for 100 bacteria to increase to 500 bacteria.
Solve problems involving Richter numbers. (Section 14.6/Objective 4)	Seismologists use the Richter scale to measure and report the magnitude of earthquakes. The equation $R = \log\dfrac{I}{I_0}$ compares the intensity I of an earthquake to a minimum or reference intensity I_0.	An earthquake in the Sandwich Islands in 2006 was about 5,011,872 times as intense as the reference intensity. Find the Richter number for that earthquake. **Solution** $R = \log\dfrac{I}{I_0}$ $= \log\dfrac{5{,}011{,}872 I_0}{I_0} = 6.7$ The Richter number for the earthquake is 6.7.

Chapter 14   Review Problem Set

For Problems 1 and 2, graph each of the functions.

1. $f(x) = 3^x - 4$ **2.** $f(x) = -3^x$

For Problems 3–6, solve each equation without using your calculator.

3. $2^x = \dfrac{1}{16}$ **4.** $16^x = \dfrac{1}{8}$

5. $3^x - 4 = 23$ **6.** $4^{3x+1} = 32$

For Problems 7–10, write each of the following in logarithmic form.

7. $8^2 = 64$ **8.** $a^b = c$

9. $3^{-2} = \dfrac{1}{9}$ **10.** $10^{-1} = \dfrac{1}{10}$

For Problems 11–14, write each of the following in exponential form.

11. $\log_5 125 = 3$ **12.** $\log_x y = z$

13. $\log_2\left(\dfrac{1}{8}\right) = -3$ **14.** $\log_{10} 10{,}000 = 4$

For Problems 15–22, evaluate each expression without using a calculator.

15. $\log_2 128$ **16.** $\log_4 64$

17. $\log 10{,}000$ **18.** $\log 0.001$

19. $\ln e^2$ **20.** $5^{\log_5 13}$

21. $\log(\log_3 3)$ **22.** $\log_2\left(\dfrac{1}{4}\right)$

For Problems 23–26, use your calculator to find each logarithm. Express answers to four decimal places.

23. $\log 73.14$ **24.** $\log 0.00235$

25. $\ln 0.014$ **26.** $\ln 114.2$

For Problems 27–30, approximate a value for each logarithm to three decimal places.

27. $\log_3 97$ **28.** $\log_8 200$

29. $\log_5 0.0065$ **30.** $\log_2 0.0036$

For Problems 31 and 32, graph each of the functions.

31. $f(x) = -1 + \log_2 x$ **32.** $f(x) = \log_2(x + 1)$

33. Express $\log_b \dfrac{x\sqrt{y}}{z^3}$ as the sum or difference of simpler logarithmic quantities. Assume that all variables represent positive real numbers.

34. Express $\log_b \dfrac{b}{cd}$ as the sum or difference of simpler logarithmic quantities. Assume that all variables represent positive real numbers.

For Problems 35–38, use your calculator to find x when given $\log x$ or $\ln x$. Express the answer to five significant figures.

35. $\ln x = 0.1724$ **36.** $\log x = 3.4215$

37. $\log x = -1.8765$ **38.** $\ln x = -2.5614$

For Problems 39–46, solve each equation without using your calculator.

39. $\log_8 x = \dfrac{2}{3}$ **40.** $\log_x 3 = \dfrac{1}{2}$

41. $\log_5 2 + \log_5(3x + 1) = 1$

42. $\log_2 x + \log_2(x - 4) = 5$

43. $\log_2(x + 5) - \log_2 x = 1$

44. $\log_3 x - \log_3(x - 5) = 2$

45. $\ln x + \ln(x + 2) = \ln 35$

46. $\log(2x) - \log(x + 3) = \log\left(\dfrac{5}{6}\right)$

For Problems 47–50, use your calculator to help solve each equation. Express solutions to the nearest hundredth.

47. $3^x = 42$ **48.** $2e^x = 14$

49. $2^{x+1} = 79$ **50.** $e^{x-2} = 37$

For Problems 51–54, (a) find f^{-1}, and (b) verify that $(f \circ f^{-1})(x) = x$ and $(f^{-1} \circ f)(x) = x$.

51. $f(x) = 4x + 5$ **52.** $f(x) = -3x - 7$

53. $f(x) = \dfrac{5}{6}x - \dfrac{1}{3}$

54. $f(x) = -2 - x^2$ for $x \geq 0$

For Problems 55 and 56, find the intervals on which the function is increasing and the intervals on which it is decreasing.

55. $f(x) = -2x^2 + 16x - 35$

56. $f(x) = 2\sqrt{x - 3}$

57. Assuming the rate of inflation is 3% per year, the equation $P = P_0(1.03)^t$ yields the predicted price P, in t years, of an item that presently costs P_0. Find the predicted price of each of the following items for the indicated years ahead.
    (a) $829 cruise in 4 years
    (b) $280 suit in 3 years
    (c) $85 bicycle in 5 years

58. Suppose it is estimated that the value of a car depreciates 25% per year for the first 5 years. The equation $A = P_0(0.75)^t$ yields the value (A) of a car after t years if the original price is P_0. Find the value (to the nearest dollar) of each of the following cars after the indicated time.
    (a) $25,000 car after 4 years
    (b) $35,000 car after 3 years
    (c) $20,000 car after 1 year

59. Suppose that $8000 is invested at 6% interest compounded monthly. How much money has accumulated at the end of 15 years?

60. If $25,000 is invested at 4% interest compounded quarterly, how much money has accumulated at the end of 12 years?

61. How long will it take $1000 to double itself if it is invested at 8% interest compounded semiannually?

62. How long will it take $1000 to be worth $3500 if it is invested at 5.5% interest compounded quarterly?

63. If $3500 is invested at 8% interest compounded continuously, how much money will accumulate in 6 years?

64. How long will it take $5000 to be worth $20,000 if it is invested at 7% interest compounded continuously?

65. Suppose that a certain radioactive substance has a half-life of 40 days. If there are presently 750 grams of the substance, how much (to the nearest gram) will remain after 100 days?

66. Suppose that a certain radioactive substance has a half-life of 20 hours. If there are presently 500 grams, how long will it take for only 100 grams to remain?

67. Suppose that the present population of a city is 50,000. Use the equation $P = P_0 e^{0.02t}$, where P_0 represents an initial population, to estimate future populations. Estimate the population of that city in 10 years, 15 years, and 20 years.

68. The number of bacteria present in a certain culture after t hours is given by the equation $Q = Q_0 e^{0.29t}$, where Q_0 represents the initial number of bacteria. How long will it take 500 bacteria to increase to 2000 bacteria?

69. An earthquake in Mexico City in 1985 had an intensity level about 125,000,000 times the reference intensity. Find the Richter number for that earthquake.

70. Calculate how many times more intense an earthquake with a Richter number of 7.8 is than an earthquake with a Richter number of 7.1.

For Problems 1–4, evaluate each expression.

1. $\log_3 \sqrt{3}$

2. $\log_2(\log_2 4)$

3. $-2 + \ln e^3$

4. $\log_2(0.5)$

For Problems 5–10, solve each equation.

5. $4^x = \dfrac{1}{64}$

6. $9^x = \dfrac{1}{27}$

7. $2^{3x-1} = 128$

8. $\log_9 x = \dfrac{5}{2}$

9. $\log x + \log(x + 48) = 2$

10. $\ln x = \ln 2 + \ln(3x - 1)$

For Problems 11–14, given that $\log_3 4 = 1.2619$ and $\log_3 5 = 1.4650$, evaluate each of the following.

11. $\log_3 100$

12. $\log_3 1.25$

13. $\log_3 \sqrt{5}$

14. $\log_3(16 \cdot 25)$

15. Find the inverse of the function
$f(x) = 3x - 6$.

16. Find the inverse of the function
$f(x) = \dfrac{2}{3}x - \dfrac{3}{5}$.

17. Determine $\log_5 632$ to four decimal places.

18. Express $3\log_b x + 2\log_b y - \log_b z$ as a single logarithm with a coefficient of 1.

19. If $3500 is invested at 7.5% interest compounded quarterly, how much money has accumulated at the end of 8 years?

20. How long will it take $5000 to be worth $12,500 if it is invested at 7% compounded annually? Express your answer to the nearest tenth of a year.

21. The number of bacteria present in a certain culture after t hours is given by $Q(t) = Q_0 e^{0.23t}$, where Q_0 represents the initial number of bacteria. How long will it take 400 bacteria to increase to 2400 bacteria? Express your answer to the nearest tenth of an hour.

22. Suppose that a certain radioactive substance has a half-life of 50 years. If there are presently 7500 grams of the substance, how much will remain after 32 years? Express your answer to the nearest gram.

For Problems 23–25, graph each of the functions.

23. $f(x) = e^x - 2$

24. $f(x) = -3^{-x}$

25. $f(x) = \log_2(x - 2)$

For Problems 1–5, evaluate each algebraic expression for the given values of the variables.

1. $-5(x - 1) - 3(2x + 4) + 3(3x - 1)$ for $x = -2$

2. $\dfrac{14a^3b^2}{7a^2b}$ for $a = -1$ and $b = 4$

3. $\dfrac{2}{n} - \dfrac{3}{2n} + \dfrac{5}{3n}$ for $n = 4$

4. $4\sqrt{2x - y} + 5\sqrt{3x + y}$ for $x = 16$ and $y = 16$

5. $\dfrac{3}{x - 2} - \dfrac{5}{x + 3}$ for $x = 3$

For Problems 6–15, perform the indicated operations and express answers in simplified form.

6. $(-5\sqrt{6})(3\sqrt{12})$

7. $(2\sqrt{x} - 3)(\sqrt{x} + 4)$

8. $(3\sqrt{2} - \sqrt{6})(\sqrt{2} + 4\sqrt{6})$

9. $(2x - 1)(x^2 + 6x - 4)$

10. $\dfrac{x^2 - x}{x + 5} \cdot \dfrac{x^2 + 5x + 4}{x^4 - x^2}$

11. $\dfrac{16x^2y}{24xy^3} \div \dfrac{9xy}{8x^2y^2}$

12. $\dfrac{x + 3}{10} + \dfrac{2x + 1}{15} - \dfrac{x - 2}{18}$

13. $\dfrac{7}{12ab} - \dfrac{11}{15a^2}$

14. $\dfrac{8}{x^2 - 4x} + \dfrac{2}{x}$

15. $(8x^3 - 6x^2 - 15x + 4) \div (4x - 1)$

For Problems 16–19, simplify each of the complex fractions.

16. $\dfrac{\dfrac{5}{x^2} - \dfrac{3}{x}}{\dfrac{1}{y} + \dfrac{2}{y^2}}$

17. $\dfrac{\dfrac{2}{x} - 3}{\dfrac{3}{y} + 4}$

18. $\dfrac{2 - \dfrac{1}{n - 2}}{3 + \dfrac{4}{n + 3}}$

19. $\dfrac{3a}{2 - \dfrac{1}{a}} - 1$

For Problems 20–25, factor each of the algebraic expressions completely.

20. $20x^2 + 7x - 6$

21. $16x^3 + 54$

22. $4x^4 - 25x^2 + 36$

23. $12x^3 - 52x^2 - 40x$

24. $xy - 6x + 3y - 18$

25. $10 + 9x - 9x^2$

For Problems 26–35, evaluate each of the numerical expressions.

26. $\left(\dfrac{2}{3}\right)^{-4}$

27. $\dfrac{3}{\left(\dfrac{4}{3}\right)^{-1}}$

28. $\sqrt[3]{-\dfrac{27}{64}}$

29. $-\sqrt{0.09}$

30. $(27)^{-4/3}$

31. $4^0 + 4^{-1} + 4^{-2}$

32. $\left(\dfrac{3^{-1}}{2^{-3}}\right)^{-2}$

33. $(2^{-3} - 3^{-2})^{-1}$

34. $\log_2 64$

35. $\log_3\left(\dfrac{1}{9}\right)$

For Problems 36–38, find the indicated products and quotients; express final answers with positive integral exponents only.

36. $(-3x^{-1}y^2)(4x^{-2}y^{-3})$

37. $\dfrac{48x^{-4}y^2}{6xy}$

38. $\left(\dfrac{27a^{-4}b^{-3}}{-3a^{-1}b^{-4}}\right)^{-1}$

For Problems 39–46, express each radical expression in simplest radical form.

39. $\sqrt{80}$

40. $-2\sqrt{54}$

41. $\sqrt{\dfrac{75}{81}}$

42. $\dfrac{4\sqrt{6}}{3\sqrt{8}}$

43. $\sqrt[3]{56}$

44. $\dfrac{\sqrt[3]{3}}{\sqrt[3]{4}}$

45. $4\sqrt{52x^3y^2}$

46. $\sqrt{\dfrac{2x}{3y}}$

For Problems 47–49, use the distributive property to help simplify each of the following:

47. $-3\sqrt{24} + 6\sqrt{54} - \sqrt{6}$

48. $\dfrac{\sqrt{8}}{3} - \dfrac{3\sqrt{18}}{4} - \dfrac{5\sqrt{50}}{2}$

49. $8\sqrt[3]{3} - 6\sqrt[3]{24} - 4\sqrt[3]{81}$

For Problems 50 and 51, rationalize the denominator and simplify.

50. $\dfrac{\sqrt{3}}{\sqrt{6} - 2\sqrt{2}}$

51. $\dfrac{3\sqrt{5} - \sqrt{3}}{2\sqrt{3} + \sqrt{7}}$

For Problems 52–54, use scientific notation to help perform the indicated operations.

52. $\dfrac{(0.00016)(300)(0.028)}{0.064}$

53. $\dfrac{0.00072}{0.0000024}$

54. $\sqrt{0.00000009}$

For Problems 55–58, find each of the indicated products or quotients and express answers in standard form.

55. $(5 - 2i)(4 + 6i)$

56. $(-3 - i)(5 - 2i)$

57. $\dfrac{5}{4i}$

58. $\dfrac{-1 + 6i}{7 - 2i}$

59. Find the slope of the line determined by the points $(2, -3)$ and $(-1, 7)$.

60. Find the slope of the line determined by the equation $4x - 7y = 9$.

61. Find the length of the line segment whose endpoints are $(4, 5)$ and $(-2, 1)$.

62. Write the equation of the line that contains the points $(3, -1)$ and $(7, 4)$.

63. Write the equation of the line that is perpendicular to the line $3x - 4y = 6$ and contains the point $(-3, -2)$.

64. Find the center and the length of a radius of the circle $x^2 + 4x + y^2 - 12y + 31 = 0$.

65. Find the coordinates of the vertex of the parabola $y = x^2 + 10x + 21$.

66. Find the length of the major axis of the ellipse $x^2 + 4y^2 = 16$.

For Problems 67–76, graph each of the functions.

67. $f(x) = -2x - 4$

68. $f(x) = -2x^2 - 2$

69. $f(x) = x^2 - 2x - 2$

70. $f(x) = \sqrt{x + 1} + 2$

71. $f(x) = 2x^2 + 8x + 9$

72. $f(x) = -|x - 2| + 1$

73. $f(x) = 2^x + 2$

74. $f(x) = \log_2(x - 2)$

75. $f(x) = -x(x + 1)(x - 2)$

76. $f(x) = \dfrac{-x}{x + 2}$

77. If $f(x) = x - 3$ and $g(x) = 2x^2 - x - 1$, find $(g \circ f)(x)$ and $(f \circ g)(x)$.

78. Find the inverse (f^{-1}) of $f(x) = 3x - 7$.

79. Find the inverse of $f(x) = -\dfrac{1}{2}x + \dfrac{2}{3}$.

80. Find the constant of variation if y varies directly as x, and $y = 2$ when $x = -\dfrac{2}{3}$.

81. If y is inversely proportional to the square of x, and $y = 4$ when $x = 3$, find y when $x = 6$.

82. The volume of a gas at a constant temperature varies inversely as the pressure. What is the volume of a gas under a pressure of 25 pounds if the gas occupies 15 cubic centimeters under a pressure of 20 pounds?

For Problems 83–110, solve each equation.

83. $3(2x - 1) - 2(5x + 1) = 4(3x + 4)$

84. $n + \dfrac{3n - 1}{9} - 4 = \dfrac{3n + 1}{3}$

85. $0.92 + 0.9(x - 0.3) = 2x - 5.95$

86. $|4x - 1| = 11$

87. $3x^2 = 7x$

88. $x^3 - 36x = 0$

89. $30x^2 + 13x - 10 = 0$

90. $8x^3 + 12x^2 - 36x = 0$

91. $x^4 + 8x^2 - 9 = 0$

92. $(n + 4)(n - 6) = 11$

93. $2 - \dfrac{3x}{x - 4} = \dfrac{14}{x + 7}$

94. $\dfrac{2n}{6n^2 + 7n - 3} - \dfrac{n - 3}{3n^2 + 11n - 4} = \dfrac{5}{2n^2 + 11n + 12}$

95. $\sqrt{3y} - y = -6$

96. $\sqrt{x + 19} - \sqrt{x + 28} = -1$

97. $(3x - 1)^2 = 45$

98. $(2x + 5)^2 = -32$

99. $2x^2 - 3x + 4 = 0$

100. $3n^2 - 6n + 2 = 0$

101. $\dfrac{5}{n - 3} - \dfrac{3}{n + 3} = 1$

102. $12x^4 - 19x^2 + 5 = 0$

103. $2x^2 + 5x + 5 = 0$

104. $(3x - 5)(4x + 1) = 0$

105. $|2x + 9| = -4$

106. $16^x = 64$

107. $\log_3 x = 4$

108. $\log_{10} x + \log_{10} 25 = 2$

109. $\ln(3x - 4) - \ln(x + 1) = \ln 2$

110. $27^{4x} = 9^{x+1}$

For Problems 111–120, solve each inequality and express solutions using interval notation.

111. $-5(y - 1) + 3 > 3y - 4 - 4y$

112. $0.06x + 0.08(250 - x) \geq 19$

113. $|5x - 2| > 13$ **114.** $|6x + 2| < 8$

115. $\dfrac{x - 2}{5} - \dfrac{3x - 1}{4} \leq \dfrac{3}{10}$

116. $(x - 2)(x + 4) \leq 0$

117. $(3x - 1)(x - 4) > 0$ **118.** $x(x + 5) < 24$

119. $\dfrac{x - 3}{x - 7} \geq 0$ **120.** $\dfrac{2x}{x + 3} > 4$

15 Systems of Equations: Matrices and Determinants

When mixing different solutions, a chemist could use a system of equations to determine how much of each solution is needed to produce a specific concentration.

© photodisc/Alamy

A 10% salt solution is to be mixed with a 20% salt solution to produce 20 gallons of a 17.5% salt solution. How many gallons of the 10% solution and how many gallons of the 20% solution should be mixed? The two equations $x + y = 20$ and $0.10x + 0.20y = 0.175(20)$ algebraically represent the conditions of the problem; x represents the number of gallons of the 10% solution, and y represents the number of gallons of the 20% solution. The two equations considered together form a system of linear equations, and the problem can be solved by solving the system of equations.

Throughout most of this chapter, we consider systems of linear equations and their applications. We will discuss various techniques for solving systems of linear equations. Then, in the last section, we consider systems that involve nonlinear equations.

Video tutorials for selected section learning objectives are available in a variety of delivery modes.

15.1   Systems of Two Linear Equations: A Brief Review

In Chapter 5 we solved systems of two linear equations in two variables using three different methods: a graphing method, a substitution method, and an elimination-by-addition method. In subsequent chapters you probably have solved some word problems using a system of two linear equations. Even so, a brief review of that material will be helpful before we introduce some additional techniques for solving systems of equations. Let's use this first section to pull together the basic ideas of Chapter 5.

Remember that any equation of the form $Ax + By = C$, where A, B, and C are real numbers (A and B not both zero), is a **linear equation** in the two variables x and y, and its graph is a straight line. Two linear equations in two variables considered together form a **system of linear equations in two variables,** as illustrated by the following examples.

$$\begin{pmatrix} x + y = 6 \\ x - y = 2 \end{pmatrix} \quad \begin{pmatrix} 3x + 2y = 1 \\ 5x - 2y = 23 \end{pmatrix} \quad \begin{pmatrix} 4x - 5y = 21 \\ -3x + y = -7 \end{pmatrix}$$

To solve a system (such as any of these three examples) means to find all of the ordered pairs that simultaneously satisfy both equations in the system. For example, if we graph the two equations $x + y = 6$ and $x - y = 2$ on the same set of axes, as in Figure 15.1, then the ordered pair associated with the point of intersection of the two lines is the **solution of the system.** Thus we say that $\{(4, 2)\}$ is the solution set of the system

$$\begin{pmatrix} x + y = 6 \\ x - y = 2 \end{pmatrix}$$

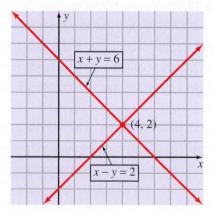

Figure 15.1

To check the solution, we substitute 4 for x and 2 for y in the two equations

$x + y = 6$ becomes $4 + 2 = 6$, a true statement
$x - y = 2$ becomes $4 - 2 = 2$, a true statement

Because the graph of a linear equation in two variables is a straight line, there are three possible situations that can occur when we are solving a system of two linear equations in two variables. These situations are shown in Figure 15.2.

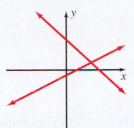
(a) Case 1: one solution

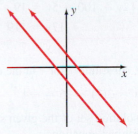
(b) Case 2: no solution

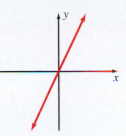
(c) Case 3: infinitely
many solutions

Figure 15.2

Case 1 The graphs of the two equations are two lines intersecting in *one* point. There is exactly one solution, and the system is called a **consistent system.**

Case 2 The graphs of the two equations are parallel lines. There is *no solution,* and the system is called an **inconsistent system.**

Case 3 The graphs of the two equations are the same line, and there arc *infinitely many solutions* of the system. Any pair of real numbers that satisfies one of the equations also satisfies the other equation, and we say that the equations are **dependent.**

Thus as we solve a system of two linear equations in two variables, we can expect one of three outcomes: The system will have *no* solutions, *one* ordered pair as a solution, or *infinitely many* ordered pairs as solutions.

The Substitution Method

Solving specific systems of equations by graphing requires accurate graphs. However, unless the solutions are integers, it is difficult to obtain exact solutions from a graph. Therefore, we will consider some other techniques for solving systems of equations.

The **substitution method,** which works especially well with systems of two equations in two unknowns, can be described as follows:

Step 1 Solve one of the equations for one variable in terms of the other. (If possible, make a choice that will avoid fractions.)

Step 2 Substitute the expression obtained in step 1 into the other equation, producing an equation in one variable.

Step 3 Solve the equation obtained in step 2.

Step 4 Use the solution obtained in step 3, along with the equation obtained in step 1, to determine the solution of the system.

Classroom Example
Solve the system:
$$\begin{pmatrix} x - 4y = -30 \\ 5x + 2y = 4 \end{pmatrix}$$

EXAMPLE 1 Solve the system $\begin{pmatrix} x - 3y = -25 \\ 4x + 5y = 19 \end{pmatrix}$.

Solution

Solve the first equation for x in terms of y to produce

$$x = 3y - 25$$

Substitute $3y - 25$ for x in the second equation and solve for y.

$$4x + 5y = 19$$
$$4(3y - 25) + 5y = 19$$

$$12y - 100 + 5y = 19$$
$$17y = 119$$
$$y = 7$$

Next, substitute 7 for y in the equation $x = 3y - 25$ to obtain

$$x = 3(7) - 25 = -4$$

The solution set of the given system is $\{(-4, 7)\}$. (You should check this solution in both of the original equations.)

EXAMPLE 2

Solve the system $\begin{pmatrix} 5x + 9y = -2 \\ 2x + 4y = -1 \end{pmatrix}$.

Solution

A glance at the system should tell us that solving either equation for either variable will produce a fractional form, so let's just use the first equation and solve for x in terms of y.

$$5x + 9y = -2$$
$$5x = -9y - 2$$
$$x = \frac{-9y - 2}{5}$$

Now we can substitute this value for x into the second equation and solve for y.

$$2x + 4y = -1$$
$$2\left(\frac{-9y - 2}{5}\right) + 4y = -1$$
$$2(-9y - 2) + 20y = -5 \qquad \text{Multiplied both sides by 5}$$
$$-18y - 4 + 20y = -5$$
$$2y - 4 = -5$$
$$2y = -1$$
$$y = -\frac{1}{2}$$

Now we can substitute $-\dfrac{1}{2}$ for y in $x = \dfrac{-9y - 2}{5}$.

$$x = \frac{-9\left(-\dfrac{1}{2}\right) - 2}{5} = \frac{\dfrac{9}{2} - 2}{5} = \frac{1}{2}$$

The solution set is $\left\{\left(\dfrac{1}{2}, -\dfrac{1}{2}\right)\right\}$.

EXAMPLE 3

Solve the system $\begin{pmatrix} 6x - 4y = 18 \\ y = \dfrac{3}{2}x - \dfrac{9}{2} \end{pmatrix}$.

Solution

The second equation is given in appropriate form for us to begin the substitution process.

Substitute $\dfrac{3}{2}x - \dfrac{9}{2}$ for y in the first equation to yield

$$6x - 4y = 18$$

$$6x - 4\left(\frac{3}{2}x - \frac{9}{2}\right) = 18$$

$$6x - 6x + 18 = 18$$

$$18 = 18$$

Our obtaining a true numerical statement ($18 = 18$) indicates that the system has infinitely many solutions. Any ordered pair that satisfies one of the equations will also satisfy the other equation. Thus in the second equation of the original system, if we let $x = k$, then $y = \frac{3}{2}k - \frac{9}{2}$. Therefore, the solution set can be expressed $\left\{\left(k, \frac{3}{2}k - \frac{9}{2}\right) \,\middle|\, k \text{ is a real number}\right\}$.

If some specific solutions are needed, they can be generated by the ordered pair $\left(k, \frac{3}{2}k - \frac{9}{2}\right)$.

For example, if we let $k = 1$, then we get $\frac{3}{2}(1) - \frac{9}{2} = -\frac{6}{2} = -3$. Thus the ordered pair $(1, -3)$ is a member of the solution set of the given system.

The Elimination-by-Addition Method

Now let's consider the **elimination-by-addition method** for solving a system of equations. This is a very important method because it is the basis for developing other techniques for solving systems that contain many equations and variables. The method involves replacing systems of equations with *simpler equivalent systems* until we obtain a system where the solutions are obvious. **Equivalent systems of equations are systems that have exactly the same solution set**. The following operations or transformations can be applied to a system of equations to produce an equivalent system.

1. Any two equations of the system can be interchanged.

2. Both sides of any equation of the system can be multiplied by any nonzero real number.

3. Any equation of the system can be replaced by the sum of that equation and a nonzero multiple of another equation.

Classroom Example

Solve the system:
$$\begin{pmatrix} 2x - 7y = 22 \\ 3x + 5y = 2 \end{pmatrix}$$

EXAMPLE 4

Solve the system $\begin{pmatrix} 3x + 5y = -9 \\ 2x - 3y = 13 \end{pmatrix}$.   (1)
    (2)

Solution

We can replace the given system with an equivalent system by multiplying equation (2) by -3.

$$\begin{pmatrix} 3x + 5y = -9 \\ -6x + 9y = -39 \end{pmatrix}$$ (3)
    (4)

Now let's replace equation (4) with an equation formed by multiplying equation (3) by 2 and adding this result to equation (4).

$$\begin{pmatrix} 3x + 5y = -9 \\ 19y = -57 \end{pmatrix}$$ (5)
    (6)

From equation (6), we can easily determine that $y = -3$. Then, substituting -3 for y in equation (5) produces

$$3x + 5(-3) = -9$$

$$3x - 15 = -9$$

$$3x = 6$$

$$x = 2$$

The solution set for the given system is $\{(2, -3)\}$.

Remark: We are using a format for the elimination-by-addition method that highlights the use of equivalent systems. In Section 15.3, this format will lead naturally to an approach using matrices. Thus it is beneficial to stress the use of equivalent systems at this time.

Classroom Example
Solve the system:
$$\left(\begin{array}{c} \dfrac{x}{6} + \dfrac{y}{3} = 3 \\ \dfrac{5x}{6} - \dfrac{y}{6} = -7 \end{array} \right)$$

EXAMPLE 5

Solve the system $\left(\begin{array}{c} \dfrac{1}{2}x + \dfrac{2}{3}y = -4 \\ \dfrac{1}{4}x - \dfrac{3}{2}y = 20 \end{array} \right)$. $\qquad$ (7)
$\qquad$ (8)

Solution

The given system can be replaced with an equivalent system by multiplying equation (7) by 6 and equation (8) by 4.

$$\left(\begin{array}{c} 3x + 4y = -24 \\ x - 6y = 80 \end{array} \right) \qquad \begin{array}{c} (9) \\ (10) \end{array}$$

Now let's exchange equations (9) and (10).

$$\left(\begin{array}{c} x - 6y = 80 \\ 3x + 4y = -24 \end{array} \right) \qquad \begin{array}{c} (11) \\ (12) \end{array}$$

We can replace equation (12) with an equation formed by multiplying equation (11) by -3 and adding this result to equation (12).

$$\left(\begin{array}{c} x - 6y = 80 \\ 22y = -264 \end{array} \right) \qquad \begin{array}{c} (13) \\ (14) \end{array}$$

From equation (14) we can determine that $y = -12$. Then substituting -12 for y in equation (13) produces

$$x - 6(-12) = 80$$
$$x + 72 = 80$$
$$x = 8$$

The solution set of the given system is $\{(8, -12)\}$. (Check this!)

Classroom Example
Solve the system:
$$\left(\begin{array}{c} 2x - y = 9 \\ 4x - 2y = 11 \end{array} \right)$$

EXAMPLE 6

Solve the system $\left(\begin{array}{c} x - 4y = 9 \\ x - 4y = 3 \end{array} \right)$. $\qquad$ (15)
$\qquad$ (16)

Solution

We can replace equation (16) with an equation formed by multiplying equation (15) by -1 and adding this result to equation (16).

$$\left(\begin{array}{c} x - 4y = 9 \\ 0 = -6 \end{array} \right) \qquad \begin{array}{c} (17) \\ (18) \end{array}$$

The statement $0 = -6$ is a contradiction, and therefore the original system is *inconsistent*; it has no solution. The solution set is $\varnothing$.

Both the elimination-by-addition method and the substitution method can be used to obtain exact solutions for any system of two linear equations in two unknowns. Sometimes it is a matter of deciding which method to use on a particular system. Some systems lend themselves to one or the other of the methods by virtue of the original format of the equations. We will illustrate this idea in a moment when we solve some word problems.

Using Systems to Solve Problems

Many word problems that we solved earlier in this text with one variable and one equation can also be solved by using a system of two linear equations in two variables. In fact, in many of these problems, you may find it more natural to use two variables and two equations.

Classroom Example
The tens digit of a two-digit number is 6 more than twice the units digit. If the sum of the digits is 6, find the number.

EXAMPLE 7

The tens digit of a two-digit number is 5 more than three times the units digit. If the sum of the digits is 9, find the number.

Solution

Let u represent the units digit, and let t represent the tens digit. The problem translates into the following system.

$$\begin{pmatrix} t = 3u + 5 \\ u + t = 9 \end{pmatrix}$$ The tens digit is 5 more than three times the units digit
   The sum of the digits is 9

Because of the form of the first equation, this system lends itself to solving by the substitution method. Substitute $3u + 5$ for t in the second equation to produce

$$u + (3u + 5) = 9$$
$$4u + 5 = 9$$
$$4u = 4$$
$$u = 1$$

Now substitute 1 for u in the equation $t = 3u + 5$ to get

$$t = 3(1) + 5 = 8$$

The tens digit is 8 and the units digit is 1, so the number is 81.

Classroom Example
Jan invested some money at 3.5% and $1100 more than that amount at 4.8%. The yearly interest from the two investments was $252. How much did Jan invest at each rate?

EXAMPLE 8

Lucinda invested $9500, part of it at 4% interest and the remainder at 6%. Her total yearly income from the two investments was $530. How much did she invest at each rate?

Solution

Let x represent the amount invested at 4% and y the amount invested at 6%. The problem translates into the following system.

$$\begin{pmatrix} x + y = 9500 \\ 0.04x + 0.06y = 530 \end{pmatrix}$$ ← The two investments total $9500
   ← The yearly interest from the two investments totals $530

Multiply the second equation by 100 to produce an equivalent system.

$$\begin{pmatrix} x + y = 9500 \\ 4x + 6y = 53{,}000 \end{pmatrix}$$

Because neither equation is solved for one variable in terms of the other, let's use the elimination-by-addition method to solve the system. The second equation can be replaced by an equation formed by multiplying the first equation by -4 and adding this result to the second equation.

$$\begin{pmatrix} x + y = 9500 \\ 2y = 15{,}000 \end{pmatrix}$$

Hence, we can determine that $y = 7500$.

Now we substitute 7500 for y in the equation $x + y = 9500$.

$$x + 7500 = 9500$$
$$x = 2000$$

Therefore, Lucinda must have invested $2000 at 4% and $7500 at 6%.

In our final example of this section, we will use a graphing utility to help solve a system of equations.

Classroom Example
$$\begin{pmatrix} 3.24x - 1.07y = -5.38 \\ 2.57x + 4.16y = 5.75 \end{pmatrix}$$

EXAMPLE 9   Solve the following system with the aid of a graphing utility:

$$\begin{pmatrix} 1.14x + 2.35y = -7.12 \\ 3.26x - 5.05y = 26.72 \end{pmatrix}$$

Solution

First, we need to solve each equation for y in terms of x. Thus the system becomes

$$\begin{pmatrix} y = \dfrac{-7.12 - 1.14x}{2.35} \\ y = \dfrac{3.26x - 26.72}{5.05} \end{pmatrix}$$

Now we can enter both of these equations into a graphing utility and obtain Figure 15.3. From this figure, it appears that the point of intersection is at approximately $x = 2$ and $y = -4$. By direct substitution into the given equations, we can verify that the point of intersection is exactly $(2, -4)$.

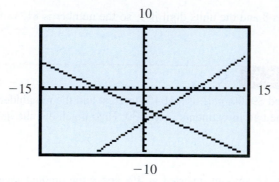

Figure 15.3

Concept Quiz 15.1

For Problems 1–8, answer true or false.

1. If we graph the equations in a system of equations, the ordered pair associated with the point of intersection of the graphs is a solution of the system.

2. When a system of two linear equations has exactly one solution, then the system is said to be consistent.

3. When the graphs for a system of two linear equations are parallel lines, then there are an infinite number of solutions.

4. Equivalent systems of equations are systems that have exactly the same solution set.

5. The substitution method for solving a system of two linear equations can only be used if one of the coefficients of the variables is 1.

6. The system $\begin{pmatrix} x + y = 4 \\ x - y = 2 \end{pmatrix}$ is a consistent system.

7. The equations of the system $\begin{pmatrix} x + y = 4 \\ x + y = 2 \end{pmatrix}$ are dependent.

8. The system $\begin{pmatrix} x + y = 4 \\ 2x + 2y = 8 \end{pmatrix}$ is an inconsistent system.

Problem Set 15.1

For Problems 1–18, solve each system by using the substitution method. **(Objective 1)**

1. $\begin{pmatrix} x + y = 16 \\ y = x + 2 \end{pmatrix}$

2. $\begin{pmatrix} 2x + 3y = -5 \\ y = 2x + 9 \end{pmatrix}$

3. $\begin{pmatrix} x = 3y - 25 \\ 4x + 5y = 19 \end{pmatrix}$

4. $\begin{pmatrix} 3x - 5y = 25 \\ x = y + 7 \end{pmatrix}$

5. $\begin{pmatrix} y = \dfrac{2}{3}x - 1 \\ 5x - 7y = 9 \end{pmatrix}$

6. $\begin{pmatrix} y = \dfrac{3}{4}x + 5 \\ 4x - 3y = -1 \end{pmatrix}$

7. $\begin{pmatrix} a = 4b + 13 \\ 3a + 6b = -33 \end{pmatrix}$

8. $\begin{pmatrix} 9a - 2b = 28 \\ b = -3a + 1 \end{pmatrix}$

9. $\begin{pmatrix} 2x - 3y = 4 \\ y = \dfrac{2}{3}x - \dfrac{4}{3} \end{pmatrix}$

10. $\begin{pmatrix} t + u = 11 \\ t = u + 7 \end{pmatrix}$

11. $\begin{pmatrix} u = t - 2 \\ t + u = 12 \end{pmatrix}$

12. $\begin{pmatrix} y = 5x - 9 \\ 5x - y = 9 \end{pmatrix}$

13. $\begin{pmatrix} 4x + 3y = -7 \\ 3x - 2y = 16 \end{pmatrix}$

14. $\begin{pmatrix} 5x - 3y = -34 \\ 2x + 7y = -30 \end{pmatrix}$

15. $\begin{pmatrix} 5x - y = 4 \\ y = 5x + 9 \end{pmatrix}$

16. $\begin{pmatrix} 2x + 3y = 3 \\ 4x - 9y = -4 \end{pmatrix}$

17. $\begin{pmatrix} 4x - 5y = 3 \\ 8x + 15y = -24 \end{pmatrix}$

18. $\begin{pmatrix} 4x + y = 9 \\ y = 15 - 4x \end{pmatrix}$

For Problems 19–34, solve each system by using the elimination-by-addition method. **(Objective 2)**

19. $\begin{pmatrix} 3x + 2y = 1 \\ 5x - 2y = 3 \end{pmatrix}$

20. $\begin{pmatrix} 4x + 3y = -22 \\ 4x - 5y = 26 \end{pmatrix}$

21. $\begin{pmatrix} x - 3y = -22 \\ 2x + 7y = 60 \end{pmatrix}$

22. $\begin{pmatrix} 6x - y = 3 \\ 5x + 3y = -9 \end{pmatrix}$

23. $\begin{pmatrix} 4x - 5y = 21 \\ 3x + 7y = -38 \end{pmatrix}$

24. $\begin{pmatrix} 5x - 3y = -34 \\ 2x + 7y = -30 \end{pmatrix}$

25. $\begin{pmatrix} 5x - 2y = 19 \\ 5x - 2y = 7 \end{pmatrix}$

26. $\begin{pmatrix} 4a + 2b = -4 \\ 6a - 5b = 18 \end{pmatrix}$

27. $\begin{pmatrix} 5a + 6b = 8 \\ 2a - 15b = 9 \end{pmatrix}$

28. $\begin{pmatrix} 7x + 2y = 11 \\ 7x + 2y = -4 \end{pmatrix}$

29. $\begin{pmatrix} \dfrac{2}{3}s + \dfrac{1}{4}t = -1 \\ \dfrac{1}{2}s - \dfrac{1}{3}t = -7 \end{pmatrix}$

30. $\begin{pmatrix} \dfrac{1}{4}s - \dfrac{2}{3}t = -3 \\ \dfrac{1}{3}s + \dfrac{1}{3}t = 7 \end{pmatrix}$

31. $\begin{pmatrix} \dfrac{x}{2} - \dfrac{2y}{5} = \dfrac{-23}{60} \\ \dfrac{2x}{3} + \dfrac{y}{4} = \dfrac{-1}{4} \end{pmatrix}$

32. $\begin{pmatrix} \dfrac{2x}{3} - \dfrac{y}{2} = \dfrac{3}{5} \\ \dfrac{x}{4} + \dfrac{y}{2} = \dfrac{7}{80} \end{pmatrix}$

33. $\begin{pmatrix} \dfrac{2x}{3} + \dfrac{y}{2} = \dfrac{1}{6} \\ 4x + 6y = -1 \end{pmatrix}$

34. $\begin{pmatrix} \dfrac{1}{2}x + \dfrac{2}{3}y = -\dfrac{3}{10} \\ 5x + 4y = -1 \end{pmatrix}$

For Problems 35–50, solve each system by using either the substitution method or the elimination-by-addition method, whichever seems more appropriate.

35. $\begin{pmatrix} 5x - y = -22 \\ 2x + 3y = -2 \end{pmatrix}$

36. $\begin{pmatrix} 4x + 5y = -41 \\ 3x - 2y = 21 \end{pmatrix}$

37. $\begin{pmatrix} x = 3y + 1 \\ x = -2y + 3 \end{pmatrix}$

38. $\begin{pmatrix} y = 4x - 24 \\ 7x + y = 42 \end{pmatrix}$

39. $\begin{pmatrix} 3x - 5y = 9 \\ 6x - 10y = -1 \end{pmatrix}$

40. $\begin{pmatrix} y = \dfrac{2}{5}x - 3 \\ 4x - 7y = 33 \end{pmatrix}$

41. $\begin{pmatrix} \dfrac{1}{2}x - \dfrac{2}{3}y = 22 \\ \dfrac{1}{2}x + \dfrac{1}{4}y = 0 \end{pmatrix}$

42. $\begin{pmatrix} \dfrac{2}{5}x - \dfrac{1}{3}y = -9 \\ \dfrac{3}{4}x + \dfrac{1}{3}y = -14 \end{pmatrix}$

43. $\begin{pmatrix} t = 2u + 2 \\ 9u - 9t = -45 \end{pmatrix}$

44. $\begin{pmatrix} 9u - 9t = 36 \\ u = 2t + 1 \end{pmatrix}$

45. $\begin{pmatrix} x + y = 1000 \\ 0.12x + 0.14y = 136 \end{pmatrix}$

46. $\begin{pmatrix} x + y = 10 \\ 0.3x + 0.7y = 4 \end{pmatrix}$

47. $\begin{pmatrix} y = 2x \\ 0.09x + 0.12y = 132 \end{pmatrix}$ **48.** $\begin{pmatrix} y = 3x \\ 0.1x + 0.11y = 64.5 \end{pmatrix}$

49. $\begin{pmatrix} x + y = 10.5 \\ 0.5x + 0.8y = 7.35 \end{pmatrix}$ **50.** $\begin{pmatrix} 2x + y = 7.75 \\ 3x + 2y = 12.5 \end{pmatrix}$

For Problems 51–68, solve each problem by using a system of equations. **(Objective 3)**

51. The sum of two numbers is 53, and their difference is 19. Find the numbers.

52. The sum of two numbers is −3, and their difference is 25. Find the numbers.

53. The measure of the larger of two complementary angles is 15° more than four times the measure of the smaller angle. Find the measures of both angles.

54. Assume that a plane is flying at a constant speed under unvarying wind conditions. Traveling against a head wind, it takes the plane 4 hours to travel 1540 miles. Traveling with a tail wind, the plane flies 1365 miles in 3 hours. Find the speed of the plane and the speed of the wind.

55. The tens digit of a two-digit number is 1 more than three times the units digit. If the sum of the digits is 9, find the number.

56. The units digit of a two-digit number is 1 less than twice the tens digit. The sum of the digits is 8. Find the number.

57. A car rental agency rents sedans at $35 a day and convertibles at $48 a day. If 32 cars were rented one day for a total of $1276, how many convertibles were rented?

58. A video store rents new release movies for $5 and favorites for $2.75. One day the number of new release movies rented was twice the number of favorites. If the total income for that day was $956.25, how many movies of each kind were rented?

59. The income from a high school band fundraiser concert was $3360. The price of a student ticket was $6 and parent tickets were sold at $10 each. Four hundred twenty tickets were sold. How many tickets of each kind were sold?

60. Michelle can enter a small business as a full partner and receive a salary of $40,000 a year and 15% of the year's profit, or she can be sales manager for a salary of $65,000 plus 5% of the year's profit. What must the year's profit be for her total earnings to be the same whether she is a full partner or a sales manager?

61. Melinda invested three times as much money at 8% yearly interest as she did at 6%. Her total yearly interest from the two investments was $1500. How much did she invest at each rate?

62. Simon invested $13,500, part of it at 4% and the rest at 7% yearly interest. His yearly income from the 7% investment was $30 less than twice the income from the 4% investment. How much did he invest at each rate?

63. One day last summer, Jim went kayaking on the Little Susitna River in Alaska. Paddling upstream against the current, he traveled 20 miles in 4 hours. Then he turned around and paddled twice as fast downstream, and with the help of the current, traveled 19 miles in 1 hour. Find the rate of the current.

64. One solution contains 30% alcohol, and a second solution contains 70% alcohol. How many liters of each solution should be mixed to make 10 liters containing 40% alcohol?

65. Santo bought 4 gallons of green latex paint and 2 gallons of primer for a total of $116. Not having enough paint to finish the project, Santo returned to the same store and bought 3 gallons of green latex paint and 1 gallon of primer for a total of $80. What is the price of a gallon of green latex paint?

66. Six cans of soda and 2 bags of potato chips cost $10.68. At the same prices, 8 cans of soda and 5 bags of potato chips cost $17.95. Find the price per can of soda and the price per bag of potato chips.

67. A cash drawer contains only five- and ten-dollar bills. There are 12 more five-dollar bills than ten-dollar bills. If the drawer contains $330, find the number of each kind of bill.

68. Brad has a collection of dimes and quarters totaling $47.50. The number of quarters is ten more than twice the number of dimes. How many coins of each kind does he have?

Thoughts Into Words

69. Give a general description of how to use the substitution method to solve a system of two linear equations in two variables.

70. Give a general description of how to use the elimination-by-addition method to solve a system of two linear equations in two variables.

71. Which method would you use to solve the system $\begin{pmatrix} 9x + 4y = 7 \\ 3x + 2y = 6 \end{pmatrix}$? Why?

72. Which method would you use to solve the system $\begin{pmatrix} 5x + 3y = 12 \\ 3x - y = 10 \end{pmatrix}$? Why?

Further Investigations

A system such as

$$\left(\begin{array}{c} \dfrac{2}{x} + \dfrac{3}{y} = \dfrac{19}{15} \\[2mm] -\dfrac{2}{x} + \dfrac{1}{y} = -\dfrac{7}{15} \end{array} \right)$$

is not a linear system, but it can be solved using the elimination-by-addition method as follows. Add the first equation to the second to produce the equivalent system

$$\left(\begin{array}{c} \dfrac{2}{x} + \dfrac{3}{y} = \dfrac{19}{15} \\[2mm] \dfrac{4}{y} = \dfrac{12}{15} \end{array} \right)$$

Now solve $\dfrac{4}{y} = \dfrac{12}{15}$ to produce $y = 5$. Substitute 5 for y in the first equation, and solve for x to produce

$$\dfrac{2}{x} + \dfrac{3}{5} = \dfrac{19}{15}$$
$$\dfrac{2}{x} = \dfrac{10}{15}$$
$$10x = 30$$
$$x = 3$$

The solution set of the original system is $\{(3, 5)\}$.

For Problems 75–80, solve each system.

73. $\left(\begin{array}{c} \dfrac{1}{x} + \dfrac{2}{y} = \dfrac{7}{12} \\[2mm] \dfrac{3}{x} - \dfrac{2}{y} = \dfrac{5}{12} \end{array} \right)$
   74. $\left(\begin{array}{c} \dfrac{3}{x} + \dfrac{2}{y} = 2 \\[2mm] \dfrac{2}{x} - \dfrac{3}{y} = \dfrac{1}{4} \end{array} \right)$

75. $\left(\begin{array}{c} \dfrac{3}{x} - \dfrac{2}{y} = \dfrac{13}{6} \\[2mm] \dfrac{2}{x} + \dfrac{3}{y} = 0 \end{array} \right)$
   76. $\left(\begin{array}{c} \dfrac{4}{x} + \dfrac{1}{y} = 11 \\[2mm] \dfrac{3}{x} - \dfrac{5}{y} = -9 \end{array} \right)$

77. $\left(\begin{array}{c} \dfrac{5}{x} - \dfrac{2}{y} = 23 \\[2mm] \dfrac{4}{x} + \dfrac{3}{y} = \dfrac{23}{2} \end{array} \right)$
   78. $\left(\begin{array}{c} \dfrac{2}{x} - \dfrac{7}{y} = \dfrac{9}{10} \\[2mm] \dfrac{5}{x} + \dfrac{4}{y} = -\dfrac{41}{20} \end{array} \right)$

79. Consider the linear system $\left(\begin{array}{c} a_1x + b_1y = c_1 \\ a_2x + b_2y = c_2 \end{array} \right)$.

   a. Prove that this system has exactly one solution if and only if $\dfrac{a_1}{a_2} \neq \dfrac{b_1}{b_2}$.

   b. Prove that this system has no solution if and only if $\dfrac{a_1}{a_2} = \dfrac{b_1}{b_2} \neq \dfrac{c_1}{c_2}$.

   c. Prove that this system has infinitely many solutions if and only if $\dfrac{a_1}{a_2} = \dfrac{b_1}{b_2} = \dfrac{c_1}{c_2}$.

80. For each of the following systems, use the results from Problem 79 to determine whether the system is consistent or inconsistent or the equations are dependent.

   a. $\left(\begin{array}{c} 5x + y = 9 \\ x - 5y = 4 \end{array} \right)$
   b. $\left(\begin{array}{c} 3x - 2y = 14 \\ 2x + 3y = 9 \end{array} \right)$

   c. $\left(\begin{array}{c} x - 7y = 4 \\ x - 7y = 9 \end{array} \right)$
   d. $\left(\begin{array}{c} 3x - 5y = 10 \\ 6x - 10y = 1 \end{array} \right)$

   e. $\left(\begin{array}{c} 3x + 6y = 2 \\[1mm] \dfrac{3}{5}x + \dfrac{6}{5}y = \dfrac{2}{5} \end{array} \right)$
   f. $\left(\begin{array}{c} \dfrac{2}{3}x - \dfrac{3}{4}y = 2 \\[1mm] \dfrac{1}{2}x + \dfrac{2}{5}y = 9 \end{array} \right)$

   g. $\left(\begin{array}{c} 7x + 9y = 14 \\ 8x - 3y = 12 \end{array} \right)$
   h. $\left(\begin{array}{c} 4x - 5y = 3 \\ 12x - 15y = 9 \end{array} \right)$

Graphing Calculator Activities

81. For each of the systems of equations in Problem 80, use your graphing calculator to help determine whether the system is consistent or inconsistent or the equations are dependent.

82. Use your graphing calculator to help determine the solution set for each of the following systems. Be sure to check your answers.

   a. $\left(\begin{array}{c} y = 3x - 1 \\ y = 9 - 2x \end{array} \right)$
   b. $\left(\begin{array}{c} 5x + y = -9 \\ 3x - 2y = 5 \end{array} \right)$

   c. $\left(\begin{array}{c} 4x - 3y = 18 \\ 5x + 6y = 3 \end{array} \right)$
   d. $\left(\begin{array}{c} 2x - y = 20 \\ 7x + y = 79 \end{array} \right)$

   e. $\left(\begin{array}{c} 13x - 12y = 37 \\ 15x + 13y = -11 \end{array} \right)$
   f. $\left(\begin{array}{c} 1.98x + 2.49y = 13.92 \\ 1.19x + 3.45y = 16.18 \end{array} \right)$

Answers to the Concept Quiz

1. True   **2.** True   **3.** False   **4.** True   **5.** False   **6.** True   **7.** False   **8.** False

15.2 Systems of Three Linear Equations in Three Variables

OBJECTIVES

1 Solve systems of three linear equations in three variables

2 Use a system of three linear equations to solve word problems

Consider a linear equation in three variables x, y, and z, such as $3x - 2y + z = 7$. Any **ordered triple** (x, y, z) that makes the equation a true numerical statement is said to be a **solution** of the equation. For example, the ordered triple $(2, 1, 3)$ is a solution because $3(2) - 2(1) + 3 = 7$. However, the ordered triple $(5, 2, 4)$ is not a solution because $3(5) - 2(2) + 4 \neq 7$. There are infinitely many solutions in the solution set.

Remark: The idea of a linear equation is generalized to include equations of more than two variables. Thus an equation such as $5x - 2y + 9z = 8$ is called a *linear equation in three variables,* the equation $5x - 7y + 2z - 11w = 1$ is called a *linear equation in four variables,* and so on.

To *solve* a system of three linear equations in three variables, such as

$$\begin{pmatrix} 3x - y + 2z = 13 \\ 4x + 2y + 5z = 30 \\ 5x - 3y - z = 3 \end{pmatrix}$$

means to find all of the ordered triples that satisfy all three equations. In other words, the solution set of the system is the intersection of the solution sets of the three equations in the system.

The graph of a linear equation in three variables is a *plane,* not a line. In fact, graphing equations in three variables requires the use of a three-dimensional coordinate system. Thus using a graphing approach to solve systems of three linear equations in three variables is not at all practical. However, a simple graphical analysis does provide us with some indication of what we can expect as we begin solving such systems.

In general, because each linear equation in three variables produces a plane, a system of three such equations produces three planes. There are various ways in which three planes can be related. For example, they may be mutually parallel; or two of the planes may be parallel, with the third intersecting the other two. (You may want to analyze all of the other possibilities for the three planes!) However, for our purposes at this time, we need to realize that from a solution set viewpoint, a system of three linear equations in three variables produces one of the following possibilities.

1. There is *one ordered triple* that satisfies all three equations. The three planes have a common point of intersection, as indicated in Figure 15.4.

2. There are *infinitely many ordered triples* in the solution set, all of which are coordinates of *points on a line* common to the three planes. This happens if the three planes have a common line of intersection (Figure 15.5a) or if two of the planes coincide, and the third plane intersects them (Figure 15.5b).

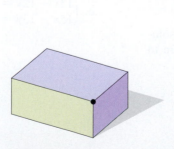

Figure 15.4

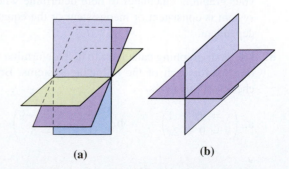

(a)   (b)

Figure 15.5

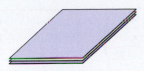

Figure 15.6

3. There are *infinitely many ordered triples* in the solution set, all of which are coordinates of *points on a plane*. This happens if the three planes coincide, as illustrated in Figure 15.6.

4. The solution set is *empty;* thus, we write ∅. This can happen in various ways, as illustrated in Figure 15.7. Note that in each situation there are no points common to all three planes.

(a) Three parallel planes

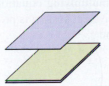

(b) Two planes coincide, and the third one is parallel to the coinciding planes.

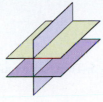

(c) Two planes are parallel and the third intersects them in parallel lines.

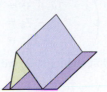

(d) No two planes are parallel, but two of them intersect in a line that is parallel to the third plane.

Figure 15.7

Now that we know what possibilities exist, let's consider finding the solution sets for some systems. Our approach will be the elimination-by-addition method, whereby systems are replaced with equivalent systems until a system is obtained that allows us to easily determine the solution set. The details of this approach will become apparent as we work a few examples.

Classroom Example
Solve the system:
$$\begin{pmatrix} x + 4y - 3z = 1 \\ 3y + 4z = 4 \\ 6z = 6 \end{pmatrix}$$

EXAMPLE 1

Solve the system $\begin{pmatrix} 4x - 3y - 2z = 5 \\ 5y + z = -11 \\ 3z = 12 \end{pmatrix}$.

    (1)
    (2)
    (3)

Solution

The form of this system makes it easy to solve. From equation (3), we obtain $z = 4$. Then, substituting 4 for z in equation (2), we get

$$5y + 4 = -11$$
$$5y = -15$$
$$y = -3$$

Finally, substituting 4 for z and -3 for y in equation (1) yields

$$4x - 3(-3) - 2(4) = 5$$
$$4x + 1 = 5$$
$$4x = 4$$
$$x = 1$$

Thus the solution set of the given system is $\{(1, -3, 4)\}$.

EXAMPLE 2

Solve the system $\begin{pmatrix} x - 2y + 3z = 22 \\ 2x - 3y - z = 5 \\ 3x + y - 5z = -32 \end{pmatrix}$.

$$\quad (4)$$
$$\quad (5)$$
$$\quad (6)$$

Solution

Equation (5) can be replaced with the equation formed by multiplying equation (4) by -2 and adding this result to equation (5). Equation (6) can be replaced with the equation formed by multiplying equation (4) by -3 and adding this result to equation (6). The following equivalent system is produced, in which equations (8) and (9) contain only the two variables y and z.

$$\begin{pmatrix} x - 2y + 3z = 22 \\ y - 7z = -39 \\ 7y - 14z = -98 \end{pmatrix}$$

$$\quad (7)$$
$$\quad (8)$$
$$\quad (9)$$

Equation (9) can be replaced with the equation formed by multiplying equation (8) by -7 and adding this result to equation (9). This produces the following equivalent system.

$$\begin{pmatrix} x - 2y + 3z = 22 \\ y - 7z = -39 \\ 35z = 175 \end{pmatrix}$$

$$\quad (10)$$
$$\quad (11)$$
$$\quad (12)$$

From equation (12) we obtain $z = 5$. Then, substituting 5 for z in equation (11), we obtain

$$y - 7(5) = -39$$
$$y - 35 = -39$$
$$y = -4$$

Finally, substituting -4 for y and 5 for z in equation (10) produces

$$x - 2(-4) + 3(5) = 22$$
$$x + 8 + 15 = 22$$
$$x + 23 = 22$$
$$x = -1$$

The solution set of the original system is $\{(-1, -4, 5)\}$. (Perhaps you should check this ordered triple in all three of the original equations.)

EXAMPLE 3

Solve the system $\begin{pmatrix} 3x - y + 2z = 13 \\ 5x - 3y - z = 3 \\ 4x + 2y + 5z = 30 \end{pmatrix}$.

$$\quad (13)$$
$$\quad (14)$$
$$\quad (15)$$

Solution

Equation (14) can be replaced with the equation formed by multiplying equation (13) by -3 and adding this result to equation (14). Equation (15) can be replaced with the equation formed by multiplying equation (13) by 2 and adding this result to equation (15). Thus we produce the following equivalent system, in which equations (17) and (18) contain only the two variables x and z.

$$\begin{pmatrix} 3x - y + 2z = 13 \\ -4x - 7z = -36 \\ 10x + 9z = 56 \end{pmatrix}$$

$$\quad (16)$$
$$\quad (17)$$
$$\quad (18)$$

Now, if we multiply equation (17) by 5 and equation (18) by 2, we get the following equivalent system.

$$\begin{pmatrix} 3x - y + 2z = 13 \\ -20x - 35z = -180 \\ 20x + 18z = 112 \end{pmatrix}$$

$$\quad (19)$$
$$\quad (20)$$
$$\quad (21)$$

Equation (21) can be replaced with the equation formed by adding equation (20) to equation (21).

$$\begin{pmatrix} 3x - y + 2z = 13 \\ -20x \quad\quad - 35z = -180 \\ \quad\quad - 17z = -68 \end{pmatrix} \quad\quad \begin{matrix} (22) \\ (23) \\ (24) \end{matrix}$$

From equation (24), we obtain $z = 4$. Then we can substitute 4 for z in equation (23).

$$-20x - 35(4) = -180$$
$$-20x - 140 = -180$$
$$-20x = -40$$
$$x = 2$$

Now we can substitute 2 for x and 4 for z in equation (22).

$$3(2) - y + 2(4) = 13$$
$$6 - y + 8 = 13$$
$$-y + 14 = 13$$
$$-y = -1$$
$$y = 1$$

The solution set of the original system is $\{(2, 1, 4)\}$.

Classroom Example

Solve the system:

$$\begin{pmatrix} x - 2y + 3z = 1 \\ 3x - 5y - 2z = 4 \\ 2x - 4y + 6z = 7 \end{pmatrix}$$

EXAMPLE 4   Solve the system $\begin{pmatrix} x + 2y - z = 4 \\ 2x + 4y - 2z = 7 \\ 3x - y + z = -1 \end{pmatrix}$. $\begin{matrix} (25) \\ (26) \\ (27) \end{matrix}$

Solution

Equation (26) can be replaced with the equation formed by multiplying equation (25) by -2 and adding the result to equation (26).

$$\begin{pmatrix} x + 2y - z = 4 \\ 0 = -1 \\ 3x - y + z = -1 \end{pmatrix} \quad\quad \begin{matrix} (28) \\ (29) \\ (30) \end{matrix}$$

Equation (29) is a contradiction; therefore there is no solution for the system of equations. The solution set of the original system is $\varnothing$.

Classroom Example

Solve the system:

$$\begin{pmatrix} 3x - 4y - z = -2 \\ 2x + 10y + 2z = -2 \\ x - 14y - 3z = 0 \end{pmatrix}$$

EXAMPLE 5   Solve the system $\begin{pmatrix} x - y + 2z = 3 \\ -2x + 3y - z = 1 \\ -x + 2y + z = 4 \end{pmatrix}$. $\begin{matrix} (31) \\ (32) \\ (33) \end{matrix}$

Solution

Equation (32) can be replaced with the equation formed by multiplying equation (31) by 2 and adding the result to equation (32). Equation (33) can be replaced with the equation formed by adding equation (31) to equation (33). The following equivalent system is produced.

$$\begin{pmatrix} x - y + 2z = 3 \\ y + 3z = 7 \\ y + 3z = 7 \end{pmatrix} \quad\quad \begin{matrix} (34) \\ (35) \\ (36) \end{matrix}$$

Equations (35) and (36) are identical. If we continue, equation (36) can be replaced with the equation formed by multiplying equation (35) by -1 and adding the result to equation (36). Then the equivalent system is produced.

$$\begin{pmatrix} x - y + 2z = 3 \\ y + 3z = 7 \\ 0 = 0 \end{pmatrix}$$

(37)
(38)
(39)

Equation (39) is an identity; therefore the original system has an infinite number of solutions. To represent the ordered triples we can solve equation (38) for y obtaining $y = -3z + 7$. We can substitute for y in equation (37) and solve that equation for x obtaining $x = -5z + 10$. Therefore, if we let $z = k$, where k is any real number, the solution set of infinitely many ordered triples can be represented by $\{(-5k + 10, -3k + 7, k) | k \text{ is a real number}\}$. We can generate a specific solution by replacing k with a number. For example, if $k = 4$, then $-5k + 10$ becomes $-5(4) + 10 = -10$, and $-3k + 7$ becomes $-3(4) + 7 = -5$. Thus the ordered triple $(-10, -5, 4)$ is a solution of the original system.

The ability to solve systems of three linear equations in three unknowns enhances our problem-solving capabilities. Let's conclude this section with a problem that we can solve using such a system.

Classroom Example
A beauty spa offers three different treatments. Each treatment requires the services of a manicurist, a hair stylist, and a masseur as indicated in the following table.

	Manicurist	Hair stylist	Masseur
Special	0.5 hour	0.5 hour	0.5 hour
Preferred	0.5 hour	0.75 hour	0.75 hour
Ultra	1 hour	1 hour	1.5 hour

The manicurist, hair stylist, and masseur have available a maximum of 40, 45, and 54 work-hours per week, respectively. How many of each treatment can the spa schedule each week so that the spa is operating at full capacity?

EXAMPLE 6

A small company that manufactures sporting equipment produces three different styles of golf shirts. Each style of shirt requires the services of three departments, as indicated by the following table.

	Style A	Style B	Style C
Cutting department	0.1 hour	0.1 hour	0.3 hour
Sewing department	0.3 hour	0.2 hour	0.4 hour
Packaging department	0.1 hour	0.2 hour	0.1 hour

The cutting, sewing, and packaging departments have available a maximum of 340, 580, and 255 work-hours per week, respectively. How many of each style of golf shirt should be produced each week so that the company is operating at full capacity?

Solution

Let a represent the number of shirts of style A produced per week, b the number of style B per week, and c the number of style C per week. Then the problem translates into the following system of equations.

$$\begin{pmatrix} 0.1a + 0.1b + 0.3c = 340 \\ 0.3a + 0.2b + 0.4c = 580 \\ 0.1a + 0.2b + 0.1c = 255 \end{pmatrix}$$
←——— Cutting department
←——— Sewing department
←——— Packaging department

Solving this system (we will leave the details for you to carry out) produces $a = 500$, $b = 650$, and $c = 750$. Thus the company should produce 500 golf shirts of style A, 650 of style B, and 750 of style C per week.

Concept Quiz 15.2

For Problems 1–8, answer true or false.

1. The graph of a linear equation in three variables is a line.

2. A system of three linear equations in three variables produces three planes when graphed.

3. Three planes can be related by intersecting in exactly two points.

4. One way three planes can be related is if two of the planes are parallel, and the third plane intersects them in parallel lines.

5. A system of three linear equations in three variables always has an infinite number of solutions.

6. A system of three linear equations in three variables can have one ordered triple as a solution.

7. The solution set of the system $\begin{pmatrix} 2x - y + 3z = 4 \\ y - z = 12 \\ 2z = 6 \end{pmatrix}$ is $\{(5, 15, 3)\}$.

8. The solution set of the system $\begin{pmatrix} x - y + z = 4 \\ x - y + z = 6 \\ 3y - 2z = 9 \end{pmatrix}$ is $\{(3, 1, 2)\}$.

Problem Set 15.2

For Problems 1–24, solve each system. (Objective 1)

1. $\begin{pmatrix} 2x - 3y + 4z = 10 \\ 5y - 2z = -16 \\ 3z = 9 \end{pmatrix}$

2. $\begin{pmatrix} -3x + 2y - 3z = -9 \\ 4x - 3z = 18 \\ 4z = -8 \end{pmatrix}$

3. $\begin{pmatrix} x + 2y - 3z = 2 \\ 3y - z = 13 \\ 3y + 5z = 25 \end{pmatrix}$

4. $\begin{pmatrix} 2x + 3y - 4z = -10 \\ 2y + 3z = 16 \\ 2y - 5z = -16 \end{pmatrix}$

5. $\begin{pmatrix} 3x + 2y - 2z = 14 \\ x - 6z = 16 \\ 2x + 5z = -2 \end{pmatrix}$

6. $\begin{pmatrix} 3x + 2y - z = -11 \\ 2x - 3y = -1 \\ 4x + 5y = -13 \end{pmatrix}$

7. $\begin{pmatrix} x - 2y + 3z = 7 \\ 2x + y + 5z = 17 \\ 3x - 4y - 2z = 1 \end{pmatrix}$

8. $\begin{pmatrix} x + 2y - z = 8 \\ x + 3y + 3z = 10 \\ -x - 3y - 3z = -10 \end{pmatrix}$

9. $\begin{pmatrix} 2x + y + 2z = 12 \\ 4x + y - z = -6 \\ -y - 5z = 10 \end{pmatrix}$

10. $\begin{pmatrix} x - 2y + z = -4 \\ 2x + 4y - 3z = -1 \\ -3x - 6y + 7z = 4 \end{pmatrix}$

11. $\begin{pmatrix} 2x - y + z = 0 \\ 3x - 2y + 4z = 11 \\ 5x + y - 6z = -32 \end{pmatrix}$

12. $\begin{pmatrix} 2x - y + 3z = -14 \\ 4x + 2y - z = 12 \\ 6x - 3y + 4z = -22 \end{pmatrix}$

13. $\begin{pmatrix} 3x + 2y - z = -11 \\ 2x - 3y + 4z = 11 \\ 5x + y - 2z = -17 \end{pmatrix}$

14. $\begin{pmatrix} 9x + 4y - z = 0 \\ 3x - 2y + 4z = 6 \\ 6x - 8y - 3z = 3 \end{pmatrix}$

15. $\begin{pmatrix} 2x + 3y - 4z = -10 \\ 4x - 5y + 3z = 2 \\ 2y + z = 8 \end{pmatrix}$

16. $\begin{pmatrix} x + 2y - 3z = 2 \\ 3x - z = -8 \\ 2x - 3y + 5z = -9 \end{pmatrix}$

17. $\begin{pmatrix} x + 3y - 2z = -4 \\ 2x + 7y + z = 5 \\ y + 5z = 13 \end{pmatrix}$

18. $\begin{pmatrix} 2x - 3y + z = -1 \\ 4x - 6y + 2z = 5 \\ x + y + 2z = 4 \end{pmatrix}$

19. $\begin{pmatrix} 3x + 2y - 2z = 14 \\ 2x - 5y + 3z = 7 \\ 4x - 3y + 7z = 5 \end{pmatrix}$

20. $\begin{pmatrix} 4x + 3y - 2z = -11 \\ 3x - 7y + 3z = 10 \\ 9x - 8y + 5z = 9 \end{pmatrix}$

21. $\begin{pmatrix} 2x - 3y + 4z = -12 \\ 4x + 2y - 3z = -13 \\ 6x - 5y + 7z = -31 \end{pmatrix}$

22. $\begin{pmatrix} 2x + 5y - 2z = -26 \\ 5x - 2y + 4z = 27 \\ 7x + 3y - 6z = -55 \end{pmatrix}$

23. $\begin{pmatrix} 5x - 3y - 6z = 22 \\ x - y + z = -3 \\ -3x + 7y - 5z = 23 \end{pmatrix}$

24. $\begin{pmatrix} 4x + 3y - 5z = -29 \\ 3x - 7y - z = -19 \\ 2x + 5y + 2z = -10 \end{pmatrix}$

For Problems 25–42, solve each problem by setting up and solving a system of three linear equations in three variables. **(Objective 2)**

25. The sum of three numbers is 20. The sum of the first and third numbers is 2 more than twice the second number. The third number minus the first yields three times the second number. Find the numbers.

26. The sum of three numbers is 40. The third number is 10 less than the sum of the first two numbers. The second number is 1 larger than the first. Find the numbers.

27. Mike bought a motorcycle helmet, jacket, and gloves for $650. The jacket costs $100 more than the helmet. The cost of the helmet and gloves together was $50 less than the cost of the jacket. How much did each item cost?

28. One binder, 2 reams of paper, and 5 spiral notebooks cost $14.82. Three binders, 1 ream of paper, and 4 spiral notebooks cost $14.32. Two binders, 3 reams of paper, and 3 spiral notebooks cost $19.82. Find the cost for each item.

29. In a certain triangle, the measure of $\angle A$ is five times the measure of $\angle B$. The sum of the measures of $\angle B$ and $\angle C$ is 60° less than the measure of $\angle A$. Find the measure of each angle.

30. Shannon purchased a skirt, blouse, and sweater for $72. The cost of the skirt and sweater was $2 more than six times the cost of the blouse. The skirt cost twice the sum of the costs of the blouse and sweater. Find the cost of each item.

31. The wages for a crew consisting of a plumber, an apprentice, and a laborer are $80 an hour. The plumber earns $20 an hour more than the sum of the wages of the apprentice and the laborer. The plumber earns five times as much as the laborer. Find the hourly wage of each.

32. A catering group that has a chef, a salad maker, and a server costs the customer $70 per hour. The salad maker

costs $5 per hour more than the server. The chef costs the same as the salad maker and the server cost together. Find the cost per hour of each.

33. A gift store is making a mixture of almonds, pecans, and peanuts, which sell for $3.50 per pound, $4 per pound, and $2 per pound, respectively. The storekeeper wants to make 20 pounds of the mix to sell at $2.70 per pound. The number of pounds of peanuts is to be three times the number of pounds of pecans. Find the number of pounds of each to be used in the mixture.

34. The organizer for a church picnic ordered coleslaw, potato salad, and beans amounting to 50 pounds. There was to be three times as much potato salad as coleslaw. The number of pounds of beans was to be six less than the number of pounds of potato salad. Find the number of pounds of each.

35. A box contains $7.15 in nickels, dimes, and quarters. There are 42 coins in all, and the sum of the numbers of nickels and dimes is two less than the number of quarters. How many coins of each kind are there?

36. A handful of 65 coins consists of pennies, nickels, and dimes. The number of nickels is four less than twice the number of pennies, and there are 13 more dimes than nickels. How many coins of each kind are there?

37. The measure of the largest angle of a triangle is twice the measure of the smallest angle. The sum of the smallest angle and the largest angle is twice the other angle. Find the measure of each angle.

38. The perimeter of a triangle is 45 centimeters. The longest side is 4 centimeters less than twice the shortest side. The sum of the lengths of the shortest and longest sides is 7 centimeters less than three times the length of the remaining side. Find the lengths of all three sides of the triangle.

39. Part of $30,000 is invested at 2%, another part at 3%, and the remainder at 4% yearly interest. The total yearly income from the three investments is $1000. The sum of the amounts invested at 2% and 3% equals the amount invested at 4%. How much is invested at each rate?

40. Different amounts are invested at 4%, 5%, and 6% yearly interest. The amount invested at 5% is $3000 more than what is invested at 4%, and the total yearly income from all three investments is $1500. A total of $29,000 is invested. Find the amount invested at each rate.

41. A small company makes three different types of bird houses. Each type requires the services of three different departments, as indicated by the following table.

	Type A	Type B	Type C
Cutting department	0.1 hour	0.2 hour	0.1 hour
Finishing department	0.4 hour	0.4 hour	0.3 hour
Assembly department	0.2 hour	0.1 hour	0.3 hour

The cutting, finishing, and assembly departments have available a maximum of 35, 95, and 62.5 work-hours per week, respectively. How many bird houses of each type should be made per week so that the company is operating at full capacity?

42. A certain diet consists of dishes A, B, and C. Each serving of A has 1 gram of fat, 2 grams of carbohydrate, and 4 grams of protein. Each serving of B has 2 grams of fat, 1 gram of carbohydrate, and 3 grams of protein. Each serving of C has 2 grams of fat, 4 grams of carbohydrate, and 3 grams of protein. The diet allows 15 grams of fat, 24 grams of carbohydrate, and 30 grams of protein. How many servings of each dish can be eaten?

Thoughts Into Words

43. Give a general description of how to solve a system of three linear equations in three variables.

44. Give a step-by-step description of how to solve the system

$$\begin{pmatrix} x - 2y + 3z = -23 \\ 5y - 2z = 32 \\ 4z = -24 \end{pmatrix}$$

45. Give a step-by-step description of how to solve the system

$$\begin{pmatrix} 3x - 2y + 7z = 9 \\ x \phantom{{}-2y} - 3z = 4 \\ 2x \phantom{{}-2y} + z = 9 \end{pmatrix}$$

Answers to the Concept Quiz

1. False   **2.** True   **3.** False   **4.** True   **5.** False   **6.** True   **7.** True   **8.** False

---

15.3   A Matrix Approach to Solving Systems

OBJECTIVES

1 Represent a system of equations as an augmented matrix

2 Transform an augmented matrix to the form $\begin{bmatrix} 1 & 0 & 0 & \vdots & a \\ 0 & 1 & 0 & \vdots & b \\ 0 & 0 & 1 & \vdots & c \end{bmatrix}$

3 Solve a system of equations by transforming the augmented matrix to reduced echelon form

In the first two sections of this chapter, we found that the substitution and elimination-by-addition techniques worked effectively with two equations and two unknowns, but they started to get a bit cumbersome with three equations and three unknowns. Therefore, we shall now begin to analyze some techniques that lend themselves to use with larger systems of equations. Furthermore, some of these techniques form the basis for using a computer to solve systems. Even though these techniques are primarily designed for large systems of equations, we shall study them in the context of small systems so that we won't get bogged down with the computational aspects of the techniques.

Matrices

A **matrix** is an array of numbers arranged in horizontal rows and vertical columns and enclosed in brackets. For example, the matrix

$$2 \text{ rows} \quad \begin{bmatrix} 2 & 3 & -1 \\ -4 & 7 & 12 \end{bmatrix}$$

$$3 \text{ columns}$$

has 2 rows and 3 columns and is called a 2×3 (this is read "two by three") matrix. Each number in a matrix is called an **element** of the matrix. Some additional examples of matrices (*matrices* is the plural of *matrix*) follow.

$$3 \times 2 \qquad 2 \times 2 \qquad 1 \times 2 \qquad 4 \times 1$$

$$\begin{bmatrix} 2 & 1 \\ 1 & -4 \\ \frac{1}{2} & \frac{2}{3} \end{bmatrix} \qquad \begin{bmatrix} 17 & 18 \\ -14 & 16 \end{bmatrix} \qquad [7 \quad 14] \qquad \begin{bmatrix} 3 \\ -2 \\ 1 \\ 19 \end{bmatrix}$$

In general, a matrix of m rows and n columns is called a matrix of **dimension $m \times n$** or **order $m \times n$.**

With every system of linear equations, we can associate a matrix that consists of the coefficients and constant terms. For example, with the system

$$\begin{pmatrix} a_1x + b_1y + c_1z = d_1 \\ a_2x + b_2y + c_2z = d_2 \\ a_3x + b_3y + c_3z = d_3 \end{pmatrix}$$

we can associate the matrix

$$\begin{bmatrix} a_1 & b_1 & c_1 & \vdots & d_1 \\ a_2 & b_2 & c_2 & \vdots & d_2 \\ a_3 & b_3 & c_3 & \vdots & d_3 \end{bmatrix}$$

which is commonly called the **augmented matrix** of the system of equations. The dashed line simply separates the coefficients from the constant terms and reminds us that we are working with an augmented matrix.

In Section 15.1, we considered the operations or transformations that can be applied to a system of equations to produce an equivalent system. Because augmented matrices are essentially abbreviated forms of systems of linear equations, there are analogous transformations that can be applied to augmented matrices. These transformations are usually referred to as **elementary row operations** and can be stated as follows:

For any augmented matrix of a system of linear equations, the following elementary row operations will produce a matrix of an equivalent system.

1. Any two rows of the matrix can be interchanged.

2. Any row of the matrix can be multiplied by a nonzero real number.

3. Any row of the matrix can be replaced by the sum of a nonzero multiple of another row plus that row.

Let's illustrate the use of augmented matrices and elementary row operations to solve a system of two linear equations in two variables.

Classroom Example
Solve the system:
$$\begin{pmatrix} 4x - y = -10 \\ 3x + 8y = -25 \end{pmatrix}$$

EXAMPLE 1

Solve the system $\begin{pmatrix} x - 3y = -17 \\ 2x + 7y = 31 \end{pmatrix}$.

Solution

The augmented matrix of the system is

$$\begin{bmatrix} 1 & -3 & \vdots & -17 \\ 2 & 7 & \vdots & 31 \end{bmatrix}$$

We would like to change this matrix to one of the form

$$\begin{bmatrix} 1 & 0 & \vdots & a \\ 0 & 1 & \vdots & b \end{bmatrix}$$

where we can easily determine that the solution is $x = a$ and $y = b$. Let's begin by adding -2 times row 1 to row 2 to produce a new row 2.

$$\begin{bmatrix} 1 & -3 & \vdots & -17 \\ 0 & 13 & \vdots & 65 \end{bmatrix}$$

Now we can multiply row 2 by $\dfrac{1}{13}$.

$$\begin{bmatrix} 1 & -3 & \vdots & -17 \\ 0 & 1 & \vdots & 5 \end{bmatrix}$$

Finally, we can add 3 times row 2 to row 1 to produce a new row 1.

$$\begin{bmatrix} 1 & 0 & \vdots & -2 \\ 0 & 1 & \vdots & 5 \end{bmatrix}$$

From this last matrix, we see that $x = -2$ and $y = 5$. In other words, the solution set of the original system is $\{(-2, 5)\}$.

■

It may seem that the matrix approach does not provide us with much extra power for solving systems of two linear equations in two unknowns. However, as the systems get larger, the compactness of the matrix approach becomes more convenient. Let's consider a system of three equations in three variables.

Classroom Example
Solve the system:
$$\begin{pmatrix} x - 2y + 3z = 9 \\ 2x - 3y + z = -2 \\ 3x + y - 5z = -8 \end{pmatrix}$$

EXAMPLE 2

Solve the system $\begin{pmatrix} x + 2y - 3z = 15 \\ -2x - 3y + z = -15 \\ 4x + 9y - 4z = 49 \end{pmatrix}$.

Solution

The augmented matrix of this system is

$$\begin{bmatrix} 1 & 2 & -3 & \vdots & 15 \\ -2 & -3 & 1 & \vdots & -15 \\ 4 & 9 & -4 & \vdots & 49 \end{bmatrix}$$

If the system has a unique solution, then we will be able to change the augmented matrix to the form

$$\begin{bmatrix} 1 & 0 & 0 & \vdots & a \\ 0 & 1 & 0 & \vdots & b \\ 0 & 0 & 1 & \vdots & c \end{bmatrix}$$

where we will be able to read the solution $x = a$, $y = b$, and $z = c$.

Add 2 times row 1 to row 2 to produce a new row 2. Likewise, add -4 times row 1 to row 3 to produce a new row 3.

$$\begin{bmatrix} 1 & 2 & -3 & \vdots & 15 \\ 0 & 1 & -5 & \vdots & 15 \\ 0 & 1 & 8 & \vdots & -11 \end{bmatrix}$$

Now add -2 times row 2 to row 1 to produce a new row 1. Also, add -1 times row 2 to row 3 to produce a new row 3.

$$\begin{bmatrix} 1 & 0 & 7 & \vdots & -15 \\ 0 & 1 & -5 & \vdots & 15 \\ 0 & 0 & 13 & \vdots & -26 \end{bmatrix}$$

Now let's multiply row 3 by $\dfrac{1}{13}$.

$$\begin{bmatrix} 1 & 0 & 7 & \vdots & -15 \\ 0 & 1 & -5 & \vdots & 15 \\ 0 & 0 & 1 & \vdots & -2 \end{bmatrix}$$

Finally, we can add -7 times row 3 to row 1 to produce a new row 1, and we can add 5 times row 3 to row 2 for a new row 2.

$$\begin{bmatrix} 1 & 0 & 0 & \vdots & -1 \\ 0 & 1 & 0 & \vdots & 5 \\ 0 & 0 & 1 & \vdots & -2 \end{bmatrix}$$

From this last matrix, we can see that the solution set of the original system is $\{(-1, 5, -2)\}$.

The final matrices of Examples 1 and 2,

$$\begin{bmatrix} 1 & 0 & \vdots & -2 \\ 0 & 1 & \vdots & 5 \end{bmatrix} \quad \text{and} \quad \begin{bmatrix} 1 & 0 & 0 & \vdots & -1 \\ 0 & 1 & 0 & \vdots & 5 \\ 0 & 0 & 1 & \vdots & -2 \end{bmatrix}$$

are said to be in **reduced echelon form.** In general, a matrix is in reduced echelon form if the following conditions are satisfied:

1. Reading from left to right, the first nonzero entry of each row is 1.

2. In the *column* containing the leftmost 1 of a row, all the remaining entries are zeros.

3. The leftmost 1 of any row is to the right of the leftmost 1 of the preceding row.

4. Rows containing only zeros are below all the rows containing nonzero entries.

Like the final matrices of Examples 1 and 2, the following are in reduced echelon form.

$$\begin{bmatrix} 1 & 2 & \vdots & -3 \\ 0 & 0 & \vdots & 0 \end{bmatrix} \quad \begin{bmatrix} 1 & 0 & -2 & \vdots & 5 \\ 0 & 1 & 4 & \vdots & 7 \\ 0 & 0 & 0 & \vdots & 0 \end{bmatrix} \quad \begin{bmatrix} 1 & 0 & 0 & 0 & \vdots & 8 \\ 0 & 1 & 0 & 0 & \vdots & -9 \\ 0 & 0 & 1 & 0 & \vdots & -2 \\ 0 & 0 & 0 & 1 & \vdots & 12 \end{bmatrix}$$

In contrast, the following matrices are *not* in reduced echelon form for the reason indicated below each matrix.

$$\begin{bmatrix} 1 & 0 & 0 & \vdots & 11 \\ 0 & 3 & 0 & \vdots & -1 \\ 0 & 0 & 1 & \vdots & -2 \end{bmatrix} \quad \begin{bmatrix} 1 & 2 & -3 & \vdots & 5 \\ 0 & 1 & 7 & \vdots & 9 \\ 0 & 0 & 1 & \vdots & -6 \end{bmatrix}$$

Violates condition 1   Violates condition 2

$$\begin{bmatrix} 1 & 0 & 0 & \vdots & 7 \\ 0 & 0 & 1 & \vdots & -8 \\ 0 & 1 & 0 & \vdots & 14 \end{bmatrix} \qquad \begin{bmatrix} 1 & 0 & 0 & 0 & \vdots & -1 \\ 0 & 0 & 0 & 0 & \vdots & 0 \\ 0 & 0 & 1 & 0 & \vdots & 7 \\ 0 & 0 & 0 & 0 & \vdots & 0 \end{bmatrix}$$

Violates condition 3   Violates condition 4

Once we have an augmented matrix in reduced echelon form, it is easy to determine the solution set of the system. Furthermore, the procedure for changing a given augmented matrix to reduced echelon form can be described in a very systematic way. For example, if an augmented matrix of a system of three linear equations in three unknowns has a unique solution, then it can be changed to reduced echelon form as follows:

1. Augmented matrix

$$\begin{bmatrix} * & * & * & * \\ * & * & * & * \\ * & * & * & * \end{bmatrix}$$

2. Get a 1 in upper left-hand corner

$$\begin{bmatrix} 1 & * & * & * \\ * & * & * & * \\ * & * & * & * \end{bmatrix}$$

3. Get zeros in first column beneath the 1

$$\begin{bmatrix} 1 & * & * & * \\ 0 & * & * & * \\ 0 & * & * & * \end{bmatrix}$$

4. Get a 1 in the second row/ second column position

$$\begin{bmatrix} 1 & * & * & * \\ 0 & 1 & * & * \\ 0 & * & * & * \end{bmatrix}$$

5. Get zeros above and below the 1 in the second column

$$\begin{bmatrix} 1 & 0 & * & * \\ 0 & 1 & * & * \\ 0 & 0 & * & * \end{bmatrix}$$

6. Get a 1 in the third row/ third column position

$$\begin{bmatrix} 1 & 0 & * & * \\ 0 & 1 & * & * \\ 0 & 0 & 1 & * \end{bmatrix}$$

7. Get zeros above the 1 in the third column

$$\begin{bmatrix} 1 & 0 & 0 & * \\ 0 & 1 & 0 & * \\ 0 & 0 & 1 & * \end{bmatrix}$$

We can identify inconsistent and dependent systems while we are changing a matrix to reduced echelon form. We will show some examples of such cases in a moment, but first let's consider another example of a system of three linear equations in three unknowns where there is a unique solution.

EXAMPLE 3

Solve the system $\begin{pmatrix} 2x + 4y - 5z = 37 \\ x + 3y - 4z = 29 \\ 5x - y + 3z = -20 \end{pmatrix}$.

Solution

The augmented matrix

$$\begin{bmatrix} 2 & 4 & -5 & \vdots & 37 \\ 1 & 3 & -4 & \vdots & 29 \\ 5 & -1 & 3 & \vdots & -20 \end{bmatrix}$$

does not have a one in the upper left-hand corner, but this can be remedied by exchanging rows 1 and 2.

$$\begin{bmatrix} 1 & 3 & -4 & \vdots & 29 \\ 2 & 4 & -5 & \vdots & 37 \\ 5 & -1 & 3 & \vdots & -20 \end{bmatrix}$$

Now we can get zeros in the first column beneath the one by adding -2 times row 1 to row 2 and by adding -5 times row 1 to row 3.

$$\left[\begin{array}{rrr:r} 1 & 3 & -4 & 29 \\ 0 & -2 & 3 & -21 \\ 0 & -16 & 23 & -165 \end{array}\right]$$

Next, we can get a one for the first nonzero entry of the second row by multiplying the second row by $-\dfrac{1}{2}$.

$$\left[\begin{array}{rrr:r} 1 & 3 & -4 & 29 \\ 0 & 1 & -\dfrac{3}{2} & \dfrac{21}{2} \\ 0 & -16 & 23 & -165 \end{array}\right]$$

Now we can get zeros above and below the one in the second column by adding -3 times row 2 to row 1 and by adding 16 times row 2 to row 3.

$$\left[\begin{array}{rrr:r} 1 & 0 & \dfrac{1}{2} & -\dfrac{5}{2} \\ 0 & 1 & -\dfrac{3}{2} & \dfrac{21}{2} \\ 0 & 0 & -1 & 3 \end{array}\right]$$

Next, we can get a one in the first nonzero entry of the third row by multiplying the third row by -1.

$$\left[\begin{array}{rrr:r} 1 & 0 & \dfrac{1}{2} & -\dfrac{5}{2} \\ 0 & 1 & -\dfrac{3}{2} & \dfrac{21}{2} \\ 0 & 0 & 1 & -3 \end{array}\right]$$

Finally, we can get zeros above the one in the third column by adding $-\dfrac{1}{2}$ times row 3 to row 1 and by adding $\dfrac{3}{2}$ times row 3 to row 2.

$$\left[\begin{array}{rrr:r} 1 & 0 & 0 & -1 \\ 0 & 1 & 0 & 6 \\ 0 & 0 & 1 & -3 \end{array}\right]$$

From this last matrix, we see that the solution set of the original system is $\{(-1, 6, -3)\}$.

Example 3 illustrates that even though the process of changing to reduced echelon form can be systematically described, it can involve some rather messy calculations. However, with the aid of a computer, such calculations are not troublesome. For our purposes in this text, the examples and problems involve systems that minimize messy calculations. This will allow us to concentrate on the procedures.

We want to call your attention to another issue in the solution of Example 3. Consider the matrix

$$\left[\begin{array}{rrr:r} 1 & 3 & -4 & 29 \\ 0 & 1 & -\dfrac{3}{2} & \dfrac{21}{2} \\ 0 & -16 & 23 & -165 \end{array}\right]$$

which is obtained about halfway through the solution. At this step, it seems evident that the calculations are getting a little messy. Therefore, instead of continuing toward the reduced echelon form, let's add 16 times row 2 to row 3 to produce a new row 3.

$$\begin{bmatrix} 1 & 3 & -4 & \vdots & 29 \\ 0 & 1 & -\dfrac{3}{2} & \vdots & \dfrac{21}{2} \\ 0 & 0 & -1 & \vdots & 3 \end{bmatrix}$$

The system represented by this matrix is

$$\begin{pmatrix} x + 3y - 4z = 29 \\ y - \dfrac{3}{2}z = \dfrac{21}{2} \\ -z = 3 \end{pmatrix}$$

and it is said to be in **triangular form.** The last equation determines the value for z; then we can use the process of back-substitution to determine the values for y and x.

Finally, let's consider two examples to illustrate what happens when we use the matrix approach on inconsistent and dependent systems.

Classroom Example
Solve the system:

$$\begin{pmatrix} x + y - z = 2 \\ 3x - 4y + 2z = 5 \\ 2x + 2y - 2z = 7 \end{pmatrix}$$

EXAMPLE 4

Solve the system $\begin{pmatrix} x - 2y + 3z = 3 \\ 5x - 9y + 4z = 2 \\ 2x - 4y + 6z = -1 \end{pmatrix}$.

Solution

The augmented matrix of the system is

$$\begin{bmatrix} 1 & -2 & 3 & \vdots & 3 \\ 5 & -9 & 4 & \vdots & 2 \\ 2 & -4 & 6 & \vdots & -1 \end{bmatrix}$$

We can get zeros below the one in the first column by adding -5 times row 1 to row 2 and by adding -2 times row 1 to row 3.

$$\begin{bmatrix} 1 & -2 & 3 & \vdots & 3 \\ 0 & 1 & -11 & \vdots & -13 \\ 0 & 0 & 0 & \vdots & -7 \end{bmatrix}$$

At this step we can stop, because the bottom row of the matrix represents the statement $0(x) + 0(y) + 0(z) = -7$, which is obviously false for all values of x, y, and z. Thus the original system is inconsistent; its solution set is $\varnothing$.

Classroom Example
Solve the system:

$$\begin{pmatrix} 4x - y + z = 5 \\ 3x + y + 2z = 4 \\ x - 2y - z = 1 \end{pmatrix}$$

EXAMPLE 5

Solve the system $\begin{pmatrix} x + 2y + 2z = 9 \\ x + 3y - 4z = 5 \\ 2x + 5y - 2z = 14 \end{pmatrix}$.

Solution

The augmented matrix of the system is

$$\begin{bmatrix} 1 & 2 & 2 & \vdots & 9 \\ 1 & 3 & -4 & \vdots & 5 \\ 2 & 5 & -2 & \vdots & 14 \end{bmatrix}$$

We can get zeros in the first column below the one in the upper left-hand corner by adding -1 times row 1 to row 2 and adding -2 times row 1 to row 3.

$$\begin{bmatrix} 1 & 2 & 2 & \vdots & 9 \\ 0 & 1 & -6 & \vdots & -4 \\ 0 & 1 & -6 & \vdots & -4 \end{bmatrix}$$

Now we can get zeros in the second column above and below the one in the second row by adding -2 times row 2 to row 1 and adding -1 times row 2 to row 3.

$$\begin{bmatrix} 1 & 0 & 14 & | & 17 \\ 0 & 1 & -6 & | & -4 \\ 0 & 0 & 0 & | & 0 \end{bmatrix}$$

The bottom row of zeros represents the statement $0(x) + 0(y) + 0(z) = 0$, which is true for all values of x, y, and z. The second row represents the statement $y - 6z = -4$, which can be rewritten $y = 6z - 4$. The top row represents the statement $x + 14z = 17$, which can be rewritten $x = -14z + 17$. Therefore, if we let $z = k$, where k is any real number, the solution set of infinitely many ordered triples can be represented by $\{(-14k + 17, 6k - 4, k) \mid k$ is a real number$\}$. Specific solutions can be generated by letting k take on a value. For example, if $k = 2$, then $6k - 4$ becomes $6(2) - 4 = 8$ and $-14k + 17$ becomes $-14(2) + 17 = -11$. Thus the ordered triple $(-11, 8, 2)$ is a member of the solution set.

Concept Quiz 15.3

For Problems 1–8, answer true or false.

1. A matrix of dimension 2×6, has 6 rows and 2 columns.

2. The augmented matrix of a system of equations is a matrix of the coefficients and constant terms of the equations.

3. Transformations that are applied to augmented matrices are called elementary column operations.

4. For any augmented matrix, two rows can be interchanged to produce an equivalent matrix.

5. For any augmented matrix, any column may be multiplied by a nonzero real number to produce an equivalent matrix.

6. The matrix $\begin{bmatrix} 1 & 0 & 0 & | & 6 \\ 0 & 1 & 1 & | & -3 \\ 0 & 0 & 1 & | & 2 \end{bmatrix}$ is in reduced echelon form.

7. The system of equations $\begin{pmatrix} x + y + 2z = 7 \\ y - 3z = 4 \\ z = -5 \end{pmatrix}$ is in triangular form.

8. Given that the matrix $\begin{bmatrix} 1 & 0 & 0 & | & 2 \\ 0 & 1 & 0 & | & -3 \\ 0 & 0 & 1 & | & 4 \end{bmatrix}$ represents a system of equations, the solution of the system is the ordered triple $(2, -3, 4)$.

Problem Set 15.3

For Problems 1–10, indicate whether each matrix is in reduced echelon form. **(Objective 1)**

1. $\begin{bmatrix} 1 & 0 & | & -4 \\ 0 & 1 & | & 14 \end{bmatrix}$

2. $\begin{bmatrix} 1 & 2 & | & 8 \\ 0 & 0 & | & 0 \end{bmatrix}$

3. $\begin{bmatrix} 1 & 0 & 2 & | & 5 \\ 0 & 1 & 3 & | & 7 \\ 0 & 0 & 0 & | & 0 \end{bmatrix}$

4. $\begin{bmatrix} 1 & 0 & 0 & | & 5 \\ 0 & 3 & 0 & | & 8 \\ 0 & 0 & 1 & | & -11 \end{bmatrix}$

5. $\begin{bmatrix} 1 & 0 & 0 & | & 17 \\ 0 & 0 & 0 & | & 0 \\ 0 & 1 & 0 & | & -14 \end{bmatrix}$

6. $\begin{bmatrix} 1 & 0 & 0 & | & -7 \\ 0 & 1 & 0 & | & 0 \\ 0 & 0 & 1 & | & 9 \end{bmatrix}$

7. $\begin{bmatrix} 1 & 1 & 0 & | & -3 \\ 0 & 1 & 2 & | & 5 \\ 0 & 0 & 1 & | & 7 \end{bmatrix}$

8. $\begin{bmatrix} 1 & 0 & 3 & | & 8 \\ 0 & 1 & 2 & | & -6 \\ 0 & 0 & 0 & | & 0 \end{bmatrix}$

9. $\begin{bmatrix} 1 & 0 & 0 & 3 & | & 4 \\ 0 & 1 & 0 & 5 & | & -3 \\ 0 & 0 & 1 & -1 & | & 7 \\ 0 & 0 & 0 & 0 & | & 0 \end{bmatrix}$

10. $\begin{bmatrix} 1 & 0 & 0 & 0 & | & 2 \\ 0 & 0 & 1 & 0 & | & 4 \\ 0 & 1 & 0 & 0 & | & -3 \\ 0 & 0 & 0 & 1 & | & 9 \end{bmatrix}$

For Problems 11–30, transform the augmented matrix to reduced echelon form to solve the system. **(Objectives 2 and 3)**

11. $\begin{pmatrix} x - 3y = 14 \\ 3x + 2y = -13 \end{pmatrix}$

12. $\begin{pmatrix} x + 5y = -18 \\ -2x + 3y = -16 \end{pmatrix}$

13. $\begin{pmatrix} 3x - 4y = 33 \\ x + 7y = -39 \end{pmatrix}$

14. $\begin{pmatrix} 2x + 7y = -55 \\ x - 4y = 25 \end{pmatrix}$

15. $\begin{pmatrix} x - 6y = -2 \\ 2x - 12y = 5 \end{pmatrix}$

16. $\begin{pmatrix} 2x - 3y = -12 \\ 3x + 2y = 8 \end{pmatrix}$

17. $\begin{pmatrix} 3x - 5y = 39 \\ 2x + 7y = -67 \end{pmatrix}$

18. $\begin{pmatrix} 3x + 9y = -1 \\ x + 3y = 10 \end{pmatrix}$

19. $\begin{pmatrix} x - 2y - 3z = -6 \\ 3x - 5y - z = 4 \\ 2x + y + 2z = 2 \end{pmatrix}$

20. $\begin{pmatrix} x + 3y - 4z = 13 \\ 2x + 7y - 3z = 11 \\ -2x - y + 2z = -8 \end{pmatrix}$

21. $\begin{pmatrix} -2x - 5y + 3z = 11 \\ x + 3y - 3z = -12 \\ 3x - 2y + 5z = 31 \end{pmatrix}$

22. $\begin{pmatrix} -3x + 2y + z = 17 \\ x - y + 5z = -2 \\ 4x - 5y - 3z = -36 \end{pmatrix}$

23. $\begin{pmatrix} x - 3y - z = 2 \\ 3x + y - 4z = -18 \\ -2x + 5y + 3z = 2 \end{pmatrix}$

24. $\begin{pmatrix} x - 4y + 3z = 16 \\ 2x + 3y - 4z = -22 \\ -3x + 11y - z = -36 \end{pmatrix}$

25. $\begin{pmatrix} x - y + 2z = 1 \\ -3x + 4y - z = 4 \\ -x + 2y + 3z = 6 \end{pmatrix}$

26. $\begin{pmatrix} x + 2y - 5z = -1 \\ 2x + 3y - 2z = 2 \\ 3x + 5y - 7z = 4 \end{pmatrix}$

27. $\begin{pmatrix} -2x + y + 5z = -5 \\ 3x + 8y - z = -34 \\ x + 2y + z = -12 \end{pmatrix}$

28. $\begin{pmatrix} 4x - 10y + 3z = -19 \\ 2x + 5y - z = -7 \\ x - 3y - 2z = -2 \end{pmatrix}$

29. $\begin{pmatrix} 2x + 3y - z = 7 \\ 3x + 4y + 5z = -2 \\ 5x + y + 3z = 13 \end{pmatrix}$ **30.** $\begin{pmatrix} 4x + 3y - z = 0 \\ 3x + 2y + 5z = 6 \\ 5x - y - 3z = 3 \end{pmatrix}$

Subscript notation is frequently used for working with larger systems of equations. For Problems 31–34, use a matrix approach to solve each system. Express the solutions as 4-tuples of the form (x_1, x_2, x_3, x_4). **(Objective 3)**

31. $\begin{pmatrix} x_1 - 3x_2 - 2x_3 + x_4 = -3 \\ -2x_1 + 7x_2 + x_3 - 2x_4 = -1 \\ 3x_1 - 7x_2 - 3x_3 + 3x_4 = -5 \\ 5x_1 + x_2 + 4x_3 - 2x_4 = 18 \end{pmatrix}$

32. $\begin{pmatrix} x_1 - 2x_2 + 2x_3 - x_4 = -2 \\ -3x_1 + 5x_2 - x_3 - 3x_4 = 2 \\ 2x_1 + 3x_2 + 3x_3 + 5x_4 = -9 \\ 4x_1 - x_2 - x_3 - 2x_4 = 8 \end{pmatrix}$

33. $\begin{pmatrix} x_1 + 3x_2 - x_3 + 2x_4 = -2 \\ 2x_1 + 7x_2 + 2x_3 - x_4 = 19 \\ -3x_1 - 8x_2 + 3x_3 + x_4 = -7 \\ 4x_1 + 11x_2 - 2x_3 - 3x_4 = 19 \end{pmatrix}$

34. $\begin{pmatrix} x_1 + 2x_2 - 3x_3 + x_4 = -2 \\ -2x_1 - 3x_2 + x_3 - x_4 = 5 \\ 4x_1 + 9x_2 - 2x_3 - 2x_4 = -28 \\ -5x_1 - 9x_2 + 2x_3 - 3x_4 = 14 \end{pmatrix}$

In Problems 35–42, each matrix is the reduced echelon matrix for a system with variables $x_1, x_2, x_3,$ and x_4. Find the solution set of each system. **(Objective 1)**

35. $\begin{bmatrix} 1 & 0 & 0 & 0 & | & -2 \\ 0 & 1 & 0 & 0 & | & 4 \\ 0 & 0 & 1 & 0 & | & -3 \\ 0 & 0 & 0 & 1 & | & 0 \end{bmatrix}$

36. $\begin{bmatrix} 1 & 0 & 0 & 0 & | & 0 \\ 0 & 1 & 0 & 0 & | & -5 \\ 0 & 0 & 1 & 0 & | & 0 \\ 0 & 0 & 0 & 1 & | & 4 \end{bmatrix}$

37. $\begin{bmatrix} 1 & 0 & 0 & 0 & | & -8 \\ 0 & 1 & 0 & 0 & | & 5 \\ 0 & 0 & 1 & 0 & | & -2 \\ 0 & 0 & 0 & 0 & | & 1 \end{bmatrix}$

38. $\begin{bmatrix} 1 & 0 & 0 & 0 & | & 2 \\ 0 & 1 & 0 & 2 & | & -3 \\ 0 & 0 & 1 & 3 & | & 4 \\ 0 & 0 & 0 & 0 & | & 0 \end{bmatrix}$

39. $\begin{bmatrix} 1 & 0 & 0 & 3 & | & 5 \\ 0 & 1 & 0 & 0 & | & -1 \\ 0 & 0 & 1 & 4 & | & 2 \\ 0 & 0 & 0 & 0 & | & 0 \end{bmatrix}$

40. $\begin{bmatrix} 1 & 3 & 0 & 2 & | & 0 \\ 0 & 0 & 1 & 0 & | & 0 \\ 0 & 0 & 0 & 0 & | & 1 \\ 0 & 0 & 0 & 0 & | & 0 \end{bmatrix}$

41. $\begin{bmatrix} 1 & 3 & 0 & 0 & | & 9 \\ 0 & 0 & 1 & 0 & | & 2 \\ 0 & 0 & 0 & 1 & | & -3 \\ 0 & 0 & 0 & 0 & | & 0 \end{bmatrix}$

42. $\begin{bmatrix} 1 & 0 & 0 & 0 & | & 7 \\ 0 & 1 & 0 & 0 & | & -3 \\ 0 & 0 & 1 & -2 & | & 5 \\ 0 & 0 & 0 & 0 & | & 0 \end{bmatrix}$

Thoughts Into Words

43. What is a matrix? What is an augmented matrix of a system of linear equations?

44. Describe how to use matrices to solve the system $\begin{pmatrix} x - 2y = 5 \\ 2x + 7y = 9 \end{pmatrix}$.

Further Investigations

For Problems 45–50, change each augmented matrix of the system to reduced echelon form and then indicate the solutions of the system.

45. $\begin{pmatrix} x - 2y + 3z = 4 \\ 3x - 5y - z = 7 \end{pmatrix}$

46. $\begin{pmatrix} x + 3y - 2z = -1 \\ 2x - 5y + 7z = 4 \end{pmatrix}$

47. $\begin{pmatrix} 2x - 4y + 3z = 8 \\ 3x + 5y - z = 7 \end{pmatrix}$

48. $\begin{pmatrix} 3x + 6y - z = 9 \\ 2x - 3y + 4z = 1 \end{pmatrix}$

49. $\begin{pmatrix} x - 2y + 4z = 9 \\ 2x - 4y + 8z = 3 \end{pmatrix}$

50. $\begin{pmatrix} x + y - 2z = -1 \\ 3x + 3y - 6z = -3 \end{pmatrix}$

Graphing Calculator Activities

51. If your graphing calculator has the capability of manipulating matrices, this is a good time to become familiar with those operations. You may need to refer to your user's manual for the key-punching instructions. To begin the familiarization process, load your calculator with the three augmented matrices in Examples 1, 2, and 3. Then, for each one, carry out the row operations as described in the text.

Answers to the Concept Quiz
1. False   **2.** True   **3.** False   **4.** True   **5.** False   **6.** False   **7.** True   **8.** True

15.4   Determinants

OBJECTIVES

1 Evaluate the determinant of a matrix

2 Compute the cofactor of an element in a matrix

3 Evaluate a determinant by expansion

4 Apply the properties of determinants to simplify the evaluation of a determinant

Before we introduce the concept of a determinant, let's agree on some convenient new notation. A **general *m* × *n*** (*m*-by-*n*) **matrix** can be represented by

$$A = \begin{bmatrix} a_{11} & a_{12} & a_{13} & \cdots & a_{1n} \\ a_{21} & a_{22} & a_{23} & \cdots & a_{2n} \\ \cdot & \cdot & \cdot & & \cdot \\ \cdot & \cdot & \cdot & & \cdot \\ \cdot & \cdot & \cdot & & \cdot \\ a_{m1} & a_{m2} & a_{m3} & \cdots & a_{mn} \end{bmatrix}$$

where the double subscripts are used to identify the number of the row and the number of the column, in that order. For example, a_{23} is the entry at the intersection of the second row and the third column. In general, the entry at the intersection of row i and column j is denoted by a_{ij}.

A **square matrix** is one that has the same number of rows as columns. Each square matrix A with real number entries can be associated with a real number called the **determinant** of the matrix, denoted by $|A|$. We will first define $|A|$ for a 2×2 matrix.

Definition 15.1

If $A = \begin{bmatrix} a_{11} & a_{12} \\ a_{21} & a_{22} \end{bmatrix}$, then

$$|A| = \begin{vmatrix} a_{11} & a_{12} \\ a_{21} & a_{22} \end{vmatrix} = a_{11}a_{22} - a_{12}a_{21}$$

Classroom Example
Find the determinant of the matrix
$\begin{bmatrix} 4 & 6 \\ -3 & 7 \end{bmatrix}$

EXAMPLE 1

If $A = \begin{bmatrix} 3 & -2 \\ 5 & 8 \end{bmatrix}$, find $|A|$.

Solution

Use Definition 15.1 to obtain

$$|A| = \begin{vmatrix} 3 & -2 \\ 5 & 8 \end{vmatrix} = 3(8) - (-2)(5)$$

$$= 24 + 10$$

$$= 34$$

Finding the determinant of a square matrix is commonly called **evaluating the determinant,** and the matrix notation is often omitted.

Classroom Example
Evaluate:
$\begin{vmatrix} 8 & 1 \\ 6 & -3 \end{vmatrix}$

EXAMPLE 2

Evaluate $\begin{vmatrix} -3 & 6 \\ 2 & 8 \end{vmatrix}$.

Solution

$$\begin{vmatrix} -3 & 6 \\ 2 & 8 \end{vmatrix} = (-3)(8) - (6)(2)$$

$$= -24 - 12$$

$$= -36$$

To find the determinants of 3×3 and larger square matrices, it is convenient to introduce some additional terminology.

Definition 15.2

If A is a 3×3 matrix, then the **minor** (denoted by M_{ij}) of the a_{ij} element is the determinant of the 2×2 matrix obtained by deleting row i and column j of A.

Classroom Example

If $A = \begin{bmatrix} 1 & 2 & 1 \\ 3 & 1 & -2 \\ 2 & 4 & 3 \end{bmatrix}$, find M_{21}.

EXAMPLE 3

If $A = \begin{bmatrix} 2 & 1 & 4 \\ -6 & 3 & -2 \\ 4 & 2 & 5 \end{bmatrix}$, find (a) M_{11} and (b) M_{23}.

Solution

(a) To find M_{11}, we first delete row 1 and column 1 of matrix A.

$$\begin{bmatrix} 2 & 1 & 4 \\ -6 & 3 & -2 \\ 4 & 2 & 5 \end{bmatrix}$$

Thus

$$M_{11} = \begin{vmatrix} 3 & -2 \\ 2 & 5 \end{vmatrix} = 3(5) - (-2)(2) = 19$$

(b) To find M_{23}, we first delete row 2 and column 3 of matrix A.

$$\begin{bmatrix} 2 & 1 & 4 \\ -6 & 3 & -2 \\ 4 & 2 & 5 \end{bmatrix}$$

Thus

$$M_{23} = \begin{vmatrix} 2 & 1 \\ 4 & 2 \end{vmatrix} = 2(2) - (1)(4) = 0$$

The following definition will also be used.

Definition 15.3

If A is a 3×3 matrix, then the **cofactor** (denoted by C_{ij}) of the element a_{ij} is defined by

$$C_{ij} = (-1)^{i+j} M_{ij}$$

According to Definition 15.3, to find the cofactor of any element a_{ij} of a square matrix A, we find the minor of a_{ij} and multiply it by 1 if $i + j$ is even, or multiply it by -1 if $i + j$ is odd.

Classroom Example

If $A = \begin{bmatrix} 4 & -2 & 1 \\ 2 & 1 & 3 \\ -3 & 2 & 1 \end{bmatrix}$, find C_{13}.

EXAMPLE 4

If $A = \begin{bmatrix} 3 & 2 & -4 \\ 1 & 5 & 4 \\ 2 & -3 & 1 \end{bmatrix}$, find C_{32}.

Solution

First, let's find M_{32} by deleting row 3 and column 2 of matrix A.

$$\begin{bmatrix} 3 & 2 & -4 \\ 1 & 5 & 4 \\ 2 & -3 & 1 \end{bmatrix}$$

Thus

$$M_{32} = \begin{vmatrix} 3 & -4 \\ 1 & 4 \end{vmatrix} = 3(4) - (-4)(1) = 16$$

Therefore,

$$C_{32} = (-1)^{3+2} M_{32} = (-1)^5 (16) = -16$$

The concept of a cofactor can be used to define the determinant of a 3×3 matrix as follows:

Definition 15.4

If $A = \begin{bmatrix} a_{11} & a_{12} & a_{13} \\ a_{21} & a_{22} & a_{23} \\ a_{31} & a_{32} & a_{33} \end{bmatrix}$, then

$$|A| = a_{11}C_{11} + a_{21}C_{21} + a_{31}C_{31}$$

Definition 15.4 simply states that the determinant of a 3×3 matrix can be found by multiplying each element of the first column by its corresponding cofactor and then adding the three results. Let's illustrate this procedure.

Classroom Example

Find $|A|$ if $A = \begin{bmatrix} 4 & -2 & 1 \\ 2 & 1 & 3 \\ -3 & 2 & -1 \end{bmatrix}$.

EXAMPLE 5

Find $|A|$ if $A = \begin{bmatrix} -2 & 1 & 4 \\ 3 & 0 & 5 \\ 1 & -4 & -6 \end{bmatrix}$.

Solution

$$\begin{aligned}
|A| &= a_{11}C_{11} + a_{21}C_{21} + a_{31}C_{31} \\
&= (-2)(-1)^{1+1}\begin{vmatrix} 0 & 5 \\ -4 & -6 \end{vmatrix} + (3)(-1)^{2+1}\begin{vmatrix} 1 & 4 \\ -4 & -6 \end{vmatrix} + (1)(-1)^{3+1}\begin{vmatrix} 1 & 4 \\ 0 & 5 \end{vmatrix} \\
&= (-2)(1)(20) + (3)(-1)(10) + (1)(1)(5) \\
&= -40 - 30 + 5 \\
&= -65
\end{aligned}$$

When we use Definition 15.4, we often say that "the determinant is being expanded about the first column." It can also be shown that **any row or column can be used to expand a**

determinant. For example, for matrix A in Example 5, the expansion of the determinant about the *second row* is as follows:

$$\begin{vmatrix} -2 & 1 & 4 \\ 3 & 0 & 5 \\ 1 & -4 & -6 \end{vmatrix} = (3)(-1)^{2+1}\begin{vmatrix} 1 & 4 \\ -4 & -6 \end{vmatrix} + (0)(-1)^{2+2}\begin{vmatrix} -2 & 4 \\ 1 & -6 \end{vmatrix} + (5)(-1)^{2+3}\begin{vmatrix} -2 & 1 \\ 1 & -4 \end{vmatrix}$$

$$= (3)(-1)(10) + (0)(1)(8) + (5)(-1)(7)$$

$$= -30 + 0 - 35$$

$$= -65$$

Note that when we expanded about the second row, the computation was simplified by the presence of a zero. In general, it is helpful to expand about the row or column that contains the most zeros.

The concepts of minor and cofactor have been defined in terms of 3×3 matrices. Analogous definitions can be given for any square matrix (that is, any $n \times n$ matrix with $n \geq 2$), and the determinant can then be expanded about any row or column. Certainly, as the matrices become larger than 3×3, the computations get more tedious. We will concentrate most of our efforts in this text on 2×2 and 3×3 matrices.

Properties of Determinants

Determinants have several interesting properties, some of which are important primarily from a theoretical standpoint. But some of the properties are also very useful when evaluating determinants. We will state these properties for square matrices in general, but we will use 2×2 or 3×3 matrices as examples. We can demonstrate some of the proofs of these properties by evaluating the determinants involved; some of the proofs for 3×3 matrices will be left for you to verify in the next problem set.

Property 15.1

If any row (or column) of a square matrix A contains only zeros, then $|A| = 0$.

If every element of a row (or column) of a square matrix A is 0, then it should be evident that expanding the determinant about that row (or column) of zeros will produce 0.

Property 15.2

If square matrix B is obtained from square matrix A by interchanging two rows (or two columns), then $|B| = -|A|$.

Property 15.2 states that **interchanging two rows (or columns) changes the sign of the determinant.** As an example of this property, suppose that

$$A = \begin{bmatrix} 2 & 5 \\ -1 & 6 \end{bmatrix}$$

and that rows 1 and 2 are interchanged to form

$$B = \begin{bmatrix} -1 & 6 \\ 2 & 5 \end{bmatrix}$$

Calculating $|A|$ and $|B|$ yields

$$|A| = \begin{vmatrix} 2 & 5 \\ -1 & 6 \end{vmatrix} = 2(6) - (5)(-1) = 17$$

and

$$|B| = \begin{vmatrix} -1 & 6 \\ 2 & 5 \end{vmatrix} = (-1)(5) - (6)(2) = -17$$

Property 15.3

If square matrix B is obtained from square matrix A by multiplying each element of any row (or column) of A by some real number k, then $|B| = k|A|$.

Property 15.3 states that **multiplying any row (or column) by a factor of k affects the value of the determinant by a factor of k.** As an example of this property, suppose that

$$A = \begin{bmatrix} 1 & -2 & 8 \\ 2 & 1 & 12 \\ 3 & 2 & -16 \end{bmatrix}$$

and that B is formed by multiplying each element of the third column by $\dfrac{1}{4}$.

$$B = \begin{bmatrix} 1 & -2 & 2 \\ 2 & 1 & 3 \\ 3 & 2 & -4 \end{bmatrix}$$

Now let's calculate $|A|$ and $|B|$ by expanding about the third column in each case.

$$|A| = \begin{vmatrix} 1 & -2 & 8 \\ 2 & 1 & 12 \\ 3 & 2 & -16 \end{vmatrix} = (8)(-1)^{1+3}\begin{vmatrix} 2 & 1 \\ 3 & 2 \end{vmatrix} + (12)(-1)^{2+3}\begin{vmatrix} 1 & -2 \\ 3 & 2 \end{vmatrix} + (-16)(-1)^{3+3}\begin{vmatrix} 1 & -2 \\ 2 & 1 \end{vmatrix}$$

$$= (8)(1)(1) + (12)(-1)(8) + (-16)(1)(5)$$

$$= -168$$

$$|B| = \begin{vmatrix} 1 & -2 & 2 \\ 2 & 1 & 3 \\ 3 & 2 & -4 \end{vmatrix} = (2)(-1)^{1+3}\begin{vmatrix} 2 & 1 \\ 3 & 2 \end{vmatrix} + (3)(-1)^{2+3}\begin{vmatrix} 1 & -2 \\ 3 & 2 \end{vmatrix} + (-4)(-1)^{3+3}\begin{vmatrix} 1 & -2 \\ 2 & 1 \end{vmatrix}$$

$$= (2)(1)(1) + (3)(-1)(8) + (-4)(1)(5)$$

$$= -42$$

We see that $|B| = \dfrac{1}{4}|A|$. This example also illustrates the computational use of Property 15.3:

We can factor out a common factor from a row or column, and then adjust the value of the determinant by that factor. For example,

$$\begin{vmatrix} 2 & 6 & 8 \\ -1 & 2 & 7 \\ 5 & 2 & 1 \end{vmatrix} = 2\begin{vmatrix} 1 & 3 & 4 \\ -1 & 2 & 7 \\ 5 & 2 & 1 \end{vmatrix}$$

Factor a 2 from
the top row

> ### Property 15.4
>
> If square matrix B is obtained from square matrix A by adding k times a row (or column) of A to another row (or column) of A, then $|B| = |A|$.

Property 15.4 states that **adding the product of k times a row (or column) to another row (or column) does not affect the value of the determinant.** As an example of this property, suppose that

$$A = \begin{bmatrix} 1 & 2 & 4 \\ 2 & 4 & 7 \\ -1 & 3 & 5 \end{bmatrix}$$

Now let's form B by replacing row 2 with the result of adding -2 times row 1 to row 2.

$$B = \begin{bmatrix} 1 & 2 & 4 \\ 0 & 0 & -1 \\ -1 & 3 & 5 \end{bmatrix}$$

Next, let's evaluate $|A|$ and $|B|$ by expanding about the second row in each case.

$$|A| = \begin{vmatrix} 1 & 2 & 4 \\ 2 & 4 & 7 \\ -1 & 3 & 5 \end{vmatrix} = (2)(-1)^{2+1}\begin{vmatrix} 2 & 4 \\ 3 & 5 \end{vmatrix} + (4)(-1)^{2+2}\begin{vmatrix} 1 & 4 \\ -1 & 5 \end{vmatrix} + (7)(-1)^{2+3}\begin{vmatrix} 1 & 2 \\ -1 & 3 \end{vmatrix}$$

$$= 2(-1)(-2) + (4)(1)(9) + (7)(-1)(5)$$

$$= 5$$

$$|B| = \begin{vmatrix} 1 & 2 & 4 \\ 0 & 0 & -1 \\ -1 & 3 & 5 \end{vmatrix} = (0)(-1)^{2+1}\begin{vmatrix} 2 & 4 \\ 3 & 5 \end{vmatrix} + (0)(-1)^{2+2}\begin{vmatrix} 1 & 4 \\ -1 & 5 \end{vmatrix} + (-1)(-1)^{2+3}\begin{vmatrix} 1 & 2 \\ -1 & 3 \end{vmatrix}$$

$$= 0 + 0 + (-1)(-1)(5)$$

$$= 5$$

Note that $|B| = |A|$. Furthermore, note that because of the zeros in the second row, evaluating $|B|$ is much easier than evaluating $|A|$. Property 15.4 can often be used to obtain some zeros before evaluating a determinant.

A word of caution is in order at this time. Be careful not to confuse Properties 15.2, 15.3, and 15.4 with the three elementary row transformations of augmented matrices that were used in Section 15.3. The statements of the two sets of properties do resemble each other, but the properties pertain to *two different concepts,* so be sure you understand the distinction between them.

One final property of determinants should be mentioned.

> ### Property 15.5
>
> If two rows (or columns) of a square matrix A are identical, then $|A| = 0$.

Property 15.5 is a direct consequence of Property 15.2. Suppose that A is a square matrix (any size) with two identical rows. Square matrix B can be formed from A by interchanging the two identical rows. Because identical rows were interchanged, $|B| = |A|$. *But* by Property 15.2, $|B| = -|A|$. For both of these statements to hold, $|A| = 0$.

Let's conclude this section by evaluating a 4×4 determinant, using Properties 15.3 and 15.4 to facilitate the computation.

Classroom Example
Evaluate:

$$\begin{vmatrix} 4 & -1 & 0 & 2 \\ 1 & -3 & 0 & 5 \\ -2 & 4 & 1 & 6 \\ 3 & -2 & 2 & -1 \end{vmatrix}$$

EXAMPLE 6

Evaluate $\begin{vmatrix} 6 & 2 & 1 & -2 \\ 9 & -1 & 4 & 1 \\ 12 & -2 & 3 & -1 \\ 0 & 0 & 9 & 3 \end{vmatrix}$.

Solution

First, let's add -3 times the fourth column to the third column.

$$\begin{vmatrix} 6 & 2 & 7 & -2 \\ 9 & -1 & 1 & 1 \\ 12 & -2 & 6 & -1 \\ 0 & 0 & 0 & 3 \end{vmatrix}$$

Now, if we expand about the fourth row, we get only one nonzero product.

$$(3)(-1)^{4+4}\begin{vmatrix} 6 & 2 & 7 \\ 9 & -1 & 1 \\ 12 & -2 & 6 \end{vmatrix}$$

Factoring a 3 out of the first column of the 3×3 determinant yields

$$(3)(-1)^{8}(3)\begin{vmatrix} 2 & 2 & 7 \\ 3 & -1 & 1 \\ 4 & -2 & 6 \end{vmatrix}$$

Next, working with the 3×3 determinant, we can first add column 3 to column 2 and then add -3 times column 3 to column 1.

$$(3)(-1)^{8}(3)\begin{vmatrix} -19 & 9 & 7 \\ 0 & 0 & 1 \\ -14 & 4 & 6 \end{vmatrix}$$

Finally, by expanding this 3×3 determinant about the second row, we obtain

$$(3)(-1)^{8}(3)(1)(-1)^{2+3}\begin{vmatrix} -19 & 9 \\ -14 & 4 \end{vmatrix}$$

Our final result is

$$(3)(-1)^{8}(3)(1)(-1)^{5}(50) = -450$$

Concept Quiz 15.4

For Problems 1–8, answer true or false.

1. A square matrix has the same number of rows and columns.

2. A determinant can be calculated for any matrix.

3. The determinant of a matrix can be zero.

4. The a_{14} element of a matrix is in the 4th row and 1st column.

5. The minor of the element a_{31} is the determinant of the matrix obtained by deleting the 3rd row and 1st column.

6. Given $A = \begin{bmatrix} 1 & -2 & 3 \\ 0 & 4 & -1 \\ 5 & -3 & 2 \end{bmatrix}$, the cofactor $C_{23} = \begin{vmatrix} 1 & -2 \\ 5 & -3 \end{vmatrix}$.

7. Interchanging two columns of a matrix has no effect on the determinant.

8. Multiplying a row of a matrix by 2 affects the determinant of that matrix by a factor of $\frac{1}{2}$.

Problem Set 15.4

For Problems 1–12, evaluate each 2×2 determinant by using Definition 15.1. (Objective 1)

1. $\begin{vmatrix} 4 & 3 \\ 2 & 7 \end{vmatrix}$ **2.** $\begin{vmatrix} 3 & 5 \\ 6 & 4 \end{vmatrix}$ **3.** $\begin{vmatrix} -3 & 2 \\ 7 & 5 \end{vmatrix}$

4. $\begin{vmatrix} 5 & 3 \\ 6 & -1 \end{vmatrix}$ **5.** $\begin{vmatrix} 2 & -3 \\ 8 & -2 \end{vmatrix}$ **6.** $\begin{vmatrix} -5 & 5 \\ -6 & 2 \end{vmatrix}$

7. $\begin{vmatrix} -2 & -3 \\ -1 & -4 \end{vmatrix}$ **8.** $\begin{vmatrix} -4 & -3 \\ -5 & -7 \end{vmatrix}$ **9.** $\begin{vmatrix} \frac{1}{2} & \frac{1}{3} \\ -3 & -6 \end{vmatrix}$

10. $\begin{vmatrix} \frac{2}{3} & \frac{3}{4} \\ 8 & 6 \end{vmatrix}$ **11.** $\begin{vmatrix} \frac{1}{2} & \frac{2}{3} \\ \frac{3}{4} & -\frac{1}{3} \end{vmatrix}$ **12.** $\begin{vmatrix} \frac{2}{3} & \frac{1}{5} \\ -\frac{1}{4} & \frac{3}{2} \end{vmatrix}$

For Problems 13–28, evaluate each 3×3 determinant. Use the properties of determinants to your advantage. (Objectives 1 and 4)

13. $\begin{vmatrix} 1 & 2 & -1 \\ 3 & 1 & 2 \\ 2 & 4 & 3 \end{vmatrix}$ **14.** $\begin{vmatrix} 1 & -2 & 1 \\ 2 & 1 & -1 \\ 3 & 2 & 4 \end{vmatrix}$

15. $\begin{vmatrix} 1 & -4 & 1 \\ 2 & 5 & -1 \\ 3 & 3 & 4 \end{vmatrix}$ **16.** $\begin{vmatrix} 3 & -2 & 1 \\ 2 & 1 & 4 \\ -1 & 3 & 5 \end{vmatrix}$

17. $\begin{vmatrix} 6 & 12 & 3 \\ -1 & 5 & 1 \\ -3 & 6 & 2 \end{vmatrix}$ **18.** $\begin{vmatrix} 2 & 35 & 5 \\ 1 & -5 & 1 \\ -4 & 15 & 2 \end{vmatrix}$

19. $\begin{vmatrix} 2 & -1 & 3 \\ 0 & 3 & 1 \\ 1 & -2 & -1 \end{vmatrix}$ **20.** $\begin{vmatrix} 2 & -17 & 3 \\ 0 & 5 & 1 \\ 1 & -3 & -1 \end{vmatrix}$

21. $\begin{vmatrix} -3 & -2 & 1 \\ 5 & 0 & 6 \\ 2 & 1 & -4 \end{vmatrix}$ **22.** $\begin{vmatrix} -5 & 1 & -1 \\ 3 & 4 & 2 \\ 0 & 2 & -3 \end{vmatrix}$

23. $\begin{vmatrix} 3 & -4 & -2 \\ 5 & -2 & 1 \\ 1 & 0 & 0 \end{vmatrix}$ **24.** $\begin{vmatrix} -6 & 5 & 3 \\ 2 & 0 & -1 \\ 4 & 0 & 7 \end{vmatrix}$

25. $\begin{vmatrix} 24 & -1 & 4 \\ 40 & 2 & 0 \\ -16 & 6 & 0 \end{vmatrix}$ **26.** $\begin{vmatrix} 2 & -1 & 3 \\ 0 & 3 & 1 \\ 4 & -8 & -4 \end{vmatrix}$

27. $\begin{vmatrix} 2 & 3 & -4 \\ 4 & 6 & -1 \\ -6 & 1 & -2 \end{vmatrix}$ **28.** $\begin{vmatrix} 1 & 2 & -3 \\ -3 & -1 & 1 \\ 4 & 5 & 4 \end{vmatrix}$

For Problems 29–32, evaluate each 4×4 determinant. Use the properties of determinants to your advantage. (Objectives 1 and 4)

29. $\begin{vmatrix} 1 & -2 & 3 & 2 \\ 2 & -1 & 0 & 4 \\ -3 & 4 & 0 & -2 \\ -1 & 1 & 1 & 5 \end{vmatrix}$ **30.** $\begin{vmatrix} 1 & 2 & 5 & 7 \\ -6 & 3 & 0 & 9 \\ -3 & 5 & 2 & 7 \\ 2 & 1 & 4 & 3 \end{vmatrix}$

31. $\begin{vmatrix} 3 & -1 & 2 & 3 \\ 1 & 0 & 2 & 1 \\ 2 & 3 & 0 & 1 \\ 5 & 2 & 4 & -5 \end{vmatrix}$ **32.** $\begin{vmatrix} 1 & 2 & 0 & 0 \\ 3 & -1 & 4 & 5 \\ -2 & 4 & 1 & 6 \\ 2 & -1 & -2 & -3 \end{vmatrix}$

For Problems 33–42, use the appropriate property of determinants from this section to justify each true statement. *Do not* evaluate the determinants. (Objective 4)

33. $(-4)\begin{vmatrix} 2 & 1 & -1 \\ 3 & 2 & 1 \\ 2 & 1 & 3 \end{vmatrix} = \begin{vmatrix} 2 & -4 & -1 \\ 3 & -8 & 1 \\ 2 & -4 & 3 \end{vmatrix}$

34. $\begin{vmatrix} 1 & -2 & 3 \\ 4 & -6 & -8 \\ 0 & 2 & 7 \end{vmatrix} = (-2)\begin{vmatrix} 1 & -2 & 3 \\ -2 & 3 & 4 \\ 0 & 2 & 7 \end{vmatrix}$

35. $\begin{vmatrix} 4 & 7 & 9 \\ 6 & -8 & 2 \\ 4 & 3 & -1 \end{vmatrix} = -\begin{vmatrix} 4 & 9 & 7 \\ 6 & 2 & -8 \\ 4 & -1 & 3 \end{vmatrix}$

36. $\begin{vmatrix} 3 & -1 & 4 \\ 5 & 2 & 7 \\ 3 & -1 & 4 \end{vmatrix} = 0$

37. $\begin{vmatrix} 1 & 3 & 4 \\ -2 & 5 & 7 \\ -3 & -1 & 2 \end{vmatrix} = \begin{vmatrix} 1 & 3 & 4 \\ -2 & 5 & 7 \\ 0 & 8 & 14 \end{vmatrix}$

38. $\begin{vmatrix} 3 & 2 & 0 \\ 1 & 4 & 1 \\ -4 & 9 & 2 \end{vmatrix} = \begin{vmatrix} 3 & 2 & -3 \\ 1 & 4 & 0 \\ -4 & 9 & 6 \end{vmatrix}$

39. $\begin{vmatrix} 6 & 2 & 2 \\ 3 & -1 & 4 \\ 9 & -3 & 6 \end{vmatrix} = 6\begin{vmatrix} 2 & 2 & 1 \\ 1 & -1 & 2 \\ 3 & -3 & 3 \end{vmatrix} = 18\begin{vmatrix} 2 & 2 & 1 \\ 1 & -1 & 2 \\ 1 & -1 & 1 \end{vmatrix}$

40. $\begin{vmatrix} 2 & 1 & -3 \\ 0 & 2 & -4 \\ -5 & 1 & 3 \end{vmatrix} = -\begin{vmatrix} 2 & 1 & -3 \\ -5 & 1 & 3 \\ 0 & 2 & -4 \end{vmatrix}$

41. $\begin{vmatrix} 2 & -3 & 2 \\ 1 & -4 & 1 \\ 7 & 8 & 7 \end{vmatrix} = 0$

42. $\begin{vmatrix} 3 & 1 & 2 \\ -4 & 5 & -1 \\ 2 & -2 & -4 \end{vmatrix} = \begin{vmatrix} 3 & 1 & 0 \\ -4 & 5 & -11 \\ 2 & -2 & 0 \end{vmatrix}$

Thoughts Into Words

43. Explain the difference between a matrix and a determinant.

44. Explain the concept of a cofactor and how it is used to help expand a determinant.

45. What does it mean to say that any row or column can be used to expand a determinant?

46. Give a step-by-step explanation of how to evaluate the determinant

$$\begin{vmatrix} 3 & 0 & 2 \\ 1 & -2 & 5 \\ 6 & 0 & 9 \end{vmatrix}$$

Further Investigations

For Problems 47–50, use

$$A = \begin{bmatrix} a_{11} & a_{12} & a_{13} \\ a_{21} & a_{22} & a_{23} \\ a_{31} & a_{32} & a_{33} \end{bmatrix}$$

as a general representation for any 3×3 matrix.

47. Verify Property 15.2 for 3×3 matrices.

48. Verify Property 15.3 for 3×3 matrices.

49. Verify Property 15.4 for 3×3 matrices.

50. Show that $|A| = a_{11}a_{22}a_{33}a_{44}$ if

$$A = \begin{bmatrix} a_{11} & a_{12} & a_{13} & a_{14} \\ 0 & a_{22} & a_{23} & a_{24} \\ 0 & 0 & a_{33} & a_{34} \\ 0 & 0 & 0 & a_{44} \end{bmatrix}$$

Graphing Calculator Activities

51. Use a calculator to check your answers for Problems 29–32.

52. Consider the following matrix:

$$A = \begin{bmatrix} 2 & 5 & 7 & 9 \\ -4 & 6 & 2 & 4 \\ 6 & 9 & 12 & 3 \\ 5 & 4 & -2 & 8 \end{bmatrix}$$

Form matrix B by interchanging rows 1 and 3 of matrix A. Now use your calculator to show that $|B| = -|A|$.

53. Consider the following matrix:

$$A = \begin{bmatrix} 2 & 1 & 7 & 6 & 8 \\ 3 & -2 & 4 & 5 & -1 \\ 6 & 7 & 9 & 12 & 13 \\ -4 & -7 & 6 & 2 & 1 \\ 9 & 8 & 12 & 14 & 17 \end{bmatrix}$$

Form matrix B by multiplying each element of the second row of matrix A by 3. Now use your calculator to show that $|B| = 3|A|$.

54. Consider the following matrix:

$$A = \begin{bmatrix} 4 & 3 & 2 & 1 & 5 & -3 \\ 5 & 2 & 7 & 8 & 6 & 3 \\ 0 & 9 & 1 & 4 & 7 & 2 \\ 4 & 3 & 2 & 1 & 5 & -3 \\ -4 & -6 & 7 & 12 & 11 & 9 \\ 5 & 8 & 6 & -3 & 2 & -1 \end{bmatrix}$$

Use your calculator to show that $|A| = 0$.

Answers to the Concept Quiz

1. True   **2.** False   **3.** True   **4.** False   **5.** True   **6.** False   **7.** False   **8.** False

15.5   Cramer's Rule

OBJECTIVES

1 Set up and evaluate the determinants necessary to use Cramer's rule

2 Use Cramer's rule to solve a system of equations

Determinants provide the basis for another method of solving linear systems. Consider the following linear system of two equations and two unknowns.

$$\begin{pmatrix} a_1x + b_1y = c_1 \\ a_2x + b_2y = c_2 \end{pmatrix}$$

The augmented matrix of this system is

$$\left[\begin{array}{cc|c} a_1 & b_1 & c_1 \\ a_2 & b_2 & c_2 \end{array} \right]$$

Using the elementary row transformation of augmented matrices, we can change this matrix to the following reduced echelon form. (The details of this are left for you to do as an exercise.)

$$\left[\begin{array}{cc|c} 1 & 0 & \dfrac{c_1b_2 - c_2b_1}{a_1b_2 - a_2b_1} \\ 0 & 1 & \dfrac{a_1c_2 - a_2c_1}{a_1b_2 - a_2b_1} \end{array} \right], \qquad a_1b_2 - a_2b_1 \neq 0$$

The solution for x and y can be expressed in determinant form as follows:

$$x = \frac{c_1b_2 - c_2b_1}{a_1b_2 - a_2b_1} = \frac{\begin{vmatrix} c_1 & b_1 \\ c_2 & b_2 \end{vmatrix}}{\begin{vmatrix} a_1 & b_1 \\ a_2 & b_2 \end{vmatrix}}$$

$$y = \frac{a_1c_2 - a_2c_1}{a_1b_2 - a_2b_1} = \frac{\begin{vmatrix} a_1 & c_1 \\ a_2 & c_2 \end{vmatrix}}{\begin{vmatrix} a_1 & b_1 \\ a_2 & b_2 \end{vmatrix}}$$

This method of using determinants to solve a system of two linear equations in two variables is called **Cramer's rule** and can be stated as follows:

Cramer's Rule (2 × 2 case)

Given the system

$$\begin{pmatrix} a_1x + b_1y = c_1 \\ a_2x + b_2y = c_2 \end{pmatrix}$$

with

$$D = \begin{vmatrix} a_1 & b_1 \\ a_2 & b_2 \end{vmatrix} \neq 0$$

$$D_x = \begin{vmatrix} c_1 & b_1 \\ c_2 & b_2 \end{vmatrix} \qquad \text{and} \qquad D_y = \begin{vmatrix} a_1 & c_1 \\ a_2 & c_2 \end{vmatrix}$$

the solution for this system is given by

$$x = \frac{D_x}{D} \qquad \text{and} \qquad y = \frac{D_y}{D}$$

Note that the elements of D are the coefficients of the variables in the given system. In D_x the coefficients of x are replaced by the corresponding constants, and in D_y the coefficients of y are replaced by the corresponding constants. Let's illustrate the use of Cramer's rule to solve some systems.

Classroom Example
Solve the system:
$$\begin{pmatrix} 3x + y = -11 \\ x + 5y = 15 \end{pmatrix}$$

EXAMPLE 1   Solve the system $\begin{pmatrix} 6x + 3y = 2 \\ 3x + 2y = -4 \end{pmatrix}$.

Solution

The system is in the proper form for us to apply Cramer's rule, so let's determine D, D_x, and D_y.

$$D = \begin{vmatrix} 6 & 3 \\ 3 & 2 \end{vmatrix} = 12 - 9 = 3$$

$$D_x = \begin{vmatrix} 2 & 3 \\ -4 & 2 \end{vmatrix} = 4 + 12 = 16$$

$$D_y = \begin{vmatrix} 6 & 2 \\ 3 & -4 \end{vmatrix} = -24 - 6 = -30$$

Therefore,

$$x = \frac{D_x}{D} = \frac{16}{3}$$

and

$$y = \frac{D_y}{D} = \frac{-30}{3} = -10$$

The solution set is $\left\{ \left(\frac{16}{3}, -10 \right) \right\}$.

Classroom Example
Solve the system:
$$\begin{pmatrix} y = -8x - 4 \\ 4x + 7y = 11 \end{pmatrix}$$

EXAMPLE 2   Solve the system $\begin{pmatrix} y = -2x - 2 \\ 4x - 5y = 17 \end{pmatrix}$.

Solution

To begin, we must change the form of the first equation so that the system fits the form given in Cramer's rule. The equation $y = -2x - 2$ can be rewritten $2x + y = -2$. The system now becomes

$$\begin{pmatrix} 2x + y = -2 \\ 4x - 5y = 17 \end{pmatrix}$$

and we can proceed to determine D, D_x, and D_y.

$$D = \begin{vmatrix} 2 & 1 \\ 4 & -5 \end{vmatrix} = -10 - 4 = -14$$

$$D_x = \begin{vmatrix} -2 & 1 \\ 17 & -5 \end{vmatrix} = 10 - 17 = -7$$

$$D_y = \begin{vmatrix} 2 & -2 \\ 4 & 17 \end{vmatrix} = 34 - (-8) = 42$$

Thus,

$$x = \frac{D_x}{D} = \frac{-7}{-14} = \frac{1}{2} \qquad \text{and} \qquad y = \frac{D_y}{D} = \frac{42}{-14} = -3$$

The solution set is $\left\{\left(\frac{1}{2}, -3\right)\right\}$, which can be verified, as always, by substituting back into the original equations.

Classroom Example
Solve the system:

$$\left(\begin{array}{l} x + \dfrac{2}{3}y = -6 \\ -\dfrac{1}{4}x + 3y = -8 \end{array} \right)$$

EXAMPLE 3

Solve the system $\left(\begin{array}{l} \dfrac{1}{2}x + \dfrac{2}{3}y = -4 \\ \dfrac{1}{4}x - \dfrac{3}{2}y = 20 \end{array} \right)$.

Solution

With such a system, either we can first produce an equivalent system with integral coefficients and then apply Cramer's rule, or we can apply the rule immediately. Let's avoid some work with fractions by multiplying the first equation by 6 and the second equation by 4 to produce the following equivalent system:

$$\left(\begin{array}{l} 3x + 4y = -24 \\ x - 6y = 80 \end{array} \right)$$

Now we can proceed as before.

$$D = \begin{vmatrix} 3 & 4 \\ 1 & -6 \end{vmatrix} = -18 - 4 = -22$$

$$D_x = \begin{vmatrix} -24 & 4 \\ 80 & -6 \end{vmatrix} = 144 - 320 = -176$$

$$D_y = \begin{vmatrix} 3 & -24 \\ 1 & 80 \end{vmatrix} = 240 - (-24) = 264$$

Therefore,

$$x = \frac{D_x}{D} = \frac{-176}{-22} = 8 \qquad \text{and} \qquad y = \frac{D_y}{D} = \frac{264}{-22} = -12$$

The solution set is $\{(8, -12)\}$.

In the statement of Cramer's rule, the condition that $D \neq 0$ was imposed. If $D = 0$ and either D_x or D_y (or both) is nonzero, then the system is inconsistent and has no solution. If $D = 0$, $D_x = 0$, and $D_y = 0$, then the equations are dependent and there are infinitely many solutions.

Cramer's Rule Extended

Without showing the details, we will simply state that Cramer's rule also applies to solving systems of three linear equations in three variables. It can be stated as follows:

Cramer's Rule (3 × 3 case)

Given the system

$$\left(\begin{array}{l} a_1x + b_1y + c_1z = d_1 \\ a_2x + b_2y + c_2z = d_2 \\ a_3x + b_3y + c_3z = d_3 \end{array} \right)$$

with

$$D = \begin{vmatrix} a_1 & b_1 & c_1 \\ a_2 & b_2 & c_2 \\ a_3 & b_3 & c_3 \end{vmatrix} \neq 0 \qquad D_x = \begin{vmatrix} d_1 & b_1 & c_1 \\ d_2 & b_2 & c_2 \\ d_3 & b_3 & c_3 \end{vmatrix}$$

$$D_y = \begin{vmatrix} a_1 & d_1 & c_1 \\ a_2 & d_2 & c_2 \\ a_3 & d_3 & c_3 \end{vmatrix} \qquad D_z = \begin{vmatrix} a_1 & b_1 & d_1 \\ a_2 & b_2 & d_2 \\ a_3 & b_3 & d_3 \end{vmatrix}$$

we have

$$x = \frac{D_x}{D}, \qquad y = \frac{D_y}{D}, \qquad \text{and} \qquad z = \frac{D_z}{D}$$

Again, note the restriction that $D \neq 0$. If $D = 0$ and at least one of D_x, D_y, and D_z is not zero, then the system is inconsistent. If D, D_x, D_y, and D_z are all zero, then the equations are dependent, and there are infinitely many solutions.

Classroom Example

Solve the system:

$$\begin{pmatrix} 2x - y + 3z = 1 \\ x + 4y - 2z = 3 \\ 3x - 2y + 3z = 5 \end{pmatrix}$$

EXAMPLE 4

Solve the system $\begin{pmatrix} x - 2y + z = -4 \\ 2x + y - z = 5 \\ 3x + 2y + 4z = 3 \end{pmatrix}$.

Solution

We will simply indicate the values of D, D_x, D_y, and D_z and leave the computations for you to check.

$$D = \begin{vmatrix} 1 & -2 & 1 \\ 2 & 1 & -1 \\ 3 & 2 & 4 \end{vmatrix} = 29 \qquad D_x = \begin{vmatrix} -4 & -2 & 1 \\ 5 & 1 & -1 \\ 3 & 2 & 4 \end{vmatrix} = 29$$

$$D_y = \begin{vmatrix} 1 & -4 & 1 \\ 2 & 5 & -1 \\ 3 & 3 & 4 \end{vmatrix} = 58 \qquad D_z = \begin{vmatrix} 1 & -2 & -4 \\ 2 & 1 & 5 \\ 3 & 2 & 3 \end{vmatrix} = -29$$

Therefore,

$$x = \frac{D_x}{D} = \frac{29}{29} = 1$$

$$y = \frac{D_y}{D} = \frac{58}{29} = 2$$

and

$$z = \frac{D_z}{D} = \frac{-29}{29} = -1$$

The solution set is $\{(1, 2, -1)\}$. (Be sure to check it!)

Classroom Example

Solve the system:

$$\begin{pmatrix} x - y + 2z = 4 \\ 3x - 2y + 4z = 6 \\ 2x - 2y + 4z = -1 \end{pmatrix}$$

EXAMPLE 5

Solve the system $\begin{pmatrix} x + 3y - z = 4 \\ 3x - 2y + z = 7 \\ 2x + 6y - 2z = 1 \end{pmatrix}$.

Solution

$$D = \begin{vmatrix} 1 & 3 & -1 \\ 3 & -2 & 1 \\ 2 & 6 & -2 \end{vmatrix} = 2 \begin{vmatrix} 1 & 3 & -1 \\ 3 & -2 & 1 \\ 1 & 3 & -1 \end{vmatrix} = 2(0) = 0$$

$$D_x = \begin{vmatrix} 4 & 3 & -1 \\ 7 & -2 & 1 \\ 1 & 6 & -2 \end{vmatrix} = -7$$

Therefore, because $D = 0$ and at least one of D_x, D_y, and D_z is not zero, the system is inconsistent. The solution set is $\varnothing$.

Example 5 illustrates why D should be determined first. Once we found that $D = 0$ and $D_x \neq 0$, we knew that the system was inconsistent, and there was no need to find D_y and D_z.

Finally, it should be noted that Cramer's rule can be extended to systems of n linear equations in n variables; however, that method is not considered to be a very efficient way of solving a large system of linear equations.

Concept Quiz 15.5

For Problems 1–7, answer true or false.

1. Cramer's rule is a method of solving a system of equations by using matrices.

2. If $D = 0$, then the system of equations is either inconsistent or the equations are dependent.

3. If $D \neq 0$, then the system of equations has either one solution or infinitely many solutions.

4. If $D = 0$ and $D_z = 4$ then the system of equations is inconsistent.

5. Cramer's rule can be extended to systems of n linear equations in n variables.

6. The solution set for the system $\begin{pmatrix} 6x - 5y = 1 \\ 4x - 7y = 2 \end{pmatrix}$ is $\left\{\left(-\dfrac{3}{22}, -\dfrac{4}{11}\right)\right\}$.

7. The solution set for the system $\begin{pmatrix} 3x - 2y + z = 11 \\ 5x + 3y = 17 \\ x + y - 2z = 6 \end{pmatrix}$ is $\left\{\left(\dfrac{15}{4}, -\dfrac{7}{12}, -\dfrac{17}{12}\right)\right\}$.

Problem Set 15.5

For Problems 1–32, use Cramer's rule to find the solution set for each system. If the equations are dependent, simply indicate that there are infinitely many solutions.
(Objectives 1 and 2)

1. $\begin{pmatrix} 2x - y = -2 \\ 3x + 2y = 11 \end{pmatrix}$

2. $\begin{pmatrix} 3x + y = -9 \\ 4x - 3y = 1 \end{pmatrix}$

3. $\begin{pmatrix} 5x + 2y = 5 \\ 3x - 4y = 29 \end{pmatrix}$

4. $\begin{pmatrix} 4x - 7y = -23 \\ 2x + 5y = -3 \end{pmatrix}$

5. $\begin{pmatrix} 5x - 4y = 14 \\ -x + 2y = -4 \end{pmatrix}$

6. $\begin{pmatrix} -x + 2y = 10 \\ 3x - y = -10 \end{pmatrix}$

7. $\begin{pmatrix} y = 2x - 4 \\ 6x - 3y = 1 \end{pmatrix}$

8. $\begin{pmatrix} -3x - 4y = 14 \\ -2x + 3y = -19 \end{pmatrix}$

9. $\begin{pmatrix} -4x + 3y = 3 \\ 4x - 6y = -5 \end{pmatrix}$

10. $\begin{pmatrix} x = 4y - 1 \\ 2x - 8y = -2 \end{pmatrix}$

11. $\begin{pmatrix} 9x - y = -2 \\ 8x + y = 4 \end{pmatrix}$

12. $\begin{pmatrix} 6x - 5y = 1 \\ 4x - 7y = 2 \end{pmatrix}$

13. $\begin{pmatrix} -\dfrac{2}{3}x + \dfrac{1}{2}y = -7 \\ \dfrac{1}{3}x - \dfrac{3}{2}y = 6 \end{pmatrix}$

14. $\begin{pmatrix} \dfrac{1}{2}x + \dfrac{2}{3}y = -6 \\ \dfrac{1}{4}x - \dfrac{1}{3}y = -1 \end{pmatrix}$

15. $\begin{pmatrix} 2x + 7y = -1 \\ x = 2 \end{pmatrix}$

16. $\begin{pmatrix} 5x - 3y = 2 \\ y = 4 \end{pmatrix}$

17. $\begin{pmatrix} x - y + 2z = -8 \\ 2x + 3y - 4z = 18 \\ -x + 2y - z = 7 \end{pmatrix}$

18. $\begin{pmatrix} x - 2y + z = 3 \\ 3x + 2y + z = -3 \\ 2x - 3y - 3z = -5 \end{pmatrix}$

19. $\begin{pmatrix} 2x - 3y + z = -7 \\ -3x + y - z = -7 \\ x - 2y - 5z = -45 \end{pmatrix}$

20. $\begin{pmatrix} 3x - y - z = 18 \\ 4x + 3y - 2z = 10 \\ -5x - 2y + 3z = -22 \end{pmatrix}$

21. $\begin{pmatrix} 4x + 5y - 2z = -14 \\ 7x - y + 2z = 42 \\ 3x + y + 4z = 28 \end{pmatrix}$

22. $\begin{pmatrix} -5x + 6y + 4z = -4 \\ -7x - 8y + 2z = -2 \\ 2x + 9y - z = 1 \end{pmatrix}$

23. $\begin{pmatrix} 2x - y + 3z = -17 \\ 3y + z = 5 \\ x - 2y - z = -3 \end{pmatrix}$

24. $\begin{pmatrix} 2x - y + 3z = -5 \\ 3x + 4y - 2z = -25 \\ -x + z = 6 \end{pmatrix}$

25. $\begin{pmatrix} x + 3y - 4z = -1 \\ 2x - y + z = 2 \\ 4x + 5y - 7z = 0 \end{pmatrix}$

26. $\begin{pmatrix} x - 2y + z = 1 \\ 3x + y - z = 2 \\ 2x - 4y + 2z = -1 \end{pmatrix}$

27. $\begin{pmatrix} 3x - 2y - 3z = -5 \\ x + 2y + 3z = -3 \\ -x + 4y - 6z = 8 \end{pmatrix}$

28. $\begin{pmatrix} 3x - 2y + z = 11 \\ 5x + 3y = 17 \\ x + y - 2z = 6 \end{pmatrix}$

29. $\begin{pmatrix} x - 2y + 3z = 1 \\ -2x + 4y - 3z = -3 \\ 5x - 6y + 6z = 10 \end{pmatrix}$

30. $\begin{pmatrix} 2x - y + 2z = -1 \\ 4x + 3y - 4z = 2 \\ x + 5y - z = 9 \end{pmatrix}$

31. $\begin{pmatrix} -x - y + 3z = -2 \\ -2x + y + 7z = 14 \\ 3x + 4y - 5z = 12 \end{pmatrix}$

32. $\begin{pmatrix} -2x + y - 3z = -4 \\ x + 5y - 4z = 13 \\ 7x - 2y - z = 37 \end{pmatrix}$

Thoughts Into Words

33. Give a step-by-step description of how you would solve the system

$$\begin{pmatrix} 2x - y + 3z = 31 \\ x - 2y - z = 8 \\ 3x + 5y + 8z = 35 \end{pmatrix}$$

34. Give a step-by-step description of how you would find the value of x in the solution for the system

$$\begin{pmatrix} x + 5y - z = -9 \\ 2x - y + z = 11 \\ -3x - 2y + 4z = 20 \end{pmatrix}$$

Further Investigations

35. A linear system in which the constant terms are all zero is called a **homogeneous system.**

    a. Verify that for a 3×3 homogeneous system, if $D \neq 0$, then $(0, 0, 0)$ is the only solution for the system.
    b. Verify that for a 3×3 homogeneous system, if $D = 0$, then the equations are dependent.

For Problems 36–39, solve each of the homogeneous systems (see Problem 35). If the equations are dependent, indicate that the system has infinitely many solutions.

36. $\begin{pmatrix} x - 2y + 5z = 0 \\ 3x + y - 2z = 0 \\ 4x - y + 3z = 0 \end{pmatrix}$

37. $\begin{pmatrix} 2x - y + z = 0 \\ 3x + 2y + 5z = 0 \\ 4x - 7y + z = 0 \end{pmatrix}$

38. $\begin{pmatrix} 3x + y - z = 0 \\ x - y + 2z = 0 \\ 4x - 5y - 2z = 0 \end{pmatrix}$

39. $\begin{pmatrix} 2x - y + 2z = 0 \\ x + 2y + z = 0 \\ x - 3y + z = 0 \end{pmatrix}$

Graphing Calculator Activities

40. Use determinants and your calculator to solve each of the following systems.

a. $\begin{pmatrix} 4x - 3y + z = 10 \\ 8x + 5y - 2z = -6 \\ -12x - 2y + 3z = -2 \end{pmatrix}$

b. $\begin{pmatrix} 2x + y - z + w = -4 \\ x + 2y + 2z - 3w = 6 \\ 3x - y - z + 2w = 0 \\ 2x + 3y + z + 4w = -5 \end{pmatrix}$

c. $\begin{pmatrix} x - 2y + z - 3w = 4 \\ 2x + 3y - z - 2w = -4 \\ 3x - 4y + 2z - 4w = 12 \\ 2x - y - 3z + 2w = -2 \end{pmatrix}$

d. $\begin{pmatrix} 1.98x + 2.49y + 3.45z = 80.10 \\ 2.15x + 3.20y + 4.19z = 97.16 \\ 1.49x + 4.49y + 2.79z = 83.92 \end{pmatrix}$

Answers to the Concept Quiz

1. False   **2.** True   **3.** False   **4.** True   **5.** True   **6.** True   **7.** True

---

15.6   Systems Involving Nonlinear Equations

OBJECTIVES

1 Graph systems of nonlinear equations

2 Solve systems of nonlinear equations

In Section 15.1 we reviewed the use of the substitution method and the elimination-by-addition method to solve a system of two linear equations. We will use both of those techniques in this section to solve systems that contain at least one nonlinear equation. Furthermore, we will use our knowledge of graphing lines, circles, parabolas, ellipses, and hyperbolas to get a visual read on the systems. This will give us a basis for predicting approximate real number solutions if there are any. In other words, we have once again arrived at a topic that vividly illustrates the merging of mathematical ideas. Let's begin by considering a system that contains one linear and one nonlinear equation.

Classroom Example
Solve the system:
$\begin{pmatrix} x^2 + y^2 = 40 \\ x + y = 4 \end{pmatrix}$

EXAMPLE 1   Solve the system $\begin{pmatrix} x^2 + y^2 = 13 \\ 3x + 2y = 0 \end{pmatrix}$.

Solution

From our previous graphing experiences, we should recognize that $x^2 + y^2 = 13$ is a circle and $3x + 2y = 0$ is a straight line. Thus the system can be pictured as in Figure 15.8. The graph indicates that the solution set of this system should consist of two ordered pairs of real numbers, which represent the points of intersection in the second and fourth quadrants.

Now let's solve the system analytically by using the *substitution method*. Change the form of $3x + 2y = 0$ to $y = -3x/2$, and then substitute $-3x/2$ for y in the other equation to produce

$$x^2 + \left(-\frac{3x}{2}\right)^2 = 13$$

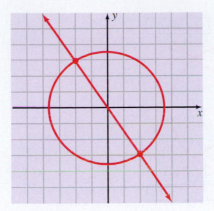

Figure 15.8

This equation can now be solved for x.

$$x^2 + \frac{9x^2}{4} = 13$$
$$4x^2 + 9x^2 = 52$$
$$13x^2 = 52$$
$$x^2 = 4$$
$$x = \pm 2$$

Substitute 2 for x and then -2 for x in the second equation of the system to produce two values for y.

$$3x + 2y = 0 \qquad\qquad 3x + 2y = 0$$
$$3(2) + 2y = 0 \qquad\qquad 3(-2) + 2y = 0$$
$$2y = -6 \qquad\qquad\qquad 2y = 6$$
$$y = -3 \qquad\qquad\qquad y = 3$$

Therefore, the solution set of the system is $\{(2, -3), (-2, 3)\}$.

Remark: Don't forget that, as always, you can check the solutions by substituting them back into the original equations. Graphing the system permits you to approximate any possible real number solutions before solving the system. Then, after solving the system, you can use the graph again to check that the answers are reasonable.

Classroom Example
Solve the system:
$$\begin{pmatrix} y - x^2 = -3 \\ y + x^2 = 2 \end{pmatrix}$$

EXAMPLE 2   Solve the system $\begin{pmatrix} x^2 + y^2 = 16 \\ y^2 - x^2 = 4 \end{pmatrix}$.

Solution

Graphing the system produces Figure 15.9. This figure indicates that there should be four ordered pairs of real numbers in the solution set of the system. Solving the system by using the *elimination method* works nicely. We can simply add the two equations, which eliminates the x's.

$$x^2 + y^2 = 16$$
$$\underline{-x^2 + y^2 = 4}$$
$$2y^2 = 20$$
$$y^2 = 10$$
$$y = \pm\sqrt{10}$$

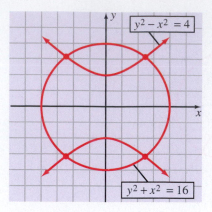

Figure 15.9

Substituting $\sqrt{10}$ for y in the first equation yields

$$x^2 + y^2 = 16$$
$$x^2 + (\sqrt{10})^2 = 16$$
$$x^2 + 10 = 16$$
$$x^2 = 6$$
$$x = \pm\sqrt{6}$$

Thus $(\sqrt{6}, \sqrt{10})$ and $(-\sqrt{6}, \sqrt{10})$ are solutions. Substituting $-\sqrt{10}$ for y in the first equation yields

$$x^2 + y^2 = 16$$
$$x^2 + (-\sqrt{10})^2 = 16$$
$$x^2 + 10 = 16$$
$$x^2 = 6$$
$$x = \pm\sqrt{6}$$

Thus $(\sqrt{6}, -\sqrt{10})$ and $(-\sqrt{6}, -\sqrt{10})$ are also solutions. The solution set is $\{(-\sqrt{6}, \sqrt{10}), (-\sqrt{6}, -\sqrt{10}), (\sqrt{6}, \sqrt{10}), (\sqrt{6}, -\sqrt{10})\}$. ∎

Sometimes a sketch of the graph of a system may not clearly indicate whether the system contains any real number solutions. The next example illustrates such a situation.

Classroom Example
Solve the system:
$$\begin{pmatrix} y = x^2 + 1 \\ 10x - 5y = 1 \end{pmatrix}$$

EXAMPLE 3 Solve the system $\begin{pmatrix} y = x^2 + 2 \\ 6x - 4y = -5 \end{pmatrix}$.

Solution

From our previous graphing experience, we recognize that $y = x^2 + 2$ is the basic parabola shifted upward 2 units, and that $6x - 4y = -5$ is a straight line (see Figure 15.10). Because of the close proximity of the curves, it is difficult to tell whether they intersect. In other words, the graph does not definitely indicate any real number solutions for the system.

Let's solve the system by using the substitution method. We can substitute $x^2 + 2$ for y in the second equation, which produces two values for x.

$$6x - 4(x^2 + 2) = -5$$
$$6x - 4x^2 - 8 = -5$$
$$-4x^2 + 6x - 3 = 0$$
$$4x^2 - 6x + 3 = 0$$

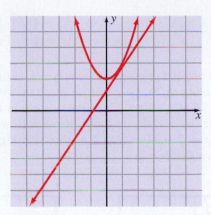

Figure 15.10

$$x = \frac{6 \pm \sqrt{36 - 48}}{8}$$

$$x = \frac{6 \pm \sqrt{-12}}{8}$$

$$x = \frac{6 \pm 2i\sqrt{3}}{8}$$

$$x = \frac{3 \pm i\sqrt{3}}{4}$$

It is now obvious that the system has no real number solutions. That is, the line and the parabola do not intersect in the real number plane. However, there will be two pairs of complex numbers in the solution set. We can substitute $(3 + i\sqrt{3})/4$ for x in the first equation.

$$y = \left(\frac{3 + i\sqrt{3}}{4}\right)^2 + 2$$

$$y = \frac{6 + 6i\sqrt{3}}{16} + 2$$

$$y = \frac{6 + 6i\sqrt{3} + 32}{16}$$

$$y = \frac{38 + 6i\sqrt{3}}{16}$$

$$y = \frac{19 + 3i\sqrt{3}}{8}$$

Likewise, we can substitute $(3 - i\sqrt{3})/4$ for x in the first equation.

$$y = \left(\frac{3 - i\sqrt{3}}{4}\right)^2 + 2$$

$$y = \frac{6 - 6i\sqrt{3}}{16} + 2$$

$$y = \frac{6 - 6i\sqrt{3} + 32}{16}$$

$$y = \frac{38 - 6i\sqrt{3}}{16}$$

$$y = \frac{19 - 3i\sqrt{3}}{8}$$

The solution set is $\left\{\left(\dfrac{3 + i\sqrt{3}}{4}, \dfrac{19 + 3i\sqrt{3}}{8}\right), \left(\dfrac{3 - i\sqrt{3}}{4}, \dfrac{19 - 3i\sqrt{3}}{8}\right)\right\}$.

In Example 3, the use of a graphing utility may not, at first, indicate whether the system has any real number solutions. Suppose that we graph the system using a viewing rectangle such that $-15 \leq x \leq 15$ and $-10 \leq y \leq 10$. As shown in the display in Figure 15.11, we cannot tell whether the line and the parabola intersect. However, if we change the viewing rectangle so that $0 \leq x \leq 2$ and $0 \leq y \leq 4$, as shown in Figure 15.12, it becomes apparent that the two graphs do not intersect.

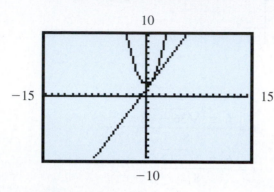

Figure 15.11   **Figure 15.12**

Concept Quiz 15.6

For Problems 1–8, answer true or false.

1. Graphing a system of equations is a method of approximating the solutions.

2. Every system of nonlinear equations has a real number solution.

3. Every nonlinear system of equations can be solved by substitution.

4. Every nonlinear system of equations can be solved by the elimination method.

5. The graph of a circle and a line will have one, two, or no points of intersection.

6. The solution set for the system $\begin{pmatrix} y = x^2 + 1 \\ y = -x^2 + 1 \end{pmatrix}$ is $\{(0, 1)\}$.

7. The solution set for the system $\begin{pmatrix} 3x^2 + 4y^2 = 12 \\ x^2 - y^2 = 9 \end{pmatrix}$ is $\{(4, \sqrt{7})\}$.

8. The solution set for the system $\begin{pmatrix} 2x^2 + y^2 = 8 \\ x^2 + y^2 = 4 \end{pmatrix}$ is the null set.

Problem Set 15.6

For Problems 1–30, (a) graph the system so that approximate real number solutions (if there are any) can be predicted, and (b) solve the system by the substitution or elimination method. **(Objectives 1 and 2)**

1. $\begin{pmatrix} x^2 + y^2 = 5 \\ x + 2y = 5 \end{pmatrix}$

2. $\begin{pmatrix} x^2 + y^2 = 13 \\ 2x + 3y = 13 \end{pmatrix}$

3. $\begin{pmatrix} x^2 + y^2 = 26 \\ x + y = -4 \end{pmatrix}$

4. $\begin{pmatrix} x^2 + y^2 = 10 \\ x + y = -2 \end{pmatrix}$

5. $\begin{pmatrix} x^2 + y^2 = 2 \\ x - y = 4 \end{pmatrix}$

6. $\begin{pmatrix} x^2 + y^2 = 3 \\ x - y = -5 \end{pmatrix}$

7. $\begin{pmatrix} y = x^2 + 6x + 7 \\ 2x + y = -5 \end{pmatrix}$

8. $\begin{pmatrix} y = x^2 - 4x + 5 \\ y - x = 1 \end{pmatrix}$

9. $\begin{pmatrix} 2x + y = -2 \\ y = x^2 + 4x + 7 \end{pmatrix}$

10. $\begin{pmatrix} 2x + y = 0 \\ y = -x^2 + 2x - 4 \end{pmatrix}$

11. $\begin{pmatrix} y = x^2 - 3 \\ x + y = -4 \end{pmatrix}$ **12.** $\begin{pmatrix} y = -x^2 + 1 \\ x + y = 2 \end{pmatrix}$

13. $\begin{pmatrix} x^2 + 2y^2 = 9 \\ x - 4y = -9 \end{pmatrix}$ **14.** $\begin{pmatrix} 2x - y = 7 \\ 3x^2 + y^2 = 21 \end{pmatrix}$

15. $\begin{pmatrix} x + y = -3 \\ x^2 + 2y^2 - 12y - 18 = 0 \end{pmatrix}$

16. $\begin{pmatrix} 4x^2 + 9y^2 = 25 \\ 2x + 3y = 7 \end{pmatrix}$ **17.** $\begin{pmatrix} x - y = 2 \\ x^2 - y^2 = 16 \end{pmatrix}$

18. $\begin{pmatrix} x^2 - 4y^2 = 16 \\ 2y - x = 2 \end{pmatrix}$ **19.** $\begin{pmatrix} y = -x^2 + 3 \\ y = x^2 + 1 \end{pmatrix}$

20. $\begin{pmatrix} y = x^2 \\ y = x^2 - 4x + 4 \end{pmatrix}$ **21.** $\begin{pmatrix} y = x^2 + 2x - 1 \\ y = x^2 + 4x + 5 \end{pmatrix}$

22. $\begin{pmatrix} y = -x^2 + 1 \\ y = x^2 - 2 \end{pmatrix}$ **23.** $\begin{pmatrix} x^2 - y^2 = 4 \\ x^2 + y^2 = 4 \end{pmatrix}$

24. $\begin{pmatrix} 2x^2 + y^2 = 8 \\ x^2 + y^2 = 4 \end{pmatrix}$ **25.** $\begin{pmatrix} 8y^2 - 9x^2 = 6 \\ 8x^2 - 3y^2 = 7 \end{pmatrix}$

26. $\begin{pmatrix} 2x^2 + y^2 = 11 \\ x^2 - y^2 = 4 \end{pmatrix}$ **27.** $\begin{pmatrix} 2x^2 - 3y^2 = -1 \\ 2x^2 + 3y^2 = 5 \end{pmatrix}$

28. $\begin{pmatrix} 4x^2 + 3y^2 = 9 \\ y^2 - 4x^2 = 7 \end{pmatrix}$ **29.** $\begin{pmatrix} xy = 3 \\ 2x + 2y = 7 \end{pmatrix}$

30. $\begin{pmatrix} x^2 + 4y^2 = 25 \\ xy = 6 \end{pmatrix}$

Thoughts Into Words

31. What happens if you try to graph the following system?

$$\begin{pmatrix} 7x^2 + 8y^2 = 36 \\ 11x^2 + 5y^2 = -4 \end{pmatrix}$$

32. For what value(s) of k will the line $x + y = k$ touch the ellipse $x^2 + 2y^2 = 6$ in one and only one point? Defend your answer.

33. The system

$$\begin{pmatrix} x^2 - 6x + y^2 - 4y + 4 = 0 \\ x^2 - 4x + y^2 + 8y - 5 = 0 \end{pmatrix}$$

represents two circles that intersect in two points. An equivalent system can be formed by replacing the second equation with the result of adding -1 times the first equation to the second equation. Thus we obtain the system

$$\begin{pmatrix} x^2 - 6x + y^2 - 4y + 4 = 0 \\ 2x + 12y - 9 = 0 \end{pmatrix}$$

Explain why the linear equation in this system is the equation of the common chord of the original two intersecting circles.

Graphing Calculator Activities

34. Graph the system of equations $\begin{pmatrix} y = x^2 + 2 \\ 6x - 4y = -5 \end{pmatrix}$, and use the TRACE and ZOOM features of your calculator to show that this system has no real number solutions.

For Problems 35–40, use a graphing calculator to approximate, to the nearest tenth, the real number solutions for each system of equations.

35. $\begin{pmatrix} y = e^x + 1 \\ y = x^3 + x^2 - 2x - 1 \end{pmatrix}$

36. $\begin{pmatrix} y = x^3 + 2x^2 - 3x + 2 \\ y = -x^3 - x^2 + 1 \end{pmatrix}$

37. $\begin{pmatrix} y = 2^x + 1 \\ y = 2^{-x} + 2 \end{pmatrix}$ **38.** $\begin{pmatrix} y = \ln(x - 1) \\ y = x^2 - 16x + 64 \end{pmatrix}$

39. $\begin{pmatrix} x = y^2 - 2y + 3 \\ x^2 + y^2 = 25 \end{pmatrix}$ **40.** $\begin{pmatrix} y^2 - x^2 = 16 \\ 2y^2 - x^2 = 8 \end{pmatrix}$

Answers to the Concept Quiz

1. True   **2.** False   **3.** True   **4.** False   **5.** True   **6.** True   **7.** False   **8.** False

Chapter 15 Summary

OBJECTIVE	SUMMARY	EXAMPLE
Review graphing systems of two equations.	Graphing a system of two linear equations in two variables produces one of the following results. **1.** The graphs of the two equations are two intersecting lines, which indicates that there is one unique solution of the system. Such a system is called a consistent system. **2.** The graphs of the two equations are two parallel lines, which indicates that there is no solution for the system. It is called an inconsistent system. **3.** The graphs of the two equations are the same line, which indicates infinitely many solutions for the system. The equations are called dependent equations.	Solve $\left(\begin{array}{c} x - 3y = 6 \\ 2x + 3y = 3 \end{array}\right)$ by graphing. **Solution** Graph the lines by determining the x and y intercepts and a check point. $x - 3y = 6$ $2x + 3y = 3$ <table><tr><th>x</th><th>y</th></tr><tr><td>0</td><td>−2</td></tr><tr><td>6</td><td>0</td></tr><tr><td>−3</td><td>−3</td></tr></table> <table><tr><th>x</th><th>y</th></tr><tr><td>0</td><td>1</td></tr><tr><td>$\frac{3}{2}$</td><td>0</td></tr><tr><td>−1</td><td>$\frac{5}{3}$</td></tr></table> 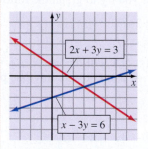 It appears that $(3, -1)$ is the solution. Checking these values in the equations, we can determine that the solution set is $\{(3, -1)\}$.
Solve systems of linear equations using substitution. **(Section 15.1/Objective 1)**	We can describe the substitution method of solving a system of equations as follows. **Step 1** Solve one of the equations for one variable in terms of the other variable if neither equation is in such a form. (If possible, make a choice that will avoid fractions.) **Step 2** Substitute the expression obtained in step 1 into the other equation to produce an equation with one variable. **Step 3** Solve the equation obtained in step 2. **Step 4** Use the solution obtained in step 3, along with the expression obtained in step 1, to determine the solution of the system.	Solve the system $\left(\begin{array}{c} 3x + y = -9 \\ 2x + 3y = 8 \end{array}\right)$. **Solution** Solving the first equation for y gives the equation $y = -3x - 9$. In the second equation, substitute $-3x - 9$ for y and solve. $2x + 3(-3x - 9) = 8$ $2x - 9x - 27 = 8$ $-7x = 35$ $x = -5$ Now, to find the value of y, substitute -5 for x in the equation $y = -3x - 9$ $y = -3(-5) - 9$ $y = 6$ The solution set of the system is $\{(-5, 6)\}$.

OBJECTIVE	SUMMARY	EXAMPLE
Solve systems of equations using the elimination-by-addition method. (Section 15.1/Objective 2)	The elimination-by-addition method involves the replacement of a system of equations with equivalent systems until a system is obtained whereby the solutions can be easily determined. The following operations or transformations can be performed on a system to produce an equivalent system: **1.** Any two equations of the system can be interchanged. **2.** Both sides of any equation of the system can be multiplied by any nonzero real number. **3.** Any equation of the system can be replaced by the *sum* of that equation and a nonzero multiple of another equation.	Solve the system $\begin{pmatrix} 2x - 5y = 31 \\ 4x + 3y = 23 \end{pmatrix}$. **Solution** Let's multiply the first equation by -2 and add the result to the second equation to eliminate the x variable. Then the equivalent system is $\begin{pmatrix} 2x - 5y = \ \ \ 31 \\ 13y = -39 \end{pmatrix}$. Now, solving the second equation for y, we obtain $y = -3$. Substitute -3 for y in either of the original equations and solve for x: $2x - 5(-3) = 31$ $2x + 15 = 31$ $2x = 16$ $x = 8$ The solution set of the system is $\{(8, -3)\}$.
Use systems of equations to solve problems. (Section 15.1/Objective 3)	Many problems that were solved earlier using only one variable may seem easier to solve by using two variables and a system of equations.	A car dealership has 220 vehicles on the lot. The number of cars on the lot is 5 less than twice the number of trucks. Find the number of cars and the number of trucks on the lot. **Solution** Letting x represent the number of cars and y the number of trucks, we obtain the following system: $\begin{pmatrix} x + y = 220 \\ x = 2y - 5 \end{pmatrix}$ Solving the system, we can determine that the dealership has 145 cars and 75 trucks on the lot.
Solve systems of three linear equations. (Section 15.2/Objective 1)	Solving a system of three linear equations in three variables produces one of the following results: **1.** There is one ordered triple that satisfies all three equations. **2.** There are infinitely many ordered triples in the solution set, all of which are coordinates of points on a line common to the planes. **3.** There are infinitely many ordered triples in the solution set, all of which are coordinates of points on a plane. **4.** The solution set is empty; it is $\varnothing$.	Solve $\begin{pmatrix} 4x + 3y - 2z = -5 \\ 2y + 3z = -7 \\ y - 3z = -8 \end{pmatrix}$. **Solution** Replacing the third equation with the sum of the second equation and the third equation yields $3y = -15$. Therefore we can determine that $y = -5$. Substituting -5 for y in the third equation gives $-5 - 3z = -8$. Solving this equation yields $z = 1$. Substituting -5 for y and 1 for z in the first equation gives $4x + 3(-5) - 2(1) = -5$. Solving this equation gives $x = 3$. The solution set for the system is $\{(3, -5, 1)\}$.

(continued)

OBJECTIVE	SUMMARY	EXAMPLE
Use systems of three linear equations to solve problems. (Section 15.2/Objective 2)	Many word problems involving three variables can be solved using a system of three linear equations.	The sum of the measures of the angles in a triangle is 180°. The largest angle is eight times the smallest angle. The sum of the smallest and the largest angle is three times the other angle. Find the measure of each angle. **Solution** Let x represent the measure of the largest angle, let y represent the measure of the middle angle, and let z represent the measure of the smallest angle. From the information in the problem, we can write the following system of equations: $$\begin{pmatrix} x + y + z = 180 \\ x = 8z \\ x + z = 3y \end{pmatrix}$$ By solving this system, we can determine that the measures of the angles of the triangle are 15°, 45°, and 120°.
Represent a system of equations as an augmented matrix. (Section 15.3/Objective 1)	A matrix is an array of numbers arranged in horizontal rows and vertical columns. A matrix of m rows and n columns is called an $m \times n$ (m-by-n) matrix.	Write an augmented matrix for the system of equations $\begin{pmatrix} 5x - 2y - z = 4 \\ 3x - y - z = 7 \\ 2x + 3y + 7z = 9 \end{pmatrix}$. **Solution** The augmented matrix of the system is $$\begin{bmatrix} 5 & -2 & -1 & \vdots & 4 \\ 3 & -1 & -1 & \vdots & 7 \\ 2 & 3 & 7 & \vdots & 9 \end{bmatrix}.$$
Solve a system of equations by transforming the augmented matrix to reduced echelon form. (Section 15.3/Objective 3)	We can change an augmented matrix of a system to reduced echelon form by applying the following elementary row operations: 1. Any two rows of the matrix can be interchanged. 2. Any row of the matrix can be multiplied by a nonzero real number. 3. Any row of the matrix can be replaced by the sum of a nonzero multiple of another row plus that row.	Solve $\begin{pmatrix} x - 2y + 3z = 4 \\ 2x + y - 4z = 3 \\ -3x + 4y - z = -2 \end{pmatrix}$. **Solution** The augmented matrix of the system $\begin{pmatrix} x - 2y + 3z = 4 \\ 2x + y - 4z = 3 \\ -3x + 4y - z = -2 \end{pmatrix}$ is $$\begin{vmatrix} 1 & -2 & 3 & \vdots & 4 \\ 2 & 1 & -4 & \vdots & 3 \\ -3 & 4 & -1 & \vdots & -2 \end{vmatrix}$$ We can change this matrix to the reduced echelon form $$\begin{vmatrix} 1 & 0 & 0 & \vdots & 4 \\ 0 & 1 & 0 & \vdots & 3 \\ 0 & 0 & 1 & \vdots & 2 \end{vmatrix}$$ where the solution set $\{(4, 3, 2)\}$ is obvious.

OBJECTIVE	SUMMARY	EXAMPLE
Evaluate the determinant of a 2×2 matrix. (Section 15.4/Objective 1)	A rectangular array of numbers is called a matrix. A square matrix has the same number of rows as columns. For a 2×2 matrix $\begin{bmatrix} a_1 & b_1 \\ a_2 & b_2 \end{bmatrix}$, the determinant of the matrix is written as $\begin{vmatrix} a_1 & b_1 \\ a_2 & b_2 \end{vmatrix}$ and defined by $\begin{vmatrix} a_1 & b_1 \\ a_2 & b_2 \end{vmatrix}$ $= a_1 b_2 - a_2 b_1$.	Find the determinant of the matrix $\begin{bmatrix} 8 & -3 \\ 5 & 2 \end{bmatrix}$. **Solution** $\begin{vmatrix} 8 & -3 \\ 5 & 2 \end{vmatrix} = 8(2) - (5)(-3) = 16 + 15 = 31$
Evaluate a determinant by expansion. (Section 15.4/Objective 3)	If $A = \begin{bmatrix} a_{11} & a_{12} & a_{13} \\ a_{21} & a_{22} & a_{23} \\ a_{31} & a_{32} & a_{33} \end{bmatrix}$, then $\lvert A \rvert = a_{11}C_{11} + a_{21}C_{21} + a_{31}C_{31}$. This definition states that the determinant of a 3×3 matrix can be found by multiplying each element of the first column by its corresponding cofactor and then adding the three results.	Evaluate $\begin{vmatrix} 2 & -1 & 3 \\ 0 & 5 & 4 \\ -3 & 6 & 1 \end{vmatrix}$ by expanding about the first column. **Solution** $\begin{vmatrix} 2 & -1 & 3 \\ 0 & 5 & 4 \\ -3 & 6 & 1 \end{vmatrix}$ $= 2(-1)^2 \begin{vmatrix} 5 & 4 \\ 6 & 1 \end{vmatrix} + 0(-1)^3 \begin{vmatrix} -1 & 3 \\ 6 & 1 \end{vmatrix}$ $+ (-3)(-1)^4 \begin{vmatrix} -1 & 3 \\ 5 & 4 \end{vmatrix}$ $= 2(-19) - 0 - 3(-19) = 19$
Use Cramer's rule to solve a 2×2 system of equations. (Section 15.5/Objective 2)	Cramer's rule for solving a system of two linear equations in two variables is stated as follows: Given the system $\begin{pmatrix} a_1 x + b_1 y = c_1 \\ a_2 x + b_2 y = c_2 \end{pmatrix}$ with $D = \begin{vmatrix} a_1 & b_1 \\ a_2 & b_2 \end{vmatrix}$ where $D \neq 0$, $D_x = \begin{vmatrix} c_1 & b_1 \\ c_2 & b_2 \end{vmatrix}, D_y = \begin{vmatrix} a_1 & c_1 \\ a_2 & c_2 \end{vmatrix}$ then $x = \dfrac{D_x}{D}$ and $y = \dfrac{D_y}{D}$.	Use Cramer's rule to solve: $\begin{pmatrix} 2x + 3y = -10 \\ x + 4y = -5 \end{pmatrix}$ **Solution** $D = \begin{vmatrix} 2 & 3 \\ 1 & 4 \end{vmatrix} = 2(4) - 1(3) = 5$ $D_x = \begin{vmatrix} -10 & 3 \\ -5 & 4 \end{vmatrix} = -10(4) - (-5)(3)$ $= -25$ $D_y = \begin{vmatrix} 2 & -10 \\ 1 & -5 \end{vmatrix} = 2(-5) - 1(-10) = 0$ $x = \dfrac{D_x}{D} = \dfrac{-25}{5} = -5$ and $y = \dfrac{D_y}{D} = \dfrac{0}{5} = 0$ The solution set is $\{(-5, 0)\}$.

(continued)

OBJECTIVE	SUMMARY	EXAMPLE
Solve a system of three linear equations in three variables by Cramer's rule. (Section 15.5/Objective 2)	Cramer's rule for solving a system of three linear equations in three variables is stated as follows: Given the system $$\begin{pmatrix} a_1x + b_1y + c_1z = d_1 \\ a_2x + b_2y + c_2z = d_2 \\ a_3x + b_3y + c_3z = d_3 \end{pmatrix} \text{ with}$$ $$D = \begin{vmatrix} a_1 & b_1 & c_1 \\ a_2 & b_2 & c_2 \\ a_3 & b_3 & c_3 \end{vmatrix} \neq 0,$$ $$D_x = \begin{vmatrix} d_1 & b_1 & c_1 \\ d_2 & b_2 & c_2 \\ d_3 & b_3 & c_3 \end{vmatrix}, \ D_y = \begin{vmatrix} a_1 & d_1 & c_1 \\ a_2 & d_2 & c_2 \\ a_3 & d_3 & c_3 \end{vmatrix},$$ $$D_z = \begin{vmatrix} a_1 & b_1 & d_1 \\ a_2 & b_2 & d_2 \\ a_3 & b_3 & d_3 \end{vmatrix} \text{ then}$$ $$x = \frac{D_x}{D}, y = \frac{D_y}{D}, \text{ and } z = \frac{D_z}{D}.$$	Use Cramer's rule to solve: $$\begin{pmatrix} 2x - y + z = 8 \\ x - 2y - 3z = -5 \\ 3x + y - 2z = -1 \end{pmatrix}$$ **Solution** Setting up and expanding the appropriate determinants we can determine that $D = 28$, $D_x = 56$, $D_y = -28$, and $D_z = 84$. Therefore, $$x = \frac{D_x}{D} = \frac{56}{28} = 2, y = \frac{D_y}{D} = \frac{-28}{28} = -1,$$ and $z = \dfrac{D_z}{D} = \dfrac{84}{28} = 3.$ The solution set is $\{(2, -1, 3)\}$.
Solve systems of nonlinear equations. (Section 15.6/Objective 2	Graphing a system of equations involving nonlinear equations is useful for predicting the number of solutions and approximating the solutions. The substitution and elimination methods can be used to find exact solutions for systems of equations involving nonlinear equations	Solve $\begin{pmatrix} y = x^2 - 2 \\ 2x + y = 1 \end{pmatrix}$. **Solution** Substituting $x^2 - 2$ for y in the second equation gives $2x + x^2 - 2 = 1$. Rewrite the equation in standard form and solve for x: $2x + x^2 - 2 = 1$ $x^2 + 2x - 3 = 0$ $(x + 3)(x - 1) = 0$ $x = -3 \quad \text{or} \quad x = 1$ To find y when $x = -3$, substitute -3 for x in the first equation: $y = (-3)^2 - 2 = 7$ To find y when $x = 1$, substitute 1 for x in the first equation: $y = 1^2 - 2 = -1$ The solution set is $\{(-3, 7), (1, -1)\}$.

Chapter 15   Review Problem Set

For Problems 1–4, solve each system by using the *substitution* method.

1. $\begin{pmatrix} 3x - y = 16 \\ 5x + 7y = -34 \end{pmatrix}$ 2. $\begin{pmatrix} 6x + 5y = -21 \\ x - 4y = 11 \end{pmatrix}$

3. $\begin{pmatrix} 2x - 3y = 12 \\ 3x + 5y = -20 \end{pmatrix}$ 4. $\begin{pmatrix} 5x + 8y = 1 \\ 4x + 7y = -2 \end{pmatrix}$

For Problems 5–8, solve each system by using the *elimination-by-addition* method.

5. $\begin{pmatrix} 4x - 3y = 34 \\ 3x + 2y = 0 \end{pmatrix}$ 6. $\begin{pmatrix} \dfrac{1}{2}x - \dfrac{2}{3}y = 1 \\ \dfrac{3}{4}x + \dfrac{1}{6}y = -1 \end{pmatrix}$

7. $\begin{pmatrix} 2x - y + 3z = -19 \\ 3x + 2y - 4z = 21 \\ 5x - 4y - z = -8 \end{pmatrix}$

8. $\begin{pmatrix} 3x + 2y - 4z = 4 \\ 5x + 3y - z = 2 \\ 4x - 2y + 3z = 11 \end{pmatrix}$

For Problems 9–12, solve each system by *changing the augmented matrix to reduced echelon form.*

9. $\begin{pmatrix} x - 3y = 17 \\ -3x + 2y = -23 \end{pmatrix}$ 10. $\begin{pmatrix} 2x + 3y = 25 \\ 3x - 5y = -29 \end{pmatrix}$

11. $\begin{pmatrix} x - 2y + z = -7 \\ 2x - 3y + 4z = -14 \\ -3x + y - 2z = 10 \end{pmatrix}$

12. $\begin{pmatrix} -2x - 7y + z = 9 \\ x + 3y - 4z = -11 \\ 4x + 5y - 3z = -11 \end{pmatrix}$

For Problems 13–16, solve each system by using *Cramer's rule.*

13. $\begin{pmatrix} 5x + 3y = -18 \\ 4x - 9y = -3 \end{pmatrix}$ 14. $\begin{pmatrix} 0.2x + 0.3y = 2.6 \\ 0.5x - 0.1y = 1.4 \end{pmatrix}$

15. $\begin{pmatrix} 2x - 3y - 3z = 25 \\ 3x + y + 2z = -5 \\ 5x - 2y - 4z = 32 \end{pmatrix}$

16. $\begin{pmatrix} 3x - y + z = -10 \\ 6x - 2y + 5z = -35 \\ 7x + 3y - 4z = 19 \end{pmatrix}$

For Problems 17–26, solve each system by using the method you think is most appropriate.

17. $\begin{pmatrix} 4x + 7y = -15 \\ 3x - 2y = 25 \end{pmatrix}$ 18. $\begin{pmatrix} \dfrac{3}{4}x - \dfrac{1}{2}y = -15 \\ \dfrac{2}{3}x + \dfrac{1}{4}y = -5 \end{pmatrix}$

19. $\begin{pmatrix} x + 4y = 3 \\ 3x - 2y = 1 \end{pmatrix}$ 20. $\begin{pmatrix} 7x - 3y = -49 \\ y = \dfrac{3}{5}x - 1 \end{pmatrix}$

21. $\begin{pmatrix} 2x + 4y = -1 \\ 3x + 6y = 5 \end{pmatrix}$

22. $\begin{pmatrix} x - y - z = 4 \\ -3x + 2y + 5z = -21 \\ 5x - 3y - 7z = 30 \end{pmatrix}$

23. $\begin{pmatrix} 2x - y + z = -7 \\ -5x + 2y - 3z = 17 \\ 3x + y + 7z = -5 \end{pmatrix}$

24. $\begin{pmatrix} 3x - 2y - 5z = 2 \\ -4x + 3y + 11z = 3 \\ 2x - y + z = -1 \end{pmatrix}$

25. $\begin{pmatrix} x - y + z = 7 \\ 2x - 3y - z = 2 \\ 3x - 4y = 9 \end{pmatrix}$

26. $\begin{pmatrix} 7x - y + z = -4 \\ -2x + 9y - 3z = -50 \\ x - 5y + 4z = 42 \end{pmatrix}$

For Problems 27–32, evaluate each determinant.

27. $\begin{vmatrix} -2 & 6 \\ 3 & 8 \end{vmatrix}$ 28. $\begin{vmatrix} 5 & -4 \\ 7 & -3 \end{vmatrix}$

29. $\begin{vmatrix} 2 & 3 & -1 \\ 3 & 4 & -5 \\ 6 & 4 & 2 \end{vmatrix}$ 30. $\begin{vmatrix} 3 & -2 & 4 \\ 1 & 0 & 6 \\ 3 & -3 & 5 \end{vmatrix}$

31. $\begin{vmatrix} 5 & 4 & 3 \\ 2 & -7 & 0 \\ 3 & -2 & 0 \end{vmatrix}$ 32. $\begin{vmatrix} 5 & -4 & 2 & 1 \\ 3 & 7 & 6 & -2 \\ 2 & 1 & -5 & 0 \\ 3 & -2 & 4 & 0 \end{vmatrix}$

For Problems 33–38, (a) graph the system, and (b) solve the system by using the substitution or elimination method.

33. $\begin{pmatrix} x^2 + y^2 = 17 \\ x - 4y = -17 \end{pmatrix}$ 34. $\begin{pmatrix} x^2 - y^2 = 8 \\ 3x - y = 8 \end{pmatrix}$

35. $\begin{pmatrix} x - y = 1 \\ y = x^2 + 4x + 1 \end{pmatrix}$ **36.** $\begin{pmatrix} 4x^2 - y^2 = 16 \\ 9x^2 + 9y^2 = 16 \end{pmatrix}$

37. $\begin{pmatrix} x^2 + 2y^2 = 8 \\ 2x^2 + 3y^2 = 12 \end{pmatrix}$ **38.** $\begin{pmatrix} y^2 - x^2 = 1 \\ 4x^2 + y^2 = 4 \end{pmatrix}$

For Problems 39–42, solve each problem by setting up and solving a system of linear equations.

39. The sum of the digits of a two-digit number is 9. If the tens digit is 3 more than twice the units digit, find the number.

40. Sara invested $2500, part of it at 10% and the rest at 12% yearly interest. The yearly income on the 12% investment was $102 more than the income on the 10% investment. How much money did she invest at each rate?

41. A box contains $17.70 in nickels, dimes, and quarters. The number of dimes is 8 less than twice the number of nickels. The number of quarters is 2 more than the sum of the number of the nickels and dimes. How many coins of each kind are there in the box?

42. The measure of the largest angle of a triangle is 10° more than four times the smallest angle. The sum of the measures of the smallest and largest angles is three times the measure of the other angle. Find the measure of each angle of the triangle.

For Problems 1–4, refer to the following systems of equations.

I. $\begin{pmatrix} 3x - 2y = 4 \\ 9x - 6y = 12 \end{pmatrix}$ **II.** $\begin{pmatrix} 5x - y = 4 \\ 3x + 7y = 9 \end{pmatrix}$

III. $\begin{pmatrix} 2x - y = 4 \\ 2x - y = -6 \end{pmatrix}$

1. For which system are the graphs parallel lines?

2. For which system are the equations dependent?

3. For which system is the solution set $\varnothing$?

4. Which system is consistent?

For Problems 5–8, evaluate each determinant.

5. $\begin{vmatrix} -2 & 4 \\ -5 & 6 \end{vmatrix}$

6. $\begin{vmatrix} \dfrac{1}{2} & \dfrac{1}{3} \\ \dfrac{3}{4} & -\dfrac{2}{3} \end{vmatrix}$

7. $\begin{vmatrix} -1 & 2 & 1 \\ 3 & 1 & -2 \\ 2 & -1 & 1 \end{vmatrix}$

8. $\begin{vmatrix} 2 & 4 & -5 \\ -4 & 3 & 0 \\ -2 & 6 & 1 \end{vmatrix}$

9. How many ordered pairs of real numbers are in the solution set for the system $\begin{pmatrix} y = 3x - 4 \\ 9x - 3y = 12 \end{pmatrix}$?

10. Solve the system $\begin{pmatrix} 3x - 2y = -14 \\ 7x + 2y = -6 \end{pmatrix}$.

11. Solve the system $\begin{pmatrix} 4x - 5y = 17 \\ y = -3x + 8 \end{pmatrix}$.

12. Find the value of x in the solution for the following system.
$$\begin{pmatrix} \dfrac{3}{4}x - \dfrac{1}{2}y = -21 \\ \dfrac{2}{3}x + \dfrac{1}{6}y = -4 \end{pmatrix}$$

13. Find the value of y in the solution for the system $\begin{pmatrix} 4x - y = 7 \\ 3x + 2y = 2 \end{pmatrix}$.

14. How many real number solutions are there for the system $\begin{pmatrix} x^2 + y^2 = 16 \\ x^2 - 4y = 8 \end{pmatrix}$?

15. Suppose that the augmented matrix of a system of three linear equations in the three variables x, y, and z can be changed to the matrix
$$\begin{bmatrix} 1 & 1 & -4 & | & 3 \\ 0 & 1 & 4 & | & 5 \\ 0 & 0 & 3 & | & 6 \end{bmatrix}$$

Find the value of x in the solution for the system.

16. Suppose that the augmented matrix of a system of three linear equations in the three variables x, y, and z can be changed to the matrix
$$\begin{bmatrix} 1 & 2 & -3 & | & 4 \\ 0 & 1 & 2 & | & 5 \\ 0 & 0 & 2 & | & -8 \end{bmatrix}$$

Find the value of y in the solution for the system.

17. How many ordered triples are there in the solution set for the following system?
$$\begin{pmatrix} x + 3y - z = 5 \\ 2x - y - z = 7 \\ 5x + 8y - 4z = 22 \end{pmatrix}$$

18. How many ordered triples are there in the solution set for the following system?
$$\begin{pmatrix} 3x - y - 2z = 1 \\ 4x + 2y + z = 5 \\ 6x - 2y - 4z = 9 \end{pmatrix}$$

19. Solve the following system:
$$\begin{pmatrix} 5x - 3y - 2z = -1 \\ 4y + 7z = 3 \\ 4z = -12 \end{pmatrix}$$

20. Solve the following system:
$$\begin{pmatrix} x - 2y + z = 0 \\ y - 3z = -1 \\ 2y + 5z = -2 \end{pmatrix}$$

21. Find the value of x in the solution for the system
$$\begin{pmatrix} x - 4y + z = 12 \\ -2x + 3y - z = -11 \\ 5x - 3y + 2z = 17 \end{pmatrix}$$

22. Find the value of y in the solution for the system
$$\begin{pmatrix} x - 3y + z = -13 \\ 3x + 5y - z = 17 \\ 5x - 2y + 2z = -13 \end{pmatrix}$$

23. Solve the system $\begin{pmatrix} x^2 + 4y^2 = 25 \\ xy = 6 \end{pmatrix}$.

24. One solution is 30% alcohol and another solution is 70% alcohol. Some of each of the two solutions is mixed to produce 8 liters of a 40% solution. How many liters of the 70% solution should be used?

25. A box contains $7.25 in nickels, dimes, and quarters. There are 43 coins, and the number of quarters is 1 more than three times the number of nickels. Find the number of quarters in the box.

Set up an equation, an inequality, or a system of equations to help solve each of the following problems.

1. A car repair bill without tax was $340. This included $145 for parts and 3 hours of labor. Find the cost per hour for the labor.

2. Find three consecutive odd integers whose sum is 57.

3. The supplement of an angle is 10° more than five times its complement. Find the measure of the angle.

4. Eric has a collection of 63 coins consisting of nickels, dimes, and quarters. The number of dimes is 6 more than the number of nickels, and the number of quarters is 1 more than twice the number of nickels. How many coins of each kind are in the collection?

5. The largest angle of a triangle is 10° more than three times the smallest angle. The other angle is 20° more than the smallest angle. Find the measure of each angle of the triangle.

6. Kaya is paid "time and a half" for each hour over 40 hours in a week. Last week she worked 46 hours and earned $455.70. What is her normal hourly rate?

7. If a DVD player costs an audio shop $300, at what price should the shop sell it to make a profit of 50% on the selling price?

8. Beth invested a certain amount of money at 8% interest and $300 more than that amount at 9%. Her total yearly interest was $316. How much did she invest at each rate?

9. Sam shot rounds of 70, 73, and 76 on the first 3 days of a golf tournament. What must he shoot on the fourth day to average 72 or less for the 4 days?

10. Eric bought a number of shares of stock for $300. A month later he sold all but 10 shares at a profit of $5 per share and regained his original investment of $300. How many shares did he originally buy and at what price per share?

11. The perimeter of a rectangle is 44 inches and its area is 112 square inches. Find the dimensions of the rectangle.

12. The cube of a number equals nine times the same number. Find the number.

13. Two motorcycles leave Daytona Beach at the same time, one traveling north and the other traveling south. At the end of 4.5 hours, they are 639 miles apart. If the rate of the motorcycle traveling north is 10 miles per hour greater than that of the other motorcycle, find their rates.

14. A 10-quart radiator contains a 50% solution of antifreeze. How much needs to be drained out and replaced with pure antifreeze to obtain a 70% antifreeze solution?

15. How long will it take $750 to double itself if it is invested at 6% simple interest?

16. How long will it take $750 to double itself if it is invested at 6% interest compounded quarterly?

17. How long will it take $750 to double itself if it is invested at 6% interest compounded continuously?

18. The perimeter of a square is 4 centimeters less than twice the perimeter of an equilateral triangle. The length of a side of the square is 2 centimeters more than the length of a side of the equilateral triangle. Find the length of a side of the equilateral triangle.

19. Heidi starts jogging at 4 miles per hour. One-half hour later Ed starts jogging on the same route at 6 miles per hour. How long will it take Ed to catch Heidi?

20. A strip of uniform width is to be cut off both sides and both ends of a sheet of paper that is 8 inches by 14 inches to reduce the size of the paper to an area of 72 square inches. Find the width of the strip.

21. A sum of $2450 is to be divided between two people in the ratio of 3 to 4. How much does each person receive?

22. Working together, Sue and Dean can complete a task in $1\frac{1}{5}$ hours. Dean can do the task by himself in 2 hours. How long would it take Sue to complete the task by herself?

23. The units digit of a two-digit number is 1 more than twice the tens digit. The sum of the digits is 10. Find the number.

24. Suppose the number of days it takes to complete a job varies inversely as the number of people assigned to the job. If it takes 12 people 8 days to do the job, how long would it take 20 people to do the job?

25. The cost of labor varies jointly as the number of workers and the number of days that they work. If it costs $3750 to have 15 people work for 5 days, how much will it cost to have 20 people work for 4 days?

26. It takes a freight train 2 hours longer to travel 300 miles than it takes an express train to travel 280 miles. The rate of the express train is 20 miles per hour greater than the rate of the freight train. Find the times and rates of both trains.

27. Suppose that we want the temperature in a room to be between 19° and 22° Celsius. What Fahrenheit temperatures should be maintained? Remember the formula $C = \frac{5}{9}(F - 32)$.

28. A country fair BBQ booth sells pork dinners for $12 and rib dinners for $15. If 270 dinners were sold for a total of $3690, how many rib dinners were sold?

29. Find two numbers such that their sum is 2 and their product is -1.

30. Larry drove 156 miles in 1 hour more than it took Nita to drive 108 miles. Nita drove at an average rate of 2 miles per hour faster than Larry. How fast did each one travel?

31. An auditorium in a local high school contains 300 seats. There are 5 fewer rows than the number of seats per row. Find the number of rows and the number of seats per row.

32. The area of a certain circle is numerically equal to twice the circumference of the circle. Find the length of a radius of the circle.

33. A class trip was to cost a total of $3000. If there had been 10 more students, it would have cost each student $25 less. How many students took the trip?

34. The difference in the lengths of the two legs of a right triangle is 2 yards. If the length of the hypotenuse is $2\sqrt{13}$ yards, find the length of each leg.

35. The length of the hypotenuse of an isosceles right triangle is 12 inches. Find the length of each leg.

36. The rental charge for 3 movies and 2 video games is $20.75. At the same prices, 5 movies and 3 video games cost $32.92. Find the price of renting a movie and the price of renting a video game.

37. Find three consecutive whole numbers such that the sum of the first plus twice the second plus three times the third is 134.

38. Ike has some nickels and dimes amounting to $2.90. The number of dimes is 1 less than twice the number of nickels. How many coins of each kind does he have?

39. Fourteen increased by twice a number is less than or equal to three times the number. Find the numbers that satisfy this relationship.

40. How many milliliters of pure acid must be added to 150 milliliters of a 30% solution of acid to obtain a 40% solution?

Answers to Odd-Numbered Problems and All Chapter Review, Chapter Test, Cumulative Review, and Cumulative Practice Test Problems

Chapter 1

Problem Set 1.1 (page 7)

1. 16 **3.** 35 **5.** 51 **7.** 72 **9.** 82 **11.** 55 **13.** 60
15. 66 **17.** 26 **19.** 2 **21.** 47 **23.** 21 **25.** 11
27. 15 **29.** 14 **31.** 79 **33.** 6 **35.** 74 **37.** 12
39. 187 **41.** 884 **43.** 9 **45.** 18 **47.** 55 **49.** 99 **51.** 72
53. 11 **55.** 48 **57.** 21 **59.** 40 **61.** 170 **63.** 164
65. 153 **71.** $36 + 12 \div (3 + 3) + 6 \cdot 2$
73. $36 + (12 \div 3 + 3) + 6 \cdot 2$

Problem Set 1.2 (page 12)

1. True **3.** False **5.** True **7.** True **9.** True
11. False **13.** True **15.** False **17.** True **19.** False
21. 3 and 8 **23.** 2 and 12 **25.** 4 and 9 **27.** 5 and 10
29. 1 and 9 **31.** Prime **33.** Prime **35.** Composite
37. Prime **39.** Composite **41.** $2 \cdot 59$ **43.** $3 \cdot 67$
45. $5 \cdot 17$ **47.** $3 \cdot 3 \cdot 13$ **49.** $3 \cdot 43$ **51.** $2 \cdot 13$
53. $2 \cdot 2 \cdot 3 \cdot 3$ **55.** $7 \cdot 7$ **57.** $2 \cdot 2 \cdot 2 \cdot 7$
59. $2 \cdot 2 \cdot 2 \cdot 3 \cdot 5$ **61.** $3 \cdot 3 \cdot 3 \cdot 5$ **63.** 4 **65.** 8
67. 9 **69.** 12 **71.** 18 **73.** 12 **75.** 24 **77.** 48 **79.** 140
81. 392 **83.** 168 **85.** 90 **89.** All other even numbers are
divisible by 2. **91.** 61 **93.** x **95.** xy

Problem Set 1.3 (page 19)

1. 2 **3.** -4 **5.** -7 **7.** 6 **9.** -6 **11.** 8 **13.** -11
15. -15 **17.** -7 **19.** -31 **21.** -19 **23.** 9 **25.** -61
27. -18 **29.** -92 **31.** -5 **33.** -13 **35.** 12 **37.** 6
39. -1 **41.** -45 **43.** -29 **45.** 27 **47.** -65 **49.** -29
51. -11 **53.** -1 **55.** -8 **57.** -13 **59.** -35 **61.** -15
63. -32 **65.** 2 **67.** -4 **69.** -31 **71.** -9 **73.** 18
75. 8 **77.** -29 **79.** -7 **81.** 15 **83.** 1 **85.** 36 **87.** -39
89. -24 **91.** 7 **93.** -1 **95.** 10 **97.** 9 **99.** -17
101. -3 **103.** -10 **105.** -3 **107.** 11 **109.** 5
111. -65 **113.** -100 **115.** -25 **117.** 130 **119.** 80
121. $-17 + 14 = -3$ **123.** $3 + (-2) + (-3) + (-5) = -7$
(7 under par) **125.** $-2 + 1 + 3 + 1 + (-2) = 1$

Problem Set 1.4 (page 25)

1. -30 **3.** -9 **5.** 7 **7.** -56 **9.** 60 **11.** -12
13. -126 **15.** 154 **17.** -9 **19.** 11 **21.** 225
23. -14 **25.** 0 **27.** 23 **29.** -19 **31.** 90 **33.** 14
35. Undefined **37.** -4 **39.** -972 **41.** -47 **43.** 18
45. 69 **47.** 4 **49.** 4 **51.** -6 **53.** 31 **55.** 4
57. 28 **59.** -7 **61.** 10 **63.** -59 **65.** 66 **67.** 7
69. 69 **71.** -7 **73.** 126 **75.** -70 **77.** 15
79. -10 **81.** -25 **83.** 77 **85.** 104 **87.** 14

89. $800(19) + 800(2) + 800(4)(-1) = 13,600$
91. $5 + 4(-3) = -7$

Problem Set 1.5 (page 33)

1. Distributive property
3. Associative property of addition
5. Commutative property of multiplication
7. Additive inverse property
9. Identity property of addition
11. Associative property of multiplication **13.** 56
15. 7 **17.** 1800 **19.** $-14,400$ **21.** -3700 **23.** 5900
25. -338 **27.** -38 **29.** 7 **31.** $-5x$ **33.** $-3m$
35. $-11y$ **37.** $-3x - 2y$ **39.** $-16a - 4b$
41. $-7xy + 3x$ **43.** $10x + 5$ **45.** $6xy - 4$
47. $-6a - 5b$ **49.** $5ab - 11a$ **51.** $8x + 36$
53. $11x + 28$ **55.** $8x + 44$ **57.** $5a + 29$ **59.** $3m + 29$
61. $-8y + 6$ **63.** -5 **65.** -40 **67.** 72 **69.** -18
71. 37 **73.** -74 **75.** 180 **77.** 34 **79.** -65
85. Equivalent **87.** Not equivalent **89.** Not equivalent

Chapter 1 Review Problem Set (page 37)

1. -3 **2.** -25 **3.** -5 **4.** -15 **5.** -1 **6.** 2 **7.** -156
8. 252 **9.** 6 **10.** -13 **11.** Prime **12.** Composite
13. Composite **14.** Composite **15.** Composite
16. $2 \cdot 2 \cdot 2 \cdot 3$ **17.** $3 \cdot 3 \cdot 7$ **18.** $3 \cdot 19$
19. $2 \cdot 2 \cdot 2 \cdot 2 \cdot 2 \cdot 2$ **20.** $2 \cdot 2 \cdot 3 \cdot 7$ **21.** 18
22. 12 **23.** 180 **24.** 945 **25.** 66 **26.** -7 **27.** -2 **28.** 4
29. -18 **30.** 12 **31.** -34 **32.** -27 **33.** -38 **34.** -93
35. 2 **36.** 3 **37.** 35 **38.** 27 **39.** 175°F
40. 20,602 feet **41.** $2(6) - 4 + 3(8) - 1 = 31$ **42.** \$3444
43. $8x$ **44.** $-5y - 9$ **45.** $-5x + 4y$ **46.** $13a - 6b$
47. $-ab - 2a$ **48.** $-3xy - y$ **49.** $10x + 74$ **50.** $2x + 7$
51. $-7x - 18$ **52.** $-3x + 12$ **53.** $-2a + 4$
54. $-2a - 4$ **55.** -59 **56.** -57 **57.** 2 **58.** 1 **59.** 12
60. 13 **61.** 22 **62.** 32 **63.** -9 **64.** 37 **65.** -39
66. -32 **67.** 9 **68.** -44

Chapter 1 Test (page 39)

1. 7 **2.** 45 **3.** 38 **4.** -11 **5.** -58 **6.** -58 **7.** 4
8. -1 **9.** -20 **10.** -7 **11.** -6°F **12.** 26 **13.** -36
14. 9 **15.** -57 **16.** -47 **17.** -4 **18.** Prime
19. $2 \cdot 2 \cdot 2 \cdot 3 \cdot 3 \cdot 5$ **20.** 12 **21.** 72
22. Associative property of addition
23. Distributive property **24.** $-13x + 6y$
25. $-13x - 21$

Chapter 2

Problem Set 2.1 (page 47)

1. $\frac{2}{3}$ 3. $\frac{2}{3}$ 5. $\frac{5}{3}$ 7. $-\frac{1}{6}$ 9. $-\frac{3}{4}$ 11. $\frac{27}{28}$ 13. $\frac{6}{11}$

15. $\frac{3x}{7y}$ 17. $\frac{2x}{5}$ 19. $-\frac{5a}{13c}$ 21. $\frac{8z}{7x}$ 23. $\frac{5b}{7}$ 25. $\frac{15}{28}$

27. $\frac{10}{21}$ 29. $\frac{3}{10}$ 31. $-\frac{4}{3}$ 33. $\frac{7}{5}$ 35. $-\frac{3}{10}$ 37. $\frac{1}{4}$

39. -27 41. $\frac{35}{27}$ 43. $\frac{8}{21}$ 45. $-\frac{5}{6y}$ 47. $2a$ 49. $\frac{2}{5}$

51. $\frac{y}{2x}$ 53. $\frac{20}{13}$ 55. $-\frac{7}{9}$ 57. $\frac{2}{9}$ 59. $\frac{2}{5}$ 61. $\frac{13}{28}$

63. $\frac{8}{5}$ 65. -4 67. $\frac{36}{49}$ 69. 1 71. $\frac{2}{3}$ 73. $\frac{20}{9}$

75. $\frac{1}{4}$ 77. $2\frac{1}{4}$ cups 79. $1\frac{3}{4}$ cups 81. $16\frac{1}{4}$ yards

85. (a) 8 (b) 12 (c) 40 (d) 42 (e) 5 (f) 7

87. (a) $\frac{11}{13}$ (b) $\frac{7}{9}$ (c) $-\frac{37}{41}$ (d) $-\frac{13}{15}$ (e) $\frac{6}{11}$ (f) $\frac{12}{17}$

(g) $\frac{7}{11}$ (h) $\frac{11}{13}$

Problem Set 2.2 (page 56)

1. $\frac{5}{7}$ 3. $\frac{5}{9}$ 5. 3 7. $\frac{2}{3}$ 9. $-\frac{1}{2}$ 11. $\frac{2}{3}$ 13. $\frac{15}{x}$

15. $\frac{2}{y}$ 17. $\frac{8}{15}$ 19. $\frac{9}{16}$ 21. $\frac{37}{30}$ 23. $\frac{59}{96}$ 25. $-\frac{19}{72}$

27. $-\frac{1}{24}$ 29. $-\frac{1}{3}$ 31. $-\frac{1}{6}$ 33. $-\frac{31}{7}$ 35. $-\frac{21}{4}$

37. $\frac{3y+4x}{xy}$ 39. $\frac{7b-2a}{ab}$ 41. $\frac{11}{2x}$ 43. $\frac{4}{3x}$ 45. $-\frac{2}{5x}$

47. $\frac{19}{6y}$ 49. $\frac{1}{24y}$ 51. $-\frac{17}{24n}$ 53. $\frac{5y+7x}{3xy}$

55. $\frac{32y+15x}{20xy}$ 57. $\frac{63y-20x}{36xy}$ 59. $\frac{-6y-5x}{4xy}$

61. $\frac{3x+2}{x}$ 63. $\frac{4x-3}{2x}$ 65. $\frac{1}{4}$ 67. $\frac{37}{30}$ 69. $\frac{1}{3}$

71. $-\frac{12}{5}$ 73. $-\frac{1}{30}$ 75. 14 77. 68 79. $\frac{7}{26}$ 81. $\frac{11}{15}x$

83. $\frac{5}{24}a$ 85. $\frac{4}{3}x$ 87. $\frac{13}{20}n$ 89. $\frac{20}{9}n$ 91. $-\frac{79}{36}n$

93. $\frac{13}{14}x + \frac{9}{8}y$ 95. $-\frac{11}{45}x - \frac{9}{20}y$ 97. $2\frac{1}{8}$ yards

99. $10\frac{3}{4}$ feet 101. $1\frac{3}{4}$ miles 103. $36\frac{2}{3}$ yards

Problem Set 2.3 (page 67)

1. Real, rational, integer, and negative
3. Real, irrational, and positive
5. Real, rational, noninteger, and positive
7. Real, rational, noninteger, and negative
9. 0.62 11. 1.45 13. 3.8 15. -3.3 17. 7.5 19. 7.8
21. -0.9 23. -7.8 25. 1.16 27. -0.272 29. -24.3
31. 44.8 33. 0.0156 35. 1.2 37. -7.4 39. 0.38
41. 7.2 43. -0.42 45. 0.76 47. 4.7 49. 4.3
51. -14.8 53. 1.3 55. $-1.2x$ 57. $3n$ 59. $0.5t$

61. $-5.8x + 2.8y$ 63. $0.1x + 1.2$ 65. $-3x - 2.3$
67. $4.6x - 8$ 69. $\frac{11}{12}$ 71. $\frac{4}{3}$ 73. 17.3 75. -97.8
77. 2.2 79. 13.75 81. 0.6 83. \$9910
85. 19.1 centimeters 87. 4.7 centimeters 89. \$6.55
91. 322. 58 miles 97. (a) $0.\overline{142857}$ (b) $0.\overline{285714}$ (c) $0.\overline{4}$
(d) $0.8\overline{3}$ (e) $0.\overline{27}$ (f) $0.08\overline{3}$

Problem Set 2.4 (page 73)

1. 64 3. 81 5. -8 7. -9 9. 16 11. $\frac{16}{81}$ 13. $-\frac{1}{8}$

15. $\frac{9}{4}$ 17. 0.027 19. -1.44 21. -47 23. -33

25. 11 27. -75 29. -60 31. 31 33. -13

35. 7 37. $\frac{79}{6}$ 39. -1 41. $9x^2$ 43. $12xy^2$

45. $-18x^4y$ 47. $15xy$ 49. $12x^4$ 51. $8a^5$

53. $-8x^2$ 55. $4y^3$ 57. $-2x^2 + 6y^2$ 59. $-\frac{11}{60}n^2$

61. $-2x^2 - 6x$ 63. $7x^2 - 3x + 8$ 65. $\frac{3y}{5}$ 67. $\frac{11}{3y}$

69. $\frac{7b^2}{17a}$ 71. $-\frac{3ac}{4}$ 73. $\frac{x^2y^2}{4}$ 75. $\frac{4x}{9}$ 77. $\frac{5}{12ab}$

79. $\frac{6y^2 + 5x}{xy^2}$ 81. $\frac{5 - 7x^2}{x^4}$ 83. $\frac{3 + 12x^2}{2x^3}$ 85. $\frac{13}{12x^2}$

87. $\frac{11b^2 - 14a^2}{a^2b^2}$ 89. $\frac{3 - 8x}{6x^3}$ 91. $\frac{3y - 4x - 5}{xy}$ 93. 79

95. $\frac{23}{36}$ 97. $\frac{25}{4}$ 99. -64 101. -25 103. -33

105. 0.45

Problem Set 2.5 (page 80)

Answers may vary somewhat for Problems 1–11.
1. The difference of a and b
3. One-third of the product of B and h
5. Two times the quantity, ℓ plus w
7. The quotient of A divided by w
9. The quantity a plus b divided by 2
11. Two more than three times y 13. $\ell + w$ 15. ab

17. $\frac{d}{t}$ 19. ℓwh 21. $y - x$ 23. $xy + 2$ 25. $7 - y^2$

27. $\frac{x - y}{4}$ 29. $10 - x$ 31. $10(n + 2)$ 33. $xy - 7$

35. $xy - 12$ 37. $35 - n$ 39. $n + 45$ 41. $y + 10$

43. $2x - 3$ 45. $10d + 25q$ 47. $\frac{d}{t}$ 49. $\frac{d}{p}$ 51. $\frac{d}{12}$

53. $n + 1$ 55. $n + 2$ 57. $3y - 2$ 59. $36y + 12f$ 61. $\frac{f}{3}$

63. $8w$ 65. $3\ell - 4$ 67. $48f + 36$ 69. $2w^2$ 71. $9s^2$

Chapter 2 Review Problem Set (page 85)

1. 64 2. -27 3. -16 4. 125 5. $-\frac{1}{4}$ 6. $\frac{9}{16}$ 7. $\frac{49}{36}$

8. 0.216 9. 0.0144 10. 0.0036 11. $-\frac{8}{27}$ 12. $\frac{1}{16}$

13. $-\dfrac{1}{64}$ **14.** $\dfrac{4}{9}$ **15.** $\dfrac{19}{24}$ **16.** $\dfrac{39}{70}$ **17.** $\dfrac{1}{15}$

18. $\dfrac{14y + 9x}{2xy}$ **19.** $\dfrac{5x - 8y}{x^2y}$ **20.** $\dfrac{7y}{20}$ **21.** $\dfrac{4x^3}{5y^2}$

22. $\dfrac{2}{7}$ **23.** 1 **24.** $\dfrac{27n^2}{28}$ **25.** $\dfrac{1}{24}$ **26.** $-\dfrac{13}{8}$ **27.** $\dfrac{7}{9}$

28. $\dfrac{29}{12}$ **29.** $\dfrac{1}{2}$ **30.** 0.67 **31.** 0.49 **32.** 2.4 **33.** -0.11

34. 1.76 **35.** 36 **36.** 1.92 **37.** $\dfrac{5}{56}x^2 + \dfrac{7}{20}y^2$

38. $-0.58ab + 0.36bc$ **39.** $\dfrac{11}{24}x$ **40.** $2.2a + 1.7b$

41. $-\dfrac{1}{10}n$ **42.** $\dfrac{41}{20}n$ **43.** $\dfrac{19}{42}$ **44.** $-\dfrac{1}{72}$ **45.** -0.75

46. -0.35 **47.** $\dfrac{1}{17}$ **48.** -8 **49.** $72 - n$

50. $p + 10d$ **51.** $\dfrac{x}{60}$ **52.** $2y - 3$ **53.** $5n + 3$

54. $36y + 12f$ **55.** $100m$ cm **56.** $5n + 10d + 25q$
57. $n - 5$ **58.** $5 - n$ **59.** $10(x - 2)$ **60.** $10x - 2$
61. $x - 3$ **62.** $\dfrac{d}{r}$ **63.** $x^2 + 9$ **64.** $(x + 9)^2$
65. $x^3 + y^3$ **66.** $xy - 4$

Chapter 2 Test (page 87)

1. (a) 81 **(b)** -64 **(c)** 0.008 **2.** $\dfrac{7}{9}$ **3.** $\dfrac{9xy}{16}$

4. -2.6 **5.** 3.04 **6.** -0.56 **7.** $\dfrac{1}{256}$ **8.** $\dfrac{2}{9}$ **9.** $-\dfrac{5}{24}$

10. $\dfrac{187}{60}$ or $3\dfrac{7}{60}$ **11.** $-\dfrac{13}{48}$ **12.** $\dfrac{4y}{5}$ **13.** $2x^2$

14. $\dfrac{4y^2 - 5x}{xy^2}$ **15.** $\dfrac{8}{3x}$ **16.** $\dfrac{35y + 27}{21y^2}$ **17.** $\dfrac{10a^2b}{9}$

18. $-x + 5xy$ **19.** $-3a^2 - 2b^2$ **20.** $\dfrac{37}{36}$ **21.** -0.48

22. $-\dfrac{31}{40}$ **23.** 2.85 **24.** $5n + 10d + 25q$ **25.** $4n - 3$

Chapters 1–2 Cumulative Review Problem Set (page 88)

1. 10 **2.** -30 **3.** 1 **4.** -26 **5.** -29 **6.** 17 **7.** $\dfrac{1}{2}$

8. $-\dfrac{7}{6}$ **9.** $\dfrac{1}{36}$ **10.** -64 **11.** 200 **12.** 0.173 **13.** -142

14. 136 **15.** $\dfrac{19}{9}$ **16.** -0.01 **17.** -2.4 **18.** $\dfrac{79}{40}$ **19.** $\dfrac{7}{50}$

20. $\dfrac{3}{5}$ **21.** $2 \cdot 3 \cdot 3 \cdot 3$ **22.** $2 \cdot 3 \cdot 13$ **23.** $7 \cdot 13$

24. $3 \cdot 3 \cdot 17$ **25.** 14 **26.** 9 **27.** 4 **28.** 6 **29.** 140

30. 200 **31.** 108 **32.** 80 **33.** $-\dfrac{1}{12}x - \dfrac{11}{12}y$ **34.** $-\dfrac{1}{15}n$

35. $-3a + 1.9b$ **36.** $-2n + 6$ **37.** $-x - 15$

38. $-9a - 13$ **39.** $\dfrac{11}{48}$ **40.** $-\dfrac{31}{36}$ **41.** $\dfrac{5 - 2y + 3x}{xy}$

42. $\dfrac{-7y + 9x}{x^2y}$ **43.** $\dfrac{2x}{3}$ **44.** $\dfrac{8a^2}{21b}$ **45.** $\dfrac{4x^2}{3y}$ **46.** $-\dfrac{27}{16}$

47. $p + 5n + 10d$ **48.** $4n - 5$ **49.** $36y + 12f + i$

50. $200x + 200y$ or $200(x + y)$

Chapter 3

Problem Set 3.1 (page 96)

1. $\{8\}$ **3.** $\{-6\}$ **5.** $\{-9\}$ **7.** $\{-6\}$ **9.** $\{13\}$ **11.** $\{48\}$

13. $\{23\}$ **15.** $\{-7\}$ **17.** $\left\{\dfrac{17}{12}\right\}$ **19.** $\left\{-\dfrac{4}{15}\right\}$

21. $\{0.27\}$ **23.** $\{-3.5\}$ **25.** $\{-17\}$ **27.** $\{-35\}$

29. $\{-8\}$ **31.** $\{-17\}$ **33.** $\left\{\dfrac{37}{5}\right\}$ **35.** $\{-3\}$ **37.** $\left\{\dfrac{13}{2}\right\}$

39. $\{144\}$ **41.** $\{24\}$ **43.** $\{-15\}$ **45.** $\{24\}$ **47.** $\{-35\}$

49. $\left\{\dfrac{3}{10}\right\}$ **51.** $\left\{-\dfrac{9}{10}\right\}$ **53.** $\left\{\dfrac{1}{2}\right\}$ **55.** $\left\{-\dfrac{1}{3}\right\}$

57. $\left\{\dfrac{27}{32}\right\}$ **59.** $\left\{-\dfrac{5}{14}\right\}$ **61.** $\left\{-\dfrac{7}{5}\right\}$ **63.** $\left\{-\dfrac{1}{12}\right\}$

65. $\left\{-\dfrac{3}{20}\right\}$ **67.** $\{0.3\}$ **69.** $\{9\}$ **71.** $\{-5\}$

Problem Set 3.2 (page 102)

1. $\{4\}$ **3.** $\{6\}$ **5.** $\{8\}$ **7.** $\{11\}$ **9.** $\left\{\dfrac{17}{6}\right\}$ **11.** $\left\{\dfrac{19}{2}\right\}$

13. $\{6\}$ **15.** $\{-1\}$ **17.** $\{-5\}$ **19.** $\{-6\}$ **21.** $\left\{\dfrac{11}{2}\right\}$

23. $\{-2\}$ **25.** $\left\{\dfrac{10}{7}\right\}$ **27.** $\{18\}$ **29.** $\left\{-\dfrac{25}{4}\right\}$ **31.** $\{-7\}$

33. $\left\{-\dfrac{24}{7}\right\}$ **35.** $\left\{\dfrac{5}{2}\right\}$ **37.** $\left\{\dfrac{4}{17}\right\}$ **39.** $\left\{-\dfrac{12}{5}\right\}$

41. 9 **43.** 22 **45.** \$18 **47.** 35 years old **49.** \$6.50
51. 6 **53.** 5 **55.** \$5.25 **57.** 6.1 inches **59.** 3
61. \$300 **63.** 4 meters **65.** 341 million **67.** 1.25 hours

Problem Set 3.3 (page 109)

1. $\{5\}$ **3.** $\{-8\}$ **5.** $\left\{\dfrac{8}{5}\right\}$ **7.** $\{-11\}$ **9.** $\left\{-\dfrac{5}{2}\right\}$

11. $\{-9\}$ **13.** $\{2\}$ **15.** $\{-3\}$ **17.** $\left\{\dfrac{13}{2}\right\}$ **19.** $\left\{\dfrac{5}{3}\right\}$

21. $\{17\}$ **23.** $\left\{-\dfrac{13}{2}\right\}$ **25.** $\left\{\dfrac{16}{3}\right\}$ **27.** $\{2\}$

29. $\left\{-\dfrac{1}{3}\right\}$ **31.** $\left\{-\dfrac{19}{10}\right\}$ **33.** $\{0\}$ **35.** $\{$All real numbers$\}$

37. $\{$All real numbers$\}$ **39.** $\varnothing$ **41.** 17 **43.** 35 and 37

45. 36, 38, and 40 **47.** $\dfrac{3}{2}$ **49.** -6 **51.** $32°$ and $58°$

53. $50°$ and $130°$ **55.** $65°$ and $75°$ **57.** \$42
59. \$9 per hour **61.** 150 men and 450 women
63. \$91 **65.** \$145

Problem Set 3.4 (page 116)

1. $\{1\}$ **3.** $\{10\}$ **5.** $\{-9\}$ **7.** $\left\{\dfrac{29}{4}\right\}$ **9.** $\left\{-\dfrac{17}{3}\right\}$

11. $\{10\}$ **13.** $\{44\}$ **15.** $\{26\}$ **17.** $\{$All reals$\}$ **19.** $\varnothing$

21. $\{3\}$ **23.** $\{-1\}$ **25.** $\{-2\}$ **27.** $\{16\}$ **29.** $\left\{\dfrac{22}{3}\right\}$

31. $\{-2\}$ **33.** $\left\{-\dfrac{1}{6}\right\}$ **35.** $\{-57\}$ **37.** $\left\{-\dfrac{7}{5}\right\}$ **39.** $\{2\}$

41. $\{-3\}$ **43.** $\left\{\dfrac{27}{10}\right\}$ **45.** $\left\{\dfrac{3}{28}\right\}$ **47.** $\left\{\dfrac{18}{5}\right\}$

49. $\left\{\dfrac{24}{7}\right\}$ **51.** $\{5\}$ **53.** $\{0\}$ **55.** $\left\{-\dfrac{51}{10}\right\}$ **57.** $\{-12\}$

59. $\{15\}$ **61.** 7 and 8   **63.** 14, 15, and 16   **65.** 6 and 11
67. 48   **69.** 14 minutes   **71.** 8 feet and 12 feet
73. 15 nickels, 20 quarters   **75.** 40 nickels, 80 dimes,
and 90 quarters   **77.** 8 dimes and 10 quarters
79. 4 crabs, 12 fish, and 6 plants   **81.** 30°   **83.** 20°, 50°, and 110°
85. 40°   **91.** Any three consecutive integers

Problem Set 3.5 (page 125)

1. True   **3.** False   **5.** False   **7.** True   **9.** True
11. $\{x\,|\,x > -2\}$ or $(-2, \infty)$

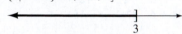

13. $\{x\,|\,x \le 3\}$ or $(-\infty, 3]$

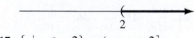

15. $\{x\,|\,x > 2\}$ or $(2, \infty)$

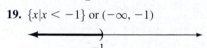

17. $\{x\,|\,x \le -2\}$ or $(-\infty, -2]$

19. $\{x\,|\,x < -1\}$ or $(-\infty, -1)$

21. $\{x\,|\,x < 2\}$ or $(-\infty, 2)$

23. $\{x\,|\,x < -20\}$ or $(-\infty, -20)$
25. $\{x\,|\,x \ge -9\}$ or $[-9, \infty)$ **27.** $\{x\,|\,x > 9\}$ or $(9, \infty)$

29. $\left\{x\,\Big|\,x < \dfrac{10}{3}\right\}$ or $\left(-\infty, \dfrac{10}{3}\right)$

31. $\{x\,|\,x < -8\}$ or $(-\infty, -8)$ **33.** $\{n\,|\,n \ge 8\}$ or $[8, \infty)$

35. $\left\{n\,\Big|\,n > -\dfrac{24}{7}\right\}$ or $\left(-\dfrac{24}{7}, \infty\right)$

37. $\{n\,|\,n > 7\}$ or $(7, \infty)$ **39.** $\{x\,|\,x > 5\}$ or $(5, \infty)$
41. $\{x\,|\,x \le 6\}$ or $(-\infty, 6]$ **43.** $\{x\,|\,x \le -21\}$ or $(-\infty, -21]$

45. $\left\{x\,\Big|\,x < \dfrac{8}{3}\right\}$ or $\left(-\infty, \dfrac{8}{3}\right)$ **47.** $\left\{x\,\Big|\,x < \dfrac{5}{4}\right\}$ or $\left(-\infty, \dfrac{5}{4}\right)$

49. $\{x\,|\,x < 1\}$ or $(-\infty, 1)$ **51.** $\{t\,|\,t \ge 4\}$ or $[4, \infty)$
53. $\{x\,|\,x > 14\}$ or $(14, \infty)$ **55.** $\left\{x\,\Big|\,x > \dfrac{3}{2}\right\}$ or $\left(\dfrac{3}{2}, \infty\right)$

57. $\left\{t\,\Big|\,t \ge \dfrac{1}{4}\right\}$ or $\left[\dfrac{1}{4}, \infty\right)$ **59.** $\left\{x\,\Big|\,x < -\dfrac{9}{4}\right\}$ or $\left(-\infty, -\dfrac{9}{4}\right)$

65. All real numbers   **67.** $\varnothing$ **69.** All real numbers   **71.** $\varnothing$

Problem Set 3.6 (page 131)

1. $\{x\,|\,x > 2\}$ or $(2, \infty)$ **3.** $\{x\,|\,x < -1\}$ or $(-\infty, -1)$

5. $\left\{x\,\Big|\,x > -\dfrac{10}{3}\right\}$ or $\left(-\dfrac{10}{3}, \infty\right)$

7. $\{n\,|\,n \ge -11\}$ or $[-11, \infty)$ **9.** $\{t\,|\,t \le 11\}$ or $(-\infty, 11]$

11. $\left\{x\,\Big|\,x > -\dfrac{11}{5}\right\}$ or $\left(-\dfrac{11}{5}, \infty\right)$ **13.** $\left\{x\,\Big|\,x < \dfrac{5}{2}\right\}$ or $\left(-\infty, \dfrac{5}{2}\right)$

15. $\{x\,|\,x \le 8\}$ or $(-\infty, 8]$ **17.** $\left\{n\,\Big|\,n > \dfrac{3}{2}\right\}$ or $\left(\dfrac{3}{2}, \infty\right)$

19. $\{y\,|\,y > -3\}$ or $(-3, \infty)$ **21.** $\left\{x\,\Big|\,x < \dfrac{5}{2}\right\}$ or $\left(-\infty, \dfrac{5}{2}\right)$

23. $\{x\,|\,x < 8\}$ or $(-\infty, 8)$ **25.** $\{x\,|\,x < 21\}$ or $(-\infty, 21)$

27. $\{x\,|\,x < 6\}$ or $(-\infty, 6)$ **29.** $\left\{n\,\Big|\,n > -\dfrac{17}{2}\right\}$ or $\left(-\dfrac{17}{2}, \infty\right)$

31. $\{n\,|\,n \le 42\}$ or $(-\infty, 42]$ **33.** $\left\{n\,\Big|\,n > -\dfrac{9}{2}\right\}$ or $\left(-\dfrac{9}{2}, \infty\right)$

35. $\left\{x\,\Big|\,x > \dfrac{4}{3}\right\}$ or $\left(\dfrac{4}{3}, \infty\right)$ **37.** $\{n\,|\,n \ge 4\}$ or $[4, \infty)$

39. $\{t\,|\,t > 300\}$ or $(300, \infty)$ **41.** $\{x\,|\,x \le 50\}$ or $(-\infty, 50]$
43. $\{x\,|\,x > 0\}$ or $(0, \infty)$ **45.** $\{x\,|\,x > 64\}$ or $(64, \infty)$

47. $\left\{n\,\Big|\,n > \dfrac{33}{5}\right\}$ or $\left(\dfrac{33}{5}, \infty\right)$

49. $\left\{x\,\Big|\,x \ge -\dfrac{16}{3}\right\}$ or $\left[-\dfrac{16}{3}, \infty\right)$

51.

53.

55.

57.

59.

61. $\varnothing$

63.

65. All reals

67. All numbers greater than 7   **69.** 15 inches
71. 158 or better   **73.** Better than 90
75. More than 250 sales   **77.** 77 or less

Chapter 3 Review Problem Set (page 137)

1. $\{-3\}$ **2.** $\{1\}$ **3.** $\left\{-\dfrac{3}{4}\right\}$ **4.** $\{9\}$ **5.** $\{-4\}$ **6.** $\left\{\dfrac{40}{3}\right\}$

7. $\left\{\dfrac{9}{4}\right\}$ **8.** $\left\{-\dfrac{15}{8}\right\}$ **9.** $\{-7\}$ **10.** $\left\{\dfrac{2}{41}\right\}$ **11.** $\left\{\dfrac{19}{7}\right\}$

12. $\left\{\dfrac{1}{2}\right\}$ **13.** $\{-32\}$ **14.** $\{-12\}$ **15.** $\{21\}$ **16.** $\{-60\}$

17. $\{10\}$ **18.** $\left\{-\dfrac{11}{4}\right\}$ **19.** $\left\{-\dfrac{8}{5}\right\}$ **20.** $\left\{\dfrac{5}{21}\right\}$

21. $\{x | x > 4\}$ or $(4, \infty)$ **22.** $\{x | x > -4\}$ or $(-4, \infty)$

23. $\{x | x \geq 13\}$ or $[13, \infty)$ **24.** $\left\{x | x \geq \dfrac{11}{2}\right\}$ or $\left[\dfrac{11}{2}, \infty\right)$

25. $\{x | x > 35\}$ or $(35, \infty)$ **26.** $\left\{x | x < \dfrac{26}{5}\right\}$ or $\left(-\infty, \dfrac{26}{5}\right)$

27. $\{n | n < 2\}$ or $(-\infty, 2)$ **28.** $\left\{n | n > \dfrac{5}{11}\right\}$ or $\left(\dfrac{5}{11}, \infty\right)$

29. $\{y | y < 24\}$ or $(-\infty, 24)$ **30.** $\{x | x > 10\}$ or $(10, \infty)$

31. $\left\{n | n < \dfrac{2}{11}\right\}$ or $\left(-\infty, \dfrac{2}{11}\right)$ **32.** $\{n | n > 33\}$ or $(33, \infty)$

33. $\{n | n \leq 120\}$ or $(-\infty, 120]$

34. $\left\{n | n \leq -\dfrac{180}{13}\right\}$ or $\left(-\infty, -\dfrac{180}{13}\right]$

35. $\left\{x | x > \dfrac{9}{2}\right\}$ or $\left(\dfrac{9}{2}, \infty\right)$ **36.** $\left\{x | x < -\dfrac{43}{3}\right\}$ or $\left(-\infty, -\dfrac{43}{3}\right)$

37.
-3 2

38.
-1 4

39. All reals

40.
1

41. 24   **42.** 7   **43.** 33   **44.** 8   **45.** 89 or better
46. 16 and 24   **47.** 18   **48.** 88 or better
49. 8 nickels and 22 dimes   **50.** 8 nickels, 25 dimes, and 50 quarters   **51.** 52°   **52.** 700 miles

Chapter 3 Test (page 139)

1. $\{2\}$ **2.** $\{3\}$ **3.** $\{-9\}$ **4.** $\{-5\}$ **5.** $\{-53\}$ **6.** $\{-18\}$

7. $\left\{-\dfrac{5}{2}\right\}$ **8.** $\left\{\dfrac{35}{18}\right\}$ **9.** $\{12\}$ **10.** $\left\{\dfrac{11}{5}\right\}$ **11.** $\{22\}$

12. $\left\{\dfrac{31}{2}\right\}$ **13.** $\{x | x < 5\}$ or $(-\infty, 5)$

14. $\{x | x \leq 1\}$ or $(-\infty, 1]$ **15.** $\{x | x \geq -9\}$ or $[-9, \infty)$

16. $\{x | x < 0\}$ or $(-\infty, 0)$ **17.** $\left\{x | x > -\dfrac{23}{2}\right\}$ or $\left(-\dfrac{23}{2}, \infty\right)$

18. $\{n | n \geq 12\}$ or $[12, \infty)$

Chapter 4

Problem Set 4.1 (page 150)

1. $\{9\}$ **3.** $\{10\}$ **5.** $\left\{\dfrac{15}{2}\right\}$ **7.** $\{-22\}$ **9.** $\{-4\}$ **11.** $\{6\}$

13. $\{-28\}$ **15.** $\{34\}$ **17.** $\{6\}$ **19.** $\left\{-\dfrac{8}{5}\right\}$ **21.** $\{7\}$

23. $\left\{\dfrac{9}{2}\right\}$ **25.** $\left\{-\dfrac{53}{2}\right\}$ **27.** $\{50\}$ **29.** $\{120\}$ **31.** $\left\{\dfrac{9}{7}\right\}$

33. $\{28\}$ **35.** $\left\{-\dfrac{5}{2}\right\}$ **37.** 55%   **39.** 60%   **41.** $16\dfrac{2}{3}\%$

19.
-2 4

20.
1 3

21. \$0.35   **22.** 15 meters, 25 meters, and 30 meters
23. 96 or better   **24.** 17 nickels, 33 dimes, and 53 quarters
25. 60°, 30°, 90°

Chapters 1–3 Cumulative Review Problem Set (page 140)

1. 3   **2.** -128 **3.** -2.4 **4.** $-\dfrac{5}{12}$ **5.** 22   **6.** 35   **7.** 16

8. -20 **9.** 20   **10.** $-\dfrac{19}{90}$ **11.** 0.09   **12.** -16

13. $-\dfrac{1}{4}x + \dfrac{4}{7}y$ **14.** $3x + 40$ **15.** $3x^2 - 2x - 6$

16. $2y^2 + 8y + 4$ **17.** $6a^2 - 2a - 9$ **18.** $-5a + 21b - 14c$
19. 12   **20.** 36   **21.** $2 \cdot 2 \cdot 3 \cdot 5 \cdot 5$

22. $2 \cdot 2 \cdot 2 \cdot 2 \cdot 3 \cdot 3$ **23.** $\dfrac{5}{32}$ **24.** $-\dfrac{27}{4}$ **25.** $\dfrac{6xy}{5}$

26. $\dfrac{5y^2 - 6x}{xy^2}$ **27.** $\dfrac{25y - 6}{15y^2}$ **28.** $-x^2$ **29.** $\dfrac{2x^3}{9}$ **30.** $\dfrac{3b^4}{a^2}$

31. $\{4\}$ **32.** $\{16\}$ **33.** $\varnothing$ **34.** $\{35\}$ **35.** $\{$All reals$\}$

36. $\left\{-\dfrac{17}{11}\right\}$ **37.** $\{x | x < 4\}$ or $(-\infty, 4)$

38. $\{x | x \geq 9\}$ or $[9, \infty)$ **39.** $\{x | x \leq 16\}$ or $(-\infty, 16]$
40. $\{x | x < -1\}$ or $(-\infty, -1)$

41.
-1 3

42.
4

43.
-1 2

44.
3

45.
-4 -1

46. All reals

47. 11 and 13   **48.** 9 on Friday and 33 on Saturday
49. 67 or fewer   **50.** 88 or higher

43. $37\dfrac{1}{2}\%$ **45.** 150%   **47.** 240%   **49.** 2.66

51. 42   **53.** 80%   **55.** 60   **57.** 115%   **59.** 90

61. 15 feet by $19\dfrac{1}{2}$ feet   **63.** 330 miles   **65.** 60 centimeters

67. 7.5 pounds   **69.** $33\dfrac{1}{3}$ pounds   **71.** 90,000

73. 137.5 grams   **75.** \$300   **77.** \$150,000
81. All real numbers except 2   **83.** $\{0\}$
85. All real numbers

Problem Set 4.2 (page 157)

1. {1.11} **3.** {6.6} **5.** {0.48} **7.** {80} **9.** {3}
11. {50} **13.** {70} **15.** {200} **17.** {450} **19.** {150}
21. {2200} **23.** $50 **25.** $3600 **27.** $20.80 **29.** 30%
31. $8.50 **33.** $12.40 **35.** $1000 **37.** 40% **39.** 8%
41. $4166.67 **43.** $1900 **45.** $785.42
49. Yes, if the profit is figured as a percent of the selling price.
51. Yes **53.** {1.625} **55.** {350} **57.** {0.06} **59.** {15.4}

Problem Set 4.3 (page 164)

1. 7 **3.** 500 **5.** 20 **7.** 48 **9.** 9 **11.** 46 centimeters
13. 15 inches **15.** 504 square feet **17.** $8 **19.** 7 inches
21. 150π square centimeters **23.** $\frac{1}{4}\pi$ square yards
25. $S = 324\pi$ square inches and $V = 972\pi$ cubic inches
27. $V = 1152\pi$ cubic feet and $S = 416\pi$ square feet
29. 12 inches **31.** 8 feet **33.** E **35.** B **37.** A **39.** C
41. F **43.** $h = \frac{V}{B}$ **45.** $B = \frac{3V}{h}$ **47.** $w = \frac{P - 2l}{2}$
49. $h = \frac{3V}{\pi r^2}$ **51.** $C = \frac{5}{9}(F - 32)$ **53.** $h = \frac{A - 2\pi r^2}{2\pi r}$
55. $x = \frac{9 - 7y}{3}$ **57.** $y = \frac{9x - 13}{6}$ **59.** $x = \frac{11y - 14}{2}$
61. $x = \frac{-y - 4}{3}$ **63.** $y = \frac{3}{2}x$ **65.** $y = \frac{ax - c}{b}$
67. $x = \frac{2y - 22}{5}$ **69.** $y = mx + b$
75. 125.6 square centimeters **77.** 245 square centimeters
79. 65 cubic inches

Problem Set 4.4 (page 170)

1. $\left\{8\frac{1}{3}\right\}$ **3.** {16} **5.** {25} **7.** {7} **9.** {24}
11. {4} **13.** $12\frac{1}{2}$ years **15.** $33\frac{1}{3}$ years
17. The width is 14 inches, and the length is 42 inches.
19. The width is 12 centimeters, and the length is
34 centimeters. **21.** 80 square inches
23. 24 feet, 31 feet, and 45 feet
25. 6 centimeters, 19 centimeters, 21 centimeters
27. 12 centimeters **29.** 7 centimeters
31. 9 hours **33.** $2\frac{1}{2}$ hours **35.** 55 miles per hour
37. 64 and 72 miles per hour **39.** 60 miles

Problem Set 4.5 (page 176)

1. {15} **3.** $\left\{\frac{20}{7}\right\}$ **5.** $\left\{\frac{15}{4}\right\}$ **7.** $\left\{\frac{5}{3}\right\}$ **9.** {2}
11. $\left\{\frac{33}{10}\right\}$ **13.** 12.5 milliliters **15.** 15 centiliters
17. $7\frac{1}{2}$ quarts of the 30% solution and $2\frac{1}{2}$ quarts of the
50% solution **19.** 5 gallons **21.** 3 quarts
23. 12 gallons **25.** 16.25%
27. The square is 6 inches by 6 inches and the rectangle is
9 inches long and 3 inches wide.

29. 40 minutes **31.** Pam is 9 and Bill is 18.
33. Jessie is 18 years old and Annilee is 12 years old.
35. $5000 at 6%; $7000 at 8%
37. $500 at 3%; $1000 at 4%; $1500 at 5%
39. $900 at 3%; $2150 at 5% **41.** $6000
43. $2166.67 at 5%; $3833.33 at 7%

Chapter 4 Review Problem Set (page 181)

1. $\left\{\frac{17}{12}\right\}$ **2.** {5} **3.** {800} **4.** {16} **5.** {73} **6.** 6
7. 25 **8.** $t = \frac{A - P}{Pr}$ **9.** $x = \frac{13 + 3y}{2}$
10. 77 square inches **11.** 6 centimeters **12.** 15 feet
13. 60% **14.** 40 and 56 **15.** 40
16. 6 meters by 17 meters **17.** $1\frac{1}{2}$ hours **18.** 20 liters
19. 15 centimeters by 40 centimeters
20. 29 yards by 10 yards **21.** 20° **22.** 30 gallons
23. $675 at 3%; $1425 at 5% **24.** $40 **25.** 35%
26. 34° and 99° **27.** 5 hours **28.** 18 gallons **29.** 26%
30. Angie is 22 years old, and her mother is 42 years old.
31. $367.50

Chapter 4 Test (page 182)

1. {−22} **2.** $\left\{-\frac{17}{18}\right\}$ **3.** {−77} **4.** $\left\{\frac{4}{3}\right\}$ **5.** {14}
6. $\left\{\frac{12}{5}\right\}$ **7.** {100} **8.** {70} **9.** {250} **10.** $\left\{\frac{11}{2}\right\}$
11. $C = \frac{5F - 160}{9}$ **12.** $x = \frac{y + 8}{2}$ **13.** $y = \frac{9x + 47}{4}$
14. 64π square centimeters **15.** 576 square inches
16. 14 yards **17.** 125% **18.** 70 **19.** $350 **20.** $52
21. 40% **22.** 875 women **23.** 10 hours
24. 4 centiliters **25.** 20 years

Chapters 1–4 Cumulative Review Problem Set (page 183)

1. $-16x$ **2.** $4a - 6$ **3.** $12x + 27$ **4.** $-5x + 1$
5. $9n - 8$ **6.** $14n - 5$ **7.** $\frac{1}{4}x$ **8.** $-\frac{1}{10}n$ **9.** $-0.1x$
10. $0.7x + 0.2$ **11.** -65 **12.** -51 **13.** 20 **14.** 32
15. $\frac{7}{8}$ **16.** $-\frac{5}{6}$ **17.** -0.28 **18.** $-\frac{1}{4}$ **19.** 5 **20.** $-\frac{1}{4}$
21. 81 **22.** -64 **23.** $\frac{8}{27}$ **24.** $-\frac{1}{32}$ **25.** $\frac{25}{36}$
26. $-\frac{1}{512}$ **27.** {−4} **28.** {−2} **29.** ∅ **30.** {−8}
31. $\left\{\frac{25}{2}\right\}$ **32.** $\left\{-\frac{4}{7}\right\}$ **33.** $\left\{\frac{34}{3}\right\}$ **34.** {200}
35. {All reals} **36.** {11} **37.** $\left\{\frac{3}{2}\right\}$ **38.** {0}
39. $\{x|x > 7\}$ or $(7, \infty)$ **40.** $\{x|x > -6\}$ or $(-6, \infty)$
41. $\left\{n|n \geq \frac{7}{5}\right\}$ or $\left[\frac{7}{5}, \infty\right)$ **42.** $\{x|x \geq 21\}$ or $[21, \infty)$
43. $\{t|t < 100\}$ or $(-\infty, 100)$ **44.** $\{x|x < -1\}$ or $(-\infty, -1)$

45. $\{n|n \geq 18\}$ or $[18, \infty)$ **46.** $\left\{x|x < \dfrac{5}{3}\right\}$ or $\left(-\infty, \dfrac{5}{3}\right)$

47. \$15,000   **48.** $45°$ and $135°$

49. 8 nickels and 17 dimes   **50.** 130 or higher

51. 12 feet and 18 feet   **52.** \$40 per pair

53. 45 miles per hour and 50 miles per hour

54. 5 liters

Chapter 5

Problem Set 5.1 (page 192)

1. Yes

3. No

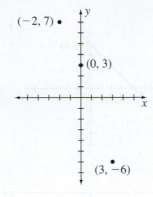

17.

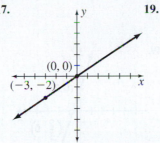

19.

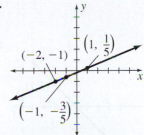

21. $y = -\dfrac{3}{7}x + \dfrac{13}{7}$ **23.** $x = 3y + 9$ **25.** $y = \dfrac{1}{5}x + \dfrac{14}{5}$

27. $x = \dfrac{1}{3}y - \dfrac{7}{3}$ **29.** $y = \dfrac{2}{3}x - \dfrac{5}{3}$

5. Yes

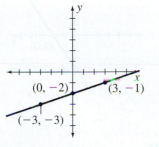

7. Yes

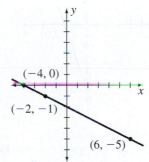

31.

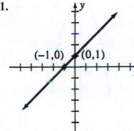

33.

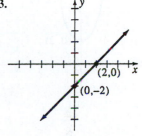

35.

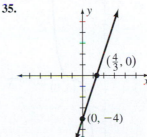

37.

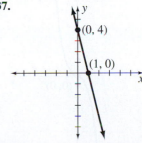

9. No

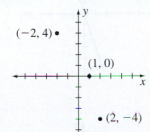

11.

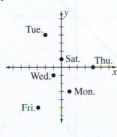

39.

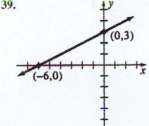

41.

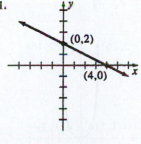

13.

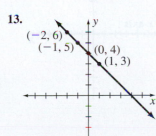

15.

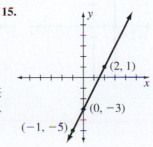

43.

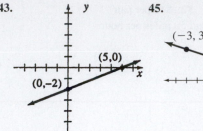

45.

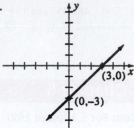

Problem Set 5.2 (page 199)

1.

3.

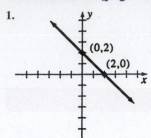

47.

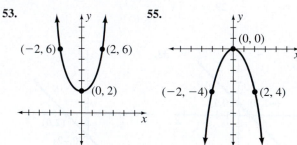

49.

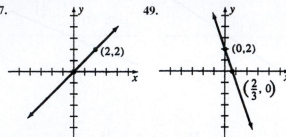

5.

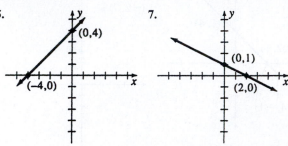

7.

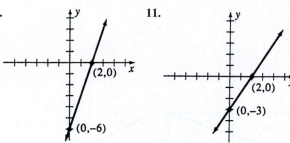

53.

55.

9.

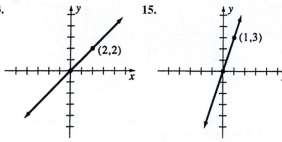

11.

57.

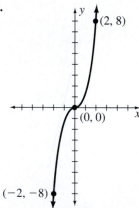

59.

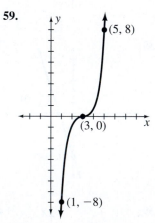

13.

15.

17.

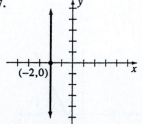

19.

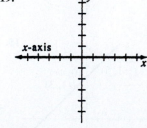

61.

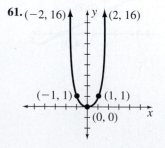

21.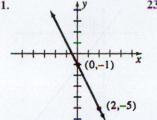

23.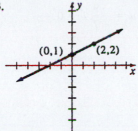

(b)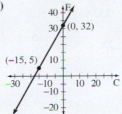

(d). $77°, 86°, -22°, -40°$

25.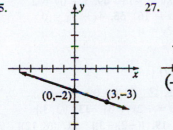

27.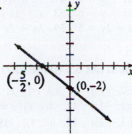

49.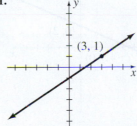

51.

Problem Set 5.3 (page 207)

1. $\dfrac{3}{4}$ **3.** $\dfrac{7}{5}$ **5.** $-\dfrac{6}{5}$ **7.** $-\dfrac{10}{3}$ **9.** $\dfrac{3}{4}$ **11.** 0 **13.** $-\dfrac{3}{2}$

15. Undefined   **17.** 1 **19.** $\dfrac{b-d}{a-c}$ or $\dfrac{d-b}{c-a}$ **21.** 4

23. -6 **25–31.** Answers will vary   **33.** Negative

35. Positive   **37.** Zero   **39.** Negative

29.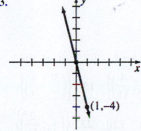

31.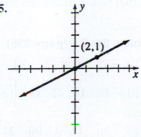

41.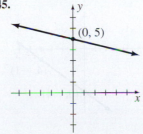

43.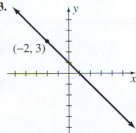

33.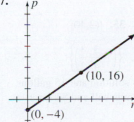

35.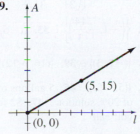

45.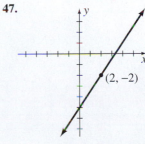

47.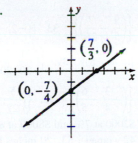

37.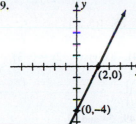

39.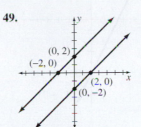

49. $-\dfrac{3}{2}$ **51.** $\dfrac{5}{4}$ **53.** $-\dfrac{1}{5}$ **55.** 2 **57.** 0 **59.** $\dfrac{2}{5}$ **61.** $\dfrac{6}{5}$

63. -3 **65.** 4 **67.** $\dfrac{2}{3}$ **69.** 5.1% **71.** 32 cm   **73.** 1.0 feet

41. $41.76; $43.20; $44.40; $46.50; $46.92

43. a.

C	0	5	10	15	20	−5	−10	−15	−20	−25
F	32	41	50	59	68	23	14	5	−4	−13

Problem Set 5.4 (page 218)

1. $2x - 3y = -5$ **3.** $x - 2y = 7$ **5.** $x + 3y = 20$

7. $0x + y = -7$ **9.** $4x + 9y = 0$ **11.** $3x - y = -16$

13. $7x - 5y = -1$ **15.** $3x + 2y = 5$ **17.** $x - y = 1$

19. $5x - 3y = 0$ **21.** $4x + 7y = 28$ **23.** $y = \frac{3}{5}x + 2$

25. $y = 2x - 1$ **27.** $y = -\frac{1}{6}x - 4$ **29.** $y = -x + \frac{5}{2}$

31. $y = -\frac{5}{9}x - \frac{1}{2}$

33. $m = -2, b = -5$

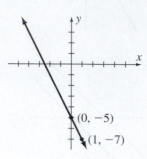

35. $m = \frac{3}{5}, b = -3$

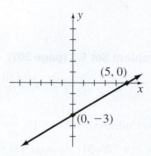

37. $m = \frac{4}{9}, b = 2$

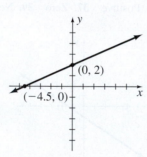

39. $m = \frac{3}{4}, b = -4$

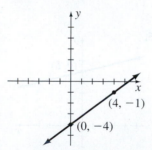

41. $m = -\frac{2}{11}, b = -1$

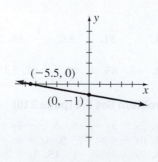

43. $m = -\frac{9}{7}, b = 0$

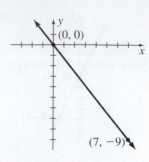

45. $2x - y = 4$ **47.** $5x + 8y = -15$ **49.** $x + 0y = 2$
51. $0x + y = 6$ **53.** $x + 5y = 16$ **55.** $4x - 7y = 0$
57. $x + 2y = 5$ **59.** $3x + 2y = 0$

Problem Set 5.5 (page 227)

1. $\{(2, -1)\}$ **3.** $\{(2, 1)\}$ **5.** $\varnothing$ **7.** $\{(0, 0)\}$
9. $\{(1, -1)\}$ **11.** $\{(x, y) \mid y = -2x + 3\}$ **13.** $\{(1, 3)\}$
15. $\{(3, -2)\}$ **17.** $\{(2, 4)\}$ **19.** $\{(-2, -3)\}$ **21.** $\{(8, 12)\}$

23. $\{(-4, -6)\}$ **25.** $\{(-9, 3)\}$ **27.** $\varnothing$ **29.** $\left\{\left(5, \frac{3}{2}\right)\right\}$

31. $\left\{\left(\frac{9}{5}, -\frac{7}{25}\right)\right\}$ **33.** $\{(-2, -4)\}$ **35.** $\{(5, 2)\}$

37. $\{(-4, -8)\}$ **39.** $\left\{\left(\frac{11}{20}, \frac{7}{20}\right)\right\}$ **41.** $\{(x, y) \mid x - 2y = -3\}$

43. $\left\{\left(-\frac{3}{4}, -\frac{6}{5}\right)\right\}$ **45.** $\left\{\left(\frac{5}{27}, -\frac{26}{27}\right)\right\}$

47. \$2000 at 7% and \$8000 at 8% **49.** 34 and 97
51. 42 women **53.** 20 inches by 27 inches
55. 60 five-dollar bills and 40 ten-dollar bills
57. 2500 student tickets and 500 nonstudent tickets

Problem Set 5.6 (page 235)

1. $\{(4, -3)\}$ **3.** $\{(-1, -3)\}$ **5.** $\{(-8, 2)\}$ **7.** $\{(-4, 0)\}$

9. $\{(1, -1)\}$ **11.** $\varnothing$ **13.** $\left\{\left(-\frac{1}{11}, \frac{4}{11}\right)\right\}$ **15.** $\left\{\left(\frac{3}{2}, -\frac{1}{3}\right)\right\}$

17. $\{(4, -9)\}$ **19.** $\{(7, 0)\}$ **21.** $\{(7, 12)\}$ **23.** $\left\{\left(\frac{7}{11}, \frac{2}{11}\right)\right\}$

25. $\varnothing$ **27.** $\left\{\left(\frac{51}{31}, -\frac{32}{31}\right)\right\}$ **29.** $\{(-2, -4)\}$

31. $\left\{\left(-1, -\frac{14}{3}\right)\right\}$ **33.** $\{(-6, 12)\}$ **35.** $\{(2, 8)\}$

37. $\{(-1, 3)\}$ **39.** $\{(16, -12)\}$ **41.** $\left\{\left(-\frac{3}{4}, \frac{3}{2}\right)\right\}$

43. $\{(5, -5)\}$ **45.** 5 gallons of 10% solution and 15 gallons
of 20% solution **47.** \$2 for a tennis ball and \$3 for a
golf ball

49. 40 double rooms and 15 single rooms **51.** $\frac{3}{4}$
53. 18 centimeters by 24 centimeters
57. **(a)** Consistent **(c)** Consistent
     (e) Dependent **(g)** Inconsistent

59. $x = \dfrac{b_1c_2 - b_2c_1}{a_2b_1 - a_1b_2}$ and $y = \dfrac{a_2c_1 - a_1c_2}{a_2b_1 - a_1b_2}$

Problem Set 5.7 (page 242)

1.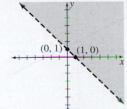

3.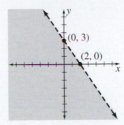

25.

27.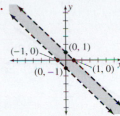

5.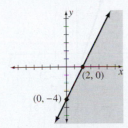

7.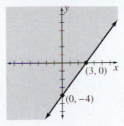

29.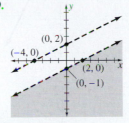

33.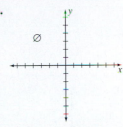

$\varnothing$

9.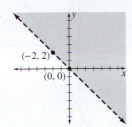

11.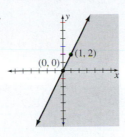

35.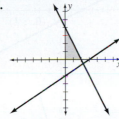

13.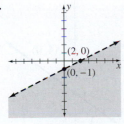

15.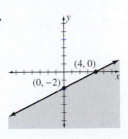

Chapter 5 Review Problem Set (page 249)

1.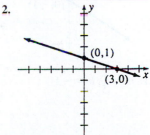

2.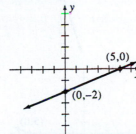

17.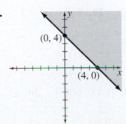

19.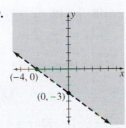

3.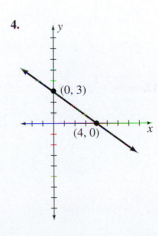

4.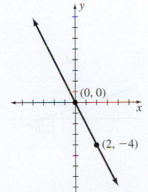

21.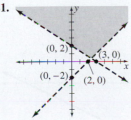

23.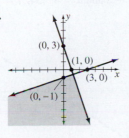

5.

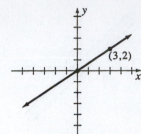

6.

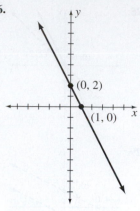

7.

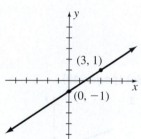

8.

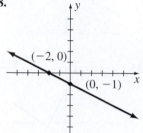

9.

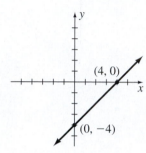

10.

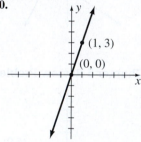

11. $m = \dfrac{2}{5}, b = -2$

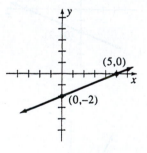

12. $m = -\dfrac{1}{3}, b = 1$

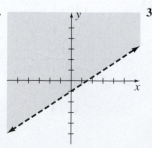

13. $m = -\dfrac{1}{2}, b = 1$

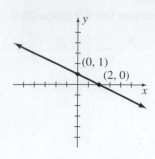

14. $m = -3, b = -2$

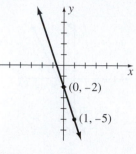

15. $m = 2, b = -4$

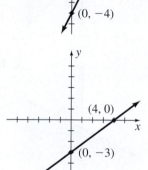

16. $m = \dfrac{3}{4}, b = -3$

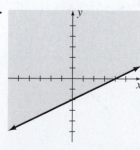

17. $-\dfrac{9}{5}$ **18.** $\dfrac{5}{6}$ **19.** $5x + 7y = -11$ **20.** $8x - 3y = 1$

21. $2x - 9y = 9$ **22.** $x = 2$ **23.** $\{(3, -2)\}$

24. $\{(7, 13)\}$ **25.** $\{(16, -5)\}$ **26.** $\left\{\left(\dfrac{41}{23}, \dfrac{19}{23}\right)\right\}$

27. $\{(10, 25)\}$ **28.** $\{(-6, -8)\}$ **29.** $\{(400, 600)\}$ **30.** $\varnothing$

31. $\left\{\left(\dfrac{5}{16}, -\dfrac{17}{16}\right)\right\}$ **32.** $t = 4$ and $u = 8$

33. $t = 8$ and $u = 4$ **34.** $t = 3$ and $u = 7$ **35.** $\{(-9, 6)\}$

36.

37.

38.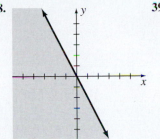

39.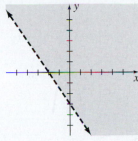

40.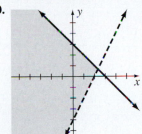

41.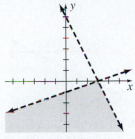

42. 38 and 75   **43.** $2500 at 6% and $3000 at 8%

44. 18 nickels and 25 dimes

45. Length of 19 inches and width of 6 inches

46. Length of 12 inches and width of 7 inches

47. 21 dimes and 11 quarters   **48.** 32° and 58°

49. 50° and 130°

50. $3.25 for a cheeseburger and $2.50 for a milkshake

51. $1.59 for orange juice and $.99 for water

Chapter 5 Test (page 251)

1. $m = -\dfrac{5}{3}, b = 5$

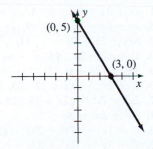

2. $m = 2, b = -4$

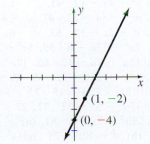

3. $m = -\dfrac{1}{2}, b = -2$

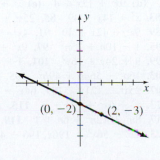

4. $m = -3, b = 0$

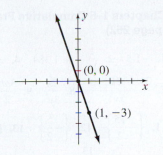

5. 8   **6.** 8   **7.** (7, 6), (11, 7) or others

8. (3, −2), (4, −5) or others   **9.** 4.6%   **10.** −2

11. $3x + 5y = 20$ **12.** $4x - 9y = 34$ **13.** $3x - 2y = 0$

14.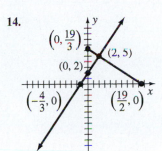

15. $\{(3, 4)\}$ **16.** $\{(-4, 6)\}$ **17.** $\{(-1, -4)\}$ **18.** $\{(3, -6)\}$

19.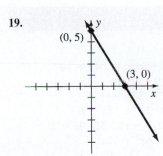

20.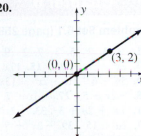

21.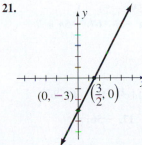

22.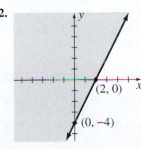

23.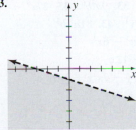

24. Paper $3.49, notebook $2.29   **25.** 13 inches

Chapters 1–5 Cumulative Practice Test (page 252)

1. -1.5 **2.** $\dfrac{1}{16}$ **3.** 1.384 **4.** $-\dfrac{16}{5}$ **5.** 0.95 **6.** -80

7. $\dfrac{3y + 2x - 4}{xy}$ **8.** $\dfrac{2x}{3y^2}$ **9.** $\dfrac{3ab}{2}$ **10.** $\{-20\}$

11. $\left\{-\dfrac{15}{2}\right\}$ **12.** $\left\{-\dfrac{5}{2}\right\}$ **13.** $\{-62\}$ **14.** $\{600\}$

15. $y = \dfrac{2x - 13}{3}$ **16.** $\{(-2, 5)\}$ **17.** $\{(5, -3)\}$

18. $\{x \mid x > -30\}$ or $(-30, \infty)$

19. $\left\{x \mid x \le -\dfrac{36}{5}\right\}$ or $\left(-\infty, -\dfrac{36}{5}\right]$

20.

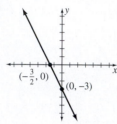

21.

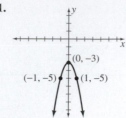

22.

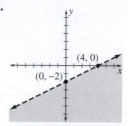

23. 36 and 99   **24.** 77 or less
25. 4 gallons of the 10% salt solution and 6 gallons of the 15% salt solution

Chapter 6

Problem Set 6.1 (page 258)

1. 3 **3.** 2 **5.** 3 **7.** 2 **9.** $8x + 11$ **11.** $4y + 10$
13. $2x^2 - 2x - 23$ **15.** $17x - 19$ **17.** $6x^2 - 5x - 4$
19. $5n - 6$ **21.** $-7x^2 - 13x - 7$ **23.** $5x + 5$
25. $-2x - 5$ **27.** $-3x + 7$ **29.** $2x^2 + 15x - 6$
31. $5n^2 + 2n + 3$ **33.** $-3x^3 + 2x^2 - 11$ **35.** $9x - 2$
37. $2a + 15$ **39.** $-2x^2 + 5$ **41.** $6x^3 + 12x^2 - 5$
43. $4x^3 - 8x^2 + 13x + 12$ **45.** $x + 11$ **47.** $-3x - 14$
49. $-x^2 - 13x - 12$ **51.** $x^2 - 11x - 8$ **53.** $-10a - 3b$
55. $-n^2 + 2n - 17$ **57.** $8x + 1$ **59.** $-5n - 1$
61. $-2a + 6$ **63.** $11x + 6$ **65.** $6x + 7$ **67.** $-5n + 7$
69. $8x + 6$ **71.** $20x^2$

Problem Set 6.2 (page 263)

1. $45x^2$ **3.** $21x^3$ **5.** $-6x^2y^2$ **7.** $14x^3y$ **9.** $-48a^3b^3$

11. $5x^4y$ **13.** $104a^3b^2c^2$ **15.** $30x^6$ **17.** $-56x^2y^3$

19. $-6a^2b^3$ **21.** $72c^3d^3$ **23.** $\dfrac{2}{5}x^3y^5$ **25.** $-\dfrac{2}{9}a^2b^5$

27. $0.28x^8$ **29.** $-6.4a^4b^2$ **31.** $4x^8$ **33.** $9a^4b^6$ **35.** $27x^6$
37. $-64x^{12}$ **39.** $81x^8y^{10}$ **41.** $16x^8y^4$ **43.** $81a^{12}b^8$
45. $x^{12}y^6$ **47.** $15x^2 + 10x$ **49.** $18x^3 - 6x^2$
51. $-28x^3 + 16x$ **53.** $2x^3 - 8x^2 + 12x$
55. $-18a^3 + 30a^2 + 42a$ **57.** $28x^3y - 7x^2y + 35xy$
59. $-9x^3y + 2x^2y + 6xy$ **61.** $13x + 22y$
63. $-2x - 9y$ **65.** $4x^3 - 3x^2 - 14x$ **67.** $-x + 14$
69. $-7x + 12$ **71.** $18x^5$ **73.** $-432x^5$ **75.** $25x^7y^8$
77. $-a^{12}b^5c^9$ **79.** $-16x^{11}y^{17}$ **81.** $7x + 5$ **83.** $3\pi x^2$
89. x^{7n} **91.** x^{6n+1} **93.** x^{6n+3} **95.** $-20x^{10n}$ **97.** $12x^{7n}$

Problem Set 6.3 (page 270)

1. $xy + 3x + 2y + 6$ **3.** $xy + x - 4y - 4$
5. $xy - 6x - 5y + 30$ **7.** $xy + xz + x + 2y + 2z + 2$
9. $6xy + 2x + 9y + 3$ **11.** $x^2 + 10x + 21$
13. $x^2 + 5x - 24$ **15.** $x^2 - 6x - 7$ **17.** $n^2 - 10n + 24$
19. $3n^2 + 19n + 6$ **21.** $15x^2 + 29x - 14$
23. $x^3 + 7x^2 + 21x + 27$ **25.** $x^3 + 3x^2 - 10x - 24$
27. $2x^3 - 7x^2 - 22x + 35$ **29.** $8a^3 - 14a^2 + 23a - 9$
31. $3a^3 + 2a^2 - 8a - 5$
33. $x^4 + 7x^3 + 17x^2 + 23x + 12$
35. $x^4 - 3x^3 - 34x^2 + 33x + 63$ **37.** $x^2 + 11x + 18$
39. $x^2 + 4x - 12$ **41.** $x^2 - 8x - 33$ **43.** $n^2 - 7n + 12$
45. $n^2 + 18n + 72$ **47.** $y^2 - 4y - 21$ **49.** $y^2 - 19y + 84$
51. $x^2 + 2x - 35$ **53.** $x^2 - 6x - 112$ **55.** $a^2 + a - 90$
57. $2a^2 + 13a + 6$ **59.** $5x^2 + 33x - 14$ **61.** $6x^2 - 11x - 7$
63. $12a^2 - 7a - 12$ **65.** $12n^2 - 28n + 15$
67. $14x^2 + 13x - 12$ **69.** $45 - 19x + 2x^2$
71. $-8x^2 + 22x - 15$ **73.** $-9x^2 + 9x + 4$
75. $72n^2 - 5n - 12$ **77.** $27 - 21x + 2x^2$
79. $20x^2 - 7x - 6$ **81.** (a) $x^2 + 8x + 16$
  (b) $x^2 - 10x + 25$ (c) $x^2 - 12x + 36$
  (d) $9x^2 + 12x + 4$ (e) $x^2 + 2x + 1$ (f) $x^2 - 6x + 9$
83. $x^2 + 14x + 49$ **85.** $25x^2 - 4$ **87.** $x^2 - 2x + 1$
89. $9x^2 + 42x + 49$ **91.** $4x^2 - 12x + 9$ **93.** $4x^2 - 9y^2$
95. $1 - 10n + 25n^2$ **97.** $9x^2 + 24xy + 16y^2$
99. $9 + 24y + 16y^2$ **101.** $1 - 49n^2$
103. $16a^2 - 56ab + 49b^2$ **105.** $x^2 + 16xy + 64y^2$
107. $25x^2 - 121y^2$ **109.** $64x^3 - x$ **111.** $-32x^3 + 2xy^2$
113. $x^3 + 6x^2 + 12x + 8$ **115.** $x^3 - 9x^2 + 27x - 27$
117. $8n^3 + 12n^2 + 6n + 1$ **119.** $27n^3 - 54n^2 + 36n - 8$
123. $4x^3 - 56x^2 + 196x$; $196 - 4x^2$

Problem Set 6.4 (page 275)

1. x^8 **3.** $2x^2$ **5.** $-8n^4$ **7.** -8 **9.** $13xy^2$ **11.** $7ab^2$
13. $18xy^4$ **15.** $-32x^5y^2$ **17.** $-8x^5y^4$ **19.** -1
21. $14ab^2c^4$ **23.** $16yz^4$ **25.** $4x^2 + 6x^3$ **27.** $3x^3 - 8x$
29. $-7n^3 + 9$ **31.** $5x^4 - 8x^3 - 12x$ **33.** $4n^5 - 8n^2 + 13$
35. $5a^6 + 8a^2$ **37.** $-3xy + 5y$ **39.** $-8ab - 10a^2b^3$
41. $-3bc + 13b^2c^4$ **43.** $-9xy^2 + 12x^2y^3$
45. $-3x^4 - 5x^2 + 7$ **47.** $-3a^2 + 7a + 13b$
49. $-1 + 5xy^2 - 7xy^5$

Problem Set 6.5 (page 279)

1. $x + 12$ **3.** $x + 2$ **5.** $x + 8$ with a remainder of 4
7. $x + 4$ with a remainder of -7 **9.** $5n + 4$
11. $8y - 3$ with a remainder of 2 **13.** $4x - 7$
15. $3x + 2$ with a remainder of -6 **17.** $2x^2 + 3x + 4$
19. $5n^2 - 4n - 3$ **21.** $n^2 + 6n - 4$ **23.** $x^2 + 3x + 9$
25. $9x^2 + 12x + 16$ **27.** $3n - 8$ with a remainder of 17
29. $3t + 2$ with a remainder of 6 **31.** $3n^2 - n - 4$
33. $4x^2 - 5x + 5$ with a remainder of -3
35. $x + 4$ with a remainder of $5x - 1$
37. $2x - 12$ with a remainder of $49x - 5$
39. $x^3 - 2x^2 + 4x - 8$

Problem Set 6.6 (page 285)

1. $\dfrac{1}{9}$ **3.** $\dfrac{1}{64}$ **5.** $\dfrac{2}{3}$ **7.** 16 **9.** 1 **11.** $-\dfrac{27}{8}$ **13.** $\dfrac{1}{4}$

15. $-\dfrac{1}{9}$ **17.** $\dfrac{27}{64}$ **19.** $\dfrac{1}{8}$ **21.** 27 **23.** 1000

25. $\dfrac{1}{1000}$ or 0.001 **27.** 18 **29.** 144 **31.** x^5 **33.** $\dfrac{1}{n^2}$

35. $\dfrac{1}{a^5}$ **37.** $8x$ **39.** $\dfrac{27}{x^4}$ **41.** $-\dfrac{15}{y^3}$ **43.** 96 **45.** x^{10}

47. $\dfrac{1}{n^4}$ **49.** $2n^2$ **51.** $-\dfrac{3}{x^4}$ **53.** 4 **55.** x^6 **57.** $\dfrac{1}{x^4}$

59. $\dfrac{1}{x^3y^4}$ **61.** $\dfrac{1}{x^6y^3}$ **63.** $\dfrac{8}{n^6}$ **65.** $\dfrac{1}{16n^6}$ **67.** $\dfrac{81}{a^8}$ **69.** $\dfrac{x^2}{25}$

71. $\dfrac{x^2y}{2}$ **73.** $\dfrac{y}{x^2}$ **75.** a^4b^8 **77.** $\dfrac{x^2}{y^6}$ **79.** x **81.** $\dfrac{1}{8x^3}$

83. $\dfrac{x^4}{4}$ **85.** $4.92(10^{11})$ **87.** $4.3(10^{-3})$ **89.** $8.9(10^4)$

91. $2.5(10^{-2})$ **93.** $1.1(10^7)$ **95.** 8000 **97.** 52,100
99. 11,400,000 **101.** 0.07 **103.** 0.000987
105. 0.00000864 **107.** 0.84 **109.** 450 **111.** 4,000,000
113. 0.0000002 **115.** 0.3 **117.** 0.000007 **119.** $875
121. 358 feet

Chapter 6 Review Problem Set (page 291)

1. $8x^2 - 13x + 2$ **2.** $3y^2 + 11y - 9$ **3.** $3x^2 + 2x - 9$
4. $-8x^2 + 18$ **5.** $11x + 8$ **6.** $-9x^2 + 8x - 20$
7. $2y^2 - 54y + 18$ **8.** $-13a - 30$ **9.** $-27a - 7$
10. $n - 2$ **11.** $-5n^2 - 2n$ **12.** $17n^2 - 14n - 16$
13. $35x^6$ **14.** $-54x^8$ **15.** $24x^3y^5$ **16.** $-6a^4b^9$
17. $8a^6b^9$ **18.** $9x^2y^4$ **19.** $35x^2 + 15x$ **20.** $-24x^3 + 3x^2$

21. $x^2 + 17x + 72$ **22.** $3x^2 + 10x + 7$ **23.** $x^2 - 3x - 10$
24. $y^2 - 13y + 36$ **25.** $14x^2 - x - 3$
26. $20a^2 - 3a - 56$ **27.** $9a^2 - 30a + 25$
28. $2x^3 + 17x^2 + 26x - 24$ **29.** $30n^2 + 19n - 5$
30. $12n^2 + 13n - 4$ **31.** $4n^2 - 1$ **32.** $16n^2 - 25$
33. $4a^2 + 28a + 49$ **34.** $9a^2 + 30a + 25$
35. $x^3 - 3x^2 + 8x - 12$ **36.** $2x^3 + 7x^2 + 10x - 7$
37. $a^3 + 15a^2 + 75a + 125$ **38.** $a^3 - 18a^2 + 108a - 216$
39. $x^4 + x^3 + 2x^2 - 7x - 5$
40. $n^4 - 5n^3 - 11n^2 - 30n - 4$ **41.** $-12x^3y^3$
42. $7a^3b^4$ **43.** $-3x^2y - 9x^4$ **44.** $10a^4b^9 - 13a^3b^7$
45. $14x^2 - 10x - 8$ **46.** $x + 4, R = -21$ **47.** $7x - 6$

48. $2x^2 + x + 4, R = 4$ **49.** 13 **50.** 25 **51.** $\dfrac{1}{16}$

52. 1 **53.** -1 **54.** 9 **55.** $\dfrac{16}{9}$ **56.** $\dfrac{1}{4}$ **57.** -8 **58.** $\dfrac{11}{18}$

59. $\dfrac{5}{4}$ **60.** $\dfrac{1}{25}$ **61.** $\dfrac{1}{x^3}$ **62.** $12x^3$ **63.** x^2 **64.** $\dfrac{1}{x^2}$

65. $8a^6$ **66.** $\dfrac{4}{n}$ **67.** $\dfrac{x^2}{y}$ **68.** $\dfrac{b^6}{a^4}$ **69.** $\dfrac{1}{2x}$ **70.** $\dfrac{1}{9n^4}$

71. $\dfrac{n^3}{8}$ **72.** $-12b$ **73.** 610 **74.** 56,000 **75.** 0.08

76. 0.00092 **77.** $(9)(10^3)$ **78.** $(4.7)(10)$ **79.** $(4.7)(10^{-2})$
80. $(2.1)(10^{-4})$ **81.** 0.48 **82.** 4.2 **83.** 2000
84. 0.00000002

Chapter 6 Test (page 293)

1. $-2x^2 - 2x + 5$ **2.** $-3x^2 - 6x + 20$ **3.** $-13x + 2$
4. $-28x^3y^5$ **5.** $12x^5y^5$ **6.** $x^2 - 7x - 18$
7. $n^2 + 7n - 98$ **8.** $40a^2 + 59a + 21$
9. $9x^2 - 42xy + 49y^2$ **10.** $2x^3 + 2x^2 - 19x - 21$
11. $81x^2 - 25y^2$ **12.** $15x^2 - 68x + 77$ **13.** $8x^2y^4$
14. $-7x + 9y$ **15.** $x^2 + 4x - 5$ **16.** $4x^2 - x + 6$

17. $\dfrac{27}{8}$ **18.** $1\dfrac{5}{16}$ **19.** 16 **20.** $-\dfrac{24}{x^2}$ **21.** $\dfrac{x^3}{4}$ **22.** $\dfrac{x^6}{y^{10}}$

23. $(2.7)(10^{-4})$ **24.** 9,200,000 **25.** 0.006

Chapters 1–6 Cumulative Review Problem Set (page 294)

1. 130 **2.** -1 **3.** 27 **4.** -16 **5.** 81 **6.** -32 **7.** $\dfrac{3}{2}$

8. 16 **9.** 36 **10.** $1\dfrac{3}{4}$ **11.** 0 **12.** $\dfrac{13}{40}$ **13.** $-\dfrac{2}{13}$

14. 5 **15.** -1 **16.** -33 **17.** $-15x^3y^7$ **18.** $12ab^7$
19. $-8x^6y^{15}$ **20.** $-6x^2y + 15xy^2$ **21.** $15x^2 - 11x + 2$
22. $21x^2 + 25x - 4$ **23.** $-2x^2 - 7x - 6$ **24.** $49 - 4y^2$
25. $3x^3 - 7x^2 - 2x + 8$ **26.** $2x^3 - 3x^2 - 13x + 20$
27. $8n^3 + 36n^2 + 54n + 27$ **28.** $1 - 6n + 12n^2 - 8n^3$
29. $2x^4 + x^3 - 4x^2 + 42x - 36$ **30.** $-4x^2y^2$
31. $14ab^2$ **32.** $7y - 8x^2 - 9x^3y^3$ **33.** $2x^2 - 4x - 7$

34. $x^2 + 6x + 4$ **35.** $-\dfrac{6}{x}$ **36.** $\dfrac{2}{x}$ **37.** $\dfrac{xy^2}{3}$ **38.** $\dfrac{z^2}{x^2y^4}$

39. 0.12 **40.** 0.0000000018 **41.** 200 **42.** $\{11\}$

43. $\{-1\}$ **44.** $\{48\}$ **45.** $\left\{-\dfrac{3}{7}\right\}$ **46.** $\{9\}$ **47.** $\{13\}$

48. $\left\{\dfrac{9}{14}\right\}$ **49.** $\{500\}$ **50.** $\{x \mid x \leq -1\}$ or $(-\infty, -1]$

51. $\{x|x > 0\}$ or $(0, \infty)$ **52.** $\left\{x|x < \dfrac{4}{5}\right\}$ or $\left(-\infty, \dfrac{4}{5}\right)$

53. $\{x|x < -2\}$ or $(-\infty, -2)$

54. $\left\{x|x \geq \dfrac{12}{7}\right\}$ or $\left[\dfrac{12}{7}, \infty\right)$

55. $\{x|x \geq 300\}$ or $[300, \infty)$

56. 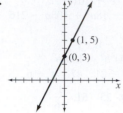 **57.**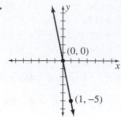

58. 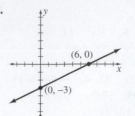 **59.**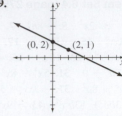

60. $x + 4y = -13$ **61.** $-\dfrac{2}{5}$ **62.** $2x + 7y = 2$ **63.** 3

64. 40 **65.** 8 dimes and 10 quarters

66. \$700 at 8% and \$800 at 9% **67.** 3 gallons

68. $3\dfrac{1}{2}$ hours **69.** The length is 15 meters, and the width is 7 meters.

Chapter 7

Problem Set 7.1 (page 302)

1. $6y$ **3.** $12xy$ **5.** $14ab^2$ **7.** $2x$ **9.** $8a^2b^2$

11. $4(2x + 3y)$ **13.** $7y(2x - 3)$ **15.** $9x(2x + 5)$

17. $6xy(2y - 5x)$ **19.** $12a^2b(3 - 5ab^3)$

21. $xy^2(16y + 25x)$ **23.** $8(8ab - 9cd)$ **25.** $9a^2b(b^3 - 3)$

27. $4x^4y(13y + 15x^2)$ **29.** $8x^2y(5y + 1)$

31. $3x(4 + 5y + 7x)$ **33.** $x(2x^2 - 3x + 4)$

35. $4y^2(11y^3 - 6y - 5)$ **37.** $7ab(2ab^2 + 5b - 7a^2)$

39. $(y + 1)(x + z)$ **41.** $(b - 4)(a - c)$

43. $(x + 3)(x + 6)$ **45.** $(x + 1)(2x - 3)$

47. $(x + y)(5 + b)$ **49.** $(x - y)(b - c)$

51. $(a + b)(c + 1)$ **53.** $(x + 5)(x + 12)$

55. $(x - 2)(x - 8)$ **57.** $(2x + 1)(x - 5)$

59. $(2n - 1)(3n - 4)$ **61.** $\{0, 8\}$ **63.** $\{-1, 0\}$

65. $\{0, 5\}$ **67.** $\left\{0, \dfrac{3}{2}\right\}$ **69.** $\left\{-\dfrac{3}{7}, 0\right\}$ **71.** $\{-5, 0\}$

73. $\left\{0, \dfrac{3}{2}\right\}$ **75.** $\{0, 7\}$ **77.** $\{0, 13\}$ **79.** $\left\{-\dfrac{5}{2}, 0\right\}$

81. $\{-5, 4\}$ **83.** $\{4, 6\}$ **85.** 0 or 9 **87.** 20 units **89.** $\dfrac{4}{\pi}$

91. The square is 3 inches by 3 inches, and the rectangle is 3 inches by 6 inches. **95. (a)** \$116 **(c)** \$750

97. $x = 0$ or $x = \dfrac{c}{b^2}$ **99.** $y = \dfrac{c}{1 + a - b}$

Problem Set 7.2 (page 308)

1. $(x - 1)(x + 1)$ **3.** $(x - 10)(x + 10)$

5. $(x - 2y)(x + 2y)$ **7.** $(3x - y)(3x + y)$

9. $(6a - 5b)(6a + 5b)$ **11.** $(1 - 2n)(1 + 2n)$

13. $5(x - 2)(x + 2)$ **15.** $8(x^2 + 4)$

17. $2(x - 3y)(x + 3y)$ **19.** $x(x - 5)(x + 5)$

21. Not factorable **23.** $9x(5x - 4y)$

25. $4(3 - x)(3 + x)$ **27.** $4a^2(a^2 + 4)$

29. $(x - 3)(x + 3)(x^2 + 9)$ **31.** $x^2(x^2 + 1)$

33. $3x(x^2 + 16)$ **35.** $5x(1 - 2x)(1 + 2x)$

37. $4(x - 4)(x + 4)$ **39.** $3xy(5x - 2y)(5x + 2y)$

41. $(2x + 3y)(2x - 3y)(4x^2 + 9y^2)$

43. $(3 + x)(3 - x)(9 + x^2)$

45. $\{-3, 3\}$ **47.** $\{-2, 2\}$ **49.** $\left\{-\dfrac{4}{3}, \dfrac{4}{3}\right\}$ **51.** $\{-11, 11\}$

53. $\left\{-\dfrac{2}{5}, \dfrac{2}{5}\right\}$ **55.** $\{-5, 5\}$ **57.** $\{-4, 0, 4\}$

59. $\{-4, 0, 4\}$ **61.** $\left\{-\dfrac{1}{3}, \dfrac{1}{3}\right\}$ **63.** $\{-10, 0, 10\}$

65. $\left\{-\dfrac{9}{8}, \dfrac{9}{8}\right\}$ **67.** $\left\{-\dfrac{1}{2}, 0, \dfrac{1}{2}\right\}$ **69.** -7 or 7

71. $-4, 0,$ or 4 **73.** 3 inches and 15 inches

75. The length is 20 centimeters and the width is 8 centimeters. **77.** 4 meters and 8 meters

79. 5 centimeters **85.** $(x - 2)(x^2 + 2x + 4)$

87. $(n + 4)(n^2 - 4n + 16)$

89. $(3a - 4b)(9a^2 + 12ab + 16b^2)$

91. $(1 + 3a)(1 - 3a + 9a^2)$

93. $(2x - y)(4x^2 + 2xy + y^2)$

95. $(3x - 2y)(9x^2 + 6xy + 4y^2)$

97. $(5x + 2y)(25x^2 - 10xy + 4y^2)$

99. $(4 + x)(16 - 4x + x^2)$

Problem Set 7.3 (page 315)

1. $(x + 4)(x + 6)$ **3.** $(x + 5)(x + 8)$ **5.** $(x - 2)(x - 9)$

7. $(n - 7)(n - 4)$ **9.** $(n + 9)(n - 3)$

11. $(n - 10)(n + 4)$ **13.** Not factorable

15. $(x - 6)(x - 12)$ **17.** $(x + 11)(x - 6)$

19. $(y - 9)(y + 8)$ **21.** $(x + 5)(x + 16)$

23. $(x + 12)(x - 6)$ **25.** Not factorable

27. $(x - 2y)(x + 5y)$ **29.** $(a - 8b)(a + 4b)$

31. $\{-7, -3\}$ **33.** $\{3, 6\}$ **35.** $\{-2, 5\}$ **37.** $\{-9, 4\}$

39. $\{-4, 10\}$ **41.** $\{-8, 7\}$ **43.** $\{2, 14\}$ **45.** $\{-12, 1\}$

47. $\{2, 8\}$ **49.** $\{-6, 4\}$ **51.** 7 and 8 or -7 and -8

53. 12 and 14 **55.** $-4, -3, -2,$ and -1 or $7, 8, 9,$ and 10
57. 4 and 7 or 0 and 3
59. The length is 9 inches, and the width is 6 inches.
61. 9 centimeters by 6 centimeters **63.** 7 rows
65. 8 feet, 15 feet, and 17 feet **67.** 6 inches and 8 inches
73. $(x^a + 8)(x^a + 5)$ **75.** $(x^a + 9)(x^a - 3)$

Problem Set 7.4 (page 321)

1. $(3x + 1)(x + 2)$ **3.** $(2x + 5)(3x + 2)$
5. $(4x - 1)(x - 6)$ **7.** $(4x - 5)(3x - 4)$
9. $(5y + 2)(y - 7)$ **11.** $2(2n - 3)(n + 8)$
13. Not factorable **15.** $(3x + 7)(6x + 1)$
17. $(7x - 2)(x - 4)$ **19.** $(4x + 7)(2x - 3)$
21. $t(3t + 2)(3t - 7)$ **23.** $(12y - 5)(y + 7)$
25. Not factorable **27.** $(7x + 3)(2x + 7)$
29. $(4x - 3)(5x - 4)$ **31.** $(4n + 3)(4n - 5)$
33. $(6x - 5)(4x - 5)$ **35.** $(2x + 9)(x + 8)$
37. $(3a + 1)(7a - 2)$ **39.** $(12a + 5)(a - 3)$
41. $3(2x + 3)(2x + 3)$ **43.** $(2x - y)(3x - y)$
45. $(4x + 3y)(5x - 2y)$ **47.** $(5x - 2)(x - 6)$
49. $(8x + 1)(x - 7)$ **51.** $\left\{-6, -\dfrac{1}{2}\right\}$
53. $\left\{-\dfrac{2}{3}, -\dfrac{1}{4}\right\}$ **55.** $\left\{\dfrac{1}{3}, 8\right\}$ **57.** $\left\{\dfrac{2}{5}, \dfrac{7}{3}\right\}$
59. $\left\{-7, \dfrac{5}{6}\right\}$ **61.** $\left\{-\dfrac{3}{8}, \dfrac{3}{2}\right\}$ **63.** $\left\{-\dfrac{2}{3}, \dfrac{4}{3}\right\}$
65. $\left\{\dfrac{2}{5}, \dfrac{5}{2}\right\}$ **67.** $\left\{-\dfrac{5}{2}, -\dfrac{2}{3}\right\}$ **69.** $\left\{-\dfrac{5}{4}, \dfrac{1}{4}\right\}$
71. $\left\{-\dfrac{3}{7}, \dfrac{7}{5}\right\}$ **73.** $\left\{\dfrac{5}{4}, 10\right\}$ **75.** $\left\{-7, \dfrac{3}{7}\right\}$
77. $\left\{-\dfrac{5}{12}, 4\right\}$ **79.** $\left\{-\dfrac{7}{2}, \dfrac{4}{9}\right\}$

Problem Set 7.5 (page 329)

1. $(x + 2)^2$ **3.** $(x - 5)^2$ **5.** $(3n + 2)^2$ **7.** $(4a - 1)^2$
9. $(2 + 9x)^2$ **11.** $(4x - 3y)^2$ **13.** $(2x + 1)(x + 8)$
15. $2x(x - 6)(x + 6)$ **17.** $(n - 12)(n + 5)$
19. Not factorable **21.** $8(x^2 + 9)$ **23.** $(3x + 5)^2$
25. $5(x + 2)(3x + 7)$ **27.** $(4x - 3)(6x + 5)$
29. $(x + 5)(y - 8)$ **31.** $(5x - y)(4x + 7y)$
33. $3(2x - 3)(4x + 9)$ **35.** $6(2x^2 + x + 5)$
37. $5(x - 2)(x + 2)(x^2 + 4)$ **39.** $(x + 6y)^2$ **41.** $\{0, 5\}$
43. $\{-3, 12\}$ **45.** $\{-2, 0, 2\}$ **47.** $\left\{-\dfrac{2}{3}, \dfrac{11}{2}\right\}$
49. $\left\{\dfrac{1}{3}, \dfrac{3}{4}\right\}$ **51.** $\{-3, -1\}$ **53.** $\{0, 6\}$ **55.** $\{-4, 0, 6\}$
57. $\left\{\dfrac{2}{5}, \dfrac{6}{5}\right\}$ **59.** $\{12, 16\}$ **61.** $\left\{-\dfrac{10}{3}, 1\right\}$ **63.** $\{0, 6\}$
65. $\left\{\dfrac{4}{3}\right\}$ **67.** $\{-5, 0\}$ **69.** $\left\{-\dfrac{4}{3}, \dfrac{5}{8}\right\}$
71. $\dfrac{5}{4}$ and 12 or -5 and -3 **73.** -1 and 1 or $-\dfrac{1}{2}$ and 2
75. 4 and 9 or $-\dfrac{24}{5}$ and $-\dfrac{43}{5}$ **77.** 6 rows and 9 chairs per row

79. One square is 6 feet by 6 feet, and the other one is 18 feet by 18 feet.
81. 11 centimeters long and 5 centimeters wide
83. The side is 17 inches long and the altitude to that side is 6 inches long.
85. $1\dfrac{1}{2}$ inches **87.** 6 inches and 12 inches

Chapter 7 Review Problem Set (page 334)

1. $(x - 2)(x - 7)$ **2.** $3x(x + 7)$
3. $(3x + 2)(3x - 2)$ **4.** $(2x - 1)(2x + 5)$ **5.** $(5x - 6)^2$
6. $n(n + 5)(n + 8)$ **7.** $(y + 12)(y - 1)$ **8.** $3xy(y + 2x)$
9. $(x + 1)(x - 1)(x^2 + 1)$ **10.** $(6n + 5)(3n - 1)$
11. Not factorable **12.** $(4x - 7)(x + 1)$
13. $3(n + 6)(n - 5)$ **14.** $x(x + y)(x - y)$
15. $(2x - y)(x + 2y)$ **16.** $2(n - 4)(2n + 5)$
17. $(x + y)(5 + a)$ **18.** $(7t - 4)(3t + 1)$
19. $2x(x + 1)(x - 1)$ **20.** $3x(x + 6)(x - 6)$
21. $(4x + 5)^2$ **22.** $(y - 3)(x - 2)$
23. $(5x + y)(3x - 2y)$ **24.** $n^2(2n - 1)(3n - 1)$
25. $\{-6, 2\}$ **26.** $\{0, 11\}$ **27.** $\left\{-4, \dfrac{5}{2}\right\}$ **28.** $\left\{-\dfrac{8}{3}, \dfrac{1}{3}\right\}$
29. $\{-2, 2\}$ **30.** $\left\{-\dfrac{5}{4}\right\}$ **31.** $\{-1, 0, 1\}$ **32.** $\left\{-\dfrac{9}{4}, -\dfrac{2}{7}\right\}$
33. $\{-7, 4\}$ **34.** $\{-5, 5\}$ **35.** $\left\{-6, \dfrac{3}{5}\right\}$ **36.** $\left\{-\dfrac{7}{2}, 1\right\}$
37. $\{-2, 0, 2\}$ **38.** $\{8, 12\}$ **39.** $\left\{-5, \dfrac{3}{4}\right\}$ **40.** $\{-2, 3\}$
41. $\left\{-\dfrac{7}{3}, \dfrac{5}{2}\right\}$ **42.** $\{-9, 6\}$ **43.** $\left\{-5, \dfrac{3}{2}\right\}$ **44.** $\left\{\dfrac{4}{3}, \dfrac{5}{2}\right\}$
45. $-\dfrac{8}{3}$ and $-\dfrac{19}{3}$ or 4 and 7
46. The length is 8 centimeters, and the width is 2 centimeters.
47. A 2-by-2-inch square and a 10-by-10-inch square
48. 8 by 15 by 17 **49.** $-\dfrac{13}{6}$ and -12 or 2 and 13
50. 7, 9, and 11 **51.** 4 shelves
52. A 5-by-5-yard square and a 5-by-40-yard rectangle
53. -18 and -17 or 17 and 18 **54.** 6 units
55. 2 meters and 7 meters **56.** 9 and 11
57. 2 centimeters **58.** 15 feet

Chapter 7 Test (page 336)

1. $(x + 5)(x - 2)$ **2.** $(x + 3)(x - 8)$
3. $2x(x + 1)(x - 1)$ **4.** $(x + 9)(x + 12)$
5. $3(2n + 1)(3n + 2)$ **6.** $(x + y)(a + 2b)$
7. $(4x - 3)(x + 5)$ **8.** $6(x^2 + 4)$
9. $2x(5x - 6)(3x - 4)$ **10.** $(7 - 2x)(4 + 3x)$
11. $\{-3, 3\}$ **12.** $\{-6, 1\}$ **13.** $\{0, 8\}$ **14.** $\left\{-\dfrac{5}{2}, \dfrac{2}{3}\right\}$
15. $\{-6, 2\}$ **16.** $\{-12, -4, 0\}$ **17.** $\{5, 9\}$ **18.** $\left\{-12, \dfrac{1}{3}\right\}$

19. $\left\{-4, \dfrac{2}{3}\right\}$ **20.** $\{-5, 0, 5\}$ **21.** $\left\{\dfrac{7}{5}\right\}$ **22.** 14 inches

23. 12 centimeters   **24.** 16 chairs per row   **25.** 12 units

Chapters 1–7 Cumulative Review Problem Set (page 337)

1. $\dfrac{5}{2}$ **2.** 6   **3.** 0.6   **4.** 20   **5.** 18   **6.** -21 **7.** $\dfrac{1}{27}$

8. $\dfrac{3}{2}$ **9.** 1   **10.** $\dfrac{12}{7}$ **11.** $-\dfrac{1}{16}$ **12.** -16 **13.** $\dfrac{25}{4}$

14. $-\dfrac{1}{27}$ **15.** $\dfrac{19}{10x}$ **16.** $\dfrac{2y}{3x}$ **17.** $-35x^5y^5$ **18.** $81a^2b^6$

19. $-15n^4 - 18n^3 + 6n^2$ **20.** $15x^2 + 17x - 4$

21. $4x^2 + 20x + 25$ **22.** $2x^3 + x^2 - 7x - 2$

23. $x^4 + x^3 - 6x^2 + x + 3$ **24.** $-6x^2 + 11x + 7$

25. $3xy - 6x^3y^3$ **26.** $7x + 4$ **27.** $3x(x^2 + 5x + 9)$

28. $(x + 10)(x - 10)$ **29.** $(5x - 2)(x - 4)$

30. $(4x + 7)(2x - 9)$ **31.** $(n + 16)(n + 9)$

32. $(x + y)(n - 2)$ **33.** $3x(x + 1)(x - 1)$

34. $2x(x - 9)(x + 6)$ **35.** $(6x - 5)^2$

36. $(3x + y)(x - 2y)$ **37.** $\left\{\dfrac{16}{3}\right\}$ **38.** $\{-11, 0\}$

39. $\left\{\dfrac{1}{14}\right\}$ **40.** $\{-1, 1\}$ **41.** $\{-6, 1\}$ **42.** $\left\{\dfrac{11}{12}\right\}$

43. $\{1, 2\}$ **44.** $\left\{-\dfrac{7}{2}, \dfrac{1}{3}\right\}$ **45.** $\left\{\dfrac{1}{2}, 8\right\}$ **46.** $\{-9, 2\}$

47. $\{x \mid x \le -1\}$ or $(-\infty, -1]$ **48.** $\{x \mid x > 13\}$ or $(13, \infty)$

49. $\left\{x \mid x \le \dfrac{3}{2}\right\}$ or $\left(-\infty, \dfrac{3}{2}\right]$ **50.** $\left\{x \mid x > \dfrac{48}{5}\right\}$ or $\left(\dfrac{48}{5}, \infty\right)$

51. $\{x \mid x \le 500\}$ or $(-\infty, 500]$

52.
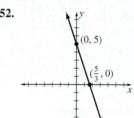

53.

54.
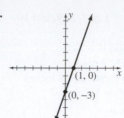

55.
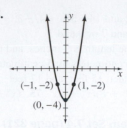

56. $\dfrac{5}{6}$ **57.** $11x + 3y = -10$ **58.** $3x - 8y = -50$

59. $2x - 9y = -12$ **60.** $\{(-6, -4)\}$ **61.** $\{(0, -3)\}$

62. $\left\{\left(-\dfrac{1}{2}, 5\right)\right\}$ **63.** $\{(-10, 6)\}$ **64.** Composite

65. 6   **66.** 72   **67.** 175%   **68.** $(2.4)(10^{-3})$

69. 3140   **70.**
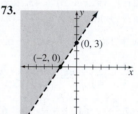

71. 16π square centimeters

72. 18   **73.**

74. 6 inches, 8 inches, and 10 inches

75. 100 milliliters   **76.** 6 feet by 8 feet

77. 2.5 hours   **78.** 7.5 centimeters

79. 27 gallons   **80.** 50° and 130°

81. 40°, 50°, and 90°

82. 40 pennies, 50 nickels, and 85 dimes

83. 14 dimes and 25 quarters

84. 97 or better   **85.** 900 girls and 750 boys

86. $2500   **87.** $1275   **88.** 1 quart

89. $1.60 for a tennis ball and $2.25 for a golf ball

Chapter 8

Problem Set 8.1 (page 346)

1. $\{4\}$ **3.** $\{-1\}$ **5.** $\{6\}$ **7.** $\left\{-\dfrac{7}{3}\right\}$

9. $\left\{\dfrac{50}{3}\right\}$ **11.** $\left\{-\dfrac{53}{2}\right\}$ **13.** $\left\{\dfrac{27}{22}\right\}$ **15.** $\{-19\}$

17. $\left\{-\dfrac{19}{16}\right\}$ **19.** $\varnothing$ **21.** $\{275\}$ **23.** $\{x \mid x$ is a real number$\}$

25. $\left\{-\dfrac{64}{79}\right\}$ **27.** $\{(2, -4)\}$ **29.** $\{(-3, -6)\}$ **31.** $\varnothing$

33. $\{(-6, 0)\}$ **35.** $\{(-8, 6)\}$ **37.** $-16, -15,$ and -14

39. 14 males and 37 females

41. $42 skirt, $48 sweater, $34 shoes

43. 14 centimeters   **45.** $25

47. 125 lb of sodium and 75 lb of chlorine

49. 8 meters and 15 meters

51. 16 miles per hour for Javier and 19 miles per hour for Domenica

Problem Set 8.2 (page 351)

1. $(-6, \infty)$ **3.** $(-5, \infty)$ **5.** $\left(-\infty, \dfrac{5}{3}\right]$ **7.** $(-36, \infty)$

9. $\left(-\infty, -\dfrac{8}{17}\right]$ **11.** $\left(-\dfrac{11}{2}, \infty\right)$ **13.** $(23, \infty)$

15. $\left(-\infty, -\dfrac{37}{3}\right]$ **17.** $\left(-\infty, -\dfrac{19}{6}\right)$ **19.** $(-\infty, 50]$

21. $(300, \infty)$ **23.** $[4, \infty)$

25. $[3, \infty)$

27. $\left(\dfrac{1}{3}, \dfrac{2}{5}\right)$

29. $(-\infty, -1) \cup \left(-\dfrac{1}{3}, \infty\right)$

31. $\left(-\dfrac{1}{4}, \dfrac{11}{4}\right)$ **33.** $[-11, 13]$ **35.** $(-1, 5)$

37. More than $200   **39.** 96 or better

41. Between $-20°C$ and $-5°C$, inclusive

43. 8.8 to 15.4, inclusive

Problem Set 8.3 (page 363)

1. $\{-7, 9\}$ **3.** $\{-1, 5\}$ **5.** $\left\{-5, \dfrac{7}{3}\right\}$ **7.** $\{-1, 5\}$

9. $\left\{\dfrac{1}{12}, \dfrac{17}{12}\right\}$ **11.** $\{0, 3\}$

13. $(-5, 5)$

15. $[-2, 2]$

17. $(-\infty, -2) \cup (2, \infty)$

19. $(-1, 3)$

21. $[-6, 2]$

23. $(-\infty, -3) \cup (-1, \infty)$

25. $(-\infty, 1] \cup [5, \infty)$

27. $(-\infty, -4) \cup (8, \infty)$

29. $(-8, 2)$ **31.** $[-4, 5]$

33. $\left(-\infty, -\dfrac{7}{2}\right] \cup \left[\dfrac{5}{2}, \infty\right)$ **35.** $(-\infty, -2) \cup (6, \infty)$

37. $\left(-\dfrac{1}{2}, \dfrac{3}{2}\right)$ **39.** $\left[-5, \dfrac{7}{5}\right]$ **41.** $[-3, 10]$ **43.** $(-5, 11)$

45. $\left(-\infty, -\dfrac{3}{2}\right) \cup \left(\dfrac{1}{2}, \infty\right)$ **47.** $(-\infty, -14] \cup [0, \infty)$

49. $[-2, 3]$ **51.** $\varnothing$ **53.** $(-\infty, \infty)$ **55.** $\left\{\dfrac{2}{5}\right\}$ **57.** $\varnothing$

59. $\varnothing$ **65.** $\left\{-2, -\dfrac{4}{3}\right\}$ **67.** $\{-2\}$ **69.** $\{0\}$

Problem Set 8.4 (page 371)

1. $9x - 3$ **3.** $4x^2 + 8x - 3$ **5.** $16x^2 - 15x + 11$

7. $2x^2 - 2xy - 5y^2$ **9.** $-\dfrac{1}{6}x^3y^4$ **11.** $30x^6$ **13.** $-18x^9$

15. $-3x^6y^6$ **17.** $81a^4b^{12}$ **19.** $-16a^4b^4$ **21.** $4x^2 + 33x + 35$

23. $9y^2 - 1$ **25.** $14x^2 + 3x - 2$ **27.** $5 + 3t - 2t^2$

29. $9t^2 + 42t + 49$ **31.** $4 - 25x^2$

33. $x^3 - 4x^2 + x + 6$ **35.** $x^3 - x^2 - 9x + 9$

37. $x^3 + x^2 - 24x + 16$ **39.** $2x^3 + 9x^2 + 2x - 30$

41. $12x^3 - 7x^2 + 25x - 6$

43. $x^4 + 5x^3 + 11x^2 + 11x + 4$

45. $64x^3 - 48x^2 + 12x - 1$

47. $125x^3 + 150x^2 + 60x + 8$

49. $x^{2n} - 16$ **51.** $x^{2a} + 4x^a - 12$

53. $6x^{2n} + x^n - 35$ **55.** $x^{4a} - 10x^{2a} + 21$

57. $4x^{2n} + 20x^n + 25$ **59.** $3x^3y^3$ **61.** $-5x^3y^2$

63. $9bc^2$ **65.** $-18xyz^4$ **67.** $-a^2b^3c^2$

69. $a^7 + 7a^6b + 21a^5b^2 + 35a^4b^3 + 35a^3b^4 + 21a^2b^5 + 7ab^6 + b^7$

71. $x^5 - 5x^4y + 10x^3y^2 - 10x^2y^3 + 5xy^4 - y^5$

73. $x^4 + 8x^3y + 24x^2y^2 + 32xy^3 + 16y^4$

75. $64a^6 - 192a^5b + 240a^4b^2 - 160a^3b^3 + 60a^2b^4 - 12ab^5 + b^6$

77. $x^{14} + 7x^{12}y + 21x^{10}y^2 + 35x^8y^3 + 35x^6y^4 + 21x^4y^5 + 7x^2y^6 + y^7$

79. $32a^5 - 240a^4b + 720a^3b^2 - 1080a^2b^3 + 810ab^4 - 243b^5$

Problem Set 8.5 (page 377)

1. $3x^3 + 6x^2$ **3.** $-6x^4 + 9x^6$ **5.** $3a^2 - 5a - 8$

7. $-13x^2 + 17x - 28$ **9.** $-3xy + 4x^2y - 8xy^2$

11. Q: $4x + 5$ **13.** Q: $t^2 + 2t - 4$ **15.** Q: $2x + 5$
R: 0   R: 0   R: 1

17. Q: $3x - 4$ **19.** Q: $5y - 1$ **21.** Q: $4a + 6$
R: $3x - 1$ R: $-8y - 2$ R: $7a - 19$

23. Q: $3x + 4$ **25.** Q: $x + 6$ **27.** Q: $4x - 3$
R: 0   R: 14   R: 2

29. Q: $x^2 - 1$ **31.** Q: $3x^3 - 4x^2 + 6x - 13$
R: 0   R: 12

33. Q: $x^2 - 2x - 3$ **35.** Q: $x^3 + 7x^2 + 21x + 56$
R: 0   R: 167

37. Q: $x^2 + 3x + 2$ **39.** Q: $x^4 + x^3 + x^2 + x + 1$
R: 0   R: 0

41. Q: $x^4 + x^3 + x^2 + x + 1$ **43.** Q: $2x^2 + 2x - 3$
R: 2   R: $\dfrac{9}{2}$

45. Q: $4x^3 + 2x^2 - 4x - 2$
R: 0

Problem Set 8.6 (page 387)

1. $2xy(3 - 4y)$ **3.** $(z + 3)(x + y)$ **5.** $(x + y)(3 + a)$

7. $(x - y)(a - b)$ **9.** $(3x + 5)(3x - 5)$

11. $(1 + 9n)(1 - 9n)$ **13.** $(x + 4 + y)(x + 4 - y)$

15. $(3s + 2t - 1)(3s - 2t + 1)$ **17.** $(x - 7)(x + 2)$

19. $(5 + x)(3 - x)$ **21.** Not factorable

23. $(3x - 5)(x - 2)$ **25.** $(5x + 1)(2x - 7)$

27. $4(x^2 + 4)$ **29.** $x(x + 3)(x - 3)$ **31.** $(3a - 7)^2$

33. $2n(n^2 + 3n + 5)$ **35.** $(5x - 3)(2x + 9)$ **37.** $(6a - 1)^2$

39. $(4x - y)(2x + y)$ **41.** Not factorable   **43.** $2n(n^2 + 7n - 10)$

45. $(x - 2)(x^2 + 2x + 4)$ **47.** $(4x + 3y)(16x^2 - 12xy + 9y^2)$

49. $4(x + 2)(x^2 - 2x + 4)$ **51.** $(a + 10)(a^2 - 10a + 100)$

53. $xy(x - y)(x^2 + xy + y^2)$ **55.** $\{-3, -1\}$

57. $\{-12, -6\}$ **59.** $\{4, 9\}$ **61.** $\{-6, 2\}$

63. $\{-1, 5\}$ **65.** $\{-13, -12\}$ **67.** $\left\{-5, \dfrac{1}{3}\right\}$

69. $\left\{-\dfrac{7}{2}, -\dfrac{2}{3}\right\}$ **71.** $\{0, 4\}$ **73.** $\left\{\dfrac{1}{6}, 2\right\}$ **75.** $\{-6, 0, 6\}$

77. $\{-4, 6\}$ **79.** $\left\{\dfrac{1}{3}\right\}$ **81.** $\{-11, 4\}$ **83.** $\{-5, 5\}$

85. $\left\{-2, 0, \dfrac{1}{2}\right\}$ **87.** $\left\{-3, 0, \dfrac{7}{4}\right\}$ **89.** 6 units **91.** 8 and 13

93. 6, 8, 10 **95.** 6 inches

97. 4 centimeters by 4 centimeters and 6 centimeters by 8 centimeters

Chapter 8 Review Problem Set (page 397)

1. $\{18\}$ **2.** $\{-14\}$ **3.** $\{0\}$ **4.** $\left\{\dfrac{1}{2}\right\}$ **5.** $\{10\}$ **6.** $\left\{\dfrac{7}{3}\right\}$

7. $\left\{\dfrac{28}{17}\right\}$ **8.** $\left\{-\dfrac{1}{38}\right\}$ **9.** $\left\{\dfrac{27}{17}\right\}$ **10.** $\left\{-\dfrac{10}{3}, 4\right\}$ **11.** $\{50\}$

12. $\left\{-\dfrac{39}{2}\right\}$ **13.** $\{200\}$ **14.** $\{-8\}$ **15.** $\{-3, 3\}$

16. $\{-6, 1\}$ **17.** $\left\{\dfrac{2}{7}\right\}$ **18.** $\left\{-\dfrac{2}{5}, \dfrac{1}{3}\right\}$ **19.** $\left\{-\dfrac{1}{3}, 3\right\}$

20. $\{-3, 0, 3\}$ **21.** $\left\{-\dfrac{4}{7}, \dfrac{2}{7}\right\}$ **22.** $\left\{-\dfrac{4}{5}, \dfrac{5}{6}\right\}$ **23.** $\{0, 1, 8\}$

24. $\left\{-10, \dfrac{1}{4}\right\}$ **25.** $x = \dfrac{2b + 2}{a}$ **26.** $x = \dfrac{c}{a - b}$

27. $x = \dfrac{pb - ma}{m - p}$ **28.** $x = \dfrac{11 + 7y}{5}$ **29.** $x = \dfrac{by + b + ac}{c}$

30. $[-5, \infty)$ **31.** $(4, \infty)$ **32.** $\left(-\dfrac{7}{3}, \infty\right)$ **33.** $\left[\dfrac{17}{2}, \infty\right)$

34. $\left(-\infty, \dfrac{1}{3}\right)$ **35.** $\left(\dfrac{53}{11}, \infty\right)$ **36.** $[6, \infty)$ **37.** $(-\infty, 100]$

38. $(-5, 6)$ **39.** $\left(-\infty, -\dfrac{11}{3}\right) \cup (3, \infty)$

40.

41.

42.

43.

44. $5x - 3$ **45.** $3x^2 + 12x - 2$ **46.** $12x^2 - x + 5$
47. $-20x^5 y^7$ **48.** $-6a^5 b^5$ **49.** $15a^4 - 10a^3 - 5a^2$
50. $24x^2 + 2xy - 15y^2$ **51.** $3x^3 + 7x^2 - 21x - 4$
52. $256x^8 y^{12}$ **53.** $9x^2 - 12xy + 4y^2$ **54.** $-8x^6 y^9 z^3$

55. $-13x^2 y$ **56.** $2x + y - 2$ **57.** $x^4 + x^3 - 18x^2 - x + 35$
58. $21 + 26x - 15x^2$ **59.** $-12a^5 b^7$ **60.** $-8a^7 b^3$
61. $7x^2 + 19x - 36$ **62.** $6x^3 - 11x^2 - 7x + 2$ **63.** $6x^{4n}$
64. $4x^2 + 20xy + 25y^2$ **65.** $x^3 - 6x^2 + 12x - 8$
66. Q: $3x^2 - 4x - 6$ **67.** Q: $5x^2 + 3x - 2$
 R: 29   R: 8
68. Q: $x^3 - 4x^2 - 7x - 1$ **69.** Q: $2x^3 + 3x^2 - 4x - 2$
 R: 0   R: 0
70. $(x + 7)(x - 4)$ **71.** $2(t + 3)(t - 3)$
72. Not factorable **73.** $(4n - 1)(3n - 1)$
74. $x^2(x^2 + 1)(x + 1)(x - 1)$ **75.** $x(x - 12)(x + 6)$
76. $2a^2 b(3a + 2b - c)$ **77.** $(x - y + 1)(x + y - 1)$
78. $4(2x^2 + 3)$ **79.** $(4x + 7)(3x - 5)$ **80.** $(4n - 5)^2$
81. $4n(n - 2)$ **82.** $3w(w^2 + 6w - 8)$ **83.** $(5x + 2y)(4x - y)$
84. $16a(a - 4)$ **85.** $3x(x + 1)(x - 6)$ **86.** $(n + 8)(n - 16)$
87. $(t + 5)(t - 5)(t^2 + 3)$ **88.** $(5x - 3)(7x + 2)$
89. $(3 - x)(5 - 3x)$ **90.** $(4n - 3)(16n^2 + 12n + 9)$
91. $2(2x + 5)(4x^2 - 10x + 25)$
92. The length is 15 meters and the width is 7 meters.
93. \$2000 at 7% and \$3000 at 8% **94.** 88 or better
95. 4, 5, and 6 **96.** \$10.50 per hour
97. 20 nickels, 50 dimes, and 75 quarters **98.** 80° **99.** \$45.60
100. 30 or more **101.** 55 miles per hour
102. Sonya for $3\dfrac{1}{4}$ hours and Rita for $4\dfrac{1}{2}$ hours

103. $6\dfrac{1}{4}$ cups **104.** 12 miles and 16 miles

105. 4 meters by 12 meters **106.** 9 rows and 16 chairs per row
107. The side is 13 feet long and the altitude is 6 feet.

Chapter 8 Test (page 399)

1. $2x - 11$ **2.** $20x^2 + 17x - 63$ **3.** $2x^3 + 11x^2 - 11x - 30$
4. $x^3 - 12x^2 y + 48xy^2 - 64y^3$
5. Q: $2x^2 - 3x - 4$ **6.** Q: $3x^3 - x^2 - 2x - 6$
 R: 0   R: 3
7. $(x - y)(x + 4)$ **8.** $3(2x + 1)(2x - 1)$

9. $\left\{\dfrac{16}{5}\right\}$ **10.** $\left\{-\dfrac{14}{5}\right\}$ **11.** $\left\{-\dfrac{3}{2}, 3\right\}$ **12.** $\{3\}$

13. $\{650\}$ **14.** $\left\{0, \dfrac{1}{4}\right\}$ **15.** $\left\{\dfrac{3}{2}\right\}$ **16.** $\{-9, 0, 2\}$

17. $\left\{-\dfrac{3}{7}, \dfrac{4}{5}\right\}$ **18.** $\left\{-\dfrac{1}{3}, 2\right\}$ **19.** $\left(-1, \dfrac{7}{3}\right)$ **20.** $(3, \infty)$

21. $(-\infty, -35]$ **22.** $\dfrac{2}{3}$ of a cup

23. 97 or better **24.** 70° **25.** 8 feet

Chapter 9

Problem Set 9.1 (page 406)

1. $\dfrac{3}{4}$ **3.** $\dfrac{5}{6}$ **5.** $-\dfrac{2}{5}$ **7.** $\dfrac{2}{7}$ **9.** $\dfrac{2x}{7}$ **11.** $\dfrac{2a}{5b}$ **13.** $-\dfrac{y}{4x}$

15. $-\dfrac{9c}{13d}$ **17.** $\dfrac{5x^2}{3y^3}$ **19.** $\dfrac{x - 2}{x}$ **21.** $\dfrac{3x + 2}{2x - 1}$ **23.** $\dfrac{a + 5}{a - 9}$

25. $\dfrac{n - 3}{5n - 1}$ **27.** $\dfrac{5x^2 + 7}{10x}$ **29.** $\dfrac{3x + 5}{4x + 1}$

31. $\dfrac{3x}{x^2 + 4x + 16}$ **33.** $\dfrac{x + 6}{3x - 1}$ **35.** $\dfrac{x(2x + 7)}{y(x + 9)}$

37. $\dfrac{y + 4}{5y - 2}$ **39.** $\dfrac{3x(x - 1)}{x^2 + 1}$ **41.** $\dfrac{2(x + 3y)}{3x(3x + y)}$

43. $\dfrac{3n - 2}{7n + 2}$ **45.** $\dfrac{4 - x}{5 + 3x}$ **47.** $\dfrac{9x^2 + 3x + 1}{2(x + 2)}$

49. $\dfrac{-2(x-1)}{x+1}$ **51.** $\dfrac{y+b}{y+c}$ **53.** $\dfrac{x+2y}{2x+y}$ **55.** $\dfrac{x+1}{x-6}$

57. $\dfrac{2s+5}{3s+1}$ **59.** -1 **61.** $-n-7$ **63.** $-\dfrac{2}{x+1}$

65. -2 **67.** $-\dfrac{n+3}{n+5}$

Problem Set 9.2 (page 412)

1. $\dfrac{1}{10}$ **3.** $-\dfrac{4}{15}$ **5.** $\dfrac{3}{16}$ **7.** $-\dfrac{5}{6}$ **9.** $-\dfrac{2}{3}$ **11.** $\dfrac{10}{11}$ **13.** $-\dfrac{5x^3}{12y^2}$

15. $\dfrac{2a^3}{3b}$ **17.** $\dfrac{3x^3}{4}$ **19.** $\dfrac{25x^3}{108y^2}$ **21.** $\dfrac{ac^2}{2b^2}$ **23.** $\dfrac{3x}{4y}$

25. $\dfrac{3(x^2+4)}{5y(x+8)}$ **27.** $\dfrac{5(a+3)}{a(a-2)}$ **29.** $\dfrac{3}{2}$ **31.** $\dfrac{3xy}{4(x+6)}$

33. $\dfrac{5(x-2y)}{7y}$ **35.** $\dfrac{5+n}{3-n}$ **37.** $\dfrac{x^2+1}{x^2-10}$ **39.** $\dfrac{6x+5}{3x+4}$

41. $\dfrac{2t^2+5}{2(t^2+1)(t+1)}$ **43.** $\dfrac{t(t+6)}{4t+5}$ **45.** $\dfrac{n+3}{n(n-2)}$

47. $\dfrac{25x^3y^3}{4(x+1)}$ **49.** $\dfrac{2(a-2b)}{a(3a-2b)}$

Problem Set 9.3 (page 419)

1. $\dfrac{13}{12}$ **3.** $\dfrac{11}{40}$ **5.** $\dfrac{19}{20}$ **7.** $\dfrac{49}{75}$ **9.** $\dfrac{17}{30}$ **11.** $-\dfrac{11}{84}$ **13.** $\dfrac{2x+4}{x-1}$

15. 4 **17.** $\dfrac{7y-10}{7y}$ **19.** $\dfrac{5x+3}{6}$ **21.** $\dfrac{12a+1}{12}$ **23.** $\dfrac{n+14}{18}$

25. $-\dfrac{11}{15}$ **27.** $\dfrac{3x-25}{30}$ **29.** $\dfrac{43}{40x}$ **31.** $\dfrac{20y-77x}{28xy}$

33. $\dfrac{16y+15x-12xy}{12xy}$ **35.** $\dfrac{21+22x}{30x^2}$ **37.** $\dfrac{10n-21}{7n^2}$

39. $\dfrac{45-6n+20n^2}{15n^2}$ **41.** $\dfrac{11x-10}{6x^2}$ **43.** $\dfrac{42t+43}{35t^3}$

45. $\dfrac{20b^2-33a^3}{96a^2b}$ **47.** $\dfrac{14-24y^3+45xy}{18xy^3}$ **49.** $\dfrac{2x^2+3x-3}{x(x-1)}$

51. $\dfrac{a^2-a-8}{a(a+4)}$ **53.** $\dfrac{-41n-55}{(4n+5)(3n+5)}$ **55.** $\dfrac{-3x+17}{(x+4)(7x-1)}$

57. $\dfrac{-x+74}{(3x-5)(2x+7)}$ **59.** $\dfrac{38x+13}{(3x-2)(4x+5)}$ **61.** $\dfrac{5x+5}{2x+5}$

63. $\dfrac{x+15}{x-5}$ **65.** $\dfrac{-2x-4}{2x+1}$ **67. (a)** -1 **(c)** 0

Problem Set 9.4 (page 428)

1. $\dfrac{7x+20}{x(x+4)}$ **3.** $\dfrac{-x-3}{x(x+7)}$ **5.** $\dfrac{6x-5}{(x+1)(x-1)}$

7. $\dfrac{1}{a+1}$ **9.** $\dfrac{5n+15}{4(n+5)(n-5)}$ **11.** $\dfrac{x^2+60}{x(x+6)}$

13. $\dfrac{11x+13}{(x+2)(x+7)(2x+1)}$ **15.** $\dfrac{-3a+1}{(a-5)(a+2)(a+9)}$

17. $\dfrac{3a^2+14a+1}{(4a-3)(2a+1)(a+4)}$ **19.** $\dfrac{3x^2+20x-111}{(x^2+3)(x+7)(x-3)}$

21. $\dfrac{x+6}{(x-3)^2}$ **23.** $\dfrac{14x-4}{(x-1)(x+1)^2}$ **25.** $\dfrac{-7y-14}{(y+8)(y-2)}$

27. $\dfrac{-2x^2-4x+3}{(x+2)(x-2)}$ **29.** $\dfrac{2x^2+14x-19}{(x+10)(x-2)}$ **31.** $\dfrac{2n+1}{n-6}$

33. $\dfrac{2x^2-32x+16}{(x+1)(2x-1)(3x-2)}$ **35.** $\dfrac{1}{(n^2+1)(n+1)}$

37. $\dfrac{-16x}{(5x-2)(x-1)}$ **39.** $\dfrac{t+1}{t-2}$ **41.** $\dfrac{2}{11}$ **43.** $-\dfrac{7}{27}$ **45.** $\dfrac{x}{4}$

47. $\dfrac{3y-2x}{4x-7}$ **49.** $\dfrac{6ab^2-5a^2}{12b^2+2a^2b}$ **51.** $\dfrac{2y-3xy}{3x+4xy}$ **53.** $\dfrac{3n+14}{5n+19}$

55. $\dfrac{5n-17}{4n-13}$ **57.** $\dfrac{-x+5y-10}{3y-10}$ **59.** $\dfrac{-x+15}{-2x-1}$

61. $\dfrac{3a^2-2a+1}{2a-1}$ **63.** $\dfrac{-x^2+6x-4}{3x-2}$

Problem Set 9.5 (page 435)

1. $\{2\}$ **3.** $\{-3\}$ **5.** $\{6\}$ **7.** $\left\{-\dfrac{85}{18}\right\}$ **9.** $\left\{\dfrac{7}{10}\right\}$

11. $\{5\}$ **13.** $\{58\}$ **15.** $\left\{\dfrac{1}{4},4\right\}$ **17.** $\left\{-\dfrac{2}{5},5\right\}$ **19.** $\{-16\}$

21. $\left\{-\dfrac{13}{3}\right\}$ **23.** $\{-3,1\}$ **25.** $\left\{-\dfrac{5}{2}\right\}$ **27.** $\{-51\}$

29. $\left\{-\dfrac{5}{3},4\right\}$ **31.** $\varnothing$ **33.** $\left\{-\dfrac{11}{8},2\right\}$ **35.** $\{-29,0\}$

37. $\{-9,3\}$ **39.** $\left\{-2,\dfrac{23}{8}\right\}$ **41.** $\left\{\dfrac{11}{23}\right\}$ **43.** $\left\{3,\dfrac{7}{2}\right\}$

45. $750 and $1000 **47.** 48° and 72°
49. $3500 **51.** $69 for Tammy and $51.75 for Laura
53. 14 feet and 6 feet **55.** 690 females and 460 males

Problem Set 9.6 (page 443)

1. $\{-21\}$ **3.** $\{-1,2\}$ **5.** $\{2\}$ **7.** $\left\{\dfrac{37}{15}\right\}$ **9.** $\{-1\}$

11. $\{-1\}$ **13.** $\left\{0,\dfrac{13}{2}\right\}$ **15.** $\left\{-2,\dfrac{19}{2}\right\}$ **17.** $\{-2\}$

19. $\left\{-\dfrac{1}{5}\right\}$ **21.** $\varnothing$ **23.** $\left\{\dfrac{7}{2}\right\}$ **25.** $\{-3\}$ **27.** $\left\{-\dfrac{7}{9}\right\}$

29. $\left\{-\dfrac{7}{6}\right\}$ **31.** $x=\dfrac{18y-4}{15}$ **33.** $y=\dfrac{-5x+22}{2}$

35. $M=\dfrac{IC}{100}$ **37.** $R=\dfrac{ST}{S+T}$ **39.** $y=\dfrac{bx-x-3b+a}{a-3}$

41. $y=\dfrac{ab-bx}{a}$ **43.** $y=\dfrac{-2x-9}{3}$

45. 50 miles per hour for Dave and 54 miles per hour for Kent
47. 60 minutes
49. 60 words per minute for Connie and 40 words per minute for Katie
51. Plane B could travel at 400 miles per hour for 5 hours and plane A at 350 miles per hour for 4 hours, or plane B could travel at 250 miles per hour for 8 hours and plane A at 200 miles per hour for 7 hours.
53. 60 minutes for Nancy and 120 minutes for Amy
55. 3 hours
57. 16 miles per hour on the way out and 12 miles per hour on the way back, or 12 miles per hour out and 8 miles per hour back

Chapter 9 Review Problem Set (page 452)

1. $\dfrac{2y}{3x^2}$ 2. $\dfrac{a-3}{a}$ 3. $\dfrac{n-5}{n-1}$ 4. $\dfrac{x^2+1}{x}$ 5. $\dfrac{2x+1}{3}$

6. $\dfrac{x^2-10}{2x^2+1}$ 7. $\dfrac{3}{22}$ 8. $\dfrac{18y+20x}{48y-9x}$ 9. $\dfrac{3x+2}{3x-2}$ 10. $\dfrac{x-1}{2x-1}$

11. $\dfrac{2x}{7y^2}$ 12. $3b$ 13. $\dfrac{n(n+5)}{n-1}$ 14. $\dfrac{x(x-3y)}{x^2+9y^2}$

15. $\dfrac{23x-6}{20}$ 16. $\dfrac{57-2n}{18n}$ 17. $\dfrac{3x^2-2x-14}{x(x+7)}$ 18. $\dfrac{2}{x-5}$

19. $\dfrac{5n-21}{(n-9)(n+4)(n-1)}$ 20. $\dfrac{6y-23}{(2y+3)(y-6)}$

21. xy^4 22. $\dfrac{3x^2y^5}{4}$ 23. $\dfrac{4x(x+1)}{x+4}$ 24. $-\dfrac{5}{3(x+2)(x+1)}$

25. $\left\{\dfrac{4}{13}\right\}$ 26. $\left\{\dfrac{3}{16}\right\}$ 27. $\varnothing$ 28. $\{-17\}$ 29. $\left\{\dfrac{2}{7}, \dfrac{7}{2}\right\}$

30. $\{22\}$ 31. $\left\{-\dfrac{6}{7}, 3\right\}$ 32. $\left\{\dfrac{3}{4}, \dfrac{5}{2}\right\}$ 33. $\left\{\dfrac{9}{7}\right\}$ 34. $\left\{-\dfrac{5}{4}\right\}$

35. $y = \dfrac{3x+27}{4}$ 36. $y = \dfrac{bx-ab}{a}$ 37. $525 and $875

38. Busboy $36; waiter $126
39. 20 minutes for Julio and 30 minutes for Dan
40. 50 miles per hour and 55 miles per hour or $8\frac{1}{3}$ miles per hour and $13\frac{1}{3}$ miles per hour
41. 9 hours 42. 80 hours 43. 13 miles per hour

Chapter 9 Test (page 453)

1. $\dfrac{13y^2}{24x}$ 2. $\dfrac{3x-1}{x(x-6)}$ 3. $\dfrac{2n-3}{n+4}$ 4. $-\dfrac{2x}{x+1}$ 5. $\dfrac{3y^2}{8}$

6. $\dfrac{a-b}{4(2a+b)}$ 7. $\dfrac{x+4}{5x-1}$ 8. $\dfrac{13x+7}{12}$ 9. $\dfrac{3x}{2}$

10. $\dfrac{10n-26}{15n}$ 11. $\dfrac{3x^2+2x-12}{x(x-6)}$ 12. $\dfrac{11-2x}{x(x-1)}$

13. $\dfrac{13n+46}{(2n+5)(n-2)(n+7)}$ 14. $\dfrac{18-2x}{8+9x}$

15. $y = \dfrac{3}{5}x - \dfrac{2}{5}$ 16. $y = \dfrac{4x+20}{3}$ 17. $\{1\}$ 18. $\left\{\dfrac{1}{10}\right\}$

19. $\{-35\}$ 20. $\{-1, 5\}$ 21. $\left\{\dfrac{5}{3}\right\}$ 22. $\left\{-\dfrac{9}{13}\right\}$ 23. $\dfrac{27}{72}$

24. 60 minutes 25. 15 miles per hour

Chapters 1–9 Cumulative Review Problem Set (page 454)

1. 16 2. -13 3. 104 4. $\dfrac{5}{2}$ 5. $7a^2+11a+1$

6. $-2x^2+9x-4$ 7. $-2x^3y^5$ 8. $16x^2y^6$ 9. $36a^7b^2$

10. $-24a^6b^4$ 11. $-18x^4+3x^3-12x^2$

12. $10x^2+xy-3y^2$ 13. $x^2+8xy+16y^2$

14. $a^3-a^2b-11ab^2+3b^3$ 15. $(x-2)(x-3)$

16. $(2x+1)(3x-4)$ 17. $2(x-1)(x-3)$

18. $3(x+8)(x-2)$ 19. $(3m-4n)(3m+4n)$

20. $(3a+2)(9a^2-6a+4)$ 21. $-\dfrac{7y^4}{x^2}$ 22. $-x$

23. $\dfrac{x(x-5)}{x-1}$ 24. $\dfrac{x+7}{x+3}$ 25. $\dfrac{9n-26}{10}$

26. $\dfrac{8x-19}{(x-2)(x+3)(x-3)}$ 27. $\dfrac{2y+3x}{6xy}$ 28. $\dfrac{m-n}{mn}$

29. $\{6\}$ 30. $\{24\}$ 31. $\left\{\dfrac{15}{7}\right\}$ 32. $\{6\}$ 33. $\left\{-2, \dfrac{10}{3}\right\}$

34. $\{-28, 12\}$ 35. $\{-8, 1\}$ 36. $\left\{-5, -\dfrac{3}{2}\right\}$

37. $\left\{-\dfrac{1}{3}, 9\right\}$ 38. $\varnothing$ 39. $P = \dfrac{A}{1+rt}$ 40. $B = \dfrac{3V}{H}$

41. $(-\infty, 2]$ 42. $(-4, 2)$ 43. $\left(-\dfrac{9}{3}, 3\right)$

44. $(-\infty, -13] \cup [7, \infty)$ 45. $(-\infty, -10] \cup (2, \infty)$
46. $5.76 47. $690 48. Width is 23 in., length is 38 in.
49. 4 hr 50. $13,600 and $54,400 51. 12 min
52. 5 in., 12 in., and 13 in. 53. $16\frac{2}{3}$ yr 54. 27 dimes and 13 quarters.

Chapter 10

Problem Set 10.1 (page 465)

1. $\dfrac{1}{27}$ 3. $-\dfrac{1}{100}$ 5. 81 7. -27 9. -8 11. 1 13. $\dfrac{9}{49}$

15. 16 17. $\dfrac{1}{1000}$ 19. $\dfrac{1}{1000}$ 21. 27 23. $\dfrac{1}{125}$ 25. $\dfrac{9}{8}$

27. $\dfrac{256}{25}$ 29. $\dfrac{2}{25}$ 31. $\dfrac{81}{4}$ 33. 81 35. $\dfrac{1}{10,000}$ 37. $\dfrac{13}{36}$

39. $\dfrac{1}{2}$ 41. $\dfrac{72}{17}$ 43. $\dfrac{1}{x^6}$ 45. $\dfrac{1}{a^3}$ 47. $\dfrac{1}{a^8}$ 49. $\dfrac{y^6}{x^2}$

51. $\dfrac{c^8}{a^4b^{12}}$ 53. $\dfrac{y^{12}}{8x^9}$ 55. $\dfrac{x^3}{y^{12}}$ 57. $\dfrac{4a^4}{9b^2}$ 59. $\dfrac{1}{x^2}$ 61. a^5b^2

63. $\dfrac{6y^3}{x}$ 65. $7b^2$ 67. $\dfrac{7x}{y^2}$ 69. $-\dfrac{12b^3}{a}$ 71. $\dfrac{x^5y^5}{5}$ 73. $\dfrac{b^{20}}{81}$

75. $\dfrac{x+1}{x^3}$ 77. $\dfrac{y-x^3}{x^3y}$ 79. $\dfrac{3b+4a^2}{a^2b}$ 81. $\dfrac{1-x^2y}{xy^2}$

83. $\dfrac{2x-3}{x^2}$ 85. $(4)(10^7)$ 87. $(3.764)(10^2)$ 89. $(3.47)(10^{-1})$

91. $(2.14)(10^{-2})$ 93. 31,400,000,000 95. 0.43

97. 0.000914 99. 0.00000005123 101. 1000 103. 1000

Problem Set 10.2 (page 475)

1. 8 3. -10 5. 3 7. -4 9. 3 11. $\dfrac{4}{5}$ 13. $-\dfrac{6}{7}$

15. $\dfrac{1}{2}$ 17. $\dfrac{3}{4}$ 19. 8 21. $3\sqrt{3}$ 23. $4\sqrt{2}$ 25. $4\sqrt{5}$

27. $4\sqrt{10}$ 29. $12\sqrt{2}$ 31. $-12\sqrt{5}$ 33. $2\sqrt{3}$

35. $3\sqrt{6}$ **37.** $-\frac{5}{3}\sqrt{7}$ **39.** $\frac{\sqrt{19}}{2}$ **41.** $\frac{3\sqrt{3}}{4}$ **43.** $\frac{5\sqrt{3}}{9}$

45. $\frac{\sqrt{14}}{7}$ **47.** $\frac{\sqrt{6}}{3}$ **49.** $\frac{\sqrt{15}}{6}$ **51.** $\frac{\sqrt{66}}{12}$ **53.** $\frac{\sqrt{6}}{3}$

55. $\sqrt{5}$ **57.** $\frac{2\sqrt{21}}{7}$ **59.** $-\frac{8\sqrt{15}}{5}$ **61.** $\frac{\sqrt{6}}{4}$ **63.** $-\frac{12}{25}$

65. $2\sqrt[3]{2}$ **67.** $6\sqrt[3]{3}$ **69.** $\frac{2\sqrt[3]{3}}{3}$ **71.** $\frac{3\sqrt[3]{2}}{2}$ **73.** $\frac{\sqrt[3]{12}}{2}$

75. 42 miles per hour; 49 miles per hour; 65 miles per hour
77. 107 square centimeters **79.** 140 square inches
85. (a) 1.414 (c) 12.490 (e) 57.000 (g) 0.374 (i) 0.930

Problem Set 10.3 (page 480)

1. $13\sqrt{2}$ **3.** $54\sqrt{3}$ **5.** $-30\sqrt{2}$ **7.** $-\sqrt{5}$ **9.** $-21\sqrt{6}$

11. $-\frac{7\sqrt{7}}{12}$ **13.** $\frac{37\sqrt{10}}{10}$ **15.** $\frac{41\sqrt{2}}{20}$ **17.** $-9\sqrt[3]{3}$

19. $10\sqrt[3]{2}$ **21.** $4\sqrt{2x}$ **23.** $5x\sqrt{3}$ **25.** $2x\sqrt{5y}$

27. $8xy^3\sqrt{xy}$ **29.** $3a^2b\sqrt{6b}$ **31.** $3x^3y^4\sqrt{7}$

33. $4a\sqrt{10a}$ **35.** $\frac{8y}{3}\sqrt{6xy}$ **37.** $\frac{\sqrt{10xy}}{5y}$ **39.** $\frac{\sqrt{15}}{6x^2}$

41. $\frac{5\sqrt{2y}}{6y}$ **43.** $\frac{\sqrt{14xy}}{4y^3}$ **45.** $\frac{3y\sqrt{2xy}}{4x}$ **47.** $\frac{2\sqrt{42ab}}{7b^2}$

49. $2\sqrt[3]{3y}$ **51.** $2x\sqrt[3]{2x}$ **53.** $2x^2y^2\sqrt[3]{7y^2}$ **55.** $\frac{\sqrt[3]{21x}}{3x}$

57. $\frac{\sqrt[3]{12x^2y}}{4x^2}$ **59.** $\frac{\sqrt[3]{4x^2y^2}}{xy^2}$ **61.** $2\sqrt{2x+3y}$

63. $4\sqrt{x+3y}$ **65.** $33\sqrt{x}$ **67.** $-30\sqrt{2x}$ **69.** $7\sqrt{3n}$

71. $-40\sqrt{ab}$ **73.** $-7x\sqrt{2x}$ **79.** (a) $5|x|\sqrt{5}$

   (b) $4x^2$ (c) $2b\sqrt{2b}$ (d) $y^2\sqrt{3y}$ (e) $12|x^3|\sqrt{2}$

   (f) $2m^4\sqrt{7}$ (g) $8|c^5|\sqrt{2}$ (h) $3d^3\sqrt{2d}$ (i) $7|x|$

   (j) $4n^{10}\sqrt{5}$ (k) $9h\sqrt{h}$

Problem Set 10.4 (page 486)

1. $6\sqrt{2}$ **3.** $18\sqrt{2}$ **5.** $-24\sqrt{10}$ **7.** $24\sqrt{6}$ **9.** 120

11. 24 **13.** $56\sqrt[3]{3}$ **15.** $\sqrt{6}+\sqrt{10}$ **17.** $6\sqrt{10}-3\sqrt{35}$

19. $24\sqrt{3}-60\sqrt{2}$ **21.** $-40-32\sqrt{15}$

23. $15\sqrt{2x}+3\sqrt{xy}$ **25.** $5xy-6x\sqrt{y}$

27. $2\sqrt{10xy}+2y\sqrt{15y}$ **29.** $-25\sqrt{6}$ **31.** $-25-3\sqrt{3}$

33. $23-9\sqrt{5}$ **35.** $6\sqrt{35}+3\sqrt{10}-4\sqrt{21}-2\sqrt{6}$

37. $8\sqrt{3}-36\sqrt{2}+6\sqrt{10}-18\sqrt{15}$

39. $11+13\sqrt{30}$ **41.** $141-51\sqrt{6}$ **43.** -10 **45.** -8

47. $2x-3y$ **49.** $10\sqrt[3]{12}+2\sqrt[3]{18}$ **51.** $12-36\sqrt[3]{2}$

53. $\frac{\sqrt{7}-1}{3}$ **55.** $\frac{-3\sqrt{2}-15}{23}$ **57.** $\frac{\sqrt{7}-\sqrt{2}}{5}$

59. $\frac{2\sqrt{5}+\sqrt{6}}{7}$ **61.** $\frac{\sqrt{15}-2\sqrt{3}}{2}$ **63.** $\frac{6\sqrt{7}+4\sqrt{6}}{13}$

65. $\sqrt{3}-\sqrt{2}$ **67.** $\frac{2\sqrt{x}-8}{x-16}$ **69.** $\frac{x+5\sqrt{x}}{x-25}$

71. $\frac{x-8\sqrt{x}+12}{x-36}$ **73.** $\frac{x-2\sqrt{xy}}{x-4y}$ **75.** $\frac{6\sqrt{xy}+9y}{4x-9y}$

Problem Set 10.5 (page 492)

1. $\{20\}$ **3.** $\varnothing$ **5.** $\left\{\frac{25}{4}\right\}$ **7.** $\left\{\frac{4}{9}\right\}$ **9.** $\{5\}$ **11.** $\left\{\frac{39}{4}\right\}$

13. $\left\{\frac{10}{3}\right\}$ **15.** $\{-1\}$ **17.** $\varnothing$ **19.** $\{1\}$ **21.** $\left\{\frac{3}{2}\right\}$ **23.** $\{3\}$

25. $\left\{\frac{61}{25}\right\}$ **27.** $\{-3, 3\}$ **29.** $\{-9, -4\}$ **31.** $\{0\}$ **33.** $\{3\}$

35. $\{4\}$ **37.** $\{-4, -3\}$ **39.** $\{12\}$ **41.** $\{25\}$ **43.** $\{29\}$

45. $\{-15\}$ **47.** $\left\{-\frac{1}{3}\right\}$ **49.** $\{-3\}$ **51.** $\{0\}$ **53.** $\{5\}$

55. $\{2, 6\}$ **57.** 56 feet; 106 feet; 148 feet
59. 3.2 feet; 5.1 feet; 7.3 feet

Problem Set 10.6 (page 496)

1. 9 **3.** 3 **5.** -2 **7.** -5 **9.** $\frac{1}{6}$ **11.** 3 **13.** 8 **15.** 81

17. -1 **19.** -32 **21.** $\frac{81}{16}$ **23.** 4 **25.** $\frac{1}{128}$ **27.** -125

29. 625 **31.** $\sqrt[3]{x^4}$ **33.** $3\sqrt{x}$ **35.** $\sqrt[3]{2y}$ **37.** $\sqrt{2x-3y}$

39. $\sqrt[3]{(2a-3b)^2}$ **41.** $\sqrt[5]{x^2y}$ **43.** $-3\sqrt[5]{xy^2}$ **45.** $5^{\frac{1}{2}}y^{\frac{1}{2}}$

47. $3y^{\frac{1}{2}}$ **49.** $x^{\frac{1}{3}}y^{\frac{2}{3}}$ **51.** $a^{\frac{1}{2}}b^{\frac{3}{4}}$ **53.** $(2x-y)^{\frac{3}{5}}$ **55.** $5xy^{\frac{1}{2}}$

57. $-(x+y)^{\frac{1}{3}}$ **59.** $12x^{\frac{13}{20}}$ **61.** $y^{\frac{5}{12}}$ **63.** $\frac{4}{x^{\frac{1}{10}}}$ **65.** $16xy^2$

67. $2x^2y$ **69.** $4x^{\frac{4}{15}}$ **71.** $\frac{4}{b^{\frac{5}{12}}}$ **73.** $\frac{36x^{\frac{4}{5}}}{49y^{\frac{4}{3}}}$ **75.** $\frac{y^{\frac{3}{2}}}{x}$ **77.** $4x^{\frac{1}{6}}$

79. $\frac{16}{a^{\frac{11}{10}}}$ **81.** $\sqrt[6]{243}$ **83.** $\sqrt[4]{216}$ **85.** $\sqrt[12]{3}$ **87.** $\sqrt{2}$

89. $\sqrt[4]{3}$ **93.** (a) 12 (c) 7 (e) 11 **95.** (a) 1024
   (c) 512 (e) 49

Chapter 10 Review Problem Set (page 503)

1. $\frac{1}{64}$ **2.** $\frac{9}{4}$ **3.** 3 **4.** 1 **5.** 27 **6.** $\frac{1}{125}$ **7.** $\frac{x^6}{y^8}$ **8.** $\frac{27a^3b^{12}}{8}$

9. $\frac{9a^4}{16b^4}$ **10.** $\frac{y^6}{125x^9}$ **11.** $\frac{x^{12}}{9}$ **12.** $\frac{1}{4y^3}$ **13.** $-10x^3$ **14.** $\frac{3b^5}{a^3}$

15. $\frac{b^3}{a^5}$ **16.** $\frac{x^4}{y}$ **17.** $-\frac{2}{x^2}$ **18.** $-\frac{2a}{b}$ **19.** $\frac{y+x^2}{x^2y}$

20. $\frac{b-2a}{a^2b}$ **21.** $\frac{2y^2+3x}{xy^2}$ **22.** $\frac{y^2+6x}{2xy^2}$ **23.** $3\sqrt{6}$

24. $4x\sqrt{3xy}$ **25.** $2\sqrt[3]{7}$ **26.** $3xy^2\sqrt[3]{4xy^2}$ **27.** $\frac{15\sqrt{6}}{4}$

28. $2y\sqrt{5xy}$ **29.** $2\sqrt{2}$ **30.** $\frac{\sqrt{15x}}{6x^2}$ **31.** $\frac{\sqrt[3]{6}}{3}$ **32.** $\frac{3\sqrt{5}}{5}$

33. $\frac{x\sqrt{21x}}{7}$ **34.** $2\sqrt{x}$ **35.** $\sqrt{5}$ **36.** $5\sqrt[3]{3}$ **37.** $\frac{29\sqrt{6}}{5}$

38. $-15\sqrt{3x}$ **39.** $24\sqrt{10}$ **40.** 60 **41.** $2x\sqrt{15y}$

42. $-18y^2\sqrt{x}$ **43.** $24\sqrt{3}-6\sqrt{14}$ **44.** $x-2\sqrt{x}-15$

45. 17 **46.** $12-8\sqrt{3}$ **47.** $6a-5\sqrt{ab}-4b$ **48.** 70

49. $\frac{2(\sqrt{7}+1)}{3}$ **50.** $\frac{2\sqrt{6}-\sqrt{15}}{3}$ **51.** $\frac{3\sqrt{5}-2\sqrt{3}}{11}$

52. $\dfrac{6\sqrt{3}+3\sqrt{5}}{7}$ **53.** $\left\{\dfrac{19}{7}\right\}$ **54.** $\{4\}$ **55.** $\{8\}$ **56.** $\varnothing$

57. $\{14\}$ **58.** $\{-10, 1\}$ **59.** $\{2\}$ **60.** $\{8\}$ **61.** 493 ft

62. 4.7 ft **63.** 32 **64.** 1 **65.** $\dfrac{4}{9}$ **66.** -64 **67.** $\dfrac{1}{9}$ **68.** $\dfrac{1}{4}$

69. 27 **70.** 8 **71.** $\sqrt[3]{xy^2}$ **72.** $\sqrt[4]{a^3}$ **73.** $4\sqrt{y}$

74. $\sqrt[3]{(x+5y)^2}$ **75.** $x^{\frac{3}{5}}y^{\frac{1}{5}}$ **76.** $4^{\frac{1}{3}}a^{\frac{2}{3}}$ **77.** $6y^{\frac{1}{2}}$

78. $(3a+b)^{\frac{5}{3}}$ **79.** $20x^{\frac{7}{10}}$ **80.** $7a^{\frac{5}{12}}$ **81.** $\dfrac{y^{\frac{4}{3}}}{x}$ **82.** $-\dfrac{6}{a^{\frac{1}{4}}}$

83. $\dfrac{1}{x^{\frac{2}{5}}}$ **84.** $-6y^{\frac{5}{12}}$ **85.** $\sqrt[4]{3^3}$ **86.** $3\sqrt[6]{3}$ **87.** $\sqrt[12]{5}$

88. $\sqrt[6]{2^5}$ **89.** $(5.4)(10^8)$ **90.** $(8.4)(10^4)$ **91.** $(3.2)(10^{-8})$

92. $(7.68)(10^{-4})$ **93.** 0.0000014 **94.** 0.000638

95. 41,200,000 **96.** 125,000 **97.** 0.000000006

98. 36,000,000,000 **99.** 6 **100.** 0.15 **101.** 0.000028

102. 0.002 **103.** 0.002 **104.** 8,000,000,000

Chapter 10 Test (page 505)

1. $\dfrac{1}{32}$ **2.** -32 **3.** $\dfrac{81}{16}$ **4.** $\dfrac{1}{4}$ **5.** $3\sqrt{7}$ **6.** $3\sqrt[3]{4}$

7. $2x^2y\sqrt{13y}$ **8.** $\dfrac{5\sqrt{6}}{6}$ **9.** $\dfrac{\sqrt{42x}}{12x^2}$ **10.** $72\sqrt{2}$ **11.** $-5\sqrt{6}$

12. $-38\sqrt{2}$ **13.** $\dfrac{3\sqrt{6}+3}{10}$ **14.** $\dfrac{9x^2y^2}{4}$ **15.** $-\dfrac{12}{a^{\frac{3}{10}}}$

16. $\dfrac{y^3+x}{xy^3}$ **17.** $-12x^{\frac{1}{4}}$ **18.** 33 **19.** 600 **20.** 0.003

21. $\left\{\dfrac{8}{3}\right\}$ **22.** $\{2\}$ **23.** $\{4\}$ **24.** $\{5\}$ **25.** $\{4, 6\}$

Chapters 1–10 Cumulative Practice Test (page 506)

1. (a) $\dfrac{12}{5}$ (b) $\dfrac{324}{121}$ (c) $-\dfrac{1}{2}$ (d) 81 **2.** $-\dfrac{4y^5}{x^4}$

3. $-\dfrac{12}{y^4}$ **4.** $6x^3+16x^2-18x+4$ **5.** $4x^2-7x-2$

6. $\{(-2, 7)\}$ **7.** (a) $6\sqrt{14}$ (b) $4\sqrt[3]{7}$ (c) $\dfrac{\sqrt{3}}{4}$

(d) $\dfrac{\sqrt{3}}{2}$ **8.** $7x+3y=-46$ **9.** 72 **10.** 429

11. $-\dfrac{7}{3}$ **12.** (a) $2\cdot2\cdot13$ (b) $2\cdot2\cdot2\cdot5$

(c) $7\cdot13$ (d) $2\cdot3\cdot13$ **13.** $\dfrac{5y-6x}{9y+5x}$ **14.** $\dfrac{2x-33}{20}$

15. $\dfrac{2x^2}{3y}$ **16.** $\dfrac{2}{3x-4}$ **17.** $\{3, 4\}$ **18.** $\{-1, 4\}$

19. $\left\{-\dfrac{7}{2}, \dfrac{5}{3}\right\}$ **20.** $\varnothing$ **21.** $\{2\}$

22. The altitude is 6 inches, and the side is 17 inches.
23. Pedro is 23 years old, and Brad is 29 years old.
24. $75 **25.** 7 inches by 11 inches

Chapter 11

Problem Set 11.1 (page 514)

1. False **3.** True **5.** True **7.** True **9.** $10+8i$
11. $-6+10i$ **13.** $-2-5i$ **15.** $-12+5i$ **17.** $-1-23i$

19. $-4-5i$ **21.** $1+3i$ **23.** $\dfrac{5}{3}-\dfrac{5}{12}i$ **25.** $-\dfrac{17}{9}+\dfrac{23}{30}i$

27. $9i$ **29.** $i\sqrt{14}$ **31.** $\dfrac{4}{5}i$ **33.** $3i\sqrt{2}$ **35.** $5i\sqrt{3}$

37. $6i\sqrt{7}$ **39.** $-8i\sqrt{5}$ **41.** $36i\sqrt{10}$ **43.** -8
45. $-\sqrt{15}$ **47.** $-3\sqrt{6}$ **49.** $-5\sqrt{3}$ **51.** $-3\sqrt{6}$

53. $4i\sqrt{3}$ **55.** $\dfrac{5}{2}$ **57.** $2\sqrt{2}$ **59.** $2i$ **61.** $-20+0i$

63. $42+0i$ **65.** $15+6i$ **67.** $-42+12i$ **69.** $7+22i$
71. $40-20i$ **73.** $-3-28i$ **75.** $-3-15i$
77. $-9+40i$ **79.** $-12+16i$ **81.** $85+0i$

83. $5+0i$ **85.** $\dfrac{3}{5}+\dfrac{3}{10}i$ **87.** $\dfrac{5}{17}-\dfrac{3}{17}i$ **89.** $2+\dfrac{2}{3}i$

91. $0-\dfrac{2}{7}i$ **93.** $\dfrac{22}{25}-\dfrac{4}{25}i$ **95.** $-\dfrac{18}{41}+\dfrac{39}{41}i$ **97.** $\dfrac{9}{2}-\dfrac{5}{2}i$

99. $\dfrac{4}{13}-\dfrac{1}{26}i$ **101.** (a) $-2-i\sqrt{3}$ (b) $\dfrac{3+i\sqrt{6}}{2}$

(c) $\dfrac{-1-3i\sqrt{2}}{2}$ (d) $-2+i\sqrt{3}$ (e) $\dfrac{10+3i\sqrt{5}}{4}$

(f) $2-2i\sqrt{3}$

Problem Set 11.2 (page 520)

1. $\{0, 9\}$ **3.** $\{-3, 0\}$ **5.** $\{-4, 0\}$ **7.** $\left\{0, \dfrac{9}{5}\right\}$ **9.** $\{-6, 5\}$

11. $\{7, 12\}$ **13.** $\left\{-8, -\dfrac{3}{2}\right\}$ **15.** $\left\{-\dfrac{7}{3}, \dfrac{2}{5}\right\}$ **17.** $\left\{\dfrac{3}{5}\right\}$

19. $\left\{-\dfrac{3}{2}, \dfrac{7}{3}\right\}$ **21.** $\{1, 4\}$ **23.** $\{8\}$ **25.** $\{12\}$ **27.** $\{\pm1\}$

29. $\{\pm6i\}$ **31.** $\{\pm\sqrt{14}\}$ **33.** $\{\pm2\sqrt{7}\}$ **35.** $\{\pm3\sqrt{2}\}$

37. $\left\{\pm\dfrac{\sqrt{14}}{2}\right\}$ **39.** $\left\{\pm\dfrac{2\sqrt{3}}{3}\right\}$ **41.** $\left\{\pm\dfrac{2i\sqrt{30}}{5}\right\}$

43. $\left\{\pm\dfrac{\sqrt{6}}{2}\right\}$ **45.** $\{-1, 5\}$ **47.** $\{-8, 2\}$ **49.** $\{-6\pm2i\}$

51. $\{1, 2\}$ **53.** $\{4\pm\sqrt{5}\}$ **55.** $\{-5\pm2\sqrt{3}\}$

57. $\left\{\dfrac{2\pm3i\sqrt{3}}{3}\right\}$ **59.** $\{-12, -2\}$ **61.** $\left\{\dfrac{2\pm\sqrt{10}}{5}\right\}$

63. $2\sqrt{13}$ centimeters **65.** $4\sqrt{5}$ inches **67.** 8 yards
69. $6\sqrt{2}$ inches **71.** $a=b=4\sqrt{2}$ meters
73. $b=3\sqrt{3}$ inches and $c=6$ inches
75. $a=7$ centimeters and $b=7\sqrt{3}$ centimeters

77. $a=\dfrac{10\sqrt{3}}{3}$ feet and $c=\dfrac{20\sqrt{3}}{3}$ feet

79. 17.9 feet   **81.** 38 meters   **83.** 53 meters
87. 10.8 centimeters   **89.** $h = s\sqrt{2}$

Problem Set 11.3 (page 526)

1. $\{-6, 10\}$ **3.** $\{4, 10\}$ **5.** $\{-5, 10\}$ **7.** $\{-8, 1\}$

9. $\left\{-\dfrac{5}{2}, 3\right\}$ **11.** $\left\{-3, \dfrac{2}{3}\right\}$ **13.** $\{-16, 10\}$

15. $\{-2 \pm \sqrt{6}\}$ **17.** $\{-3 \pm 2\sqrt{3}\}$ **19.** $\{5 \pm \sqrt{26}\}$

21. $\{4 \pm i\}$ **23.** $\{-6 \pm 3\sqrt{3}\}$ **25.** $\{-1 \pm i\sqrt{5}\}$

27. $\left\{\dfrac{-3 \pm \sqrt{17}}{2}\right\}$ **29.** $\left\{\dfrac{-5 \pm \sqrt{21}}{2}\right\}$ **31.** $\left\{\dfrac{7 \pm \sqrt{37}}{2}\right\}$

33. $\left\{\dfrac{-2 \pm \sqrt{10}}{2}\right\}$ **35.** $\left\{\dfrac{3 \pm i\sqrt{6}}{3}\right\}$ **37.** $\left\{\dfrac{-5 \pm \sqrt{37}}{6}\right\}$

39. $\{-12, 4\}$ **41.** $\left\{\dfrac{4 \pm \sqrt{10}}{2}\right\}$ **43.** $\left\{-\dfrac{9}{2}, \dfrac{1}{3}\right\}$

45. $\{-3, 8\}$ **47.** $\{3 \pm 2\sqrt{3}\}$ **49.** $\left\{\dfrac{3 \pm i\sqrt{3}}{3}\right\}$

51. $\{-20, 12\}$ **53.** $\left\{-1, -\dfrac{2}{3}\right\}$ **55.** $\left\{\dfrac{1}{2}, \dfrac{3}{2}\right\}$

57. $\{-6 \pm 2\sqrt{10}\}$ **59.** $\left\{\dfrac{-1 \pm \sqrt{3}}{2}\right\}$

61. $\left\{\dfrac{-b \pm \sqrt{b^2 - 4ac}}{2a}\right\}$ **65.** $x = \dfrac{a\sqrt{b^2 - y^2}}{b}$

67. $r = \dfrac{\sqrt{A\pi}}{\pi}$ **69.** $\{2a, 3a\}$ **71.** $\left\{\dfrac{a}{2}, -\dfrac{2a}{3}\right\}$ **73.** $\left\{\dfrac{2b}{3}\right\}$

Problem Set 11.4 (page 535)

1. $\dfrac{1 \pm \sqrt{5}}{2}$ **3.** $-2 \pm \sqrt{3}$ **5.** $\dfrac{2 \pm \sqrt{2}}{3}$

7. $\dfrac{-2 \pm \sqrt{3}}{2}$ **9.** $\dfrac{-3 \pm 2\sqrt{3}}{2}$ **11.** $\{-1 \pm \sqrt{2}\}$

13. $\left\{\dfrac{-5 \pm \sqrt{37}}{2}\right\}$ **15.** $\{4 \pm 2\sqrt{5}\}$ **17.** $\left\{\dfrac{-5 \pm i\sqrt{7}}{2}\right\}$

19. $\{8, 10\}$ **21.** $\left\{\dfrac{9 \pm \sqrt{61}}{2}\right\}$ **23.** $\left\{\dfrac{-1 \pm \sqrt{33}}{4}\right\}$

25. $\left\{\dfrac{-1 \pm i\sqrt{3}}{4}\right\}$ **27.** $\left\{\dfrac{4 \pm \sqrt{10}}{3}\right\}$ **29.** $\left\{-1, \dfrac{5}{2}\right\}$

31. $\left\{-5, -\dfrac{4}{3}\right\}$ **33.** $\left\{\dfrac{5}{6}\right\}$ **35.** $\left\{\dfrac{1 \pm \sqrt{13}}{4}\right\}$ **37.** $\left\{0, \dfrac{13}{5}\right\}$

39. $\left\{\pm\dfrac{\sqrt{15}}{3}\right\}$ **41.** $\left\{\dfrac{-1 \pm \sqrt{73}}{12}\right\}$ **43.** $\{-18, -14\}$

45. $\left\{\dfrac{11}{4}, \dfrac{10}{3}\right\}$ **47.** $\left\{\dfrac{2 \pm i\sqrt{2}}{2}\right\}$ **49.** $\left\{\dfrac{1 \pm \sqrt{7}}{6}\right\}$

51. Two real solutions; $\{-7, 3\}$ **53.** One real solution; $\left\{\dfrac{1}{3}\right\}$

55. Two complex solutions; $\left\{\dfrac{7 \pm i\sqrt{3}}{2}\right\}$

57. Two real solutions; $\left\{-\dfrac{4}{3}, \dfrac{1}{5}\right\}$

59. Two real solutions; $\left\{\dfrac{-2 \pm \sqrt{10}}{3}\right\}$ **65.** $\{-1.381, 17.381\}$

67. $\{-13.426, 3.426\}$ **69.** $\{-8.653, -0.347\}$
71. $\{0.119, 1.681\}$ **73.** $\{-0.708, 4.708\}$ **75.** $k = 4$ or $k = -4$

Problem Set 11.5 (page 541)

1. $\{2 \pm \sqrt{10}\}$ **3.** $\left\{-9, \dfrac{4}{3}\right\}$ **5.** $\{9 \pm 3\sqrt{10}\}$

7. $\left\{\dfrac{3 \pm i\sqrt{23}}{4}\right\}$ **9.** $\{-15, -9\}$ **11.** $\{-8, 1\}$

13. $\left\{\dfrac{2 \pm i\sqrt{10}}{2}\right\}$ **15.** $\{9 \pm \sqrt{66}\}$ **17.** $\left\{-\dfrac{5}{4}, \dfrac{2}{5}\right\}$

19. $\left\{\dfrac{-1 \pm \sqrt{2}}{2}\right\}$ **21.** $\left\{\dfrac{3}{4}, 4\right\}$ **23.** $\left\{\dfrac{11 \pm \sqrt{109}}{2}\right\}$

25. $\left\{\dfrac{3}{7}, 4\right\}$ **27.** $\left\{\dfrac{7 \pm \sqrt{129}}{10}\right\}$ **29.** $\left\{-\dfrac{10}{7}, 3\right\}$

31. $\{1 \pm \sqrt{34}\}$ **33.** $\{\pm\sqrt{6}, \pm 2\sqrt{3}\}$

35. $\left\{\pm 3, \pm\dfrac{2\sqrt{6}}{3}\right\}$ **37.** $\left\{\pm\dfrac{i\sqrt{15}}{3}, \pm 2i\right\}$

39. $\left\{\pm\dfrac{\sqrt{14}}{2}, \pm\dfrac{2\sqrt{3}}{3}\right\}$ **41.** 8 and 9   **43.** 9 and 12

45. $5 + \sqrt{3}$ and $5 - \sqrt{3}$ **47.** 3 and 6

49. 9 inches and 12 inches   **51.** 1 meter

53. 8 inches by 14 inches   **55.** 20 miles per hour for Lorraine and 25 miles per hour for Charlotte, or 45 miles per hour for Lorraine and 50 miles per hour for Charlotte

57. 55 miles per hour   **59.** 6 hours for Tom and 8 hours for Terry

61. 2 hours   **63.** 8 students   **65.** 50 numbers   **67.** 9%   **69.** 6%

75. $\{9, 36\}$ **77.** $\{1\}$ **79.** $\left\{-\dfrac{8}{27}, \dfrac{27}{8}\right\}$ **81.** $\left\{-4, \dfrac{3}{5}\right\}$

83. $\{4\}$ **85.** $\{\pm 4\sqrt{2}\}$ **87.** $\left\{\dfrac{3}{2}, \dfrac{5}{2}\right\}$ **89.** $\{-1, 3\}$

Problem Set 11.6 (page 549)

1. $(-\infty, -2) \cup (1, \infty)$

3. $(-4, -1)$

5. $\left(-\infty, -\dfrac{7}{3}\right] \cup \left[\dfrac{1}{2}, \infty\right)$

7. $\left[-2, \dfrac{3}{4}\right]$

9. $(-1, 1) \cup (3, \infty)$

11. $(-\infty, -2] \cup [0, 4]$

13. $(-7, 5)$ **15.** $(-\infty, 4) \cup (7, \infty)$ **17.** $\left[-5, \dfrac{2}{3}\right]$

19. $\left(-\infty, -\dfrac{5}{2}\right] \cup \left[-\dfrac{1}{4}, \infty\right)$ **21.** $\left(-\infty, -\dfrac{4}{5}\right) \cup (8, \infty)$

23. $(-\infty, \infty)$ **25.** $\left\{-\dfrac{5}{2}\right\}$ **27.** $(-1, 3) \cup (3, \infty)$

29. $(-\infty, -2) \cup (2, \infty)$ **31.** $(-6, 6)$ **33.** $(-\infty, \infty)$
35. $(-\infty, 0] \cup [2, \infty)$ **37.** $(-4, 0) \cup (0, \infty)$

39. $(-\infty, -1) \cup (2, \infty)$ **41.** $(-2, 3)$ **43.** $(-\infty, 0) \cup \left[\dfrac{1}{2}, \infty\right)$

45. $(-\infty, 1) \cup [2, \infty)$ **47.** $(-6, -3)$ **49.** $(-\infty, 5) \cup [9, \infty)$

51. $\left(-\infty, \dfrac{4}{3}\right) \cup (3, \infty)$ **53.** $(-4, 6]$ **55.** $(-\infty, 2)$

Chapter 11 Review Problem Set (page 555)

1. $2 - 2i$ **2.** $-3 - i$ **3.** $8 - 8i$ **4.** $-2 + 4i$ **5.** $2i\sqrt{2}$
6. $5i$ **7.** $12i$ **8.** $6i\sqrt{2}$ **9.** $-2\sqrt{3}$ **10.** $6i$ **11.** $\sqrt{7}$
12. $i\sqrt{3}$ **13.** $30 + 15i$ **14.** $86 - 2i$ **15.** $-32 + 4i$

16. $25 + 0i$ **17.** $\dfrac{9}{20} + \dfrac{13}{20}i$ **18.** $-\dfrac{3}{29} + \dfrac{7}{29}i$ **19.** $2 - \dfrac{3}{2}i$

20. $-5 - 6i$ **21.** $\{-8, 0\}$ **22.** $\{0, 6\}$ **23.** $\{-4, 7\}$

24. $\left\{-\dfrac{3}{2}, 1\right\}$ **25.** $\{\pm 3\sqrt{5}\}$ **26.** $3 \pm 3i\sqrt{2}$

27. $\left\{\dfrac{-3 \pm 2\sqrt{6}}{2}\right\}$ **28.** $\{\pm 3\sqrt{3}\}$ **29.** $\{-9 \pm \sqrt{91}\}$

30. $\{-3 \pm i\sqrt{11}\}$ **31.** $\{5 \pm 2\sqrt{6}\}$ **32.** $\left\{\dfrac{-5 \pm \sqrt{33}}{2}\right\}$

33. $\{-3 \pm \sqrt{5}\}$ **34.** $\{-2 \pm i\sqrt{2}\}$

35. $\left\{\dfrac{1 \pm i\sqrt{11}}{3}\right\}$ **36.** $\left\{\dfrac{1 \pm \sqrt{61}}{10}\right\}$

37. One real solution with a multiplicity of 2
38. Two nonreal complex solutions
39. Two unequal real solutions   **40.** Two unequal real solutions

41. $\{0, 17\}$ **42.** $\{-4, 8\}$ **43.** $\left\{\dfrac{1 \pm 8i}{2}\right\}$ **44.** $\{-3, 7\}$

45. $\{-1 \pm \sqrt{10}\}$ **46.** $\{3 \pm 5i\}$ **47.** $\{25\}$ **48.** $\left\{-4, \dfrac{2}{3}\right\}$

49. $\{-10, 20\}$ **50.** $\left\{\dfrac{-1 \pm \sqrt{61}}{6}\right\}$ **51.** $\left\{\dfrac{1 \pm i\sqrt{11}}{2}\right\}$

52. $\left\{\dfrac{5 \pm i\sqrt{23}}{4}\right\}$ **53.** $\left\{\dfrac{-2 \pm \sqrt{14}}{2}\right\}$ **54.** $\{-9, 4\}$

55. $\{-2 \pm i\sqrt{5}\}$ **56.** $\{-6, 12\}$ **57.** $\{1 \pm \sqrt{10}\}$

58. $\left\{\pm\dfrac{\sqrt{14}}{2}, \pm 2\sqrt{2}\right\}$ **59.** $\left\{\dfrac{-3 \pm \sqrt{97}}{2}\right\}$

60. 34.6 ft and 40.0 ft   **61.** 47.2 m   **62.** $4\sqrt{2}$ in.
63. $3 + \sqrt{7}$ and $3 - \sqrt{7}$ **64.** 8 hr
65. Andre: 45 mph and Sandy: 52 mph
66. 8 units   **67.** 8 and 10   **68.** 7 in. by 12 in.
69. 4 hr for Reena and 6 hr for Billy   **70.** 10 m

71. $(-\infty, -5) \cup (2, \infty)$ **72.** $\left[-\dfrac{7}{2}, 3\right]$ **73.** $\left[-\dfrac{1}{2}, \dfrac{1}{2}\right]$

74. $(-\infty, 2) \cup (5, \infty)$ **75.** $(-\infty, -6) \cup [4, \infty)$

76. $\left(-\dfrac{5}{2}, -1\right)$ **77.** $(-9, 4)$ **78.** $\left[-\dfrac{1}{3}, 1\right)$

Chapter 11 Test (page 557)

1. $39 - 2i$ **2.** $-\dfrac{6}{25} - \dfrac{17}{25}i$ **3.** $\{0, 7\}$ **4.** $\{-1, 7\}$

5. $\{-6, 3\}$ **6.** $\{1 - \sqrt{2}, 1 + \sqrt{2}\}$ **7.** $\left\{\dfrac{1 - 2i}{5}, \dfrac{1 + 2i}{5}\right\}$

8. $\{-16, -14\}$ **9.** $\left\{\dfrac{1 - 6i}{3}, \dfrac{1 + 6i}{3}\right\}$ **10.** $\left\{-\dfrac{7}{4}, \dfrac{6}{5}\right\}$

11. $\left\{-3, \dfrac{19}{6}\right\}$ **12.** $\left\{-\dfrac{10}{3}, 4\right\}$ **13.** $\{-2, 2, -4i, 4i\}$

14. $\left\{-\dfrac{3}{4}, 1\right\}$ **15.** $\left\{\dfrac{1 - \sqrt{10}}{3}, \dfrac{1 + \sqrt{10}}{3}\right\}$

16. Two equal real solutions   **17.** Two nonreal complex solutions

18. $[-6, 9]$ **19.** $(-\infty, -2) \cup \left(\dfrac{1}{3}, \infty\right)$ **20.** $[-10, -6)$

21. 20.8 ft   **22.** 29 m   **23.** 3 hr   **24.** $6\dfrac{1}{2}$ in.   **25.** $3 + \sqrt{5}$

Chapters 1–11 Cumulative Review Problem Set (page 558)

1. $\dfrac{64}{15}$ **2.** $\dfrac{11}{3}$ **3.** $\dfrac{1}{6}$ **4.** $-\dfrac{44}{5}$ **5.** -7 **6.** $-24a^4b^5$

7. $2x^3 + 5x^2 - 7x - 12$ **8.** $\dfrac{3x^2y^2}{8}$ **9.** $\dfrac{a(a + 1)}{2a - 1}$

10. $\dfrac{-x + 14}{18}$ **11.** $\dfrac{5x + 19}{x(x + 3)}$ **12.** $\dfrac{2}{n + 8}$

13. $\dfrac{x - 14}{(5x - 2)(x + 1)(x - 4)}$ **14.** $y^2 - 5y + 6$

15. $x^2 - 3x - 2$ **16.** $20 + 7\sqrt{10}$ **17.** $2x - 2\sqrt{xy} - 12y$

18. $-\dfrac{3}{8}$ **19.** $-\dfrac{2}{3}$ **20.** 0.2 **21.** $\dfrac{1}{2}$ **22.** $\dfrac{13}{9}$ **23.** -27

24. $\dfrac{16}{9}$ **25.** $\dfrac{8}{27}$ **26.** $3x(x + 3)(x^2 - 3x + 9)$

27. $(6x - 5)(x + 4)$ **28.** $(4 + 7x)(3 - 2x)$
29. $(3x + 2)(3x - 2)(x^2 + 8)$ **30.** $(2x - y)(a - b)$

31. $(3x - 2y)(9x^2 + 6xy + 4y^2)$ **32.** $\left\{-\dfrac{12}{7}\right\}$

33. $\{150\}$ **34.** $\{25\}$ **35.** $\{0\}$ **36.** $\{-2, 2\}$ **37.** $\{-7\}$

38. $\left\{-6, \dfrac{4}{3}\right\}$ **39.** $\left\{\dfrac{5}{4}\right\}$ **40.** $\{3\}$ **41.** $\left\{\dfrac{4}{5}, 1\right\}$

42. $\left\{-\dfrac{10}{3}, 4\right\}$ **43.** $\left\{\dfrac{1}{4}, \dfrac{2}{3}\right\}$ **44.** $\left\{-\dfrac{3}{2}, 3\right\}$ **45.** $\left\{\dfrac{1}{5}\right\}$

46. $\left\{\dfrac{5}{7}\right\}$ **47.** $\left\{-2, 2, \pm\dfrac{i\sqrt{6}}{6}\right\}$ **48.** $\{0, \pm i\}$ **49.** $\{-6, 19\}$

50. $\left\{-\dfrac{3}{4}, \dfrac{2}{3}\right\}$ **51.** $\{1 \pm 5i\}$ **52.** $\{1, 3\}$ **53.** $\left\{-4, \dfrac{1}{3}\right\}$

54. $\{-2 \pm 4i\}$ **55.** $\left\{\dfrac{1 \pm \sqrt{33}}{4}\right\}$ **56.** $(-\infty, -2]$

57. $\left(-\infty, \dfrac{19}{5}\right)$ **58.** $\left(\dfrac{1}{4}, \infty\right)$ **59.** $(-2, 3)$

60. $\left(-\infty, -\dfrac{13}{3}\right) \cup (3, \infty)$ **61.** $(-\infty, 29]$ **62.** $[-2, 4]$

63. $(-\infty, -5) \cup \left(\dfrac{1}{3}, \infty\right)$ **64.** $(-\infty, -2] \cup (7, \infty)$

65. $(-3, 4)$ **66.** $\dfrac{13}{3}$ **67.** $\dfrac{3}{5}$ **68.** $3x + 11y = -10$

69. $5x + 2y = 9$ **70.** $\{(-1, -2)\}$ **71.** $\{(8, -9)\}$

72. $\{(4, 6)\}$ **73.** $-\dfrac{30y^2}{x^2}$ **74. (a)** $12\sqrt{6}$ **(b)** $6\sqrt[3]{2}$

(c) $\dfrac{3\sqrt{15}}{20}$ **(d)** $2xy^2\sqrt{6xy}$ **75. (a)** $(7.652)(10^3)$

(b) $(2.6)(10^{-5})$ **(c)** $(1.414)(10^0)$ **(d)** $(1)(10^3)$

76. 6 liters   **77.** \$900 and \$1350

78. 12 inches by 17 inches   **79.** 5 hours

80. 7 golf balls   **81.** 12 minutes   **82.** 7%

83. 15 chairs per row   **84.** 140 shares

85. 60° and 80°   **86.** 70°   **87.** More than 8.5%

88. \$0.25 per lemon and \$0.33 per orange   **89.** \$90

Chapter 12

Problem Set 12.1 (page 575)

1. 15   **3.** $\sqrt{13}$ **5.** $3\sqrt{2}$ **7.** $3\sqrt{5}$ **9.** 6   **11.** $3\sqrt{10}$

13. The lengths of the sides are $10, 5\sqrt{5},$ and 5. Because $10^2 + 5^2 = (5\sqrt{5})^2$, it is a right triangle.

15. The distances between $(3, 6)$ and $(7, 12)$, between $(7, 12)$ and $(11, 18)$, and between $(11, 18)$ and $(15, 24)$ are all $2\sqrt{13}$ units.

17. $\dfrac{4}{3}$ **19.** $-\dfrac{7}{3}$ **21.** -2 **23.** $\dfrac{3}{5}$ **25.** 0   **27.** $\dfrac{1}{2}$ **29.** 7

31. -2 **33.** $(-3, -1); (3, 1); (3, -1)$

35. $(7, 2); (-7, -2); (-7, 2)$ **37.** $(5, 0); (-5, 0); (-5, 0)$

39. x axis   **41.** y axis   **43.** x axis, y axis, and origin

45. x axis   **47.** None   **49.** Origin   **51.** y axis   **53.** All three

55. x axis   **57.** y axis

59.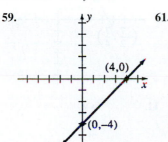

61.

63.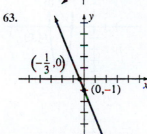

65.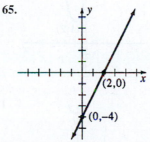

67.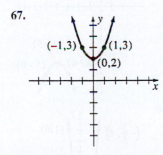

69.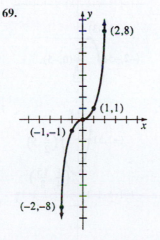

71.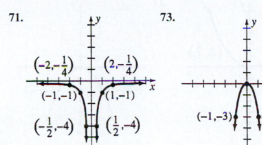

73.

75.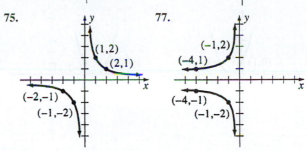

77.

79.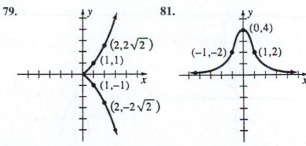

81.

83.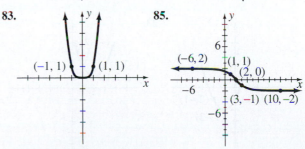

85.

Problem Set 12.2 (page 585)

1.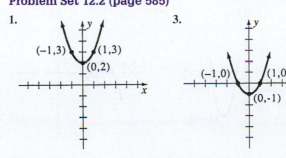

3.

5.
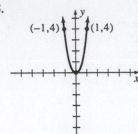
$(-1,4)$ $(1,4)$

7.
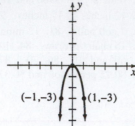
$(-1,-3)$ $(1,-3)$

29.
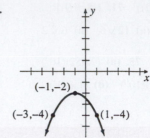
$(-1,-2)$
$(-3,-4)$ $(1,-4)$

9.
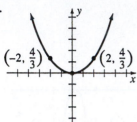
$\left(-2,\dfrac{4}{3}\right)$ $\left(2,\dfrac{4}{3}\right)$

11.
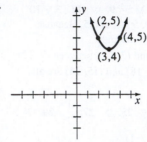
$(-2,-2)$ $(2,-2)$

Problem Set 12.3 (page 592)

1.
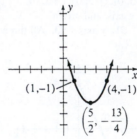
$(2,5)$
$(4,5)$
$(3,4)$

3.
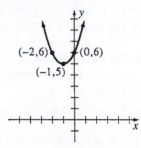
$(-2,6)$ $(0,6)$
$(-1,5)$

13.
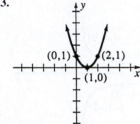
$(0,1)$ $(2,1)$
$(1,0)$

15.
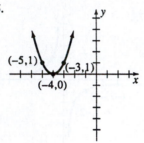
$(-5,1)$ $(-3,1)$
$(-4,0)$

5.
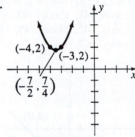
$(1,-1)$ $(4,-1)$
$\left(\dfrac{5}{2},-\dfrac{13}{4}\right)$

7.
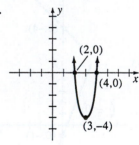
$(-4,2)$ $(-3,2)$
$\left(-\dfrac{7}{2},\dfrac{7}{4}\right)$

17.
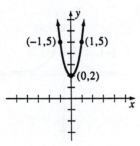
$(-1,5)$ $(1,5)$
$(0,2)$

19.
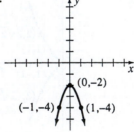
$(0,-2)$
$(-1,-4)$ $(1,-4)$

9.
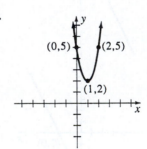
$(0,5)$ $(2,5)$
$(1,2)$

11.
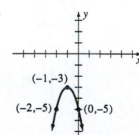
$(2,0)$
$(4,0)$
$(3,-4)$

21.
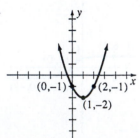
$(0,-1)$ $(2,-1)$
$(1,-2)$

23.
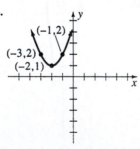
$(-1,2)$
$(-3,2)$
$(-2,1)$

13.
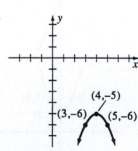
$(-1,-3)$
$(-2,-5)$ $(0,-5)$

15.
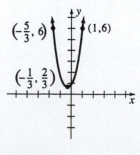
$(4,-5)$
$(3,-6)$ $(5,-6)$

25.
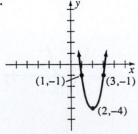
$(1,-1)$ $(3,-1)$
$(2,-4)$

27.
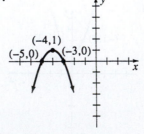
$(-4,1)$
$(-5,0)$ $(-3,0)$

17.
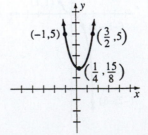
$(-1,5)$ $\left(\dfrac{3}{2},5\right)$
$\left(\dfrac{1}{4},\dfrac{15}{8}\right)$

19.
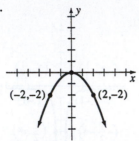
$\left(-\dfrac{5}{3},6\right)$ $(1,6)$
$\left(-\dfrac{1}{3},\dfrac{2}{3}\right)$

21.

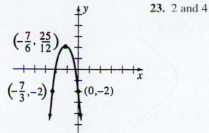

23. 2 and 4

3.

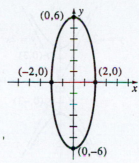

5.

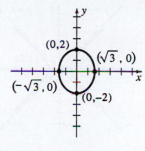

25. $\dfrac{5 - \sqrt{13}}{2}$ and $\dfrac{5 + \sqrt{13}}{2}$ **27.** No x intercepts

29. -2 and $-\dfrac{1}{3}$ **31.** No x intercepts   **33.** $\dfrac{3}{2}$ **35.** $(1, 3), r = 4$

37. $(-3, -5), r = 4$ **39.** $(0, 0), r = \sqrt{10}$ **41.** $(8, -3), r = \sqrt{2}$

43. $(-3, 4), r = 5$ **45.** $\left(-\dfrac{1}{2}, 4\right), r = 2\sqrt{2}$

7.

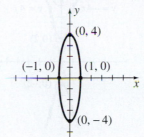

9.

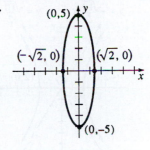

47.

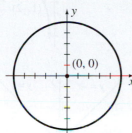

49.

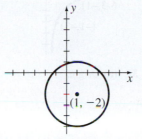

11.

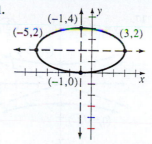

51.

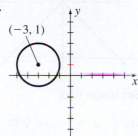

53.

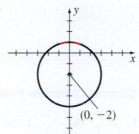

13.

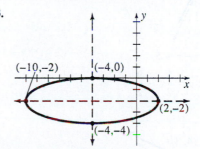

55.

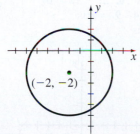

15.

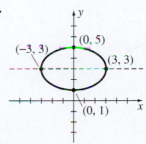

57. $x^2 + y^2 - 6x - 10y + 9 = 0$ **59.** $x^2 + y^2 + 8x - 2y - 47 = 0$
61. $x^2 + y^2 + 4x + 12y + 22 = 0$ **63.** $x^2 + y^2 - 20 = 0$
65. $x^2 + y^2 - 10x + 16y - 7 = 0$
67. $x^2 + y^2 - 8y = 0$ **69.** $x^2 + y^2 + 8x - 6y = 0$
75. (a) $(1, 4), r = 3$ (c) $(-6, -4), r = 8$ (e) $(0, 6), r = 9$

Problem Set 12.4 (page 597)

1.

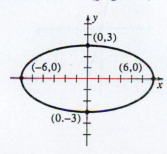

Problem Set 12.5 (page 604)

1. $(4, 0), (-4, 0),\quad y = \dfrac{1}{3}x$ and $y = -\dfrac{1}{3}x$

3. $(0, 6), (0, -6),\ y = 3x$ and $y = -3x$

5. $\left(\dfrac{2}{5}, 0\right), \left(-\dfrac{2}{5}, 0\right),\ y = \dfrac{5}{3}x$ and $y = -\dfrac{5}{3}x$

7. 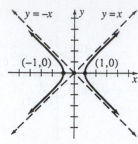 **9.**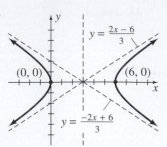

11. $y = -\dfrac{\sqrt{10}}{2}x$ $\quad y = \dfrac{\sqrt{10}}{2}x$ 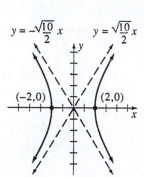 **13.**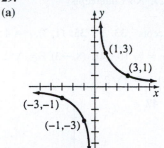

15. 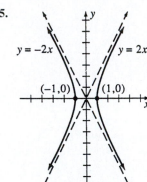 **17.**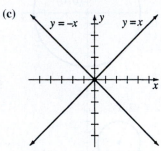

19. $y = x - 1$ and $y = -x - 5$

21. $y = -\dfrac{3}{2}x - \dfrac{3}{2}$ and $y = \dfrac{3}{2}x - \dfrac{9}{2}$

23.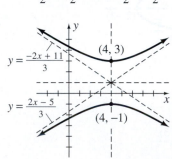

25.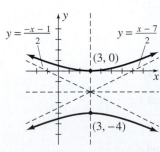

27.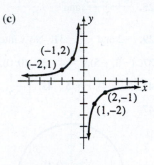

29.
(a)   (c)

31. (a) Origin   (c)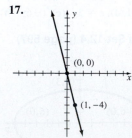

Chapter 12 Review Problem Set (page 610)

1. (a) $\dfrac{6}{5}$ (b) $-\dfrac{2}{3}$ **2.** (a) -4 (b) $\dfrac{2}{7}$ **3.** 5, 10, and $\sqrt{97}$
4. $7x + 4y = 1$ **5.** $3x + 7y = 28$ **6.** $2x - 3y = 16$
7. $x - 2y = -8$ **8.** $2x - 3y = 14$ **9.** y axis
10. x axis **11.** Origin **12.** y axis
13. x axis, y axis, and origin

14. 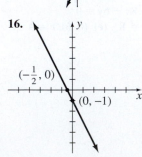 **15.**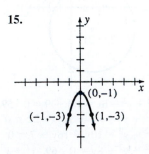

16. **17.**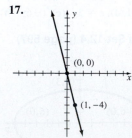

18.

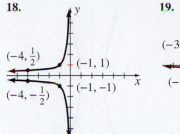

$(-4, \frac{1}{2})$ $(-1, 1)$
$(-4, -\frac{1}{2})$ $(-1, -1)$

19.

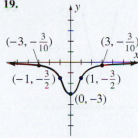

$(-3, -\frac{3}{10})$ $(3, -\frac{3}{10})$
$(-1, -\frac{3}{2})$ $(1, -\frac{3}{2})$
$(0, -3)$

54.

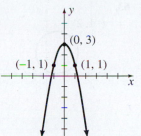

$(0, 3)$
$(-1, 1)$ $(1, 1)$

55.

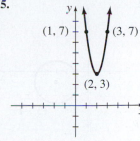

$(1, 7)$ $(3, 7)$
$(2, 3)$

20. 316.8 feet   **21.** 8 inches   **22.** $(0, 6)$ **23.** $(0, -8)$
24. $(-3, -1)$ **25.** $(-1, 0)$ **26.** $(7, 5)$ **27.** $(6, -8)$
28. $(-4, -9)$ **29.** $(-2, -3)$ **30.** $x^2 + y^2 - 36 = 0$
31. $x^2 + y^2 - 4x + 12y + 15 = 0$
32. $x^2 + y^2 + 8x + 16y + 68 = 0$ **33.** $x^2 + y^2 - 10y = 0$
34. $(-7, 4)$ and $r = 7$ **35.** $(-8, 0)$ and $r = 5$
36. $(6, -8)$ and $r = 10$ **37.** $(0, 0)$ and $r = 2\sqrt{6}$
38. Major $= 16$ and minor $= 4$ **39.** Major $= 8$ and minor $= 6$
40. Major $= 10$ and minor $= 4$
41. Major $= 2\sqrt{14}$ and minor $= 4$
42. Major $= 6$ and minor $= 2$ **43.** Major $= 8$ and minor $= 4$
44. Major $= 6$ and minor $= 4$

45. Major $= 8$ and minor $= 6$ **46.** $y = \pm\frac{1}{3}x$ **47.** $y = \pm 2x$

48. $y = \pm\frac{5}{3}x$ **49.** $y = \pm\frac{1}{2}x$

50. $y = -\frac{5}{2}x - 2$ and $y = \frac{5}{2}x + 8$

51. $y = -\frac{1}{6}x + \frac{25}{6}$ and $y = \frac{1}{6}x + \frac{23}{6}$

52.

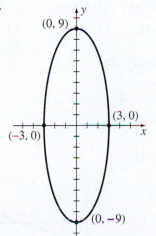

$(0, 9)$
$(-3, 0)$ $(3, 0)$
$(0, -9)$

53.

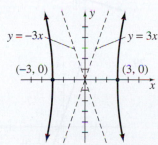

$y = -3x$ $y = 3x$
$(-3, 0)$ $(3, 0)$

56.

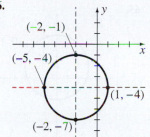

$(-2, -1)$
$(-5, -4)$
$(1, -4)$
$(-2, -7)$

57.

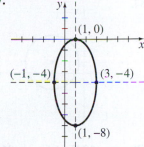

$(1, 0)$
$(-1, -4)$ $(3, -4)$
$(1, -8)$

58.

$(-3, 3)$
$y = 2x + 3$ $y = -2x - 9$
$(-3, -9)$

59.

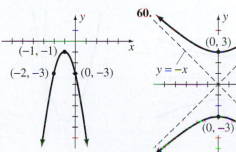

$(-1, -1)$
$(-2, -3)$ $(0, -3)$

60.

$(0, 3)$
$y = -x$ $y = x$
$(0, -3)$

61.

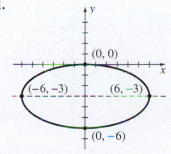

$(0, 0)$
$(-6, -3)$ $(6, -3)$
$(0, -6)$

62.

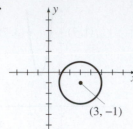

63.

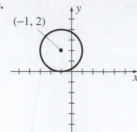

23.

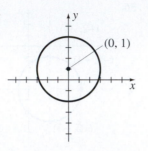

24.

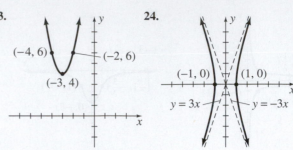

64.

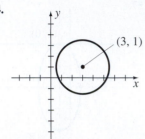

65.

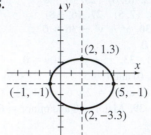

25.

Chapter 12 Test (page 611)

1. $-\dfrac{6}{5}$ **2.** $\dfrac{3}{7}$ **3.** $\sqrt{58}$ **4.** -5 and $\dfrac{1}{2}$

5. $-\dfrac{5}{3}$ and $\dfrac{3}{2}$ **6.** y axis **7.** x axis **8.** 480 feet

9. 43 centimeters **10.** $(3, 0)$
11. $(-4, -2)$ **12.** $x^2 - 4x + y^2 - 16y + 59 = 0$
13. $(6, -4)$ and $r = 7$ **14.** 8 units
15. 4 units **16.** $y = \pm 4x$

17.

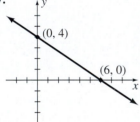

18.

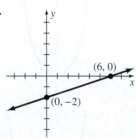

19.

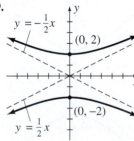

20.

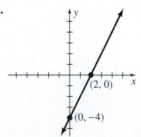

21.

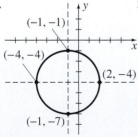

22.

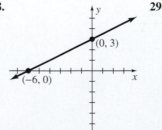

Chapters 1–12 Cumulative Review Problem Set (page 612)

1. **(a)** $-\dfrac{2}{3}$ **(b)** -32 **(c)** $\dfrac{1}{4}$ **(d)** $\dfrac{15}{2}$

2. **(a)** $4\sqrt{6}$ **(b)** $2\sqrt[3]{5}$ **(c)** $\dfrac{2\sqrt{6}}{9}$ **(d)** $\dfrac{2(\sqrt{7} + \sqrt{2})}{5}$

3. $-24x^4y^5$ **4.** $6x^3 - 5x^2 - 8x - 2$ **5.** $x^3 + 2x^2 - x - 4$

6. $\dfrac{-6x + 11}{12}$ **7.** $14 + \sqrt{16}$

8. **(a)** Origin **(b)** x axis **(c)** y axis **(d)** y axis

9. -21 **10.** -192 **11.** 36 **12.** $\dfrac{1}{4}$ **13.** $4x - 3y = -13$

14. $\{2\}$ **15.** $\left\{-\dfrac{53}{10}\right\}$ **16.** $\{3000\}$ **17.** $\left\{-\dfrac{5}{3}, 5\right\}$

18. $\left\{\dfrac{1 \pm \sqrt{13}}{6}\right\}$ **19.** $[-1, 4]$ **20.** $(-\infty, 2) \cup (3, \infty)$

21. $(-1, 6)$ **22.** $34°$ and $56°$ **23.** 12 dimes and 25 quarters
24. 3 hours **25.** \$1.79

26.

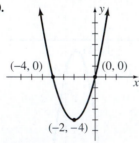

27.

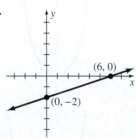

28.

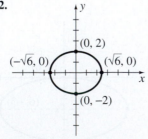

29.

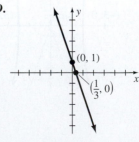

30.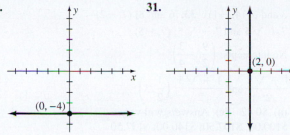

31. (2, 0)

(0, −4)

41.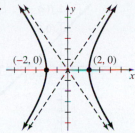

(0, 5)
(−2, 0) (2, 0)
(0, −5)

42.

(−2, 0) (2, 0)

32. $7\sqrt{2}$ **33.** $\sqrt{41}$ **34.** $3x - 5y = 6$ **35.** $x + y = 3$
36. $2x + y = -1$ **37.** $3x + y = 1$
38. $C(-4, -2), r = 5$ units **39.** $(1, 5)$ **40.** $y = \pm\frac{4}{3}x$

43.

(1, 4)

Chapter 13

Problem Set 13.1 (page 622)

1. $D = \{1, 2, 3, 4\}, R = \{5, 8, 11, 14\}$. It is a function.
3. $D = \{0, 1\}, R = \{-2\sqrt{6}, -5, 5, 2\sqrt{6}\}$. It is not a function.
5. $D = \{1, 2, 3, 4, 5\}, R = \{2, 5, 10, 17, 26\}$. It is a function.
7. $D = \{$all reals$\}$ or $D: (-\infty, \infty)$, $R = \{$all reals$\}$ or
   $R: (-\infty, \infty)$. It is a function.
9. $D = \{$all reals$\}$ or $D: (-\infty, \infty)$, $R = \{y \,|\, y \geq 0\}$ or $R: [0, \infty)$.
   It is a function. **11.** $\{$all reals$\}$ or $(-\infty, \infty)$
13. $\{x \,|\, x \neq 1\}$ or $(-\infty, 1) \cup (1, \infty)$
15. $\left\{x \,|\, x \neq \frac{3}{4}\right\}$ or $\left(-\infty, \frac{3}{4}\right) \cup \left(\frac{3}{4}, \infty\right)$
17. $\{x \,|\, x \neq -1, x \neq 4\}$ or $(-\infty, -1) \cup (-1, 4) \cup (4, \infty)$
19. $\{x \,|\, x \neq -8, x \neq 5\}$ or $(-\infty, -8) \cup (-8, 5) \cup (5, \infty)$
21. $\{x \,|\, x \neq -6, x \neq 0\}$ or $(-\infty, -6) \cup (-6, 0) \cup (0, \infty)$
23. $\{$All reals$\}$ or $(-\infty, \infty)$
25. $\{t \,|\, t \neq -2, t \neq 2\}$ or $(-\infty, -2) \cup (-2, 2) \cup (2, \infty)$
27. $\{x \,|\, x \geq -4\}$ or $[-4, \infty)$ **29.** $\left\{s \,|\, s \geq \frac{5}{4}\right\}$ or $\left[\frac{5}{4}, \infty\right)$
31. $\{x \,|\, x \leq -4 \text{ or } x \geq 4\}$ or $(-\infty, -4] \cup [4, \infty)$
33. $\{x \,|\, x \leq -3 \text{ or } x \geq 6\}$ or $(-\infty, -3] \cup [6, \infty)$
35. $\{x \,|\, -1 \leq x \leq 1\}$ or $[-1, 1]$
37. $f(0) = -2; f(2) = 8; f(-1) = -7; f(-4) = -22$
39. $f(-2) = -\frac{7}{4}; f(0) = -\frac{3}{4}; f\left(\frac{1}{2}\right) = -\frac{1}{2}; f\left(\frac{2}{3}\right) = -\frac{5}{12}$
41. $g(-1) = 0; g(2) = -9; g(-3) = 26; g(4) = 5$
43. $h(-2) = -2; h(-3) = -11; h(4) = -32; h(5) = -51$
45. $f(3) = \sqrt{7}; f(4) = 3; f(10) = \sqrt{21}; f(12) = 5$
47. $f(1) = -1; f(-1) = -2; f(3) = -\frac{2}{3}; f(-6) = \frac{4}{3}$
49. $f(a) = a^2 - 7a; f(a - 3) = a^2 - 13a + 30;$
   $f(a + h) = a^2 + 2ah + h^2 - 7a - 7h$

51. $f(-a) = 2a^2 + a - 1; f(a + 1) = 2a^2 + 3a;$
   $f(a + h) = 2a^2 + 4ah + 2h^2 - a - h - 1$
53. $f(-a) = -a^2 + 2a - 7; f(-a - 2) =$
   $-a^2 - 2a - 7; f(a + 7) = -a^2 - 16a - 70$
55. $f(-2) = 27; f(3) = 42; g(-4) = -37; g(6) = -17$
57. $f(-2) = 5; f(3) = 8; g(-4) = -3; g(5) = -4$
59. -3 **61.** $-2a - h$ **63.** $4a - 1 + 2h$
65. $-8a - 7 - 4h$
67. $h(1) = 48; h(2) = 64; h(3) = 48; h(4) = 0$
69. $C(75) = \$74; C(150) = \$98; C(225) = \$122; C(650) = \$258$
71. $I(0.04) = \$20; I(0.06) = \$30; I(0.075) = \$37.50; I(0.09) = \$45$

Problem Set 13.2 (page 633)

1.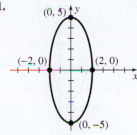

$f(x)$

(2, 0)
(0, −4)

3.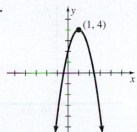

$f(x)$

(−1, −2) (1, −2)

5.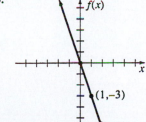

$f(x)$

(1, −3)

7.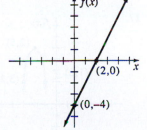

$f(x)$

(−1, −2)
(−2, −3) (0, −3)

9.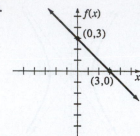

11.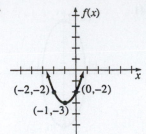

13.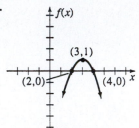

15.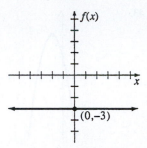

17.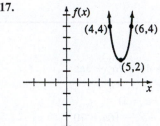

19.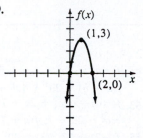

21.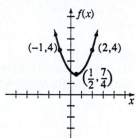

23.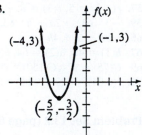

25.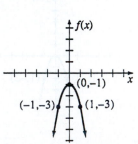

27.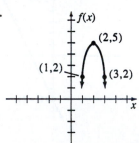

29.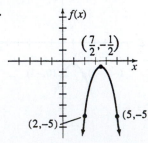

31. 3 and 5; $(4, -1)$ **33.** 6 and 8; $(7, -2)$

35. $7 + \sqrt{5}$ and $7 - \sqrt{5}$; $(7, -5)$

37. No intercepts; $\left(\dfrac{9}{2}, -\dfrac{3}{4}\right)$

39. $\dfrac{1 - \sqrt{5}}{2}$ and $\dfrac{1 + \sqrt{5}}{2}$; $\left(\dfrac{1}{2}, 5\right)$

41. **(a)** $0.42 **(c)** Answers will vary.

43. $100.00; $107.50; $140.00; $157.50

45. $2.10; $4.55; $20.72; $29.40; $33.88

47. $f(p) = 0.8p$; $7.60; $12; $60; $10; $600

49. 80 items **51.** 5 and 25 **53.** 60 meters by 60 meters

55. 1100 subscribers at $13.75 per month

Problem Set 13.3 (page 642)

1. B **3.** A **5.** F

7.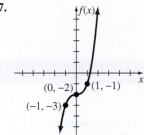

9.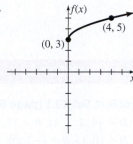

11.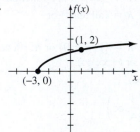

13.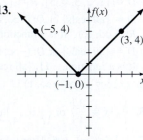

15.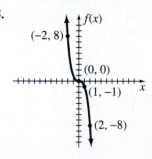

17.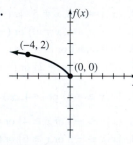

19.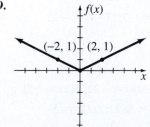

21.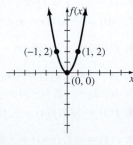

23.

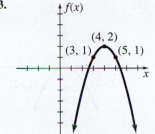

(4, 2)
(3, 1) (5, 1)

25.

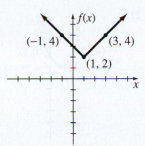

(−1, 4)   (3, 4)
(1, 2)

47.

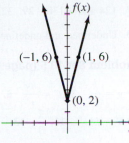

(−4, 3)
(−5, 0) (−3, 0)

49.

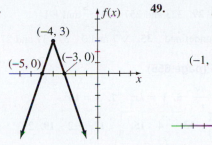

(−1, 6)   (1, 6)
(0, 2)

27.

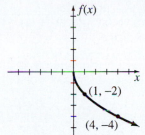

(1, −2)
(4, −4)

29.

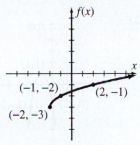

(−1, −2)   (2, −1)
(−2, −3)

51. (a)

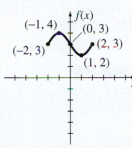

(−1, 4)   (0, 3)
(−2, 3)   (2, 3)
(1, 2)

51. (c)

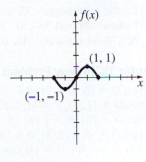

(1, 1)
(−1, −1)

31.

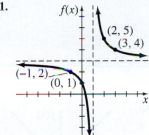

(2, 5)
(3, 4)
(−1, 2)
(0, 1)

33.

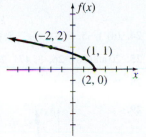

(−2, 2)
(1, 1)
(2, 0)

Problem Set 13.4 (page 649)

1. $(f \circ g)(x) = 15x - 3, D = \{\text{all reals}\}$ or $D: (-\infty, \infty)$;
$(g \circ f)(x) = 15x - 1, D = \{\text{all reals}\}$ or $D: (-\infty, \infty)$

3. $(f \circ g)(x) = -14x - 7, D = \{\text{all reals}\}$ or $D: (-\infty, \infty)$;
$(g \circ f)(x) = -14x + 11, D = \{\text{all reals}\}$ or $D: (-\infty, \infty)$

5. $(f \circ g)(x) = 3x^2 + 11, D = \{\text{all reals}\}$ or $D: (-\infty, \infty)$;
$(g \circ f)(x) = 9x^2 + 12x + 7, D = \{\text{all reals}\}$ or $D: (-\infty, \infty)$

7. $(f \circ g)(x) = 2x^2 - 11x + 17, D = \{\text{all reals}\}$ or $D: (-\infty, \infty)$;
$(g \circ f)(x) = -2x^2 + x + 1, D = \{\text{all reals}\}$ or $D: (-\infty, \infty)$

9. $(f \circ g)(x) = \dfrac{3}{4x - 9}, D = \left\{x \,|\, x \neq \dfrac{9}{4}\right\}$ or
$D: \left(-\infty, \dfrac{9}{4}\right) \cup \left(\dfrac{9}{4}, \infty\right); (g \circ f)(x) = \dfrac{12 - 9x}{x}$,
$D = \{x \,|\, x \neq 0\}$ or $D: (-\infty, 0) \cup (0, \infty)$

11. $(f \circ g)(x) = \sqrt{5x + 4}, D = \left\{x \,|\, x \geq -\dfrac{4}{5}\right\}$ or $D: \left[-\dfrac{4}{5}, \infty\right)$;
$(g \circ f)(x) = 5\sqrt{x + 1} + 3, D = \{x \,|\, x \geq -1\}$ or $D: [-1, \infty)$

13. $(f \circ g)(x) = x - 4, D = \{x \,|\, x \neq 4\}$ or $D: (-\infty, 4) \cup (4, \infty)$;
$(g \circ f)(x) = \dfrac{x}{1 - 4x}, D = \left\{x \,|\, x \neq 0 \text{ and } x \neq \dfrac{1}{4}\right\}$ or
$D: (-\infty, 0) \cup \left(0, \dfrac{1}{4}\right) \cup \left(\dfrac{1}{4}, \infty\right)$

15. $(f \circ g)(x) = \dfrac{2\sqrt{x}}{x}, D = \{x \,|\, x > 0\}$ or $D: (0, \infty)$;
$(g \circ f)(x) = \dfrac{4\sqrt{x}}{x}, D = \{x \,|\, x > 0\}$ or $D: (0, \infty)$

17. $(f \circ g)(x) = \dfrac{3x + 3}{2}, D = \{x \,|\, x \neq -1\}$ or
$D: (-\infty, -1) \cup (-1, \infty)$;
$(g \circ f)(x) = \dfrac{2x}{2x + 3}, D = \left\{x \,|\, x \neq 0 \text{ and } x \neq -\dfrac{3}{2}\right\}$ or
$D: \left(-\infty, -\dfrac{3}{2}\right) \cup \left(-\dfrac{3}{2}, 0\right) \cup (0, \infty)$

35.

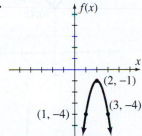

(2, −1)
(1, −4)   (3, −4)

37.

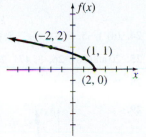

(3, 2)
(2, −1)
(1, −4)

39.

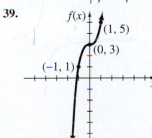

(1, 5)
(0, 3)
(−1, 1)

41.

(−3, 4)
(−2, 2)   (1, 0)

43.

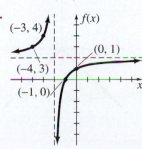

(−3, 4)
(0, 1)
(−4, 3)
(−1, 0)

45.

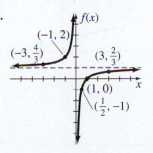

(−1, 2)
$\left(-3, \dfrac{4}{3}\right)$ $\left(3, \dfrac{2}{3}\right)$
(1, 0)
$\left(\dfrac{1}{2}, -1\right)$

27. 124 and -130 **29.** 323 and 257   **31.** $\dfrac{1}{2}$ and -1

33. Undefined and undefined   **35.** $\sqrt{7}$ and 0   **37.** 37 and 27

Problem Set 13.5 (page 655)

1. $y = \dfrac{k}{x^2}$ **3.** $C = \dfrac{kg}{t^3}$ **5.** $V = kr^3$ **7.** $S = ke^2$

9. $V = khr^2$ **11.** $\dfrac{2}{3}$ **13.** -4 **15.** $\dfrac{1}{3}$ **17.** -2 **19.** 2

21. 5   **23.** 9   **25.** 9   **27.** $\dfrac{1}{6}$ **29.** 112

31. 12 cubic centimeters   **33.** 28   **35.** 2 seconds
37. 12 ohms   **39.** (a) \$87.50   (b) \$273   (c) \$560
41. 3560.76 cubic meters   **43.** 0.048

Chapter 13 Review Problem Set (page 663)

1. Yes; $D = \{6, 7, 8, 9\}, R = \{2, 3, 4, 5\}$
2. No; $D = \{1, 2\}, R = \{1, 3, 5, 7\}$
3. No; $D = \{0\}, R = \{-6, -5, -4, -3\}$
4. Yes; $D = \{0, 1, 2, 3\}, R = \{8\}$
5. $f(2) = -1; f(-3) = 14; f(a) = a^2 - 2a - 1$
6. $f(2) = 0; f(-1) = -6; f(3a) = \dfrac{6a - 4}{3a + 2}$
7. $D = \{\text{all reals}\}$
8. $D = \{x \mid x \neq 5\}$ or $D: (-\infty, 5) \cup (5, \infty)$
9. $D = \{x \mid x \neq 0, x \neq -4\}$ or
   $D: (-\infty, -4) \cup (-4, 0) \cup (0, \infty)$
10. $D = \{x \mid x \leq -5 \text{ or } x \geq 5\}$ or
   $D: (-\infty, -5] \cup [5, \infty)$
11. 6   **12.** $4a + 1 + 2h$ **13.** \$0.72
14. $f(x) = 0.7x$; \$45.50; \$33.60; \$10.85

15. 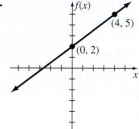   **16.**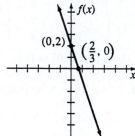

17. 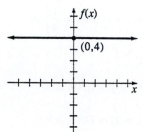   **18.**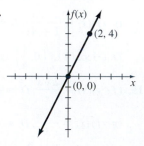

19. $f(x) = 6x - 100$; 110   **20.** $f(x) = \dfrac{10}{3}x + 150$; \$300

21. 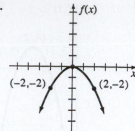   **22.**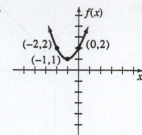

23.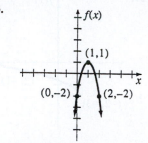

24. (a) $(-5, -28); x = -5$ (b) $\left(-\dfrac{7}{2}, \dfrac{67}{2}\right); x = -\dfrac{7}{2}$

25. 20 and 20   **26.** 3, 47   **27.** Width, 15 yd; length, 30 yd

28. 25 students

29. 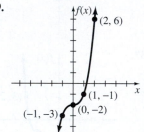   **30.**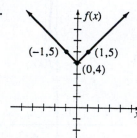

31. 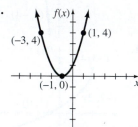   **32.**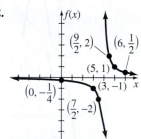

33. 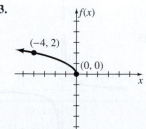   **34.**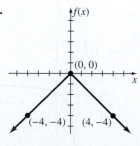

35.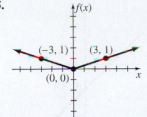

36.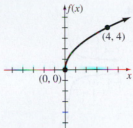

18.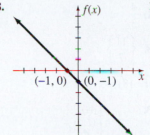

19.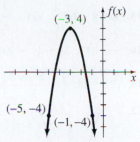

37.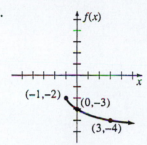

38.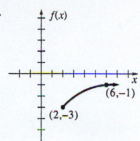

20.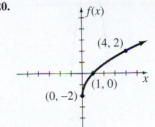

21.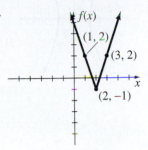

39.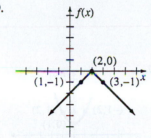

40.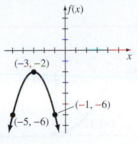

22.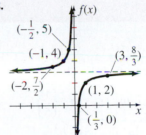

23.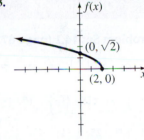

24.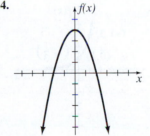

25.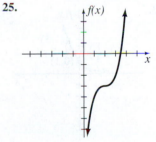

41. $(f \circ g)(x) = 6x - 11; (g \circ f)(x) = 6x - 13$

42. $(f \circ g)(x) = x^2 - 2x - 1; (g \circ f)(x) = x^2 - 10x + 27$

43. $(f \circ g)(x) = 4x^2 - 20x + 20; (g \circ f)(x) = -2x^2 + 15$

44. 179   **45.** -6 **46.** \$375   **47.** 600 in.^{2} **48.** 10 min

49. 128 lb   **50.** $k = 9$ **51.** $y = 120$

Chapter 13 Test (page 665)

1. $D = \left\{ x \mid x \neq -4 \text{ and } x \neq \dfrac{1}{2} \right\}$ or

$D: (-\infty, -4) \cup \left(-4, \dfrac{1}{2}\right) \cup \left(\dfrac{1}{2}, \infty\right)$

2. $D = \left\{ x \mid x \leq \dfrac{5}{3} \right\}$ or $D: \left(-\infty, \dfrac{5}{3}\right]$ **3.** $\dfrac{11}{6}$ **4.** 11

5. $(-6, 3)$ **6.** $6a + 3h + 2$ **7.** $(f \circ g)(x) = -21x - 2$

8. $(g \circ f)(x) = 8x^2 + 38x + 48$ **9.** $(f \circ g)(x) = \dfrac{3x}{2 - 2x}$

10. $f(x) = 1.50x + 5$ **11.** 6 and 54   **12.** -4 **13.** 15   **14.** \$96

15. The graph of $f(x) = (x - 6)^3 - 4$ is the graph of
$f(x) = x^3$ translated 6 units to the right and 4 units downward.

16. The graph of $f(x) = -|x| + 8$ is the graph of $f(x) = |x|$
reflected across the x axis and translated 8 units upward.

17. The graph of $f(x) = -\sqrt{x + 5} + 7$ is the graph of
$f(x) = \sqrt{x}$ reflected across the x axis and translated 5 units
to the left and 7 units upward.

Chapters 1–13 Cumulative Practice Test (page 666)

1. 12   **2. (a)** x axis   **(b)** $f(x)$ axis, or y axis
(c) Origin   **(d)** Origin

3. -9 and $\dfrac{1}{2}$ **4.** $(-1, -2)$ **5.** 32

6. (a) $4xy^2\sqrt{2x}$ **(b)** $2xy\sqrt[3]{4y}$ **(c)** $\dfrac{\sqrt{2}}{3}$

(d) $\dfrac{6 - \sqrt{6}}{15}$ **7.** $\dfrac{11}{10}$ **8.** $6x^3 + 5x^2 - 19x - 20$

9. $\dfrac{21}{20x}$ **10.** $2x^3 - 3x^2 + 4x - 5$ **11.** $x^2 + 2x + 4$

12. 0.001414   **13.** 115%   **14.** $-14 - 32i$

15. (a) 8 **(b)** $\dfrac{4}{5}$ **(c)** $-\dfrac{1}{8}$ **(d)** $\dfrac{4}{3}$

16. $-\dfrac{2}{3}$ **17.** $3x + 4y = 20$ **18.** $\left\{\dfrac{11}{7}\right\}$

19. $\{2\}$ **20.** $\left\{\dfrac{1 \pm i\sqrt{2}}{3}\right\}$ **21.** $\{8\}$

22.

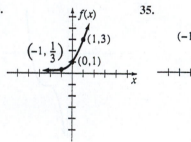

23.

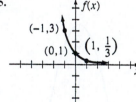

24.

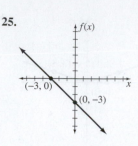

25.

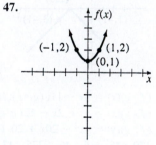

Chapter 14

Problem Set 14.1 (page 673)

1. $\{6\}$ **3.** $\left\{\dfrac{3}{2}\right\}$ **5.** $\{7\}$ **7.** $\{5\}$ **9.** $\{1\}$ **11.** $\{1\}$

13. $\{-3\}$ **15.** $\left\{\dfrac{3}{2}\right\}$ **17.** $\{-3\}$ **19.** $\{1\}$ **21.** $\left\{\dfrac{1}{5}\right\}$

23. $\{0\}$ **25.** $\{-1\}$ **27.** $\left\{\dfrac{5}{2}\right\}$ **29.** $\{3\}$ **31.** $\left\{\dfrac{1}{2}\right\}$

33.

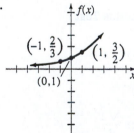

35.

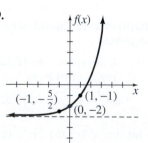

37.

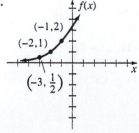

39.

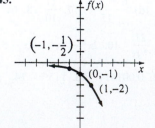

41.

43.

45.

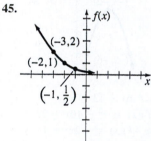

47.

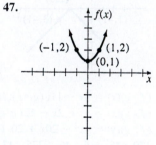

49.

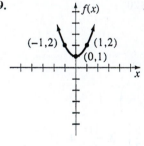

51.

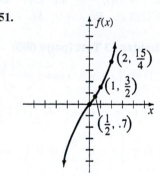

Problem Set 14.2 (page 682)

1. (a) $1.55 **(b)** $4.17 **(c)** $2.33 **(d)** $2.28
(e) $21,900 **(f)** $246,342 **(g)** $658
3. $283.70 **5.** $659.74 **7.** $1251.16 **9.** $2234.77
11. $9676.41 **13.** $13,814.17 **15.** $567.63
17. $1422.36 **19.** $5715.30 **21.** $14,366.56
23. $26,656.96

25.

	1 yr	5 yr	10 yr	20 yr
Compounded annually	$1060	1338	1791	3207
Compounded semiannually	1061	1344	1806	3262
Compounded quarterly	1061	1347	1814	3291
Compounded monthly	1062	1349	1819	3310
Compounded continuously	1062	1350	1822	3320

27. Nora will have $0.24 more.   **29.** 9.54%
31. 50 grams; 37 grams

33. 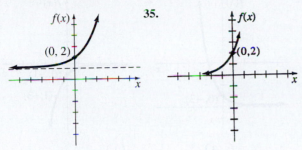   **35.**

37.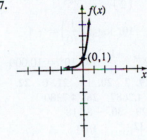

39. 2226; 3320; 7389   **41.** 2000
43. (a) 82,888   (c) 96,302   **45.** (a) 6.5 pounds per square inch
   (c) 13.6 pounds per square inch

Problem Set 14.3 (page 692)

1. Yes   **3.** No   **5.** Yes   **7.** Yes   **9.** Yes   **11.** No   **13.** No
15. Domain of f: $\{1, 2, 5\}$
   Range of f: $\{5, 9, 21\}$
   $f^{-1} = \{(5, 1), (9, 2), (21, 5)\}$
   Domain of f^{-1}: $\{5, 9, 21\}$
   Range of f^{-1}: $\{1, 2, 5\}$
17. Domain of f: $\{0, 2, -1, -2\}$
   Range of f: $\{0, 8, -1, -8\}$
   f^{-1}: $\{(0, 0), (8, 2), (-1, -1), (-8, -2)\}$
   Domain of f^{-1}: $\{0, 8, -1, -8\}$
   Range of f^{-1}: $\{0, 2, -1, -2\}$
27. No   **29.** Yes   **31.** No   **33.** Yes   **35.** Yes
37. $f^{-1}(x) = x + 4$
39. $f^{-1}(x) = \dfrac{-x - 4}{3}$ **41.** $f^{-1}(x) = \dfrac{12x + 10}{9}$
43. $f^{-1}(x) = -\dfrac{3}{2}x$ **45.** $f^{-1}(x) = x^2$ for $x \geq 0$
47. $f^{-1}(x) = \sqrt{x - 4}$ for $x \geq 4$
49. $f^{-1}(x) = \dfrac{1}{x - 1}$ for $x > 1$
51. $f^{-1}(x) = \dfrac{1}{3}x$ **53.** $f^{-1}(x) = \dfrac{x - 1}{2}$

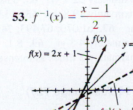

55. $f^{-1}(x) = \dfrac{x + 2}{x}$ for $x > 0$

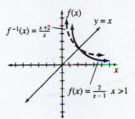

57. $f^{-1}(x) = \sqrt{x + 4}$ for $x \geq -4$

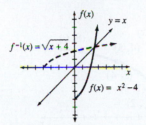

59. Increasing on $[0, \infty)$ and decreasing on $(-\infty, 0]$
61. Decreasing on $(-\infty, \infty)$
63. Increasing on $(-\infty, -2]$ and decreasing on $[-2, \infty)$
65. Increasing on $(-\infty, -4]$ and decreasing on $[-4, \infty)$

71. (a) $f^{-1}(x) = \dfrac{x + 9}{3}$ (c) $f^{-1}(x) = -x + 1$
   (e) $f^{-1}(x) = -\dfrac{1}{5}x$

Problem Set 14.4 (page 701)

1. $\log_2 128 = 7$ **3.** $\log_5 125 = 3$ **5.** $\log_{10} 1000 = 3$
7. $\log_2\left(\dfrac{1}{4}\right) = -2$ **9.** $\log_{10} 0.1 = -1$ **11.** $3^4 = 81$
13. $4^3 = 64$ **15.** $10^4 = 10,000$ **17.** $2^{-4} = \dfrac{1}{16}$
19. $10^{-3} = 0.001$ **21.** 4   **23.** 4   **25.** 3   **27.** $\dfrac{1}{2}$ **29.** 0
31. -1 **33.** 5   **35.** -5 **37.** 1   **39.** 0   **41.** $\{49\}$
43. $\{16\}$ **45.** $\{27\}$ **47.** $\left\{\dfrac{1}{8}\right\}$ **49.** $\{4\}$ **51.** 5.1293
53. 6.9657   **55.** 1.4037   **57.** 7.4512   **59.** 6.3219
61. -0.3791 **63.** 0.5766   **65.** 2.1531   **67.** 0.3949
69. $\log_b x + \log_b y + \log_b z$ **71.** $\log_b y - \log_b z$
73. $3\log_b y + 4\log_b z$ **75.** $\dfrac{1}{2}\log_b x + \dfrac{1}{3}\log_b y - 4\log_b z$
77. $\dfrac{2}{3}\log_b x + \dfrac{1}{3}\log_b z$ **79.** $\dfrac{3}{2}\log_b x - \dfrac{1}{2}\log_b y$
81. $\left\{\dfrac{9}{4}\right\}$ **83.** $\{25\}$ **85.** $\{4\}$ **87.** $\left\{\dfrac{19}{8}\right\}$ **89.** $\{9\}$
91. $\{1\}$ **93.** $\left\{-\dfrac{22}{5}\right\}$ **95.** $\varnothing$ **97.** $\{-1\}$

Problem Set 14.5 (page 707)

1. 0.8597   **3.** 1.7179   **5.** 3.5071   **7.** -0.1373 **9.** -3.4685
11. 411.43   **13.** 90,095   **15.** 79.543   **17.** 0.048440
19. 0.0064150   **21.** 1.6094   **23.** 3.4843   **25.** 6.0638
27. -0.7765 **29.** -3.4609 **31.** 1.6034   **33.** 3.1346
35. 108.56   **37.** 0.48268   **39.** 0.035994

41.

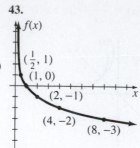

43.

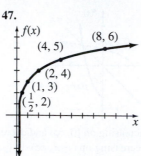

45.

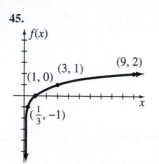

47.

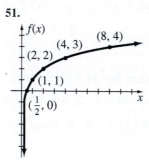

49.

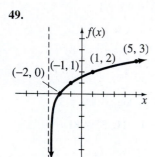

51.

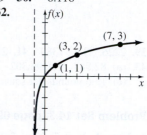

53.

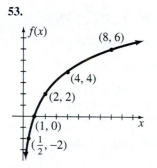

55. 0.36 **57.** 0.73 **59.** 23.10 **61.** 7.93

Problem Set 14.6 (page 716)

1. {3.15} **3.** {2.20} **5.** {4.18} **7.** {0.12} **9.** {1.69}

11. {4.57} **13.** {2.46} **15.** {4} **17.** $\left\{\dfrac{19}{47}\right\}$

19. {1} **21.** {8} **23.** 4.524 **25.** -0.860 **27.** 3.105
29. -2.902 **31.** 5.989 **33.** 7.2 years **35.** 11.6 years
37. 6.8 hours **39.** 1.5 hours **41.** 34.7 years **43.** 6.7
45. Approximately 8 times

Chapter 14 Review Problem Set (page 725)

1.

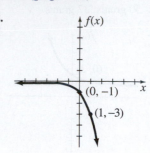

2.

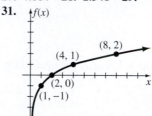

3. {-4} **4.** $\left\{-\dfrac{3}{4}\right\}$ **5.** {3} **6.** $\left\{\dfrac{1}{2}\right\}$ **7.** $\log_8 64 = 2$

8. $\log_a c = b$ **9.** $\log_3\left(\dfrac{1}{9}\right) = -2$ **10.** $\log_{10}\left(\dfrac{1}{10}\right) = -1$

11. $5^3 = 125$ **12.** $x^z = y$ **13.** $2^{-3} = \dfrac{1}{8}$ **14.** $10^4 = 10,000$

15. 7 **16.** 3 **17.** 4 **18.** -3 **19.** 2 **20.** 13 **21.** 0 **22.** -2
23. 1.8642 **24.** -2.6289 **25.** -4.2687 **26.** 4.7380
27. 4.164 **28.** 2.548 **29.** -3.129 **30.** -8.118

31.

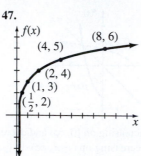

32.

33. $\log_b x + \dfrac{1}{2}\log_b y - 3\log_b z$ **34.** $1 - \log_b c - \log_b d$

35. {1.1882} **36.** {2639.4} **37.** {0.013289}

38. {0.077197} **39.** {4} **40.** {9} **41.** $\left\{\dfrac{1}{2}\right\}$ **42.** {8}

43. {5} **44.** $\left\{\dfrac{45}{8}\right\}$ **45.** {5} **46.** $\left\{\dfrac{15}{7}\right\}$ **47.** {3.40}

48. {1.95} **49.** {5.30} **50.** {5.61}

51. $f^{-1}(x) = \dfrac{x - 5}{4}$ **52.** $f^{-1}(x) = \dfrac{-x - 7}{3}$

53. $f^{-1}(x) = \dfrac{6x + 2}{5}$ **54.** $f^{-1}(x) = \sqrt{-2 - x}$ for $x \le 2$

55. Increasing on $(-\infty, 4]$ and decreasing on $[4, \infty)$
56. Increasing on $[3, \infty)$
57. **(a)** \$933.05 **(b)** \$305.96 **(c)** \$98.54
58. **(a)** \$7910 **(b)** \$14,766 **(c)** \$15,000 **59.** \$19,632.75
60. \$40,305.65 **61.** Approximately 8.8 yr
62. Approximately 22.9 yr **63.** \$5656.26
64. Approximately 19.8 yr **65.** 133 g
66. 46.4 hr **67.** 61,070; 67,493; 74,591
68. Approximately 4.8 yr **69.** 8.1
70. Approximately 5 times as intense

Chapter 14 Test (page 727)

1. $\dfrac{1}{2}$ **2.** 1 **3.** 1 **4.** -1 **5.** {-3} **6.** $\left\{-\dfrac{3}{2}\right\}$ **7.** $\left\{\dfrac{8}{3}\right\}$

8. {243} **9.** {2} **10.** $\left\{\dfrac{2}{5}\right\}$ **11.** 4.1919 **12.** 0.2031

13. 0.7325   **14.** 5.4538   **15.** $f^{-1}(x) = \dfrac{x+6}{3}$

16. $f^{-1}(x) = \dfrac{3}{2}x + \dfrac{9}{10}$ **17.** 4.0069   **18.** $\log_b\left(\dfrac{x^3 y^2}{z}\right)$

19. $6342.08

20. 13.5 years   **21.** 7.8 hours   **22.** 4813 grams

23.   **24.**

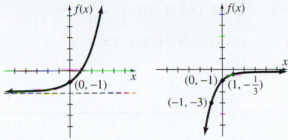

25.

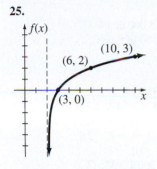

Chapters 1–14 Cumulative Review Problem Set (page 728)

1. -6 **2.** -8 **3.** $\dfrac{13}{24}$ **4.** 56   **5.** $\dfrac{13}{6}$ **6.** $-90\sqrt{2}$

7. $2x + 5\sqrt{x} - 12$ **8.** $-18 + 22\sqrt{3}$ **9.** $2x^3 + 11x^2 - 14x + 4$

10. $\dfrac{x+4}{x(x+5)}$ **11.** $\dfrac{16x^2}{27y}$ **12.** $\dfrac{16x + 43}{90}$ **13.** $\dfrac{35a - 44b}{60a^2 b}$

14. $\dfrac{2}{x-4}$ **15.** $2x^2 - x - 4$ **16.** $\dfrac{5y^2 - 3xy^2}{x^2 y + 2x^2}$ **17.** $\dfrac{2y - 3xy}{3x + 4xy}$

18. $\dfrac{(2n-5)(n+3)}{(n-2)(3n+13)}$ **19.** $\dfrac{3a^2 - 2a + 1}{2a - 1}$ **20.** $(5x - 2)(4x + 3)$

21. $2(2x+3)(4x^2 - 6x + 9)$ **22.** $(2x+3)(2x-3)(x+2)(x-2)$

23. $4x(3x+2)(x-5)$ **24.** $(y-6)(x+3)$ **25.** $(5 - 3x)(2 + 3x)$

26. $\dfrac{81}{16}$ **27.** 4   **28.** $-\dfrac{3}{4}$ **29.** -0.3 **30.** $\dfrac{1}{81}$ **31.** $\dfrac{21}{16}$

32. $\dfrac{9}{64}$ **33.** 72   **34.** 6   **35.** -2 **36.** $\dfrac{-12}{x^3 y}$ **37.** $\dfrac{8y}{x^5}$

38. $-\dfrac{a^3}{9b}$ **39.** $4\sqrt{5}$ **40.** $-6\sqrt{6}$ **41.** $\dfrac{5\sqrt{3}}{9}$ **42.** $\dfrac{2\sqrt{3}}{3}$

43. $2\sqrt[3]{7}$ **44.** $\dfrac{\sqrt[3]{6}}{2}$ **45.** $8xy\sqrt{13x}$ **46.** $\dfrac{\sqrt{6xy}}{3y}$ **47.** $11\sqrt{6}$

48. $-\dfrac{169\sqrt{2}}{12}$ **49.** $-16\sqrt[3]{3}$ **50.** $\dfrac{-3\sqrt{2} - 2\sqrt{6}}{2}$

51. $\dfrac{6\sqrt{15} - 3\sqrt{35} - 6 + \sqrt{21}}{5}$ **52.** 0.021   **53.** 300

54. 0.0003   **55.** $32 + 22i$ **56.** $-17 + i$ **57.** $0 - \dfrac{5}{4}i$

58. $-\dfrac{19}{53} + \dfrac{40}{53}i$ **59.** $-\dfrac{10}{3}$ **60.** $\dfrac{4}{7}$ **61.** $2\sqrt{13}$

62. $5x - 4y = 19$ **63.** $4x + 3y = -18$

64. $(-2, 6)$ and $r = 3$ **65.** $(-5, -4)$ **66.** 8 units

67.   **68.**

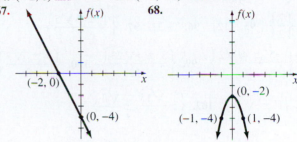

69.   **70.**

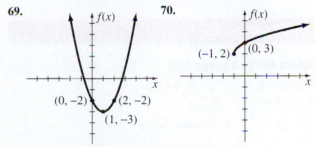

71.   **72.**

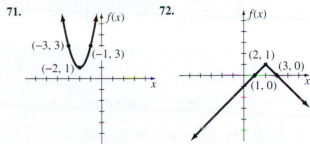

73.   **74.**

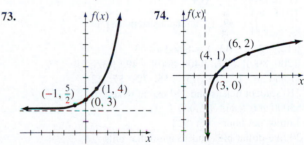

75.   **76.**

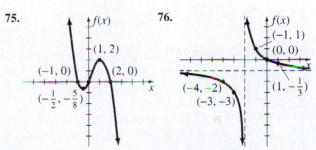

77. $(g \circ f)(x) = 2x^2 - 13x + 20$; $(f \circ g)(x) = 2x^2 - x - 4$

78. $f^{-1}(x) = \dfrac{x+7}{3}$ **79.** $f^{-1}(x) = -2x + \dfrac{4}{3}$

80. $k = -3$ **81.** $y = 1$ **82.** 12 cubic centimeters

83. $\left\{-\dfrac{21}{16}\right\}$ **84.** $\left\{\dfrac{40}{3}\right\}$ **85.** $\{6\}$ **86.** $\left\{-\dfrac{5}{2}, 3\right\}$

87. $\left\{0, \dfrac{7}{3}\right\}$ **88.** $\{-6, 0, 6\}$ **89.** $\left\{-\dfrac{5}{6}, \dfrac{2}{5}\right\}$

90. $\left\{-3, 0, \dfrac{3}{2}\right\}$ **91.** $\{\pm 1, \pm 3i\}$ **92.** $\{-5, 7\}$ **93.** $\{-29, 0\}$

94. $\left\{\dfrac{7}{2}\right\}$ **95.** $\{12\}$ **96.** $\{-3\}$ **97.** $\left\{\dfrac{1 \pm 3\sqrt{5}}{3}\right\}$

98. $\left\{\dfrac{-5 \pm 4i\sqrt{2}}{2}\right\}$ **99.** $\left\{\dfrac{3 \pm i\sqrt{23}}{4}\right\}$ **100.** $\left\{\dfrac{3 \pm \sqrt{3}}{3}\right\}$

101. $\{1 \pm \sqrt{34}\}$ **102.** $\left\{\pm\dfrac{\sqrt{15}}{2}, \pm\dfrac{\sqrt{3}}{3}\right\}$

103. $\left\{\dfrac{-5 \pm i\sqrt{15}}{4}\right\}$ **104.** $\left\{-\dfrac{1}{4}, \dfrac{5}{3}\right\}$ **105.** $\varnothing$ **106.** $\left\{\dfrac{3}{2}\right\}$

107. $\{81\}$ **108.** $\{4\}$ **109.** $\{6\}$ **110.** $\left\{\dfrac{1}{5}\right\}$ **111.** $(-\infty, 3)$

112. $(-\infty, 50]$ **113.** $\left(-\infty, -\dfrac{11}{5}\right) \cup (3, \infty)$ **114.** $\left(-\dfrac{5}{3}, 1\right)$

115. $\left[-\dfrac{9}{11}, \infty\right)$ **116.** $[-4, 2]$ **117.** $\left(-\infty, \dfrac{1}{3}\right) \cup (4, \infty)$

118. $(-8, 3)$ **119.** $(-\infty, 3] \cup (7, \infty)$ **120.** $(-6, -3)$

Chapter 15

Problem Set 15.1 (page 739)

1. $\{(7, 9)\}$ **3.** $\{(-4, 7)\}$ **5.** $\{(6, 3)\}$

7. $a = -3$ and $b = -4$

9. $\left\{\left(k, \dfrac{2}{3}k - \dfrac{4}{3}\right)\right\}$, a dependent system

11. $u = 5$ and $t = 7$ **13.** $\{(2, -5)\}$

15. $\varnothing$, an inconsistent system   **17.** $\left\{\left(-\dfrac{3}{4}, -\dfrac{6}{5}\right)\right\}$

19. $\left\{\left(\dfrac{1}{2}, -\dfrac{1}{4}\right)\right\}$ **21.** $\{(2, 8)\}$ **23.** $\{(-1, -5)\}$

25. $\varnothing$, an inconsistent system

27. $a = 2$ and $b = -\dfrac{1}{3}$ **29.** $s = -6$ and $t = 12$

31. $\left\{\left(-\dfrac{1}{2}, \dfrac{1}{3}\right)\right\}$ **33.** $\left\{\left(\dfrac{3}{4}, -\dfrac{2}{3}\right)\right\}$ **35.** $\{(-4, 2)\}$

37. $\left\{\left(\dfrac{11}{5}, \dfrac{2}{5}\right)\right\}$ **39.** $\varnothing$, an inconsistent system

41. $\{(12, -24)\}$ **43.** $t = 8$ and $u = 3$

45. $\{(200, 800)\}$ **47.** $\{(400, 800)\}$ **49.** $\{(3.5, 7)\}$

51. 17 and 36   **53.** $15°, 75°$ **55.** 72   **57.** 12

59. 210 student tickets and 210 parent tickets

61. $5000 at 6% and $15,000 at 8%

63. 3 miles per hour   **65.** $22

67. 30 five-dollar bills and 18 ten-dollar bills

73. $\{(4, 6)\}$ **75.** $\{(2, -3)\}$ **77.** $\left\{\left(\dfrac{1}{4}, -\dfrac{2}{3}\right)\right\}$

Problem Set 15.2 (page 747)

1. $\{(-4, -2, 3)\}$ **3.** $\{(-2, 5, 2)\}$ **5.** $\{(4, -1, -2)\}$

7. $\{(3, 1, 2)\}$ **9.** $\varnothing$ **11.** $\{(-1, 3, 5)\}$

13. $\{(-2, -1, 3)\}$ **15.** $\{(0, 2, 4)\}$

17. $\{(17k - 43, -5k + 13, k) \mid k$ is a real number$\}$

19. $\{(4, -1, -2)\}$ **21.** $\{(-4, 0, -1)\}$

23. $\{(2, 2, -3)\}$ **25.** $-2, 6, 16$

27. Helmet = $250; jacket = $350; gloves = $50

29. $\angle A = 120°$; $\angle B = 24°$; $\angle C = 36°$

31. Plumber = $50 per hour; apprentice = $20 per hour; laborer = $10 per hour

33. 4 pounds of pecans, 4 pounds of almonds, and 12 pounds of peanuts

35. 7 nickels, 13 dimes, and 22 quarters

37. $40°, 60°$, and $80°$

39. $5000 at 2%, $10,000 at 3%, $15,000 at 4%

41. 50 of type A, 75 of type B, and 150 of type C

Problem Set 15.3 (page 756)

1. Yes   **3.** Yes   **5.** No   **7.** No   **9.** Yes

11. $\{(-1, -5)\}$ **13.** $\{(3, -6)\}$ **15.** $\varnothing$ **17.** $\{(-2, -9)\}$

19. $\{(-1, -2, 3)\}$ **21.** $\{(3, -1, 4)\}$ **23.** $\{(0, -2, 4)\}$

25. $\{(-7k + 8, -5k + 7, k)\}$ **27.** $\{(-4, -3, -2)\}$

29. $\{(4, -1, -2)\}$ **31.** $\{(1, -1, 2, -3)\}$

33. $\{(2, 1, 3, -2)\}$ **35.** $\{(-2, 4, -3, 0)\}$ **37.** $\varnothing$

39. $\{(-3k + 5, -1, -4k + 2, k)\}$

41. $\{(-3k + 9, k, 2, -3)\}$

45. $\{(17k - 6, 10k - 5, k)\}$

47. $\left\{\left(-\dfrac{1}{2}k + \dfrac{34}{11}, \dfrac{1}{2}k - \dfrac{5}{11}, k\right)\right\}$ **49.** $\varnothing$

Problem Set 15.4 (page 766)

1. 22   **3.** -29 **5.** 20   **7.** 5   **9.** -2 **11.** $-\dfrac{2}{3}$

13. -25 **15.** 58   **17.** 39   **19.** -12

21. -41 **23.** -8 **25.** 1088

27. -140 **29.** 81   **31.** 146   **33.** Property 15.3

35. Property 15.2   **37.** Property 15.4

39. Property 15.3   **41.** Property 15.5

Problem Set 15.5 (page 772)

1. $\{(1, 4)\}$ **3.** $\{(3, -5)\}$ **5.** $\{(2, -1)\}$ **7.** $\varnothing$

9. $\left\{\left(-\dfrac{1}{4}, \dfrac{2}{3}\right)\right\}$ **11.** $\left\{\left(\dfrac{2}{17}, \dfrac{52}{17}\right)\right\}$ **13.** $\{(9, -2)\}$

15. $\left\{\left(2, -\dfrac{5}{7}\right)\right\}$ **17.** $\{(0, 2, -3)\}$ **19.** $\{(2, 6, 7)\}$

21. $\{(4, -4, 5)\}$ **23.** $\{(-1, 3, -4)\}$

25. Infinitely many solutions

27. $\left\{\left(-2, \dfrac{1}{2}, -\dfrac{2}{3}\right)\right\}$ **29.** $\left\{\left(3, \dfrac{1}{2}, -\dfrac{1}{3}\right)\right\}$

31. $(-4, 6, 0)$ **37.** $(0, 0, 0)$

39. Infinitely many solutions

Problem Set 15.6 (page 778)

1. $\{(1, 2)\}$ **3.** $\{(1, -5), (-5, 1)\}$

5. $\{(2 + i\sqrt{3}, -2 + i\sqrt{3}), (2 - i\sqrt{3}, -2 - i\sqrt{3})\}$

7. $\{(-6, 7), (-2, -1)\}$ **9.** $\{(-3, 4)\}$

11. $\left\{\left(\dfrac{-1 + i\sqrt{3}}{2}, \dfrac{-7 - i\sqrt{3}}{2}\right), \left(\dfrac{-1 - i\sqrt{3}}{2}, \dfrac{-7 + i\sqrt{3}}{2}\right)\right\}$

13. $\{(-1, 2)\}$ **15.** $\{(-6, 3), (-2, -1)\}$

17. $\{(5, 3)\}$ **19.** $\{(1, 2), (-1, 2)\}$ **21.** $\{(-3, 2)\}$

23. $\{(2, 0), (-2, 0)\}$

25. $\{(\sqrt{2}, \sqrt{3}), (\sqrt{2}, -\sqrt{3}), (-\sqrt{2}, \sqrt{3}), (-\sqrt{2}, -\sqrt{3})\}$

27. $\{(1, 1), (1, -1), (-1, 1), (-1, -1)\}$

29. $\left\{\left(2, \dfrac{3}{2}\right), \left(\dfrac{3}{2}, 2\right)\right\}$

Chapter 15 Review Problem Set (page 785)

1. $\{(3, -7)\}$ **2.** $\{(-1, -3)\}$ **3.** $\{(0, -4)\}$

4. $\left\{\left(\dfrac{23}{3}, -\dfrac{14}{3}\right)\right\}$ **5.** $\{(4, -6)\}$ **6.** $\left\{\left(-\dfrac{6}{7}, -\dfrac{15}{7}\right)\right\}$

7. $\{(-1, 2, -5)\}$ **8.** $\{(2, -3, -1)\}$ **9.** $\{(5, -4)\}$

10. $\{(2, 7)\}$ **11.** $\{(-2, 2, -1)\}$ **12.** $\{(0, -1, 2)\}$

13. $\{(-3, -1)\}$ **14.** $\{(4, 6)\}$ **15.** $\{(2, -3, -4)\}$

16. $\{(-1, 2, -5)\}$ **17.** $\{(5, -5)\}$ **18.** $\{(-12, 12)\}$

19. $\left\{\left(\dfrac{5}{7}, \dfrac{4}{7}\right)\right\}$ **20.** $\{(-10, -7)\}$ **21.** $\varnothing$

22. $\{(1, 1, -4)\}$ **23.** $\{(-4, 0, 1)\}$ **24.** $\varnothing$

25. $\{(-4k + 19, -3k + 12, k) | k \text{ is a real number}\}$

26. $\{(-2, -4, 6)\}$ **27.** -34 **28.** 13   **29.** -40 **30.** 16

31. 51   **32.** 125   **33.** $\{(-1, 4)\}$

34. $\{(3, 1)\}$ **35.** $\{(-1, -2), (-2, -3)\}$

36. $\left\{\left(\dfrac{4\sqrt{2}}{3}, \dfrac{4}{3}i\right), \left(\dfrac{4\sqrt{2}}{3}, -\dfrac{4}{3}i\right), \left(-\dfrac{4\sqrt{2}}{3}, \dfrac{4}{3}i\right),\right.$
$\left.\left(-\dfrac{4\sqrt{2}}{3}, -\dfrac{4}{3}i\right)\right\}$

37. $\{(0, 2), (0, -2)\}$

38. $\left\{\left(\dfrac{\sqrt{15}}{5}, \dfrac{2\sqrt{10}}{5}\right), \left(\dfrac{\sqrt{15}}{5}, -\dfrac{2\sqrt{10}}{5}\right), \left(-\dfrac{\sqrt{15}}{5}, \dfrac{2\sqrt{10}}{5}\right),\right.$
$\left.\left(-\dfrac{\sqrt{15}}{5}, -\dfrac{2\sqrt{10}}{5}\right)\right\}$

39. 72   **40.** $900 at 10% and $1600 at 12%

41. 20 nickels, 32 dimes, and 54 quarters

42. $25°, 45°,$ and $110°$

Chapter 15 Test (page 787)

1. III   **2.** I   **3.** III   **4.** II   **5.** 8   **6.** $-\dfrac{7}{12}$ **7.** -18

8. 112   **9.** Infinitely many   **10.** $\{(-2, 4)\}$ **11.** $\{(3, -1)\}$

12. $x = -12$ **13.** $y = -\dfrac{13}{11}$ **14.** Two   **15.** $x = 14$

16. $y = 13$ **17.** Infinitely many   **18.** None

19. $\left\{\left(\dfrac{11}{5}, 6, -3\right)\right\}$ **20.** $\{(-2, -1, 0)\}$

21. $x = 1$ **22.** $y = 4$

23. $\left\{(3, 2), (-3, -2), \left(4, \dfrac{3}{2}\right), \left(-4, -\dfrac{3}{2}\right)\right\}$

24. 2 liters   **25.** 22 quarters

Chapters 1–15 Cumulative Word Problem Review Set (page 788)

1. $65 per hour   **2.** 17, 19, and 21   **3.** 70°

4. 14 nickels, 20 dimes, and 29 quarters

5. $30°, 50°,$ and $100°$ **6.** $9.30 per hour   **7.** $600

8. $1700 at 8% and $2000 at 9%   **9.** 69 or less

10. 30 shares at $10 per share

11. 8 inches by 14 inches   **12.** $-3, 0,$ or 3

13. 66 miles per hour and 76 miles per hour

14. 4 quarts   **15.** $16\dfrac{2}{3}$ years

16. Approximately 11.64 years

17. Approximately 11.55 years   **18.** 6 centimeters

19. 1 hour   **20.** 1 inch   **21.** $1050 and $1400

22. 3 hours   **23.** 37   **24.** $4\dfrac{4}{5}$ days   **25.** $4000

26. Freight train: 6 hours at 50 miles per hour; express train: 4 hours at 70 miles per hour

27. $66.2° < F < 71.6°$ **28.** 150   **29.** $1 + \sqrt{2}$ and $1 - \sqrt{2}$

30. Larry: 52 miles per hour; Nita: 54 miles per hour

31. 15 rows and 20 seats per row

32. 4 units   **33.** 30 students

34. 4 yards and 6 yards   **35.** $6\sqrt{2}$ inches

36. $3.59 per movie and $4.99 per game

37. 21, 22, and 23   **38.** 12 nickels and 23 dimes

39. $[14, \infty)$ **40.** 25 milliliters

Index